The Ray Society

INSTITUTED 1844

This volume is No. 161 of the series
and commemorates the 150th Anniversary
of the Foundation of the Ray Society

LONDON
1994

Registered Charity Number 208082

Revised and updated edition
of Volume No. 144

VERA FRETTER & ALASTAIR GRAHAM

British
Prosobranch
Molluscs

THEIR FUNCTIONAL ANATOMY
AND ECOLOGY

Made and printed in Great Britain by
Henry Ling Limited
The Dorset Press, Dorchester, Dorset

Set in Monotype Optima

© *The Ray Society, 1994*

ISBN 0 903874 23 7

Sold by The Ray Society,
Registered Charity Number 208082
c/o Intercept Ltd., P.O. Box 716
Andover, Hants., SP10 1YG

VERA FRETTER & ALASTAIR GRAHAM

British Prosobranch Molluscs

THEIR FUNCTIONAL ANATOMY
AND ECOLOGY

Made and printed in Great Britain by
Henry Ling Limited
The Dorset Press, Dorchester, Dorset

Set in Monotype Optima

© *The Ray Society, 1994*

ISBN 0 903874 23 7

Sold by The Ray Society,
Registered Charity Number 208082
c/o Intercept Ltd., P.O. Box 716
Andover, Hants., SP10 1YG

BRITISH PROSOBRANCH MOLLUSCS
THEIR FUNCTIONAL ANATOMY AND ECOLOGY

I T is with some temerity that we offer an updated version of this book, in the hope that it may still prove valuable to those interested in the anatomy and ecology of prosobranch gastropods, especially of such species as are found in Britain and the British seas. We are encouraged in this hope since the recent gradual but steady reduction in the teaching of invertebrate zoology has left many with a background inadequate for any advanced study which they might wish to undertake.

Since 1962, when the first edition appeared, the flow of publications dealing with proso-branchs has become so great that it did not seem practicable for us to deal effectively with it all. We have therefore felt constrained to modify the contents in several ways: (1) by the deletion of certain parts either excellently dealt with in other recent publications, or peripheral to the main theme. These have not been replaced. (2) By rewriting parts where new facts and ideas rendered the original accounts inadequate, out-of-date, or wrong. Part I contains the original text minus deletions; Part II contains the new chapters. Details of these revisions are given in the next section 'Notes on the preparation and use of this book'.

We are very indebted to those who have read drafts of new chapters, for their interest in the work and for their comments, which have pointed out omissions and places where the treatment was inadequate or wrong. For their willingness to undertake this task we thank E. B. Andrews, A. D. Ansell, R. S. K. Barnes, J. A. Crothers, J. A. Kitching, C. Little, M. C. Pilkington, W. F. Ponder, D. G. Raffaelli, E. R. Trueman, and A. Warén. We are also grateful to Prof. K. Simkiss and Dr A. R. Jones for the facilities of their department, and Dr K. Harrison and Dr N. J. Evans of the Ray Society for help in solving some of the many problems which arose during the preparation and publication of this book.

NOTE ADDED DECEMBER, 1992

Vera Fretter died on 15 October 1992. Although both she and I had read and commented on what had been written by the other, there were certain areas where her special interests and knowledge made her the obvious author. These primarily related to reproduction—the anatomy and functioning of genital ducts, the different types of spawn, and what could be learned about larvae and their life. She was also an experienced shore collector whose intimate knowledge of the ways of intertidal prosobranchs added a great deal to our accounts of their ecology. She and I both believed that a book of this sort should be well and richly illustrated: her gifts as an artist contributed much to this.

Alastair Graham

It may be helpful to the reader to be aware of the following facts.

Chapter 1, Introduction, offers an introduction to the entire subsequent text, both Part I and Part II. It also includes definitions of many of the higher taxa used in prosobranch classification. References given in this chapter are included in the first bibliography.

Part I (chapters 2 to 22) includes most of the text of the first edition but the following sections have been deleted and have not been replaced: that treating of molluscan radiation outside the gastropods; that dealing with the physiology and biochemistry of shell secretion; the chapter on the parasites of prosobranchs; and the appendices on classification and on habitats and distribution. Other parts of the original text have been extended, or replaced by accounts taking cognizance of new facts and ideas: these appear in Part II and refer to toxoglossan alimentary anatomy and feeding, the endocrine, excretory and vascular systems, much of the ecology, and prosobranch classification and relationships. Where such extensive rewriting was not necessary some minor additions or connections are enclosed in square brackets [ ].

Part II (chapters 23 to 35) includes new material based on papers which have appeared since about 1961 up to 1990, occasionally later. Most chapters include a list of cross references to other chapters in this Part.

References in Part I appear in the first bibliography; those in Part II in the second. A very few, judged necessary for understanding, appear in both.

The reader will find basic facts in the chapters of Part I, and these may be supplemented and updated by reference to Part II. To facilitate this, the list of figures and the index both cover the entire text.

Names used in this book

The names of genera and species which were used in the first edition of this book were those current at the time that it was written. Since then revisionary work on taxonomy has changed many. It therefore seems proper to offer an alphabetical list of the names employed in the first edition (which are retained in the first part of the following pages) together with a list of synonyms where such changes have taken place. These include the names used in the second part of the following pages, together with those to be found in some other publications dealing with molluscs. It should be emphasized that this list makes no claims to be giving the 'correct' names of genera and species, only those used by various authors. The first name in each entry is that used in the first edition.

Aclis supranitida : *A. minor*
Acmaea tessulata : *A. testudinalis* : *Tectura testudinalis*
Acmaea virginea : *Tectura virginea*

Alvania beani : includes *A. calathus* and *A. reticulata*
Alvania carinata : *A. striatula*
Alvania crassa : *Manzonia crassa*
Aporrhais pespelicani : *A. pespelecani*
Balcis alba : *Melanella alba*
Balcis devians : *Vitreolina philippii*
Barleeia rubra : *B. unifasciata*
Bythinella scholtzi : *Marstoniopsis scholtzi*
Caecum imperforatum : *C. trachea*
Calliostoma papillosum : *C. granulatum*
Cantharidus clelandi : *Jujubinus clelandi* : *Clelandella clelandi*
Cantharidus exasperatus : *Jujubinus exasperatus*
Cantharidua striatus : *Jujubinus striatus*
Carinaria mediterranea : *C. lamarcki*
Cerithiopsis jeffreysi : *C. pulchella*
Chrysallida spiralis : *Partulida spiralis*
Cingula alderi : *Obtusella intersecta*
Cingula cingillus : *C. trifasciata*
Cingula proxima : *Ceratia proxima* : *Onoba proxima*
Cingula semicostata : *Onoba semicostata* : may also include *Onoba aculeus*
Cingula semistriata : *Alvania semistriata* : *Putilla semistriata*
Cingula vitrea : *Hyala vitrea* : *Onoba vitrea*
Cingulopsis fulgida : *Eatonina fulgida*
Clathrus clathrus : *Scala clathrus* : *Epitonium clathrus*
Clathrus turtonis : *Scala turtonis* : *Epitonium turtonis*
Diodora apertura : *D. graeca*
Emarginula conica : *E. rosea*
Emarginula reticulata : *E. fissura*
Eulima trifasciata : *E. bilineata*
Eulimella macandrei : *E. scillae*
Eulimella nitidissima : *Ebala nitidissima*
Hymeniacidon sanguinea : *H. perleve*
Ianthina : *Janthina*
Lacuna vincta : *L. divaricata*
Laminaria cloustoni : *L. hyperborea*
Littorina littoralis : *L. obtusata* (may cover *L. aestuarii* and/or *L. mariae*)
Littorina neritoides : *Melarhaphe neritoides*
Littorina saxatilis : may include any or all of the segregates *arcana, neglecta. nigrolineata, rudis,*
    tenebrosa
Lora elegans : *Oenopota elegans*
Lora turricula : *Oenopota turricula*
Mangelia coarctata : *Cytharella coarctata*
Menestho clavula : *Liostomia clavula*
Menestho divisa : *Evalea divisa* : *Ondina divisa*
Menestho obliqua : *Evalea obliqua* : *Ondina obliqua*
Nassarius incrassatus : *Hinia incrassata*
Nassarius reticulatus : *Hinia reticulata*

Natica alderi : N. poliana : Lunatia alderi : Polinices alderi
Natica catena : Lunatia catena : Polinices catena
Natica pallida : Lunatia pallida : Polinices pallida
Ocenebra aciculata : Ocinebrina aciculata : Tritonalia aciculata
Odostomia eulimoides : Brachystomia eulimoides
Odostomia lukisi : Brachystomia lukisi
Odostomia scalaris : Brachystomia rissoides
Patella aspera : P. ulyssiponensis
Patella intermedia : P. depressa
Patina pellucida : Helcion pellucidum
Pelseneeria stylifera : Stilifer turtoni
Philbertia asperrima : Raphitoma asperrima
Philbertia gracilis : Comarmondia gracilis
Philbertia leufroyi : Raphitoma leufroyi
Philbertia linearis : Raphitoma linearis
Philbertia teres : Teretia teres : Raphitoma anceps
Potamopyrgus jenkinsi : P. antipodarum : P. badia
Rissoa inconspicua : Pusillina inconspicua
Rissoa lilacina : may include R. porifera and/or R. rufilabrum
Rissoa membranacea : Rissostomia membranacea
Rissoa parva : R. interrupta : Turboella parva : T. interrupta
Rissoa sarsi : R. albella : Pusillina sarsi
Triphora perversa : T. adversa (may also refer to T. pallescens and/or T. similior)
Tritonalia aciculata : Ocenebra aciculata : Ocinebrina aciculata
Trophon muricatus : Trophonopsis muricatus (may include T. barvicensis)
Turbonilla elegantissima : T. lactea
Turbonilla jeffreysi : T. scalaris
Turbonilla fenestrata : Tragula fenestrata

For a long time malacologists have been in the habit of using the ending -acea to indicate a superfamily. The International Commission on Zoological Nomenclature has, however, recently recommended that the ending should be -oidea, rather than -acea. Though this is only a recommendation (ICZN 29A) rather than a requirement, today's recommendations tend to become tomorrow's requirements; we have therefore written superfamily names throughout Part II to conform with the recommendation.

LIST OF ILLUSTRATIONS

LIST OF ILLUSTRATIONS

xix

INTRODUCTION

T HE Mollusca are one of the great groups of the animal kingdom. If we measure the
success of a stock of animals by paying attention to the number of individuals, the size
to which they grow and the variety of different modes of life to which they have
adapted themselves, then it is clear that not more than three or four of the numerous
groups of animals known to the zoologist have proved outstandingly successful. Undoubtedly the most markedly so are the arthropods—the crustaceans, insects, spiders, centipedes,
extinct trilobites and similar creatures; another successful group—which we inevitably, tend
to regard as the highest of all—is the vertebrates, in which fishes, amphibians, reptiles, birds
and mammals are included; and a third is the phylum Mollusca, to which belong all the
animals which are dealt with in this book and many more.

The word Mollusca, which is derived from the Latin *mollis*, meaning soft, may seem at first
sight a curious name for animals one of the outstanding characteristics of which is a hard,
calcareous shell, sometimes, as in the case of the common mussels and cockles, completely
encasing the body. This apparent misnomer originated historically in that it was first used by
the French naturalist Cuvier in 1798 for the squids and cuttlefish, a group of the Mollusca
amongst living forms of which the shell is always (except in the pearly nautilus) either reduced
and covered by the soft flesh of the animal or is altogether wanting. Since then, the relationship of the more familiar snails, slugs and bivalves to the squids and cuttles has been made
certain and the name extended to them as well. Despite its seeming unsuitability, however, it
was a fortunate accident which led to the group being named so as to emphasize the
importance of the soft parts rather than of the shell, because although the latter is of
importance to the palaeontologist—is, in fact, the only part of the animal which is usually
fossilized and so preserved for his inspection—it is the soft parts which are of outstanding
interest to the zoologist, whether he be anatomist, physiologist, or, simply, naturalist.
Paradoxically it is the soft parts which at the start produce and mould the shell within which
they are later to lie; it is the soft parts which trim away unwanted or obstructive portions of
shell which inhibit growth or development, and it is, finally, the extraordinary plasticity of the
soft body of the molluscs which has allowed them to become adapted for life in a great
number of different ecological niches.

The ancestral mollusc must have appeared on earth many millions of years ago, too early
for any fossil record of it to have been preserved, because by the time the oldest fossiliferous
rocks were laid down, evolution had already produced molluscs of many different patterns.
What this ancestral type may have looked like, therefore, can only be deduced from a
study of these fossils and of present day forms according to the laws of comparative morphology and embryology, guides of a certain degree of trustworthiness but not by any means
infallible. It would seem, on their basis, that it was marine in its habitat, creeping over the
surface of the shores or sublittoral regions of an archaic sea, and had evolved from a group
related to the same stock as that from which arthropods and annelids are derived and
originally arising from a turbellarian-nemertine ancestry.

The turbellarian-nemertine group consists of unsegmented, acoelomate animals which, in the case of the flatworms, glide, partly by ciliary and partly by muscular action, over the substratum. The body is flat and thin in a dorsoventral direction so that the viscera have to be packed away in the limited space between the upper and lower surfaces. This shape is connected with the fact that it has neither blood system nor special respiratory organs, and the alimentary canal of the larger polyclads and triclads branches repeatedly and reaches into all parts of the body, supplying them with food; in the smaller rhabdocoels it may remain a simple sac. The leaflike shape allows oxygen to diffuse into its deepest parts. Although the brain and the major sense organs are located at one end, which normally goes first and so is the head, the mouth is often placed on the ventral surface some way behind; but there are again many smaller forms in the Rhabdocoelida in which it lies anteriorly. There is no other opening to the gut at all.

These anatomical features limit the turbellarians to a low level of metabolism and this is to some extent exaggerated by functional peculiarities. The lack of an anus prevents a continuous stream of food from passing along the gut and imposes an alternation of inward and outward movements, and it appears from the investigations of Westblad (1922) that during digestion the walls of the gut fuse with one another across the cavity so as to convert it into a solid spongy mass of tissue within the spaces of which the food is digested. When the process is completed the vacuolated syncytial mass breaks down to form a hollow tube with epithelial walls once more and the indigestible residue of the meal is left in the lumen for egestion. This essentially intracellular digestive phase clearly imposes a kind of paralysis on the digestive system and, along with the lack of respiratory organs and vascular system, limits rather strictly the size and complexity which the turbellarians have been able to reach in their evolutionary history.

The outermost layer of the turbellarian is a richly glandular and ciliated epidermis, resting on a layer of connective tissue. Many of the glands associated with the skin lie embedded in this connective tissue and discharge their secretion through long necks which lie between the epithelial cells. In the connective tissue, too, run muscle strands, some circular in direction, some longitudinal, some crossing from the upper to the lower epidermis and collectively forming a complex network of fibres differing a little in the details of its arrangement from one part of the body to another. The connective tissue also forms the matrix (parenchyma) in which the gut, the nervous system and the reproductive organs are embedded. The nervous system is centred in two cerebral ganglia placed in the head as dorsal enlargements on a ring of nervous tissue encircling the gut in those forms which have the mouth placed anteriorly, but not related to it in others. To them run sensory nerves from receptors, tactile, chemical and visual, located at the anterior end of the body, and from them arise cords which pass posteriorly. These are better developed on the ventral side, in accordance with the biological principle of neurobiotaxis, which asserts that nerve cells migrate inside an animal towards the region from which they receive most stimulation, and a nerve-net ramifies in relation to them and the musculature of the body underneath the epidermis. The movement of the animal, effected and controlled largely by this nerve-net under the direction of the cerebral ganglia, involves the simultaneous discharge of secretion from the numerous gland cells to produce a layer of mucus over which the animal glides, the rhythmic contraction of the muscles of the body and the beating of the cilia of the epidermal cells. In different circumstances the effect of the cilia may predominate or the effect of the muscles, but the secretion of the cutaneous glands always appears to be important.

The excretory organs are 'flame cells' or protonephridia, running in the parenchymatous connective tissue and discharging to the exterior.

The turbellarians, like all the platyhelminths, are hermaphrodite, and the fertilized eggs undergo spiral cleavage to give rise (in those marine forms in which a free-swimming stage occurs) to a larva of trochophoral pattern. When fully formed this is an ovoid body with a number of ciliated lobes projecting from the equatorial region by means of which it swims. A mouth opens on the ventral side leading to a blind pouch, which is the archenteron or beginning of the gut. Between body surface and gut wall is a cavity, the primary body cavity or blastocoel, within which lies a number of cells which are the rudiments of the parenchymatous material found in the adult animal. Some of these cells are derived from the ectodermal cells of the body wall and are therefore ectomesoderm. In animals which exhibit spiral cleavage true mesoderm, such as is formed from the teloblasts of annelids and is responsible for the formation of the coelom, is derived normally from cell 4d (p. 384). In the turbellarians and nemertines, however, though this gives rise to endoderm, some mesenchyme, some muscles and the genital organs, no coelom sacs are formed and as a consequence the turbellarians and nemertines are acoelomate. The ducts by which their genital products are conveyed to the exterior may, nevertheless, be regarded as coelomoducts and the cavity within them as coelomic, but no part of the cavity between epidermis and gut wall is of this nature. Their excretory protonephridia, too, are to be regarded as straightforward ectodermal ingrowths without any coelomic connexions.

The nemertine members of the platyhelminth assemblage of invertebrates have advanced upon the turbellarians in a number of respects, all of which may be associated with a more active mode of life. These advances give rise to animals with bodies which mimic those of annelid worms, though they are not metamerically segmented, and have probably arisen because of the adoption by both groups of similar ways of living. The body has become worm-shaped and the alimentary canal has acquired an anus, placed terminally at the posterior end, and digestion has become partly—if not wholly—an extracellular process. Both these changes mean that a continuous stream of food can pass through the gut without the pauses for digestion and egestion which have to occur in the turbellarian.

The organization of the body wall has altered, probably in connexion with the more active wriggling and creeping by means of which nemertines move, although it still retains a ciliated and glandular epidermis. Within this there now lie, however, definite layers of circular and longitudinal muscles. The brain has enlarged, as might be expected to occur in a more active animal and two of the longitudinal nerves, one on each side of the body, are of much greater importance than the others. They are still in relation to a well developed nerve-net in a subepidermal position. The departure of the body from a leaf-like shape and the adoption of a cylindrical one have been permitted by the invention of a vascular system, neither extensive nor probably very efficient, but certainly permitting some transport of food and respiratory gases. The nemertines have become distinctly specialized in their invention of a proboscis as a special food catching organ which is armed, at least in some cases, with chitinous stylets for piercing and gripping the body of their prey.

Many changes were involved in the transformation of an animal with this organization into any to which the label mollusc could be given unreservedly. According to the recent extensive and detailed researches of Salvini-Plawén (much of his earlier work summarized in his review of 1985) some stages in this evolutionary history may be paralleled in the organization of living genera. In what are presumably the historically earlier stages a number of basic molluscan features were established in what was still essentially a planarian type of body, though specializations adapting the animals to their particular mode of life were also present. Later forms show a structural pattern more advanced and, from its resemblance to that familiar from the study of living forms and their adaptations, recognizably molluscan.

If we follow Salvini-Plawén (1985) the first molluscs (fig. 1) still resembled flatworms in their external features, being more or less flattened, and creeping on their ventral surface. At the posterior end of the body there had developed a depression (the mantle or pallial cavity), its walls formed from flap-like extensions of the body wall (the mantle skirt), and containing that collection of structures known as the pallial complex: a pair of ctenidia (or at least respiratory areas), paired chemosensory organs, the paired openings of excretory and genital ducts, and a median anus, showing that the animals had escaped from the limitations of the blind platyhelminth gut. With this particular group of contents this is clearly a molluscan mantle cavity, but, unlike that of modern forms, it shows only a slight tendency to extend forwards along the sides of the body. Anteriorly the ventral surface, like that of a flatworm, formed a ciliated and richly glandular surface on which the animal could move. There was no distinction here of head from foot, no tentacular outgrowths, and the mouth opened to this surface near, or perhaps at, its anterior end, leading to a buccal cavity armed with a series of radular teeth. Dorsally the body was covered with a fur of spicules of aragonite, and was without a continuous shell.

Animals at a comparable grade of organization—though it is difficult to be sure how much of it is truly primitive and how much simply adaptive—are represented amongst living molluscs by a group of animals called caudofoveates which are largely unknown to the general zoologist. Most burrow into soft marine substrata and in adaptation for this mode of life have become worm-like and circular in transverse section, losing any creeping sole that may have been present in their ancestors. The cup-shaped mantle cavity is terminal, has no anterior extensions, and its opening, surrounded by mantle skirt on all sides, lies flush with the surface of the substratum. This restriction of the cavity prevents the entry of particulate matter, an event avoided by all burrowing molluscs.

At a later stage in evolution, with adaptation to a more active mode of life, the pallial grooves extending the mantle cavity anteriorly grew progressively further forwards under the cover of a projecting mantle skirt, until right and left grooves met anteriorly, so forming a complete peripheral pallial groove and isolating the central area of the ventral surface, which formed the head-foot. Later head and foot became separated by a groove, allowing each to develop its own specializations. Accompanying these changes was a series of others affecting internal anatomy, of which only a few need be mentioned here. The radula developed a more complex and diversified series of teeth which adapted it for dealing with a variety of foods, though these were probably still, as in more primitive types, bacterial and algal mats and small, soft macrophytes. Associated with this the glandular equipment of the gut (salivary, oesophageal, and the main gastric digestive glands) enlarged, as did also the kidneys, which were not obvious in simpler forms, and the intestine elongated. The result was a marked increase in the volume of the viscera relative to that of the head-foot, leading to a swelling of the dorsal part of the body in which they were lodged, forming a visceral mass or hump, so that the animals departed for good from the ancestral flattened shape. To protect this mass there developed a continuous calcareous shell, replacing the earlier spicular covering except sometimes in a few localized parts of the body.

The animal which has been created from a hypothetical flatworm-like ancestor lived a littoral or sublittoral life, its shell giving a measure of protection against waves and predators. It breathed by means of the gills in the mantle cavity; it crept about on the sole of the foot,

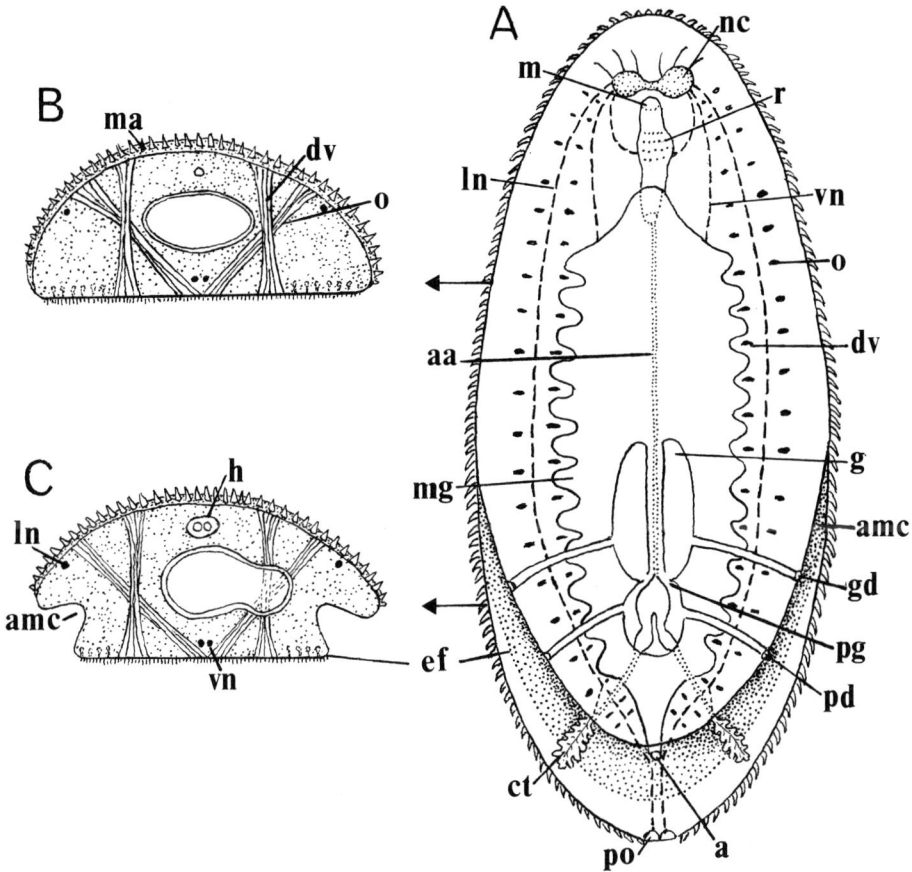

FIG. 1.—A, Dorsal view of possible molluscan archetype, modified from Salvini–Plawén (1985). B, C, sections of A at levels indicated.

a, anus; aa, anterior aorta; amc, anterior extension of mantle cavity; ct, ctenidium; dv, dorsoventral muscle bundle; ef, edge of foot; g, gonad; gd, gonoduct opening to mantle cavity; h, heart in pericardial cavity; ln, lateral nerve cord; ma, mantle with cuticle and irregularly arranged calcareous spicules; mg, pouched mid gut; nc, nerve centre; o, oblique muscle bundle; pd, pericardial duct opening to mantle cavity, perhaps with excretory tissue; pg, pericardial connection with gonad; po, primitive osphradium; r, radula; vn, ventral nerve cord.

which was sufficiently broad and flat to prevent dislodgement by waves and currents; and it fed on minute plant or detrital material raked into the gut by the radula. It now possessed all the essential basic characters of the molluscs, which may be enumerated thus:

(1) a body of triploblastic but acoelomate organization;
(2) the body divisible into head, foot, and visceral mass;
(3) the visceral mass covered by a shell secreted by the mantle, that is, the skin of the visceral mass;
(4) a radula in the buccal cavity;

(5) a mantle cavity under a fold of mantle and shell in which gills lay and to which the anus, excretory and genital organs discharged.

Perhaps the nearest, though still a distant, approach to the ancestral gastropod amongst living molluscs is the monoplacophoran *Neopilina*, whose discovery was announced by Lemche in 1957. Its anatomy was described by Lemche & Wingstrand (1959) and a revised account given by Wingstrand (1985). Related animals have since been recorded from many places and several habitats (Clarke & Menzies, 1959; McLean, 1979; Warén, 1988) and studied alive (Lowenstam, 1978). The organization of *Neopilina* proved to be such that it could not be classified in any of the groups of living molluscs then known but only in the class Monoplacophora Knight, 1952, at that time represented only by fossils.

The main characters of *Neopilina* and its living relatives are a limpet-like shell showing no spiral coiling (Wingstrand, 1985), which covers the animal completely. There is a small central foot, not very muscular, surrounded by a broad pallial groove in which lie the head (anteriorly), the anus (posteriorly), and five pairs of gills (laterally) each with an axis bearing some lamellae. The point which caught the attention of malacologists when this description was first published was the suggestion of metamerism given by the gills. The idea was strengthened by some internal features, in particular the five pairs of muscles, with their accompanying nerves, running from the shell to the foot, and the apparently similar arrangement of the excretory organs. Not only did these features suggest that molluscs were metamerically segmented, though short-bodied, but the interpretation of five pairs of spaces (said to connect with the excretory organs) as coelom sacs suggested that they were also coelomate, and made a link with the annelid-arthropod line (already indicated by their shared spiral cleavage pattern) even closer than had been supposed. Wingstrand's (1985) reworking of the *Neopilina* material, however, showed that the so-called coelom sacs were parts of the oesophageal glands, and left some doubt about the anatomy of the excretory system. Together with other critical comments, summarized by Salvini-Plawén (1985), these findings greatly weakened the arguments for metamerism, and it is fair to say that while the origin of the phylum from some stock at more or less the platyhelminth level is now generally accepted, it must have been at a low level amongst the proto-articulates and quite distinct from the line leading to annelids and arthropods. The apparent metamerism of molluscs is to be explained as incidental to other evolutionary events or anatomical arrangements and not as a basic characteristic.

The discovery of living monoplacophorans, however, has aroused interest in the animals and the part which they may have played as a basal stage in the evolution of the various molluscan classes, a topic explored further by Stasek (1972). Whilst the role of monoplacophorans as molluscan pioneers may be acceptable, argument still persists as to whether the type of monoplacophoran ancestral to gastropods was a tergomyan (a limpet-like animal, possibly bilaterally symmetrical, with several pairs of dorsoventral shell muscle bundles) or a cyclomyan (with a dorsal coiled shell and only one pair of shell muscles). Many of the proposed theories of the origin of torsion favour the latter (e.g. Pojeta & Runnegar, 1976) whilst others (e.g. Ghiselin, 1966) start from the former.

In the evolution of prosobranchs there is one important departure from the body plan just described which is both essential for their mode of life and as a prerequisite for that peculiar characteristic which they exhibit, torsion, the anticlockwise rotation through 180° of the visceral hump on the head-foot so that the mantle cavity and the contained pallial complex moved from a position over the back end of the foot along the animal's right side to an anterior one over the head. The reasons for this change remain arguable and there are

B

A

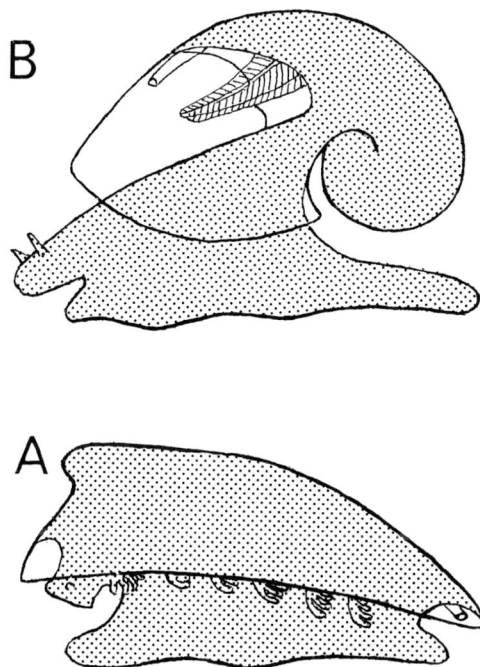

FIG. 2.—Diagrams to show the relationship between head-foot and visceral mass in A, a tergomyan monoplacophoran; B, a prosobranch gastropod. Note the narrowness of the connection between head-foot and
visceral mass in B. In both diagrams the roof of the lateral part of the mantle cavity is not shown.

probably several—shelter for the head-foot, balance of the heavy, shelled visceral hump on
the head-foot as the animal creeps, better ventilation of the respiratory organs, or some larval
advantage. In modern gastropods the mechanism by which torsion is brought about may be
an asymmetry of muscles running from shell to head-foot, or growth, or phases relying on first
one and then the other of these activities. However, unless the whole change has been
brought about by growth, a slow process with no very obvious advantage to be gained from
intermediate positions, one feature of organization is required which is not present in that
described above: this is a relatively narrow stalk linking the visceral mass dorsally with the
head-foot ventrally (fig. 2B), which can absorb (and will exhibit) the twist introduced by
torsion. This is a permanent anatomical feature of many adult prosobranchs and its temporary formation is to some extent within the power of at least some monoplacophorans, as
illustrated by Lowenstam (1978) in *Vema* (fig. 2A). It is, however, almost universally present in
prosobranch larvae, even if it is obliterated by later growth. This may, indeed, be the main
reason why torsion is usually timed to be initiated during this stage of the life history.

The situation regarding the origin of gastropods has been modified by our recently much
increased knowledge of molluscan fossils dating from early Cambrian times. These are mostly
small, the shell only 1–5 mm long, but those known nevertheless include representatives of
three classes: evidently differentiation of the phylum had begun in Precambrian times, when

the animals were possibly still smaller. The implications of such small size for the anatomical organization of the animals (in particular the pallial arrangements and the need for gills) and for the shell (whether it was mineralized or not, whether needing to be balanced over a head-foot or merely a protective dorsal cover for a flattish body) have not yet been properly worked out though the work of Haszprunar (1992) has contributed significantly to them. And since the position of the mantle cavity, the need to balance a mineralized shell, and possible asymmetry of the shell muscles have appeared to many malacologists to lie at the heart of the torsion problem, the varied and numerous theories which have attempted to explain its occurrence must all still be regarded as very much on approval. It is doubtful, for example, whether any theory attributing its origin to larval stages (Garstang, 1928; Crofts, 1937; Underwood, 1972) has demonstrated real larval as distinct from possible adult benefits, and whether the animals which first underwent torsion had problems with shell balance which it solved. Some 'problems' said to have been solved by torsion may not have been problems at all to such small organisms.

Though the anterior mantle cavity is the most conspicuous part of the prosobranch pallial groove the rest of it is not lost: it remains as a shallow groove underneath a brief mantle skirt encircling the lateral and posterior faces of the stalk linking visceral mass and head-foot. This is now mainly occupied by the muscles running from the one part to the other. One might, indeed, speculate that the helicoid spiralling of the visceral hump which characterizes most prosobranchs was advantageous not just because it allowed a compact arrangement of the growing volume of the viscera which increased size and activity demanded, but also because it did so in such a way as to facilitate the development of such a stalk.

Further evolution of the prosobranch line brought into being the fundamental features of modern gastropod organization. The new position of the mantle cavity allowed the withdrawal of the whole head-foot into its shelter, where it was protected by the shell against predators and disadvantageous environmental conditions, and by the development on the posterior end of the foot, the last part to be pulled into the shelter of the shell, of a horny plate, the operculum. Enlargement of the cavity has allowed it to act thus as a compensation sac, and has also allowed, or accompanied, the increase in size of the respiratory organs which was essential both for greater activity and for the needs of a body which was growing larger. The head, no longer tied to the main locomotor organ, showed a marked process of cephalization, becoming larger, more muscular, freer in its movements, and provided with tentacles carrying tactile, olfactory, and visual sense organs for testing the environment. Part of its increased size reflected increase in the complexity of the radula and the odontophore on which it was carried, associated with the developing need for more, and more nutritious, food. Motility was improved by the development of a foot more powerful both in its movements and in its powers of adhesion, allowing the animals to live in rougher and more exposed habitats than had previously been possible.

In addition to torsion another organizational feature has played a great part in the evolution of gastropods: this is the spiral coiling of the visceral hump, a process, it must be emphasized, totally distinct from torsion. Some such coiling was present in pretorsional forms and it may have been an ancient character of the phylum since some is visible in other classes. Once the two features coexisted, however, they seem to have interacted and together were, at least in part, responsible for much of the later evolutionary change. Coiling of the visceral hump may have arisen simply as a device reducing its surface in relation to its volume, so easing locomotion. At this stage, with equal growth on the two sides of the body, the result would have been a plane spiral, as is exhibited by the extinct bellerophonts and, amongst living forms, many planorboid pulmonates (though in this case it is certainly a secondary condition). At a later stage, probably because it improved carriage of the shell on the head-foot after torsion, growth of the two sides of the mantle skirt and visceral mass became unequal, giving the helicoid spiral shell so typical of living prosobranchs.

Normally it is the post-torsional left side which grows more rapidly. Because of this differential growth rate there is much more space available for the accommodation of such organs as gill, osphradium, and the like, which lie on the left than there is for their partners on the right, and this results in the right gill and other organs being often smaller than those on the other side of the body. The more tightly the visceral mass coiled the more compact the gastropod body became, but the greater was the compression of the right side which this entailed; the greatest compression compatible with proper functioning, in fact, becomes possible only with complete loss of the organs on the right. There can be distinguished two grades of prosobranch mollusc exhibiting torsion: a more primitive, in which right and left sets of pallial organs are both present, though there may be differences in size between the two (the Archaeogastropoda or Diotocardia) and a derived grade in which the bulk of the right set has disappeared (the Caenogastropoda or Monotocardia).

Though this account of the loss of organs on the right side has received some general acceptance it cannot be a full explanation because not all the organs on the right side disappear, and some on the left do. In all gastropods there is only one gonad, the post-torsional right, and it is the left that has gone. The archaeogastropods retain and, indeed, expand the right kidney whilst the post-torsional left is reduced, sometimes markedly. It therefore looks as if in addition to the effects of reduction of the right side another factor is involved, the need to ration space within the visceral mass, where kidneys and gonads were primitively placed. The reasons which determined that this was achieved by loss of the left gonad and reduction of the left kidney were probably functional. The right kidney in early gastropods was the main excretory organ, the left already differentiated for other activities; the former, therefore, had to remain in the visceral mass, the latter was moved into the mantle skirt where, indeed, it functioned better [see chapter 27]. The right gonad was that which persisted because of its relationship with the only kidney remaining in the visceral mass.

Though it is not possible to be dogmatic as to whether the primitive gastropods were more limpet-like or more topshell-like (and both shapes abound in the archaeogastropods) it seems likely that limpet organization may well be a secondarily simplified condition adapted for the rock-clinging mode of life which these animals exhibit. Certainly the topshell pattern is the one from which caenogastropod organization has been derived.

Caenogastropods include an immense variety of species many of which are littoral in their habitat, but they live a more active life than the archaeogastropods. Most are vegetarians, like winkles, but some, like neogastropods, are carnivores. Of British forms a few, like *Melarhaphe neritoides*, live so high on the beach that they are practically terrestrial and two species, *Pomatias elegans* and *Acicula fusca*, are completely so, being found in hedge bottoms and amongst dead leaves on chalky ground in such places as the North Downs and the Chiltern Hills. A few become at home in fresh water and may, like *Bithynia tentaculata* and *Potamopyrgus jenkinsi*, abound in appropriate places. Many marine forms, particularly neo-gastropods, have become sublittoral. More surprisingly, a few have become adapted for pelagic life, like the violet snail *Janthina* which keeps itself at the surface of the sea by means of a float of air bubbles entangled in mucus from the same glands of the foot as produce the slime over which more ordinary snails creep; or like the heteropods *Atlanta*, *Carinaria*, and *Pterotrachea*, which reduce the visceral mass and shell to negligible proportions and inflate the rest of the body to form a swollen, gelatinous mass which allows them to remain at the surface of the sea.

The groups of gastropod molluscs which have just been mentioned, the Diotocardia (= Archaeogastropoda) and Monotocardia (= Caenogastropoda), show the full consequences of the process of torsion. As a result their gills, which in their untwisted ancestors lay behind the heart, projecting into a backwardly facing mantle cavity, now face forwards and lie anterior to the heart. For this reason the two groups are united in a single primary division of the class Gastropoda to which the name Prosobranchia has been given; because torsion also introduced a twist into the visceral loop of the nervous system this group is also known as the Streptoneura. Throughout this prosobranch group runs a tendency to relinquish the rock-clinging habit and to adopt a freer mode of life. A still more mobile life would undoubtedly be possible were the shell to be lost and the animal become slug-like. Clearly this cannot happen without some alternative protection being given to the visceral mass: this has been effected in some gastropods by the expansion and hollowing of the foot to provide a cavity into which the visceral mass might be sunk, so that it no longer appeared as a projection from the animal's back. At the same time as these changes were taking place a third also occurred: a process by which some approach to the original untwisted disposition of the parts of the body was achieved. This may be described as 'detorsion' though to what extent it has involved a true detorsion or merely an adjustment of parts through differential growth may be debated. Which of all these events is the primary change and which merely its concomitants is difficult to say: each probably contributes its own significant share to the success of the total change. The final result of this evolutionary sequence is an unshelled slug-like animal, at least superficially bilaterally symmetrical, though different kinds of animal show various degrees of reduction of shell, incorporation of the visceral mass in the foot, and of 'detorsion'. These gastropods, because the 'detorsion' which they have undergone has restored the original relationship of gill and heart, are called the Opisthobranchia. They have all, without doubt, been derived from some monotocardian ancestor, a statement which it is possible to make because of the fact that all, whilst embryonic or larval, present a condition akin to that produced by torsion which is undone in the later stages of development, and also because they do not possess any pallial organ belonging to the post-torsional right, a state of affairs found only in monotocardians.

To complete this summary of the adaptive radiation of the gastropods it is necessary to mention a third group, the Pulmonata, also derived from a monotocardian ancestry and related to opisthobranchs, a fact reflected in the inclusion of both in the single taxon Euthyneura, a name indicative of the loss (though by different routes) of the twist in the visceral loop. The facies of the pulmonate body is that of the monotocardian: in most a large visceral hump covered by a stout calcareous shell is still present, and there has been no detorsion, so that the mantle cavity still lies above the animal's head and faces more or less forwards. The cavity (or perhaps a new outgrowth from it) has been converted to a lung and the original gill completely lost; the members of this group, which includes the common garden and most freshwater snails, breathe air. The shell is retained as a protection against desiccation and osmotic changes, but, as in the other gastropod evolutionary lines, there has been repeated in pulmonates the same trend towards a naked body, a visceral hump, not coiled, and pulled into a foot blown out to make room for it, and a slug-like shape: this gives rise to one of the most successful of all molluscan groups, the land slugs.

Boss (1971) estimated the number of different kinds of living molluscs known at 5×10^4, concluding that 80% of these were gastropods and many of these prosobranchs. This number includes recent finds of animals, many with novel patterns of structure, broadening still further our conception of what this amazingly versatile group of animals has done in the course of its radiation. The number of genera recognized by taxonomists, however, has probably risen relatively faster than the number of newly discovered prosobranchs as a result of the current tendency, sometimes validly based on more detailed knowledge, sometimes not, to destroy broad generic groupings and raise previous subgenera to generic level. Thus in 1865 Jeffreys, in his classical work 'British Conchology', could place all of the fifteen species of topshell which he described within the one genus *Trochus* Linné, 1758: a current check-list would allocate them to the six genera *Margarites* Gray, 1847, *Solariella* Wood, 1842, *Gibbula* Risso, 1828, *Monodonta* Lamarck, 1801, *Jujubinus* Monterosato, 1884, and *Calliostoma* Swainson, 1840, and modern taxonomists restrict the genus *Trochus* to some Indo-Pacific species. Since all these genera, *Jujubinus* apart, had long been established before Jeffreys wrote, his inclusion of British topshells in a single genus reflects his attitude to classification rather than ignorance, but it is a good illustration of today's multiplication of names at the generic level. To a certain extent a similar realization that what were regarded as single, though perhaps rather variable species were in reality groups of closely related but distinct taxa has increased the number of recognized species. The best known local example of this is in the genus *Littorina*. If once again we turn to Jeffreys (1865) we find four species listed, where today at least double that number would be accepted. As with *Trochus* many of these species had already been described before Jeffreys wrote: he therefore appears as a confirmed 'lumper'. The same is true of his treatment of *Patella*, of which he lists only the one species, though at least three malacologists had previously concluded that there were three species of *Patella* in the north-eastern Atlantic.

Several other events have helped to raise the number of taxa. On the one hand there is the increased exploration of areas such as the deep sea, previously reachable only in a very limited, haphazard, and indirect way by means of the laborious and time-consuming method of dredging, which brought all animals dead to the surface. Now manned submersibles allow detailed collecting and direct observation of the living animal in its proper habitat. Scuba diving has encouraged a comparable change in the collecting and observing of animals in the shallow depths in which it is practicable. A much more detailed examination of shells (still often all that is available of a particular species) has become possible with use of the scanning electron microscope and made differences and similarities, particularly in the systematically valuable protoconch, obvious enough for any eye to see. Further, the increased availability of some animals, little and inadequately studied before, has made their organization better known. Lastly, a more precise comparison of species is offered by the biochemical approach, through which at least the enzyme equipment of species is known and by which, in the long run, their DNA or the amino acid sequences of their proteins may be defined.

A marked change has taken place in recent years in the thrust of ecological work, of which two facets in particular may be mentioned. Whereas the main effort of the ecologist was once primarily, and in what was at that time a burgeoning discipline properly, directed at the autecology of the species in which he was interested, it would now be aimed at the synecology of the community of which that species was part, and its autecology re-interpreted or extended in the light of what that study revealed. Prey-predator relationships, animal-plant interactions, the sharing of resources within a given area by species which might otherwise be directly competitive, the life history and reproductive methods best adapted for each animal's success, demographic analysis of populations, along with the reverberating effects of all the findings throughout the community, are now prominent fields of ecological study. A second change is in the attention which ecologists now pay to the physiological activities of the animals which they study, to the flow of energy through trophic systems, and the share of this that each species possesses as determined by measurements of its energy budget. Since these studies are productive of meaningful results only when large numbers of observations can be made, nearly all work of this nature has been carried out on abundant animals, and particularly on those prosobranchs readily obtainable between tide-marks, such as limpets, winkles and dog whelks. Where such numbers are unavailable ecology still remains autecology.

The study of anatomy has produced less change than has the ecologists' approach to their work, so that the anatomical accounts written earlier are still largely valid, though recent work has shown that some require revision. The extension of experimental methods to the study of such systems as, in particular, those concerned with locomotion, endocrine secretion, circulation, and excretion, has brought much more information to light and rendered earlier accounts out of date. To a certain extent, too, study of the ultrastructure of some cells has produced more definite ideas of the function of some parts of the prosobranch body. Though studies of such sense organs as the osphradium have led to new ideas about prosobranch relationships, those of the central nervous system, largely because of the small size of its cells in comparison with those of opisthobranch and pulmonate gastropods, have not been so fruitful of ideas about its working as in those animals. Limitations of small size, too, have severely restricted work on the prosobranch circulatory and some other systems.

'British Prosobranch Molluscs', published in 1962, was being written at a time when the classification of gastropods most commonly used was that of Thiele (1929–1931) as later modified by Wenz (1938), and at a time when discussion of molluscan phylogeny had been all but non-existent for about half a century. The few years immediately before its appearance, however, contained two events which brought that period to an end. The first of these was Lemche's announcement of the discovery of Neopilina in 1957 and two years later his account with Wingstrand of its anatomy and the implications of that for the origin and evolution of molluscs. This provoked a vastly increased interest in molluscan origins, in their lower groups and in the relationships of these to the higher molluscan groups. It also began a drive which has continued with growing force for palaeontologists to interpret their fossil molluscs as living, functional organisms and not merely as convenient ways of dating rocks. Later discoveries of other deep-sea molluscs, comparable to that of Neopilina though less sensational, have sustained and augmented the interest in lower molluscan groups and in primitive gastropods.

The second event was the publication in 1960 of the first molluscan part of the 'Treatise on Invertebrate Paleontology'. Here Cox (1960b) attempted the first recent revision of higher gastropod classification, having previously published (1960a) some of the thoughts which had influenced him. It was a modest revision indeed, with the union of Thiele's Mesogastropoda and Wenz's Neogastropoda into the Caenogastropoda as its major change. Taylor & Sohl's (1962) classification, published as the first paper in the first part of the newly established journal Malacologia, presumably for the guidance of future contributors, was still largely the Thiele-Wenz classification. From then onwards, however, events have moved more rapidly. The classification of Golikov & Starobogatov (1975), reflecting some ideas from 'British Prosobranch Molluscs', was a much more novel and fundamental revision, introducing several new features as taxobases. Since then a flow of new ideas both about the phylogeny and evolution of the molluscs as a whole and of the gastropods in particular has come from Kosuge (1964, 1966), Climo (1975), the Austrian school [Salvini-Plawén, (1972, 1980, 1985);

Salvini-Plawén & Haszprunar (1987); Haszprunar (1985a, 1985b, 1985c, 1988a, 1988b); Rath (1988)], from Lindberg (1986, 1988), and from Ponder & Warén (1988). These workers have forced some totally new concepts, many based on the use of new characters, upon the attention of malacologists as regards the history and evolution of the animals which they study.

To some extent the grasp of these ideas has been made more difficult by the (perhaps over-hasty) invention and use of a host of new (and forgotten old) names for taxa higher than family level. As some of these, together with some possibly unfamiliar synonyms, are freely used in the following pages it may be helpful to give brief descriptions of them here in an alphabetical list. Some indication of possible relationships is given in fig. 3, and these are discussed in chapter 36.

ALLOGASTROPODA Haszprunar, 1985. A name for the rump of Habe and Kosuge's group Heterogastropoda after removal of those superfamilies which Haszprunar regarded as true prosobranchs (Ctenoglossa). The rump contains four Recent superfamilies: Architectoni-coidea, Pyramidelloidea, Omalogyroidea, and Rissoelloidea. These are still regarded as prosobranchs by Haszprunar (1988a) but are classified as an order of Heterobranchia (= Euthyneura) by Ponder & Warén (1988). Their placing of Pyramidelloidea should certainly be taken as correct, though that of the others may still be *sub judice*.

APOGASTROPODA Haszprunar, 1985. Haszprunar's name for a group containing all those prosobranchs which are not archaeogastropods. It is therefore equivalent to Caenogastropoda as used by Cox (1960a, 1960b). Haszprunar, however, restricts the term Caenogastropoda to the main prosobranch evolutionary line, assigning the Architaenio-glossa to the archaeogastropods, the Valvatidae to the Ectobranchia, and the others to the Allogastropoda. Ponder and Warén (1988) differ from Haszprunar in retaining Caeno-gastropoda as a larger group which includes Architaenioglossa but they transfer both Ectobranchia and Allogastropoda to the euthyneurans.

ARCHAEOGASTROPODA Thiele, 1925. Originally an order of the subclass Prosobranchia including all which have paired pallial organs, plus some others which do not. The latter include the docoglossan limpets and the nerites and their relatives. In this use it corresponds to Diotocardia Mörch, 1863, not all of which (docoglossans and nerites again) have two auricles. The term is now perhaps best used with a lower case initial letter as an adjective referring broadly to prosobranchs at a primitive level of organization rather than as the name of a taxon. Salvini-Plawén & Haszprunar (1987) and Haszprunar (1988a, 1988b), however, regard it as a useful paraphyletic taxonomic name and its familiarity may well ensure its use in this capacity for a long time. Their criterion for identifying an archaeogastropod is that it should possess a hypoathroid nervous system, that is one in which the pleural and pedal ganglia are closely associated and lie ventral to the oesophagus, remote from the cerebrals. On this view architaenioglossans must be treated as archaeogastropods, though they resemble them only in this one respect.

ASPIDOBRANCHIA Schweigger, 1820. An order of prosobranchs typically with two ctenidia, each with a double row of lamellae; approximately equivalent to Archaeogastropoda, to Diotocardia, and contrasted with Pectinibranchia.

CAENOGASTROPODA Cox, 1960*. Proposed as the name of an order of prosobranchs combining Mesogastropoda Thiele, 1925 and Stenoglossa Bouvier, 1887 (= Neogastropoda

*Though frequently quoted as 1959 [e.g. 'Treatise on Invertebrate Paleontology' (published 1960), Part I, p. I152; Ponder & Warén, 1988, p. 290] the publication date of *Proc. malac. Soc. Lond.* vol. 33, part 6, in which Cox's paper appeared (pp. 239–304), was February, 1960.

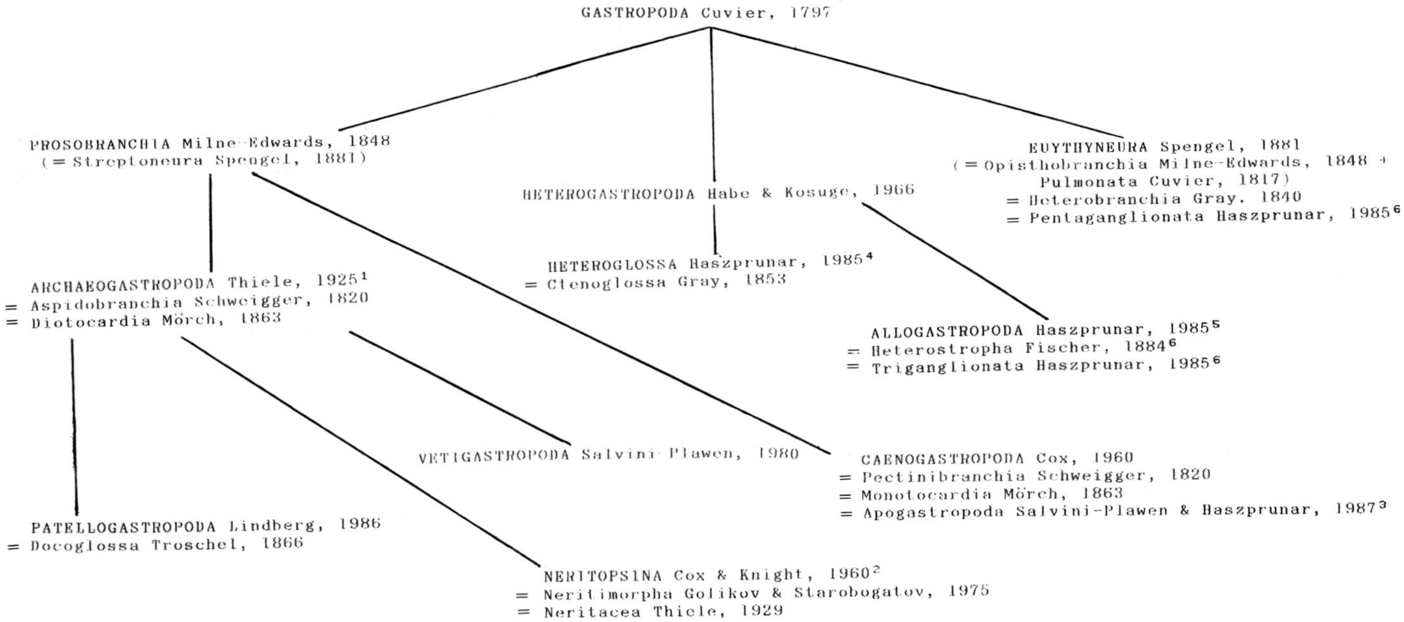

GASTROPODA Cuvier, 1797

PROSOBRANCHIA Milne-Edwards, 1848
(= Streptoneura Spengel, 1881)

EUTHYNEURA Spengel, 1881
(= Opisthobranchia Milne-Edwards, 1848 +
Pulmonata Cuvier, 1817)
= Heterobranchia Gray, 1840
= Pentaganglionata Haszprunar, 1985[6]

HETEROGASTROPODA Habe & Kosuge, 1966

HETEROGLOSSA Haszprunar, 1985[4]
= Ctenoglossa Gray, 1853

ALLOGASTROPODA Haszprunar, 1985[5]
= Heterostropha Fischer, 1884[6]
= Triganglionata Haszprunar, 1985[6]

CAENOGASTROPODA Cox, 1960
= Pectinibranchia Schweigger, 1820
= Monotocardia Mörch, 1863
= Apogastropoda Salvini-Plawen & Haszprunar, 1987[3]

ARCHAEOGASTROPODA Thiele, 1925[1]
= Aspidobranchia Schweigger, 1820
= Diotocardia Mörch, 1863

VETIGASTROPODA Salvini-Plawen, 1980

NERITOPSINA Cox & Knight, 1960[2]
= Neritimorpha Golikov & Starobogatov, 1975
= Neritacea Thiele, 1929

PATELLOGASTROPODA Lindberg, 1986
= Docoglossa Troschel, 1866

FIG. 3.—Diagram to show possible relationships of prosobranch higher level taxa.

[1]Includes Cocculiniformia Haszprunar, 1987; Architaenioglossa is placed here by Haszprunar (1988a) but is regarded as a caenogastropod group by Ponder & Warén (1988), as also earlier by Haszprunar.

[2]This grouping is sometimes regarded as of equal rank with Archaeogastropoda and Caenogastropoda.

[3]This is Salvini-Plawén & Haszprunar's name for all prosobranchs not archaeogastropods. It includes Ectobranchia Fischer, 1884 (=Valvatacea) though these are excluded from the caenogastropods by Haszprunar (1988a) and Ponder & Warén (1988) on account of their opisthobranch affinities.

[4]United with caenogastropods both by Haszprunar (1988a) and Ponder & Warén (1988).

[5]Originally regarded by Haszprunar (1985b) as a group of Heterobranchia, but later (1988a) as a prosobranch group above archaeogastropod level (i.e. apogastropod) distinct from caenogastropods; placed by Ponder & Warén (1988) in Heterostropha, an order in the subclass Heterobranchia (=Euthyneura).

[6]Triganglionata and Allogastropoda were names used by Haszprunar (1985b, 1985c) for the same group of heterogastropods, those showing annectant characters with euthyneurans. They were accepted as true euthyneurans by Ponder & Warén (1988) and placed in the Heterostropha.

[7]Pentaganglionata is Haszprunar's (1985c) name for a taxon which includes opisthobranchs and pulmonates.

Wenz, 1938) on the grounds of the similarity or identity of many of their features. In Cox's meaning it originally achieved only limited success with taxonomists, but is now beginning to be widely used and it remains a useful adjective.

CTENOGLOSSA Gray, 1853. A name revived by Haszprunar (1988a) for the group of superfamilies (Triphoroidea, Janthinoidea, Eulimoidea, Cerithiopsoidea) removed from Heterogastropoda Habe and Kosuge, 1966 on the grounds that they are true caenogastropods. Earlier, Haszprunar (1985a) had placed them in a group Heteroglossa.

DIOTOCARDIA Mörch, 1863. Prosobranchs with two auricles, equivalent to Archaeogastropoda and to Aspidobranchia.

DOCOGLOSSA Troschel, 1866. Those prosobranchs with a docoglossate radula. The name Patellogastropoda Lindberg, 1986 is to be preferred since a docoglossate radula occurs in groups other than prosobranch gastropods.

ECTOBRANCHIA Fischer, 1884. A group containing only the superfamily Valvatoidea, placed therein after assessment of its systematic position by Rath (1988). It is classified within the broad group Apogastropods by Haszprunar (1988a) though distinct from both caenogastropods and allogastropods; it is regarded as a superfamily within the order Heterostropha of the subclass Heterobranchia by Ponder and Warén (1988). Both schemes emphasize the isolated position of the valvatids.

EUTHYNEURA Spengel, 1881. One of the two subclasses, the other being Streptoneura, into which Spengel divided the Gastropoda on the basis of the arrangement of the visceral loop of the nervous system. Approximately equal to Opisthobranchia Milne-Edwards, 1848 plus Pulmonata Cuvier, 1817.

HETEROBRANCHIA Gray, 1840. A term used by Gray in apposition to Ctenobranchia Schweigger, 1820, which was Schweigger's name for the subclass Prosobranchia. Gray's Heterobranchia included opisthobranchs and pulmonates and was intended to emphasize their common ancestry. For long replaced in common use by Spengel's term Euthyneura it has been recently revived, with some modification, by Haszprunar (1985c) and is used by Ponder and Warén (1988) for Euthyneura (including Heterostropha) though Haszprunar (1988a) prefers to use the term Pentaganglionata for opisthobranchs and pulmonates in the strict sense, and to use Allogastropoda for the Heterostropha.

HETEROGASTROPODA Habe and Kosuge, 1966. A group created to contain those monotocardian prosobranchs with a radula neither taenioglossate, rachiglossate nor toxoglossate, with an acrembolic proboscis, and often with pallial glands producing purple secretions. This is now generally taken to be a collection of animals showing convergent organization rather than as a group of related species, and is split into two other taxa.

HETEROGLOSSA Haszprunar, 1985. A group of families removed from Heterogastropoda Habe and Kosuge, 1966 on the ground that they are undoubted prosobranchs. Not, however, used by Salvini-Plawén and Haszprunar (1987) in their classification of that group, nor, more recently, by Haszprunar (1988a, 1988c); replaced by Ctenoglossa.

HETEROSTROPHA Fischer, 1884. The group of lower heterobranch (= euthyneuran) families characterized by heterostrophic shells, corresponding approximately to Allogastropoda Haszprunar, 1985.

MESOGASTROPODA Thiele, 1925. The central group of the subclass Prosobranchia, members of which have lost the most primitive features still retained by archaeogastropods, possess a taenioglossate radula, and are without the specializations of neogastropods.

MONOTOCARDIA Mörch, 1863. Prosobranchs with only one auricle; equivalent to Caenogastropoda and to Pectinibranchia.

NERITOPSINA Cox, 1960 (= NERITIMORPHA Golikov and Starobogatov, 1975 =NERITACEA Thiele, 1929). A group of prosobranchs of unusual morphology, often classi-fied as archaeogastropods but, if so, to be regarded as an evolutionary line distinct at that level from both vetigastropods and patellogastropods. Sometimes regarded as a distinct order.

PATELLOGASTROPODA Lindberg, 1986. An order crested to include the docoglossan prosobranch limpets, previously taken to be a superfamily (Patellacea or Patelloidea) of Archaeogastropoda though they are distinct from the other groups in that order in a large number of ways.

PECTINIBRANCHIA de Blainville, 1814. An order of prosobranchs possessing a single (left) ctenidium with only a right series of lamellae, approximately equivalent to Caenogastropoda and Monotocardia; contrasted with Aspidobranchia.

PENTAGANGLIONATA Haszprunar, 1985. A new name for Euthyneura, including opisthobranchs and pulmonates but excluding Heterostropha.

PROSOBRANCHIA Milne-Edwards, 1848. Those gastropods with anterior mantle cavity, ctenidia anterior to the heart, and a twisted visceral loop. Equivalent to Streptoneura Spengel, 1881.

RACHIGLOSSA Gray, 1853. Those neogastropods with a rachiglossan radula, neither cancellarioidean nor conoidean.

STENOGLOSSA Bouvier, 1887. Used by Thiele (1929) as the name of an order containing the most advanced prosobranchs; replaced by Neogastropoda Wenz, 1938 by most current writers.

STREPTONEURA Spengel, 1881. One of the two subclasses, the other being Euthyneura, into which Spengel divided the Gastropoda on the basis of the arrangement of the visceral loop of the nervous system. Approximately equivalent to Prosobranchia Milne-Edwards, 1848.

TOXOGLOSSA Troschel, 1847. Prosobranchs with a toxoglossan radula, all placed in the superfamily Conoidea.

TRIGANGLIONATA Haszprunar, 1985. That group of Heterobranchia which contains the allogastropods; contrasted with Pentaganglionata. Ponder and Warén (1988) use the term Heterostropha in its place.

VETIGASTROPODA Salvini-Plawén, 1980. Created as a suborder within the order Archaeogastropoda to include the superfamilies Fissurelloidea, Scissurelloidea, Pleurotomarioidea, Trochoidea and some others. Approximately equivalent to the group Rhipidoglossa Mörch, 1865.

CHAPTER 2

THE ANATOMY OF *LITTORINA* TO ILLUSTRATE
PROSOBRANCH ORGANIZATION

I N order that the anatomy of a prosobranch mollusc may be understood it seems best to describe that of one animal in particular with which comparison may later be made. For this purpose we have selected the common edible winkle *Littorina littorea*. The reasons which have led to the choice of this particular mollusc are these: it is an extremely common animal which may easily be obtained alive in any part of the country; it represents a central, rather unmodified type of prosobranch gastropod, and it is of a size which allows, with reasonable care and the help of no more than a simple lens, the checking of most of its organization by dissection.

If some living winkles are put into sea water and watched they will soon emerge from their shells and start creeping over the surface of the vessel into which they have been placed. It will be seen that the animal moves in a gliding fashion on the under surface of a wedge-shaped part of its body, which is broadly truncated at the front and tapers to a rounded point behind. This is the foot (fig. 4, f), confluent anteriorly with the head, which is a more or less cylindrical structure projecting forwards above its front end. Towards its posterior end the head carries a pair of laterally placed tentacles (t), one on each side. These are delicate, tapering, finger-shaped structures which are mobile and contractile and are kept in frequent movement as the animal creeps. At the base of each, on the outer side, is a cushion-like bulge on which a dark spot with a lighter halo around it may be seen. This is the eye stalk, which has become fused to the outer side of the base of the tentacle, and the dark spot on it (e) is the eye. The tentacle, which is tactile and olfactory, is thus the seat of three major senses. At the end of the snout (sn) projecting in front of the tentacles lies the mouth, bordered by lips (fig. 6, ol) which are complete dorsally, but interrupted in the mid-ventral line.

On the dorsal surface of the foot posteriorly will be seen an oval, plate-like structure. This may be difficult to see because of the way in which the shell lies over it, but if the shell be gently pushed aside the animal will continue to creep and permit the hinder half of the foot to be seen. The disc of material which lies here is the operculum (op, figs. 4, 5), which serves to block the opening of the shell after the mollusc has retracted within (fig. 8). In the winkle it is made of conchiolin, and grows as the animal grows by the addition of strips of material along the edge which faces forwards on the extended foot. Here the operculum dips into a groove (oge, fig. 7) which runs transversely across the dorsal surface of the foot and it is in this that new material is secreted. The under surface of the operculum is attached to the skin of the special lobe of the dorsal part of the foot which carries it, which is called the operculigerous disc (opd), and to part of its under surface there runs a slip of muscle (columellar muscle) the other end of which is fastened to the columella of the shell; by contraction of this the operculum will be pulled against the mouth of the shell, which it fits accurately. The edge of the operculum which lies anteriorly as the winkle crawls comes against the columella of the shell (col) when the animal retracts: it is therefore called the columellar edge (ce). Similarly

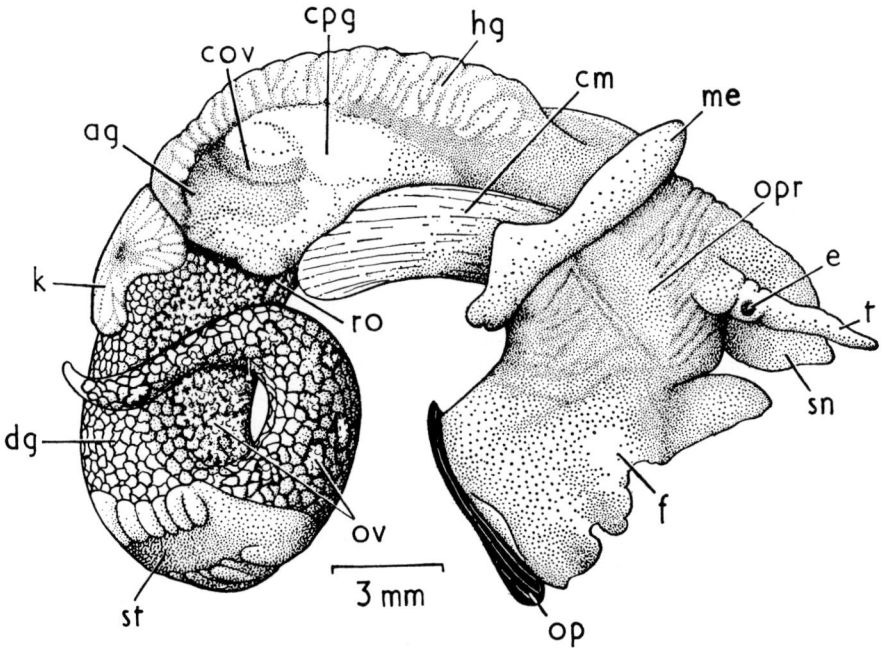

FIG. 4.—*Littorina littorea:* female removed from shell and seen from the right side.
ag, albumen gland; cm, columellar muscle; cov, covering gland; cpg, capsule gland; dg, digestive gland; e, eye on eye stalk; f, foot; hg, hypobranchial gland; k, kidney; me, mantle edge; op, operculum; opr, ovipositor; ov, ovary; ro, renal section of oviduct; sn, snout; st, stomach; t, tentacle.

the edge of the operculum which is posterior when the winkle is extended comes against the outer lip of the mouth of the shell when the snail is retracted. It is therefore called the labial edge (lee). The side of the operculum on the animal's right when extended comes to lie against the upper part of the mouth of the shell in the retracted state, and that on the winkle's left against the lower part; these are known as the sutural (sus) and siphonal (sis) edges respectively.

The visceral hump of the animal is completely covered by the shell and the two structures have the same shape, though the shell is more capacious than the visceral mass since it accommodates not only that part of the body of the animal but the head and foot as well when these are retracted. The shell is made of calcium carbonate in a matrix of the protein conchiolin. The calcareous part of the shell is arranged in different layers, and the whole is covered by a layer of conchiolin known as the periostracum. In some gastropods it is very obvious, but this is not the case in *Littorina littorea.* The periostracum and the outer layer of calcareous matter can normally be added to only at the mouth of the shell, but the innermost layers are secreted by the entire surface of the visceral mass. This is necessary in order that the older parts of the shell, which were produced when the animal was young and small, should be thick enough and strong enough to protect the body of the winkle when it has grown larger and older. In theory there is perhaps no limit to the size to which gastropods may grow, but in practice there is a size which is not exceeded by each species and when this

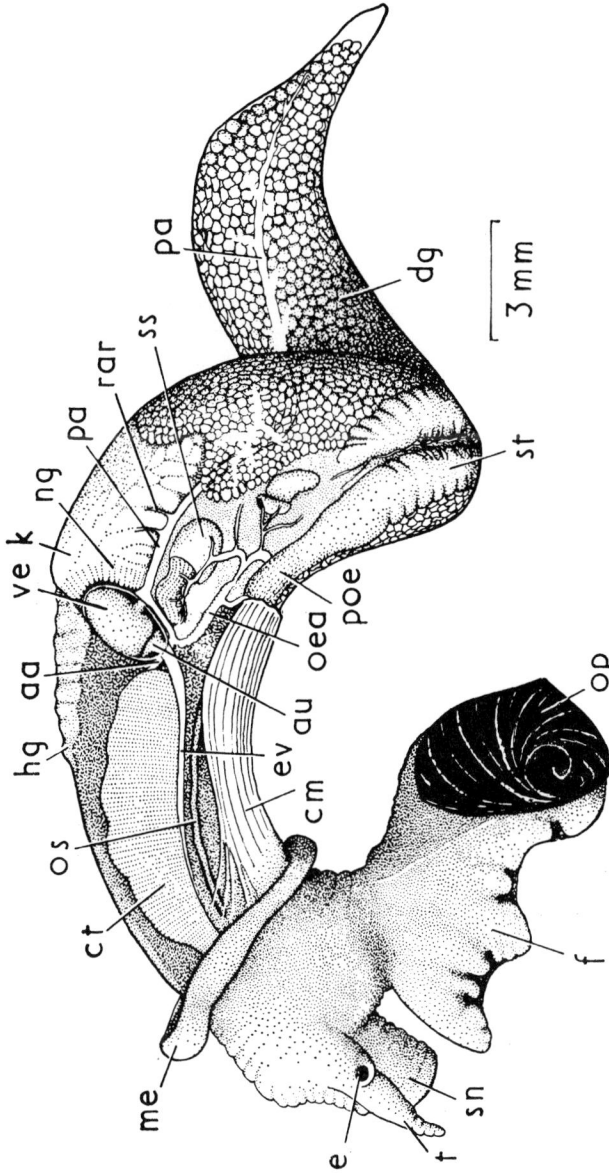

FIG. 5.—*Littorina littorea*: animal removed from shell and seen from the left side.
aa, anterior aorta; au, auricle; cm, columellar muscle; ct, ctenidium; dg, digestive gland; e, eye on eye stalk; ev, efferent branchial vessel; f, foot; hg, hypobranchial gland; k, kidney; me, mantle edge; ng, nephridial gland; oea, oesophageal artery; op, operculum; os, osphradium; pa, posterior aorta; poe, posterior oesophagus; rar, renal artery; sn, snout; ss, style sac region of stomach leading forward to intestine; st, stomach; t, tentacle; ve, ventricle.

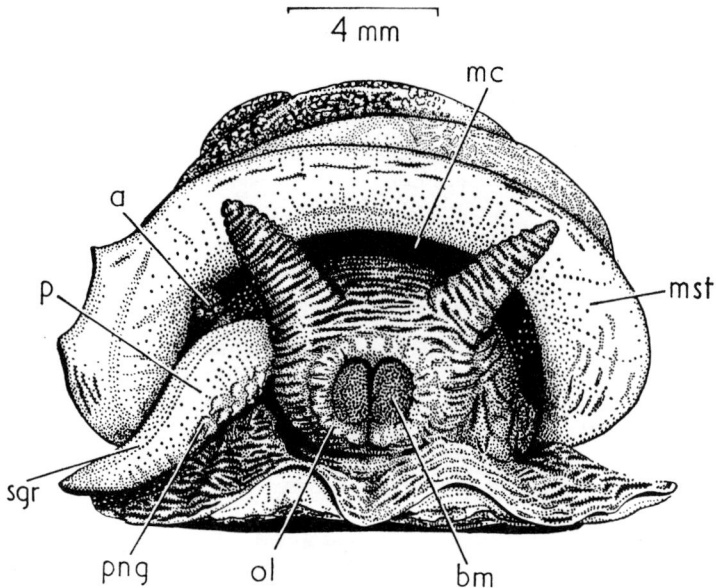

FIG. 6.—*Littorina littorea:* male, removed from shell, in anterior view. The mouth is half open.
a, anus; bm, odontophore; mc, mantle cavity; mst, mantle skirt; ol, outer lip; p, penis; png, penial glands; sgr, seminal groove.

is reached straightforward growth of the shell ceases, though other growth—in particular general thickening and the formation of teeth around the mouth—may continue for a while.

The shell of *Littorina littorea* (fig. 8) is more or less conical in shape and consists in essence of a tube of gradually expanding diameter coiled in a helicoid spiral, the direction of coiling when the shell is viewed from above the apex being right-handed, that is, in the same direction as the movement of the hands of a clock. The tube is closed at its inner end, but is open at its outer end, and this opening is the mouth, through which the body of the winkle can be protruded for movement and feeding. The growth of the tube is so arranged that each successive turn of the spiral (a whorl) is applied to the outer surface of the previous one, which it partly conceals (fig. 9). In any one shell, therefore, there will be one turn of the spiral— the largest—the outer side of which is completely visible. This turn ends at the mouth of the shell and is the youngest (most recently secreted) part. In it, when the winkle is retracted, will be found the animal's head and foot, that is, all the body except the visceral hump: for this reason it is called the body whorl (bw, and see fig. 37). The remaining whorls of the spiral, of each of which only a part is visible, constitute the spire of the shell. The line of contact where two whorls meet is called a suture (sts).

As the animal lies in its shell the outer part of each whorl corresponds to the dorsal surface of the body, and the inner to the ventral. The lower side of each turn of the shell corresponds to the left side of the visceral mass and the upper or apical to the right—these, it will be remembered, must have started originally the other way round and have been brought into this new position by the process of torsion. As a result of the dextral coiling of the visceral

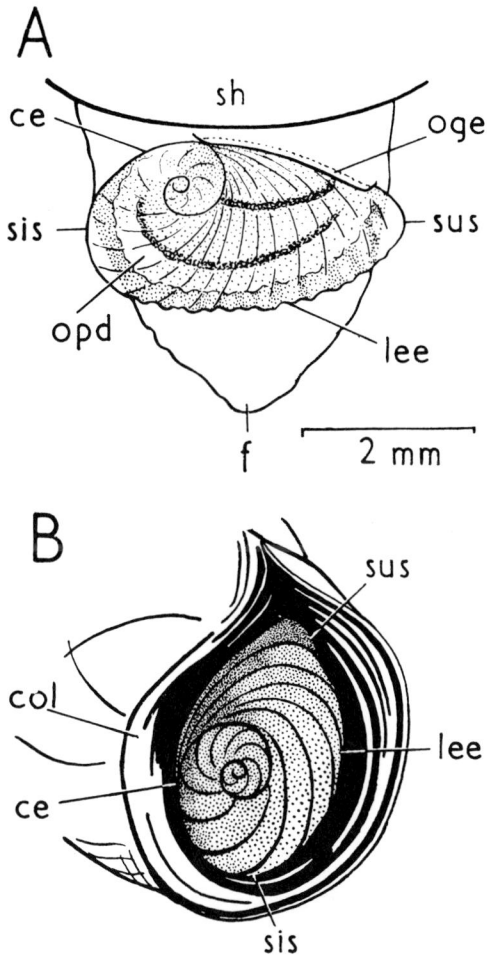

FIG. 7.—*Littorina littorea:* A, dorsal view of posterior end of foot of creeping winkle to show relationships of operculum; B, mouth of shell to show relationships of operculum when winkle is retracted.

ce, columellar edge of operculum; col, columella of shell; f, foot; lee, labial edge of operculum; oge, opercular groove of foot; opd, operculigerous disc seen through operculum; sh, shell; sis, siphonal side of operculum; sus, sutural side of operculum.

hump, much more space is available on the left-hand side of that part of the body than there is on the right, and this is reflected in the arrangement of the organs which occur there, as will be seen below.

Where the inner sides of the spirally coiled whorls are brought into contact with one another there results a more or less solid central pillar round which the whorls of the shell rotate. This is the columella (col). If the contact between the concave sides of successive whorls is very intimate then it will be solid; if less so, the columella may be hollow with a small

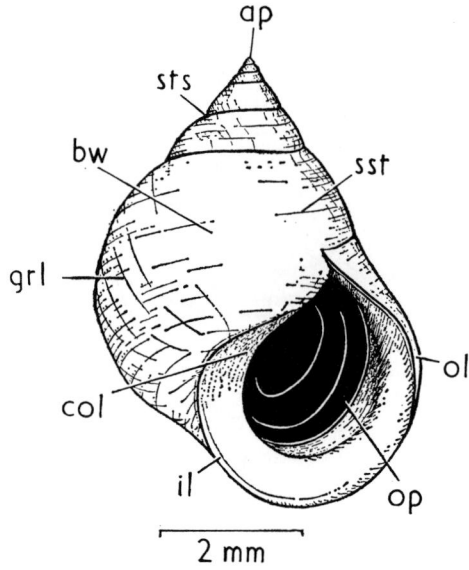

FIG. 8.—*Littorina littorea:* shell, apertural view.
ap, apex; bw, body whorl; col, columella; grl, growth line; il, inner lip; ol, outer lip; op, operculum; sst, spiral stria; sts, suture.

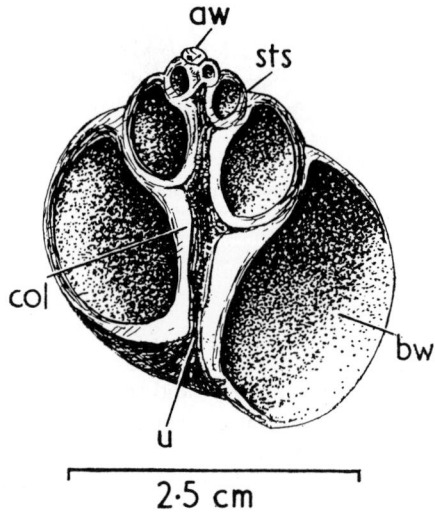

FIG. 9.—*Natica catena:* shell, halved vertically.
aw, apical whorl of shell; bw, body whorl; col, columella; sts, suture; u, open umbilicus.

part of external space lying in its centre and opening to the outside at its base. This opening is the umbilicus (u).

The mouth of the shell is formed of an outer lip (fig. 8, ol) on the animal's dorsal side and an inner lip (il)—often inconspicuous against the columella—on its ventral.

The shell is added to as the animal grows, and except in a few cases, of which the winkle is not one, loses no part during its life-time. The initial shell which was present during larval life, the protoconch, forms the extreme summit of the spire and often is different in appearance, or texture, or architecture, from the shell secreted by the metamorphosed animal: in most winkles it will have been worn away. The shell usually exhibits changes in rate of growth marked by lines (grl) representing successive positions of the mouth: these lie parallel to the columellar axis and are described as vertical. In addition, the shell of the edible winkle exhibits a number of lines which run spiralwise down the shell and are obviously due to slight differences in the rate of secretion or to irregularities of the outer lip itself. These lines are called spiral striae (sst). In the case of the edible winkle the shell is rather smooth and neither striae nor vertical lines are pronounced. The body whorl is large, as it must be when such a large body has to be enclosed, but the rest of the shell, the spire, is rather short, tapering smoothly to the apex. The sutures are visible but are not deeply marked and the upper whorls have flat sides so that the spire is rather regularly conical. The mouth is large, somewhat compressed from side to side and the outer lip tends to run up the body whorl so as to approach it tangentially. There is no umbilicus. The whole shell is solidly built, is of a dark colour, with brown streaks usually clearly visible near the lips; there is usually a white patch on the columellar side.

Young shells have the outer lip with a slightly crenulated margin, but with sexual maturity it becomes much thicker, and this, and possibly the effect of wear on a shell the rate of growth of which has been reduced, causes it to be smooth.

When a living edible winkle is observed moving, the head and foot are extended outside the shell, but the visceral mass does not leave its shelter: the only other part visible is a fold of tissue resting against the inside of the mouth of the shell. This is the edge of the mantle skirt (mst, fig. 6), which forms a fold round the body of the animal at the point of union of visceral mass (on one side) and the head and foot (on the other. Between the mantle skirt and the underlying body lies a space, which is the mantle cavity (mc, fig. 6), and the depth of this will be the same as the length of the mantle skirt. Further observation of the living animal will show that, except when it is completely withdrawn into the shell, a current of water passes into the cavity on the left side, and leaves it on the right.

To examine the extent and contents of the mantle cavity after the winkle has been removed from its shell it should be opened by a cut through the mantle skirt which starts medianly at the anterior end and bears to the animal's left posteriorly. If the two halves of the mantle skirt are then pulled apart the inside of the mantle cavity will be visible (see fig. 10). It is low in a dorsoventral direction (see fig. 11), broad at the mouth and tapering to a narrow inner end. Its floor is formed by the dorsal integument of the head and anterior part of the visceral mass and its roof by the mantle skirt. All the structures which lie partly in the head-foot and partly in the visceral mass make their way from the one part of the body to the other by a narrow connexion ventral to the innermost end of the mantle cavity. Except at its free edge the mantle skirt is everywhere delicate, but all the organs of importance in the mantle cavity lie on it, and are developments from it.

At the left side lies the gill or ctenidium (ct, figs. 5, 10), which has the form of an elongated axis lying along the mantle skirt in an antero-posterior direction from the innermost part of

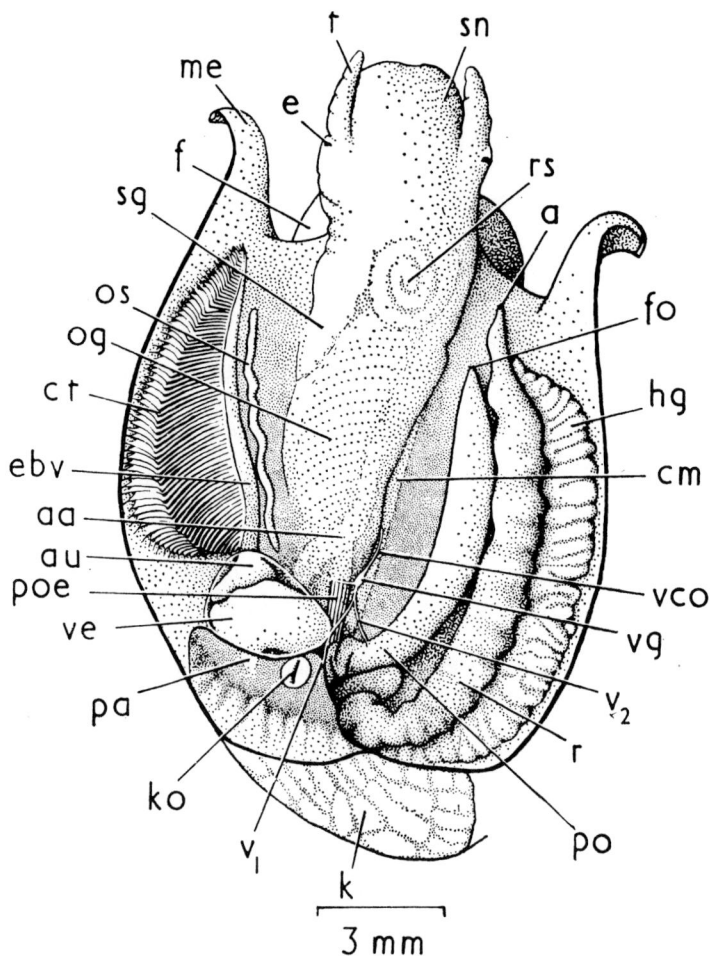

FIG. 10.—*Littorina littorea:* animal removed from shell and mantle cavity opened mid-dorsally to display its contents. Some other structures are seen by transparency.

a, anus; aa, anterior aorta; au, auricle; cm, columellar muscle; ct, ctenidium; e, eye on eye stalk; ebv, efferent branchial vessel; f, foot; fo, female opening; hg, hypobranchial gland; k, kidney; ko, kidney opening; me, mantle edge; og, oesophageal gland; os, osphradium; pa, posterior aorta; po, pallial oviduct; poe, posterior oesophagus; r, rectum; rs, radular sac; sg, salivary gland; sn, snout; t, tentacle; v_1, nerve to heart and kidney; v_2, genital nerve; vco, visceral connective; ve, ventricle; vg, visceral ganglion.

the cavity almost to its mouth. Attached to the axis is a series of branchial leaflets (blf, fig. 11), 50–60 in number. Each of these has a triangular shape, with one side of the triangle fused to the mantle skirt, and they all hang into the mantle cavity like a series of pages from a book. Along the axis of the ctenidium lies a blood vessel, the efferent branchial vessel (ebv, figs. 10, 11), which may be traced back from the posterior end of the ctenidial axis to the heart (au, ve, figs. 5, 10). Blood reaches the branchial leaflets by way of small afferent branchial vessels (av,

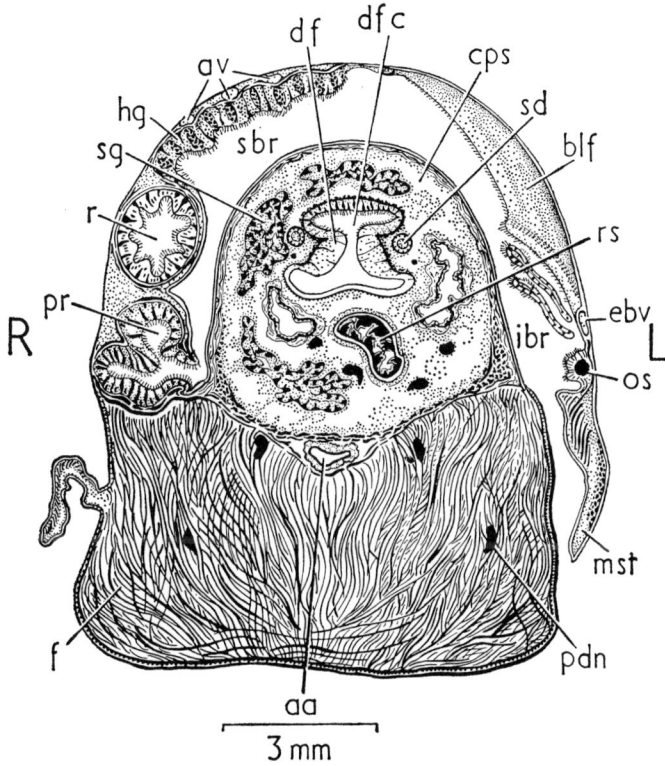

FIG. 11.—*Littorina littorea:* transverse section through male at level of middle of mantle cavity. R, L, mark right and left sides.

aa, anterior aorta; av, afferent branchial vessels; blf, branchial leaflet; cps, cephalopedal blood sinus; df, dorsal folds of oesophagus; dfc, dorsal food channel of oesophagus; ebv, efferent branchial vessel; f, foot; hg, hypobranchial gland; ibr, infrabranchial part of mantle cavity; mst, mantle skirt; os, osphradium; pdn, pedal nerve; pr, prostate gland; r, rectum; rs, radular sac; sbr, suprabranchial part of mantle cavity; sd, salivary duct; sg, salivary gland.

fig. 11) which enter the right edge of the attached side of each leaflet from the mantle skirt: they are too small to be visible except with high magnification. The gill lamellae contain a network of blood spaces through which the blood travels from afferent to efferent vessel. Each lamella is ciliated and the leaflets are collectively responsible for driving water from the infrabranchial space (ibr, fig. 11) which lies below and to the left of the gill, into the supra-branchial space (sbr) above it and to the right. This is, in fact, the driving force of the current which enters the mantle cavity on the left and leaves it on the right and which has already been referred to.

Parallel to the ctenidial axis and along its left side runs a narrow pigmented ridge. This is the osphradium (os, figs. 10, 11) one of the animal's major sense organs, and it is believed to test the water entering the mantle cavity, although the precise nature of the testing is still not known. It may be a straightforward chemical testing, or one discriminating the amount of

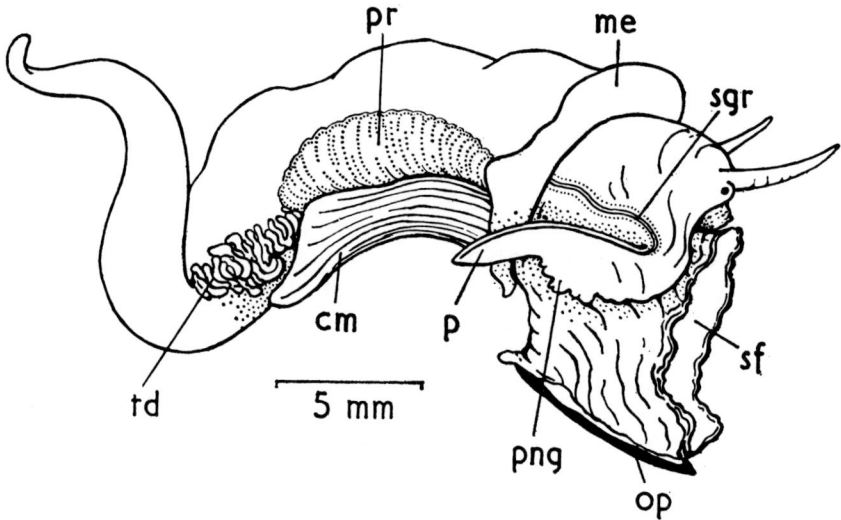

FIG. 12.—*Littorina littorea:* male, removed from shell, to show genital duct.
cm, columellar muscle; me, mantle edge; op, operculum; p, penis; png, penial glands; pr, prostate; sf, sole of foot; sgr, sperm groove; td, testicular duct acting as vesicula seminalis.

suspended particulate matter in the water washing its surface, or perhaps both these things at once. The osphradium is certainly well placed for this.

To the right of the ctenidium lies a stretch of mantle skirt which shows little specialization immediately alongside the gill. Microscopic examination, however, reveals numerous mucous cells here and towards the right, where the epithelium is puckered and ridged, their number increases greatly. (The puckering occurs only in the edible winkle, not in other British species of *Littorina*.) This area is the hypobranchial or mucous gland (hg) and it produces secretion for trapping and cementing particulate matter sucked into the mantle cavity in the respiratory water current, prior to its expulsion on the right.

Still more to the right will be observed a tube running parallel to the right edge of the hypobranchial gland. This is the rectum (r), which emerges from the visceral hump at the inner end of the mantle cavity and runs along the mantle skirt to open at the anus (a) placed near the mouth of the cavity. To the right of the rectum lies the terminal part of the reproductive duct. This duct differs according to whether the winkle is male or female, and whether the animal is examined during the breeding season or not. Outside the breeding season the reproductive apparatus becomes much reduced, to be re-formed when the next reproductive period approaches. In males the duct from the testis opens to the mantle cavity by a pore situated deeply within it on the right of the rectum. This discharges to a ciliated groove which runs forward along the floor of the mantle cavity until it reaches a position posterior and ventral to the right cephalic tentacle, where it run on to the penis (p, figs. 6, 12, 13), a large, curved, paddle-shaped structure. Within the mantle cavity the groove traverses the centre of a rich glandular field which is the prostate gland (pr, figs. 12, 13).

In females conditions differ in that the section of the duct which lies on the mantle skirt is not open to the mantle cavity as in males, but is a closed tube and because of this the female

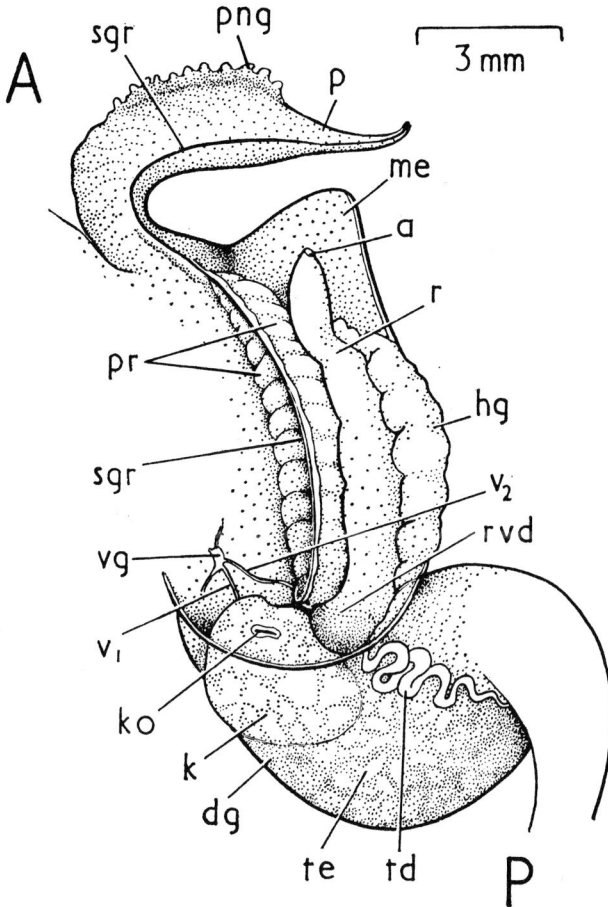

FIG. 13.—*Littorina littorea:* dissection to show male reproductive system. A, P, mark anterior and posterior ends. a, anus; dg, digestive gland; hg, hypobranchial gland; k, kidney; ko, kidney opening to mantle cavity; me, mantle edge; p, penis; png, penial glands; pr, prostate; r, rectum; rvd, renal section of vas deferens; sgr, sperm groove; td, testicular duct acting as vesicula seminalis; te, testis; v_1, renopericardial nerve; v_2, rectal and genital nerve; vg, visceral ganglion.

aperture lies near the anus. This section of the female duct is very glandular, especially in the breeding season, when the capsule gland (cpg, fig. 4), which secretes the bulk of the wall of the capsule within which the eggs are laid, forms a large, chalk-white mass lying ventral to the rectum in the inner half of the mantle cavity. Alongside it on the outer (shell) side are two other associated glands: one of these is the albumen gland (ag), which is of a translucent buff colour and secretes the albuminous fluid in which each egg is embedded in the capsule, and the other is a special area of the capsule gland, apricot in colour (cov), which produces the covering around the egg and its albumen (ec, fig. 193C, F). There is, of course, no penis in the female, but there is a structure occupying a corresponding position on the right side of the head, in the form of an unpigmented glandular tract running down the side towards the foot. This is the ovipositor (opr, fig. 4), a function of which is to carry the egg capsule out of the mantle cavity and launch it on its pelagic life. The precise use of the glands is not known: they must be, in part, simply lubricating, but they have also been held to form the tough outer layer of the capsule wall.

The only other structure to be directly connected with the mantle cavity is the kidney (k, figs. 4, 5, 10), a pinkish or brownish structure occupying a basal position on the visceral hump towards the left side, and discharging to the innermost part of the mantle cavity by a slit-like opening (ko, fig. 10) with conspicuous lips.

A certain amount of the internal anatomy of the body of the winkle is visible by transparency through the skin. This is particularly true of organs which lie in the posterior part of the head and in the visceral mass where the integumentary coverings are delicate. Through the unpigmented floor of the mantle cavity may be seen a number of structures which are drawn in fig. 10. These are mostly connected with the anterior end of the alimentary tract and of them the oesophagus with the oesophageal glands (og) is most conspicuous, overlaid anteriorly on the right by the spirally coiled radular sac (rs) and on the left by salivary glands (sg). The anterior or cephalic aorta (aa) is also visible by transparency running forward and to the right from the heart. At the innermost end of the floor of the mantle cavity there may also be noticed the visceral ganglia (vg) placed at the posterior end of the visceral loop of the nervous system (see below) and nerves may be traced from the ganglia to the rectum, heart and kidney (v_1, v_2).

The visceral hump is covered by a thin and transparent body wall, the pallium or mantle, through which most of the underlying viscera may easily be seen. They are shown in figs. 4, 5 which are drawings of edible winkles removed from their shells. When viewed from the left (fig. 5) the basal part of the visceral hump is seen to be occupied by the pericardial cavity containing the heart (au, ve), which receives the efferent branchial vessel (ev) from the ctenidial axis, and which sends off an anterior aorta (aa) to the head and foot and a posterior aorta to the organs lying in the visceral hump (pa). Behind the pericardial cavity (i.e. further towards the apex of the spire) is the kidney (k) and that part of it which borders the pericardial cavity may be picked out from the rest by its lighter colour as the nephridial gland (ng). The remainder of the visceral hump is mainly occupied by the tubules of the digestive gland (dg), dark brown in colour. Should the animal be breeding, however, greater or lesser areas of gonad will lie over the digestive gland, especially on the columellar (concave) side of the whorls, and obscure it from sight. The gonad (te, fig. 13; ov, fig. 4) is grey or grey-green in colour in males, but yellow or pinkish in females. On the basal, convex region of the visceral mass parts of the stomach (st, figs. 4, 5) are visible, and quite ventrally, underneath the mantle cavity, is the white band-shaped mass of the columellar muscle (cm), originating on a short length of the columella of the shell and running into the head, the foot, on to the operculum and into the mantle skirt to pull these parts into the shell for protection when the animal is disturbed.

We may now turn to an account of the internal anatomy of the winkle. This may easily be investigated by opening the body by a cut through the mid-line of the floor of the mantle cavity. The cut should be extended forwards to the tip of the snout and backwards towards the inner end of the mantle cavity, though care must be taken in the region of the visceral ganglia. The pericardial cavity may be opened together with the kidney, but it is unprofitable to attempt the dissection of the upper visceral mass unless particular information regarding such structures as lie there is being sought.

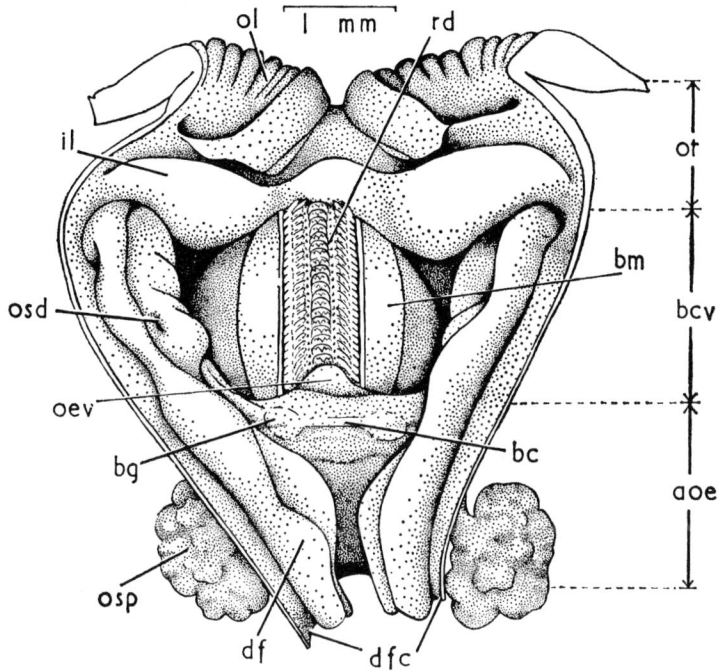

FIG. 14.—*Littorina littorea:* buccal cavity and anterior oesophagus opened mid-dorsally.
aoe, anterior oesophagus; bc, buccal commissure, seen by transparency; bcv, buccal cavity; bg, buccal ganglion, seen by transparency; bm, odontophore; df, dorsal fold of oesophagus; dfc, dorsal food channel; il, inner lip; oev, oesophageal valve; ol, outer lip; osd, opening of salivary duct; osp, oesophageal pouch; ot, oral tube; rd, radula on upper surface of buccal mass.

The most conspicuous object revealed in the anterior part of the body is the buccal mass, an elaborate muscular apparatus tied to the body wall by muscles and tendons. It surrounds the buccal cavity which receives anteriorly the oral tube leading from the mouth and which, posteriorly, opens to the oesophagus and, beneath that, to the radular sac. The oral tube (ot, fig. 14) is very short and bounded internally by a transverse ridge projecting into the cavity from its ventral and lateral surfaces, but not dorsally: this is the inner lip (il). Two prominent folds project from the roof of the buccal cavity, one on either side of the mid-line, which continue down the oesophagus. These are the dorsal folds (df) and the channel between them the dorsal food channel (dfc). The floor of the buccal cavity is raised into a tongue-like prominence covered with shining cuticle, fused to the floor of the cavity posteriorly and free at its anterior tip. This is the odontophore (bm, figs. 6, 14), a complex mass of muscles attached in part to skeletal structures called cartilages, from their histological resemblance to the cartilage of vertebrates, and in part to the body wall. It is a very mobile structure which is pushed in and out of the mouth, turned up and down and twisted right and left as the winkle feeds. Over the median part of its mid-dorsal surface runs the radula (rd, fig. 14), a belt of cuticular material bearing teeth in regular transverse and longitudinal rows. These teeth are formed at the inner end of the radular sac and are gradually moved forward along the sac and

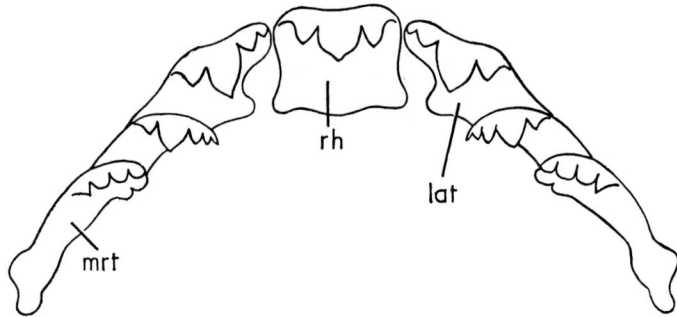

FIG. 15.—*Littorina littorea:* one transverse row of radular teeth.
lat, lateral tooth; mrt, marginal tooth; rh, rachidian tooth.

on to the surface of the odontophore as those already there are lost or broken in use. The radular sac ends blindly and lies coiled in a spiral dorsal to the oesophagus on the right side of the body. In length it may be as much as 50 mm or about twice the shell length. Each transverse row (fig. 15) contains a definite number of teeth of several shapes and sizes: these comprise a single medianly placed central or rachidian tooth with 3 large cusps on it (rh), flanked on each side by 3 rather similarly shaped teeth each with a number of cusps on its edge. The most median of these teeth, which is next to the rachidian, is called a lateral tooth (lat); the two others are marginal teeth (mrt). The arrangement of each transverse radular row is identical with that of its neighbours in front and behind and it can be represented by a formula

2 marginal—1 lateral—1 rachidian—1 lateral—2 marginal

or, more simply,

2—1—1—1—2.

All the cusps on all the teeth are curved so that they point inwards. As the odontophore is moved in and out of the mouth the radula is moved backward and forwards over the angled edge at the tip of the odontophore. Anterior to this edge the marginal and lateral teeth are spread sideways; posterior to it they are folded in towards the mid-line. As the radula is moved backwards and forwards the teeth spring from the one position to the other—on the inward movement moving towards the middle line. As they carry out this movement the recurved cusps rake detritus or algae into the mid-line where it gets caught on the cusps of the median rachidian teeth. These form a conveyor belt carrying the food into the gut. The process is lubricated by saliva and the two salivary glands (sg, fig. 17) which lie dorsal to the oesophagus, on the animal's left, discharge their secretion by ducts (sd, figs. 11, 17) which enter the buccal cavity at points (osd, fig. 14) on the latero-dorsal walls just lateral to the dorsal folds. The secretion is mainly mucus.

The next section of the alimentary tract is the oesophagus, which leads from the buccal cavity (in the head) to the stomach (in the visceral mass). It therefore traverses the region of the body which has been affected by torsion and it will be as well to consider what the effects of this have been before describing what is actually visible in a dissected winkle.

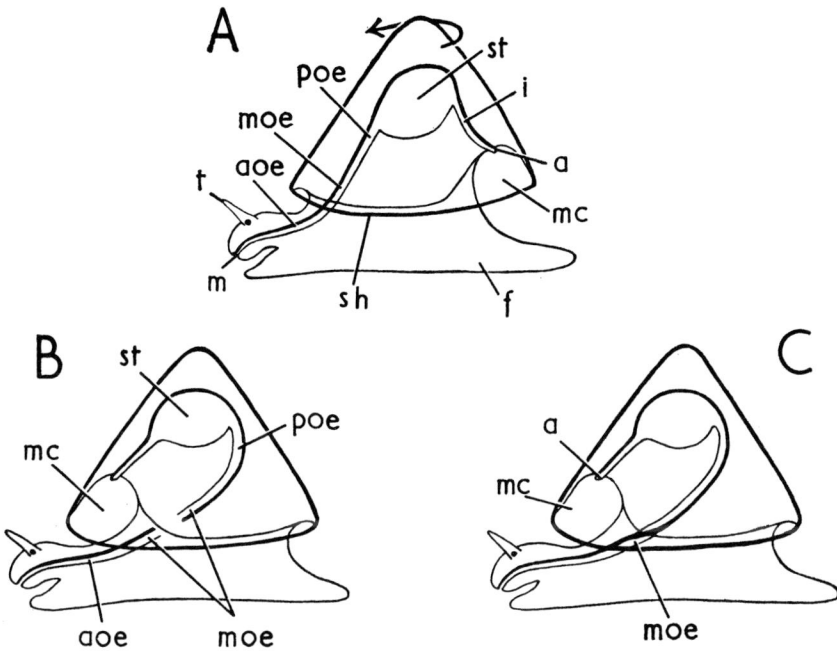

FIG. 16.—Diagrams to show the effect of torsion upon the gut. A, pretorsional stage; C, after torsion; B, for explanation see text. The morphologically dorsal surface of the gut is marked by a thicker line.

a, anus; aoe, anterior oesophagus; f, foot; i, intestine; m, mouth; mc, mantle cavity; moe, mid-oesophagus; poe, posterior oesophagus; sh, shell; st, stomach; t, tentacle.

The progastropod (fig. 16A) possessed a visceral hump with the mantle cavity excavated from its posterior face. The anus (a) discharged into this cavity and was medianly placed. Between mouth and anus the oesophagus (aoe, moe, poe) ran dorsally to the stomach (st), which lay near the summit of the visceral mass, and the intestine (i) descended again to the more ventral anus. The gut was therefore U-shaped, the U being inverted and the mouth and anus marking the ends of its limbs. When torsion occurred the visceral mass rotated in an anticlockwise direction through 180° on an axis which we may suppose passed vertically through its centre, so that the posteriorly placed anus was brought forward. If, as in fig. 16B and C, we draw this new arrangement in its simplest form, it will be realized that, in order to keep the dorsal side of the oesophagus in connexion with the dorsal side of the stomach and intestine, it is necessary to introduce a twist in the course of the gut. Thus in the basal part of the visceral hump the original dorsal surface comes to lie topographically ventral, and the original ventral surface dorsal. This twist occurs in the oesophagus of the prosobranch mollusc, and as the oesophagus happens to be a rather elaborate part of the gut with clearly differentiated dorsal and ventral sides, the twisting is particularly easy to see.

The first part of the oesophagus, which may be called the anterior oesophagus (aoe, figs. 14, 16, 17), lies anterior to the region of torsion and is therefore normally disposed. It begins at the posterior end of the buccal cavity and dorsally, apart from the continuations of the two folds and food channel, there is nothing remarkable about its structure. Two oesophageal pouches

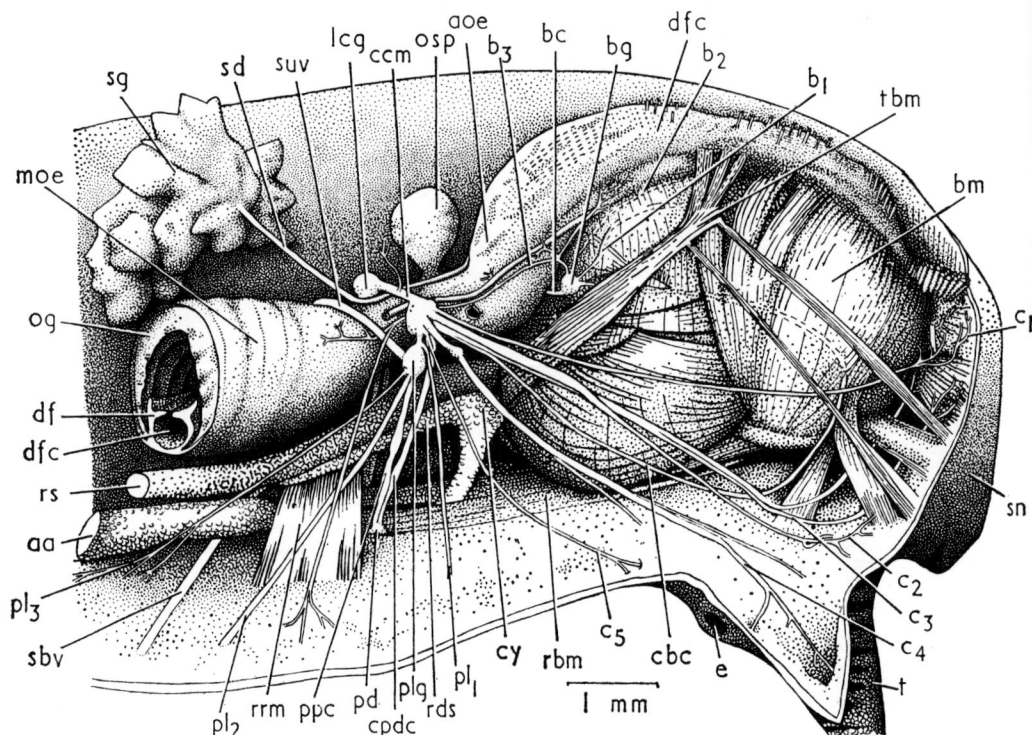

FIG. 17.—*Littorina littorea:* dissection of head from the right side. The head has been opened by a cut in a right dorsolateral position and the right body wall and tentacle pulled ventrally. The oesophageal pouch and salivary gland on the right side have been removed.

aa, anterior aorta; aoe, anterior oesophagus; b_1, b_2, b_3, buccal nerves; bc, buccal commissure; bg, buccal ganglion; bm, buccal mass; c_1, dorsal labial nerve; c_2, lateral labial nerve; c_3, ventral labial nerve; c_4, tentacular nerve; c_5, optic nerve; cbc, cerebrobuccal connective; ccm, cerebral commissure; cpdc, cerebropedal connective; cy, cephalic aorta; df, dorsal folds of oesophagus now twisted to mid-ventral position; dfc, dorsal food channel; e, eye; lcg, left cerebral ganglion; moe, mid-oesophagus; og, oesophageal gland; osp, oesophageal pouch; pl_1, pl_2, pl_3, pleural nerves; pd, pedal ganglion; plg, right pleural ganglion; ppc, pleuropedal connective; rbm, retractor muscles of buccal mass; rds, radular sinus; rrm, retractor muscles of radular sac; rs, radular sac in radular sinus; sbv, sub-oesophageal part of visceral loop; sd, salivary duct; sg, salivary gland; sn, snout; suv, supra-oesophageal part of visceral loop; t, tentacle; tbm, tensor muscles of buccal mass.

(osp, figs. 14, 17), however, open from its lateral walls anteriorly, one on each side. They are hollow, spherical structures with lobulated walls, of unknown significance. At the posterior end of this section the gut is surrounded by a nerve ring (fig. 17).

The next part is the mid-oesophagus (moe, figs. 16, 17) and it is this which is twisted by torsion. The twist is such that the mid-dorsal line of the gut curves over to the left and eventually on to the underside, whilst the morphological mid-ventral part curves up the right side until it lies above. The course of the twist is easy to follow because the morphologically mid-dorsal part of the oesophagus bears the food channel and its two edging folds and these can easily be traced from an anatomically mid-dorsal position at the anterior end of this section to a mid-ventral position at the posterior end. The mid-ventral line is similarly marked by other structures to be mentioned below and its curvature may also be easily followed (fig. 17).

The mid-oesophagus is an elongated spindle-shaped structure lying between the nerve ring in front and the posterior end of the mantle cavity, where it narrows to join the posterior oesophagus (poe, fig. 10). The swelling of this section of the alimentary tract is due to the presence of oesophageal glands in the lateral and ventral walls. These are flung into a series of folds so that they present the appearance of a series of compartments one behind the other, but all opening towards the central axis of the oesophagus. The epithelium which lines the mid-oesophagus is ciliated. The direction of the ciliary beat is backwards on all the surfaces of the food channel, but in the lateral glandular areas is arranged so as to bring their secretion into the food channel, where it will be mixed with the food passed back from the mouth. This secretion has been shown to contain digestive enzymes, so that at the posterior end of the mid-oesophagus there is a mixture of mucus, food particles and digestive enzymes ready to be passed into the next section of the gut. This is the posterior oesophagus, a rather narrow tube with numerous longitudinal ridges and no special glandular equipment, which runs up the visceral mass to open into the stomach.

The stomach (st, figs. 4, 5, 18) lies entirely in the visceral mass, mainly on the outer (convex) side towards the animal's left and is an elongated pouch extending through more than one whorl of the spiral. The oesophagus opens into it about half way between its two ends (OA, fig. 18), and the intestine (I) leaves the lower end, the upper end being blind. Internally the cavity of the upper half of the stomach is almost divided into two parts by a tall longitudinal fold (FF) which ends near the summit. The oesophagus opens into the chamber on one side of this fold, whilst the ducts of the digestive gland (DD), three in number, and the intestine are connected to the chamber on the opposite side. This arrangement means that the stomach is provided with a caecum, along which the food which is received from the oesophagus has to pass before it is exposed to secretions from the digestive gland. During this time, however, the food is in contact with the secretions from the oesophageal glands and perhaps from the salivary glands, and doubtless, there-fore, some digestion occurs. In the other chamber of the stomach the food is exposed to secretion from the digestive gland, and muscular action will mix food and enzyme and cause a reflux of digested food and, perhaps, of particulate matter into the ducts and tubules of the digestive gland for absorption. The distal part of the stomach, leading forwards to the intestine, may be distinguished as the style sac. Along its length run two ridges, the major (T1) and minor (T2) typhlosoles, with the intestinal groove (G1) between them. This originates at the point where the ducts of the digestive gland open into the stomach. The indigestible residue of a meal is forced into the style sac of the stomach where it comes under the influence of two sets of ciliary currents, one, on the typhlosoles, which tends to drive the mass towards the intestine; the other, on the remaining areas of the walls, which rotates the contents. As a consequence the indigestible remains are gradually compacted with mucus into a rod of faecal material and passed out of the style sac into the intestine. There will also be a certain amount of waste material which is derived from the digestive gland: this may be particulate indigestible matter which has accidentally got into the ducts of the gland along with matter for absorption, or it may be the indigestible remains of particulate food ingested during intracellular digestion (if such occurs in *Littorina littorea*), or it may be true excretory matter which is discharged from the digestive gland cells. Whatever its origin, it will escape from the ducts of the digestive

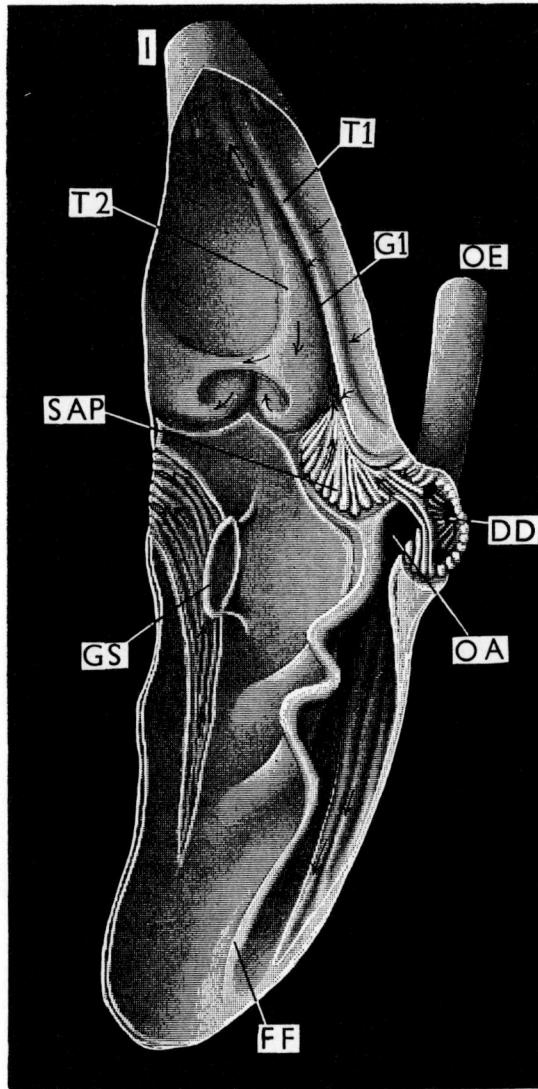

FIG. 18.—*Littorina littorea*: stomach, opened longitudinally. Arrows show direction of ciliary currents.
    DD, opening of ducts from digestive gland; FF, longitudinal fold; G_1, intestinal groove; GS, gastric shield; I, intestine; OA, opening of oesophagus to stomach; OE, oesophagus; SAP, posterior sorting area; T_1, major typhlosole; T_2, minor typhlosole.

gland into the intestinal groove and so make its way along the style sac and become incorporated in the rod of faecal matter which enters the intestine.

    The intestine is a tube of smaller diameter than the style sac of the stomach, with numerous longitudinal folds running along its walls. It runs from the left side of the body, where the

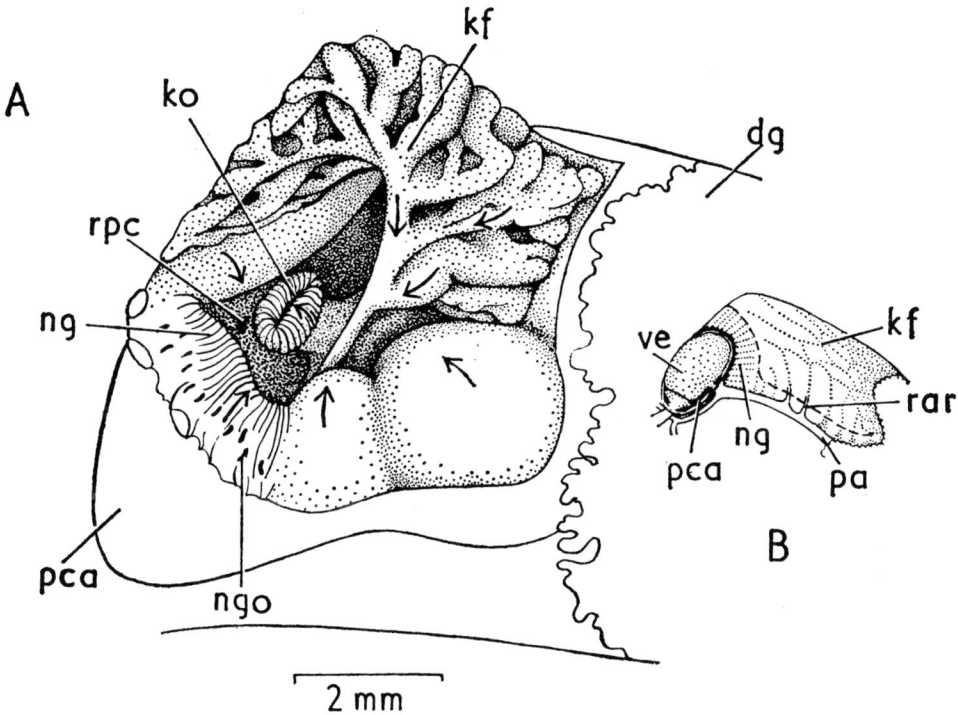

FIG. 19.—*Littorina littorea:* A, view of the inside of the kidney sac which has been opened along the broken line marked in the inset fig. B, which has been copied from fig. 5. Arrows show the direction of ciliary currents.
dg, digestive gland; kf, folded surface of kidney; ko, opening of kidney to mantle cavity; ng, nephridial gland; ngo, opening from kidney sac to tubules of nephridial gland; pa, posterior aorta; pca, pericardial cavity; rar, renal artery; rpc, opening of renopericardial canal to main kidney sac; ve, ventricle.

stomach is placed, across to the right in a C-shaped loop and there enters the right part of the mantle skirt, along which it passes anteriorly to open at the anus. The rod of faecal matter which it receives from the style sac becomes segmented into oval faecal pellets which are still further compacted by intestinal secretions, so that when discharged they do not easily disintegrate within the mantle cavity.

The kidney (k, figs. 5, 10) lies on the left side of the visceral hump abutting against the upper end of the mantle cavity to which it opens by a slit-like aperture (ko, fig. 10). Internally it communicates by means of a very fine ciliated canal (rpc, fig. 19) with the pericardial cavity. The kidney is a roomy sac with walls differentiated for varied purposes at different places. On the right there projects into the cavity a series of lobulated folds (kf), mainly covered by an excretory epithelium resting on blood spaces, but with some ciliated cells which produce currents leading excretory matter towards the outer opening, which is guarded by a sphincter muscle and may be opened by radially running dilator muscles. Over its lips beats a strong outward ciliary current.

The left anterior wall forms part of the nephridial gland (ng, fig. 5) which borders the pericardial cavity. Here a series of blind diverticula opens (ngo, fig. 19) from the main cavity of the kidney and pushes into blood spaces, so that a close intermingling is achieved. Some of the cells of this nephridial gland are excretory, others are again ciliated and drive material extracted from the blood into the kidney for excretion. The details of the way in which excretory and osmoregulatory functions are shared by these two parts of the kidney is not certain.

The vascular system of *Littorina littorea* comprises the heart, lying within the pericardial cavity, and a number of vessels and spaces. Arteries distribute the blood to the main parts of the body, branching over and over again until they are minute, and end by pouring their contents into a series of blood spaces around the main organs and in between the muscles and connective tissue layers of the body. From this series of spaces other vessels take the blood to the excretory and respiratory organs, whence it is returned to the heart. The vascular system of the winkle differs from the more familiar vertebrate pattern in a number of important respects: the heart receives, primarily, oxygenated blood to circulate to the body and is, therefore, a systemic heart whereas that of the vertebrate is originally a branchial heart pumping deoxygenated blood to respiratory organs for oxygenation; there is no capillary system between the arteries and veins but a series of indefinite haemal spaces, though these are not necessarily of dimensions very different from capillaries. All the organs are directly bathed in blood, and the circulation must be slower than in the vertebrate and less definite in its course. Many of the mollusc's movements, moreover, are hydraulic in mechanism, and it seems likely that their accomplishment will bring about vascular disturbance just as compression of our own veins does. Most gastropods have various sheets of muscle and connective tissue subdividing the haemocoelic spaces so that they have greater control over the distribution of blood in their body; these are not well developed in the winkle, except for one major one which separates the haemocoel within the visceral mass from that in the combined head and foot. This rums across the body at the level of the posterior oesophagus just anterior to the visceral ganglia.

In *Littorina littorea* the heart (au, ve, figs. 5, 10) lies in the pericardial cavity on the left side near the base of the visceral mass, anterior and ventral to the kidney. The anterior wall of the pericardium is in contact with the innermost end of the mantle cavity. The heart is composed of two chambers, an auricle and a ventricle. The auricle (au) lies anterior and slightly ventral to the ventricle (ve) which is set across the body. The two communicate by an aperture guarded by a single tongue-shaped valve (avv, fig. 20), so arranged as to permit free flow of blood from auricle to ventricle but not in the reverse direction. The two chambers are covered by an outer epithelium (vep, fig. 22), but there is no endothelial lining. Both have bundles of muscle (hm) forming a series of strands, much less elaborate in the auricle than in the ventricle. The latter opens posteriorly and ventrally to a single vessel, which is sometimes called the bulbus or truncus arteriosus (ob, fig. 20), and which splits almost at once into two, a larger anterior aorta (aa) taking blood forwards, and a smaller posterior aorta (pa) which takes blood into the upper coils of the visceral mass. These channels have muscular walls.

In most individuals the arteries have around their walls a thick investment of connective tissue laden with granules of calcium carbonate which makes them show up a brilliant white in a fresh animal. The course of the major arteries is therefore readily observable.

The anterior aorta (aa, figs. 5, 10, 17, 21, 23) runs forward under the floor of the pericardial cavity on its way to the head and foot, lying dorsal to and left of the posterior oesophagus at this point. It gives off a branch, the oesophageal or style sac artery (oea, figs. 5, 21), which runs posteriorly, dorsal to the oesophagus, into connective tissue filling the space between the oesophageal and intestinal limbs of the stomach, to both of which structures it sends

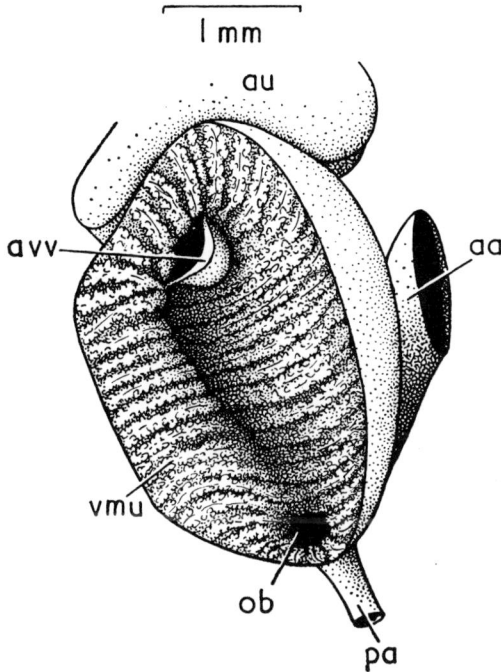

FIG. 20.—*Littorina littorea:* ventricle, opened.
     aa, anterior aorta; au, auricle; avv, auriculo-ventricular valve; ob, opening from ventricle to bulbus; pa, posterior aorta; vmu, muscles in wall of ventricle.

branches. The main aorta now penetrates into the cephalopedal haemocoel, lying dorsal to the mid-oesophagus (aa, fig. 23), and, like that structure, it shows here the effects of torsion, crossing the mid-oesophagus dorsally so that it comes to lie on its right, and then curving underneath until it lies in the mid-ventral line at the level of the anterior end of the mid-oesophagus, directly below the radular sac (aa, fig. 17). At this point the anterior aorta passes ventrally through the nerve ring into the foot through a gap in the sheet of transverse muscle which separates the cephalic and pedal haemocoelic cavities; it then divides into a main pedal artery (pda, fig. 21A) and a main cephalic artery (car). The pedal artery turns backwards and branches within the muscular masses of the foot emptying the blood finally into the haemocoelic spaces of that organ (cps, fig. 21B). The cephalic artery climbs dorsally back into the head, gives off a radular artery which enlarges to form a sinus (rds) around the radular sac, and then passes into the buccal mass to supply blood to that complex—blood which then escapes to the cephalopedal haemocoel (cps).

     The posterior or visceral aorta (pa, figs. 5, 21A) leaves the bulbus at its dorsal end and passes up the visceral mass along the left border of the kidney, giving off a number of renal arteries (rar) as it does so. The aorta then passes deeply into the central parts of the visceral mass, emerges on to the surface on the columellar side and runs in this position to its summit, giving off a large number of arteries to the intestine, the stomach and, principally, to the digestive

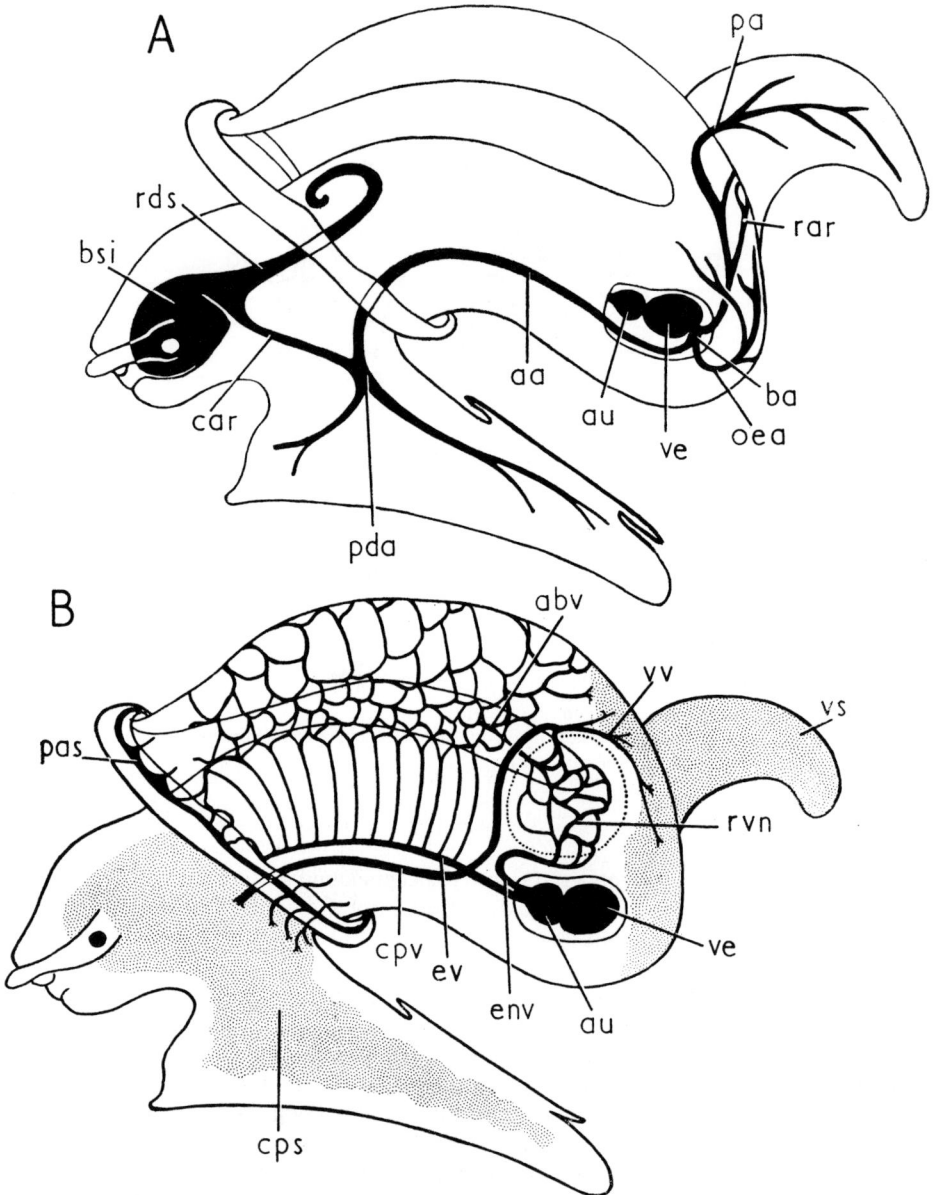

FIG. 21.—*Littorina littorea:* diagrammatic representations of A, the main parts of the arterial system and B, the main parts of the venous system as seen in animals viewed from the left side. The main blood sinuses are stippled. The proportions of the parts have been distorted for the sake of clarity.

aa, anterior aorta; abv, afferent branchial vessels; au, auricle; ba, bulb of aorta; bsi, buccal sinus; car, cephalic artery; cps, cephalopedal blood sinus; cpv, cephalopedal vein; env, efferent vessel from nephridial gland; ev, efferent branchial vessel; oea, oesophageal artery; pa, posterior aorta; pas, pallial blood sinus; pda, pedal artery; rar, renal artery; rds, radular sinus; rvn, renal vessels; ve, ventricle; vs, visceral sinus; vv, visceral vein.

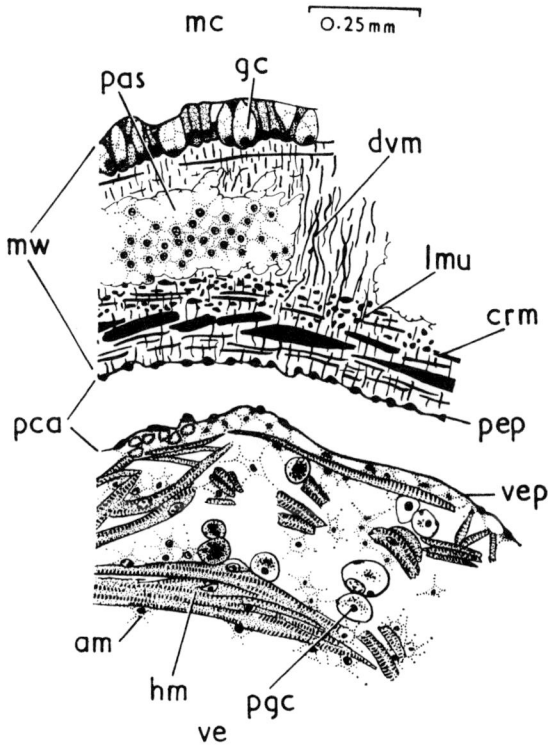

FIG. 22.—*Littorina littorea*: transverse section showing wall of ventricle and of pericardial cavity and inner epithelium of mantle skirt.
am, amoebocyte; crm, circular muscle; dvm, dorsoventral muscle; gc, gland cell in epithelium lining mantle cavity; hm, striped muscle of ventricle; lmu, longitudinal muscle; mc, mantle cavity; mw, wall of mantle cavity = inner part of mantle skirt; pas, pallial blood space; pca, pericardial cavity; pep, epithelium lining pericardial cavity; pgc, pigment cell; ve, cavity of ventricle; vep, outer epithelium of ventricle.

gland and gonad. All these vessels pass blood into a large visceral haemocoel (vs, fig. 21B) by which these organs are surrounded.

Though the distribution of blood from the heart to the body is thus achieved in a fairly direct way, the venous return is more complex, especially when it is realized that both the kidney and the respiratory organ are incorporated in it. From the visceral mass blood may pass directly into haemal spaces in the mantle skirt (pas) mainly by way of a large vessel lying around and on the right of the rectum, and the same is also true for blood in the cephalic and pedal haemocoels, which are connected broadly to the mantle skirt on the left. Nevertheless there is a second, more important, route by which blood may return—by way of the kidney. A short cephalopedal vein (cpv) returning blood from the main cephalopedal sinus (cps) and visceral veins (vv) returning blood from the visceral sinus (vs) unite near the right side of the kidney and liberate their blood by way of one main renal vessel into the numerous blood vessels (rvn) which lie in the wall of the kidney sac. From here the blood may pass to the nephridial gland; in *Littorina*, this is then carried by an efferent vein (env) direct to the auricle,

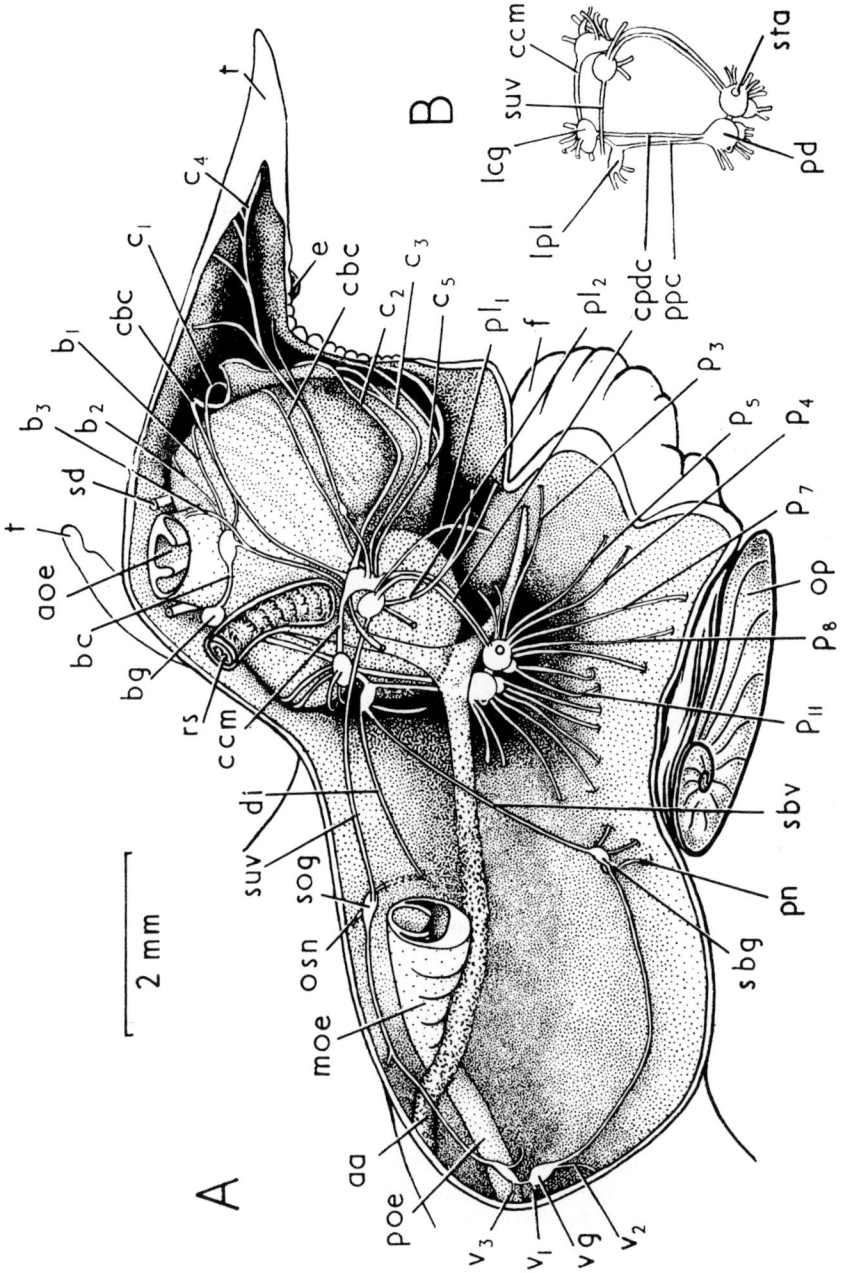

although in some other prosobranchs it may go to pallial blood spaces lying in the mantle skirt. Blood may also pass directly into these pallial spaces from the visceral sinus and from the lateral parts of the cephalopedal sinus.

The blood which has reached the haemal spaces in the mantle skirt (pas, fig. 21B) (by whatever route) will undoubtedly undergo some degree of oxygenation there, but it is gradually passed into channels which form the afferent branchial vessels (abv) leading into the right edges of the ctenidial leaflets. The blood passes through spaces within the gill lamellae and finally escapes from these into the efferent branchial vessel in the ctenidial axis (ev) along which it flows to the auricle of the heart.

The blood is a colourless or slightly bluish liquid, the colour being due to the blood pigment haemocyanin. It contains numerous amoebocytes (am, fig. 22). The muscle of the ventricular wall is striated (hm). Many cells laden with pigmented material, possibly excretory in nature, cling to the muscle fibres inside the heart: it is these which make both the auricle and the ventricle look brownish in colour (pgc).

The nervous system of the gastropod consists of a number of ganglia, interlinked, and giving off nerves to various organs. The ganglia lie mainly in the head, in a group around the oesophagus, but some are placed in the visceral hump to co-ordinate the activities of the viscera. Originally the entire system was symmetrical but as a consequence of torsion and also of the asymmetry introduced by the spiral coiling of the visceral hump the final arrangement is usually quite asymmetrical.

Around the oesophagus is a ring of six ganglia (figs. 17, 23). These are on each side: the cerebral, most dorsally; the pedal, most ventrally; and the pleural intermediately. The three ganglia on each side are linked to each other and the right and left cerebrals and the right and left pedals are linked across the mid-line of the body. Longitudinal connexions between different ganglia are called connectives; connexions between corresponding ganglia of the right and left sides are called commissures. Thus there are cerebral and pedal commissures in the nerve ring and there are cerebropleural, cerebropedal and pleuropedal connectives on each side. To these ganglia there come many nerves from sense organs such as the eyes and the statocysts, the tentacles and other structures on the head, foot and mantle skirt, and

---

FIG. 23.—*Littorina littorea*: A, dorsolateral dissection of head, foot and base of visceral mass to show the nervous system. The gut has been cut at the level of the anterior oesophagus; the greater part of the radular sac has been removed. Both anterior oesophagus and radular sac have been pulled forwards through the nerve ring and turned dorsally, and the two salivary glands have been cut out. The mid-oesophagus has been displaced posteriorly. B, diagram to show ganglia in nerve ring.

aa, anterior aorta dividing anteriorly into pedal and cephalic arteries; aoe, anterior oesophagus; b_1, b_2, buccal nerves; b_3, anterior branch of oesophageal nerve, the posterior branch being cut behind the nerve ring; bc, buccal commissure; bg, buccal ganglion; c_1, dorsal labial nerve; c_2, lateral labial nerve; c_3, ventral labial nerve; c_4, tentacular nerve with ganglion; c_5, optic nerve; cbc, cerebrobuccal connective; ccm, cerebral commissure; cpdc, cerebropedal connective; di, dialyneury on left linking supra-oesophageal and left pleural ganglia; e, eye; f, foot; lcg, left cerebral ganglion; lpl, left pleural ganglion; moe, mid-oesophagus; op, operculum; osn, osphradial and branchial nerve; p_3, p_4, p_5, p_7, p_8, p_{11}, pedal nerves; pd, left pedal ganglion; pl_1, pl_2, pleural nerves; pn, pallial nerves; poe, posterior oesophagus; ppc, pleuropedal connective; rs, radular sac in radular sinus; sbg, sub-oesophageal ganglion; sbv, sub-oesophageal part of visceral loop; sd, salivary duct (cut); sog, supra-oesophageal ganglion; sta, statocyst; suv, supra-oesophageal part of visceral loop; t, tentacle; v_1, visceral nerve to heart and kidney; v_2, visceral nerve to rectum and genital duct; v_3, columellar nerve; vg, visceral ganglion.

nerves go to muscles in the same situations. The muscles of the odontophore, which form an extraordinarily complex apparatus calling for delicately controlled manipulation, are provided with two special buccal ganglia, connected with the cerebrals. These are placed at the point where the anterior oesophagus and the radular sac separate from the buccal cavity and are linked by a buccal commissure which runs between the anterior oesophagus above and the radular sac below.

Another group of ganglia is associated with the innervation of the mantle skirt and the viscera. Connectives between these form the visceral loop which runs from the pleural ganglion on each side of the nerve ring in the head into the visceral mass, the two parts connecting there so as to form a complete nerve loop. On this are borne five or six ganglia, the pleurals anteriorly, the single or double visceral (or abdominal) ganglion which lies at the innermost end of the mantle cavity, and the parietal ganglia, which lie between the pleurals anteriorly and the viscerals behind. Originally, in the progastropod, the entire loop lay ventral to the gut, but it will be realized that this part of the nervous system, like the gut and the anterior aorta, has to make its way between the head, which has not been disturbed by torsion, and the visceral hump, which has, and it will therefore be affected by the process. This has the effect of transferring ganglia which were originally on the right side of the body to the left, and vice versa, and of turning part of the system upside down just as happens to the mid-oesophagus. As a consequence of this the visceral loop of the nervous system comes to lie dorsal to the gut behind the region of torsion, though not anteriorly, and it is also flung into a figure-of-eight shape where the nerves cross from their original right and left position in the head to their new left and right positions in the visceral hump. This twisted state of the visceral loop is called streptoneury.

Once torsion has taken place the visceral loop is twisted at a point which lies between the two pleural ganglia in the nerve ring anteriorly and the two parietal ganglia posteriorly. The pleural ganglia, being cephalic in position, retain their original situations; the parietals, however, have come to lie on opposite sides of the body, the original right on the left and vice versa. The course of the visceral loop is now, therefore, as follows: right pleural ganglion to the (original right) parietal (now left of the oesophagus), to the (original right) visceral (now left of the oesophagus), to the (original left) visceral ganglion (now right of the oesophagus), to the (original left) parietal ganglion (now right of the oesophagus), and it is completed by a connexion to the left pleural ganglion in the nerve ring. Because torsion also involves the turning upside down of the posterior half of the visceral loop the connexion between right pleural and parietal ganglia now crosses the oesophagus dorsally, whereas the connexion between left pleural and parietal ganglia lies ventral to the oesophagus. The two visceral ganglia and the commissure which links them are now dorsal to the oesophagus. It is largely because of these new relationships to the alimentary canal that the parietal ganglia are given the names by which they are generally known in gastropods—the original right parietal, now lying left of the oesophagus and connected to the right pleural by a connective which lies dorsal to the oesophagus is called the supra-oesophageal (or, sometimes, but less happily, the supra-intestinal) ganglion; the original left parietal, now lying right of the oesophagus and connected to the left pleural by a connective which lies ventral to the oesophagus is called the sub-oesophageal (or sub-intestinal) ganglion.

The supra- and sub-oesophageal ganglia are concerned in the innervation of the mantle skirt and the organs which are situated there, such as the gill and the osphradium, for which the supra-oesophageal ganglion in particular is important: there may, indeed, be special ganglia, osphradial and branchial, connected to it. The pleural ganglia themselves may be involved in some control over the lateral parts of the mantle lying contiguous to and joined to the side of the foot and the head, which are controlled by the pedal and cerebral ganglia respectively. Since the head-foot has not undergone torsion, whereas the visceral mass—and so the mantle skirt—has, this would mean that neighbouring areas of mantle skirt would be controlled by ganglia the nervous pathways between which might be quite lengthy. As an example we may consider the control of part of the mantle edge on the animal's left side. The anterior part of this might be innervated by fibres originating in the left pleural ganglion (and would be in working relationship with pedal structures controlled by the left pedal ganglion), the posterior part by fibres connected to the supra-oesophageal ganglion. To ensure proper nervous co-ordination it is clear that links between the left pleural and the supra-oesophageal ganglia—and possibly the left pedal ganglion—are desirable. There is, however, no direct initial contact with these two nerve centres and the shortest route by which a connexion can in fact be established is by way of the left cerebropleural connective, left cerebral ganglion, cerebral commissure, right cerebral ganglion, right cerebropleural connective, right pleural ganglion and the connective between that and the supra-oesophageal ganglion, a route which almost certainly involves at least four synaptic junctions, or by way of the visceral ganglion, which is no shorter, and less likely. Gastropods are not noted for the rapidity of their reactions, but they need more efficient nervous pathways than this, and many of them modify the original plan of the visceral loop so as to bring about direct connexions between the right pleural ganglion and the sub-oesophageal ganglion on the one hand, and between the left pleural and supra-oesophageal ganglia on the other. Occasionally connexions with the pedal ganglia are also established. These are called zygoses or zygoneuries and they occur more frequently on the right side of the nerve ring (between right pleural and sub-oesophageal ganglia) than on the left. A second way in which the gastropod may attempt to co-ordinate the control of these pallial organs is by means of what is known as a dialyneury; this is a peripheral fusion between the pallial nerves emanating from the two ganglia involved and it would seem to achieve a result which is possibly physiologically similar to what a zygoneury does.

In *Littorina littorea* the nervous system is illustrated in figs. 17, 23, 24, 25. The cerebral ganglia (lcg, figs. 17, 23) lie behind the buccal mass, one on either side of the oesophagus, which is narrow at this point. Each is connected to the other cerebral ganglion by the cerebral commissure (ccm, figs. 17, 23, 24) which passes dorsal to the anterior oesophagus, and to the pleural (lpl, plg, figs. 17, 23B and pedal (pd) ganglia by connectives. The pleural ganglia lie close behind and a little ventral to the cerebrals, so that the cerebropleural connectives (cpc, fig. 24) are short. The pleural ganglia are slightly smaller than the cerebrals, and, like them, are connected to the pedals by a pleuropedal connective (ppc, figs. 17, 23) which runs alongside the cerebropedal (cpdc). The pedal ganglia lie side by side ventral to the anterior aorta in the blood spaces of the foot. The connectives to the dorsal ganglia are therefore long.

From each cerebral ganglion (fig. 24) there originate four connexions to other ganglia: (a) the cerebral commissure (ccm); (b) the cerebropleural connective (cpc); (c) the cerebropedal connective (cpdc); and (d) the cerebrobuccal connective (cbc). This runs forward from the ganglion with one of the nerves going to the superficial muscles of the buccal mass.

In addition, each cerebral ganglion gives off 5 nerves (figs. 17, 24). These are:

c1 A nerve of moderate size to the dorsal integument and muscles of the snout; this leaves the anterior border of the ganglion medially and somewhat ventrally, near the origin of the cerebral commissure.

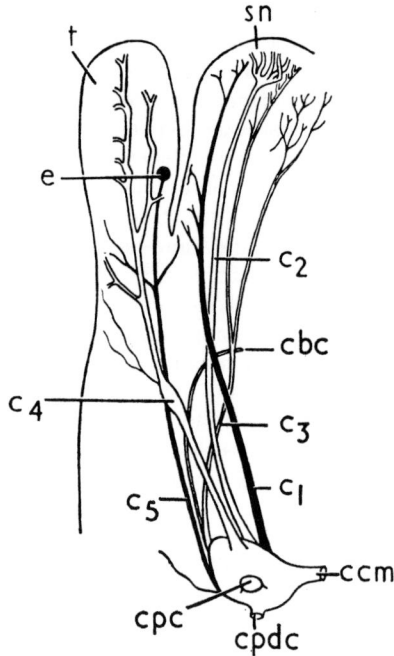

FIG. 24.—*Littorina littorea:* left cerebral ganglion and nerves (modified from Leyon, 1947).
c_1, dorsal labial and integumentary nerve; c_2, lateral labial and integumentary nerve; c_3, ventral labial and integumentary nerve; c_4, tentacular nerve with ganglion; c_5, optic nerve; cbc, cerebrobuccal connective; ccm, cerebral commissure; cpc, cerebropleural connective; cpdc, cerebropedal connective; e, eye; sn, snout; t, tentacle.

$c2$ A stout nerve to the lateral integument and muscles of the snout; this leaves the ventral side of the anterior border of the ganglion lateral to the origin of nerve $c1$.

$c3$ A large nerve to the ventral integument and muscles of the snout, which divides into two branches of equal size about half way to the skin of the snout; it arises from the ventral surface laterally and its basal section is bound with part of the cerebrobuccal connective by tissue.

$c4$ The tentacular nerve: this bears a small ganglionic enlargement a short distance from the cerebral ganglion; it originates from the dorsal surface of the ganglion and gives off numerous branches to the skin of the tentacles. It is olfactory in function.

$c5$ The optic nerve; a nerve of moderate size which arises from the ventral surface of the ganglion lateral and posterior to the nerve $c4$. It goes to the eye and the skin in its neighbourhood.

The pleural ganglia (plg, fig. 17; lpl, fig. 23B) lie close to the cerebrals, but slightly more posterior and ventral. Each gives off 3 connexions to other ganglia: (a) the cerebropleural; (b) the pleuropedal (ppc); and (c) the pleuroparietal connectives. The right pleural ganglion (plg, fig. 17) gives rise to a connective (suv, figs. 17, 23) which runs (dorsal to the oesophagus) to the supra-oesophageal ganglion (sog, fig. 23) lying on the left; the left pleural ganglion gives rise to

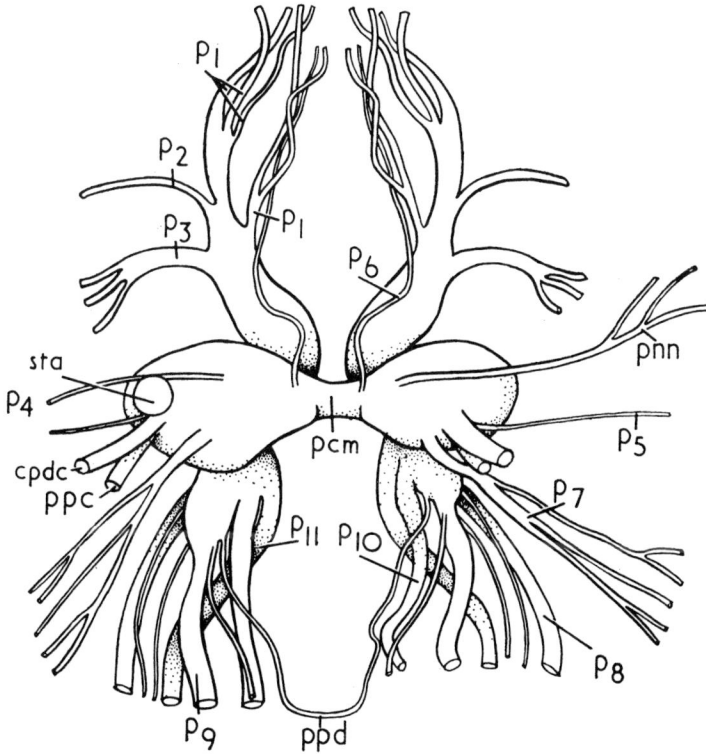

FIG. 25.—*Littorina littorea:* posterodorsal view of pedal ganglia and nerves.
cpdc, cerebropedal connective; p_1, p_2, p_3, p_4, p_5, p_6, p_7, p_8, p_9, p_{10}, p_{11}, pedal nerves; pcm, pedal commissure; pnn, penial nerve; ppc, pleuropedal connective; ppd, posterior pedal commissure; sta, statocyst.

a connective (sbv) which runs (ventral to the oesophagus) to the sub-oesophageal ganglion (sbg) lying on the right. There is no zygoneury in *Littorina littorea*.

The right pleural ganglion gives rise to 3 nerves ($pl1$, $pl2$, $pl3$, fig. 17) which innervate tissue at the right mantle edge, where it joins the side of the cephalopedal mass. The left pleural ganglion innervates a corresponding region on the left, but in this case one of the nerves joins another which has come from the supra-oesophageal ganglion so as to give rise to a dialyneury (di, fig. 23) on the animal's left.

From each pedal ganglion (pd, fig. 23; see also fig. 25) arise 3 connexions to other ganglia: (a) a very short pedal commissure (pcm); (b) a much longer pleuropedal connective (ppc); and (c) a cerebropedal connective (cpdc). Each also gives rise to 11 nerves to various parts f the foot. These tend to arise in groups from lobelike expansions of the ganglion. They are:

$\left.\begin{array}{l} p1 \\ p2 \\ p3 \end{array}\right\}$ A group of small nerves arising from a median lobe posterior to the pedal commissure: they run to the median part of the anterior border of the foot.

$\left.\begin{array}{l} p4 \\ p5 \\ p6 \\ p7 \end{array}\right\{$ This group of nerves originates mainly from the lateral margin of the ganglion. (5) divides into 3 branches; and (6) and (7) each into 2. The anterior ones run to the antero-lateral border of the foot, the posterior ones to the posterior part of the sole. (4) links with a nerve from the sub-oesophageal ganglion so as to complete a dialyneury on the animal's right. (4) also innervates the penis.

$\left.\begin{array}{l} p8 \\ p9 \\ p10 \\ p11 \end{array}\right\}$ This group of stout nerves arises in a posterior lobe of the ganglion. They run posteriorly to the back of the sole and to the operculigerous disc.

The buccal ganglia (bg, figs. 17, 23) are placed lateral to the gut at the point where oesophagus and radular sac originate from the buccal cavity. Each is the source of 2 connexions to other ganglia: (a) the buccal commissure (bc); and (b) the cerebrobuccal connective (cbc). Each ganglion also sends 2 nerves (b1, b2) which pass forwards into the muscles of the buccal mass and a third (b3) which curves backwards along the oesophagus running alongside the salivary duct. It sends branches to this duct, to the dorsal food channel in the roof of the buccal cavity, and continues under the nerve ring. Here it again divides, one half passing under the supra-oesophageal connective and following the oesophagus as far as the stomach; the other passes above the connective and goes to the body wall, where it innervates the muscles suspending the oesophagus.

The visceral loop contains ganglia from each of which a number of nerves takes origin. From the supra-oesophageal, connectives run forwards to the right pleural and posteriorly to the visceral ganglia; this ganglion is also the origin of pallial nerves and of the branchial and osphradial nerves. The sub-oesophageal ganglion is connected to the left pleural and the visceral ganglia and gives off 3 nerves (pn, fig. 23A) to the mantle edge on the right and to the anus and reproductive aperture. Occasionally a circumpallial nerve is formed, running round the edge of the mantle skirt, by the fusion of the pallial nerves from these 2 ganglia. The 2 ganglia lie on the floor of the haemocoelic space in which the oesophagus and anterior aorta run and are partly buried in the muscles which form it. The connectives which link them to the visceral ganglia are similarly situated so that the posterior part of the visceral loop is less easy to see and dissect than the anterior half.

The visceral ganglia (vg, figs. 10, 13, 23) are represented by a single structure lying below the integument forming the floor of the mantle cavity almost at its innermost end. A smaller ganglion is sometimes visible close to this on the course of the connective which passes over the posterior oesophagus and anterior aorta on its way to the supra-oesophageal ganglion. Two main nerves originate from the larger ganglion: one goes to the heart and excretory organ (v1), whilst the other is a genital nerve (v2) passing to the oviducal region in the female and to the male pore in the opposite sex. From the smaller ganglion nerves pass to the columellar muscle (v3).

The most important sense organs of the winkle are probably local tactile and similar sensory structures in the skin, but in addition to these the animal possesses at least three other major sense organs—the eyes, the statocysts, and the osphradium.

The eyes (e, figs. 4, 5) are situated at the base of the tentacles, on their outer side. Strictly speaking they are on stalks which have become secondarily fused to the tentacles. Each eye is a hollow vesicle, formed in development by intucking of the skin, and within the cavity of the organ is a spherical cuticular lens. The retina is formed of light perceptive cells shielded by black pigmented ones. The skin over the eye is clear and forms a kind of conjunctiva.

How good these may be as image-forming organs is not known: the winkle certainly reacts to a shadow falling on the eyes by a contraction into its shell.

The statocysts (sta, fig. 25) lie near the pedal ganglia, though not, in *Littorina littorea,* as close to their surface as is often the case. That on the left is further forwards than that on the right, and both are surrounded by connective tissue containing granules of calcium carbonate. Each statocyst has the form of a small, spherical sac, completely closed in the adult state and filled with fluid. In this is found a statolith, a spherical mass of calcium carbonate embedded in an organic matrix and often showing concentric shells, perhaps due to rhythms in the animal's growth. The statoliths fall under the influence of gravity on to the sensory epithelium which lines the cavity.

The osphradium (os, fig. 10) is usually regarded as a receptor for chemical stimuli, although some writers believe that it may be also, or, instead, sensitive to the amount of particulate matter which is carried in suspension in the water to which it is exposed. In *Littorina littorea* it forms a ridge projecting from the mantle skirt into the mantle cavity and lying alongside and parallel to the ctenidial axis. It contains numerous sensory cells and ganglionic enlargements of the osphradial nerves. It is sometimes said that the nerve fibres which run to the osphradium start in the cerebral ganglia although they reach their destination in nerves which originate in the supra-oesophageal. This is known to be true of the nerves supplying the statocysts: although running in nerves from the pedal ganglia the fibres have their real origin in the cerebral ganglia. The cerebral ganglia thus have direct information not only from the snout, the tentacles and the eyes, but also from the statocysts and, possibly, from the osphradium as well, on which to base the behaviour which the winkle will exhibit in any particular situation.

Edible winkles are either male or female. The former may easily be distinguished, at least during the breeding season, by the presence of a penis (p, fig. 6) lying on the side of the head behind the right tentacle. Females have a smooth, unpigmented patch of skin in the same region, the ovipositor (opr, fig. 4). When the animals are not breeding both the penis and the ovipositor are greatly reduced in size and conspicuousness.

In males the testis (te, fig. 13) is a large, diffuse, branching organ which lies in the upper parts of the visceral mass, mainly more superficial than the digestive gland, but thrusting itself, wherever space is available, between the lobules of that gland. It is commonly a greyish, grey-green or grey-brown colour. The tubules of the testis join along the columellar (inner) side of the visceral mass to produce a duct which is formed from the walls of the gonad itself: this is the testicular duct (td, figs. 12, 13) which runs towards the mantle cavity. Its course, at first straight, becomes more convoluted as the cavity is approached, and this part is used as a seminal vesicle for the storage of spermatozoa. During the breeding season it appears for this reason as a chalk-white tube. The duct then narrows (rvd, fig. 13) and opens to the inner end of the mantle cavity at the male pore. The narrower section of the male duct is presumably derived embryologically from the kidney of the right side and it may therefore be distinguished as the renal section of the vas deferens.

From the male opening a ciliated sperm groove (sgr) runs forward on the floor of the mantle cavity on to the side of the head and to the base of the penis (p). Along the pallial stretch of this groove its walls are tall and hypertrophied by the development of folds of glandular tissue which produce a well developed prostate gland (pr).

The penis is conical, slightly flattened from side to side and carries the sperm groove to its summit along the dorsal edge. When at rest it forms a sickle-shaped structure lying behind the right tentacle and hidden within the mantle cavity, but at times of copulation it becomes engorged with blood and greatly extended. On the ventral border of the penis, though not

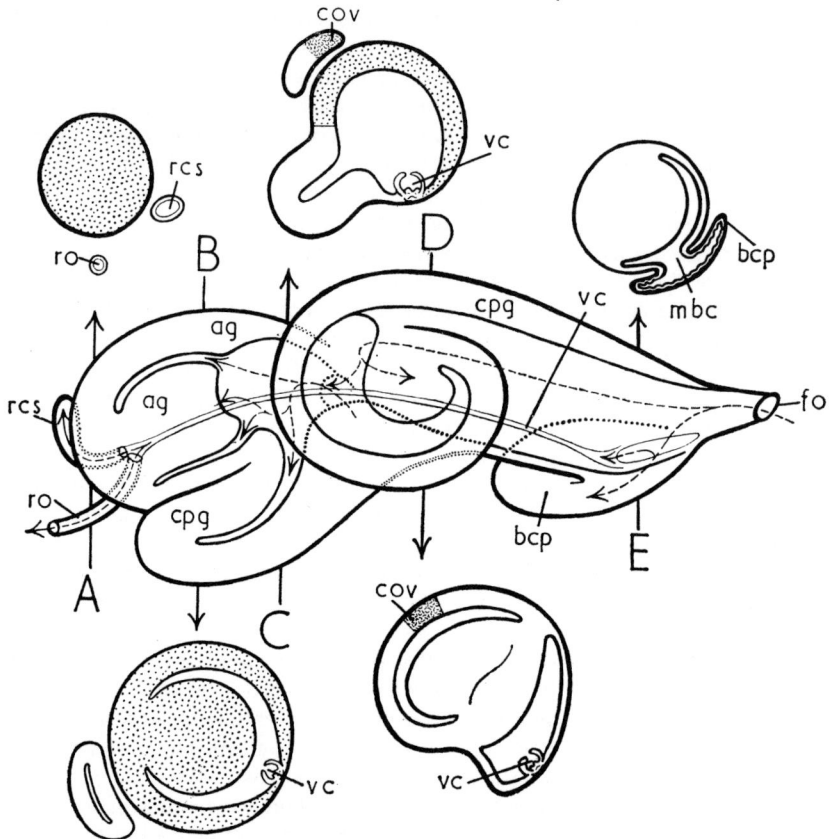

FIG. 26.—*Littorina littorea*: diagram of the female genital tract represented as a transparent object seen from the right side. A, B, C, D and E are transverse sections across the duct at the levels indicated. The arrows within the main diagram are intended to show channels of communication between sections of the duct: they are not related to the movements of germ cells or secretions. In the sections the albumen gland has been coarsely stippled; the covering gland has been finely stippled; the capsule gland is unshaded. To prevent obscurity the covering gland has not been shown in the main diagram.

ag, albumen gland; bcp, bursa copulatrix; cov, covering gland; cpg, capsule gland; fo, female opening to mantle cavity; mbc, mouth of bursa copulatrix; rcs, receptaculum seminis; ro, renal section of oviduct; vc, ventral channel.

extending right to the tip, lies a row of mamilliform structures which give the penis a serrated edge: these are glands (png, figs. 6, 12, 13) the secretion of which is said to hold the penis in position during copulation when only the elongated tip passes through the female pore and enters the bursa copulatrix, into which the spermatozoa are discharged.

The reproductive organs of the female winkles (figs. 4, 26) are more elaborate than those of the male. This is due to the facts that not only do they have to provide for the elaboration of female gametes and lead them to the exterior, but must also make provision for their fertilization in the course of this journey and for enclosing them, plus food, in a protective capsule within which the earliest developmental stages are undergone.

The ovary (ov, fig. 4) lies in the visceral hump in a position similar to that occupied by the testis in males. It may be recognized by its colour, which is yellow to pink or violet, and is very similar to that of the kidney. The ovary extends into the first part of the oviduct, which is therefore an ovarian duct; this runs to the inner end of the mantle cavity along a straight path. At one point along this part of the duct a side branch opens to the pericardial cavity. This is the gonopericardial duct, and it is said to be the inner part of the right kidney. The part of the oviduct between the ovary and the gonopericardial duct is therefore gonadial in origin, whereas the next section between the gonopericardial duct and the mantle cavity is the renal section of the oviduct (ro, figs. 4, 26). The female winkle differs from the male in retaining the original connexion between the renal section of the genital duct and the pericardial cavity, possibly because ova are large and immobile whilst—were it retained in the male—small, motile spermatozoa might travel along such a connexion and block it.

The female winkle also differs from the male in that the original opening of the renal section of the genital duct to the mantle cavity is not retained. Instead the oviduct opens at this point into a further section of duct which has been formed by the folding off of the wall of the mantle skirt to form a tube. In this way a third section of genital duct has been produced in the female, which may be distinguished as the pallial section of the oviduct. This lies alongside, and to the right of, the rectum and has brought the female pore near the anus, close to the mouth instead of in the depths of the mantle cavity.

The pallial section of the female genital tract is composed of a series of glands, which become extremely large and prominent at the height of the breeding season, and, in addition, of chambers for the reception and storage of spermatozoa received in copulation.

The renal section of the oviduct opens ventrally to the upper end of the pallial section. From here to near the female pore there stretches a ciliated groove which marks the morphological ventral side of this part of the female genital tract. This is the ventral channel (vc, fig. 26). At its upper end opens a small blind tubule, the receptaculum seminis (rcs), whilst between its lower end and the external aperture there opens to it a large blind sac, the bursa copulatrix (bcp), which extends below and to the left of the rest of the duct. These three parts of the female apparatus are concerned primarily with the reception and storage of spermatozoa. During copulation the penis is placed into the bursa copulatrix and spermatozoa and prostatic secretion are discharged. The spermatozoa then swim along the ventral channel of the female duct to the receptaculum seminis, which is usually visible at the upper end of the mass of glandular tissue as a refringent white streak due to the spermatozoa within it. They are arranged with their heads stuck to or embedded in the surface of the cells lining the receptaculum and they may receive nourishment from these.

The upper end of the ventral channel is thus a region which eggs, descending from the ovary, have to traverse in close proximity to a store of spermatozoa. It is, therefore, the most likely site of fertilization, and we may imagine a stream of fertilized eggs passing into the glandular mass which the female duct forms at this point. Perhaps because of the presence of the rectum immediately to the left of the female tract almost all its glandular development occurs on the right side; this is the explanation of the fact that the ventral channel lies on the left side.

The albumen gland (ag) is the most posterior section of the pallial duct. It has a translucent buff appearance in the fresh state and secretes material available for the feeding of the developing egg. Anteriorly it is followed by a second gland which is opaque white in the living animal. This is the capsule gland which at first lies ventrolateral to the albumen gland, but nearer the genital aperture expands to form a great mass (cpg). Across the right side of this gland runs a strip of orange or orange-pink tissue which secretes a kind of shell around the albumen which covers the fertilized egg. This 'shell' is usually called the egg covering and the gland may therefore be referred to as the covering gland (cov). The three glands co-operate to produce the egg capsule of the edible winkle.

A number of eggs, usually 3–5, enters the upper end of the ventral channel from the upper oviduct. There they are fertilized and passed forward. Their presence stimulates the albumen gland to secrete so that they become embedded in a mass of albumen from this source. At this point the eggs may be travelling in a group, but it is equally possible that they may be progressing down the duct singly, because the next thing which happens is that each, with its coat of albumen round it, becomes enclosed in an egg covering secreted by the covering gland. After this, however, the eggs become grouped, because it is in groups of 1–5 that they are to be found enclosed in the egg capsule, the material for which is produced by the capsule gland.

The capsule is of a disc shape, about 1 mm diameter or rather less, with a swelling in the centre in which the eggs are accommodated, each with its own covering and supply of albumen. The peripheral parts of the capsule form a flat flange to the central swelling and the whole has a somewhat hat-like shape. The outermost skin of the capsule is much tougher than the inner layers, although it retains its glass-like clarity. The hardening has been said to be due to the effect of the secretion of glands on the ovipositor. The ovipositor (opr, fig. 4) is a broad belt of tissue, slightly raised above the general level of the skin of the head, but made most obvious by the fact that it is almost completely unpigmented. It is ciliated and undoubtedly helps in carrying the capsules forward out of the mantle cavity. The capsules are pelagic, and give rise to planktonic veliger larvae. They are laid 2–12 hrs after copulation has occurred, usually at a high tide period and by night, until an estimated total of about 500 capsules has been produced. There may then be a pause after which laying may be resumed. The breeding season is long, January to July (Millport), or, at Plymouth, between November and May, the males ripening a little before the females. The breeding season has a peak in spring, during which each female may lay several lots of capsules, so that the total number produced in the whole breeding period may total about 5,000 per individual (Tattersall, 1920).

After laying, the capsules sink slowly, the rate being determined by the number of eggs they contain: this allows them to be well distributed by surface currents. The cleavage of the egg is completed after the first day and the larvae hatch on the sixth day as free-swimming veligers. The larva (fig. 227) has a yellowish spiral shell of one and a half turns and no definite architectural markings apart from faint spiral striae. Around the head the body is drawn out into two semicircular velar lobes (from which it is named) edged by strongly beating cilia, which are its main locomotor organs. The anterior parts of these lobes are marked with dark purple pigment, by means of which it may be recognized as the veliger of *Littorina littorea*. These are also visible as two semicircular patches showing through the shell when the animal is retracted. The larva swims in the plankton for a period of about two weeks by which time the shell has grown to two whorls and darkened in colour to a brownish hue. At metamorphosis the velum is lost and as a consequence the young winkle falls to the bottom and assumes a crawling mode of life.

CHAPTER 3

THE SHELL

T HE body of a mollusc is typically enclosed within a shell, and to many this dead product of the animal's secretory activity is more familiar than the living organism itself. Because of their beauty of shape and colour, molluscan shells have for long been collected with eagerness and at much expense; indeed, until a recent period within the last 200 years a 'cabinet' of shells gathered from all the world over was regarded as one of the main features of a naturalist's collection (Dean, 1936). Shells were sought from many sources and beautiful rarities like *Conus gloria-maris* sometimes changed hands at very high prices (see Cooke, 1895, p. 121). Today, collections of this nature are no longer regarded so highly, though many people still derive great pleasure from the contemplation of small private stocks or more extensive ones in public ownership in museums.

The molluscan shell is secreted by the mantle, or pallium, that is, by the epithelial covering of the visceral hump and mantle skirt and it has, therefore, the same shape as that part of the animal's body. Growth proceeds by the addition of new rings of conchiolin impregnated with calcium carbonate to the edge of the shell, and by the addition of calcium carbonate over the whole of its inner surface. The marginal accretion represents the formation of new shelly substance to accommodate the increased volume of the visceral hump of a larger animal; the addition to the inside thickens it so as to give the extra strength which is called for in a larger structure. Normally, the calcareous material secreted at the pallial edge is in the crystalline form of prisms of calcite lying normal to the surface of the shell, whereas the inner part is made of layers of crystals of aragonite parallel to the shell surface. In lower prosobranchs this part is often known as nacre, or mother-of-pearl, and much of the lustrous appearance of molluscan shells, especially their blue, or blue-green refulgence, depends upon the refraction of light by this inner layer. The relative extents of these layers and their precise microstructure vary within the prosobranch series and have been described by Bøggild (1930) in some detail. External to the whole mass of calcareous matter there lies a layer of conchiolin not impregnated with mineral material: this is the epidermis or periostracum. Sometimes, perhaps usually in prosobranchs, it is so delicate as to be all but invisible; sometimes, as in *Capulus* (fig. 27), it forms a distinct layer over the rest of the shell and imparts its own colour to the whole; sometimes, as in *Trichotropis* (fig. 94), *Velutina* or young *Viviparus* (fig. 56), it not only covers the shell but is drawn out into hairlike processes standing erect on the surface; usually it becomes worn off the older parts of a shell, and it is often absent from all but the youngest. The thickness of the periostracal layer is related to the habitat in which a mollusc lives, being greatest in animals from freshwaters or high latitudes and least in those coming from the warmer seas. Where it is lost in terrestrial or freshwater animals erosion often follows, so that it has a clear protective function.

Since it is generally assumed that the visceral hump of the ancestral gastropod was conical or cap-shaped, it follows that the shell which it produced was similar. With evolutionary advance, however, the visceral hump elongated and became coiled into a helicoid spiral,

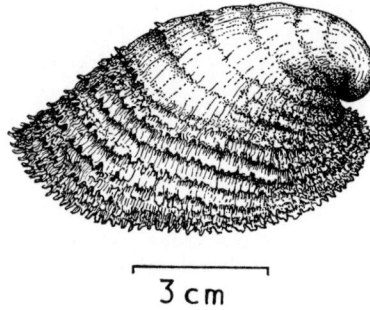

3 cm

FIG. 27.—*Capulus ungaricus:* the Hungarian cap shell. Note the thickened and spiny periostracal layer.

with the necessary consequence that the shell became similarly elongated and coiled: it can, in fact, be regarded as a long tube, approximately circular in transverse section and of gradually increasing diameter, which is rolled into a helicoid spiral coil. This, it has been suggested by Cox (1955), should be known as the helicocone; it is sometimes known as the conchospiral. The mathematical properties of the plane logarithmic or equiangular spiral are discussed by D'Arcy Thompson (1942). We may imagine a tube coiled so that the centre of each turn (or whorl) lies in a plane (fig. 28A). This would produce a shell which is disc-like in shape, actually biconcave because of the fact that the helicocone increases in diameter from the central part outwards (fig. 28B). In the vast majority of prosobranchs, however, the coil of the helicocone is not in one plane but is such as would be produced by winding it round the surface of a cone from apex to base: it is this shape which is a helicoid spiral, and a shell with that configuration is known as conispiral (fig. 28C, D). According to the size of the apical angle of the cone round which we imagine the helicocone to grow a great number of shapes may be produced, with the planispiral shell as that limiting case where the apical angle equals 180°. Normally the direction of coiling is right-handed, or in a clockwise direction when viewed from the origin: this gives rise to what is known as a dextral shell. Occasionally, as an abnormality in most prosobranchs, but regularly in a few species when adult (and in all opisthobranch larvae), the direction of coiling is reversed, when a sinistral shell results. The only British prosobranch which is normally sinistral is *Triphora perversa,* but Ancey (1906) recorded the occurrence of occasional sinistral specimens in the following species: *Viviparus viviparus, Valvata piscinalis, Theodoxus fluviatilis, Acicula fusca, Buccinum undatum, Nucella lapillus, Neptunea antiqua, Littorina littorea* and *L. saxatilis.* In these cases the entire symmetry of the animal is reversed, a case of *situs inversus;* in some, however, like *Lanistes* (an African ampullariid), the shell is sinistral but the animal retains the normal disposition of the parts found in dextral prosobranchs. This has been explained (Simroth, 1896–1907) as being due to reversal of the direction in which the apex of the shell points. Instead of the spire coiling downwards to the mouth it coils downwards from the mouth and so gives an apparent sinistrality. These shells are sometimes known as hyperstrophic. If we start with a planispiral shell (fig. 28A, B) showing dextral coiling we may imagine this to be converted into a conispiral shell in either of two ways. The first of these involves pressing a conical surface up from below against the centre of the planispiral shell so as to elevate the initial turns of the spiral above the original plane in which they lay, though leaving the latest formed turn (the mouth of the shell) in that position (fig. 28C, D). This gives a type of conispiral shell known as orthostrophic and is the usual prosobranch arrangement. The second method involves the pressing of the central turns of the shell downwards below the original plane but, once again, leaving the last turn there (fig. 28E, F) and so producing the type of shell known as hyperstrophic. In both these cases the direction of coiling has obviously not been affected, yet if the shells produced by the two methods are looked at in the conventional way—with the apex upwards and the mouth facing the observer—the orthostrophic shell will appear dextral and the hyperstrophic sinistral. The term 'ultra-dextral' has been used to describe this kind of sinistral coiling, which can only be distinguished from true sinistrality by examination of the soft parts of the animal, or of the operculum. The direction of coiling is presumably genetically controlled. This has been proved for the pulmonate *Lymnaea peregra* (Boycott, Diver, Garstang & Turner, 1930), but is also strongly suggested by the local occurrence of purely sinistral populations of normally dextral shells, such as that of *Neptunea antiqua* at St Jean de Luz, France.

Normally, each turn of the spiral as it forms is laid down in contact with the wall of the whorl which preceded it, so that the shell forms a solid object. In a few genera of prosobranchs such as *Vermetus, Tenagodus* (= *Siliquaria*), *Caecum* and some trichotropids the successive turns of the spiral are (either from the start or later) not in contact, giving an open spiral: these are known as evolute shells. In others, like *Clathrus* (fig. 141), *Pomatias, Valvata, Cirsotrema* and its allies (Clench & Turner, 1950) the successive whorls just touch one another, and the transverse section of each whorl is more or less circular. In many genera, however, each whorl overlaps part of that immediately preceding and successive ones are tightly pressed against one another; when this happens, the cross section of each whorl may no longer be circular. The line along which two successive turns of shell meet is known as the suture (fig. 37A), and where contact is just made this will be strictly linear. In animals where the side of one whorl overlaps another, contact is along a surface and the suture line which is visible externally is the outer edge of that surface. In cross section whorls may be circular, in which case the suture lines will be deeply sunk and the whorls bulge outwards between them: this gives rise to a ventricose shell (fig. 29A), examples of which are *Ampullarius, Viviparus* (fig. 175A), *Clathrus* (fig. 141), *Valvata* (fig. 302), *Littorina saxatilis, Hydrobia ventrosa*. Alternatively, the outer side of the whorl may be flattened, when there is no (or only little) dip to the suture lines: this gives rise to a flat-sided shell (fig. 29C, D), examples of which are to be found in the trochids, *Littorina littorea, Hydrobia ulvae, Cingula cingillus, Balcis alba, Turbonilla elegantissima*. Other shapes of shell tube can give rise to an angulated or turreted shell (fig. 29B), when the upper part of each whorl projects outwards below the suture line. Shells of this shape are seen in *Alvania carinata. Trichotropis borealis, Trophon barvicensis* and *Lora turricula*.

Further major factors in giving a shell its characteristic appearance are also dependent upon the relationship of each whorl to its older and younger neighbours. As the shell grows, it can be regarded as rotating in three-dimensional space around the protoconch, which acts as origin. With each rotation the diameter of the whorl becomes larger. If the rate of increase is constant then a surface tangential to the whorls is a regular cone, and this type of shell is conical, as in a trochid. This term may be used whether the sides of the whorls are flat or ventricose. It frequently happens that the rate of increase is not regular, but becomes either steadily greater, or steadily less, as the mollusc gets older. Shells of this type may be called conoidal, in that their shape is nearly conical. In the former group the sides of the cone are concave, in the latter group convex: Cox (1955) suggested that those with concave sides be called coeloconoid (fig. 29D) (e.g. *Calliostoma zizyphinum, Littorina littorea, Aclis supranitida, Pelseneeria stylifera, Neptunea antiqua*), whilst those with convex sides be known as cyrtoconoid (fig. 29C) (e.g. *Gibbula cineraria, G. umbilicalis, Rissoa parva, Bittium reticulatum, Triphora*

A

B

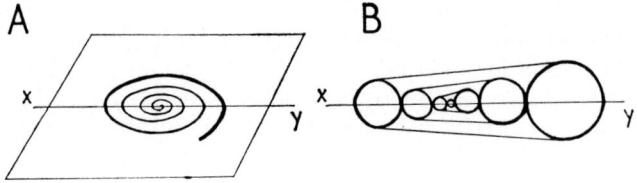

C

D

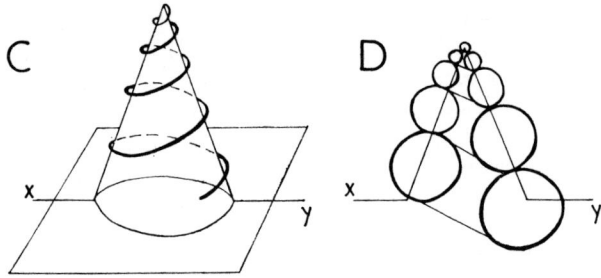

E

F

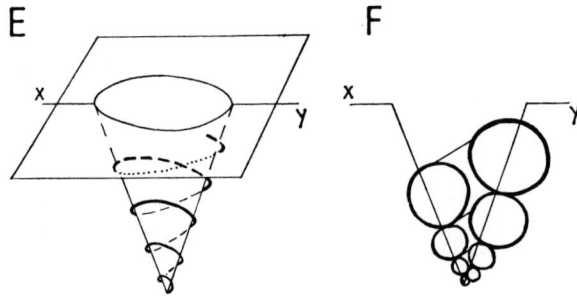

G

H

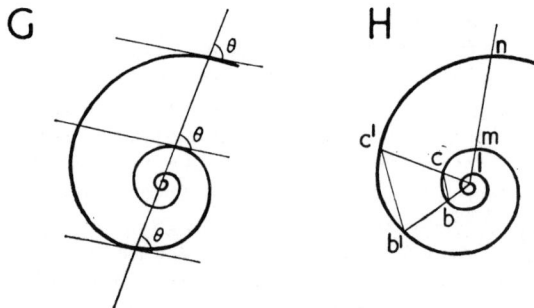

perversa, Cerithiopsis tubercularis, Eulima trifasciata). The term 'extraconical' is sometimes used instead of coeloconoid. Successive whorls may also vary in respect of the extent to which an older one is overlapped by a younger: if the amount of overlap is very slight then, necessarily, a rather tall shell results, the height of which is due to the fact that it is formed by the sum of the separate heights of the constituent whorls. If the amount of overlap is great, then a much squatter shell results until, at the limit, all whorls lie together in one and the same plane. In the former case tangents drawn to the sides of the helicocone meet beyond the apex at a small angle; in the latter case the angle is greater than a right angle and may be in the neighbourhood of 180°. This angle is usually known as the apical angle and is represented by α. *Turritella communis* (fig. 40A) is an example of a prosobranch with a small apical angle (about 17°), whereas *Conus* spp. (fig. 40B), *Ianthina* spp. (fig. 291), *Valvata* spp. (fig. 302) and *Velutina* spp. exemplify shells with a rather wide apical angle. In the final case, where the coiled tube of the shell lies symmetrically in a single plane the shell will obviously have the form of a biconcave disc, because of the fact that the diameter of the tube grows steadily greater with age. This type of shell, as pointed out by Cox (1955), has been called planispiral or isostrophic. No British prosobranch has a shell which is strictly isostrophic, however, except perhaps *Omalogyra atomus* (fig. 198B) and *Valvata cristata*, the best known examples being found in the bellerophonts. It may be noted here that the sinistral shell of *Lanistes* (see p. 52) which has been explained as an 'ultra-dextral' one—a bad term as it is not the dextrality of the shell which is excessive, but the apical angle—is, in fact, a shell in which the apical angle is obtuse, and in this genus exceeds 270°.

In certain cases where the helicocone is comparable to a screw of low pitch, its early turns are visible at one pole, dipping to the apex of the shell, but the last turn conceals all the others at the other pole of the shell: this type of shell is known as involute (fig. 30). No British prosobranch has a truly involute shell, though those of *Velutina velutina* and some species of *Lacuna* are very nearly so; the opisthobranch *Cylichna truncata* bears a genuine involute shell. Still another relationship of the whorls of a shell is that in which the youngest turn grows so as to conceal all those that have gone before: this condition is known as convolute and may be illustrated by the shells of *Trivia* spp. (fig. 31) or *Simnia patula* (fig. 288).

In the various formations of the helicocone which have been described above attention has been directed primarily to the degree of overlapping of the whorls. The general appearance

FIG. 28.—Diagrams to illustrate the coiling of gastropod shells. A represents a plane on which a spiral line marks the centre of a helicocone coiling dextrally; B represents a vertical section through such a shell along the line xy. This type of shell is planispiral.

C represents a plane on which a spiral line marks the centre of a helicocone coiling dextrally; a cone has been pressed upwards through the plane from below, its apex coinciding with the centre of the spiral. D represents a vertical section through such a shell along the line xy. This type of shell is conispiral and orthostrophic.

E represents a plane on which a spiral line marks the centre of a helicocone coiling dextrally; a cone has been pressed downwards through the plane from above, its apex coinciding with the centre of the spiral. F represents a vertical section through such a shell along the line xy. This type of shell is conispiral and hyperstrophic. Note that although the coiling of the spiral is unchanged it appears to be reversed when the shell is viewed in the conventional way.

G is a logarithmic or equiangular spiral; note that the angles θ between a line drawn through the central point of the spiral and tangents at the points where it intersects the spiral, are constant. H is a logarithmic or equiangular spiral in which two triangles linking the origin with the points bc and b'c' respectively are inscribed, by joining corresponding points in successive turns; note that these figures are similar. The line lmn, which is a radius vector, is cut by the spiral into lengths which are in geometrical proportion.

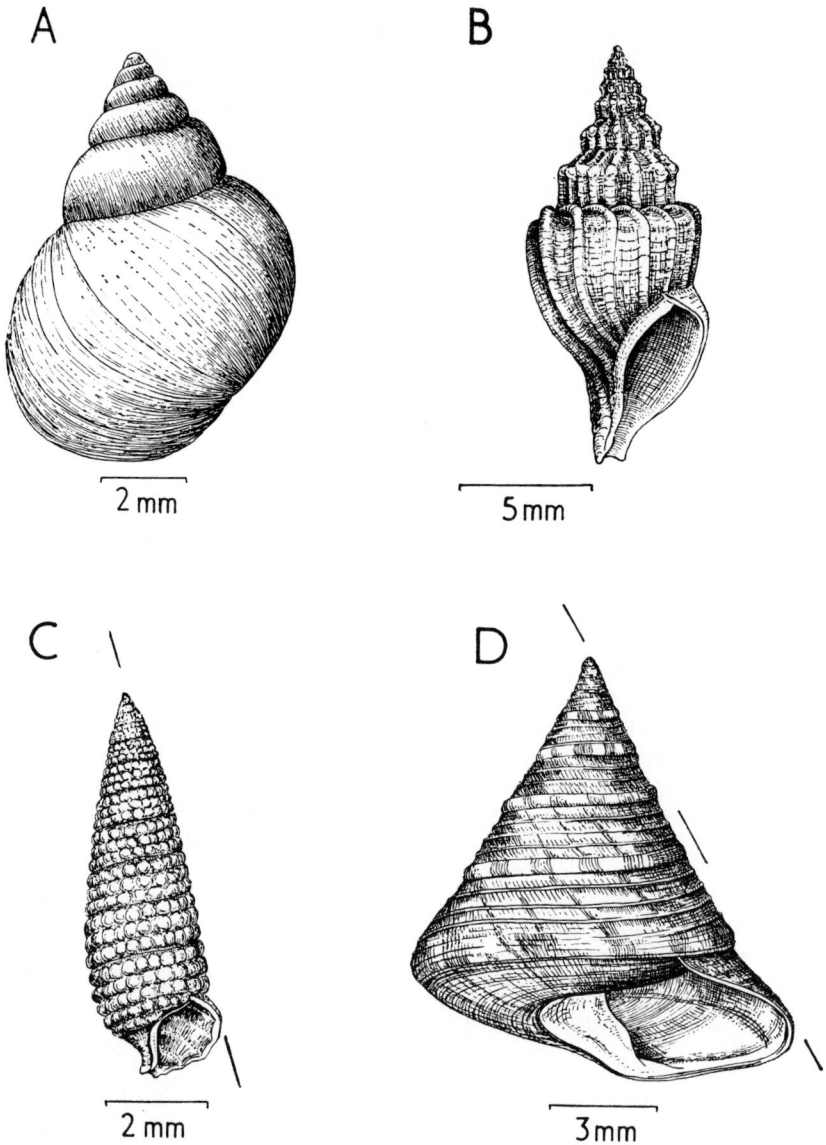

FIG. 29.—To illustrate shell shape. A, *Littorina saxatilis,* to show a ventricose shell; B, *Lora turricula,* a turreted shell; C, *Cerithiopsis tubercularis,* a cyrtoconoid shell; D, *Calliostoma zizpyhinum,* a coeloconoid shell. In C and D lines are drawn from the apex to form a tangent to the outer lip in order to reveal the convexity of the spire in C and its concavity in D.

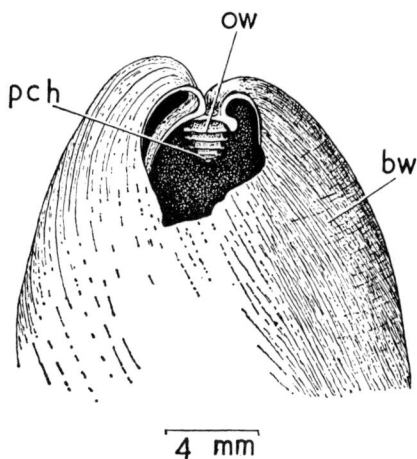

FIG. 30.—*Bulla ampulla:* an involute shell broken open to show the spire covered over by the body whorl.
bw, body whorl; ow, broken edge of older whorl now covered by body whorl; pch, position of protoconch.

of a gastropod shell, however, is also much affected by another aspect of this to which no reference has yet been made: this is the relation of the whorls to the axis round which the turns of the spiral revolve. As the helicocone grows from the protoconch its configuration may be regarded as due to the movement of the generating circle of mantle skirt which secretes it (a) around the origin in a spiral, and (b) along an axis at right angles to the plane of the spiral. In certain cases the inner wall of the whorls may all rest on the central axis around which the shell lies, forming a solid spiral; in others these walls come to lie at greater and greater distances from it so that a hollow spiral results. In shells conforming to the former plan it is naturally not possible to see the inner walls of the whorls except by breaking the shell: in the latter group they may be looked at from the base of the shell, where there is an opening which allows one to look up the centre of the series of turns comprising it (fig. 274). This opening is the umbilicus, or, more accurately, the inferior umbilicus. In some shells (e.g. *Architectonica, Valvata cristata, Skeneopsis planorbis*) almost the whole extent of the inner sides of the whorls of the shell is visible through a wide umbilical opening (figs. 32A, 282), but this is distinctly unusual: these shells are phaneromphalous. In most prosobranchs the shell is hemiomphalous with the umbilicus reduced to a chink (*Gibbula magus* (fig. 32C), *Clathrus, Monodonta lineata* (fig. 273), *Lacuna* spp., *Viviparus*) or crack (*Natica,* fig. 9) or it is lost altogether (cryptomphalous, anomphalous) because of the tightness of the spiral coiling of the shell, or by its closure with secreted shelly material known as callus.

In involute and isostrophic shells there may be not only an inferior umbilicus produced as described above, but also a superior or apical umbilicus produced by the later coils uprising (because of their greater diameter) around earlier ones (fig. 30).

The inner walls of the rotating tube which makes up the helicocone may, as just mentioned, come into contact with one another where there is no umbilicus, and produce a central pillar around which the different whorls spiral. This central axis of the shell is the columella and to it is attached the great columellar muscle by contraction of which the head and foot of the animal may be withdrawn into the shell. In most species the surface of the

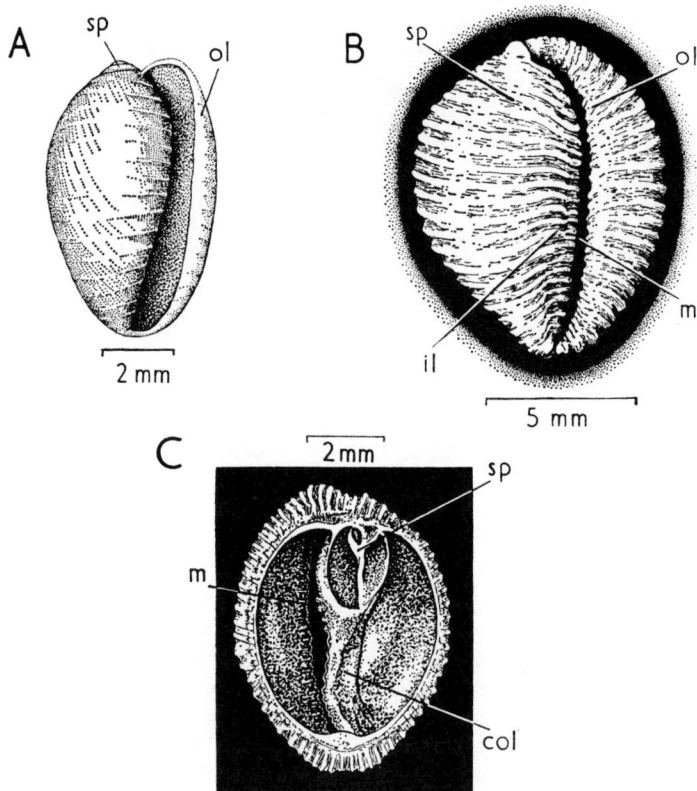

FIG. 31.—*Trivia monacha:* A, shell of young animal showing exposed spire; B, shell of adult with concealed spire (convolute); C, shell of adult halved to show internal spire.
col, columella; il, inner lip; m, mouth of shell; ol, outer lip; sp, spire.

columella is smooth, but in some, such as *Monodonta* and *Odostomia,* there is a tooth on the columella, i.e. a tooth-shaped projection on its lower end, visible when the intact shell is examined (tcl, fig. 273). This will, obviously, be the end of a ridge which runs the length of the columella and provides extra surface for the origin of the columellar muscle. In some species (Harpidae) more than one such fold may occur, when the columella may present a plaited appearance. These 'teeth' or plications are probably due to secretion of material over the surface of the columella by a mantle which has outgrown the space available for it and which has therefore become folded. Folds will be most marked when the mollusc is withdrawn into the shell and this will in turn depend in part upon whether the origin of the columellar muscle (which pulls upon the mantle edge) is remote—far up the columella—or near. It will also be related to the pull of the muscle and is a more likely development where that is excessive, as in animals with a long proboscis.

At the lower end of the helicocone lies the mouth or aperture (fig. 37A) of the shell, through which the body of the gastropod extends during activity. The edge of the mouth, the growing zone of the shell, is known as the peristome and may be considered as composed of an outer lip or labrum (lying away from the axis of the shell) and an inner or columellar lip or labium

C

A

1 cm

B

1   cm

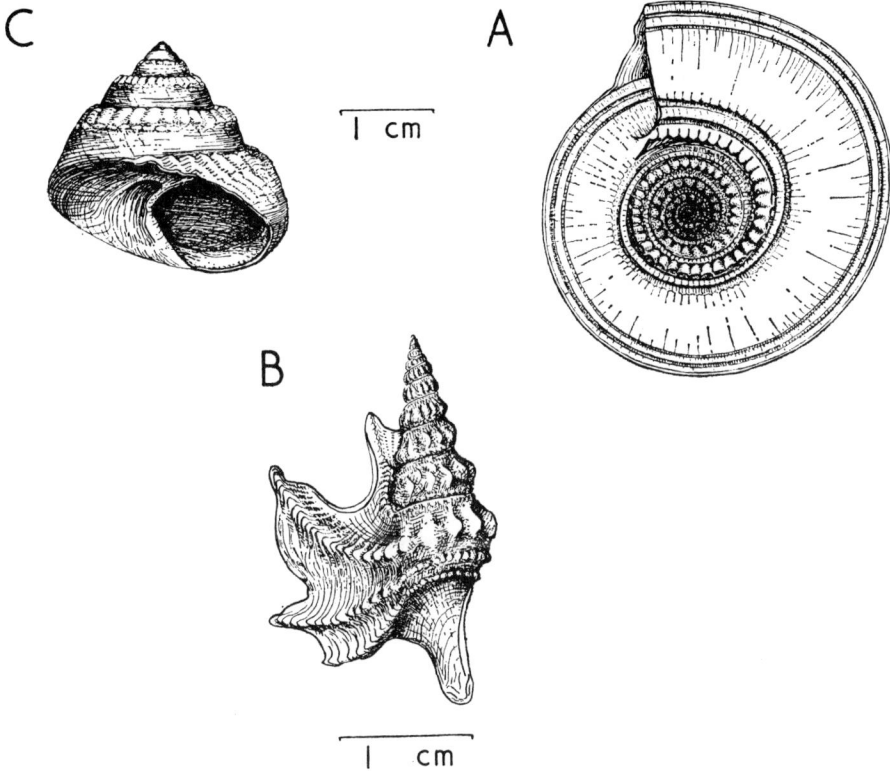

FIG. 32.—Shells: A, *Architectonica* sp., seen from below to show wide umbilicus; B, *Aporrhais pespelicani*, to show expanded outer lip; C, *Gibbula magus*, to show adapical ribs, and reduced umbilicus.

(lying close to the axis of the shell). The degree of development of the margin of the mouth varies greatly from one kind of gastropod to another; rarely indeed does it project as a free edge all round as it does in *Pomatias elegans*. More commonly the right side of the aperture (as viewed in dextral shells seen in apertural view) is complete, forming the outer lip, but the left, forming the inner or columellar lip, is much less well developed, and may be reduced to little more than a callus or glaze detectable on the outside of the previous whorl of the shell. Indeed, in some animals, even this sign of inner lip appears to be absent and the peristome can then be regarded as composed of a U-shaped edge, the circle being completed by the wall of the previous whorl applied to the arms of the U. This is an area known frequently as the parietal wall—as Cox (1955) commented, a most unfortunate term which he suggested might be replaced by the term parietal region. That part of the inner lip which runs over the surface of the previous whorl is also known as the parietal lip.

The peristome may be, and in some primitive shells is, a plane structure. With advance in evolution, however, this may no longer be true and the peristome exhibits a complex tridimensional curvature; some of the changes leading to this are given below and it is clear that they have a functional basis and may be understood in terms of the animal's way of living. Other changes, however, which cannot so far be interpreted in any such way also affect the peristome, although these seem somehow to be concerned with the change from primitiveness to an advanced state. One of these, mentioned by Davies (1939), is the angle which the plane of the peristome makes with the axis of the helicocone round which it rotates: in shells like those of *Calliostoma, Gibbula* (fig. 37B) or *Haliotis* the peristomial plane cuts the central axis at an angle of about 70°, whereas in most higher prosobranchs, mesogastropods and stenoglossans (fig. 37C) alike, the angle becomes very small. Davies remarked that this change in angle seems to be a necessary prerequisite for such further evolution in the peristome as the formation of siphonal canals, though the underlying functional significance is not apparent [p. 567.]

In the majority of prosobranchs the peristome is entire, that is, it forms a smooth curve from one end to the other. In a few of these animals, however, this curve is interrupted either by an ingrowth or by an outgrowth corresponding to similar changes in the shape of the edge of the underlying mantle skirt. An embayment there gives rise to a marginal slit at the lip of the shell as in *Emarginula* (fig. 33), *Scissurella* (fig. 248A), young specimens of *Puncturella* and *Diodora* and at times, *Haliotis* (see below, p. 76).

An outgrowth of the mantle edge occurs locally on the left side in prosobranchs of the families Cerithiidae, Cerithiopsidae, Triphoridae, Lamellariidae, Eratoidae, Muricidae, Buccinidae, Nassariidae, Fasciolariidae and Turridae and the shell is expanded correspondingly. In the first three families (fig. 29C) the degree of outgrowth is slight, forming a spoon-shaped bulge on the mantle edge and an expansion of similar shape on the peristome where the outer lip runs on to the lower end of the columella. In the other families this expansion of the mantle edge forms a long outgrowth which is kept rolled into a tubular form and which is mobile and muscular (s, fig. 34): it is an inhalant siphon and allows the gastropod to draw water into the mantle cavity either whilst it burrows, or from a spot remote from that which its movement or its feeding may be fouling. In these siphonate forms the shell is usually drawn out into a siphonal tube for the accommodation of the pallial siphon (sit, fig. 36A), though this has not happened in eratoids (fig. 35) and lamellariids. The length of the siphonal or anterior canal of the shell and of the pallial siphon do not necessarily correspond—*Buccinum undatum* (fig. 269) and *Nassarius reticulatus* (fig. 266) for example, have long siphons and short canals, whereas in some species of *Trophon,* and more particularly in some foreign species of *Murex* (fig. 36A), the siphonal canal may be very long and the siphon not protrude much from it. The siphonal canal is normally open along one side like the siphon, but may, as in *Ocenebra erinacea* (figs. 39, 263), be closed by fusion of the edges.

At the opposite end of the aperture, where the outer lip abuts adapically against the surface of the body whorl, a posterior notch or canal is occasionally developed as an outgrowth of the peristome (e.g. some conoideans). This accommodates the right side of the mantle edge where the rectum ends and so permits the faecal products to be directed away from the body on discharge. This corresponds functionally to the slit or hole of zeugobranchs but has been separately evolved.

It will be useful at this stage to introduce some other terms used in the topographical description of the gastropod shell before turning to the question of what ornamentation its surface may bear (fig. 37A). In the majority of shells the constituent whorls are visible, but all except the youngest are partially concealed by overlap of later formed turns. When the mollusc retracts itself within the shelter of the shell the most recently formed whorl naturally contains its head and foot, whereas older whorls never contain anything except the visceral mass: for this reason the youngest whorl, the whole external surface of which is visible, is called the body whorl. It extends back from the aperture for one complete turn of the

FIG. 33.—Shells. Top left, *Emarginula reticulata,* × 11. Top right, *Diodora apertura,* × 4. Bottom right, *Patina pellucida pellucida,* from fronds of *Laminaria,* × 7. Bottom left, *Patina pellucida laevis,* from holdfast of *Laminaria,* × 6.

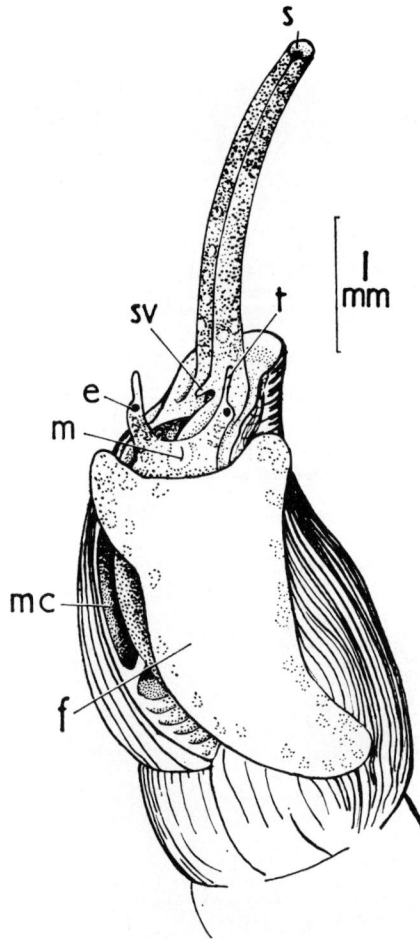

FIG. 34.—*Mangelia powisiana:* alive, in ventral view, with extended siphon.
e, eye; f, foot; m, mouth; mc, mantle cavity; s, siphon; sv, valve at base of siphon; t, tentacle.

helicocone, and its height is to be measured by the distance between the lowest point on
the apertural rim and the point nearest the apex of the shell (its extreme adapical point) at the
level at which the body whorl begins. The remainder of the shell is known as the spire. Its
height is to be measured by the distance between the apex and the lowest visible point (the
extreme abapical point) on its youngest whorl. The basis of measurement of the whorls in
these two cases is thus different, but is justifiable on the fact that much of the general
appearance of a shell is linked with the ratio between the height of the body whorl and the
height of the spire and the latter is, in turn, largely dependent upon the degree to which each
of the whorls is covered by its successor in the helicocone. A further measurement of the
shell which helps to express its general appearance is the degree of swelling exhibited by the
helicocone, that is, the diameter of its constituent tube. This may be measured by the size of

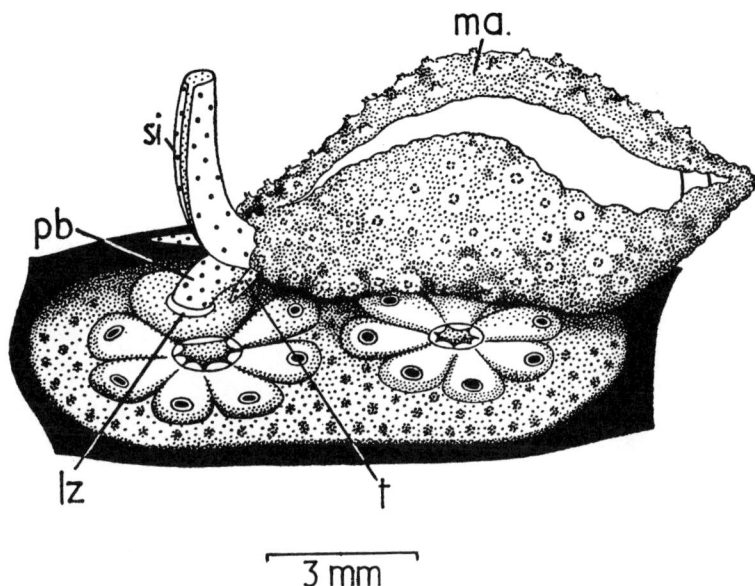

FIG. 35.—*Erato voluta:* feeding on *Botryllus.*
   lz, lip of mouth of zooid; ma, mantle extended over outer surface of shell; pb, proboscis thrust into oral
siphon of zooid of *Botryllus;* si, siphon; t, tentacle.

the broadest part of the body whorl, which is known as its periphery, and which is clearly equal to twice the greatest radius of the body whorl. This measurement will also be related to the apical angle α.

   The initial shell of a gastropod is secreted during embryonic life by the deposition of conchiolin and, slightly later, calcareous matter over the apical part of the visceral mass, and this is gradually expanded during development by the addition of rings to its margin. This part of the shell is usually devoid of ornament and is often clearly separated from later additions (*Epitonium* sp. fig. 231). It is called the embryonic shell or protoconch I. If the embryos hatch as juveniles this shell abuts the initial part of the adult shell or teleoconch, produced by the same process of marginal secretion but with its own distinctive pattern of ornament. If, however, there is a larval stage, especially if it is prolonged, the two are separated by an area of shell formed during larval life, the larval shell or protoconch II; again this may have its own pattern of ornament or colour. In some gastropods the early parts of the teleoconch are, like the protoconch, smooth surfaced, and so long as the mantle skirt which secretes the shell grows in length at a rate which equals its rate of growth in breadth, this smooth condition will persist and give rise to a shell such as that seen in *Natica* (Grabau, 1928) or *Tricolia* (fig. 39). In many prosobranchs, however, the two rates of growth do not remain equal and as the animal becomes older the growth of the mantle in breadth is greater than its growth in length. The mantle edge therefore becomes disproportionately broad for the size of aperture in which it lies and can only be accommodated there by being puckered. This folding of the mantle edge will be reflected in a folding of the shell which is secreted by it and a series of crenulations running along the length of the helicocone is the consequence. These, if small, are known as

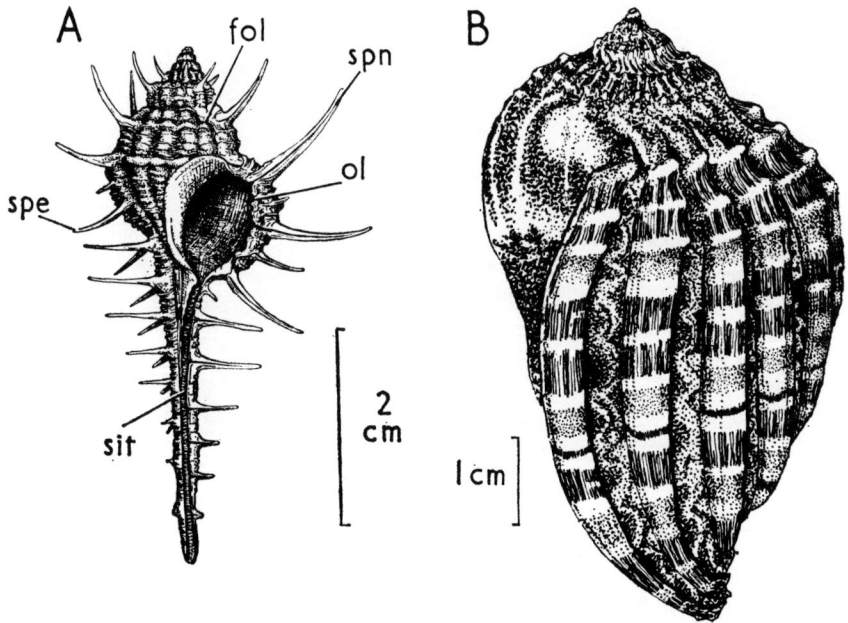

FIG. 36.—Shells: A, *Murex tenuispina;* B, *Harpa major.*
    fol, former position of outer lip; ol, outer lip; sit, siphonal tube; spe, spine which will be removed because it
    obstructs aperture when growth next occurs; spn, spine on outer lip.

striae and because they share in the general spiral motion of the helicocone they are usually called spiral striae (figs. 37A, 38). Provided that the ratio (growth of the mantle in breadth/ growth of the mantle in length) remains constant during the lifetime of the mollusc these striae will increase in number at a constant rate, new striae being intercalated between older ones on the younger whorls of the shell. In some cases, after an initial disparity of growth rates has produced a number of striae, the rates become equalized and no further striae appear. In most shells the first stria to appear lies half-way between the adapical and abapical margins of the whorl, the next two subdivide its adapical and abapical halves and later series appear in between these. Shells which show this type of spiral ornamentation in varying degrees of perfection are: various species of rissoid like *Cingula alderi, C. proxima, C. semistriata* (fig. 285B); the species of *Littorina; Eulimella nitidissima,* and *Neptunea antiqua.*

In a further series of prosobranchs the spiral striae do not remain of equal prominence but some become exaggerated in height and appear more prominent than others. This is particularly noticeable in the shells of some trochids, such as *Calliostoma zizyphinum* (fig. 29D), and it is also to be seen in *Alvania carinata, Turritella communis* (fig. 40A), *Aclis minor* and *Trichotropis borealis.* One particular form taken by this exaggeration of spiral striae is that in which the first formed one becomes simultaneously exaggerated and angulated. This may often occur along with a flattening of the rest of the surface of the whorl, especially of that part which lies adapical of the angulated stria; the wall of the whorl below the main stria may also flatten on occasion, but does so much more seldom. This arrangement suggests that the exaggerated stria results from the presence of one large fold of the edge of the mantle skirt which

A

apex

whorl

suture

spiral stria

height
of
spire

rib

inner lip

parietal
lip

aperture

height
of
body
whorl

columellar
lip

outer lip

siphonal
canal

B

C

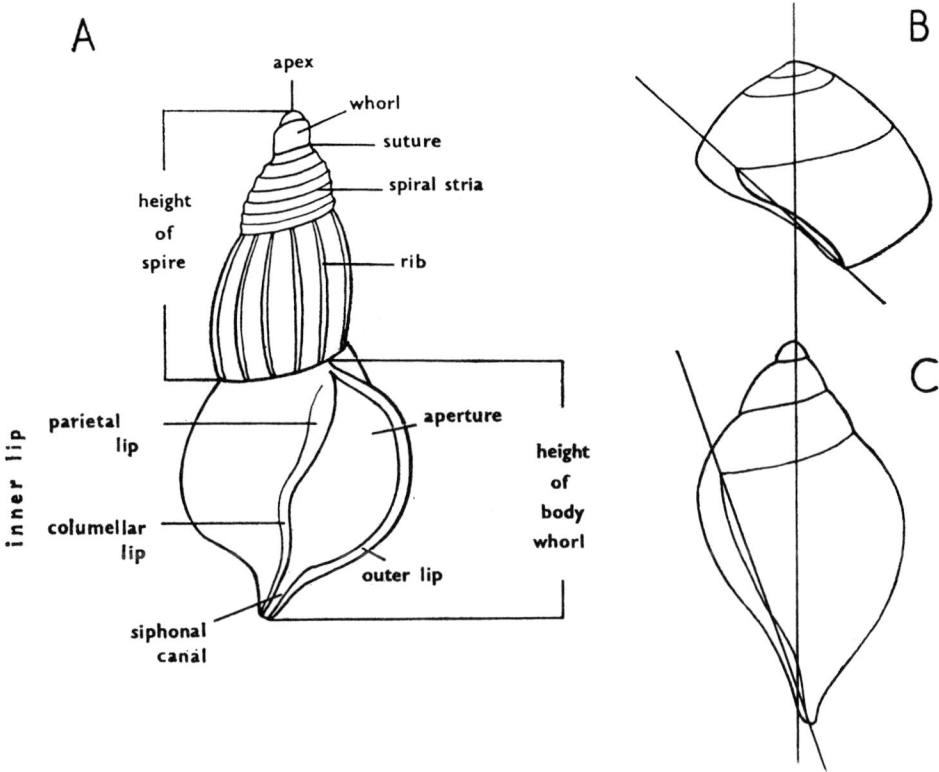

FIG. 37.—A, diagram of gastropod shell to explain some of the terms used in its description; B, shell of a trochid compared with C, shell of *Nucella*, to show difference in the angle made by the outer lip and vertical axis.

accommodates all the surplus breadth. The result is an angulated shell, showing a shoulder projecting around the adapical part of each whorl, with or without other spiral striae abapically. The appearance of the shell varies greatly with (a) the position on the whorl—adapical or abapical—of the angulated stria and (b) the relationship between this and the degree of overlap of two successive whorls. When the angle is low down on the whorl and is placed at the suture line with the next whorl a straight-sided trochoid shell results (fig. 29D). If the angle made by the stria is obtuse the effect on the body whorl is such that the shell tapers at the base (e.g. *Turritella communis*, fig. 40A); if it is acute the base may be flat, as in *Calliostoma*, or even concave. The angle may, however, be placed high up on the whorl. If this is, again, coincident with the suture line between whorls then a straight-sided spire results, but this time the spire is short in relation to the length of the rest of the shell as in *Conus* (figs. 40B, 41A); if the succeeding whorl embraces the older coil below the angle then an angulated or turreted shell results, the length of its spire varying with the degree of whorl left exposed at each turn and also with the angle made by the shell adapical of the angle and the axis. If the shoulder slopes then the spire will be tall (fig. 29B); if the shoulder angle comes to lie at right angles to the axis of the shell then a pagoda-like outline may be produced. Not many British

FIG. 38.—Shells. Top left, *Nucella lapillus,* abapertural view, × 2. Top right, *Nucella lapillus,* mature, apertural view, × 2. Bottom left, *Nassarius reticulatus,* mature, apertural view, × 2. Bottom right, *Nassarius incrassatus,* young, apertural view, × 4.

FIG. 39.—Shells. Top left, *Tricolia pullus;* note the calcareous operculum, × 10. Top right, *Ocenebra erinacea,* × 2. Bottom left, *Rissoa parva,* ribbed shell, abapertural view, showing characteristic comma-shaped pigment streak on upper part of outer lip, × 18. Bottom right, *Rissoa parva* var. *interrupta,* smooth shell, apertural view. × 18.

A

B

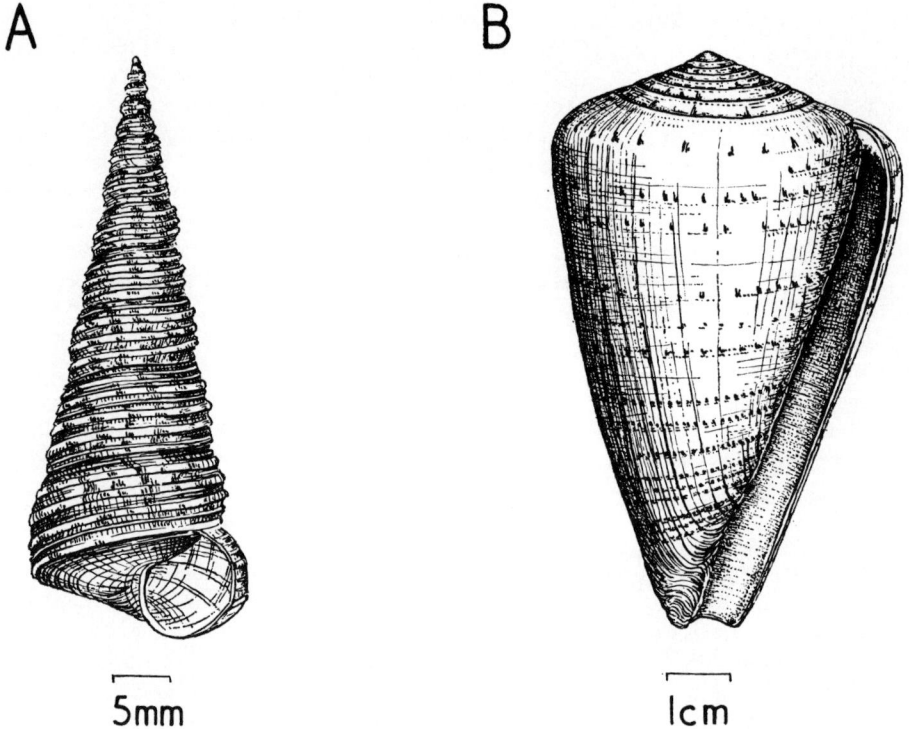

5mm

Icm

FIG. 40.—Shells: A, *Turritella communis;* B, *Conus* sp.

shells can be used to illustrate these varieties of angulated shells, but a few show them complicated either by the presence of subsidiary striae in the abapical half of each whorl or by the presence of other types of ornamentation to be described below. Shells of this type are exemplified by *Alvania carinata, Aporrhais pespelicani* (fig. 32B), *Trophon barvicensis.*

The spiral striae running along the course of the helicocone represent a continuous element in the architectural ornamentation of the shell. There is, in addition, a discontinuous element represented by structures which lie across the helicocone and which, therefore, are elongated in a direction more or less parallel to the axis of the shell. If they are precisely parallel to it they may be described as orthocline, if they are inclined at their adapical end towards the direction of growth of the helicocone they are called prosocline, if in the opposite sense, opisthocline. Prosocline structures are of much commoner occurrence than opisthocline.

The occurrence of this element in shell architecture is due to the fact that the fundamental growth process is discontinuous and may be made more so by its reaction to environmental variation, seasonal or otherwise. The secretion of new shell material always involves the addition of a strip of substance to the peristome, and the outer surface of the shell is almost always marked by a series of growth lines where these strips adjoin one another (e.g. *Natica,* where they constitute the only element of this nature in the shell). Superimposed upon the more or less regular secretion of these pieces of new shell may be a seasonal effect, due, perhaps, to cessation of secretion during hibernation from cold, or during aestivation to avoid desiccation, and this may produce marks by means of which the age of the animal may be determined. Quite apart from seasonal rhythms, however, periodic swelling of the mantle skirt seems to occur, especially in marine species, and to result in the formation of a greatly thickened piece of shell placed at intervals along the length of the helicocone. These thickenings are known as ribs or costae (fig. 37) and are usually spaced with complete regularity along the whorls of the shell, suggesting that the thickening of the mantle edge, or the hypertrophied secretory activity which is responsible for their occurrence, is a rhythmical event, although it is difficult to see any underlying reason for it as it is clearly not related to seasonal change and there are some species, e.g. *Rissoa parva* (fig. 39) in which shells of two types, ribbed and ribless, occur. In almost every individual shell in which ribs are normally found one rib will be found to coincide with the outer lip of the shell aperture: this is known as the labial rib or varix. It is extremely rare to find a shell of a species in which these occur without a labial rib, that is to say, in the period between the secretion of one rib and the next. From this it may be deduced that the growth of the helicocone is not regular, but spasmodic; that the growth between one rib and the next takes place rapidly and lasts a short time—only 2 days in the case of *Murex* spp. according to Abbott (1954)—which is why animals in this state are only rarely found. The secretion of ribs, therefore, may not represent so much an increased rate of secretion as a decreased rate of growth, so that the shelly material which is produced piles up as the mantle skirt remains stationary. Some prosobranchs found in this country which show shells bearing ribs (with or without spiral striae) are: *Rissoa parva, R. inconspicua, Alvania crassa, Turbonilla elegantissima, Nassarius incrassatus* (fig. 38) and *Mangelia coarctata*.

Although it is possible to explain the formation of labial and other ribs as due either to an increased rate of secretion of shelly material recurring periodically, growth remaining constant, or, alternatively, to a decreased rate of growth recurring periodically whilst secretion remains constant, there are other events occurring in the formation of some shells which are more complex and seem to involve simultaneous change in both rate of growth and rate of secretion. These are structures such as varices and spines, which are advanced types of shell ornamentation evolved from ribs. They are exhibited by few British gastropods—indeed none shows either type of decoration in anything except a rudimentary form, and it is necessary to refer to exotic shells in quoting typical examples of both. Just as the secretion of shelly substance at the outer lip of the shell produces a labial rib (the position of the secretory mantle edge being constant), so a varix arises when the mantle edge turns outwards while still secreting. This produces an out-turned outer lip projecting more or less at right angles to the rest of the body whorl, perhaps ending smoothly, or in a number of points, or in a number of spines. Examples of this may be seen in *Clathrus* sp. (fig. 141) and in *Aporrhais pespelicani* (fig. 32B) amongst British shells and (particularly clearly) in such others as *Pterocera* and various species of the genus *Murex* (fig. 36A). In *Aporrhais* and *Pterocera* the expansion of the outer lip occurs only in the fully grown shell of which it is a permanent feature, but like the formation of ribs, the formation of out-turned lips may occur at regular intervals during the life-time of the gastropod and the outer surface of the whorls will therefore be marked by a series of out-turned flanges, with or without spines, indicating previous positions occupied by the outer lip. It is these structures which are termed varices. They may be seen beautifully in *Clathrus* spp. and (on a small scale) in young specimens of *Nucella lapillus* (fig. 262) and in adults which have been living in deep water where the shell has not been worn smooth, and on shells of *Ocenebra erinacea*. *Trophon barvicensis* shows them, equally small, but partly decorated by spines. For the best examples of this type of structure, however, it is again necessary to quote from animals which are not found in this country: *Harpa* spp. (fig. 36B) show well formed, but simple varices, whilst species of *Murex* such as *M. tenuispina* (fig. 36A) or *M. palma-rosae* show them in a much more elaborate and spiny form. In the formation of these it is clear that a very considerable growth in the mantle edge has occurred, leading to a great increase in the amount of calcareous matter produced.

The secretion of varices or of spine rows which run across the whole breadth of a whorl, from its adapical to its abapical border, is clearly going to complicate the formation of the shell when, in the process of growth, the next turn of the helicocone brings the inner lip to lie alongside the former outer lip. In this position the varix will block much of the aperture and interfere with the movements of the animal in and out of the shell: in some cases indeed (*Murex tenuispina*) the length of the spines is so great that they project completely across the aperture and beyond the outer lip. In these circumstances it is necessary for the spines or the varix to be removed from that part of the older whorl about to be overgrown by the younger. How this is undertaken is not known, but it is presumably by the same process as erosion of shell (though on a much less impressive scale) is known to be achieved in other instances (see p. 72).

If a shell exhibiting ribs or varices or spine rows is examined (fig. 141) it will be noticed that successive varices are not arranged in rows the projection of which on an axial plane would be parallel to the axis of the shell, but that they run in spirals which coil in a direction opposite to that in which the shell itself is turning. This indicates a steady decrease in the proportional amount of shell which is laid down between the 'quiescent' periods—or, alternatively expressed, the periods during which the animal makes new shell become progressively shorter. This has been regarded by Grabau (1902) as an indication that Minot's law of senescence applies to these animals, which show signs of ageing from the very start of their existence. This may be so, but it is well to remember that the size of ribs, varices, spines and other outgrowths gets progressively greater as the size of a shell increases and it may be that the total physiological activity of the animal, in the secretion of shell plus ornamentation, increases steadily throughout its lifetime.

The two types of structure described above, spiral striae and ribs, are the bases of ornamentation of the gastropod shell: the former is in Davies' phrase (1939) a 'space rhythm', the latter a 'time-rhythm'. Either may occur alone, but more frequently both kinds are present and the definitive structure of the shell is achieved by means of the interaction of the two types of architecture. In many cases there is produced in this way a simple network or reticulation over the surface of the shell; sometimes, as in many of the Stenoglossa, the spiral striae run clearly over the surface of the ribs of the shell, sometimes, as in *Chrysallida* and *Turbonilla,* the ribs run equally obviously over the striae, but in many others the two reinforce one another at the nodes of the reticulation and produce a surface elevated into bosses or short tubercles where ribs and striae cross. This is well seen in many British shells: *Cantharidus striatus, Alvania cancellata, Bittium reticulatum, Triphora perversa, Cerithiopsis tubercularis* (fig. 29C), *Nassarius reticulatus* (fig. 38) and *Philbertia asperrima* will serve as examples. In other shells such as those in which spines occur it will be discovered that these also are located at points where ribs and striae cross (e.g. *Murex* spp.) and the points on the expanded lip of *Aporrhais* (fig. 32B) and *Pterocera* are similarly related to striae. This reticulation or cancellation of the surface is also related to the colour pattern of the shell (see p. 79).

In the examples which have been discussed so far the decorative elements of the shell surface have been supposed to affect the entire extent of the whorls. This, however, is by no means necessarily so, and it is common to find that the adapical part is decorated in a different way from the abapical half. This will, in most cases, be detectable only on the body whorl because that is normally the only coil of which the abapical half is visible. A common difference of this sort is the disappearance of ribs over the abapical part of the whorl so that they form small projections limited to the adapical region. Frequently, too, their interaction with striae is graded from a maximum in the adapical region to a minimum abapically. This tendency is particularly noticeable in angulated shells where the primary stria is exaggerated, and if this interacts with ribs (which may also be exaggerated adapically) there is produced a series of prominent tubercles or spines on the keel of the whorl.

Some examples of British prosobranchs showing a limitation of ribs to the adapical part of the whorls are: *Gibbula magus* (fig. 32C), *Rissoa parva* (fig. 39), *Cingula semicostata* (fig. 285A). *Buccinum undatum* and many other Stenoglossa show ribs on the shell which become less and less prominent towards the abapical parts of the whorls, finally disappearing. *Lora trevelliana, L. turricula* (fig. 29B), *Trophon barvicensis,* are native shells showing an approach to a turreted outline, but few British shells show this distinctly; for examples of these one must again have recourse to prosobranchs from abroad, e.g. *Tectarius pagoda, Melanatria fluminea, Melania amarula, Cerithium nodulosum,* and some cones, like *Conus marmoreus.*

In addition to the striae, ribs, spines and other kinds of external processes which may appear on the shell of a prosobranch gastropod a certain amount of similar growth is also found to occur internally, though normally to a much lesser extent. This is in addition to the secretion of aragonite which goes on over the whole of the internal surface, and, in most cases, is limited to the inner region of the lips and to mature animals. So long as the animal is immature and further shell growth is likely to occur the outer lip remains thin and sharp-edged. When maturity is reached, growth of the helicocone ceases at least in some cases and the only further deposition of calcareous matter leads to the formation of a thicker lip to the aperture of the shell and to the appearance of bosses, tubercles or teeth projecting into the mouth. This takes place in a very distinct way in the dog whelk *Nucella lapillus,* young specimens of which (fig. 262) are found to have a thin lip devoid of internal processes, whereas mature specimens (fig. 38) show a thick, blunt-edged outer lip carrying a number of rounded projections internally. Those on the inside of the outer lip are often related to the striae on its external surface, as if the same overactive parts of the mantle skirt were responsible for both sets of structures (*Nassarius,* fig. 38). The appearance of these ingrowths at maturity is not, perhaps, widespread amongst prosobranchs though they are visible in many Stenoglossa and in *Trivia,* but the thickening of the outer lip at maturity is common. [See p. 575.]

Changes in the shape of the aperture with age and in the development of a labial rib have been correlated with stages in the maturity of the reproductive system of *Rissoa parva* by Gostan (1958). He has shown a correspondence between the appearance of the shell and sexual development summarized in Table 1.

TABLE 1

Shell	Reproductive system
No ribs	Rudimentary
1st rib; angulated outer lip, no labial rib	Gonad and duct unlobed, solid Gonad lobed, duct hollow
Ribs; rounded outer lip, no labial rib	Sexes recognizable
Ribs; labial rib	Mature

Other changes may also occur with the onset of maturity. In animals of the genus *Trivia* (and of the Cypraeidae) the young shell shows a short spire with a large, smooth body whorl. When mature, however, the pattern of coiling changes and the shell becomes convolute, the mature body whorl enveloping all those of greater age (fig. 31). In addition to this the eratoids and cypraeids differ from other prosobranchs in that the mantle skirt extends beyond the edge of the shell and may completely cover the entire external surface (figs. 35, 206). Like any other part of the mantle this secretes shell, but adds it to the outer surface of the body whorl, not the inner. As a result this becomes overlaid by a polished sheet of material which obscures much of the irregularity normally seen on the outer surface of a prosobranch shell, and results in the smooth, glossy surface familiar in the shell of all cowries, although in *Trivia* it bears striae. Pelseneer (1932) called this the epiostracum. The same overgrowth of the mantle, but followed by fusion of the right and left halves, gives the wholly internal shell of *Lamellaria* (fig. 286), which is also glossy.

In addition to these internal growths on the lips or columella which, according to Dall (1889), take place because the aperture has become too large for the animal living within, but which might equally be due to folding of the mantle for the very reverse reason, similar ingrowths may occur throughout the inner surface of the helicocone. Alternatively septa may be laid down to separate the uppermost whorls of the spire, into which the visceral hump no longer extends, from the lower ones which it occupies, or within these whorls so as to exclude a peripheral part of their volume and leave only a central part alongside the columella for the visceral mass to lie in. These changes are particularly common in shells with tall spires. Ingrowths from the columella encroaching on the cavity of the whorls are particularly well known in prosobranchs of the family Cymatiidae (e.g. *Distortrix*), whilst the formation of septa is seen in the shell of terebrids, melaniids, vermetids and cerithiids (fig. 41A). In *Caecum* (fig. 41B) and a few other genera such as *Truncatella* (fig. 306A, B) a septum is laid down, when the animal has grown to a certain size, in such a way as to close off completely the upper coils of the shell, which are then broken off and lost. This truncated shell is said to be decollated. In the case of *Caecum*, the shell of which is coiled in an open spiral, this occurs twice or several times and gives the animal the appearance of living inside a shell which is a slightly arched tube, bearing little resemblance to the coiled spiral shell of the typical prosobranch.

It is well known that the molluscan mantle has not only the power of secreting shelly substance but is equally capable of removing what has already been secreted: examples of this have already been given in dealing with shells possessing varices and spines. In addition to removing such obstructive growths the mantle may also remove internal parts of the shell, either as part of a remodelling, or wholly, so as to give the animal greater freedom of movement within the shell. These alterations must, of course, not interfere radically with the mechanical strength of the shell, although they may in certain cases be relatively extensive, as in the pulmonates of the family Ellobiidae described by Crosse & Fischer (1882) and by Morton (1955), in which the whole of the older part of the columella and the internal parts of the upper whorls are absorbed. The same thing has been shown to occur in many neritids (Woodward, 1892) and in some cones, though in these animals the internal partitions are not removed, but merely reduced in thickness.

The above description of the formation and decoration of the prosobranch shell has been written primarily with reference to the shape of shell which is most commonly found in that group of gastropods, the spirally coiled helicocone. There are, however, a number of special cases to be considered, in which the shell appears—at least superficially—to possess not this

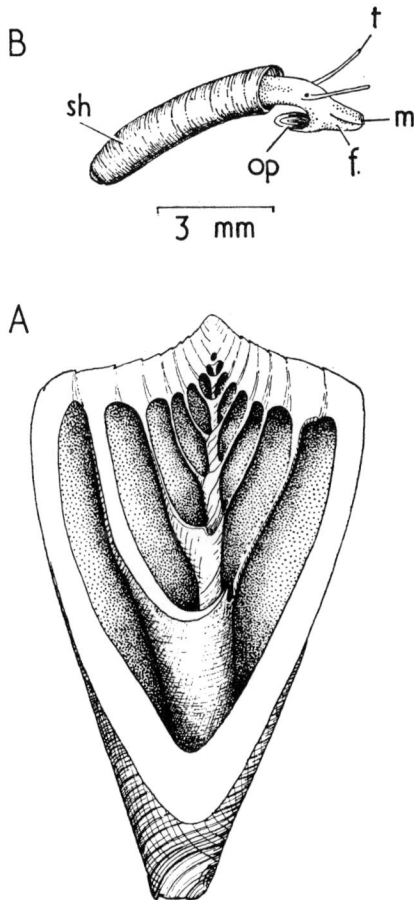

FIG. 41.—A, *Conus* sp., shell in longitudinal section to show internal septa; B, *Caecum glabrum,* from life. f, foot; m, mouth; op, operculum; sh, shell; t, tentacle.

shape, but some other. One of these has already been mentioned in referring to the genus *Caecum,* the shell of which is a short curved tube, related to the more usual prosobranch shell as explained above. Many prosobranchs have conical shells with, in the adult condition, little or no apparent sign of spiral coiling, and the question may be asked how these shells have been produced. They vary amongst themselves, too, in that some (*Patella, Propilidium, Patina* (fig. 33), *Acmaea*) are simple cones, others (*Calyptraea, Crepidula* (fig. 42A)) have internal partitions, whilst still others have apical holes (*Diodora* (fig. 33), *Puncturella*) or marginal slits (*Emarginula,* fig. 33). The first and last of these have been derived from a spirally wound helicocone of which they represent the body whorl, the whole of the spire having been lost. This shape is due to the fact that increase in the radius of the helicocone occurs very rapidly, so that the initial turns of the spiral are minute and disappear easily. In young limpets of any one of these genera (fig. 43A, D) the spire of the shell is visible as a small coil at the apex of the

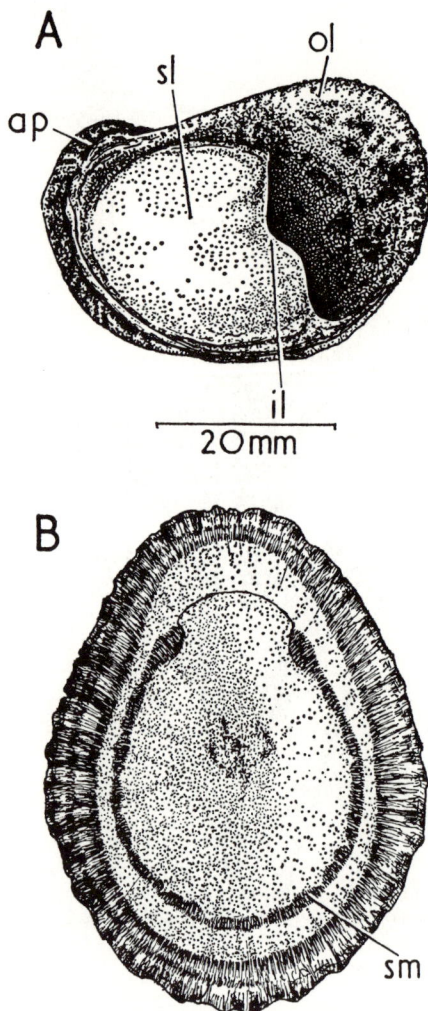

FIG. 42.—Shells of limpets in ventral view: A, *Crepidula fornicata*, the slipper limpet; B, *Patella vulgata*, the common limpet.
ap, apex of shell; il, inner lip of shell; ol, outer lip of shell; sl, shelf; sm, attachment of shell muscle.

expanded terminal part in which the body of the animal is housed, but it has no internal cavity, which has been early sealed off by shelly substance, and it is usually rapidly eroded so as to be no longer visible. Since the conical shell of a limpet represents the body whorl of the helicocone with the aperture clamped against the substratum, the spiral striae are represented by the ribs which radiate from its apex, and the growth lines by lines parallel to the mouth of the shell. As in more typical prosobranchs these frequently interact to give a reticulated pattern over the surface, particularly well seen in *Emarginula* spp. (fig. 33) and in

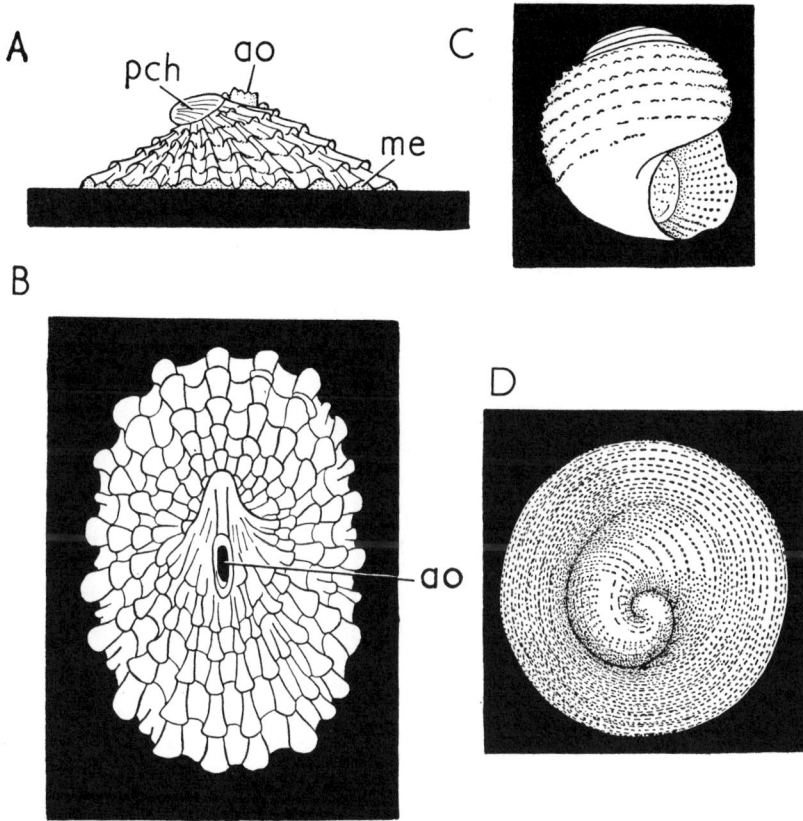

FIG. 43.—Shells of recently metamorphosed animals. A, *Diodora apertura*, from life, in side view from the right; B, *Diodora apertura*, shell of the same specimen in apical view; C, *Alvania punctura*, from life; D, *Calyptraea chinensis*, apical view showing protoconch.

ao, apical opening; me, mantle edge; pch, protoconch.

Diodora apertura (fig. 33), less clearly in *Patella* spp. and hardly at all in such smooth-shelled animals as *Patina* (fig. 33), *Acmaea* or *Calyptraea*.

In *Emarginula* the slit which extends from the margin is due to the mantle underneath being similarly split, the opening overlying the anus and excretory apertures and allowing escape of the respiratory water current from the mantle cavity. With growth of the shell the slit becomes filled in with shelly material at its upper end, its previous positions being marked by this, forming a feature on the shell known as the slit band. A similar appearance is to be found on the shells of *Scissurella* (fig. 248) and the pleurotomariids, which have the usual spiral shape. In *Puncturella* and *Diodora* (fig. 43A, B) the young shell looks like an *Emarginula*, with a marginal slit, but with further growth the mantle edges come together again at the lower end of the slit, re-unite and so convert what was an emargination into a perforation through the mantle skirt. In *Puncturella* this lies on the anterior face of the shell, but in *Diodora* it comes to occupy the summit of the cone. In both it serves as an exit for the pallial water stream. In

Haliotis (fig. 84) a series of such holes forms with growth of the shell, of which the 5–6 youngest are open, older ones being sealed internally by nacre. In *Puncturella* and *Diodora* only one hole exists. As it grows in size with growth of the shell it is clear that its margins must be resorbed by the mantle which lines it.

The remaining limpets, *Calyptraea* (fig. 43D) and *Crepidula* (fig. 42A), have shells which are derived from the ordinary spiral gastropod shell in a different way, giving a simple conical shell in the former genus (Chinaman's hat) but one retaining a distinct spiral coiling in the latter (slipper limpet). Both are easily distinguished from the shells of other limpets by the occurrence of a shelf or septum partially subdividing the internal cavity (fig. 42A). This is a specialization due to the behaviour of the mantle during development, well described by Werner (1955). The unhatched larval stage of *Crepidula fornicata* possesses a shell (fig. 44A) which is formed partly of a single conical piece secreted by the shell gland, and partly of a number of pieces added to this by secretion from the edge of the mantle skirt. The result is a shell coiled spirally in a plane, the mouth, before torsion, overhanging the posterior end of the animal and the apex the anterior end (exogastric coiling), after torsion these positions being reversed (endogastric coiling). At this stage the shell is symmetrical, its plane of symmetry being coincident with the sagittal plane of the body. The veliger hatches and takes up a pelagic existence, and during this shell growth continues, but becomes asymmetrical. The right side of the mantle skirt starts to grow much more rapidly than the left, the growth rate of which is very slow, and the shell secreted by the mantle also becomes asymmetrical. The result is the formation of a shell (fig. 44B) which is greatly expanded laterally but of low height, possessing a relatively enormous mouth and resembling that of *Haliotis* in general shape. At this stage the left half of the mantle skirt has come to lie ventrally and the whole of the dorsal, right and left parts of the mouth of the shell are lined by the expanded right half. As a consequence of the restricted space available to it because of this change in orientation, the left half of the mantle expands posteriorly underneath the older part of the shell in that area. It does not cease from the production of shelly material in this position, however, and so gives rise to a small ledge of calcareous material running round the posterior part of the body whorl at the level to which its edge reaches (fig. 44C). This is the so-called pallial line (ple) on the shell, but it is, in fact, the beginning of a pronounced calcareous ledge which is gradually built on to the posterior part of the shell and which comes to link itself anteriorly with the right and left edges of the greatly expanded mouth of the shell derived from the right half of the mantle (fig. 44D). In this way the beginnings of a limpet-like shell are achieved, the bulk of it from the activity of the right pallial half, the smaller posterior half from the shell edge ('rim' of Moritz, 1939) produced by the posterior tip of the back-turned left pallial half ('accessory mantle fold' of Moritz, 1939). When this stage is reached the two parts of the mantle become fused and are no longer distinguishable one from the other.

When this fusion of the different parts of the mantle has occurred it follows that the posterior part of the shell, on the underside, is now covered by mantle both inside and out—the former by the proximal part of the mantle skirt, the latter by the posteriorly reflected distal part—and at the mouth of the shell these two layers join one another at an angle approaching 360°. Over all this surface calcareous matter continues to be secreted and results in the complete covering over of the original spire of the shell, so as to render it invisible from below, and also, at the point where the mantle is bent backwards, in the formation of a sheet of calcareous material gradually growing forwards towards the anterior end. This is the septum (sl), which gives the shell of all members of the family Calyptraeidae a characteristic

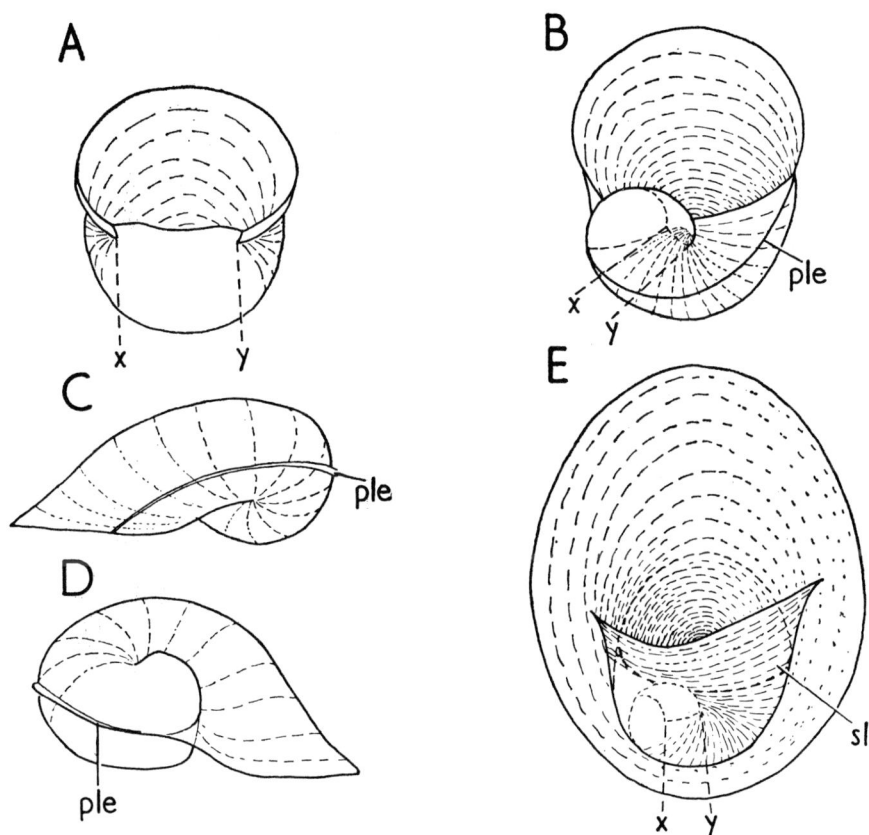

FIG. 44.—*Crepidula fornicata*: development of adult shell. A, shell of veliger stage ready for hatching, from below; B, shell of older veliger, from below; C, shell of veliconcha stage, from left; D, shell of veliconcha stage from right; E, shell of young metamorphosed animal, from below. After Werner.

ple, mantle line (the beginning of the septum); sl, septum; x and y mark corresponding points in figures A, B and E.

appearance and which also provides, on its underside, a surface for the insertion of the pedal muscles by means of which attachment to the substratum is secured.

During larval life and early metamorphosed life the ornamentation of the shell may change abruptly at several points, each of these, in general, representing the addition of an extra element of ornamentation. One of these steps represents the final stage of decoration and this stage will then be retained for the remainder of the animal's lifetime, perhaps complicated at maturity by the formation of teeth or spines as cessation of ordinary growth takes place. Examples of this are to be seen in *Cingula vitrea* (= *Onoba vitrea* of Thorson, 1946) where the shell shows first an unornamented protoconch which changes abruptly to a shell showing spiral striae. In *Alvania punctura* (fig. 43C) a protoconch with a few spiral striae passes suddenly to a shell showing spiral striae on its abapical half but with small raised spots adapically; this changes to one with delicate ribs which finally interact with the striae to give a shell with a reticulate ornament. In *Turritella communis* the protoconch develops as an unsculptured shell, but later whorls show the development of spiral striae some of which become exaggerated in still younger parts of the shell. *Bittium reticulatum* shows the same growth sequence, but adds to it a final stage in which ribs develop and combine with the striae, some of which are exaggerated, to produce the series of tubercles which characterize the shell of this species. A particularly clear case is that of *Lora trevelliana* (= *Bela trevelyana* of Thorson, 1946): in this species the initial part of the shell has a few dots on it, and the succeeding whorl develops spiral lines some of which eventually become transformed into exaggerated striae. At the beginning of the third turn ribs appear and give rise to a reticulate pattern on its surface. Further examples could be multiplied, but it is clear that there is a regular sequence in the development of sculpture on the shell of a gastropod. This has been claimed (e.g. by Grabau, 1928) to be a recapitulation of ancestral shell characteristics, but it is not safe to interpret the shell structure in this way. In so far as some aspects of patterning are essential prerequisites for the development of others, however (e.g. the presence of both spiral striae and ribs for the appearance of a reticulated or cancelled sculpture) these stages do have a certain ordinal significance and the sculpture may be of value as a sign of either primitiveness or advancement. Grabau's stages are as follows: smooth, round-whorled stage (naticoid); primitive ribbed, spiral stage; angulated, with nodes where ribs and striae cross; keeled stage; secondary round-whorled stage (by loss of sculpture); spinous stage; smooth stage (by loss of spines); loose spiral stage (by loss of power to coil, e.g. *Vermetus*). These later stages show gerontism in the disappearance of the sculptural characteristics of the shell.

The gastropod shell is frequently coloured. In some the predominant colour is that of an overlying periostracal layer, but in addition to that—and in animals where the periostracum is not evident—the calcareous material is usually pigmented by the inclusion of coloured substance. This is found in the outer layer of the shell only, which is, it will be recalled, the part secreted by the edge of the mantle skirt. This does not mean that the inner surface of the gastropod shell is inevitably colourless because, as is well known, the inner side of a shell like *Haliotis* or a top shell shows beautiful mother-of-pearl tints. However, this effect is physical in origin and it is the colouring of the shell by means of pigments of chemical nature which is confined to its outer layer. The chemical nature and possible origin of these substances are discussed on p. 126: here it is their relationship to the sculpturing of the surface of the shell which is of interest.

Like the ornament on the shell, its colour often shows specialization in space and time; and since its mere presence must mirror the activity of chromogenic cells in the mantle skirt, there must be a spatial differentiation of that part of the body with or without a superimposed temporal rhythm. If no pigmentation at all is produced at the mantle edge then the resulting shell is white, although it may be sufficiently translucent to acquire a spurious colour from the underlying viscera. If pigment is incorporated into the shell by adding secretion from pigment glands to the calcareous secretion, and if these glands lie uniformly along the length of the mantle skirt, then a uniformly coloured shell will result, the tint depending upon the nature and amount of pigment secreted. Uniformly coloured shells of this sort may be found, for example, in *Littorina littoralis* and *Barleeia rubra*. They are, however, unusual, and it is much more common to find a pattern in the colouring of the shell, which implies a secretion of pigment which is not uniform. If manufacture of pigment is localized at points or stretches of mantle edge which secrete continuously then the result is a series of spiral lines or bands of pigment as in *Eulima glabra*, *E. trifasciata*, *Lacuna parva*, *Cingula cingillus* and some specimens of *Nucella lapillus* (fig. 38). Should the secretion of pigment be rhythmical and occur in outbursts of activity separated by rest periods, then these bands will be interrupted and appear as a line of dots or blocks of pigment as may be seen in *Alvania punctura* and *Natica catena* (fig. 296A).

In many other prosobranchs the ability to manufacture pigment for incorporation in the substance of the shell is present along the entire length of the mantle edge but only intermittently. As a result axial lines of colour may be produced in the shell, and if spiral lines are also present there may be produced a criss-cross of coloured lines precisely comparable to the reticulation of spiral striae and ribs in its sculpturing. Coloured 'ribs' of this sort are well seen in many trochids, *Rissoa guerini, Turritella communis* and *Neptunea antiqua,* whilst a reticulation due to interaction of the continuous spiral and discontinuous axial elements may be seen in some toxoglossans. Occasionally these axial lines fail to be complete, more particularly at the periphery of the whorls, an event which gives the impression that they are crossed by a light spiral band as in *Rissoa parva* (fig. 39). As with true ribs or spine rows these coloured lines or reticulations are set spirally down the whorls of the spire.

The two types of pigmentation referred to seem at first sight to be brought about by two different kinds of event in the edge of the mantle skirt. There are, however, a number of shells which show a type of pigment marking which links the one with the other. In *Tricolia pullus* (fig. 39), for example, the shell often bears a number of zigzag or V-shaped lines lying with the point of the V pointing up the helicocone towards the apex. This kind of mark suggests that at a particular instant of time a group of cells in the edge of the mantle skirt, previously quiescent or incapable of manufacturing pigment, suddenly burst into activity; and further, that a wave of secretion then passed along the edge of the mantle skirt away from this point, all cells relapsing into quiescence after a short burst of secretion. If the spread of the wave of metabolic activity is slow in relation to the rate of shell growth a long and narrow V mark results; if the reverse relationship holds, a wide V will be formed; if spread is more rapid in one direction than another then the V will be asymmetrical; whilst if the rate of propagation of the disturbed metabolism is one subject to acceleration or deceleration then a mark which is not rectilinear but curved will be the visible consequence. If the activity starts and stops abruptly the lines will be sharp; if either is slow then the lines may fade gradually into the background. This kind of marking is of frequent occurrence and may be seen on many trochids (fig. 29D) as well as on *Tricolia pullus* (fig. 39), *Rissoa parva, Theodoxus fluviatilis,* some specimens of *Littorina littoralis, Natica alderi,* and—perhaps the best known examples—such exotic shells as those of *Harpa major* (fig. 36B), *Cypraea zic-zac, Oliva porphyria* and *Conus geographus.* Where these zigzag lines may be traced across the entire breadth of a whorl, as in some specimens of *Tricolia pullus,* and as in some of the lines on *Harpa major,* they presumably represent a single transverse line which differs from the straight line seen in *Viviparus* (fig. 175A) only in that the ability to secrete pigments has not been developed at all points in the mantle edge simultaneously. In many cases, however, as in some shells of *Tricolia pullus* or those of *Oliva porphyria,* the jumble of short zigzags is so complicated and they are mixed with so many spots, blotches and short curves that it is difficult to decide whether the decoration results from the disorganization of axial lines or from the spreading of attempts to produce spiral lines.

The pigmentation of the shell in relation to its sculpture has been well examined by Wrigley (1932, 1934, 1942 and general discussion in 1948). From his work it is clear that there is, in general, an inverse relationship between pigment formation and excess secretion of calcareous matter. Where tubercles, spines, varices, teeth and similar out-growths occur, the secretion of pigment is reduced and these parts stand out as light coloured projections against the generally darker background of the rest of the shell. Pigment abounds in the areas between them. This is well seen in many shells: *Gibbula magus, Clathrus* spp., on ribbed specimens of *Rissoa parva, Aporrhais pespelicani, Bittium reticulatum, Cerithiopsis tubercularis, Nassarius* spp. and many toxoglossans. The physiological basis underlying this is obscure and it may be that better understanding would be obtained of the relationship were it thought of, not as a negative correlation between secretion of pigment and the formation of calcareous outgrowths, but as a positive correlation between pigment formation and the growth of the mantle skirt; for it will be recalled from the discussion above that the formation of tubercles, ribs and the like occurs during a period when shell and pallial growth has been arrested, whereas the areas between these—which are, as Wrigley points out, the pigmented parts—coincide with periods during which growth is maximal. If the formation of pigment were in any way connected with the increased metabolism which affects the mantle skirt during these periods, then it might be possible to assume that that is the cause of the material being added to the shell then, but not at other times.

Although much of the general structure, decoration and sculpture of the gastropod shell appears to be wholly unrelated to the environment in which the animal lives, this is not entirely so, as indicated by Berner (1942). He pointed out that *Murex brandaris* has long spines on muddy bottoms but only short ones on sand and rocks. This may, however, be a spurious adaptation in that wear of the spines is less in the one case than in the other, just as it is possible to find shells of *Nucella lapillus* with an imbricated surface when they have grown in deep waters and on softer substrata, whereas those living on intertidal rocks suffer enough abrasion to remove these. The only difference between the case of *Nucella* and that of *Murex* is that the latter is susceptible of interpretation as an adaptation, the former is not. Other examples of the effect of the environment producing adaptations—or apparent adaptations—are variations in the height of *Patella* shells at different levels of the beach (see p. 467) and a thickening of the shells of *Nucella lapillus, Nassarius reticulatus* and some others in brackish habitats.

Because both the shell which covers the visceral hump of a prosobranch and the operculum which lies on the dorsal surface of its foot are made of a basis of conchiolin, more or less impregnated with salts of calcium, many malacologists have tried to show that the two structures are merely parts of one single shell. This view was originally put forward by the French naturalist Adanson (1757), and was supported by his compatriot Dugès in 1829, by the British worker Gray in 1850 and may still be encountered in conchological literature, as, for example, in Fleischmann (1932) and Pruvot-Fol (1954). On this theory the only univalves would be the cephalopods (unless the aptychus of ammonoids represents a second valve): gastropods and lamellibranchs would agree in being bivalved and the columellar muscle of the former could be equated with the adductors of the latter. The chitons remain distinct with a multivalvular shell difficult to homologize with that of the other molluscan classes, though Simroth (1896–1907) made the statement that he would like to homologize the gastropod shell with the anterior valves of a chiton and the operculum with the most posterior were it not for the fact that the last chiton valve is morphologically dorsal to the anus, whereas the gastropod operculum lies ventral to it.

De Blainville and Lamarck denied Adanson's proposition that the operculum and shell were homologous, on the ground that the one was a pallial secretion and the other pedal, but Dugès overcame this objection by supposing that the operculum was in fact secreted by the mantle skirt when the mollusc had withdrawn into the shell and was only secondarily associated with the foot. Doubtless this theory was the more easily tenable in view of the fact that the epiphragm produced by hibernating pulmonates, which looks and acts very much like a temporary operculum, is largely pallial in origin and is produced by stylommatophoran snails when retracted.

No real light was shed on the origin of the operculum and the different types which could be found in prosobranchs until the work of Houssay (1884), the majority of whose findings are current today, though a more recent re-investigation by Kessel (1942) has extended our knowledge considerably and corrected some mistakes of Houssay's. There is room, how-ever, for a proper histochemical investigation of the secretion and regeneration of the prosobranch operculum [p. 589.]

Not all prosobranchs have an operculum when adult, but it seems to be ubiquitous in their embryonic and larval stages. Most of those with a final limpet shape lose their operculum at metamorphosis (*Haliotis, Emarginula, Puncturella, Diodora, Acmaea, Lepeta, Propilidium, Patina, Patella, Calyptraea, Crepidula,* and *Capulus*), as do, too, a number of pelagic forms like *Ianthina* and the heteropods (Franc, 1949); in some limpets, such as *Phenacolepas* and some forms from hydrothermal vents, the operculum is not lost but greatly reduced in size and it may become internal. In these cases the loss of the operculum can easily be related to the animals' mode of life, but the reason for its loss in a further series is obscure. In *Lamellaria,* with an internal shell, an operculum is clearly pointless and its absence understandable, whilst in the case of *Trivia* the narrowness of the mouth of the shell may perhaps render an operculum superfluous, but it is difficult to suggest any reason why some genera of stenoglossan (*Philber-tia, Mangelia*) should have no operculum, whilst others—of almost identical structure and habits so far as is known—should possess one (*Lora, Haedropleura, Typhlomangelia*). The absence of an operculum in the eulimids may be because of their parasitic habits, or because they have relations with the opisthobranchs within which group a trend towards loss of the operculum regularly occurs, even when the shell persists.

Houssay distinguished two types of operculum on a basis of whether they showed a spiral or a non-spiral construction. Kessel also distinguished two types of operculum, but based these on the material of which the were made—horny and calcareous. As all opercula turn out to be fundamentally similar, neither classification is wholly happy, but perhaps the division into spiral and other types is the better. A description of the operculum of a single species of prosobranch, *Littorina littorea,* has already been given (p. 17, fig. 7) and it will be recalled that the bulk of the operculum has its origin in a groove (oge) lying transversely across the dorsal surface of the foot. Here secretion goes on steadily throughout the lifetime of the mollusc, but although the foot is a bilaterally symmetrical part of the body the secretion of opercular material proceeds at a greater rate on the right side than on the left. The result of this is that the operculum is gradually pushed round in a clockwise direction (when examined in dorsal view) on the opercular disc on the foot. To its underside are attached muscle fibres belonging to the columellar muscle: as clockwise rotation of the growing operculum occurs these fibres must migrate, though the mechanism by which this is brought about seems to be completely unknown. As in other similar situations the muscle fibres end on epithelial cells through which tonofibrillae run to transmit the pull from muscle to operculum. Houssay (1884) believed epithelial cells were absent at the attachment of the operculum, but this was shown to be wrong by Fischer (1940a).

A further consequence of the asymmetrical addition of opercular material to its columellar edge (ce) and of the fact that the breadth of the added strip increases with growth of the mollusc, is that the operculum has a spiral structure with a nucleus representing the beginning of its formation. In an animal such as *Littorina* this lies towards the left side or siphonal edge (sis). The spiral, like the spiral of the shell, is an equiangular spiral and the rates of growth of the two structures are so adjusted that the operculum is always of approximately the same size as the mouth of the shell. Whereas the shell coils dextrally, however, the operculum coils sinistrally and vice versa, and this remains true even in hyperstrophic shells.

The conchiolin or 'horny' substance comprises the bulk of the material out of which the operculum of *Littorina* is made, but not all. A second substance is applied underneath the horny material: it is shiny and for this reason is known as the varnish or gloss. The activity of the cells in the opercular groove accounts for the production of the operculum in *Littorina* and probably most of the prosobranchs in which a similar, horny operculum is found. Complication is limited to a stratification of alternate layers of conchiolin and varnish in older animals. This is brought about by the rotation which the operculum undergoes. When the posterior edge, which has an outer layer of conchiolin and an under layer of varnish, has rotated forwards through 180° and so overlies the opercular groove once more, new conchiolin will be secreted under the preexisting gloss and a new varnish layer applied under that in turn, so that a sandwich structure of alternating conchiolin and varnish is produced.

In the cases so far mentioned the opercular groove stretches across a large part of the dorsal pedal surface. There appears, according to Kessel (1942), to be a trend leading to the shortening of this groove so that it becomes only a short slit on the right side of the foot. This is common in trochids (fig. 46A) and has the effect of producing an operculum of spiral pattern, but with a large number of turns instead of the small number found in animals like *Littorina* (fig. 46B). These opercula are known as polygyrous, whereas those of *Littorina* pattern are called oligogyrous. Apart from this difference in the size of opercular groove and the effects which stem from it the two kinds of operculum are identical.

The edges of the opercula described above, except for the columellar one which is buried in the opercular groove, are free and therefore flexible to some extent, since the conchiolin out of which they are made is not rigid. This means that some variation in the extent to which the animal withdraws into the shell is permissible, the opercular edge bending outwards as the animal withdraws further. With animals which possess a siphonal canal, it has been claimed that a more precise fitting of the opercular edge to the shell mouth is required and that this is achieved by an operculum which is not of spiral construction but which has a marginal or terminal nucleus and which grows in size by the addition of new material to the opposite edges. This style of operculum is that exhibited by the Stenoglossa, and is well shown by *Nucella* (fig. 45B) and *Nassarius* (fig. 46D). On its external face it presents a series of markings lying around the terminal nucleus (n) which represent growth lines. On its underside there shows a series of concentric marks related to an eccentric nucleus but bearing no relationship to the growth marks on the opposite side. In addition, the underside shows areas of different texture, one part (the greater) being roughened and representing the area of attachment of the columellar muscle, the other (limited to a strip around the labial margin drawn out to fine lines at the siphonal and sutural ends) being glossy and marking where varnish has been secreted. Over this area the concentric marks have been covered and lie under the gloss. Sections show that the conchiolin which forms the outer layer is secreted in an opercular groove as before, but that this does not cause a rotation of the operculum on the opercular disc, but merely a transverse gliding so ensuring that the projection which fits into the siphonal canal of the shell is not displaced. Gloss, therefore, never comes to be added all round the conchiolin but only along its labial edge which lies at the posterior end of the opercular disc, and the alternation of layers of conchiolin and varnish which occurs in spiral

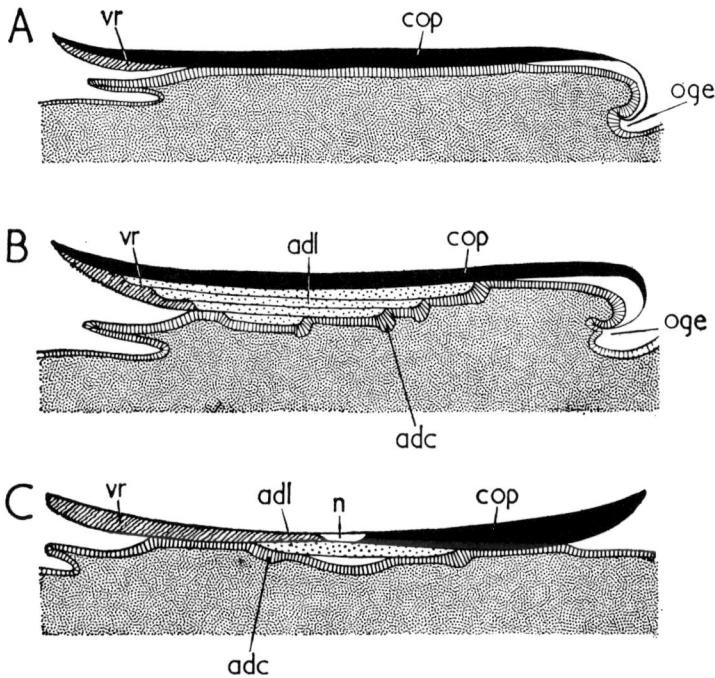

FIG. 45.—Diagrammatic sagittal sections of opercula: A, *Littorina littoralis*; B, *Nucella lapillus*; C, *Viviparus viviparus*. After Houssay and Kessel. Conchiolin layer is shown in solid black, varnish with diagonal hatching, adventitious layers stippled.

adc, cells secreting adventitious layers; adl, adventitious layer; cop, conchiolin layer; n, nucleus; oge, opercular groove; vr, varnish.

opercula is therefore never encountered in this type. It may also be seen from an examination of sections that the concentric markings on the underside of the operculum are due to the presence of plates of conchiolin lying below the main one. These are known as adventitious layers (adl, fig. 45B) and they presumably add strength to the operculum over the area to which the columellar muscle is attached. They are secreted by circles of gland cells in the epithelium of the opercular disc (adc). In the literature these are usually known as 'chitinogenous' cells, but this is a bad name and derives from times when anything of a cuticular nature was regarded as 'chitin' by the zoologist.

A certain number of opercula contain calcareous material as well as conchiolin. The amount of this may be relatively small as in *Pomatias elegans*, or considerable as in *Theodoxus fluviatilis*, or may constitute so prominent a part of the operculum as to leave it in doubt as to whether any other constituent can occur. This is the type of operculum found in turbinids in general, and in *Tricolia pullus* (fig. 39), the only British member of the family, which is immediately identifiable on this single feature. The two first of these are, in effect, simple spiral opercula in which the matter which is secreted is partly conchiolin, partly calcareous, as in the shell, but in the turbinids, according to Kessel, the secretion of the operculum follows a rather different course. The calcareous matter is produced, not by the opercular groove, but

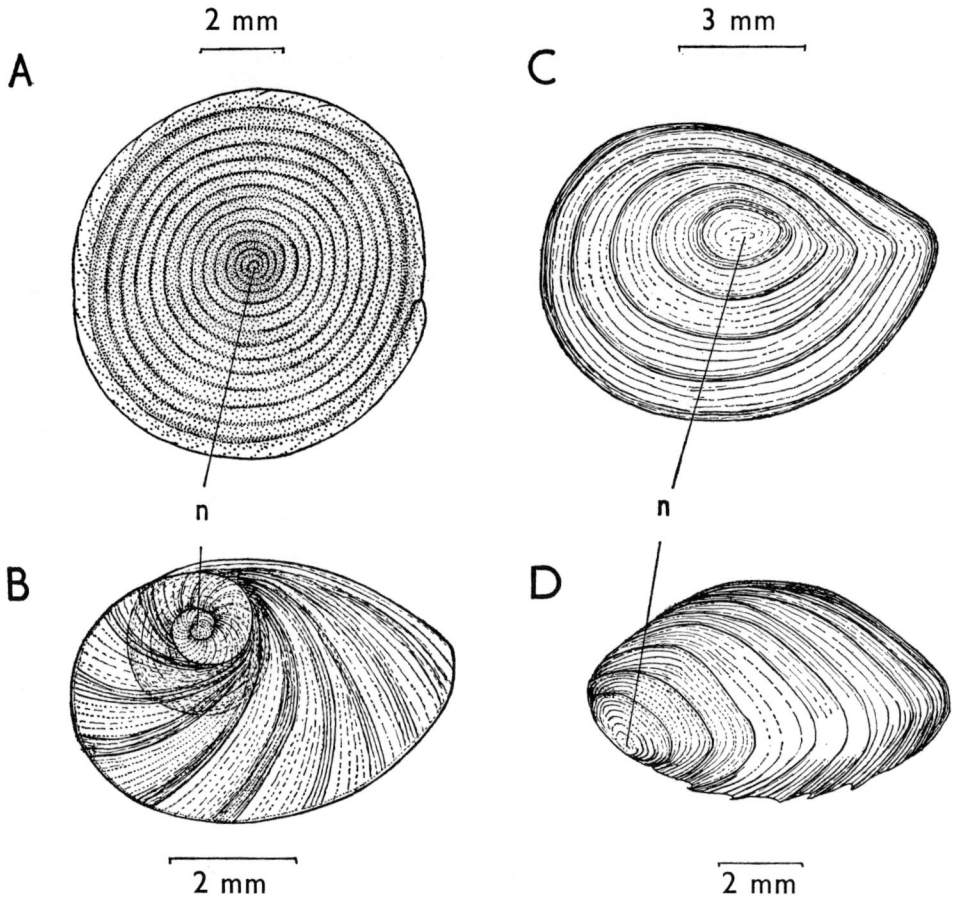

FIG. 46.—Opercula. A, *Gibbula cineraria*, polygyrous spiral type; B, *Littorina littorea*, oligogyrous spiral type; C, *Viviparus viviparus*, concentric type; D, *Nassarius reticulatus*, with marginal nucleus. n, nucleus.

by the expanded margin of the opercular disc which forms the posterior lip of the groove, a structure which he calls the *Deckelsaum* or opercular flange. The conchiolin part of the operculum comes in the normal way from the opercular groove, but is applied underneath the calcareous operculum by an eversion of the opercular groove brought about by blood pressure and taking place in a rhythmical way. The whole structure rotates like any spiral operculum and this may give rise to a spiral keel of calcareous matter on the outer surface of the operculum as in *Turbo rugosus*. In other cases (*Tricolia pullus*) the outer surface may be completely smooth, whilst in others again (*Turbo sarmaticus*) the outside may be irregularly roughened. In all of these a varnish underlies the calcareous matter.

Some other opercula appear to be built on a pattern which is different from those already described. These are the concentric opercula, of which by far the best known is that of *Viviparus* (figs. 45C, 46C). Superficially, this type appears to be made of concentric rings of conchiolin which may (in other genera) often be impregnated with salts of calcium. When examined in sections, however, the apparent simplicity vanishes and the operculum is found to consist of several different components, each of which occupies its own characteristic fraction of the whole area. In *Viviparus* the central areole turns out to be made of an embryonic nucleus (n); anterior to that the operculum is made of conchiolin (cop), posterior to it of varnish (vr), whilst underneath there will probably be found adventitious layers of further conchiolin (adl). In the early embryos of *Viviparus* an opercular groove occurs as a shallow invagination along the anterior edge of the opercular disc, secreting conchiolin, whilst at the posterior end is a gland secreting varnish: at this stage, therefore, conditions in *Viviparus* are precisely comparable to those in other prosobranchs. Later, however, the groove everts, its secretory epithelium forming a line across the anterior border of the opercular disc, and from this stage the edge of the operculum projects freely all around the opercular disc. With growth this group of secretory cells transforms into the anterior half of a circle the posterior half of which is formed by the cells which manufacture the varnish, so that from this time on, growth of the operculum depends upon the expansion of 2 semi-circles, one of varnish, the other of conchiolin. The structure of the operculum is further complicated by the presence of cells secreting adventitious conchiolin. These are at first placed together centrally, underneath the operculum, but with time the group of cells expands and becomes converted into a circle of ever-increasing radius (adc), the central cells losing the ability to secrete, and ultimately becoming only loosely attached to the operculum. The circle of secreting cells lays down the material of the adventitious layer as it expands under the overlying operculum. From this it follows that the relative dispositions of varnish and adventitious layers vary in the two types of operculum which have them: in opercula with marginal and terminal nuclei (e.g. *Nucella*) the varnish underlies the adventitious layers, but in concentric opercula with a central nucleus the adventitious layer lies under the varnish. Hubendick (1948) has shown that the rings in the operculum of *Viviparus* are related to overwintering but because other factors may produce them they cannot be used to measure the age of a snail with certainty.

Like the shell, the operculum may be repaired when damaged, or even replaced should it be altogether lost. The ability to do this seems to vary from species to species, since Techow (1910) got no regeneration in *Viviparus viviparus,* though Hankó (1913) obtained positive and similar results with *Murex brandaris* and *Nassarius mutabilis*. Not unnaturally, snails from which the operculum (or parts of it) had been removed spent much time withdrawn into the shell and secreted quantities of mucus; the operation was not, however, fatal except in one instance. The exposed surface of the opercular disc was completely re-clothed with epithelial cells 10 days after the removal of the operculum; many of the cells were immigrants from the surrounding surface of the foot, but some were derived from the circles of secreting cells responsible for producing the adventitious layers found in the opercula of these animals. Dedifferentiation of these cells occurred so that the wound was covered with a uniform epithelium of unspecialized cells. This condition persisted for about a further 3 weeks, when the cells transformed to a columnar secretory type filled with granules which passed out and produced a cuticular covering. This was repeated so that a multilayered operculum was formed. The production of cuticle started centrally and gradually passed peripherally until the entire operculum had been replaced, the new structure being thickest centrally and tapering to its margin. It showed no differentiation of structure apart from being laminated, and neither varnish nor calcareous material was present. The entire process—which was carried out in a Mediterranean spring—took from 21 March to 5 June.

CHAPTER 4

THE MANTLE CAVITY

THE mantle cavity is one of the features diagnostic of the molluscs, and throughout the group retains such a standard pattern as to indicate that it must have been arranged in much the same way in the original members of the phylum, whatever changes may have taken place as adaptive radiation occurred. The cavity (fig. 47) is fundamentally a pocket on the posterior face of the visceral mass, which comprises its anterior wall and floor. The roof and sides are formed of a fold of body wall hanging down from the more dorsal parts of the visceral hump, forming a structure called the mantle skirt (mst). Within the mantle cavity lies an assemblage of structures which collectively constitute the pallial complex. In the primitive mollusc these comprised (a) the anus; (b) 2 excretory openings; (c) 2 genital apertures; (d) right and left osphradia; (e) right and left gills; and (f) right and left hypobranchial glands. These are disposed in a relatively constant way. The anus lies medianly towards the inner end of the cavity, which primitively was perhaps not a very deep one, and it was flanked on either side by the excretory openings. To the outer side of these lay the gills and lateral to these again, or possibly on them, were the osphradia. The inner epithelium of the mantle skirt between the gill and the anus was extremely glandular and formed the hypobranchial glands. The position of the genital openings can be stated less certainly as they may have been situated on either the median or the lateral side of the gills: evidence suggests the former since they lie median to the gills in chitons, bivalves and cephalopods.

In the gastropods the mantle cavity has been rotated by the process of torsion which the members of that class undergo during development so that it lies anteriorly on the visceral mass. Strictly, it is not confined to that area, because the mantle skirt extends as a frill round the whole of the stalk by which the visceral mass is connected to the head and foot. At all points except directly behind the head, however, the mantle skirt is very brief and the cavity which lies underneath is correspondingly shallow. If we exclude the very drastic modifications which have taken place within the class in adaptation to different ways of life, the sole other difference between the pallial complex of gastropods and that of other molluscs is that the gonads do not discharge by openings separate from those of the kidneys.

The mantle cavity of the earliest gastropods was perhaps not very deep (see fig. 47), and the 3 openings (anus (a) and kidney apertures (eo)) lay at the innermost end, whilst the gills (ct) arose at a similar level but projected forwards towards its mouth. Each gill, known as a ctenidium because of its comb-like structure, was composed of an axis (cta) carrying a double row of leaflets, set alternately. The axis of each ctenidium stretched obliquely forwards across the lateral parts of the cavity, so that one row of leaflets lay below and towards the mid-line whilst the other lay above and more laterally. Along the inner part of its length it was possibly fastened by a membrane (called a suspensory membrane) to the side wall of the mantle cavity laterally (efm), and to the mantle skirt in the neighbourhood of the mid-line medianly (afm). The distal, anterior part of the ctenidial axis, however, projected freely into the mantle cavity and was not supported by membranes. Yonge (1947) supposed that a supporting

A

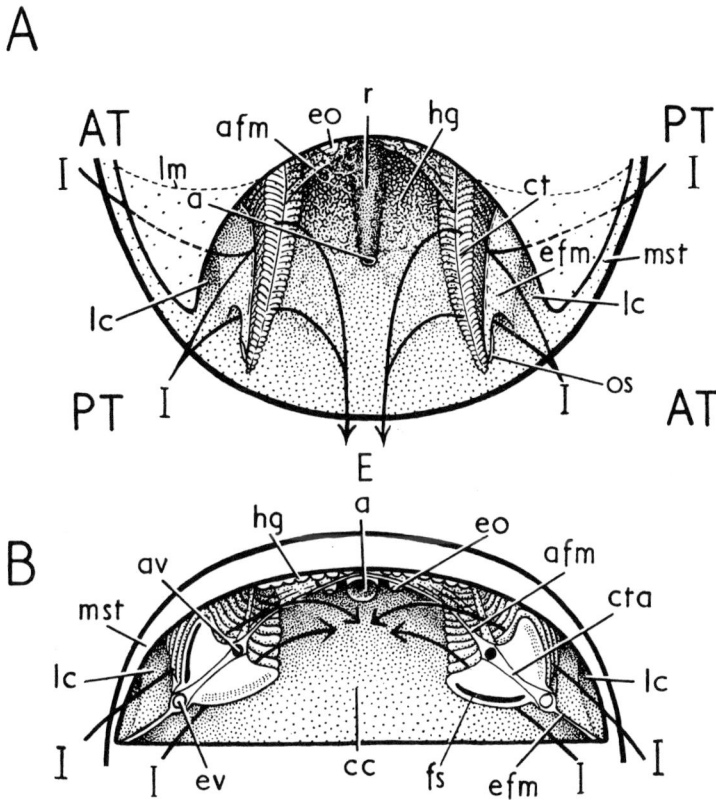

FIG. 47.—A. Diagrammatic representation of the contents of the mantle cavity of a gastropod and a possible pretorsional ancestor seen by transparency through the mantle skirt. The reference letters AT, PT, on the left mark the anterior and posterior ends of the pretorsional stage; those on the right refer to a gastropod which has undergone torsion. Arrows show direction of water currents.

B. View of the contents of the inner half of the mantle cavity of a primitive gastropod. The outer half has been cut away and the deeper part is seen through the cut. Arrows show direction of water currents. It is debatable as to whether skeletal rods were present in the gill lamellae and whether the animal was bilaterally symmetrical.

a, anus; afm, afferent membrane of ctenidium; av, afferent branchial vessel; cc, central compartment of mantle cavity; ct, ctenidium; cta, ctenidial axis; E, exhalant; efm, efferent membrane of ctenidium; eo, excretory opening; ev, efferent branchial vessel; fs, skeletal rod in ctenidial leaflet; hg, hypobranchial gland; I, inhalant; lc, dorsolateral compartment of mantle cavity; lm, inner limit of mantle cavity; mst, mantle skirt; os, osphradium; r, rectum.

endoskeleton (fs, figs. 47, 49) in the form of chitinous rods was often developed in the axis and in the leaflets it bore. Because of the way in which the ctenidia were attached to the walls of the mantle cavity this was divided in its deeper parts into 3 compartments—a left dorso-lateral one (lc, figs. 47, 48A, 54) between the axis of the left gill and the left wall of the mantle cavity; a right dorsolateral compartment with similar relations to the right gill; and a central compartment (cc) into which discharged the anus and the right and left kidney apertures, one of which was also the outlet for reproductive cells. In the left compartment lay the left row of leaflets belonging to the left gill; in the right lay the right row of leaflets of the right gill; in the central lay both the other rows. For functional reasons explained below the membrane on the dorsal (median) side (afm) of the axis usually extended less far forward than the membrane connecting the gill axis to the floor of the mantle cavity (efm).

The hypobranchial glands (hg, lhg, rhg) developed particularly on the inner wall of the mantle skirt in the deeper parts of the central compartment, but extended on to the gill membranes and anal papilla.

Since the respiratory organs—delicate structures which could not be directly exposed on the surface of the body of the gastropod—were housed within the mantle cavity, they had also to be ventilated. It was, in fact, largely the ciliated surfaces of the ctenidia themselves which became responsible for this task. Since, in addition, there had to be an escape route for the faecal and excretory matter from the anus and kidney openings, it was convenient to use the outgoing stream of water from the ctenidia for carrying this material out of the mantle cavity, though contraction of the shell muscle, pulling the visceral mass down on to the head and foot, would also forcibly eject water from the mantle cavity carrying this material with it.

In all, or almost all, respiratory organs the circulation of the blood is so arranged that the efferent blood vessels lie on the side of the respiratory surface which is bathed by the incoming stream of water: in this way the blood leaving the respiratory organ is in equilibrium with the water of highest oxygen content and minimal carbon dioxide tension. The gastropods are no exception to this rule though its efficiency has been measured only by Hazelhoff (1938), who showed that *Haliotis* removed 56% of the oxygen from the incoming water and *Murex brandaris* 38%. Not all of this passed necessarily though the ctenidium. In the primitive gastropods the incoming current of water (I, fig. 47) entered the mantle cavity lateroventrally and left (E) in a more median and dorsal position, the 2 ctenidia lying across the stream. The efferent blood vessels of the ctenidia (ev) were situated on their ventrolateral sides and the afferent ones (av) on the upper side. Water passed between the ctenidial leaflets and accumulated in the dorsal half of the mantle cavity. As a consequence of this arrangement of the water currents, any suspensory membrane of the ctenidial axis which fastened it to the floor of the mantle cavity could be quite elongated without interfering with the flow of water, and the osphradia (os), lying partly on the anterior edge of these membranes and partly on the ventral edges of the free ctenidial axis, were well placed to test the water which was coming into the mantle cavity. Because of this, too, any supporting gill endoskeleton (fs) developed along the ventral part of each ctenidial axis and its lateral leaflets. On the upper side of each ctenidium, where the water emerged into the main mantle cavity from the inter-lamellar spaces, an extensive dorsal supporting membrane would probably interfere with the circulation of the water, especially in view of the fact that it was this water which had to remove waste (from kidneys or gut) and reproductive products. Any upper suspensory membrane (afm) of the early gastropod gill was, therefore, reduced or even lost. Thus water tended to pass from the right and left lateral compartments of the mantle cavity into the central one and so augmented the volume available for washing waste away.

In the progastropod the mantle cavity lay mainly in the visceral hump posteriorly (fig. 47). Water was sucked in laterally (I) and somewhat ventrally by the ciliated epithelium of the ctenidial filaments and escaped medianly and somewhat dorsally (E). In the early gastropods torsion brought this whole apparatus forwards. In this situation water entered and left as before so that the outgoing stream, though normally clearing the head because it was leaving the mantle cavity dorsally rather than ventrally, was liable to carry the faecal and excretory water stream into its close proximity. This condition is represented in fig. 48A and is such an insanitary arrangement that it has been superseded in all living groups of gastropods. The ciliary currents in the mantle cavity of gastropods have been particularly investigated by Yonge (1938, 1947). On the assumption that what he described for living prosobranchs can be applied to the extinct ancestral types now under consideration then the picture of what went on within their mantle cavity is this.

Each gill filament may be taken as corresponding to that of *Haliotis* (fig. 49) and be regarded as a triangular leaflet fastened to the ctenidial axis (cta) by its base. On each of the flat surfaces of the filament is a delicate epithelium occasionally ciliated over the major part, but richly so over a strip overlying the endoskeletal support (fs) near the lower (efferent) margin. These are the lateral cilia (ltc), and it is they which are responsible for driving water from one side of the gill to the other and so for creating the current through the mantle cavity.

The water enters through a narrow slit into the relatively spacious mantle cavity (fig. 48A). When this happens the speed of the current decreases and the larger particles suspended in it fall on to the floor, where they encounter a ciliated epithelium which carries them to the mouth of the cavity and outside (see legend to fig.). This is the current called current A by Yonge (1938). Medium-sized particles do not drop on to the floor of the mantle cavity until a later stage and so tend to fall nearer the mid-line, whence currents also carry them to the mouth of the cavity (B). The most minute particles, however, stay in suspension until they reach the gill filaments, over the surface of which stretches a sheet of mucus on which they are trapped. Frontal cilia (fc, fig. 49) on the efferent edge of the leaflet and abfrontal cilia (aci) on the afferent edge then sweep the mucus and embedded particles by a variety of routes on to the mantle skirt and the surface of the hypobranchial gland (hg, fig. 48A). The secretion of this cements them together so that they remain in the main exhalant stream (C) in which they are swept out of the mantle cavity along with excretory and faecal matter, and, in the appropriate season, the reproductive cells.

In the original disposition of the body of the primitive gastropod, as indicated above, this exhalant stream, laden with waste, left the mantle cavity anteriorly in the mid-line over the animal's head. Such an arrangement is not encountered in any living gastropod, and it may therefore be assumed that its apparent insanitary effects had some degree of reality. In all living gastropods alternative means of disposing of the waste matter have been introduced, which have the effect of diverting the stream away from the head, and which in some animals also involve the separation of the sanitary and the respiratory aspects of the activity of the cavity.

The first of these devices which may be mentioned is that used by Zeugobranchia (e.g. *Pleurotomaria, Haliotis, Puncturella, Diodora, Emarginula*); in some cases it permits of an almost perfect external bilateral symmetry, more perfect, perhaps, than that achieved by any other prosobranch. In these animals one may imagine with Garstang (1928) that the impact of the exhalant stream of water, laden with metabolites, against the edge of the mantle skirt, has slowed down its rate of growth so that a bay is formed in the mid-point anteriorly, directly over the head. This bay in the mantle skirt has consequential effects upon the shell, with the result that a slit-like emargination appears (fig. 33) at this point, and it is through this slit in the mantle skirt and shell that the outgoing stream of water now passes. The edge of the mantle skirt may be turned outwards a little through the slit in the shell so as to form a kind of incomplete siphon (fig. 43A), which may be used to direct the water upwards, or sideways—in any direction except towards the head. This is the state of affairs in *Pleurotomaria* and *Scissurella* (fig. 248B). A further evolutionary step separates the opening of the mantle cavity

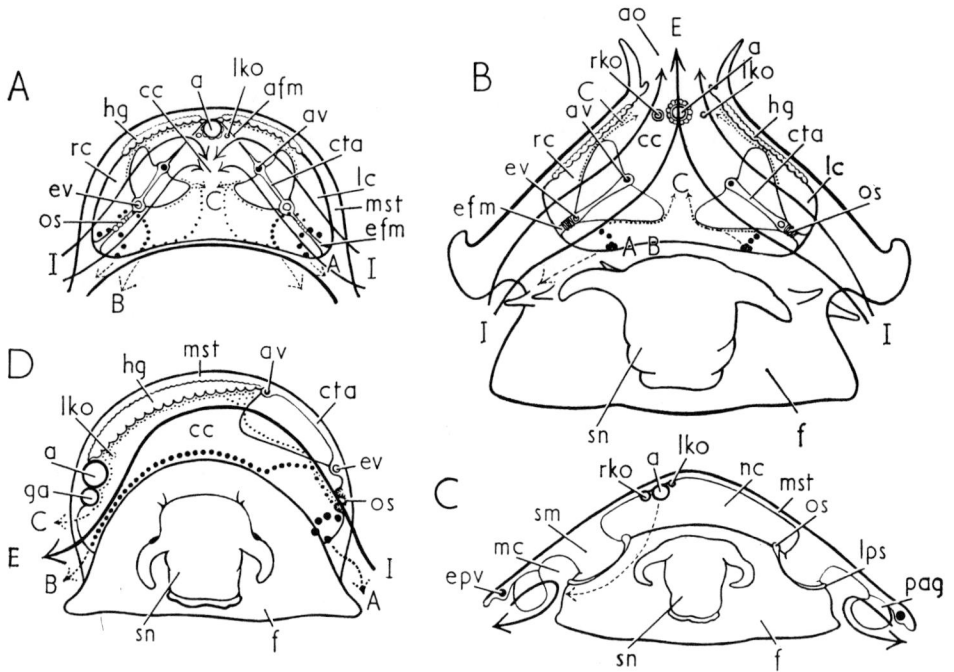

FIG. 48.—Diagrammatic transverse sections to show water and ciliary currents in the mantle cavity of A, a
hypothetical, primitive gastropod; B, *Diodora;* C, *Patella;* D, a mesogastropod. Continuous arrows show the
direction of water currents; broken arrows show the direction of ciliary currents A, B and C, and the particles
affected by them are indicated by dots of three sizes. The presence of ctenidial membranes is debatable.
    a, anus; afm, afferent membrane of ctenidium; ao, apical opening of mantle cavity; av, afferent branchial
vessel; cc, central compartment of mantle cavity; cta, ctenidial axis; E, exhalant; efm, efferent membrane of
ctenidium; epv, efferent pallial vessel; ev, efferent branchial vessel; f, foot; ga, genital opening; hg, hypobranchial gland; I, inhalant; lc, left dorsolateral compartment of mantle cavity; lko, left kidney opening; lps,
lateral pallial streak; mc, mantle cavity; mst, mantle skirt; nc, nuchal cavity; os, osphradium; pag, pallial gill; rc,
right dorsolateral compartment of mantle cavity; rko, right kidney and genital opening; sm, shell muscle; sn,
snout.

into two parts by the fusion of the anterior edges of the mouth of the slit. When this has
occurred there are two unconnected pallial openings, the main one over the animal's head
and neck, which is now an inhalant aperture (for water currents), and the separated slit
through which the outgoing current passes. At first located near the anterior margin of the
shell (*Emarginula* (figs. 33, 245B), *Puncturella* (fig. 246)) it may be moved by secondary growth of
the shell so that it comes to lie further back, when the outgoing current may be directed
completely away from the head (*Diodora,* fig. 43A). Despite the changes in pallial arrangements shown by zeugobranchs particulate matter may still be removed from the mantle
cavity by the main inhalant aperture either by ciliary currents A and B, which cannot be
separately distinguished, or by the animal pulling the shell down on the head and foot by
contraction of the shell muscle, a movement which has the effect of washing water and
particles forcibly out of the cavity. The arrangements in this case are shown in fig. 48B.

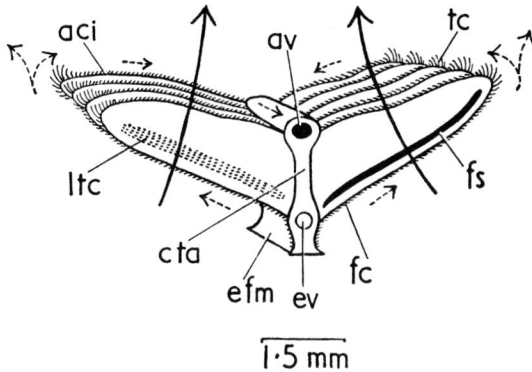

FIG. 49.—*Haliotis tuberculata:* stereogram of portion of ctenidium showing a short length of axis bearing leaflets. The nearest leaflet on the right has been cut open longitudinally. Continuous arrows show the direction of water currents; broken arrows show the direction of ciliary currents.

aci, abfrontal cilia; av, afferent branchial vessel; cta, ctenidial axis; efm, efferent membrane of ctenidium; ev, efferent branchial vessel; fc, frontal cilia; fs, endoskeletal support of branchial leaflet; ltc, lateral cilia; tc, terminal cilia.

As explained elsewhere (p. 52), the gastropod visceral mass is normally wound in a right-handed spiral. This involves differential growth along its two sides, the left hypertrophying, the right atrophying. This tendency is marked in some of the zeugobranchs (*Pleurotomaria, Scissurellidae*), is much less marked in others (*Haliotis*) and has been completely lost with the assumption of a limpet-like shape by the Fissurellidae. This evolutionary trend is perhaps to be linked with the fact that the animals have found a workable means of ventilating the mantle cavity by apical or sub-apical openings in the shell. It presupposes, however, the adoption of a particular ecological niche in that it decreases the value of the shell as a protection against desiccation.

Alternative solutions of the problem of the ventilation of the mantle cavity have been found by such gastropods as are not occupants of this niche. One of the simplest of these is that adopted by the majority of members of the Patellacea (=Docoglossa). In these (e.g. *Patina, Patella*, figs. 258, 48c, 50) the mantle cavity has given up its respiratory function in that the ctenidia have been lost. The main site of exchange of respiratory gases is now provided by secondary gills (pag) which are developed on the pallial skirt near its edge. Superficially these resemble the ctenidial leaflets, but they are new structures, and the cilia on their surface create a local current of water which enters dorsally and leaves ventrally. The main anterior part of the mantle cavity (the nuchal cavity (nc), lying over the animal's neck) is now concerned only with accommodating the head when that is retracted, and in housing the anus (a), the osphradia (os) and the openings of the two kidneys (lko, rko). Hardly any current of water is maintained through it in the adult and faeces and other material are led out by ciliary means (fig. 259B). Presumably it is for this reason that the faeces (fig. 126H) are highly compacted by the elaborately coiled intestine which these animals possess (fig. 124). All members of the Patellacea have apparently lost the hypobranchial gland—presumably functionally replaced by the gland cells of the intestine (see p. 222). *Acmaea* (=*Patelloida*) (figs. 51, 250, 251) still retains a single ctenidium (ct) (although some workers believe this to be a new pseudo-ctenidium) and therefore resembles the animals of the next section [see p. 700].

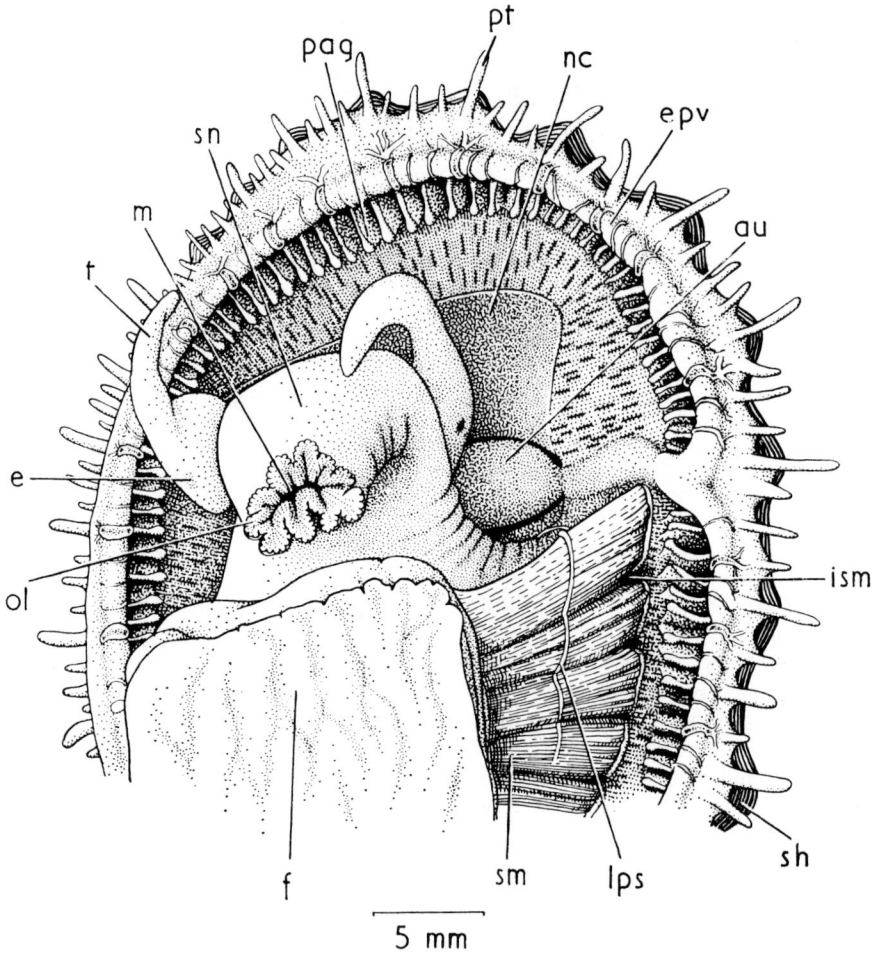

FIG. 50.—*Patella vulgata:* anterior half in ventral view.
    au. auricle; e, eye; epv, efferent pallial vessel; f, foot; ism, interstice between bundles of shell muscle; lps, lateral pallial streak; m, mouth; nc, nuchal cavity; ol, outer lip; pag, pallial gill; pt, pallial tentacle; sh, shell; sm, shell muscle; sn, snout; t, tentacle.

    Most prosobranchs (Monotocardia) have altered the arrangements within their mantle cavity in relation to the different growth rates of the right and left sides of the visceral mass already referred to. In a primitive gastropod the exhalant part of the mantle cavity is marked by the anus and it lies in a median position, inhalant streams of water converging on it from right and left. Could the animal push the mid-line over to one side, so to speak, the anus and exhalant stream would then be lateral. This is precisely what the spiral coiling allows: the hypertrophy of the organs of the animal's (topographical) left side makes them occupy almost the entire breadth of the mantle cavity, pushes across the left kidney opening and the

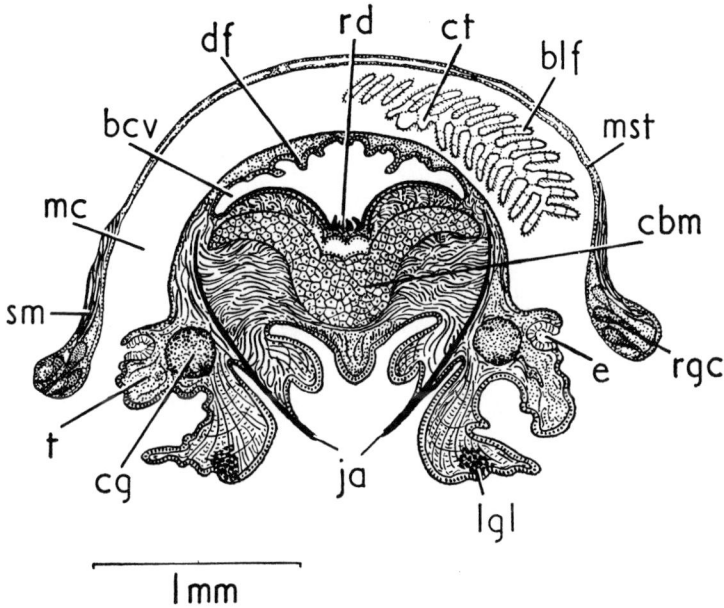

FIG. 51.—*Acmaea virginea:* transverse section at level of base of tentacles.

bcv, buccal cavity; blf, branchial leaflet; cbm, cartilage of buccal mass; cg, cerebral ganglion; ct, ctenidium; df, dorsal fold of buccal cavity; e, eye; ja, jaws; lgl, labial glands; mc, mantle cavity; mst, mantle skirt; rd, radula; rgc, repugnatorial glands of edge of mantle skirt; sm, shell muscle; t, tentacle.

anus, which is morphologically median, so that they come to lie at the extreme topographical right of the mantle cavity, whilst the organs of the right side atrophy almost completely. The original, double, lateral to median current in the mantle cavity is thus replaced by a single transverse current which enters the mantle cavity on the left, washes across the animal's back and emerges, with its waste, on the right (fig. 48D).

Associated with this change in the general disposition of the pallial complex of mesogastropods go changes in its detailed arrangement. The general trend to replace double water currents by a single one involves the suppression of one whole row of ctenidial leaflets.

Only the left ctenidium survives in Monotocardia, and only the right series of ctenidial leaflets remains attached to the axis in most species. In the Trochacea and Neritacea, which are diotocardian but have only one gill, the left, there exists a state of affairs which shows how this may have arisen. The ctenidium is bipectinate with right and left rows of leaflets attached to the axis. In the Neritacea (fig. 52) it (ct) projects rather freely into the mantle cavity and its suspensory membranes (afm, efm) are weakly developed, especially on the dorsal side, so that water can flow freely above it and below. The right ctenidium is reduced to a small vascular knob (ho), the so-called 'organe creux' of Lenssen (1899). In most Trochacea, however (figs. 53, 54), only the anterior third of the ctenidium is free, the posterior part being tied above to the mantle skirt by an afferent membrane (afm) and to the wall of the mantle cavity below by an efferent one (efm). The free tip is stiffened by an internal skeleton (fs).

In this region and in the middle third of the gill two sets of leaflets are carried on the axis. In the anterior third these project into the main mantle cavity; in the middle third only those on

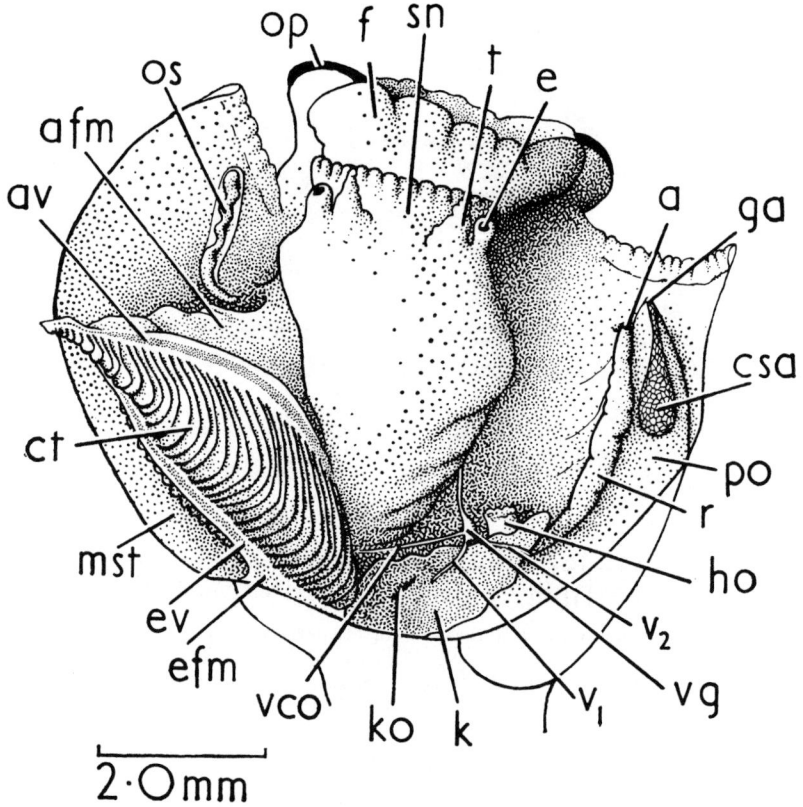

FIG. 52.—*Theodoxus fluviatilis:* dissected to display the contents of the mantle cavity.
   a, anus; afm, afferent membrane of ctenidium; av, afferent branchial vessel; csa, crystal sac on oviduct; ct, ctenidium; e, eye; efm, efferent membrane of ctenidium; ev, efferent branchial vessel; f, foot; ga, genital opening; ho, 'organe creux' of Lenssen, vestigial right ctenidium; k, kidney; ko, kidney opening; mst, mantle skirt; op, operculum; os, osphradium; po, pallial oviduct; r, rectum; sn, snout; t, tentacle; v_1, renopericardial nerve; v_2, genital nerve; vco, part of visceral loop; vg, visceral ganglion.

the right side of the axis can do this, those on the left being cut off from it by the suspensory membranes. They therefore project into a small cavity (lc, fig. 54) cut off from the main one, and obviously difficult to ventilate effectively, though Clark (1958) has been able to show that the same kind of circulation exists within it as within the main pallial chamber. The left leaflets rapidly become small and die out towards the base of the gill where the cavity in which they lie also disappears. In its most posterior third, therefore, the trochacean gill consists of an axis fused broadside to the mantle skirt and carrying a single row of leaflets which project from its right side into the general mantle cavity. This is, in fact, the condition in which the ctenidium exists throughout its entire length in all the monotocardian gastropods which Thiele (1929–35) placed in the Mesogastropoda and Stenoglossa. As a result the ctenidial axis appears to be attached to the mantle skirt not only by its dorsal and ventral edges, but along its whole

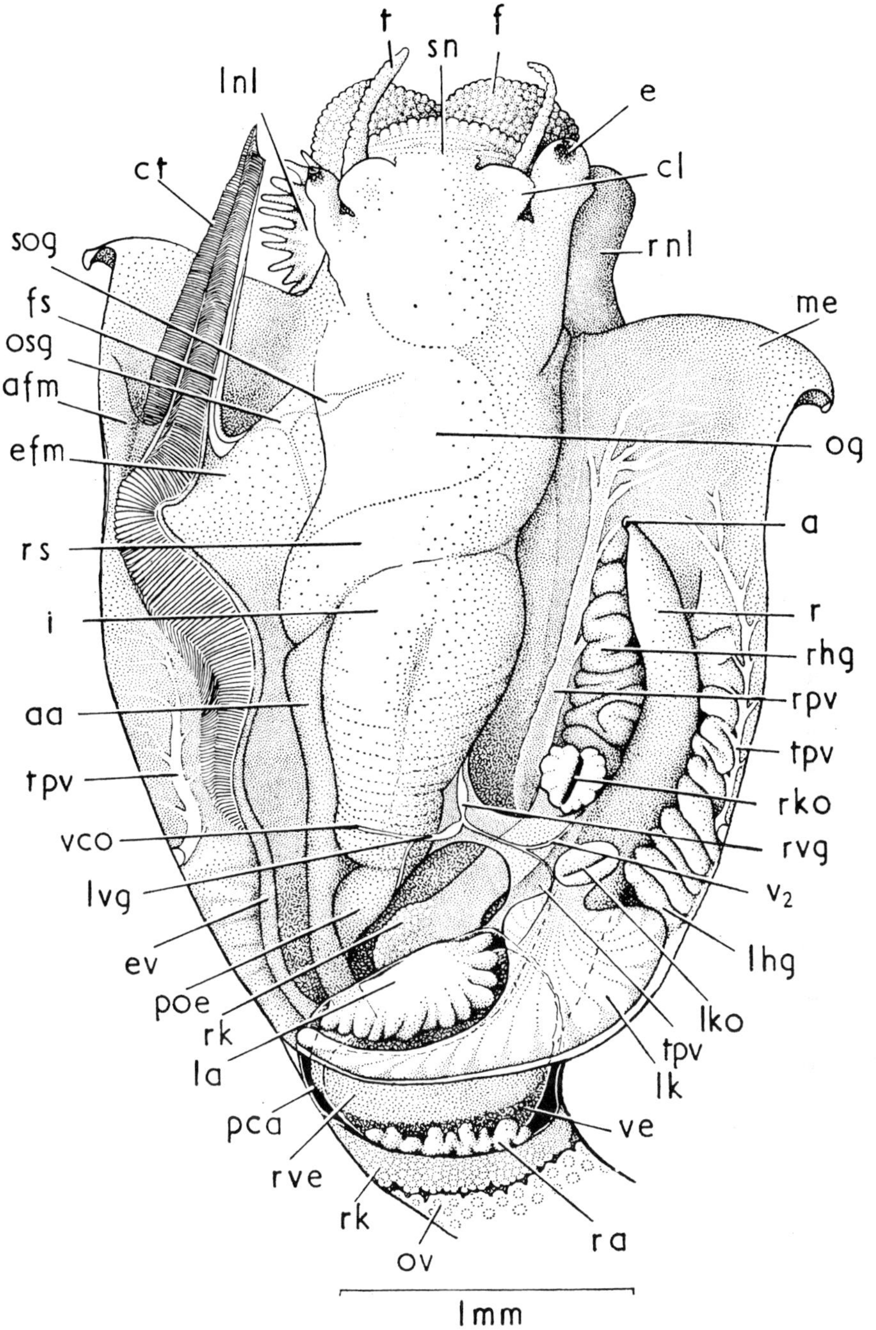

FIG. 53.—*Monodonta lineata*: dissected to display the contents of the mantle cavity.

a, anus; aa, anterior aorta; afm, afferent branchial membrane; cl, cephalic lappet; ct, ctenidium; e, eye on eye stalk; efm, efferent membrane of ctenidium; ev, efferent branchial vessel; f, foot; fs, skeleton of gill; i, intestine, by transparency; la, left auricle; lhg, left hypobranchial gland; lk, left kidney or papillary sac; lko, left kidney opening; lnl, left neck lobe; lvg, left visceral ganglion; me, mantle edge; og, oesophageal glands, by transparency; osg, osphradial ganglion, by transparency; ov, ovary; pca, pericardial cavity; poe, posterior oesophagus, by transparency; r, rectum; ra, right auricle; rhg, right hypobranchial gland; rk, right kidney; rko, right kidney and genital opening; rnl, right neck lobe; rpv, right pallial vein; rs, radular sac, by transparency; rve, rectum in ventricle; rvg, right visceral ganglion; sn, snout; sog, supra-oesophageal ganglion; t, tentacle; tpv, transverse pallial vein; v$_2$, genital nerve; vco, part of visceral loop; ve, ventricle.

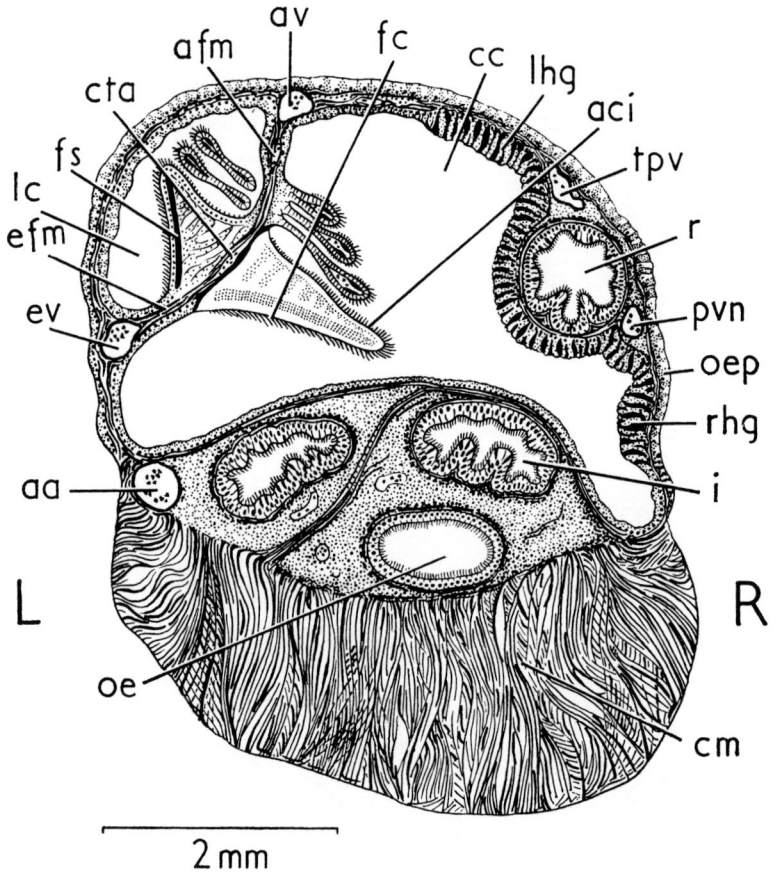

FIG. 54.—*Monodonta lineata:* transverse section at level of middle third of mantle cavity.
aa, anterior aorta; aci, abfrontal cilia; afm, afferent membrane of ctenidium; av, afferent branchial vessel; cc, central compartment of mantle cavity; cm, columellar muscle; cta, ctenidial axis; efm, efferent membrane of ctenidium; ev, efferent branchial vessel; fc, frontal cilia; fs, skeleton of branchial leaflet; i, intestine; L, left; lc, left compartment of mantle cavity; lhg, left hypobranchial gland; oe, oesophagus; oep, outer epithelium of mantle skirt; pvn, pallial vein; R, right; r, rectum; rhg, right hypobranchial gland; tpv, transverse pallial vein.

breadth and the triangular leaflets which it bears are apparently joined to the mantle skirt along the whole of one side.

A second change which is to be associated, but less certainly, with the trend towards more effective ventilation and cleansing of the mantle cavity, is the altered position of some or all of the apertures which lie within it. In most of the lower Diotocardia (figs. 249, 250, 251, 258) the anus lies well back from the edge of the mantle skirt near the inner end of the mantle cavity and it is flanked by small papillae on which the right and left kidney openings are placed, the former acting also as a genital pore. In the Zeugobranchia (fig. 249) these are in close proximity to the slit or hole in the mantle through which the exhalant current of water passes.

They lie in corresponding positions in the Patellacea (figs. 250, 251, 258) where there is no hole. In the Trochacea (fig. 53) and Neritacea (fig. 52), however, a change has occurred which is met with again in the mesogastropods and stenoglossans: the anus in the first group, and the genital duct (extending from the right kidney aperture) in addition in the others, now open to the mantle cavity close to its mouth so that faeces and gametes are no longer shed into the deeper parts. The purely renal opening stays as a pore in the deepest part of the mantle cavity.

How this extension of rectum and genital duct has been achieved can be deduced from the fact that in some species the genital duct is merely a ciliated gutter stretching along the mantle skirt (sgr, fig. 13), and from the fact that even when it forms a closed duct traces of its origin from a gutter are occasionally discernible (fig. 173J). It is therefore possible that the anus has been brought forward to its new position in the same way, though traces of this are not discoverable.

Another factor to be considered is this: with the alteration in growth rates between right and left sides the extension of the mantle skirt is maximal on the left, minimal on the right, and of intermediate extent between. The altered position of the anus may be merely relative and the real change a deepening of the left half of the cavity. However it has been brought about, the forward position of the anus and the genital pore has the effect at one and the same time of keeping faeces from contaminating the mantle cavity, leading germ cells to the edge and facilitating copulation, which might otherwise prove too difficult a process.

The various changes which have accompanied and permitted the evolution of a ciliary method of food collecting in prosobranch gastropods have been well discussed by Yonge (1938) and Werner (1952, 1959). [This particular trend was once believed to be limited to sedentary monotocardians but has since been found in some diotocardians (see p. 626).] The reason for its absence in zeugobranchs is most likely to be found in the disposition of the currents within the mantle cavity—so long as there are two sets of these, right and left, converging in the mid-line, it will prove impossible for the material which they carry in suspension to be collected into a place where the gastropod may use it. It is only when the water current is the transverse stream of the mesogastropod that this happens.

As explained above, suspended material which enters the mantle cavity of a mesogastropod is normally sorted into three lots (fig. 48D), each of which is treated differently by the animal. The basis of the sorting is size and weight, the largest and heaviest particles settling out of the water current soonest, therefore to the left side of the mantle cavity, the smallest ones being trapped only when they reach the gill surface. Those of the first group are led out of the mantle cavity by a tract along the left side of the animal's neck which takes them to the side of the foot, from which they fall off or are swept away. This, it will be recalled was called current A by Yonge. The others, medium-sized, which fall on to the floor of the inner part of the mantle cavity, or the smallest, on the gills, are carried by currents B and C respectively— on the floor of the mantle cavity in the case of B, over the hypobranchial gland in the case of C—and ultimately reach the right side of the cavity and are carried out.

In the evolution of a ciliary feeding mechanism the main step which had to be taken was the diversion of at least some of the material carried by these three currents, A, B and C, to the animal's mouth, a diversion which clearly had to occur before they joined the faecal and excretory stream. One of the simplest of the alterations is found in *Bithynia tentaculata* (Lilly, 1953; Schäfer, 1953a, b), though it is combined with a specialized filter. This is formed of a mucous net secreted from the hypobranchial gland and stretched from that structure to the

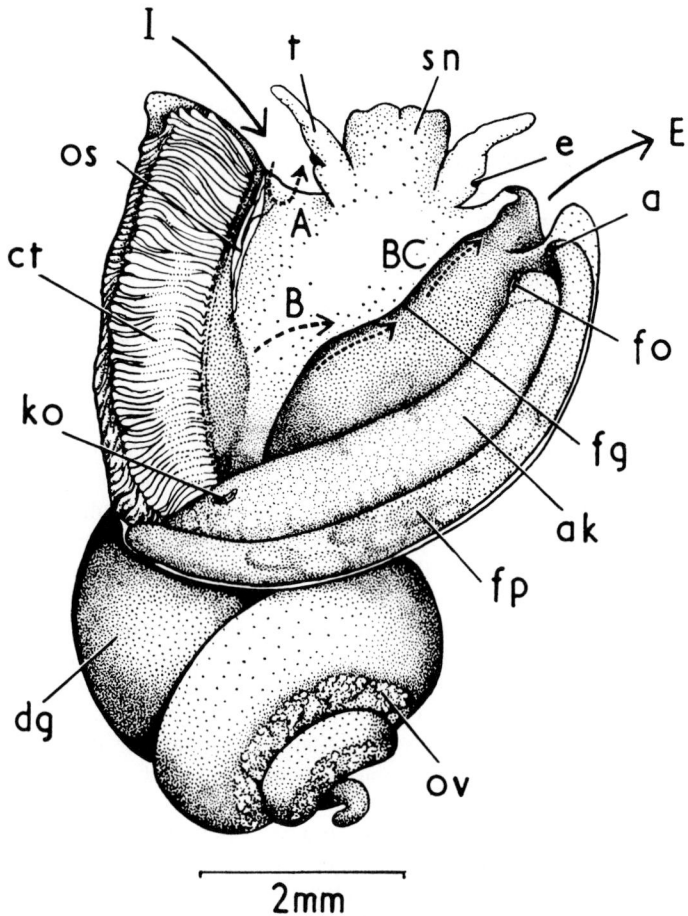

FIG. 55.—*Bithynia tentaculata:* dissection to show contents of mantle cavity. Continuous arrows show the course of water currents; broken arrows show the course of ciliary currents A, B and C (for explanation see text).

a, anus; ak, anterior diverticulum of kidney overlying pallial oviduct; ct, ctenidium; dg, digestive gland; E, exhalant; e, eye; fg, food groove; fo, female opening; fp, faecal pellets in rectum; I, inhalant; ko, kidney opening; os, osphradium; ov, ovary; sn, snout; t, tentacle.

gill. Material caught here is carried down the right side of the mantle cavity in a distinct ciliated groove (fg, fig. 55). The material which emerges from the front end of this passes to the propodial region in front of the snout (sn), from where it may be raked into the gut by the radula. Material from current A is not collected and it appears as if material from current C were diverted so as to join B (BC). This is a regular feature of ciliary feeding prosobranchs (fig. 61) and in other genera involves the elongation of the ctenidial filaments until their free tips rest on the floor of the mantle cavity. This is thus split into two sections, an inhalant chamber on the left (ibr), and an exhalant chamber on the right (sbr). The two are separated by the

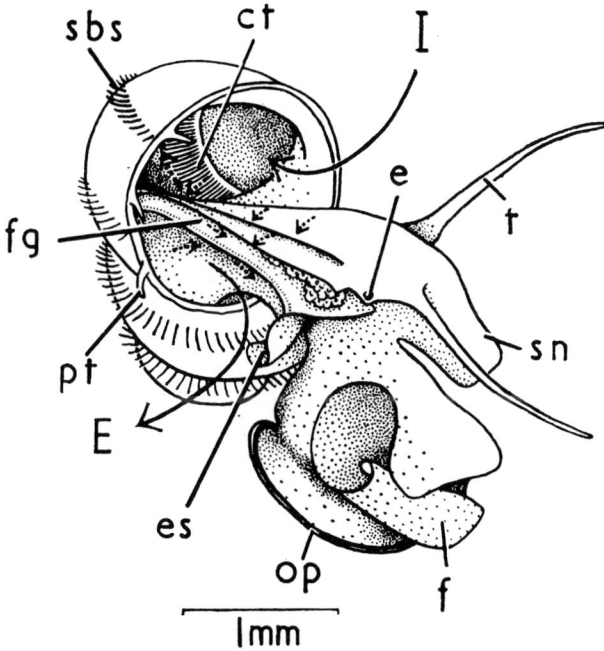

FIG. 56.—*Viviparus contectus:* young specimen. Continuous arrows show the course of water currents; broken arrows show the course of ciliary currents. Based on Cook.
ct, ctenidium; E, exhalant; e, eye; es, exhalant siphon; f, foot; fg, food groove with accumulation of food at anterior end; I, inhalant; op, operculum; pt, pallial tentacle; sbs, bristle on shell; sn, snout; t, tentacle.

ctenidium which forms an incomplete barrier through which water is driven by the beat of the lateral cilia. Over the surface of the ctenidium there is now secreted a net of mucus (bf) from goblet cells lying in the ctenidial epithelium and this net, continuously renewed, is continuously driven down the leaflet edge by the frontal cilia (fc, fig. 62), entrapping suspended particles from the water which passes through its interstices as it does so. In this way there is produced by the prosobranch the two essential parts of any ciliary food-collecting mechanism—(a) something to create a water current, and (b) something to strain suspended particulate matter from it.

With further elaboration of the food-collecting apparatus a number of changes appears. One of the most widespread is the conversion of the tract on the right of the mantle cavity, along which the food particles are led to the mouth, into a deep gutter. A groove is visible in *Bithynia, Calyptraea, Crepidula,* but has become converted into a distinct gutter in *Viviparus* (fg, fig. 56) and reaches its maximal development in *Turritella* (fg, fig. 57) where it forms a great trough, with high, upraised sides, which runs across the whole of the floor of the mantle cavity to a point just under the right cephalic tentacle. These devices help to keep food and faeces separate. A second change affects the size and efficiency of the ciliated surface which creates the current: by increasing the length of the mantle cavity it is possible to house a longer ctenidium and so create a greater current; and by elongating the ctenidial leaflets the same end is achieved. These changes may be noted by comparing the mantle cavity and ctenidium

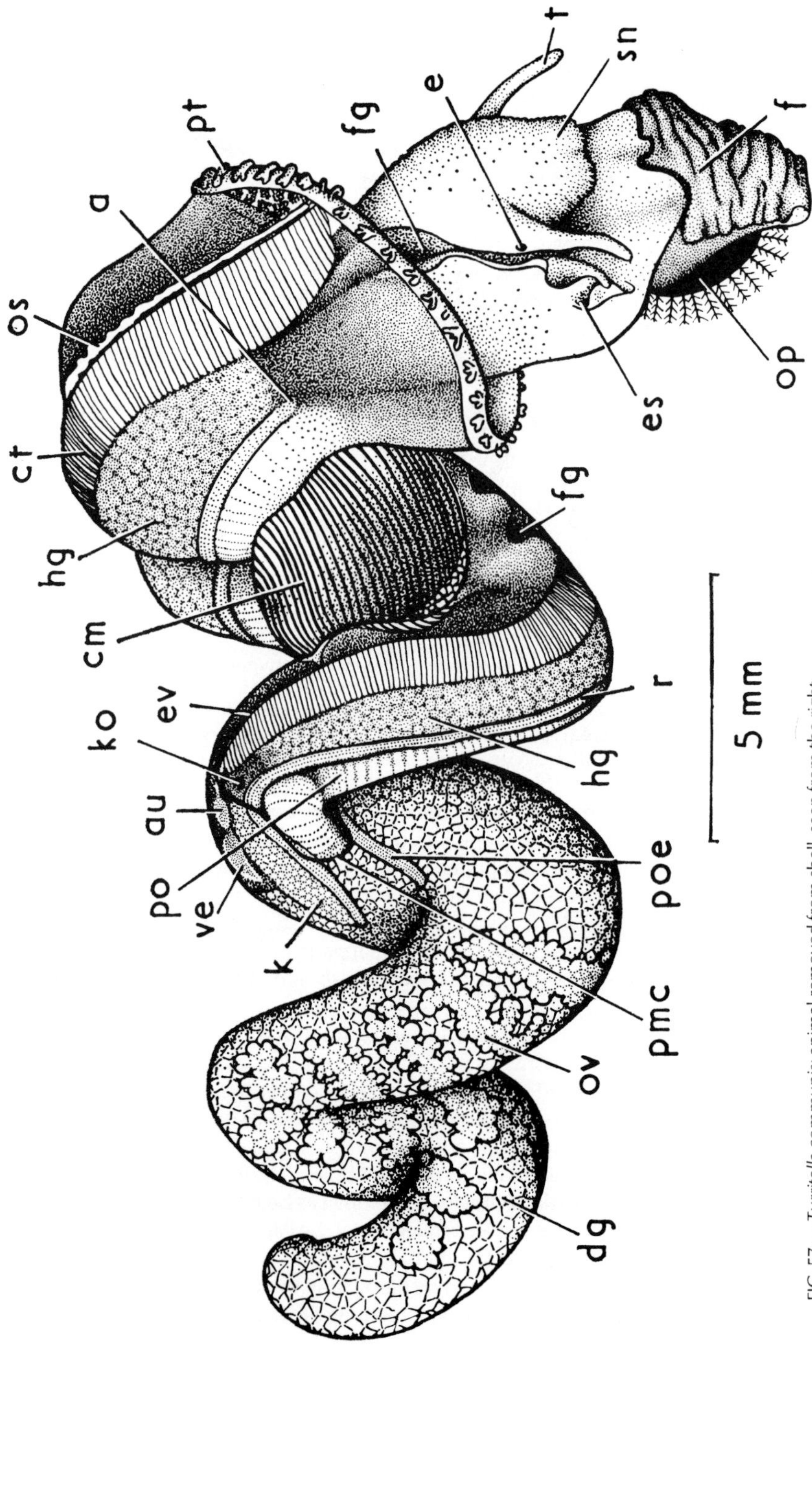

FIG. 57.— *Turritella communis*: animal removed from shell, seen from the right.
a, position of anus; au, auricle; cm, columellar muscle; ct, ctenidium, by transparency; dg, digestive gland; e, eye; es, exhalant siphon; ev, efferent branchial vessel; f, foot; fg, food groove; hg, hypobranchial gland; k, kidney; ko, kidney opening, by transparency; op, operculum; os, osphradium; ov, ovary; pmc, posterior limit of mantle cavity; po, pallial oviduct; poe, posterior oesophagus; pt, pallial tentacle; r, rectum; sn, snout; t, tentacle; ve, ventricle.

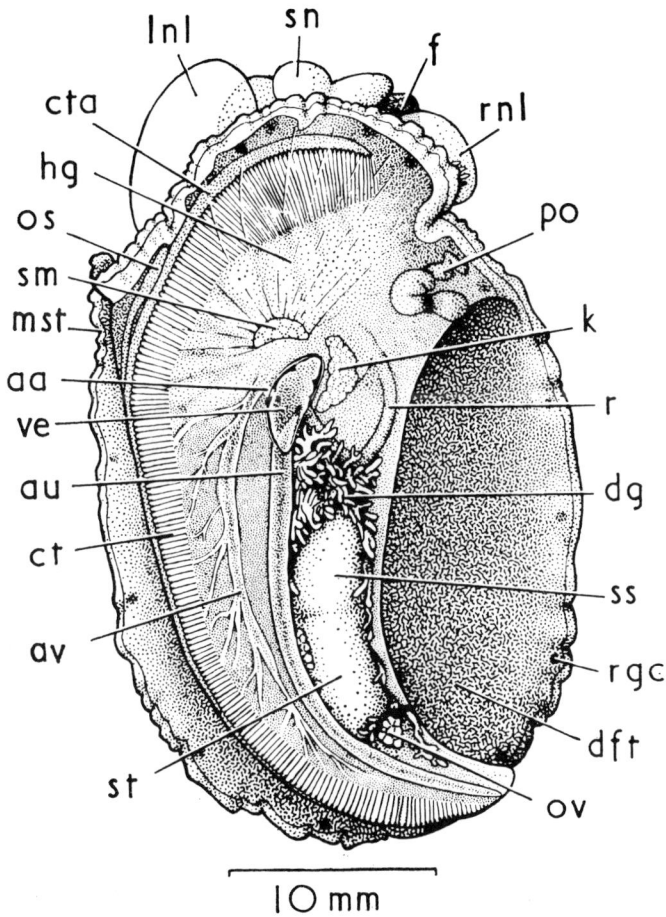

FIG. 58.—*Crepidula fornicata*: dorsal view of animal removed from shell.

aa, anterior aorta; au, auricle; av, afferent branchial vessel; ct, ctenidium; cta, ctenidial axis; dft, dorsal surface of foot; dg, digestive gland; f, foot; hg, hypobranchial gland; k, kidney; lnl, left neck lobe; mst, mantle skirt; os, osphradium; ov, ovary; po, pallial oviduct; r, rectum; rgc, repugnatorial gland; rnl, right neck lobe; sm, shell muscle; sn, snout; ss, style sac region of stomach; st, stomach; ve, ventricle.

of *Bithynia* (fig. 55), which is not particularly specialized, with those of *Calyptraea, Turritella* (fig. 57), and (especially) *Crepidula* (fig. 58) which are.

Increased gill area produces a greater water current and allows the animal to sieve a greater volume in a given time, but for effective food trapping, this necessitates a greater production of mucus. This is provided, at least in part, from the increased gill surface itself, but in some of the more elaborate forms this appears to be inadequate and has to be supplemented by the development of totally new sites. The most important of these is the 'endo-style' of the Calyptraeacea (fig. 59; ens, fig. 61), first described by Orton (1912a, b; 1913b), and

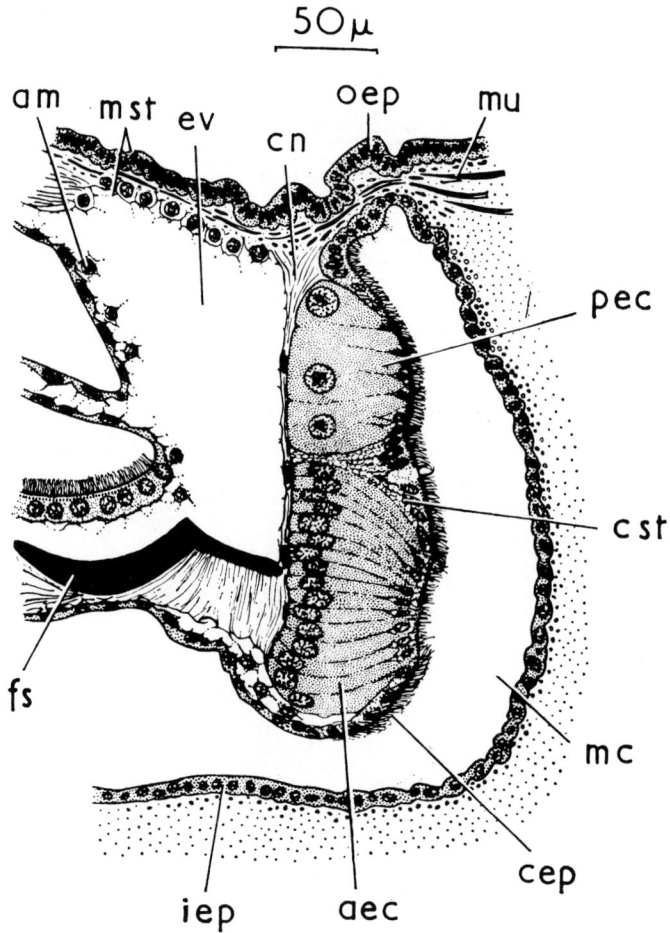

FIG. 59.—*Crepidula fornicata:* transverse section of ctenidial axis to show endostyle.
    aec, anterior gland cells of endostyle; am, amoebocyte; cep, ciliated epithelium of ctenidial axis; cn, connective tissue; cst, strip of ciliated cells between two parts of endostyle; ev, efferent branchial vessel; fs, skeleton of branchial filament; iep, inner pallial epithelium; mc, mantle cavity; mst, mantle skirt; mu, muscle fibre; oep, outer pallial epithelium; pec, posterior gland cells of endostyle.

so called because its function is a precise parallel to that of the endostyle of the protochordates, in that it provides mucus out of which a sheet is made, stretched (in these animals) over the gill slits to sieve the water current which escapes from the pharynx to the atrial cavity. The histology of the endostyle has been described in detail by Werner (1953), who showed that it is composed of two longitudinal strips of gland cells (aec, pec, fig. 59) alternating with ciliated cells. Between these groups of gland cells runs a tract composed predominantly of ciliated cells (cst) within which an occasional goblet cell may be found. The whole structure overlies the efferent branchial vessel (ev) the cavity of which is kept open partly by the gill skeleton (fs).

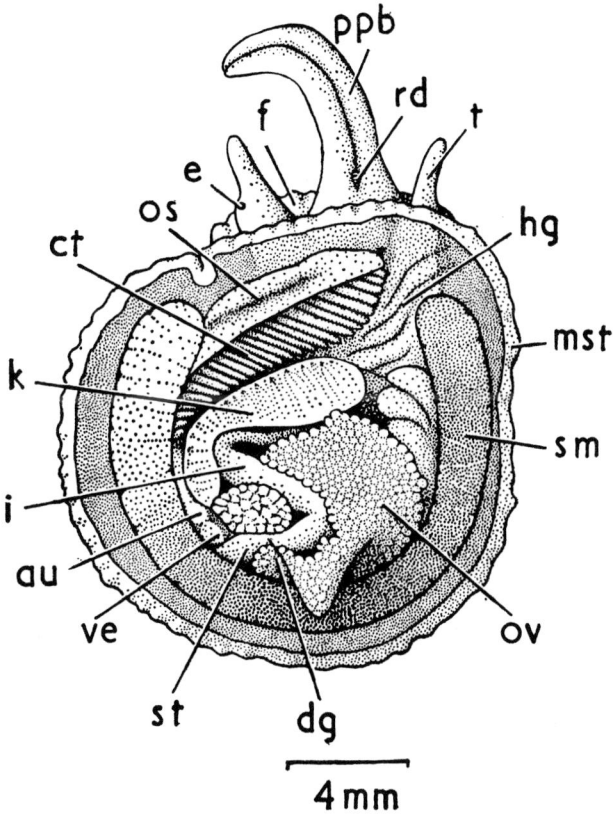

FIG. 60.—*Capulus ungaricus:* dorsal view of animal after removal from shell.
   au, auricle; ct, ctenidium; dg, digestive gland; e, eye; f, foot; hg, hypobranchial gland; i, intestine; k, kidney; mst, mantle skirt; os, osphradium; ov, ovary; ppb, proboscis; rd, radula; sm, shell muscle; st, stomach; t, tentacle; ve, ventricle.

The cilia beat forwards, that is across the endostyle, so that its secretion is carried on to the ventral face of the ctenidial filaments.

A third change which may affect the mantle cavity in the evolutionary trend towards ciliary food collecting is the inclusion of Yonge's current A amongst the sources of particulate food: this has been managed only in the more advanced types. In *Capulus* (fig. 60) the material collected from the right side of the mantle cavity is assembled on the propodium (f), anterior to the mouth, on which the proboscis (pb) rests. To this, too, comes the material which is collected from the floor of the mantle cavity on the left by current A, and food from both sources is taken to the mouth by the proboscis.

The most complex of the devices for exploiting current A, however, is that described by Werner (1953) for *Calyptraea, Crepidula* and some other members of the Calyptraeidae. In these ciliary feeders may be seen the apex of the trend amongst gastropods (figs. 58, 59, 61, 62, 63): the mantle cavity is very long, the gill extensive, its filaments elongate, an endostyle is

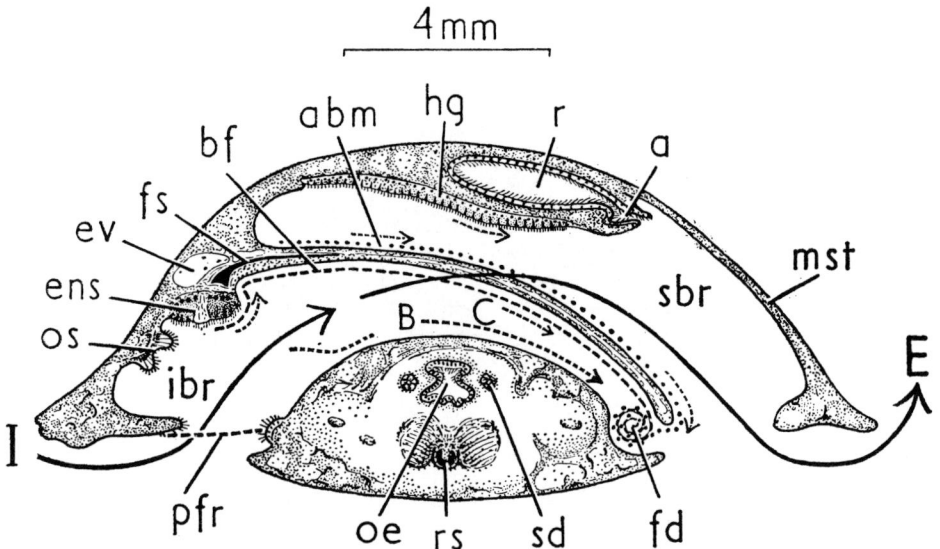

FIG. 61.—*Crepidula fornicata*: transverse section to show currents in the mantle cavity. Continuous arrows show direction of water currents; broken arrows show direction of ciliary currents, B, C (for explanation see text).

a, anus; abm, abfrontal mucous stream; bf, branchial mucous filter collecting fine particles; E, exhalant; ens, endostyle; ev, efferent branchial vessel; fd, food and mucous string in food groove; fs, skeleton in branchial filament; hg, hypobranchial gland; I, inhalant; ibr, infrabranchial part of mantle cavity; mst, mantle skirt; oe, oesophagus; os, osphradium; pfr, pallial mucous filter collecting coarse particles; r, rectum; rs, radular sac; sbr, suprabranchial part of mantle cavity; sd, salivary duct.

present and there is, in addition, what is called by Werner a pallial filter (pfr, figs. 61, 63) as well as the branchial one (bf, figs. 61, 62). This lies across the mouth of the mantle cavity on the left of the head as a web of mucous threads, secreted by glands at the point where the mantle skirt joins the head-foot, which strain the bigger particles out of the incoming water current and transport them to a food pocket in the anterior part of the mantle skirt. From there they may be withdrawn by the radula.

In the calyptraeids, Werner (1953) has described the elaborate mechanisms responsible for the collection and transportation of the food caught on the branchial filter. This, as shown in figs. 61, 62 (bf), is a sheet of mucus secreted from the endostyle (ens), supported by the gill filaments and carried towards their tip by the beating of the frontal cilia (fc). The bulk of the small food particles entering the mantle cavity in the inhalant water current is caught in this filter; a few may succeed in passing between the filaments into the suprabranchial space (sbr) but these then fall either on the abfrontal mucous membrane (abm) and are carried by the abfrontal cilia (aci) to the tip of the filament, or on to the hypobranchial gland (hg). It is only the small fraction of the total suspended matter in the water entering the mantle cavity making up the last of these groups which eludes the food-collecting apparatus. The remainder moves to the tips of the filaments which, in these animals, rest on the food groove on the floor of the mantle cavity on the right. Here the detailed arrangement of the cilia on gill tip and food groove rolls the mucus and its attached food particles into a cylinder rather like a Swiss roll,

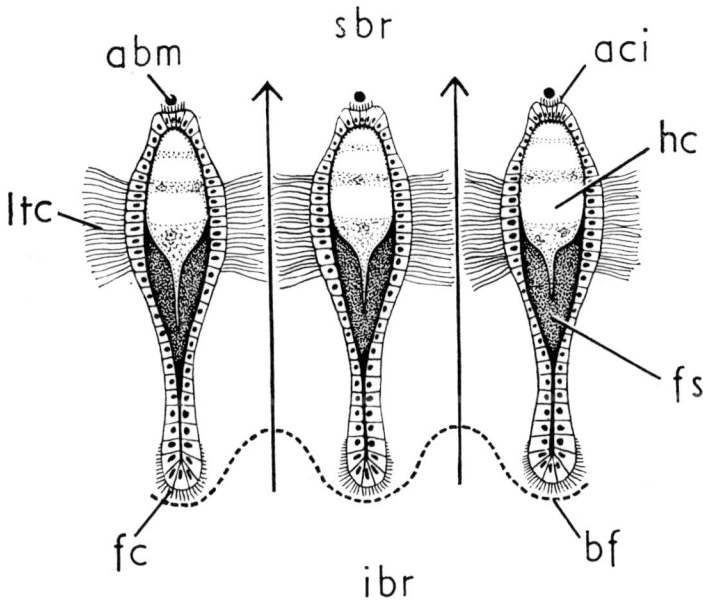

FIG. 62.—*Crepidula fornicata*: transverse section of 3 gill filaments to show the associated mucous food trap. Continuous arrows show direction of water currents. Based on Werner.
    abm, abfrontal mucous stream; aci, abfrontal cilia; bf, branchial mucous filter collecting fine particles; fc, frontal cilia; fs, skeleton in branchial filament; hc, haemocoelic space; ibr, infrabranchial part of mantle cavity; ltc, lateral cilia; sbr, suprabranchial part of mantle cavity.

with the mucus corresponding to the sponge, and the food particles to the jam (fd). This is carried forwards along the groove (fg, fig. 63), along the base of the neck lobe (nl) on the right, ventral to the tentacle (t) to its end by the mouth (efg).

    There are a few other features of the mantle cavity of ciliary feeding prosobranchs, and other modifications of their structure, which may be conveniently referred to here. Most of these animals are sedentary and live on bottoms which are hard: if too much suspended matter were to enter the mantle cavity there is danger of the filter becoming clogged and certainly its efficiency being decreased even if it is not actually damaged. A few ciliary-feeding gastropods, nevertheless, do live on muddy bottoms. Of these the best known is perhaps *Turritella* (fig. 64) in which the mantle edge has been thickened, especially on the left side, and drawn out into pinnately branched tentacles (pt). These project dorsally a little way over the mouth of the shell, but their greatest length is directed over the inhalant aperture of the mantle cavity, across which they stretch as a coarse filter. By means of this device the amount of suspended particulate matter which does find its way into the mantle cavity is kept within manageable limits. *Crepidula fornicata* and *Calyptraea chinensis* may also live in silty situations, but these animals are attached by the sucker-like foot and contrive to raise themselves to a position where there is no danger of mud fouling their feeding mechanism by settling on shells and stones.

    All particles which are caught in the mantle cavity are trapped in mucus and transported towards the mouth embedded in this material. The radula licks lengths of it into the buccal

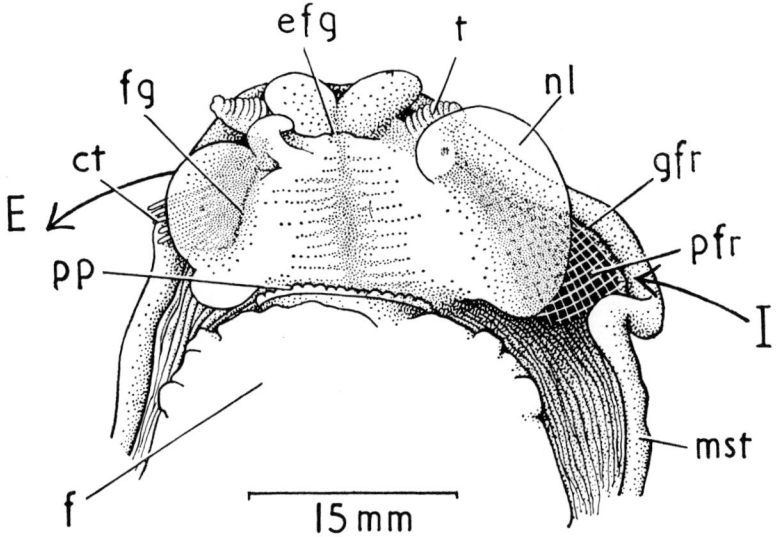

FIG. 63.—*Crepidula fornicata:* ventral view of the anterior half of the body. Arrows show direction of water currents. Details added from Werner.

    ct, ctenidium; E, exhalant; efg, anterior end of food groove; f, foot; fg, food groove, by transparency; gfr, groove along which food and mucus from pallial filter travel to food pouch; I, inhalant; mst, mantle skirt; nl, neck lobe; pfr, pallial mucous filter; pp, propodium; t, tentacle.

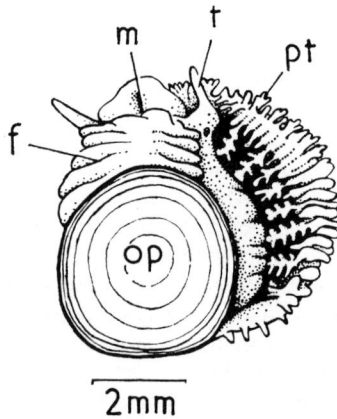

FIG. 64.—*Turritella communis:* to show details of the inhalant pallial aperture.

    f, foot; m, mouth; op, operculum simplified; pt, pallial tentacle; t, tentacle.

cavity and may, in prosobranchs which feed entirely by this means, never be used for manipulating any more abrasive substance. As a consequence wear on the radular teeth is minimal, as has been shown by Peile (1937). The salivary glands, too, which usually produce the mucus required for the lubrication of the radular movements, find themselves replaced functionally by the glands of the mantle cavity and in molluscs like *Crepidula* form two tubular structures, and in *Capulus* or *Turritella* two tufts of acini discharging to the buccal cavity, much smaller than the voluminous glands of prosobranchs which feed in other ways.

These changes in the various parts of the pallial complex and the associated changes in the first part of the alimentary canal [see also p. 596] are, in general, parallel to those which have also occurred in the lamellibranchs, and point clearly to the way in which the structure of this second class of molluscs has been reached.

Amongst the prosobranchs a ciliary feeding process has been described for the following animals found in the British Isles:

Viviparus contectus (Cook, 1949, *V. viviparus* in error), *Bithynia tentaculata* (Schäfer 1953a; Lilly, 1953), *Turritella communis* (Graham, 1938), *Capulus ungaricus* (Yonge, 1938), *Calyptraea chinensis* (Orton, 1912b; Werner, 1953), *Crepidula fornicata* (Orton, 1912a, b; Werner, 1953).

Viviparus and *Bithynia* are freshwater animals (see pp. 553–8). The former tends to bury itself in muddy detritus at the bottom of the streams in which it lives, and feeds, at least partly, on what it collects from its mantle cavity. It seems to be able to live in this habitat without the specialized filters required by *Turritella*. *Bithynia*, whilst normally feeding on detritus and small algae collected by the radula, supplements this by eating what it collects in the mantle cavity. This seems to be true of specimens living in certain habitats only, and observers of the animal (Starmühlner, 1952; Lilly, 1953; Werner, 1953) agree that it cannot be its primary feeding method.

Turritella (Yonge, 1946) is found in mud and sometimes muddy gravel at a depth of a few fathoms, living just below the surface but maintaining connexion with the overlying water by means of two openings. One of these is inhalant and is made by the foot; the other is exhalant and is due to the outflowing current from the mantle cavity. This is given momentum enough to make and keep open the hole by the fact that it is ejected from an exhalant siphon formed from the right side of the pallial skirt. Through this waste water and faecal pellets are ejected clear of the inhalant opening.

Crepidula and *Calyptraea* are to be found, often in enormous numbers, the latter on stones in the lower parts of the beach and below, the former in clumps of individuals attached to one another in chains (see p. 343). *Capulus* is also sublittoral and has been shown by Sharman (1956) to be particularly common on shells of *Pecten*, living on the edge of the valve, which its presence distorts, and supplementing what it collects by means of the water current through the mantle cavity by thrusting its proboscis into the mantle cavity of the bivalve and appropriating what it can of that animal's food (see p. 475).

All these ciliary feeding gastropods, it will be noted, have become exceedingly sedentary in their mode of life and this is, indeed, an essential corollary of the ciliary feeding habit. Schäfer (1953b) has made some quantitative estimations of the efficiency of ciliary feeding in *Bithynia tentaculata* and shown that one individual filters about 400 ml of water in a day and extracts about 8 mg of suspended matter from a suspension containing 20 mg/l. More effective feeders must extract still greater quantities.

CHAPTER 5

THE SKIN

T HE skin of an animal is invariably an interesting part of the body in that it has to provide protection for the living protoplasm within against the rigours of the environment without and at the same time allow the animal to become aware of its surroundings so that it may move successfully through them. The more thoroughly does the skin provide protection—and in the case of freshwater or terrestrial animals the need for this may be great—the more difficult will it be for it to allow communication with the outside world. These two cutaneous functions are therefore fundamentally antagonistic and some compromise becomes necessary. In many molluscs the matter becomes more acute in that a great extent of the exposed surface of the body is covered by a shell and is therefore frankly protective and almost devoid of sense organs, save in a few special cases like the chitons and perhaps the extinct tryblidiaceans. This has the effect of concentrating sense organs on other parts.

In gastropod molluscs, not only is this true, but in addition another area, the sole of the foot, becomes preoccupied with locomotor activity, whilst elsewhere, especially in the mantle cavity, other parts are involved in the transport of such material as genital products, excretory matter, faeces and other waste. The skin of gastropods is therefore concerned not only with the ordinary functions of protection and sensation, but with a number of other activities which are special to the group.

The fundamental structure of the skin is an epidermal epithelium resting on a mat of connective tissue through which run muscle fibres. No gastropod has become terrestrial in the sense that some arthropods and vertebrates have become land dwellers: the epidermis never becomes waterproofed either by the growth of a cuticle or the keratinization of the surface layers as it does in these two groups, and molluscs are terrestrial only in the narrow sense that they are not aquatic, and they survive on land only by avoiding truly xerotic conditions and by being active only in restricted habitats of high humidity. The modifications of structure shown by the skin are therefore mainly in relation to its glandular equipment, the height of the cells and their equipment with cilia. In ordinary circumstances (fig. 65A) the epidermis is a low columnar epithelium (epc), the cells of which are ciliated, and it includes a number of gland cells of the goblet cell type (gc). From this state the skin may depart towards a taller ciliated epithelium (as on the sole of the foot (fig. 66)) or one which is almost squamous (as on the lining of the mantle cavity and visceral hump (fig. 81B)) and the proportion of gland cells may be either high or low. In places, too, cilia may be absent (fig. 65C).

The gland cells which are present in the more simple cases (sole of the foot; skin of the head) are goblet cells (gc, fig. 65A, C), most of which secrete mucus. The secretion is stained metachromatically by toluidine blue and by alcian blue and other mucous stains. It is therefore comparable to the epithelial mucins of many other groups of animals and is probably an acid mucopolysaccharide. Some other types of secretion are also produced by a second type of goblet cell; these are not so frequent in the sole of the foot as in epithelia in other situations. In this case the staining reactions suggest that the secretion is protein.

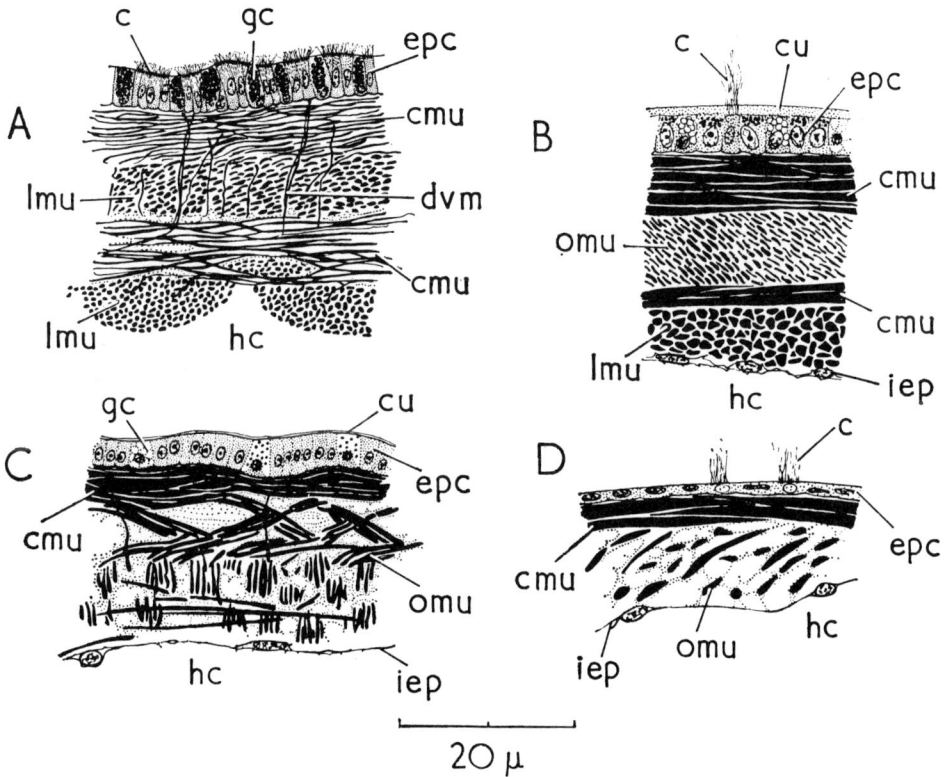

FIG. 65.—Structure of the body wall as shown by transverse sections of dorsal surface of head.
A, *Gibbula cineraria*; B, *Emarginula reticulata*; C, *Diodora apertura*; D, *Calyptraea chinensis*.
c, cilia; cmu, circular muscles; cu, cuticle; dvm, dorsoventral muscles; epc, epidermis; gc, gland cell; hc, haemocoelic cavity; iep, inner epithelium; lmu, longitudinal muscles; omu, oblique muscles.

On the head, over most of the floor of the mantle cavity and on the upper surface of the foot the gland cells usually lie in the thickness of the epithelium. In other situations, however, where the amount of secretion which is called for is greater, they often become sunk into the underlying connective tissue (sgc, fig. 66) and open to the exterior by long necks lying between the epidermal cells. In this way the local concentration of glands may be greatly increased. Gland cells arranged in this way may be found on the sole and sides of the foot, in the shell gland at the edge of the mantle skirt (p. 126 and fig. 80), often in the roof of the mantle cavity and in a number of other, special, situations. These cells are alleged to differ from corresponding cells in other phyla in being derived from connective tissue cells, at least in some species.

There are three areas of skin which call for lengthier treatment because of the particularly important part which they play in the life of a prosobranch mollusc. These are the skin over the foot, that over the visceral hump and that at the mantle edge.

The foot is not only the main locomotor organ by means of which the gastropod creeps over the surface of the substratum, but it is also involved in a number of other activities since

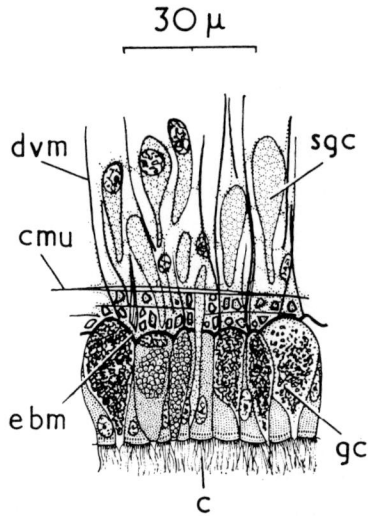

FIG. 66.—*Calyptraea chinensis:* vertical section through epidermis and underlying tissue of sole of foot.
c, cilia; cmu, circular muscles; dvm, dorsoventral muscles; ebm, basement membrane; gc, gland cell in epidermis; sgc, gland cell in subepithelial connective tissue.

it bears the operculum, is, on occasion, sensory, is often concerned with the manipulation of egg capsules and almost always forms the penis in such males as have one. In Neritidae, [some hydrothermal vent limpets] and Viviparidae the penis is cephalic (p. 318).

Primitively the molluscan foot is arranged in three sections lying one behind the other and separated from one another by transverse grooves. These are the propodium, the meso-podium and the metapodium. This arrangement is very rarely seen in its entirety, some of the more primitive heteropods being the only gastropods in which it is clearly visible. Except in a few cases, of which the Naticidae are the most obvious (fig. 296), the propodium loses its identity by merging into the front end of the mesopodium (pp, msp, fig. 67), from which is formed the flat surface on which the gastropod creeps, and the metapodium (mpt), which is the posterior lobe on which the operculum (op) is carried, becomes similarly fused to its dorsal surface.

The foot is equipped with a number of glands most of which are primarily concerned with lubricating the locomotion of the animal, but it also contains others which have nothing to do with this activity. The descriptions given by Lang (1896), Thiele (1897) and Pelseneer (1906a) are inaccurate because of the failure of these workers to realize that these glands are not all locomotory. The following classification of the glands present in the foot of prosobranchs is modified from Touraine (1952). [But also see p. 586.]

The *anterior pedal mucous gland* (apg, fig. 68) is a collection of gland cells opening to the sole of the foot at the anterior end of the mesopodium. The openings of the glands may be to the general surface of the foot so that its anterior margin shows no macroscopic evidence of the presence of this gland (*Calyptraea chinensis; Crepidula fornicata* (fig. 63; trochids, e.g. *Cantharidus clelandi* (fig. 69)); but in the majority of species the gland cells discharge into a transverse furrow which runs across the anterior end of the sole, separating the mesopodium

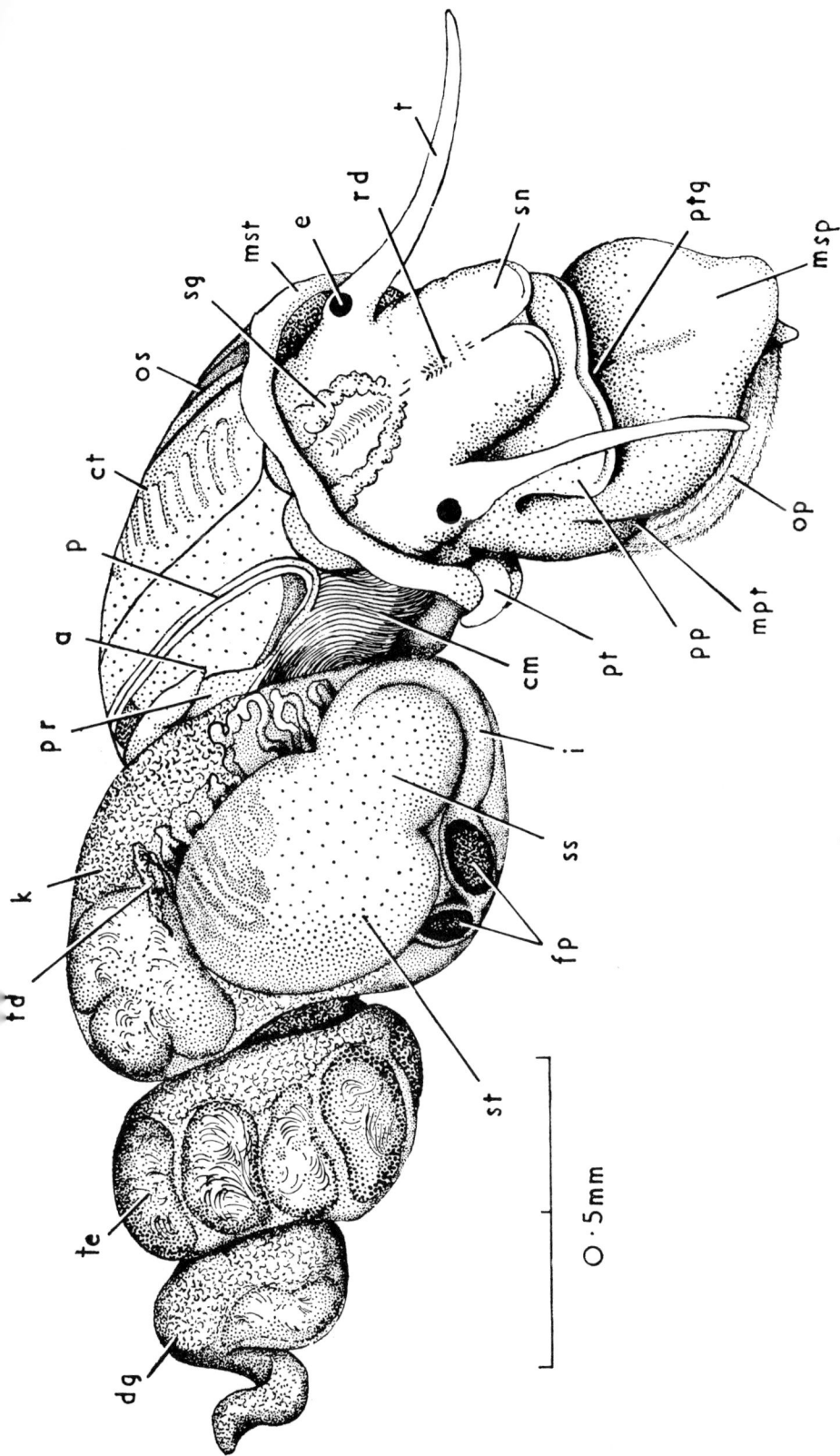

FIG. 67.—*Cingula semicostata*: animal removed from shell, seen from the right.
a, anus; cm, columellar muscle; ct, ctenidium; dg, digestive gland; e, eye; fp, faecal pellets in rectum; i, intestine; k, kidney; mpt, metapodium; mst, mantle skirt; op, operculum; os, osphradium; p, penis with penial duct; pp, propodium; pr, prostate; pt, pallial tentacle; ptg, mouth of posterior pedal gland; rd, radula, by transparency; sg, salivary gland, by transparency; sn, snout; ss, style sac region of stomach; st, stomach; t, tentacle; td, testicular duct; te, testis.

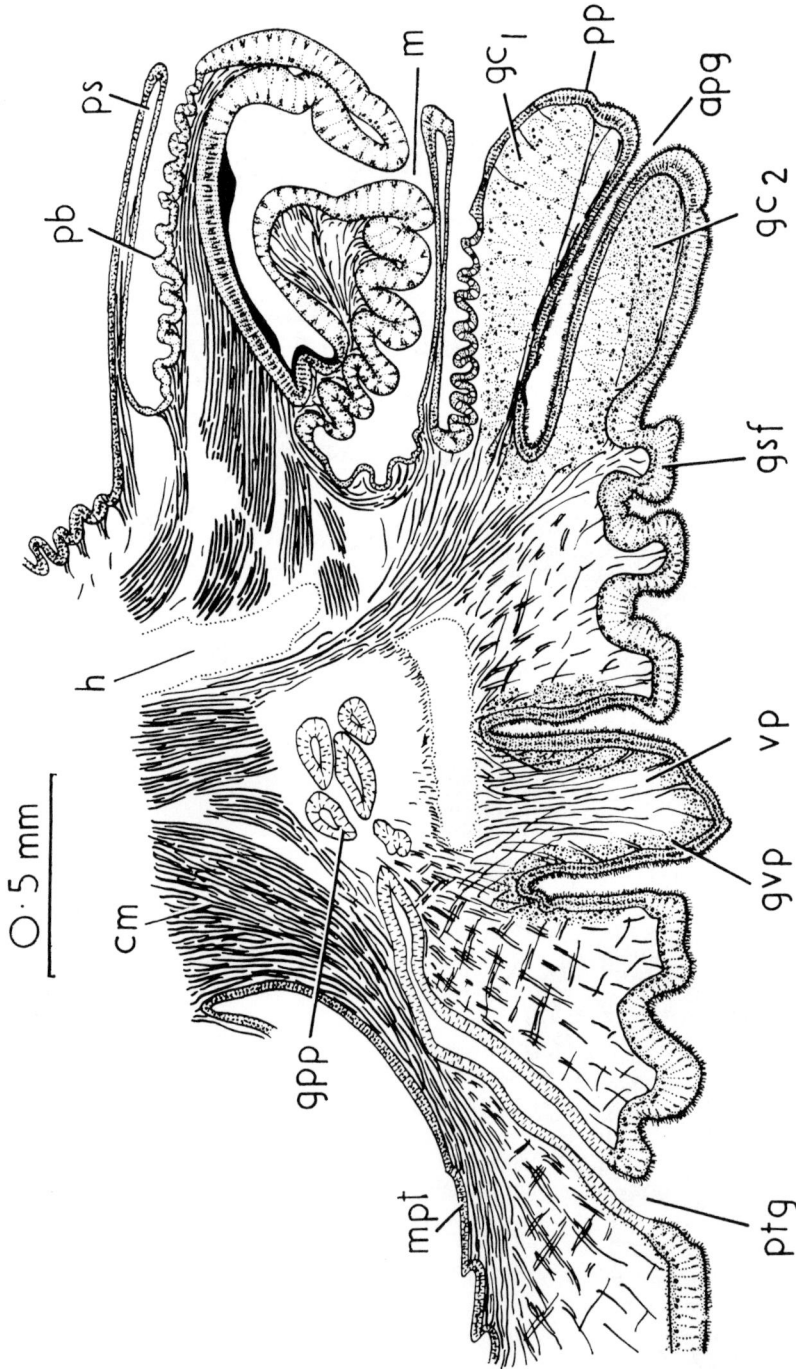

FIG. 68.—*Trivia* sp.: parasagittal section through head and foot to show pedal glands and proboscis. The animal had been narcotized before fixing: otherwise the papilla of the ventral pedal would have been retracted.

apg, mouth of anterior pedal mucous gland; cm, columellar muscle; gc_1, first type of gland in anterior pedal mucous gland; gc_2, second type of gland in anterior pedal mucous gland; gpp, gland cells of posterior pedal mucous gland; gsf, gland cells of sole gland; gvp, gland cells of ventral pedal gland; h, haemocoelic space; m, mouth; mpt, metapodium; pb, proboscis; pp, propodium; ps, proboscis sheath; ptg, mouth of posterior pedal mucous gland; vp, papilla of ventral pedal gland, partly protruded.

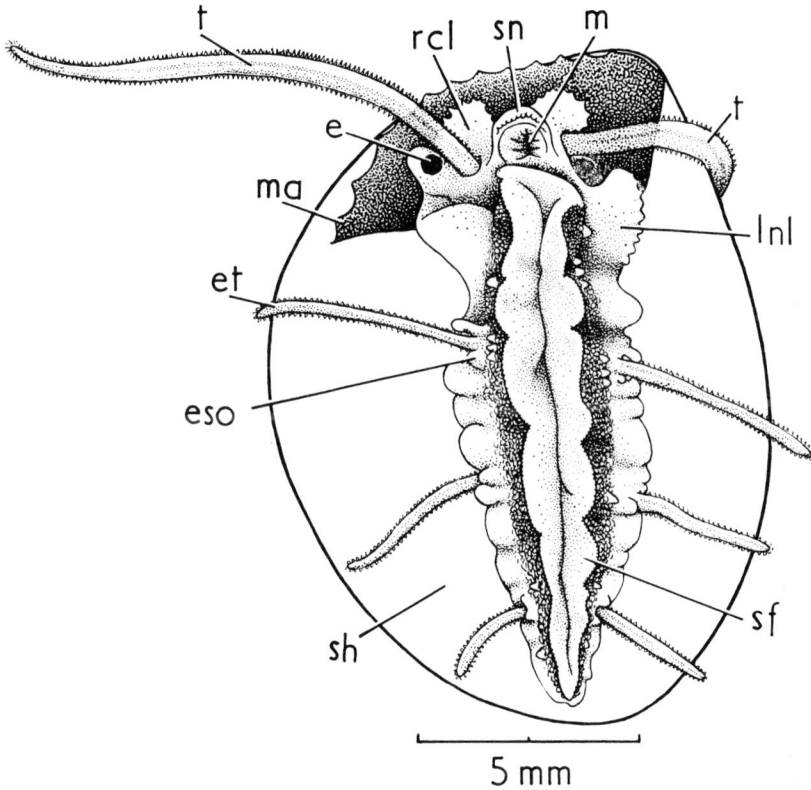

FIG. 69.—*Cantharidus clelandi:* ventral view; the shell is left unshaded.
e, eye; eso, epipodial sense organ; et, epipodial tentacles; lnl, left neck lobe; m, mouth; ma, mantle; rcl, right cephalic lappet; sf, sole of foot; sh, shell; sn, snout; t, tentacle.

behind from the reduced propodium in front (pp, fig. 68). In *Calliostoma* the groove occurs but gland cells are confined to its lips. In some cases this furrow is the mouth of an invagination which may run some way into the deeper parts of the foot in a longitudinal direction. This is known as the sagittal canal of the gland and it is slightly developed in Lacunidae, Cerithiopsidae and the Stenoglossa (fig. 261) but much more in the eratoids (fig. 68) and eulimids where its secretion may help in attaching the mollusc to its prey. In *Pomatias elegans* the canal runs into tubules lying deep in the haemocoel of the anterior part of the foot, and it is also well developed in *Acicula fusca* (Creek, 1953)—presumably a reflection of the greater need for mucus in the locomotion of a terrestrial as compared with an aquatic animal and leading on to the great development of this gland in the terrestrial pulmonates.

Although it is referred to as a 'mucous' gland it is not the same mucous material which is produced here as in the ordinary goblet cells of the skin. Only some of its cells produce acid mucopolysaccharide: most of the cells do not show metachromasia with toluidine blue nor do they stain with alcian blue, whereas they do stain so as to suggest that their secretion might be mucoprotein, neutral mucopolysaccharide or even not 'mucus' at all. Possibly in relation to this difference in the staining of the cells of the gland it is to be noted that most of them are not simple goblet cells, but are arranged in nests in subepithelial spaces. The epithelium which lines the transverse furrow or the sagittal canal is not normally glandular, but is composed of columnar ciliated cells, between which the true gland cells discharge. The anterior pedal gland is the normal source of most of the mucus over which an aquatic prosobranch creeps.

The *sole gland* is the name which may be given to the collection of glands which pour secretion on to the surface of the sole on which the animal moves. The name is given by Touraine (1952) only to those cells which are subepithelial in position, and he does not use it to include those goblet cells which lie within the epithelium. It seems somewhat illogical to separate these two collections of gland cells on this basis alone, even if, as is mentioned below, they may be to some extent chemically different, and it is therefore proposed to use the term 'sole gland' for the sum of all the gland cells opening here, whatever their position in the foot. As might be expected, there is a relationship between the development of the sole gland and the degree of locomotor activity which the animals exhibit. In the Zeugobranchia and Patellacea, most of which have adopted a limpet-like mode of life and do not wander far from a 'home', there are only rare glands in the epithelium of the pedal sole and most of those are subepithelial, mainly single cells scattered evenly over the surface, though concentrated in anterior and posterior masses in *Emarginula,* and laterally in *Patella.* Their secretion appears to be in some cases (the Patellacea) an acid mucopolysaccharide comparable to the secretion of goblet cells; in others (the Zeugobranchia) a mucoprotein or neutral mucopolysaccharide.

Many prosobranchs possess goblet cells in the epithelium of the sole of the foot (gsf, fig. 68) in addition to the subepithelial ones, and they seem to be exclusively given over to the secretion of an acid mucopolysaccharide and any other kind of secretion has its origin in the deeper cells. The number of goblet cells in the epithelium may be quite low (Zeugobranchia, Patellacea, *Theodoxus fluviatilis*), moderate (*Calyptraea chinensis* (fig. 66), *Crepidula fornicata, Natica catena*), or very high (*Trivia* (fig. 68), *Nassarius reticulatus*). The trochids, *Bittium* and *Littorina* spp., on the other hand, have no goblet cells at all in this position. The subepithelial cells of the monotocardians do not usually secrete acid mucopolysaccharides, and in most animals are scattered in an irregular way throughout the interstices of the muscular feltwork which lies under the pedal epithelium. In some, however (*Littorina littorea* is a good example), these cells are arranged in bundles.

Occasionally the sole gland is constructed to a pattern; thus in *Lacuna* spp. the gland cells are confined to two lateral bands and the centre of the foot has none. Special arrangements are also to be found in *Ianthina* in relation to the habit this animal has of enclosing air in bubbles of mucus to form a float, a process which has been described by Fraenkel (1927b). Across the anterior edge of the foot in this animal runs a cleft marking the position of the anterior pedal mucous gland. Along the centre of the anterior half of the mesopodium (msp, fig. 291) runs a groove into which the sole gland discharges mucus; this groove is called the funnel and posteriorly it ends in a transverse fold covered with smaller longitudinal grooves and rich in gland cells. It is here that air, trapped from the atmosphere, is covered with mucus to make the bubbles of the float (see p. 528).

The *posterior pedal mucous gland* opens by a pore in the middle of the sole of the foot (ptg, figs. 68, 70), and there frequently runs a groove from that point to the posterior end (gv, fig. 70) along which its secretion flows. It does not occur in more than a limited number of proso-branchs which are especially those of small size like the rissoids, *Rissoella, Skeneopsis,*

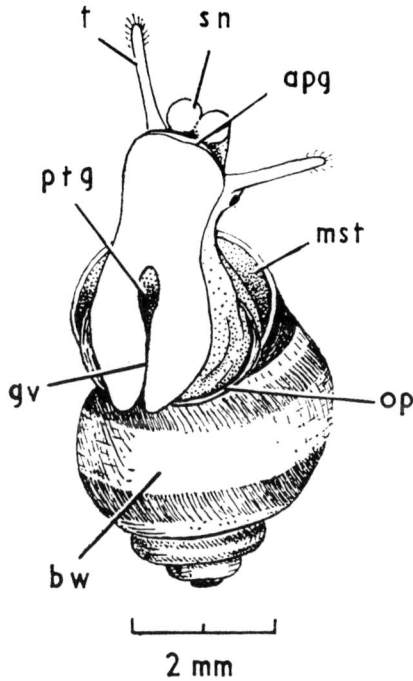

FIG. 70.—*Cingulopsis fulgida:* ventral view.
apg, mouth of anterior pedal mucous gland; bw, body whorl of shell; gv, groove from pedal gland conducting mucus to posterior tip of foot; mst, mantle skirt; op, operculum; ptg, mouth of posterior pedal gland; sn, snout; t, tentacle.

Omalogyra, Cerithiopsidae, triphorids and eratoids. The animals which have it tend to use its secretion, which assumes the form of a thread, in much the same way as a spider uses the thread of silk which it spins, and they climb up and down it from one water level to another. It may also act as an accessory lubricating gland in general locomotion, or as a source of adhesive mucus preventing the mollusc from being dislodged by water currents. The pore leads by way of a duct lined by a ciliated epithelium to a number of secretory tubules (gpp, fig. 68) which lie in the pedal haemocoel at a rather deep level. In some species (rissoids, *Omalogyra atomus, Skeneopsis planorbis* (ptg, fig. 176B)) these tubules extend dorsally into the head and even into the visceral hump, lying around the nerve ring and alongside the oesophagus. In most cases the tubules are lined by goblet cells regularly alternating with supporting cells and their secretion is identical with that from the epithelial goblet cells of the pedal sole.

It may be that this posterior pedal mucous gland is homologous with the posterior pedal gland found in some slugs, the secretion of which also emerges as a thread and which can be used as a means of climbing or descending from one level to another.

The last gland of widespread occurrence in the foot is the *ventral pedal gland,* which occurs, in the female sex only, in some of the higher Mesogastropoda (Lamellariacea) and in the Stenoglossa. It has no relation to the locomotor activities of the animals, but is concerned with the attachment of egg capsules to the substratum on which they are laid. In such lower prosobranchs as attach their eggs to a substratum this is effected by the foot. The use of the gland for this purpose is an advance from this.

The establishment of the true function of this gland was long delayed. Pelseneer (1910) suggested that it was the place in which the egg capsule was actually secreted around eggs and albumen received from the female genital duct. That this could not be the true story was shown by Kostitzine (1940) who discovered a capsule, more or less fully formed, within the genital duct of a female whelk (*Buccinum undatum*). At the same time doubt was also thrown on Pelseneer's account of the function of the gland by Fischer's discovery (1940b) that there were no mucous cells in the ventral pedal gland of *Nucella lapillus*, although parts of the wall of the capsule are made of that material. He suggested that the ventral pedal gland was responsible for imparting its final shape to the capsule rather than for its production. Pelseneer's view, however, was re-affirmed by Franc (1941b) who observed the eggs of *Thais haemastoma* emerge from the genital pore encased only in a transparent, lens-shaped membrane and enter the ventral pedal gland. The mollusc remained with this pressed against the substratum for about 15–20 min at the end of which time the animal rose and left a fully formed egg case *in situ*. It had, in fact, already been shown by Ankel (1929), of whose work the French writers were apparently not aware, that the ventral pedal gland of *Nassa mutabilis* shapes a capsule which is received from the genital duct and attaches it to the substratum, and this was also confirmed for a variety of Stenoglossa by Fretter (1941).

In Stenoglossa the gland lies within the muscles and haemocoel of the foot, opening by a pore in the mid-line slightly nearer the front than the hinder end. The pore leads to a deep cavity with folded walls, which is lined by a ciliated columnar epithelium. In the epithelium lie gland cells, but the bulk of these is to be found in clusters of cells placed under the epithelium. The secretion of the cells is sometimes partly mucus, partly some other substance. Muscle fibres form a network around and between the groups of cells and radial fibres also run outward from the folds. The shape of the cavity of the gland resembles that of the egg capsule which emerges from it, and there is no doubt that the one is caused by the other.

The Lamellariacea amongst the mesogastropods possess ventral pedal glands, and these are constructed on a plan rather different from that on which the stenoglossan gland is built. This may be correlated with the slightly different uses to which the gland is put in the two cases. In Stenoglossa the gland is used for shaping the wall of the egg capsule while it is still soft into its final form and for pressing it against stone or weed or whatever substratum is appropriate for its attachment. During this process some of the material of the capsule wall is moulded by the muscles around the gland, and by the general pressure exerted by the body of the mollusc, into a small attachment plate which fits any irregularities in the substratum closely and allows of a good grip. It is doubtful whether the gland cells add anything to the capsule: their secretions are more concerned with lubricating the fixation process, or hardening the materials of which the capsule is composed.

The Lamellariacea deposit their egg capsules in a different way, embedding them in holes excavated in the tissues of compound ascidians. These cavities are bitten out by the radula and the egg capsules pushed into them. The capsules are transferred from the genital duct to the cavity of the ventral pedal gland and ejected from it into their resting place within the ascidian. For this reason the ventral pedal gland of these mesogastropods is not a simple cavity as in Stenoglossa but is provided with a central region (vp, fig. 68) which is protrusible and acts as a ramrod for pushing the capsule home and perhaps shaping its mouth. The gland, like those of stenoglossans, is lined by a ciliated columnar epithelium which contains

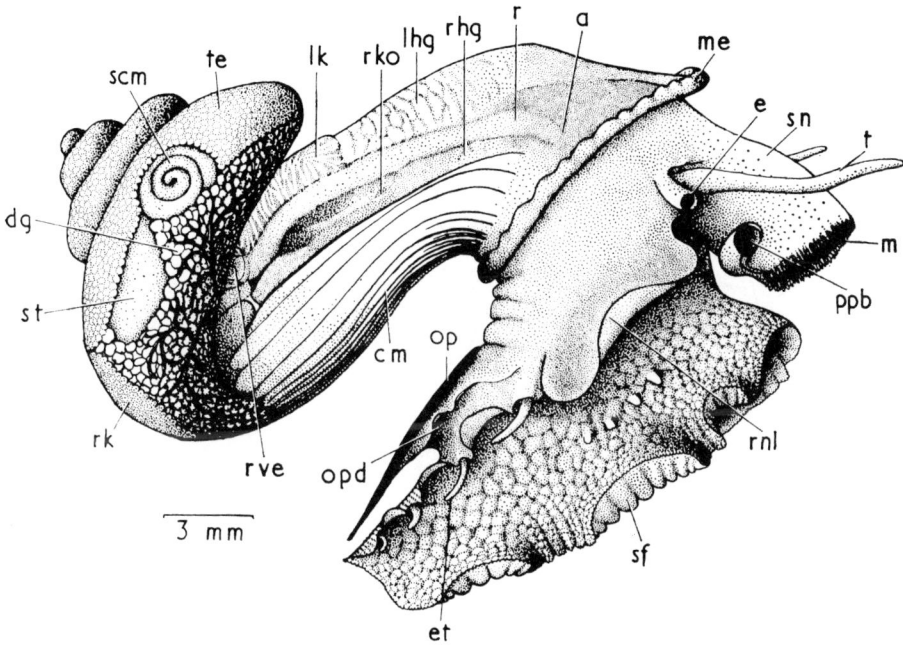

FIG. 71.—*Calliostoma zizyphinum:* male removed from shell and seen from right side; some organs are seen by transparency.

a, anus; cm, columellar muscle; dg, digestive gland; e, eye; et, epipodial tentacle; lhg, left hypobranchial gland; lk, left kidney or papillary sac; m, mouth; me, mantle edge; op, operculum; opd, operculigerous disc; ppb, pseudoproboscis; r, rectum; rhg, reduced right hypobranchial gland; rk, right kidney; rko, opening of right kidney (urinogenital); rnl, right neck lobe; rve, rectum within ventricle; scm, spiral caecum of stomach; sf, sole of foot; sn, snout; t, tentacle; te, testis.

goblet cells and has groups of gland cells in the subjacent tissue (gvp), although the secretion is not mucus as in members of that group. The protrusible papilla contains similar glands, but is not ciliated, and the whole structure is strongly muscular.

The gastropod foot has other glandular structures associated with it in addition to those that have just been mentioned, but these are located on the sides rather than upon the sole, and are not normally concerned with the locomotor activities of the animals. Such a structure is the lateral glandular streak (fig. 72) which lies along the side of the foot at the level of the shell muscle anteriorly, in young specimens of *Patella*, and which recurs in a number of other genera. The secretion from the lateral pedal glands of prosobranchs, whether irregularly scattered over the general surface or not, is partly mucous and partly protein. The function of these glands is probably, in the broad sense, repugnatorial, a first indication that it may be so being provided by the fact that they tend to be located in exposed situations, and that corresponding glands are particularly abundant at the mantle edge of those gastropods the shape of which qualifies them for the name of 'limpet'; they also abound all over the skin of those gastropods which have become naked by reduction or loss of the shell.

In the Zeugobranchia and Patellacea the cells on the side of the foot are mainly large subepithelial cells secreting material which is protein or mucous in nature; cells of a similar

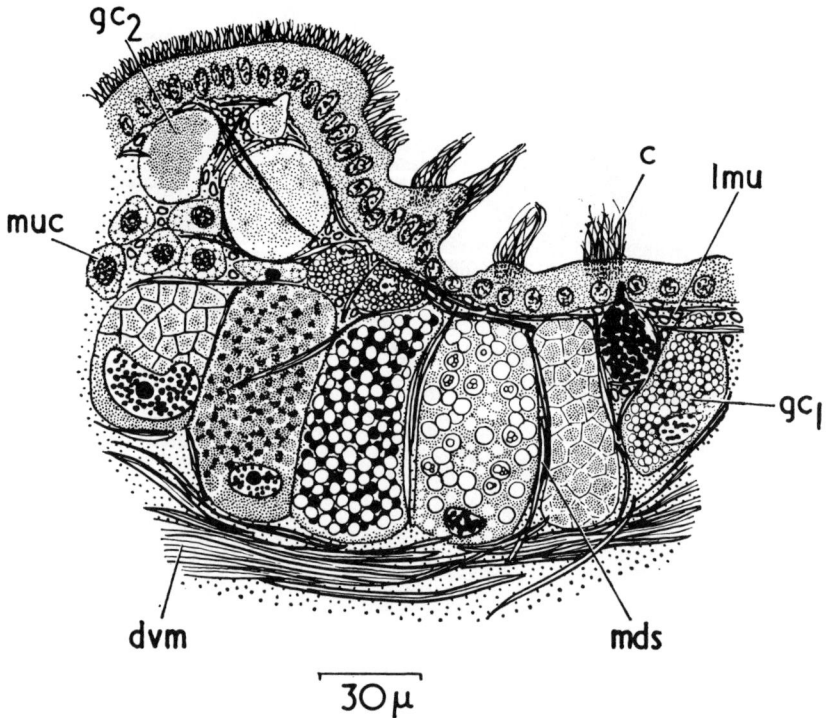

FIG. 72.—*Odostomia unidentata:* transverse section through lateral glandular streak.
c, cilia; dvm, dorsoventral muscles of foot; gc_1, gc_2, gland cells of two types; lmu, longitudinal muscles of foot; mds, muscles for discharge of secretion; muc, mucous cell.

sort do not occur on the mantle edge in most genera. In *Acmaea,* however, the mantle edge becomes laden with an agglomeration of gland cells of three types (figs. 73, 74). The first (rg_1, gc_1) is a large cell, deeply sunk in the connective tissue and haemocoelic spaces of that region and discharging by a long neck between the epithelial cells of the edge. The contents of the cells reflect the light and appear white in living animals, giving a white edge around the mantle skirt. Their secretion emerges from the mouth of the gland as a viscid, white thread which does not disperse quickly. The second type of gland (gc_2) is similarly situated and smaller, so lying closer to the mantle edge. It is not visible in the living state but in sections has homogeneous contents which stain readily. The third type of gland cell (rg_3, gc_3) has bright red contents and may therefore be seen in the living animal. It is the smallest of the three and the gland cells appear in life as small red streaks running between the necks of the cells of the first type. The nature of the secretion and the cause of the colour are not known.

The mantle edge of molluscs of the genera *Crepidula* and *Calyptraea* contains multicellular glands (rgc, fig. 75). These lie at regular intervals round the edge of the mantle skirt and may be seen in the living animal as a series of whitish objects buried in its thickness (rgc, fig. 58). From the gland, which is more or less spherical, a narrow duct runs to open (drg, fig. 75) at the surface of the mantle skirt on the ventral surface of the middle pallial fold (p. 126). The

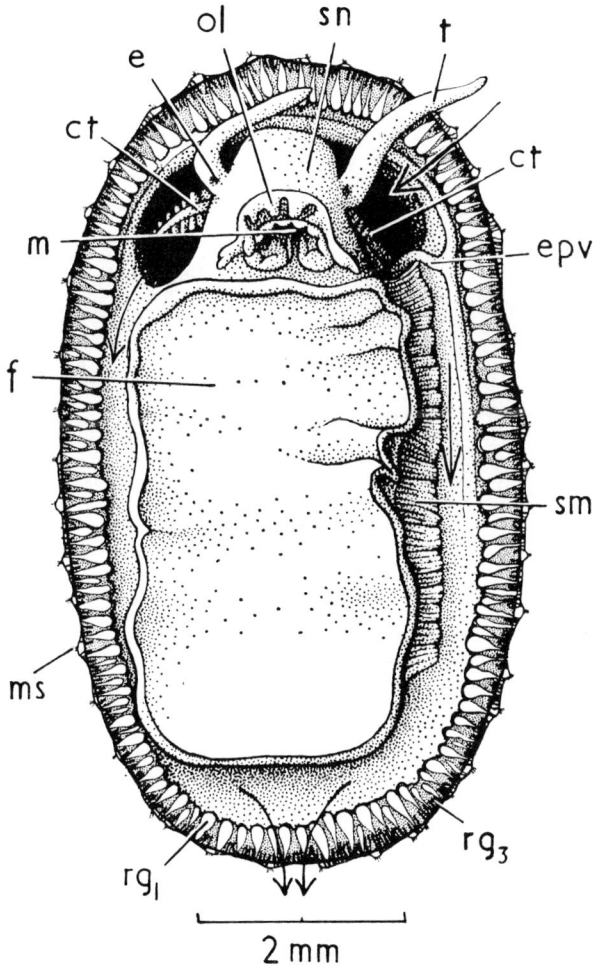

FIG. 73.—*Acmaea virginea:* ventral view.
    ct, ctenidium in mantle cavity; e, eye; epv, efferent pallial vessel; f, foot; m, mouth; ms, cilia on sensory
processes of mantle edge; ol, outer lip; rg_1, large repugnatorial glands; rg_3, small repugnatorial glands; sm,
shell muscle; sn, snout; t, tentacle.

cells in the gland are large and their secretion is protein, not mucus. The body of the gland
and the duct are surrounded by muscle fibres which probably help in the discharge of the
secretion. As in *Acmaea* and some opisthobranchs in which similar glands occur this happens
only after rather drastic stimulation and gives rise to persistent threads. Similar glands
are found in *Onchidella celtica, Acteon tornatilis, Haminea* and other opisthobranchs, the
secretion of which has been shown to be distinctly toxic to other animals, and it therefore
seems proper to assume that in such prosobranchs as have them these glands are repugnatorial too. It may be, of course, that whilst the skin glands of an opisthobranch like *Acteon*

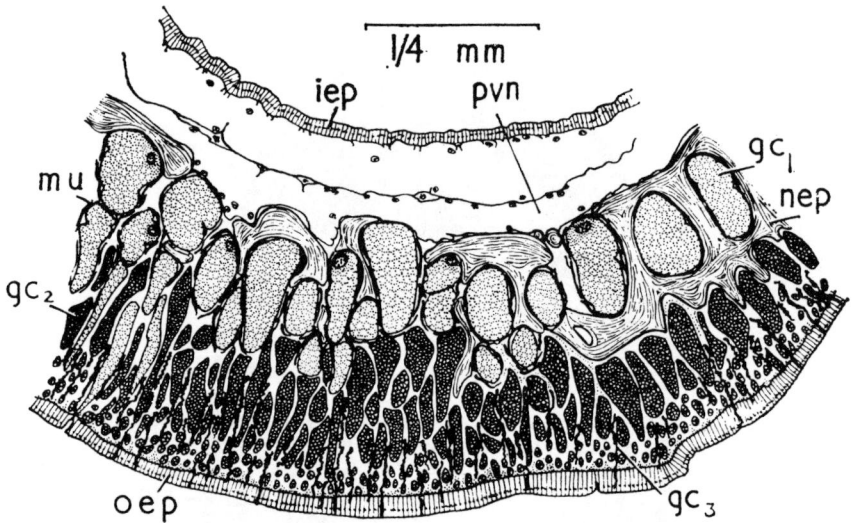

FIG. 74.—*Acmaea virginea:* tangential section of part of edge of the mantle skirt showing the nerve plexus, gland cells and muscle fibres around gland cells. The sensory organs on the outer pallial epithelium do not occur at this level.

gc_1, gc_2, gc_3, different types of pallial gland cells; iep, inner pallial epithelium; mu, muscle fibre; nep, nerve plexus; oep, outer pallial epithelium; pvn, pallial vein.

are positively dangerous to a small predator, the skin glands in other gastropods are simply the cause of a bad taste, and the glands which occur on the sides of the foot of *Diodora* may be better called antiseptic rather than truly repugnatorial.

Another conspicuous glandular field on the surface of the prosobranch body is the hypobranchial gland. A considerable amount of effort has been put into attempts to prove that it reduces a toxic secretion and Dubois (1909) has gone so far as to claim that dog whelks are helped to overcome their prey by its use, a claim recently repeated by Clench (1947). A similar claim has been made for the purple secretion of the hypobranchial gland of *Ianthina* feeding on *Velella* (Hardy, 1956). It is true that extracts of the whole gland of stenoglossans have been shown by Dubois (1909) and Jullien (1948) to be poisonous to some animals when injected, but the latter has also shown in experiments with fish that the secretion of the hypobranchial gland does not diffuse freely into the water round an uninjured dog whelk and that although this does occur when the dog whelk has been injured the concentration of secretion never rises to a level which is sufficiently high to act as an external poison. More recently Erspamer (1952) and Whittaker & Michaelson (1954) have identified this poisonous substance in *Murex* spp. and in *Urosalpinx* and *Nucella* respectively as urocanylcholine. Fischer (1925) claimed that the hypobranchial secretion in *Nucella* has sexual significance in that it leads to aggregations of animals attracted by its odour, but the most reasonable assumption as to its significance in the life of the prosobranch is that, as described on p. 89, it is concerned solely with the production of a glairy slime which will cement particles together as they are being swept out of the mantle cavity, and simultaneously provide a conveyor belt moved by the beating cilia. Mucous cells, mainly of the goblet cell type, abound for this reason on most

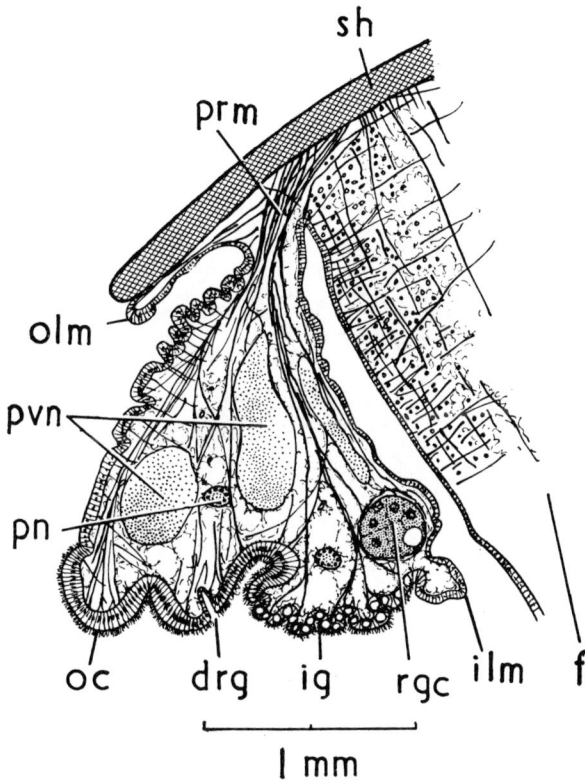

FIG. 75.—*Calyptraea chinensis:* transverse section through mantle edge.
drg, opening of duct from repugnatorial gland; f, foot; ig, inner glandular zone of middle pallial lobe; ilm, inner pallial lobe; oc, outer ciliated zone of middle pallial lobe; olm, outer pallial lobe; pn, pallial nerve; prm, pallial retractor muscle; pvn, pallial blood vessels; rgc, repugnatorial gland; sh, shell.

parts of the wall of the mantle cavity and are especially abundant on the ctenidial leaflets, but the hypobranchial gland is a particularly rich source of this and other secretions.

In some of the Diotocardia the hypobranchial gland is double and lies right and left of the rectum (hg, fig. 76). The two parts are equally developed in some Zeugobranchia (fig. 76), but vary considerably within the Trochacea. In this group the right half is somewhat reduced in *Gibbula, Monodonta* (rhg, fig. 53) and most species of *Cantharidus,* but in *C. clelandi* the left half is reduced and in *Margarites* (hg, fig. 160) there is only a right gland present. In *Tricolia pullus* (fig. 77) a remnant of the left gland occurs along with a well developed right one. Clark (1958) has attempted to rationalize this distribution by relating it to the course of the rectum: a right gland develops only where the rectum curves to the left (see also p. 294). Hypobranchial glands seem to have been completely lost in the Patellacea (fig. 48C). The Neritacea (at least *Theodoxus fluviatilis* (fig. 52)) possess no undoubted hypobranchial gland but on the extreme right of the mantle cavity posteriorly a few glandular tubules open. These are lined by an epithelium which secretes mucus and they have been claimed (Thiele, 1897;

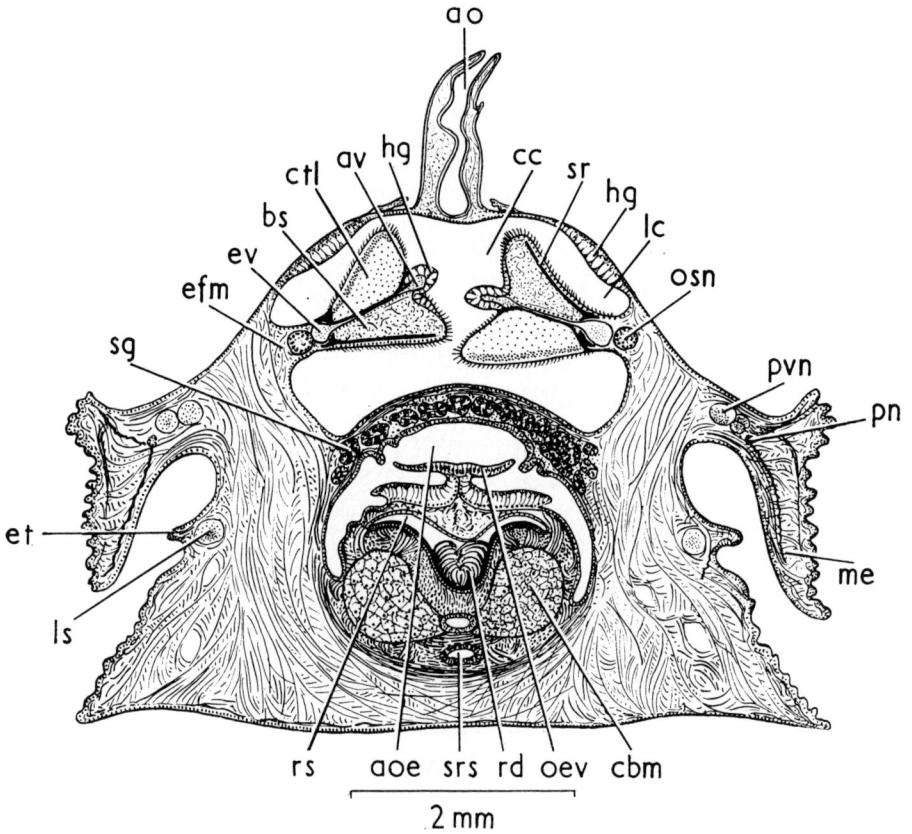

FIG. 76.—*Diodora apertura:* transverse section at level of apical mantle opening.
ao, apical opening of mantle cavity; aoe, anterior oesophagus; av, afferent vessel of ctenidium; bs, blood space in ctenidial leaflet; cbm, cartilage of buccal mass; cc, central compartment of mantle cavity; ctl, ctenidial leaflet; efm, efferent membrane; et, epipodial tentacle; ev, efferent vessel of ctenidium; hg, hypobranchial gland; lc, dorsolateral compartment of mantle cavity; ls, lateral pedal sinus; me, mantle edge; oev, oesophageal valve; osn, osphradial nerve; pn, pallial nerve; pvn, pallial vein; rd, radula; rs, radular sac; sg, salivary gland; sr, skeletal rod in ctenidial leaflet; srs, subradular space.

Bourne, 1908) as the homologue of the right hypobranchial gland. Mucous cells in the mantle skirt between the ctenidium and the rectum represent a reduced left hypobranchial gland.

The monotocardians have all apparently lost the right hypobranchial gland but retain that of the left side more or less well developed. It occupies the field between the attachment of the ctenidial axis on the left and the rectum on the right on the inner side of the mantle skirt. In some the degree of development is slight (*Crepidula,* fig. 58; *Turritella,* fig. 57), in others it is sufficiently well developed to fling this part of the epithelium into irregular low folds (*Littorina,* fig. 10; *Aporrhais,* fig. 171; *Nassarius,* fig. 115) whilst in some genera (*Buccinum*) the development is considerable and the gland is a conspicuous folded structure on the roof of the mantle cavity.

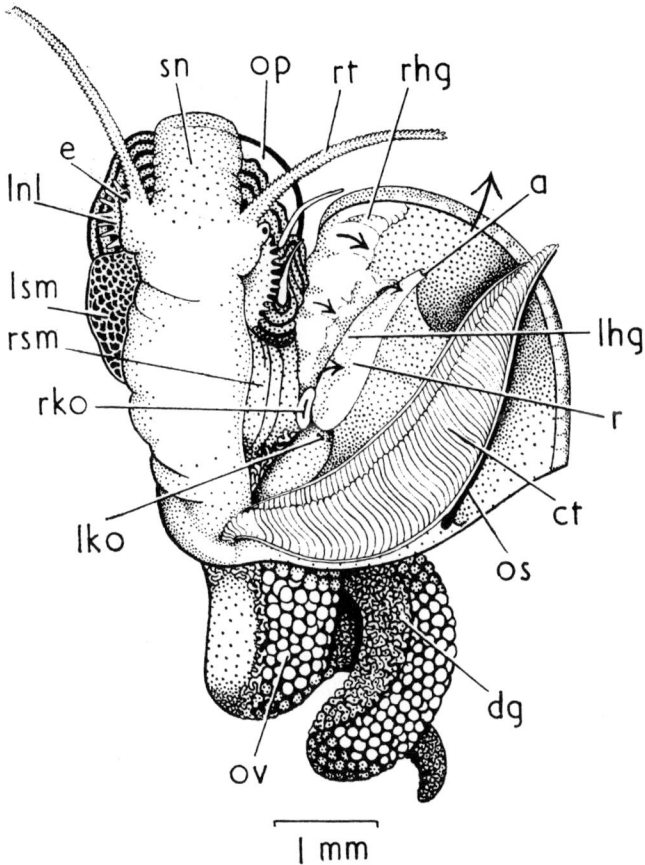

FIG. 77.—*Tricolia pullus:* dissection to show contents of mantle cavity. Arrows show direction of ciliary currents. a, anus; ct, ctenidium; dg, digestive gland; e, eye; lhg, left hypobranchial gland; lko, opening of left kidney; lnl, left neck lobe; lsm, left shell muscle; op, operculum; os, osphradium; ov, ovary; r, rectum; rhg, right hypobranchial gland; rko, opening of right kidney; rsm, right shell muscle; rt, right tentacle; sn, snout.

The nature of the secretion of the hypobranchial gland has not been investigated with modern histochemical methods to the extent that that of some other glands in gastropods has been. Some of its cells secrete mucus apparently identical with that produced by goblet cells elsewhere in the skin, but these may not be the commonest kind; it is because of their presence, however, and because of the fact that the secretion in general resembles mucus, that this gland is occasionally called the mucous gland.

In *Diodora,* the hypobranchial gland of which has been examined histochemically by Gabe (1951*b*), mucous cells are relatively few and are limited to the part of the gland lying on the upper surface of the ctenidial axis and to the anal papilla. Elsewhere the cells of the (fixed) gland usually appear very empty although occasional granules of secretion occur; they give

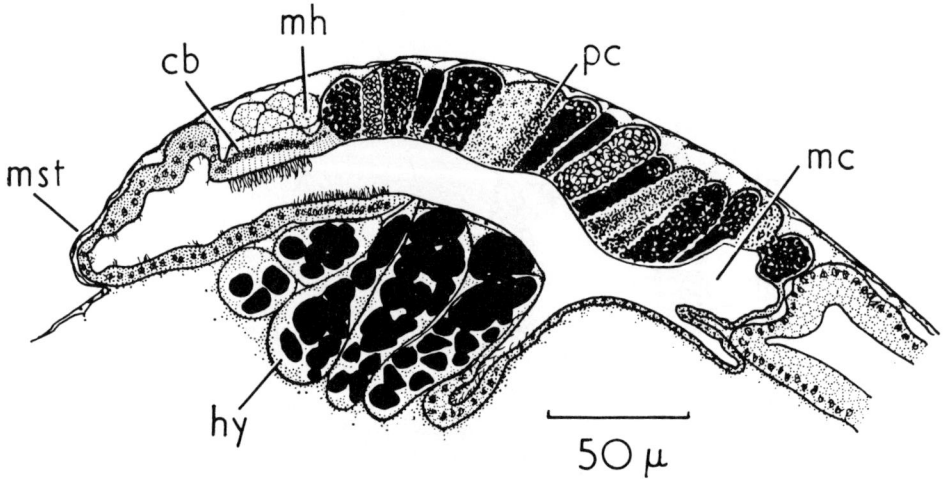

FIG. 78.—*Turbonilla jeffreysi*: section of inner part of mantle cavity.
cb, ciliated strip; hy, hypertrophied gland cells on floor of mantle cavity; mc, mantle cavity; mh, mucous cell of hypobranchial gland; mst, mantle skirt; pc, protein cell of hypobranchial gland.

no clear staining reaction and Gabe, like previous workers, came to no clear conclusion as to what they do secrete. *Emarginula* and *Puncturella* show glands with identical staining reactions. In *Scissurella,* where the left gland is much smaller than the right, mucous cells occur along with a second type with minutely granular contents (Bourne, 1910). These gland cells alternate regularly with slender supporting cells, some of which are ciliated. In *Diodora* and its relatives the binding properties of the secretion of the hypobranchial gland are possibly less than in many monotocardian prosobranchs because of the more thorough ventilation of the mantle cavity allowed by the apical hole in the shell. Crofts (1929) reported the presence of two types of cell in the hypobranchial gland of *Haliotis*: both of these were said to secrete mucus.

In many gastropods true mucous cells are relatively more abundant than they are in the zeugobranchs and they form in some species the major constituent of the hypobranchial gland. This is true of the Calyptraeacea, Lamellariacea and Cypraeacea. In the last two of these groups the hypobranchial gland is very extensive, in *Velutina* covering the entire surface of the mantle cavity right of the gill. In most mesogastropods, however, other cells secreting protein material are added to the mucous cells and in some genera (*Turbonilla*, fig. 78; pyramidellids; *Omalogyra*; *Skeneopsis*), these are the common type of cell in the hypobranchial gland and mucous cells occur very sparsely, although it is as well to add that most of these animals are probably opisthobranchs or have strong opisthobranch affinities. In these genera hypertrophy of the protein gland cells (pc, hy) is apt to occur; they become extraordinarily large and have, indeed, on occasion been mistaken for eggs lying within the mantle cavity. What is involved in the changing proportions of the secretions from the hypobranchial gland is by no means clear. It is worth noting, however, that the hypobranchial secretion tends to contain more mucus in those gastropods which have a ctenidium in the mantle cavity, whereas it tends not to be mucus in animals where this has been—for whatever reason—replaced by ciliated strips (cb). In these (e.g. *Turbonilla*) such mucous cells (mh) as do occur lie in close relation to the ciliated strips and the remainder of the hypo-branchial gland has none. It may be that the current of water maintained by ciliated strips through the mantle cavity is less powerful than that maintained by a ctenidium and requires a different kind of lubricant and transport medium for the particles which have to be moved. The nature of the varying hypobranchial secretions is a point for further investigation.

In a number of prosobranchs the secretion of the hypobranchial gland is coloured. This occurs, for example, in *Cirsotrema*, *Nucella lapillus*, *Ocenebra erinacea* and probably all muricaceans, *Clathrus* spp. and *Ianthina janthina*. In all these animals the secretion is purple in its final form, but in the stenoglossans it is a greenish-yellow colour when first liberated. This gradually turns to a red hue and then deepens to purple. The change is dependent on the presence of oxygen and light and will not occur in their absence, although this appears to happen in *Murex trunculus* (Dubois, 1909). Chemically the purple pigment (the Royal or Tyrian Purple of the ancients, made in classical times from species of *Murex*) is 4-4'-dibromindigo or 6-6'-dibromindigo, related to natural indigo or woad and presumably manufactured by the mollusc from a tryptophane source. The production of the coloured secretion has been fully investigated by Letellier (1889, 1890) and Dubois (1902a, b); 1903a, b, Erspamer (1947) and Bouchilloux & Roche (1955). Letellier and Dubois suggested that a colourless chromogen was acted upon in the light by an enzyme which Dubois called purpurase, and converted into dibromindigo. Erspamer showed that the source of the substances involved was the median part of the hypobranchial gland, the enzyme being produced in the anterior third of the gland and the chromogen (or a prochromogen) in the posterior part. Like Dubois before him, Erspamer found that whilst the chromogens varied from species to species, the enzyme was identical in all. The study of the secretion has been taken further by Bouchilloux & Roche using chromatography: they have shown the presence of three prochromogens in *Murex, M. trunculus* having prochromogen 1 and prochromogen 2; *M. brandaris* prochromogen 3. The prochromogens consist of a sulphate group linked to an indoxylic group, which contains bromine in prochromogens 2 and 3, and is probably 6-bromoindoxyl, but has no bromine and is probably indoxyl in the case of prochromogen 1. The origin of the indoxyl compounds is not known. Prochromogen 1 of *M. trunculus* is converted by the removal of the sulphate radicle by a sulphatase called purpurase into a blue pigment which is indigo itself; pro-chromogen 2, in the same species, is similarly converted into a red-mauve pigment which is bromindigo and is very similar to the 6-6'-dibromindigo of *M. brandaris* made from pro-chromogen 3. The pigments of *Ocenebra erinacea* and *Nucella lapillus* have not been similarly investigated but are no doubt closely related. Little work has been done on purple pigments from other prosobranch sources. The familiar purple pigment which can be obtained from animals of the genus *Aplysia* is chemically quite different from the prosobranch purples, being a haem derivative.

One major activity of the molluscan integument is to give rise to a shell and it is now recognized that this is secreted by the mantle (Réaumur, 1711; Tullberg, 1881) and is not the result of a direct transformation of the animal's tissue as supposed by von Nathusius-Königsborn (1877). This activity is confined to the skin which forms the outer covering of the visceral mass or hump, though in exceptional cases something corresponding to a shell may be produced from skin over other parts of the body (*Teredo, Hipponyx*). If this does happen, however, the structure of the secreted material is invariably simpler than true shell.

The secretion of the shell is undertaken by special areas of the mantle. The shell is composed, in a general way, of an organic base, conchiolin, which is impregnated with inorganic salts of calcium, predominantly calcium carbonate, and which may exist as a separate layer, not so impregnated, over the outer surface, where it is known as the periostracum. The secretion of the organic base and of the initial calcareous material of the shell takes place normally only at the free edge of the mantle skirt, and it is the gradual growth which occurs along this line which generates the shell. The total secretion at this place, however, does not produce more than a thin shell and later deposition of extra calcareous matter thickens and strengthens it. This secondary growth involves the addition of calcareous salts and conchiolin over the whole of the inner surface of the shell and is undertaken by the whole outer pallial epithelium. In many cases the crystalline character of the calcareous matter produced at these two sites is different, the edge of the mantle skirt producing prisms, the general pallial epithelium, on the other hand, laying down plates which alternate with conchiolin.

In most lamellibranchs, and in some gastropods like the Zeugobranchia (fig. 79), the free edge of the mantle skirt is flung into three folds, the outer (olm), middle (mlm) and inner (ilm) pallial folds. In lamellibranchs the periostracum originates from the groove between the outer and middle folds and there has been debate as to whether the site of secretion is the lateral aspect of the middle fold, or the medial aspect of the outer. Older workers (Tullberg, 1881; de Villepoix, 1892; List, 1902; Manigault, 1939) have mainly chosen the former situation; more recent ones (Kessel, 1944; Brown, 1952) have favoured the latter. The calcareous matter, on the other hand, is produced from the lateral surface of the outer fold and from the rest of the pallial surface. There is, therefore, at the edge of the mantle folds a small gap between the newly secreted periostracum and the ventralmost edge of the calcareous part of the shell. In gastropods the same general relationships hold, but the proportionate sizes of the folds are different and the production of periostracum often much less. The most important change in the relative sizes of the pallial folds of gastropods in the present content is the reduction of the outer pallial fold which has the effect of bringing the groove in which the periostracum is secreted into close proximity to the site of secretion of the calcareous matter of the shell, and the two processes appear to go on more or less side by side.

Conchiolin is the name given to the non-calcareous matrix of the shell. It is perhaps a complex of substances rather than one substance in the strict chemical sense; it appears to be protein (Friza, 1932) and to vary from animal to animal and even from part to part of the shell (Beedham, 1958b). The protein also appears to be tanned (Trueman, 1949). Bevelander & Benzer (1948), working with oysters, have shown that a great mass of glands, which lie with ducts opening to the groove where the periostracum is formed, have protein and polysaccharide contents. These may, therefore, be assumed to be the source of the conchiolin of the shell. In gastropods they are much less conspicuous and, in some genera, it is difficult to find any glands at all in this position. In most, they occur in small groups of cells embedded in the connective tissue of the mantle edge and discharging to its margin (fig. 80).

[Secretion of the periostracal and the other layers of the shell has been extensively investigated recently with the help of electron microscopy. The results of this work have been expertly summarized by workers who have been leaders in the field (Wilbur & Saleuddin, 1983).]

Most molluscan shells are coloured, and inspection shows that this is limited to the outer layer of the shell, such colour as occurs in the inner layers being of physical origin. This location of the colour therefore implies the presence, at the mantle edge, of glands which secrete pigments at the time and near the place where the calcareous material of the outer layer is being formed. The nature of the pigment which is produced in such situation is often

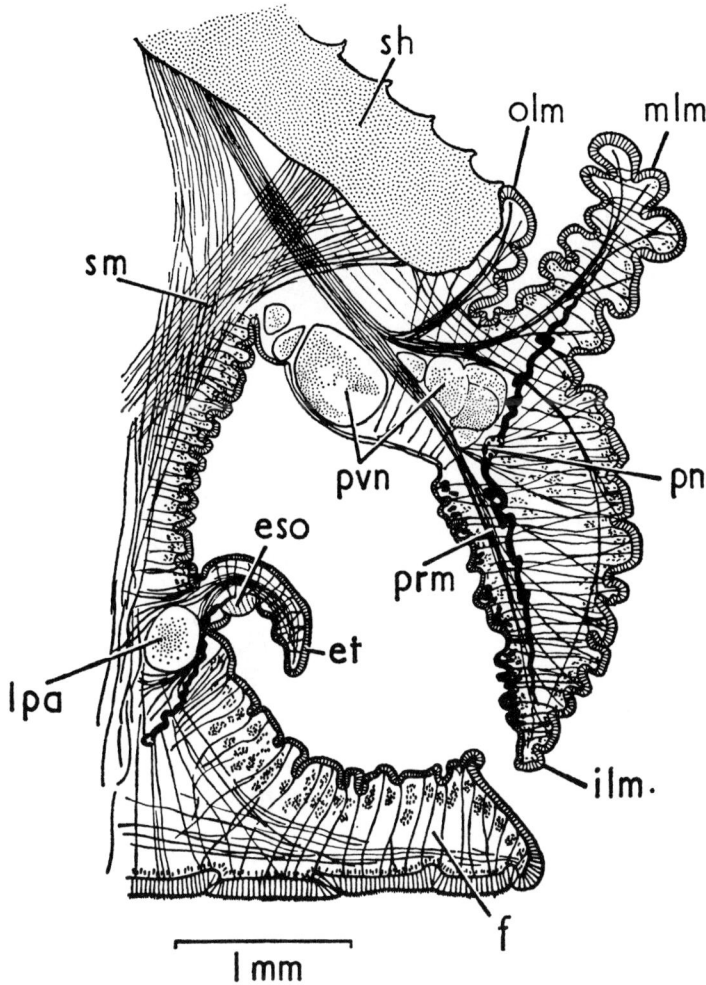

FIG. 79.—*Diodora apertura:* transverse section through lateral part of foot and mantle edge.
eso, epipodial sense organ; et, epipodial tentacle; f, foot; ilm, inner pallial lobe; lpa, lateral pedal artery; mlm, middle pallial lobe; olm, outer pallial lobe; pn, pallial nerve; prm, pallial retractor muscle; pvn, pallial vein; sh, shell; sm, shell muscle.

unknown, but has been investigated in a number of animals. In the following account the findings of Comfort (1951) are mainly relied upon.

There is a difference between the lower prosobranchs and the higher in that the shell pigments of the latter are bound to protein material incorporated in the shell, whereas those of the diotocardians are not and so may be obtained in solution by dissolving the shell. The bound pigments have not been properly investigated chemically as no methods for their extraction have yet been elaborated. They may be chromoproteins with melanin groups

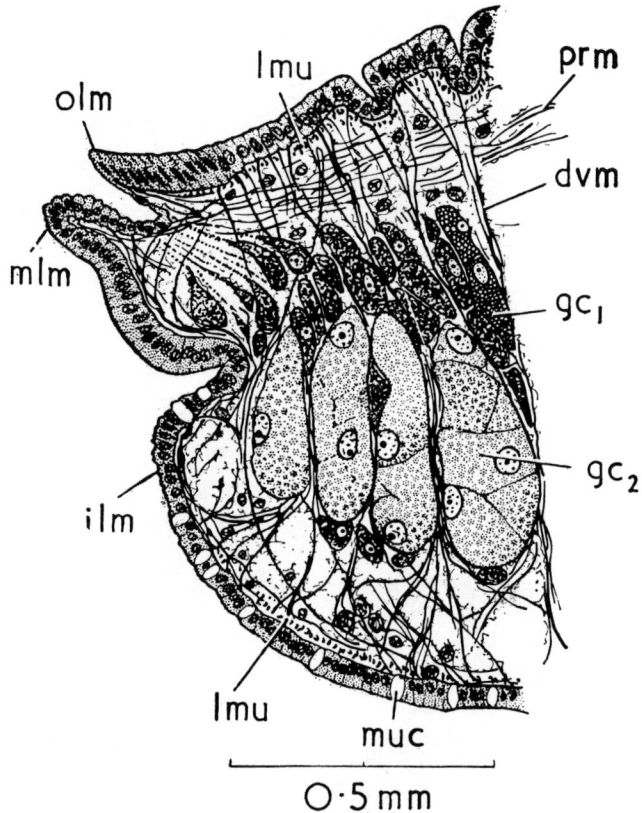

FIG. 80.—*Siphonaria* sp.: transverse section of mantle edge.
    dvm, dorsoventral muscle; gc_1, gc_2, cells of shell gland; ilm, inner pallial lobe; lmu, longitudinal muscle; mlm, middle pallial lobe; muc, mucous cell; olm, outer pallial lobe; prm, pallial retractor muscle.

incorporated. The only pigments which have been demonstrated in these shells without doubt are pyrroles, though it has been suggested that indigo-like pigments and melanins also occur, and there are numerous others, still chemically unknown.

Indigo and 6-6'-dibromindigo have already been discussed in relation to the hypobranchial gland of Stenoglossa. Some of these animals, on occasion, display violet tints in their shell, and the assumption is readily made that these are chemically identical or related. There is no proof, however, that this is so, and the statement that ianthinine, the pigment which gives the purple colour to the shell and hypobranchial secretion of *Ianthina* spp., is indigoid (Moseley, 1877) appears unlikely to be true.

Melanins are likely substances to be responsible for various shades of yellow, through brown, to black, in the pigmentation of shells as in other parts of the body, but they have never been demonstrated in any prosobranch.

Pigments of pyrrole type are apparently the most abundant source of colour of chemical origin in the shells of the lower prosobranch. These may be of two types: (a) where the pyrrole

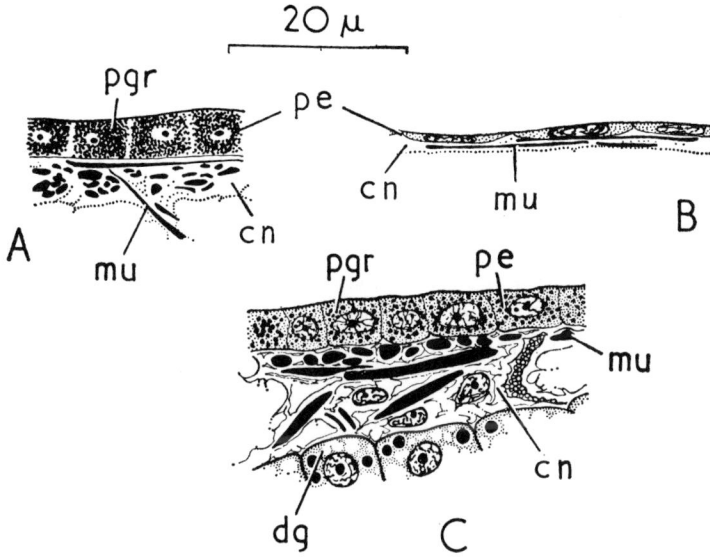

FIG. 81.—Transverse sections of mantle from surface of visceral hump: A, *Patella vulgata;* B, *Emarginula reticulata;* c, *Nassarius reticulatus.*

cn, connective tissue; dg, digestive gland; mu, muscle fibre; pe, mantle epithelium; pgr, pigment granule.

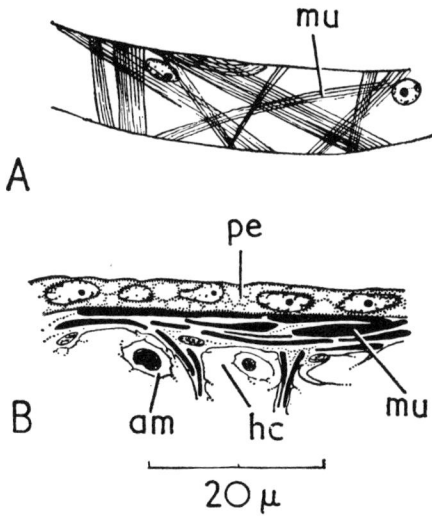

FIG. 82.—*Ocenebra erinacea:* mantle from surface of visceral hump; A, surface view; B, transverse section.

am, amoebocyte; hc, blood space; mu, muscle fibre; pe, mantle epithelium.

rings form a ring: these are porphyrins; and (b) where the pyrrole rings form a chain: these are bilins. The latter are much less common than the former, but are responsible for the green colour of certain shell (*Turbo, Haliotis*). The porphyins are the usual source of colour in the shells of the lower prosobranchs, the principal one which is found being uroporphyrin I, which occurs in a pure state, but in marine forms only. Another porphyrin encountered in many shells is conchoporphyrin. According to Comfort (1951) these porphyrins are to be found in the following British species of gastropods: *Acmaea virginea, Gibbula magus, G. cineraria, Cantharidus striatus, Monodonta lineata, Velutina velutina, Erato voluta, Trivia monacha, Acteon tornatilis.*

This list suggests a general distribution in the Trochacea and Lamellariacea, and, although investigations are not numerous, they suggest that porphyrins are also found in the shells of Zeugobranchia, Cypraeacea and of *Tricolia*. They do not occur in *Patella*.

The origin of the porphyrins is not known. In other animals porphyrins are usually held to be chlorophyll derivatives, and if this were true of gastropods, their deposition in the shell could be regarded as fundamentally an excretion of material otherwise unmanageable. There is, however, no evidence of a biochemical kind that this is indeed the case so that the source of the pigment remains unknown. That dietary factors can, on occasion, affect the colour of the shell by altering the kind of pigment which is laid down in it, has nevertheless, been shown to be true of some prosobranchs. Dr D. R. Crofts tells us that month-old *Haliotis* fed on red weeds develop red shells and Moore (1936) has brought evidence to suggest that the steno-glossan *Nucella lapillus* makes shell which is predominantly grey in colour, though yellow may also occur, where the diet is of barnacles; where the diet includes mussels, however, brown or even purple pigment is deposited in the shell. If an animal be reared first on the one food and then on the other a sharp change in colour will be recorded in the shell at the point corresponding to the changed diet. This is, however, unlikely to be the full story [see p. 573]. More recently Ino (1949) has shown that the type of food eaten influences the colour of the shell of the turbinid *Turbo cornutus*. If these animals eat calcareous algae then the shell is coloured; if they are not allowed to eat this type of food their shell is white, and there are incidental changes in spinyness. Blinded animals behave in the same way as intact ones, showing that there is no visual response such as might be part of a procryptic coloration.

CHAPTER 6

THE MUSCULAR SYSTEM

T HE body wall of the gastropod is dermomuscular (fig. 83) and shares in the locomotor activities of the animal as in annelids, the haemocoelic fluid acting as an internal skeleton. Whereas the arrangement of the muscles in the annelid is regular, with outer layers of fibres running in a circular direction and inner layers running longitudinally, the two sets forming antagonistic pairs, this is not so obvious in gastropods, where greater irregularity of arrangement is found (figs. 65, 83). In annelids the coelomic fluid is separated by metamerically arranged septa into compartments and the septa are usually numerous. In gastropod molluscs septa are also developed with the same effect of limiting the movement of the fluid in the body cavity, but they are few in number, never the simple transverse partitions that they are in annelid worms and not the walls of coelomic sacs. As might be expected, the degree of development of the muscles in the body wall can be correlated with the use to which each area is put: it is minimal in the mantle (figs. 81, 82), underneath the shell, where movement is negligible, greater in the mantle skirt (figs. 75, 79, 80), still better developed over the surface of the head (fig. 83B), the sides of the foot and the floor of the mantle cavity, and reaches the highest development in the muscles developed in the sole of the foot in connexion with locomotion (figs. 68, 286). All these muscles are histologically plain, except in *Scissurella*.

In addition to the muscles of the body wall described above the gastropod possesses muscles which are attached to the inner surface of the shell which acts as an exoskeleton. The inner ends of these shell muscles have, therefore, a fixed origin unlike the muscles of the body wall. They are, presumably, simply special groups of dermal muscles which have secondarily acquired this connexion. The most important is that known as the columellar muscle. It originates on the columella of the shell in those prosobranchs where this is spirally coiled, runs down the concave side of the visceral hump to a ventral position slightly to the right side of the body (cm, figs. 4, 5, 264, 265, 324). From there it runs into the head-foot. In that region it splits into bundles which spread into the head, into the anterior and posterior halves of the foot and on to the inner surface of the operculum (cm, figs. 68, 133, 261). The part which enters the head sends branches to the skin and snout, the tentacles, the buccal mass and radular sac (ort, pmr, fig. 107); the sections which enter the foot fan out into bundles which are inserted on the inner surface of the pedal epithelium, and the opercular branch is linked to the underside of that structure. When the columellar muscle contracts all these parts of the body are pulled within the shelter of the shell. Bozler (1930) has shown that in *Helix pomatia* this muscle can contract to a tenth of its relaxed length.

A few fibres from the columellar muscle fan from its dorsal surface into the right side of the mantle skirt in the region of the anal and reproductive apertures, and a few separate from it on the left side to run into the mantle skirt near the anterior end of the osphradium and ctenidium (figs. 5, 324). In limpets there is present a series of muscles lying at right angles to its anterior edge and originating from the shell a little way in from its margin: these are retractor

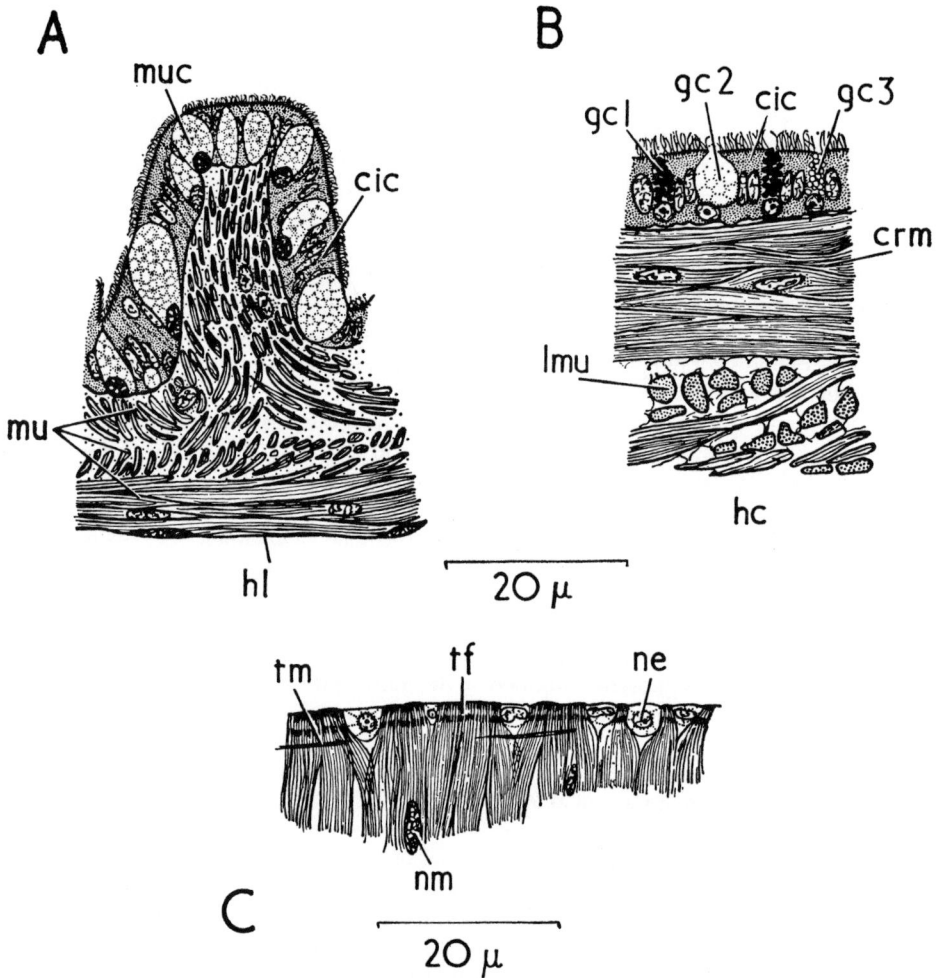

FIG. 83.—A & B, *Nucella lapillus:* histology of body wall, A, basal part of introvert; B, body wall dorsa to introvert; C, *Emarginula reticulata:* histology of insertion of shell muscle on to shell, part of a transverse section.

cic, ciliated cell; crm, circular muscle; gc_1, gc_2, gc_3, three different types of gland cell; hc, haemocoel; hl, layer of cells lining haemocoel; lmu, longitudinal muscle; mu, muscle fibres; muc, mucous cell; ne, nucleus of pallial epithelial cell; nm, nucleus of muscle cell of shell muscle; tf, tonofibrillae of epithelial cell; tm, transverse muscle fibre.

muscles of the pallial edge (prm, figs. 75, 79). In other prosobranchs, however, it is the columellar branches which retract the mantle skirt, and this is pulled completely away from the shell edge when the snail withdraws into its shelter.

At other situations within the body of the gastropod other muscles are to be found. The gut, the reproductive ducts and some of the major arteries have plain muscles in their walls which propel their contents. Sphincter muscles lie around the external kidney opening (rsp,

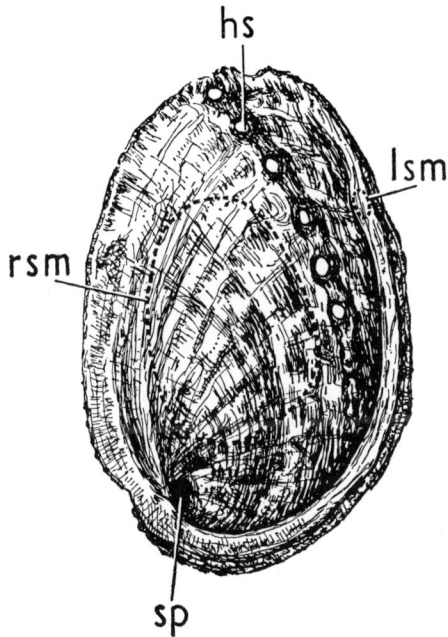

FIG. 84.—*Haliotis tuberculata:* shell, seen from below.
hs, hole in shell; lsm, impression of left shell muscle; rsm, impression of right shell muscle; sp, spire.

fig. 329) and at various places on the gut and reproductive tracts. Two situations, however, deserve special mention in connexion with the muscular system, in that the fibres which occur there are not histologically plain, but striated, though probably with a spiral pattern rather than the more complex structure of vertebrate striped muscle (Hanson & Lowy, 1957). These are the ventricle (fig. 22) and the buccal mass, both places in which more rapid and powerful contraction is called for. The arrangement of the musculature in these places is dealt with elsewhere (pp. 172, 651). In the case of the muscles of the odontophore, skeletal attachment is provided by the cartilage-like material which occurs there.

Some prosobranchs have not spiral shells, and in these the columellar muscle tends to be replaced by one (sm, fig. 246) running from the inner surface of the shell (lsm, rsm, fig. 84; sm, fig. 42B) into the foot and head, to the same destinations as before except that in these limpet-like animals the operculum has been lost. In most of these prosobranchs the muscle has a horseshoe-shaped origin on the shell, the open part of the shoe facing anteriorly and marking the situation of the mantle cavity (sm, figs. 60, 245, 246, 249, 258). From its origin fibres spread downwards into the head and the foot (sm, fig. 260), not so much to bring about the withdrawal of these parts of the body into the shell as to pull the shell down to the substratum. The fibres of the shell muscle do not run for any distance in distinct bundles, but radiate into fans which spread into head and foot. From both right and left origins of the shell muscle fibres spread to the entire breadth of the foot so as to produce a decussation in the mid-line of the body. As in other prosobranchs, the radular sac and the cartilages of the buccal mass are connected to the shell by slips of muscle. Slightly different arrangements exist in connexion with the retractor muscles of the mantle edge: in limpets these muscles run into the shell muscle and originate from the shell in the same region as that, either immediately on its lateral margin (*Emarginula,* Patellacea (prm, fig. 260)) or, after crossing over some of the more lateral fibres of the shell muscle, intermingled with its fibres (*Diodora,* prm, fig. 79).

The columellar muscle of prosobranchs is attached to the shell in a series of bundles, each with a separate origin and insertion. The shell muscle of zeugobranchs shows no such separation; in patellaceans, however, it is arranged in a series of bundles separated by narrow clefts, and this pattern is often visible in the scar left by the muscle on the shell. Thiem (1917*b*) has shown that these clefts allow for the passage of blood from the venous spaces in the foot and amongst the viscera into the blood spaces of the mantle skirt, where at least partial oxygenation may occur. From the mantle edge the blood is collected into pallial veins and sent to the auricle. The degree of oxygenation which occurs here, or the percentage of blood which is exposed to the water here, depends partly upon the degree of development of pallial gills and partly upon whether ctenidia are present in the mantle cavity. In Zeugobranchia such as *Diodora* (fig. 249), *Emarginula* (fig. 245) and *Puncturella* (fig. 246) two ctenidia are present, no particular development of pallial gills occurs, and, although some division of the shell muscle is to be noted, this is not great, and presumably the bulk of the respiratory exchange occurs within the main part of the mantle cavity over the head. The muscle is little divided in some members of the Patellacea e.g. *Lottia, Patina* and *Patella:* these have either no ctenidia but well developed pallial gills (the Cyclobranchia, e.g. *Patina* and *Patella* (fig. 50)), or one ctenidium and less well developed pallial gills (*Lottia*); because of the high degree of development of the pallial gills oxygenation is easily effected and only a small number of afferent vessels is required. The shell muscle is therefore little split. In other members of the Patellacea, however, pallial oxygenation is less easily brought about because of the absence of special gills on the mantle edge (*Acmaea,* fig. 73) and although the animal possesses a ctenidium in the nuchal cavity this may not always be capable of effective functioning. Pallial respiration is therefore essential, but not easy, and has to be facilitated by bringing the blood through the shell muscle in many small vessels. The same is also true for the abranchiate Lepetidae which, perhaps because they are small, have neither ctenidia nor pallial gills (fig. 254). In all these animals the shell muscle is much interrupted for the passage of afferent pallial veins.

It is now necessary to enquire what relationship exists between the single columellar muscle which is found in most prosobranch molluscs and the single horseshoe-shaped shell muscle which is found in the limpet-like forms. Knight (1947) has shown that the extinct Bellerophontacea, regarded by some as the earliest known prosobranchs, had two muscles inserted on the columella of the shell, which ran across the body of the animal from side to side, one at its right extremity and the other at its left. The two muscles were equally developed and the animal apparently completely bilaterally symmetrical. Two columellar muscles may still be found in some living prosobranchs, but in all these the shell is coiled in a helicoid spire, which introduces a lateral asymmetry, instead of in a plane spire as was the case in the bellerophonts. Associated with the asymmetry of the shell is an asymmetry of the columellar muscles and that on the (post-torsional) right is always larger than that on the left. Two unequal muscles of this nature are to be found in the following archeogastropods: Haliotidae (lsm, rsm, fig. 84), Scissurellidae (lsm, rsm, fig. 247), *Tricolia* (lsm, rsm, fig. 77) and possibly other turbinids, Neritidae, Helicinidae, and in the mesogastropods *Rissoella, Lamellaria* (lsm, rsm, fig. 85), *Trivia* and *Velutina.* In the genus *Hydrocena* (Neritacea), according to Thiele (1929–35), the columellar muscle is single in the head and foot, but splits into right

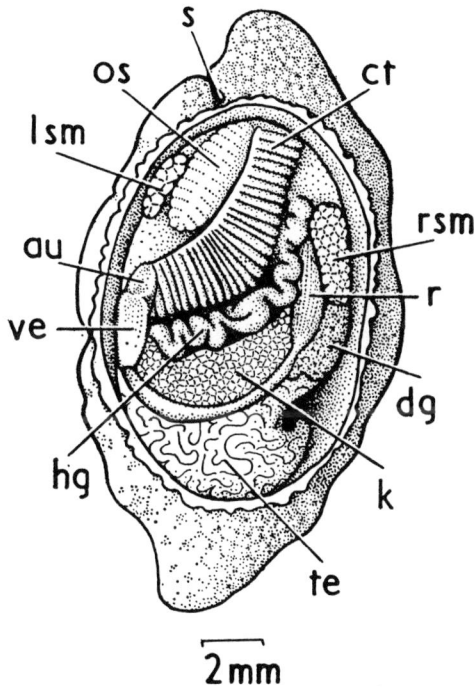

FIG. 85.—*Lamellaria perspicua:* dissection to show animal after removal of the shell and the mantle overlying it. au, auricle; ct, ctenidium seen through mantle; dg, digestive gland; hg, hypobranchial gland; k, kidney; lsm, left shell muscle; os, osphradium; r, rectum; rsm, right shell muscle; s, siphonal notch at edge of mantle; te, testis; ve, ventricle.

and left halves which embrace the visceral mass before being inserted on the shell: this might represent a partially double state. Where only a single columellar muscle exists it is the right member of the pair which persists and the left which has disappeared. The double muscle is of importance in many prosobranchs in helping to bring about torsion and may persist into the adult of primitive forms and of those monotocardians in which the body whorl is expanded.

The shell muscle of familiar patelliform gastropods never shows any trace of doubleness in the adult animal. Nevertheless it is usually believed that it has been produced by two muscles, one on each side, increasing in size and gradually extending their origin on the shell backwards until they join to form the definitive horseshoe-shaped mass. Suggestions as to how this may have happened are given by the two elongated muscles of the limpet-like *Septaria* (Bourne, 1908), [by the elongated shell muscles described in some recently discovered hydrothermal vent limpets (Fretter, 1988, 1989)] and is also supported by Crofts' account (1955) of their development in *Patella*.

Since it is by means of the contraction of the fibres of the columellar or shell muscles that a prosobranch mollusc retracts within the shelter of the shell, and withdraws the edge of the mantle skirt before that is trapped between the lip of the shell and the operculum or substratum, it is necessary for the fibres to have a firm attachment to the substance of the shell. Over the area where attachment occurs the calcification of the shell is interfered with and, as a consequence, it appears as a slight depression on the surface of the shell—the muscle scar (sm, fig. 42B). The fibres of the muscle do not make direct contact with the material of the shell, however, nor do they penetrate into its substance (fig. 83C). The muscle fibres end at the inner surface of the mantle epithelium which covers the whole area of attachment just as it lines the whole of the rest of the shell. At this point, however, the mantle cells are low, squamous cells and contain fibrillae (tonofibrillae) (tf, fig. 83C) which are continuous with the myofibrillae of the muscle cells. These tonofibrillae run across the pallial epithelium to the outer surface of the cells where they are securely fastened to the inner surface of the shell. Recent work by Hubendick (1958) on the pulmonate *Acroloxus lacustris* has shown the details of this arrangement and suggests that it is probably generally applicable to all molluscs. Here the muscles end in a dense layer of connective tissue underlying the epidermal cells, the bases of which interlock and interdigitate with it. Neighbouring epidermal cells similarly interlock, so as to make the epithelium and underlying tissue a cohesive whole. The most remarkable detail of this structure, however, is the presence of vast numbers of finger-like processes projecting from the outer surface of the cells and dipping into depressions on the inner surface of the shell. These form a brush border of what Hubendick calls microvilli. They have a diameter of 0·05–0·1 μm at their base, a length of about 2·5 μm and number 25–100 per sq μm (25–100 $\times$ 10^6/mm^2). The tonofibrillae enter these microvilli.

That the attachment of muscle to shell is by no means easily broken is illustrated by the well known ability of animals like limpets to cling firmly to their home, especially after they have been stimulated by an unsuccessful attempt to dislodge them. This ability allows limpets to live in places exposed to currents so strong that other gastropods cannot withstand them (see, for examples, Lilly *et al.*, 1953). The force which a limpet is able to withstand before being dislodged has been known since the time of Réaumur (1711) and has often been investigated since (see Aubin, 1892; Lawrence-Hamilton, 1892; Menke, 1911; Krumbach, 1918; Loppens, 1922; Thomas, 1948): various investigations have suggested that a weight of 6–7 kg must be applied to a limpet before it looses its hold on the rock. How this is achieved is not clear, except that it is not due to the foot as a whole acting as a sucker actuated by atmospheric pressure, as Woodward (1875) believed, although this may be part of the mechanism. The foot may act as a whole, or part of it may grip whilst the remainder is free, and injuring the sole by radial cuts does not affect its ability to grip the substratum. It is clear that muscular contraction is involved since, if a limpet attached to a sheet of glass is watched, it will be seen that the area of the foot increases when an unsuccessful attempt to remove the animal is made; but it is often assumed that some kind of glandular cement may help, partly by giving increased contact between foot and substratum, partly by the very stickiness of the secretion. As Ankel (1947) has suggested, the release of the grip on the substratum would then be a more complicated matter than if muscular contraction alone were the cause, because whereas a simple relaxation would stop the grip in this case, the sticky secretion would have to disperse or be dispersed in the other. Alkalinity of sea water might be sufficient to disperse an acid secretion or at least destroy its effect, and this has been found to be an actual deterrent to the adhesion of limpets (Réaumur, Menke) and chitons (Hoffmann, 1938). Evidence of the secretion of such a substance has been offered by Ankel (1947) in connexion with *Gibbula cineraria*, a mollusc which wanders over surfaces and feeds by rasping diatoms and detritus as it goes. As it does this it leaves no trail of slime, yet, if knocked off a glass plate which is then stained, an impression of the sole of the foot will be left at the place where the mollusc was dislodged, in the form of many oval drops of mucus representing the mouths of glands, from which long fibrous threads spread in all directions. This secretion must be produced in response to the force used in removing the animal from the glass and Ankel regards it as a reflex. If this be accepted as a major activity of the glands on the prosobranch foot, it may be that it is an important function of much of the sole gland, especially in the more sedentary types.

The main locomotor organ of the gastropod is the foot, and it is characteristic of the class that this has become flattened ventrally to form a sole by the activity of which the animals creep over the ground on which they dwell. It is only in exceptional cases (e.g. some heteropods) where the animal has become free-swimming, that this is no longer true.

It has an elaborate system of muscles. The extrinsic musculature is the columellar muscle, but in addition to this the foot has many muscles confined to itself, longitudinal, transverse and dorsoventral bundles running in a rather irregular manner to produce a feltwork. Amongst these muscles lie glands, connective tissue cells, nerves and blood vessels. The vessels are largest near the central parts of the foot where they constitute part of an extensive sinus fed with blood from the pedal branch of the anterior aorta. As a result the musculature tends to be thicker laterally and the whole is roughly bilaterally symmetrical. Dorsally the musculature thins markedly and is restricted to the sides of the foot; centrally at this level viscera may be found. The foot is therefore divisible into a dorsal visceral part and a ventral muscular part in its middle region. Anteriorly and, in particular, posteriorly the whole thickness of the foot may be filled with the muscular network, and it is this which is used as the animal's means of locomotion.

If a snail of any sort be allowed to creep over the surface of a glass plate it may be inverted and the sole of the foot examined through the glass. A series of waves of local contraction and relaxation of the pedal musculature will then be seen to travel along the sole, their number and their direction of travel depending upon the kind of mollusc which is being examined (fig. 86). Vlès (1907) distinguished direct waves and retrograde waves according to whether their direction of travel was in the same sense as that in which the mollusc moved or the reverse. Direct waves therefore travel along the sole from the hind end forwards, retrograde waves from the front end backwards. The waves may be monotaxic when they occupy the entire breadth of the foot, ditaxic if there is a double series out of phase with one another, each occupying half the foot, or tetrataxic in which no less than four sets of waves travel along the sole.

Carlson (1905a), Parker (1911) and Copeland (1922) have shown that these rhythmical waves are the cause of the animals' locomotion forwards, but not only failed to show how the passage of waves over the pedal surface brought this about, but were also uncertain whether the waves, which appear as dark bands moving over the sole, were furrows or ridges on its surface and whether a given point on the foot was in continuous though fluctuating movement, or whether its motion was intermittent. Van Rijnberk (1919) and ten Cate (1922) believed the waves were protrusions from the surface, whereas Olmsted (1917) and Bonse (1935) regarded them as grooves; Biedermann (1905) may be mentioned as one who thought of the movement as continuous, whereas Bonse believed that a forward movement of a given point occurred only as a wave passed over it. Once this had happened, he thought that it then remained motionless relative to the substratum until the next wave in the series arrived.

More recent work by Lissmann (1945, 1946) has provided answers to most of these questions and the following is largely based on his account. In *Helix pomatia* a monotaxic

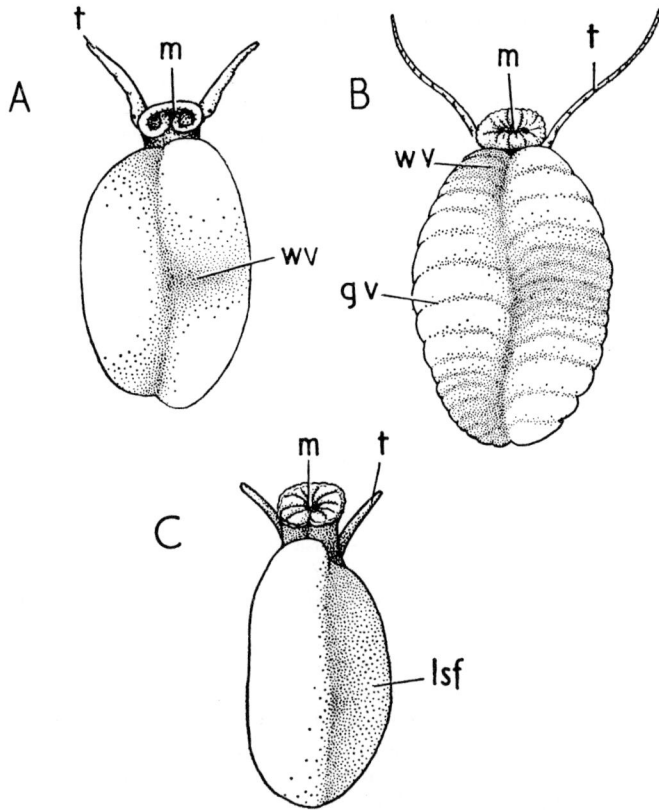

FIG. 86.—Ventral views of 3 prosobranchs to show locomotor waves on the sole of the foot; A, *Littorina saxatilis;* B, *Gibbula cineraria;* C, *Pomatias elegans.* In all cases the animal is progressing towards the top of the page, but in A and B the ditaxic waves travel to the posterior end of the foot. In *Pomatias,* which steps alternately with right and left halves of the foot, the left half is shown raised from the substratum.

gv, transverse groove on foot, characteristic of trochids; lsf, left half of sole of foot raised from ground; m, mouth; t, tentacle; wv, wave of contraction on foot.

series of direct waves passes along the sole of the foot: by tracing the position of such natural reference points on the sole as mucous glands or of marked points in cinephotographic records, Lissmann was able to show that the waves passing forwards over the foot are waves of contraction and that normally a point moves only during the passage of such a wave. Between waves a particular point on the sole is characteristically at rest, but may in certain cases slip a little backwards under the influence of the wave of contraction next behind, or be dragged a little forward under the residual effect of the wave of contraction ahead. He was also able to show that coincident with the wave of contraction of the longitudinal muscles, visibly marked by a dark band on the sole, the surface of the foot was locally lifted clear of the substratum, an effect perhaps produced by contraction of the dorsoventral musculature. The snail is thus resting, at any given moment, on a series of transverse ridges formed on the sole of the foot, just as a long-bodied arthropod is held up on numerous legs.

The same kind of statement may be made about other gastropods. In *Haliotis,* which exhibits ditaxic direct waves, or *Littorina* (fig. 86A) with retrograde waves, there tend to be three interlinked areas of the sole—two on one side and one on the other—on which the animal is supported, as in an insect. In *Pomatias* (fig. 86C) a slightly different mechanism appears to exist, in that the two halves of the foot work independently of one another and out of phase, so that the animal appears to shuffle along first with the one side of the foot and then with the other, the tip of the snout also being involved and acting as a prop or third support during the stepping. If one half of the foot be watched it will be seen that the stepping starts by a lifting of the lateral margin and then of the whole breadth at the posterior end. This lifting is accompanied by a contraction of the longitudinal muscles which shortens the posterior half. The lifting and shortening continue until the whole half-foot is off the ground. Once this has been achieved it is then replaced, the posterior end regaining contact first, at a position in advance of that from which it was lifted because of the contraction of the longitudinal muscles. From this point forwards the foot gradually comes to rest on the ground, elongating as it does so, with the result that the anterior end is one step ahead of the starting position. This process is then repeated by the other half of the foot.

The study of *Pomatias* is useful in that it provides evidence as to how the forward movement of the snail is achieved by the events which occur in the foot. It will be noted that in each half-foot three factors are important in bringing movement about. At first the half-foot moves so as to bring the posterior part forwards, gripping the substratum in the meantime with the anterior end; later the posterior part grips the substratum and the anterior end is protracted. The three essential parts of the process are therefore the mechanisms responsible for adhesion to the ground, for contracting and for elongating the sole. The first of these is probably in part the close fit which is possible between the soft under surface of the foot and the substratum and, in part, the secretion of slime from the pedal glands; the contraction of the foot is brought about by contraction of the musculature, but the elongation is due to a mechanism still not clear but probably involving the hydrostatic pressure of the blood. What causes increased blood pressure is not certain.

The locomotion of *Helix,* though more like that of the majority of prosobranchs, is less easy to analyse. The alternating waves of contraction and relaxation which sweep along the foot have the result of keeping a state of tension in the posterior part so that it is pulled forwards, whilst a thrust exerted on the anterior end pushes it forwards; a central point acts to some extent as an anchorage. Lissmann points out that as the fastening of the snail foot to the ground is achieved predominantly by the extent of the area of contact this is best done by applying the relaxed parts to the substratum, whereas in worms, which grip by setae, the areas of contact may be contracted areas. Movement of the sole therefore coincides with longitudinal contraction and therefore with direct waves. It is interesting to note, nevertheless, that some prosobranchs do progress forwards by means of retrograde waves, notably all the common species of *Littorina.*

The use of the foot in two halves, as in *Pomatias,* is a common occurrence in many rissoaceans and is reflected, presumably, in the division of the foot into right and left halves by a median groove. In these animals, too, the anterior half of the foot tends to be separated from the posterior half by a constriction and this again seems to reflect a degree of independence in the use of the two parts, the anterior part operating by itself when the snail is negotiating corners or passing on to the surface film. These creatures, particularly the hydrobiids, use their proboscis as an accessory locomotor device just as *Pomatias* does; this is very noticeable when the animals are creeping out of water and they are denied the support which it would give. The same habit is emphasized in snails of the genus *Truncatella* (Clench & Turner, 1948, and see p. 548) which are therefore known as 'looping snails'.

In a certain number of prosobranchs no muscular waves can be detected on the surface of the foot as the animal moves and it seems to glide over the substratum, presumably by means of the cilia with which the sole of the foot is covered. This may be seen in *Hydrobia ulvae* when it is under water and moving horizontally over a smooth surface; when moving vertically waves become apparent in the posterior half of the foot. The smaller *Nassarius* species, *incrassatus,* can similarly move without any visible contractions, as observed for the related *Alectrion* by Copeland (1919) and for *Polinices* (Copeland, 1922). This is probably true for many small gastropods as noted by Gersch (1934) for *Skeneopsis,* where, when the animal stops creeping, the pedal cilia are found to have stopped beating, an observation which suggests that they may be under nervous control.

If the movement of a prosobranch is indeed dependent upon waves of contraction passing along the pedal musculature from one end to the other or upon metachronal waves affecting the pedal cilia, then reversal of the direction of motion would seem to involve a reversal of the nervous impulses upon which these waves depend and of the direction of the effective ciliary beat. This is a somewhat difficult thing to envisage and it has been commonly stated that prosobranchs cannot in fact move backwards. Ten Cate (1923), for example, put slugs into blindly ending tubes of so narrow a bore that they were unable to turn round in them; in these circumstances the slugs moved towards the blind end of the tubes and stayed stuck there. Nevertheless Gersch (1934) devised situations in which animals of *Gibbula cineraria* held between glass plates were compelled to turn if they were going to make any further progress. This they managed by producing the normal direct waves on one half of the foot and a series of retrograde waves on the other, so as to produce a rotatory effect. He also reported that *Patina pellucida* which, like other limpets, has direct ditaxic waves, can move backwards and that when it does so the foot shows retrograde waves moving over its sole. This reversal of activity in the muscles, which would apparently involve a reversed polarity in the pedal nerve net, must be centrally controlled from the pedal ganglia, but the nervous control of the locomotion of gastropods is very little understood.

The shape of the foot has been investigated in relation to the habitat and way of life by Rotarides (1934), who gave figures for the ratio between the height and breadth of the foot in a number of gastropods (Table 2).

TABLE 2

Ratio between height and breadth of foot. 1 = breadth in each case			
Patella caerulea	1 : 0·4	*Calyptraea chinensis*	1 : 0·7
Haliotis tuberculata	1 : 0·8	*Murex brandaris*	1 : 0·8
Littorina neritoides	1 : 0·7	*Nassa mutabilis*	1 : 0·3
Pomatias elegans	1 : 1		

All of these except *Pomatias* are aquatic. In *Pomatias* the equality of the height and breadth agrees with the figures obtained from the terrestrial pulmonates and is presumably related in some way to the need for supporting a heavier body in an aerial environment. Rotarides also believed that in animals with a heavier body the distinction between the visceral and the muscular parts of the foot is more pronounced. This is exaggerated where the animal moves over soft bottoms, where, presumably, the difficulty of movement becomes greater; this may also be partly overcome by having a broad sole (e.g. *Buccinum, Natica*). A circular foot tends to occur in limpets, where the problem of holding on to the substratum is often made worse by the roughness of the water in which the animals live; the adhesive power of the foot is increased by a number of factors of which the broad attachment to the shell and the insertion of the muscle splayed over the whole foot are perhaps the most important.

Simroth (1882) gave a few relative rates for the movement of prosobranchs: *Viviparus viviparus* 2–3, *Bithynia tentaculata* 2, *Pomatias elegans* 0·7. He pointed out that the difference between *Pomatias* and the two others reflected the fact that it has to carry its own full weight: it may also reflect different locomotor mechanisms. Stephenson (1924) recorded a speed of 5–6 yards/min in *Haliotis*, but this is maintained only momentarily.

CHAPTER 7

THE ALIMENTARY SYSTEM—1

THE alimentary canal of the gastropod mollusc consists of a stomodaeal section of ectodermal origin, a central stretch derived from the embryonic endoderm and of a terminal proctodaeal length which has been formed from the ectoderm of the underside of the mantle skirt. The mouth lies on the animal's head, terminally, and the anus discharges into the mantle cavity, primitively in the mid-line, but in most cases more or less displaced to the right side. In a few genera of Docoglossa the anus lies at the innermost end of the mantle cavity, but in most prosobranchs the rectum runs along the mantle skirt and opens closer to the mouth of the cavity.

The mouth leads into a short tubular part of the gut, of variable length, the oral tube. This enlarges into a more spherical cavity, the buccal cavity, from the posterior end of which arise two tubes lying one directly over the other. The lower one is the radular sac, the dorsal one the oesophagus. Into the buccal cavity discharge buccal glands, mostly goblet cells, but there are in addition one or two pairs of larger salivary glands provided with ducts. All the apparatus up to the beginning of the oesophagus is of ectodermal origin and so is stomodaeal and may be distinguished as the fore-gut, and it may be that in some prosobranchs the oesophagus is also stomodaeal.

The oesophagus leaves the posterior end of the buccal cavity dorsally and runs to the stomach, which is placed near the base of the visceral mass, commonly rather to the left. The anterior part of the oesophagus may carry lateral outpouchings, but the mid-oesophagus is primitively the seat of extensive secretory epithelium, especially in its lateral walls, which constitutes the oesophageal glands. The posterior oesophagus leads, primitively, to the upper end of the stomach as it lies in the visceral mass.

The stomach of the prosobranch gastropod (fig. 87) is a complex structure into which open the ducts of the digestive gland, which is arranged in two lobes, one on the right and the other on the left, the two usually being unequally developed. In the more archaic prosobranchs the stomach is pear-shaped with the swollen base of the pear towards the upper end of the visceral mass and receiving the oesophagus (oe) and the ducts of the digestive gland (dd). It is often drawn out near its upper end into a spirally coiled caecum (scm). Internally its walls bear in part a ciliated epithelium (sap) responsible for the movement—and often the sorting—of the particulate matter within, in part a cuticle which is frequently raised into a tooth-like eminence, the gastric shield (gs), helping in the trituration of food. One ciliated groove, which is called the intestinal groove (g_1), is a permanent feature of this part of the stomach, lying alongside a fold which is called the major typhlosole (t_1). Both these structures are often related to the duct from the larger liver lobe and run into the second part of the stomach (corresponding to the narrow end of the pear) which lies lower in the visceral mass. This is a cylindrical or conical section connected to the main part of the stomach dorsally and leading into the intestine ventrally, and known as the style sac (ss). The intestinal groove and major typhlosole run along its walls in the majority of animals and continue into the intestine (i). This

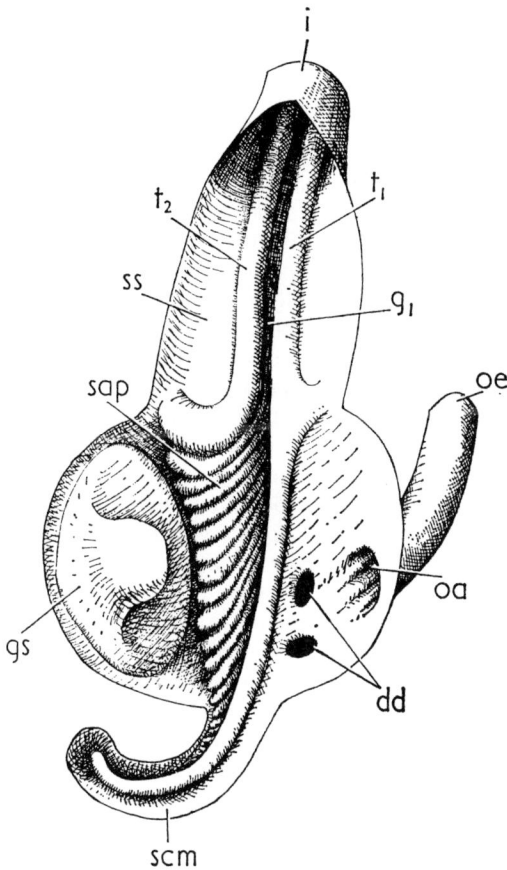

FIG. 87.—Diagram of the stomach of a generalized prosobranch, opened mid-dorsally.
dd, openings of ducts of digestive gland; g_1, intestinal groove; gs, gastric shield; i, intestine; oa, oesophageal aperture; oe, oesophagus; sap, posterior sorting area; scm, spiral caecum; ss, style sac region; t_1, major typhlosole; t_2, minor typhlosole.

runs forward, looping to the neighbourhood of the kidney and pericardial cavity—in the archaeogastropods, indeed, passing through the ventricle—and thereafter passing into the mantle skirt. After a more or less extended course in this the intestine opens at the anus at the summit of a short papilla. Occasionally a rectal or anal gland may occur. It is certain that some part of this intestinal region is of ectodermal, proctodaeal origin; it may be only the small papilla hanging freely into the mantle cavity at the summit of which the anus opens, or it may be the entire stretch which is embedded in the mantle skirt.

The mouth lies on the underside of the snout. In most gastropods this means that because of the downward curvature of the snout, it lies in a horizontal plane just clearing the substratum over which the animal moves; in some of the higher prosobranchs, however, the snout is straighter and the mouth then lies obliquely. In Stenoglossa and in such higher mesogastropods as possess a proboscis, the aperture visible on the underside of the head is not the true mouth but the opening of a proboscis sac within which the proboscis lies. The formation of a proboscis has occurred in animals which feed on material not immediately accessible: it occurs in prosobranchs which bore into other animals to find their food (*Natica, Doliacea, Muricacea*); in those which are carrion feeders and must use their proboscis to reach or probe into the body of the prey (*Buccinacea*); in those which seek access to the most nutritious parts of the body of their prey by way of natural apertures (Lamellariacea, Cypraeacea, Cerithiacea); by those with still more specialized ways of feeding (Toxoglossa— see p. 614; Pyramidellidae—see p. 240). The formation of a proboscis has not taken place among all prosobranchs in exactly the same way. When a proboscis is fully everted the true mouth lies at its tip and the apparent mouth—really that of the proboscis sac—disappears. The proboscis is then seen in its true morphological relations, as an elongation of that part of the snout lying anterior to the tentacles, an exaggerated example of what Amaudrut (1898) called snout formation or pretentacular elongation: all forms agree in this. It is in respect of the way in which this elongated snout is disposed of within the body on retraction that they differ. The simpler case appears to be the acrembolic type (fig. 88C, D), which is a simple introvert or inturned part of what is truly outer body surface. Here the retractor muscles (rmp) are inserted primarily at the tip of the proboscis and to a lesser extent along its side walls; when retraction occurs it is the apex of the proboscis which is pulled in first so that it comes to be the most deeply invaginated part and the whole process is like turning a stocking outside in by putting one's hand in and pulling the toe towards the top. This type of proboscis is found in the families Scalidae, Aclididae, Eulimidae, Pyramidellidae, Naticidae, Lamellariidae, Eratoidae. It may be regarded as an exaggeration of a tendency in all prosobranchs to withdraw the mouth into the snout to a certain extent when subject to noxious stimulation.

In more advanced prosobranchs (Cassididae, Doliidae, Columbellidae and Tritoniidae amongst mesogastropods, and Stenoglossa) the proboscis is of the type called pleurembolic (fig. 88A, B). Here the retractor muscles (rmp) are inserted on the sides of the proboscis mainly towards the base and on their contraction it is only this basal part which is invaginated to form the proboscis sac, the distal half being pulled into the shelter of that without turning inside out. Since only about half the length of the proboscis, when retracted, has to be accommodated within the head of the gastropod with this type of proboscis, it is clear that this is a device more economical of space than the acrembolic type in which the entire invaginable length has to be stored. Or, from another point of view, the pleurembolic type of proboscis may be relatively longer than the acrembolic type when a given volume of storage space is used. It is also mechanically more efficient [see also p. 612].

The mouth is a rounded or slit-like opening on the end of the snout in the primitive gastropods, usually elongated in a transverse direction, a fact more easily seen in the preserved than in the living animal. Around the actual opening lies a ridged outer lip, the ridges marking the outer ends of longitudinal folds in the tube to which the mouth leads. Frequently this ring-like tip is interrupted in the mid-ventral line so that it becomes horseshoe-shaped (*Patella* (fig. 50), *Littorina* (fig. 6), *Lacuna* (fig. 89)), but this is due to an inturning of the lip at that point rather than to a true interruption to its course. In many Trochidae, on the other hand, an eversion of the outer lip is to be found at this spot and a gutter-like process projects from the mid-ventral line of the mouth. It is short and tied to the mid-ventral wall of the snout in *Gibbula cineraria* (vlp, fig. 90), still shorter in *G. umbilicalis*, well developed and turned to the right neck lobe in *Calliostoma zizyphinum* (ppb, fig. 71) and *C. papillosum* (ppb, fig. 91), but absent in *Monodonta lineata*. In the Acmaeidae and Lepetidae the outer lip is elongated into a

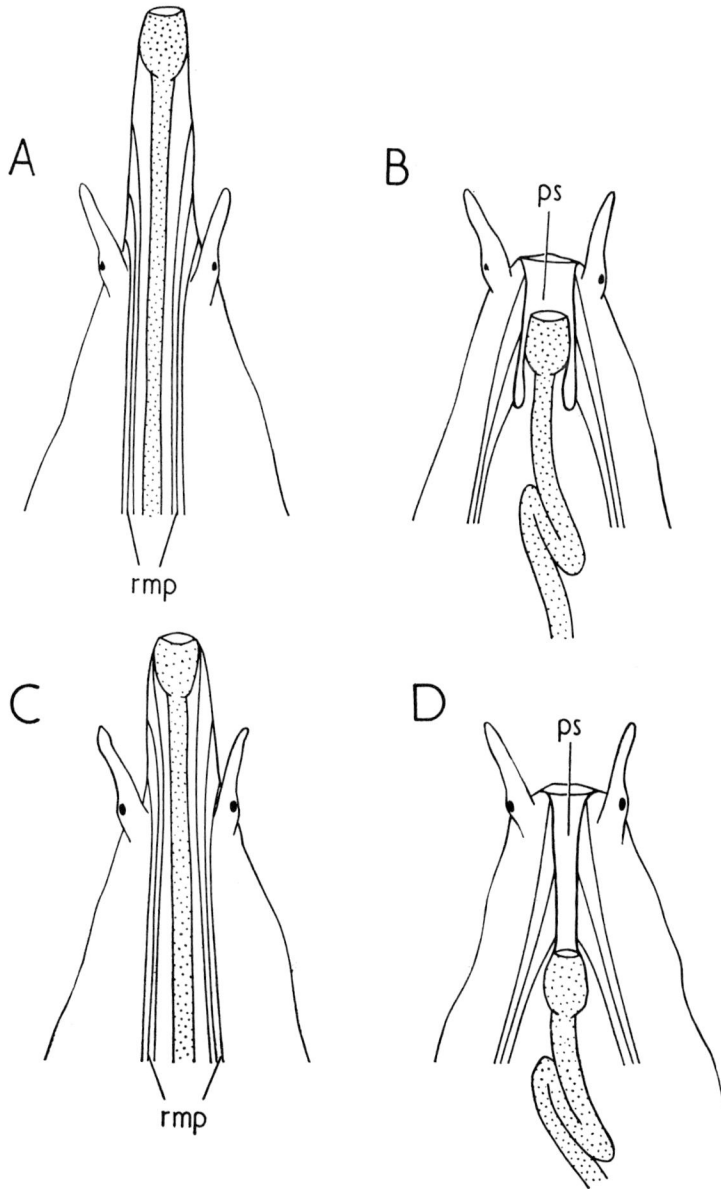

FIG. 88.—Diagrams showing proboscis structure: A, pleurembolic type extended; B, pleurembolic type retracted; C, acrembolic type extended; D, acrembolic type retracted. The gut is stippled.
ps, proboscis sheath; rmp, retractor muscles of proboscis.

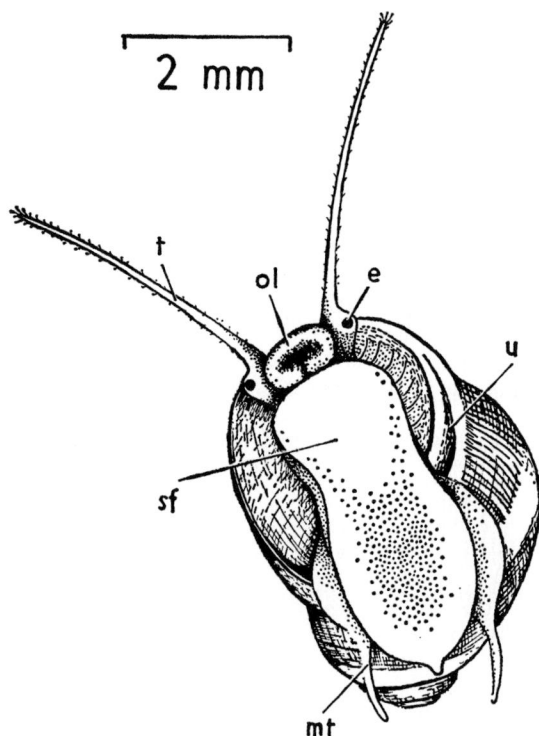

FIG. 89.—*Lacuna vincta:* young animal creeping, ventral view.
e, eye; mt, metapodial tentacle; ol, outer lip; sf, sole of foot; t, tentacle; u, umbilicus.

thin frill extending outward around the mouth and incomplete ventrally (*Acmaea tessulata:* ool, fig. 92; *A. virginea:* ol, fig. 73), and is limited to a dorsal hood-like veil in *Propilidium exiguum* (fig. 254). Thiem (1917*b*) has shown that this extended outer lip is rich in sensory structures in parts (ool, fig. 92), but cuticularized elsewhere (mol). In other prosobranchs the lip is bordered directly by the skin of the snout.

Where the opening on the surface of the head is not a true mouth but leads to a proboscis sac (e.g. *Lamellaria,* fig. 93, m; *Balcis,* fig. 139B), the front of the head is usually tapering and the slit longitudinal in direction without any special development of sensory lips around it. In these circumstances, too, the true mouth tends to be a simple rounded aperture.

The mouth may on occasion show other arrangements as, for example, in *Trichotropis borealis* and *Capulus ungaricus.* Here the ventral and ventrolateral parts of the lips are drawn out to form a proboscis (ppb, figs. 60, 90, 94), in *Capulus* almost equal in length to that of the shell, with a groove extending along the whole of its dorsal side at the proximal end of which the mouth is placed and rather resembling the proboscis of the trochids. It is used in connexion with the peculiar feeding arrangements of these animals. In *Trichotropis borealis* the proboscis appears to be kept permanently turned towards the right part of the mouth of the mantle cavity (Graham, 1954a), as was first thought to be true of *Capulus* (Orton, 1912b). Yonge (1938) has shown that the proboscis of *Capulus* is actually freely movable.

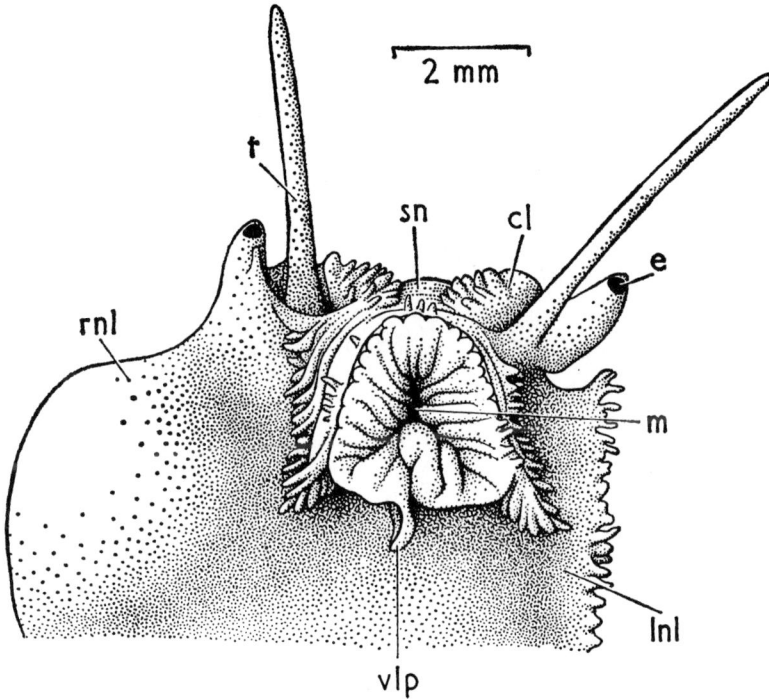

FIG. 90.—*Gibbula cineraria:* head, ventral view.
cl, cephalic lappet; e, eye; lnl, left neck lobe; m, mouth; rnl, right neck lobe; sn, snout; t, tentacle; vlp, ventral lip drawn out into median projection.

The mouth leads to the first part of the alimentary tract, a tube of short length, the oral tube (ot, fig. 14). Along it run continuations of the folds which marked the outer lip, except on the ventral side. It is lined by an epithelium bearing a cuticle and its walls are muscular and may be partly glandular. It leads into an expanded chamber which is the buccal cavity (bcv). A number of characteristic structures lie here. Embedded in the dorsal or lateral walls of the cavity at its anterior end may be a pair of jaws (ja, fig. 100), or they may be replaced by a single median one. Behind the level of the jaws there extends into the buccal cavity from the ventrolateral walls a pair of inner lips (il, fig. 14), which may be drawn across the buccal cavity like a curtain so as to separate it more or less completely from the oral tube. Projecting into the buccal cavity from the posterior floor, like a tongue, is the odontophore (bm), over the dorsomedian surface of which runs the radula (rd). The odontophore is free dorsally, laterally, and ventrally near its tip. The space between the ventral surface of the odontophore and the floor of the buccal cavity is the sublingual space or cavity; the space between the sides of the odontophore and the lateral walls of the buccal cavity—often roomy—may sometimes be referred to as buccal pouches. These parts of the buccal cavity are partly cuticularized but especially on the inner lips and in parts of the lateral pouches are also partly glandular; the surface of the odontophore, however, is almost wholly covered with a thick cuticle and presents, therefore, a glistening appearance on dissection.

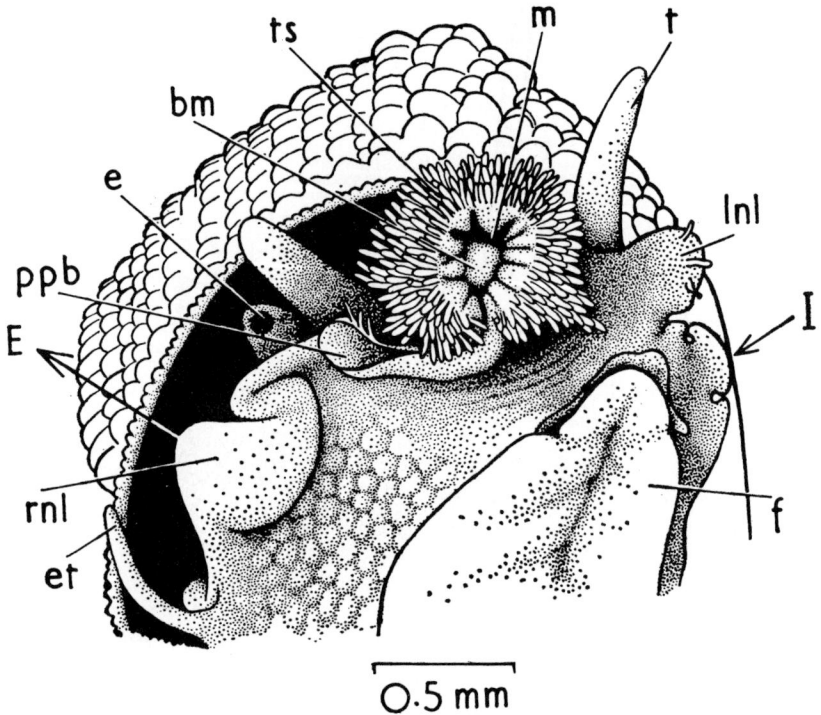

FIG. 91.—*Calliostoma papillosum:* anterior part of body, ventral view.
bm, anterior tip of buccal mass; e, eye; E, direction of exhalant water current; et, epipodial tentacle; f, foot; I, direction of inhalant water current; lnl, left neck lobe; m, mouth; ppb, pseudoproboscis formed from ventral lip; rnl, right neck lobe; t, tentacle; ts, tentacles on snout.

The roof of the buccal cavity is more elaborately built than any other part. Its most conspicuous feature is a pair of longitudinally directed folds which lie on either side of the mid-dorsal line (df, fig. 14); they begin some little way behind the jaw and run posteriorly out of the buccal cavity into the oesophagus, forming the lateral walls of a trough-like space which occupies the dorsomedian wall. These are the dorsal folds of the fore-gut enclosing the dorsal food channel (dfc). Frequently each fold appears double where it runs on the buccal roof into which it fades away anteriorly. In contrast to the rest of the buccal walls the epithelium which lines these folds and the channel is, in most prosobranchs, richly and strongly ciliated, the cilia on the folds beating into the channel and those in the channel maintaining a strong current backwards into the oesophagus.

Arising from the lateral wall of the buccal cavity on each side, ventral to the dorsal folds, is a sheet of tissue which sweeps horizontally across the buccal cavity to meet its partner from the opposite wall. This forms a septum (oev) which is at one and the same time the floor of the oesophagus and the roof of the radular sac. Ventral to this septum the buccal cavity is continued into a narrow tube, the radular sac, which extends backwards ventral to the oesophagus to a varying distance, ending blindly. Within it is secreted the radula (see p. 167). The septum is sometimes said to mark the inner limit of the stomodaeal region which is then

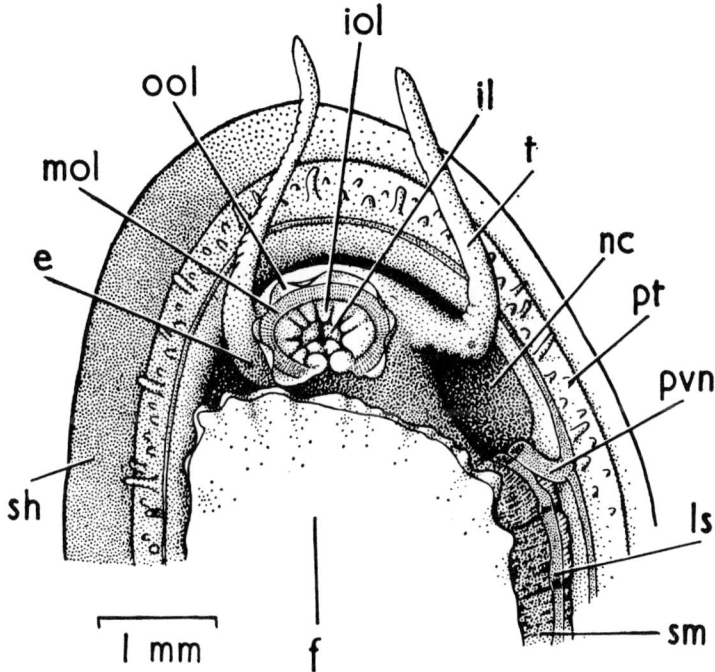

FIG. 92.—*Acmaea tessulata:* anterior part of body, ventral view.

e, eye; f, foot; il, inner lip; iol, inner zone of outer lip; ls, lateral blood sinus; mol, middle zone of outer lip; nc, nuchal cavity ; ool, outer zone of outer lip; pt, pallial tentacle; pvn, pallial vein; sh, shell; sm, shell muscle; t, tentacle.

coincident with the inner end of the buccal cavity, the oesophagus behind it being of endodermal origin, but many observers regard the whole gut as far as the posterior end of the oesophagus as being ectodermal (see p. 397). It is right to add at this point that in the view of Nisbet (1953), the line separating buccal cavity and oesophagus should be drawn, not vertically across the gut of the animal as has just been suggested, but obliquely, so that whilst its ventral end would coincide with the mouth of the radular sac its dorsal end would cut the roof of the buccal cavity at the anterior end of the dorsal folds and dorsal food channel, posterior to the point of entry of the salivary ducts. This would separate the whole of the food conducting apparatus (food channel and folds) from the buccal cavity. This may be a better arrangement anatomically, and is certainly so from the functional point of view, than that which divides it partly into a buccal section and partly into an oesophageal half.

In connexion with the buccal cavity is a number of glands which may be dealt with before the complexities of the buccal mass are treated. Most of these glands are unicellular goblet cells lying in the buccal epithelium secreting mucus. They are particularly abundant on the inner lips and in the lateral parts of the buccal wall, but may occur in some animals in other positions. Thus in *Diodora* and *Haliotis* there is a concentration at the base of the jaw (Ziegenhorn & Thiem, 1926) in the roof of the buccal cavity, and *Theodoxus* (Whitaker, 1951)

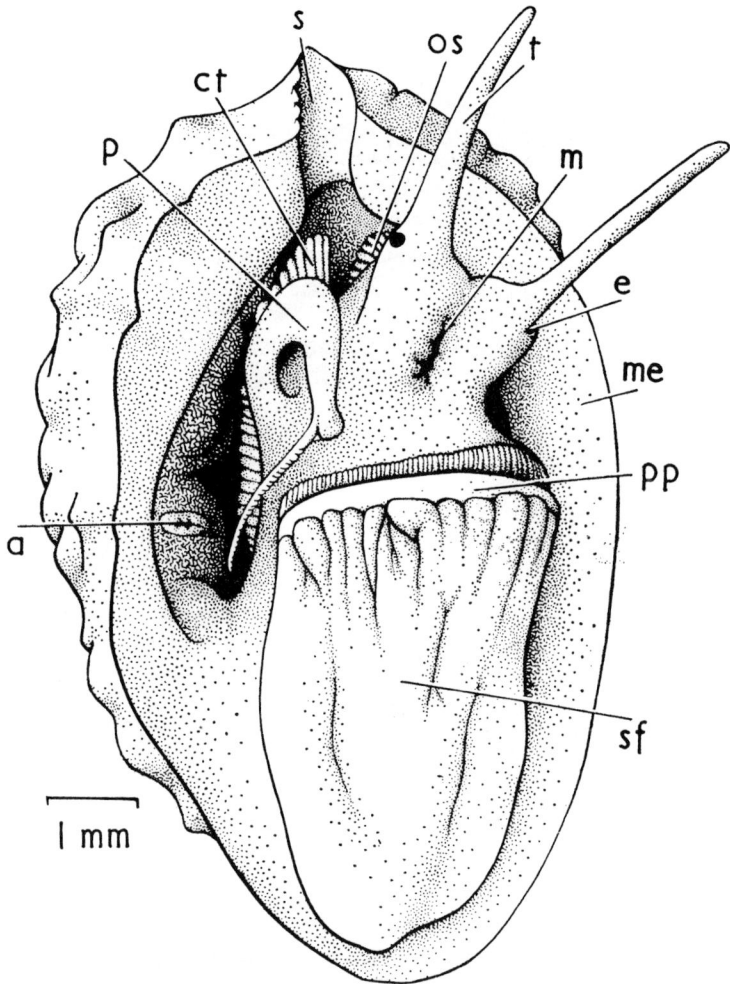

FIG. 93.—*Lamellaria perspicua*: male in ventral view.
a, anus; ct, ctenidium; e, eye; m, mouth; me, mantle edge; os, osphradium; p, penis, with flagellum; pp, propodium; s, siphon; sf, sole of foot; t, tentacle.

and *Septaria* (Bourne, 1908) show a pair of blind tubules of a glandular nature discharging to the inner end of the space ventral to the odontophore.

The most important of the glands connected to the buccal cavity, however, are the salivary glands, which discharge to its roof just anterior and lateral to the ends of the dorsal folds. In the more primitive groups such as the Zeugobranchia these glands (sg, fig. 95) are not conspicuous, but form a single pouch or a tuft of a few short tubes discharging by a common opening, which hardly merits the name of duct, to the buccal cavity. All the epithelial tissue, including that of the duct, is secretory, apart from occasional ciliated cells wedged between

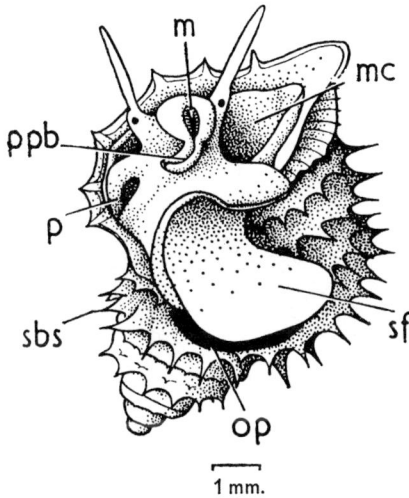

FIG. 94.—*Trichotropis borealis:* male, in ventral view.
     m, mouth; mc, mantle cavity; op, operculum; p, penis; ppb, pseudoproboscis formed from ventral lip; sbs, bristles on shell; sf, sole of foot.

the gland cells (fig. 96D). Most of the cells secrete mucus, though some secrete other substances: in neither case is the secretion digestive and in these lower prosobranchs the gland appears to be concerned solely with producing a lubricant for the feeding processes and an adhesive for the food particles. In the Trochacea the salivary glands resemble those of the Zeugobranchia. In the Patellacea (sg, fig. 253) and most mesogastropods, however, the salivary glands are much increased in complexity and have become bulky masses of secretory tissue. The glands are now found in the posterior part of the head, or even in the basal part of the visceral hump, and are connected to the buccal cavity by ducts which run along the roof of the oesophagus and buccal cavity lateral to the attachment of the dorsal folds, thus avoiding the congestion which they would produce if they remained in the snout. In most cases the ducts run through the nerve ring. In the Patellidae each gland has two ducts running parallel to one another and opening close together. In the Acmaeidae two histologically distinct types of so-called salivary glands occur: the ordinary ones which open to the roof of the buccal cavity by ducts alongside the dorsal folds (the 'pharyngeal salivary glands' of Thiem (1917b)), and a number of tubules which open separately to the buccal pouches or even to the oesophagus (these are not present in *Acmaea virginea*). The double pair of *Patella* and its relatives probably represents a doubling of the first type rather than two different sorts, since their histology is identical. In all these cases the glandular tissue is confined to the main mass of the gland and the duct is a purely conducting tube, leading the secretion to the buccal cavity, and lined, in relation to this function, with a ciliated epithelium which, together with secretion pressure, appears to be the main means of transporting the secretion, muscle fibres being notably rare. The secretory cells, as in zeugobranchs, alternate with ciliated cells and produce predominantly mucous substance although basic protein secretions are also of frequent occurrence (fig. 96C). Few prosobranchs secrete digestive enzymes in their saliva, which appears to be primarily a lubricant for the food-collecting and swallowing activities of

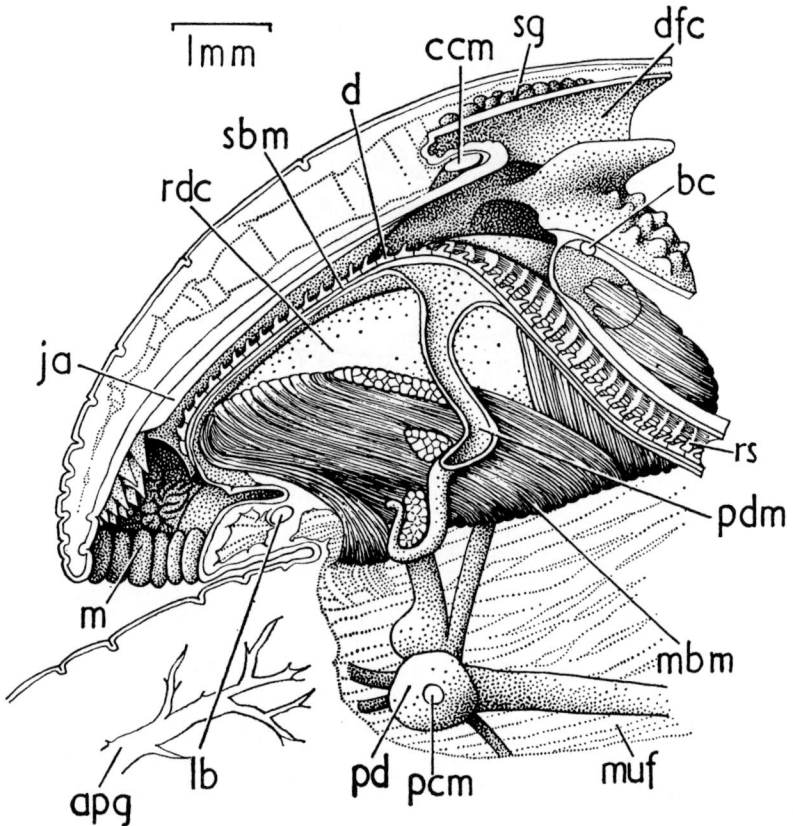

FIG. 95.—*Diodora apertura*: right sagittal half of head and part of foot, seen from median side.
    apg, anterior pedal gland; bc, buccal commissure; ccm, cerebral commissure; d, radular tooth; dfc, dorsal food channel; ja, jaw; lb, labial commissure; m, mouth; mbm, muscles of odontophore; muf, muscle fibres; pcm, pedal commissure; pd, pedal ganglion; pdm, diverticulum extending ventrally between two halves of odontophore; rdc, radular cartilage; rs, radular sac; sbm, subradular membrane; sg, salivary gland.

the animal. Hirsch (1915) and Mansour-Bek (1934) recorded the occurrence of a variety of proteolytic enzymes in the glands of species of *Murex*, and Jenkins (1955) recorded amylase in those of *Littorina littorea*. Welsh (1956) has shown that the salivary glands of *Buccinum* produce 5-hydroxytryptamine, and Fänge (1957, 1958) has recorded a similar substance in those of *Cassidaria* and an acetylcholine-like substance, possibly neurine, in those of *Neptunea antiqua*. He suggested that in addition to its well known carrion feeding, this animal may also use its toxic saliva to permit the capture of active prey.

    In the Stenoglossa (Muricacea, Buccinacea) the salivary glands lie in a position which is different from their location in less advanced prosobranchs. In the diotocardians (fig. 95) the glands (sg) lie usually over the buccal roof and posterior to the point at which the cerebral commissure (ccm) runs. In the monotocardians the point of entry of the salivary duct into the buccal cavity shows that the fundamental morphology of the glands is the same, but as a

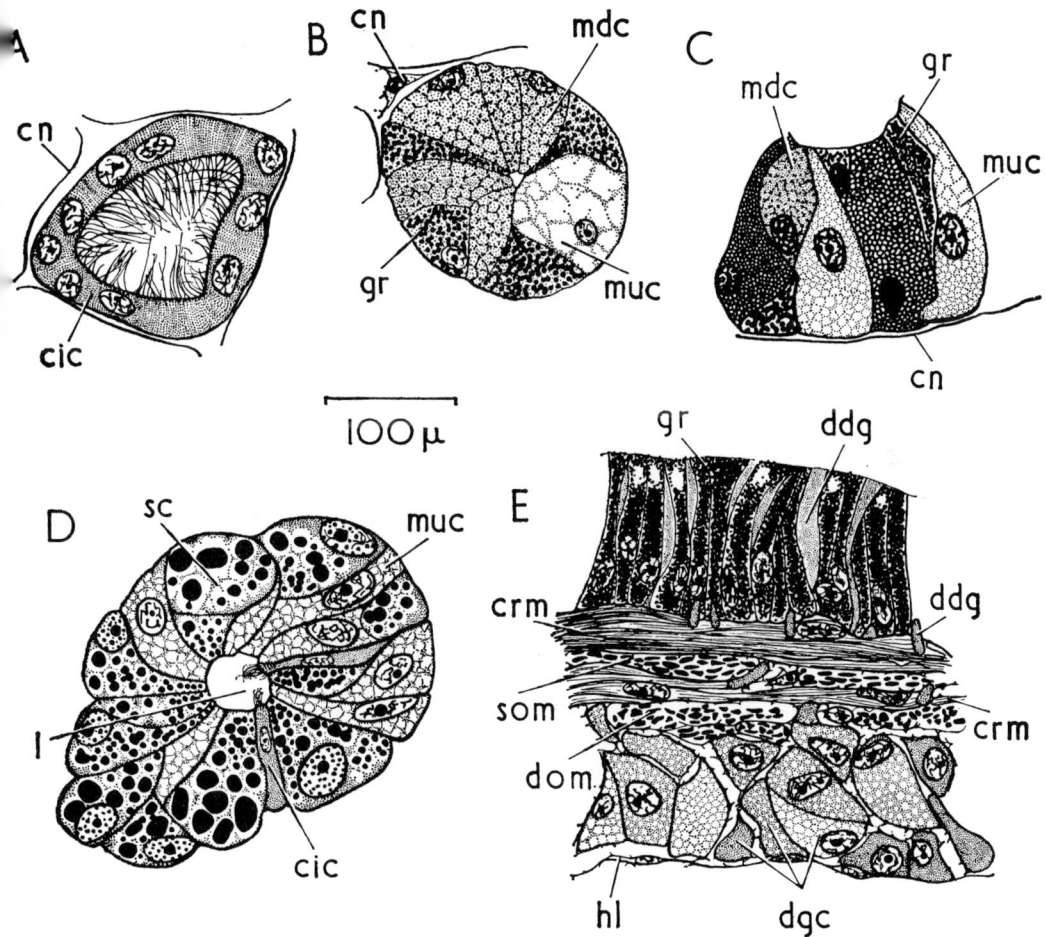

FIG. 96.—Histology of salivary glands. A, *Nassarius reticulatus,* transverse section of salivary duct; B, *Nassarius reticulatus,* section of lobe of gland; C, *Diodora apertura,* section of salivary cells; D, *Patella vulgata,* section of lobe of gland; E, *Nucella lapillus,* section of part of accessory gland.

cic, ciliated cell; cn, connective tissue; crm, circular muscle layer; ddg, duct of deep gland cell; dgc, deep gland cell; dom, deep layer of oblique muscles; gr, granule of secretion; hl, limit of haemocoelic space; l, lumen; mdc, mucoid cell; muc, mucous cell; sc, secreting cell; som, superficial layer of oblique muscles.

consequence of the backward shift of both gland and cerebral commissure the salivary duct (sd, fig. 17) runs through the nerve ring. The Stenoglossa (fig. 97) have elongated the anterior part of the gut in the course of making a proboscis (pb) and this elongation lies anterior to the nerve ring. As a consequence of differential growth rates involved in this process the salivary glands (sg) have been pulled anteriorly through the nerve ring and now lie in front of the cerebral commissure and so their ducts do not pass through it. Despite this, because of the need to exclude bulky glands from a slender, retractile proboscis, considerable elongation of the ducts has occurred and the glands lie at the inner end of the proboscis in

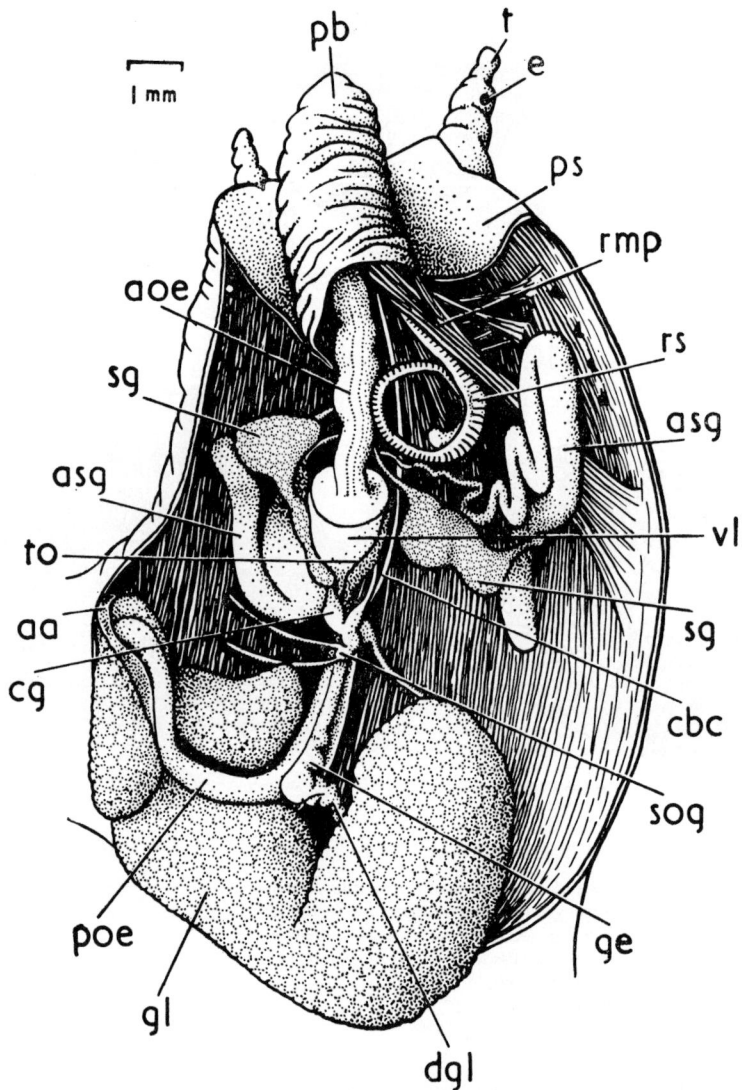

FIG. 97.—*Nucella lapillus:* dissection to show anterior part of gut. The proboscis sheath has been opened to show the proboscis and the anterior part of the body cavity dissected.

aa, anterior aorta; aoe, anterior oesophagus; asg, accessory salivary gland; cbc, cerebrobuccal connective; cg, cerebral ganglion; dgl, duct of the gland of Leiblein; e, eye; ge, 'glande framboisée'; gl, gland of Leiblein; pb, proboscis; poe, posterior oesophagus; ps, proboscis sheath; rmp, retractor muscle of proboscis; rs, radular sac; sg, salivary gland; sog, supra-oesophageal ganglion; t, tentacle; to, line marking site of separation of glands from rest of oesophagus; vl, valve of Leiblein.

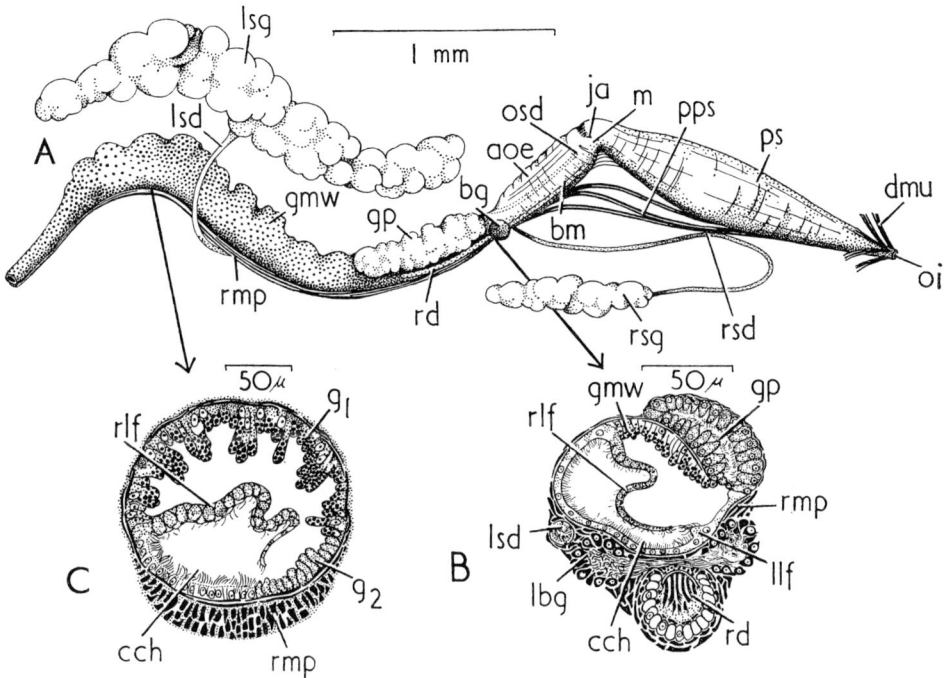

FIG. 98.—*Cerithiopsis tubercularis:* A, anterior part of alimentary canal removed from animal and viewed from the right; B, transverse section at level indicated by arrow; C, transverse section at level indicated by arrow. aoe, anterior oesophagus; bg, right buccal ganglion; bm, buccal mass; cch, ciliated channel; dmu, dilator muscles; g$_1$, first type of gland cell of mid-oesophagus; g$_2$, second type of gland cell of mid-oesophagus; gmw, glandular wall of mid-oesophagus; gp, glandular diverticulum of oesophagus; ja, jaw; lbg, left buccal ganglion; llf, left longitudinal fold; lsd, left salivary duct; lsg, left salivary gland; m, position of mouth; oi, opening of introvert; osd, opening of salivary duct; pps, protractor muscles of proboscis; ps, wall of introvert; rd, radular sac; rlf, right longitudinal fold; rmp, retractor muscles of proboscis; rsd, right salivary duct; rsg, right salivary gland.

the neighbourhood of the nerve ring. The production of a proboscis in the mesogastropods seems to have been accompanied merely by an elongation in the length of the salivary duct allowing the gland to remain alongside the nerve ring in the body of the mollusc when the proboscis is everted (eratoids: Fretter, 1951a). In *Cerithiopsis tubercularis* (Fretter, 1951b) the slimness of the animal has caused the right (rsg, fig. 98) and left (lsg) glands to be staggered to permit better packing, as has occurred with the internal organs of snakes. The arrangement of the salivary glands in *Triphora perversa* is discussed below (p. 208).

The size and shape of the salivary glands vary within the prosobranchs to a considerable extent. The normal structure of the gland is acinous but it may depart from this and tubular glands occur in a number of families (hydrobiids, rissoids (fig. 67), assimineids, Ptenoglossa (fig. 99), Calyptraeidae, Pyramidellidae (fig. 133)). The size of the gland may be related to the amount of lubrication called for, and where this is slight, the glands may be quite small. This is noticeable where the lubricant is provided by other glands such as those in the mantle cavity, as occurs in ciliary feeders: then the salivary glands may be very small (*Capulus, Turritella*). In

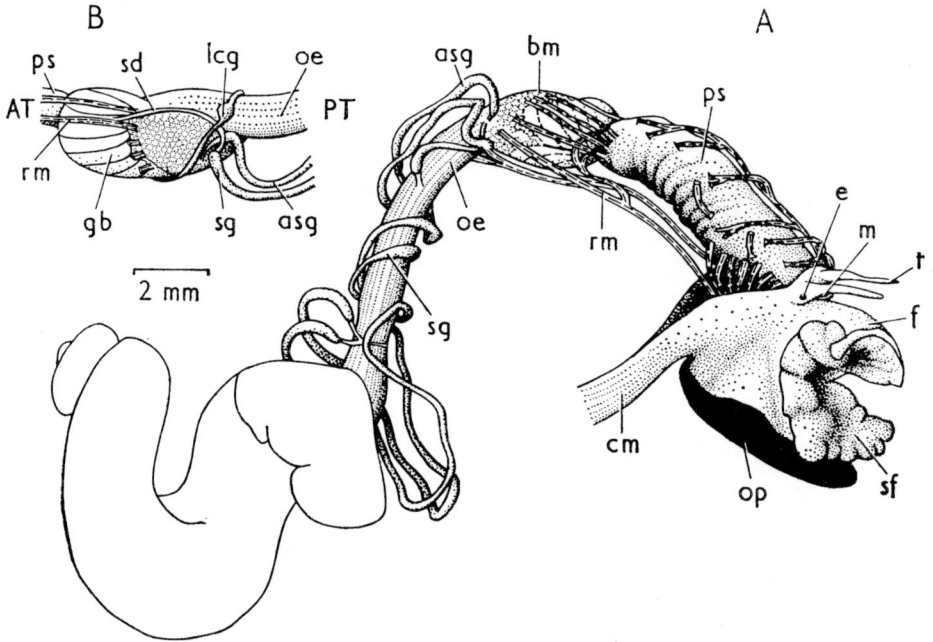

FIG. 99.—*Clathrus clathrus:* A, anterior part of alimentary canal removed from animal and viewed from the right; B, buccal region, seen from the left.

asg, accessory salivary gland; AT, anterior; bm, buccal mass; cm, columellar muscle; e, eye; f, foot; gb, glands in wall of buccal cavity; lcg, left cerebral ganglion; m, mouth of introvert; oe, oesophagus; op, operculum; ps, wall of introvert; PT, posterior; rm, retractor muscle; sd, salivary duct; sf, sole of foot; sg, salivary gland; t, tentacle.

other animals the glands may be wholly absent (Neritacea, Eulimidae) and in these animals secretion of lubricating substances is probably sufficient at other places to compensate for their loss, though it is not always certain where these places are. On the other hand, the salivary glands have become hypertrophied in some prosobranchs: they are particularly large in Ampullariidae and in some families of the Doliacea. In *Dolium* itself, Weber (1927) has shown that this is due to the fact that they secrete a moderate concentration of sulphuric acid in the saliva, which is used to poison the prey which the animal attacks. The gland is covered with a layer of muscle tied to the foot and on contraction this squeezes the gland against the solid background provided by this part of the body and so ejects the acid saliva through the long duct and into the body of the prey under considerable pressure. *Dolium* and *Cassis* both show an accessory salivary gland on the course of the duct near its origin from the main gland; its significance is not known [see p. 626].

In a few prosobranchs special arrangements affect the organization of the salivary glands. In the family Scalidae, for example, *Clathrus clathrus* (fig. 99A) and other species possess a pair of long, tubular salivary glands (sg) tied to the oesophagus at their inner ends by muscle strands. They narrow to form ducts at the level of the nerve ring. The ducts (sd, figs. 99B, 100) penetrate this and then run forwards alongside the buccal mass and oral tube. Here they

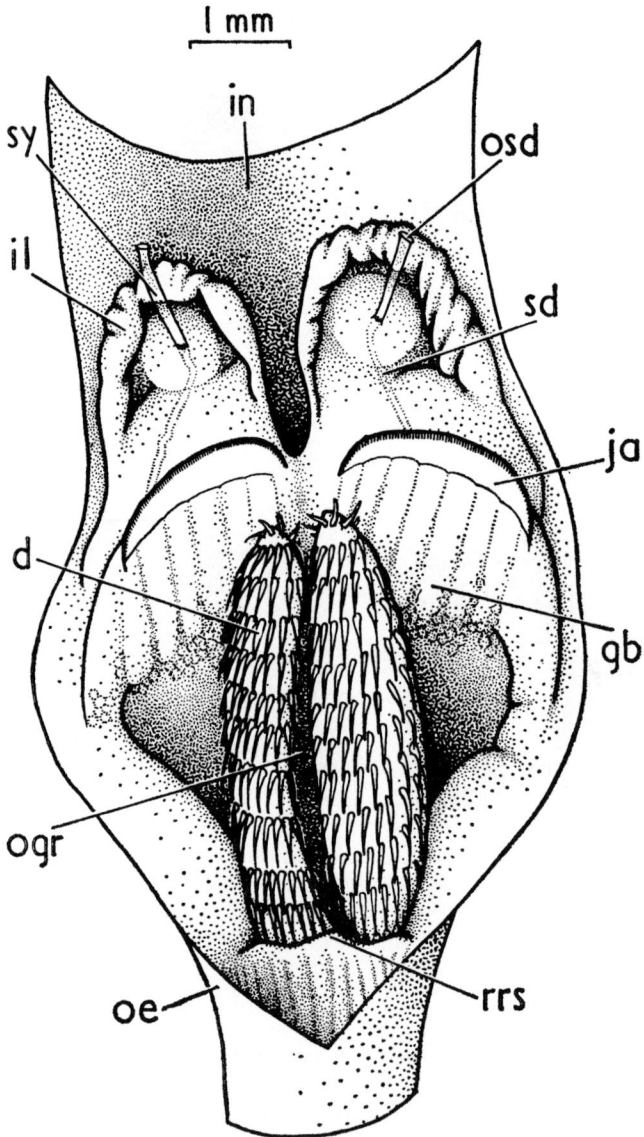

FIG. 100.—*Clathrus clathrus:* dissection showing the base of the introvert, the buccal cavity and the anterior part of the oesophagus opened mid-dorsally.

d, radular tooth; gb, glands in wall of buccal cavity; il, inner lip; in, introvert; ja, jaw; oe, oesophagus; ogr, groove on mid-dorsal surface of odontophore; osd, opening of salivary duct; rrs, roof of radular sac; sd, salivary duct; sy, stylet on which salivary duct opens.

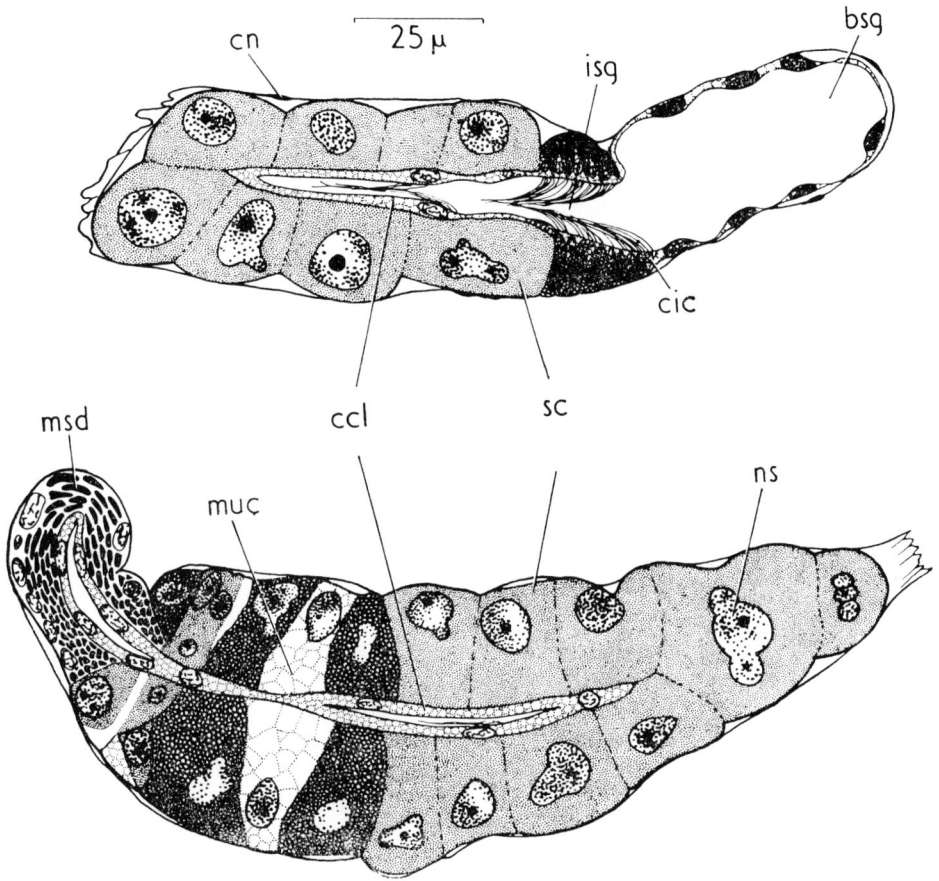

FIG. 101.—*Odostomia lukisi:* longitudinal sections through the salivary gland; the upper section cuts the proximal half of the gland, the lower the distal half and the initial part of the duct.

bsg, bladder of salivary gland; ccl, ciliated epithelium of lumen; cic, ciliated cell; cn, connective tissue; isg, intermediate section of salivary gland; msd, circular muscles of salivary duct; muc, mucous cell; ns, nucleus of secreting cell; sc, secreting cell.

expand somewhat to form reservoirs which open to the exterior on the inner lips (il), each of which carries a mammiform swelling on its inner surface. Projecting from the centre of this is a hollow chitinous stylet (sy) which bears the opening of the salivary duct at its summit (osd). As the scalids possess an acrembolic proboscis this apparatus is normally hidden at the base of the introvert (in), but will project when the proboscis is everted. Its precise use is unknown, as the feeding of these animals has not been observed, but it seems likely that it forms part of a device for overwhelming prey, perhaps by the injection of a poisonous salivary fluid.

A somewhat similar device is employed by the pyramidellids, although its anatomical basis is different. In these animals the salivary glands (sg, fig. 133) are, again, tubular, the innermost part of the tube being a thin-walled bladder (bsg). Next to that is a small section (isg, fig. 101) in

which small gland cells alternate with ciliated cells (cic). This section is brief and the bulk of the rest of the wall of the tube is lined by ciliated cells of a squamous shape (ccl) overlying large gland cells (sc) the secretion of which enters the gland between the ciliated cells. The gland produces very little mucus and the nature of its secretion is not known. A very muscular duct (msd) leads the secretion to the buccal cavity. Here (fig. 102), in the mid-dorsal line, the two ducts (sd) unite and open into the central cavity of a hollow stylet (sy) which in this case, however, is the animal's jaw, which has been rolled into a hollow cylinder with a tip tapered like a hypodermic needle. When the mollusc feeds, this is thrust through the skin of the prey and saliva injected. As pyramidellids appear to feed primarily on fluid sucked from the body of their prey, the salivary secretion may well be an anticoagulant.

Along with the salivary glands which have just been described a number of prosobranch gastropods possess other glands of a salivary type. Some of these are merely exaggerations of the glands already mentioned in the walls of the buccal cavity. Others, however, are additional structures and of these, the most important are perhaps the tubular salivary glands of Stenoglossa (asg, fig. 97). These occur with the ordinary acinous type alongside which they lie as white tubules, somewhat bent. They narrow anteriorly into ducts which do not traverse the nerve ring but run forwards to the underside of the proboscis, where they unite to form a median duct of very fine diameter which runs to open on the mid-ventral line of the mouth. The gland (fig. 96E) is lined by a columnar epithelium containing spherules of secretion (gr) which seem to be of a protein nature. This epithelium rests on a thick layer of muscle made of two sheets of fibres running diagonally and at right angles to one another and arranged in outer and inner sets (crm, som, dom). External to this are gland cells (dgc), three or four layers thick, which send long necks (ddg) through the muscle coat and between the epithelial cells, to discharge secretion to the central space. The function of these gland cells is not known. They do not occur in Buccinacea, which do not bore into the shells of other molluscs, whereas they do occur in most Muricacea, which are active borers. They have been recorded in *Oliva* (Küttler, 1913) but it is not certainly known to what extent they occur in other volutaceans nor what the feeding habits of the animals may be. They do not seem to produce an acid which might help in boring (Graham, 1941) nor a 'calcase' such as has been suggested by Ankel (1937a) might help in the process. It is possible that their secretion may have some toxic effect on their prey, or contribute towards an external digestion of it (though Graham (1941) failed to find either proteolytic or amylolytic enzymes in their secretion in *Nucella*) or— most probably—they may be simply a source of further lubrication for the radula during the boring process, the opening of the duct being well placed for this function [see p. 625].

This second pair of salivary gland of the Muricacea leads to a discussion of the poison gland of another group of Stenoglossa, the Toxoglossa, which are well known for their ability to inflict a poisonous bite on their prey. Four species of the genus *Conus, C. geographus, C. tulipa, C. aulicus* and *C. textile*, all from the Indopacific region, are known to be actively poisonous and to have harmed and killed man. Most toxoglossans, including the Mediterranean species (*C. mediterraneus*) and the British species of *Philbertia* and *Mangelia* described by Robinson (1955), have the same apparatus (fig. 103) and, presumably, the same toxic powers, though their small size prevents them from being troublesome to man [see pp. 614–20].

In the roof of the buccal cavity of many prosobranchs, anteriorly, lie the jaws or jaw. These are primitively placed right and left of the mid-line, but in some species they may become approximated and fuse to a single piece. When double, they may be used in a scissor-like way to shear pieces off the prey, as in *Natica*, but more usually—and always when single—they form a stiff edge against which pieces of food may perhaps be pressed by the odontophore

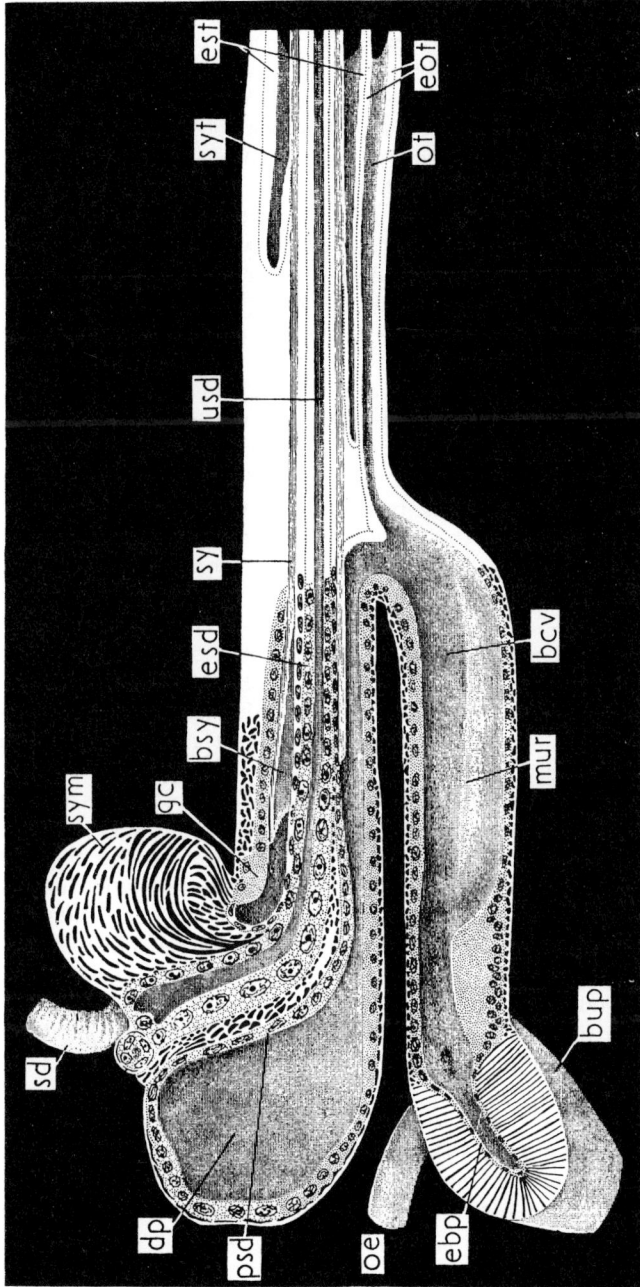

FIG. 102.—*Odostomia*: stereogram of a sagittal half of the base of the stylet and associated structures. bcv, buccal cavity; bsy, base of stylet; bup, buccal pump; dp, dorsal pouch of buccal cavity; ebp, epithelium of buccal pump; esd, epithelium of salivary duct; est, epithelium of oral tube; eot, epithelium of oral tube; gc, gland cell; mur, mucous ridge; oe, oesophagus; ot, oral tube; psd, projection containing salivary ducts; sd, salivary duct; sy, stylet; sym, muscles moving stylet; syt, stylet tube; usd, united salivary ducts.

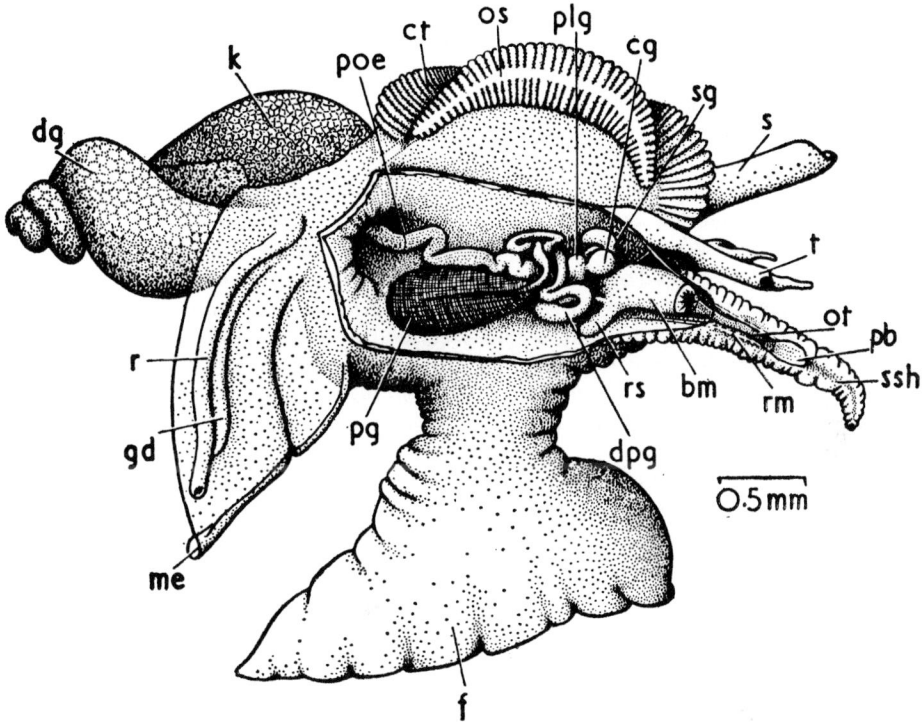

FIG. 103.—*Mangelia powisiana:* animal removed from shell; the mantle skirt has been cut by a median longitudinal incision and the halves deflected and the anterior part of the body dissected. The proboscis has been
pulled out from the proboscis sheath and its tip and the oral tube are seen by transparency through the walls
of the suspensory sheath.

bm, buccal mass; cg, right cerebral ganglion; ct, ctenidium; dg, digestive gland; dpg, poison gland; f, foot;
gd, genital duct; k, kidney; me, mantle edge; os, osphradium; ot, oral tube; pb, proboscis; pg, muscular sac;
plg, right pleural ganglion; poe, posterior oesophagus; r, rectum; rm, retractor muscle of proboscis; rs, radular
sac; s, siphon; sg, salivary gland; ssh, suspensory sheath; t, tentacle.

and bitten or broken. The name 'jaw' unfortunately suggests this type of biting action, but as
pointed out by Starmühlner (1952) and Nisbet (1953) the jaw is much more important in
manipulating the radula, or in preventing food escaping from the buccal cavity than in biting.
Each jaw is an exaggeration of the cuticular covering present on the walls of much of the
buccal cavity and is secreted by the cells there. Chemically it appears to be chitinous. Each
cell secretes numerous threads of material which may be seen, in fixed and stained sections,
as cilia-like structures extending from the outer surface of the cell across a narrow space
(presumably an artefact) to the jaw. Each cell thus produces a piece of material which joins
on to those secreted by neighbouring cells, though retaining its independence to a sufficient
degree to make the jaw look as if composed of numerous rods. This arrangement (or
differential wear at the free edge of the jaw) often produces a serrated edge. The whole
structure is deeply embedded in a groove running along or across the anterodorsal wall of the
buccal cavity. Muscles run from the base to the body wall and buccal mass which may

protract or retract it. In Docoglossa the (single) jaw does not show either the rod-like structure or the serrated edge.

In a number of prosobranchs, in correlation with a mode of feeding which does not call for any kind of biting, jaws have vanished. This is true of the Trochidae, for the most part, the Neritidae, Lacunidae, Pomatiasidae, Acmidae, Assimineidae (where they are vestigial), Eulimidae, the parasitic forms, Calyptraeidae, and all Stenoglossa, though vestiges may appear in some of this last group. The heteropods have also lost jaws as part of their adaptation to pelagic life. In a few groups (Docoglossa, Ampullariidae, Cerithiopsidae, Scalidae and Pyramidellidae) the jaws have united to a single piece, in the pyramidellids a tubular structure penetrated by the single salivary duct. In other families the jaws are paired.

The most characteristic structure of the molluscan buccal cavity is the radula, one of the hallmarks of the phylum. It is a chitinous ribbon continually added to at its inner end, which is placed at the posterior end of the radular sac, a blind diverticulum from the posterior end of the odontophore, and stretching forward from there over the dorsal surface of the buccal mass on the floor of the buccal cavity. The ribbon bears teeth placed regularly alongside one another in transverse rows and regularly behind one another in longitudinal series and the number of these and the shape of the teeth differ from species to species, though remaining fairly constant within one species. As a consequence of this and the fact that they are imperishable and may be extracted from dried bodies, the radula is an important organ from the taxonomic point of view. From the functional viewpoint, however, there are many important questions arising in connexion with it to which it is still impossible to give any adequate answer.

Each row of teeth on the radula normally repeats precisely the number and shape of the teeth in the rows in front of it and behind. In it there is usually an odd number of teeth due to the fact that the row consists of a single tooth, centrally placed, with a series of others on either side, those on the right being the mirror image of those on the left. The middle tooth is the rachidian or central; those on either side of it are broadly known as the laterals. These usually diminish in size from the mid-line laterally, sometimes in an even gradation of similar teeth, but, more commonly, a group of teeth nearer the rachidian (the lateral or intermediate teeth) is distinguishable from a group lying further away (the marginal teeth). The biggest lateral tooth may be called the dominant. Since the number of teeth in a row is a specific character it is customary to represent it by means of a formula in which either figures representing the number of teeth of each kind in a single row are given, or figures for the rachidian and the right half-row, the left half being assumed to be the mirror image of the right. Thus we may write either $30 + 4 + 1 + 4 + 30$ or $R + 4 + 30$. Occasionally figures (1, 2, 3 . . .) under the tooth number give their relative sizes. If the number of teeth of one kind is high and variable it may be denoted by n; if very high or virtually uncountable by ∞. D may be used to show the dominant; * a tooth which has no cusp. A few examples of uses of this sort are given here:

$$\infty + D + 4 + R + 4 + D + \infty \quad \text{(Rhipidoglossa)}$$
$$3 + R + 3 \quad \text{(Taenioglossa)}$$
$$n + 0 + n \quad \text{(Dorid)}$$
$$R \quad \text{(Ascoglossa)}$$

The various patterns of radular structure which are encountered among the prosobranchs are these:

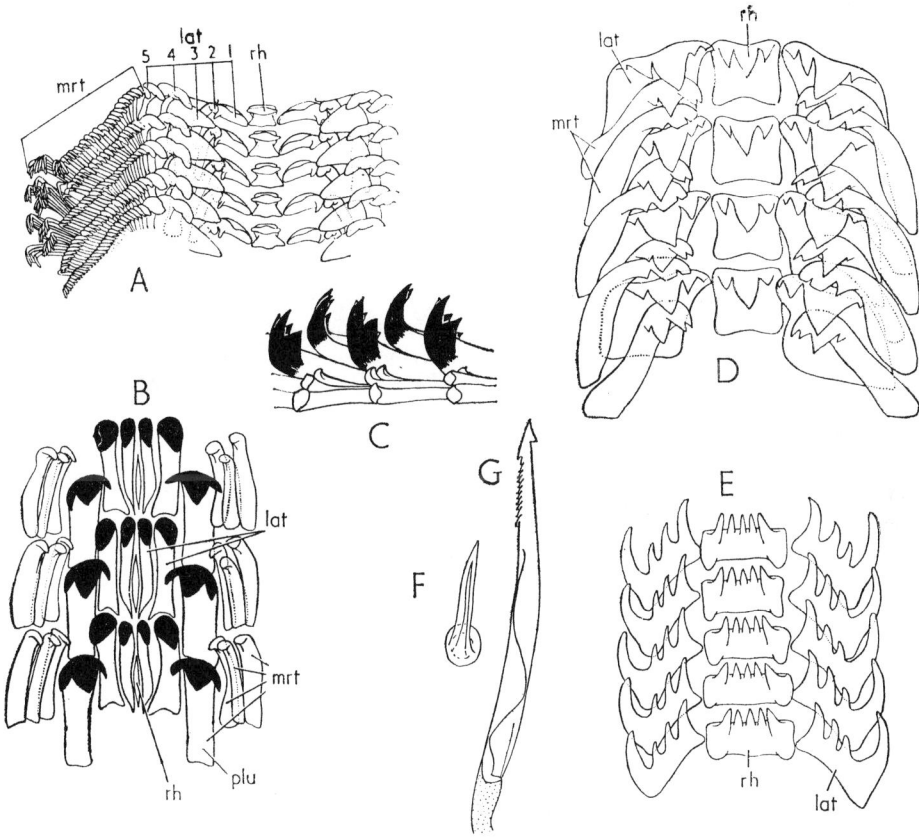

FIG. 104.—Radulae. A, rhipidoglossan (*Haliotis*), dorsal view; B, docoglossan (*Patella*), dorsal view; C, docoglossan (*Patella*), side view (rh, not seen); D, taenioglossan (*Littorina*), dorsal view; E, rachiglossan (*Buccinum*), dorsal view; F, toxoglossan (*Mangelia*), single tooth; G, toxoglossan (*Conus*), single tooth.

To facilitate comparison A, B, D are drawn with the anterior end above; in E it is below. lat, lateral tooth; mrt, marginal tooth; plu, pluricuspid tooth; rh, rachidian or median tooth.

1. Rhipidoglossan (from the Greek *rhips, rhipidos,* a fan) (fig. 104A): formula

$$\infty + 1 + 4 + R + 4 + 1 + \infty \quad \text{or} \quad \infty + 1 + D + 3 + R + 3 + D + 1 + \infty.$$

The rachidian tooth (rh) is large and often not cusped. On either side of it lies a fan of smaller teeth which can be divided into five laterals (lat) with the outermost dominating; beyond comes a vast array of needle-like marginals (mrt) the most median of which is sufficiently differentiated to be counted as unlike the rest. The half rows on one side alternate with those on the other. A rhipidoglossate radula is met with in the Zeugobranchia, Trochacea and Neritacea. It is in all probability the most primitive type of gastropod radula.

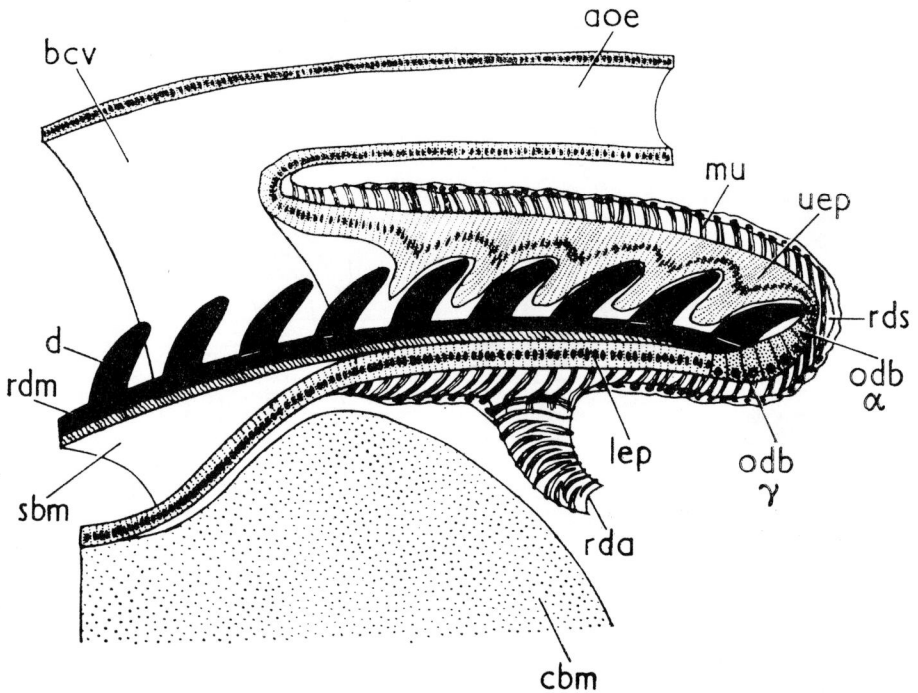

FIG. 105.—Diagrammatic longitudinal section of the radular sac and of the radula with related parts of the alimentary canal.

aoe, anterior oesophagus; bcv, buccal cavity; cbm, cartilage of buccal mass; d, radular tooth; lep, lower epithelium of radular sac; mu, muscles in wall of radular sinus; odbα, α odontoblasts; odbγ, γ odontoblasts; rda, radular artery; rdm, radular membrane; rds, radular sinus; sbm, subradular membrane; uep, upper epithelium of radular sac.

2. Docoglossan (from Greek *dokos,* a beam or spear) (fig. 104B, C): formula

$$3 + D + 2 + R + 2 + D + 3.$$

R may be absent (Acmaeidae) and is usually small and unpigmented (rh) and therefore not easily seen in those families in which it occurs. The marginals (mrt) are also uncoloured except for the dominant (plu) which is pigmented black like the laterals (lat). The teeth are unicuspid except the dominant which is pluricuspid, possibly due to fusion, though this is not evident in the Lepetidae. This type of radula occurs in the Patellacea.

3. Taenioglossan (from Greek *tainia,* a band) (fig. 104D): formula

$$3 + R + 3 \quad \text{or} \quad 2 + 1 + R + 1 + 2.$$

The teeth which lie on either side of the rachidian (rh) are usually distinguishable into two marginal (mrt) and one lateral (lat) on each side, but it is very doubtful whether this division corresponds to the similar one in rhipidoglossate types: it is more likely that the teeth known here as laterals and marginals correspond only to the lateral teeth of the rhipidoglossate forms. On this interpretation the teeth corresponding to the marginal teeth of the rhipido-glossate radula are regarded as not present at all. It may be, however, that fusion of the marginal teeth has occurred to give a reduced number. Radulae of taenioglossate pattern occur in all the families of monotocardian prosobranchs which are placed by Thiele in the group Mesogastropoda, except for those which have become secondarily devoid of a radula, and the few families which are ptenoglossan.

4. Ptenoglossan (from Greek *ptenos*, feathered) (fig. 100): formula

$$n + 0 + n.$$

This type of radula is clearly modified from a taenioglossate type and is the main support for the supposition that the few lateral teeth of that pattern have been evolved by a fusion of the more numerous teeth of an ancestral rhipidoglossan type. All the teeth are of similar, rather simple, hook-like shape and give rise to a very efficient grasping organ. This type of radula occurs in Ianthinidae and Scalidae amongst British prosobranchs.

5. Rachiglossan (from Greek *rhachis*, a spine) (fig. 104E): formula

$$1 + R + 1.$$

This narrow type of radula is found in the Muricacea and Buccinacea amongst British members of the Stenoglossa.

6. Toxoglossan (from Greek *toxon*, a bow or *toxotes*, an archer) (fig. 104F, G): formula

$$D + 0 + D \text{ or } 1 + 0 + 0 + 0 + 1.$$

The 2 teeth in each row alternate and are in use only one at a time. They are of complex structure [see p. 614].

These six types of radulae show a gradual evolution from the polyodont rhipidoglossan, with an extremely large number of teeth in each row, to a number of patterns with few teeth per row (oligodont) reaching as a climax the toxoglossan where only a half row, containing a single tooth, is in use at any particular time. This evolutionary change, it appears, has taken place at least twice, since a study of the functioning of the radula shows that whilst it is possible to put the rhipidoglossan—taenioglossan—rachiglossan—toxoglossan radulae together, the docoglossan type differs fundamentally from these and must be looked upon as a separate line of change [see p. 608]. The ptenoglossan type would appear as a regression from the taenioglossan towards the ancestral rhipidoglossan type, at least in part, whilst the rhipidoglossan radula is itself reminiscent of the broad ribbon of little specialized teeth which is to be found in the buccal cavity of chitons and of the basal families of the tectibranchs (e.g. Acteonidae).

The shape of the radular teeth is directly related to the kind of food which the animals eat and the way in which it is manipulated, just as the dentition of a mammal and its food and feeding habits are correlated. The length of the radula appears also to be related to the amount of work which has to be done in feeding: where wear is extensive a long radula is present, where wear is less (because softer food is eaten) the length of the radula is much reduced. In the latter case the radular sac forms only a small bulge behind the muscles of the buccal mass; in the former case it may loop backwards and forwards, as it does in *Patella*, amongst the viscera, or be coiled in a spirally wound heap, as in *Littorina*. Some figures of the length of the radula are given in Table 3.

TABLE 3

Animal	Radula length / Shell length	Radula / $\sqrt[3]{}$ Shell volume	Author
Patella vulgata	1·30-2·30 Mean 1·75	—	Fischer-Piette, 1935*b*
Patella vulgata	1·13-2·00 Mean 1·51	—	Eslick, 1940
Patella vulgata	1·20-2·20 Mean 1·75	—	Fischer-Piette, 1941a
Patella vulgata	1·40-2·00 Mean 1·60	3·20-4·80 Mean 3·90	Fischer-Piette, 1948
Patella vulgata	1·80	—	Pelseneer, 1935
Patella vulgata	1·69	—	Evans, 1953
Patella aspera	0·95-1·40 Mean 1·15	—	Fischer-Piette, 1935*b*
Patella aspera	0·93-1·25 Mean 1·05	—	Eslick, 1940
Patella aspera	0·80-1·40 Mean 1·15	—	Fischer-Piette, 1941a
Patella aspera	1·00-1·10 Mean 1·05	2·50-3·50 Mean 2·90	Fischer-Piette, 1948
Patella aspera	1·11	—	Evans, 1953
Patella intermedia	1·60-2·50 Mean 2·10	—	Fischer-Piette, 1935*b*
Patella intermedia	1·60-2·70 Mean 2·10	—	Fischer-Piette, 1941a
Patella intermedia	1·40-2·00 Mean 1·60	4·10-5·70 Mean 4·71	Fischer-Piette, 1948
Patella intermedia	1·88	—	Evans, 1953
Gibbula cineraria	1·31	—	Pelseneer, 1935
Littorina littorea	2·00	—	Pelseneer, 1935
Littorina saxatilis	2·33	—	Pelseneer, 1935
Buccinum undatum	1·00	—	Pelseneer, 1935
Nucella lapillus	0·30	—	Pelseneer, 1935

In some exotic prosobranchs the radula may be still longer, e.g. in *Tectarius pagoda* it is said to be not less than seven times as long as the animal's shell. Although the length of the radula has always been assumed, since the days of Simroth (1896–1907), to be correlated with the degree of hard usage to which it is put, no proof of this has ever been given, nor is there any clear idea of how the effect would be brought about. Peile (1937) has certainly shown that the effects of wear are visible on radular teeth, but no one has convincingly shown for any one species that differences in the food, or in the substratum from which the food is collected, make differences to the length of the radula.

In the young animal the radula appears as a secretion from the inner end of the radular sac, a finger-shaped caecum which grows out from the posterior end of the buccal cavity and is usually slightly bifid at its tip. Its floor opens on to the dorsal surface of the odontophore, its roof turns over to the floor of the oesophagus. Both are lined by a columnar epithelium which presents the same histological appearance as that into which it is continued in the rest of the gut: that is to say, the upper epithelium has no special characteristics, but the lower secretes a cuticle like that overlying the buccal mass. This is known (for reasons which will appear later) as the subradular or elastic membrane. The upper wall of the sac becomes folded inwards so as to bulge like a typhlosole into its cavity and, except near its mouth, the lumen of the sac is wholly or almost wholly obliterated by this, so that it is shaped like a broad U in transverse section. The gutter which runs externally along the roof of the sac is filled in with a characteristic spongy connective tissue and the whole structure is surrounded by a blood sinus (rds, fig. 105) connected to the cephalic branch of the anterior aorta. The outer walls of this are muscular and various muscle strands originating in the walls of the head, or the columellar muscle or the buccal mass are inserted upon it. This outer wall is the radular sheath.

The copious blood supply and the connective tissue rich in stored nutritive materials seem to be related to the fact that throughout the lifetime of a prosobranch a continuous secretion of radular teeth is occurring in which the greater part of the walls of the sac would appear to be involved, only the tissues nearest to the buccal cavity not actively helping in this process. The cells in the sac which are responsible for the initiation of tooth production are the odontoblasts, which lie at the innermost end of the radular sac, rather more ventrally than dorsally, so that the terminal wall of the radular sac is clothed by a backward extension of the same epithelium as covers the roof. In any one animal these odontoblasts are of various size, the cells known as α and β cells, which are most posterior in position, being small and the γ cells, the biggest, being more anterior (fig. 105, β cells not labelled). They also vary in size and number from group to group, there being many small odontoblasts in prosobranch gastropods, but only a small group of relatively giant cells (somewhat unnecessarily distinguished as odontophytes) in pulmonates and opisthobranchs: for this reason almost all work upon the secretion of radular teeth has been carried out on the latter two groups, although there is no reason to suppose that the process is any different from that occurring in prosobranchs. The odontoblasts appear to be arranged in groups each of which is responsible for the production of one tooth.

There have long been two views as to the way in which the odontoblasts produce a radular tooth, which parallel the two views which have been held regarding the manufacture of the molluscan shell. Trinchese (1878) and Pruvot-Fol (1925, 1926) have both assumed that each group of odontoblasts is directly transformed into a tooth and thereby extinguished as a productive unit. The next tooth in that series must, therefore, be due to the activity of a new group of odontoblasts, as will every successive one. This continual production of new nests of odontoblasts they see in a periodic transformation of cells from the upper epithelium of the radular sac which is assumed to migrate over the terminal wall of the sac and change rhythmically into odontoblasts, the rhythm coinciding with that of tooth production. Most investigators of this subject, however, have regarded the formation of a radular tooth as merely an exaggerated local secretion of cuticular material such as goes on to a lesser extent in many other sites in the body of the animal. These investigators still differ among themselves, nevertheless, on the matter of the renewal of odontoblasts; Rössler (1885), Beck (1912),

Spek (1921) and Prenant (1925) believed that each group of odontoblasts persists throughout the life of the mollusc and produces all the teeth; Rücker (1883), Rottmann (1901), Schnabel (1903) and Sollas (1907) believed that a group of odontoblasts is exhausted by the secretion of one tooth, as did Pruvot-Fol (1925), and therefore believed in their rhythmical replacement as described above. Bloch (1896) also believed in their replacement, but not in the same regular way after the secretion of each tooth. Although we have no personal experience of this problem this seems to us the less likely happening.

The secretory process may, perhaps, resemble that which produces chaetae in annelids. This has been described in a number of genera by Schepotieff (1903, 1904), in *Nereis* by Pruvot (1913), *Myzostomum* by Jägersten (1937), *Sabellaria* by Ebling (1945) and, in oligochaets, by Bourne (1894) and Vandebroek (1936). The most interesting of these studies is by Ebling, who showed how the secretion of a chaeta is carried out within the chaetal sac primarily by the activity of one cell lying at its base. To the primary core produced by this cell the other cells on the opposite wall of the sac add a secondary, superficial layer so that the final chaeta has a double structure. As the chaetae mature they migrate up the chaetal sac along its ventral wall, and since their primary cells travel with them it is clear that there must be a continuous production of new chaetoblasts at the base of the sac, apparently from the same cell population on the dorsal wall as gives rise to the cells adding the secondary material to the primary core of the chaetae.

The resemblances between what happens here and what may be happening in the gastropod are very close and it may be that in the mollusc, as in the annelid, a migration of cells from the dorsal wall of the sac in one direction keeps up the supply of odontoblasts for the manufacture of successive new teeth, whilst the same cells, moving in the opposite sense, lie over the radular teeth and add secondary material to them as they migrate from the radular sac into the buccal cavity.

The secretion of a single tooth has been described for several euthyneurous gastropods (fig. 105). It involves a group of odontoblasts (odb) lying ventrally at the inner end of the radular sac. These decrease in size from the large γ cells anteriorly to the small α cells posteriorly. This gradient of size coincides with gradient in the amount of secretion produced, the α cells producing minimal quantities, the γ cells maximal. As a result of this a wedge-shaped mass of secretion accumulates over the odontoblasts, its tip pointing posteriorly and overlying the α cells. A further difference between the large γ cells on the one hand and the smaller β and α cells on the other is that secretion is continuous in the former but periodic in the latter. When the secretory phase has finished in the α and β cells the continued secretion of material by the γ cells causes the material lying over the α and β cells to be pulled away from their surface, and when the next phase of secretion occurs in these cells a discrete mass of secretion is formed which is not connected to the mass previously produced in that situation. The γ cells, however, secrete continuously and the sum of the activity of all the cells is a ribbon of material (the product of the γ cells) from which, at regular intervals, rise up the tapered blocks which have come from the α and β cells. The continuous ribbon is called the radular membrane (rdm), the blocks (from α and β cell) are the teeth (d). Jones, McCance & Shackleton (1935) have shown that the body of the radula in *Patella* is made partly of a protein and partly of a polysaccharide strengthened by salts of iron and silicon. Although the salts may vary from animal to animal the body of the radula probably remains substantially the same.

With successive waves of tooth production the teeth attached to the radular membrane are inevitably pushed forward on the floor of the radular sac away from the site of secretion.

In this position they meet and over-ride the cuticle which lies there and which is an extension of that covering the buccal mass, so that the two cuticular layers, one above (the radular membrane), the other below (the cuticle) are in contact, with the upper appearing to slip forward over the lower. This arrangement persists so long as the radula grows, and because of its position the lower cuticular layer is, in this situation, called the subradular membrane (sbm). Amaudrut (1898) called it the elastic membrane. Like so many cuticular secretions in gastropods it looks fibrous, this structure being apparently due to the escape of the secretion in thread-like masses from the upper surface of the cell.

This is the classical interpretation of the fact that the membrane to which the radular teeth are attached shows a distinct double layering with different staining properties. Recent work by Runham (personal communication), however, suggests that there is only a single cuticle the upper part of which is tanned where the radula lies; the source of the tanning material is probably the overlying epithelium. This interpretation would imply that not only the radular teeth and the 'radular' membrane, but also the whole of the 'subradular' membrane would be gradually moving forwards from the posterior towards the anterior end of the odontophore.

The teeth recently produced by the activity of the odontoblasts do not present the same appearance to the eye as do those which lie nearer the buccal cavity, and it is clear that between their initial formation and their appearance on the buccal mass they change. This alteration, it is agreed by all workers on the subject except Rottmann (1901) and Schnabel (1903), is due to the activity of the cells forming the epithelium of the dorsal wall of the radular sac (uep). After the teeth have broken away from the groups of odontoblasts and have moved forward in the radular sac, their upper surface is in contact with the epithelium on the roof of the radular sac. The activity of these cells adds to the composition of the teeth and is responsible for the final form which they assume and their ultimate chemical constitution and physical consistency (Prenant, 1924a). It is at this stage, probably, that are added the salts of iron and silicon shown by Jones, McCance & Shackleton (1935) to be responsible for much of the hardness of the teeth in Patella. Although this much may be said about the secretion of the gastropod radula, many details of the process remain unknown and it is clear that this is a problem which urgently calls for proper histochemical investigation.

When the radula is first formed it inevitably lies within the radular sac. As the animal grows it is found to lie over the dorsal surface of the odontophore. A continuous secretion of new teeth occurs at the inner end of the radular sac: those teeth which lie at the tip of the odontophore show signs of wear and can be seen to be torn off during feeding. Small individuals have a small odontophore and would seem to require small-sized teeth, whilst fully grown animals have a larger odontophore and, presumably, larger teeth. All these things would point to a steady forward movement of the radula out of the sac on to the odonto-phore during the lifetime of the mollusc, and to a gradual increase in the size of the teeth present on it. That an increase in the size of the teeth with age occurs is undoubted: it can be seen by looking at a preparation of the radula of the cephalopod Eledone, but in most prosobranchs (e.g. Patella) the teeth have to be carefully measured before this is detected, even in animals with a long radula of many rows. Nevertheless in a young Patella, Pruvot-Fol (1926) recorded that the teeth are only one-seventh the length of the teeth of a typical adult: from this it may be deduced that between youth and maturity a very large number of rows of teeth are used, broken off and replaced by others which have been formed within the radular sac.

The question therefore arises as to how these teeth move out of the radular sac on to the odontophore, and this is a question to which it must be admitted there is no generally agreed answer. A number of causes have been supposed to bring it about. First is a slipping forwards of the radular membrane, plus the teeth it bears. This may be the way in which the movement takes place but it can hardly be supposed to occur spontaneously: some pull from the front or push from behind must set it in motion. Both these things have, in fact, been imagined as acting: the forces exerted on the front of the radula as the animal feeds have been suggested by Rücker (1883) and others as pulling the radula forwards over the odontophore and out of the radular sac, and a variety of other muscular tractions on the radula in its sac have been supposed to act in the same way. Similarly the secretion of new teeth at the inner end of the radula will push those already formed forwards. To what extent this force exists is not known, but Hoffmann (1932) has suggested that its effect will be felt no further forward than the innermost 3 rows of teeth. Migration of the radular teeth forwards out of the radular sac on to the odontophore would seem to involve a gliding or slipping of the teeth on the radular membrane over the underlying subradular membrane, or of teeth and membrane together over the epithelium, and although measurements such as those given by Pruvot-Fol (1926) for *Patella* (mentioned above) would seem to have this as a necessary consequence, no proof of this movement has ever been obtained.

One further factor is undoubtedly involved in the change of position of the radular teeth—the relative rates of growth in the buccal cavity and radular sac. If the buccal cavity extends backwards more rapidly than the radular sac grows, then the posterior wall of the buccal cavity will uncover a stretch of radula previously enclosed within the sac, and so produce an apparent forward movement. This is known to occur (Pruvot-Fol, 1926), but it must be confessed that to say that the forward movement of the radula is due to such an extent to this factor and to such an extent to another is an impossible task.

The radular teeth may be immovably fixed to the membrane of which they form part, or they may be movably articulated with it. The former condition is characteristic of all rachidian teeth and of all teeth in a docoglossate radula: the latter applies to the lateral and marginal teeth of the other types. It is important in connexion with the use of these teeth in feeding, and will be described in the section dealing with that (p. 179). The teeth are also adapted in shape and size for the use to which they are put by their owner and this too will be best illustrated when their use is considered.

Before this can be done, however, attention must be turned to the odontophore, which is an elaborate apparatus of great complexity evolved for the manipulation of the radula. It is obviously composed of a right and a left half connected across the mid-line where, on its dorsal side, runs a groove in which lies the radula. The whole of this, and the bulk of the rest of the surface is covered with a thick and shining cuticle, though this is thinner towards the ventral side, where it may disappear and be replaced by glandular epithelium. Internally the odontophore is usually supported by 'cartilages', pieces of tough, resilient material which earn this name because of the resemblance which they exhibit to vertebrate cartilage from the histological point of view. The number of cartilages varies from 5 pairs (*Patella*) to 2 (most prosobranchs) or even a single bilobed piece (*Ianthina*). They are wrapped in layers of tough connective tissue and not only contribute towards the maintenance of the shape of the odontophore—which they are able to do because of the turgor of the cells out of which they are composed, like a notochord—but also give origin to a great number of muscles which are concerned with the movement of the odontophore and of the radula upon it.

The histology of these has been described most carefully by Nowikoff (1912) in *Patella caerulea*, *Diodora apertura* and *Haliotis tuberculata*. In the two last genera and in the posterior cartilages of *Patella*—as indeed in the majority of prosobranchs—the cartilages are made of highly vacuolated cells containing a peripherally placed nucleus and a few granules. Each cell secretes matrix which is frequently fibrous. The cells lying centrally in the cartilage mass are polyhedral and large, those lying peripherally are small and flattened and merge into the perichondrium, a dense mass of fibrous material. In the anterior cartilages of *Patella* the cells are less vacuolated, richer in cytoplasm and granules and the matrix is alveolar: these differences may be related to differences in the stresses to which the cartilages are exposed.

The muscles of the buccal mass may be arranged in series according to the functions which they subserve: there are protractors, which will project the odontophore towards, or through, the mouth; retractors, which will bring about the reverse effect; intrinsic muscles, which affect the shape; and a series of tensor muscles which have the effect of immobilizing it in any particular position. In addition to these muscles a further series of radular muscles is attached to the radula and to the radular sheath.

Some account of the musculature of the buccal mass has usually been given by most investigators of different forms of prosobranchs, though few of these except Amaudrut (1898) have been particularly thorough. More recently a small number of more complete investigations has been made, especially by Carriker (1943) for *Urosalpinx,* and by Nisbet (1953) for *Monodonta.* The second of these is particularly important in that it approaches the problem from a physiological as well as an anatomical viewpoint and so gives a more complete picture of the functioning of the buccal mass than does the work of any other author. The muscles appear cross-striated but this may perhaps be another case of 'double striation' or helical arrangement of myofibrils rather than the true striation of vertebrates (Hanson & Lowy, 1957). They also contain haemoglobin (Lankester, 1872) and for this reason often appear as a red mass shining through the body wall, and Ball & Meyerhof (1940) have shown in *Busycon* the presence of a complete cytochrome system. Fänge & Mattisson (1958) have described a high concentration of mitochondria in the buccal muscles of *Buccinum undatum* and have shown that the oxygen consumption is very high and that their respiratory enzymes are not easily poisoned by cyanide or carbon monoxide. These characteristics would permit a high level of activity even if conditions are not particularly favourable.

CHAPTER 8

T HE musculature which is associated with the movements of the buccal mass and radula of *Monodonta* is complex and Nisbet (1953) distinguished no less than 33 different muscles. In addition some strands of stout connective tissue, which he called tendons, are involved. These immobilize certain areas. Three are important: the mid-ventrally placed nuchal tendon (nt, figs. 106, 107) which runs from the inner end of the sublingual pouch to the musculature of the foot (f), and the dorsal buccal tendon (dbt) on each side, which runs from the base of each jaw outwards to the musculature of the side of the head. As a result of the presence of these tendons the three points from which they run remain more or less unchanged during the various feeding movements. In addition a pair of ventrolateral buccal tendons steadies the lateral buccal areas.

The feeding process may be divided into a cycle of operations each of which involves the primary activity of a group of muscles and the synergic action of many others. The different phases of the cycle (which has also been described for the related *Gibbula cineraria* by Ankel (1938a) and his pupil Eigenbrodt (1941)) were given by Nisbet as follows: (1) the opening of the mouth, along with which occurs an uncovering of the cuticularized edges of the two jaws; (2) the protraction of the odontophore and, simultaneously, a backward movement of the subradular membrane over the tip and sides of the anterior end of the odontophore; (3) the opening of the radula; (4) the closing of the radula; (5) the retraction of the odontophore; and (6) the shutting of the mouth. These actions— which are not separated in performance by the animal but flow together and overlap in one continuous movement—may now be investigated one by one.

(1) *Opening of the mouth* (fig. 107A, B, C). The tip of the snout of *Monodonta* shows, at rest, a circular surface crossed by a vertical cleft which is the closed mouth, lying between the lateral lips. Diagonal extensions from the mouth dorsally mark off a dorsal lip (dlp) or pre-mandibular fold and similar extensions ventrally mark off a ventral lip (vlp). These are mobile and move during the opening of the mouth, but the outer rim of the circular end of the snout is not involved in this process. When the mouth opens the dorsal and lateral lips move swiftly, the ventral lip less so. These movements are produced by synchronous contraction of (i) the mandibular protractor, (ii) the mandibular retractor and (iii) the inner ventral buccal protractor muscles.

The mandibular protractor muscle (mpm) has its origin primarily in the oral rim of the snout dorsally, but secondarily in the dorsal lip (dlp). From these places it runs to be inserted on the dorsal (anterior) surface of the jaw (ja) towards its base. Its contraction will either protract the jaw (if the retractor of the jaw synchronously relaxes) or will retract the dorsal lip (if the jaw is immobilized by simultaneous contraction of its retractor).

The mandibular retractor muscle (mr) runs from an origin in the dorsal wall of the snout to the dorsal (anterior) surface of the jaw, being inserted near the lower free edge. Some of its

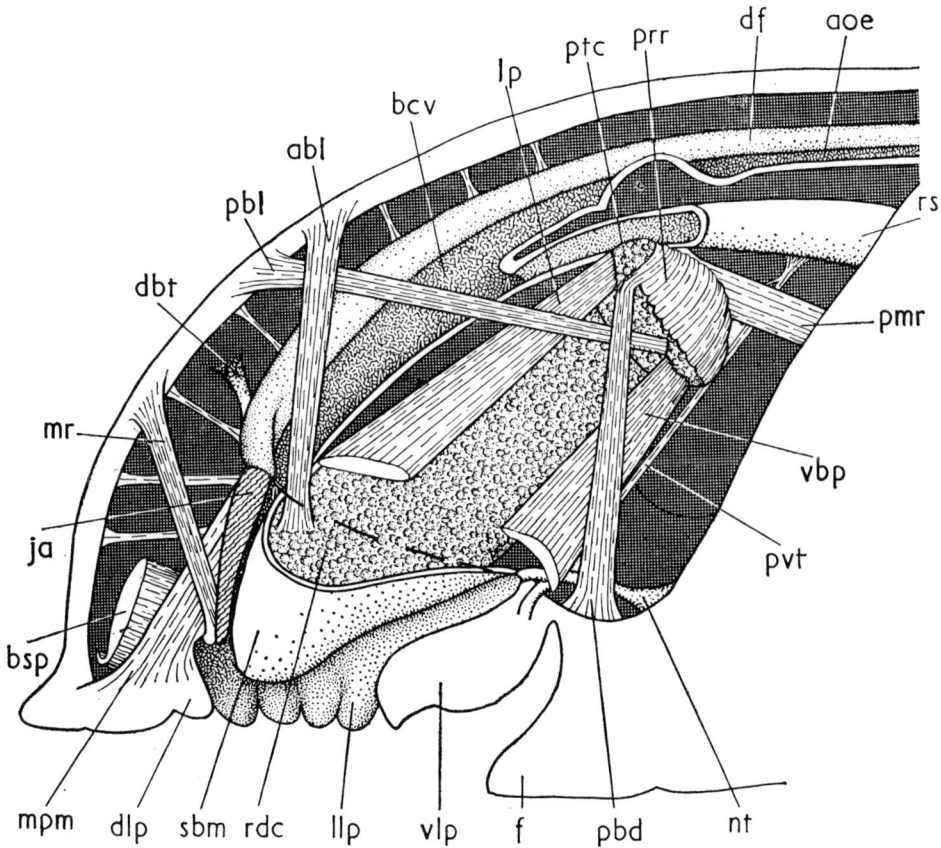

FIG. 106.—*Monodonta lineata*: diagram of lateral dissection of the anterior part of the head and foot, showing the anterior part of the alimentary tract and some of the muscles associated with it. Radular teeth are omitted. All cut surfaces are left white; the haemocoel is cross-hatched; the jaw is diagonally hatched and the cavity of the gut stippled. After Nisbet.

abl, anterior buccal levator muscle; aoe, anterior oesophagus; bcv, buccal cavity; bsp, buccal sphincter muscle; dbt, dorsal buccal tendon; df, dorsal fold; dlp, dorsal lip; f, foot; ja, jaw; lp, lateral protractor muscle; llp, lateral lip; mpm, mandibular protractor muscle; mr, mandibular retractor muscle; nt, nuchal tendon; pbd, posterior buccal depressor muscle; pbl, posterior buccal levator muscle; pmr, postmedian retractor muscle of the radula; prr, posterior retractor muscle of the radula; ptc, posterior radular cartilage; pvt, posterior ventral radular tensor muscle; rdc, anterior radular cartilage; rs, radular sac; sbm, subradular membrane; vbp, ventral buccal protractor muscle; vlp, ventral lip.

fibres run to the more fleshy parts of the jaw. When it contracts it pulls these away so as to expose the edge and (if the mandibular protractor simultaneously relaxes) will retract the jaw. The simultaneous contraction of mandibular protractor and retractor which occurs at this stage of the feeding cycle means that the jaw is steadied and the dorsal lips withdrawn.

The inner ventral buccal protractors (ivp) run from the lateral and ventral lips dorsally and posteriorly along the underside of the odontophore on each side, and end by being attached to the ventral end of each posterior cartilage. In so far as the opening of the mouth is affected by these muscles it is the attachment to the cartilage which acts as the origin of the muscle and its endings within the lips which behave as its insertion.

(2) *Protraction of the odontophore and of the subradular membrane* (fig. 107C, D). These actions are occurring whilst the mouth is being opened: their separation here is purely to simplify description. If an animal feeding on a glass plate be watched it will be seen that the radula appears in the mouth before that is fully opened, having been brought to this position by a forward movement of the odontophore within the buccal cavity. This movement continues until the tip of the odontophore, and the radula which lies over it, are pressed against the substratum. As this happens the radula opens and is in the open state when actually brought into contact with the substratum.

The forward movement of the odontophore will be discussed in this section and the factors which open the radula in the next.

The forward movement of the odontophore is achieved by means of the lateral and ventral buccal protractor muscles, and the posterior buccal levator and depressor muscles.

The lateral protractors (lp) have each a triple origin and are inserted on the dorsal third of each posterior cartilage (ptc). The most dorsal of the 3 sections into which this muscle is divisible runs forward from the cartilage into the lateral wall of the buccal cavity, passing obliquely round that to its dorsal side. There it spreads forward and is attached to the wall of the head (fig. 107B). The second section lies ventral to the first and its fibres, running forwards in the lateral wall of the buccal cavity, join those of the outer buccal constrictor muscle (obi, see below) which are also running there (fig. 107A). The third section, still more ventral and deeper, joins the inner buccal constrictor (ibm, see below) in exactly similar fashion (fig. 107E). These points are all relatively fixed, with the result that when the protractor contracts the posterior cartilage is pulled forward, and, as it is cupped against the posterior surface of the anterior cartilage, and joined to it by groups of muscles, this forward pull is transmitted to the anterior cartilage and the whole mass moves forward.

The same effect is achieved by the simultaneous contraction of the ventral buccal protractor (vbp, fig. 106). One slip of this has already been described in connexion with the withdrawal of the lateral and ventral lips during the opening of the mouth; the bulk of the muscle, however, helps to protract the odontophore. The fibres of this major part are inserted like those of its smaller section, on the posterior cartilage (ptc) ventrally. From there they run forward and fan out to give a number of slips which end on the walls of the head, laterally and ventrally.

The posterior buccal levator muscle (pbl) lies on each side. It has its origin dorsolaterally on the wall of the head and from there passes round the side of the buccal cavity to be inserted on the ventral part of the posterior cartilage (ptc). Since its course is obliquely backwards and downwards it not only raises the posterior end of the odontophore but has a protractor effect. The posterior buccal depressor (pbd) is, likewise, a paired muscle which originates in the musculature of the ventral body wall in the neck region and runs thence dorsally to the posterior cartilage. Some fibres are inserted on the cartilage but most join the posterior retractor muscle of the radula (prr). Like the levator this muscle has a protractor action in addition to its depressor one and simultaneous contraction of both levator and depressor will cause a moderate degree of protraction alone.

The simultaneous contraction of the ventral and lateral protractors and the posterior levators and depressors will produce a drive of the odontophore forwards and downwards

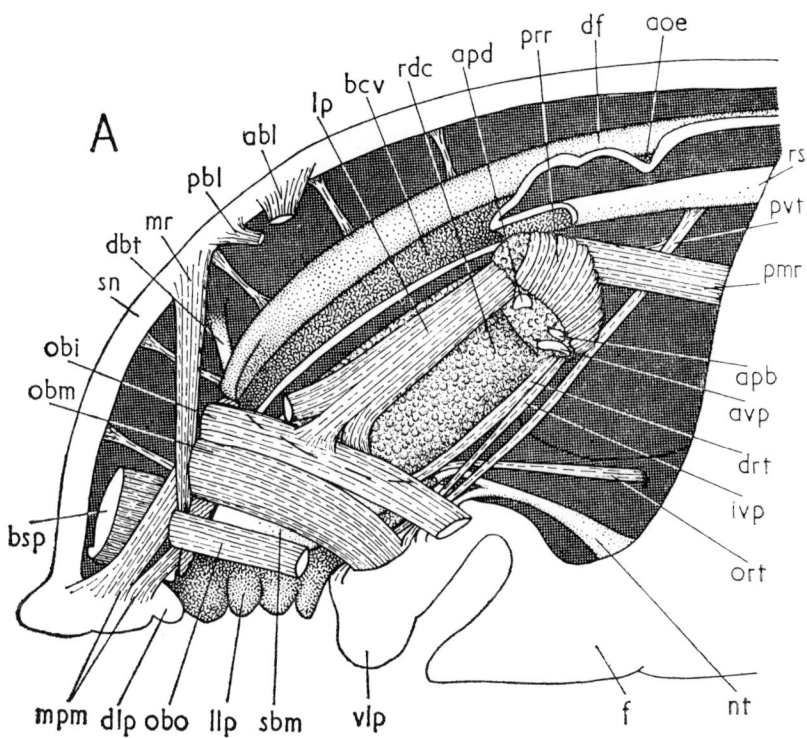

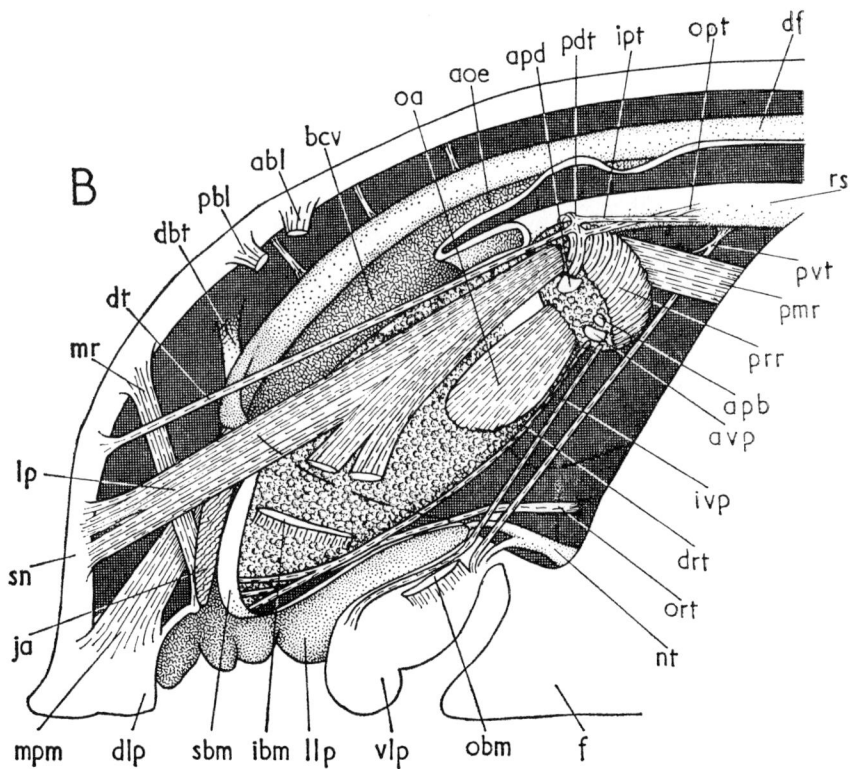

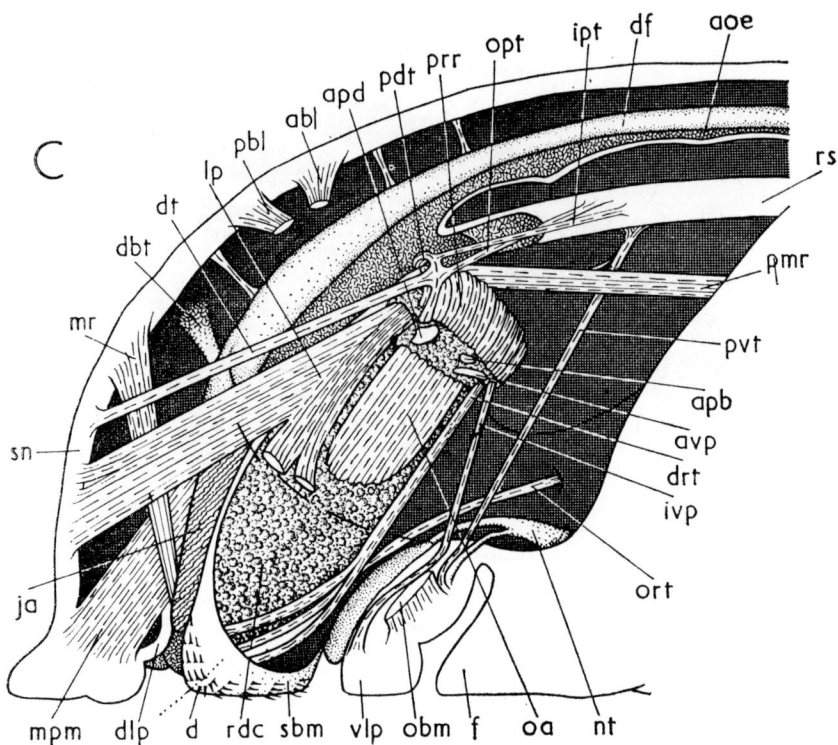

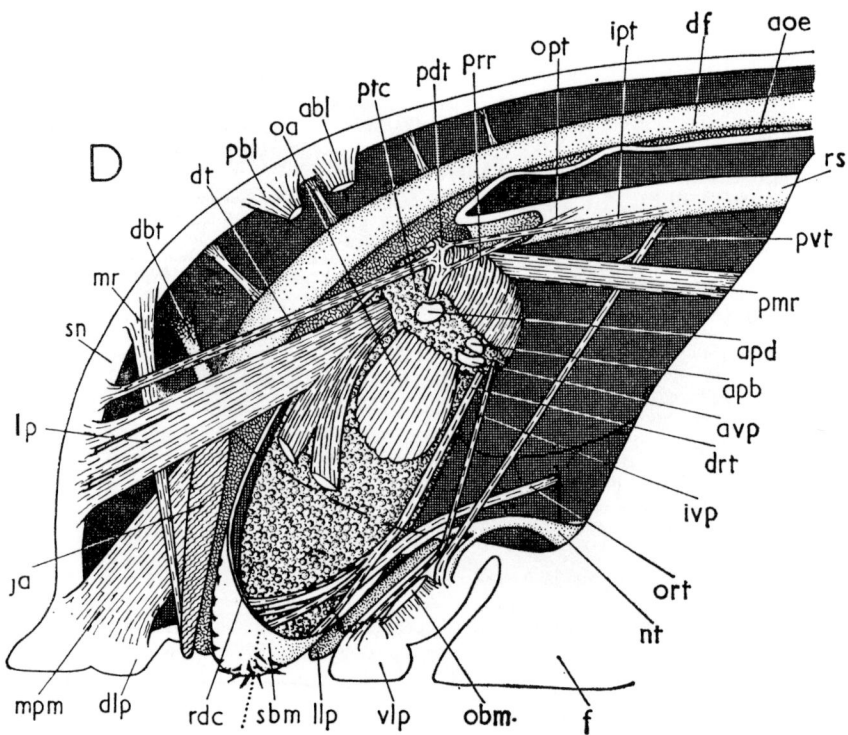

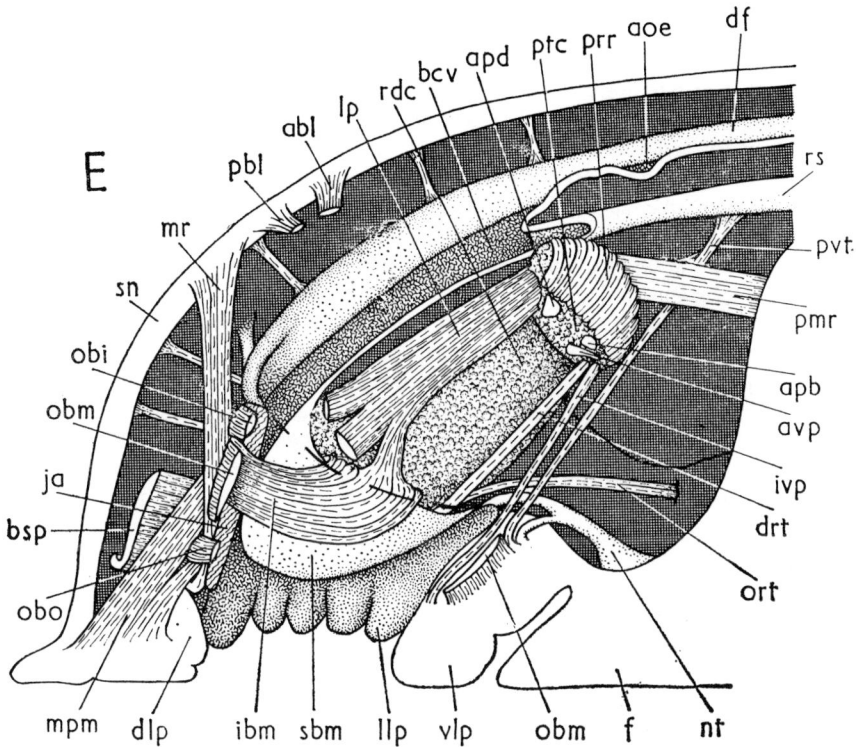

FIG. 107.—*Monodonta lineata:* a series of diagrams of lateral dissections of the anterior part of the head and foot showing the cycle of operations in the feeding process. A, the mouth is shut and the odontophore retracted in the position of rest. B, the mouth is opening, the dorsal lip is withdrawn and the odontophore is beginning to protract. C, the mouth is open, the dorsal lip withdrawn, the odontophore is protracted and the subradular membrane pulled forwards; note the contact between the jaws and the dorsal surface of the buccal mass, and the different position taken up by the radular teeth according to whether they lie outside the bending plane (to the right of the dotted line) or within it (to its left). D, the mouth is open, the dorsal lip withdrawn, and the radula has been closed by the backward movement of the anterior radular cartilages and therefore of the bending plane (marked by the dotted line). E, the mouth is open, the dorsal lip protracted and the odontophore is almost fully retracted. Radular teeth are shown diagrammatically in C and D only. All cut surfaces have been left white, the haemocoelic cavities are cross-hatched, the jaw diagonally hatched and the cavity of the gut and the buccal tendons stippled. The broken line stretching across the odontophore from the dorsal end of the jaw to the inner end of the subradular part of the buccal cavity marks the boundary of the buccal cavity laterally and is a fixed line maintained by the dorsal buccal and nuchal tendons. Based on Nisbet.

abl, anterior buccal levator muscle, cut; aoe, anterior oesophagus; apb, attachment of posterior buccal levator muscle to posterior cartilage; apd, attachment of posterior buccal depressor muscle to posterior cartilage; avp, attachment of ventral protractor muscle to posterior cartilage; bcv, buccal cavity; bsp, buccal sphincter muscle; d, radular tooth just external to the bending plane; dbt, dorsal buccal tendon; df, dorsal fold; dlp, dorsal lip; drt, direct radular tensor muscle; dt, dorsal buccal tensor muscle; f, foot; ibm, inner buccal constrictor muscle; ipt, inner posterior tensor muscle of the radula; ivp, inner ventral buccal protractor muscle; ja, jaw; lp, lateral protractor muscle; llp, lateral lip; mpm, mandibular protractor muscle; mr, mandibular retractor muscle; nt, nuchal tendon; oa, outer approximator muscle of the cartilages; obi, inner part of outer buccal constrictor muscle; obm, median part of outer buccal constrictor muscle; obo, outer part of outer buccal constrictor muscle; opt, outer posterior tensor muscle of the radula; ort, oblique radular tensor muscle; pbl, posterior buccal levator muscle, cut; pdt, posterodorsal tensor muscle of the cartilages; pmr, postmedian retractor muscle of the radula; prr, posterior retractor muscle of radula; ptc, posterior radular cartilage; pvt, posterior ventral radular tensor muscle; rdc, anterior radular cartilage; rs, radular sac; sbm, subradular membrane; sn, wall of snout; vlp, ventral lip.

which brings its tip up to and through the mouth, and with it will be brought the radula and subradular membrane. The odontophore, however, is not like a piston, wholly unconnected to the walls of the cylinder within which it moves, but at certain points the subradular membrane, which forms its outer covering, is reflected on to the walls of the buccal cavity. This occurs, in particular, at the inner end of the sublingual pouch underneath the odonto-phore, and at the inner ends of the lateral buccal pouches at the sides of the radula. These lines are kept more or less fixed in position by the tendons running to the body wall. As a consequence of this the forward movement of the cartilages within the odontophore causes the subradular membrane to be pulled backwards and downwards to a certain extent because of its connexions with the buccal epithelium at these places. The movement of the subradular membrane is possible because it is invaginated within a deep groove in the mid-dorsal line of the odontophore where it bears the radula, and this invagination is pulled out and the groove obliterated. The effects which this movement of the subradular membrane has upon the radula are dealt with in the next section. In addition to the purely mechanical effects of the tendons in causing this movement, it is actively promoted by the contraction of muscles, the most important of which are the tensor muscles shown in fig. 107B, C, D. The direct tensors (drt) are relatively thin bands originating in the anterior surface of the posterior cartilage close to the origin of the ventral protractors (avp). Each passes forwards ventrally towards the tip of the anterior cartilage (rdc) where it splits into two or three bands which curve dorsally round the front of the cartilage and are inserted dorsally, in a somewhat recurved way, on to the subradular membrane (sbm). The oblique tensors (ort) are inserted alongside the direct ones and the first part of their course is almost identical; instead of originating on the posterior cartilage, however, these muscles are tied to the columellar muscle, of which, indeed, they are best regarded as slips.

Whilst this anterior movement of the odontophore and the simultaneous movement of the subradular membrane are occurring a further action takes place which has the effect of pressing the jaw against the dorsal side of the odontophore as it moves forward within the buccal cavity. This presses the jaw against the groove in which the radula lies and so appears to antagonize the effect of the tendons and tensor muscles in pulling this open: it at least prevents it from being effective until a level more anterior than the tip of the jaw has been reached. This pressure of the jaw against the advancing odonto-phore is produced by the contraction of the outer and inner buccal constrictor muscles (fig. 107A, E).

The outer buccal constrictor muscle is in reality triple, the three parts being placed successively behind one another. The outermost (obo) has its insertion in connective tissue between the 2 jaws mid-dorsally and thence curves round the wall of the buccal cavity in a ventral direction, ending on the lateral borders of the mouth. The innermost of the three (obi) begins in the same area of connective tissue, but overlies the base and not the apex of the jaw, and then curves ventrally in a similar fashion. With its fibres are amalgamated those of the middle part of the lateral protractor (lp) muscle (fig. 107A). The middle section of the outer buccal constrictor (obm) is unpaired and runs sphincter-wise round the wall of the buccal cavity at about the level of the middle of the jaw. It is not a complete ring of muscle, having a gap mid-ventrally. On contraction these muscles pull the jaw backwards.

The inner buccal constrictor muscle (ibm, fig. 107E) has a similar course to the outer but lies at a deeper level. Each muscle is fastened to the cartilaginous material of which the jaw is made and then runs round the lateral wall of the buccal cavity to the point where that joins the side of the odontophore. It then bends sharply forwards under the subradular membrane and ends on the outer surface of the anterior cartilage. To it are joined the fibres of the most ventral part of the lateral protractor.

(3) *Opening of the radula.* As explained above (p. 162) the radula consists of a continuous ribbon of cuticular material carrying rows of teeth. This extends forwards out of the mouth of the radular sac over the mid-dorsal line of the odontophore on top of the subradular membrane, which is the name given to the cuticle of the odontophore itself When the odontophore is in the retracted position the subradular membrane dips into a longitudinal groove which lies over the gap between the two pairs of cartilages. As an accompaniment of the movement of protraction this groove is pulled open so that the radula comes to lie flush with the general surface of the odontophore instead of sunk in a groove. This is due to three separate factors: (i) the pull on the subradular membrane maintained by the fixed posterior walls of the buccal cavity laterally and ventrally; (ii) the direct action of the tensor muscles, and (iii) the hydrostatic pressure of the blood within the haemocoelic spaces of the odontophore creating turgor pressure.

In addition to the stretching of the subradular membrane which abolishes the gully within which the radula lay, there is an actual backward and downward movement of the subradular membrane sideways and over the tip of the cartilages within the odontophore. At the point where the radula passes over the cartilages a change occurs in the position of the teeth relative to the surface of the odontophore: internal to this point (i.e. on the buccal cavity side) the teeth lie flat, one over the other, approximately parallel to the surface. As they move forwards over the points of the anterior cartilages the teeth erect themselves, the rachidian and lateral teeth rotating forwards through an angle of about 90° so as to stand almost vertically erect on the surface of the odontophore, and the marginal teeth swing sideways and forwards through an arc of about 90° so as to form a series of fan-like structures on either side of the central tract of rachidian and lateral teeth. As the radula is pulled forwards this movement will be seen to affect each row of teeth as it turns over the line marked by the anterior horns of the anterior cartilages. As a result the teeth lying at the tip of the odontophore and on its underside are in the opened condition and rest over a broad area of substratum. This behaviour of the radular teeth was first adequately described by Ankel (1936*b*), who named the level at which it occurred the 'Knickkante', a term which may be translated as the bending plane.

(4) *Closing of the radula* (fig. 107D). Just as the movement of the radula over the bending plane formed by the stationary anterior horns of the anterior cartilages produces a change in the orientation of the radular teeth, so may the same result be achieved by a movement of the anterior cartilages underneath the stationary radula. In fact, relative movement of cartilages and radula at the bending plane, however produced, will result in a change in the position of the radular teeth. When the movement transfers teeth from within the bending plane to the outside, then an erection and outward rotation of the teeth will occur; when the movement is in the opposite sense then an infolding and flattening results.

It is interesting to note that these movements of the radular teeth, which are the critical events in the feeding action of the radula, are due not so much to the pull of muscles on the structures which do the actual movement, as to the reaction of these structures to changing tensions in the cuticular sheet to which they are attached. These are achieved partly by the pull of muscles on the cuticular material at other places, partly by tensions created in it by stretching over cartilages and partly by tension created by blood pressure. This effect is reminiscent of similar movements of cuticular structures in arthropods (e.g. insect wings, butterfly proboscis) and may be a characteristic feature of this kind of structure wherever it occurs.

Opening of the radula and erection of its teeth occur in *Monodonta* primarily because the subradular membrane is pulled forwards over the bending plane by muscular action; but the closing of the radular teeth into the resting position is achieved by a movement of the bending plane from an anterodorsal to a posteroventral position. This is brought about by a downwards and backwards movement of the anterior tips of the anterior cartilages, so that they swing across the opened mouth from the neighbourhood of the dorsal lip to the neighbourhood of the ventral lip. As this happens one row of radular teeth after another closes, the rachidian and lateral teeth collapsing on to the radular membrane and successive rows of marginal teeth sweeping inwards towards the middle line and folding down in the same way, brushing any particulate matter which may be lying in their path into a median heap as they do so. This action constitutes the main food-collecting action of the radula.

The movement of the cartilages within the odontophore is due to the intrinsic muscles of the odontophore only to a very slight extent: most of it is brought about by a movement of the jaws which press on to the upper surface of the odontophore and so cause motion of the cartilages within. The jaws are pulled ventrally by the action of 3 muscles: (i) the inner buccal constrictor (ibm, fig. 107E); (ii) the outer buccal constrictor (obi, obm, obo, fig. 107A); and, (iii) the buccal sphincter (bsp, fig. 107A, E). The first two of these muscles have already been described and the movement which they bring about at this point in the feeding cycle is clearly only a continuation and enhancement of the effect described in section (2). The buccal sphincter muscle forms a band passing round the anterior wall of the buccal cavity over the muscles of the jaws, the fibres decussating posteriorly. It emerges from the side walls of the neck region on the right and left and probably properly belongs to the columellar muscle. It therefore is shaped like an open figure-of-eight with the open limbs pointing backwards, and its contraction will tend to close the mouth and to press the jaws against the odontophore. Although the bulk of the movement of the anterior cartilages is effected by pressure from the jaws, for which these muscles are responsible, some of the movement is due to intrinsic muscles of the odontophore, the outer approximators of the cartilages (oa, fig. 107B, C, D). These form a sheet of muscle, on each side, of which the fibres originate along the anterior edge of the posterior cartilage (ptc) and run forwards for varying distances to be inserted on the anterior cartilages (rdc). Since the posterior cartilages are held firmly at this stage of the feeding cycle by tonus of the protractor muscles, contraction of the less powerful approximator muscles will cause movement of the anterior cartilages on the posterior ones and, in fact, a backward rotation.

(5) *Retraction of the odontophore* (fig. 107E). After the posterior movement of the odonto- phoral cartilages just described the radular teeth have scraped particulate matter off the surface against which they have been applied and in their in-swinging closing motion have grasped it and swept it towards the lateral and rachidian teeth. The next movement of the feeding cycle involves the withdrawal of the odontophore into the buccal cavity along with whatever the radular teeth have plucked from the substratum. This movement is due to the contraction of the retractor muscles of the radula, the paired posterior retractors (prr) and the paired postmedian retractors (pmr).

Each posterior retractor muscle of the radula has its origin on the outer face of the posterior cartilage. Each is partly continued into the posterior buccal depressor (pbd, fig. 106) at its dorsal end and into the ventral protractor (vbp) at its ventral end. From their origin the fibres of the muscle extend round the posterior face of the cartilage and then run forwards to be inserted on the lateral and ventral walls of the radular sac (rs). Contraction of the muscle will pull the radular sac back until the level of the muscle insertions has reached the level of the posterior face of the posterior cartilage. This effect, however, depends upon the contraction of the postmedian radular retractors anchoring the odontophore to the columellar muscle: if this does not occur then the role of the muscle is changed and its contraction causes a partial protraction of the odontophore.

The postmedian retractors of the radula are united to one another to form a single band, except at two points along their course, the first anteriorly, where the right and left muscles separate to allow another muscle to pass between them, the second posteriorly, where, after entering the columellar muscle, the two run to separate origins amongst the fibres of that muscle. The insertions of the muscle are on the ventral surface of the radular sac, posterior to those of the posterior retractors. Their action is to provide a posterior anchor to the whole buccal apparatus and to withdraw the radula, and so the subradular membrane and the odontophore.

(6) *Shutting the mouth.* This is not, in general, an action which is due to direct muscular action. The movements of the jaw, however, which have been described in section (4) have, as one of their results, the gradual shutting of the mouth and, as soon as the odontophore has been retracted, relaxation of the muscles which had opened the mouth allows it to close. The closed mouth is the resting state and therefore does not involve extensive muscular contraction.

In feeding, the cycle of operations just described would be repeated in a regular way as the animal creeps over a suitable substratum, and loose detritus would be brushed off the surface by the rotating marginal radular teeth which act somewhat after the style of the brushes of an electric polisher. This material is then pulled into the buccal cavity on the retraction of the odontophore. It will be noticed that the marginals are the only teeth that move extensively over the substratum and, therefore, that they are the only ones which can brush it. Since this is an important part of the collection of food the teeth must not be pressed too tightly against the substratum in case the whole apparatus jams. This implies that the central and lateral teeth, when erected, do not scratch deeply into the substratum: in fact, they show little signs of movement in this stage of the feeding cycle and are more concerned with pulling into the buccal cavity the material gathered by the marginal teeth than with collecting food material themselves. As a consequence of the weakness of the jaws, too, these are not used to any extent as a means of biting: their real importance has to do with holding the radula shut during protraction and moving the cartilages after it has been applied to the substratum, and these are crucial features of the feeding process. On all these grounds it is clear that the radula of *Monodonta* (and, indeed, of other rhipidoglossans) is a sweeping or brushing radula rather than a scraping or gnawing type, and, because of this, the animals must be microphagous, feeding on protophyte and protozoan material and detritus of all sorts. This is recognized in the name which Ankel (1938a) has applied to the rhipidoglossan radula— Randbursten-radula—which may be translated as a 'border-brush radula'.

Before any survey of the other types of prosobranch buccal masses is made it is worth while remarking one other feature of that of *Monodonta* and, therefore, presumably of all other rhipidoglossan prosobranchs. As remarked above, the odontophore of *Monodonta* is used in such a way that the marginal teeth must be free enough to move and yet close enough to the substratum to sweep fine particles from it. A certain amount of mobility is also necessary to allow for irregularities in the surface against which the rotating teeth might get caught. The feeding operations therefore call for nice manipulation of the odontophore. This appears to be controlled by a further series of muscles shown in fig. 107B, C, D. These are fine strands of muscle, not powerful enough to be capable of moving the odontophore, but associated with structures which, according to Nisbet (1953), are proprioceptive sense organs. This series consists of 4 muscles, 3 of which are paired, all interconnected, and related to the cartilages, the wall of the snout and the radular sac so as to form a kind of sling suspending the buccal mass in the head. The alleged proprioceptors are placed mainly at the points where the unions and intersections of the members of the sling occur, and they are, therefore, in a position where they may, perhaps, measure the relative tensions in the parts of the sling. Since these are determined largely by the main protractors and retractors this system acts as a kind of monitor system, giving the animal information of the precise state of affairs in the buccal mass and allowing a delicate control which would permit the radula to be applied to the substratum with the correct degree of pressure for the collection of food.

The muscles which act thus are the posterodorsal tensor, the inner and outer posterior radular tensors and the dorsal buccal tensors, of which all are paired except the first.

The posterodorsal tensor (pdt, fig. 107B, C, D) is a thin band of muscle which runs transversely over the dorsal surface of the buccal mass above the radular sac (rs), but below the oesophagus (aoe), from the outer surface of one posterior cartilage (ptc) to the outer surface of the other. It can therefore directly control the degree of approximation of these two cartilages, or, by measuring tension generated in its fibres by the separation of the cartilages, reflexly stimulate other more powerful muscles to control this separation.

The inner and outer posterior radular tensors (ipt, opt) run posteriorly to an insertion on the walls of the radular sac from an anterior origin. In the case of the inner muscle the origin is on the posterodorsal tensor (pdt); in the case of the outer muscle it is on the dorsal surface of the posterior cartilage (ptc). These muscles will, therefore, be stretched on protraction of the odontophore and can signal to the central nervous system the extent of this process.

The dorsal buccal tensor (dt) is also inserted on the posterodorsal tensor (pdt) and runs forward from there to an origin on the anterior wall of the snout. It antagonizes the posterior radular tensors in that it is stretched when the odontophore is retracted and may, therefore, provide a measure of this process. During protraction and retraction tonic contraction and relaxation keep all the members of this system taut and appear to allow the animal to gauge the position of the whole buccal apparatus, a matter of significance to any animal with the type of radula possessed by *Monodonta*.

It is unfortunate that, although descriptions of the buccal mass in other rhipidoglossan prosobranchs have been published (e.g. by Woodward (1901a) for *Pleurotomaria*, and Crofts (1929) for *Haliotis*) these do not include a sufficiently detailed account of the musculature to say whether corresponding tensor muscles occur. Woodward's account of the musculature of *Pleurotomaria* allows it to be said that muscles homologous with the lateral and ventral protractors of *Monodonta* occur as well as those corresponding to the main retractors, from which it may be concluded that protraction and retraction of the odontophore are achieved similarly in both animals. No description of the musculature in the walls of the buccal cavity is given, which makes it impossible to decide whether there is a movement of the subradular membrane over the cartilages or of the cartilages within the subradular membrane such as to cause opening and closing of the radular teeth, nor is there any mention of muscles which might behave as tensors.

On turning to the account given by Crofts of *Haliotis* it is noticeable that there is a greater resemblance to *Monodonta*. Protractor muscles and retractor muscles appear to correspond in both molluscs, though a greater number of protractors occurs in *Haliotis*. These are related to a third pair of cartilages which are placed near the mouth. More important, a tensor system of muscles is described and a number of buccal muscles, from the arrangement of which it seems likely that the buccal mass and radula of *Haliotis* function in a way very similar to those of *Monodonta*. Although no account has been given of the details of the buccal mass in *Gibbula* spp. it is, nevertheless, likely that in these gastropods, too, the mode of functioning is similar to that of *Monodonta*. This is based partly on the fact that Nisbet (1953) recorded *Gibbula* as being structurally similar to *Monodonta*, partly on Eigenbrodt's description (1941) of the movements of the radula of *Gibbula*, and partly on the photographs of the results of the radular action given by Ankel (1938a). This last author investigated the appearance of a sheet of glass covered with algal growth after it had been grazed by a variety of molluscs, including *Gibbula*. Where feeding has occurred the radular teeth scrape the algae off the surface of the glass and so expose it and from the pattern which the feeding leaves the action of the various teeth involved may be deduced. In the case of *Gibbula* feeding leaves a broad track, broader than the radula in the buccal cavity, which is marked by innumerable fine scratches each of which is the arc of a circle running more or less transversely across the track. Few marks are visible in the central part of the track apart from the central ends of these arcs. From this it may be deduced (1) that the marginal teeth open outwards, making the radula much broader when set against the substratum than in the buccal cavity; (2) that the substratum is scratched by the rotary movement of these teeth; and (3) that the rachidian and lateral teeth either do not touch the substratum at all or do so very lightly—though it is possible that whatever marks they may make on it are later obliterated by the raking action of the marginals.

One feature of the odontophore of some rhipidoglossan prosobranchs (the family Fissurellidae) remains to be mentioned. This is a caecum which lies in the haemocoelic space in the mid-line of the odontophore between the two sets of radular cartilages and their associated muscles (pdm, fig. 95). The caecum is lined by an epithelium which is an extension of that over the general surface of the odontophore (fig. 108), but the cells are modified for secretion. The nucleus lies basally in dense cytoplasm whilst the distal part of the cell is vacuolated and contains granules which may be seen discharged in the cavity of the caecum. The walls are not muscular. The significance of this structure is unknown, and what part it plays—if any—in the functioning of the odontophore is equally uncertain. It must, presumably, be formed at a late stage of development in the animals which possess it, as the subradular membrane is complete over its mouth. That this membrane is complete, in fact, speaks for the idea that it must be secreted towards the posterior end of the buccal mass and move forward from there, perhaps sharing in the radular movement. In this region, indeed, radular and subradular membranes could be regarded as merely the differentiated inner and outer layers of one and the same structure.

The prosobranchs which are not rhipidoglossan have radulae which appear to function in a different way, and may therefore have buccal masses in which the musculature is differently arranged. The first of these groups which may be dealt with is the Taenioglossa, a group which has what Ankel (1938a) called a 'spreading-tooth radula' or 'splay-tooth radula' and which corresponds, approximately, to the group Mesogastropoda of Thiele. The conditions in this type (fig. 104D) are readily derivable from the rhipidoglossan. The radula is characterized by the presence of a bending plane marked by the anterior horns of the cartilages of the

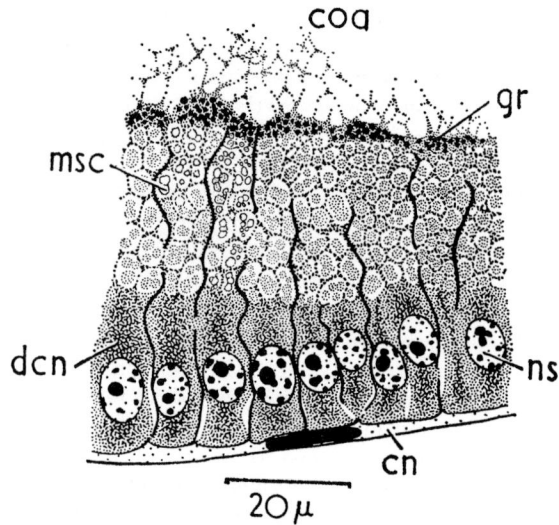

FIG. 108.—*Diodora apertura:* part of the wall of the buccal caecum.
cn, connective tissue; coa, coagulum of secretion in lumen; dcn, dense cytoplasm; gr, granule of secretion; msc, mass of secretion within vacuole; ns, nucleus of secreting cell.

buccal mass and at this the teeth behave as in the rhipidoglossan radula. The marginal teeth, however, are much fewer, though they execute the same kind of outward sweep on passing forwards over the bending plane and an inward sweep on being withdrawn, and so are less efficient as sweeping agents. This is, perhaps, compensated by one important difference, however, discernible on comparing the trace of a taenioglossan radula with that of a rhipidoglossan, in that the impress of the rachidian and lateral teeth is clearly visible on the substratum. It appears, therefore, that the marginal teeth of the taenioglossan radula may still brush loose material towards the centre of the radula, but that this is not such an important part of the feeding mechanism as it is in *Monodonta:* it may be that as its importance diminished so did the number of marginal teeth. Its diminution has accompanied, and, possibly, been caused by the activity of the more centrally placed teeth which may gather material directly from the substratum, or loosen it for the marginals to brush together, or both. In many of these gastropods, too, a cuticularized jaw exists and the action of the rachidian and lateral teeth against this may allow something resembling a bite to be taken.

The form and musculature of the buccal mass has been rather superficially examined in a number of taenioglossan prosobranchs, usually without any reference to the way in which the apparatus functions. The most important of these works is possibly that in which Johansson (1939) described the buccal musculature of *Littorina* and some rissoids, more particularly since the working of the radula of *Littorina* has been carefully described by Ankel (1937c). The buccal mass of some rissoaceans has also been described by Bregenzer (1916) and Krull (1935), and Nisbet (1953) has included references to *Littorina*. The working of the radula and the feeding movements have also been described in *Viviparus* by Eigenbrodt (1941) who stated that they are similar to those seen in *Littorina*. The musculature has also been described by Starmühlner (1952) and this therefore allows anatomy and function to be correlated.

The buccal mass of *Viviparus* is small and globular, with an inflated dorsal food channel extending along its dorsal surface (dfc, fig. 109). The sides of the food channel overlap the sides of the odontophore and the duct of the salivary gland (sd) runs between these two structures with lobules of glandular tissue lying around a central lumen. The dorsal and lateral walls of the food channel are connected to the side and dorsal walls of the snout by a large number of small strands of dilator muscle (bd), which may also help in protraction. One larger muscle (pr) runs from the base of the dorsal folds to the body wall, passing between the cerebropedal connective in front and the pleuropedal connective behind as it does so: it acts as a main retractor and simultaneously dilates the oesophagus.

The buccal mass itself is anchored to the body wall by a number of muscles. These include the following paired muscles.

(1) Lateral buccal protractors (lp): these originate on the lateral walls of the snout well forward and run, as several converging strands, to be inserted on the posterior edge of the cartilage which forms the support of the odontophore on each side.

(2) Ventral protractors (vbp): these also originate on the lateral walls of the snout on each side, ventral to the origin of the previous muscle, and pass to an insertion near the anterior end of each cartilage, somewhat ventrally. The action of these two muscles is to protrude the odontophore. The same effect is partly achieved by a third pair of muscles which also originate from the wall of the snout. These are the protractor muscles of the subradular membrane (psm). Their origin lies on the snout wall anterior and ventral to those of the other protractors: from there they run to the lateral aspect of the cartilage, pass dorsally to its upper end and then medially, where they are inserted on the under surface of the subradular membrane (sbm). The effect of their isolated contraction is (a) to pull the odontophore forward and (b) to stretch the subradular membrane laterally, which helps to open the radular teeth. A few fibres apparently belonging to this muscle run only between subradular membrane and the dorsal part of the cartilage: these would limit the extent of stretching of the membrane and enhance the protraction effect.

The combined effect of all these muscles, which appear to act synergically, is to pull the odontophore forwards from the buccal cavity, through the oral tube, so that it may be applied to the surface of the ground on which the animal is feeding. At the same time, because the posteriorly inserted protractors have a dorsal origin anteriorly and the anteriorly inserted ones have a more ventral origin, the odontophore is also upended so that its anterior tip is applied to the substratum more or less at right angles. This has already been noted by Eigenbrodt (1941).

Another set of muscles is concerned with a movement of the subradular membrane outwards over the cartilages, which occurs simultaneously with the combined rotation and protraction already described. This is due to the action of muscles inserted ventrally on the subradular membrane and originating on the cartilages of the odontophore and the radular sac itself. The latter consists apparently of a median muscle, the median radular protractor, which is inserted on the mid-ventral part of the subradular membrane near the tip of the buccal mass and which runs posteriorly between the two cartilages to an origin on the underside of the radular sac near its posterior end. The former comprise a pair of muscles, the lateral radular protractors, inserted on the subradular membrane one to either side of the median protractor and running back alongside that muscle to the ventral end of the cartilage, to which they are fastened posteriorly. When these muscles contract the subradular membrane will be pulled out of the buccal cavity and down over the apex of the odonto-phore; as it does so it is tensed by the spreading effect exerted by the protractor of the subradular membrane and passes, in this stretched state, over the bending plane marked by the anterior horns of the cartilages. This has the effect of causing the radular teeth to rotate outwards as they are protruded.

The tip of the odontophore is thus brought into contact with the substratum after the opening of the mouth in such a way that the horns of the odontophore lie at the level of the dorsal lip; the radular teeth which are in contact with the ground have, therefore, been drawn over that and so are in the outspread position.

The next event in the feeding cycle, as recorded by Eigenbrodt (1941), is the passage of the horns of the cartilages across the mouth from the level of the dorsal lip to that of the ventral lip. Since they move behind the radula this moves the bending plane behind the teeth which are exposed and in contact with the substratum. Nisbet (1953) observed that a to and fro movement of the cartilages behind the radular teeth may occur so that they open and close several times in succession, rasping the substratum each time. Finally, however, a backward movement occurs corresponding to the single one described by Eigenbrodt and, simul-taneously with this, the radular teeth are drawn into the buccal cavity so that whatever material they may have grasped as they shut at the instant the bending plane moved behind them, is ingested. The two components of this movement appear to be due to another series of muscles, as follows:

(a) Posterior retractor of the radula (prr): this comprises a pair of muscles which originate in a mass of muscle lying at the anterior end of the cartilages and in the roof of the buccal cavity at the same level. From this origin they pass backwards along the outer face of each cartilage, rather ventrally, curve medianly round the posterior end and thence run forwards through the space between the cartilages, and end by inserting on the wall of the radular sac and on the subradular membrane underneath the radula.

(b) Postmedian retractor of the radula (pmr): this is an apparently single muscle posteriorly but is composed of separate right and left halves anteriorly, where it runs within the buccal mass. The muscle originates in the mid-line of the foot where it separates from the bundles of columellar muscle which lie there: it then runs to the ventral side of the posterior end of the radular sac and bifurcates into right and left halves before penetrating the sheath of circular muscle which lies around the sac. Within this the two parts of the muscle run forwards underneath the radular sac and end by becoming inserted on the ventral surface of that structure. This muscle acts not only as a means for withdrawing the radula over the horns of the cartilage, but also as a retractor of the entire odontophore. It is by virtue of the tone of this muscle, too—steadying the radular sac—that contraction of the median protractor of the radula is able to bring about its full effect.

(c) The retractor (pr) which runs from the skin of the snout to the posterior end of the cartilage on each side, and which also sends branches to the oesophageal wall.

The mechanism by which movement of the cartilages within the odontophore is brought about is not clear in Viviparus. It may be that their backward drive is managed by the same means as in Monodonta, i.e. by pressure exerted on them by musculature in the roof of the buccal cavity in the neighbourhood of the jaws, but it is not easy to find muscles which would translate their tips anteriorly again and so permit the to and fro rocking observed by Nisbet. The muscles which do seem able to do this (not confirmed by observation) are the inner buccal constrictors (ibc), a pair which have their attachments to the roof of the buccal cavity far forwards at one end and to the outer face of the anterior end of the cartilages at the other.

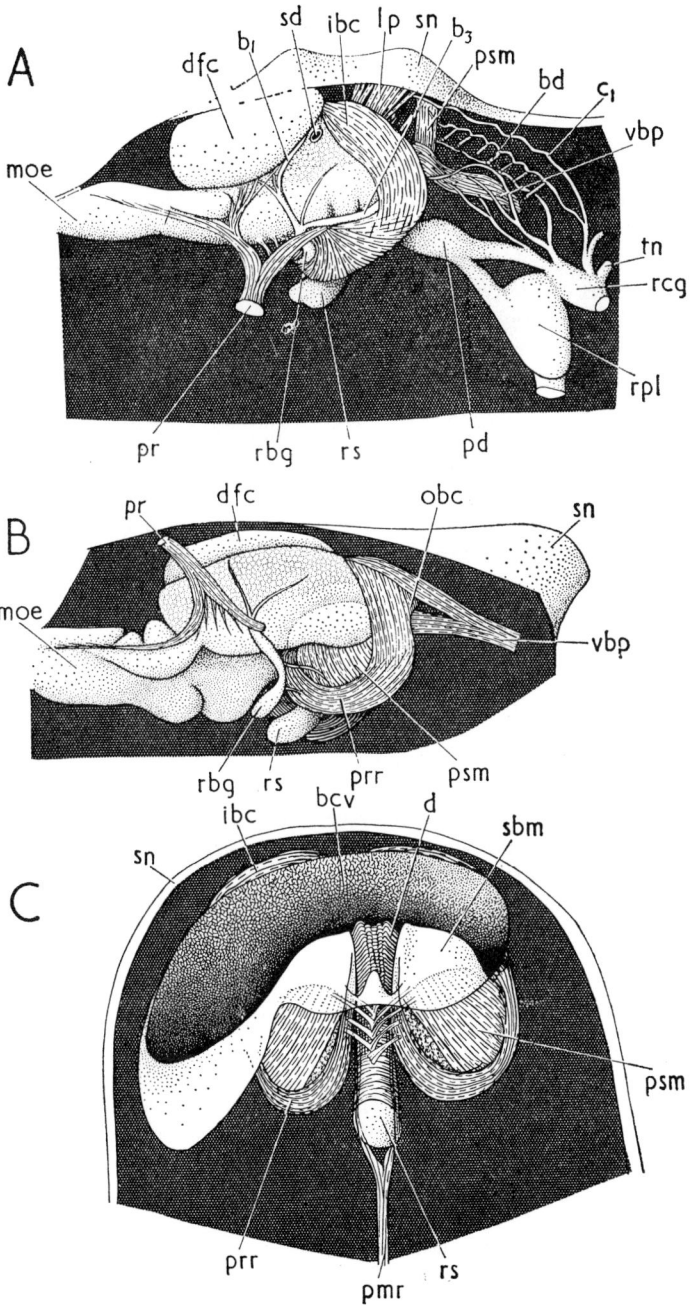

On the assumption that the roof of the buccal cavity can be immobilized by contraction of other muscles, these would then pull the horns of the cartilages dorsally when they contracted. Anterior to these, other muscles, the outer buccal constrictors (obc), will press the jaws against the odontophore and so effect the backward movement of the horns of the cartilages in the same manner as in *Monodonta*.

The following table (p. 189), in which Starmühlner's names (1952) are italicized, suggests homologies between the muscles of *Monodonta* and those of *Viviparus*. To what extent these homologies are justifiable is uncertain in view of our relative ignorance of their comparative anatomy and our total ignorance of their development.

A study of this table shows that a number of homologies can apparently be established between the muscles of the two gastropods under discussion, with a general identity of function as well as of arrangement. Since Nisbet has described no less than 33 separate muscles in *Monodonta* it is clear that the taenioglossan has undergone a considerable simplification of its buccal mass and many muscles have vanished: as suggested above, this may perhaps be correlated with the fact that the rhipidoglossan has to poise its odontophore rather accurately over the substratum in order that the marginal teeth may function properly, and to do this the system of tensor muscles is necessary. With the more coarse type of activity displayed by the taenioglossan, which sets its odontophore down on to the substratum without the same degree of adjustment, these are unnecessary and have gone.

The evolution from rhipidoglossan to taenioglossan radula has involved a number of other changes: the odontophore has shortened and is supported by only a single pair of cartilages, whilst the marginal teeth are reduced in number. These changes are continued, and to some extent increased, in the Stenoglossa, a group which receives its name from the fact that marginal teeth are not found on the radula at all, and the number of lateral teeth is reduced to one on each side or none. Each row of teeth in the radula of a rachiglossan (fig. 104E) consists of only three teeth, a rachidian flanked by a single lateral tooth on each side. The radula behaves like the rhipidoglossan and taenioglossan types in that it possesses a bending plane and movement of the teeth outwards over the bending plane (or of the bending plane inwards under the teeth) will cause the teeth to erect themselves instead of lying flat on the surface of the subradular membrane, whilst the lateral teeth also curve outwards. Movement of the radula or of the bending plane in the opposite direction causes the reverse effects, and, in particular, a biting or gripping action as the lateral teeth curve inwards towards the

---

FIG. 109.—*Viviparus viviparus*: dissections of head to show buccal mass. A, dorsolaterally from the right; the nerve ring has been cut mid-dorsally and the right half pulled ventrally to the right; the oeosophageal dilator muscle on the right has been cut and pulled ventrally and away from the nerve ring; B, similar view; nerve ring completely removed, the oesophageal dilator muscle lies in its normal position but has been cut from the body wall and the superficial muscles over the posterior end of the radular cartilages dissected away to expose the posterior retractor of the radula and the protractor of the subradular membrane muscles; C, dorsal view, the buccal cavity and the posterior half of the buccal mass opened.

b$_1$, b$_2$, buccal nerves; bcv, buccal cavity; bd, buccal dilator muscle, cut; c$_1$, dorsal snout nerve; d, radular tooth; dfc, dorsal food channel; ibc, inner buccal constrictor muscle; lp, lateral protractor muscle; moe, mid-oesophagus; obc, outer buccal constrictor muscle; pd, pedal ganglion; pmr, post-median retractor muscle of the radula; pr, posterior retractor and oesophageal dilator muscle, cut; prr, posterior radular retractor muscle; psm, protractor muscle of the subradular membrane; rbg, right buccal ganglion; rcg, right cerebral ganglion; rpl, right pleural ganglion; rs, radular sac; sbm, subradular membrane; sd, opening of salivary duct to buccal cavity; sn, snout; tn, tentacular nerve; vbp, ventral buccal protractor muscle.

TABLE 4

Monodonta	Use	Viviparus	Use
Lateral protractor	Protracts odontophore	Lateral protractor *Protractor of odontophore*	Protracts odontophore and rotates ventrally
Ventral protractor	Protracts odontophore	Ventral protractor *Protractor of odontophore*	Protracts odontophore and rotates ventrally
?	—	Protractor of the subradular membrane *Radular tensor* and *radular folder*	Protracts odontophore and spreads subradular membrane
?	—	Branch of above	Protracts odontophore
?	—	Median radular protractor *Radular protractor*	Pulls radular outwards
Direct anterior radular tensors	Pull radula outwards	Lateral radular protractors *Radular protractor*	Pull radula outwards
Posterior retractor of radula	Retracts radula	Lateral radular protractors *Elevator of buccal mass*	Retracts radula
Postmedian retractor of radula	Retracts radula and odontophore	Postmedian retractor of radula *Radular retractor*	Retracts radula and odontophore
Inner buccal constrictor	Moves cartilages back and constricts buccal cavity	Inner buccal constrictor *Buccal constrictor*	Moves cartilages back and constricts buccal cavity
Outer buccal constrictor	Moves cartilages back and constricts buccal cavity	Outer buccal constrictor *Buccal constrictor*	Moves cartilages back and constricts buccal cavity
Buccal sphincter	Closes oral tube	Buccal sphincter *Buccal constrictor*	Closes oral tube
Buccal dilators	Dilate buccal cavity	*Buccal dilators*	Dilate buccal cavity
Oesophageal dilators + ?	Dilate dorsal food channel	Oesophageal dilators and *retractor of buccal mass*	Dilate dorsal food channel
Retractor of transverse fold	Retracts oesophageal valve and transverse fold	Retractor of oesophageal valve	Retracts oesophageal valve and subradular membrane below
Ventral approximator of cartilages	Pulls cartilages together ventrally	*Ventral approximator of cartilages*	Pulls cartilages together ventrally, separates them dorsally

midline. This biting effect is exaggerated by the fact that the radula lies outspread on the under surface of the buccal mass but in a deep groove running over the dorsal surface. The rotation which occurs as each tooth swings over the bending plane is therefore caused partly by the movement from the erect to the supine position and partly by infolding of the subradular and radular membranes on passing from the outspread condition into the groove. It is, therefore, a wide arc, and as the teeth are provided with sharp fangs they pierce and pull on the food which is being eaten.

The movement of the radula of a gastropod in relation to the underlying cartilages was likened by Huxley (1853) to the movement of a rope over a pulley, whereas Geddes in 1879 regarded the two structures as so closely united that movement in the one could only be caused by movement in the other. If there is any group in which the second kind of movement occurs it is not the Rachiglossa: they are, contrariwise, a group in which all the movement of the radula appears to be of the type described by Huxley, and to be achieved by the radular retractor and protractor muscles pulling the subradular membrane backwards and forwards over the bending plane created by the anterior horns of the cartilages, and the ties between subradular membrane and cartilages are few and loose to permit this to occur. There appears to be little or no activity due to the movement of the cartilages behind the subradular membrane.

The buccal mass has been described in a number of Stenoglossa—in *Buccinum* by Dakin (1912) and Brock (1936), in *Sycotypus* (= *Busycon*) by Herrick (1906), by Wilsmann in *Buccinum undatum* (1942), and by Carriker in *Urosalpinx* (1943). The muscles mentioned by Carriker lie in and around the buccal mass, a relatively small structure which lies at the apex of the proboscis, whether that be withdrawn or extended. Most of the surface of the buccal mass and of the buccal cavity and anterior oesophagus is moored to the inner surface of the proboscis by numerous slender bundles of muscle called buccal or oesophageal tensors by Carriker: these also act as dilators of the buccal cavity and dorsal food channel. The buccal mass is kept under tension by a sheet of circular muscle running around it and the buccal cavity at a rather anterior level: this will constrict the buccal cavity and press the jaws down over the dorsal surface of the odontophore. It may, therefore, help to fold the radular teeth away into the dorsal groove as they are withdrawn into the buccal cavity, or prevent their precocious uprising as they move in the opposite direction.

Protraction of the odontophore is brought about by two pairs of muscles which originate on the lips around the mouth and are inserted, one pair dorsally, the other pair ventrally, at the posterior end of each buccal cartilage. Retraction is due to the action of two other pairs of muscles which run back from the odontophore to more posterior origins. One of these pairs originates on the proboscis wall ventrally and laterally and is inserted on the other muscles of the odontophore; these are the lateral retractors of the odontophore. The second pair, the odontophoral retractors, originates, according to Carriker, on the walls of the cephalic haemocoel, well behind the base of the proboscis: this may indicate a real attachment on the columellar muscle. From there the muscles run to the ventral part of the buccal cartilages posteriorly, but a slip passes further forward and is treated by Carriker as a separate, powerful muscle, the dorsal subradular membrane retractor. A slip from the lateral odontophoral retractor on each side is also involved in the formation of this muscle, which is inserted on the subradular membrane on the dorsal side of the odontophore. Some of its fibres, however, originate on the posterior ends of the buccal cartilages. The action of this muscle is to pull the subradular membrane dorsally and posteriorly into the buccal cavity over the bending plane. It is helped in this by the contraction of another muscle on each side, the lateral retractor of the subradular membrane, which runs from the posterior end and outer side of the cartilage on to the subradular membrane over the odontophore. This complex of muscles, therefore, separately or simultaneously retracts the odontophore and the subradular membrane over it and so is responsible for the effective feeding action of the radula.

The outward movement of the subradular membrane over the bending plane, which is responsible for the opening of the teeth of the radula, is primarily due to the action of a pair of ventral retractors of the subradular membrane, to give them Carriker's name. These originate on the buccal cartilages posteriorly and run forwards, ventral to each cartilage, to be inserted on the subradular membrane near the posterior end of the sublingual cavity. They are probably aided in their work by an increased tension in the subradular membrane which is produced by the contraction of the divaricator muscles of the cartilages. These run from the lateral wall of each cartilage out to the side walls of the proboscis: their contraction will clearly stretch and tauten the subradular membrane over the dorsal surface of the odonto-phore, will tend to open the groove in which the radula lies there and so make it more susceptible to the forces which open it as it passes over the bending plane.

The other muscles which are mentioned by Carriker in his account of *Urosalpinx* are oral sphincters, oral retractors and tensors of the radular sac. The first of these forms a ring of circular muscle round the mouth, the second a cone of muscular strips which radiate from the lips back to the wall of the snout, and the third, which run from the innermost end of the radular sac to the buccal artery, serve as a means of securing the sac in the proboscis haemocoel during protraction and retraction.

The apparent homologies of these muscles with those of *Buccinum* (Wilsmann, 1942), *Viviparus* (Starmühlner, 1952) and *Monodonta* (Nisbet, 1953) are given in Table 5.

On comparing this list of muscles active in *Urosalpinx* with those in the taenioglossan and rhipidoglossan it will be seen that their modification can be linked with the modification in the use of the radula. Broadly speaking, the stenoglossan has selected a few of the varied movements of which the odontophore of the rhipidoglossan is capable and has developed an effective feeding mechanism out of them and the taenioglossan has done the same. But whereas in the taenioglossan it is the rotary phase of the total pattern of rhipidoglossan activity which has been exaggerated, it is the movement of the subradular membrane over stationary cartilages which is the most important feature of the working of the stenoglossan odontophore, and it is, therefore, the muscles which allow this to occur which are best developed. Since it is the retraction of the subradular membrane which is the effective working stroke in the feeding cycle it is the muscles which bring this about (dorsal and lateral retractors of the subradular membrane) which are the best developed and most powerful in the odontophore. It is difficult to see what muscles in *Viviparus* or *Monodonta* correspond to these, partly because the closure of the radular teeth is brought about by a completely different mechanism in these animals, partly because the entry of the food into the buccal cavity is managed by the retraction of the whole odontophore rather than by that of the radula alone, and partly because there is so little evidence, comparative or embryological, on which to build firm homologies.

In carnivores the musculature must be normally greater than in herbivores and in biters transverse forces must be greater than longitudinal ones. In prosobranchs which gnaw or scrape with the radula longitudinal forces predominate and this is particularly evident in a second evolutionary trend in radular function which may be traced in another group of the prosobranchs, the Docoglossa. Their radula is distinct from the rhipidoglossan type in a number of ways: the rachidian tooth is minute or absent, the marginals do not exceed three

TABLE 5

Urosalpinx	Buccinum	Viviparus	Monodonta
Buccal tensor	(Buccal dilators)	Buccal dilators	Buccal dilators
Oesophageal tensors	?	Oesophageal dilators	Oesophageal dilators
Buccal circular muscles	Buccal circular muscles	Buccal constrictor	Buccal constrictors
Lateral odontophoral retractor	Retractor pharyngis?	?	?
Odontophoral retractors	Median retractor of radula	Postmedian retractor of radula?	Postmedian retractor of radula?
Dorsal odontophoral protractors	Posterior jugal muscle	Lateral buccal protractor	Lateral buccal protractor
Ventral odontophoral protractor	Anterior jugal muscle	Ventral buccal protractor	Ventral buccal protractor
Oral sphincter	?	Buccal sphincter	Buccal sphincter
Oral retractors	?	?	?
Radular sac tensors	?	?	Posteior ventral radular tensors?
Ventral subradular membrane retractors	Radular retractors	Lateral radular protractors	Direct anterior radular tensors
Dorsal subradular membrane retractors	Ventral tensors	?	Ventrolateral tensor of buccal membrane?
Lateral subradular membrane retractors	Lateral tensors	?	?
Divaricators of cartilages	?	Protractor of the subradular membrane?	?
Transverse muscle	Horizontal muscle	Approximator of cartilages	Ventral and outer (?) approximators of cartilages
?	Dorsal protractor of radula	?	?

in number and are not prominent, and, most important, a bending plane is either totally absent or its effect is so slight as not to be important.

The working of the radula has been described for *Patella* and *Patina* by Ankel (1938a) and again (for *Patella*) by Eigenbrodt (1941). A number of previous workers have also described in less detail how limpets feed (Davis & Fleure, 1903; Orton, 1913b) and the former of these and Geddes (1879) and Amaudrut (1898) have given brief accounts of the cartilages and associated musculature. There are at least 4 pairs of cartilages, anterior, posterior, anterolateral and ventrolateral. Of these the first two are held firmly together by tough membrane and, as they articulate with one another by flat surfaces, it does not seem likely that much movement between them is possible. The third and fourth pairs lie anteriorly, lateral to the anterior cartilages and are not in contact either with them or with one another: instead they are attached by muscle to a variety of structures in the head.

The musculature of *Patella* (Graham, 1964) is much more complex than previous descriptions would suggest. It is also noticeably more powerful than that of the various prosobranchs so far dealt with, this being due to two main causes, a hypertrophy of the muscles which interconnect the cartilages and of those which protract the odontophore and radula. The consequence of the first of these changes is that the odontophore is protruded as a very stiff and inflexible pad, and the consequence of the second is that it is protruded with very great power. This, in turn, may be related to the fact that since there is no effective bending plane in the radula (although some movement of tooth on radular membrane does occur at the apex of the odontophore), the action of the teeth in a docoglossan radula is fundamentally different from that of any of the other types which have been so far dealt with. In these, each radular tooth, with the possible exception of the rachidian, scrapes the substratum only as it pivots through an arc of a circle on its base, as that moves over the bending plane. In other words, the bending plane and the working place coincide. In the docoglossan, on the other hand, the tooth and the odontophore move together without any pivoting of the one on the other. In the first case, therefore, the muscular equipment need be powerful enough only to move the subradular membrane over the horns of the cartilages (or vice versa) with sufficient force to move against the substratum the radular teeth of the one or two rows that lie over the bending plane: contact between teeth and substratum is limited to the immediate neighbourhood of that plane, as may be seen by reference to the feeding tracks left by these animals. In the docoglossan, however, a large part of the underside of the protracting odontophore, with numerous rows of teeth, is applied to the substratum and the whole structure forced across it like a rasp: in this action the working place lies (morphologically) anterior to the bending plane and it clearly calls for much more muscular power; this is responsible for the hypertrophy of the buccal musculature mentioned above.

A further important difference exists between the docoglossan and the other prosobranchs in respect of the use of their odontophore. It will be recalled that the effective food-gathering part of the feeding cycle in *Monodonta* occurred at the end of the protraction of the odontophore and radula, which had the effect of laying the radula down on the substratum with its teeth open. Then the backward movement of the cartilages and the retraction of the radula caused the rotation of the marginal teeth which brushed the substratum and their withdrawal into the buccal cavity. The same kind of movement is the essential part of the feeding of a taenioglossan, and it is the retraction of the radula which is again the basis of feeding in a stenoglossan. The docoglossan differs from this, as shown by Ankel (1938a) and Eigenbrodt (1941) in that the important part of the feeding cycle is the phase of protraction: the odontophore is everted through the mouth so that it strikes the substratum near the ventral lip, with a number of radular rows applied to the substratum. The whole structure is then moved forwards across the substratum to the region of the dorsal lip, the tips of the teeth scoring the substratum with a number of parallel lines (showing absence of rotation) as this happens. This movement then merges into the beginning of retraction, with the subradular membrane being pulled into the buccal cavity close to the dorsal lips and the jaw. The muscles which require to be well developed for carrying out such an action are, therefore, those which stiffen the entire odontophore against the vibration and shocks of a forceful application of the radula to the substratum, and those which will drive the odontophore forwards despite the friction engendered by so many teeth rasping the substratum on which the animal is feeding. This need explains the great development of the protractor muscles, of the muscles binding cartilage to cartilage and, also, the very large size of the odontophore.

Some other peculiarities of the radula and feeding of Docoglossa may be noted here and correlated with what has just been pointed out. The radular teeth in many Docoglossa, especially the laterals (fig. 104B, C), are often provided with dark, pigmented denticles: this appears to indicate hardness. All the teeth differ from those of other prosobranchs in having a broad base of attachment to the radular membrane instead of a more or less linear one: this is clearly linked with the lack of rotatory movement. Finally it is obvious that this kind of use must expose the teeth to a very high degree of wear and is probably the reason why limpets possess such long radulae. All the Docoglossa appear to be able to apply considerable pressure on the substratum on which they feed: *Patina* easily excavates great caves under the stipe of *Laminaria* in which to shelter and feed, whilst *Patella* leaves clear rake-like markings where it has been feeding on the surface of even moderately hard rocks.

The functional morphology of the buccal mass of the ptenoglossan *Ianthina* has been described by Graham (1965) and the general mode of functioning of its type of radula is known. Ptenoglossa is the name given to a small group of prosobranchs which includes the families Ianthinidae and Scalidae. The ianthinids capture and swallow whole specimens of the siphonophores *Velella, Porpita,* and *Physalia;* scalids are known (Ankel, 1938a; Thorson, 1958) to attack coelenterates too, but as they usually, or perhaps always, attack anemones it seems unlikely that they do more than bite small pieces from their prey. [See p. 621.]

In *Ianthina* (and *Clathrus* (fig. 100) is similar) the mouth leads through an oral tube past a sphincter muscle into a capacious buccal cavity in the lateral walls of which are embedded jaws and on the floor of which lies a large odontophore occupying the greater part of the available space. Dorsally, over the odontophore, it opens widely into the oesophagus, which is a simple folded tube without any trace of the dorsal folds or other features usually found in the prosobranch. No clear radular sac is visible and the radula appears to be secreted in the posterior part of the buccal cavity rather than in a separate sac.

The odontophore is supported by a single piece of cartilage on each side which is composed of highly vacuolated cells and extends across the mid-line from side to side. There is also cartilage developed in the roof of the buccal cavity and in its floor, below the sublingual pouch. The odontophore is deeply cleft mid-dorsally and rises into two halves placed right and left of the mid-line; the two halves also extend forwards from the anterior end of the bridge which links them together ventrally. The whole structure thus forms a trough-like structure with steep sides and an incomplete floor, which projects forwards into the buccal cavity from its posterior wall and can, when necessary, be everted through the oral tube and mouth, the two halves then diverging. All its surfaces bear radular teeth, all of which are alike—curved, fang-like structures set on the radular ribbon by an elongated, narrow base in such a way that all the teeth point inwards, the arrangement being reminiscent of the teeth on the jaws of a non-poisonous snake. Indeed, the whole feeding mechanism is comparable to that of a python or similar creature. The teeth are movable on the radular ribbon in such a way as to bend inwards, but they cannot bend outwards: as the mollusc feeds, therefore, and protracts the odontophore, the teeth swing inwards and lie flat on its surface so as to slip under the body of the prey; when the odontophore is retracted, however, the teeth erect themselves and pull the prey into the buccal cavity. Any outward movement on the part of the prey, too, will be stopped by the recurved points of the teeth, whereas inward movement will be facilitated by their downfolding. The grip is intensified by the approximation of the two halves, like a hand grasping. It may be that, although this is a competent enough piece of apparatus for dealing with relatively inactive prey such as the coelenterates upon which *Ianthina* mainly feeds, a more lively kind of animal would be able to make its escape. If this were so, then the need for paralysing the prey, which is presumably the effect of the saliva injected by scalids, becomes obvious.

The form of the radular teeth is usually closely related to the use to which they are put, and to the type of food eaten. It has been discussed by Cooke (1895, 1920) and Peile (1937) amongst others. The typical radular tooth is recurved at its apex, where it is frequently denticulate, so as to form an efficient rasping tool and also a structure adapted for the retention of the material rasped from the substratum, like a hand scraping up sand or snow. Since the precise use of the teeth varies from animal to animal and from one type of radula to another the matter may be approached on that basis.

The Docoglossa, as has been shown above, differ from other prosobranchs in that their radula is moved from behind forwards over the substratum. Since it is the teeth lying on the underside of the odontophore which are brought into contact with the ground, this forward movement means that their recurved tips dig into the substratum and rasp off threads or particles which can then be carried over the tip of the odontophore into the buccal cavity. Loose particles may also be gathered. *Patina* appears to be able to rasp weeds effectively in this way, but *Patella* and *Acmaea,* to judge from their gut contents, seem rather to take unicellular algae than parts rasped from larger plants. Davis & Fleure (1903) suggested these types of food for *Patella,* but added a method of feeding in which pieces of weed are held in the lips whilst the radula rasps them, and they believed that, whilst the anterior end of the radula is rasping on the substratum, the next inner piece is triturating the food already gathered against the jaw. Peile (1937) recorded little sign of use on the most anterior teeth, but assumed, since the teeth detach readily from the radular ribbon when it is being manipulated, that worn teeth fall off readily in nature, and are replaced by fresh ones moving forward. No wear on the teeth in from the tip is to be seen, so that no support from this source is available for the idea that the limpet chews its food between radula and jaw. From another point of view it seems unlikely that this action, if it occurs at all, is extensive—the occurrence of uncrushed fragments of vegetable material in the gut and even in the faeces.

In the case of the rhipidoglossan type of radula working place and bending plane coincide and the bulk of the action of the radular teeth is due to the sweeping action which they have on the substratum, whilst the rachidian and lateral teeth appear to be more important as a conveyor belt for the material gathered by the other teeth than as collectors of food themselves. These facts also may be related to the form of the teeth, the marginals being long bristle-like structures, very numerous, and well adapted for the use to which they are put. Gwatkin (1914) was one of the earliest to note this, talking of a 'whirlpool motion' of the radula when the small trochid *Margarites helicinus* feeds. It may also be observed in the other trochids and in the turbinid *Tricolia pullus.* All these appear to feed mainly by collecting diatoms and detrital matter from the substratum. Peile (1937) noted that the tip of the radula is formed from a few rows containing marginal teeth only, from which the more median teeth have been lost. He suggested that these marginals may not only gather up small particles, but also exclude large ones and act as a filter allowing water to escape, but not the particles.

As discussed above, it is not known to what extent the lateral and median teeth of a rhipido-glossan are involved in the collection of food: some evidence bearing on this may be obtained from a study of the radula, in that examination shows a blunting of the most anterior lateral teeth, suggesting that they hit and move over the substratum. This is also borne out by Crofts' account (1929) of feeding in *Haliotis,* which, she said, feeds on minute encrusting algae and, also, by rasping fragments from the surface of weeds by means of the narrow lateral teeth. She also, like Davis & Fleure, supposed that the radula can chew food against the jaw once it has been rasped from the substratum, but wear from this cause (if it occurs) is too slight to be detectable. Other rhipidoglossans also show blunting of the lateral teeth of the radula, so that it appears probable that some scraping of the surface of the substratum is carried out in addition to the brushing due to the marginals.

Most taenioglossans have a feeding method fundamentally similar to that of rhipido-glossans, with the brushing minimized and the scraping exaggerated. This may be responsible for the signs of wear exhibited by the teeth of *Littorina* and for the length of the radula in species of that genus. In some Taenioglossa the primary effect of the movement of the radular teeth which occurs as they pass over the bending plane appears to be grasping rather than rasping or scraping and in this case one would naturally expect to find less signs of wear. This lack of wear has been shown to occur in *Pila* (Prashad, 1932), *Viviparus conectus, V. viviparus, Bithynia tentaculata* and *Pomatias elegans* (Peile, 1937) and, in at least some of these animals (*Viviparus, Bithynia*), may be correlated with the fact that they are ciliary feeders, and the radular teeth are used for no harder work than grasping the mucous string laden with food particles which is elaborated in the mantle cavity. The same is also true of *Crepidula fornicata, Calyptraea chinensis* and *Capulus ungaricus,* which are also ciliary food collectors, and in which little wear is detectable on the radula.

On the other hand the radula of taenioglossans may show signs of excessive wear as in *Natica,* a fact which puzzled Peile, since at the time at which he was writing this gastropod was supposed to bore through the shells of the bivalves which it ate by means of acid secreted from a gland placed below and behind the mouth (Schiemenz, 1891). Since then the work of Ziegelmeier (1954) and Carriker (1959) has shown that all the removal of shell is apparently due to rasping by the teeth on the radula and the degree of wear which these exhibit is not surprising.

For the same reason, the teeth of those rachiglossans which feed after boring through the shells of other molluscs, show much damage and those in the most anterior rows of the radula of *Nucella, Ocenebra* and *Urosalpinx* are often completely devoid of cusps, so intensely have they been ground down. In other groups of Rachiglossa, however, the fact that the animals are carnivores which do not bore or are carrion feeders (Buccinacea) is reflected in the lesser degree of wear exhibited by the teeth, which are used as grasping and pulling organs rather than as rasps. In these the cusps are not worn down in the smooth, rounded way found in *Nucella,* for example, such as regular wear on a hard substance would produce, but, if not complete, show irregular fractures, such as would happen on accidental contact with hard material.

Toxoglossan radular teeth, in their most highly developed form as in cones and many turrids, are quite unlike those met with in any other prosobranchs (figs. 104F, G; 317), and their function is as different from those in other groups of prosobranchs as their shape is aberrant. (Their evolution and role in feeding, together with the accompanying ancillary modifications of the alimentary tract, are described in chapter 25 and chapter 26.)

CHAPTER 9

THE ALIMENTARY SYSTEM—3

T HE part of the gut which lies behind the mouth of the radular sac is the oesophagus. Whilst this landmark conveniently sets off its ventral anterior limit, its dorsal one is by no means so clear and is perhaps best regarded as lying still further forwards over the odontophore at a point just behind the opening of the salivary ducts. The posterior limit of the oesophagus is its entry to the stomach which lies in the visceral mass. It is this section of the gut, therefore, which, in a prosobranch, has to traverse the part of the body in which the effects of torsion are obvious; as the dorsal and ventral sides of the oesophagus carry characteristic structures, it is easy to follow the twist which has been imposed by torsion. Graham (1939), investigating the oesophagus in a variety of prosobranchs, has suggested that it is divisible into 3 sections. These are: (1) the anterior oesophagus, lying directly behind the posterior end of the buccal cavity, but in front of the region affected by torsion; (2) the mid-oesophagus, which includes the whole length of the part involved in torsion and usually sections anterior and posterior to that; and (3) the posterior oesophagus, which runs to the stomach. Since this division coincides with the separation of a mid-region characterized by the development of much glandular tissue not present in either of the other two parts, it has much to recommend it.

The dorsal wall of the anterior oesophagus and of the mid-oesophagus in an animal like *Patella* (Graham, 1932) or a trochid (Randles, 1905; Nisbet, 1953), is marked by the fact that along it run two great folds separated by a deep channel. These have been called the dorsal folds (df, fig. 110B) by Graham and the channel between the dorsal food channel (dfc). In some prosobranchs the dorsal folds show signs of being bifid. Along the ventral wall of the mid-oesophagus in these animals runs an apparently single fold of tissue (vf), but in the anterior oesophagus this can be seen to arise as two separate ridges lying right and left of the mid-line and converging and uniting posteriorly. In the mid-oesophagus of a limpet or an animal like *Calliostoma,* too (fig. 110C), the ventral fold can be shown to be double, so retaining traces of the two folds of the anterior oesophagus. In that part of the gut the anterior end of the diverging halves of the fold and the lip of the radular sac demarcate a triangular area which is often raised into a flap-like structure drawn into an anteriorly directed point. This may be called the oesophageal valve (oev, fig. 14) and it acts like the mammalian epiglottis in preventing the entry of food to the radular sac; it may also prevent regurgitation of food from the oesophagus when the animal is feeding, an occurrence which protraction of the odontophore might facilitate.

In the anterior oesophagus the lateral walls, which lie between the oesophageal valve ventrally and the dorsal fold dorsally, are often drawn out into capacious pouches (osp, fig. 14), which may communicate widely with the rest of the cavity (*Patella,* Zeugobranchia, Trochidae) or be connected to it only by a narrow mouth, as is the case in winkles. However this may be, these side walls are not adapted for the production of digestive juices, although they may contain plentiful goblet cells secreting mucus. Presumably they provide a certain amount of loose wall and space for the accommodation of food during feeding, and for easing the movements of the odontophore and radula.

A mid-ventral pouch found in *Erato voluta* by Fretter (1951a) should perhaps be regarded as comparable to these paired oesophageal pouches. It opens to the oesophagus behind the nerve ring and extends forward from there through the nerve ring and along the proboscis (mud, fig. 111). As in other oesophageal pouches its walls are lined by mucous cells alternating with ciliated supporting cells.

In the mid-oesophagus, however, the lateral walls are expanded to form a pouch-like extension of the gut cavity on either side and their surface is further increased in area by the development of finger-like outgrowths from the inner surface (Zeugobranchia, fig. 110A; Trochacea, fig. 110C) or by the formation of lamellae (Patellacea, figs. 110B, 253; Littorinacea, fig. 110E; Cypraeacea; Lamellariacea; Doliacea). Both kinds of out-growth and the wall between are covered by a glandular epithelium not found in other parts of the gut, which is responsible for secreting digestive enzymes. It forms a gland, the oesophageal gland, which, since the saliva is usually devoid of digestive action, is normally the first digestive secretion which the food meets on its course along the alimentary canal.

By comparison with the two parts of the oesophagus which lie in front of it, the posterior oesophagus is relatively simple. The two dorsal folds and the ventral fold of the anterior and mid-oesophagus die out, or merge into other folds, and the glandular epithelium of the lateral walls disappears. Consequently the diameter of the oesophagus is much reduced and the posterior oesophagus is a narrow tube with an internal surface raised into a series of longi-tudinal folds none of which is more prominent than any other, and all of which contain only mucous goblet cells.

The most interesting part of this oesophageal region is the mid-oesophagus, which, because of the dilatation caused by the lateral glands is a conspicuous object, easily seen on dissection, and often known as the crop. Within the prosobranchs it has undergone an extensive evolutionary change and, as Amaudrut (1898) was the first to point out, it is an easy structure to investigate from the point of view of comparative anatomy since its relationships with other structures in this area, also affected by torsion, are almost constant. The effects of torsion on the mid-oesophagus itself are plain to see (fig. 112): the dorsal folds and food-channel (dfc'), which mark the morphological mid-dorsal line, curve over the left side until they come to lie in the topographical mid-ventral line. Simultaneously the fold which lies mid-ventrally at the anterior end of the mid-oesophagus curves round the right side until it lies in the topographical mid-dorsal line. The glandular areas (og) also curve across so as to lie on the opposite side of the body from that to which they actually belong. The anterior limit of the area affected by torsion is marked by the supra-oesophageal part of the visceral loop, which curves over the oesophagus from the right pleural ganglion (rpl) on its way to the supra-oesophageal ganglion (sog) on the left, while the posterior limit of the area involved in the twist is similarly marked by the anterior aorta, which curves over the oesophagus from a dorsal position on the left, where it has emerged from the heart (h), to a ventral position on the right, where it breaks into cephalic and pedal branches. Whilst both these structures curve over the mid-oesophagus the precise part of that organ where this happens varies from animal to animal: it may be far forward, close to the posterior end of the anterior oesophagus, as in *Calliostoma* and *Pomatias*, or it may be relatively close to the posterior end of the mid-oesophagus, as in *Gibbula* or the Stenoglossa.

From the functional point of view the mid-oesophagus divides itself into two clear parts: the glandular lateral outpouchings and the rest, which is so arranged that the ventral fold (vf,

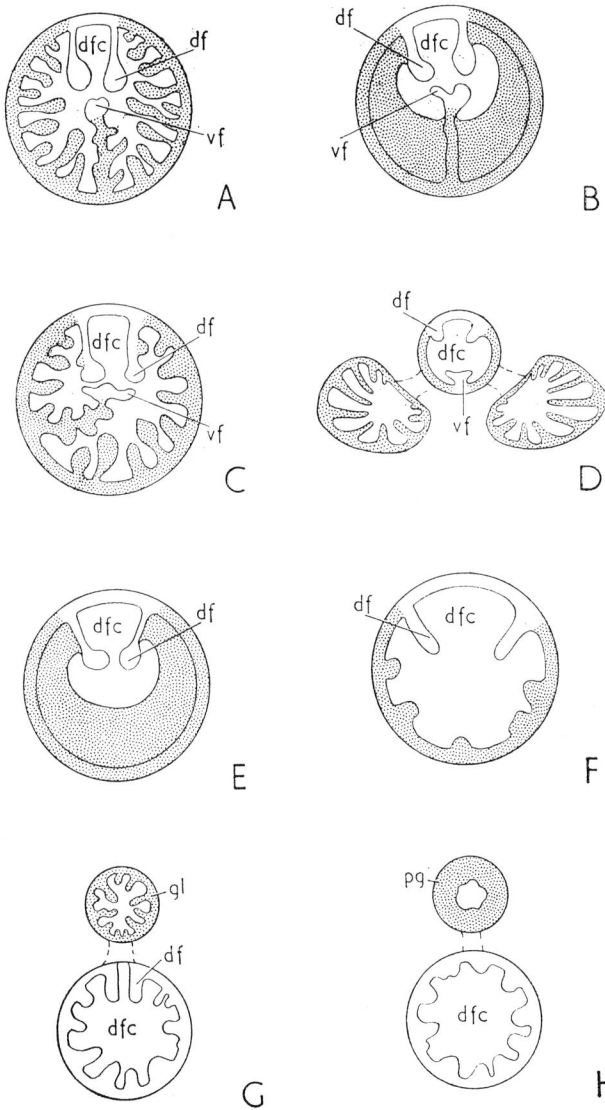

FIG. 110.—A series of diagrammatic transverse sections of the mid-oesophagus of prosobranchs. A, zeugo-branch; B, patellacean; C, trochacean; D, *Theodoxus fluviatilis*: E, *Littorina littorea,* and also typical of most mesogastropods; F, mesogastropods with a crystalline style; G, rachiglossan; H, toxoglossan. In these sections the dorsal folds, the ventral fold and the other walls of the dorsal food channel are not stippled; the glandular epithelium of the oesophageal gland and of structures homologous with it, is. Broken lines indicate connexions between glands and the main part of the oesophagus not occurring at the level of the sections.

df, dorsal fold; dfc, dorsal food channel; gl, gland of Leiblein of rachiglossan; pg, poison gland of toxoglossan; vf, ventral fold.

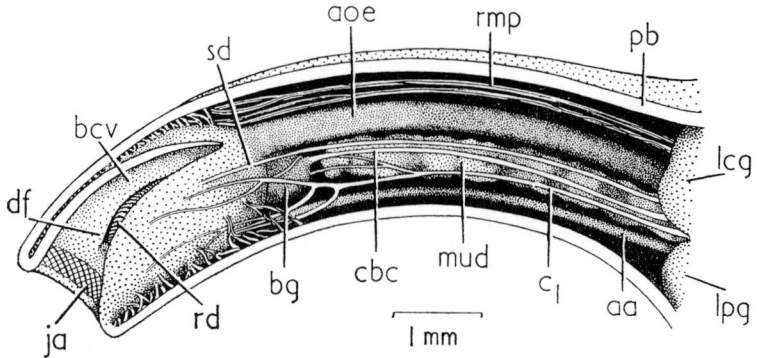

FIG. 111.—*Erato voluta*: longitudinal section through the proboscis, partly extended. The left wall of the proboscis has been removed and part of the left lateral wall of the buccal cavity. The haemocoel is black.
aa, anterior aorta; aoe, anterior oesophagus; bcv, buccal cavity; bg, buccal ganglion; c_1, cerebral nerve to muscles of buccal mass; cbc, cerebrobuccal connective; df, dorsal fold; ja, jaw; lcg, left cerebral ganglion; lpg, left pedal ganglion; mud, mucous diverticulum of anterior oesophagus; pb, proboscis; rd, radula; rmp, retractor muscles of proboscis; sd, salivary duct.

fig. 110A, B, C) fits over the dorsal food channel (dfc) between the dorsal folds (df) and converts that trough into a virtually closed channel. The histology of these structures is related to the different parts which they play in this process. Thus the epithelium which lines the dorsal food channel, the median faces and summits of the dorsal folds and the summit of the mid-ventral fold is a tall columnar ciliated epithelium rich in mucous cells. The direction of the beat of the cilia on the walls of the food channel is towards the stomach, as it is also on the summit of the ventral fold. Since that structure forms, in effect, the floor of the channel into which the food is passed from the radula, this arrangement makes the dorsal food channel into a food-carrying tube transporting it back to the stomach, and secreting mucus to bind the food particles together and to lubricate their travel.

On the summit of the dorsal folds, and for a very short part of their lateral wall over which the same type of epithelium spreads, the direction of the ciliary beat is transverse—out of the lateral pouches of the mid-oesophagus into the food channel. This brings whatever substances may be produced there into the food channel where they may be mixed with the food. In the lateral pouches, or on the villi, a wholly different type of epithelium is found, in which 2 types of cell occur, one glandular, the other ciliated. The glandular cells are low in height and, in *Patella* (fig. 113A), where the pouches contain lamellae, they alternate more or less regularly with ciliated cells (cic). Their cytoplasm is vacuolated (gc), lobed superficially and almost always contains spherules of secretory material (msc). In *Gibbula*, where the pouches contain villi, the glandular cells appear on the sides of the villi, intermingled with ciliated cells as in *Patella* and resembling the gland cells of that animal. The apex of the villus is devoid of ciliated cells but is covered with an epithelium of cells looking like the glandular ones but with a firm, almost cuticularized surface, which does not support the idea that they are secretory. In all these animals the direction of the beat of the cilia in the lateral glandular pouches is such as to drive material either towards the dorsal or the ventral folds, where they can be directed into the food channel. Since many workers have shown that the secretion of these pouches

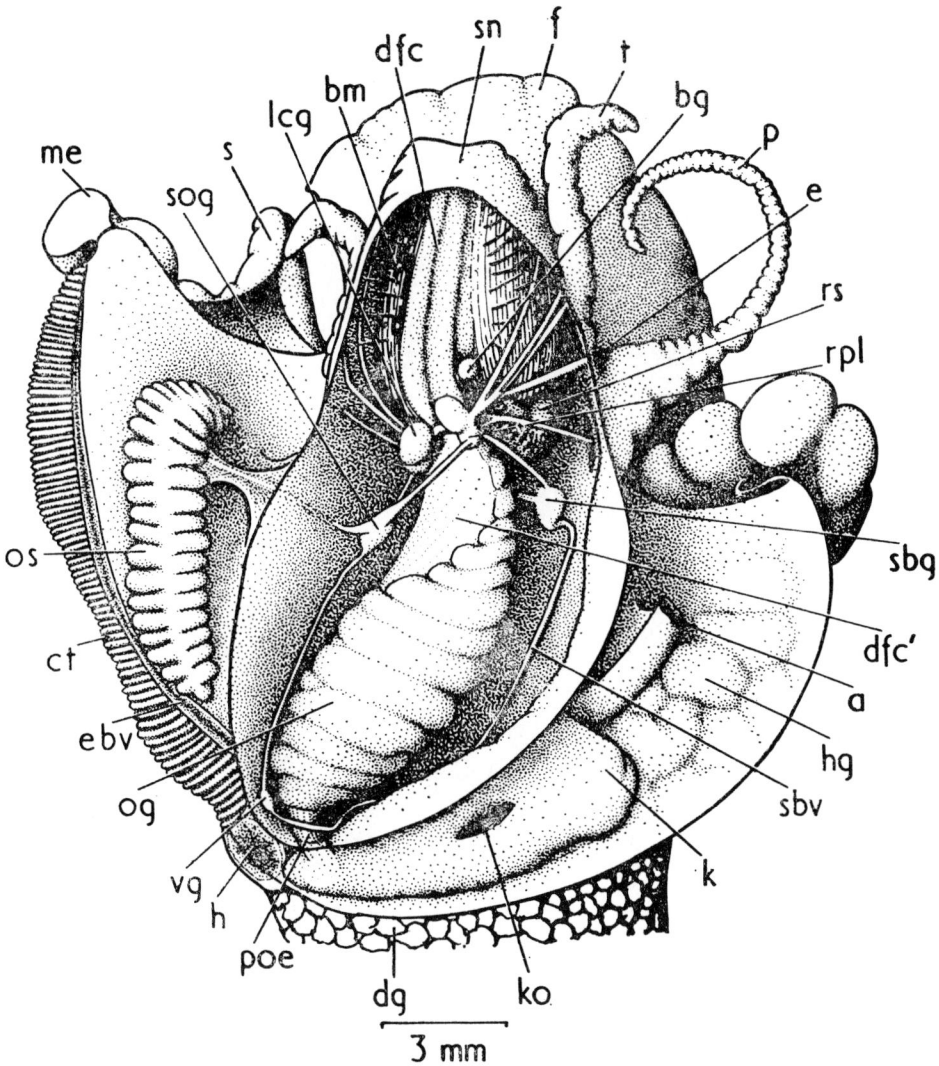

FIG. 112.—*Trivia monacha*: dissection of anterior part of body. The mantle skirt has been cut medianly and the two halves deflected laterally.

a, anus; bg, buccal ganglion; bm, buccal mass; ct, ctenidium; dfc, dorsal food channel of buccal cavity seen by transparency; dfc', dorsal food channel in mid-oesophagus seen by transparency; dg, digestive gland; e, eye; ebv, efferent branchial vessel; f, foot; h, heart; hg, hypobranchial gland; k, kidney; ko, opening of kidney; lcg, left cerebral ganglion; me, mantle edge; og, oesophageal gland; os, osphradium; p, penis; poe, posterior oesophagus; rpl, right pleural ganglion; rs, radular sac; s, siphon; sbg, sub-oesophageal ganglion; sbv, sub-oesophageal part of viceral loop; sn, snout; sog, supra-oesophageal ganglion; t, tentacle; vg, visceral ganglion.

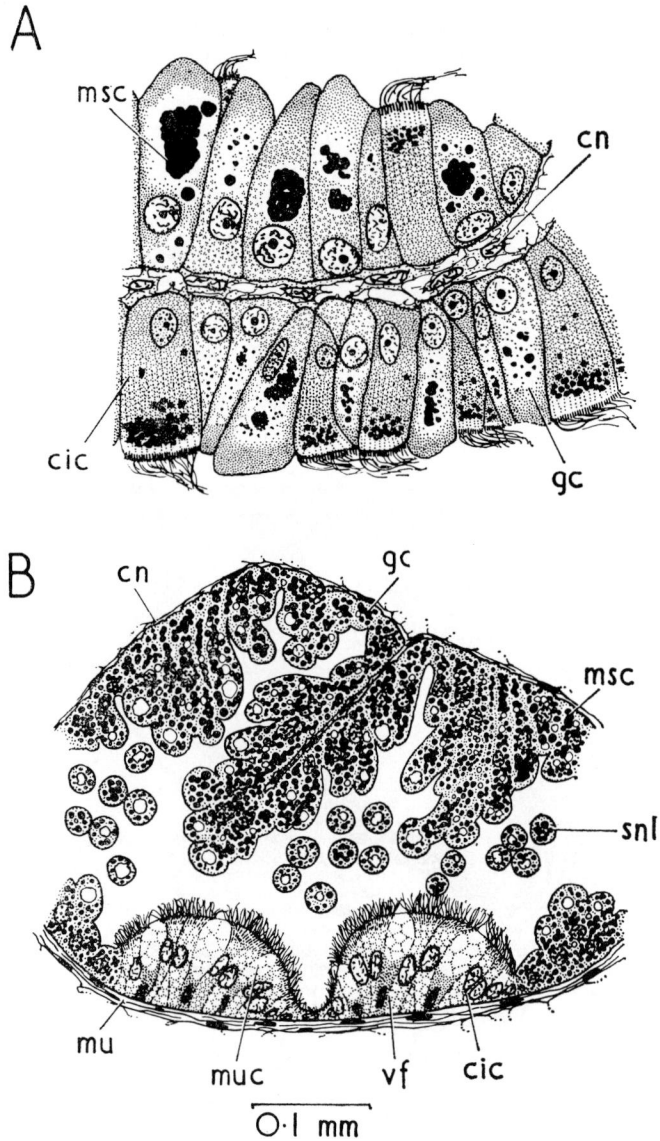

FIG. 113.—Histology of oesophageal glands. A, *Patella vulgata;* B, *Nucella lapillus,* gland of Leiblein.
cic, ciliated cell; cn, connective tissue; gc, gland cell; msc, mass of secretion; mu, muscle fibre; muc, mucous cell; snl, secretion in lumen of gland, nipped off from tip of gland cell; vf, ventral fold.

contains digestive enzymes the whole apparatus allows for the transport of food from the buccal cavity back to the stomach and for the simultaneous mixing of the food and the enzymatic secretion of the lateral pouches during the transport.

As has already been described, the mid-oesophagus has two folds running along its dorsal wall and a double, or potentially double, fold running along its mid-ventral line. The double-ness of this fold and the way in which it is formed by the approximation and more or less complete fusion of two folds in the anterior oesophagus suggest that the primitive organiz-ation of the ventral half of the oesophagus mirrored that of its dorsal half, and that a ventral food channel separated a right glandular mass from a left one just as does a dorsal channel. No living prosobranch, however, possesses an oesophageal structure of this sort, though in development Fischer (1892) has shown that the oesophagus is marked by lateral glandular strips separated by non-glandular conducting strips in the mid-dorsal and mid-ventral lines. All the diotocardian prosobranchs (the Neritacea excepted) and many mesogastropods show a mid-oesophagus with, at most, a double fold along the mid-ventral line to mark where a ventral food channel may once have stood. In many of the monotocardians, however, even this is lost and in *Littorina*, for example, the right and left glandular areas are connected without interruption across the mid-ventral line (fig. 110E).

Considerable evolutionary change has affected the oesophageal glands in the Steno-glossa, and these seem to have undergone a completely different kind of alteration in the rachiglossan members of the group from what has affected them in the Toxoglossa. In the Rachiglossa much of the evolution must be set down to two things: (1) the elongation of the snout to form a proboscis, and (2) the production of a gland set off from the main course of the gut and connected to it by a duct instead of lying along it as in a mesogastropod—perhaps a more efficient arrangement.

The formation of a proboscis not only calls for an elongation of the body wall in gastropods (a group in which the proboscis is an introvert) but inevitably also for an elongation of the gut which lies within the body wall. In the Rachiglossa (figs. 114, 115) it is easy to see that the part of the gut which has been elongated is almost entirely derived from the anterior oesophagus: the buccal cavity bears the normal relationships of a prosobranch buccal cavity and shows no signs of increased size, and, at the level of the nerve ring, lies the part affected by torsion and therefore mid-oesophageal in origin. Between these two levels, stretching along the whole length of the proboscis, lies a part of the gut (aoe) which is posterior to the mouth of the radular sac, anterior to the point of torsion and marked by a small number of longitudinal folds of which two are outstandingly large and spread forwards dorsal to the odontophore Ciliary currents stream backwards on and between these two folds dorsally and the ducts of the normal salivary glands (sd) lie along their bases and open to the buccal cavity at their anterior ends. Between them, on the ventral side of the gut, lies a shallow groove which opens anteriorly on to a triangular area forming the roof of the radular sac. From these relationships it is evident that the two main folds are the dorsal folds, that the broader dorsal channel between them is the dorsal food channel, and that that section of the gut which has contributed the bulk of the elongation involved in the formation of the proboscis is the anterior oesophagus. When the proboscis is retracted it is flung into an S-bend; when the proboscis extends it straightens.

This part of the gut ends at the base of the proboscis by expanding into a pear-shaped structure which is known as the valve of Leiblein (vl, figs. 97, 114, 115, 116), the broad base of the pear being anterior and its narrower end tapering posteriorly. It narrows to a part of the gut which is surrounded by the nerve ring almost immediately behind which (since the rachiglossan nervous system is partly concentrated) the supra-oesophageal nerve crosses the gut from right to left and the anterior aorta (aa, fig. 114) from left to right, the two almost in contact. The oesophagus, after traversing the nerve ring, expands a little in diameter and in

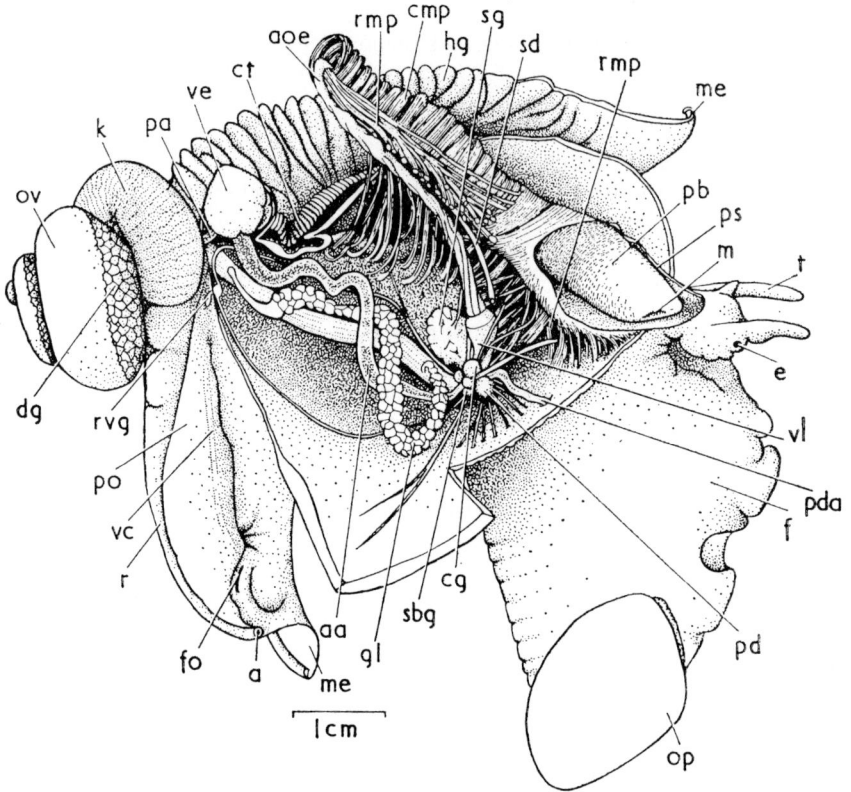

FIG. 114.—*Buccinum undatum:* dissection of anterior part of body from the right side. The mantle skirt has been
cut medianly and deflected above and below, and a window cut in the proboscis sheath.

a, anus; aa, anterior aorta; aoe, anterior oesophagus; cg, right cerebral ganglion; cmp, circular muscles of
proboscis; ct, ctenidium; dg, digestive gland; e, eye; f, foot; fo, female opening; gl, gland of Leiblein; hg,
hypobranchial gland; k, kidney; m, mouth; me, mantle edge; op, operculum; ov, ovary; pa, posterior aorta;
pb, proboscis lying in proboscis sheath; pd, pedal ganglion; pda, pedal artery; po, pallial oviduct; ps, cut wall
of proboscis sheath; r, rectum; rmp, retractor muscles of proboscis; rvg, right visceral ganglion; sbg, sub-
oesophageal ganglion; sd, right salivary duct, cut; sg, left salivary gland; t, tentacle; vc, ventral channel within
capsule gland, seen by transparency; ve, ventricle; vl, valve of Leiblein.

the Muricacea, but not the other groups, bears a mass of lobed glandular tissue (ge, figs. 97,
116) called the 'glande framboisée' by Amaudrut (1898) and the 'median unpaired fore-gut
gland' by Haller (1888). Immediately behind this, in all groups, a duct (dgl, figs. 97, 115) leaves
the oesophagus and runs to a brown or yellow mass of glandular tissue which is wrapped
around the oesophagus and the anterior aorta. This is the gland of Leiblein (gl, figs. 97, 114,
115) and it has been shown, by numerous workers on different animals, to be a source of
powerful digestive enzymes. From the point of entry of the duct from the gland of Leiblein a
relatively narrow tube, with walls flung into several longitudinal folds, runs back to the
stomach. This is the posterior oesophagus (poe, fig. 97) and the problem which calls for
discussion is whether all that lies between the anterior oesophagus in front and the posterior

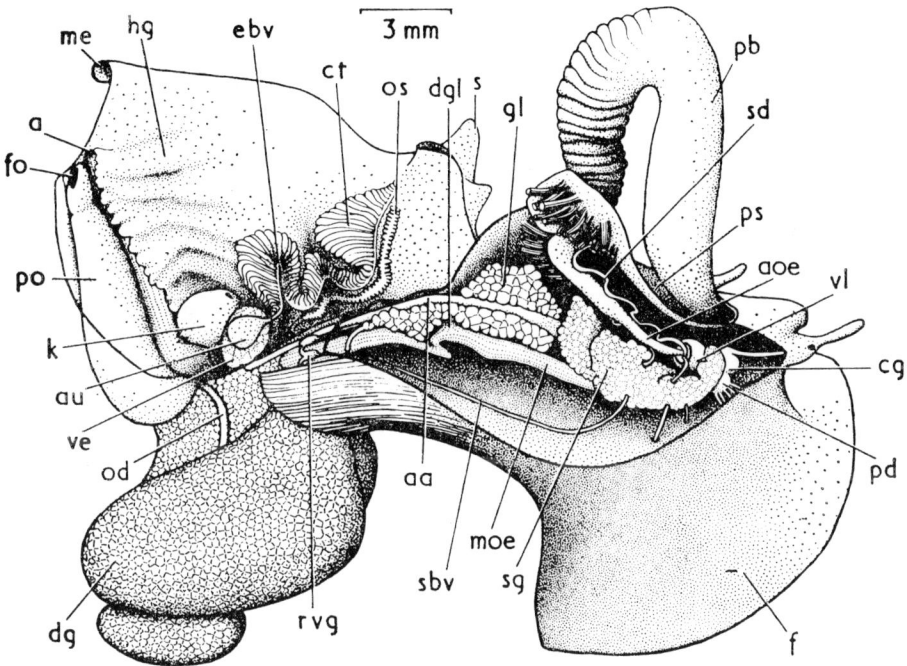

FIG. 115.—*Nassarius reticulatus;* dissection of anterior part of body from the right side. The mantle skirt has been cut along the right side and raised dorsally and the proboscis sheath opened and the proboscis pulled out of it.

　　a, anus; aa, anterior aorta; aoe, anterior oesophagus; au, auricle; cg, right cerebral ganglion; ct, ctenidium; dg, digestive gland; dgl, duct of the gland of Leiblein; ebv, efferent branchial vessel; f, foot; fo, female opening; gl, gland of Leiblein; hg, hypobranchial gland; k, kidney; me, mantle edge; moe, mid-oesophagus; od, ovarian duct; os, osphradium; pb, proboscis; pd, right pedal ganglion; po, pallial oviduct; ps, proboscis sheath; rvg, right visceral ganglion; s, siphon; sbv, sub-oesophageal part of visceral loop; sd, right salivary duct; sg, salivary gland; ve, ventricle; vl, valve of Leiblein.

oesophagus behind is homologous with the mid-oesophageal region of the lower proso-branchs; and, if it is, how the transformation which it has undergone in the stenoglossans has been effected.

　　If the oesophageal region of *Nucella lapillus* be slit longitudinally the appearance which is obtained is seen in fig. 116. The two dorsal folds (df) continue into the valve of Leiblein and run across it, spiralling from a lateroventral position at its anterior end, round the right side, to a mid-dorsal position at its posterior end. As they do so they draw close together so that they lie alongside one another at the posterior end of the valve, with only the narrowest of clefts between. Now here is certainly an appearance which suggests torsion, but at first sight it seems that it cannot actually be torsion since we are talking of folds twisting dorsally round the right side of the oesophagus, whereas in *Patella,* a trochid or *Littorina* they twist ventrally round the left side. Nevertheless the connexions of these folds anteriorly, their histology (tall columnar ciliated cells with numerous intermingled mucous cells) and the direction of the ciliary currents over them, all agree with the idea that they are the homologues of the dorsal

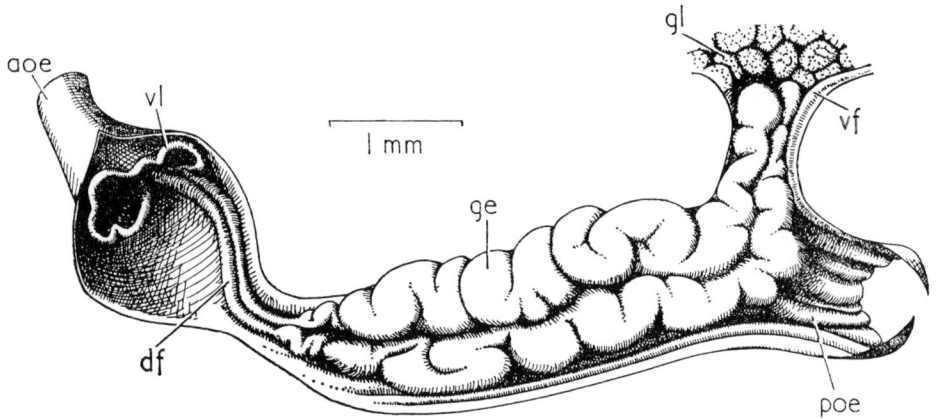

FIG. 116.—*Nucella lapillus:* dissection of the mid-oesophagus. The anterior oesophagus lies to the left, the posterior oesophagus to the right. The mid-oesophagus between has been opened by a cut along the morphologically median dorsal line, which has been extended along the duct of the gland of Leiblein into the gland itself. Since this part of the gut is involved in torsion this cut starts topographically mid-dorsally at the anterior end and twists to a topographically mid-ventral position posteriorly. The gut has been straightened so that this twist is not seen.

aoe, anterior oesophagus; df, dorsal fold; ge, 'glande framboisée'; gl, gland of Leiblein; poe, posterior oesophagus; vf, ventral fold; vl, valve of Leiblein.

folds of other prosobranchs. They have, however, migrated ventrally down the side wall of the oesophagus until they now lie ventrolaterally or even ventrally. In so doing they have expanded the dorsal food channel and constricted the lateral and ventral walls of the oesophagus. The course which they now follow is therefore, quite properly, one that corresponds to the course taken along the mid-oesophagus of *Patella* or *Gibbula* by the ventral fold.

If the folds are traced behind the valve of Leiblein, along the narrow stretch which runs from there through the nerve ring and back to the duct of the gland of Leiblein, they will be found to run along the dorsal side of the oesophagus and become greatly folded and expanded. It is, in fact, these expanded convolutions of the dorsal folds which produce the 'glande framboisée' (ge) in this situation. Finally the folds enter the duct of the gland of Leiblein (gl) and die away near the point where the glandular tissue of that organ begins. On the walls of the gland, however, starting near its posterior tip, runs a double fold (vf), which extends along the opposite wall of the duct of the gland from that on which the dorsal folds lie, and then curves backward into the posterior oesophagus (poe). This may well be interpreted as a structure homologous with the ventral fold of the more typical mid-oesophagus.

If we now turn attention to the cleft which separates the two dorsal folds in their new, ventral position we ought to find something which corresponds to the ventral half of the normal mid-oesophagus. What we do find, in fact, is a strip of oesophageal wall lined by an epithelium of squamous cells, contrasting completely with that which forms the lining of the remainder of the oesophagus. The cells are featureless and show no trace of glandular activity, or, indeed, of being more than a membranous seal to the gut cavity stretched between the dorsal folds.

In all probability, the correct interpretation which should be placed upon this part of the alimentary tract (fig. 110G) is this. In the formation of the rachiglossan proboscis, as has been seen, great elongation of the anterior part of the gut is necessary, largely accomplished by extension of the anterior oesophagus, but needing still further alteration in the arrangement of the parts. That the elongation of the anterior oesophagus has not been adequate to keep pace with the growth rate of the proboscis is shown, for example, by the fact that the (ordinary) salivary glands now lie anterior to the nerve ring, their ducts not passing through that structure: the ducts have grown less fast than the body wall and so the glands have been pulled forwards. Similarly, part of the mid-oesophagus, represented by the valve of Leiblein, has been pulled forwards through the nerve ring by the traction exerted on it by the growing body wall: this can be recognized as mid-oesophageal by the torsion of the gut shown in that position. Now the mid-oesophagus is normally a dilated region of the gut by virtue of the glandular pouches attached to it, a configuration which does not lend itself to movement through a relatively narrow gap like the nerve ring. The rachiglossan has therefore undergone an evolution in the course of which the lateral glandular pouches of the mid-oesophagus and the ventral fold which runs between right and left halves, have been torn off backwards from the food channel so that they are now connected to the mid-oesophagus only at its posterior end. To fill the hole in the oesophageal wall the dorsal folds have migrated ventrally until they come into contact with one another in the mid-ventral line, but, even so, there is left between them a narrow gap which is filled in with a scar tissue represented by the thin epithelium described above. It is extremely unlikely, partly because of the close contact maintained between the two dorsal folds, partly because of the way in which the cilia on the dorsal folds maintain their original direction of beating—out of the lateral parts of the oesophagus into the food channel and therefore out of the cleft between the folds—that food particles ever reach this epithelium, which might easily be broken by their impact.

One further elaboration of this region of the alimentary canal in *Nucella* remains to be described. At the anterior end of the valve of Leiblein each dorsal fold is expanded into a flap-like structure which hangs freely into the cavity of the gut (vl, fig. 116), and into a pad of mucous cells lying at the base of the flap. The two flaps normally project backwards into the lumen of the pharynx, which is dilated at this spot largely to accommodate them. Their free surfaces are fringed with extremely long cilia which beat, however, very languidly. This structure acts as a valve, reacting partly mechanically in preventing regurgitation of food from the more posterior parts of the gut during the elongation of the proboscis, and probably also chemically, since Brock (1936) has shown that in the whelk *Buccinum undatum* stomach contents or secretion from the digestive gland stimulate the flaps to come together and prevent forward movement. Since the valve of Leiblein is a more or less constant feature of the rachiglossan gut it is probable that it acts in the same way in each animal.

The conditions which have just been described apply in particular to *Nucella*, but would be equally true for any member of the Muricacea. A certain number of differences are noticeable in the Buccinacea and Volutacea. In the Buccinacea the valve of Leiblein is reduced or even absent (*Galeodes, Semifusus, Busycon*) and so are the dorsal folds, which never give rise to a 'glande framboisée'. More important, perhaps, is the fact that the part of the mid-oesophagus which shows the effects of torsion is not the part which forms the valve of Leiblein, but the length behind that, which runs through the nerve ring and back to the opening of the duct of the gland. This is marked in *Neptunea* and *Buccinum* by a groove, lined with scar tissue, which represents the original line of attachment of the gland and which twists from the mid-ventral line up the right side to a dorsal position; in *Nassarius* even this vestige has been lost.

In the Muricacea the gland of Leiblein (gl, fig. 97) is a large and solid mass of glandular tissue lying rather compactly behind the nerve ring and shown by Hirsch (1915) (and probably by Mansour-Bek, 1934) to secrete digestive enzymes in the Mediterranean *Murex trunculus*. In the Buccinacea it appears (*Buccinum,* gl, fig. 114; *Nassarius,* gl, fig. 115) much less solid, its walls more transparent, and it extends backwards along the posterior oesophagus as a thin finger-shaped caecum the walls of which are not glandular at all. This tendency of the organ to become less glandular appears to extend to other members of the Buccinacea and the Volutacea: Vanstone (1894) described *Semifusus* spp. as having only a small caecum to represent the gland and *Galeodes melongena* as having lost it altogether, whilst Woodward (1901b) described *Volutocorbis abyssicola* as having only a sac-like gland. This would seem to suggest an evolutionary trend within the Rachiglossa towards the suppression of the gland of Leiblein, possibly associated with the assumption of greater activity by the digestive gland.

From the histological point of view the mid-oesophagus of these stenoglossan proso-branchs shows an advance on that of lower forms. The main tubular part of the gut is lined by a tall, ciliated, columnar epithelium with numerous mucous cells interspersed, especially in the valve of Leiblein and where the dorsal folds are elaborated in the Muricacea to form the 'glande framboisée'. The epithelium rests on a wall with a little circular and longitudinal muscle in it, but the cilia appear to be responsible for much of the movement of food. The gland (fig. 113B), in the forms where it is well developed, has a central cavity into which a large number of partitions project. In the partitions and in the outer walls lie a few muscle fibres (mu), but there is no muscular capsule around it. The epithelial cells which clothe the inner walls are mainly club-shaped and bulge into the cavity, and their distal tips appear to be nipped off when secretion occurs (snl). The nuclei lie centrally, and the cytoplasm is usually stuffed with large numbers of protein spherules (msc). The cells are very fragile and burst readily on handling. Occasional mucous cells are intermingled with the other gland cells. Sometimes the gland cells have a striated distal border which may indicate a change of phase. Franc (1952) has shown the gland to be rich in alkaline phosphatase in *Tritonalia* (= *Ocenebra*) *aciculata*.

The second group of the Stenoglossa is the Toxoglossa, a group which has achieved some notoriety by containing the animals known as cone shells, capable of inflicting a severe, poisonous bite even on man. The poison gland of these animals has been taken to be the homologue of the gland of Leiblein (fig. 111H), and therefore of the oesophageal pouches of lower prosobranchs, since the days of Amaudrut (1898). [For recent work see chapters 25, 26.]

There still remain to be dealt with a few cases where the mid-oesophageal region of the gut has undergone some special modification. The most puzzling example of this is shown by the sinistral prosobranch *Triphora perversa* described by Fretter (1951b) (fig. 117). In this animal, which has a long proboscis correlated with its habit of feeding on sponges amongst other things, there are two large salivary glands which open by a single duct into the ventral part of the oesophagus. The part of the duct near the glands has secretory cells in it like those found in the glands themselves, but the distal part, near the oesophagus, contains other types of gland cells different from anything in the salivary gland or the oesophagus. The oesophagus itself is a tube of small diameter, lined by a ciliated columnar epithelium: no sign of dorsal folds, ventral folds, lateral glandular pouches or of torsion can be detected. Overlying it, however, and connected to it by a duct, is a gland (gp), the internal cavity of which is lined by a glandular epithelium flung into septate folds (tfd) and which is therefore reminiscent of many a mid-oesophageal gland. The duct leaves the posterior end of the gland and opens to the oesophagus not far from the point where that enters the stomach (st).

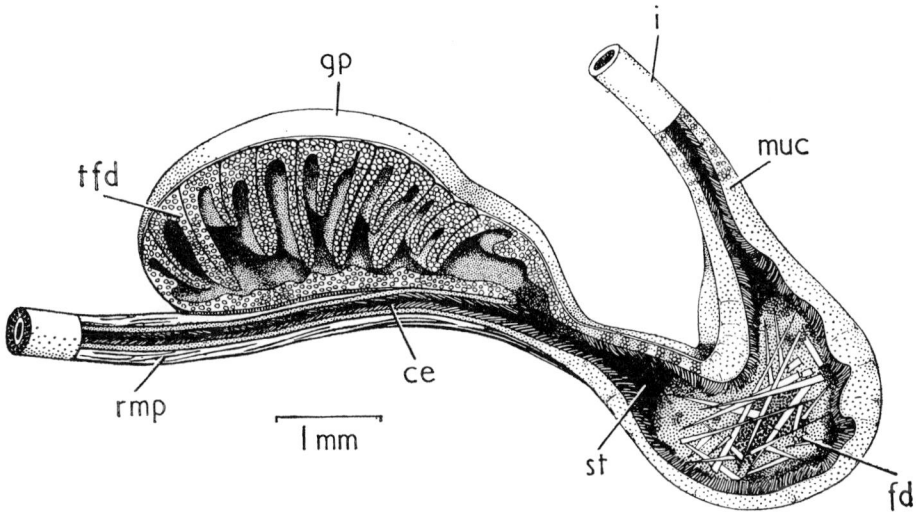

FIG. 117.—*Triphora perversa:* diagrammatic longitudinal section through the posterior oesophagus, stomach and part of the intestine.
    ce, ciliated epithelium of oesophagus; fd, food in stomach; gp, glandular pouch; i, intestine; muc, mucous cells of intestine; rmp, retractor muscles of proboscis; st, stomach; tfd, transverse fold.

It may be that here, again, a separation of the glandular equipment of the mid-oesophagus has occurred in relation to the formation of a lengthy proboscis, but, as in Toxoglossa, this would appear to have occurred in *Triphora* without leaving any scar along the line of separation between oesophagus and gland. Perhaps, therefore, it is a new structure which has been evolved in *Triphora* for its own special functions. That would, to a certain extent, fit in with the unusual arrangements which the salivary glands exhibit. Fretter, however, offered an explanation of the salivary arrangement which involves the glandular equipment of the mid-oesophagus, and, if this is accepted then the posterior oesophageal pouch must have originated *de novo* or a split of the oesophageal glands into anterior and posterior halves must be postulated. In the lower prosobranchs the salivary ducts run dorsolaterally along the wall of the anterior part of the oesophagus at the insertion of the dorsal folds (sd, fig. 17). If the latter be supposed to migrate round the oesophageal wall ventrally—as they have already been shown to do in the rachiglossan stenoglossans—then they and the two salivary ducts will come to be approximated mid-ventrally. It is possible that in this way a fusion would be brought about, so that a single salivary duct would open mid-ventrally into the anterior part of the oesophagus, as is actually observed. Conceivably, too, some of the remaining glandular tissue of strictly oesophageal origin could be incorporated into the terminal part of this single duct so as to produce the type of gland cell recorded as occurring here but nowhere else.

Finally, it must be noted that a considerable number of prosobranchs appear to have no glandular structures of any sort in connexion with the oesophagus. This point has been investigated by Graham (1939) and it appears to be generally true that, whenever a crystalline style is found in the stomach, the oesophageal glands are either so vestigial as to be functionless, or have vanished altogether (fig. 110F). [See p. 596.]

The British prosobranchs which are known to possess a style are these: all Rissoidae; *Hydrobia ulvae; H. ventrosa; Potamopyrgus jenkinsi; Bythinella scholtzi; Pseudamnicola confusa; Bithynia tentaculata; B. leachi; Assiminea grayana; Pomatias elegans; Turritella communis; Aporrhais pespelicani; Crepidula fornicata; Calyptraea chinensis; Capulus ungaricus.*

A few other prosobranchs which have no crystalline style, but are, in fact, either pronounced carnivores or parasites, have also lost all trace of oesophageal glands. These include animals such as *Ianthina, Clathrus, Balcis, Eulima* and the pyramidellids; as all these show opisthobranch affinities this loss of oesophageal glands is probably a matter of inheritance rather than of adaptation.

The posterior oesophagus, the section of gut which follows the mid-oesophagus is, in contrast to that length of the alimentary canal, a simple tube carrying food and such enzymes as are produced by the salivary and oesophageal glands towards the stomach. The only development of this region which calls for comment is the pouch-like caecum attached to it in *Buccinum* and *Neptunea.* This appears to be a simple expansion of the gut, presumably providing some extra space which may help to counteract any tendency towards regurgitation of food in the event of elongation of the proboscis.

The next section of the alimentary canal is the stomach, which lies in the visceral hump usually towards the left side. Much of it is invisible in surface view because of the way in which it is covered by other viscera, of which kidney, digestive gland and gonad are the most important. As has been indicated above (p. 31) the alimentary canal of gastropods is primitively U-shaped when seen from the side, rising from the mouth to the stomach in the visceral hump and then falling again to the point where the anus discharges to the mantle cavity. In the hypothetical pre-torsional stage the mantle cavity was posteriorly placed, the entry of the oesophagus to the stomach was anterior and the origin of the intestine from it posterior. Even after torsion has displaced the mantle cavity and anus to an anterior position the more primitive groups (fig. 118A) still maintain this orientation of the parts and the oesophagus opens (oa) at the topographically posterior end of the stomach and the intestine (i) leaves the topographically anterior end. Because of the twist which occurs in the oesophageal region of the gut the dorsal surface of the stomach is the morphologically dorsal side and its ventral surface is morphologically ventral, as reference to fig. 16 will show.

A large number of more or less extensive descriptions of the anatomy of the stomach of various prosobranchs have been published, the value of which is diminished by the fact that they have often been opened in different ways in different states of contraction and almost always described in inadequate detail. The following descriptions rest mainly on the comparative account given by Graham (1949). The most primitive type of stomach which is apparently present in the prosobranchs is that of the Trochacea (see Owen, 1958): that of *Monodonta* is shown in fig. 119A. It is an ovoid sac embedded in digestive gland, lying near the base of the visceral hump, on the animal's left. It is divisible into two parts, a more posteriorly placed globular part, and a more cylindrical anterior, the style sac (SS). The oesophagus (OE) opens near the posterior end of the globular portion on to a groove on the right gastric wall. Into this groove discharges a single duct from the digestive gland (DD). The groove, bounded by a marked fold (T_1) on its left, runs into the mouth of a spirally coiled caecum (SCM) which extends from the posterior end of the stomach amongst the tubules of the digestive gland. Almost the whole of the remainder of the wall of the posterior globular part of the stomach is covered by a thick cuticle, the gastric shield, which is raised near the middle of the left side into a prominent boss (GS). Along its ventral edge the gastric shield borders a fold (FF) and this, in turn, forms the boundary of a strip of stomach wall (SAP) edged by the fold T_1 already

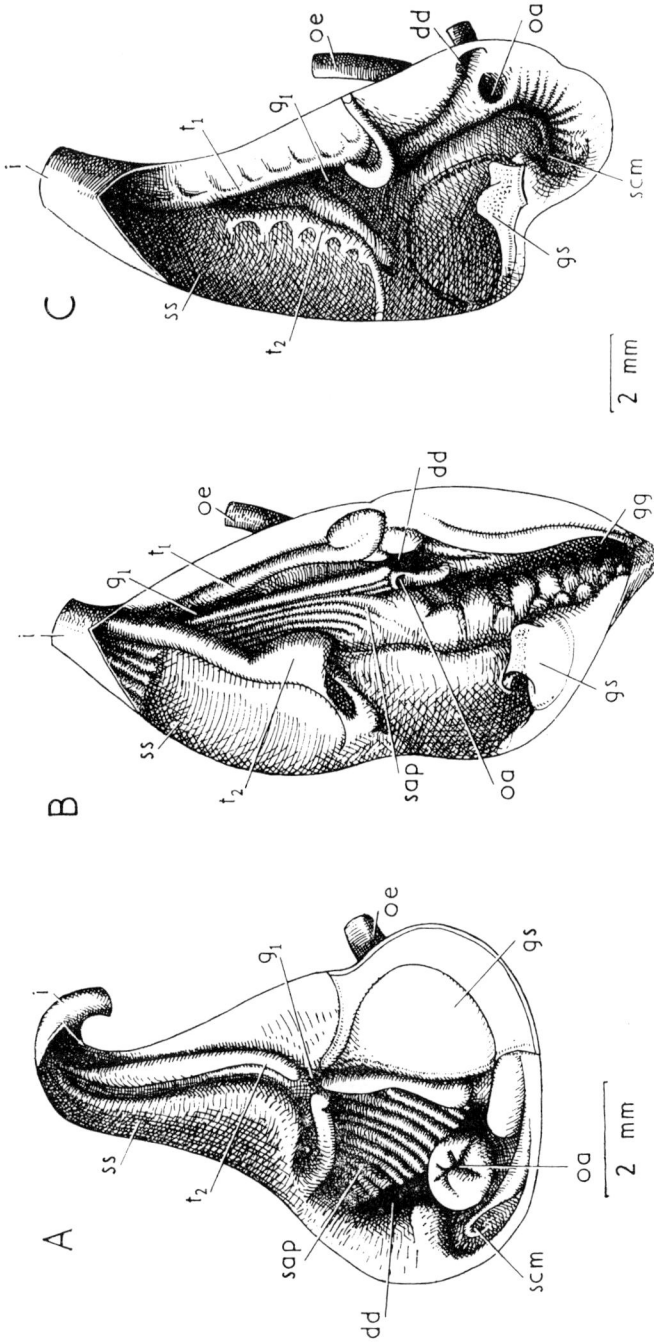

FIG. 118.—Dissections of stomachs, opened by a dorsal longitudinal cut. A, *Diodora apertura*; B, *Theodoxus fluviatilis*; C, *Bithynia tentaculata*. dd, opening of ducts of digestive gland; g₁, intestinal groove; gg, posterior groove; gs, gastric shield; i, intestine; oa, oesophageal aperture; oe, oesophagus; sap, sorting area; scm, spiral caecum, reduced; ss, style sac region; t₁, major typhlosole; t₂, minor typhlosole.

referred to and, like it, disappearing into the mouth of the spiral caecum. This strip is crossed by a large number of parallel grooves and ridges and extends anteriorly into the base of the cylindrical part of the stomach. Examination of this region in a living animal shows that diverse, vigorous, ciliary currents beat over its surface, and that it is, in fact, an area where particles are sorted, mainly on a basis of size. It has been called the posterior sorting area of the stomach. The folds which run across it do not make actual contact with the major typhlosole (T_1), but are prevented from so doing by a groove, the intestinal groove (G_1), which lies between and separates the two structures. Along this a fast ciliary stream sets towards the intestinal end of the stomach. Posteriorly, like the sorting area and the groove into which the oesophagus and the duct from the digestive gland open, the intestinal groove disappears into the mouth of the spiral caecum. If that be slit open (fig. 119B, C) the relations of the structures within it may be seen, and it is obvious that the major typhlosole (T_1) extends to the very tip of the caecum, where it ends, whereas the intestinal groove (G_1) curves round its tip and so becomes continuous with the groove connected with the oesophagus and digestive gland— they are, in fact, one and the same groove.

The only other feature which calls for mention in the posterior part of the stomach of a trochid is another groove which emerges from the caecum on the side opposite to that on which the ingoing and outcoming intestinal groove runs: this curves across to the base of the boss on the gastric shield and cilia on it beat in that direction.

The intestinal end of the stomach (SS) is cylindrical or conical in shape, broader where it is fitted to the posteriorly placed globular part, narrower anteriorly where it runs into the intestine. Along its ventral wall the intestinal groove extends from the posterior part of the stomach, with an extension of the major typhlosole (T_1) on its right. On the left it is bordered by a second typhlosole (T_2), the minor, which resembles the major but does not pass into the posterior half of the stomach. At the base of the style sac these typhlosoles diverge some-what, and in the triangular space so formed are wedged the tip of the sorting area and a special area rich in mucous glands.

On the basis of this description and a comparison between the trochid stomach and that of other molluscs it is possible to say that the generalized prosobranch stomach (fig. 87) may be expected to show the following features. The oesophagus (oe) opens on the right side near the posterior end, the intestine (i) emerges from the anterior end of the style sac (ss), and the duct, or ducts, from the digestive gland (dd) open near the oesophageal aperture (oa). From the posterior end of the stomach there extends a spiral caecum (scm). Internally the funda-mental features of the stomach wall appear to be: (1) the extensive cuticularized area, the gastric shield (gs), lying over the dorsal and left walls; (2) the intestinal groove (g_1) along the major typhlosole (t_1), the latter extending across the whole length of the stomach from the intestine to the apex of the spiral caecum, the former curving round that and running to the duct of the digestive gland; (3) the sorting area (sap) which lies in relation to the intestinal groove between it and the gastric shield; and (4) the minor typhlosole (t_2) which forms a second lip to the intestinal groove in the style sac.

Within the prosobranch gastropods it is possible to trace a relatively small number of evolutionary trends which affect the organization of the stomach. One of these is the gradual disappearance of the spiral caecum. Although this structure is clearly recognizable in most cephalopod and lamellibranch molluscs and therefore appears to be a fundamental feature of molluscan gastric organization, it is, in fact, only the Trochacea and the families of Zeugobranchia with spirally coiled shells (Pleurotomariidae, Haliotidae, Scissurellidae) which possess it in a well formed state amongst the Prosobranchia. In the remaining Zeugobranchia,

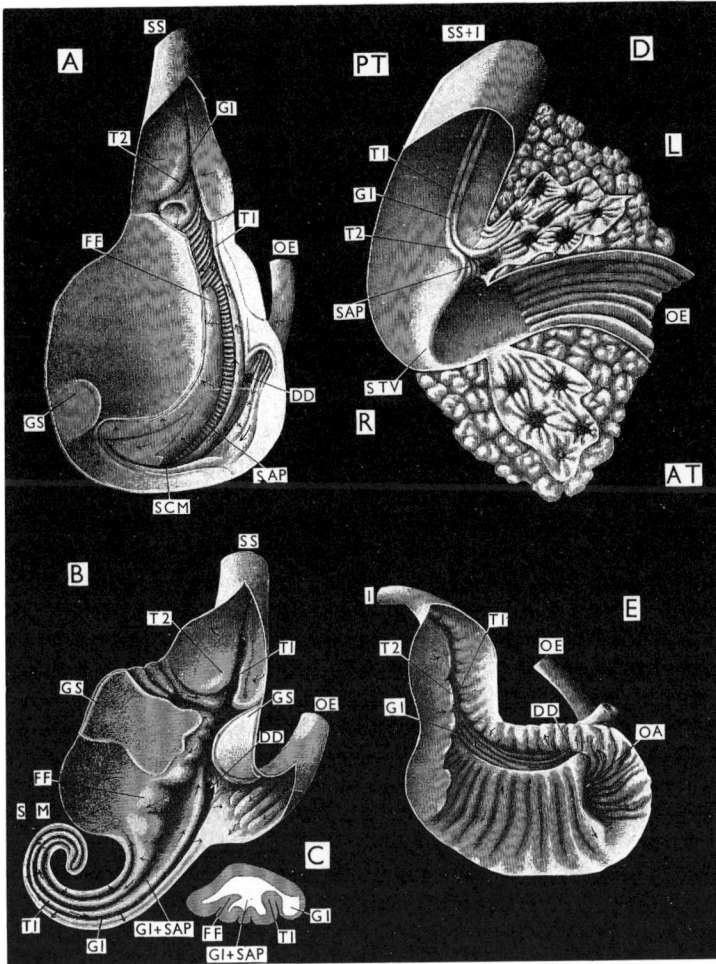

FIG. 119.—Dissections of stomachs, opened by a dorsal longitudinal cut. A, *Monodonta lineata*; B, *Calliostoma zizyphinum*; C, *Calliostoma zizyphinum*, T.S. spiral caecum; D, *Patella vulgata*; E, *Nucella lapillus*. Arrows show the course of ciliary currents. All × about 13.

AT, anterior end; DD, opening of duct of digestive gland; FF, fold emerging from spiral caecum; G_1, intestinal groove; G_1 + SAP, area representing fusion of intestinal groove and sorting area; GS, gastric shield; I, intestine; L, left; OA, oesophageal aperture; OE, oesophagus; PT, posterior end; R, right; SAP, sorting area; SCM, SM, spiral caecum; SS, style sac region; SS + I, part of gut formed from style sac or intestine—boundaries not detectable; STV, fold in stomach acting as valve; T_1, major typhlosole; T_2, minor typhlosole.

e.g. *Diodora* (fig. 118A), *Emarginula* (Graham, 1939), the stomach shows the usual division into a globular posterior portion and a narrower anterior style sac, with the oesophagus opening into the former on a prominent papilla (oa) flanked by large ducts (dd), right and left, from the digestive gland. From these, sorting area (sap), intestinal groove (g_1) and major typhlosole run in characteristic fashion towards the style sac (ss), but behind there lies only a slight depression (scm) which could be held to represent the spiral caecum. A greater representation of the caecum may perhaps be recognizable in the Neritacea, if the conditions in *Theodoxus fluviatilis* (fig. 118B) are at all typical of the group. In this animal the oesophagus opens on the right (oa), into the posterior chamber, between the openings of two ducts from the digestive gland (dd) and from these apertures a pronounced groove (gg) leads to the posterior apex of the stomach. This is edged on both sides by folds, one of which runs along the base of the gastric shield (gs), which is not an extensive cuticularized area but covers only a limited part of the stomach wall, though it is raised into a curved crest in its middle. The fold which borders the other side of the groove is not obviously connected with the major typhlosole (t_1), which appears to originate at the mouth of one of the ducts from the digestive gland. From the posterior apex of the stomach a groove leads towards the base of the gastric shield, and more anteriorly a grooved and ridged sorting area (sap) lies at the base of the style sac, wedged between the two typhlosoles which run there, and in this area, too, lies a pouch-like depression from which mucus is secreted. This stomach obviously bears resemblances to that of *Monodonta* and, in particular, the slightly protuberant upper apex has a number of points in common with the spiral caecum of the trochid, though it is much less well developed and not at all spirally wound. In similar fashion it is possible to suggest that small pockets which lie at the upper end of the stomach of *Bithynia tentaculata* (fig. 118C), *Turritella communis* (fig. 120C), *Aporrhais pespelicani*, *Bittium reticulatum*, *Calyptraea chinensis* (fig. 120B) and *Crepidula fornicata* (fig. 120A) are vestiges of the elaborate caecum of the trochids. In most of the prosobranch stomachs, however, not even this persists, and it is not possible to see any trace of it in the Docoglossa (fig. 119D), Architaenioglossa, Valvatacea, Littorinacea, Rissoacea, the higher groups of the mesogastropods or Stenoglossa (fig. 119E). The explanation of this is partly functional, as will be seen below.

A second evolutionary trend which is discernible in the prosobranchs is the gradual migration of the oesophageal opening from its morphologically anterior but topographically posterior position to a point which is topographically anterior. This involves a gradual shifting along the ventral side of the stomach and brings it much closer to the base of the style sac, so that the oesophagus and the style sac appear to be coming off the posterior part of the stomach together, and, at least in some cases, that part then looks like a caecum extending up the visceral hump. The changed position of the oesophageal opening is probably due as much to the mechanical traction of the more anterior parts of the alimentary canal as to any other cause, but, as a discussion of function will show later, it is also tied up with the altered importance of the spiral caecum and may be partly responsible for the disappearance of that structure in most prosobranchs, though it is difficult to disentangle cause and effect in this connexion. Associated with these changes is an alteration in the importance of the major typhlosole: with the reduced size or even absence of the caecum there is correlated a tendency for the typhlosole to stop near the base of the style sac, and this is, indeed, almost a necessary condition for the migration of the oesophageal opening towards the same point. This is indicated even in the trochids because the stomach of *Calliostoma zizyphinum* (fig. 119B) shows the oesophagus moving forwards and a breaking of the major typhlosole (T_1) into 2 sections, an anterior one which is confined to the style sac (SS) and a posterior one which is restricted to the caecum (SCM). It is across the break in the typhlosole which has been made in this way that the oesophagus, and, to some extent, the ducts of the digestive gland too, migrate forwards.

A third change in the stomach of prosobranchs is definitely connected with a change in feeding habits. The more archaic members of the sub-class are all herbivorous and mainly

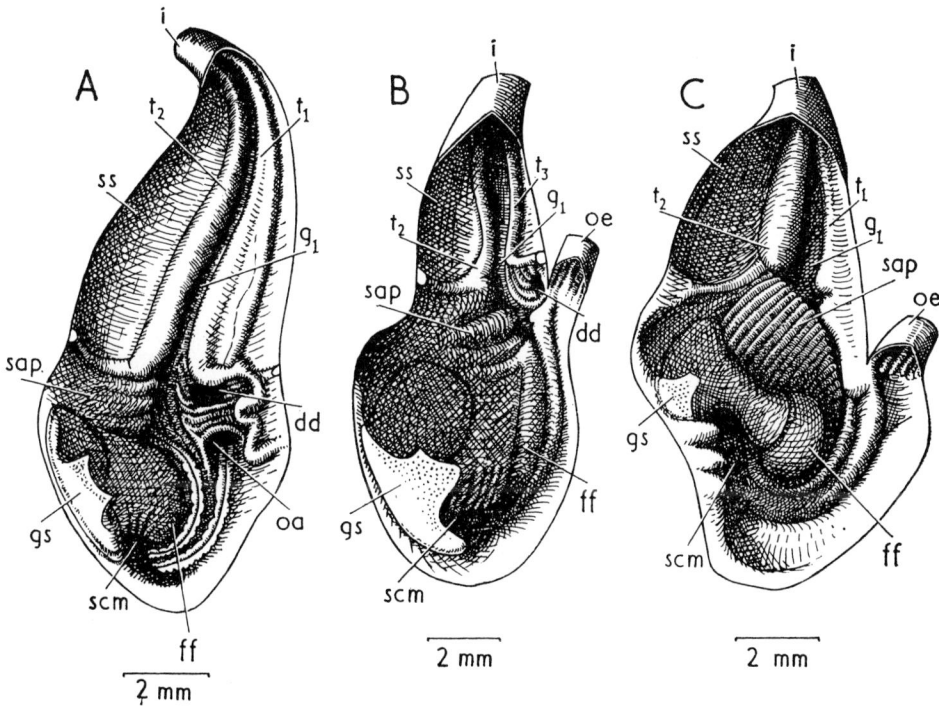

FIG. 120.—Dissections of stomachs, opened by a dorsal longitudinal cut. A, *Crepidula fornicata;* B, *Calyptraea chinensis;* C, *Turritella communis.*

dd, opening of duct of digestive gland; ff, fold emerging from spiral caecum, g₁, intestinal groove; gs, gastric shield; i, intestine; oa, oesophageal aperture; oe, oesophagus; sap, sorting area; scm, spiral caecum, much reduced; ss, style sac region; t₁, major typhlosole; t₂, minor typhlosole; t₃, fold on major typhlosole.

microphagous, collecting algae and detritus of all kinds by means of their radula; a few may, in addition, bite or scrape small pieces off larger plants by means of the radula acting against the jaw. Most of the higher monotocardians, however, are carnivores, and the problems of digestion are different. There may, too, be a greater emphasis on extra-cellular digestion in the carnivorous forms and a greater emphasis on the sorting of material for ingestion and intra-cellular digestion by the cells of the digestive gland in the case of the microphagous herbivores. Whichever of these may be the effective cause there is a pronounced simplification of the stomach in the higher prosobranchs and it is reduced to little more than a sac to which oesophagus, intestine and the ducts of the digestive gland open, with only vestiges of intestinal groove and typhlosoles and often no trace at all of caecum, sorting area or gastric shield. This is seen, for example, in the stomach of *Nucella lapillus* (fig. 119E) (Graham, 1949), *Natica catena* (fig. 121A) and *Buccinum undatum* (Brock, 1936).

The same kind of evolution, more unexpectedly, has occurred in the Patellacea (fig. 119D), where the morphologically anterior part of the stomach can be regarded as almost wholly lost, a mere trace sandwiched between the well developed oesophagus (OE) on the one hand and the equally prominent style sac (SS) on the other. Two typhlosoles (T₁, T₂) run along the latter with a vestigial sorting area (SAP) between, but all the other features of this part of

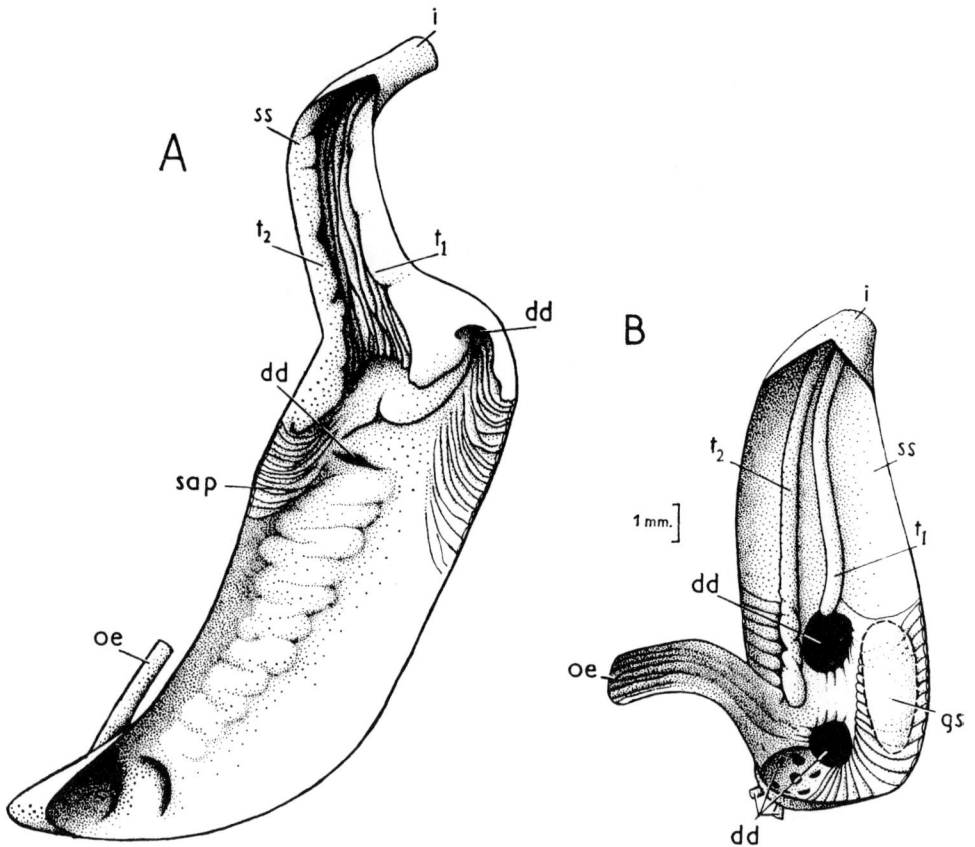

FIG. 121.—Dissections of stomachs, opened by a dorsal longitudinal cut. A, *Natica catena;* B, *Capulus ungaricus*. dd, opening of ducts of digestive gland; gs, gastric shield; i, intestine; oe, oesophagus; sap, sorting area; ss, style sac region; t_1, major typhlosole; t_2, minor typhlosole.

the prosobranch stomach have disappeared. It is difficult to suggest a convincing reason for their loss in these animals.

Throughout, the stomach is lined by a columnar epithelium which, in certain areas, contains mucous goblet cells. Other types of gland are not common. The height of the epithelium is variable and it is principally to this that may be attributed the lesser folds and grooves which are so prominent over its inner surface. The cells are not noteworthy in any way except in so far as they may be either ciliated (cic, fig. 122E) or cuticularized (gs, fig. 122D). In the latter case they exhibit the same structure as in the jaw—a fibrous layer interposed between the distal surface of the cell and the overlying cuticle, suggesting that the secretion escapes from pores in the cell surface and that the individual threads coalesce to form a sheet over the epithelium.

The style sac is marked off from the rest of the stomach not only anatomically but also histologically. The typhlosoles, major and minor, bear tall, columnar, ciliated cells interspersed with which are gland cells, mainly mucous. The rest of the sac is lined by an extremely characteristic epithelium the like of which is not found in any other situation. The cells are cubical or columnar and are filled with rather dense cytoplasm in which a large, rounded nucleus is centrally placed. The distal surface is densely clothed with long, close-set cilia each with a prominent basal granule. From the basal granules intracellular fibrillae converge fanwise to the side of the nucleus and can, on occasion, be traced even as far as the basal surface of the cell.

In gastropods with a crystalline style it appears to be the gland cells on the typhlosolar regions of the style sac which are responsible for the secretion of the substance out of which the style is made, just as they are in lamellibranchs (Nelson, 1918). The evolutionary step which has occurred in so many bivalves whereby the typhlosoles fuse across the intestinal groove and so separate that (as an intestine) from the main cavity of the style sac, has occurred much less commonly in prosobranchs and is found only in *Tornus* (Woodward, 1899), *Typhobia* (Robson, 1922) and *Pterocera* (Yonge, 1932). It is well to remember in this connexion, however, that whereas all bivalves (save the protobranchs) possess a crystalline style, only a very small number of gastropods do, and it is probable that more advanced structure will be found only in the former class.

Opening into the stomach lies the digestive gland which fills the greater part of the visceral hump except, possibly, during the breeding season when the gonad is large. The digestive gland is the organ most clearly visible on the surface of the visceral mass when a prosobranch is removed from its shell. Superficially examined, it appears as a vast mass of branching tubules; more carefully seen, it is found to be composed of 2 lobes of unequal size each connected to the stomach by a single duct. Of these lobes that on the left is by far the larger. Much modification of the ducts may occur: in many Trochacea and Docoglossa, for example, the 2 ducts from the 2 lobes unite before reaching the stomach so that only a single aperture appears on the stomach wall. In others the reverse kind of change seems to have occurred and each lobe may open to the stomach by several apertures as if multiple ducts were present. This is presumably due to the opening out of the main ducts on to the walls of the stomach.

The tubules of which the gland is composed extend into the visceral haemocoel and so are bathed in blood (fig. 122). Only a very thin layer of connective tissue (cn) appears to separate the digestive cells from the blood (hc), and, in places, even this may be absent. A few muscle fibres are normally present around them; cells containing glycogen may be found in the connective tissue. The amount of material which is stored here depends not only upon such obvious factors as the degree of starvation of the animal but also upon the season and the sexual state. Linke (1934b) has shown that in a prosobranch such as *Littorina littorea* (fig. 123) the bulk of the visceral hump is occupied by digestive gland (dg) and gonad (t) during the breeding season with a minimal amount of connective tissue (cn) containing reserve food separating the tubules of these 2 organs. During the resting period which intervenes between two breeding periods, however, much of the reproductive system is broken down and, at these times, the visceral hump is primarily made up of digestive gland and a voluminous connective tissue rich in stored foodstuffs in which only vestiges of gonad may be traced.

In most prosobranchs, at least the main ducts of the digestive gland are lined by a ciliated, columnar epithelium similar to that lining the stomach (ste, fig. 122E): it may be that it is an eversion of the stomach wall which has given rise to them. The tubules of the gland, however, are lined by a totally different type of epithelium within which at least two different types of cell may always be distinguished. One of these appears to correspond, broadly, to the

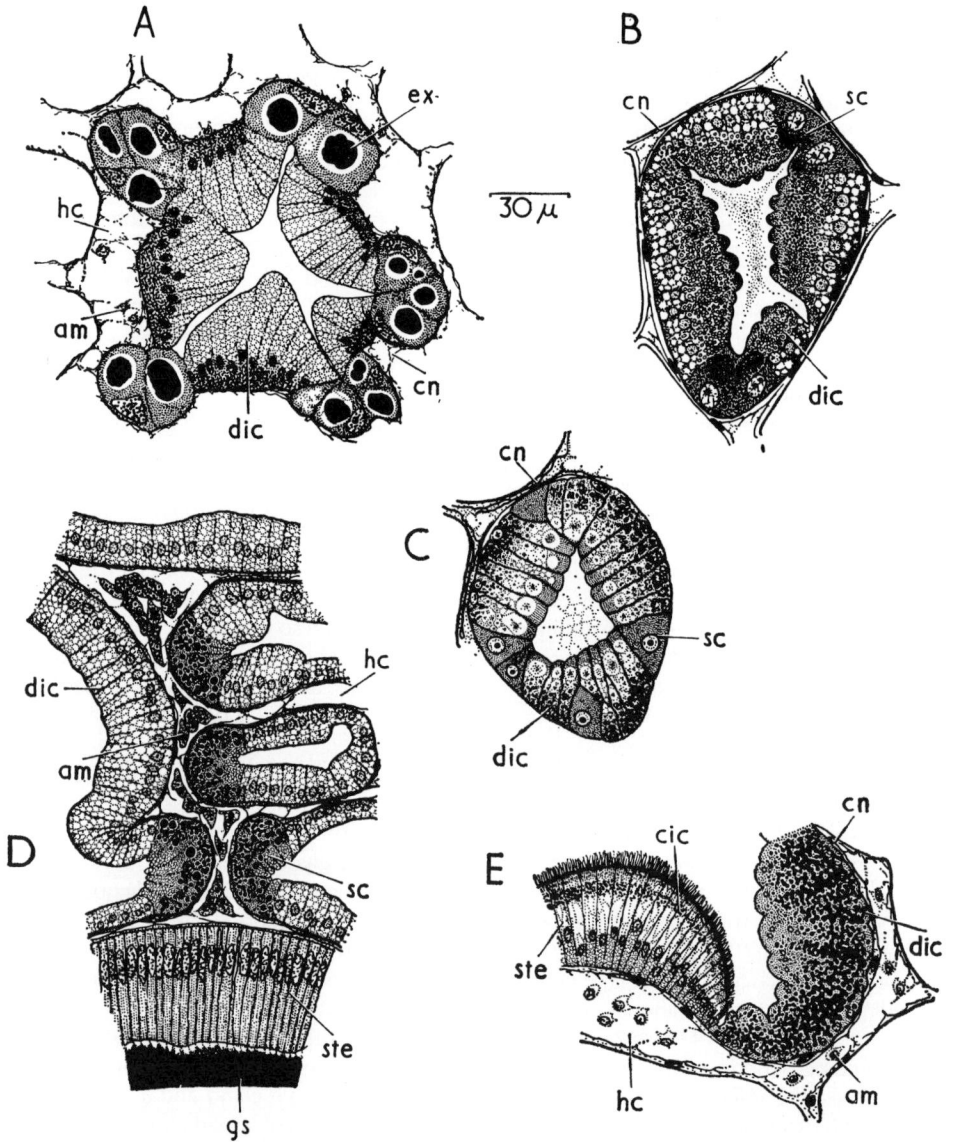

FIG. 122.—Histology of the digestive gland. A, *Bithynia tentaculata;* B, *Lacuna vincta;* C, *Patella vulgata;* D, *Gibbula cineraria;* E, *Natica catena.* Fig. D contains a piece of cuticularized gastric wall, and fig. E a piece of ciliated duct.

am, amoebocyte; cic, ciliated cell; cn, connective tissue; dic, digestive cell; ex, excretory cell; gs, cuticle of gastric shield; hc, haemocoelic space; sc, secretory cell; ste, epithelium of stomach wall.

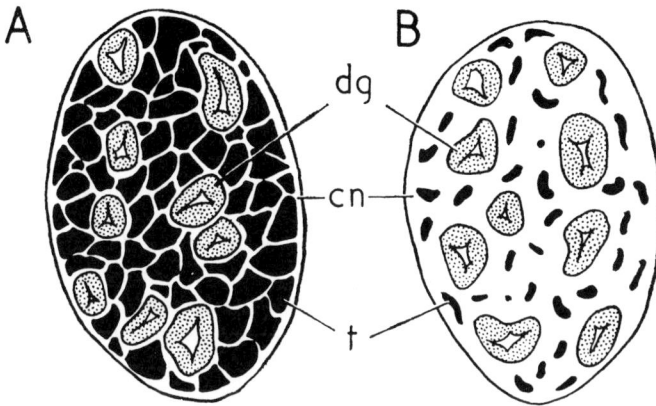

FIG. 123.—*Littorina littorea*: diagrammatic transverse sections of the visceral hump. A, during the breeding season; B, in non-breeding period. After Linke.
cn, connective tissue; dg, digestive gland; t, testis.

digestive cell described by Yonge (1925*b*) in the digestive gland of bivalves, but the second sort is different from the 'young' cells which occur in the crypts of the tubules in members of that class.

The histology of the digestive gland varies considerably from animal to animal amongst the prosobranchs. This may be partly due to genuine specific differences related to such questions as the animal's food and whether ingestion of particulate food is carried on, but some of the differences may be due to a rhythmical cycle of activity related to secretion, ingestion and the like, such as has been described for *Helix pomatia* by Krijgsman (1925, 1928).

The diotocardian prosobranchs all appear to be microphagous herbivores, occasionally supplementing this diet by rasping pieces off larger plants, though others, like *Diodora* and *Emarginula,* have the habit of feeding on sponges. It is not surprising, therefore, that there is a broad resemblance between the digestive glands of these animals (fig. 122C, D). The commonest cell is the digestive cell (dic), a tall, columnar cell with highly vacuolated cytoplasm which rarely stains at all intensely. The distal border is often denser than the rest of the cell, may show signs of striations normal to the surface and, in life, is often ciliated, though the cilia either drop off or are withdrawn on fixation and so do not appear in sections (see Owen, 1956). It may, on the other hand, be lobed, and suggest the nipping off of part of the cell with secretion: the nucleus may be central or basal and has not a prominent nucleolus. These cells can, in some animals (*Patella, Acmaea*) be shown to ingest particulate food into vacuoles and to digest it intracellularly, and to take it up from solution in most prosobranchs. The second type of cell (sc) is less numerous and tends to occur in groups in the crypts located at the angles of the tubules. In it the cytoplasm is dense, though it may be somewhat vacuolated, and always stains darkly. It is not recorded by workers in this field as taking up food, either in solution or in particles, from the lumen of the gut. The cell is triangular in section, with a broad surface set along the base of the epithelium and abutting against the blood space beyond, and tapers to a fine point where it reaches the cavity of the tubule. The nucleus is basal, large, and contains a prominent nucleolus, whilst darkly staining spherules often abound in the cytoplasm. In trochids and some other groups these cells have often been seen to project into the intertubular haemocoelic spaces. They seem, therefore, adapted for the uptake of material from the blood by way of their expanded bases and for elaborating it into some secretion which would be then shed to the lumen of the tubule. Both types of cell seem able to manufacture secretion of some sort: presumably these are unlike and in view of the greater number of the first type of cell it seems more likely that they would be the source of any digestive enzymes that might be secreted.

Whilst this appears to be the general histological structure of the digestive gland not only in the diotocardians but in most prosobranchs, some depart from it. In the rissoids, hydrobiids, *Littorina,* naticids, *Bittium,* and calyptraeids there frequently occur large yellowish concretions, usually spherical but frequently quite irregular in shape. Their real nature is unknown, but as they may often be found, apparently unaltered, in the faeces, they appear to be excretory matter of some sort. In sections of the digestive gland of these animals (fig. 122A) these concretions are found to occur in the cells which lie in the crypts of the gland and, therefore, to correspond to the glandular cells of a limpet which have a special relation to the vascular system: this is especially well seen in *Bittium* where these cells occupy almost the whole of the peripheral wall of the tubules abutting against the visceral haemocoel, and in freshwater prosobranchs where they project markedly into the surrounding haemocoelic spaces. In these circumstances the material of which the yellow spherules is composed is more likely to be truly excretory than faecal matter derived from the indigestible residue of what has been ingested by the first type of cell in the gland. This is the more likely, too, in that ingestion of particulate food is not a certain event in the digestive physiology of some of the molluscs that produce such spherules (p. 225).

In some higher mesogastropods which have become carnivorous and in the Stenoglossa the digestive gland shows an apparently more elaborate histology in that a third kind of cell seems to occur. This may be a second kind of gland cell, introduced because the original gland cell has become preoccupied with excretory activity, or it may be simply a secretory phase of the ordinary digestive cell which is out of step with its neighbours.

The stomach is connected to the anus by the intestine and the terminal portion of this is distinguished as the rectum, though the boundary between the two is somewhat arbitrarily drawn. The rectum usually shows considerable longitudinal folds on its walls, whereas the intestine is normally smooth. The intestine leaves the stomach at the distal end of the style sac, the point being marked by the disappearance of the ciliated epithelium characteristic of that part, by a decrease in diameter and, in some animals, by a slight sphincter muscle. The intestinal groove and the 2 typhlosoles, major and minor, normally continue along the intestine, dying away after greater or lesser distances, but occasionally continuing as far as the anus.

The length of the intestine varies considerably: in *Diodora* it is relatively short—in *D. apertura* measuring only 4 times the length of the shell—whereas in the Docoglossa it is vastly longer—in *Patella vulgata* equalling 8 times the shell length (fig. 124). One reason for this difference in intestinal development appears to be the need for consolidation of the faecal material before it is passed into the mantle cavity. (See p. 89.) The dangers of fouling this are evident, but may be minimized by the elaboration of the faecal matter into pellets as it travels along the intestine from stomach to anus. The length of the intestine may, perhaps, be regarded as proportional to the urgency of this requirement in relation to the kind of faecal matter produced, for few signs of other activity on its part have ever been recorded [see p. 223].The brevity of the intestine in *Diodora* would then be correlated with the presence of an apical pallial aperture through which the exhalant current from the mantle cavity escapes,

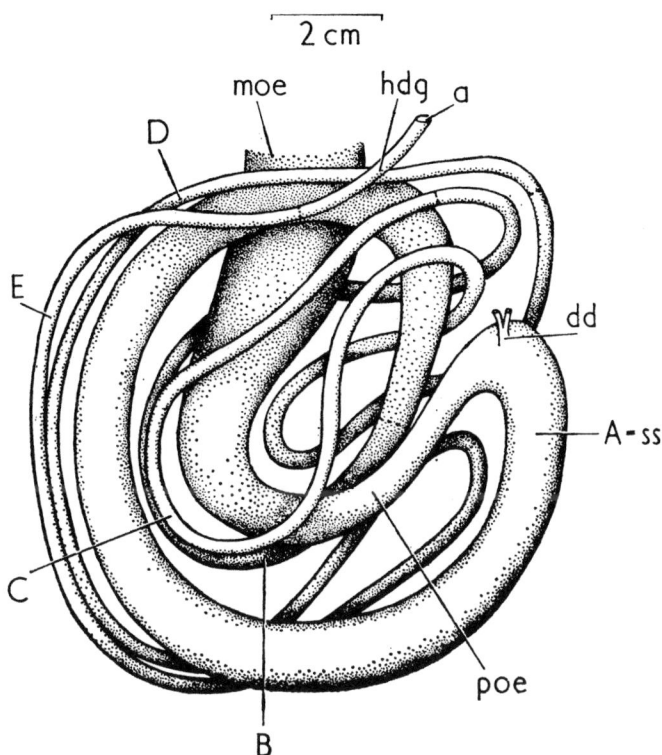

FIG. 124.—*Patella vulgata:* diagram of the course of the alimentary canal lying within the visceral mass.
a, anus; dd, ducts of the digestive gland; hdg, hind gut or rectum; moe, mid-oesophagus; poe, posterior oesophagus; A = ss, proximal section of intestine homologous with style sac of stomach; B, C, D, E, sections of intestine.

washing over the anus as it does so, and carrying faecal matter from contact—or likelihood of contact—with the ctenidia.

In a general way most herbivorous prosobranchs have a longer intestine than do carnivorous forms. In *Trivia, Lamellaria, Ianthina, Balcis, Nucella, Buccinum* and *Mangelia,* for example, the intestine runs straight from style sac to anus and is not long. In *Diodora* it forms a simple loop between stomach and anus, in *Emarginula* a large loop, with a smaller loop superimposed near the anus. In the trochids the intestine runs forward from the stomach, loops tightly on itself once and then passes forwards towards the head (i, fig. 162); from this position it bends back towards the stomach and curves on to the mantle roof to reach the anus (a). In the herbivorous monotocardian prosobranchs it is more or less possible to discern an underlying plan for the intestinal coiling: in many, on emerging from the style sac the intestine is flung into an S-bend, the initial loop towards the animal's right, by the kidney, the second towards the left and the neighbourhood of the pericardial cavity from where it runs on to the mantle skirt. This pattern of intestinal coiling is found, for example, in littorinids, lacunids, rissoids, calyptraeids, capulids, turritellids, and aporrhaids. In the Docoglossa—at least the larger ones—the intestine becomes much more elaborately coiled and forms, in *Patella vulgata,* 6 loops around the visceral dome before running to the anus (fig. 124).

In the Zeugobranchia, Trochacea and Neritacea the intestine runs through the ventricle of the heart before running on to the mantle skirt to open at the anus (rve, figs. 53, 71). During development it starts ventral to the heart but the ventricle wraps round it as growth occurs. This region is generally referred to as the rectum, but, in forms where penetration of the ventricle does not occur, there is very little reason for separating one part from another. The whole intestinal length is clothed with a columnar epithelium the cells of which vary in height and so often give rise to shallow longitudinal folds. These, especially in the rectum, are often raised on connective tissue masses so that folds of considerable dimensions run along the gut. The cells are usually ciliated and gland cells abound, becoming more frequent in the neighbourhood of the anus, though a zone rich in this type of cell may occur near the style sac as well (e.g. *Trichotropis*).

In only a few species of prosobranch has the intestine been at all carefully investigated from the histological viewpoint. Gabe (1951*b*) has investigated it in *Diodora* and shown that several segments are discernible. The epithelium rests on a basement membrane of collagen and on a double muscle layer, of circular fibres internally and longitudinal and oblique mixed externally. This is thin where the rectum lies within the ventricle but becomes thicker towards the anus. At the level of the heart the epithelium is low (15–20 μm) and its cells contain neither iron-containing granules nor alkaline phosphatase. The chondriome is poorly developed and gland cells are sparse. A second region may be distinguished lying between the heart and the style sac and a third between the heart and the anus. The former is covered by epithelial cells taller than those in the heart (35–40 μm) and ciliated. The base of the cytoplasm is rich in mitochondria whilst the more distal parts contain numerous yellow granules which are rich in iron. The anal region has gland cells and alkaline phosphatase in the epithelium.

Graham (1932) divided the intestinal region of the limpet *Patella vulgata* (fig. 124) into 5 sections on a histological basis, designated by reference letters A, B, C, D and E. Of these section A corresponds, in part, to the style sac of the stomach of other prosobranchs. To the remaining 4, however, there falls to be added a lengthy hind-gut, so that between stomach and anus, there are still traceable 5 sections which are histologically differentiated, though all are alike in being composed of a columnar epithelium resting on a connective tissue basis through which run layers of inner circular and outer longitudinal muscles. Section A, which may be compared with the style sac of the stomach, nevertheless shows some signs of difference from the usual appearance of that region, in that the component cells are tall and narrow (28–30 μm), with only short cilia. They contain, distally, numerous yellow-green pigment granules, and seem to secrete some mucoid substance. Section B is narrower in diameter and marked by shallow longitudinal folds. Its epithelium (18 μm) resembles that of section A in most respects, but contains fat droplets. Section C, which forms 2 of the 6 loops into which the intestine is flung, is lined by an epithelium (16–20 μm) mainly similar to that of previous sections, save for the absence of fat droplets, but, in addition, it contains gland cells which are wedged between the bases of the ciliated cells and connected to the lumen of the gut by long, slender necks. Their cytoplasm is packed with numerous minute granules which are very refringent in fresh material and stain so as to suggest that they are protein. In section D these glands also occur, but so does a second sort, the clavate gland, which produces large spherules of secretion with the same refringency and staining properties as the much more minute granules of the basal glands. Section E is, in general, reminiscent of section C, with basal glands alone, but its cilia are very short and its pigmentation greater. The hind-gut, on the other hand, is almost unpigmented, and its walls are flung into several longitudinal folds, two of which largely subdivide the cavity into dorsal and ventral channels, the faeces always, for some reason, using the latter. Mucous cells occur here.

In some prosobranchs a rectal or anal gland is to be found. The occurrence of this has been wrongly attributed to some Zeugobranchia: in *Diodora* and *Emarginula,* but not *Haliotis* nor *Scissurella* apparently, a long tube runs alongside the intestine and rectum, with which it communicates near the anus. This was described first by Haller (1884) who called it a genital duct, but later Pelseneer (1906a) regarded it as an anal gland. It is, in fact, a siphon, and it rejoins the intestine just distal to the point where that leaves the style sac. Investigation shows that it is in reality the intestinal groove which has been separated from the main intestinal lumen by the fusion of the 2 typhlosoles over it. It retains a ciliated epithelium, but what its significance in the life of the animal may be is still unknown, though it always appears empty in sections. It may be that it offers an escape for intestinal fluid if the animal has to pull the shell suddenly over the viscera, but the resistance which would be offered to the flow of fluid by a tube of such inconsiderable dimensions makes its use in this way rather improbable.

A genuine anal gland in the form of a caecal outgrowth from the rectum in the neighbourhood of the anus, is known to occur in *Murex, Ocenebra, Nucella* (rgl, fig. 148), *Urosalpinx* and *Trophon* and may well occur in other, or all, genera of muricacean stenoglossans. According to Pelseneer (1906a) an anal gland also occurs in the Naticidae and Simroth (1896–1907) stated that *Puncturella* possesses one. These, however, are doubtful: Fretter (1946a) failed to find any gland in *Natica catena,* whilst that alleged to occur in *Puncturella* may well be the end of the siphon found in other conical-shelled zeugobranchs and described above.

In *Nucella lapillus* (fig. 125) the anal gland has the form of a group of caeca which unite with one another to form a duct leading to the rectum just within the anus. In young *Nucella* the gland is a simple diverticulum from the rectum lying between the rectum and mantle and extending back to the level of the posterior end of the mantle cavity. The gland is surrounded by the same blood sinus as lies around the rectum. Whilst the animal is young the epithelium is a simple, ciliated, columnar epithelium like that covering the rectum. Later as outpouching of the walls occurs, the cells begin to develop small brown granules: these two processes continue, to give rise to the dark brown or even black gland of the adult. The cells remain ciliated even when full of granules, the cilia beating gently towards the duct. The granules come to lie in vacuoles and finally break away and are lost. Experimental work shows that injected materials such as trypan blue or iron saccharate are picked out of the blood and expelled through the cells of the gland. [See p. 604.]

In a certain number of other prosobranchs a simpler type of rectal or anal gland occurs in the form of an enlargement of the terminal part of the intestinal groove or as a pouch on the side of the end of the rectum. The former condition occurs in some trochids and has been described by Fretter (1955b) and by Deshpande (1957); the latter is found in *Scissurella crispata.* The degree of development of the gland of the trochids varies, being greatest in *Margarites helicinus,* moderate in *Gibbula umbilicalis,* small in *G. cineraria* and absent elsewhere. In all the animals in which it occurs this type of rectal gland seems to be different in function from that of the stenoglossans and is to be regarded as a lubricant of the terminal part of the gut rather than as an excretory organ.

Little absorption of the products of digestion seems to take place through the intestinal wall: nevertheless it is improbable that the intestinal contents are not altered in some way during their sojourn in this part of the gut, as is true of the intestine of *Helix,* where no true

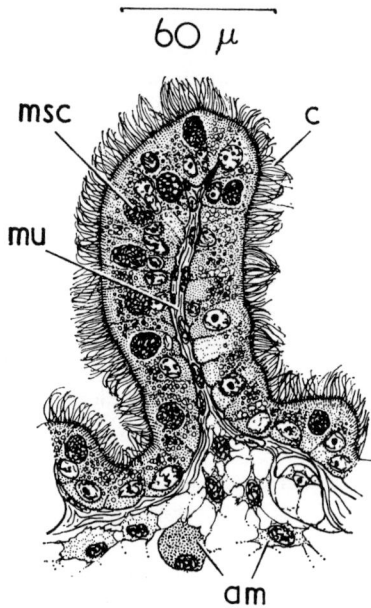

FIG. 125.—*Nucella lapillus:* histology of anal gland.
   am, amoebocytes; c, cilia; msc, mass of secretion within cell; mu, muscle fibre.

absorption occurs, but through the walls of which soluble substances may diffuse (Jordan & Lam, 1918; Hörstadius-Kjellström & Hörstadius, 1940), even those with molecules as large as disaccharide sugars (Jordan & Begemann, 1921). Gabe & Prenant (1949) have made similar suggestions for chitons, and Fretter (1952), using radioactive phosphorus and iodine, showed that salts could diffuse through the intestinal wall of snails and slugs.

   The formation of faeces begins in the stomach, the style sac region of that organ (ss, fig. 87) having this as its primary function. Within it there may be distinguished 2 spaces incompletely separated from one another by the 2 typhlosoles, major and minor (t_1, t_2), which run along it. Of these much the smaller is the intestinal groove (g_1), which arises at the main opening of the digestive diverticula into the stomach (dd) and runs thence to the style sac; the main channel of the style sac, however, is merely an analward extension of the principal part of the stomach, to which the oesophagus and digestive gland ducts open and on the walls of which lies the gastric shield (gs). Food is led into this chamber mixed, in most prosobranchs, with digestive enzymes derived from the oesophageal glands, and to this mixture is added further enzymatic material in the secretion of the digestive gland. The mixture of food and enzyme is then acted upon by the ciliary currents and by the muscular wall of the stomach, which is extensively protected against abrasion in the lower gastropods by the cuticular gastric shield. This may be correlated with the tendency of these animals to be (at least incidentally) detritus feeders; in the higher prosobranchs, which tend to be more exclusive carnivores, the gastric shield is reduced or lost as the need for its protection is minimal. When digestion has occurred the same processes of muscular squeezing and ciliary streaming press a solution of digested food material out of the mass in the stomach into the ducts and tubules of the digestive gland, where it is taken up by the absorbing digestive cells (dic, fig. 122). In certain cases (diotocardians and some lower monotocardians) minute particulate matter which enters the gland may be phagocytosed by the cells and digested intracellularly in vacuoles. These processes, however, leave in the main chamber of the stomach a mass of indigestible material which is gradually moved into the style sac. In the style sac it comes under the influence of two sets of ciliary currents, the main one (on the greater part of the walls) rotating the mass, the others (on the typhlosoles) moving it along the style sac towards the anus, the combined effect being a rotatory movement in the direction of the intestine. On the typhlosoles and in an area at the point where the style sac springs from the main part of the stomach mucous cells abound and the material in the style sac is gradually rolled and cemented into a rod of firm gelatinous consistency. In many prosobranchs provision is made for the return to the main cavity of the stomach of particles which fail to become incorporated in this rod by means of a ciliary current on the typhlosoles running in that direction. The occurrence of this current is a matter of some theoretical importance in considering the origin of the crystalline style (see below). The rod manufactured in the style sac moves out of that part of the gut into the intestine where it may undergo further cementing with secretion from intestinal glands, further compacting and perhaps a final segmentation into pellets.

In addition to material which enters the style sac from the main part of the stomach there is a second stream which passes into the intestinal groove and comes predominantly from the ducts of the digestive gland. It is composed partly of particulate matter which has failed to be incorporated in the main mass in the stomach and which has been passed into the ducts of the digestive gland, partly of a certain amount of true excretory matter which has been extracted from the blood by the excretory cells of the digestive gland, and sometimes of the indigestible residue of food undergoing intracellular digestion. These 3 kinds of material become intermingled in the intestinal groove and pass along the style sac to enter the intestine, where they become associated with the faecal rod emerging from the main part of the sac. The faecal rod in the intestine, therefore, is made up of 2 kinds of material: the bulk from the stomach, the second from the digestive gland. Using the terminology suggested by Carriker (1946), the former may be called the stomach string, and the latter the liver string. They are frequently different in consistency (the liver string containing only microscopic particles) and in colour, so that when both become associated in the intestine it is still possible to distinguish the parts of the faecal rods or pellets to which they give rise.

The stomach string may have no particular form apart from that imparted to it by the compacting and rolling which it has undergone in the stomach and style sac. In those proso-branchs which possess a spirally coiled gastric caecum, however (Haliotidae, Scissurellidae, Trochidae, Turbinidae), the presence of this structure affects the organization of the stomach string. In these animals, as shown by Graham (1949), the string of material which enters the stomach from the oesophagus is passed over the entrance of the ducts of the digestive gland into the mouth of the caecum (see fig. 119A, B). It travels along this to the apex of the coil and then back to the main cavity of the stomach where it is added as a thread-like structure to the end of the mass of material which fills that space. This treatment ensures a very thorough admixture of food and enzymes of both oesophageal and other origin. Although some of this structure is modified by the treatment which the food mass receives in the style sac it can usually still be recognized in the faeces even after discharge.

As a consequence of the differential origin of the faecal material and of the various treatments which it receives in different parts of the gut, the faecal rods or pellets which leave

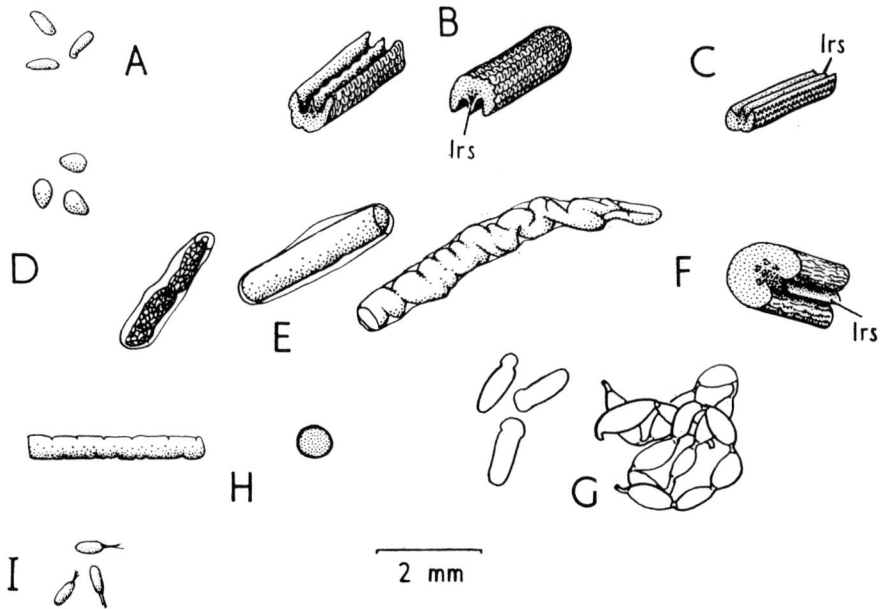

FIG. 126.—Faeces. A, *Acmaea virginea;* B, *Gibbula cineraria;* C, *Gibbula magus;* D, *Turritella communis;* E, *Calliostoma zizyphinum;* F, *Monodonta lineata;* G, *Natica alderi;* H, *Patella vulgata* in side view and T.S.; I, *Bittium reticulatum.*
         lrs, liver string.

the anus are sometimes objects of considerable complexity, the architecture of which was first shown by Moore (1931–32) to have some value as a specific character. These differences are most marked in the lower gastropods, the faeces in the higher being often simple oval pellets without surface markings, enclosed in a superficial mucous layer. Of these faecal masses Moore (1939) regarded the simple rod of *Patella vulgata* (fig. 126H) as the most primitive: here there are only segmental markings and the faecal matter lies within a skin derived from intestinal glands (Graham, 1932). The rod, which may be up to 1·7 cm long, segments fairly readily along the line of the markings but does not easily undergo further breakdown. Moore suggested that the extremely common pellets of higher gastropods (*Turritella,* fig. 126D; *Bittium,* fig. 126 I) have been derived from the segmentation of a rod such as this. The fissurellids produce somewhat similar rod-shaped faeces, but these are flattened in section. Much more elaborate are the faeces of trochids: these consist of a rod (e.g. *Monodonta lineata,* fig. 126F) composed of a stomach string, U-shaped in section, with a liver string (lrs) plugging the opening of the U and running like a keel along one side of the rod. In the intestine this lay in the intestinal groove. The stomach string shows a regular gradation of material, the superficial layers always being composed of the finest particles and the largest invariably occurring centrally in the area where stomach and liver strings are in contact: this may perhaps be related to the rotatory movement in the style sac to which is due the initial shaping of the rod. The rod also shows a series of sinuous ripple-like markings over its surface. Similar faecal rods occur in other trochids (*Gibbula cineraria,* fig. 126B; *G. magus,* fig. 126C). *Calliostoma zizyphinum,* however, differs in that the faeces produced comprise an irregular rod containing sand, calcareous or vegetable detritus embedded merely in a mass of mucus (fig. 126E).

As mentioned above, a certain number of prosobranch gastropods possess the structure known as a crystalline style. The way in which this has arisen may now be briefly dealt with. When the anatomy of the stomach of these animals is investigated the style is found to lie in a structure which is so directly comparable with the style sac region of other gastropods in respect of its anatomical relationships, its histology and the direction of the ciliary currents along its walls as to leave little doubt that the one is homologous with the other. Inside each lies a cylindrical structure composed largely of mucus, but in the one group it is the beginning of a faecal rod travelling to the intestine, and in the other has adsorbed enzymes which are being carried to the stomach for release. If the style sacs are homologous then it is worth asking whether their contents are not equally so: Yonge (1939) and Graham (1939, 1949) have both concluded that this is so and that the crystalline style is, indeed, merely a transformation of the faecal rod. In addition to the points already mentioned in support of their homology there may also be mentioned the facts that food particles are often embedded in the mucus of the style—just as they are in the faecal rod—and that the bulk of the mucus in both cases is derived from gland cells on the typhlosoles. It appears that in the evolution of a few groups of monotocardians, as in the evolution of the lamellibranchs, the adoption of a rigorous microphagy emphasized the phase of intracellular digestion within the digestive gland at the expense of the extracellular digestion which occurred within the stomach. When this happened the stomach string became relatively free of food particles and detritus because these now all passed to the digestive gland and, therefore, became incorporated in the liver string. The stomach string was thus susceptible to a transformation into a mucous rod with adsorbed enzymes and gave rise to the structure which we know as the crystalline style. The direction of movement which it underwent changed from a rotatory posterior to a rotatory anterior one, ciliary currents which could bring this about being already present, and the end of the rod which lay alongside the gastric shield in the main part of the stomach was no longer the place at which the rod was being formed, but the site of its solution in the gastric contents to liberate the enzymes which it contained. These appear to be simple amylases, though the style has been less thoroughly explored from this point of view than that of lamellibranchs, where it is known to contain glycogenase and oxidase as well (Yonge, 1926). Oncomelania (=Hypsobia) nosophora has a cellulase in its style (Winkler & Wagner, 1959). In addition to the liberation of enzymes in the gastric cavity the solution of the style helps to control the pH of the gut (being normally the most acid substance present) and its rotation helps the transport of food through the stomach.

So far as other digestive enzymes in the gut of prosobranch gastropods are concerned there is not a great deal of information in the literature, the choice of those animals which have been investigated being obviously dictated by the need for adequate supplies of tissue or gut juice for this type of physiological work. Thus the animals used have been mainly Patella vulgata, Haliotis spp., Natica spp., Viviparus spp., Vermetus novae-hollandiae, Pterocera crocata, Pterotrachea spp. and some of the larger stenoglossans belonging to the genera Murex, Nucella, Buccinum, Neptunea and Busycon.

In the diotocardians amongst these it is likely that the saliva is solely lubricatory and that the salivary glands secrete no enzymes (Patella: Graham, 1932). On the other hand the oesophageal and digestive glands in these animals seem to secrete a digestive fluid capable of attacking proteins, fats and carbohydrates (Patella: Roaf, 1906, 1908; Rosén, 1937; Haliotis: Albrecht, 1921, 1923). Graham (1932) found no evidence of the secretion of digestive enzymes from the digestive gland of limpets though he did record the secretion of an amylolytic enzyme from the oesophageal gland. All writers, however, are agreed that the intestine is not responsible for secreting enzymes and that the uptake of digested food occurs in the digestive gland, within the cells of which intracellular digestion may also take place. Although little work has been done to demonstrate it, it is likely that this pattern of digestive activity is the common one amongst the herbivorous and microphagous prosobranchs (except where a crystalline style is found). The enzymes, apart from a lipase extracted from the digestive gland of *Viviparus viviparus* by Rosén (1932), have not been purified so that their properties are only vaguely known: they appear to work best at pH 5–6, which is the pH of the parts of the gut in which they naturally occur (Yonge, 1925a). Rosén (1937) has shown that the proteinase of the digestive gland of *Patella vulgata* requires to be activated by hydrogen sulphide or similar reducing agent and that it is inhibited by iodoacetic acid: this suggests that the enzyme is probably a cathepsin, and may, therefore, be intracellular.

In herbivores with a crystalline style free proteolytic enzymes do not normally occur because they would digest the style itself (Yonge, 1930); in them, however, free carbohydrases and lipases may be found (*Vermetus novae-hollandiae*: Yonge, 1932; *Pterocera crocata*: Yonge, 1932) derived from the salivary or digestive glands. *Pterocera* and *Strombus* also secrete a cellulase, though the site of its manufacture is not known, and in this respect they differ from all the other prosobranchs. Dodgson & Spencer (1954) reported the occurrence of sulphatases in a number of prosobranchs (*Patella*, *Monodonta*, *Calliostoma* and *Littorina littorea* in particular) which may help in the digestion of polysaccharide sulphates in the algal food which they ingest.

In carnivorous prosobranchs the investigation of digestive enzymes has been more closely pursued. Amongst the mesogastropods the most complete account is the one given by Hirsch (1915) for two species of *Natica*, *N. hebraea* and *N. millepunctata*. Here enzymes are absent from the salivary glands but a proteinase occurs in the secretion of the oesophageal glands whilst the digestive gland secretes proteolytic, amylolytic and lipolytic enzymes. The secretion of these is timed so that the digestive gland is, on the whole, active after food has been caught but does not secrete in a starved animal. This is particularly noticeable in the case of the proteolytic enzymes which might otherwise attack the gut wall. More thorough examinations of the digestive enzymes—some using modern methods for purification— have been made in the case of some Stenoglossa. Amongst the first of these to be investigated was the American *Busycon canaliculatum* by Mendel & Bradley (1905a, b, 1906). These workers found a proteolytic enzyme in the saliva, another (or the same) and a diastase in the gastric contents, whilst an extract of the digestive gland contained enzymes capable of attacking starch, glycogen, sugars and some proteins. Roaf (1906) obtained similar results with the digestive glands of *Nucella lapillus* and *Neptunea antiqua* and gave the further information that the enzymes were most active in acid media. Hirsch (1915) gave an account of digestion in *Murex trunculus*. Here the salivary glands ('kleine Vorderdarmdruse') secrete a proteolytic enzyme but not an amylolytic one; extracts of the gland of Leiblein ('grosse Vorderdarmdruse'), on the other hand, were found to digest starch but apparently to be without action on other types of food; the fluid in the stomach and extracts of the digestive gland, however, contained enzymes which were effective on proteins, carbohydrates and fats. Frequently, as in *Natica*, these were absent from a starved animal and were secreted only after feeding. A more elaborate investigation of the related species *M. anguliferus* was made by Mansour-Bek (1934). Here the saliva was shown to contain a proteinase most active at pH 8·2 and also a dipeptidase; the same enzymes were found in extracts of the gland of Leiblein, in the stomach fluid and in extracts of the digestive gland. The gland of Leiblein and the digestive gland also manufacture a carboxypolypeptidase and an aminopolypeptidase. In the fluids in which they normally occur these have an optimum pH at 7·6 or 8·2; this is unchanged when purified. The proteinase does not require either zookinase or enterokinase for activation, which seems to suggest that it is not an enzyme of the cathepsin type and, if a trypsin, is different from the familiar type of vertebrates. The cyclical production of enzyme in step with the feeding activity of the mollusc may partly be related to the activity of the enzyme.

Brock (1936) and Mansour-Bek (1934) have given an account of the enzymes—and, in the case of the former, of much of the physiology of the alimentary tract—of the common whelk, *Buccinum undatum*. The salivary glands in this animal secrete saliva containing enzymes attacking peptone and glycylglycine at a neutral pH (water extract) or at pH 8·0–8·2 (glycerol extract); no enzyme of carbohydrase type occurs in their secretion nor in that of the gland of Leiblein, which secretes proteolytic enzymes. The fluid in the stomach, on the other hand, contains an amylase, a lipase and enzymes capable of attacking proteins and their breakdown products: these are presumably derived from the digestive gland since they may also be detected in extracts of that structure. These enzymes work optimally around neutrality, the natural pH of the stomach being slightly lower.

In all these carnivorous prosobranchs extracellular digestion in the stomach appears to be the rule, followed by an absorption of the products of digestion by the digestive gland. This also seems to be true of many of the herbivores though it may be supplemented by a phagocytosis of particulate matter in the digestive gland of these animals. In the zeugobranchs and in the Patellacea intracellular digestion seems to be relatively more important still. As in all groups of the animal kingdom diet and enzymes are correlated, proteolytic enzymes being more marked in carnivores and those of a diastatic type predominating in herbivorous animals. In contrast to the lamellibranchs and to some opisthobranch gastropods wandering amoeboid or phagocytic cells seem to play little part in the digestive processes of prosobranchs.

CHAPTER 10

FEEDING

T HE molluscs have exhibited such ability to adapt themselves to life in so many different types of habitat that it is not surprising that they have learned how to feed in a variety of different ways. Adaptive radiation in the Mollusca, as in other phyla, has involved adaptation to feeding niches. The gastropods appear to be equipped fundamentally with apparatus permitting them to take in particles of food which are either inactive (like detrital deposits) or which are scraped from the surface of a plant or animal and the way in which different types of gastropod are enabled to do this by means of their radula has been described above. Perhaps the only point which requires mention at this stage is the way in which they graze the vegetation, moving forwards and oscillating from side to side as they progress (ftr, fig. 197B). This has been called 'pendulum feeding' by Ankel (1938a). Several species have also evolved a ciliary means of collecting their food, out of the series of ciliary currents which is primarily concerned with the maintenance of a water stream through the mantle cavity and the transport of such particulate material as accidentally enters in that way: this, too, has been dealt with above (p. 97). There still remain, however, other proso-branchs with feeding mechanisms which do not belong to either of these types, or with feeding mechanisms of such complexity as to merit fuller treatment. Of these, those which have to extract their food from the deeper parts of the bodies of other animals, those which bore through shells of other molluscs to get to their food, and those which pierce the bodies of their prey to suck blood or other fluid are the most important.

Of the first group the members of the Lamellariacea and Cypraeacea are the most import-ant, feeding on sedentary animals of a variety of sorts, mainly tunicates, though *Simnia patula* eats the coelenterates *Alcyonium digitatum* and *Eunicella verrucosa,* and *Velutina plicatilis* eats *Tubularia indivisa* (Ankel, 1936a). Molluscs of the genera *Lamellaria, Erato* and *Trivia* all eat colonial tunicates and use the proboscis to do so. *Erato* is the most selective of these in that it thrusts the proboscis through the oral aperture of a zooid of the tunicate *Botryllus* or *Botrylloides* and so reaches directly to the more nutritious and tasty parts of the body (fig. 35); *Trivia,* on the other hand, which also eats botryllid and didemnid ascidians, tears and devours the test so as to expose these parts, which it then eats, though it cannot digest the poly-saccharide material out of which the test is made. *Velutina velutina* (fig. 127) attacks the solitary ascidians *Ascidia* and *Phallusia* (Ankel, 1936a) and also *Styela coriacea* (Diehl, 1956). It lives on or near the ascidians, resembling the last in colour and surface texture, and appears to feed like *Lamellaria* and *Trivia,* biting holes in the tunicate and rasping with the proboscis.

Two groups of prosobranchs feed on animals the bodies of which are enclosed within shells. These are the naticids and the rachiglossans belonging to the group Muricacea, including the British genera *Nucella,* the dog whelk, *Ocenebra,* the rough tingle, and *Urosalpinx,* the American whelk tingle or oyster drill. The naticids eat almost entirely bivalves, especially *Donax, Tellina, Macoma* and *Mactra,* but *Mya, Abra, Spisula, Venus, Aloidis* and *Nucula* are known to be eaten, and probably any shell of appropriate shape and size will be

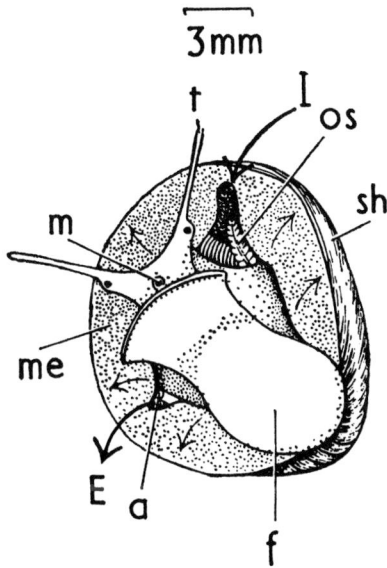

FIG. 127.—*Velutina velutina:* animal in ventral view. Arrows show direction of ciliary currents.
a, anus; f, foot; E, exhalant water current; I, inhalant water current; m, mouth; me, mantle edge; os, osphradium; sh, shell; t, tentacle.

attacked, though *Natica* certainly seems to prefer bivalves with thin and incompletely closing shells to those with thick shells which shut firmly. *Nucella lapillus* feeds largely on mussels and limpets, or on barnacles; it is able to force barnacle shells apart by muscular action of the proboscis without boring. *Ocenebra erinacea* attacks *Paphia, Cardium, Venus* and oysters, though it will also bore through gastropod shells. *Urosalpinx* eats the young of bivalves, especially oysters and *Venus,* in the spat stage, to which it is very destructive. It has also been recorded (Galtsoff, Prytherch & Engle, 1937) as eating large numbers of barnacles, preferring them to all other kinds of food. Hancock's observations (1954) at Burnham-on-Crouch, however, showed *Urosalpinx* boring shells of molluscs which were covered with barnacles, apparently preferring the less accessible mollusc to the more easily entered crustacean: he cautions, however, that the barnacles belonged to the genus *Elminius* and were therefore only recently established in Essex waters and so may not yet have been appreciated as possible food by the oyster drill. There is, in fact, a considerable body of evidence (Orton, 1929a; Orton, 1950b; Hancock, 1957) that *Ocenebra*—and presumably the other genera as well—have feeding habits and preferences and take some time to appreciate the value of strange sources of food.

In all the cases described above (except barnacles) the body of the prey is reached only after a hole has been drilled in its protective shell, and valves of gastropods and lamellibranchs with holes of this nature may often be picked up on beaches (fig. 128).

The mechanism of boring in these gastropods, especially naticids, has been repeatedly debated. The recent work of Carriker (see chapter 26) has shown that the drilling is done by a combination of chemical and mechanical activity. *Nucella lapillus* produces a cylindrical hole

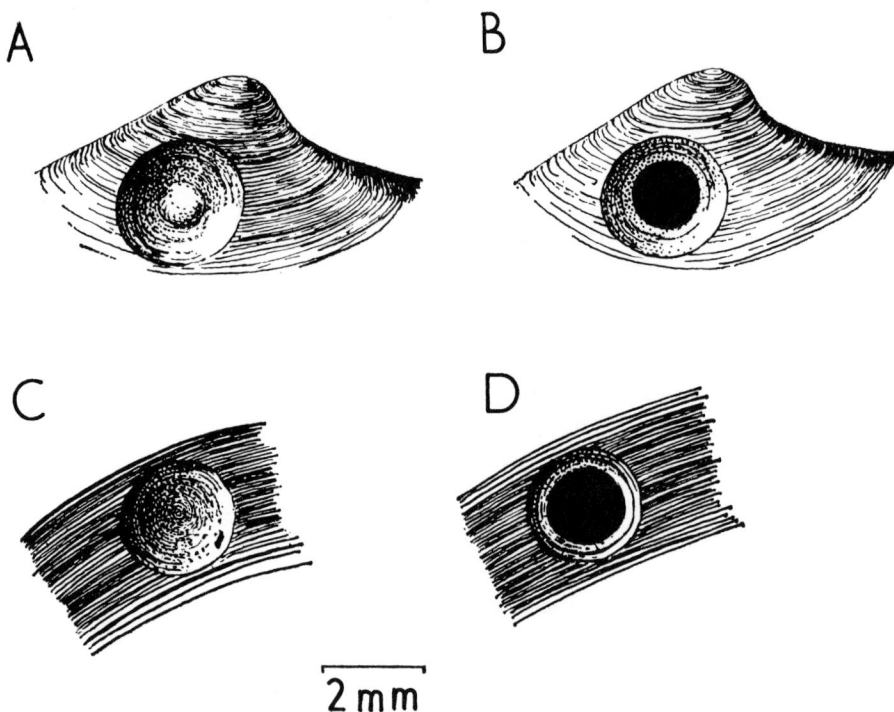

FIG. 128.—Shells bored by prosobranchs. A, shell of bivalve incompletely bored by *Natica:* note the boss in the centre; B, shell of bivalve perforated by *Natica;* C, shell of bivalve incompletely bored by *Nucella:* note absence of boss; D, shell perforated by *Nucella.*

up to 1·75 mm deep in 10 hrs' boring of a shell of a *Patella* or after 2 days' work on the shell of *Mytilus.* In the case of *Ocenebra erinacea* figures given by Fischer (1865) suggest a similar rate of working—3–4 hrs to bore an 'average' oyster, but more recent records by Orton (1927) and Piéron (1933) imply a much lower rate of penetration: Orton gave an average of 134 hrs for boring and eating an oyster, or 75–100 hrs for boring alone. Piéron's figures are:

> 7·5 hrs per 0·1 mm of shell bored (*Cardium*).
> 8·3 hrs per 0·1 mm of shell bored (*Cardium*).
> 8·3–13·3 hrs per 0·1 mm of shell bored (*Paphia*).

Hancock (1957) gave figures of 144 hrs for boring and eating a mussel, 168 for *Paphia* or a small oyster, and 216–240 hrs for a medium sized oyster—in June at 16–20°C. The holes which are made by this mollusc are very like those of *Nucella* but tend to be shallower (1–1·25 mm deep).

Piéron has also shown that holes made by *Ocenebra* and by *Nucella* and left unfinished have a flat base (fig. 128C). This state of affairs is commoner in thick than thin shells, as if the task of boring had exhausted the mollusc.

Mechanical boring has been advocated not only by Pelseneer (1925) and Graham (1941) among recent writers but also by Jensen (1951) and Korringa (1952). Carriker (1955, 1959) is the

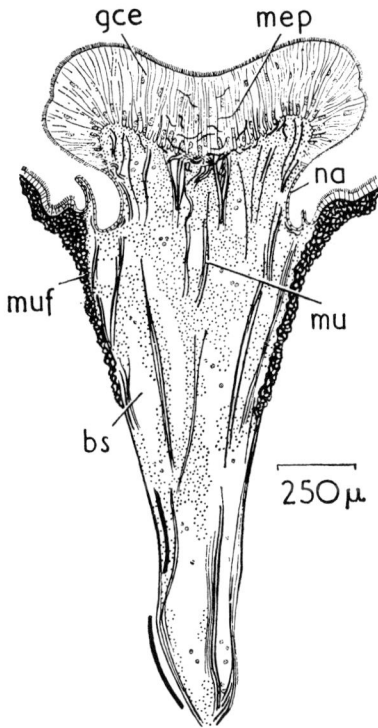

FIG. 129.—*Urosalpinx cinerea:* diagrammatic longitudinal section of the everted accessory boring organ.
bs, blood space; gce, glandular and ciliated epithelium; mep, muscle fibres penetrating between epithelial cells; mu, muscles controlling movement of organ; muf, other muscles of foot; na, epithelium on neck of boring organ.

only recent worker to propose that some chemical activity is also involved in the process. This he has shown to be due to the accessory boring organ, a sucker-like structure (fig. 129) which lies in the mid-ventral line of the foot a little posterior to the anterior edge. This structure was first described by Fretter (1941) and was later dealt with by Carriker (1943) who then called it the accessory proboscis. It normally lies withdrawn into a sac lying in the pedal tissue; when everted, however, it swells into a large, rounded projection of diameter comparable with that of the true proboscis. It is covered by a tall epithelium containing alternate gland and ciliated cells (gce), the former secreting spherules of some material which is not mucus. Fretter's ideas as to the use to which this structure was put centred mainly on its functioning as a sucker which would help to steady the body of the predator on that of the prey during the boring process, a need which was increased by the fact that much of the anterior end of the foot is used to steady the true proboscis during boring and feeding rather than for gripping the substratum: she also found no evidence of its secretion having any action on calcareous material. Carriker (1955), on the other hand, believed that this structure has a chemical effect on the substance of the shell which, without dissolving it, nevertheless makes it easier for the radula to remove it by rasping. He has brought impressive evidence

(1959) in support of this chemomechanical theory of boring with experiments involving the amputation of the proboscis and of the accessory boring organ in the muricids *Urosalpinx cinerea* and *Eupleura caudata*. Both structures are regenerated with surprising rapidity, but only those animals with both organs bore, suggesting their co-operation in this process. He has also been able to show that the accessory boring organ, like that of *Natica* (Ankel, 1937a), will etch calcareous shells when closely applied to them.

When *Nucella* is about to feed it attaches itself to the body of the prey by means of the foot, the anterior end of which is contracted and turned dorsally off the prey. The median part of the edge is particularly contracted and comes to form a groove along which the proboscis extends. The anterior corners of the foot then curve dorsally and meet above the proboscis which is therefore completely embraced and held by the foot.

The behaviour of the radula of a feeding *Nucella* is almost impossible to observe because of the way in which the proboscis is hidden; nevertheless it can be exposed by gentle removal of the foot, when it will be seen that the tip of the odontophore is being continually pushed out of the mouth so that the radular teeth can rasp at the surface of the shell. After executing a number of rasping strokes the mollusc rests a little and it is during this period that the accessory boring organ is placed in the hole. When boring is begun again it will be noticed that the odontophore has been rotated to one side so that the direction of rasping is altered. So far as can be seen the rotation is predominantly one of the odontophore within the proboscis rather than of the proboscis itself. As the rasping proceeds a straight-sided cavity is produced with finely polished walls (fig. 128D), examination of which fails to reveal the marks made by the radular teeth. There is no doubt about the reality of the rasping, however, because if a dog whelk be disturbed in the middle of its drilling, killed, and its gut opened, innumerable sickle-shaped flakes of crystalline material can be found there, which prove to be calcium carbonate.

Carriker (1955) gave a similar account of the boring of oysters by *Urosalpinx,* using a method which allowed of observation of the drilling with a binocular microscope. At first the radular teeth made little impression on the shell but after the process had proceeded for a little the gastropod altered its position on the bivalve, creeping forwards until the accessory boring organ lay over the point at which the proboscis had been active. When this movement has occurred it was everted into the mark made by the proboscis and left thus for a little time. It is then retracted, the animal backs, the proboscis everted and drilling begun again. This rhythm persists throughout boring. Precisely what is going on during the period when the accessory boring organ is applied to the shell it is not possible to say: there is clearly no real solution of the calcareous matter as the chips (in *Nucella*) from the gut are still soluble with evolution of CO_2 in acid. If, as seems possible and as Carriker thinks, some secretion from the accessory organ occurs, then the function of this would seem to be that of softening the shell material for easier treatment by the radula. The same might also be true of similar structures in other members of the Muricacea. Carriker (1959) has suggested that the organ might produce, not an acid, but a chelating agent or, possibly, an enzyme attacking the conchiolin matrix of the shell. The role of other glands, such as the accessory salivary glands, in boring is still unclear. [See pp. 623, 625.]

Much more argument has surrounded the question of how naticids bore the shells of the bivalves which they eat. Some authors have assumed that this was done by chemical means:the first of these was Réaumur (1711), but Schiemenz (1891) was the first to support the idea with any proper evidence. He argued that the radular teeth were not hard enough, nor the proboscis mobile enough, to bore a cylindrical hole, but he was, perhaps, primarily urged towards the chemical theory of boring by the discovery of a hemispherical boss lying under the ventral lip, which he called a 'boring gland' and which, he believed, produced acid for making a hole in the shell. The secretion of the gland, he showed, reddened litmus. Other workers have upheld this theory. Hirsch (1915) showed that the diameter of the gland and the diameter of the hole bored were identical and therefore assumed that the one had been made by the other. Boettger (1930) and Ankel (1937a) also supported this idea, the latter showing that glands removed from *Natica* etched the gloss from the surface of a shell of *Trivia*. Repeating the experiment later Ankel (1938a) got no similar result and therefore suggested the presence of an enzyme, 'calcase' (which is very reminiscent of what Carriker (1955) says of *Urosalpinx*), although what precise effect this might have on the material of which the shell is composed is left vague. Giglioli (1949) has also supported this theory.

Another group of workers has suggested that boring is mechanical and is accomplished entirely by radular action. Fischer (1922) investigated the 'boring gland' of *Natica* and found it composed mainly of muscle, without obvious gland cells, and its surface had no effect on litmus. For these and other reasons he argued that it could have no part in the process of boring. Pelseneer (1925) and Loppens (1926) came to the same conclusion. Jensen (1951) discovered egg cases of *Raia* sp. and *Sipho* sp. bored by a mollusc which was probably either *Natica affinis* or *N. pallida*, though the identifications are admittedly not proved. These show clear markings of radular teeth on their edges and round the periphery, suggesting a mechanical drilling by the radula, though it may, of course, be true that naticids could bore such horny objects with the radula but must have recourse to chemical agents when attempting to drill the calcareous shell of another mollusc. Turner (1953) has also taken this view. Wheatley (1947) has gone so far as to suggest that naticids can devour certain types of prey without boring the shell at all. [See p. 713.]

The most full description of the boring of shells by naticids has been given recently by Ziegelmeier (1954), who concluded that the hole is made entirely by mechanical means. He was able to reach this conclusion because he managed to overcome the very great difficulty of seeing what the mollusc was doing to its prey during the feeding process by keeping starved specimens of *Natica* in aquaria. This difficulty, which had defeated previous observers, arises partly from the fact that *Natica* wraps its prey in its foot during the process of boring (to such an extent that Pelseneer believed that the bivalve died of suffocation) and also because it will bore only when buried in sand.

Natica feeds mainly on *Donax*, *Tellina* and *Mactra* (Piéron, 1933) but Ziegelmeier found that it would also bore holes in the shells of numerous other tellinids and of *Abra alba*, *Spisula solida*, *S. subtruncata*, *Venus gallina*, *Mya arenaria*, *Nucula nucleus*, *N. nitida* and *Aloidis gibba*. Giglioli (1949) also recorded the boring of gastropods and gave examples of cannibalism. The bivalve is gripped in the foot during boring and is almost invisible. Because of the way in which it must be held, boring is usually confined to a limited number of positions and, sometimes, because their asymmetry allows a better grip on the prey, the bivalves are bored through one valve more frequently than the other. Ziegelmeier investigated about 200 borings, almost all of which were made in the mid-region of the valve where that was broadest, and very few near its angles. Stinson's observations (1946) support this, but Belding (1930) noted that most shells had been bored towards the posterior end of the valves. *Aloidis* is regularly bored through this right valve, which is thicker than the left, but offers *Natica* a better grip. Piéron gave the following figures for shells of *Donax* entered by *Natica*:

TABLE 6

| | Site of boring | | | | | |
Animal	Right valve	Left valve	Ant. third	Mid third	Post. third	Near hinge
Donax (Atlantic)	54·3%	45·7	22·3	73·4	4·3	29·1
Donax (Mediterranean)	57·0%	43·7	29·5	60·0	10·5	27·0

These figures suggest that there are privileged positions for boring, almost certainly related to the way in which the prey is held, and are in agreement with earlier figures given by Boettger (1930). Schiemenz (1891) stated that 63·6% of valves bored by *Natica* were left valves; corresponding figures given by Pelseneer (1925) are 43·75% and by Piéron (1933), 43–45%, and these may also have the same explanation. According to Pelseneer (1925) and Verlaine (1936) the site of boring is often related to the position of the underlying gonad. Verlaine has published evidence suggesting that *Natica alderi* learns how to locate its hole so as to reach the gonad: thus of a series of shells of *Tellina* (=*Macoma*) *balthica* bored by that prosobranch the dimensions of the hole (and therefore the age of the animal which bored it) and its situation in relation to the gonad are as follows:

TABLE 7

Size of hole	% over gonad	% at edge of gonad	% not over gonad
0·5 mm or less	45	18	36
0·5–1 mm	78	9	13
1–1·5 mm	91	9	0
1·5 mm or more	91	9	0

The hole made by *Natica* on boring is recognizably different from that made by the muricacean borers (fig. 128A, B), in that the lips show an initial abrasion and the walls are curved; incomplete ones are easily recognized by the presence of a small upgrowth arising from the centre of a concave base. Ziegelmeier's account (1954) of the boring process shows how this arises.

The bivalve is held by the propodium (fig. 297). As in the muricaceans the mid-anterior region of the propodium retracts deeply so as to form a groove into which the proboscis everts and its tips then curve dorsally to hold this steady. The proboscis is rotated through 90° to right or left and applied to the shell surface against which the radula works; as successive radular strokes follow one another the proboscis gradually rotates back from the twisted to the normal position (fig. 130). After this a rest period ensues, of 2–3 min if the mollusc has just

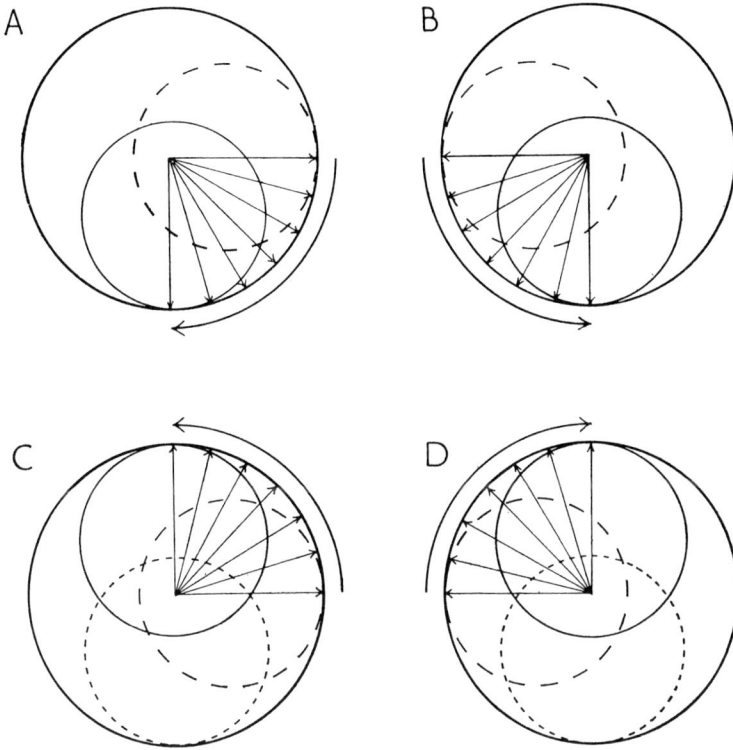

FIG. 130.—Diagrams to show the method of using the radula employed by *Natica* whilst boring. The borehole is dealt with in bursts of activity separated by rests; boring is done sector by sector represented here as the four quadrants of a circle, A, B, C and D. In each diagram the large circle represents the edge of the borehole, the small broken circle the initial position taken up by the tip of the proboscis applied to the shell, the small complete circle its final position; intermediate positions are not shown. In C and D the dotted circles represent the resting position of the proboscis. Within the small circles continuous arrows represent the pathways followed by the radula during each burst of rasping. Note that the final position of the proboscis is always straight, the initial one always twisted; the continuous arrow outside the large circle gives the direction of movement during untwisting. From Ziegelmeier.

started boring, of about 5 min if it has been active for some time. During the resting period the proboscis is lifted off the shell and the accessory boring organ which lies under the ventral lip is brought close to the hole. Ziegelmeier never saw it enter it. The proboscis is then applied to the hole in a twisted position so as to allow the radula to scratch at another sector. The proboscis is always twisted so that the first scratchings are from right to left or vice versa at the beginning of a working period, and it always untwists itself so that the radula is scraping in an anteroposterior direction at the end. The central knob is obviously left because it lies at a spot where the radular action is least effective.

During the process of boring Ziegelmeier was able to see white shell material passing down the oesophagus. After the boring is completed (at the rate of 0·1 mm/4 hrs) the prosobranch begins to feed—a process which may last anything up to 60 hrs, and which may involve the animal in the ingestion of its own weight of bivalve flesh. During this defaecation occurs and

TABLE 8

Animal	Bivalves/day	Bivalves/month	Investigator
Polynices triseriata (young)	0·34	10·2	Stinson (1946)
P. triseriata (adult)	0·4	12·0	Stinson (1946)
P. heros (adult)	0·22	6·6	Wheatley (1947)
P. triseriata (adult)	0·07	2·1	Wheatley (1947)

up to 12 isolated, white pellets are passed, which are composed mainly of shell fragments. Later a chain of pellets is seen to escape from the anus which are dark green in colour and slimy to the touch. They are oval and measure about 1 mm long by 0·2–0·3 mm in diameter. Some figures for the amount of food eaten by naticids are given in Table 8 taken from Giglioli (1949). They refer to the American species *Polynices heros* and *P. triseriata*.

Ziegelmeier has supplemented the information given by Fischer (1922) on the histological nature of the accessory boring organ. He described it as covered by a cuticle the thickness of which is greatest centrally, and which appears unperforated by pores through which the secretion of such gland cells as are present could escape, and it appears to have no effect on calcareous material with which it is brought into contact. Ziegelmeier, indeed, concluded that it may well be primarily a tactile sense organ giving information necessary for boring in view of the fact that this process is invariably carried out by *Natica* as it lies buried in the sand. In this respect Ziegelmeier's observations (made on *Natica* (*Lunatia*) *nitida*) are not in agreement with ours on *N. catena* (fig. 131) in which it is quite apparent that secretion from the gland cells (msc) can pass through the cuticle, which is more accurately interpreted as a rodlet border (rb).

In view of the similarity which exists between the structure on the ventral lip of *Natica* and that on the sole of the foot of muricids it is difficult to believe that the boring process is not fundamentally similar, or even identical, in the two groups of prosobranchs. Carriker's latest observations and experiments (1959) make it seem highly probable that a combination of chemical and mechanical attack allows boring in muricids and it seems likely that the same alternation occurs in naticids as well. Ziegelmeier noted the same rhythm of radular activity and 'rest' during which the accessory organ is brought near to the hole which is being made: it seems probable that during this phase some chemical attack is made which depends, like that of muricids, on the secretion of the accessory boring organ.

The Buccinacea among the rachiglossan Stenoglossa are provided with a long proboscis but do not bore, being primarily feeders on living and (less frequently) dead but fresh animals, especially polychaetes (figs 132, 269). Some of them, nevertheless, can open the shells of bivalves in order to feed on the animal within. This has been described by Colton (1908) and by Carriker (1951) for the American genus *Busycon* (= *Sycotypus*). The prosobranch appears to detect the lamellibranch by means of the exhalant current of water emerging from the dorsal siphon; it then climbs on to the shell and grips it in the foot so that the ventral edges of the 2 valves lie under the outer lip of its own shell. By pulling this down with a slow contraction of the columellar muscle the lip is forced between the 2 valves until a piece of

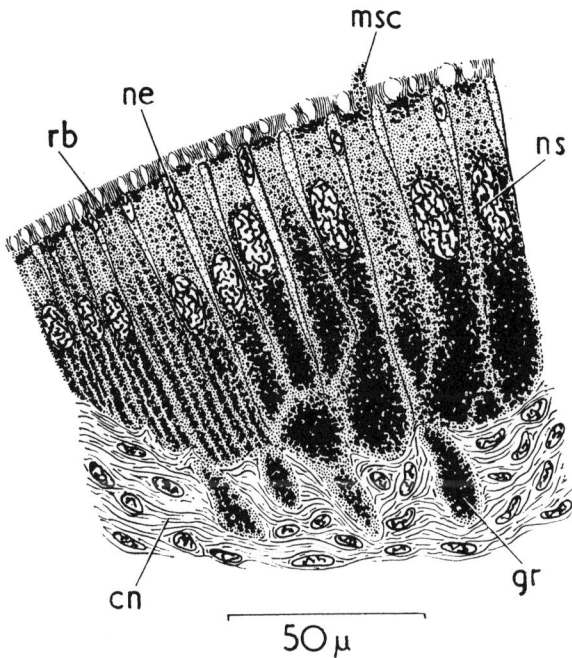

FIG. 131.—*Natica catena*: part of a vertical section through the accessory boring organ; the more lateral part of the organ is to the left.

cn, connective tissue; gr, granules of secretion; msc, mass of secretion escaping through rodlet border; ne, nucleus of supporting epithelial cell; ns, nucleus of secreting cell; rb, rodlet border.

one of them is broken off. When this hole is large enough the proboscis is everted, passed through it and the process of feeding begun. *Busycon* eats bivalves such as *Mytilus, Venus, Mya* or *Ensis* at the rate of 4–5/week; oysters are eaten at a rate of 0·84/week in winter, but 2–7/week in spring. *Venus mercenaria* is eaten at a rate of 0·86 in winter, but 0·35 in spring, suggesting that the animal's taste or appetite varies from season to season. *Buccinum undatum*, finding bivalves with their shell open, will prevent the valves from closing by wedging them apart with the anterior end of its own shell and then will start to feed on the helpless prey. Dakin (1912) said that they incapacitate it by first attacking the adductor, but this is contrary to the general behaviour of these animals. [See p. 483.]

The food consumption of the boring muricaceans has also been measured in a number of instances. Hancock (1957) has carried out experiments with *Ocenebra erinacea* which show that feeding is most intense during July and August off the coast of Essex and does not occur at all when the temperature falls to less than 10–11°C, which would appear to involve the animals in a starvation period extending from the beginning of December to April in the latitude of the British Isles. *Ocenebra*, however, is known to be particularly sensitive to low temperatures and related genera may not be so much affected: *Buccinum* is certainly not. Cole (1942) gave a feeding rate of 0·165 oyster spat per tingle per day. The same worker found that *Urosalpinx cinerea* devoured oyster spat at an average rate of 0·438 per day or 2·92 per week. The figures given by Carriker (1955) are higher. Animals measuring 2–4 mm ate 33·6

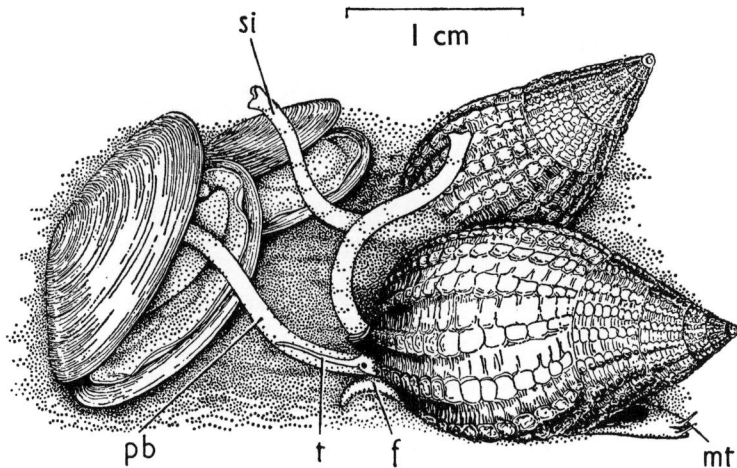

FIG. 132.—*Nassarius reticulatus:* eating dead bivalves of the species *Tellina crassa.*
    f, foot; mt, metapodial tentacle; pb, proboscis thrust into bivalve; si, extended siphon of the gastropod; t, tentacle.

oyster spat per week, whereas those 21 mm in height ate 0·05 oysters (53 mm long) per week. Like *Ocenebra, Urosalpinx* is affected by temperature and appears to starve during the coldest months of the year (Cole, 1942) and to feed voraciously during the warmer months and especially whilst spawning.

In contrast to the prosobranchs which have just been dealt with in the preceding section stand those gastropods which have developed a method of sucking liquid food into their gut from the body of their prey. This has been achieved mainly by those placed in the family Pyramidellidae, which are dealt with here although they have recently been shown to belong to the group Opisthobranchia rather than to the prosobranchs. To this end their gut has been highly modified since it requires not only pumping machinery for the sucking of the blood but also piercing apparatus to make the fluid available. The genera *Odostomia, Chrysallida* and *Turbonilla* have all undergone a very similar process of specialization.

The animals are provided with an elongated acrembolic proboscis (figs. 133, 134; see also fig. 102) normally withdrawn to form an introvert. The elongation of the gut which is involved in the production of this affects the oral tube (ot), the buccal cavity (bcv) and the oesophagus (oe), all of which are consequently much longer than in an ordinary gastropod. At the apex the proboscis is converted into a sucker with opening and closing muscles by means of which it can be anchored to the body of the prey after elongation. In the exact centre of the sucker in the genus *Turbonilla* lies the only aperture which occurs, the mouth; in the genera *Odostomia* and *Chrysallida* an opening is also to be found centrally placed on this sucker (lsa) but, in addition there is a second, a short, crescentic slit, lying on the surface of the sucker about half-way between the centre and the ventral edge. This second opening is the mouth (m), which leads into the oral tube, whereas the opening in the centre of the sucker leads into another tube which is separate from the oral tube, except at its extreme inner end, and which lodges a hollow, cylindrical rod of cuticular material tapered to a fine point at its outer end. This is the stylet (sy), and the cavity in which it lies is the stylet tube (syt). At its inner end the stylet is secreted from the dorsal wall of the buccal cavity as a cuticular layer the lateral margins of which converge ventrally and unite so as to form a tube. It is homologous with the jaw of other prosobranchs (Fretter & Graham, 1949). Into it runs a single tube formed from the union of the 2 ducts from the salivary glands (sd), and the whole apparatus can be moved by a complex series of muscles (sym) fastened to its base. The stylet can be thrust out of the mouth of the stylet tube and forms a hypodermic needle which can be forced through the skin of the prey and along which saliva can then be injected.

At the inner end of the oral tube, near the point at which (in *Odostomia* and *Chrysallida*) that part of the gut and the stylet tube are connected, the one with the other, arises the buccal cavity. As it passes inwards this splits into a dorsal part (dp) in which the stylet is secreted, and into which the tube formed by the union of the salivary ducts projects (sd), but which otherwise leads nowhere, and a ventral part (bcv), which is the continuation of the main channel of the alimentary tract. This, in its turn, can be divided into an anterior part and a posterior caecal part, with the oesophagus (oe) coming off between. The anterior part is not marked by any particular histological feature beyond an abundance of mucous cells; the posterior pouch, however, forms a buccal pump (bup). Its walls are thick, but little of this is due to the epithelial lining, which is formed of featureless, squamous cells only 2 μm high (ebp, fig. 102). These rest directly on the muscle of the walls, which is composed of radial fibres inserted on the proximal faces of the epithelial cells and originating on a deeper layer of connective tissue, and of transverse sheets of fibres which run rather more than half-way across the walls of the tube. The radial fibres will clearly distend the lumen of the pouch and the transverse fibres will constrict it. All the fibrillae within the muscle cells show marked striation. Similar muscle fibrillae are to be seen lying around the ducts of the salivary glands and are presumably used to force the salivary secretion along the ducts and through the fine central canal of the stylet into the body of the prey.

The salivary glands (sg, fig. 133; and see fig. 101) are tubular structures, anchored to the body wall by muscular strands and differentiated along their length into areas with varying function. The innermost part is sac-like (bsg), with thin walls, and appears to act as a store for the secretion; next outermost lies a brief region lined by ciliated cells and containing small gland cells. The next region occupies the greater length of the gland and is the major source of saliva; its walls are covered by a squamous epithelium of ciliated cells under which lie the secreting cells in the form of large, cubical masses. Few of these secrete mucus: most produce a protein.

The process of feeding in pyramidellids has been described by a number of workers and it has long been known that these animals are ectoparasites, though it is only recently that the details of the feeding act have been made known. Pelseneer (1914) suggested that *Odostomia scalaris*, which normally lives in close association with *Mytilus edulis* (fig. 137), feeds by extending its proboscis into the mantle cavity of the mollusc and sucking mucus or similar material, food or pseudo-faeces, possibly by the aid of secretion of salivary enzymes which might effect some degree of external digestion. Rasmussen (1944) published a figure (without comment) which showed two animals of this species feeding on mussels, and it was left to Ankel (1948) and Fretter & Graham (1949) to show that pyramidellids were blood and tissue suckers. Ankel and Fretter & Graham made observations mainly on species of *Odostomia* feeding on the polychaete worm *Pomatoceros*, *O. plicata* in Ankel's case and *O. lukisi* and *O. unidentata* in the other. These animals lurk (fig. 134) near the opening of a worm tube waiting for the polychaete to expand its crown of tentacles (tw). When this has happened the mollusc will then evert the proboscis (pb), which moves with a slightly spiral movement when it gets

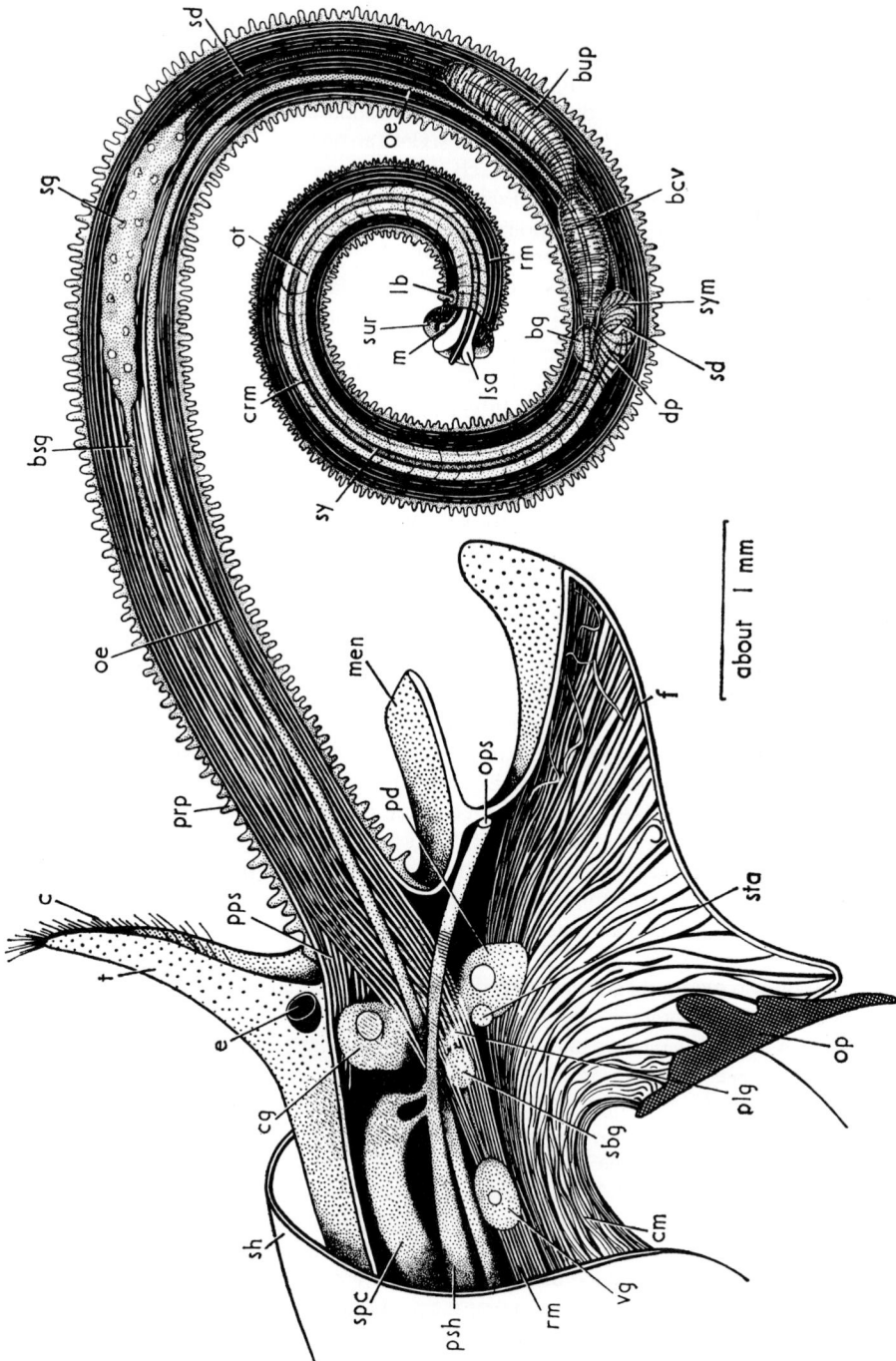

FIG. 133.—*Odostomia unidentata*: sagittal half of the anterior end of a specimen which is protruding head and foot from the shell and has its proboscis extended. This is shown in a conventional position. The haemocoel is black.

bcv, buccal cavity; bg, buccal ganglion; bsg, bladder of salivary gland; bup, buccal pump; cg, cerebral ganglion; c, cilia; cm, columellar muscle; crm, circular muscles; dp, dorsal pouch of buccal cavity; e, eye; f, foot; lb, labial commissure; lsa, lip of stylet aperture; m, mouth; men, mentum; oe, oesophagus; op, operculum; ops, opening of penial sheath; ot, oral tube; pd, pedal ganglion; plg, pleural ganglion; pps, protractor muscles of proboscis; prp, papilla on epithelium of proboscis; psh, penial sheath; rm, retractor muscle; sbg, sub-oesophageal ganglion; sd, salivary duct; sg, salivary gland; sh, shell; spc, sperm sac; sta, statocyst; sur, sucker; sy, stylet in stylet tube; sym, muscles moving stylet; t, tentacle; vg, visceral ganglion.

TABLE 9

Pyramidellid	Host
Chysallida obtusa	Ostrea edulis
Chrysallida spiralis	Sabellaria spp.
Chrysallida seminuda	Crepidula fornicata
Odostomia unidentata	Pomatoceros triqueter
Odostomia conoidea	Astropecten irregularis
Odostomia lukisi	Pomatoceros triqueter
Odostomia plicata	Pomatoceros triqueter
Odostomia scalaris	Mytilus edulis (small)
Odostomia eulimoides	Pecten maximus
	Chlamys opercularis
	Ostrea edulis
	(Turritella)
Odostomia trifida	Mya arenaria
Turbonilla jeffreysi	Some coelenterate,
	probably Halecium sp.
Turbonilla elegantissima	Audouinia tentaculata
	Amphitrite gracilis
Evalea diaphana	Phascolion strombi
Odostomia turrita	Homarus

close to the worm as if it were carrying out exploratory movements. The proboscis is brought close to the tentacle with great caution until it rests on one of the filaments. At this instant the worm may react slightly by jerking the tentacle, but normally it permits the contact with the proboscis without any movement. The sucker at the end of the proboscis then slides along the tentacle, appearing to search for an appropriate spot, which is usually on the inner side. When this is assured the sucker grips, the stylet is driven outwards so as to penetrate the body of the worm and vigorous pumping movements of the buccal apparatus suck fluid into the mollusc's gut.

Fretter & Graham (1949) have suggested that each species of pyramidellid is normally associated with a particular host (figs. 135, 136, 137) and does not usually occur apart from the neighbourhood of the host. Although Cole & Hancock (1955) thought that pyramidellids are not so precise in their feeding habits as was originally suggested, it still seems that each species is predominantly associated with one particular host. The associations which have so far been recorded are listed in Table 9.

Other gastropods besides the Pyramidellidae have become parasitic, but few of these are British and few of them have been sufficiently adequately investigated for it to be known how they feed. Of the British forms Pelseneeria and some of the eulimids (or melanellids) may be mentioned. These have a stout proboscis which is passed deep into the tissues of the host, unlike the superficial attachment of the pyramidellid, and appears to be able to attach the mollusc to its host without the help of the foot.

Pelseneeria lives on echinoids and appears to digest the epidermis by means of enzymes secreted over the body of the host. Their source is unknown. The eulimids are also associated with echinoderms and Balcis devians (= Eulima distorta) has been found on Mesothuria intestinalis, Echinus esculentus, Strongylocentrotus drobachiensis according to Pelseneer

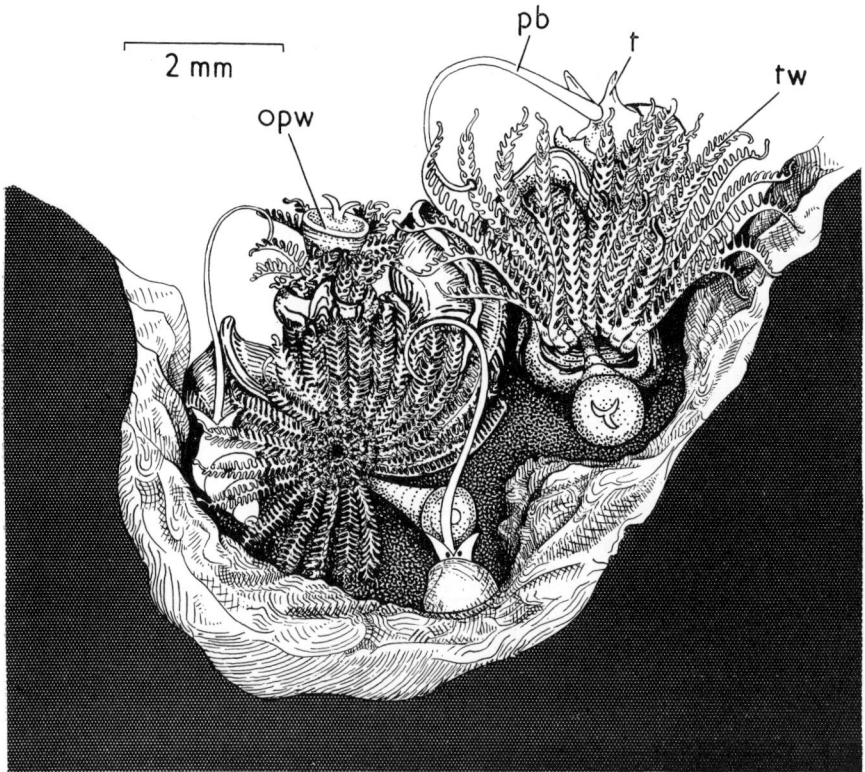

FIG. 134.—*Odostomia unidentata:* feeding on the tubicolous polychaete worm, *Pomatoceros triqueter.*
opw, operculum of worm; pb, proboscis of gastropod; t, tentacle of gastropod; tw, tentacle of worm.

(1928) and also on *Antedon bifida* according to Fretter (1955a) (fig. 139). The echinoderms which are parasitized by other species of eulimid such as *Balcis alba* are not known, though this species occurs abundantly with *Spatangus purpureus* at Plymouth. The alimentary tract of these animals has been extraordinarily modified in connexion with their parasitic mode of life and has been described to a certain extent by Koehler & Vaney (1912) for *Eulima equestris* and some other species, by Risbec (1954) for *E. acutissima* and an unidentified species, and by Fretter (1955a) for *Balcis devians* and *B. alba.*

The foot of *Balcis alba* and *B. devians* is well developed compared with that of the more specialized molluscan parasites of echinoderms, some of which are embedded in the tissues of the host and have lost the foot as they have no need to move about in search of food. The opercular lobe (opl, fig. 289) is large and the pedal glands which secrete mucus have become much enlarged (apg, ptg, figs. 139A, 140) the anterior having hypertrophied to a greater extent than the posterior, and spread backwards in the general body cavity, alongside the oesophagus, as far as the base of the visceral mass. The glands are also enlarged in other parasitic prosobranchs such as *Eulima equestris, Pelseneeria* and *Mucronalia* (Vaney, 1913)

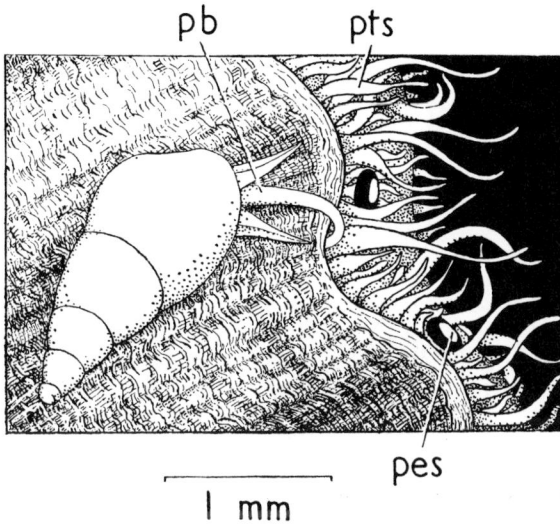

FIG. 135.—*Odostomia eulimoides:* feeding on the scallop, *Pecten maximus.*
pb, proboscis of gastropod; pes, pallial eye of scallop; pts, pallial tentacle of scallop.

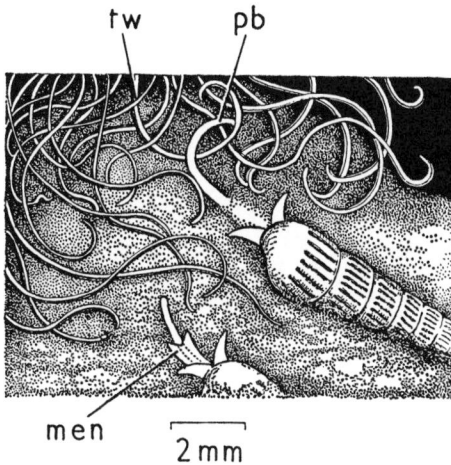

FIG. 136.—*Turbonilla elegantissima:* feeding on the polychaete worm, *Cirratulus cirratus.*
men, mentum; pb, proboscis of gastropod; tw, tentacle of worm.

and in *Megadenus* (Rosen, 1910), but not in *Stilifer* (Hirase, 1932) nor in the endoparasitic types in which, indeed, the foot is wholly lost. When *B. devians* is feeding the foot is contracted so that the glandular tissue is bunched together, with the openings of the 2 glands not far from one another nor from the wound made by the proboscis. The secretion may help to keep the parasite in position whilst it is fixed by the proboscis, though the possibility of it producing some toxic or digestive effect on the tissues of the host has not yet been explored.

2 mm

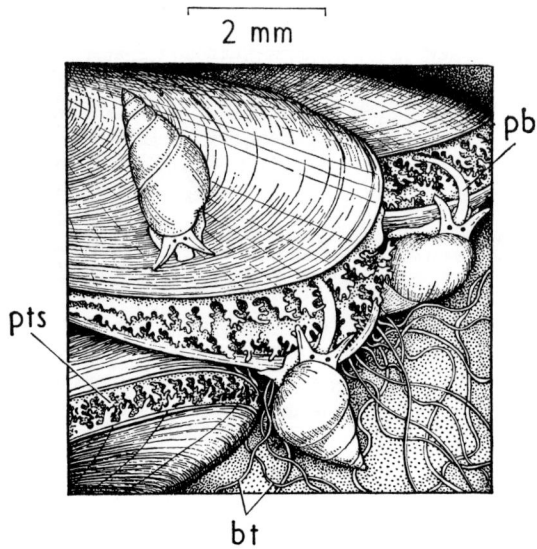

FIG. 137.—*Odostomia scalaris:* feeding on small specimens of the common mussel, *Mytilus edulis.*
bt, byssus threads of mussel; pb, proboscis of gastropod; pts, pallial tentacle of mussel.

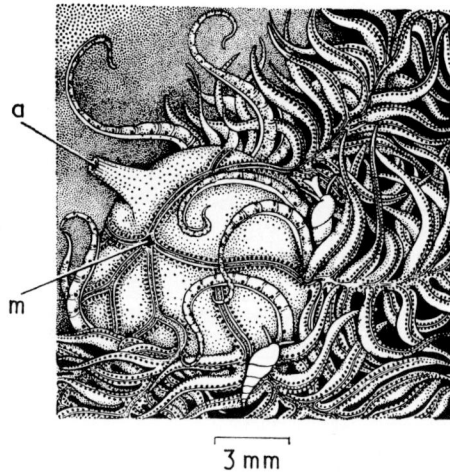

3 mm

FIG. 138.—*Balcis devians:* on the surface of the crinoid, *Antedon bifida.*
a, anus of crinoid; m, mouth of crinoid.

Balcis devians and *B. alba* possess a long acrembolic proboscis, withdrawn into the haemocoel when not in use (fig. 139). When protruded it can excavate the tissues of the host to a depth equivalent to the length of the shell or more. In *B. alba* it is proportionately even longer, for the oesophagus, which is drawn through the introvert on its protrusion is many

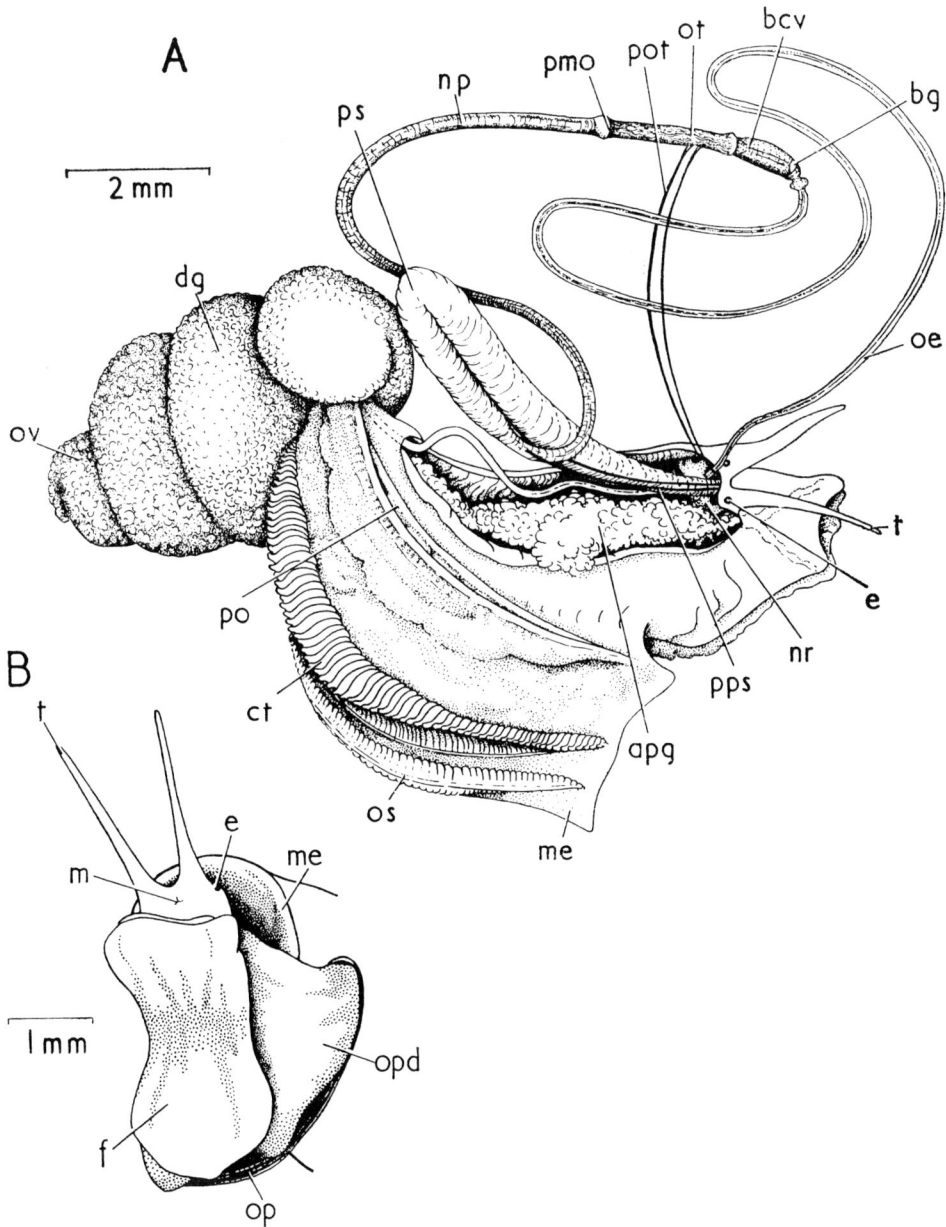

FIG. 139.—*Balcis alba*: A, dissection of animal removed from shell. The mantle skirt has been cut along the left side and folded to the right, the anterior haemocoelic space has been opened and the anterior part of the gut lifted out and displayed; B, the whole animal in ventral view.

apg, anterior pedal gland; bcv, buccal cavity; bg, buccal ganglion; ct, ctenidium; dg, digestive gland; e, eye; f, foot; m, mouth; me, mantle edge; np, narrow part of proboscis; nr, nerve ring; oe, oesophagus; op, operculum; opd, operculigerous disc; os, osphradium; ot, oral tube; ov, ovary; pmo, position of true mouth; po, open pallial oviduct; pot, protractor muscle of oral tube; pps, protractor muscle of proboscis; ps, proboscis sheath; t, tentacle.

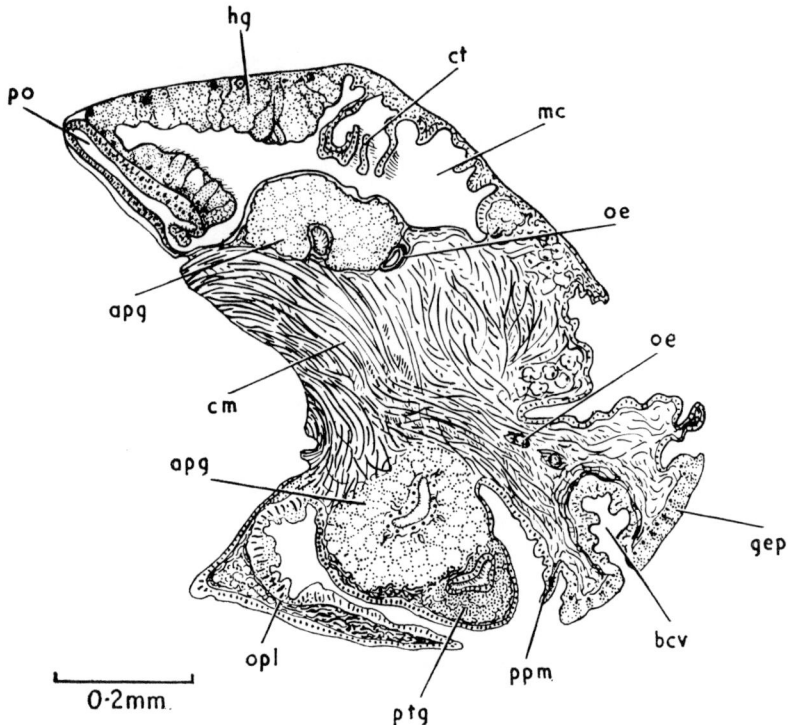

FIG. 140.—*Balcis devians*: oblique transverse section passing through the extended proboscis, the pedal glands and the mantle cavity.

apg, anterior pedal gland; bcv, buccal cavity; cm, columellar muscle; ct, ctenidium; gep, glandular epithelium; hg, hypobranchial gland; mc, mantle cavity; oe, oesophagus; opl, operculigerous lobe of foot; po, pallial oviduct; ppm, pseudopallium; ptg, posterior pedal gland.

times the length of the shell (fig. 139A), suggesting that it passes well into the body of the host, perhaps seeking the gonad. At its inner end the proboscis (ps) opens into a narrower tube (np), rather abruptly in *Balcis alba,* but gradually in the unidentified species of *Eulima* described by Risbec (1954), and this leads to a distinct boundary which apparently marks the position of the true mouth (pmo). Behind this, specimens of *Eulima* sp. (Risbec, 1954) exhibit 2 sections of gut, one, anterior, which is clearly buccal, the second, posterior, which must be the oesophagus. As the animal has no radula, no salivary glands, no oesophageal glands and as Risbec does not describe the position of any buccal ganglia, it is difficult to relate this featureless tube to the alimentary tract of less specialized animals. The work of Fretter (1955a), however, permits this. In *Balcis devians* and *B. alba* 3 sections are discernible in this stretch, two relatively short, but the third, which passes through the nerve ring (nr), greatly elongated. There are, once again, no salivary glands, nor radula nor oesophageal glands to help in identifying parts, but at the posterior limit of the second section are to be found the buccal ganglia (bg). This, therefore, allows the second section to be identified as the buccal cavity (bcv) and the greatly elongated third part as the oesophagus (oe), whilst the first part, immediately behind the true mouth, must be the oral tube (ot).

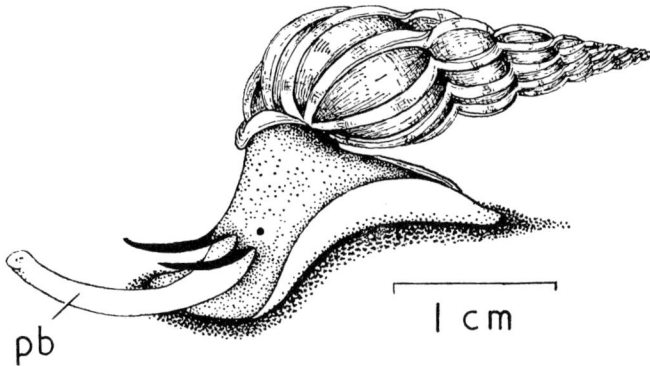

FIG. 141.—*Clathrus clathrus:* animal crawling with partly extended proboscis.
pb, proboscis.

Histologically all these sections of the gut are very much alike and strikingly devoid of gland cells. The oesophagus is ciliated, the other parts not, and all are extremely muscular. It is difficult to tell how much of this apparatus can be everted into the body of the prey: certainly not the oesophagus which passes through the nerve ring and is there, and in that neighbour-hood, securely attached to surrounding tissue. It is probable, however, that the true mouth comes to lie at the summit of the proboscis when that is fully extended. Feeding must clearly be by sucking and by eating such particulate material as the muscular region round the mouth can manage to tear from the prey. From the minute size of the oesophagus (oe) where it passes though the nerve ring (nr), it seems likely that these parasitic eulimids are primarily feeders on fluids.

From the nerve ring the oesophagus runs to a stomach embedded in the visceral mass. As in pyramidellids its separation from the digestive gland is not clear and it appears more as an excavation in that than a more normal stomach. The cells in the digestive gland are much laden with darkly staining spherules, again as in the pyramidellids.

Rosen's statement (1910) that *Eulima polita* (=*Balcis alba*) is free-living, not parasitic, and has a radula and an oesophagus with attached glands is clearly based on a misidentification.

A few other prosobranchs perhaps deserve a mention at this point, mainly because the anatomy of their alimentary tract suggests unusual modes of feeding, though these are, in most cases, still not known. These animals are sometimes placed together in a group called the Ptenoglossa, and include, among British genera, *Ianthina, Cirsotrema, Clathrus, Graphis, Aclis, Pherusina* and *Cima*. Of this list *Ianthina* and *Clathrus* are the only animals not almost wholly unknown except for their shells. The anatomy of *Clathrus* has been described by Bouvier (1886) and Thiele (1928) and that of *Ianthina* by Thiele (1928) and Laursen (1953). Thiele (1928) gave a few details of *Aclis* and Dall (1889) described its shell as heterostrophic, which suggests a link with the pyramidellids, but apart from this our knowledge of this group of little prosobranchs is extremely meagre.

In *Ianthina* a small proboscis is developed (pb, fig. 291), the tip of which is normally inturned. The buccal mass and its covering of recurved teeth, all alike, have already been described. Thiele (1928) described 2 pairs of salivary glands on each side discharging by a common duct opening far forward on the dorsal wall of the buccal cavity: in *I. janthina*, however, the 2 salivary glands on each side open to the buccal cavity by separate ducts, one, from a posterior pair almost mid-dorsally, the second, from an anterior pair, very laterally and anterior to the cutting edge of the jaw. What degree of variation occurs within the genus is not known. The glands themselves are tubular. In the related genus *Recluzia* there are also 2 pairs of salivary glands the ducts from which lead to a pair of cuticular stylets surrounded by a sheath of muscle. These run along the lateral walls of the buccal cavity and open near the mouth. This is reminiscent of the genus *Clathrus* (figs. 99, 100), in which again there are 2 pairs of tubular salivary glands, dorsally and ventrally placed. Of these the dorsal pair (asg, fig. 99) open to the dorsal wall of the buccal cavity a short distance in front of the nerve ring, in a position not dissimilar to that occupied by salivary ducts in most prosobranchs. The second, ventral pair (sg), however, send ducts which pass through the nerve ring, travel in the lateral wall of the buccal cavity, expand into reservoirs and finally open (osd, fig. 100) through cuticular stylets (sy) which arise from the centre of mammiform swellings lying, apparently external to the mouth, in a lateral position. The mouth, however, lies at the base of a long proboscis (pb, fig. 141) of acrembolic type and as no one, apparently, has ever seen *Clathrus* feeding, the precise arrangement of these parts in the everted condition is not known. The buccal mass is very like that of *Ianthina* and the mollusc has 2 sharp-edged jaws (ja, fig. 100) in the neighbourhood of which the buccal wall appears to contain still other masses of glandular tissue (gb). The introvert is very muscular and, in its apical region, rich in gland cells; it is lined by a cuticle apically, but is ciliated at its base.

Beyond the fact that *Clathrus* is clearly adapted for a carnivorous mode of life and, perhaps, for the ingestion of entire animals, it is not possible to say much about its feeding, which remains one of the most interesting points in this connexion yet to be cleared up. The jaws perhaps permit us to conclude that *Clathrus* bites pieces off the body of the prey and hauls them in with the radula. Ankel (1936a) said that this mollusc eats coelenterates, and Thorson (1958) described a related American species as feeding by sucking fluid from sea anemones [see p. 621].

Thiele (1928) stated that *Aclis* is related to *Clathrus* and has a long, retractile proboscis, a pair of jaws and a radula with needle-like teeth. So far as the other animals in the group are concerned nothing appears to be certainly known of their structure or habits, a gap which would undoubtedly be interesting to fill.

[The food of British prosobranchs (so far as known) is given in chapter 26.]

CHAPTER 11

THE VASCULAR SYSTEM

THE vascular system of the gastropod molluscs consists of a central contractile heart which is linked to the various parts of the body by a series of distributing vessels, the arteries, and by a series of collecting vessels, the veins. The arteries and the veins are connected to each other partly by means of small channels which almost deserve to be regarded as capillaries, but partly also by large cavities, the venous sinuses. These differ from the other parts of the vascular system in that in at least some places they are not lined by any special endothelium, and the blood comes into direct contact with the cells of the organs amongst which they lie. The main arteries and veins, and, on occasion, some of these sinuses, are lined with endothelium.

The heart is a systemic heart, that is, it is placed so as to pump blood through the arteries to the body, collecting it from the respiratory organs to do so. The kidney and its blood vessels occupy a particularly important place in the circulation, being intercalated between the main organs and the respiratory capillary bed. There is thus a renal portal system with afferent renal veins, draining blood from the body to the kidney, and efferent renal veins taking blood from the kidney to the respiratory organs.

The general plan of the circulatory system in most prosobranchs (fig. 21) is as follows. Blood leaves the heart by one or other of 2 main arteries, an anterior aorta and a posterior (or visceral) aorta. The anterior aorta runs forward to end in a series of sinuses in the head and foot; the posterior aorta runs up the visceral hump, usually on the outer, convex, side of the spiral and opens into visceral sinuses. From those sinuses which are placed in the head and the foot the blood is collected into a vessel which lies near the visceral ganglia and in relation to the kidney. Into it there also comes blood from the visceral sinuses, by way of a vessel running along the inner, concave, side of the visceral hump. All this blood is drained into the kidney whence most of it is taken to the mantle skirt, distributed to the ctenidium, and so returned to the heart. A part of it, however, passes from the kidney to the nephridial gland (see p. 639) and is then drained, not into the roof of the mantle cavity, but directly to the heart, so that it does not pass through the ctenidium.

In the diotocardian prosobranchs the course of the circulation is affected by the duplicity of the pallial organs which occurs in these animals and also by the asymmetry of the kidneys which they show. It has been described for a number of animals: for *Haliotis*, by Milne-Edwards (1846), Wegmann (1884) and Crofts (1929), for *Acmaea* by Willcox (1898) and Thiem (1917b), for fissurellids by Ziegenhorn & Thiem (1926), for *Patella* by Milne-Edwards (1846), Wegmann (1887), Haller (1894) and Boutan (1900), and for trochids by Robert (1900), Nisbet (1953) and Deshpande (1957).

In *Monodonta* and other trochids the heart lies in a pericardial cavity placed transversely across the basal part of the visceral mass (la, ra, ve, figs. 53, 324). It is composed of a ventricle (ve, fig. 142A), through the centre of which runs the rectum (r), flanked anteriorly by a left auricle (la), and posteriorly by a right auricle (ra). From the left end of the ventricle there

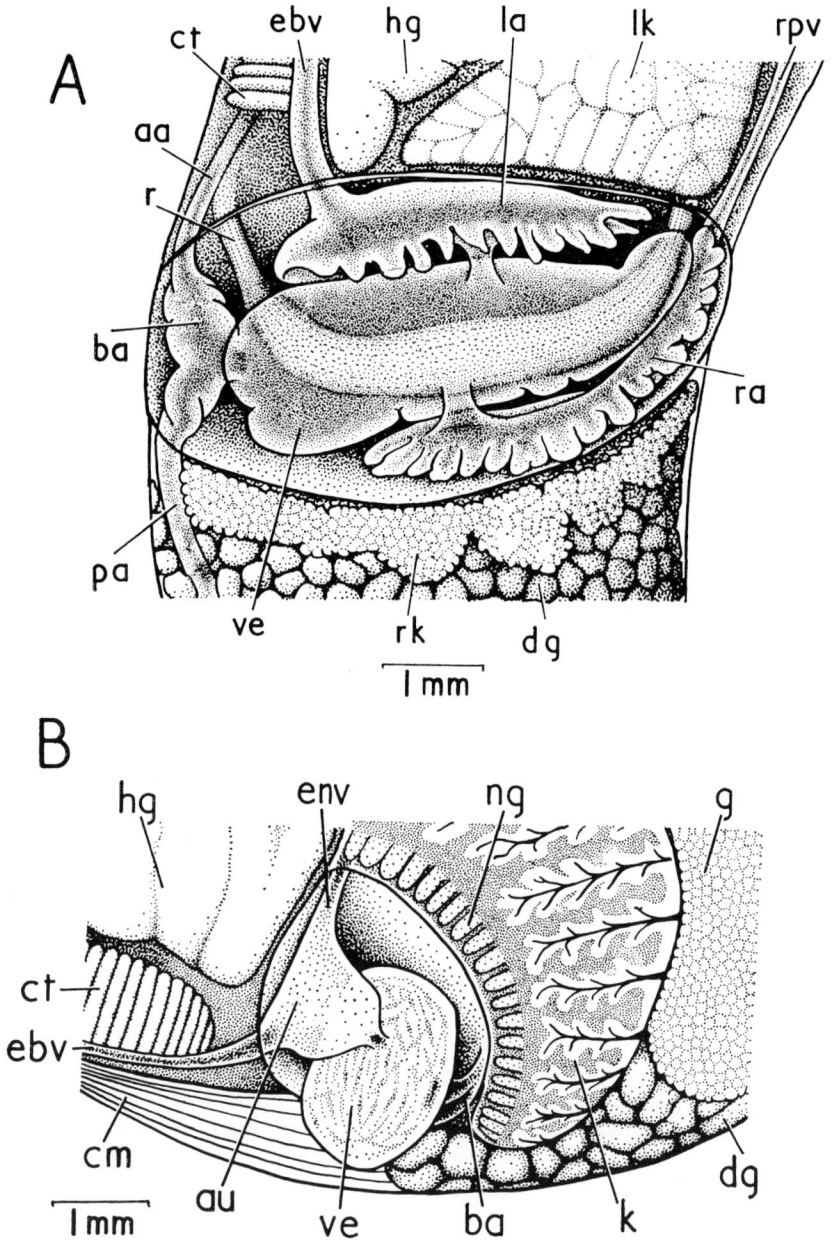

FIG. 142.—A, *Monodonta lineata:* heart and surrounding organs. Note that the roof of the pericardial cavity has been removed and that the anterior end is at the top of the figure. B, *Nucella lapillus:* heart and surrounding organs. Note that the roof of the pericardial cavity has been removed, that the ventricle is displaced and that the anterior is to the left.

aa, anterior aorta; au, auricle; ba, bulbus aortae; cm, columellar muscle; ct, ctenidium; dg, digestive gland; ebv, efferent branchial vein; env, efferent vein of nephridial gland; g, gonad; hg, hypobranchial gland; k, kidney of monotocardian; la, left auricle; lk, left kidney or papillary sac; ng, nephridial gland; pa, posterior aorta; r, rectum; ra, right auricle; rk, right kidney; rpv, right pallial vein; ve, ventricle.

emerges a single vessel, which may be called the bulb of the aorta (using Boutan's term) (ba). As soon as it has left the pericardial cavity this divides to form 2 vessels which are the anterior (aa) and posterior (pa) aortae.

The posterior aorta (pa, fig. 324B) climbs the visceral mass giving off vessels to such sections of the alimentary and reproductive systems as are to be found in that part of the body. These end by opening to the visceral haemocoelic spaces lying amongst the acini of the gonad (ov) and digestive gland (dg). The anterior aorta (aa, fig. 53) passes forwards, on the left side of the body, at the level of attachment of the mantle skirt, ventral to the line of the efferent branchial vessel (ev). It penetrates the muscular wall of the body and so comes to lie alongside a forwardly directed loop of intestine (i) placed in this area. At the anterior end of this loop the artery penetrates a muscular partition, the transverse septum (trs, figs. 143, 144, 149), which forms a boundary between the haemocoelic space in which the intestine lies and other anterior ones in the head. Having penetrated this, the aorta runs across the body, closely adherent to, or even embedded in the septum, until it reaches the right side, where it again turns and resumes its forward course, enlarging in diameter as it does so, and comes to lie underneath the buccal mass, in a space called (by Nisbet, 1953, whose nomenclature is being followed here) the cephalopedal sinus (cps, fig. 143). This forms a distributive centre from which blood passes to a further series of structures lying at the anterior end of the body: (1) an anterior pedal artery (apa) passes downwards and forward in the mid-line, supplying the anterior part of the foot; (2) paired lateral pedal arteries (lpa) pass backwards one on either side of the foot; (3) a connexion (ppa) takes blood to a cavity placed directly under the cephalopedal sinus, the pleuropedal sinus, so called because it surrounds the pleuropedal ganglionic mass, from which blood spaces extend backwards through the foot alongside the pedal nerve cords; (4) similar spaces pass dorsally alongside the cerebropleural (cpl) and cerebropedal (cpdc) connectives, pouring blood into a large haemocoelic cavity, the dorsal cephalic sinus, situated in the dorsal part of the head; (5) extensions pass forwards to the snout, the eye stalks and the tentacles; (6), (7) and (8) 3 routes by which blood can pass into the sinuses of the buccal mass. These are: a dorsal buccal sinus lying dorsally in the buccal mass and extending into the oesophageal valve (the sheet of tissue lying between the anterior oesophagus above and the radular sac below) and a ventral buccal sinus filling most of the space between the more ventrally placed muscles and cartilages of the odontophore.

Blood from these different destinations ultimately drains into a small number of sinuses lying around, or in relation to, the oesophagus and anterior loop of intestine already described. All the blood from the buccal sinuses is gathered into one called the ventral cephalic sinus (vcs). This lies around the buccal sinuses, between the outer surface of the buccal mass and the lateral walls of the head, and it extends backwards around the radular sac (rs) to where that ends anterior to the transverse septum. Blood from this sinus, however, cannot escape directly backwards at this point, because of this and other septa, but must pass dorsally into another sinus, the dorsal cephalic sinus (dcs), already referred to; into this also drains all the blood from the other haemocoelic spaces in the head. The dorsal cephalic sinus lies morphologically dorsal to the mid-oesophagus, and extends with it as far as the transverse septum, where it ends. The transverse septum is thus an important landmark, coinciding not only with the posterior, blind ends of the 2 cephalic blood sinuses, dorsal and ventral, but also marking the posterior end of the radular sac and, incidentally, the point at which the mid-oesophagus runs into the posterior oesophagus. Since the dorsal cephalic sinus overlies the mid-oesophagus it is also, like that part of the gut, twisted to the left as it passes backwards, and

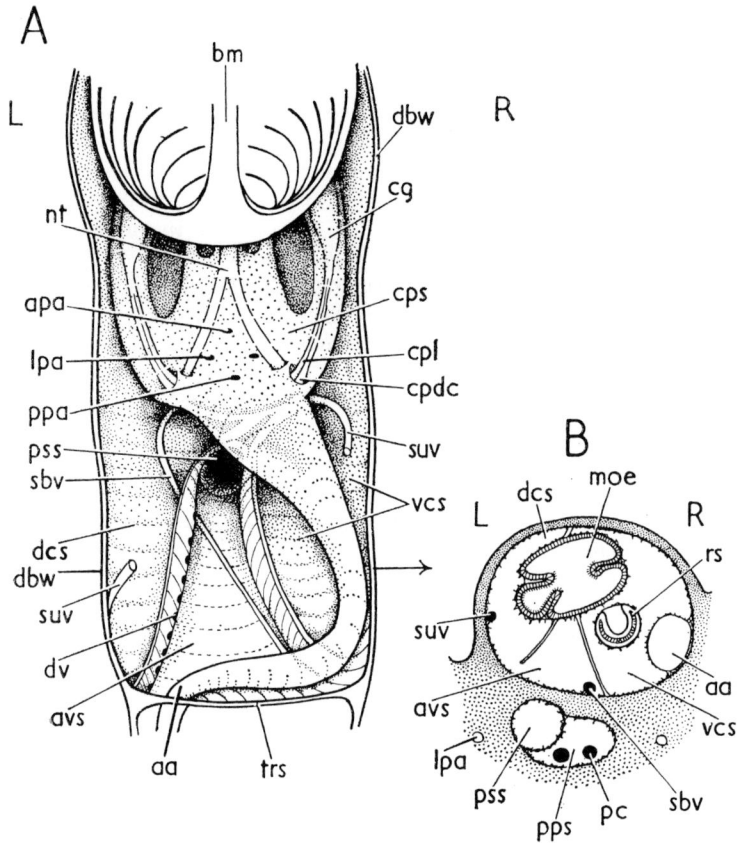

FIG. 143.—*Monodonta lineata:* A, semidiagrammatic representation of the anterior part of the body lying between the transverse septum (trs) posteriorly and the buccal mass (bm) in front; the buccal mass has been turned forwards to expose structures beneath and the radular sac and oesophagus have been removed. Some nerves and muscular partitions have also been partly taken away opening some haemocoelic spaces. The ventral cephalic sinus (vcs) is following the twist of the mid-oesophagus round the right body wall from a ventral position anteriorly to a dorsal position near the level of the transverse septum. The dorsal cephalic sinus (dcs) is similarly curving from a dorsal position anteriorly to a ventral position posteriorly where it connects with the anterior ventral visceral sinus (avs) by means of holes in a muscular partition (dv). B, transverse section at level marked by arrow in A.

aa, anterior aorta; apa, opening to anterior pedal artery; avs, anterior ventral visceral sinus; bm, buccal mass; cg, cerebral ganglion; cpdc, cerebropedal connective; cpl, cerebropleural connective; cps, cephalopedal sinus; dbw, dorsal body wall cut; dcs, dorsal cephalic sinus; dv, connexions between the dorsal cephalic and anterior ventral visceral sinuses; lpa, opening to lateral pedal artery; moe, mid-oesophagus; nt, nuchal tendon; pc, pedal cord; ppa, opening to pleuropedal sinus; pps, pleuropedal sinus; pss, opening of pedal sinus to the anterior ventral visceral sinus; rs, radular sac; sbv, sub-oesophageal part of visceral loop cut; suv, supra-oesophageal part of visceral loop cut; trs, transverse septum; vcs, ventral cephalic sinus.

divided into 3 parts by double sheets of muscle and connective tissue; these run between the bases of the dorsal folds and the overlying body wall, so as to produce lateral haemocoelic compartments over the lateral glandular areas and a median channel over the food groove,

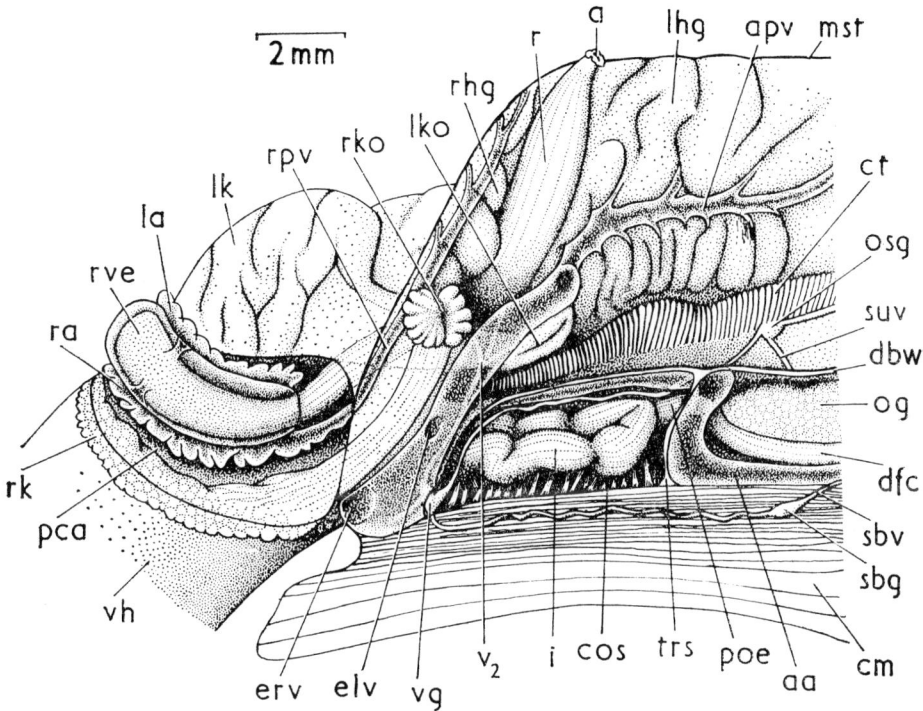

FIG. 144.—*Monodonta lineata:* dissection to show the relationships of the vascular system to other structures at the inner end of the mantle cavity. The pericardial cavity has been opened and exposes the heart and the posterior part of the right kidney. The body wall on the right side has been removed and some dissection done to display the venous sinus and posterior part of the visceral loop. The mantle skirt has been cut along the extreme right side and lifted dorsally.

a, anus; aa, anterior aorta; apv, anterior pallial vein; cm, columellar muscle; cos, collecting sinus; ct, ctenidium; dbw, cut edge of dorsal body wall; dfc, dorsal food channel of oesophagus; elv, opening of vein from left kidney to transverse pallial vein; erv, opening of vein from right kidney to transverse pallial vein; i, anterior loop of intestine; la, left auricle; lhg, left hypobranchial gland; lk, left kidney; lko, opening of left kidney; mst, cut edge of mantle skirt; og, oesophageal gland; osg, osphradium overlying osphradial ganglion; pca, pericardial cavity; poe, posterior oesophagus; r, rectum; ra, right auricle; rhg, right hypobranchial gland; rk, right kidney; rko, opening of right kidney; rpv, right pallial vein; rve, rectum within ventricle; sbg, sub-oesophageal ganglion; sbv, sub-oesophageal part of visceral loop; suv, supra-oesophageal part of visceral loop; trs, transverse septum; v_2, rectal and genital nerve (the guide line ends on the nerve within the transverse pallial vein at the point where it divides giving branches to the two kidney openings and the rectum, and at a point where the duct of the left kidney may be seen by transparency under the vein); vg, visceral ganglion lying at the point where the collecting sinus opens to the transverse pallial vein; vh, visceral hump.

which may be called the median dorsal channel of the dorsal cephalic sinus. The muscular curtains which separate it from the lateral compartments are incomplete and allow blood to enter the dorsal channel, which it does because there is no other outlet from the lateral spaces. The median channel thus collects the blood which is returning from all parts of the head, leads it backwards and ventrally, following the torsion of the gut, and finally spills it into a sinus which extends posteriorly below the ventral edge of the transverse septum and ventral to the oesophagus and anterior intestinal loop, though dorsal to the columellar muscle. This is called the anterior ventral visceral sinus (avs) into the anterior end of which a large pedal sinus (pss) returns blood from the foot. It is thus a collector of all blood emerging from the head and foot.

Behind the transverse septum the roof of the anterior ventral visceral sinus (avs) is formed of a stout sheet of muscle, continuous with the transverse septum anteriorly and with a similar vertical sheet posteriorly, the vertical septum, which separates all the cephalic and pedal blood spaces so far mentioned from the major visceral blood sinuses. This sheet lies horizontally under the oesophagus and anterior intestinal loop, and is there a complete septum, but on the right it is perforated by a large number of openings through which blood passes dorsally from the anterior ventral visceral sinus into a space called the collecting sinus by Nisbet (1953) (cos, fig. 144).

Posteriorly on the right, the lateral body wall, the vertical and horizontal septa and the dorsal body wall come together to form the corner of body cavity into which the collecting sinus runs, and it is at this point that there arises a large and important vein which passes on to the mantle skirt; this is the transverse pallial vein, inside which the visceral ganglia (vg) lie and through which some of the visceral nerves (v_2) run. It is formed by the union of the collecting sinus (cos) from the cephalopedal regions in front, and a large vein (erv) which emerges from the right kidney (rk). The T-junction where the 3 vessels converge is traversed by strands of muscle and tough connective tissue which may permit some control of the blood flow at this point in the circulation. The transverse pallial vein, so formed, now runs on to the mantle skirt, following a forward course more or less ventral to the rectum (r), though actually crossing that structure diagonally from right to left, and flanked as it does so by the ducts and openings of the right (rko) and left (lko) kidneys (papillary sac). It therefore occupies a position which is morphologically median, a fact of some importance from the point of view of comparative anatomy. Into it, at this point, flow 3 vessels: one, according to Deshpande (1957), from the rectum, another from the left kidney (elv), and the third from the region of the right hypobranchial gland, along which it flows parallel to, and to the right of, the rectum. Small vessels leave its main channel and spread between the lobes of the gland and others ramify in the mantle skirt in front of the anterior end of the gland. The transverse pallial vein itself crosses left of the rectum and then turns forward to run through the left hypobranchial gland, where it is sometimes known as the anterior pallial vein (apv); it gives off branches which pass between the lobes of the gland. Some of these are small, others are vessels of considerable size, but all pass to the left towards the base of the ctenidium where they flow into a longitudinal vessel which runs along the dorsal surface of the ctenidial axis and from which spring the numerous capillaries leading blood into the branchial leaflets (ctl, fig. 145). This longitudinal vessel is, therefore, clearly the afferent branchial vein (abv).

Whilst there can be no doubt that blood is passing into the transverse pallial vein from the right renal vein—that this is, in other words, a renal efferent vessel—there can be no similar certainty about the vessel connected to the papillary sac (the left kidney), in view of the fact that the vascular system of that kidney is also directly connected to the left auricle. It may, therefore, be possible for blood to flow from the transverse pallial vein to the left auricle either by way of the ctenidium and the efferent branchial vessel, or by way of the left kidney. At first sight it may seem that a similar doubt must also exist regarding the direction in which blood flows along the anterior pallial vein, since this might be either a tributary or distributary of the transverse pallial. This matter appears to be resolved by the occurrence of another vein, the efferent or right pallial vein (rpv, figs. 142A, 144), which runs along the mantle skirt at its

ctl

hg

abv

0.5 mm

tpv

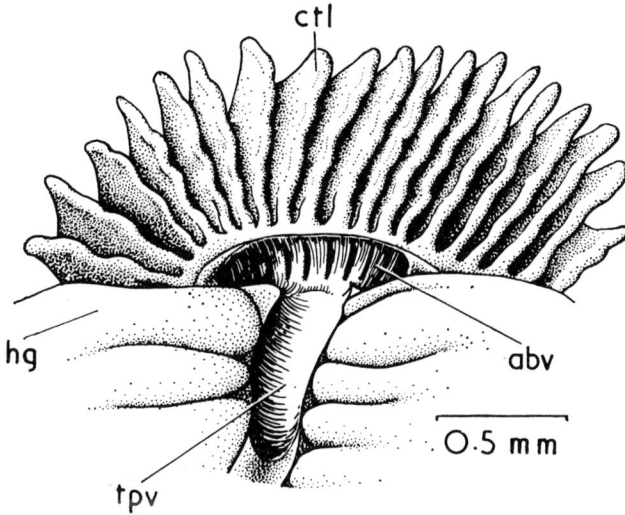

FIG. 145.—*Calliostoma zizyphinum:* a dissection to show the relationship of the transverse pallial vein to the afferent branchial vein.
abv, afferent branchial vein showing openings to blood spaces in ctenidial leaflets; ctl, ctenidial leaflet; hg, hypobranchial gland; tpv, transverse pallial vein.

extreme right margin, ventral to the right hypobranchial gland (rhg) and just dorsal to the right edge of the columellar muscle (cm). It connects anteriorly with the branches of the anterior pallial vein (apv) and carries blood to the right auricle of the heart (ra). In view of this it seems reasonable to conclude that the anterior pallial vein is, in fact, distributing blood from the transverse pallial vein to the right part of the mantle skirt which will then be collected into the efferent pallial vein.

Now the efferent pallial vein has been regarded by several writers (e.g. Thiele, 1897) as the homologue of the right efferent ctenidial vessel, the right gill having vanished, though its efferent has persisted. This view would be strengthened by the description just given, from which it would appear that the anterior pallial vein is a part of the transverse pallial vein system distributing blood by way of the right hypobranchial gland to a right ctenidium and by way of the left hypobranchial gland to the left ctenidium. The basal part of the transverse pallial vein is a (morphologically) median vessel bringing blood to what was originally a symmetrically arranged respiratory area in the roof of the mantle cavity.

The only remaining part of the vascular system to be described in *Monodonta* (fig. 146B) is the venous return of blood from the sinuses in the visceral mass. This is also achieved by way of the kidneys, and is, in fact, the main source of the blood which enters the transverse pallial vein by way of the right efferent renal vein (erv) and, perhaps, by the vein connecting the transverse pallial vein and the left kidney as well.

The historical origin of the system which has just been described would appear to lie in a gastropod in which there were two equally developed hypobranchial and ctenidial areas in the mantle roof. It is, therefore, interesting to compare the arrangement of the vascular system in trochids with what is found in the other diotocardians. Of these the zeugobranchs are, from this aspect, clearly the more important and the members of the Patellacea not so

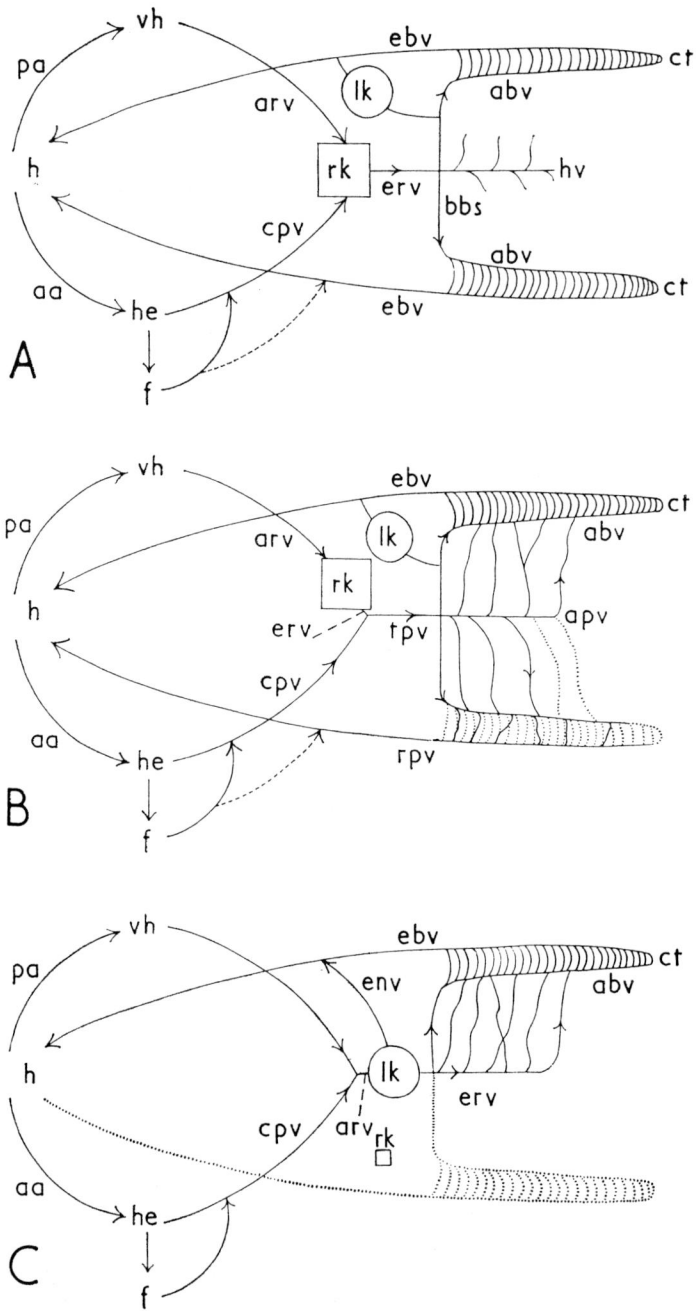

FIG. 146.—Diagrams of the vascular system of: A, *Haliotis*, based on Crofts; B, a trochid; C, a monotocardian. Solid lines indicate vessels occurring; pecked lines mark vessels which probably exist; dotted lines are vessels which have been lost.

aa, anterior aorta; abv, afferent branchial vein; apv, anterior pallial vein; arv, afferent renal vein; bbs, basibranchial sinus; cpv, cephalopedal vein; ct, ctenidium; ebv, efferent branchial vein; env, efferent vessel of nephridial gland; erv, efferent renal vein; f, foot; h, heart; he, head; hv, hypobranchial vessels; lk, left kidney; pa, posterior aorta; rk, right kidney; rpv, right pallial vein; tpv, transverse pallial vein; vh, visceral hump.

significant. In *Haliotis* (fig. 146A) the anterior aorta (aa) takes blood to the anterior part of the body (he, f), whence it is collected into venous sinuses: unfortunately, these have not been described in detail comparable to what is known about *Monodonta,* so that a closer comparison is not possible. They are said, however, to open into the 'abdominal' visceral sinuses, from which they reach afferent veins leading to the right kidney (rk): if this is so, there is then a major difference between *Haliotis* and the trochids, because in the former all blood would traverse the right kidney, whereas in the latter the blood from the head-foot by-passes the excretory organ and passes straight to the mantle skirt, and only blood from the viscera traverses the kidney.

From the right kidney blood passes along the efferent renal vessel (erv) to one known as the basibranchial sinus (bbs) which lies transversely across the mantle skirt at its attachment to the visceral mass, between the bases of the ctenidia (ct). It lies ventral to the rectum, crossing it from side to side in what is a morphologically median situation, and in addition to its main factor (the right renal efferent) it also receives vessels related to the rectum and the left kidney or papillary sac. From its ends arise the afferent ctenidial vessels (abv) running one along the axis of each gill. There seems no difficulty, after the recital of these facts, in homologizing the transverse pallial vein of trochids with this so-called basibranchial sinus of *Haliotis* (as suggested by Bernard, 1890), and the conditions in this animal would also support the idea that the efferent pallial vein is the homologue of the right efferent ctenidial vessel.

Whereas *Haliotis* shows a departure from the trochids as regards the situation of the right kidney in the circulatory system the fissurellids appear to occupy an intermediate position, according to Spillmann (1905) and Ziegenhorn & Thiem (1926). In *Fissurella crassa* and *Glyphis graeca* (= *Diodora apertura*) the anterior aorta carries blood, as usual, to the head, foot and anterior viscera, and the posterior aorta carries blood to the visceral mass. Whereas all the blood from the visceral mass is carried to the right kidney, as in *Haliotis* and the trochids, and all the blood from the cephalic and pedal region passes straight to the basibranchial sinus as in trochids, some blood from the anterior viscera goes to the right kidney as in *Haliotis*.

In these zeugobranchs the left kidney seems all but a vestige embedded in the anterior pericardial wall. According to a diagram given by Crofts (1929) it appears to receive only a supply of oxygenated blood from an efferent ctenidial vessel, and this, after traversing its tissues, is then returned to the basibranchial sinus and sent to the gills. [For recent work see chapter 27.]

In view of the reduction or absence of ctenidia in the Patellacea and their replacement by secondary pallial outgrowths, it is not surprising to find that the plan of the vascular system in these animals has been modified. In *Acmaea,* in which one ctenidium still persists, some blood is sent directly there and so to the heart, but there are other routes by which blood may reach the auricle (Willcox, 1898; Thiem, 1917b). From the head most of the blood is collected into sinuses related to the nerve ganglia, the pleural sinus, the anterior neural sinus and the posterior neural sinus. These connect with pedal veins into which also passes the blood from the major spaces of the foot, and this volume of blood is then passed through gaps in the shell muscle into the mantle edge and collected into a vein running around the free edge of the mantle skirt which leads to the auricle (epv, fig. 73). Although there are no true pallial gills set around the mantle edge in these animals there must, nevertheless, be a considerable degree of oxygenation of the blood possible in this situation and it is likely that this blood is as well oxygenated as that which passes through the ctenidium. In fact, according to Thiem (1917b), there is a correlation between the degree of folding in the mantle edge, the number of vessels passing through gaps in the shell muscle and the presence or absence of a ctenidium: thus in scurriids, with a well developed ctenidium and well developed folds on the mantle edge, these breaks are few; in acmaeids, which have no folds but still retain the ctenidium, they are numerous, whilst in the lepetids, which have neither marginal pallial folds nor ctenidium, the gaps through the shell muscle, and therefore the break-up of the blood flow into a large number of small vessels permitting ready oxygenation, are very frequent.

In addition to the two routes by means of which venous blood may be returned to the heart—both of which are so arranged as to introduce respiratory surfaces *en route*—the acmaeids also show a third way, but this does not include any significant respiratory organ. It drains blood from some of the anterior viscera (such as salivary glands, digestive gland and parts of the intestine) which abut against the posterior wall of the nuchal cavity, and passes it by means of channels which lie in the floor of the mantle cavity to the heart. Little, if any, respiratory exchange appears to be likely to occur in this situation, but the vessels are small and the amount of blood transmitted by them is probably not great. Some of the blood from these organs, however, and also apparently all from the posterior part of the visceral mass, drains through the right kidney and is then added to the stream which is passing through the shell muscle to be oxygenated in the mantle edge prior to its return to the heart. A minute fraction of this is said (Thiem, 1917*b*) to make its way through the left instead of the right kidney.

It is clear from this description of the acmaeid circulatory system that the ctenidium plays a significantly less important part than it did in either *Haliotis* or the trochids, and, also, that the right kidney is not incorporated into the circulation in such a strategic way as in these animals. It is not surprising, therefore, to find that the cyclobranchs, the patellids, have been able to lose their ctenidia wholly, and compensate for their loss by an expansion of the pallial branchial equipment. In *Patella* (Milne-Edwards, 1846; Wegmann, 1887; Boutan, 1900; Davis & Fleure, 1903) the heart (au, ve, figs. 253, 258) lies at the anterior end of the visceral mass well to the left, near the anterior end of the shell muscle (sm). Three apparent chambers lie parallel to one another across the breadth of the pericardial cavity: the auricle (au) most anteriorly, the ventricle (ve) in the centre and a bulbus aortae (ba) (considered by Wegmann (1887) to be a subdivision of the ventricle) posteriorly. From the left side of the bulbus arises a vessel, often called the genital artery, which runs posteriorly over the visceral hump supplying the gut and gonad: this is clearly the posterior aorta. From its right side leaves a vessel which curves ventrally, under the rectum, and then runs forwards into the head, where it opens to a large cephalic blood space. This, in turn, gives rise to anterior and posterior pedal arteries and to arteries which supply the more ventrally placed viscera and those which lie to the animal's right. The vessel which originates from the right side of the bulbus corresponds to the anterior aorta of the other types described. From all these parts the blood eventually makes its way into visceral sinuses lying in the visceral mass and it is from these that the venous return starts. The blood passes through interstices in the shell muscle (ism, fig. 50) into an afferent branchial vein which runs round the base of the mantle skirt on the outer side of the shell muscle and is interrupted only anterior to the head. From this the blood is passed through the spaces in the gill leaflets (pag) into an efferent branchial vein (epv) running round the complete circle of the mantle skirt near its outer edge. At one point on the left side, near the anterior end of the shell muscle, this ring vessel gives rise to a large trunk which runs to the auricle (au). Two subsidiary routes may be added to this principal one just described, exactly comparable to those found in acmaeids. Some blood from the visceral sinuses passes through the right kidney and is led from it to the afferent branchial vein by an efferent renal vessel which crosses the rectum near the anus, crosses the kidney duct and makes its way on to the mantle skirt at the anterior end of the right half of the shell muscle. The relationships of this vessel suggest that it may well be the homologue of the basibranchial sinus of *Haliotis* and the transverse pallial vein of the trochids. What part the reduced left kidney plays in the circulation is obscure, but it may contribute some blood to this vessel. Lastly, there exists a pathway by which blood from visceral spaces can pass into the floor of the nuchal cavity, from which it is drained directly into the auricle. The quantity of blood which does this is not great, and it seems unlikely that much oxygenation can occur in such a position in a limpet.

It is now necessary to turn to the remaining prosobranchs, the mesogastropods and stenoglossans (fig. 146C). In these, as the name Monotocardia implies, there is only a single auricle in the heart (au, fig. 142B), which corresponds to the left auricle of the diotocardians and, like it, receives blood from the left ctenidium (ct) and left kidney (k, fig. 142B; lk, fig. 146C). With the disappearance of the right auricle the right or efferent pallial vein disappears too, and the whole of the right vascular complex of the roof of the mantle cavity is lost. The right kidney (rk) remains only as an element in the organization of the genital duct, and as the glandular sections of this appear to take the place of, and may have arisen from the right hypobranchial gland and have been folded off the surface of the mantle skirt, there is nothing to prevent the rectum from coming to lie well over to the right side of the body. This allows for the expansion of the left ctenidium and left hypobranchial gland in the mantle skirt and for the increased volume of the left kidney as that replaces the right as the main excretory organ. Associated with this—perhaps a necessary prerequisite of it—the rectum no longer penetrates the ventricle as it does in so many diotocardians, a fact which allows its migration to the right to occur readily; and the heart now lies more along the longitudinal axis of the body on the left, in proximity to the base of the gill, rather than across it. This allows the base of the efferent branchial vessel (ebv) to be quite short and the auricle often gives the appearance of being attached almost to the last ctenidial filament.

In *Littorina littorea* a bulbus aortae (ba, fig. 147A) lies in the posterior part of the pericardial cavity and anterior and posterior aortae (pa) diverge from it, the former passing to the head and the latter climbing up the visceral hump. Both end by passing blood to cephalic, pedal or visceral haemocoelic spaces. The anterior aorta is affected by torsion in the mid-oesophageal region and passes from the left to the right ventral side of the oesophagus before dividing, near the nerve ring, into cephalic and pedal branches (see fig. 17). As in the trochids, the blood from the sinuses to which these lead is collected into a venous sinus which lies alongside the posterior half of the mid-oesophagus. This sinus is separated by a vertical septum on its posterior side from the visceral parts of the body, but there is a gap in the septum ventrally on the right (ocp, fig. 147B). Through this, which is controlled by muscle, blood passes into a vein which runs vertically towards the mantle skirt and in which lie the visceral ganglia (vg). Across it run a number of strands of muscle and stout connective tissue (cn) which suggest valvular control of the blood flow, and nerves to the viscera leave the ganglia and travel along it. Into this vessel there opens another which can be traced along the concave side of the visceral hump, running between the posterior oesophagus below and the digestive gland and genital duct above. This is a visceral vein (vv) into which blood collects from all the numerous irregular haemocoelic channels which ramify among and between the lobules of gonad and digestive gland. The union of these vessels in the neighbourhood of the visceral ganglia collects almost all the blood from head, foot and visceral mass into a single channel. This proves to be an afferent vein of the left kidney (arv, fig. 147A): it may be traced into the wall of the kidney, entering ventrally and towards the left, close to the bulbus aortae (ba). From there it runs dorsally, passing near the external aperture (ko) and projecting broadly into the cavity

A

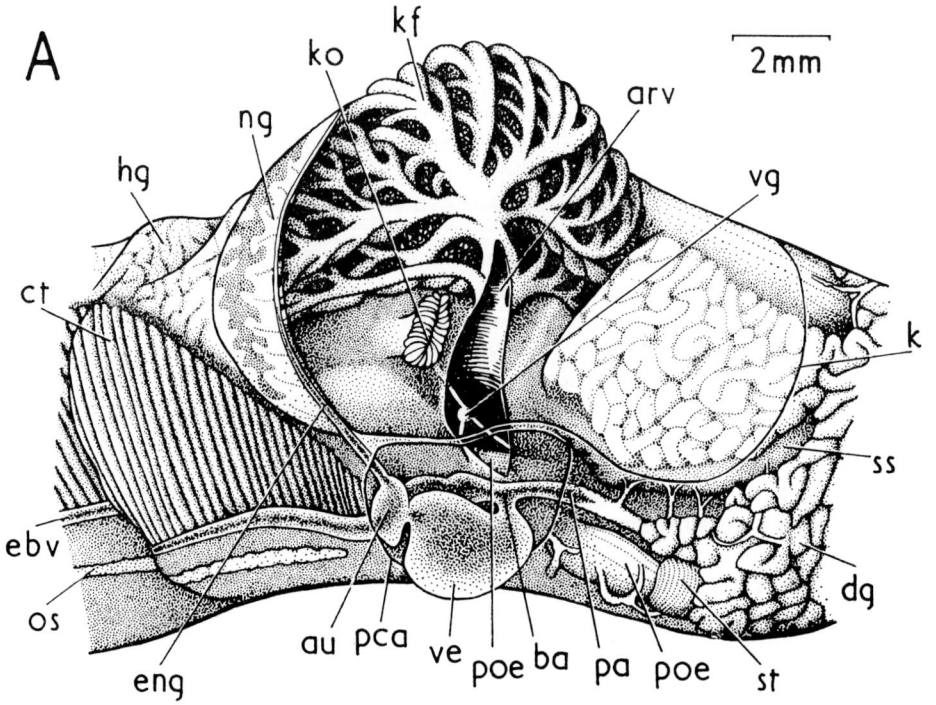

B

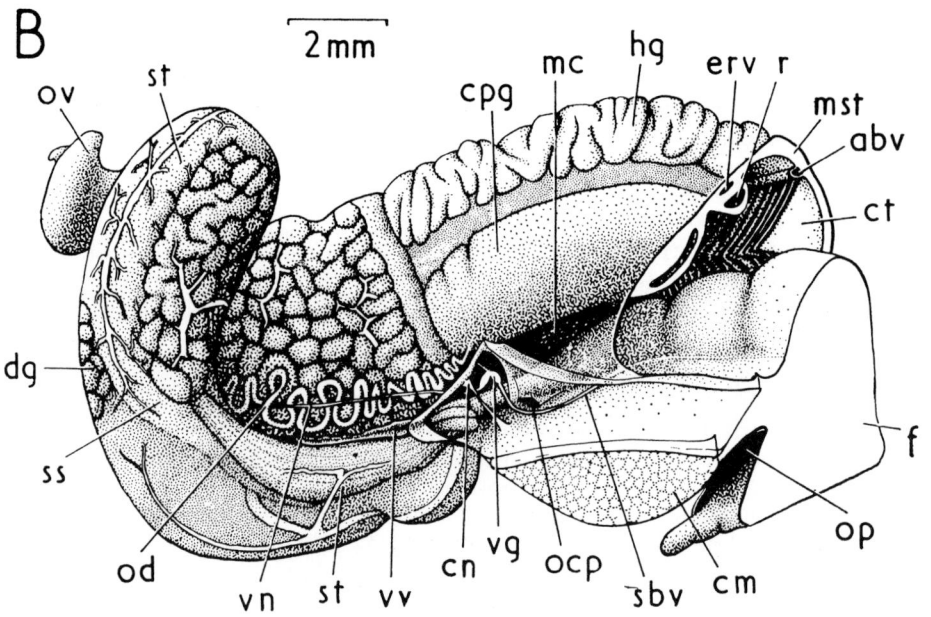

of the kidney sac. Nearing the dorsal wall it starts to branch, subdividing in pinnate fashion to supply the folds of excretory tissue (kf) which run over the inner and dorsal walls of the kidney, and breaking down into vessels of almost capillary dimensions.

From these renal capillaries the bulk of the blood is collected into a prominent vein which emerges from the kidney dorsally and to the right and runs forwards on the mantle skirt more or less dorsal to the rectum and genital duct (erv, figs. 147B, 148). It gives rise to a series of vessels which pass to the left in the roof of the mantle cavity between the lobes of the hypobranchial gland, which are, indeed, largely due to their presence. They ramify and anastomose among themselves and from this network arise the different channels passing into the ctenidial leaflets.

This plan of the vascular system seems to be general not only amongst the mesogastropods but also in the Stenoglossa (*Buccinum*: Dakin, 1912; *Nucella*: present authors). It presents a clear resemblance to the circulatory system of the diotocardians with a number of equally marked differences, of which the most important appear to be that all blood (except a relatively small part which makes its way from visceral hump or foot directly into vessels lying in the peripheral parts of the mantle skirt) passes through the kidney on its way to the heart; and that this kidney is the left rather than the right. These matters raise the question of the homologies of the vessels in the neighbourhood of the kidney.

When a specimen of *Monodonta* is dissected from the right side in the region around the vertical and transverse septa the view shown in fig. 144 is obtained: the visceral ganglia (vg) lie at the point where the collecting sinus, the renal vein and the transverse pallial vein all join. If the same dissection be carried out on a monotocardian (e.g. *Littorina* or *Nucella*) the appearance obtained will be that shown in fig. 147B. The two arrangements are so nearly identical in the relationships between blood vessels and ganglia, muscles and mantle skirt that it is difficult to avoid the conclusion that homologous vessels form the T-junction in each case. Yet the functions of these vessels and their relationships to the kidney are not the same in the 2 animals: the vessel emerging from the direction of the visceral hump is an efferent renal vein in the case of the trochid (erv, fig. 144), but a visceral vein in the monotocardian (vv, fig. 147B), whilst the vessel passing dorsally is running on to the mantle skirt in the diotocardian but is an afferent vein of the kidney in the monotocardian. This seems to suggest that the homologies

---

FIG. 147.—*Littorina littorea*: A, dissection from the left to show vessels in relation to the kidney. The pericardial cavity has been opened to expose the heart. The kidney sac has been opened and the afferent renal vein slit from the point where it is formed ventrally by union of veins from head-foot and visceral mass, to the point where it branches into the kidney folds. B, dissection from the right to show vessels in relation to kidney. The animal is represented with the anterior end removed. A venous space above the columellar muscle is opened as well as the base of the afferent renal vein.

abv, afferent branchial vein; arv, afferent renal vein; au, auricle; ba, bulbus aortae; cm, columellar muscle; cn, connective tissue strut; cpg, capsule gland; ct, ctenidium; dg, digestive gland; ebv, efferent branchial vein; eng, efferent vessel of nephridial gland; erv, efferent renal vein overlying rectum; f, foot; hg, hypobranchial gland; k, cut edge of kidney sac; kf, folded surface of kidney; ko, opening of kidney to mantle cavity; mc, mantle cavity; mst, cut edge of mantle skirt; ng, nephridial gland; ocp, opening of cephalopedal vein; od, ovarian duct; op, operculum; os, osphradium; ov, ovary; pa, posterior aorta; pca, cut edge of pericardium; poe, posterior oesophagus; r, rectum; sbv, sub-oesophageal part of visceral loop; ss, style sac region of stomach; st, stomach; ve, ventricle; vg, visceral ganglion lying in blood space where the cephalopedal vein, visceral vein and afferent renal vein unite, and giving off the visceral loop anteroventrally, a nerve to the kidney opening anterodorsally and a visceral nerve posteriorly; vn, visceral nerve; vv, visceral vein.

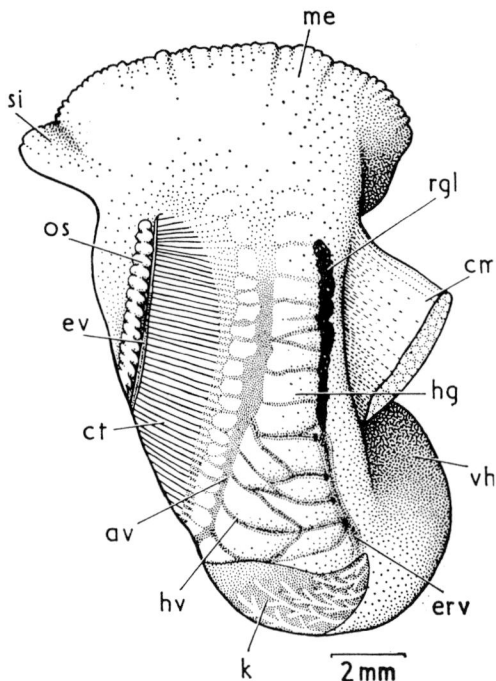

FIG. 148.—*Nucella lapillus:* dorsal view of mantle skirt and base of visceral hump.
av, afferent vessel of ctenidium; cm, columellar muscle; ct, ctenidium; erv, efferent renal vein; ev, efferent vessel of ctenidium; hg, hypobranchial gland; hv, hypobranchial vessel; k, kidney; me, mantle edge; os, osphradium; rgl, rectal gland; si, siphon; vh, visceral hump.

are spurious. Nevertheless it is likely that this is not so: the anatomical relationships between vascular and other structures are too intimately identical for there to be real doubt as to this, and the explanation of the changes in the flow of blood in these channels must be looked for elsewhere, and, in particular, in the fact that in the one case the kidney is the right kidney and in the other it is the left. Reference to the series of diagrams in fig. 146 will make this clear. In fig. 146A the vascular system of *Haliotis* is shown with the right kidney (rk) placed so as to intercept the main blood flow from both anterior and posterior halves of the body. The efferent vessel from this (erv) leads to the basibranchial sinus (bbs) and so to the gills (ct), and it also connects with a rectal vessel (hv). It is not easy to determine how the hypobranchial glands receive their blood supply, but they seem to be related to a pallial vein which leads to visceral sinuses and so to the right kidney. The left kidney or papillary sac is connected to the basibranchial sinus and left efferent ctenidial vessel, but the route which the blood takes through these vessels can only be surmised.

Fig. 146B shows the arrangement in the trochids, in which it seems that the system is fundamentally similar except in 3 main points: (1) the right kidney (rk) invades the vascular system draining only the visceral hump (arv) and does not penetrate that which is running back from the head and foot (cpv); (2) the transverse pallial vein (= basibranchial sinus) sends blood to the afferent ctenidial vessel on the left through the hypobranchial gland; and (3) with

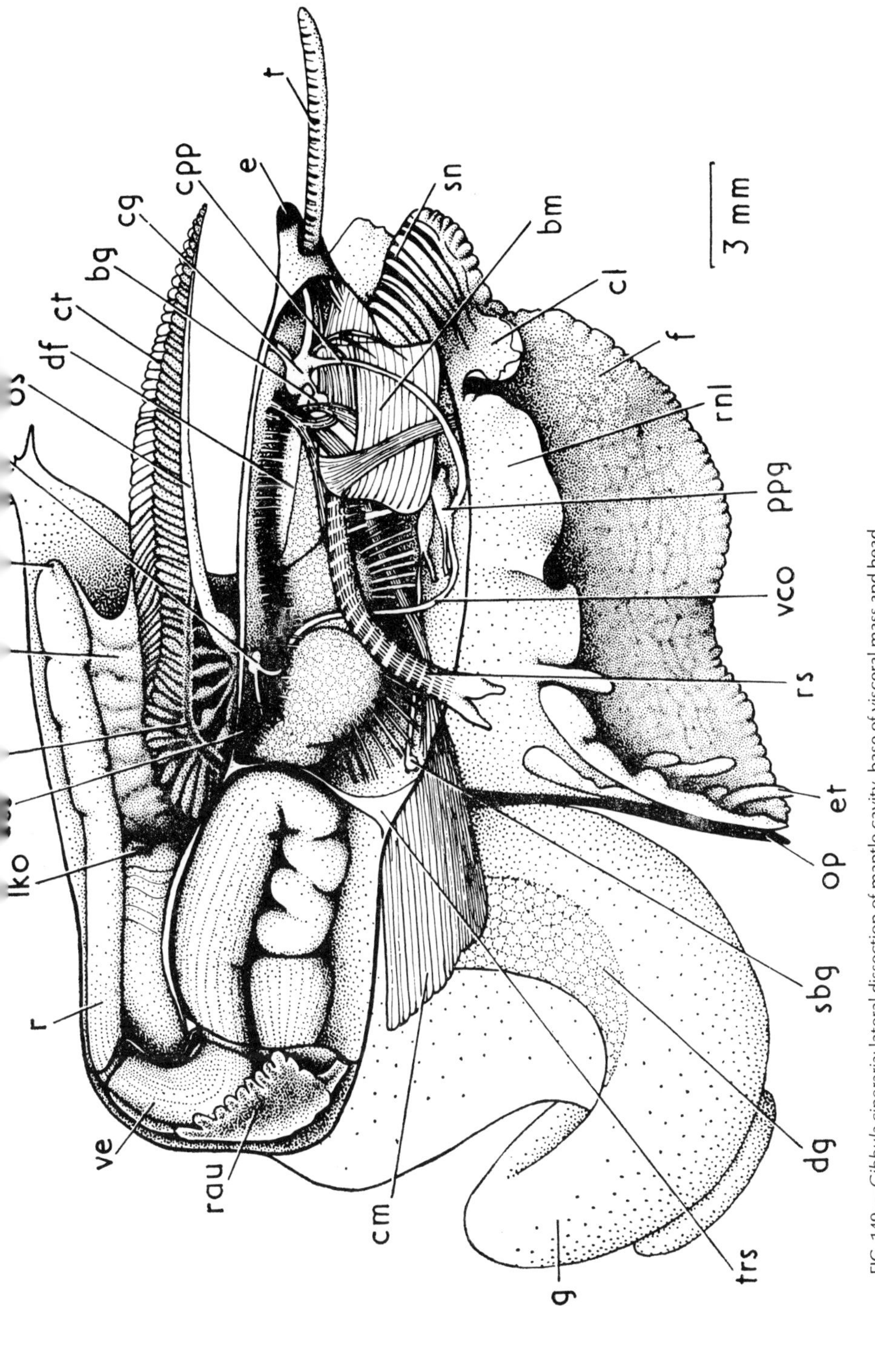

FIG. 149.—*Gibbula cineraria*: lateral dissection of mantle cavity, base of visceral mass and head.
a, anus; bg, buccal ganglion; bm, muscles of buccal mass; cg, cerebral ganglia; cl, cephalic lappet; cm, columellar muscle; cpp, cerebropleuropedal connective; ct, ctenidium; cta, ctenidial axis; dcs, dorsal cephalic sinus; df, line of dorsal fold on oesophageal wall; dg, digestive gland; e, eye; et, epipodial tentacle; f, foot; g, gonad; lhg, left hypobranchial gland; lko, opening of left kidney (papillary sac); op, operculum; os, osphradium; ppg, pleuropedal ganglion; r, rectum; rau, right auricle; rnl, right neck lobe; rs, radular sac; sbg, sub-oesophageal ganglion; sn, snout; sog, supra-oesophageal ganglion; t, tentacle; trs, transverse septum; vco, visceral connective; ve, ventricle.

the disappearance of the right ctenidium the blood from the transverse pallial vein on that side makes its way directly through the anastomoses of the right hypobranchial gland to the efferent pallial vein (= right efferent ctenidial vessel (rpv)). As a result of the new relationship between the right kidney and vascular system the anterior end of the visceral venous return now appears as a renal efferent (erv). The left kidney (lk) is still, as in *Haliotis,* somewhat ambiguously sited between transverse pallial vein and left efferent ctenidial blood stream (in this case the connexion is with the auricle and not the efferent vessel).

In the monotocardians a fundamental alteration has occurred in that the right kidney (rk) is reduced and the left (lk) expanded to form the main excretory organ. In its hypertrophy this kidney, too, has come to invade the afferent ctenidial flow of blood from head, foot and visceral hump, but the right kidney is no longer related to it. It is not an unlikely assumption that contact between the expanding left kidney and the vascular system came about at a place different from that at which the right kidney and bloodstream had commingled, and if we suppose that it occurred (as shown in fig. 146C) in front of the junction between the converging vessels from head-foot (cpv) and visceral hump then both of these would unite to form a single afferent renal vein (arv) which is made from the ventral part of the vessel known respectively as transverse pallial vein or basibranchial sinus in the two diotocardians. In these circumstances the vessel bringing blood from the head and foot will look the same in trochid and monotocardian (cos, fig. 144; ocp, fig. 147B), but the vessel returning blood from the visceral mass will be an efferent renal vessel in the trochid (erv, fig. 144), but, because of the disappearance of the right kidney, a simple visceral vein in the other (vv, fig. 147B). The circulation on the mantle skirt has also been modified, and not merely by the simple abolition of the entire right moiety. The main vessel emerging from the kidney runs dorsal to the rectum (erv, fig. 148) and sends branches (hv) through the hypobranchial gland towards the ctenidium, sometimes forming an afferent ctenidial channel (av) at or near the base of the gill leaflets, sometimes, however, apparently running directly into these without any longitudinal vessel intervening. From its situation over the rectum the main vessel looks as if it might be interpreted as a hypertrophied rectal vein, but it might also be regarded as the main left half of the transverse pallial vein with a new orientation, and it appears unprofitable to attempt to discriminate between these two points of view.

It will be recalled that the left kidney of the diotocardians apparently receives its blood supply from the basibranchial sinus and returns it to the left efferent ctenidial vessel (*Haliotis*) or is related to the transverse pallial vein and auricle (*Monodonta*). This connexion is found to persist in the higher prosobranchs, though in a restricted sense. If the kidney of a mesogastropod or stenoglossan be examined externally on the surface of the visceral mass, the appearance represented in fig. 142B will be seen. Over most of the surface run the insertions of the main folds of the walls on which lies the excretory tissue and into which run the branches of the afferent renal vein. Along the side of the kidney sac which borders the pericardial cavity, however, the appearance of the kidney is quite different: this marks the position of a specialized part usually known as the 'nephridial gland' (ng)—an unfortunate name in that the kidney of molluscs is technically a coelomoduct and not a nephridium, but perhaps acceptable until knowledge of the functional importance of this part of the kidney provides a basis for a better one.

The nephridial gland consists of a mass of blood spaces into which projects a large number of blind tubules from the main kidney sac. These tubules are of small diameter and are lined by cells some of which are ciliated: it seems likely that the liquid from the main kidney sac and the blood can be brought into intimate contact in this part of the kidney and it may, therefore, act as an osmoregulatory organ. The blood reaches the haemal spaces of the nephridial gland from the main afferent renal vessel, but it is drained away into a separate efferent from that which collects the blood from the rest of the kidney. This vessel, the efferent of the nephridial gland (env, figs. 142B, 146C), passes directly to the auricle and so by-passes the circulation of the mantle skirt. This is clearly reminiscent of the blood supply of the left kidney of the diotocardians; that it has persisted into the monotocardians offers a proof, from the unexpected angle of its vascular supply, that the sole functional kidney of the higher gastropods is indeed the left.

The heart is innervated by nerves from the pleurovisceral loop or from the visceral ganglion, almost invariably on the right side (crn, fig. 165). In the more primitive genera the auricles are said to receive an innervation which is different from that reaching the ventricle, the nerve to which passes along the aorta (Zeugobranchia and Trochacea: Haller, 1884, 1894; Patellacea and Neritacea: Bouvier, 1887). In higher animals the different parts of the heart are all innervated by branches of the same nerve (*Natica*: Carlson, 1905*b*; *Ianthina*, *Hipponyx*: Suzuki, 1934, 1935). As in vertebrates, these nerves have the function of altering the rate at which the heart is beating, the propagation of the beat within the heart itself being myogenic. Which part acts as pace-maker is not known, although Zubkov (1934), like others before him, attributed this function to a small ganglion placed in the region of the junction between ventricle and bulbus aortae in a number of genera. On the other hand, any part of the heart appears capable of initiating the beat in *Haliotis*. The function of the cardiac nerves in prosobranchs appears to be both inhibitory and excitatory.

CHAPTER 12

THE NERVOUS SYSTEM AND SENSE ORGANS

T HE molluscan nervous system consists essentially of a circum-oesophageal ring from which two longitudinal nerve cords pass posteriorly on each side. One of these pairs is ventral and lies embedded in the foot; the other is dorsal and is related to the visceral hump and mantle cavity. The former pair constitutes the pedal cords, the latter the pleural or pallial nerve cords. In most primitive molluscs it is not possible to distinguish ganglia on the nerve cords: nerve cells are, in fact, generally distributed along them, forming a peripheral zone, whilst the central area constitutes a neuropile, composed of fibres only. The pedal cords are connected to one another across the mid-line of the body by numerous transverse nerves so that this part of the nervous system resembles a ladder in appearance. It therefore bears a superficial resemblance to the ventral nerve cords of many arthropods and annelids—apart from the absence of ganglia—and it has been assumed by some workers that this indicates affinity with these groups. In many species of chitons the pedal and pallial cords of the same side are also frequently connected by anastomoses, but the right and left pallial cords are not directly connected to one another except at the extreme posterior end of the animal, where they unite dorsal to the rectum.

The circum-oesophageal ring at the anterior end of the body of a chiton is connected to the pedal and pallial cords laterally, is completed dorsally by a cord running over the initial part of the gut and by a similar but more slender connexion ventrally. The dorsal part of this ring is known as the cerebral commissure, the ventral part as the labial commissure.

In gastropods the nervous system is constructed on the same general plan but is complicated by the effects of the torsion which these molluscs undergo. The parts of the system which lie in the head and foot of the animal are not involved in this process, so that the relationship of the pedal cords and cephalic sections is not affected, but the pallial cords, which run from the head into the visceral hump, and are known, in the gastropods, as the visceral loop, are. The process of torsion (as explained on, pp. 6, 31) involves the rotation in an anticlockwise direction of the visceral hump on the head-foot through 180°. This has the effect of twisting the visceral loop into the shape of a figure-of-eight and of entangling it with the alimentary canal in such a way that one half now crosses from the right anteriorly to the left posteriorly dorsal to the oesophagus, and the other half crosses from the left anteriorly to the right posteriorly ventral to the oesophagus. The two parts unite posteriorly dorsal to the posterior oesophagus, but as this has turned over during the process of torsion, this topographically dorsal position indicates that the posterior end of the visceral loop is morphologically ventral to the alimentary canal. In this respect the gastropods agree with the other molluscs, and it is the chitons which are anomalous in having the orientation of these parts reversed.

In addition to showing the effects of torsion, the gastropods differ from chitons in that an aggregation of nerve cells to form ganglia has occurred, although the rate at which this takes place varies within the group and from place to place within the nervous system. In all, however, 7 pairs of principal ganglia appear at various situations on the nerve cords of a gastropod, and various other ganglia may also be encountered in particular prosobranchs. These are usually developed in proximity to one of the major sense organs.

The evolution of the nervous system within the prosobranchs is most easily followed by examination of a series of examples. Of these the most primitive is *Haliotis* (fig. 150), the nervous system of which has been described by de Lacaze-Duthiers (1859), Haller (1884, 1886), Bouvier (1887) and Crofts (1929).

It is one of the characteristic features of the lower prosobranchs that the haemocoelic cavity separating the body wall of the head from the buccal mass is extremely narrow, a fact which makes the dissection of the nervous system in that area difficult, for it is precisely there that the cerebral ganglia (cg) lie and the commissure (ccm) which links them one with the other. The ganglia lie laterally and merge without obvious boundary into the commissure; all are pressed against the buccal musculature so as to be concave on the inner side. Ventrally and anteriorly each ganglion is extended into a nerve trunk (lgn) running underneath the extreme anterior end of the oral tube, the 2 trunks linking up with each other in the mid-ventral line to complete a circum-oesophageal nerve ring. The ventral half of this constitutes the labial commissure and it contains numerous nerve cells, although these are not aggregated into visibly differentiated labial ganglia. The connexion between these ganglionic areas and the cerebral ganglia is known as the labial connective. From its cerebral end a further connective arises, which passes between different layers of buccal muscles to the cleft between the origin of the radular sac below and that of the anterior oesophagus above: here each swells into a small buccal ganglion (bg), the right and left ganglia being connected by a buccal commissure. The terms stomatogastric ganglion and commissure may also be applied to these structures.

Postero-ventrally, on each side, the cerebral ganglion is extended into 2 trunks which run backwards and ventrally out of the head into the anterior parts of the foot, where they connect with a very large mass of nervous material which proves to contain 4 ganglionic centres. Two of these lie more ventrally and take the form of elongated strands, running one on either side of the mid-line, throughout the greater length of the foot: these are the pedal ganglion cords (pdc). The two are connected to one another by 15–36 transverse commissures (pcm) at more or less regular intervals along their entire length, though the commissure at the anterior end is considerably larger than any of the others. At this level two other ganglia lie dorsal to the pedal ganglia: these are the pleural ganglia, sometimes known as the pallial ganglia. On each side the nerves from the cerebral ganglion connect, one with the pleural ganglion, one with the pedal ganglion of the same side. The former trunk is called the cerebropleural connective (cpc), the latter the cerebropedal connective (cpdc).

The main part of the nervous system of *Haliotis*—what may be regarded as its central rather than peripheral nervous system—is completed by nerve trunks leading to ganglia which are related to the pleural ganglia and which constitute the visceral loop. From the right pleural ganglion arises a trunk (the supra-oesophageal connective (suv)) which runs dorsally over the mid-oesophagus to the animal's left side, where it connects with a ganglion placed near the end of the efferent membrane of the left gill. This is the left branchial ganglion (lg). A similar trunk passes to a right branchial ganglion (rg) from the left pleural ganglion with the difference that this time the trunk passes ventral to the mid-oesophagus and is, for that reason, known as the sub-oesophageal connective (sbv). The visceral loop is completed by nerve trunks which run posteriorly from close to the branchial ganglia to unite, dorsal to the posterior oesophagus, at a point nearly level with the posterior end of the mantle cavity: at

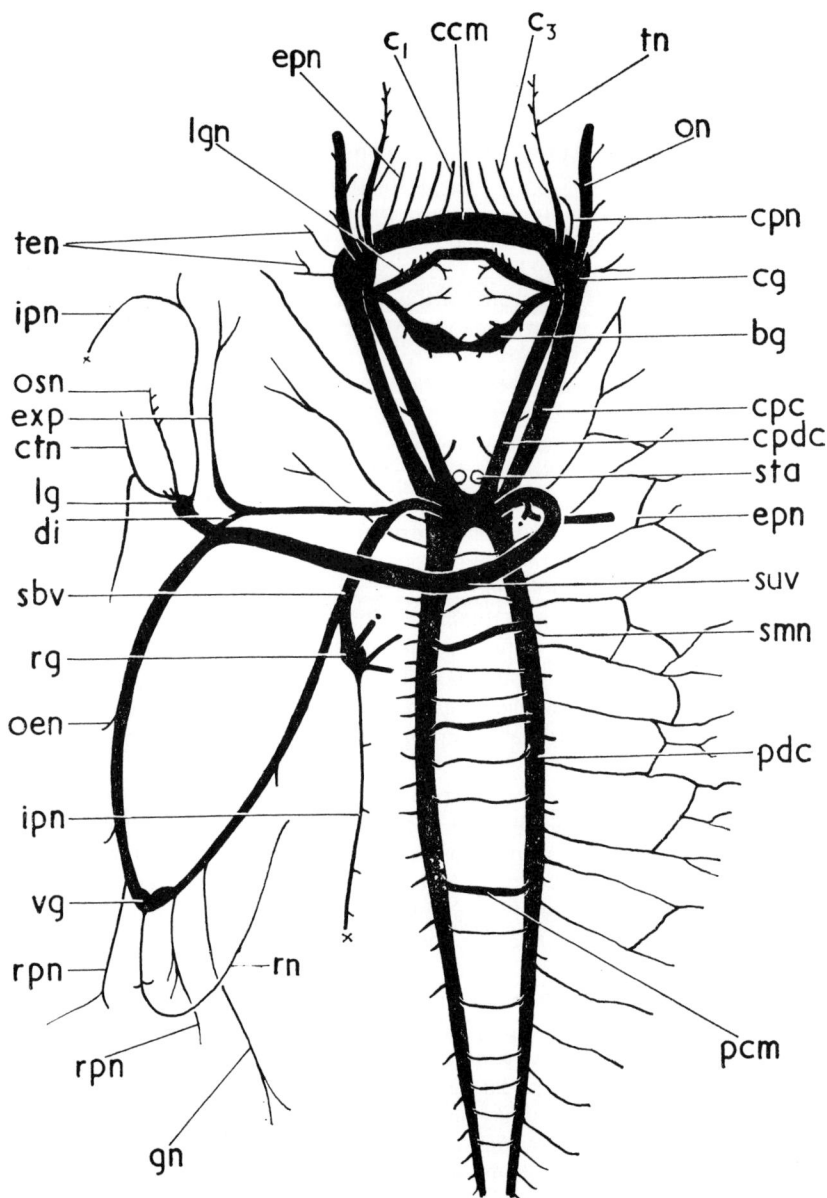

FIG. 150.—*Haliotis tuberculata:* diagram of the nervous system. Based on Crofts.

bg, buccal ganglion; c_1, c_3, cerebral nerves; ccm, cerebral commissure; cg, cerebral ganglion; cpc, cerebropleural connective; cpdc, cerebropedal connective; cpn, cephalic pleat nerve; ctn, ctenidial nerve; di, dialyneury; epn, epipodial nerve; exp, external pallial nerve; gn, gonadial nerve; ipn, internal pallial nerve; lg, left branchial ganglion; lgn, labial commissure; oen, oesophageal nerve; on, optic nerve; osn, osphradial nerve; pcm, pedal commissure; pdc, pedal cord; rg, right branchial ganglion; rn, rectal nerve; rpn, renopericardial nerve; sbv, sub-oesophageal part of visceral loop; smn, nerve to shell muscle; sta, statocyst; suv, supra-oesophageal part of visceral loop; ten, tegumentary nerve; tn, tentacular nerve; vg, visceral ganglion. The points where a right dialyneury has been cut are marked . . ; those where a circumpallial nerve has been cut are marked x x.

this point lies (somewhat asymmetrically) a single ganglion. This is known as the visceral or abdominal ganglion (vg). In *Haliotis* it is diffuse, somewhat irregular, and might be thought of as formed by the union or partial separation of right and left concentrations of nerve cells.

There are, therefore, in *Haliotis*, 6 pairs of ganglia with a seventh median ganglion which may have a double structure. These are connected by transverse connexions which link ganglia carrying the same name, e.g. buccal and buccal, pedal and pedal. Such connexions are known as commissures. The only ganglia not linked by commissures are the pleural and branchial. The ganglia are also linked longitudinally and in these cases the links have ganglia carrying different names at their ends, e.g. cerebral and pleural, pleural and visceral. This type of nerve trunk is known as a connective.

Each ganglion innervates a special area of the animal's body. This may be most conveniently indicated by listing the nerves arising from each.

Cerebral ganglion and commissure	(1)	Tentacular nerve (tn) to sense organs in the tentacle. It is often ganglionated at its base.
	(2)	Optic nerve (on) to eye and also to skin and muscles of eye stalk. Small ganglia may lie on its branches.
	(3)	Statocyst nerve to statocyst (sta).
	(4)	A group of nerves (c_1, c_3), 6–8 in number, to the dorsal and lateral lips.
	(5)	Epipodial nerve (epn) to anterior part of epipodium on head between tentacle and snout.
	(6)	Cephalic pleat nerve (cpn) (Crofts, 1929) to the cephalic pleat or lappet, a transverse fold of skin across the dorsal surface of the head between the tentacles.
	(7)	Tegumentary nerves (ten), 2 on each side, to the skin on the side of the head.
Labial ganglion	(1)	Nerves, 4 on each side, to the ventral lips.
	(2)	Nerves to muscles of the buccal mass.
Buccal ganglion	(1)	(Arising more accurately from the buccal commissure.) Nerve pair to muscles of radular sheath.
	(2)	(Arising also from the buccal commissure.) Nerve pair to radular membrane.
	(3)	Nerve to oesophageal valve.
	(4)	Nerve to radular sheath and oesophageal pouches.
	(5)	(From the buccal connective.) Nerve to dorsal wall of buccal cavity, where right and left nerves anastomose.
	(6)	(From the buccal connective.) Nerve to buccal pouch, salivary gland, wall of buccal cavity and oesophageal gland.
Pleural ganglion	(1)	The external pallial nerve (exp) from each ganglion runs laterally and dorsally to enter the corresponding mantle lobe, running forwards to its anterior end, innervating its edge. Each nerve forms an anastomosis (di) with the corresponding branchial ganglion, a condition known as a dialyneury.

Pedal ganglion   (1) Epipodial nerves (epn), 2 from the cerebropedal connective and
                             many from the pedal ganglion to innervate the epipodial tentacles.
                        (2) Nerves, numerous, to shell muscles (smn).
                        (3) Anterior pedal nerve to anterior part of foot ventrally.
                        (4) Posterior pedal nerves, numerous, to central and posterior parts of
                             foot, ventrally.

Branchial ganglion   (1) Osphradial nerve (osn) to osphradium.
                        (2) Ctenidial nerve (ctn) to gill.
                        (3) Internal pallial nerve (ipn) to corresponding half of mantle skirt.
                        (4) Tegumentary nerve to body wall nearby.
                        (5) (From left part of visceral loop.) Oesophageal nerve (oen).
                        (6) (From each side of visceral loop.) Reno-pericardial nerve (rpn).
                        (7) (From right part of visceral loop.) Gonadial nerve (gn).

Visceral ganglion   (1) Rectal nerve (rn).

The nervous system of *Haliotis* is like that of most of the more primitive prosobranchs. The points which emphasize its primitiveness are: (1) the slight degree of separation of ganglia from nerve trunks; (2) the elongation and forward position of the cerebral ganglia; (3) the length of the commissure between the cerebral ganglia; (4) the presence of labial ganglia and of a labial commissure; (5) the indirect connexion of buccal ganglia to cerebral ganglia by way of the labial commissure; (6) the origin of the statocyst nerves directly from the cerebral ganglia; (7) the ventral position of the pleural ganglia, which appear almost as if they were dorsal lobes of the pedal ganglia; (8) the extreme elongation of the pedal ganglion cords; (9) the numerous pedal commissures giving a ladder-like nervous system in the foot, reminiscent of the double ventral nerve cord with transverse connexions of annelids and arthropods; (10) the absence of ganglia on the visceral loop apart from the abdominal; (11) the origin of many nerves from the visceral loop; (12) the fact that the only connexions between the branchial ganglia and the pleural ganglia of the same topographical side occur indirectly, as anastomoses between the peripheral ends of nerves arising from these centres, i.e. are of the type known as a dialyneury.

Though these are primitive features in the organization of the nervous system of prosobranchs, several point to an advance over the condition found in chitons, yet indicating a distinct relationship with that group; others, like the dialyneury, are novelties probably due to the imposition of torsion. A dialyneury, for example, allows—at least in theory—a better co-ordination of pallial control in that there is no longer any necessity for messages from one part of the pallial edge to be relayed to the left pleural—left cerebral—right cerebral—right pleural—supra-oesophageal ganglia in order to reach a contiguous part of mantle innervated from the last of these nerve centres. To what extent the connexions which occur in *Haliotis* in fact permit this to occur is not known, but this union is the beginning of one evolutionary trend which persists throughout the prosobranchs and leads to a general concentration and closer union of nerve centres. A special aspect of this main change affects the pedal and pleural ganglia, the former of which become gradually shorter and concentrated anteriorly, whilst the pleurals, which in *Haliotis* appear almost as appendages of the pedal ganglia, migrate away from these in a dorsal direction until they lie alongside the cerebral ganglia. The type of nervous system which has pleural and pedal ganglia contiguous, ventral to the gut and both linked by long connectives to the cerebral ganglia, is known as hypoathroid (fig. 153A); it is characteristic of the Archaeogastropoda but is also to be found in some of the lower mesogastropods, mainly Architaenioglossa.

The nervous system of *Patella vulgata* (fig. 151) has been described by Bouvier (1887), Gibson (1887), Pelseneer (1898–99) and Davis & Fleure (1903). It shows the continuation of a number of trends incipiently visible in *Haliotis* or other rhipidoglossans. The cerebral ganglia (cg) are moderately expanded swellings lying well forward in the head and connected to one another by a long commissure (ccm) dorsal to the gut. From their posteroventral ends 2 connectives (cpc, cpdc) run ventrally and back to the pedal (pd) and pleural (plg) ganglia which lie, one alongside the other on each side, underneath the buccal mass in the haemocoel of the anterior part of the foot. The 2 pedal ganglia lie near the mid-line, the 2 pleurals are more laterally placed, and there is a short but distinct pleuropedal connective uniting the two. The pedal ganglia are drawn out into cords (pdc) running back along the foot, but there are only a few commissures (pcm) linking these, one at the extreme anterior end, the others at irregular intervals along their length.

The cerebral ganglia are also related to the labial (lgn) and buccal ganglia (bg). The former lie close under the skin, in the wall of the oral tube, ventrolaterally, and united across the mid-line by a labial commissure. The buccal (or stomatogastric) commissure lies in the usual place, dorsal to the radular sac but ventral to the anterior oesophagus, and carries, right and left of the mid-line, slight swellings which are the buccal ganglia: these connect to the labial ganglia.

The visceral loop of *Patella* is not a very conspicuous part of the nervous system, perhaps because of the general reduction in importance of the mantle cavity and its contents in the life of the animal. As a result of the shortening of that part of the body, the visceral ganglion (vg) is placed nearly in the same transverse plane as the pleurals, with the result that the figure-of-eight of the visceral loop lies almost vertically. Because of the brevity of the visceral loop no nerves leave it: all come instead from the associated ganglia, of which *Patella* possesses three. A supra-oesophageal ganglion (sog) of moderate size lies on the visceral loop at the point where the nerve to the left osphradium leaves, a much smaller sub-oesophageal ganglion (sbg) (hardly, indeed, deserving the name) lies at the corresponding point on the right, and there is a single visceral ganglion (vg) related to the rectum. These supra- and sub-oesophageal ganglia co-exist with the homologues of the branchial (now called osphradial) ganglia of *Haliotis* so that they are real evolutionary novelties: they may be referred to collectively as parietal ganglia.

According to Thiem (1917b) the peripheral nerves conform to a more or less standard pattern in the Docoglossa (=Patellacea). Six nerves leave each cerebral ganglion in addition to the commissure and connectives, innervating the tentacle (tn), the eye (on), and the dorsal and lateral walls of the snout (lbn). The labial ganglia give nerves to the inner lips. The buccal ganglia similarly give nerves to muscles of the odontophore and the dorsal folds of the oesophagus. From the pleural ganglion on each side originates a variable number of pallial nerves (pn). The supply of nerves to the viscera is achieved on a pattern rather different from what has been described for *Haliotis*: the supra- and sub-oesophageal ganglia give off only one nerve, which runs to the corresponding osphradial (or branchial) ganglion (osg). Although the osphradium itself is of considerably less importance in most prosobranchs, its ganglion is relatively more important since from the left osphradial ganglion come off nerves to the osphradium, the gill (if present), pericardium and left kidney, whilst from the right osphradial ganglion arise nerves going to the right kidney and the salivary glands. The visceral ganglion

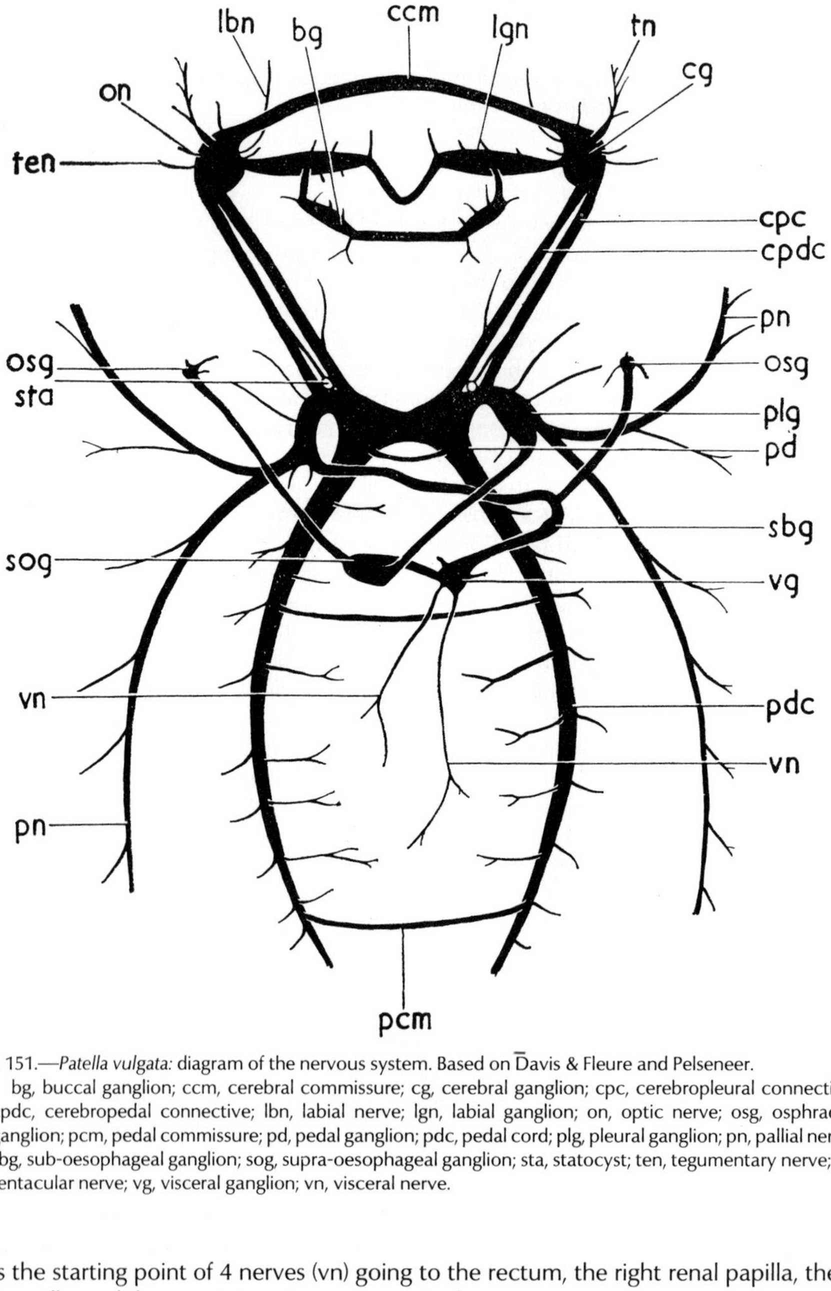

FIG. 151.—*Patella vulgata:* diagram of the nervous system. Based on Davis & Fleure and Pelseneer.
bg, buccal ganglion; ccm, cerebral commissure; cg, cerebral ganglion; cpc, cerebropleural connective; cpdc, cerebropedal connective; lbn, labial nerve; lgn, labial ganglion; on, optic nerve; osg, osphradial ganglion; pcm, pedal commissure; pd, pedal ganglion; pdc, pedal cord; plg, pleural ganglion; pn, pallial nerve; sbg, sub-oesophageal ganglion; sog, supra-oesophageal ganglion; sta, statocyst; ten, tegumentary nerve; tn, tentacular nerve; vg, visceral ganglion; vn, visceral nerve.

(vg) is the starting point of 4 nerves (vn) going to the rectum, the right renal papilla, the left renal papilla, and the remaining viscera respectively.

In *Acmaea* (fig. 152) the tendency towards a lessening of the visceral loop is carried to greater lengths than in the Patellidae (Willcox, 1898), and it is no longer twisted into a

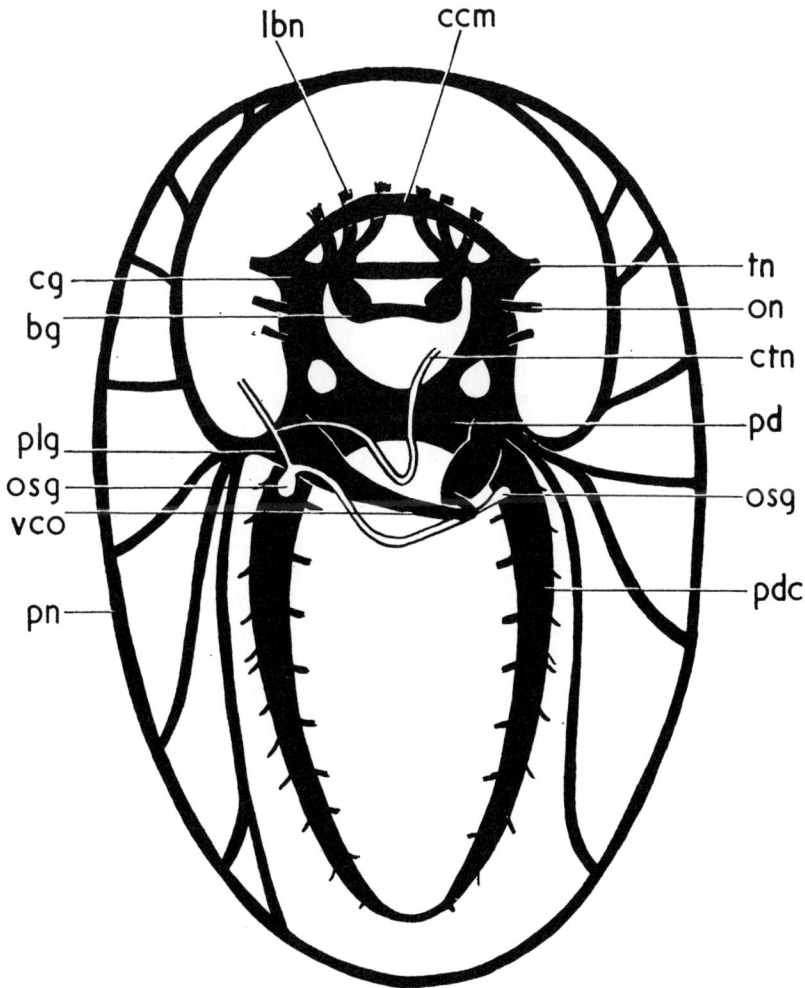

FIG. 152.—*Acmaea fragilis:* diagram of the nervous system. Based on Wilcox.
bg, buccal ganglion; ccm, cerebral commissure; cg, cerebral ganglion; ctn, ctenidial nerve; lbn, labial nerve; on, optic nerve; osg, osphradial ganglion; pd, pedal ganglion; pdc, pedal cord; plg, pleural ganglion; pn, pallial nerve (circumpallial); tn, tentacular nerve; vco, visceral loop.

figure-of-eight, except in rare specimens. Instead, it is a slightly asymmetrical nerve trunk (vco) running from one pleural ganglion to the other. There are no visible ganglia on it, though cells present in it may represent the visceral ganglion. It gives rise to 2 nerves, one going to each osphradial ganglion, that on the left being much the larger and innervating the ctenidium (ctn).

The Trochacea exhibit a condition in the nervous system very like the zeugobranchs, the main difference being the development of a supra-oesophageal ganglion on the course of the visceral loop (sog, fig. 53); a corresponding sub-oesophageal ganglion is less frequently developed (sbg, fig. 144), probably because of the loss of the right gill. A left osphradial ganglion (osg) also occurs from which the osphradium and ctenidium are innervated.

The remaining group of the Archaeogastropoda with which we are here concerned, the Neritacea, departs markedly from the others in respect of the state of its nervous system and shows in this, as in nearly all other features of its organization, a more advanced state. This is particularly true of the visceral loop where, in *Theodoxus,* the part which runs from the right pleural ganglion to the supra-oesophageal is so thin as to suggest that it is all but functionless, a theory borne out by its absence in some members of the group. With the disappearance of the supra-oesophageal connective from the visceral loop the supra-oesophageal ganglion and the closely associated left osphradial ganglion would be isolated from the rest of the anterior nervous system if a new connexion had not been established with the left pleural ganglion. Such a direct link between a parietal and a pleural ganglion is known as a zygoneury, and it clearly confers the same functional advantage as a dialyneury but in a more effective manner. Because of the disappearance of the gill and osphradium on the right, the sub-oesophageal ganglion is not tied to a particular level in the body as is the supra-oesophageal and it is therefore free to migrate forwards until it comes close to the left pleural ganglion, to which it is attached by the sub-oesophageal connective. At the same time a zygoneury develops to connect to the right pleural ganglion and its final position is between the 2 pleurals. In other respects the central nervous system of a neritid is less modified, the pleural ganglia still lying alongside the pedals, which are long swollen cords with numerous commissures, the whole effect being ladder-like. The cerebral ganglia are fusiform, with a long commissure, and a labial commissure is present to which the buccal ganglia connect. The cerebropleural and cerebropedal connectives join so as to appear as a double nerve trunk. Two visceral ganglia lie at the posterior end of the visceral loop.

In the distribution of the peripheral nerves two points are deserving of mention. Of these one is the aberrant innervation of the osphradium and ctenidium, which is by way of a nerve which leaves the left pleural ganglion rather than from the supra-oesophageal. This arrangement might appear to indicate that the osphradium and ctenidium of the Neritacea were new structures and not homologous with similar formations in other prosobranchs, as has, indeed, been held by Thiele (1929–35). This, however, seems to be a rather extravagant conclusion to reach in view of the similarities in location and structure which exist between neritacean and other prosobranch gills and osphradia, and a much simpler explanation would be that the point of separation of the branchial and osphradial nerve has migrated from the supra-oesophageal ganglion along the zygoneury which connects it with the left pleural ganglion so that it now appears to originate in that centre. (This might also involve a transfer of control from right to left pleural ganglion.) The branchial and osphradial nerve connects with nerves from the supra-oesophageal ganglion so that a dialyneury occurs in addition to the zygoneury already described.

In most mesogastropods a number of changes occurs in the nervous system so that it shows a more advanced state than that of archaeogastropods. These are not very evident in the lowest group of the Mesogastropods, the Architaenioglossa, in some of which an almost unmodified hypoathroid arrangement persists. Most mesogastropods, however, show a tendency for the following changes: (1) the cerebral ganglia move dorsally and lie closer together; (2) for this reason the cerebral commissure becomes shorter; (3) the cerebral ganglia tend to migrate posteriorly to the level of the anterior end of the mid-oesophagus, and so behind the buccal mass: this allows greater concentration; (4) the labial commissure tends to disappear; (5) the labial ganglia disappear; (6) as a consequence of this the buccal ganglia are the only ones in relation to the anterior end of the gut connected to the cerebrals; because of their original link with the labial ganglia, the cerebrobuccal connective traces a circuitous route from the one ganglion to the other, and, though appearing as a single nerve, must be regarded as morphologically made up partly of a cerebrolabial connective and partly of a labiobuccal one; (7) the pleural ganglia become wholly separated from the pedals and tend to migrate dorsally towards the cerebrals; (8) the pleuropedal connective therefore elongates; (9) the pedal ganglia tend to become concentrated anteriorly; (10) the pedal commissures become reduced, most mesogastropods having only one, though a second, much more slender one, often occurs behind this; (11) the posterior pedal nerves elongate as the pedal ganglia concentrate anteriorly; (12) the statocyst nerve runs from the cerebral ganglion through the cerebropedal connective and therefore appears to originate from the pedal ganglion.

Of these changes the most important are those affecting the three major ganglia, cerebral, pleural and pedal. They have the general effect of bringing them all into proximity in the neighbourhood of the point where anterior and mid-oesophagus join behind the buccal mass; this forms a nerve ring at a point where the gut is often narrow enough to ensure close contact between the ganglia and so allow a real nerve centre to arise. Although no gastropod appears to have made much of the chance of mental evolution which this situation offers, it is precisely the same situation which the cephalopods have exploited with conspicuous success.

Two main stages may be recognized in the evolutionary trend which has just been outlined. One is that in which the pleural ganglia have become well separated from the pedals with which they had been previously associated, but have not yet migrated to the proximity of the cerebrals: they still lie ventral to the gut. This is a condition known as dystenoid (fig. 153B) and it is exhibited by *Viviparus*, for example. In most mesogastropods, however, cerebral and pleural ganglia lie alongside one another—in some cases contiguous, in others even fused—dorsal to the gut and connected to the pedal ganglia, which always retain their primitive ventral situation, by connectives of equal length. This is known as an epiathroid nervous system (fig. 154A).

The nervous system of *Viviparus* has been described by Bouvier (1887). The cerebral ganglia lie in juxtaposition above the oesophagus and behind the buccal mass, embedded to a certain extent in the salivary glands. From them nerves pass to the usual destinations—eye, tentacle (which on the right side in males is modified to act as a penis), dorsal and lateral skin of the snout and the dorsal and lateral lips. Most nerves of the last group emerge from a particularly prominent anteroventral prolongation of the ganglion, which is known as the labial lobe. That this is in fact a labial ganglion conjoined with the cerebral is suggested by the fact that one of the nerves originating from it is the labial commissure, which still persists in *Viviparus*. The buccal ganglia are also related to these lobes. The pleural ganglia lie ventral to the cerebrals, but quite apart from the pedals, on which 2 connectives converge on each side, one cerebropedal, the other pleuropedal. These are sufficiently apart dorsally, neverthe-less, for one of the extrinsic muscles of the buccal mass to pass between them. The pedal ganglia are still primitive and have the form of elongate cords with 3 commissures linking right to left along the length of the foot.

So far as the rest of the nervous system is concerned there is little call for comment. As in many of the lower gastropods a sub-oesophageal ganglion is absent, though a supra-oesophageal one is present giving off a nerve to the osphradium. The visceral loop near the

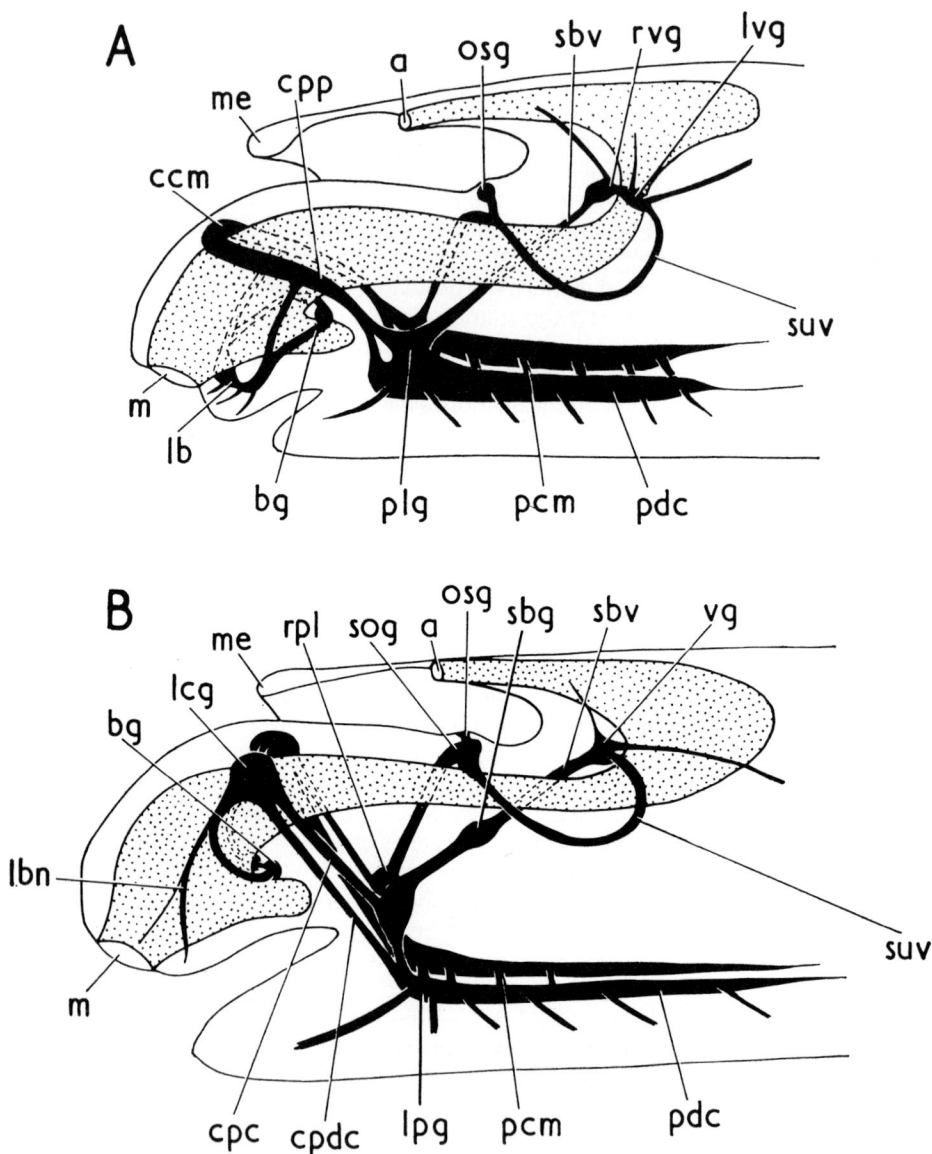

FIG. 153.—Lateral views of the anterior end of a prosobranch to show A, diagram of hypoathroid nervous system (e.g. rhipidoglossan); B, diagram of dystenoid nervous system (e.g. *Viviparus*).

a, anus; bg, buccal ganglion; ccm, cerebral commissure; cpc, cerebropleural connective; cpdc, cerebro-pedal connective; cpp, cerebropleuropedal connective; lb, labial ganglion; lbn, labial nerve; lcg, left cerebral ganglion; lpg, left pedal ganglion; lvg, left (or left part of) visceral ganglion; m, mouth; me, mantle edge; osg, osphradial ganglion; pcm, pedal commissure; pdc, pedal cord; plg, pleural ganglion; rpl, right pleural ganglion; rvg, right (or right part of) visceral ganglion; sbg, sub-oesophageal ganglion; sbv, sub-oesophageal part of visceral loop; sog, supra-oesophageal ganglion; suv, supra-oesophageal part of visceral loop; vg, visceral ganglion.

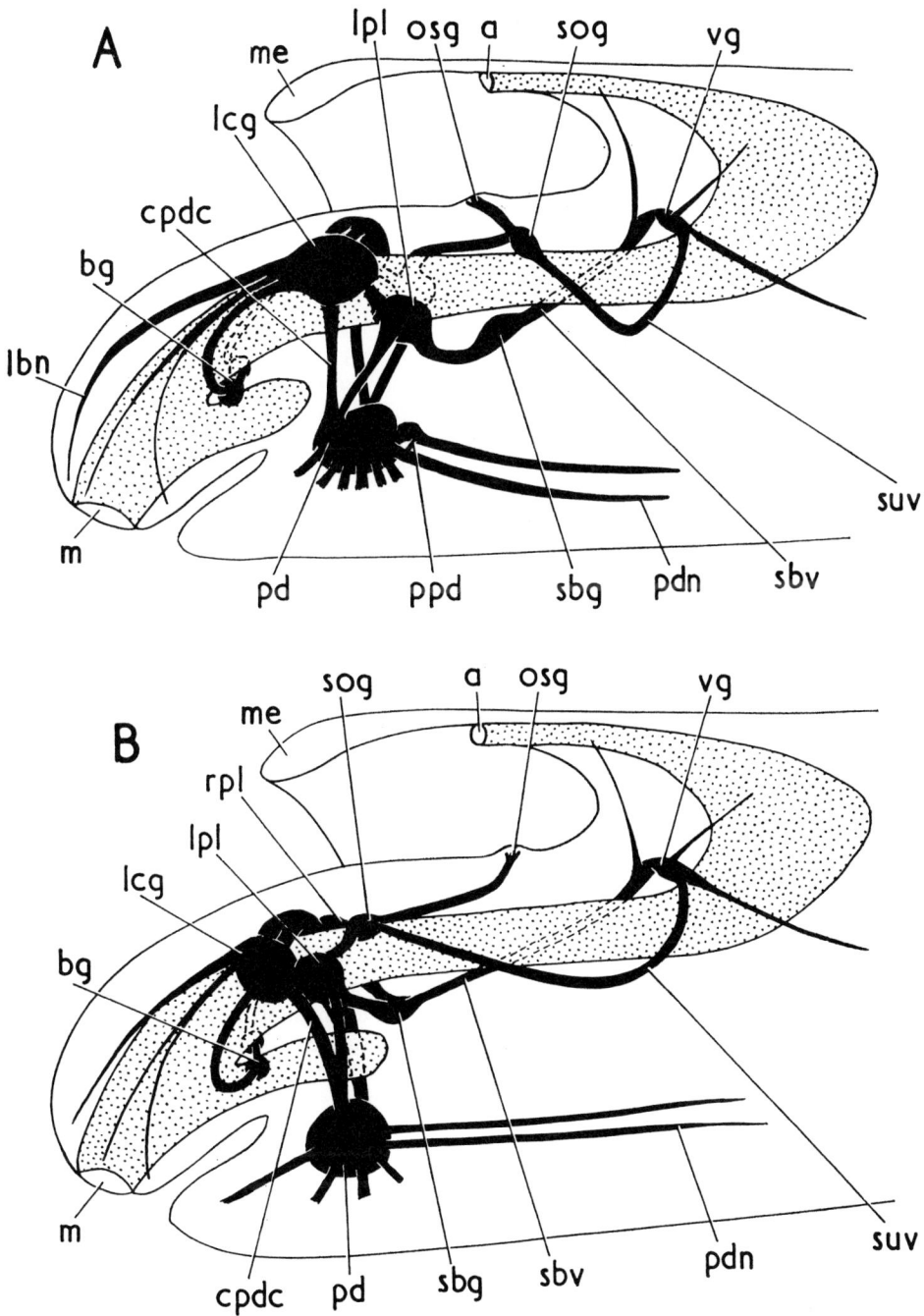

FIG. 154.—Lateral views of the anterior end of a prosobranch to show A, diagram of epiathroid nervous system (e.g. most monotocardians); B, diagram of concentrated nervous system (e.g. higher monotocardians).

a, anus; bg, buccal ganglion; cpdc, cerebropedal connective; lcg, left cerebral ganglion; lbn, labial nerve; lpl, left pleural ganglion; m, mouth; me, mantle edge; osg, osphradial ganglion; pd, pedal ganglion; pdn, pedal nerve; ppd, postpedal commissure; rpl, right pleural ganglion; sbg, sub-oesophageal ganglion; sbv, sub-oesophageal part of visceral loop; sog, supra-oesophageal ganglion; suv, supra-oesophageal part of visceral loop; vg, visceral ganglion.

supra-oesophageal ganglion is the origin of the branchial nerve, and pallial nerves leave from the region on the right where a sub-oesophageal ganglion might have occurred. On both sides of the body these anastomose with pallial nerves from the corresponding pleural ganglia so that right and left dialyneuries exist.

A further advance still is found in *Littorina*, the nervous system of which has already been described in detail (p. 41). In this animal both cerebral and pleural ganglia lie dorsal to the gut behind the buccal mass, although the pedals are still placed in the haemocoel of the foot and are connected to the others by long connectives (figs. 17, 23). They are now ovoid nervous masses, no trace of the long pedal cords remaining, except for lobed outgrowths of the main ganglia from which the nerves arise. Of the original multiple commissures only two remain, one thick and prominent, the other a very tenuous connexion linking the lobes from which the posterior pedal nerves spring, and possibly of little functional importance (pcm, ppd, fig. 25).

The labial ganglia and the labial commissure have been lost: only the buccal ganglia survive in the innervation of the anterior gut walls (bg, fig. 17). The ventral lips, originally innervated from the labial ganglia, now receive their nerve supply from the cerebral ganglia which also innervate the rest of the snout, the tentacle, the eye and, by fibres running in the cerebropedal connective, the statocyst.

The visceral loop carries supra-oesophageal, sub-oesophageal and a bilobed visceral ganglion. The first of these innervates the osphradium and gill, the sub-oesophageal supplies the right half of the mantle and on both sides dialyneuries unite these nerves and others originating in pleural and pedal ganglia (di, fig. 23). No zygoneury has been formed.

A number of stages representing various degrees of concentration of the ganglia of the oesophageal region and of the visceral loop may be distinguished amongst prosobranchs, the first being the condition already described for *Littorina*, in which the ganglia of the visceral loop occupy their normal position and no zygoneury has been developed on either side of the body. This may be encountered in the families Lacunidae, Pomatiasidae and Hydrobiidae, and is, therefore, characteristic of the Littorinacea and of the lower Rissoacea.

A number of other prosobranchs show a similar arrangement of the ganglia, but in them zygoneuries have been evolved so that the parietal ganglia are directly linked to the pleural ganglia of both sides. This occurs in the Ianthinidae and in the Aporrhaidae. It is also met with in the Capulidae, but in these animals the parietal ganglia have undergone a degree of forward migration, though they are still some way from the pleurals. The supra-oesophageal ganglion, indeed, retains its position left of the oesophagus and the sub-oesophageal ganglion lies on the right of that structure. The Ianthinidae differ from the others in this group in having the pleural ganglia fused to the corresponding cerebral ganglia.

The third stage which may be discriminated is derived directly from the second by a forward movement of the sub-oesophageal ganglion so that it comes to lie between the right and left pleurals underneath the gut. In this position it has in fact become a new member of the circum-oesophageal nerve ring established by the other ganglia. The supra-oesophageal ganglion maintains its typical situation. This group of prosobranchs contains the genera *Bittium, Clathrus, Natica, Trichotropis* and *Turritella*.

In a fourth group (fig 154B) the forward migration of the sub-oesophageal ganglion (sbg) has also involved the supra-oesophageal ganglion (sog) so that both these structures come to lie alongside the corresponding pleural ganglion (lpl, rpl) and the visceral loop appears to pass from supra-oesophageal ganglion on the right to sub-oesophageal ganglion on the left with only the visceral ganglion (or, more usually, ganglia) (vg) along its length. This state of affairs is encountered in the higher members of the Rissoacea (*Alvania, Cingula, Rissoa, Tornus*),

Calyptraea (Weise, 1924) and *Crepidula, Lamellaria* and *Velutina,* in the Stenoglossa (fig. 163) and in the parasitic animals *Balcis, Eulima* and the pyramidellids. In most of these the visceral loop remains long and streptoneurous, but in some (*Cingula,* the pyramidellids) the visceral ganglion has moved far forwards and in the case of the pyramidellids the loop has also become euthyneurous. In some of the genera mentioned in this list (*Crepidula* and *Calyptraea)* the cerebral and pleural ganglia on each side have fused.

A final stage in the evolution of the nervous system of the mesogastropods is exemplified by a number of animals in which some fusion of ganglia on nerve ring and visceral loop has occurred. This may also be accompanied by fusion of cerebrals and pleurals so that, on occasion, complex nervous masses of multiple origin may arise. The animals which are known to possess nervous systems of this sort are *Valvata, Assiminea, Paludinella, Omalogyra, Skeneopsis* and some species of *Eulima.* As these are all small, or extremely small, it appears likely that the fusion of the ganglia is as much an adaptation to this as an attempt to produce a more highly co-ordinated nervous system.

In *Valvata* (Bouvier, 1887; Bernard, 1888) the cerebral and pleural ganglia on each side are fused, and the 2 pedals are closely approximated, although they still lie ventrally in the foot. In the visceral loop the supra-oesophageal ganglion has moved forward and fused with the right pleural and the sub-oesophageal has migrated until it lies alongside the left pleural. The single visceral ganglion has retained its posterior position on the visceral loop and there is one other ganglion placed near the point where a connexion is established between a pallial nerve originating in the supra-oesophageal ganglion and another coming from the left pleural ganglion: this is probably an osphradial ganglion. A similar dialyneury, but no ganglion, occurs on the animal's right.

In *Assiminea* and *Paludinella* (Thiele, 1929–35) a similar fusion between the right cerebral, right pleural and supra-oesophageal ganglia has taken place but the left has retained a greater degree of separation, the left pleural and the sub-oesophageal ganglion lying between the cerebral above and the pedal ganglion below. In *Omalogyra atomus* (fig. 176A) the nervous system retains its primitive disposition except that on each side the cerebral and pleural ganglia have fused. No visceral ganglion can be made out in the nervous system of this minute animal, but it is probable that this is a consequence of the difficulty of working with such a small creature and it may well be present. A similar nervous system occurs in *Skeneopsis planorbis,* but in the species of *Rissoella, opalina* and *diaphana,* a greater degree of fusion of ganglia has taken place and the visceral loop seems to have been drawn forward and probably untwisted in the process. Two large ganglia lie above the buccal region and 2 below, whilst buccal ganglia lie alongside the root of the radular sac. The 2 ventral ganglia have the statocysts lying beside them and are clearly pedal; the dorsal ones, however, have been formed by the fusion of ganglia and their morphological value remains uncertain.

The sensory equipment of the prosobranchs cannot be compared in its acuity with that of the cephalopod molluscs or with that of many other types of animal. Their bodies, however, are well supplied with sense organs of various sorts, of which eyes, osphradia, tentacles and statocysts are the most important distance receptors and numerous other organs set in various parts of the body the major local ones. The whole surface of the body is naturally sensitive to contact and chemical stimulation, but the tentacles and lips of the head, the sides of the foot and the mantle edge, including the siphon where one is developed, appear to be more sensitive than the surface in general.

The tentacles are obviously the seat of a well developed tactile sense and are used by the animal as a main guide to its movements throughout the environment in which it lives. They are kept in continuous movement, the main direction being up and down, and their motion is timed with the locomotor waves which are moving along the foot. In view of the probability that the cephalic tentacles are simply the most anterior of a series set along head and foot this is not surprising. In many prosobranchs the sensitivity of the tentacles is increased by the presence of innumerable immobile cilia set over their surface: these are particularly obvious in the trochids, rissoids and pyramidellids, in the last group of which they have been shown to be compound cirri (Fretter & Graham, 1949). The cilia are connected to sensory cells which in turn are related to the terminations of branches of the tentacular nerve. In the trochids they have been examined by Burdon-Jones & Desai (in lit.) who refer to them as brush organs. They are composed of a group of 3–5 bipolar neurones, the T-shaped outer ends of which underlie a ring of immobile cilia. Burdon-Jones & Desai regard them as touch-taste receptors. They are particularly common on the cephalic and epipodial tentacles, on the papillae at the base of the latter, on the epipodial ridge and on the fringed edge of the left neck lobe. In these situations they appear to be of importance in testing the nature of the substratum and the turbidity of the water entering the mantle cavity. The poor visual equipment of the ordinary prosobranch (Willem, 1892c) makes the tactile sensitivity of these structures important, and probably similar but less well developed sensory structures are located at all other parts of the body. In the archaeogastropods, for example, special sensory organs of unknown signifi-cance are placed at the base of the epipodial tentacles (eso, fig. 79); these may be compared to those on the mantle edge of Acmaea (ms, fig. 73).

The tentacles, however, are also supposed to be the seat of an olfactory sense. This is additionally ascribed to another sense organ, the osphradium or Spengel's organ as it was originally called, located in the mantle cavity. Ankel (1936a), for example, has shown how specimens of Nassarius are able to find their way to food when only the siphon projects out of the sand in which they are buried, so that the siphon and the mantle cavity are the only parts of the body exposed to stimulation. This mechanism and the differentiation between smell and taste has been studied by Henschel (1932).

The osphradium is an organ located in the mantle cavity either on or near the ctenidium. Like that structure, it is double in the diotocardians, single in the monotocardians, and it tends to be lost when the ctenidium is lost. The basommatophoran pulmonates are notable examples of a group in which the ctenidium has gone but the osphradium persists, but there are some examples of the same thing in prosobranchs, e.g. Pomatias (Garnault, 1887) and Assiminea grayana. In the nuchal cavity of Patella there are 2 orange papillae which are normally called osphradia, and there is certainly a nerve centre associated with each. In addition to this, however, Spengel (1881) found blood spaces, and for this reason regarded the structure as the remains of the whole osphradium-ctenidium complex. Bernard (1890), re-investigating its structure, confirmed Spengel's description without, however, committing himself to Spengel's homologies.

Bernard (1890) has given a good survey of the gross morphology of the osphradium throughout the prosobranch series and this has been supplemented by Yonge (1947). In the diotocardians the usual appearance of the osphradium is a ridge placed on the ctenidial axis, in Emarginula and Diodora lying both dorsally and ventrally. In the trochids, on the other hand, the osphradium forms a kidney-shaped elevation at the place where the free portion of the ctenidial axis joins the mantle skirt, on the anterior border of the efferent membrane (osg, fig. 144), and it occupies a similar position in Haliotis. In the other diotocardians the osphradium is not closely related to the ctenidium but lies near it on the roof or floor of the mantle cavity, as in the Neritacea where it is formed by a ridge on the mantle skirt. In the docoglossans some uncertainty surrounds the organ, especially in view of the loss of gills which these animals often exhibit. Thus in *Acmaea* Bernard (1890) described as osphradia a pair of tubercles lying, the left by the base of the ctenidium, the right by the reproductive aperture, but was unable to find sensory epithelium on their surface. Willcox (1898), on the other hand, thought that she had found sensory cells. Thiem (1917a, b) described osphradia in several acmaeids but again failed to find either sense cells or nerves from the underlying ganglia: he also failed to find the tubercles in *A. virginea*. He regarded these structures as partially ctenidial in origin like those of *Patella*. Yonge (1947) believed that neither of these tubercles is an osphradium at all and located this structure on the left side in a patch of sensory cells anterior to the tubercle: no corresponding structure, however, lies in front of the right tubercle. The osphradia of *Patella* have already been mentioned and similar structures—except for the lack of vascular spaces (Bernard, 1890)—occur in *Patina*.

In the majority of the mesogastropods the osphradium has the form of a linear ridge lying parallel to the ctenidial axis and to its left. This is to be found, for example, in *Littorina* spp. (figs. 10, 11), in *Bithynia tentaculata* (fig. 55), in the rissoids, in *Pomatias elegans* (fig. 305), in *Bittium reticulatum* (fig. 182B), in *Aporrhais pespelicani* (fig. 171), in *Crepidula fornicata*; in *Trivia* (fig. 164) and in *Natica catena* (fig. 172A) the osphradium enlarges somewhat to occupy a triangular area in the former genus. There is an indication of a considerable increase in the size of the organ in the Stenoglossa, where the ridge has developed a double series of lateral foldings, giving the whole organ something of the appearance of a bipectinate ctenidium (fig. 267). For this reason the osphradium has sometimes been known as the 'fausse branchie'.

The osphradium consists of an epithelium of sensory cells mixed with pigment cells overlying a ganglion. The ganglion is related to the branchial nerve and the supra-oesophageal and right pleural ganglia. The precise relationships of the nerve vary within the prosobranchs and three different arrangements may be distinguished. In the majority of rhipidoglossans a nerve runs from the supra-oesophageal part of the visceral loop and expands into a branchial ganglion. This sends nerves to the gill, the osphradium and to local parts of the mantle skirt. In the Docoglossa and the remaining rhipidoglossans (Neritacea) there is no branchial ganglion and the osphradium receives its nerve directly from the supra-oesophageal ganglion, from the same nerve, indeed, as goes to the ctenidium. In the monotocardians, also, there is no branchial ganglion and the osphradial branch of the branchial nerve gradually assumes so much more importance than that to the gill that it is better designated the osphradial nerve. [For recent work see Chapters 26 and 36.]

Although the function of the osphradium has usually been regarded as olfactory, or gustatory, and the organ the seat of a chemical sense since it is placed across or along the stream of water entering the mantle cavity (fig. 268), it is worthwhile noting that it is also sited at a place where it would be affected by the amount of particulate matter in the incoming water. For this reason Hulbert & Yonge (1937) have held that it is used by the animal as an indicator of the amount of suspended matter rather than as an organ of more strictly chemical sense.

The second organ of sense which acts as a distance receptor is the eye, found in almost all gastropods, the only exceptions being a few blind ones which come from habitats to which light does not penetrate in sufficient quantities to make the possession of a photo-sensitive organ useful and in a few other cases (e.g. *Ianthina*) where this explanation does not hold, and where, indeed, a plausible reason for its absence or ineffectiveness is hard to find. The gastropod eye must be regarded mainly as an organ acting as a simple detector of light and the direction from which it comes, since its image-forming powers seem to be extremely poor

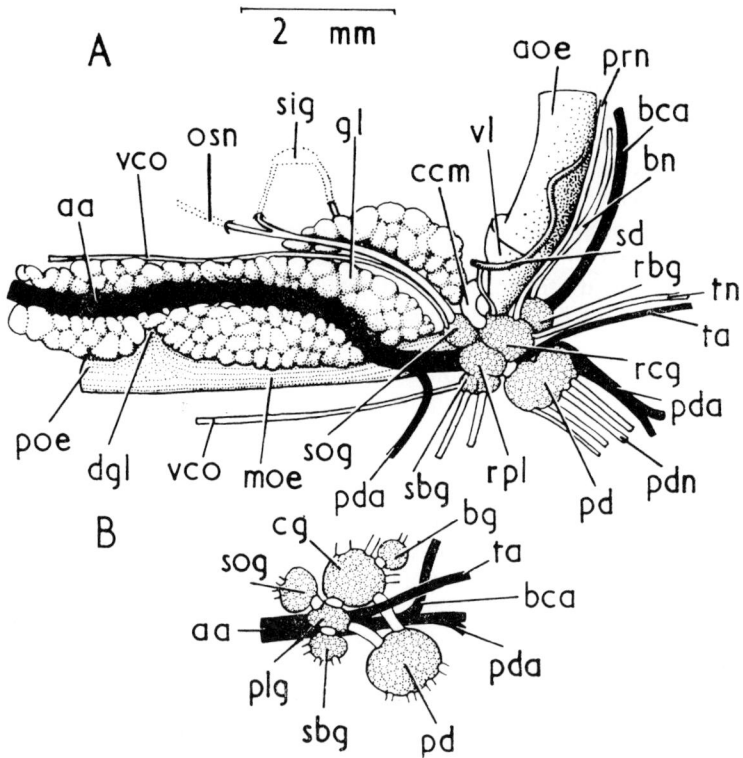

FIG. 155.—*Nassarius reticulatus:* A, the nerve ring and associated structures; B, nerve ring stretched to show the course of related blood vessels.

aa, anterior aorta; aoe, anterior oesophagus; bca, buccal artery; bg, buccal ganglion; bn, buccal nerve; ccm, cerebral commissure; cg, cerebral ganglion; dgl, duct of gland of Leiblein; gl, gland of Leiblein; moe, mid-oesophagus; osn, osphradial nerve; pd, pedal ganglion; pda, pedal artery; pdn, pedal nerve; plg, pleural ganglion; poe, posterior oesophagus; prn, proboscis nerve; rbg, right buccal ganglion; rcg, right cerebral ganglion; rpl, right pleural ganglion; sbg, sub-oesophageal ganglion; sd, salivary duct, the salivary gland removed; sig, siphonal ganglion; sog, supra-oesophageal ganglion; ta, tentacular artery; tn, tentacular nerve; vco, visceral loop; vl, valve of Leiblein.

(Willem, 1892a, b, c). The eye is normally situated on an eye stalk which lies on the side of the head immediately posterior to the cephalic tentacle. Although originally separate from this (e.g. *Haliotis*)—and still partially so in the Trochacea (figs. 53, 69, 90)—the 2 structures become so closely fused in most prosobranchs that the eye usually appears to be placed on a small bulge at the base of the tentacle (figs. 10, 57). In a small number of animals the eyes have migrated to a deeper situation and in the pyramidellids they have also migrated towards the mid-line, so that they now lie between the 2 tentacles (figs. 133, 134, 290).

In the lower families of the diotocardian prosobranchs (Haliotidae, Fissurellidae, Patellidae) the eye is an open vesicle with a moderately wide mouth (e.g. *Patella*, fig. 156A); in the Trochacea whilst still open, the vesicle has a constricted aperture (fig. 156B). In the Neritacea and all other prosobranchs (e.g. *Trichotropis*, fig. 156C) the vesicle has closed off from the

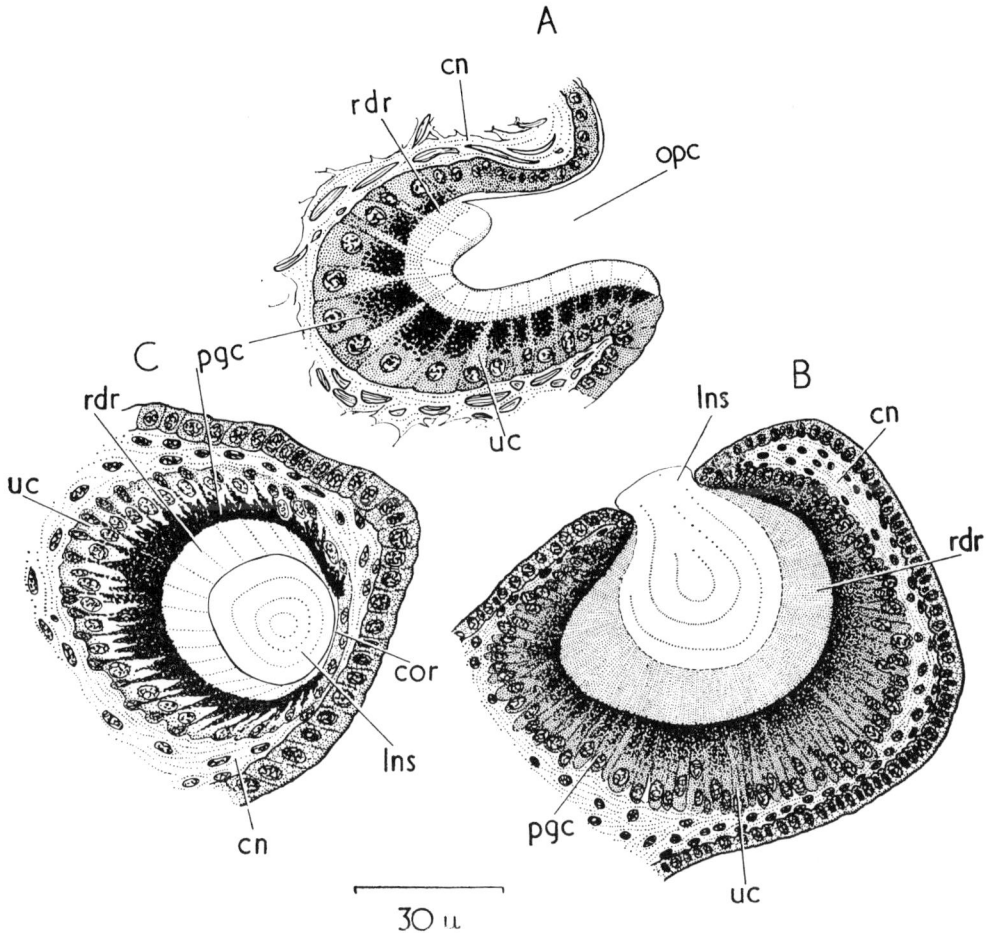

FIG. 156.—Eyes in vertical section. A, *Patella vulgata;* B, *Cantharidus clelandi;* C, *Trichotropis borealis.*
cn, connective tissue; cor, cornea; lns, lens; opc, opening of optic cup; pgc, pigmented cell of retina; rdr, rod of retinal cell; uc, unpigmented cell of retina.

surface of the skin and has no connexion with the outside environment. In the case of the open eye of the typical diotocardian the cavity is filled, wholly or partly, with a secreted mass of material acting, perhaps, as a lens or refracting medium (lns, fig. 156B). Its detailed structure is a matter of debate: according to Hilger (1885) and most other writers on the subject the cavity of the eye is filled with a homogeneous mass of emplema (to give it Grenacher's name), whereas Patten (1886b) found it divisible into a hard, outer, biconvex lens and an inner vitreous in *Haliotis,* which was the only gastropod with which he worked. In the Patellacea, however, the cavity of the eye (opc) is usually thought to be filled with sea water. In this case the eye might be regarded as either a more primitive stage from which the eye of the other diotocardians with its emplema might be derived, or a degeneration from this by loss of emplema. Hilger (1885), however, thought that he had some evidence suggesting the presence of a cuticle over the open mouth of the optic vesicle, which would bring the patellacean eye more into line with that of the other diotocardians: no other worker has suggested the occurrence of this.

In the prosobranchs with closed eyes the homogeneous vitreous appears to be limited to the Neritacea, and the great majority have a spherical lens lying under the outermost part of the optic vesicle (lns, fig. 156C) and separated by a small quantity of vitreous from the retina: it has, however, been suggested that this gap between lens and retina is an artefact and that no vitreous is naturally present.

Although there is this considerable degree of variation in the general structure of the eye within the prosobranchs, the histology of the wall of the optic vesicle appears, on the other hand, to be relatively uniform throughout the class. The bulk of the epithelium lining the vesicle makes up the light-percipient layer or retina, a small part underneath the skin in closed eyes forming a cornea (cor, fig. 156C) made of thin, unpigmented and translucent cells like those of the overlying epidermis. Little progress was at first made in understanding the histological nature of the retina because—inevitably—it was originally attempted to explain it on the same basis as that of the vertebrate eye, and it was only when Hensen (1865) showed that its development was wholly different that some success in understanding it was reached. The most important investigators of its structure have been Fraisse (1881), Hilger (1885), Grenacher (1886), Patten (1886b), Carrière (1889), Hesse (1900) and Bäcker (1903), and their findings as to retinal structure are almost identical. All are agreed that in the retina there occur two types of cell, one containing pigment (pgc), the other without (uc), but they differ amongst themselves as to the function that the two types subserve in the actual sensory act. The pigmented cells and unpigmented cells alike are supposed to be sensitive to light in the more primitive forms although the latter are usually held to be the more important; in higher prosobranchs the unpigmented cells come to be the only sensory ones and the pigmented cells have no direct visual importance, though they are presumably useful accessory structures. This conclusion is reached mainly because it is only the unpigmented cells which bear rods distally and are connected by neurofibrillae to the underlying nerve fibres. The pigmented cells, which may share in sensitivity in lower forms, become, in the more specialized eyes, only supporting cells for the unpigmented, though they may share in the production of the cuticular material of the lens and even of the rods. Bäcker (1903), for example, has shown that each rod consists of a stainable axis embedded in cuticle: he regarded the axis as being essentially a surface prolongation of the unpigmented cell and the cuticular material in which it is embedded as a supporting and perhaps also an optical device secreted by the pigmented cells.

The heteropods are free-swimming active predators that catch their prey by sight, and this is possible because of the important development which their eyes have undergone. These have been carefully investigated by Hesse (1900) in *Carinaria mediterranea*. The eyes are large—3·8 mm in an animal 15 cm in length—and are shaped like a tapered cylinder (fig. 157A, B) with its long axis parallel to the long axis of the body and its narrower end directed forwards. This end bulges somewhat because of the presence of a spherical lens (lns). The cylinder is circular in cross-section in its anterior half but is flattened dorsoventrally in its posterior part. The optic nerve (on) runs from its base to the cerebral ganglion.

The wall of this eye is lined by an epithelium composed mainly of pigmented cells. These are absent over the lens, where a clear transparent cornea (cor) occurs and also over the greater part of the dorsal surface. A few small unpigmented spots (us) are also present in the

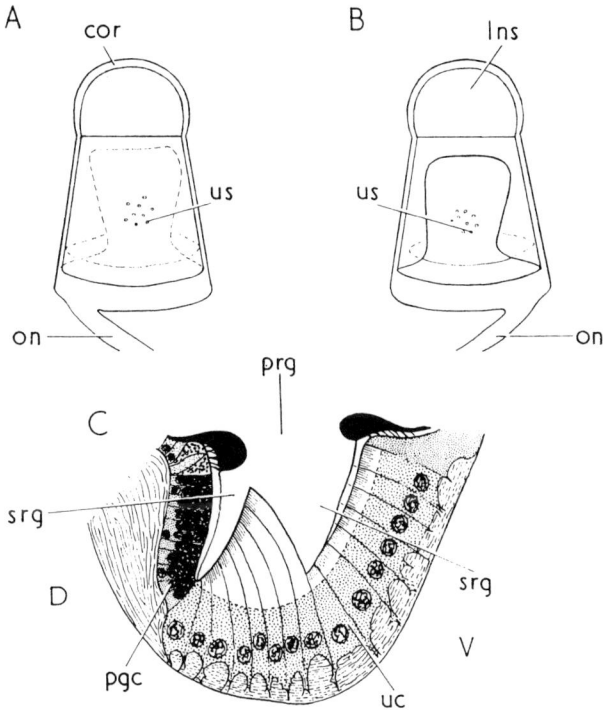

FIG. 157.—Heteropod eye: A, in ventral view; B, in dorsal view; C, vertical section of retina. After Hesse.
cor, cornea; D, dorsal; lns, lens; on, optic nerve; pgc, pigment cell; prg, primary retinal groove; srg, secondary retinal groove; uc, unpigmented retinal cell; us, unpigmented spot in pigmented wall of eye; V, ventral.

ventral wall. Since the heteropods normally swim upside down this means that the window in the pigmented layer is pointing down into the depths of the water.

The retina (fig. 157C) is a narrow band of cells, running across the base of the eye from right to left. It lies at the base of a groove (prg) and is itself scored by 2 deep grooves (srg). The dorsal wall of the groove is made of pigment cells (pgc) extending from the main cavity of the eye, its ventral wall by retinal cells (uc). The dorsal half of the retina is composed of cells of varying height, the ventral ones being tallest, the dorsal ones smallest. As a result the surface of these cells, like that of the ventral ones, looks across the retinal groove towards the pigment cells and is therefore parallel to the long axis of the eye. The images of the objects being viewed fall along these surfaces and the retina is therefore provided with two surfaces which give the animal information about the distance of the object which is being viewed. No accommodation is required—no mechanism to achieve it exists—since the arrangement of the sensitive retinal surfaces gives this information directly. This retina gives the animal a view of what lies ahead, and the mollusc arranges to sight potential prey in this way before making its final pounce upon it, an action for which the accurate judgement of distance is necessary. An accessory source of visual information, however, is provided by other sensitive cells placed further forwards in the eye: these are, in fact, the cause of the unpigmented spots on the ventral wall of the eye opposite the window in the dorsal wall. These record movement in the water below and the objects to which they react might then be inspected in more detail if the heteropod turned and looked straight at them through the front of the eye and the main part of the retina.

In prosobranchs, as in other molluscs, there also occurs the type of sense organ known as a statocyst. This name and the idea underlying it—that the organ is one for equilibration—were first published by Delage (1887) in relation to what had previously been called an auditory capsule or otocyst (Lacaze-Duthiers, 1872). The experimental work of Tschachotin (1908) gave firm support to the idea that these structures were indeed organs of balance rather than of hearing.

The statocysts arise as ectodermal invaginations which sink inwards and close off, though occasional hollows in the statocyst nerve may perhaps be vestiges of the original invagination (Buddenbrock, 1915). They form vesicles which in most prosobranchs lie in the proximity of the pedal ganglia, though the precise situation is somewhat variable. They were originally thought to be innervated by nerves from the pedal ganglia, but Lacaze-Duthiers (1872) showed that the fibres really arose in the cerebral ganglia and ran thence, sometimes direct to the statocyst (*Pomatias elegans, Capulus ungaricus, Viviparus viviparus, Theodoxus fluviatilis*), sometimes branching off to the statocyst from the cerebropedal connective (*Patella vulgata, Natica catena*) but in most species run through the pedal ganglia and so mimic pedal nerves in appearance. The statocyst nerve arises from the posterior part of the corresponding cerebral ganglion alongside the area from which the optic nerve springs.

As in the case of the eyes much attention has been paid to the statocysts of heteropods, where they are large and accessible, but the organs appear to be very uniform throughout the prosobranchs. The following account is based on Ilyin (1900), Tschachotin (1908) and Pfeil (1922). Each statocyst is a spherical cavity about 0·25 mm in diameter lined by an epithelium containing 2 types of cell, giant and syncytial. The giant cells are few in number and vary a little (11–13 in *Helix pomatia*) but whatever number occurs in the right statocyst will also occur in the left. They consist of a central area which contains the nucleus and which occupies the whole depth of the epithelium; from this pseudopodia-like extensions reach out in all directions over the inner surface of the statocyst. These meet, tip to tip, similar extensions from other giant cells. The greater part of the inner surface of the statocyst is lined by the giant cells. The second type, the syncytial cell, forms the rest of the wall; it therefore underlies the arms radiating from the bodies of the giant cells and fills up the whole thickness of the epithelium in the lacunae between them.

The whole inner surface, whatever type of cell it may be formed from, is ciliated, and the wall rests on an outer layer of connective tissue. Beyond the fact that the statocyst is undoubtedly an organ of equilibration, which was convincingly shown by Tschachotin (1908), the functioning of the statocyst is not known. Pfeil (1922), however, believed that it is the giant cells which are the essential sensory cells and that the syncytial base on which they rest is merely for their support. The cilia are very short and do not seem to have sense hairs of different length mixed up with them.

The cavity of the statocyst is filled with fluid, secreted from the cells which line it, and in this float calcareous particles, either one statolith or several smaller statoconia. Though stato-conia appear on the whole to be characteristic of the more primitive prosobranchs and the single statolith of the more advanced types there is no very clear systematic base to this difference. Statoconia occur in all the Archaeogastropoda and in the Viviparidae, Valvatidae and some other families of mesogastropods, mostly with a distribution in freshwater. The other mesogastropod families and the Stenoglossa have a single statolith in each statocyst.

The nervous system of many animals, in addition to its more ordinary role of rapid correlation of sensory and motor activity, undertakes in addition the long term control of many processes by means of the secretion of hormones or neurohormones. In vertebrates and insects this is a well known phenomenon, but, though less spectacular, it also occurs in molluscs and in particular in the gastropods. The first record of a hormone-like substance secreted in a mollusc was made in *Aplysia* and *Pleurobranchaea* by Scharrer (1935) who observed droplets or granules in cells of most of the ganglia which also entered the nerves. The number of animals in which this was found to occur grew rapidly, mainly by the work of Gabe (1953a, b), until Gersch (1959) was able to state that the total number of gastropod species known to elaborate neurohormones was 71, of which 23 were prosobranchs, 46 opisthobranchs and 2 pulmonates. These last are *Ferrissia* (Lever, 1957) and *Helix* (unpublished work by Jungstand quoted by Gersch). As in many other groups of animals the neurosecretory cells are likely to have an annual cycle of activity though this has so far only been exposed for those found in *Viviparus* (unpublished work by Gorff quoted by Gersch), where granules can be found in some ganglion cells and traced from there into certain nerves where they lie amongst the nerve fibres. They abound in summer and have a minimum in winter.

The precise physiological role played by neurosecretory substances in gastropods is not clear, although some physiological effects brought about by them are quite obvious. Their occurrence has mainly been tested for by measuring their effects on the heart of *Aplysia* or *Helix* or, more recently, *Venus,* or on the crop of *Aplysia,* all of which organs are commonly stimulated to increased amplitude and rate of contraction by extracts of ganglia or nerves (Scharrer, 1937; Gabe, 1954; Welsh, 1953, 1957). Analysis by paper chromatography showed that the substance responsible for this was not adrenaline, noradrenaline nor 5-hydroxytryptamine, and might perhaps be a mixture of substances. In *Helix*, Meng (1958) has shown that there are 2 neurohormones, one causing an acceleration and the other a retardation of the heart beat: the former he showed to be 5-hydroxytryptamine, the latter acetylcholine.

CHAPTER 13

THE REPRODUCTIVE SYSTEM—1

M OLLUSCS are considered to be acoelomate, lacking coelomic sacs, though possessing certain body cavities which in other groups might be interpreted as coelomic. In present-day molluscs such cavities relate to the heart, the gonads, and the kidneys, though their interlinking by ducts suggests that they represent parts of a single space, as they are in fact in development. It is assumed that in the ancestral molluscs the pericardial cavity may have been paired as it is in *Neopilina* but that the two parts soon fused. Each gonad communicated with the mantle cavity by a gonoduct and posteriorly with the pericardial cavity (fig. 1). From each side of this cavity a duct ran, parallel and posterior to the gonoduct, to open to the mantle cavity: these pericardial ducts may have had an excretory function. In the ancestral prosobranch it is probable that the single persisting gonad, the right, had lost its direct connexion with the mantle cavity and now discharged to the right renopericardial duct which, like the left, had expanded to form a kidney (fig. 323B). Such an arrangement now holds in *Diodora*, *Puncturella* and the trochids (fig. 323D, E) but in the Patellacea (Goodrich, 1895) and in *Pleurotomaria* (Woodward, 1901a) it is with the kidney (fig. 323C). The connexion with the kidney may be by a simple longitudinal slit as in *Patella*, or it may be by several openings as Thiem (1917b) described for monobranchs (e.g. *Acmaea*), which break through only when the gonad is ripe. There is some doubt as to the precise state of affairs in *Haliotis* since according to Haller (1894), Fleure (1903), Totzauer (1902) and Meyer (1913) there is a right renopericardial duct into which the gonad opens, but according to Perrier (1889), Erlanger (1892) and Crofts (1929) this duct is absent in the adult and the gonad discharges directly to the right kidney. It may be possible to reconcile these contradictory statements on the assumption that a renopericardial duct does in fact form at an early stage as found in the early post-veliger by Crofts (1937) in *Haliotis* and receives the gonadial duct, and that in a certain number of individuals the pericardial end persists, but in another series it aborts, leaving the gonad and kidney in apparently direct connexion.

The position of the gonad varies. In diotocardians with a helicoid spiral shell (trochids (ov, fig. 324), *Pleurotomaria* and *Scissurella* (te, fig. 247)) it occupies a position similar to that in the monotocardians and, with the digestive gland, comprises the visceral coils, lying mainly on the columellar side. The spire is reduced to one coil in *Haliotis* and the gonad is in a corresponding position. In patelliform genera (*Acmaea, Lepeta, Patella, Patina, Puncturella, Propilidium*) it lies between the visceral mass and foot (ov, fig. 253) except anteriorly, and when ripe spreads around the periphery of the visceral mass sometimes more particularly on the right side (g, figs. 249, 258).

Most diotocardians are littoral. The sexes are separate and there is no copulation; the eggs are fertilized after they leave the female. In *Patina pellucida, Patella caerulea* (Ankel, 1936a), *P. lusitanica* (von Medem, 1945) and *Gibbula tumida* (Gersch, 1936) the male and female are close together during the emission of the gametes, and this is probably true for other species. The gonad is relatively much greater in volume than in species which copulate. In female

Patella vulgata the ovary is 1 : 3·75 the body weight, whereas in the monotocardian *Lamellaria perspicua* it is only 1 : 21·6. *Haliotis tuberculata* will shed about 20,000 eggs at one spawning, emitting them through all the pores of the shell, the greatest quantity passing through the second and third oldest, and sperm from the male may make the water turbid for a distance of 3 ft (Crofts, 1929). The males are the first to spawn thus stimulating the females to shed their eggs (Boutan, 1892). This phenomenon of a ripe male inducing females to spawn is of widespread occurrence in marine invertebrates, and trochids have been observed to behave likewise (Robert, 1902), but Gersch (1936) denied it for *Gibbula cineraria,* though he admitted that spawning, once started, spreads rapidly through a whole population.

In some invertebrates the fusion of egg and sperm is brought about by substances secreted by each of these cells which help to overcome the hazards of external fertilization. Both gametes have been shown to secrete 2 types of fertilization substances, or gamones. Gynogamone I, secreted by the eggs, accelerates the movements of the sperm so that they reach the egg more rapidly. Once the sperm has reached the egg it is agglutinated by gynogamone II, which has little or no effect on sperm of other species. The sperm produce androgamone I which slows down their movement before they are liberated. After liberation they are stimulated by gynogamone I and may move up a gradient of this gamone to the egg. Later androgamone I cancels the effect of gynogamone I and so prevents the sperm being attracted to a fertilized egg. The sperm also produce androgamone II which causes local solution of the egg jelly and membrane and so facilitates penetration. These 4 gamones have been reported in *Haliotis, Diodora* and *Patella* (von Medem, 1945) and even egg cases have been found to be as effective as the whole egg in attracting spermatozoa, probably due to contamination with gynogamone substance.

Although some diotocardians shed their eggs singly to become planktonic, others embed them in a gelatinous secretion forming fixed egg masses. In the former case they develop to free trochophore or veliger larvae, but those in protective coverings are not freed until they have developed to the crawling stage. The eggs have investments which are produced by the ovary. In *Patella* and *Patina* the egg membrane is surrounded only by a gelatinous covering which swells when the eggs are discharged into the water. According to Lebour (1937) the eggs of *Gibbula cineraria, G. umbilicalis, Monodonta* and *Tricolia* have an albuminous layer within the gelatinous sheath, though we are unable to identify this and believe that the apparent double covering is merely a line of contact between denser inner jelly and a thinner more superficial layer, all of which is secreted by the egg within the ovary (jo, fig. 158A). All these forms lay their eggs singly. In the top shells of the genera *Monodonta* (fig. 53) and *Gibbula* the urinogenital aperture of the female (rko) is, unlike that of the left kidney (lko), provided with glandular rosette-shaped lips which are yellow or bright orange in the living animal. They are not developed in the male. In *Tricolia* (rko, fig. 77) such lips are present in both sexes, but are larger in females. The secretion from the lips is wholly mucus and may augment secretion from the hypobranchial gland in entangling the egg stream within the mantle cavity. According to Gersch (1936) the hypobranchial gland secretes most actively during the breeding season, and he concluded that it provides the embedding medium for the egg masses in *Gibbula tumida.*

Within the family Acmaeidae there is surprising variation in the method of spawning. *Acmaea virginea* sheds the sex cells (fig. 159A) singly into the plankton, but Willcox (1905a) recorded that in *A. tessulata* the eggs are embedded in very thin mucus in which they lie one cell deep and at regular intervals apart; the mucus is secreted by the sole of the foot. She has also observed a kind of copulation in this species, in which the male mounts the left side of

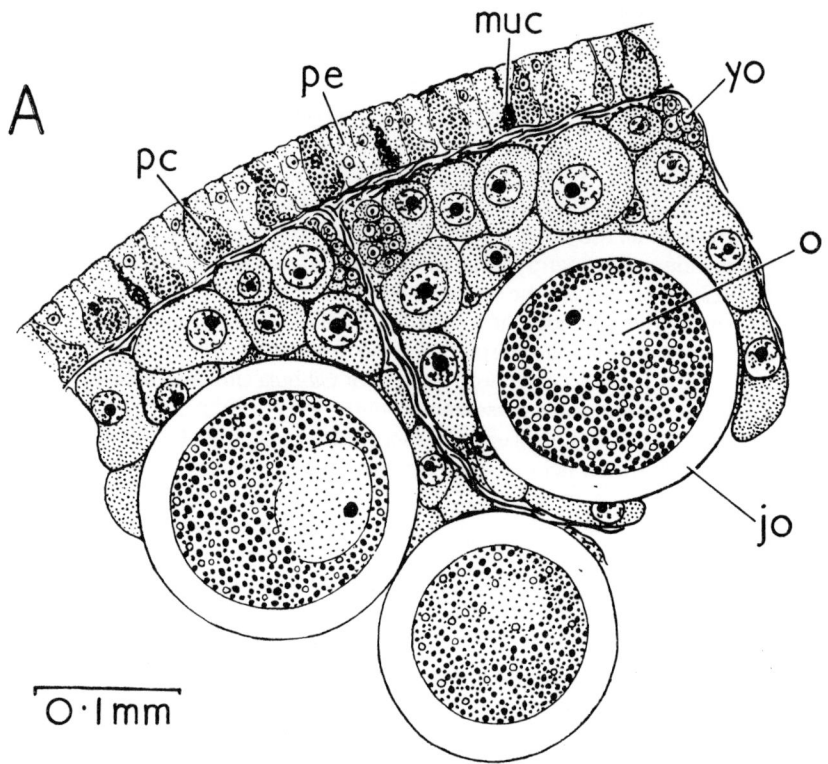

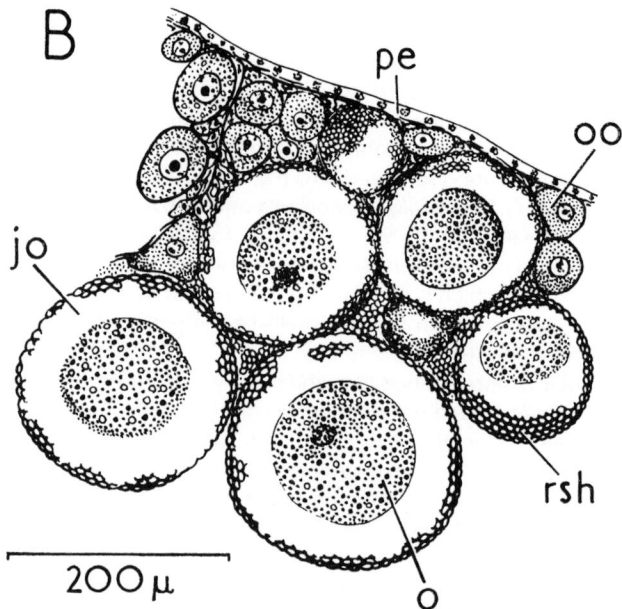

FIG. 158.—Transverse section through superficial part of upper visceral hump showing outer pallial epithelium and adjacent ovary: A, *Monodonta lineata*; B, *Diodora apertura*.

jo, jelly round egg; muc, mucous cell; o, ovum; oo, oocyte; pc, protein secreting cell; pe, pallial epithelium; rsh, reticulated shell; yo, young oocyte.

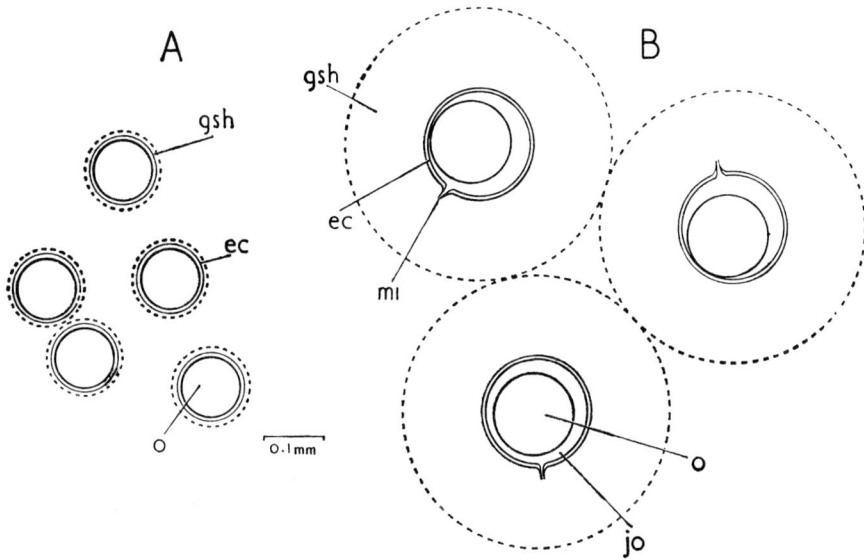

FIG. 159.—A, *Acmaea virginea*: eggs as they leave the urinogenital aperture. B, *Diodora apertura*: three adjacent eggs from egg mass.
   ec, egg covering; gsh, gelatinous sheath of egg; jo, jelly round egg; mi, micropyle; o, ovum.

the shell of the female to discharge sperm. These are carried to the vicinity of the female opening by cilia on the underside of the mantle skirt. Finally, Thorson (1935) has shown that *Acmaea rubella* incubates its eggs.

   Diodora apertura, *Cantharidus exasperatus*, *C. striatus*, *Calliostoma zizyphinum* and *Margarites helicinus* produce spawn of differing shapes. In *Diodora* (fig. 159B) it consists of a layer of eggs one cell thick and several inches across, with each egg joined to its neighbour and the whole firmly fixed to the substratum. Around the urinogenital aperture there are no glandular lips which might produce a cementing fluid, and, according to Boutan (1886), this comes from an accessory gland on the wall of the urinogenital duct. However, sections of the mature ovary (fig. 158B) show that each ovum secretes a shell (rsh) which appears reticulate in fixed material, and an inner jelly coat (jo). The shell, in particular, swells when the egg is discharged (gsh, fig. 159B) to form a gelatinous sheath of considerable thickness, and the eggs may adhere together by these sheaths alone. When the eggs leave the urinogenital aperture they pass to the anterior end of the mantle cavity in a continuous stream and are spread on the under surface of a stone by the foot. Sperm from the male are said to pass through the apical opening of the mantle and shell and fertilize the eggs as they are laid. Medem (1945), however, stated that in *Fissurella* (= *Diodora*) *nubecula* there are spermatophores formed by testicular epithelium surrounding packets of sperm and that the eggs are fertilized in the ovary, though he gave no indication as to how the spermatophores reach the female. This is the only example of spermatophores recorded from the diotocardians with the exception of some Neritacea (not the British *Theodoxus*), and they have a penis.

   Spawn masses of *Margarites helicinus* (Fretter, 1955b) (fig. 160A) are irregular clumps of 100 or more eggs deposited on seaweed and on the undersides of stones. Each clump adheres by

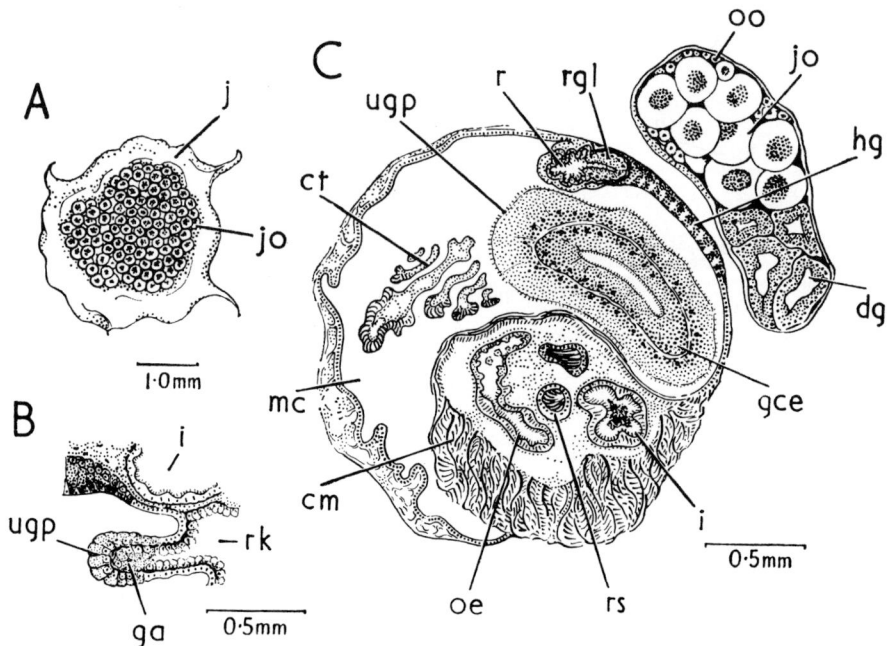

FIG. 160.—*Margarites helicinus*: A, single spawn mass; B, longitudinal section through the urinogenital papilla of a male animal; C, transverse section through a female at the level of the inner part of the mantle cavity; the section also cuts one turn of the visceral hump.

cm, columellar muscle; ct, ctenidium; dg, digestive gland; ga, urinogenital duct; gce, glandular and ciliated epithelium; hg, hypobranchial gland; i, intestine; j, jelly of egg mass; jo, jelly round egg; mc, mantle cavity; oe, oesophagus; oo, oocyte; r, rectum; rgl, rectal gland; rk, right kidney; rs, radular sac; ugp, urinogenital papilla, projecting into mantle cavity.

the outer envelope which is moulded by the foot of the female and produced into anchoring threads. The eggs have gelatinous coats (jo, fig. 160A, C) developed in the ovary and the outer secretion in which each spawn mass is embedded (j, fig. 160A) comes from a gland associated with the urinogenital papilla (ugp, figs. 160C, 161). This, in the female, projects freely into the mantle cavity, with no attachment to the mantle skirt or body wall and the epithelium of tall gland cells which lines it also covers its outer surfaces. In males the papilla (ugp, fig. 160B) is very much smaller.

Calliostoma zizyphinum produces an egg ribbon many times longer than broad, attached at one end. It contains several hundred eggs each about 0·28 mm in diameter, which is about twice the diameter of the eggs of *Gibbula cineraria* and *G. umbilicalis*. This species also shows sexual dimorphism. In the male the left and right kidney apertures lie level with one another at the posterior end of the mantle cavity, but in the female the right one (rko, fig. 162) is considerably further forwards since a glandular section is added which, unlike that of *Margarites*, is in the thickness of the mantle skirt and is assumedly derived from a closed off portion of the mantle. It appears to replace, or even be derived from the posterior part of the right hypobranchial gland; this is reduced in female *Calliostoma* though not in males, and is fully developed in female *Margarites*. The glandular section of the duct in *Calliostoma* (ugp) is

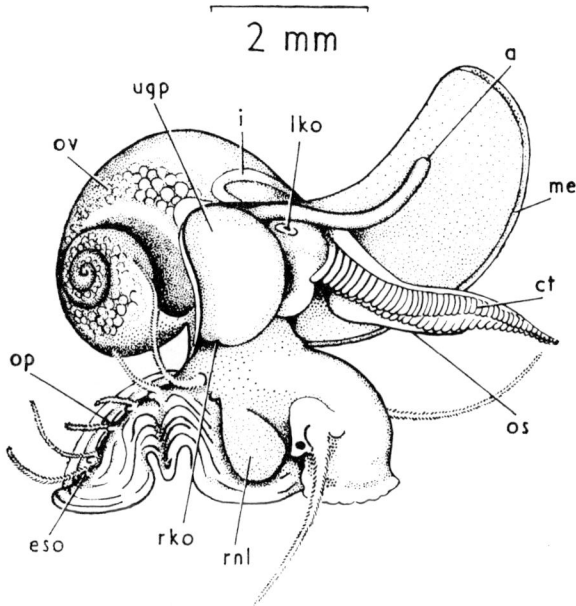

FIG. 161.—*Margarites helicinus:* female, dissected to show contents of the mantle cavity from the right.
a, anus; ct, bipectinate ctenidium; eso, epipodial sense organ; i, intestine; lko, opening of left kidney; me, mantle edge; op, operculum; os, osphradium; ov, ovary; rko, opening of right kidney; rnl, right neck lobe; ugp, urinogenital papilla.

built on the same plan as the pallial oviduct of the monotocardians—lined by columnar ciliated epithelium, the lateral walls deep and thickened by tightly packed bundles of sub-epithelial glands, the dorsal and ventral walls narrow and comparatively thin. The secretion from the duct is mucus and as the eggs pass to the urinogenital aperture, each covered by a gelatinous sheath, a further fluid is poured over them and binds them into an egg ribbon. Both rectum and pallial oviduct lie on the right side of the mantle cavity where, in the lower aspidobranchs, there is a second ctenidium. Thus in *Calliostoma* the female genital tract is made up of (a) the ovarian duct which discharges the eggs into the kidney, (b) part of the right kidney and its duct, and (c) a glandular duct—urinogenital in function—derived from the mantle. This triple origin of the genital duct is the general plan on which that of higher gastropods, both male and female, is built.

The eggs of monotocardians have more elaborate investments from the pallial oviduct which in these prosobranchs traverses the whole length of the mantle cavity on the right side, running parallel with the rectum (fig. 163A). The proximal glandular area (ag) provides the albumen in which the eggs are embedded, and the shell which may surround this. From the more anterior parts (rlc, llg) are secreted the protective outer coverings of the egg mass, which vary in thickness and consistency. Within these coverings the egg develops to the veliger or crawling stage; in no species is there a free trochophore as in the more primitive diotocardians. The pallial oviduct is also elaborated for the reception and storage of sperm since the eggs are fertilized in its upper part; there may be a bursa copulatrix (bcp, fig. 163B) into which the seminal fluid is deposited, and a receptaculum seminis in which the sperm are

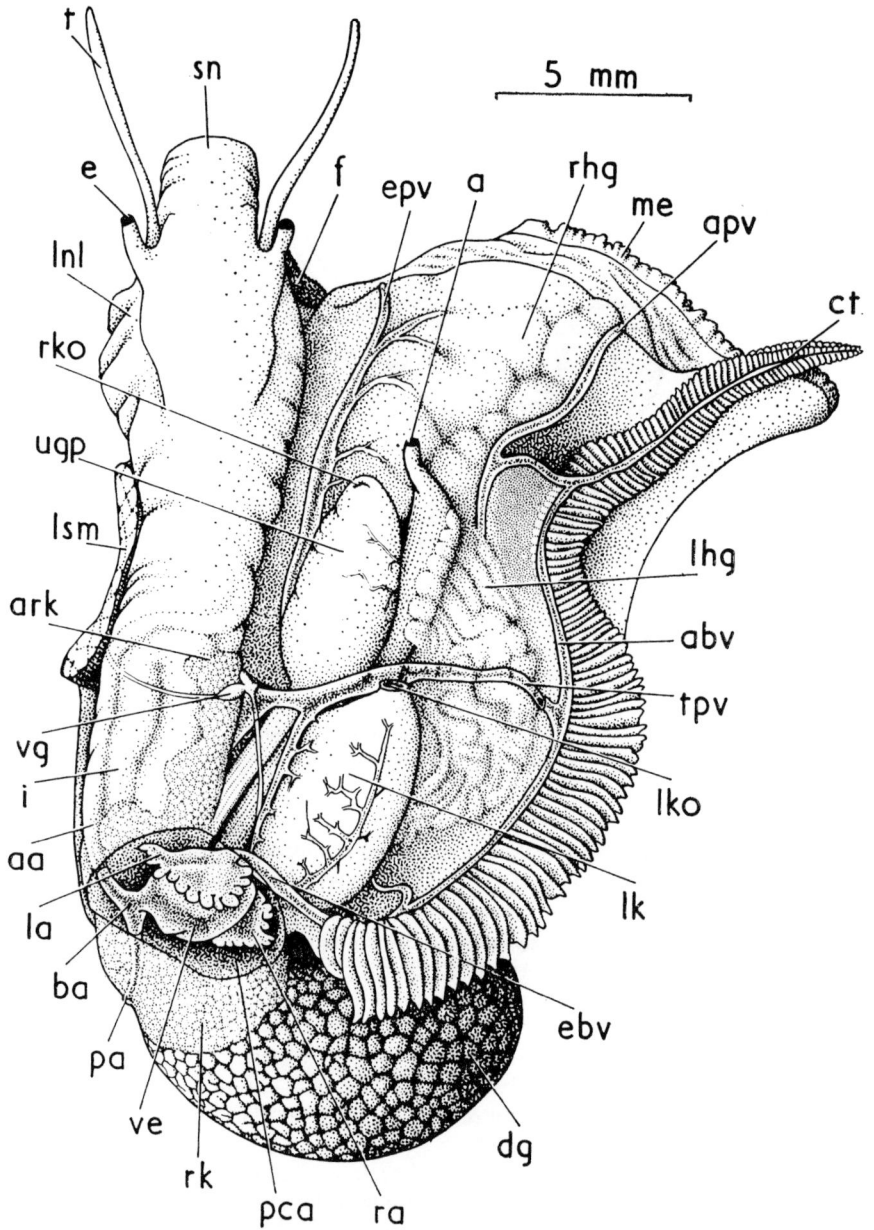

FIG. 162.—*Calliostoma zizyphinum:* female dissected to show the vascular system in the mantle skirt and the contents of the mantle cavity. Some other organs are seen by transparency.

a, anus; aa, anterior aorta; abv, afferent branchial vein; apv, anterior pallial vein; ark, anterior lobe of right kidney; ba, bulbus aortae; ct, ctenidium; dg, digestive gland; e, eye; ebv, efferent branchial vein; epv, efferent pallial vein; f, foot; i, intestine; la, left auricle; lhg, left hypobranchial gland; lk, left kidney (=papillary sac); lko, opening of left kidney; lnl, left neck lobe; lsm, left shell muscle; me, mantle edge; pa, posterior aorta; pca, pericardial cavity opened; ra, right auricle; rhg, right hypobranchial gland; rk, right kidney; rko opening of right kidney (urinogenital); sn, snout; t, tentacle; tpv, transverse pallial vein; ugp, urinogenital papilla; ve, ventricle; vg, visceral ganglion.

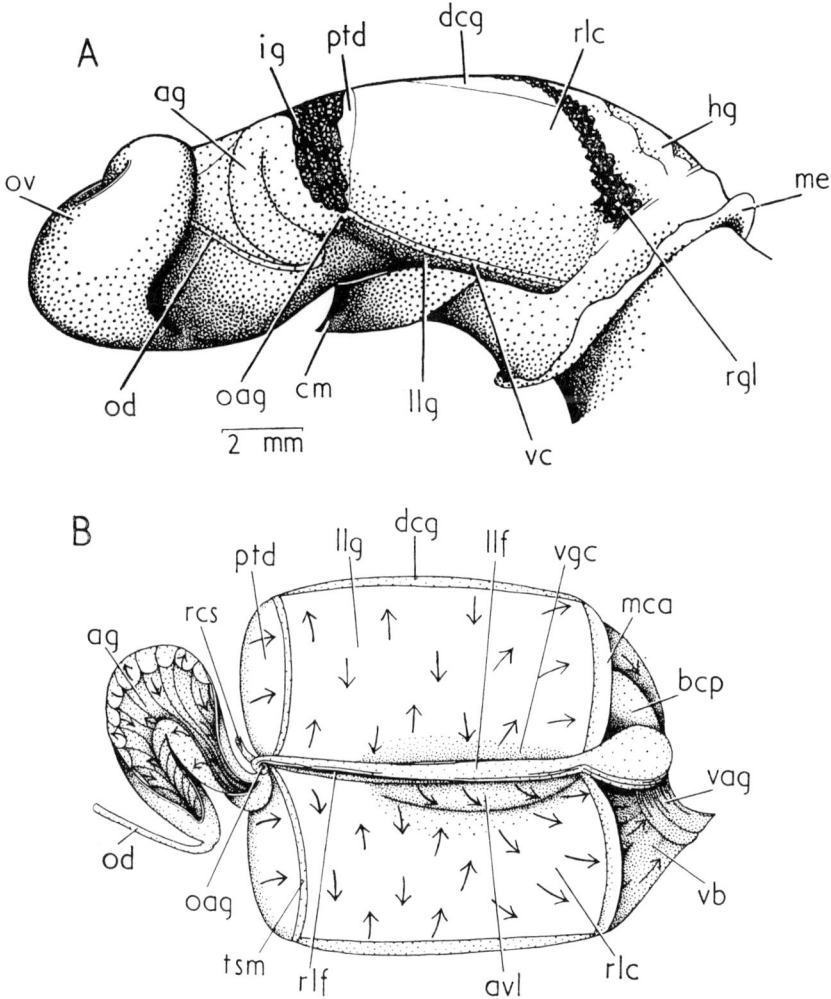

FIG. 163.—*Nucella lapillus:* A, visceral hump seen from right side after removal of the shell; B, female genital duct. The muscular vestibule and the capsule gland have been opened by a dorsal longitudinal incision, and the albumen gland by a longitudinal incision along the right side. Arrows show ciliary currents. The bursa copulatrix and the anterior end of the ventral channel are distended with sperm.

ag, albumen gland; avl, anteroventral lobe of capsule gland; bcp, bursa copulatrix; cm, columellar muscle; dcg, gland cells of dorsal wall of capsule gland; hg, hypobranchial gland; ig, ingesting gland; llf, left longitudinal fold; llg, left lobe of capsule gland; mca, mucous cells of the anterior border of capsule gland; me, mantle edge; oag, opening of albumen gland to capsule gland; od, ovarian duct; ov, ovary; ptd, posterior tip of capsule gland; rcs, receptaculum seminis, into which the ingesting gland opens; rgl, rectal gland; rlc, right lobe of capsule gland; rlf, right longitudinal fold; tsm, transverse muscular strip; vag, vagina; vb, vestibule; vc, ventral channel of capsule gland; vgc, anteroventral gland cells of capsule gland.

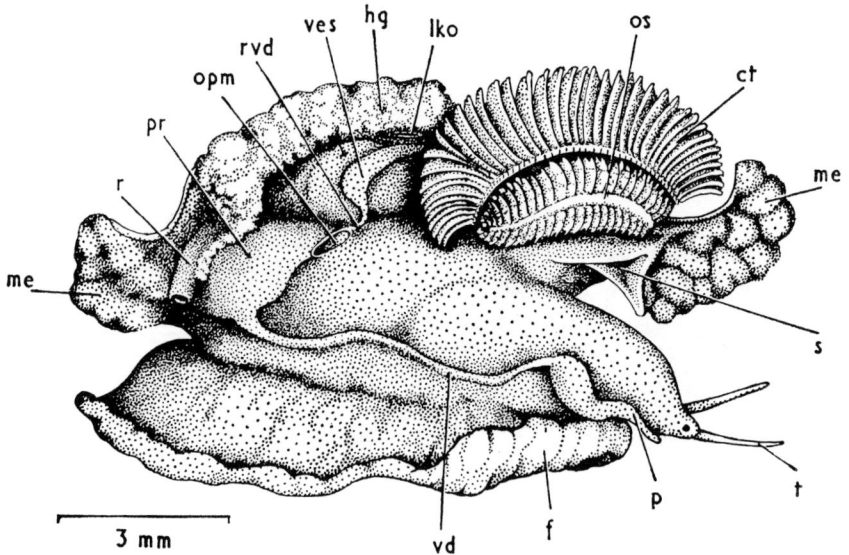

FIG. 164.—*Trivia monacha:* male, dissected to show contents of mantle cavity from the right.
ct, ctenidium; f, foot; hg, hypobranchial gland; lko, opening of left kidney; me, mantle edge; opm, opening of prostate to mantle cavity; os, osphradium; p, penis; pr, prostate; r, rectum; rvd, renal vas deferens; s, siphon; t, tentacle; vd, vas deferens; ves, vesicula seminalis.

stored. The complexity of the male genital duct is associated with the habit of internal fertilization (fig. 164). In most species sperm are transferred to the oviduct by a penis (p), which, save in a few, is pedal in origin and situated behind the right cephalic tentacle, and the vas deferens (vd) opens at or near its tip. The pallial section of the vas deferens may be enlarged and glandular forming a prostate (pr).

The posterior region of the genital duct of monotocardians runs along the columellar side of the visceral hump (od, fig. 163A) and connects the gonad with the pallial duct. It is thin-walled and comparatively narrow. In females of some species, the distal part of this section is in communication with the pericardium by a gonopericardial duct (gpd, fig. 165); in males there may be a vestige of this. The position and structure of this duct and our knowledge of its development in *Viviparus* confirm its homology with the post-torsional right renopericardial duct of Diotocardia. The region of the gonoduct with which it connects is, therefore, derived from the right kidney or its duct, and will be referred to as the renal genital duct (rvo). In female *Littorina* (Linke, 1933) and in the stenoglossans (Fretter, 1941) this region is histologically similar to the gonopericardial duct and is quite short: both have a ciliated epithelium in which the cells are densely packed, and a subepithelial muscular coat. The gonopericardial duct opens into the pericardium by a ciliated funnel around which the musculature is pronounced, and the funnel may be closed off from the pericardial cavity. The initial part of the genital duct which leads from the gonad to the renal section is comparatively long and shows evidence of a gonadial origin, since Linke (1933) has shown that in *Littorina* its epithelium is similar to the undifferentiated cells of the gonad. For this reason it will be referred to as the gonadial duct. In males its lower part stores ripe sperm (ves, fig. 164; td, fig. 166), In Neritacea there is some modification of the connexion between the genital and pericardial

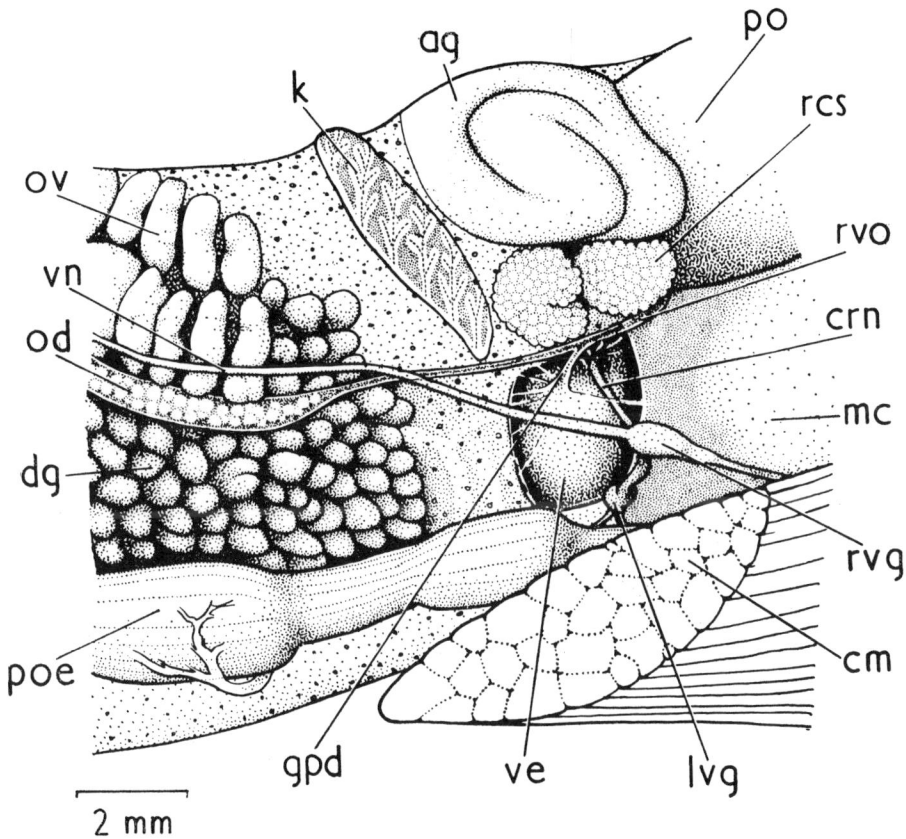

FIG. 165.—*Nucella lapillus:* female, part of base of visceral hump from the right side dissected to expose the pericardial cavity, which has not, however, been opened, the ovarian and gonopericardial ducts.

ag, albumen gland displaced to expose receptaculum seminis; cm, columellar muscle; crn, nerve to heart and kidney; dg, digestive gland; gpd, gonopericardial duct; k, kidney; lvg, left visceral ganglion; mc, mantle cavity seen through mantle skirt; od, ovarian duct; ov, ovary; po, pallial oviduct; poe, posterior oesophagus; rcs, receptaculum seminis; rvg, right visceral ganglion; rvo, renal section of oviduct; ve, ventricle; vn, visceral nerve.

cavities. In this group these cavities are more extensive than in other gastropods and more spacious than the haemocoelic spaces, which tend to be filled with an abundant development of connective tissue. One large cavity, the pericardium, surrounds the heart and is continued into a wide passage from which the renopericardial canal branches; a further large expansion, the gonadial cavity communicates with the oviduct by a conspicuous funnel. This connexion has not been traced in the male. Bourne (1908) stated that there can be no doubt that the gonad has been derived from the wall of the right part of the gonadial cavity.

The origin of the various parts of the reproductive system can be followed in the embryology of one of the most highly organized taenioglossans, which retains in its development an unexpectedly primitive condition. The classical work of Erlanger (1891*b*; 1894) on *Paludina* (= *Viviparus*) *vivipara* showed that there are 3 evaginations from the primary mesodermal sac, which itself will form the pericardial cavity. Two of these are the rudiments of the right and left kidneys, each becoming gradually constricted from the pericardium, though retaining a connexion by way of the renopericardial canal. They lie against the ectoderm, which very soon forms duct-like prolongations of the mantle cavity one towards each: the right one coalesces with the original right kidney (which will become the single kidney of the adult) and forms its duct, whilst the left is arrested in its growth and the corresponding kidney disappears. Both kidney rudiments and their pallial apertures are involved in 180° torsion. The third outgrowth of the pericardium forms later close to the origin of the original left kidney; it is nipped off from the pericardial epithelium and forms a vesicle which is the rudiment of the gonad. At the same time an ingrowth from the mantle cavity, the arrested duct of the kidney which has disappeared, grows towards the gonad and finally fuses with it to form its duct.

A further study of the development of *Viviparus viviparus* was made by Drummond (1903) who agreed with Erlanger's account with one significant exception. She found that both the original left kidney and its duct, which is formed from the mantle, are present at the time when the gonad is formed and from that time decrease in importance. The gonad is at first solid and connected to the kidney by a thickening of the pericardial wall. At a later stage it becomes hollow and its lumen communicates with that of the original left kidney in the region of the thickening, close to the renopericardial aperture. The genital products must therefore pass through the post-torsional right kidney and its duct—though these are relatively insignificant in size as compared with the gonadial and pallial sections of the mature oviduct.

The ducts of the right and left kidneys as shown in the embryology of *Viviparus* are extremely short invaginations from the innermost end of the mantle cavity, yet in the adult the genital aperture and the opening of the ureter are at its mouth. Johansson (1950) has studied a later stage in the development of the genital and renal ducts and has shown that the two invaginations arise from the inner end of a longitudinal groove in the mantle skirt. He thought that the pallial ureter had originated in close association with the pallial genital duct in one of two ways: either the posterior end of the pallial genital groove enclosed the opening of the left invagination of the mantle cavity as well as the right and then closed and divided to form pallial genital duct and ureter, or the ureter has arisen from an external groove along the pallial genital duct. This groove would widen posteriorly to embrace the opening of the invagination forming the initial part of the ureter and then close to form the long pallial ureter of the adult. However, our present interest is in the pallial genital duct, which first appears as a longitudinal groove in the mantle skirt, later closing so that no indication of the open condition remains. Such a change from an open channel to a closed duct probably occurred in the course of evolution of the genital system of the monotocardians. It occurs in the life history of each individual of the protandrous hermaphrodites *Calyptraea*, *Crepidula* and *Capulus*. In the male phase the vas deferens is a narrow, ciliated groove traversing the length of the mantle cavity and passing anteriorly to the penis (sgr, fig. 177A), whilst in the transition to the female phase the pallial part of the groove alone persists; it closes and the walls hypertrophy to form the albumen and capsule glands which bulge into the mantle cavity (po, fig. 190).

The question as to the ultimate fate of the right kidney (apart from its duct) still remains unanswered. Both are arrested in their development, no renal epithelium is differentiated and they form only a very small part of the whole genital tract. Linke (1933) suggested that the receptaculum seminis of Monotocardia, which in many species is placed at the inner end of the pallial duct, may be derived from the duct of the kidney, and although Thiele (1929–35) would agree with this for the *Littorina* spp. (rcs, fig. 26) he suggested that in many other prosobranchs its position is incompatible with this.

The structure of the genital system and the formation and deposition of egg capsules are as well known for *Nucella lapillus* as for any prosobranch, and, since the dog whelk is a very familiar intertidal species passing its whole life history on the shore, its genital system (except the gonad) will be described in some detail. It selects a moist and shady surface of rock for a spawning ground and some 30 or more individuals may congregate there, not feeding for the time. In such a group uncovered by the tide some individuals may be seen copulating whilst others are depositing egg capsules. Experiments show that copulation is repeated at intervals between which a few capsules are laid.

There are no easily visible sexual characters by which the sexes of *Nucella* are distinguishable externally except the penis, which lies behind the right tentacle, though this is difficult to see as it is concealed in the mantle cavity. Pelseneer (1926a) recognized a size difference and stated that in general the broadest specimens are females. If the shell be removed from a mature male (fig. 166) the testis (te) is seen in the upper coils of the visceral mass as numerous tubules lying over and between the lobes of the digestive gland (dg). The tubules join one another and form a common duct, the testicular duct (td), which passes forward along the columellar side in a superficial position. This duct acts as a vesicula seminalis and during breeding its epithelium will ingest and digest the effete, or perhaps superfluous, spermatozoa. At the anterior end of the vesicula seminalis a sphincter closes the entrance to a short, straight, ciliated duct running beneath the intestine and pericardium to the prostate (pr) at the posterior end of the mantle cavity. The duct is surrounded by a layer of circular muscles. It is the renal vas deferens. There is no vestige of a gonopericardial canal as in the closely related *Ocenebra erinacea* in which the renal vas deferens gives off towards the pericardium a short diverticulum, the blind end of which is connected with a slight prominence on the pericardial wall by a band of dense connective tissue and muscle fibres.

The renal vas deferens leads through an entrance guarded by a sphincter to the narrow ventral wall of the prostate, which runs parallel with the rectum (r) to the edge of the mantle skirt. Its lateral walls are thickened by the profuse development of glands which are grouped in clusters beneath the ciliated epithelium, so that in transverse section the lumen appears as a vertical slit. The prostate is closed off from the mantle cavity (mc) except at the posterior extremity near the renal vas deferens, where there is a minute aperture between two flaps of tissue, which are ventral extensions of the epithelium lining the gland and are continuous with the inner epithelium of the mantle skirt. In front of the opening these flaps fuse with one another, but the line of fusion remains distinct as two opposing strips of epithelial cells (fig. 173J). From the anterior end of the gland a narrow duct (vd) passes along the right side of the head to the penis (p) which lies behind the right cephalic tentacle. Along the whole length of the duct can be seen the two strips of epithelium which fused to close it, evidence of its derivation from an open seminal groove such as is found in *Littorina* (sgr, fig. 12). Only one type of secreting cell occurs in the prostate gland, opening to the lumen between the ciliated cells and providing a fluid in which the sperm are discharged. The cilia cause an anteriorly directed current, which probably has little effect on the forward passage of seminal fluid, which is more forcibly moved by the peristalsis of the anterior section of the duct leading to the penis. This duct is narrow and runs in the body wall along a slight ridge; it is lined by columnar ciliated cells and surrounded by a thick layer of circular muscles.

The penis (p, fig. 166) is flattened, and its duct is not centrally placed but lies towards the posterior edge. The histological structure of the epithelium is similar to that of the preceding section of the vas deferens. At its tip the two layers of fused epithelium which close the duct separate to form the penial aperture. A layer of circular muscles lies beneath the epithelium

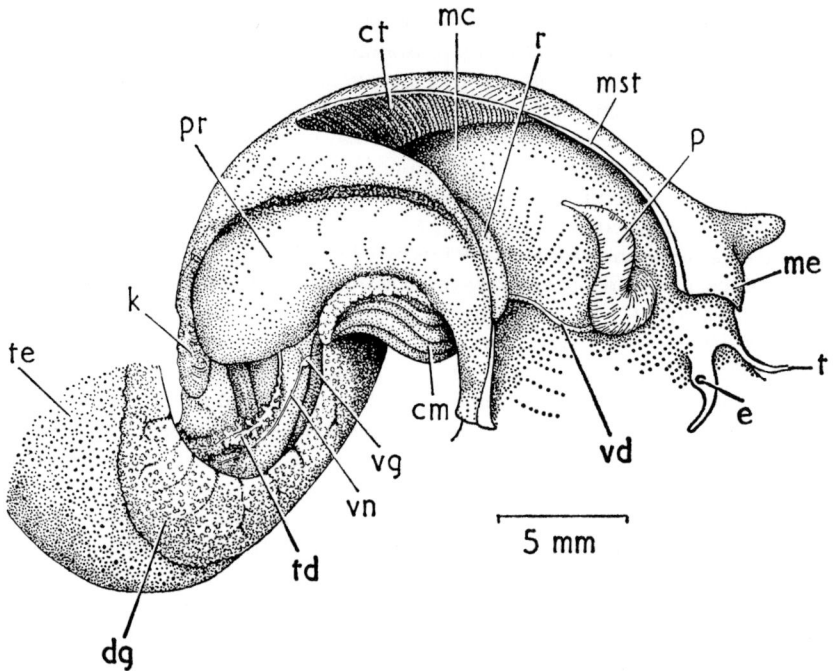

FIG. 166.—*Nucella lapillus:* male, removed from shell and partly dissected to show genital duct.
　　cm, columellar muscle; ct, ctenidium; dg, digestive gland; e, eye; k, kidney; mc, mantle cavity opened; me, mantle edge; mst, cut edge of mantle skirt; p, penis; pr, prostate gland; r, rectum; t, tentacle; td, testis duct acting as vesicula seminalis; te, testis; vd, vas deferens leading from prostate to penis; vg, visceral ganglion; vn, visceral nerve.

of the duct and extends beneath the fused strips of epithelium. Throughout the thickness of the penis vascular spaces are numerous, and between them run muscle fibres which are in association with a thick coat of muscles beneath the outer epithelium; the penial nerve, which arises from the right pedal ganglion, is centrally situated.

　　In the female the ovary (ov, figs. 163A, 165) spreads over the surface of the digestive gland in the visceral mass and when mature it may attain a quarter of the total body weight. The majority of the eggs which it produces will provide the food within the egg capsule for the embryos, which in this way obtain sufficient nourishment to complete development and emerge in the crawling stage. After a period of spawning the ovary passes to a period of rest and yellow-brown inclusions appear which indicate the resorption of unused oocytes (Kostitzine, 1934); the ripe ones disintegrate and are attacked by phagocytes (absorption of unused ova is also recorded for *Turritella communis* by Pérez & Kostitzine (1930)). From the ovary a thin-walled ovarian duct (od, fig. 165) leads forwards and ventrally on the right side of the viscera, and can be seen through the integument anteriorly. It passes beneath the kidney (k) where the glandular lining of the duct is replaced by a columnar, ciliated epithelium thrown into longitudinal folds by variations in the thickness of the underlying connective tissue. This is the renal section (rvo), which leads to the albumen gland (ag) at the posterior end of the mantle cavity. As it approaches this gland it is joined by a gonopericardial duct (gpd) of similar histological structure. The gonopericardial duct opens to the pericardium by an inconspicuous ciliated funnel, and around this opening the musculature is well developed so that the passage can be closed. A sphincter also surrounds the opening of the renal oviduct into the albumen gland.

The pallial oviduct (fig. 163A) lies in the thickness of the mantle skirt on its extreme right and bulges into the mantle cavity; in young females (1·2–1·5 mm long) it is seen to arise as a longitudinal gutter of the pallial wall (Kostitzine, 1949). A mature individual removed from its shell and viewed from the right side shows the albumen gland (ag) as a U-shaped loop at the posterior end of the duct, with the concavity of the gland directed ventrally, and the distal limb opening to the much larger capsule gland. Between the two glands is a mass of deep brown tubules which will be referred to as the ingesting gland (ig), and its duct as the receptaculum seminis. The capsule gland leads forward to the anterior end of the mantle cavity and, on dissection, it can be seen as an opaque white or yellowish mass divided into right and left lobes (rlc, llg). These are joined dorsally and ventrally by a comparatively thin and narrow wall forming dorsal and ventral sutures (vc), so that in transverse section the lumen of the gland has the appearance of a dorsoventral slit. From the narrow ventral wall, where gland cells are absent, arise two longitudinal folds of tissue which form a channel between them. This can be shut from the lumen of the capsule gland so that it becomes a functionally closed duct. It leads posteriorly to the duct of the ingesting gland, the receptaculum seminis (rcs, fig. 163B), and anteriorly to a bursa copulatrix (bc). The bursa is a pouch into which the penis discharges sperm and prostatic fluid, and the receptaculum seminis an area near the site of fertilization where the sperm may be stored for a longer period, orientated and nourished by the female. The albumen gland (ag) opens into the posterior ventral wall of the capsule gland on the right side of the ventral channel (oag). Anteriorly the capsule gland leads to a muscular vestibule (vb) through which the ventral channel passes to the bursa copulatrix. When filled with sperm the bursa bulges into the vestibule. The short vagina (vag), into which the penis is inserted to deposit sperm into the bursa, and from which the fully formed egg capsules pass from the vestibule to the exterior, opens on the right anterior extremity of the mantle cavity, ventral to the anus; the opening is surrounded by a sphincter.

The pallial oviduct has a ciliated epithelium (ci, fig. 167A), except in the ventral channel in which the epithelium is columnar and unciliated. In the albumen and capsule glands the walls are thickened by subepithelial gland cells grouped in clusters, the ducts from each cluster running parallel with each other to open between the ciliated cells. In the albumen gland there are 3 kinds of secreting cells and their secretions combine to form the albuminous fluid in which the eggs are embedded.

If the capsule gland is cut open mid-dorsally (fig. 163B) it will be seen that the right lateral wall has a ventral longitudinal cleft separating an anteroventral lobe (avl) from the rest of this glandular area. This lobe, together with two adjacent longitudinal strips of tissue (vgc), is more translucent than the surrounding area and of a slightly yellowish hue. The subepithelial gland cells are all similar and produce 2 different secretions which occur in the cells as 2 different types of spherules, one protein and the other muco-protein. An area along the dorsal wall (dcg) of the capsule gland has a similar appearance in the living state, and the gland cells here also produce 2 secretions. Near the posterior end of the capsule gland 2 narrow strips of tissue (tsm), one on either side and arising near the opening of the albumen gland, separate right and left posterior tips (ptd) from the main mass of the gland. Beneath the ciliated epithelium of the strips is a layer of circular muscles and a few muscles radiate outwards: it is this musculature which distinguishes the strips and constricts them from the surrounding

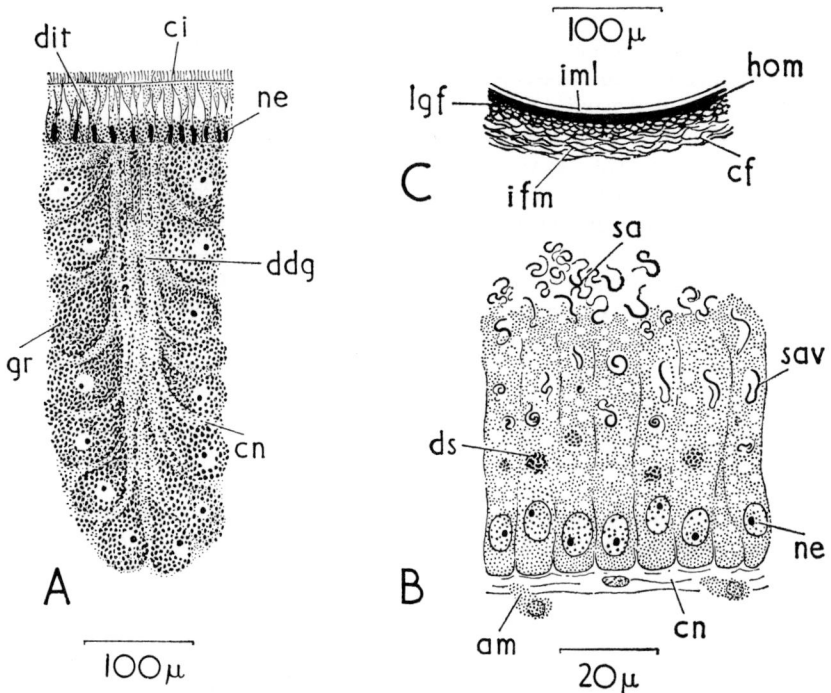

FIG. 167.—*Nucella lapillus:* parts of transverse sections through A, the wall of the capsule gland, showing single cluster of gland cells; B, the ingesting gland; C, the wall of an egg capsule.

am, amoebocyte; cf, circular fibres; ci, cilia; cn, connective tissue; ddg, duct of deep gland cell; dit, distal tip of duct of subepithelial gland cell filled with mucoid protoplasm; ds, sperm undergoing digestion; gr, granular protein secretion; hom, homogeneous layer; ifm, interfibrillar mucoid matrix; iml, inner mucous layer of capsule; lgf, longitudinal fibres; ne, nucleus of ciliated cell; sa, spermatozoa; sav, spermatozoon in vacuole of cytoplasm.

wall. Both the strips and the posterior tips of the gland are made up of mucous cells, and similar cells border the anterior extremity of each lobe (mca). The gland cells (fig. 167A) constituting the main mass of the capsule gland contain colourless granules of an irregularly oval shape, and again produce 2 types of secretion, an amorphous mucoid secretion (dit) and a protein secretion in granules (gr).

Hence all the gland cells, except those of the posterior tip and the anterior border of each lobe of the capsule gland, produce a double secretion and it is the intertwining of these which is responsible for the fibrous structure of the wall of the egg capsule.

The vestibule, vagina and bursa copulatrix are clothed by a columnar, ciliated epithelium and are extremely distensible. The walls of the bursa are thrown into folds of various depths and sperm are everywhere attached to the epithelium, while more may fill the lumen; they have been deposited here by the penis. Excess sperm may be found in the ventral channel closely packed together, with their heads embedded in the distal cytoplasm of the epithelium, and their tails projecting into the lumen. Somehow or other they reach the receptaculum seminis (rcs, fig. 163B), the walls of which are surrounded by a very thick coat of circular muscles and have masses of sperm invariably attached to their columnar epithelial lining. The receptaculum leads into the ingesting gland (ig, fig. 163A) which is composed of blind tubules and has the appearance of a digestive gland (fig. 167B). In the lumina of the tubules large numbers of unorientated sperm (sa) may be present, but some lie motionless against the free surface of the epithelium as if trapped. These sperm are engulfed by the cells, where they lie in vacuoles (sav) in the cytoplasm and are digested (ds): it seems probable that the products of their digestion give the gland its brown colour. Blood spaces with amoebocytes (am) surround the gland, from which the amoebocytes appear to carry waste. This gland may serve as a mechanism for ridding the animal of unwanted sperm and, perhaps, of deriving nourishment from them.

The method of functioning of the pallial oviduct may best be considered after a brief description of the egg capsule. This is a vase-shaped structure about 8 mm high (fig. 234A). It is circular in transverse section and broadest in the middle where its diameter is about 2 mm. At one end the capsule tapers to a short stalk and then expands into a basal disc (b) which is firmly anchored by cement to the substratum. At the opposite end there is a circular aperture which is filled with a plug of mucus (pl). Two longitudinal lines of thickening (su) can be traced over the smooth surface, and these are placed so as to divide the wall into two approximately equal lobes. Distally the suture of one side meets that of the other through the plug which is thus subdivided. The wall consists of 3 layers (Ankel, 1937b) in addition to a thin mucous sheet (iml, fig. 167C) which covers it internally, separating it from the albumen and eggs. The three comprise an inner layer of a homogeneous transparent substance (hom), probably conchiolin, and middle and outer layers of fibres orientated in definite ways and separated by distinct spaces. The inner fibres appear to run in a longitudinal direction (lgf) and the outer ones in a circular direction (cf). Sections of newly formed capsules show the interfibrillar spaces filled with a mucous or mucoid substance (ifm), but when the capsule has weathered for some time this tends to contract leaving the spaces between the fibres observed by Ankel (1937b). In the stem and basal disc the fibres are more irregularly arranged. [See p. 665.] The capsule contains several hundred eggs embedded in an albuminous fluid: only about 15–30 hatch, the remainder being devoured by their fellows (Lebour, 1937).

If an egg capsule is found within the capsule gland the plug is seen to be at the upper end, near the posterior mucous tips, and the base anterior. From this it can be deduced that the plug is secreted by the posterior mucous tips and the capsule wall by the rest of the gland. The method of formation of the capsule would appear to be as follows. Several hundred eggs pass down the gonadial duct and distend the albumen gland, where they are embedded in albuminous fluid. The exact site of fertilization is unknown: if sperm from the receptaculum and the ventral channel are passed into the albumen gland on the relaxation of the sphincter muscle which guards its opening, then fertilization will occur in the lumen. Otherwise spermatozoa may be poured on to the eggs as they enter the capsule gland. Meanwhile the cavity of the capsule gland becomes filled with a mass of secretion. Its cells produce first the protein matter and then, in increasing quantities, the mucoid substance, and the result is that the lumen is filled with fluid which is almost pure protein in the centre, but consists, in its outer layers, of an emulsion in which the mucoid material is dispersed in a continuous phase of protein. Because of the steadily increasing quantities of mucus and of the accompanying decrease in the production of protein, the size of the mucoid droplets steadily increases from the centre to the periphery of the mass, and the strands of protein which separate the drops gradually decrease in size. The predominantly transverse ciliary currents on the walls of the gland rotate the mass and draw out the drops of mucoid material into streaks parallel to the transverse axis of the gland, and the protein material separating them is drawn out into strands elongated in the same direction. When the duct between the albumen gland and capsule gland is opened, the albumen and the eggs are forcibly passed into the central portion of the secretion lying in the latter, so as to invaginate this into a vase-shaped structure with a round hole at the inner end into which the plug will later be fitted. This process of invagination deposits the eggs and their accompanying albumen in the centre of the mass of secretion occupying the capsule gland, which is composed of unmixed protein material, and from this is formed the innermost layer of the capsule. It has also the effect of drawing the outer emulsion of protein and mucoid secretion into sheets around the inner homogeneous layer in which the direction of elongation of the mucoid droplets now lies parallel to the direction of movement of the eggs, that is, parallel to the long axis of the gland: from this results the longitudinal direction of the strands of the inner part of the fibrous coat. The outer part of this layer, being still exposed to the ciliary currents on the wall of the gland, retains the original alignment of the drops and so gives rise to the outer part of the fibrous coat in which the fibres are circular in direction, except for material which gives rise to the basal disc. This is accumulated at the anterior end of the capsule gland where the direction of the ciliary beat is mainly anterior. With the disappearance, on exposure, of the mucous dispersed phase the space occupied by it is left as a series of lacunae separating what now appear as strands of fibrous material. The mucus which lines the wall internally is secretion from the posterior tips, and was dragged along with the mass of eggs when it passed into the capsule gland. These tips continue to pour out secretion while the wall of the capsule is being elaborated by the more anterior parts of the genital duct. The mucus forms an accumulation which is fitted into the hole in the upper part of the capsule apparently by the muscular action of the transverse strips which border the mucous tips anteriorly. These press the upper edge of the wall of the capsule on to the mass of mucus so that the cavity within is securely closed. The suture which divides the plug into 2 equal halves demarcates the limit of the secretion produced by each posterior tip.

The egg capsule is passed from the oviduct along a temporary groove of the foot to the sole and is inserted in the ventral pedal gland which lies immediately behind the accessory boring organ. It is held in an approximately vertical position with the plug innermost and the gland embraces the capsule tightly so that only the base protrudes from its opening. Whilst held in this position the wall of the capsule is compressed and moulded to the final, smooth, vase-shaped outline. The stalk is constricted from the basal region and the latter finally pressed out to form a disc and fixed to the substratum by the sole of the foot. Mucus is secreted by the subepithelial cells of the gland to act as a lubricant during the fashioning process. Finally the foot is lifted off the capsule and the wall of conchiolin hardens still further in contact with sea water.

Variation in the structure of the reproductive ducts within the monotocardians is chiefly concerned with the degree of closure of the pallial section and differences in its glandular equipment, the structure of the penis, and, in the female, the number and position of the sperm pouches. Owing to its comparative simplicity in structure the male system exhibits fewer differences from species to species than does the female and will be considered first.

In some Monotocardia the testis produces more than one type of sperm which develop in the same tubules: eupyrene sperm (fig. 168) which fertilize the eggs, and have nuclei containing the haploid number of chromosomes, and others which are atypical and may be hyperpyrene (with more than the haploid number), dyspyrene or oligopyrene (with only part of the haploid set), or apyrene (enucleate). In the diotocardian gastropods with external fertilization the eupyrene sperm (fig. 168) are similar in structure to those of the more primitive Metazoa

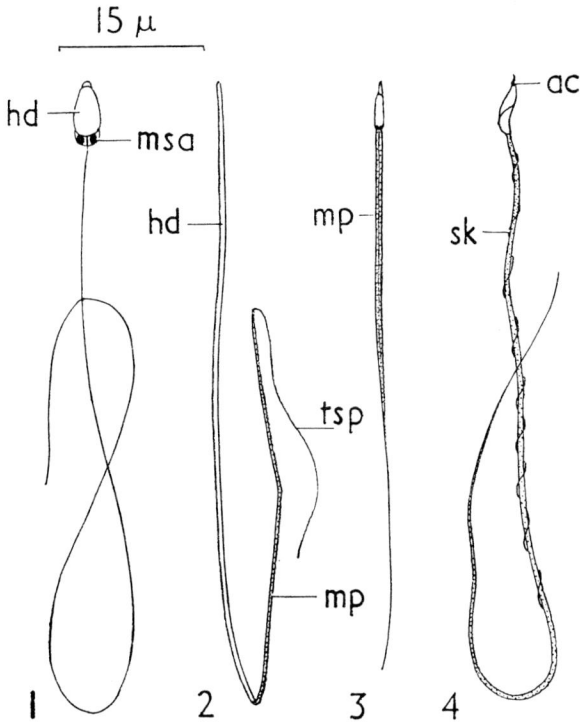

FIG. 168.—Eupyrene sperm: 1, *Gibbula cineraria*; 2, *Cingula semistriata*; 3, *Bittium reticulatum*; 4, a pyramidellid. After Franzén.
    ac, acrosome; hd, head of sperm; mp, middle piece of sperm; msa, mitochondrial sphere in middle piece of sperm; sk, spiral keel on sperm; tsp, tail of sperm.

(Franzén, 1956). Each consists of a short head (hd) containing the nucleus, which is typically oval in outline, a middle piece of 4 or 5 mitochondrial spheres (msa) usually lying transversely across the base of the head and embedded in an insignificant mass of protoplasm, and a tail composed of a long filament. Except for a terminal region the tail filament is covered by a thin layer of undifferentiated cytoplasm. In the Monotocardia the eupyrene sperm differ from this. In them the mitochondria collect around the base of the nucleus in 4 large spheres as in the Diotocardia, but later they disperse to form a cylindrical covering over a part of the axial filament, and a more or less elongated middle piece is formed (mp). The electron microscope has revealed that in *Viviparus viviparus* the cylindrical covering comprises 4 tape-like fibres, spirally wound, and that the axis of both middle piece and tail is composed of about 7 fibres (Hanson, Randall & Bayley, 1952). A variable amount of change also occurs in the nucleus: in some species (*Bittium reticulatum*, fig. 168, 3) it remains relatively short, forming about 1/20 of the total length of the sperm (Franzén, 1955), whereas in others it lengthens so that both head and middle piece are filiform, and in *Cingula semistriata* (fig. 168, 2) the nucleus comprises about half the length of the sperm. The length of the tail varies considerably: it may be longer than the middle piece, as in *Clathrus clathrus* and *Crepidula fornicata*, or shorter as in *Viviparus viviparus*. These modifications of the eupyrene sperm of the Monotocardia would appear to be associated with internal fertilization, for they are found alike in species in which copulation occurs, in which spermatozeugmata occur (*Clathrus clathrus, Ianthina* sp. (Ankel, 1926)) and species with neither penis nor spermatozeugmata (*Turritella communis* (Bataillon, 1921), *Bittium reticulatum* (Franzén, 1955)), the only common feature being internal fertilization. Within the female the sperm are stored in a limited area. They are packed parallel with one another with no apparent space between, their heads embedded in the epithelium of the storage area. From this epithelium they may derive nourishment. Undoubtedly the shape of the monotocardian spermatozoon facilitates close packing, and other differences from the sperm of Diotocardia may also be associated with this period of suspended activity within the female duct. In the receptaculum seminis of *Theodoxus fluviatilis* the orientation of the sperm is unusual for it is the tails and not the heads which are embedded in the epithelium.

In opisthobranchs and pulmonates the structure of the mature spermatozoon is still further modified. The boundary between middle piece and tail is indistinct for these two regions consist of a central axis around which a thin membrane of cytoplasm is coiled in a spiral, and there is usually a spiral ridge on the surface of the nucleus. The pyramidellids (fig. 168, 4) have this type of sperm with a spiral keel (sk), giving further evidence of their opisthobranch affinity. Also in their sperm the nucleus is short as it is in a number of primitive opisthobranchs, including *Acteon tornatilis,* whereas in other opisthobranchs the nucleus is elongated, though short heads also occur in some nudibranch sperm (Franzén, 1955). The primitive features of the diotocardian spermatozoon cannot be traced in the stages of spermatogenesis of euthyneurous gastropods and pulmonates.

The spermatocytes which give rise to the atypical forms of spermatozoa show abnormal growth and the maturation division is distorted or absent (Schitz, 1920a, b); sooner or later the chromatin, at least in part, passes into solution and the centrioles undergo multiple division. In *Bithynia tentaculata* 2 atypical types of sperm have been recognized (Ankel, 1924, 1933), hyperpyrene and oligopyrene; in other species of *Bithynia* only one has been described. In the Scalidae (Ankel, 1926, 1930a) the atypical spermatocyte grows to resemble a young oocyte and at the time of solution of the chomatin about 2,000 centrioles appear which elongate to form fibres; these give rise to a plate extended posteriorly into a tail. From one cell 50 μm in diameter a structure 900 μm is developed which acts as a nurse to thousands of eupyrene sperm attached along the tail in spiral rows. Such compound structures, which Ankel called spermatozeugmata, may be found in the vesicula seminalis; when transferred to sea water each swims by the undulating movements of the anterior plate. It may be that in this way the functional sperm can travel longer distances. Similar spermatozeugmata are developed in *Ianthina* and *Cerithiopsis* (fig. 169). In *Cerithiopsis* the eupyrene sperm are arranged along the tail of the atypical spermatozoon in longitudinal rows. In other prosobranchs the atypical sperm do not act as nurses to the eupyrene forms and resemble them more closely in structure and size—in fact in *Bithynia tentaculata* the two are similar in appearance (Ankel, 1933). In *Viviparus* the atypical sperms are vermiform and motile, having 8–16 tails apiece (Meves, 1903; Hanson, Randall & Bayley, 1952); each has one chromosome, and the head lies like a thimble over the anterior end of the long and rather thick middle piece which is enclosed in a sheath containing a polysaccharide. The proximal centrioles lie in the apex of the head, and the distal ones at the base of the brush tail. The two types of spermatozoa, eupyrene and apyrene, occur in about equal numbers in the vesicula seminalis and are transferred to the receptaculum of the female in this proportion. There is then a gradual reduction in the number of apyrene sperm since they have the shorter life (Ankel, 1925). In *Nassarius* and *Fusus* (Pelseneer, 1935) the oligopyrene sperm are fusiform with little power of movement. In *Nucella lapillus* the two types are identical to look at when ripe though they have undergone different maturation processes (Portmann, 1931*b*) so that the atypical form has an abnormal chromatin content.

Sperm dimorphism occurs in a number of families (Viviparidae, Hydrobiidae (Bithyniinae), Turritellidae, Aporrhaidae, Cerithiidae, Scalidae, Calyptraeidae, Capulidae, Lamellariidae, Eratoidae, Muricidae, Buccinidae, Nassariidae and Fasciolariidae) yet its significance remains obscure. It has been suggested that the typical sperm of *Viviparus* provide nourishment for the functional ones (Hanson, Randall & Bayley, 1952) since the sheath of the middle piece disintegrates in the female duct liberating the polysaccharide. However, in other species their food reserves may be required for more important activity than this, for it is tempting to think that spermatozeugmata fulfil the function of a penis where that is absent (Cerithiopsidae, Scalidae, Ianthinidae). In *Ianthina janthina* there is neither penis in the male nor bursa copulatrix and receptaculum seminis in the female, and degenerating spermatozeugmata liberating clouds of eupyrene sperm have been seen in all parts of the female duct from the mucous gland to the ovary (Graham, 1954*b*). This pelagic prosobranch may be a protandrous hermaphrodite which passes through more than one breeding cycle in its life history and until our knowledge is more complete the possibility of self-fertilization cannot be overlooked. As it has neither active locomotion nor penis cross-fertilization is difficult, but the gregarious habit, keeping the animals in close contact, and the powerfully swimming spermatozeugmata probably permit it to occur. Moreover Wilson & Wilson (1956) described the liberation of aggregates of spermatozeugmata from a single animal in captivity which supports the suggestion that the eupyrene sperm are transported from a male to a female individual. As yet no spermatozeugma has been found in the female reproductive ducts of *Cerithiopsis* or the protandrous hermaphrodite *Clathrus*; in contrast to *Ianthina* the pallial oviduct is open and has a receptaculum seminis.

There is a record of apyrene sperm in the cytoplasm of eggs of *Aporrhais pespelicani* 20 minutes after the egg had been fertilized by normal spermatozoa, though they were then ejected (Kuschakewitsch, 1910). In his experiments on artificial fertilization of this prosobranch Kuschakewitsch found that only the eupyrene sperm orientated themselves with respect to the egg; the entrance of an atypical spermatozoon seems fortuitous.

It is of interest to note that in the Turbellaria there is more than one type of spermatozoon, all functional, though a given species has only one kind. In some species each has a filamentous head and 2 flagella (*Dendrocoelum lacteum*), in others each sperm has 2 projecting bristles, one on each side (*Macrostomum*), whilst in the alloiocoel genus *Plagiostomum* the spermatozoon is broad with wing-like lateral extensions reminiscent of the atypical sperm of prosobranchs.

The origin of apyrene spermatozoa and spermatozeugmata may perhaps be traced to cells which were described by Ankel (1930*c*) in species of *Littorina*. These cells, derived from the germinal epithelium of the testis, measure about 10 μm in diameter, are loaded with yolk granules and the eupyrene sperm bore into them with their heads. Sometimes the whole body moves by co-ordinated beating of the sperm tails, but usually they are totally quiescent. In the ducts of the male this is the way in which the sperm are usually found: it is only after transfer to the bursa copulatrix of the female that the two separate, the one degenerating and the sperm re-attaching themselves by their head to the epithelium of the bursa. A somewhat similar association between sperm and nutritive cells occurs in the bivalves *Montacuta ferruginosa* and *M. substriata* (Oldfield, 1959). [For further information on sperm types see p. 665.]

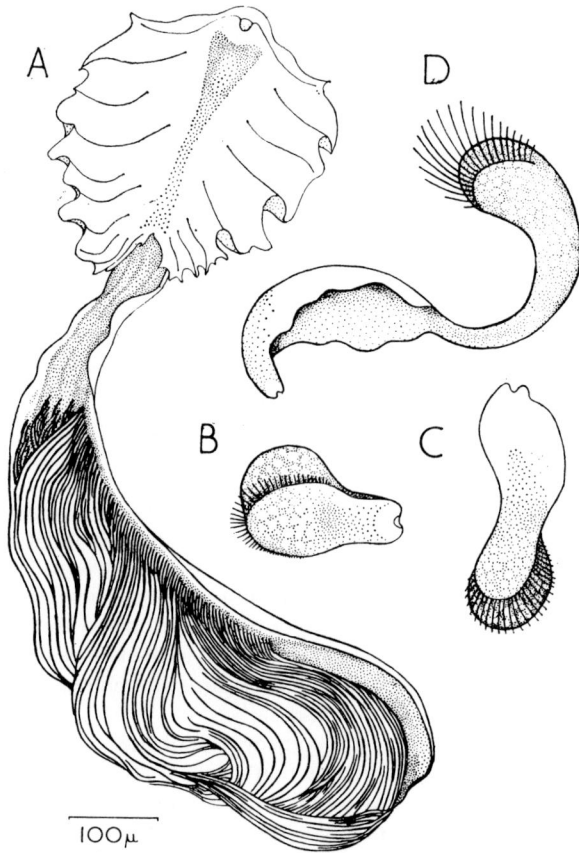

FIG. 169.—*Cerithiopsis tubercularis:* A, spermatozeugma with longitudinal rows of eupyrene spermatozoa on its tail; B, C, D, developing spermatozeugmata.

The epithelium of the vesicula seminalis in *Littorina* ingests a certain number of the sperm and nurse cells, and digests them. Ingestion continues during the whole period of sexual activity, suggesting that the production of sperm is in excess of requirements. Perhaps only senile sperm are removed in this way, and by such a selective mechanism an effective stock is maintained within the duct; or the absorption may be haphazard, and merely a means of safeguarding the duct against blockage. A similar ingestion of sperm by the epithelium of the vesicula seminalis occurs in the Stenoglossa (Fretter, 1941).

In the egg capsules of some prosobranchs there may be found abortive ova known as food eggs, which are eaten by the normal embryos. They are ova arrested in development and commonly unfertilized (Staiger, 1951) as in *Pisania maculosa* (fam. Buccinidae), *Fasciolaria tulipa* and *F. lignaria* (fam. Fasciolariidae), or fertilized as in *Nucella lapillus, Buccinum undatum* and *Murex trunculus,* though with no subsequent syngamy of the male and female pronuclei. In both *F. lignaria* and *M. trunculus* these ova undergo cleavage resulting in groups of cells with haploid nuclei, and Portmann (1925) described their segmentation in *N. lapillus*. All these rachiglossans have two types of sperm, and it has been suggested that the cause of sterility of the food eggs is their atypical fertilization by oligopyrene sperm (Portmann, 1927, 1931a). In a study of *F. tulipa* Hyman (1923) (not realizing that the majority of the food eggs in this species are unfertilized) supported this theory on the grounds that the proportion of eggs to nurse cells (1 : 59) is remarkably close to that of eupyrene to oligopyrene sperm (1 : 50). However, this cannot be the only cause of the origin of sterile ova, for in *Natica catena* only one type of spermatozoon has been traced yet numbers of food eggs are found in the egg capsule (Ankel, 1930b). This theory fails to take into account possible differential survival of sperm types within the female. To what extent atypical sperm survive there is not known; in *Crepidula onyx* they do not enter the receptaculum seminis (Coe, 1942), but disintegrate in the cavity of the pallial oviduct. A second suggestion as to the origin of food eggs we owe to Glaser (1906) who drew a parallel, later emphasized by Burger & Thornton (1935), between the production of two types of ova and the two types of sperm and considered that the cause of sterility was inherent in the behaviour of the ova. This view has been upheld by Staiger (1950c). He has shown that in *Pisania maculosa* only 2% of the eggs undergo normal development; 90% are not fertilized (even though they must pass the site of fertilization on their passage through the oviduct), suggesting that the mechanism by which germ cells are brought together does not operate in their case. The remainder which are fertilized and will become food eggs have their development arrested at 4 different stages along the course of development of the normal eggs; in one of these polyspermy occurs. These facts indicate that within one species the sterile ova do not form a homogeneous population. They are perhaps genetically determined and their sterility due to a heterozygotic factor system. This suggestion is supported by the constant proportion of all types of eggs in capsules of any individual and in broader limits within any one species. For instance in *Murex trunculus* all ova are fertilized by eupyrene sperm and only 4% develop normally. Of the others 6% undergo no cleavage and the others divide whilst the male pronucleus remains in the vegetative area; it never reaches the animal pole and the parthenogenetical development is soon arrested. In *Buccinum undatum* it is seldom that an egg is unfertilized and no atypical sperm are involved. Abnormalities in eggs appear after fertilization so that the suggestion that there may be two types of eupyrene sperm only one of which would stimulate the female nucleus, is still a plausible one. About 75% of the eggs in a capsule stop development at the metaphase of the first maturation division and then degenerate according to Staiger (1951); the remaining 25% complete maturation but only in 1% does syngamy occur and normal development follow.

Polyspermy is usually pathological, resulting in irregular cleavage. In *Fasciolaria lignaria* it is relatively common and in one capsule up to 77 eupyrene sperm have been found in a single egg whilst the majority of the food eggs were unfertilized (Staiger, 1950c). Eggs entered by two sperm are known, however, to undergo normal development since in *Murex trunculus* (Staiger, 1951) one spermatozoon may remain in the yolk, paralleling the behaviour of that entering a food egg, and the other fertilize the ovum.

In a number of mesogastropods the vas deferens anterior to the renal section is an open groove throughout its length to the tip of the penis (fig. 173B, C); the species in which this occurs are not necessarily related. An open duct is found in *Littorina littorea* (sgr, figs. 12, 13), *L. littoralis* and *L. saxatilis* (sgr, fig. 170A) (Linke, 1933), yet in the closely related *Cremnoconchus syhadrensis* (Linke, 1935b), a freshwater Indian form, and the terrestrial *Pomatias elegans*, the duct is closed. It is open in the advanced genera *Aporrhais* (sgr, fig. 171A), *Balcis* and *Natica* (sgr, fig. 172) and in the male phase of the protandrous hermaphrodites *Calyptraea, Crepidula*

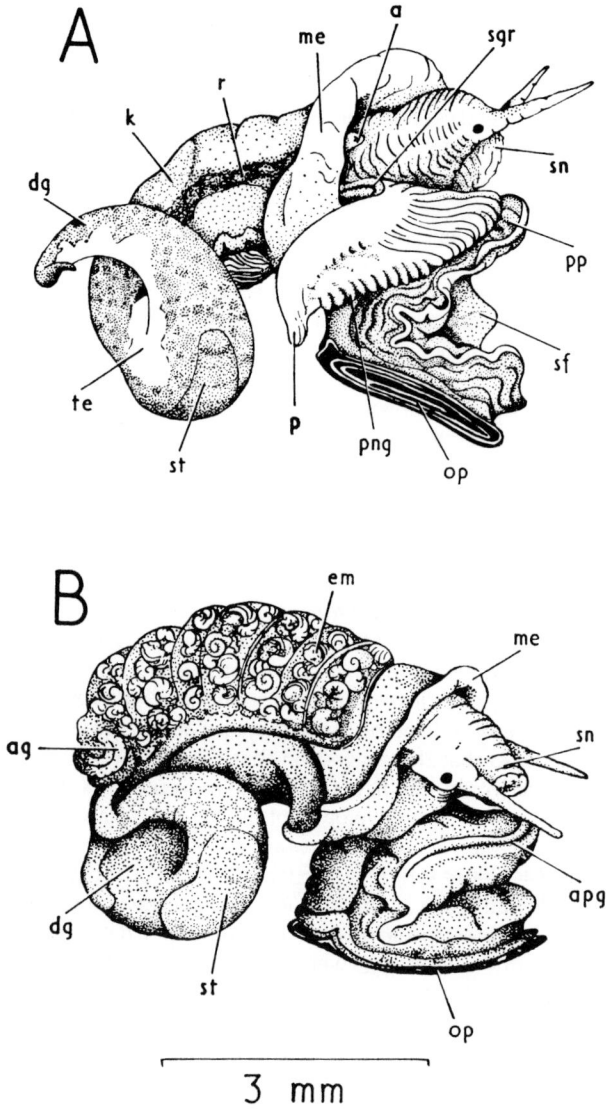

FIG. 170.—*Littorina saxatilis:* A, male removed from shell, seen from right side; B, female, with young in brood pouch, removed from shell and seen from right side.

a, anus; ag, membrane or covering gland; apg, opening of anterior pedal gland; dg, digestive gland; em, embryos in brood pouch which is subdivided into compartments by septa; k, kidney; me, mantle edge; op, operculum; p, penis; png, penial glands; pp, propodium; r, rectum with faecal pellets; sf, sole of foot; sgr, seminal groove; sn, snout; st, stomach; te, testis.

TABLE 10

To show the varying development of the commonest type of food egg in the Stenoglossa (after Staiger).

F, fertilization; L, laying; M, metaphase of first cleavage division; M_1, M_2, first and second maturation divisions; S, union of pronuclei.

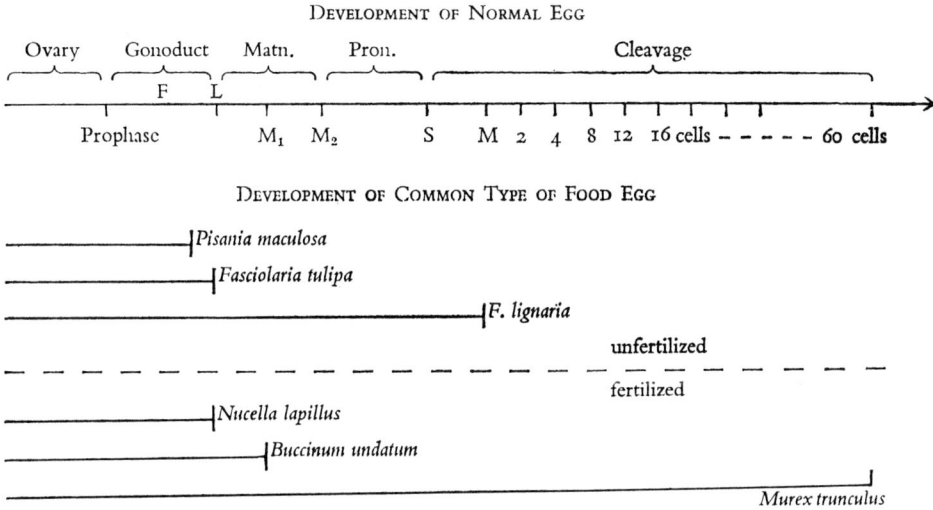

DEVELOPMENT OF NORMAL EGG

Ovary	Gonoduct	Matn.	Pron.		Cleavage
F	L				
Prophase		M_1 M_2	S	M 2 4	8 12 16 cells – – – – – 60 cells

DEVELOPMENT OF COMMON TYPE OF FOOD EGG

Pisania maculosa

Fasciolaria tulipa

F. lignaria

unfertilized

fertilized

Nucella lapillus

Buccinum undatum

Murex trunculus

and *Capulus*. The groove is characteristically ciliated throughout its length. The pallial section may be narrow (*Calyptraea* (pvd, fig. 173B), *Crepidula, Capulus, Balcis*) or it may be bordered by a glandular prostate of considerable size (*Littorina* (pr, fig. 173C), *Aporrhais* (pr, fig. 171A), *Natica*). Other genera in which an open duct occurs lack a penis and the vas deferens ends at the mouth of the mantle cavity: in all these there is a large prostate gland. *Cerithiopsis tubercularis, Bittium reticulatum* and *Clathrus clathrus* have an open prostate and no copulatory organ. In *Cerithiopsis* and *Clathrus* spermatozeugmata occur: it may be that the oligopyrene sperm obviate the possession of a penis by carrying the eupyrene sperm to the female.

In the evolution of the male genital duct an open pallial groove for the passage of sperm through the mantle cavity may have been the first part to appear: the sperm would enter the female by way of the inhalant respiratory current. However, the living gastropods in which this condition obtains are in other respects advanced, so that the simplicity of their genital system is likely to be secondary and may be correlated with the shape of the shell (p. 328). Small size may also affect the reproductive ducts and may be associated with the lack of a penis in *Cingulopsis*, its very large size in *Skeneopsis* and its unusual position and structure in *Omalogyra* and the pyramidellids.

In the majority of monotocardians the vas deferens is closed and there is a penis. The pallial region of the duct may have an outlet to the inner end of the mantle cavity (dm, opm, figs. 172B; 173D, F, I) which is slit-like or at the end of a short muscular duct, and, perhaps, represents the incomplete closure of an originally open groove (*Lacuna vincta, Cingula cingillus, C. semicostata, C. semistriata, Rissoa inconspicua, R. parva, R. lilacina, Alvania punctura, Barleeia rubra, Circulus striatus, Erato voluta, Trivia monacha* (opm, fig. 164), *T. arctica,*

Lamellaria perspicua, Nucella lapillus, Ocenebra erinacea, Buccinum undatum, Nassarius reticulatus). In the rachiglossans *Ocenebra erinacea* and *Nucella lapillus* the longitudinal line of closure may be traced throughout the length of the male duct to the penial aperture, and the opening of the prostate to the posterior end of the mantle cavity is seen as a region in which these epithelia have not fused. In some species of rissoids (*Cingula semicostata, C. semistriata,* fig. 173D) only the regions of the duct anterior to the prostate show evidence of having been derived from an open groove.

There is typically a prostate along the pallial region of the male duct. It may form a wide sac-like portion and have subepithelial glands thickening the deep lateral walls (*Hydrobia ulvae, Erato voluta, Trivia monacha, T. arctica, Nucella lapillus, Ocenebra erinacea* (pr, fig. 173J)) or the gland cells may lie in the epithelium only as in the small rissoids (*Cingula semicostata, C. semistriata, Alvania punctura, Rissoa inconspicua*) and in *Barleeia rubra, Assiminea grayana, Cingulopsis fulgida* and *Pomatias elegans.* In *Pomatias* and *Hydrobia ulvae* the posterior end of the gland is embedded in the tissues of the visceral mass. Alternatively, the prostate may be a narrow tube of about the same diameter as the rest of the male duct and extending beyond the mouth of the mantle cavity; the glands may be in the epithelium (*Lamellaria perspicua* and *Nassarius reticulatus*) or subepithelial (*Buccinum undatum*). In the freshwater architaenioglossan *Viviparus* the prostate extends to the base of the penis and is conspicuous in its anterior part owing to its bright orange colour; the duct has a very thick layer of muscles throughout its length. In *Cingula cingillus, Rissoa parva* and *R. lilacina* the pallial section of the vas deferens is not glandular, though in *Cingula* it is relatively broad. In the two species of the genus *Rissoa* the penial duct is broad and has gland cells alternating with ciliated cells like the prostate of other forms. This is unusual, for the penial duct is typically narrow and muscular like the section of the duct which runs in the body wall and links the pallial vas deferens with the penis.

Although in the mesogastropods there is an evolutionary trend towards the complete closure of the male duct, the persistence of the opening into the mantle cavity in so many forms suggests that it plays some important role. The opening is typically into the posterior end of the cavity, though in *Hydrobia ulvae* (dm, fig. 173E) it is further forwards and near the anterior part of the gland. Sperm are liberated from the vesicula seminalis only at the time of copulation and are transferred to the female in prostatic secretion. The flow of seminal fluid towards the penial aperture is brought about by the relaxation of muscles at the distal end of the vesicula seminalis, which cease their sphincter-like action and release sperm to the vas deferens where peristalsis forces them anteriorly; the pressure set up in the duct must be considerable. If during copulation the male is disturbed, the withdrawal of the penis and rapid contraction into the shell may only be possible with an escape of seminal fluid from the vas deferens to relieve the pressure there. Although some escape may be made through the penial opening there is an obvious advantage in having a more posterior outlet as well. However, one species, *Neptunea antiqua* (Johansson, 1942), shows the loss of this outlet, for although a diverticulum towards the mantle cavity is developed in a position similar to the duct in *Buccinum undatum* (dm, fig. 172B), it has no opening there. In *Cingula cingillus* the vesicula seminalis enters the broad pallial region of the male duct by way of a long papilla which projects into the lumen from the dorsal wall and there is a second communication between these two regions situated more posteriorly near the outlet of the pallial duct to the mantle cavity. It is this second opening which in emergency allows the escape of sperm to the mantle cavity in order to relieve the pressure in the vas deferens.

In *Cingulopsis fulgida* (fig. 173G) there is neither vas deferens anterior to the mantle edge nor penis (Fretter, 1953). The pallial duct is closed and forms a prostate: it has no accessory

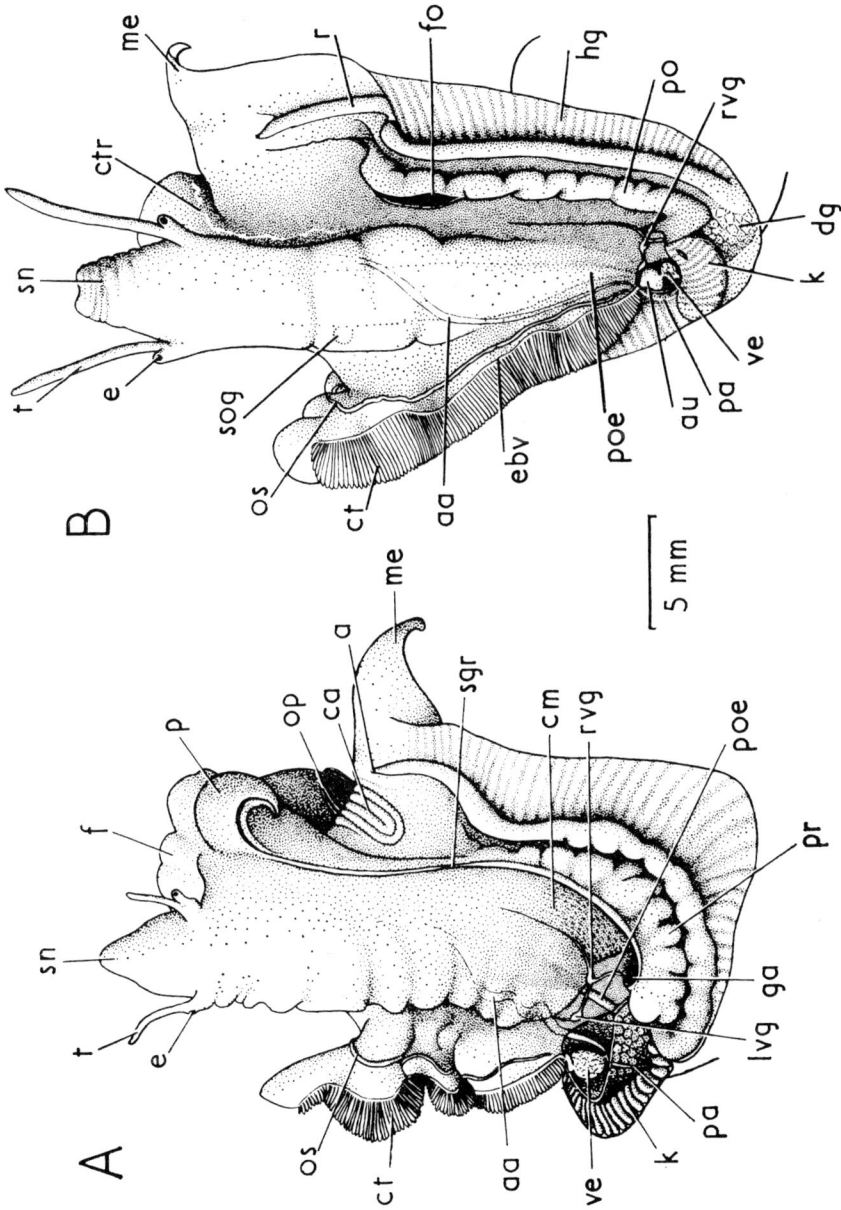

FIG. 171.—*Aporrhais perspelicani*: A, male, removed from shell, with the mantle cavity and pericardial cavity opened to display the contents; B, female, similarly dissected.

a, anus; aa, anterior aorta; au, auricle; ca, ciliated area; cm, columellar muscle; ct, ctenidium; ctr, exhalant tract; dg, digestive gland; e, eye; ebv, efferent branchial vein; f, foot; fo, female opening; ga, genital aperture, opening of renal vas deferens to prostate; hg, hypobranchial gland; k, kidney; lvg, left visceral ganglion; me, mantle edge; op, operculum; os, osphradium; p, penis; pa, posterior aorta; po, pallial oviduct; poe, posterior oesophagus; pr, prostate gland; r, rectum; rvg, right visceral ganglion; sgr, seminal groove; sn, snout; sog, supra-oesophageal ganglion, seen by transparency; t, tentacle; ve, ventricle.

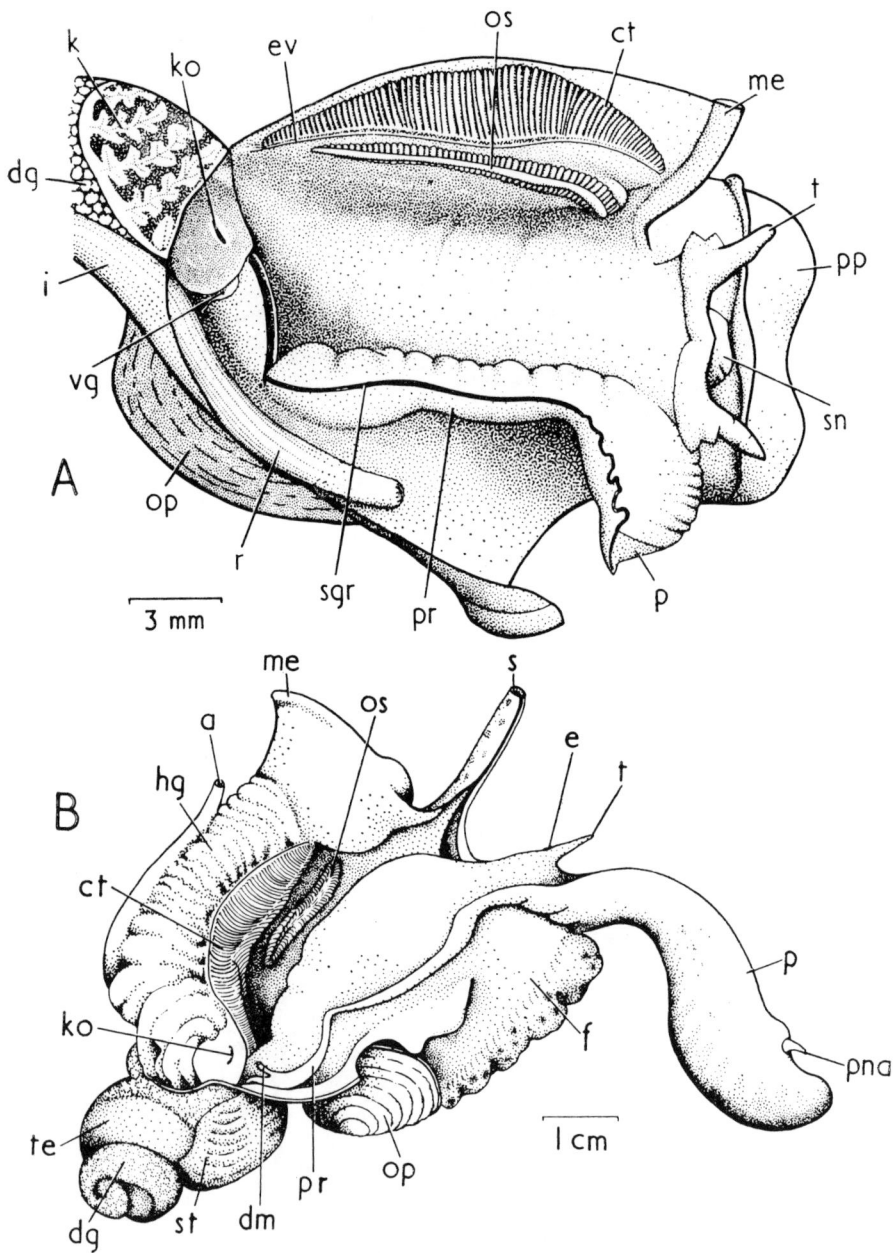

FIG. 172.—A, *Natica catena:* male, removed from shell, with the mantle cavity opened to show its contents; viewed dorsally. B, *Buccinum undatum:* male, similarly dissected, but seen somewhat from the right side.

a, anus; ct, ctenidium; dg, digestive gland; dm, diverticulum of male duct opening to mantle cavity; e, eye; ev, efferent branchial vessel; f, foot; hg, hypobranchial gland; i, intestine; k, kidney; ko, kidney opening; me, mantle edge; op, operculum; os, osphradium; p, penis; pna, penial aperture; pp, propodium; pr, prostate gland; r, rectum; si, siphon; sgr, seminal groove; sn, snout; st, stomach; t, tentacle; te, testis; vg, visceral ganglion.

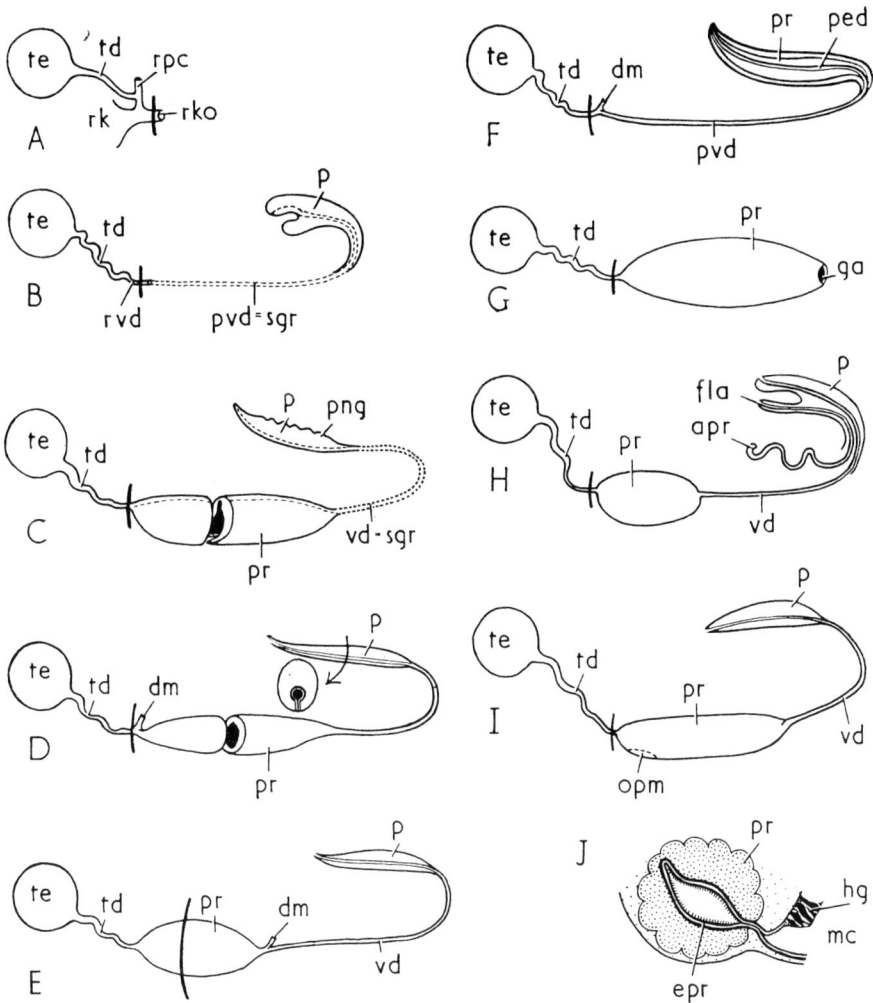

FIG. 173.—Comparative diagrams of the male genital ducts of: A, *Calliostoma zizyphinum*; B, *Calyptraea chinensis*; C, *Littorina littorea*; D, *Cingula semistriata* (with transverse section of penis to show closure of duct by fusion of epithelia); E, *Hydrobia ulvae*; F, *Rissoa parva*; G, *Cingulopsis fulgida*; H, *Bithynia tentaculata*; I, *Ocenebra erinacea*; J, *Ocenebra erinacea*: diagrammatic transverse section of prostate to show closure of gland by fusion of epithelia. The limit of the mantle cavity is indicated by a thick line.

apr, accessory prostate gland; dm, diverticulum of male duct opening to mantle cavity; epr, epithelium through which prostatic glands open; fla, flagellum; ga, genital aperture; hg, hypobranchial gland; mc, mantle cavity; opm, opening of prostate to mantle cavity; p, penis; ped, penial duct; png, penial glands; pr, prostate gland; pvd, pallial vas deferens; rk, right kidney; rko, right kidney opening; rpc, renopericardial canal; rvd, renal vas deferens; sgr, seminal groove; td, testis duct; te, testis; vd, vas deferens.

opening to the mantle cavity, and presumably this is correlated with the fact that the duct opens at the anterior end of the cavity and ends there. Similarly in *Theodoxus fluviatilis,* in the small terrestrial prosobranch *Acicula fusca* and in *Simnia patula* the pallial duct is closed and forms a prostate, and the male opening at its anterior end is the only outlet. A penis is present, however, and in the short space intervening between genital aperture and penis the flow of seminal fluid in *Theodoxus* appears to be guided by a furrow in the overlying mantle to the deep cuticularized groove which runs up the outer edge of the dorsoventrally flattened penis to its tip; in *Acicula* and *Simnia* there is a ciliated groove along the right side of the head which unites the opening of the prostate with the penial groove.

Some species show an advance in the organization of the male duct in having the prostate separate from the duct along which the sperm pass, and a hypertrophy of this gland is a characteristic of some freshwater species. In *Bithynia tentaculata* the gland (pr, fig. 174A) is composed of numerous tubules clustered around the posterior part of the pallial vas deferens and opening at intervals into it; they are joined into a more or less solid mass by connective tissue (Lilly, 1953). In the hermaphrodite *Valvata piscinalis,* the prostate (pr, fig. 179B) is the largest accessory gland of the reproductive system (Cleland, 1954): it spreads into the visceral coils posterior to the mantle cavity and its anterior parts penetrate the haemocoel alongside the muscles of the foot. The gland consists of small tubules which join one another to form a broad duct and near the opening of the duct the vas deferens has a similar glandular lining. *Velutina velutina* (fig. 186C), also hermaphrodite, has 3 large accessory glands separated from the main course of the genital duct and one of these (pr) is associated with the vas deferens.

Copulation occurs in the majority of monotocardians. The penis is typically of pedal origin, innervated by the right pedal ganglion and situated on the head behind the right tentacle; this is also true for *Pomatias elegans* (Creek, 1951) despite statements to the contrary. There are, however, exceptions to this, for in *Bithynia tentaculata* it appears to be pallial, being innervated from the sub-oesophageal ganglion, and in the Viviparidae (fig. 175) it seems to be cephalic since the vas deferens runs through the right tentacle to a finger-like process which normally lies folded back in a pouch (ops) on the right side and opens at its tip (ga). This is in agreement with other peculiarities met in the Architaenioglossa and perhaps suggests that they have evolved from a different stock from the other monotocardians. In *Omalogyra* and *Odostomia* (fig. 179C, D) the copulatory organ is a muscular invaginable tube, so different from the common pattern that it will be dealt with independently. Apart from these animals the penis is not invaginable: when at rest it is hidden in the mantle cavity. It is essentially a muscular and vascular process: muscles lie round the seminal duct or groove and make up most of the surrounding tissue, and between them run abundant blood spaces. When the penis is erected blood engorges these spaces and may enlarge it to twice its normal size, or more. The shape of the penis varies. It may be conical, broad at the base and tapering to a pointed tip (*Capulus* and *Trivia* (p, fig. 164)) and there may be glands around the base and tip secreting to the surface (*Crepidula*). Or the penis may be laterally compressed as in *Buccinum* (p, fig. 172B), or relatively larger and thinner as in the minute forms *Skeneopsis* (fig. 176B) and the rissoids (fig. 67) in which it extends to the posterior end of the mantle cavity of the male when folded back at rest, and at copulation approaches the posterior end of the mantle cavity of the female. It must pass through the length of the pallial oviduct to reach the bursa copulatrix in the rissoids, but in *Skeneopsis* it follows a longitudinal groove (ch, fig. 186B) between the shell and the mantle on the right side of the female, to a sperm pouch (rcs) which has its opening there (mrs). In *Buccinum* the external opening of the vas deferens is at the apex

A

i mm

osg
os
oe
ct
me
ko
vg
e
apr
fla
t
ve au
apo
r
pna
ves
m
dg
pnd
rvd
es
p
te
fp
r
a
me
pr
ak
vd

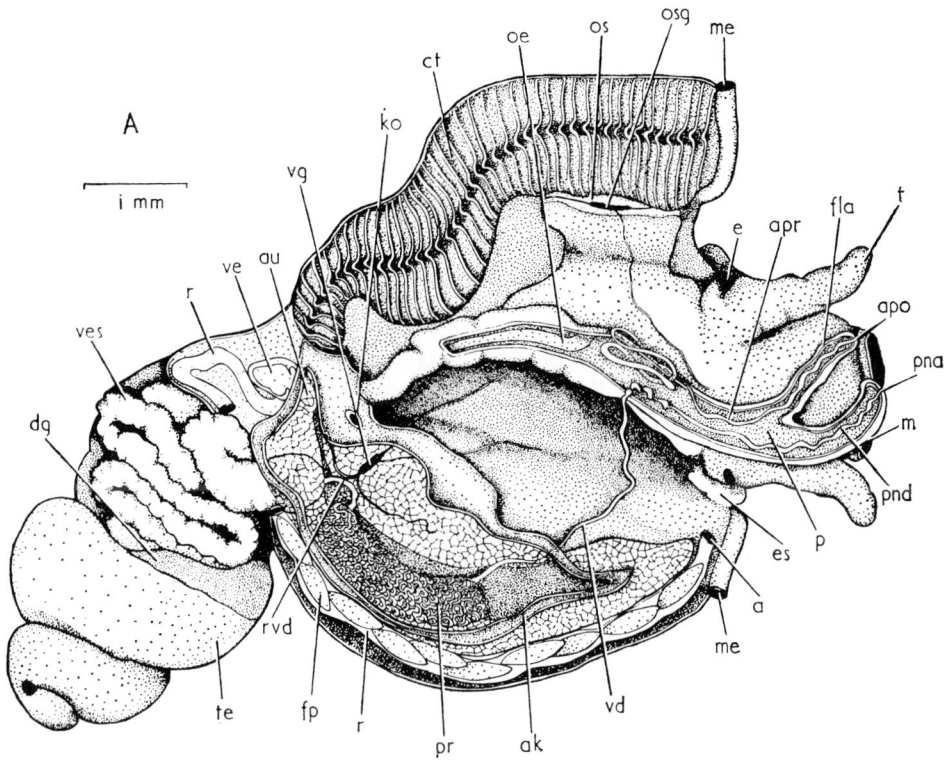

of a short papilla (pna, fig. 172B) situated subterminally on the anterior edge of the penis: at copulation the penis is not inserted far into the female duct for the seminal fluid is deposited in a bursa copulatrix near the genital aperture (Fretter, 1941).

Other devices for filling sperm pouches at the upper end of the pallial oviduct are associated with some modification of penial structure. In *Lamellaria perspicua* (p, fig. 93) the penis has a stout basal region, which is laterally compressed, and a slender apical region which is set at right angles to the base and arises from it subterminally, the origin being set in a groove. At copulation the basal part lies in the pallial oviduct, which it equals in length, and the slender part extends to the receptacular ducts (dr, rcs, fig. 185C). The free edges of the penial groove may embrace the base of the slender part and steady it. In *Calyptraea chinensis* (Giese, 1915) there is a different modification: the penis (p, fig. 173B), which may measure two-thirds the body length, has a distal lobe which is distended at copulation to grip the wall of the upper end of the pallial oviduct, and the vas deferens, an open groove, broadens beyond the origin of the lobe to a cup-shaped depression which is directed towards the group of openings of the receptaculum seminis (rcs, fig. 190).

The outer epithelium of the penis is frequently glandular, producing secretions which may help to lubricate its movement in the female and secure its position during copulation. The glands may lie in the epithelium and be very abundant, as in *Barleeia rubra*, which has,

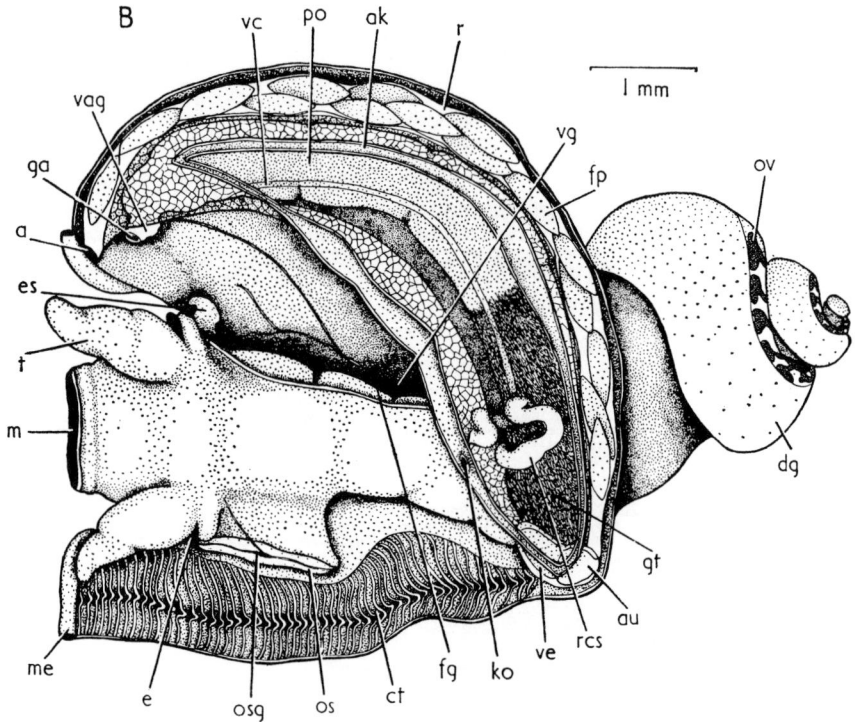

FIG. 174.—*Bithynia tentaculata:* A, male, dissected to show reproductive organs. The mantle cavity has been opened dorsally and the kidney overlying the vas deferens divided longitudinally and its halves pulled apart to expose that structure; the penis and its flagellum dissected to show the penial duct and accessory prostate gland. B, female, with the mantle cavity and anterior part of the kidney opened dorsally, as in the male. After Lilly.

a, anus; ak, lumen of anterior extension of kidney; apo, opening of accessory prostate gland; apr, accessory prostate gland; au, auricle; ct, ctenidium; dg, digestive gland; e, eye; es, exhalant siphon; fg, food groove; fla, flagellum of penis; fp, faecal pellet in rectum; ga, genital aperture; gt, glandular tubules of pallial oviduct; ko, kidney aperture; m, mouth; me, mantle edge; oe, oesophagus; os, osphradium; osg, osphradial ganglion; ov, ovary; p, penis; pna, penial aperture; pnd, penial duct; po, pallial oviduct; pr, prostate gland; r, rectum; rcs, receptaculum seminis; rvd, renal vas deferens; t, tentacle; te, testis; vag, vagina; vc, ventral channel of pallial oviduct; vd, vas deferens leading from prostate to penis; ve, ventricle; ves, vesicula seminalis; vg, visceral ganglion.

perhaps, an unusual method of copulation (p. 333), or some may be grouped in large subepithelial bundles which project from the surface as prominent papillae in *Caecum glabrum* (Götze, 1938). In this minute form (fig. 41B) there is one large adhesive gland on the underside of the penis and a number of smaller ones elsewhere; in each the secreting cells open separately to the outside, though the openings are clustered together. *Littorina littorea* (Linke, 1933) has longitudinal rows of compound glands along the anteroventral wall of the broad, flat penis (png, figs. 12, 13), which appear as mamilliform projections. The deep seminal groove runs along the opposite wall. At copulation only the tip of the penis enters the female duct and the broad middle region lies in the mantle cavity of the female, secured, apparently, by secretion from these glands. Mucus from cells in the penial epithelium may help to

A

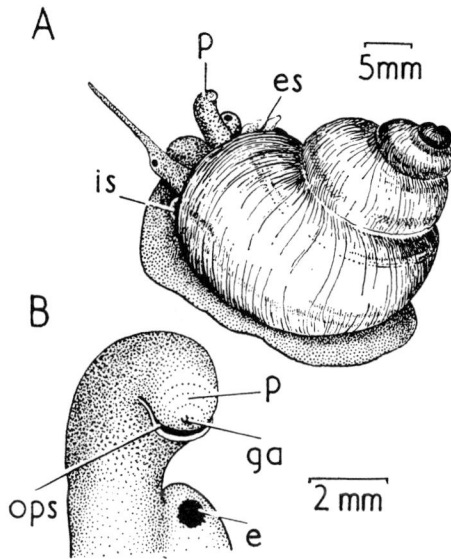

B

FIG. 175.—*Viviparus viviparus*, male: A, whole animal; B, details of penis.
e, eye; es, exhalant siphon; ga, genital aperture; is, inhalant siphon; ops, opening of penis sheath; p, penis.

lubricate the passage of the organ. Compound adhesive glands are also developed in the penis of *L. littoralis* and *L. saxatilis* (png, fig. 170A). Each opens to the surface by a ciliated duct leading from a reservoir which is lined by gland cells and ciliated cells. The wall of the reservoir has a coat of circular muscles penetrated by the ducts of 10–25 tubular glands which are grouped around it. The extremely long penis of *Assiminea grayana* has a longitudinal row of about 8 muscular projections along its anteroventral wall. They are not associated with glands, but may serve the same purpose of securing a hold on the female during copulation; the pallial oviduct has no vaginal channel.

There is some question as to the function of the penial gland of *Bithynia tentaculata* (apr, figs. 173H, 174A). The penis (p) is bifid, with a median flagellum (fla). An accessory gland opens at the tip of the flagellum (apo, fig. 174A): it consists of a single tubule coiling in the tissues of the penis and extending posteriorly into the haemocoel. When the penis is erected the flagellum diverges from it so that it is doubtful whether at copulation its gland discharges into the female duct; the secretion may be poured on to the wall of the mantle cavity and hold the penis in position. This may also be true for the pelagic prosobranch *Carinaria* which has a penial appendage bearing the opening of a gland (fig. 293B) and in which copulation may be difficult owing to its mode of life.

In a number of opisthobranchs, especially the Ascoglossa, the penis is armed with spines or hooks which are cuticular secretions used to stimulate the partner. Only in one prosobranch, the ectoparasite *Pelseneeria stylifera* (Ankel, 1936a) are such spines known to occur.

One of the outstanding characteristics of the hermaphrodite reproductive system of the pyramidellids (Fretter & Graham, 1949) is the position of the penis and its associated structures which lie in the haemocoel, apart from the rest of the genital system; the penis is contained in a sheath which opens beneath the mentum and passes back through the nerve ring (ops, figs. 179D, 133). Normally in a gastropod the passage through the nerve ring is traversed only by the digestive tube and its associated glands, muscles and blood vessels. Perhaps the penis was carried into this position by the development of the long, acrembolic proboscis which characterizes the pyramidellids. The opening of the penial sheath is connected with the opening of the pallial genital duct by a ciliated tract which runs back along the right side of the head, ventral to the base of the tentacle, on to the floor of the mantle cavity to end at the genital aperture (sgr). The sheath lies dorsal to the gut and may be flung into one or two broad loops when the animal is withdrawn into its shell. It is broad posteriorly and narrows to its opening, and on the dorsal wall, in front of the tip of the retracted penis, opens the short duct of a large muscular sac, the sperm sac (spc). A duct, the vas deferens (vd), arises near the opening of the sac and runs posteriorly in the thickness of the dorsal wall of the sheath towards the base of the penis. In *Odostomia* and *Chrysallida* the vas deferens, a closed duct, passes through the penis to its tip, and in the surrounding tissues are large blood spaces: when the penis is erected it appears as a long whip-like structure, turgid with blood and tapering to a fine point. During copulation the individual which acts as male creeps on to the surface of the shell of its partner, everts the penis and bends it ventrally so that it passes into the mantle cavity and through the genital aperture of the female. The withdrawal of the penis is brought about by the contraction of its longitudinal muscles and those of the sheath, and the consequent expulsion of blood into the general haemocoel. The fibrillae of these muscles are striated. The sperm sac must be filled prior to copulation, for it is difficult to see how sperm could reach it once the penis is protruded. The common genital aperture (ga) lies on the floor of the mantle cavity to the right, and about one-third of the total depth of the cavity inwards from its mouth. The path of the spermatozoa can be followed from this opening along the ciliated tract (sgr) to the opening of the penial sheath, and through the anterior ciliated part of the sheath into the sperm sac. During copulation contraction of the walls of the sac will force its contents along the vas deferens to the penis. In *Turbonilla elegantissima* (Fretter, 1951c) the opening of the hermaphrodite duct is on the propodium in front of the right tentacle and to the right of the mentum; in fact it is anterior to the opening of the penial sheath. The penis has no duct and either no blood spaces or only insignificant ones. In the retracted state (p, fig. 179E) it has the appearance of a long muscular scoop attached posteriorly to the wall of the sheath. The vas deferens (vd) arises from the mouth of the sperm sac (spc) as a groove which is morphologically open, though physiologically closed by apposition of the lips. It runs along the penial sheath to the posterior end of the penis, then opens to the concave surface of this scoop-shaped organ. Seminal fluid is propelled through vas deferens and penis by local muscular contractions.

Omalogyra atomus (fig. 176A) is one of the most minute of British molluscs and its reproductive system is specialized. It is hermaphrodite (Fretter, 1948) and during the summer months, when the *Ulva* on which it feeds is abundant, one generation follows another very rapidly. An individual collected at this time of the year has no structure which could act as a copulatory organ, and it is probable that self-fertilization is then practised. Only immature forms tide over the winter and these come to maturity in the spring. They differ from the summer forms in certain structural details. In the reproductive system the male organs are the first to mature. Through the lumen of the vas deferens runs a muscular tube (p, figs. 176A, 179C), open and directed inwardly, its anterior end arising from a large sac (bcp) connected to the anterior hermaphrodite portion of the pallial duct; the sac may be homologous with the bursa copulatrix of other forms. It is assumed that this tube is the copulatory organ: sperm liberated from the vesicula seminalis (ves, td) and prostatic secretion from the wall of the

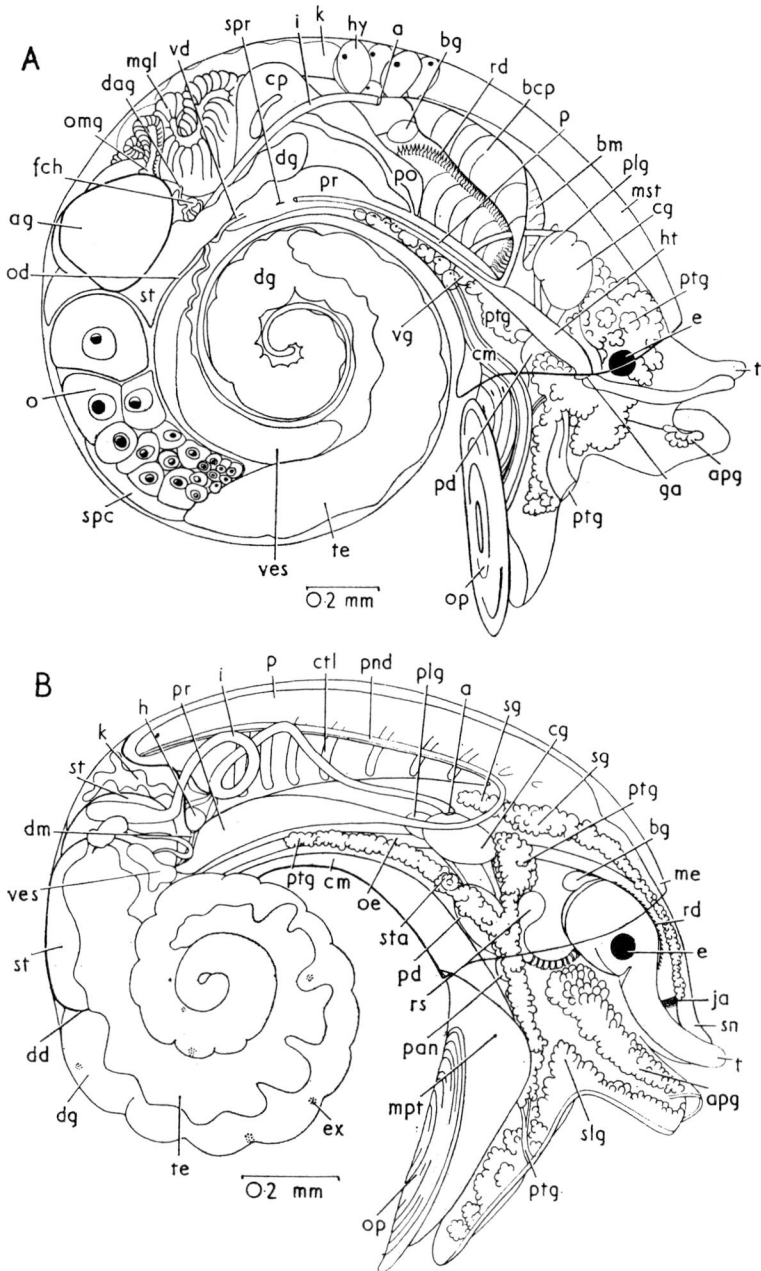

FIG. 176.—A, *Omalogyra atomus;* B, *Skeneopsis planorbis.* The animals are seen as transparent objects.

a, anus; ag, albumen gland; apg, anterior pedal gland; bcp, bursa copulatrix; bg, buccal ganglion; bm, buccal mass; cg, cerebral ganglion; cm, columellar muscle; cp, capsule gland; ctl, ctenidial leaflet; dag, duct of albumen gland; dd, opening of duct of digestive gland to stomach; dg, digestive gland; dm, diverticulum of male duct opening to mantle cavity; e, eye; ex, excretory cell in digestive gland; fch, fertilization chamber; ga, genital aperture; h, heart; ht, hermaphrodite duct; hy, hypertrophied glands opening to mantle cavity near anus; i, intestine; ja, jaw; k, kidney; me, mantle edge; mgl, mucous gland; mpt, metapodium; mst, mantle skirt; o, ovum in ovary; od, ovarian duct; oe, oesophagus; omg, opening of albumen gland to mucous gland; op, operculum; p, penis; pan, parapedal ganglion; pd, pedal ganglion; plg, pleural ganglion; pnd, penial duct po, pallial oviduct; pr, prostate gland; ptg, posterior pedal gland (the labels placed on the mouth of the gland and on its anterior and posterior lobes); rd, radula; rs, radular sac; sg, salivary gland; slg, sole gland; sn, snout; spc, sperm sac; spr, opening of sperm sac into prostate gland; st, stomach; sta, statocyst; t, tentacle; te, testis; vd, pallial vas deferens; ves, seminal vesicle; vg, visceral ganglion.

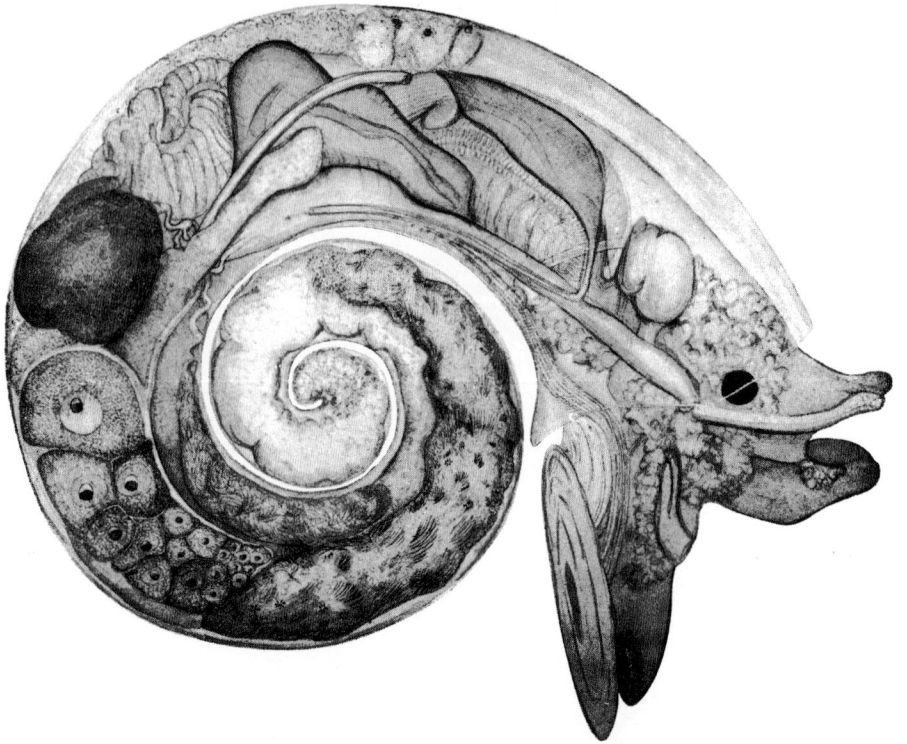

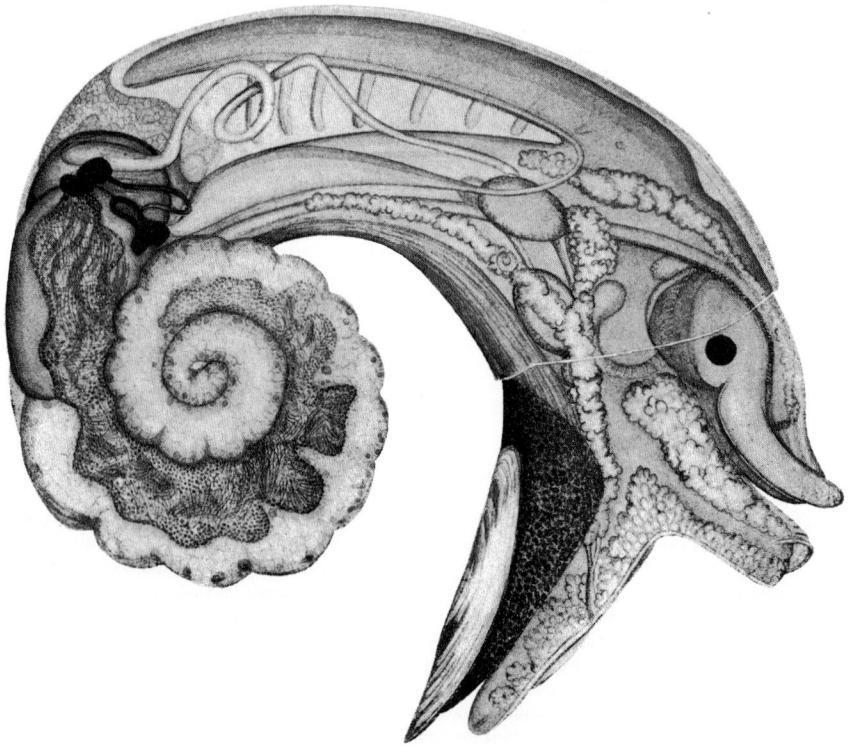

posterior half of the vas deferens (pr) could be sucked through the penial tube into the muscular sac (bcp); the direction of the tube could then be reversed so that it protrudes through the genital aperture and into the hermaphrodite duct of the copulating partner. The muscular wall of the sac would contract and so transfer the seminal fluid. This suggested method of copulation is unusual and may be correlated with the small size of the mollusc as well as the isostrophic shell. There is a second muscular sac (spc) which is connected to the posterior end of the prostate by a narrow, ciliated duct; it is only present in animals collected during the spring months. The sac frequently contains sperm and prostatic secretion in large quantities, and the sperm appear as though they are disintegrating. Perhaps its function is to clear the pallial vas deferens of sperm and prostatic fluid which have failed to enter the copulatory organ. A similar method of clearing the genital duct of unwanted material is found in the hermaphrodite opisthobranchs, in which the bursa frequently fulfils this purpose.

Woodward (1899) has described the reproductive system of the rare monotocardian *Adeorbis* (= *Tornus*) *subcarinatus* and stated that the male has neither penis nor accessory gland and that the vas deferens opens 'high up and close to the external opening of the kidney'. The oviduct, however, has a pallial section. Unless *Adeorbis* is an exception to all other monotocardians which have been studied, internal fertilization occurs; yet we have no suggestion as to the way in which it is brought about.

There would appear to exist a connexion between the state of development of the gonad and the state of development of the secondary sexual organs in at least some prosobranch molluscs. This is particularly well seen in *Littorina*. *L. littorea* breeds at Plymouth from November to May, chiefly in February and March, and during these months when the gonad is ripe the penis, prostate and pallial oviduct are at a maximal size. The reduction of the gonad during late spring and early summer is, however, concurrent with a reduction in size of these organs. During July and August no male has a functional penis: it is of insignificant size or lost, the glands are reduced to only the pits of the chief follicles, the muscles are thin and short—in fact the tissues assume an embryonic appearance which is retained until the gonad enlarges and matures in the autumn. Linke (1934b) has shown that at the end of the breeding season material from the destruction of the genital system is stored in connective tissue cells (fig. 123). A permanent reduction in the size of the gonad may be brought about by trematode infection, and as the gonad is destroyed the secondary sexual structures are reduced to vestiges. This also holds for *L. saxatilis*. In this viviparous species the annual rhythm of sexual activity appears less marked, since the breeding seasons of the varieties (see p. 503), each with its own resting period, tend to overlap (James, personal communication).

A similar correlation between the condition of the gonad, penis and pallial genital duct occurs in the protandrous hermaphrodites *Calyptraea, Crepidula* (fig. 177) and *Capulus*. At the transition from male to female stage the reduction of the penis (p) is synchronized with the reduction of the testis, and as the ovary develops the walls of the pallial duct, which is formed from the closure of the narrow seminal groove, hypertrophy to form the albumen and capsule glands. Such events suggest a hormonal control, though this has been little investigated. Rohlack (1959) has, indeed, found an oestrogen (not identical with that of vertebrates) in the ovary of *Littorina littorea*, though not in any other part of the body. Despite this she suggested that the sex cycle is probably controlled by temperature: perhaps both factors interact. Like Linke (1934a) she was unable to carry out successful injections since experimental animals invariably died. No androgens were found in males. [See p. 657.]

CHAPTER 14

THE REPRODUCTIVE SYSTEM—2

I N the more primitive oviparous British monotocardian gastropods (Lacunidae, Hydrobiidae, Rissoidae) the oviduct is closed throughout its length and the genital aperture is near the mouth of the mantle cavity alongside the anus (fig. 178). There is no evidence in the adults of these, or of more advanced groups, that the pallial region has been formed by the fusion of the lips of a longitudinal groove on the mantle skirt as has been described for the prostate of *Ocenebra* (fig. 173J) and *Nucella*. However, sections of *Calyptraea chinensis* show that on the ventral wall of the pallial oviduct there is a narrow longitudinal strip of cubical cells which are neither ciliated, nor like the epithelium elsewhere, nor under-lain by glands. This is the line of closure in the transformation of the male stage with an open seminal groove to the female with a closed duct. In the majority of the mesogastropods this ventral wall is relatively thin and its musculature well developed: in some animals the penis slides along it in passing from the genital aperture to the inner end of the pallial duct (rissoids, *Capulus, Calyptraea, Crepidula, Lamellaria*), and in others this ventral path is followed by the sperm which, deposited in the lower regions of the duct, are making their way to the receptaculum seminis. When used as a path for sperm the ventral channel may be elabor-ated. In *Littorina* spp. and in the Rachiglossa the bursa copulatrix and receptaculum seminis are at opposite ends of the channel which is arched over on either side by a longitudinal fold so that the sperm travel in a physiologically closed duct, separated from the reproductive glands.

Johansson (1948a) has shown that in the development of *Hydrobia ulvae* the closure of the pallial duct is delayed and in young forms it opens widely to the mantle cavity by a ventral slit. At the free edge of the median lip of this opening there is a ciliated groove or gutter which extends beyond the slit anteriorly, and runs posteriorly along the internal line of closure of the duct to form the vaginal channel. It leads to the bursa copulatrix at the inner end of the pallial duct. The receptaculum seminis is immediately beyond the bursa and the channel can be traced beyond its opening as far as the gonopericardial duct. From this Johansson concluded that the gonopericardial duct in this, and in other species in which it occurs (Johansson, 1956a), marks the division between the mesodermal and ectodermal parts of the genital duct; a similar conclusion was reached by Krull (1935). The renal oviduct will, theoretically, include a section of each part, the mesodermal region being derived from the rudiment of the right kidney, and the ectodermal from an invagination of the mantle cavity which forms its duct. With the ontogenetic and phylogenetic modifications of the genital duct the exact bound-aries of this section in the various species of monotocardians are no longer traceable with certainty in the adult, nor are they of any real significance.

In some genera a gonopericardial duct occurs in the female, while in the male there may be a vestige in the form of a dense strand of connective tissue passing from the vas deferens to the pericardium (*Littorina* spp., *Ocenebra erinacea, Nassarius reticulatus*). In the protandrous hermaphrodites *Calyptraea* and *Crepidula* the duct first makes its appearance as a strand of

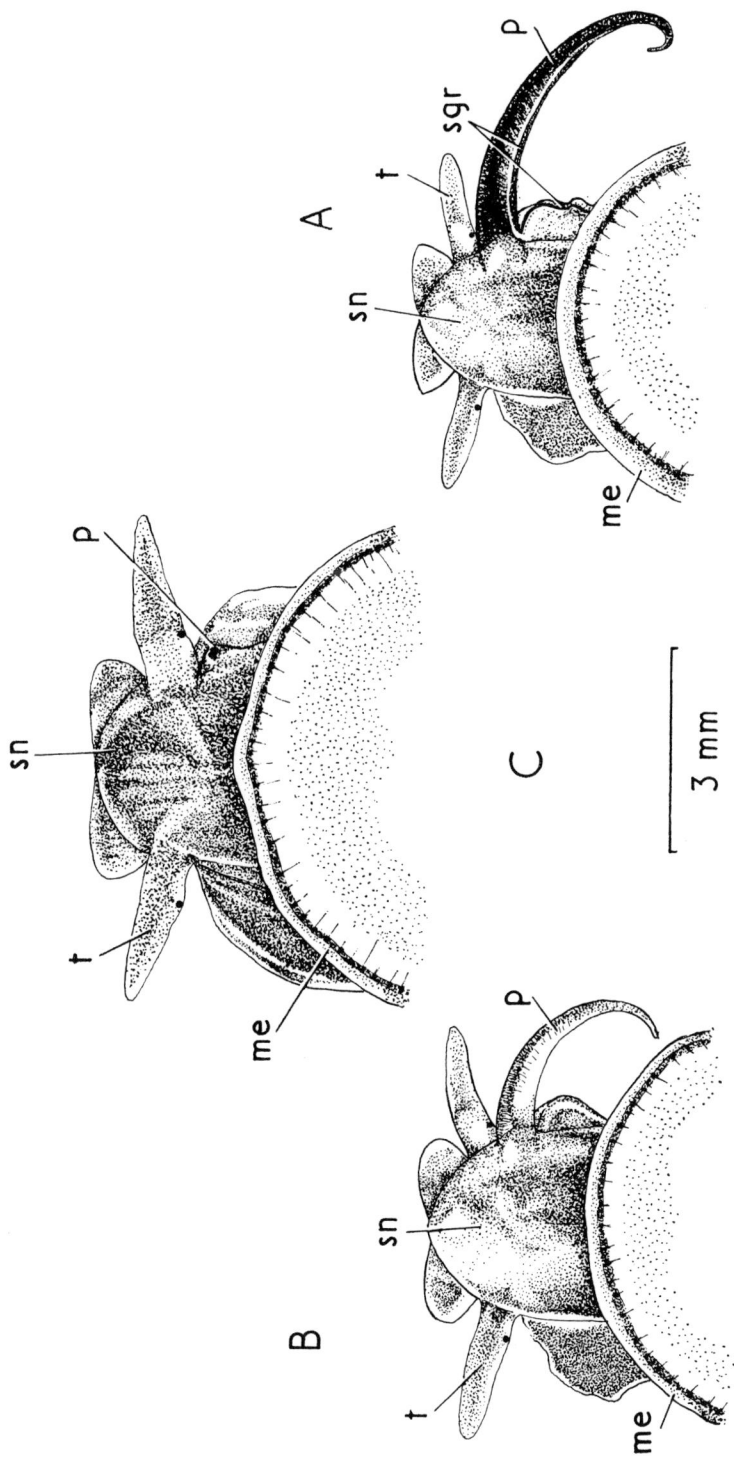

FIG. 177.—*Crepidula fornicata*: dorsal view of anterior halves of 3 animals to show changes in penis with change in sex. A, functional male; B, transitional stage; C, functional female.

me, mantle edge; p, penis; sgr, seminal groove; sn, snout; t, tentacle.

3 mm

connective tissue as the female organs develop (Giese, 1915) and later it cavitates. The occurrence of a gonopericardial duct cannot be regarded as giving any indication of phylogenetic relationship since it is found in the Stenoglossa (gpd, fig. 165), which are specialized in other respects, and is lost in many of the more primitive mesogastropods: in *Bithynia tentaculata* its presence has been mentioned by some workers (Krull, 1935) and denied by others (Lilly, 1953), suggesting that it is not always developed. The persistence of the duct in females shows that at least it is no handicap in the functioning of the genital system, and its formation at the transition from male to female stage of *Calyptraea* and *Crepidula* suggests that it may serve some essential role. There is an unusual condition in *Circulus striatus* where the kidney (k, fig. 180A) extends between the pericardium and the genital duct (ag), and this duct opens into the kidney and not into the pericardium (gk). A similar link between the female genital duct and the kidney occurs in *Truncatella subcylindrica* (fig. 306C).

Johansson (1948a) stated that the ciliated groove, which in *Hydrobia* is incorporated in the ventral wall of the pallial duct to form the ventral channel, is not completely enclosed by the duct in *Rissoa violacea* (=*R. lilacina*) and *R. membranacea*: it is present in immature and mature individuals as an outer ventral gutter against the median wall of the capsule gland, and, as a consequence, this gland has no clearly defined ventral channel. However, the gutter must be enclosed posteriorly as Johansson regarded the bursa copulatrix as an outgrowth from it, and similarly he regarded the vagina as an enclosed anterior part. Moreover in *Cingula cingillus*, *C. semicostata*, *C. semistriata*, *Rissoa parva*, *R. lilacina*, *Alvania beani*, *A. abyssicola*, *A. punctura* a thin-walled vaginal channel is well defined along the capsule gland but there is no longitudinal fold of tissue which separates it, even incompletely, from the lumen of the gland as in *Hydrobia ulvae* and *Potamopyrgus jenkinsi*.

Most species in which the prostate gland is open throughout its length and no penis is developed, have an open pallial duct in the female (*Turritella* (figs. 178M, 181), *Bittium* (fig. 182), *Cerithiopsis*, *Clathrus*). It is in the form of a glandular groove with deep lateral walls, joined by a narrow dorsal wall and open ventrally. It may be physiologically closed with an anterior aperture, and along the ventral free edge of the median wall there may be a ciliated groove (thought to be homologous with that described for *Hydrobia*) which in *Bittium reticulatum* is most pronounced and leads to the bursa (bcp, rcs, fig. 182) and the receptaculum. Johansson (1947) referred to this groove as a sperm-collecting gutter. This open condition of the genital ducts and the lack of a penis, is presumably an advantage; in other respects the species which have it are advanced. In some genera it appears to be correlated with a long, narrow mantle cavity containing a relatively large ctenidium. In *Turritella*, *Bittium*, *Cerithiopsis* and *Clathrus* the shell is a close spiral with a small apical angle and small mouth; the mantle cavity is deep, following the course of the tight spiral of the shell, and it may have reached a minimal breadth for maintaining proper ventilation. Should it be still further restricted by the presence of a penis in the male, its breadth might be brought below the limit for efficiency; in support of this it is to be noted that the female glands swell relatively less than in species in which the duct is closed. Perhaps it is for this reason that the sperm are transferred to the female by a method involving an open pallial duct. Moreover, in such tightly coiled spirals there will be a greater degree of shortening of the right side of the body and so less space for the right half of the pallial complex, which is, in the monotocardian, the genital duct and rectum. *Turritella communis* is a ciliary feeder living in muddy situations and is specialized in accordance with this. The mantle cavity is isolated from the environment by a curtain of tentacles which hangs from the mantle skirt across the pallial opening (pt, fig. 64) and so prevents silting of the cavity: should copulation occur, the introduction of the penis into the female aperture would impair

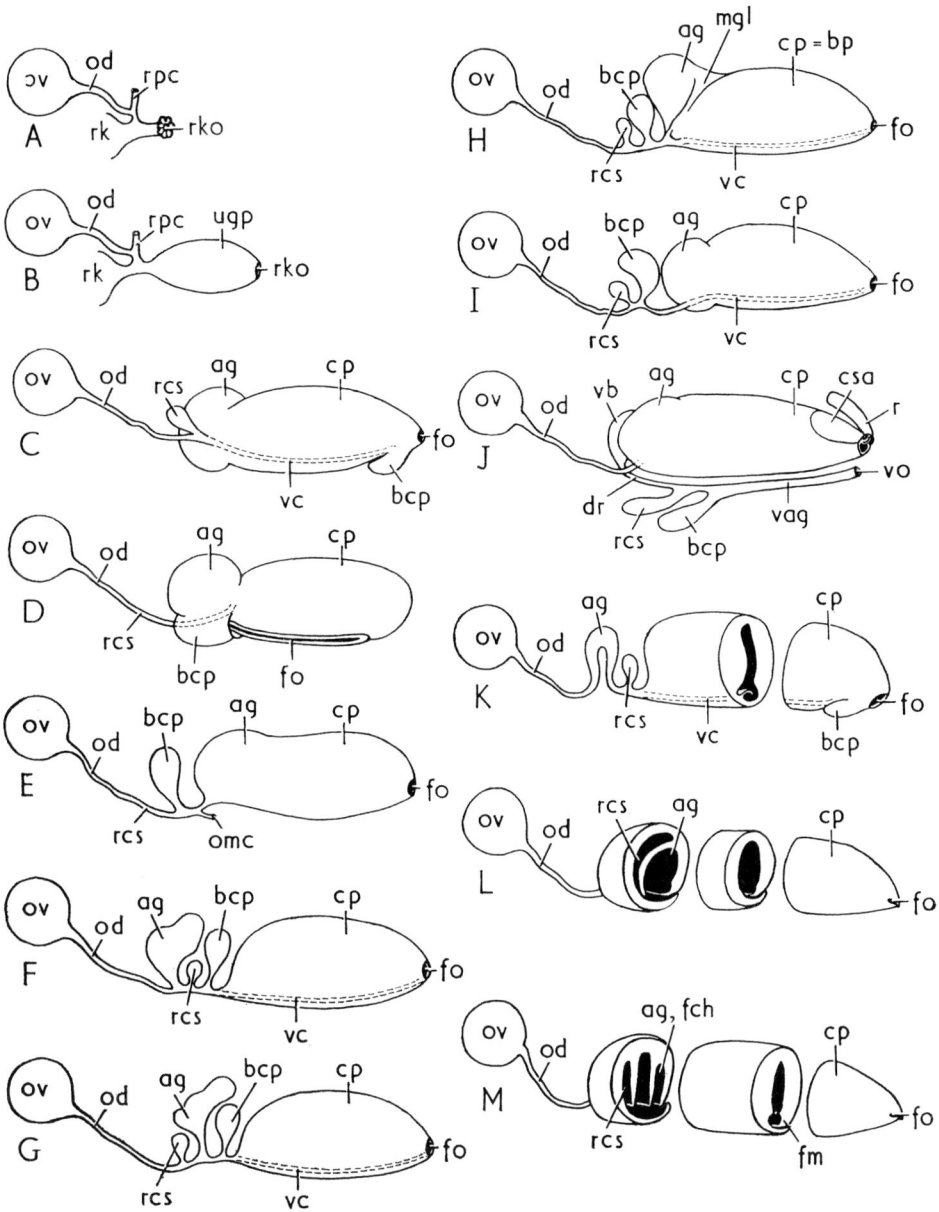

FIG. 178.—Comparative diagrams of the female genital ducts of: A, *Gibbula* and *Monodonta*; B, *Calliostoma*; C, *Littorina littorea*; D, *Pomatias elegans*; E, *Acicula fusca*; F, *Cingula semicostata, C. semistriata*; G, *Cingula cingillus*; H, *Potamopyrgus jenkinsi*; I, *Hydrobia ulvae*; J, *Theodoxus fluviatilis*; K, *Nucella lapillus*; L, *Clathrus clathrus*; M, *Turritella communis*.

ag, albumen gland; bcp, bursa copulatrix; bp, brood pouch; cp, capsule gland; csa, crystal sac; dr, duct of receptaculum; fch, fertilization chamber; fm, flange bordering edge of left wall of pallial oviduct; fo, female opening; mgl, mucous gland; od, oviduct; omc, opening to mantle cavity; ov, ovary; r, rectum; rcs, receptaculum seminis; rk, right kidney; rko, right kidney opening surrounded by a glandular lip in A; rpc, renopericardial canal; ugp, urinogenital papilla; vag, vagina; vb, vestibule; vc, ventral channel; vo, vaginal opening.

this isolation. With an open pallial duct exposing the receptaculum seminis at its upper end the sperm may be transferred to the female by the inhalant water current. The animals are gregarious and spawning of large numbers probably occurs simultaneously.

There are other advanced mesogastropods with a more spacious mantle cavity in which there is no penis and the pallial genital ducts are open (e.g. *Fagotia esperi* (Soós, 1936)). The presence of this apparently primitive condition in these species may be correlated with some unknown factor in their mode of life, or perhaps an alteration in shell shape is recent and the structure of the genital duct not yet changed to conform to the new type of shell.

The British eratoids, *Erato voluta* (fig. 35), *Trivia monacha* and *T. arctica* (figs. 185A, B), and the terrestrial prosobranch *Pomatias elegans* (fig. 178D) have a comparatively short and deep pallial section to the oviduct, with the capsule gland opening ventrally to the mantle cavity along the greater part of its length. In the male there is a penis and the vas deferens is closed except, perhaps, for a small outlet to the mantle cavity posteriorly. The egg capsules are large and spherical (figs. 197, 206) and embedded in some kind of substratum when they are laid. The ventral opening of the capsule gland may be correlated with the size and shape of the capsules, and the mantle cavity is broad so that it would not be incommoded by their passing through it. A similar ventral aperture occurs in the hermaphrodite *Velutina velutina* (fig. 186C) which also has large egg capsules and a broad mantle cavity; these capsules are embedded in the tissues of the compound ascidian *Styela coriacea* according to Diehl (1956).

Two kinds of sperm pouch are developed on the female duct (fig. 178) communicating (with few exceptions) with the ventral channel and apparently developed as outgrowths from it: the bursa copulatrix (bcp) which receives sperm and prostatic fluid, and the receptaculum seminis (rcs) to which the sperm pass from the bursa. In the receptaculum the sperm are normally orientated, lying closely packed, each with the tip of its head embedded in the cytoplasm of an epithelial cell and its tail directed into the lumen, and here they have a longer stay than in the bursa, only leaving the pouch at the time of fertilization. Occasionally, but not usually (Stenoglossa (Fretter, 1941)), sperm are also orientated in the bursa. The position of the sperm pouches varies and also their number. The receptacular pouch is typically at the inner end of the pallial duct, not far from the site of fertilization: it may lie between the albumen and capsule glands (fig. 178F, K), or posterior to both (fig. 178G), or, more unusually, it may arise from the albumen gland (*Trivia monacha, T. arctica, Lamellaria perspicua* (rcs, fig. 185A, B, C)). Although in some species the bursa copulatrix is immediately anterior to the receptaculum, in others it is at the opposite end of the ventral channel and near the genital aperture (*Littorina* (fig. 178C); Stenoglossa (fig. 178K)). Johansson (1953) considered the proximal position of the bursa to be the more primitive and associated originally with an open pallial groove which received sperm by way of the inhalant stream of water, the distal pouch having arisen after the closure of the pallial duct and the development of a penis. In the 2 sperm pouches of *Bithynia tentaculata*, one at each end of the pallial duct, spermatozoa mingled with prostatic fluid have been identified (Krull, 1935; Lilly, 1953) suggesting that each acts as a bursa and that the depth to which the penis discharges its contents may vary. An even more unusual condition is found in *Aporrhais pespelicani* which has 3 bursae in the anterior half of the pallial duct. They are embedded in the thickness of the lateral walls and open near the slit-like genital aperture (Johansson, 1948*b*); ingoing ciliary currents mark the entrance to each. There is still another pouch which opens mid-ventrally not far from the posterior lip of the genital opening with which it is connected by a groove. The effective beat of the cilia along this groove is away from the pouch, which, Johansson suggested, is a receptaculum seminis.

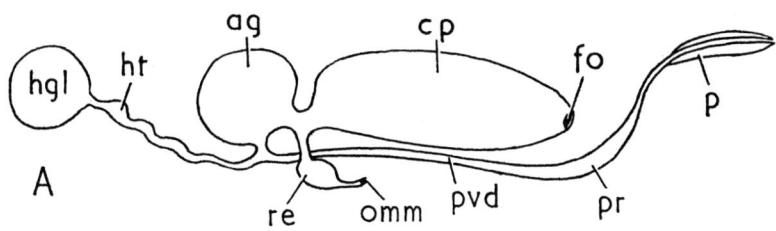

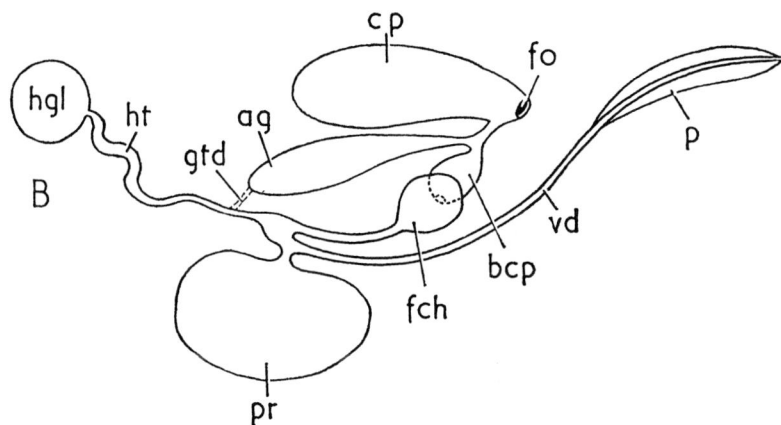

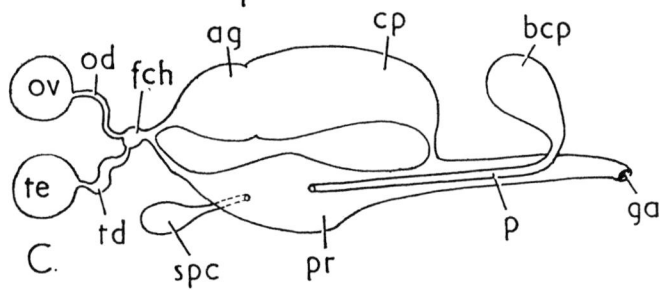

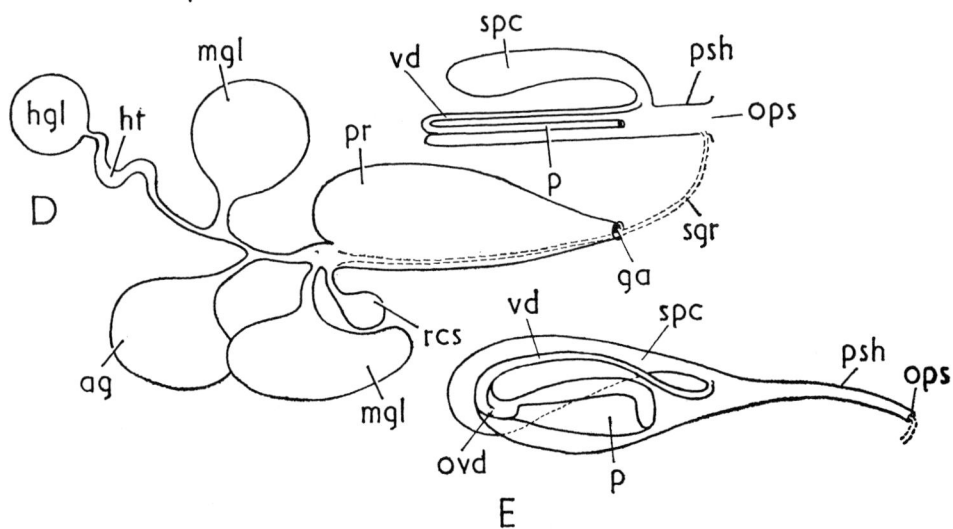

In some species sperm are stored in a region of the gonoduct which is proximal to the pallial glands, in which case the receptacular pouch may be reduced, as in *Alvania subsoluta* (Johansson, 1956a), or lost, as in the Indian pond snail *Cremnoconchus*, the terrestrial *Pomatias elegans* and *Lacuna pallidula* (Gallien & Larambergue, 1938). In *Pomatias* sperm are orientated over the entire surface of this region of the oviduct (rcs, fig. 178D), whereas in *Cremnoconchus* and *Lacuna* its wall is constricted longitudinally to form one channel for the eggs and another for the orientated sperm. The small terrestrial prosobranch *Acicula fusca* has a single pouch at the inner end of the pallial duct (bcp, fig. 178E) where orientated sperm are always present in large numbers, sometimes with prostatic fluid. It has been termed the bursa copulatrix (Creek, 1953) though its functions would appear to be combined with those of a receptaculum. A certain number of sperm also accumulate in the region of the oviduct immediately posterior to the pouch. Here the epithelial cells are large except in two or three places where they form longitudinal grooves in which sperm are orientated. The large cells lining the duct between the rows of sperm are either mucous glands or ingesting cells which take up sperm and apparently digest them in the cytoplasm. Creek suggested that the sperm accumulated in the duct may be the excess of those which left the sperm pouch at the time ripe eggs were liberated from the ovary, and, trapped in the duct, they orientate themselves and remain healthy for some days. This appears to be the course of events in *Trivia monacha* in which sperm with a similar history are accumulated in a bunch of branching diverticula of the renal oviduct (br, fig. 185A); they are orientated in the lower parts, whilst at the blind ends of the tubules they lose their orientation, their heads shorten and they are ultimately devoured and digested by amoebocytes. In *Bithynia tentaculata* (Lilly, 1953) orientated sperm are frequently found in the glandular tubules of the posterior part of the pallial glands (gt, fig. 174B), as well as in the coiled muscular part of the gonoduct (rcs) which precedes both these glands and the proximal bursa. This coiled region resembles the receptacular part of the duct of *Alvania abyssicola* and is presumably of ectodermal origin, representing an invagination of the mantle cavity (Johansson, 1956a).

In a few species the vaginal channel is separated anteriorly from the pallial oviduct so that the female genital system has two openings to the mantle cavity. In *Theodoxus fluviatilis* (figs. 178J, 187) these openings are alongside one another, and from the vaginal duct (vag) arise the bursa copulatrix (bcp), which receives the penis, sperm and prostatic secretion, and the receptaculum seminis (rcs), in which the sperm are orientated with their tails attached to the epithelium. A narrow, coiled duct leaves the stalk of the receptaculum and passes to a fertilization chamber at the upper end of the pallial duct; it is the only connexion between the vaginal channel and the oviduct. In *Cingulopsis fulgida* (fig. 183) the separated channel forms

---

FIG. 179.—Comparative diagrams of the hermaphrodite genital ducts of: A, *Rissoella diaphana;* B, *Valvata piscinalis;* C, *Omalogyra atomus;* D, *Odostomia* spp.; E, penis and related structures of *Turbonilla elegantissima.*

ag, albumen gland; bcp, bursa copulatrix; cp, capsule gland; fch, fertilization chamber; fo, female opening; ga, genital aperture; gtd, position of Garnault's duct; hgl, hermaphrodite gland; ht, hermaphrodite duct acting as vesicula seminalis; mgl, mucous gland; od, ovarian duct; omm, opening of muscular sac to mantle cavity; ops, opening of penial sheath; ov, ovary; ovd, opening of vas deferens to base of spoon-shaped penis; p, penis; pr, prostate; psh, penial sheath; pvd, pallial vas deferens; rcs, receptaculum seminis; re, muscular pouch, probably homologous with receptaculum seminis; sgr, seminal groove; spc, sperm sac; td, testis duct acting as vesicula seminalis; te, testis; vd, vas deferens.

a narrow muscular duct (mdt) which is shorter than the oviduct so that it opens well within the mantle cavity. It leads to the upper end of the pallial duct where there is a single receptacular pouch with orientated sperm (rcs) and a muscular sac which, during the breeding season, has been found to contain a fluid with staining properties similar to those of prostatic secretion (as, sas)—it is probably a bursa copulatrix. There is no penis in the male and the method by which sperm reach the female can only be surmised: seminal fluid directed to the mantle cavity with the inhalant flow of water may be drawn up the accessory duct, and at some point on its course the sperm are separated from the prostatic secretion. This may occur in the muscular sac, though no sperm have been found with its contents.

In *Acicula fusca* (fig. 178E) the pallial oviduct has no thin-walled vaginal channel, but anterior to the origin of the sperm pouch the oviduct opens to the mantle cavity by a very short duct (omc), which may represent a much reduced vaginal channel separate from the duct. Creek (1953) suggested, however, that it is too short to be used by the penis since the mantle cavity is deep and the penis small; but she has not observed the method of copulation. Nor has it been observed in *Barleeia rubra* in which a diverticulum of the mantle cavity opens into the upper end of the pallial oviduct, near the double receptaculum seminis. In this species there is no vaginal channel and it is thought that during copulation the penis passes through the mantle cavity to the upper end of the pallial duct to deposit seminal fluid; this might account for the unusual abundance of gland cells in the epithelium covering the penis.

Johansson (1953) regarded the posterior opening of the oviduct to the mantle cavity as a primitive character. It may be assumed that in the gastropods the pallial ducts in both sexes were originally open grooves and internal fertilization may have been practised before the penis evolved. When during the course of evolution the oviducal groove closed, a proximal opening for the entry of sperm may have been retained at least until the penis developed. This condition is fulfilled in *Triphora perversa* (Johansson, 1953) in which the male has an open pallial duct and no penis, and the female a closed duct except for a posterior opening leading to the bursa copulatrix.

Other species in which no copulation takes place have an open pallial oviduct and the sperm pouches are typically in the thickness of the lateral walls and open ventrally to the mantle cavity. There may be both bursa and receptaculum, or receptaculum only. At the inner end of the pallial duct of *Turritella communis* (figs. 178M, 181) are 2 pouches, one on either side of the longitudinal glandular tract. The pouch lying against the right wall, near the columellar muscle, is the smaller and acts as a receptaculum seminis (rcs). Ciliated ridges pass from its opening to the mantle cavity and the effective beat of the cilia is away from the pouch, so that the sperm must swim against the ciliary current to enter it. The left pouch (af) is embedded in the thickness of the left wall and has a fairly extensive opening along the ventral edge; anteriorly the lips of this opening fuse and are continuous with the flange bordering the free edge of the wall. The lips are very mobile and can envelop the openings of the renal oviduct and receptaculum seminis. This second pouch is glandular and the eggs may be fertilized here and surrounded by albumen, though Johansson (1946) regarded it as a bursa copulatrix and the flange bordering the free edge of the wall as a sperm-collecting gutter (as in *Bittium reticulatum*). Each capsule consists of a group of eggs embedded in albumen and surrounded by a wall which is secreted as the spawn mass passes anteriorly. *Clathrus clathrus,* also with an open pallial duct and no copulatory organ, has only one sperm pouch, a receptaculum, which is posterior and embedded in the thickness of the right wall, but in contrast, in *Bittium reticulatum* (fig. 182), there are 3 sperm pouches. The bursa (bcp) in *Bittium* is anterior to a receptaculum (rcs) in the left wall of the duct, and along the free edge of the

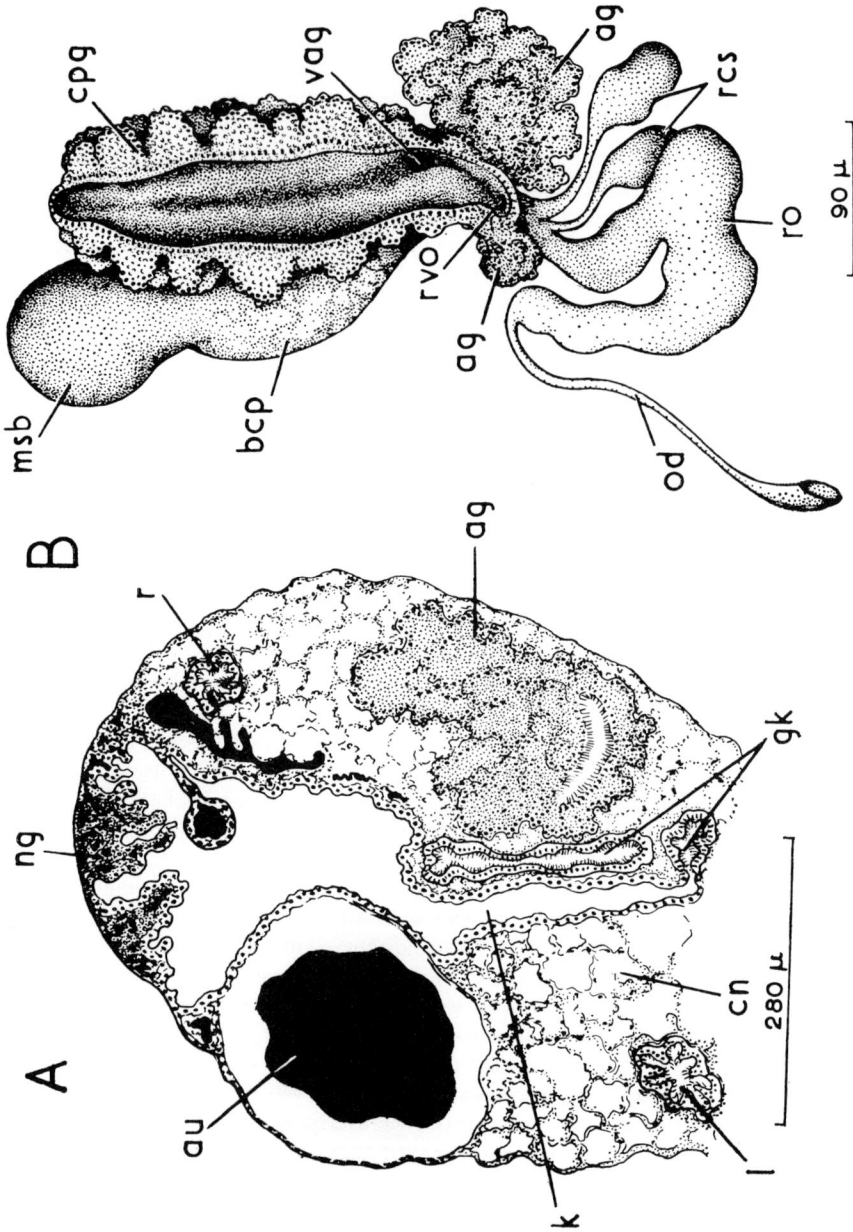

FIG. 180.—*Circulus striatus*: A, transverse section of female showing the connexion between the genital duct and the kidney; B, dissection of the female reproductive system. The capsule gland has been opened by a longitudinal cut, the ovarian duct transected near the ovary, and some of the albumen gland removed.

ag, albumen gland; au, auricle; bcp, bursa copulatrix; cn, connective tissue; cpg, capsule gland; gk, renal oviduct and its opening into kidney; k, kidney; l, intestine; msb, muscular sac at end of bursa copulatrix; ng, nephridial gland; od, ovarian duct; r, rectum; rcs, receptaculum seminis; ro, renal oviduct; rvo, opening of renal oviduct; vag, vagina.

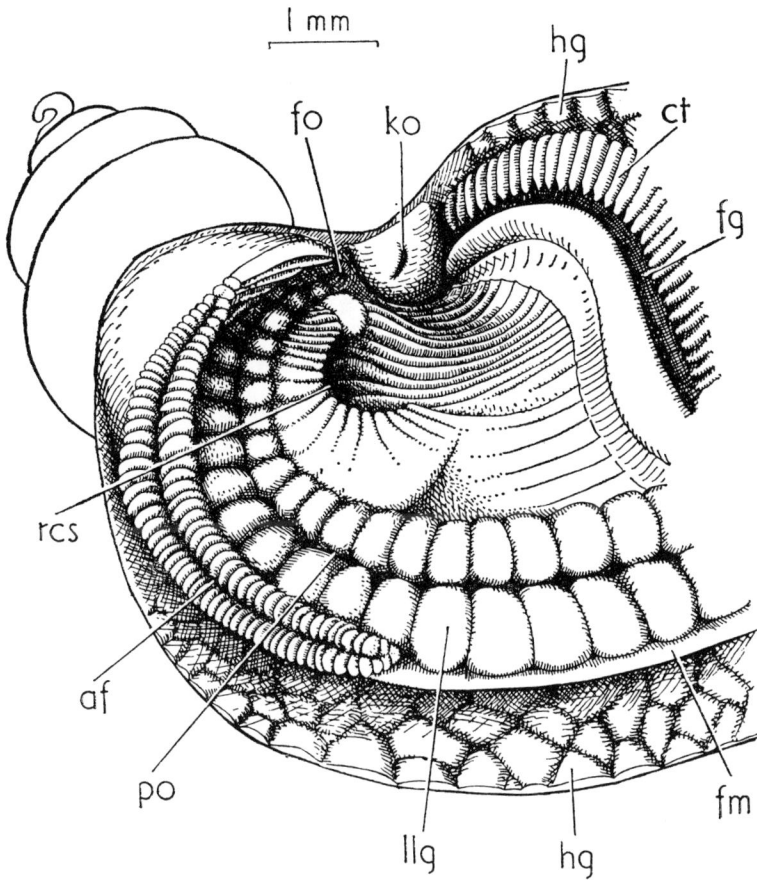

FIG. 181.—*Turritella communis*: upper part of the pallial oviduct.

    af, albumen gland and fertilization pouch; ct, ctenidium; fg, food groove; fm, flange bordering edge of left wall of pallial oviduct; fo, female aperture; hg, hypobranchial gland; ko, kidney opening; llg, left lobe of capsule gland; po, pallial oviduct; rcs, receptaculum seminis (see text).

wall leading to its opening is a pronounced ciliated groove. The third pouch is alongside the right wall of the oviduct and its opening is opposite that of the bursa (rcs); Johansson (1956*b*) found orientated sperm in the lumen, and suggested that this pouch is the homologue of the receptaculum of *Turritella* and that it is being functionally replaced by a new receptaculum in the opposite wall of the duct.

    Balcis alba has an open pallial oviduct (po, fig. 139) and there is a penis in the male. At the inner end of the mantle cavity the glandular oviduct is closed for a very short distance before it narrows and receives the muscular duct of a receptaculum seminis. The mantle cavity of *Balcis* is comparatively deep and curved, and it is doubtful whether the penis can reach the remote receptaculum, though it may deposit the sperm near its duct. The pallial oviduct is

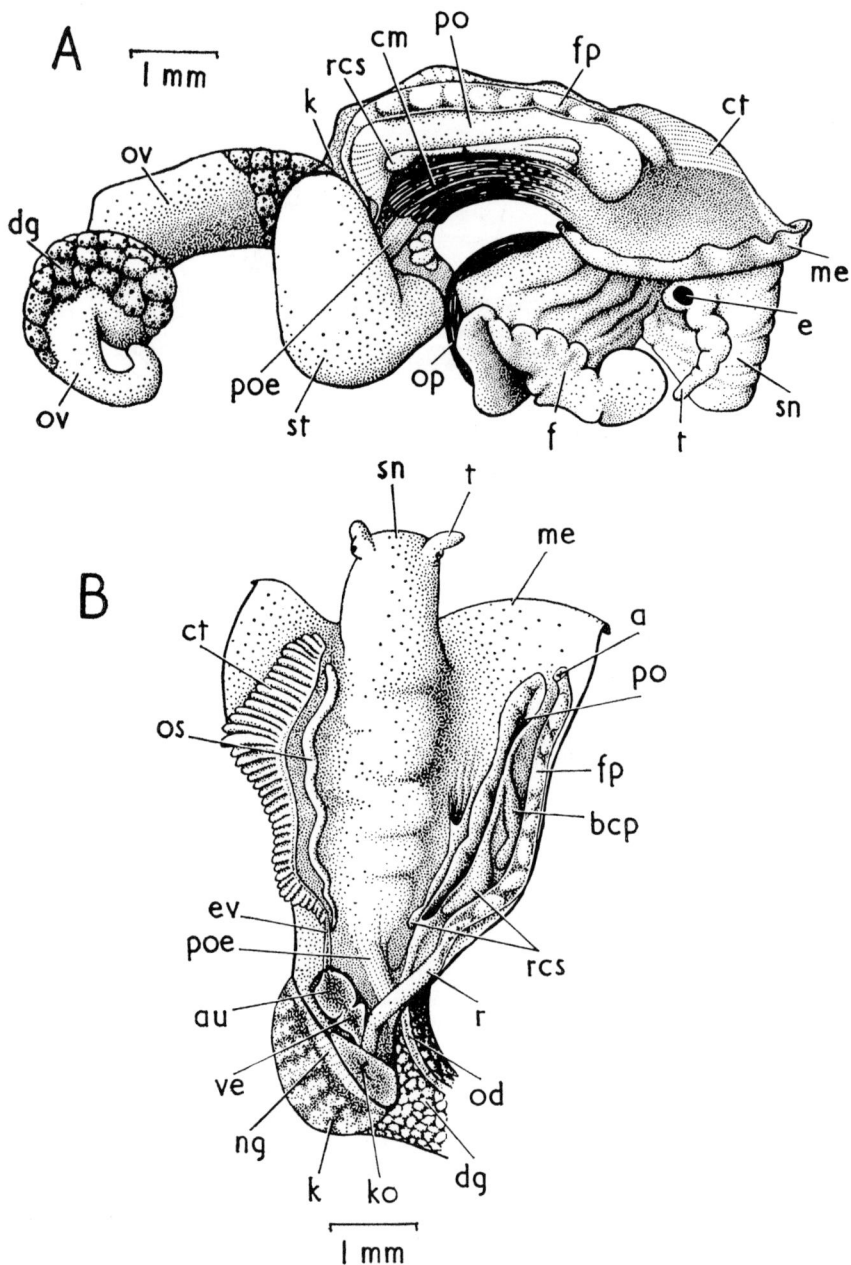

FIG. 182.—*Bittium reticulatum*; A, female removed from shell and seen from the right; B, female removed from shell and mantle cavity opened to show its contents.

a, anus; au, auricle; bcp, bursa copulatrix; cm, columellar muscle; ct, ctenidium; dg, digestive gland; e, eye; ev, efferent branchial vein; f, foot; fp, faecal pellet in rectum; k, kidney; ko, kidney opening; me, mantle edge; ng, nephridial gland; od, ovarian duct; op, operculum; os, osphradium; ov, ovary; po, pallial oviduct; poe, posterior oesophagus; r, rectum; rcs, receptaculum seminis (see text); sn, snout; st, stomach; t, tentacle; ve, ventricle.

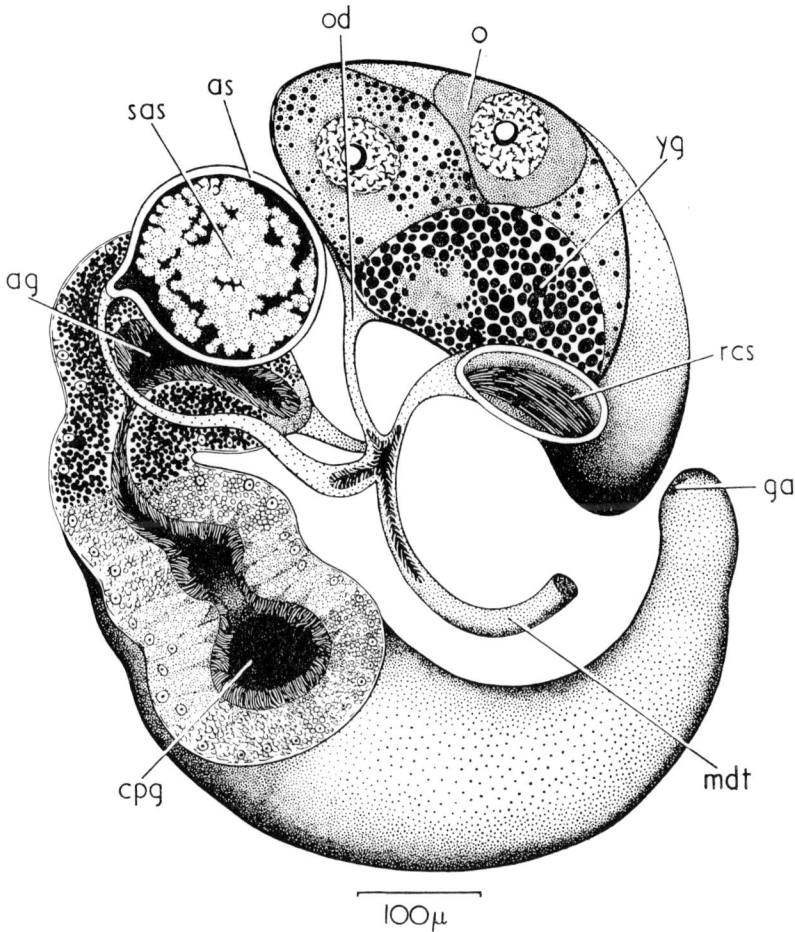

FIG. 183.—*Cingulopsis fulgida:* semi-diagrammatic representation of the female reproductive system, with some parts opened.

ag, albumen gland; as, accessory sac, perhaps homologous with bursa copulatrix; cpg, capsule gland; ga, genital aperture; mdt, muscular duct; o, ovum in ovary; od, ovarian duct; rcs, receptaculum seminis; sas, secretion in accessory sac; yg, yolk granule.

also open in *B. devians;* in this species, however, no males have been described (Fretter, 1955a).

Only a single sperm pouch, a receptaculum seminis with orientated sperm, is present in a number of monotocardians with closed oviducts; it is a dorsal outgrowth of the initial part of the pallial duct with no direct connexion with the ventral channel. As far as is known this pouch receives the sperm directly from the penis, which penetrates far into the oviduct. It may be a simple sac as in *Capulus* (Giese, 1915), *Erato voluta* and *Trivia monacha* (rcs, fig. 185A), or subdivided into diverticula, 6 in *Calyptraea* (Giese, 1915) (fig. 190), *Lamellaria* (fig. 185C) and

Trivia arctica (fig. 185B), 3 in *Crepidula* (Giese, 1915) and *Trichotropis* (Graham, 1954a). The duct of each diverticulum is muscular and may help the uptake of sperm during copulation and their later ejection for fertilization. There is no bursa in the toxoglossan *Mangelia* (Robinson, 1955) and the penis is extremely long and probably penetrates along the ventral channel to the proximal end of the capsule gland, where there is a receptaculum seminis and sperm-ingesting gland.

Internal fertilization and the manufacture of egg capsules in the closed pallial duct may lead to an accumulation of waste, and in some species there are means of ridding the duct of this. Mention has already been made of ingestion of spermatozoa in the female duct of *Acicula, Trivia monacha, Nucella* and *Mangelia*. In *Nucella* copulation and capsule formation may continue for some months, necessitating the maintenance of an efficient stock of sperm, and in accordance with this the ingesting gland is large. It is similarly developed in the other Stenoglossa as an outgrowth from the receptaculum seminis, though it may be smaller in those with a more restricted breeding period. The receptaculum (the duct of the ingesting gland (ig, rcs, fig. 163)) may be surprisingly short and surrounded by a thick muscular coat. In *Ocenebra* and *Buccinum* yolk granules have been found in the lumen of the gland and in the resorptive cells. The waste yolk, which is considerable in *Buccinum,* must originate from eggs which, for some reason, have not been included in a capsule. Sometimes amoebocytes may be seen in the bursa copulatrix of this whelk ingesting prostatic secretion which has been left there after the sperm have passed to the receptaculum. *Cerithiopsis tubercularis* has a sperm-ingesting gland associated with the receptaculum seminis (fig. 184). At the extreme posterior end of the open pallial oviduct is a small cul-de-sac which receives ventrally the renal oviduct and dorsally accommodates spermatozoa which have entered the duct by way of the inhalant water current. The sperm are orientated in a shallow pouch lined by ciliated cells and a duct leads dorsally from the pouch and bifurcates distally. Each branch (dr) opens to a receptaculum which is lined by relatively enormous cells, except around the entrance where there is a patch of low columnar epithelium with orientated sperm (sa). In the cytoplasm of the epithelium elsewhere there are tangled balls of sperm which are undergoing digestion (ds): the length of time these have been in the female is unknown. In some prosobranchs the sperm in the receptaculum appear to remain healthy for several weeks. In *Viviparus,* Ankel (1925) found sperm in the pouch 5 months after copulation; some, though only a few, were oligopyrene sperm which have about half the length of life of the eupyrene sperm. Sperm which make their way to the albumen gland in *Viviparus* live even longer—11 months has been estimated for the eupyrene and 7 for the oligopyrene—and Ankel (1925) suggested that the longer life is due to the fact that they are less crowded there than in the receptaculum.

The minute hermaphrodite *Rissoella diaphana* (fig. 186A) has another method by which waste material is cleared from the genital duct. There is a muscular sac (re) which communicates with the ventral wall of the posterior end of the capsule gland by a short duct, and has an outlet to the mantle cavity by way of a longer duct (omm). After an egg capsule has been laid the pouch may be enlarged by enormous quantities of unwanted secretions and spermatozoa; at other times it is empty and quite minute. Presumably the contents are emptied to the mantle cavity, for there is no indication that the mollusc puts them to any profitable use. The relationship of this pouch to the other parts of the genital duct suggests that it is homologous with a receptaculum seminis. The oviduct has no pouch in which orientated sperm are stored.

The glandular equipment of the pallial duct of the monotocardian gastropod provides the nutritive and protective coverings for the eggs. The inner surface of the duct is lined by a

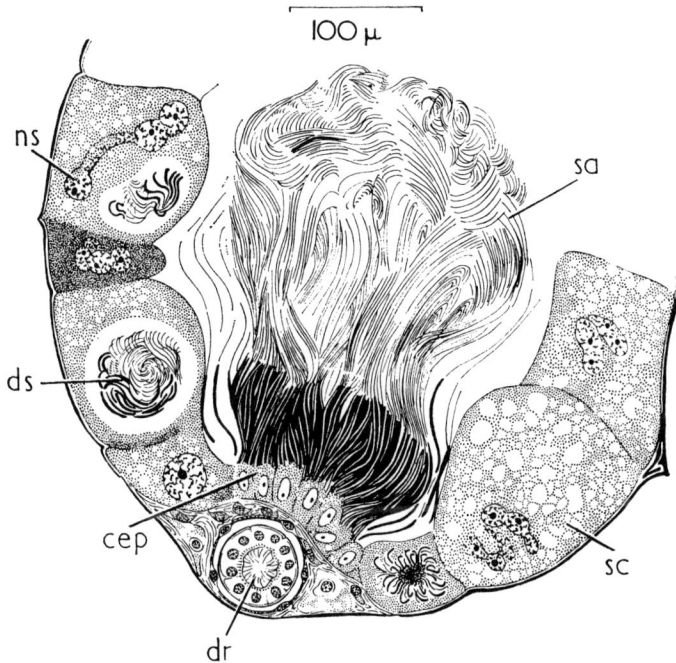

FIG. 184.—*Cerithiopsis tubercularis:* part of transverse section through receptaculum seminis, showing absorption of sperm.

cep, ciliated epithelium; dr, duct of receptaculum seminis; ds, sperm undergoing digestion; ns, nucleus of secreting cell; sa, orientated spermatozoa; sc, secreting cell.

ciliated epithelium and the secreting cells may be glands placed in the epithelial layer only (fig. 186A), or (frequently) grouped in clusters concentrated beneath the deep lateral walls (fig. 167A). The cilia can do little more than distribute the secretion over the surface of the duct, and the movement of the egg mass is brought about by muscles in the connective tissue surrounding the duct. Glands of the posterior part of the pallial duct secrete the albumen, whilst anteriorly the secreting cells form a capsule gland. In some species each egg with its supply of albumen is surrounded by a shell which is secreted by a special area of the wall of the duct near the anterior end of the albumen gland.

The simplest arrangement of glands is in the open pallial oviduct, which has an uninterrupted development of secreting cells from the inner end of the mantle cavity to its anterior limit, and the sperm pouches are placed typically against the lateral walls or in their thickness. In the closed duct, however, the outgrowth of sperm pouches from the ventral channel may separate a proximal from a distal glandular area, as in some rissoids and the Rachiglossa (fig. 178F, G, K). Moreover there is a tendency for the proximal glandular development to project posteriorly into the visceral mass. In some species a further complication arises when some of the accessory glands separate from the main course of the oviduct; in hermaphrodite forms this may also apply to the prostate (*Valvata* (pr, fig. 179B), *Velutina* (pr, fig. 186C)). The minute *Caecum glabrum* has separate sexes and alongside the oviduct are 2 glands, one on either

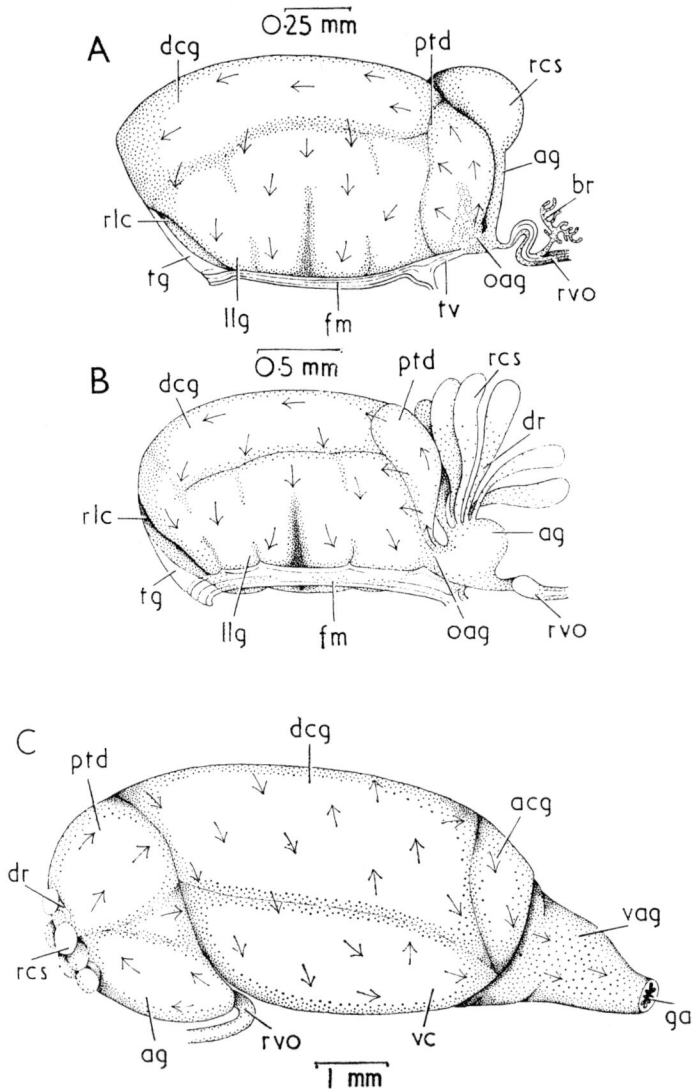

FIG. 185.—Renal and pallial oviducts, seen from the right, of A, *Trivia monacha;* B, *T. arctica;* C, *Lamellaria perspicua.*

acg, anterior lobe of capsule gland; ag, albumen gland; br, branching diverticula on renal oviduct; dcg, dorsal lobe of capsule gland; dr, duct of receptaculum seminis; fm, flange of tissue forming part of mouth of genital duct; ga, genital opening; llg, left lobe of capsule gland; oag, opening of albumen gland to capsule gland; ptd, posterior tip of capsule gland; rcs, receptaculum seminis; rlc, right lobe of capsule gland; rvo, renal oviduct; tg, thin anterior wall of capsule gland; tv, thin ventral wall of capsule gland; vag, vagina; vc, ventral channel.

side. They have been termed the uterus and nidamental glands by Götze (1938), and it is suggested that the former provides the albuminous covering for the eggs. It has no connexion with the oviduct except at the genital aperture, which consequently appears as a double opening. The opening of the nidamental gland is alongside but independent, and this gland, which probably produces the outer covering of the capsule, discharges not to the oviduct but to the mantle cavity. Such an unusual arrangement is presumably associated with the small size of the snail. A less extreme example illustrating the separation of the various elements of the pallial duct is found in female *Natica,* in which one gland lies along the course of the oviduct, and a large one is separated from it and opens to the base of the vagina alongside the oviduct and the bursa copulatrix.

Although in the majority of gastropods the sexes are separate and unchanged throughout the life of the individual, some species are consecutive hermaphrodites, first male and then female, and others are simultaneous hermaphrodites. The approach of the female phase in protandrous hermaphrodites involves not only a change in the gonad, but, in the monotocardians, the loss of the penis (if one is developed) and modifications of the pallial duct so that in older individuals there is no indication of the previous sex. When limpets of the genus *Patella* are examined they are found to be either male or female, yet there is evidence that in *P. vulgata* sex change occurs (Orton, 1920) in at least 90% of the population. Perhaps most if not all individuals are male when they first mature and change to female at the age of 1 year or more (Orton, 1928a). A study of this species from a range of localities in the British Isles (Orton, Southward & Dodd, 1956) has shown that the smaller limpets, between 16 and 25 mm in shell length, are at least 90% male. In those with a shell length of about 40 mm the sexes are approximately equal, while 60–70% of specimens having a shell length of 60 mm are female; in still larger specimens an even higher proportion of females occurs. Sex reversal (which takes place in a number of prosobranch molluscs) would seem the most reasonable explanation for this change in sex proportion with growth. There are no secondary sexual characters to involve considerable anatomical changes and since between periods of spawning the gonad has a long resting phase when sex is not determinable, a change occurring then would pass unnoticed. The Mediterranean limpet, *P. caerulea,* is also a protandrous hermaphrodite and the change from male to female gonad occurs during the resting period (Bacci, 1947; Pellegrini, 1948). In British patellids other than *P. vulgata* sex change is far less probable: 30–40% females are already present in the smallest size groups of maturing *P. aspera,* and in *P. intermedia* every size group with mature gonads shows 70% males—a high proportion of males may perhaps be correlated with external fertilization in a sedentary animal. [See p. 697.]

Occasionally limpets with a hermaphrodite gonad are found. Dodd (1956) examined 43,257 individuals of *P. vulgata* and 5 were hermaphrodite, a percentage of 0·012; a similar percentage was obtained for *P. intermedia* and 10 times more for *P. aspera.* He concluded that such hermaphroditism is accidental and has no connexion with change of sex.

The other diotocardians in which hermaphrodites have been reported are *Acmaea fragilis* (Willcox, 1898), and *A. rubella* (Thorson, 1935), which are not British, and *Puncturella noachina* (Rammelmeyer, 1925). Willcox (1898) examined 13 specimens of *A. fragilis* from New Zealand and found 5 female, 6 male and 2 hermaphrodites. She concluded that all individuals are at first male and then pass through a brief hermaphrodite phase before changing to females. Thiem (1917b), however, held that all monobranchs have separate sexes and that the sex of an individual is constant. In *Puncturella noachina* Rammelmeyer (1925) found spermatogenesis occurring in the gonad (of about 20 individuals) which had ripe eggs and assumed

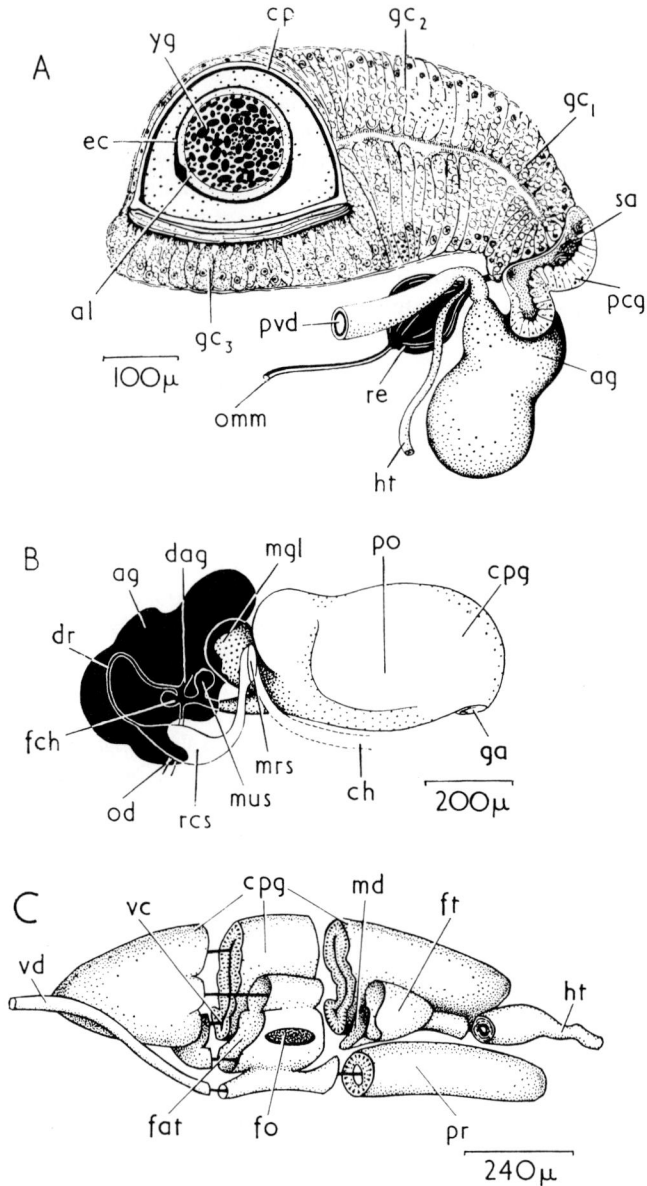

FIG. 186.—Genital system of A, *Rissoella diaphana;* B, *Skeneopsis planorbis;* C, *Velutina velutina;* the first includes a section across an egg capsule within the capsule gland.

ag, albumen gland; al, albumen; ch, channel along inner surface of mantle; cp, egg capsule; cpg, capsule gland; dag, duct of albumen gland; dr, duct of receptaculum seminis; ec, egg covering; fat, female atrium; fch, fertilization chamber; fo, female opening; ft, female duct; ga, genital opening; gc_1, gland cells secreting albumen-like substance; gc_2, gland cells secreting mucin-conchiolin; gc_3, gland cells which thicken base of egg capsule; ht, hermaphrodite duct; md, male duct; mgl, mucous gland; mrs, mouth of receptaculum seminis; mus, muscular pouch; od, ovarian duct; omm, opening of muscular sac to mantle cavity; pcg, posterior lobe of capsule gland; po, pallial oviduct; pr, prostate gland; pvd, pallial vas deferens; rcs, receptaculum seminis; re, muscular sac, probably homologue of receptaculum seminis; sa, spermatozoa; vc, ventral channel; vd, vas deferens; yg, yolk granules in egg.

that the species is hermaphrodite, though presumably differing from *Patella vulgata* and *Acmaea fragilis* in being a simultaneous hermaphrodite.

In the monotocardians consecutive hermaphroditism occurs in the Scalidae, the Ianthinidae (Ankel, 1926), the Capulidae and the Calyptraeidae. It has been studied particularly well in *Crepidula* perhaps because of the economic importance of this genus, its abundance and the unique association of the individuals. These limpets form groups of up to 12 animals each clinging to the shell of the one beneath in such a way that the right lips of the shells, and therefore the genital apertures, approximate and the whole chain bends to the right (fig. 189). The groups continue from year to year, newly arrived young attaching themselves to the tip of the group as the old individuals die at the bottom. On an average one individual is added to the chain each year (Walne, 1956). The whole chain may be fixed to the substratum by its oldest member or the oldest shell may be empty and the chain unattached. These chains have usually been regarded as an association for breeding but recent work by Wilczynski (1955) has suggested that they may be more important in creating a strong feeding current from which all members will benefit. The limpets take 4–5 years to reach full size and most individuals live for 8–9 years (Walne, 1956). Only the youngest are able to move freely, for after about 2 years the power of locomotion is gradually lost.

Conklin (1897) was the first to show that *Crepidula fornicata* is a protandrous hermaphrodite and this was later verified by Orton (1909). The oldest animals are females, the youngest males. At the time when an individual settles the gonad has both spermatogonia and oogonia, but in the male phase the oocytes remain small with the nuclei apparently devoid of chromatin. The penis becomes functional only when the gonad has produced enough sperm to fill the vesicula seminalis (Coe, 1948). The gonadial duct at this stage acts as a vesicula seminalis and at copulation the sperm are freed to the renal vas deferens and so to the posterior end of the mantle cavity. From here a ciliated groove passes forwards to the outer edge of the food groove, along which it runs anteriorly, and then crosses to approach the base of the penis. The penis is long and tapering with glands around the base and at the tip; when in use it is passed into the mantle cavity of the underlying female and inserted into the pallial oviduct. Association of young *Crepidula* with the female prolongs the male phase and stimulates spermatogenesis (Coe, 1944). In the older members of the chain, the females, the genital duct differs from that of the male in having a gonopericardial duct between the renal oviduct and the pericardial cavity, and a closed pallial duct with glandular walls. At the upper end of the pallial duct are tubules comprising the receptaculum seminis into which sperm are deposited by the penis. There is no bursa copulatrix. Sperm in the tubules of the receptaculum remain functional for more than a year (Coe, 1942) and in a single female of the species *onyx* the tubules store sufficient sperm to fertilize the 50,000–200,000 eggs produced within that time. The penis in the female phase is reduced to a vestige and is lost completely in older females (fig. 177). The early stages in its reduction can be followed in one or two individuals in the middle of a chain which are undergoing sex inversion (fig. 189): regression of the penis begins as spermatogenesis stops and continues as sperm in the gonad degenerate, and as follicle cells appear and multiply and eggs develop. Before the female becomes functional all sperm are cytolyzed, but the reduced penis may sometimes be seen in an animal incubating a spawn mass (fig. 204A). The pallial oviduct is formed during the transitional period by the closing of that part of the sperm groove which runs through the mantle cavity and by the hypertrophy of its walls. The isolation of a male during the breeding season will lead to the onset of the female phase (Coe, 1944), though usually the first female stage is reached in the third summer (Cole, 1956).

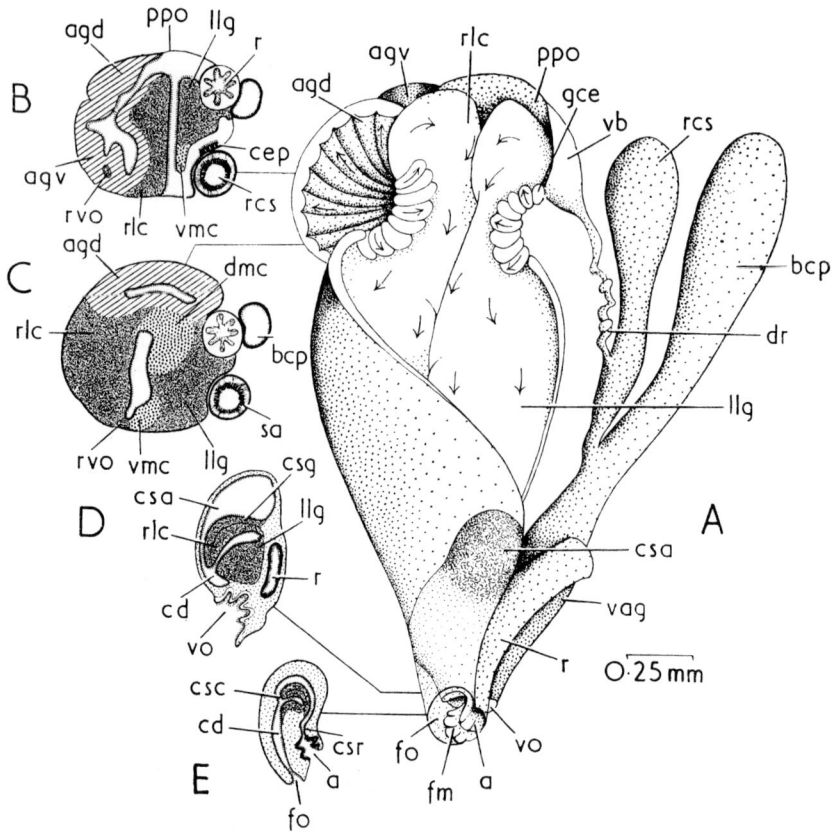

FIG. 187.—*Theodoxus fluviatilis*: A, pallial oviduct, seen from the dorsal surface, posterior end above. The bursa copulatrix and the receptaculum seminis have been dissected to one side and parts of the albumen and capsule glands opened. Arrows show the course of ciliary currents. B, C, D, E, transverse sections across the duct at the levels marked; in these the bursa and receptaculum are shown in their natural positions.

a, anus; agd, dorsal lobe of albumen gland; agv, ventral lobe of albumen gland; bcp, bursa copulatrix; cd, duct of capsule gland; cep, strip of tall columnar ciliated cells; csa, crystal sac; csc, opening of crystal sac to duct of capsule gland; csg, glandular ventral wall of crystal sac; csr, opening of crystal sac to rectum; dmc, dorsal strip of mucoid gland cells; dr, duct of receptaculum seminis; fm, fold acting as valve; fo, opening of pallial oviduct; gce, glandular and ciliated strip spreading ventrally from albumen gland; llg, left lobe of capsule gland; ppo, posterior wall of pallial oviduct; r, rectum; rcs, receptaculum seminis; rlc, right lobe of capsule gland; rvo, renal oviduct; sa, orientated spermatozoa; vag, vagina; vb, vestibule; vmc, ventral strip of mucoid gland cells; vo, opening of vagina.

The work of Gould (1919, 1947) and Coe (1938a, b, 1944, 1948) suggests that the transition from male to female occurs at different times in different individuals and indicates that sex change is influenced by other animals in the chain and by external factors. When immature limpets are cultured in association with mature females, the great majority assume the functional male phase and Gould (1919) held that the intimate association of the young individual with an older one is necessary for the male phase to develop. He presented evidence (1919, 1952) to show that the formation and maintenance of the male phase is influenced by a substance or substances secreted into the water by mature females, and this view was at first rejected by Coe (1944), who stated that there is no direct evidence that either the development of the male phase, or its prolongation by association with the female in any species of *Crepidula* depends upon chemical secretion or hormone which might pass through the water from the body of the female to the male. Coe also suggested (1938b) that a male individual, during association with a female, receives stimuli through the sense organs which influence its male characters, presumably by hormones secreted by its own body. However, in a later paper (1953) he supported Gould's conclusion that a female liberates a substance influencing the sexual activity of young individuals and this, he considered, immobilizes them in her vicinity; this influence is strictly specific. Not all young males react in the same way to the mating stimulus, for some leave their mating positions after a few weeks and move to neighbouring objects where they begin transformation to the female phase. Should an immature individual settle on a rock instead of joining a colony, the male phase may be relatively brief and followed rapidly by the transitional and then the female phase, or the animal may pass directly from the non-sexual to the female condition. Alternatively the solitary limpet may remain in the male state until an immature individual settles on it and then its sex changes (Coe, 1948). These variations may be due to congenital differences in the tendency of individuals towards maleness or femaleness (Gould, 1952; Coe, 1953; Montalenti, 1960). For *C. plana* it has been shown that the omission of the male phase may not be due to lack of opportunity for association with older animals, but may be brought about by starvation (Gould, 1947; Coe, 1948). In the Japanese slipper limpets, *C. aculeata* and *C. walshi,* small animals settling beside large females become male at a size about half that at which isolated animals do (Ishiki, 1936). Solitary males become hermaphrodite but those associated with females remain male and may grow as large as the female. When several small limpets of *C. walshi* settle together one rapidly becomes female, the others male.

Should individuals comprising a chain be separated the oldest are unable to attach themselves again to any object, perhaps because the mouth of the shell has grown to fit over another *Crepidula* shell or a specific rock surface and is useless in fitting against a surface of different shape. Those individuals die of starvation or through attack by enemies. The younger members of the chain, however, tend to collect in new groups with females at the bottom and young males at the top (Coe, 1938b). If this is achieved without attempted copulation or during periods when the animals are not sexually active then it would seem to imply that these limpets display a recognition of sex which in other prosobranchs is pronounced only during periods of copulation, but this is perhaps not surprising in animals which are permanently associated.

The sex cycle of *Calyptraea chinensis* differs from that of *Crepidula* in several respects. In *Calyptraea* sex change always occurs at a particular stage in the life cycle (Pellegrini, 1949) and not, as in *Crepidula,* at different times in different animals (fig. 191); moreover, each individual passes through a functional male phase. There appears to be no self-fertilization (Wyatt, 1957). Males and females are associated only during the breeding season when the smaller male is carried by the female; at other times they are isolated. Bacci (1951) working on *Calyptraea chinensis* at Naples, where breeding occurs between December and May, showed that in the developing gonad of a young individual and in the youngest functional male which has recently paired, oocytes are absent. In larger specimens up to 6 mm in length, which may be isolated or in copulation, oocytes are developing, for during the functional male phase the gonad becomes hermaphrodite. During its first breeding period each individual is a functional male only. This may last 3–4 months and then, as the gonad changes, the excess sperm are resorbed and the penis is reduced and finally lost. The individual may have a second and third season as a female, and meanwhile growth continues and it may attain 17 mm in shell length. The changes in sex appear to be unaffected by external stimuli. In the male phase the pallial genital tract is similar to that of Crepidula in that there is an open seminal groove, which runs from the inner end of the mantle cavity to near the tip of the penis. With the loss of the penis that part of the seminal groove which extends beyond the opening of the mantle cavity is lost, too, whereas that part within the cavity closes to form the pallial oviduct, and narrow tubules, usually 6, grow out from the dorsal wall at the upper end of the duct; their openings lie close together and can be closed by muscles in their walls, and the blind end of each swells to form a vesicle for sperm (rcs, fig. 190). These structures constitute the receptaculum seminis. In the female phase a gonopericardial duct is present.

The sex cycle of Capulus ungaricus and Trichotropis borealis (Graham, 1954a) resembles that of Calyptraea chinensis, though there are differences in the structure of the genital duct (Giese, 1915). In the functional male of Capulus the renal vas deferens opens anteriorly to a small pouch from which the seminal groove passes to the mouth of the mantle cavity, and then along the right side of the head and along the penis to its tip. The pouch was regarded by Giese as the initial part of the uterus and even in the male phase a single dorsal vesicle opens from it, the future receptaculum seminis. The penis has no glands and tapers to a pointed tip. Very young males were not examined by Giese to see whether the upper part of the uterus and the receptaculum were already developed. In the male phase of Trichotropis there are rudiments of the female accessory glands which may, perhaps, have some prostatic function (Graham, 1954a). In neither species has a gonopericardial duct been found. [For more recent work on hormonal control in calyptraeids see chapter 28.]

Much less is known about the change in sex in the various members of the Scalidae and Ianthinidae. Ankel (1936a) stated that in Scala (= Clathrus) clathrus in the Gulf of Naples each breeding season brings a sex reversal, the female phase of the gonad overlapping with the male so that oogenesis begins in the superficial ends of tubules in which spermatogenesis can still be followed in the lumina and ducts. Laursen (1953) suggested that in Ianthina there is a similar sex change repeated several times throughout the life history, not perhaps in each breeding season, though this is not confirmed by Ankel (1926, 1930a) for I. pallida (=I. bicolor of Ankel; see Laursen, 1953), nor by Graham (1954b) for I. janthina. These workers found that the smaller animals are males and the larger females and concluded that sex change occurs only once in the life of the individual. In both genera the change in the genital duct between male and female phase is relatively slight. There is neither penis in the male, nor bursa copulatrix in the female, nor a receptaculum in Ianthina. In Clathrus the pallial duct is open in both sexes and the female differs from the male in having a receptacular pouch at its inner end. In Ianthina janthina (Graham, 1954b) the genital duct is closed throughout its length. The pallial section in the male (fig. 188A) consists of a caecal posterior part with numerous tubular glands which are not mucous (pgl), and an anterior prostatic part in which the walls are covered with an epithelium mainly of mucous cells and are flung into irregular foldings (fpr). All parts of the male duct contain spermatozeugmata and the glands may provide them with nutrient material. However, Graham suggested that some part of this glandular equipment foreshadows the female condition (fig. 188B) in which the caecal posterior section of the duct is enlarged to form the albumen gland (ag) and there is hypertrophy of the anterior section (mgl) so that the walls, almost entirely covered by large mucus-secreting cells, are irregularly folded to form 3 pouches on the right side. This species is viviparous and free-swimming veligers are liberated (Wilson & Wilson, 1956). Spermatozeugmata liberating eupyrene sperm may be seen in all parts of the female duct. Fertilization appears to take place in the ovary, which is filled with developing eggs and young, and spermatozeugmata in all stages of degeneration. The genital duct contains older embryos and the mucous glands produce a histotrophe for their nourishment. The site of fertilization is unusual, though Ankel (1930a) had already suggested that in *I. pallida* fertilization probably occurs as soon as the eggs have escaped from the germinal epithelium and in the very tubules in which they have matured. Two British species of *Ianthina*, *I. exigua* Lamarck and *I. pallida* Thompson are oviparous, and according to Laursen embryos of oviparous species are nourished by mucus from the egg capsule, so that in all species mucus of oviducal origin is the pabulum on which the embryos are reared.

There are 4 British prosobranchs which are viviparous—*Littorina saxatilis* (fig. 170B), *Potamopyrgus jenkinsi* (fig. 301), *Viviparus viviparus* (fig. 215) and *V. contectus*—and the pallial duct is modified to form a brood pouch in which the embryos are retained until their development is complete, although Smidt (1944) stated that the young of *P. jenkinsi* may hatch with a very small velum and have a brief pelagic stage. Both albumen and shell glands are developed in the pallial oviduct of the viviparous species, but the capsule gland is replaced by a thin-walled brood pouch; before the embryos enter it each is provided with albumen and a shell. Along the ventral wall of the pouch in *Littorina* and *Viviparus* (Ankel, 1936a) are 2 longitudinal folds limiting a sperm channel which leads to the receptaculum, and a similar channel is developed in *Potamopyrgus*. The brood pouch is more complex in *L. saxatilis* and may accommodate up to 900 eggs according to Linke (1933), though in winkles from the Channel Pelseneer (1911) found only two-thirds this number, and at Whitstable the larger individuals have only about one-third (Berry, 1956) at the height of their breeding season; very young individuals may have less than 40 eggs. What varieties of the species were examined by these workers is not known, nor is it known whether the varieties differ from one another in this respect. Thorson (1946) suggested that one factor affecting productivity is the salinity of the water in which the animals are living. Transverse folds arise from the longitudinal folds and subdivide the lumen of the pallial oviduct of *L. saxatilis* into smaller pouches in which the embryos are grouped. Groups of about 20 are at the same stage of development with the youngest ones nearest the upper end; the embryos move forward as their development proceeds. The wall of the pouch is vascular and its epithelium is ciliated and has mucous and protein-secreting cells. These glands produce a fluid which surrounds the embryos and is kept in circulation by the ciliary currents. There is no evidence that the embryos obtain nourishment from the fluid or directly from the maternal tissues; if they are removed from the brood pouch they will develop in sea water; moreover, the shell is impermeable to proteins (Linke, 1933). The albumen, however, supplements the yolk provided in the egg. It becomes less viscous during development and the veliger rotates in its albuminous covering and gradually devours it (fig. 214). The pronounced vascularization of the brood pouch is presumably correlated with the respiratory requirements of the brood. A holotrichous ciliate *Protophrya* is frequent in the mantle cavity of *L. saxatilis* and enters the pallial oviduct where it may be seen creeping over the surface of the egg shells.

At Cullercoats, on the Northumberland coast, Seshappa (1947) found specimens of a thin-shelled variety of *L. saxatilis* otherwise agreeing with the variety *patula* of Jeffreys (1865). The reproductive system is unusual in that a capsule gland replaces the brood pouch, and individuals reared at the Cullercoats laboratory laid gelatinous egg masses. [See p. 681.]

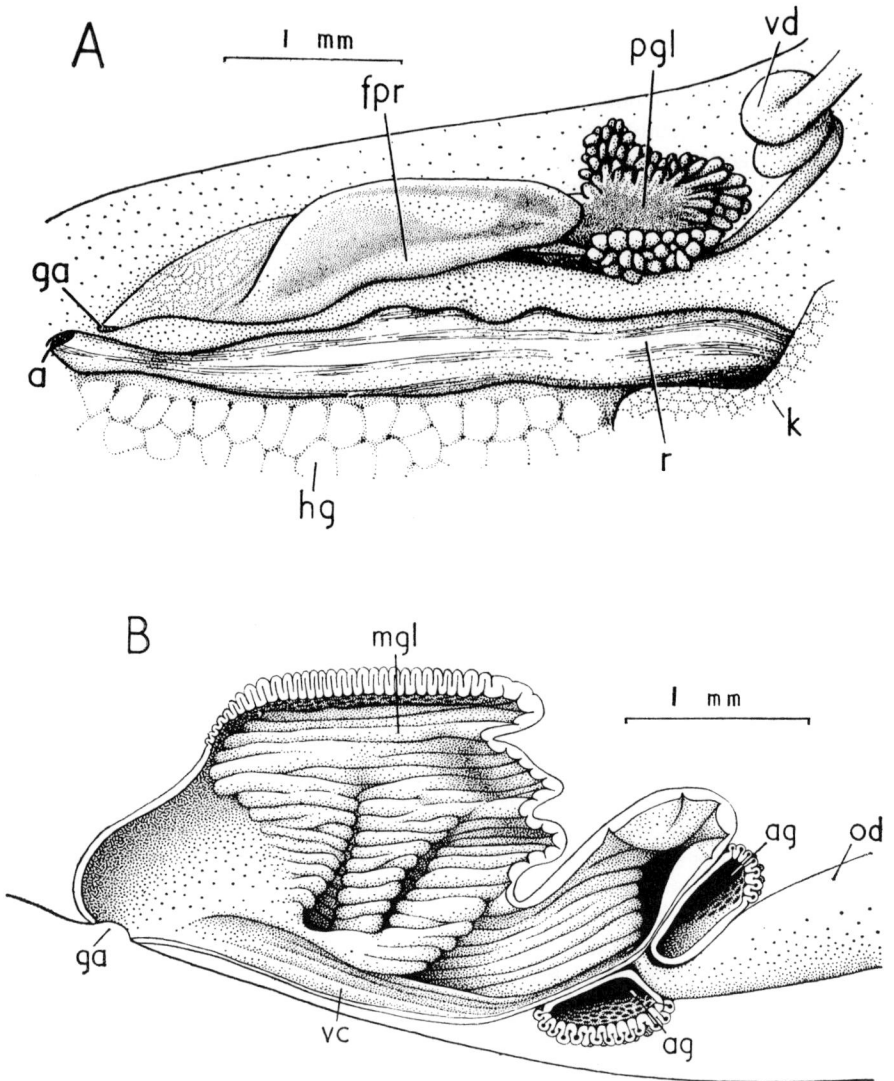

FIG. 188.—*Ianthina janthina*: A, dissection to show genital duct of animal in male stage, seen in ventral view from the mantle cavity; B, dissection of genital duct of animal in female stage.

a, anus; ag, albumen gland; fpr, fold within prostate gland alongside ventral channel; ga, genital aperture; hg, hypobranchial gland; k, kidney; mgl, mucous gland; od, ovarian duct; pgl, pigmented gland on male duct; r, rectum; vc, ventral channel; vd, vas deferens.

The two other viviparous species, which are freshwater forms, have fewer embryos—35–40 in a fully grown *Potamopyrgus* and up to 96 (Heywood, personal communication) in *Viviparus viviparus*—and in these forms the brood pouch is not subdivided into compartments nor is there evidence that the embryos obtain nourishment other than yolk and

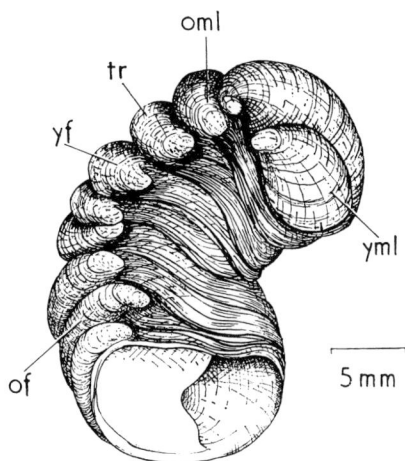

FIG. 189.—*Crepidula fornicata:* chain.
of, old female; oml, old male; tr, transitional form; yf, young female; yml, young male.

albumen. Until recently no male of *P. jenkinsi* had been found and the species was known to be parthenogenetic. Boycott (1919) isolated very young individuals which, at maturity, reproduced, and since there was no evidence of a hermaphrodite gland he concluded that the species is parthenogenetic, the only parthenogenetic British prosobranch. These results have been verified by others (Quick, 1920). The discovery of a single male of the variety *carinata* (Patil, 1958) sheds new light on the origin of parthenogenesis in the species (p. 551). The reproductive system agrees in general plan with that of *Hydrobia ulvae*, but has no accessory opening to the mantle cavity. The specimen was of a size equivalent to the normal parthenogenetic form, the testis contained ripe sperm, though it was small, and the prostate was of moderate size and embedded in the visceral mass. [See p. 663.]

The 2 pouches opening into the oviduct of *P. jenkinsi* (bcp, rcs, fig. 178H) correspond in position to the sperm pouches of related forms; they have not been found to be used as such, though at least one of them has some functional importance. This is the pouch corresponding to the bursa copulatrix. It is relatively enormous in size and in all specimens with embryos in the brood pouch is filled with secretion. This secretion has staining properties which suggest that it is albuminous, and there is also a mucous component which must be derived from a gland adjacent to the albumen gland. Each egg in the brood pouch is surrounded by albumen and externally by a thin mucous coat which forms the shell. It receives both secretions from glands at the posterior end of the brood pouch (ag, mgl); the mucous gland surrounds the posterior wall of the pouch (bp). It is assumed that secretions not used in the formation of the egg capsules are sucked into the bursa copulatrix which thus acts as a waste dump. This may be a subsidiary function of the bursa in related species, though they produce egg capsules, and unwanted secretions from the duct may be extruded when the capsules are spawned. The ultimate fate of the waste in *P. jenkinsi* is unknown: it may be voided to the mantle cavity by way of the ventral channel which is partially separated from the brood pouch by an epithelial fold.

The species of monotocardians which are simultaneous hermaphrodites are *Valvata piscinalis* (fig. 179B), *Rissoella diaphana* (figs. 179A, 186A), *R. opalina*, *Pelseneeria stylifera*, *Velutina velutina* (fig. 186C) and perhaps *Omalogyra atomus* (figs. 176A, 179C). Sperm and ova are either produced in the same acini of the gonad (*Rissoella* spp.), or in different acini (*Valvata, Velutina*), or the gonad is bilobed with an ovary and a testis (*Omalogyra*). The lay-out of the hermaphrodite ducts varies considerably from species to species, suggesting that her-maphroditism has arisen independently in the different groups to which these animals belong. It is simplest in *Rissoella*. The hermaphrodite duct, on reaching the posterior end of the body whorl, divides into 2 branches which diverge, and one leads to the pallial vas deferens (pvd, fig. 179A) and the other to the pallial oviduct (ag). The plan of each of the 2 pallial ducts resembles that of small monotocardians in which the sexes are separate: the vas deferens runs forwards beneath the pallial oviduct, becomes glandular, forming a prostate, towards the mouth of the mantle cavity and passes up the right side of the head and through the penis to its tip; no glands occur in the penial duct. The pallial oviduct comprises an albumen gland and a capsule gland, and the duct of a muscular sac opens ventrally to the wall which separates them. The female aperture is at the mouth of the mantle cavity. Excess secretion and sperm from the pallial oviduct collect in the muscular sac and are discharged to the mantle cavity by way of a short duct (omm). The hermaphrodite duct of *Velutina* is, in plan, a modification of the genital duct of *Trivia*. In both sexes in *Trivia* (fig. 185A, B) the pallial region forms a pouch of considerable size which is glandular, with a pronounced develop-ment of muscles, and opens by a ventral longitudinal slit to the mantle cavity. The corres-ponding pallial region in *Velutina* (fig. 186C) is a similar pouch, and its opening is the female aperture (fo), but the glandular elements are stripped from the walls leaving a columnar ciliated epithelium rich in mucous cells, and a thick muscle coat. There is a tubular gland opening into the posterior wall, ventral to the opening of the hermaphrodite duct, and a large composite gland running parallel with the pouch and extending beyond it both anteriorly and posteriorly. This gland has a ventral channel, with secreting cells unlike those elsewhere, and with an independent opening to the anterior wall of the pouch, adjoining the opening from the dorsal part of the same gland. The posterior tubular gland is probably a prostate (pr) since from its opening a short, deep, ciliated groove, into which is directed a groove from the hermaphrodite duct, passes anteriorly to the opening of the vas deferens. The narrow vas deferens (vd) runs up the right side of the head and through the penis to its tip. The egg capsule in *Trivia* is moulded by the ventral pedal gland of the female. This gland is not developed in *Velutina,* and in this species the capsule may be fashioned by the muscular pouch of the hermaphrodite duct. In *Mangelia* and perhaps other toxoglossans the outer wall of the capsule is shaped by a terminal section of the oviduct and not by a ventral pedal gland. Despite this all capsules have the same accuracy of form.

The genital system of *Valvata piscinalis* and *Omalogyra atomus* is more complex. In *Valvata* (fig. 179B) there are 2 genital openings—the male at the tip of the penis and the female near the mouth of the mantle cavity. As in *Rissoella,* the hermaphrodite duct (ht), which acts as a vesicula seminalis, bifurcates anteriorly to give oviduct and vas deferens. The latter has a straight course to the tip of the penis, and, not far from its origin, receives the opening of an enormous prostate gland (pr) which is separated from the duct and spreads in the haemocoel. The arrangement of the oviduct and its glands is not so simple. The initial part of the duct is narrow, and it then swells rather abruptly into a large pouch (fch), the fertilization chamber (Cleland, 1954), constricts again, and enlarges into a second pouch (bcp), the bursa copulatrix. It passes from the bursa as a narrow channel which, after receiving the duct from the

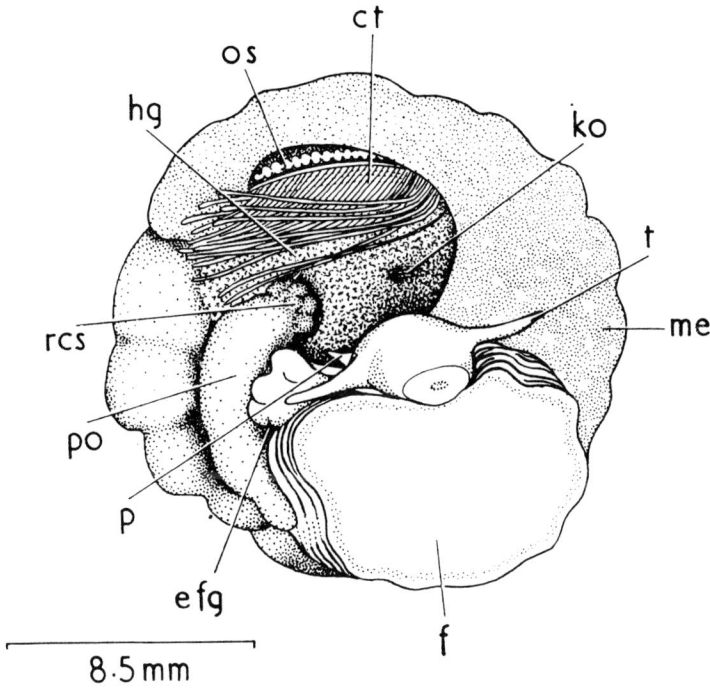

FIG. 190.—*Calyptraea chinensis:* female, in ventral view, dissected to show receptaculum seminis.
ct, ctenidium; efg, expanded anterior end of food groove; f, foot; hg, hypobranchial gland; ko, kidney opening; me, mantle edge; os, osphradium; p, remains of penis; po, pallial oviduct; rcs, receptaculum seminis; t, tentacle.

albumen gland (ag), opens into the anterior end of the capsule gland (cp), not far from the genital aperture (fo). It has been suggested by Johansson (1955) that this course, along which the eggs travel to the capsule gland, represents the vaginal channel of less specialized monotocardians, which has separated from the accessory female glands. He described a narrow connexion (gtd) between the albumen gland and the hermaphrodite duct (it was previously found by Garnault) and suggested that the separation of the channel starts from that point. The connexion is apparently inessential since it is not always present.

Omalogyra atomus (figs. 176A, 179C) stands apart from the other hermaphrodite forms in having the gonad divided into ovary (ov) and testis (te) and only one genital opening near the mouth of the mantle cavity on the right side. The only structure which could act as a copulatory organ is a whip-like muscular tube (p) (not present in summer individuals, p. 324) which may be protruded through the genital opening. From the genital aperture (ga) the hermaphrodite duct passes back for a short distance and bifurcates into vas deferens and oviduct which are both glandular, forming the prostate gland (pr), and the capsule and albumen glands (cp, ag) respectively. Posteriorly each duct narrows before opening to the fertilization chamber (fch), which also receives the testicular (td) and ovarian (od) ducts. Immature individuals tide over the winter, and animals collected in spring show the vas deferens as the principal genital duct. At this stage the opening of the pallial oviduct into the

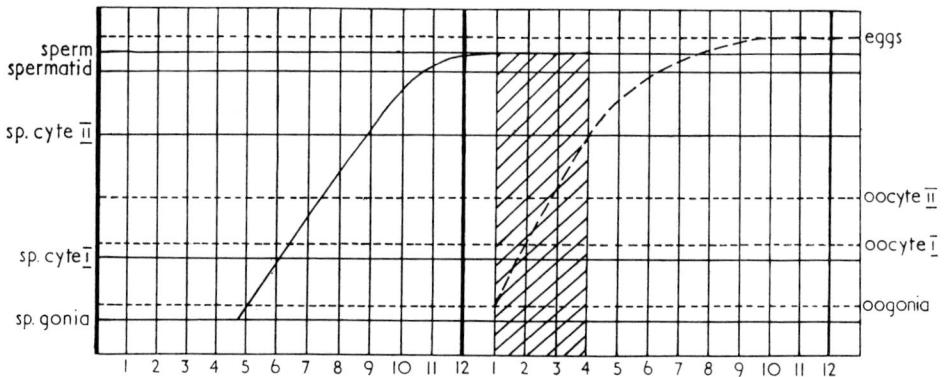

FIG. 191.—Graph showing the reproductive cycle of *Calyptraea chinensis*. Figures on the abscissa represent months of the year; on the ordinate the solid line represents the development of spermatozoa, the pecked line, of eggs. The hatched area represents the transition from male to female. After Pellegrini.

hermaphrodite duct is far too small to allow the passage of an egg capsule—in other words the animal is purely male. It is assumed that later the reproductive system undergoes the necessary changes associated with the adoption of the female phase. This is found in all summer individuals. The capsule gland is then voluminous and lies on the right side of the vas deferens and not above it as in spring forms, and it is broadly open to the hermaphrodite duct anteriorly; the opening of the vas deferens into this duct is minute and surrounded by a sphincter; there is no sperm sac or penial tube and the bursa has a long muscular duct, but has itself an insignificant musculature. The histology of the bursa in the female phase resembles that of the sperm sac (spc) in the male phase (p. 325), and the function of these 2 structures appears to be similar, for in the bursa waste secretion from the genital ducts may accumulate and later be disposed of; the accumulation is greatest after an egg capsule has been deposited. This anatomical femaleness does not prevent the formation of apparently ripe spermatozoa in the testis, and their passage into such parts of the male system as are present, and although the lack of a penis and associated structures would appear inevitably to preclude copulation and cross-fertilization, it may not be incompatible with successful self-fertilization.

The pyramidellids, which are superficially prosobranch in appearance, show evidence of their opisthobranch affinity in the structure of the reproductive organs (fig. 179D). The gonad is hermaphrodite (hgl) and sperm and ova are produced in the same tubules. There is a single genital duct (ht) which extends from the gonad to the genital aperture (ga). This is at the anterior end of the mantle cavity in *Odostomia* and *Chrysallida* and on the propodium in *Turbonilla elegantissima*. The duct is divided into an upper thin-walled region and a broader pallial part which is glandular and ciliated in its proximal half, and ciliated and muscular distally. The gland cells of the pallial duct are epithelial in position, and they appear to be functional only during early sexual maturity when the male system, but not the female, is mature; they presumably produce a prostatic secretion. Running along the median wall of the pallial duct as far as its inner end is a ciliated gutter which communicates with a single sperm pouch (rcs). After copulation spermatozoa fill this pouch, surround its opening to the pallial duct and are concentrated along the ventral gutter. Secretions concerned with the

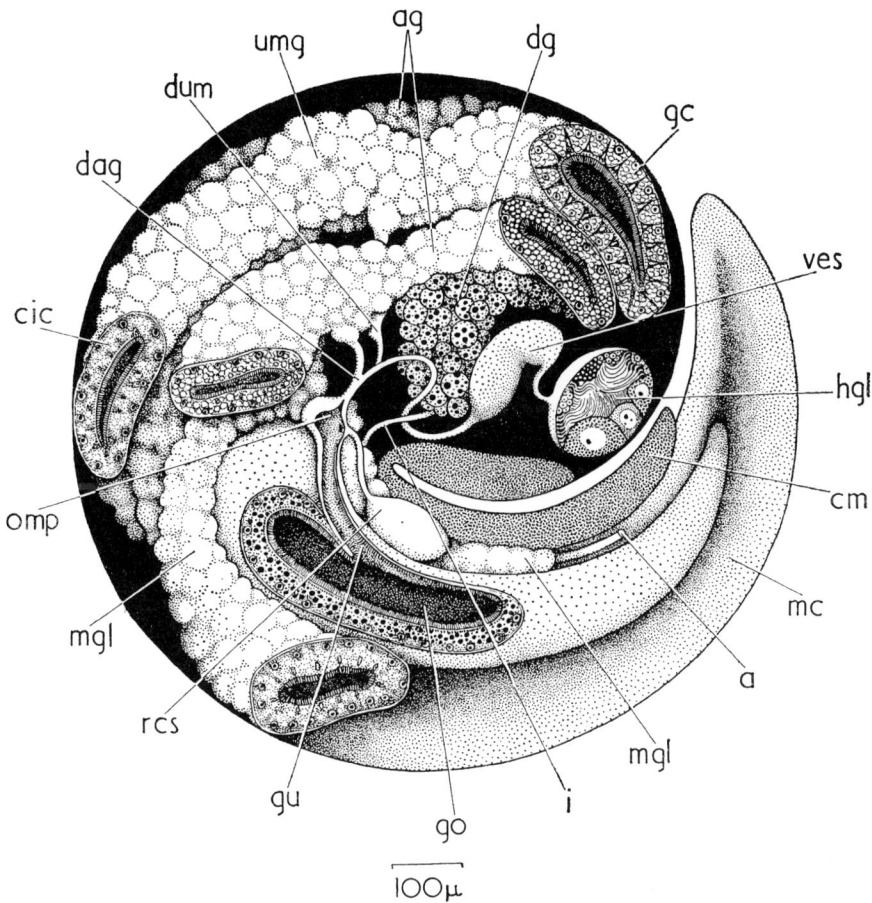

FIG. 192.—*Odostomia lukisi*: the visceral hump has been cut across at the level of the reproductive ducts and the cut surface is viewed from above; the haemocoel is black.

a, anus; ag, albumen gland; cic, ciliated cell; cm, columellar muscle; dag, duct of albumen gland; dg, digestive gland; dum, duct of upper mucous gland; gc, gland cell; go, glandular part of pallial hermaphrodite duct; gu, gutter along glandular part of pallial hermaphrodite duct; hgl, hermaphrodite gland; i, intestine; mc, mantle cavity; mgl, lower mucous gland; omp, opening of lower mucous gland into glandular part of pallial hermaphrodite duct; rcs, receptaculum seminis; umg, upper mucous gland; ves, seminal vesicle.

formation of egg capsules come from glands which are separated from the hermaphrodite duct and open into its upper thin-walled part (ag, mgl). As in opisthobranchs the ova receive their albuminous covering as they pass along the spermoviduct and do not enter the albumen gland, and large mucous glands provide a jelly in which the eggs are embedded. Other opisthobranch characters are evident in the open seminal groove (sgr) leading from the common genital aperture to the penis, which is invaginable, and the sperm sac, which is attached to the penial sheath and is filled with spermatozoa before copulation (p. 324). This general plan resembles the arrangement of the male organs in the opisthobranch *Philine aperta,* though the position of the penis in the pyramidellid is unique (perhaps associated with the evolution of the long acrembolic proboscis), and the structure of this organ resembles more closely that of *Omalogyra atomus.*

Within the prosobranch gastropods all types of sexuality occur. The majority are separately sexed. Of the hermaphrodites some are simultaneous hermaphrodites, others consecutive hermaphrodites and in one species, *Clathrus clathrus,* sex reversal appears to take place at each breeding season. Hermaphroditism prevails in the other two orders of the class, the Opisthobranchia and Pulmonata, so that of more than 20,000 species of living gastropods only about half are unisexual. It is partly on this account that there have been varied speculations as to the sex of the archimollusc and archigastropod. Simroth (1896–1907) suggested that the archigastropod was hermaphrodite and, as in Turbellaria, the gonad was comprised of male and female follicles. The Diotocardia with protandrous species support this idea of a hermaphrodite ancestor, but Simroth regarded the lack of copulation and accessory glands in this group as secondary and not primary, as is more commonly believed. On the contrary, Pelseneer (1895) held that the archimollusc was not hermaphrodite, and he thought of the most archaic forms of the different groups of molluscs as unisexual, with hermaphroditism as a sign of advancement. Giese (1915) agreed with him. A study of the reproductive ducts of the hydrobiids influenced Krull (1935) in supporting Simroth's view. The subdivision of the pallial oviduct into the main glandular egg-conducting area and the ventral ciliated groove he likened to the bipartite division of the spermoviduct of monaulic pulmonates. He regarded the pulmonates as the most primitive and suggested that the Monotocardia and the Neritidae evolved from them on the one hand and the opistho-branchs on the other, with the more archaic gastropods, the Diotocardia, branching from the stem arising from the hermaphrodite Urgastropod before the pulmonates evolved. These rather unorthodox ideas are open to much criticism; as Hubendick (1945) has pointed out, too much importance is attached to a single character—the spermoviduct—in assessing the affinity between such major groups of the Mollusca. However, in view of the fact that hermaphroditism occurs in all classes of the phylum with the exception of the most highly specialized, the cephalopods, it is tempting to believe that the ancestral mollusc was hermaphrodite, and although this condition may not have persisted into most modern diotocardians, the tendency to produce a hermaphrodite condition is undoubtedly present.

CHAPTER 15

SPAWN

THE eggs of diotocardians are each surrounded by a vitelline membrane and a sphere of albumen, both produced by the ovum, and in species which form spawn masses they are bound together in a gelatinous secretion. This is produced either by the ovum (*Diodora*) (gsh, fig. 159B), or by the enlarged urinogenital papilla (*Margarites;* ugp, fig. 161), or by a short pallial duct (*Cantharidus, Calliostoma;* ugp, fig. 162). It appears to form no barrier to external fertilization but this must occur close to the time of spawning, for the secretion hardens considerably when exposed to sea water. The egg mass is manipulated by the foot which fastens it to the substratum. No monotocardian is known in which the ovum produces all the investments which surround it—in fact there is no suggestion except in the family Lacunidae (Hertling, 1928) that it produces anything but the vitelline membrane. Their eggs have more elaborate coverings derived from the pallial oviduct, which extends through the length of the mantle cavity; these last longer and the eggs are slower to develop than those of diotocardians. They are provided with albumen which is used as food for the developing embryos and which in *Viviparus* contains mineral salts but neither carbohydrates nor fat (Charin, 1926). The albumen is not homogeneous; the part immediately surrounding the embryo is typically semifluid and clear, the remainder more viscous and usually granular, changing in consistency as development proceeds. In the land snail *Succinea putris* (George & Jura, 1958) there is an outer insoluble skin of albumen which gives positive tests for arginine, tyrosine, -SS- groups and polysaccharides, and which presumably constitutes the last food of the embryo before it escapes. The inner albumen comprises polysaccharides and simple proteins soluble in water. These snails have no free larva and need a considerable supply of food. All but a few species of monotocardians produce egg capsules with a resistant wall which may allow passage of salts and water. This is unlikely to be true of freshwater forms which must maintain osmotic independence, except *Viviparus,* where the capsules remain in the oviduct and have been shown by Charin (1926) to be permeable to water and salts, and perhaps in the parthenogenetic *Potamopyrgus jenkinsi,* which is also viviparous. The wall frequently shows a fibrillar appearance due to conchiolin, and a suture separating it into 2 equal halves, which reflect the bilobed structure of the oviduct. The manufacture of a capsule, which occurs after the eggs have been fertilized, may take some hours. Their shape and size vary from species to species, though the differences are often trivial, and may be due to the varying methods by which they are manipulated after leaving the oviduct. Usually they are secured to a hold-fast. Only in 2 British species (*Littorina littorea, L. neritoides*) is the capsule liberated to the plankton, and in these it is passed to an ovipositor (opr, fig. 281A) where it is moulded to its definite shape and freed to the sea. In some species a fixed gelatinous spawn mass is produced in which the eggs, each in an albuminous covering, are isolated from one another.

The diversity in the type of spawn produced by species of one family may be as great as that of species from unrelated families. The littorinid species commonly found on rocky shores may be cited as an example. They have contrasting spawn and breeding habits which are associated with the differing provision made for the young and the suppression of the larval phase in the life history (Tattersall, 1920). Two have planktonic capsules and veliger larvae, others fasten gelatinous spawn to the substratum and the young emerge in the crawling stage; still others are ovoviviparous. Their zonation is especially interesting. *Littorina neritoides* is the highest of them all, living in crevices of exposed rocks often above levels ever reached by the sea, for which reason it was thought for some time to be viviparous. However, its method of reproduction by planktonic capsules was discovered almost simultaneously by Linke (1935a) and Lebour (1935c). The capsule (fig. 193D, E) has the shape of a biconvex disc measuring about 0·18 mm in diameter and with a height of about half this. It contains a single egg (de) surrounded by albumen (al) and then an egg covering (ec). Lysaght (1941) has shown that in the Plymouth area *Littorina neritoides* spawns only when it is submerged. She suggested that the snails living well above the tide do not migrate downwards, so that except in severe storms spawning can only take place at the fortnightly spring tides. However their spawning period is during the winter and spring (September–April) when the level of the tide may be increased by rough weather. Observations made by us on the Gower peninsula, S Wales, seem to indicate a genuine downward movement of about 8 ft in March at spring-tide time carried out by very ripe animals, though no opportunity of checking the occurrence of planktonic capsules at the same time was available. Egg capsules may frequently be taken from the pallial oviduct, fully formed though softer than capsules in the plankton, suggesting that they are ready to be released when conditions are appropriate. It is of interest to find that the fortnightly spawning rhythm is also evident in individuals which live in high rock pools and are always submerged. After a planktonic life the metamorphosing larvae may be cast up with the tide, or settle lower on the shore and migrate to some extent to dry rocks, but the way in which the young reach the habitat of the adult is unknown. Lysaght (1941) has suggested that the larvae are able to settle only on exposed rock faces devoid of fucoids. [See p. 674.]

The common periwinkle, *Littorina littorea*, also has pelagic capsules, different in shape, though similar in construction (Caullery & Pelseneer, 1910). It breeds chiefly in the spring, though at Plymouth capsules may be found in the plankton during any month (Lebour, 1947). The capsule (fig. 193C, F) resembles in shape a British soldier's steel helmet. It measures about 1 mm across and contains 1–5 eggs (de) which are pink in colour; up to 9 eggs have been recorded from a single capsule by Linke (1933), though so large a number is rare. The shape is moulded by the ovipositor which, according to Linke (1933), produces the tough external membrane with the circular brim (om). A gelatinous fluid (gf) fills the space between this membrane and the egg covering (ec). The supply of albumen around each egg (al) is small and unimportant as food. The capsules are liberated an hour or two after copulation and then intermittently for a month or more, the single copulation sufficing (Tattersall, 1920). Hatching, which may occur on the sixth day in normal sea water (Tattersall, 1920), is due to increased osmotic pressure within the capsule causing a rupture in the wall (Linke, 1933). The eggs develop in water of salinity 20‰ or over, though the lower salinities slow down development (Hayes, 1927b).

The viviparous periwinkle, *Littorina saxatilis*, breeds throughout the year at Plymouth (Lebour, 1937), though at Whitstable there is a decline in activity (p. 503) from June to mid-August, when the ovary is reduced (Berry, 1956). This species extends several feet above HWM, though not so high as *Littorina neritoides*, and is also found as low as half-tide level. The eggs (fig. 214A), which are retained in the brood pouch during development, are each surrounded by a larger supply of albumen (fa, va) than in the preceding species, and then an

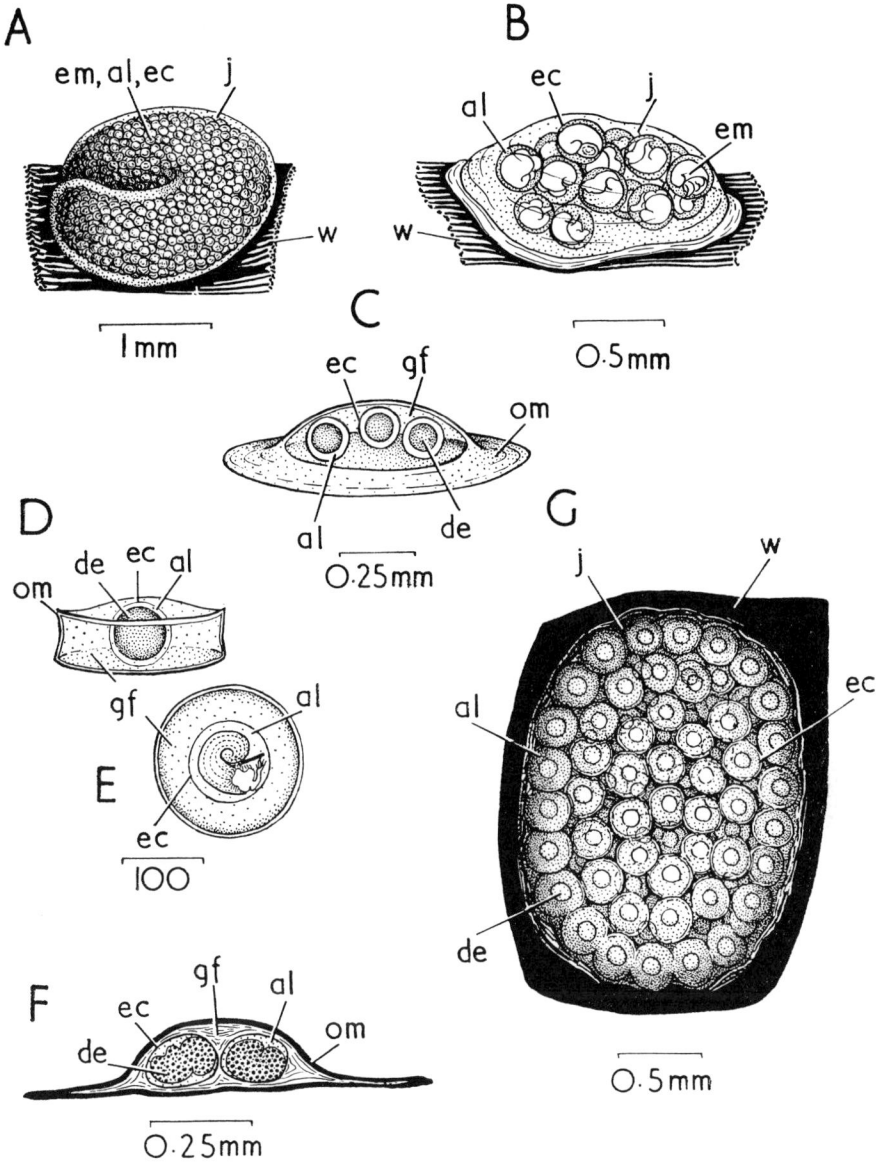

FIG. 193.—Spawn. A, *Lacuna vincta*; B, *Lacuna pallidula* to show embryos nearly hatching; C, *Littorina littorea*; D, *Littorina neritoides*, from the side; E, the same from above; F, *Littorina littorea*, in vertical section; G, *Lacuna pallidula* newly laid, from above.

al, albumen; de, developing egg; ec, egg covering; em, embryo; gf, gelatinous fluid; j, jelly of egg mass; om, outer membrane; w, weed.

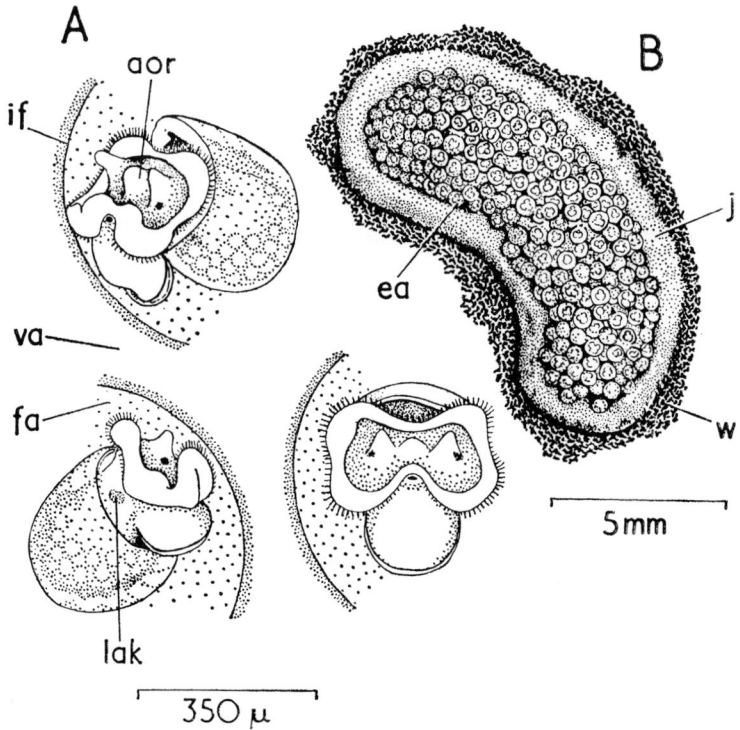

FIG. 194.—*Littorina littoralis:* A, developing embryos; B, spawn on weed.
   aor, alleged apical organ; ea, egg with albumen and egg covering; fa, fluid albumen; if, interface between
fluid and viscous albumen; j, jelly of egg mass; lak, larval kidney; va, viscous albumen; w, weed.

egg covering (ec). The albumen is eaten as food and in the early stages of development is fluid only in the vicinity of the embryo. Its jelly-like consistency at the periphery changes as development proceeds. The veliger (fig. 214B) rotates in the fluid and at a later stage (fig. 214C, D) when the albumen has gone, the embryo creeps around the egg covering. Finally it rasps the wall of its cell with the radula and makes a hole through which it escapes (Linke, 1933). Freed individuals move rapidly amongst the embryos in the lower part of the oviduct; they escape through the genital aperture and live in the same rock crevices as the adults.

   The flat periwinkle, *Littorina littoralis,* lives only where large algae are growing, especially *Ascophyllum, Fucus vesiculosus* and *F. serratus,* and deposits its gelatinous egg mass (fig. 194B) on damp, unexposed fronds, especially of *Fucus.* It usually spawns at night, a few hours after copulation. The spawn is flat and may be kidney-shaped in outline, measuring 7 × 3 mm (Linke, 1933; Lebour, 1937), or it may be oval or circular. It contains 90–150 eggs, each with albumen and egg covering as in the other species, arranged regularly in 2–3 layers one above the other and embedded in a jelly (j) which is hard throughout. The jelly prevents the spawn from drying and protects the embryos from infection; only if the spawn mass is in the later stages of development and unhealthy does it allow entry to ciliates and nematode worms. The young take 2–3 weeks to hatch, passing through the veliger stage (fig. 194A) within the egg covering and biting their way out with the radula. By this time the surrounding jelly has swollen with the uptake of water and the young can make their escape. The adults can live in water of low salinity (10–15‰) but Hertling (1928) has shown that the development of the egg ceases at a salinity of 25‰ or below.

The spawn of *Littorina littoralis* appears to have been frequently confused with that of *Lacuna pallidula* (fig. 193G) which is also laid on *Fucus* or at times on *Laminaria* fronds, and from which the young hatch in the crawling stage. In fact it is not at all clear that the confusion has been dispelled. Hertling & Ankel (1927) differentiated between the spawn of the 2 species on the grounds of shape, consistency of jelly and the arrangement of the eggs. They stated that the spawn of *Littorina littoralis* is kidney-shaped in outline, the jelly in which the eggs are embedded hard throughout, and the eggs (each surrounded by albumen and egg covering) arranged in 2–3 layers and never squashed against one another, whereas the egg mass of *Lacuna* has the shape of an inverted watch glass, is approximately circular in outline, and the jelly rather soft except for a stiff rind. The number of eggs in each mass may be less than or about the same as in *Littorina*: Hertling & Ankel (1927) quoted 60–140, Pelseneer (1911) 110–125, and Thorson (1946) gave a number as low as 13. Thorson found, however, that all the spawn masses in The Sound are smaller and have a much lower content of embryos than those described from more Atlantic localities, where the size varies from a length of 3·9–5·3 mm and a breadth of 3·3–4·5 mm (Lebour, 1937). *Littorina littoralis* does not always lay masses of spawn with a kidney-shaped outline. Very frequently they are circular or oval, approximating in both shape and size to those allegedly characteristic of *Lacuna pallidula,* and this is the main cause of confusion between the 2 species. The 2 may be easily differentiated in later stages because in *Lacuna* the foot of the embryo, just before hatching, shows the enlarged metapodial lobes (mt, fig. 278A) on either side (Pelseneer, 1911) which allow it to be identified without doubt. [See p. 691.]

In adult *Lacuna pallidula* there is pronounced sexual dimorphism (fig. 195) and the female (fl) may attain more than twice the size of the male (ml). Gallien & Larambergue (1938), working at Wimereux, found that the females attain a shell length of 12 mm and the males 5·8 mm; the weight of the female may be 10 times that of the male (Thorson, 1946). The species is an annual though the females live longer than the males which disappear after copulation. The breeding period is late winter and spring and may extend into the summer. At this time the animals move from deeper water to the inter-tidal zone and they may be seen in pairs, the male attached to the shell of the female towards the opening of the mantle cavity on the right side; this appears to be the position for copulation, as in many other species.

The spawn of only one other British species of *Lacuna* has been described, that of *L. vincta*. It forms the familiar yellowish rings (fig. 193A) which may be found from January to early summer on various weeds, green, brown and red; Lebour (1937) stated that other egg masses of similar shape which are pink or green almost certainly belong to another species. The rings are slightly spiral and each is made of a gelatinous substance with 1,000–1,200 eggs embedded in it. These eggs are only half the size of those of *L. pallidula,* or less than half, the diameter ranging from 103–128 μm. Each is surrounded by a small sphere of albumen and then an egg covering. Hertling (1928) held that the albumen is not used as food, but to raise the osmotic pressure, and that it brings about the swelling of the egg covering which doubles in diameter during development and finally bursts to liberate the embryos. The young hatch as veligers after a development lasting 2–3 weeks, the rate of development being influenced by the temperature (Hertling, 1931). The jelly in which the eggs are embedded is no barrier to the entry of salts and the egg covering allows the escape of CO_2 and the passage of water and

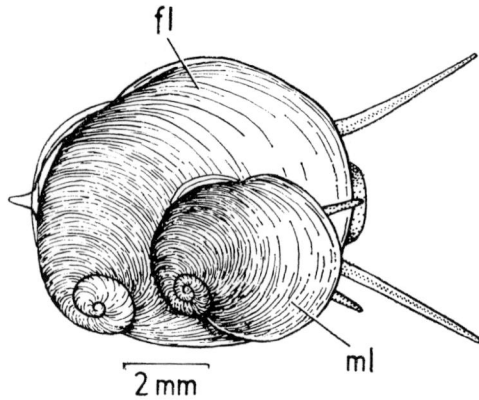

FIG. 195.—*Lacuna pallidula:* male and female in pairing position, from above.
fl, female; ml, male.

salts in both directions (Hertling, 1928). To begin with the jelly is viscous, but gradually it swells with the uptake of water and allows the veligers to leave. Hertling (1928) has distinguished a very fine inner membrane close to the egg of *L. vincta* which, he suggested, is secreted by the egg itself. It is visible up to the veliger stage, separating the fluid albumen surrounding the embryo from the outer albumen which is more viscous. The presence of such a membrane in *L. pallidula* is uncertain, though an interface between thick and thin albumen simulating a membrane can be identified in *Littorina littoralis* (if, fig. 194A). If Hertling's suggestion is correct, it would mean that at least in some members of the Lacunidae the ovum, as in the archaeogastropods, secretes coverings other than the vitelline membrane.

There is only one other species of British prosobranch, *Bittium reticulatum,* which produces a coiled, gelatinous egg mass (fig. 196C). As in *Lacuna vincta* it contains several hundred eggs each with a supply of albumen and an egg covering separating it from its neighbours. However the eggs of *Bittium* are much smaller (about 60 μm diameter and the supply of albumen correspondingly reduced. There is little possibility of confusing the spawn of the 2 species since in *Bittium* it is a narrow cord coiled in an anticlockwise direction to form a tight spiral which is fixed to rock, stone or weed, and it measures only about half the diameter of the spawn of *L. vincta.* The young escape as veligers with horn-coloured shells.

The importance of albumen as a supply of food for the developing embryo is emphasized in the terrestrial prosobranch, *Pomatias elegans,* which occurs in large numbers in parts of southern England where a thin layer of comparatively alkaline soil (pH 7·5–7·9) covers calcareous rocks. *Pomatias* is a member of the Stirps Littorinacea (Thiele, 1929–35) and may be regarded as a culmination of the trend towards a terrestrial habit which is displayed by the littorinids. The spherical egg capsules (fig. 197), which measure about 2 mm in diameter, are deposited just beneath the surface of the soil on warm humid days between March and September, the largest number being laid in May and June. The female covers each with a layer of soil particles (snd) about 0·5 mm thick as soon as it is laid, so that they are well camouflaged (Creek, 1951). The capsule contains a single egg (de) measuring about 0·14 mm in diameter. This is similar to the diameter of the eggs of *Littorina littorea* (0·13 mm, Linke, 1933) which, in contrast, has a pelagic stage in its development. The egg of *Pomatias* contains little

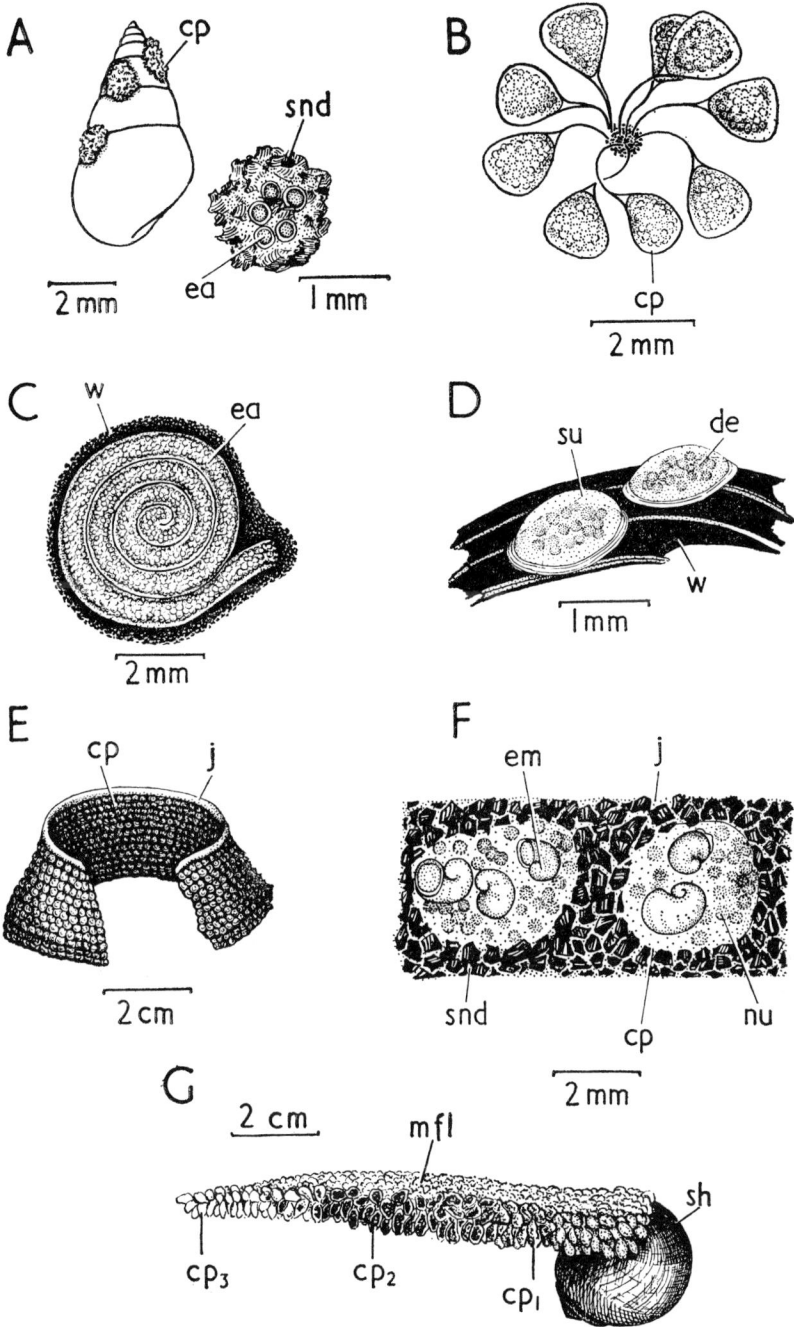

FIG. 196.—Spawn. A, *Hydrobia ulvae*, spawn attached to shell on left, and one spawn mass on right; B, *Turritella communis*; C, *Bittium reticulatum*; D, *Rissoa parva*; E, *Natica catena*; F, *Natica catena*, section showing 2 egg cases and surrounding jelly; G, *Ianthina exigua*.

cp, egg capsule; cp_1, youngest capsules; cp_2, capsules containing shelled embryos; cp_3, empty capsules; de, developing egg; ea, developing egg with albumen and egg covering; em, embryo; j, jelly of egg mass; mfl, mucous float; nu, food egg; sh, shell; snd, sand grains; su, suture of egg capsule; w, weed.

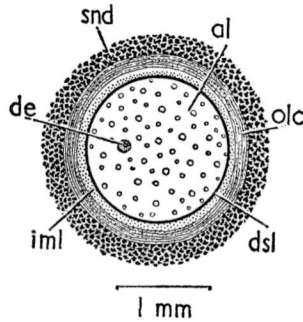

FIG. 197.—*Pomatias elegans:* egg capsule in section.
al, albumen; de, developing egg; dsl, deeply staining layer round albumen; iml, inner mucous layer of capsule; olc, outer fibrous layer of capsule; snd, soil particles.

yolk, but a large supply of albumen (al) which is used as food, and the cephalic vesicle (cv, fig. 224) provides the embryo with a special means of absorbing this. Between the albumen and the covering of earth, and of about the same thickness as the latter, is a transparent coat made up of an inner mucoid layer (iml) and an outer layer of concentrically arranged fibres of conchiolin (olc) embedded in a mucoid matrix; this construction is similar to that of the resistant walls of egg capsules of some Stenoglossa (*Ocenebra, Nucella, Buccinum*) (fig. 208). In the Stenoglossa there is a special gland in the foot of the female which finally shapes and deposits the capsule. In *Pomatias* this is done by the surface of the sole which rolls the capsule in the soil. Creek (1951) suggested that the mucus which holds together the soil particles comes from the anterior pedal mucous gland. The embryo tales long to develop, about 3 months, and when it hatches its supply of albumen is exhausted and its volume, as compared with that of the first cleavage stage, has increased fifteen-fold. However the albumen and yolk do not together supply the full requirements for development since the embryo absorbs water and salts from the soil; moreover Creek (1951) has shown that there is some factor in the natural soil water which is necessary. This is in contrast to the terrestrial snail *Helix* for there is no evidence that its egg absorbs water or anything else from the environment but oxygen; the egg coverings are not impermeable, but water in excess of the animal's requirements is provided (Needham, 1938).

The spawn of the only other British terrestrial prosobranch, *Acicula fusca,* is unknown.

Amongst the smallest egg capsules of marine prosobranchs, yet by far the most abundant on many coasts, are those of the various species of rissoaceans (Table 11). They are usually lens-shaped (fig. 196D), circular in outline, with a flattened area attached to the substratum and the free surface convex. Occasionally they are ovoid (*Cingula semicostata*) or are nearly spherical (*Barleeia rubra*) and with only a small area of attachment. The capsule contains 1, 2 or several eggs (up to 100 have been recorded in *Rissoa guerini*) which all float together in an albuminous fluid. The limiting wall is fairly thick and tough, and usually so transparent that the contents of the capsule can be seen through it; frequently it shows a fibrillar structure. When only 1 or 2 eggs occur together, the young emerge in the crawling stage (*Cingulopsis fulgida, Cingula semicostata, C. cingillus, Barleeia rubra*), but with larger numbers of smaller eggs the embryos are freed as veligers. The veligers may have a long planktotrophic life, and because of their immense numbers form an important constituent of inshore plankton.

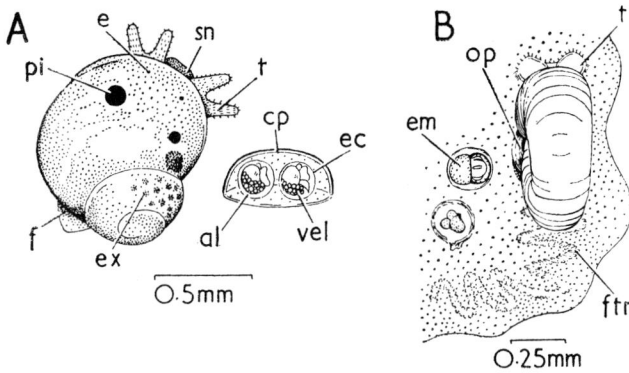

FIG. 198.—A, *Rissoella opalina,* whole animal seen from above, with one egg capsule alongside; B, *Omalogyra atomus,* whole animal on weed, seen from above, with 2 egg capsules.

al, albumen; cp, egg capsule; e, eye; ec, egg covering; em, embryo; ex, pigmented spherules in excretory cells of digestive gland; f, foot; ftr, feeding track; op, operculum; pi, group of pigmented cells at anus; sn, snout; t, tentacle; vel, veliger.

Moreover, since their breeding times vary, young and old stages of different species may be seen in the plankton at all times of the year (Table 11). Many rissoids inhabit regions between tide marks and may be found in every type of habitat on rocky shores, extending seawards from about mid-tide level. On some shores, especially in the SW, *Rissoa parva* is the most numerous species. Its capsules are laid on weed, often in rock pools on the lower part of the beach. Each contains 6–50 eggs which hatch in about 10 days as veliger larvae. At the time of hatching a suture, which appears as a distinct fine line over the convex surface of the wall, ruptures and forms the escape hole. The wall of the capsules of *R. guerini* and *Barleeia rubra* breaks in a similar way, but in *Cingula semicostata* the wall has concentric striae radiating from a centre near the surface of attachment and this is the site of emergence of the young (Rasmussen, 1951), and the lens-shaped capsule of *R. membranacea* has an oval hole in the centre of the convex surface covered by a membrane which ruptures when the larvae are ready to escape (Smidt, 1938).

There are a few minute mesogastropods classified near the rissoids, inhabitants of rock pools, which have egg capsules similar to the rissoid type. They breed during the late spring and summer months. In *Skeneopsis planorbis* and *Omalogyra atomus* (Fretter, 1948) the capsules (fig. 198B) are spherical or ovoid, slightly flattened along the surface of attachment to an algal frond or filament. The outer wall is relatively thick and made of conchiolin threads embedded in a mucoid matrix. Through it can be seen the single egg (sometimes 2 in *Skeneopsis*) embedded in albumen. The eggs are heavily yolked and the young pass through the veliger stage within the capsule. In *Skeneopsis* the diameter of the capsule is about 0·4 mm. In *Omalogyra* it is half this, though the development of the egg is more rapid and may be completed in 10 days, whereas in *Skeneopsis* it may take twice as long. The 2 species of *Rissoella* are found in the same rock pools. They both have hemispherical capsules attached along the flattened base to weed. The capsule of *R. diaphana* (fig. 186A) has 1 or 2 eggs and that of *R. opalina* (fig. 198A), which is larger (breadth 0·65–0·4 mm, height 0·5 mm), 2 or rarely 3. Unlike the rissoids each egg has a supply of albumen and its own covering (ec) and they

TABLE 11
Egg Capsules and Seasonal Breeding of some Marine Rissoaceans (partly based on Lebour, 1934a)

Species	Months during which breeding has been recorded at Plymouth												Egg capsules
	J	F	M	A	Ma	Ju	Jl	Au	S	O	N	D	
Cingulopsis fulgida					——	——	——	——	——	——			Capsule attached to weed, often corallines, lens-shaped (0·32 mm across), contains one egg, no free larvae.
Cingula semicostata	——	——	——	——							—		Capsules in muddy gravel under stones, ovoid (0·48–0·64 × 0·32–0·42 mm), very tough, thick wall, small area of attachment, contains one egg, no free larvae.
C. semistriata				——	——	——	——	——					Capsule attached to weed or other substratum, lens-shaped (0·56–0·64 mm across and 0·24 mm high), contains 12–22 eggs, free veliger larvae.
C. cingillus				——	——	——							Capsule laid in narrow crack or crevice of rock, attached, lens-shaped (0·64–0·72 mm across), contains one and sometimes up to four eggs, no free larvae.
Alvania punctura							——	——	——	——			Sublittoral, capsule laid in captivity on shell or weed, lens-shaped (0·32–0·48 mm across), contains 12–14 eggs, free veliger larvae.
Rissoa sarsi	——	——	——	——							—		Capsule lens-shaped, attached (0·48 mm across), contains 10–14 eggs, free veliger larvae.
R. inconspicua						——	——	——					Sublittoral. Capsule laid in captivity on weed, debris or shell of another, lens-shaped (0·48–0·64 mm across), contains 6–9 eggs, free veliger larvae.
R. parva	——	——	——	——			——	——	——				Capsule attached to weed, lens-shaped (0·64 mm across), contains 6–50 eggs, free veliger larvae.
R. guerini		——	——	——			——	——	——	——			Capsule attached to weed, lens-shaped (0·96–1·4 mm across), contains 80–100 eggs, free veliger larvae.
R. membranacea	——	——	——								—		Capsules attached to *Zostera*, lens-shaped (1·4–1·6 mm across), contains 40–60 eggs, free veliger larvae. With loss of *Zostera* through disease this rissoid has disappeared.
Barleeia rubra				——	——	——	——	——	——				Capsules attached to weed, spherical, small area of attachment, contains one egg, no free veligers.

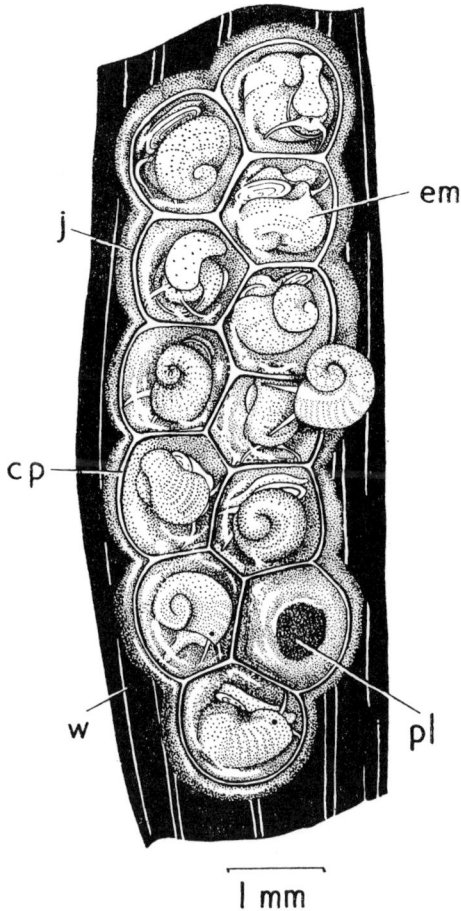

I mm

FIG. 199.—*Bithynia tentaculata:* spawn on weed.
cp, wall of egg capsule; em, embryo in capsule; j, jelly of egg mass; pl, plug hole through which embryo has just escaped; w, weed.

float in a fluid which fills the capsule. There is an intracapsular veliger stage (vel) and development is completed in about a fortnight.

The British prosobranchs which inhabit fresh water are members of Thiele's (1929–35) Stirps Rissoacea or are more primitive forms. They comprise *Viviparus, Bithynia, Valvata* and *Bythinella scholtzi,* which are found exclusively in fresh water, and *Theodoxus* and *Potamopyrgus jenkinsi* which live also in brackish water. *Viviparus* and *P. jenkinsi* are viviparous and the others lay egg capsules and there is no planktonic larval phase. The capsules of *Bithynia* (fig. 199) and *Valvata* (fig. 200) are both laid on weed during the late spring and summer months. Those of *Bithynia* are usually attached to the under surface of leaves and are unmistakable, for a number of approximately hemispherical capsules are accurately

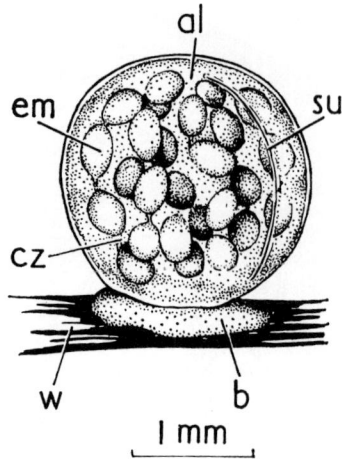

FIG. 200.—*Valvata piscinalis:* egg capsule on weed.
al, outer albuminous fluid; b, base of capsule; cz, chalaza; em, outer mucoid coat surrounding albumen in which embryo is embedded; su, suture; w, weed.

arranged side by side in 2–3 parallel rows, those of one row alternating and interlocking with those of the other. The capsules adhere to the weed (w) by the outer coat and are moulded and pressed into place by the sole of the foot; there is no special pedal gland for this purpose. Each is filled with albumen and contains a single egg, orange in colour, which can be seen through the transparent wall. The wall is composed of threads of a protein, probably conchiolin, intermingled with a mucoid secretion, except for the centre of the convex surface where there is a disc-shaped, mucoid plug. The plug separates from the capsule when the snail is ready to hatch (pl). The number of capsules in a single spawn mass varies considerably: occasionally there may be as few as 4 in each row, though Moquin-Tandon (1855) gave a maximum of 70 capsules and Nekrassow (1928) found 98. However, such large numbers are infrequent. The young escape after an embryonic period of about 2–3 weeks, depending on temperature. The egg mass of *Valvata* is of a very different form. The capsule of *V. piscinalis* is spherical, approximately 1 mm in diameter, and is attached to weed by a cement (b). Around part of the wall opposite to the point of attachment is a suture (su) (Cleland, 1954) along which the capsule bursts when the young hatch. There are several eggs in each capsule: 9–19 appears to be a common number, though Germain (1930) mentioned 10–60. Each egg embedded in albumen forms an ellipsoidal mass enclosed within a very thin mucoid covering (em). The case around an egg is at first continuous with that of the next and is constricted to a thread (cz) between the two. In this way all eggs within a capsule may be joined into a continuous string which is coiled in an outer fluid (al), though the connexions soon break. A single snail may lay about 10 capsules with a total of about 150 eggs (Frömming, 1956). The embryonic period lasts 15–30 days. During this time the mucoid covering around each egg is ruptured and it would appear that the swelling of the contents of the capsule eventually bursts the wall and liberates the young (Heywood, personal communication). The capsules of *V. cristata* described by Nekrassow (1928) differ in shape and in the number of eggs each

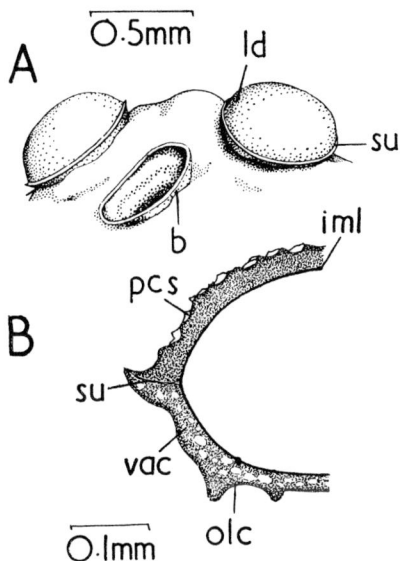

FIG. 201.—*Theodoxus fluviatilis*: A, egg capsules on a stone; the young animal has escaped from one and the base remains; B, vertical section of part of wall of egg capsule.

b, base of capsule; iml, inner mucous layer of capsule; ld, lid of capsule; olc, viscous outer layer of capsule; pcs, particles from crystal sac; su, suture; vac, vacuoles filled with mucus.

contains. There may be up to 4 eggs enclosed in a sac-shaped capsule attached at the base and according to Frömming the embryonic period lasts 30–40 days.

Theodoxus fluviatilis lives on stones, sunken wood and aquatic plants in canals, rivers, lakes and also in brackish water. The spawn (fig. 201), fixed to the substratum or to the shell of another snail of the same species, is a flattened sphere made up of approximately equal halves sutured together (su) around the equator (Andrews, 1935). The lid (ld) breaks off when the young escape. The walls are of tough conchiolin, white to straw colour, lined internally by a homogeneous membrane enclosing an albuminous fluid in which the eggs float. The upper half, or lid, is strengthened by a surface layer of sand grains and diatom cases (pcs), which are poured on to it and adhere to its outer sticky covering as the capsule passes from the capsule gland to the genital aperture. These particles come from the crystal sac which lies against the terminal part of the pallial oviduct (csa, fig. 52). The sac collects them from the rectum for storage and later they pass to the terminal region of the oviduct with which the sac communicates (Fretter, 1946b). Bondeson (1940) has shown that capsules laid by individuals living in fresh water are larger and contain more eggs than those laid in brackish water. Thus a capsule from brackish water may have a diameter of 800 μm and contain 50–60 eggs, whereas one from freshwater may have a diameter of 1,000–1,200 μm and contain 140–150 eggs. In each capsule one egg cleaves more readily than its neighbours and this precocious embryo uses the others as food and develops to the creeping stage before it hatches. Embryos developing in freshwater and with a larger supply of food attain a larger size (751–910 μm) than those developing in brackish water (650–775 μm). An interesting correlation between ecological conditions, egg size and life history occurs in another brackish water gastropod, *Odostomia scalaris,* a pyramidellid, which is also marine. It is an ectoparasite of *Mytilus* and is found with its host in habitats of varying salinities. The egg masses (fig. 202C, D, E) are fixed to the shell of the bivalve or other substratum. Each consists of an irregularly shaped mass of jelly which may measure 2–4 mm across and about 2 mm high and contain 40 or more eggs, their size varying with the number. Each egg (de) lies at the centre of an oval mass of transparent albumen (al) which is covered by a thin, though tough, egg covering (ec), and the eggs are linked together by a continuation of the covering from one to the next (cz). Thus the covering may be considered as a long tube coiled irregularly in the jelly, expanded at regular intervals by the eggs and constricted between them to a fine strand. (The spawn of some specimens of *Odostomia eulimoides* (fig. 202A, B) and of the primitive opisthobranch *Onchidella celtica* is similar, though another pyramidellid, *Eulimella nitidissima,* lays ovoid gelatinous egg masses with only about 6–7 eggs in each, and they are not linked together by a covering (Rasmussen, 1944).) Pelseneer (1914) was the first to study the development of *Odostomia scalaris* and he described the egg mass as containing an average of 50 oval eggs, each about 0·38 mm in the longest diameter, which develop in 25 days to miniatures of the adult. He gives no mention of the locality from which the specimens came. Later Rasmussen (1944) described the egg masses and embryonic development of specimens from the innermost part of the Isefjord, Denmark, where the salinity is 20‰. The eggs, however, are smaller (0·18–0·20 mm diameter) than those described by Pelseneer and about 500 occur in one mass. The young are freed as larvae after a development of 6·5 days at 19°C. The length of the free-swimming stage is unknown, but the small velum indicates that it is short. At Frederikssund in the Roskildefjord, where the salinity is lower (16·1‰), Rasmussen (1951) found the individual eggs to be even larger than those described by Pelseneer and in their development the planktonic phase omitted. It seems probable that Pelseneer's specimens were also from a more brackish water, and under these conditions the pelagic phase in development is omitted.

Hydrobia ulvae is another species which inhabits brackish water and is capable of withstanding wide fluctuations in salinity (p. 544); in favourable habitats such as mud flats and salt marshes it is present in myriads. Spawning is spread over several months. On some parts of the Tamar estuary, Plymouth, the main period is autumn and on others, such as St John's Lake, it extends to February and March (Rothschild, 1941a). For the Swansea area Quick (1920) mentioned May–August, but suggested that it extends to other months. The capsules are fastened to shells of other individuals of the same species, male and female, perhaps for lack of other firm substrata in the habitat. As many as 22 have been recorded on one shell (Linke, 1939), though this is exceptional and the average is 4 even where the population is dense. Each (cp, fig. 196A) has the shape of a typical rissoid capsule, fastened by the flattened base and with the free surface convex. This surface is covered with sand grains (snd). The capsule contains a number of eggs floating in an albuminous fluid (ea): Lebour (1937) gave the number as 3–7, Smidt (1951) 4–16, and Quick (1920) as many as 25. Smidt (1944) held that the pelagic larval stage is suppressed, on the grounds that although the adults are very abundant in Copenhagen harbour he did not find the larval stage there, though he admitted that there is a brief distributive phase before settlement. Other workers, however (Henking, 1894; Quick, 1920; Lebour, 1937; Linke, 1939; Thorson, 1946), have described a free veliger, although Lebour (1937) suggested that its planktonic life is short. Thorson (1946), obviously surprised at the apparent lack of knowledge of the life history of this common species, studied hauls from the Southern Harbour, Copenhagen, and found numbers of larvae which he identified without doubt (comparing them with veligers hatched in captivity) as belonging to *H. ulvae.*

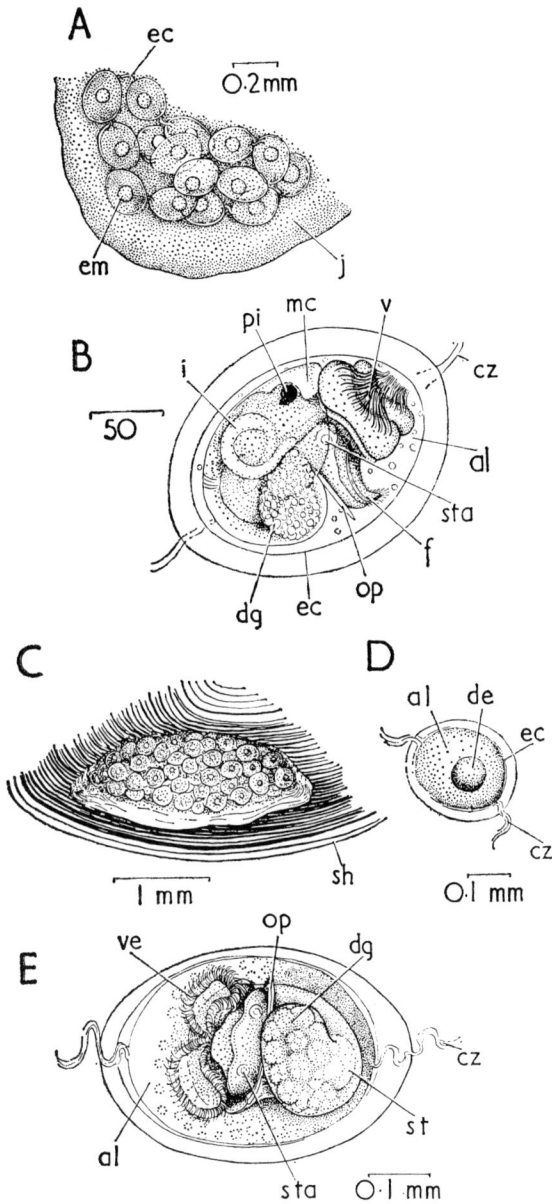

FIG. 202.—*Odostomia eulimoides:* A, part of egg mass showing eggs in an irregular string embedded in clear jelly; B, part of egg string to show veliger larva just before hatching.

O. scalaris: C, egg mass on shell of *Mytilus:* D, a single egg at early stage of development; E, part of egg string to show veliger larva just before hatching.

al, albumen; cz, chalaza; de, developing egg; dg, digestive gland; ec, egg covering; em, embryo; f, foot; i, intestine; j, jelly of egg mass; mc, mantle cavity; op, operculum; pi, pigment patch; sh, shell of *Mytilus;* st, stomach; sta, statocyst; v, ve, velum.

All had a small and scarcely bilobed velum and the organs of the body relatively advanced, indicating a short pelagic phase. However, the numbers he obtained were nothing to what might have been expected, for in certain areas of the bottom more than 1,000 adults/m sq may be found, all with egg capsules on their shells in summer months. Perhaps there is more to be discovered about the life history of this animal and the possibility still exists that the larval stage may be free-swimming or suppressed according to circumstances (assuming that it has been correctly identified in the first place). Even if set free, larvae may avoid capture by staying near the bottom, and all evidence seems to point to the conclusion that the planktonic phase is at most a brief, and probably a non-feeding stage of the life history. [See chapter 29].

It is well known that in the related species *Hydrobia ventrosa,* which lives amongst algae and on mud in brackish water, the larval stage is suppressed. The globular capsules, covered with grains of sand, are fixed to stones; each contains a single egg which hatches as a young snail. These capsules are similar to those of *Assiminea grayana* which are passed from the oviduct as the snail creeps over the mud on which it lives. They, however, are unattached, but, owing to the sticky consistency of their walls, may become grouped in clumps and be camouflaged by adhering particles. It might be expected that the veliger stage of this species (which inhabits brackish water) would not be free: a free larva, however, has been described from capsules kept in an aquarium (Sander, 1950). Development has not been studied in natural conditions.

When the egg capsule of the mesogastropod leaves the genital duct it is, in most species, manipulated by the foot which may give it a final moulding while the wall is still pliable. The wall hardens in contact with the water and it may be that the foot applies secretion to hasten this. In certain species the wall is reinforced by particulate matter which is pressed into it while soft. In *Theodoxus* this is done before the capsule leaves the duct; in *Hydrobia ulvae* and *H. ventrosa* it is after this and presumably occurs when the foot fixes the capsule. The minute egg cases of the marine prosobranch *Aporrhais pespelicani* are deposited singly, or 2 or 3 together, in the muddy sand in which the adults live and they are obscured by the particles which accidently adhere to the thick outer wall. The capsules are spherical, about 0·24 mm in diameter, and each contains only 1 egg.

Two genera the members of which live on sandy shores and produce large egg masses, strengthened and camouflaged by the sand or mud, are *Clathrus* and *Natica.* Vestergaard (1935) was the first to give an account of the spawn of *Clathrus* (fig. 203). The adults of *Clathrus clathrus* spawned in an aquarium which had sand in the bottom, and she described a winding string of triangular capsules 3 cm long and each capsule 2 mm high. The larvae hatched 9 days later through an aperture at the apical end of the capsule. Spawn from the Salstone in the Salcombe estuary, Devon, which is described by Lebour (1937), differs from this, for the capsules are covered with fine mud instead of sand. They are polygonal and very irregular, only occasionally being triangular. Moreover, each string measured 5–6 cm and the individual capsules 2–4 mm across at the widest part. Lebour stated that several such spawn masses were found lying on mud near green weed and that one adult was found beneath the surface of the mud with the egg string attached to it by the slimy thread which links the capsules together. It may be that spawning takes place beneath the surface as in *Natica.* On some occasions *Clathrus* has been seen on the surface of sand trailing its egg string along. One string obtained by Prof. L. A. Harvey in the Scilly Isles (fig. 203) exceeded 23 cm in length and was made up of 170 capsules closely packed, each pyramidal in shape and about 1·5 mm across the base. The walls of the capsules (cp) were covered with sand grains

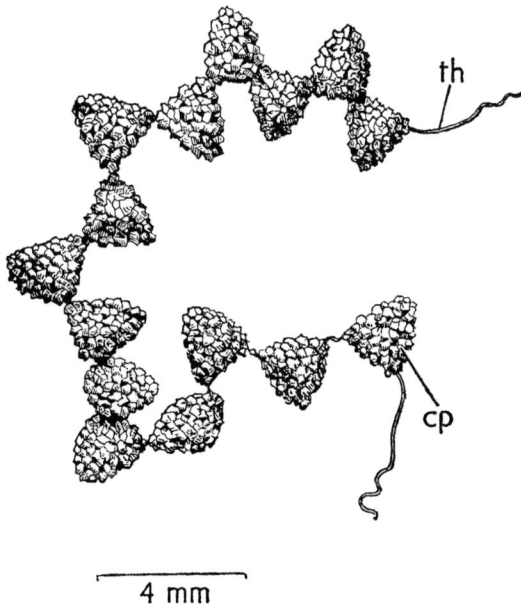

FIG. 203.—*Clathrus* sp.: spawn.
cp, egg capsule; th, thread linking capsules.

securely cemented together and the thread of secretion (th), probably conchiolin, which held them was naked for a short distance at either end. It seems, therefore, that *Clathrus* will use either mud or sand particles in the manufacture of the spawn and that the size of the capsules, and more especially their number, may vary from one string to another. Thorson (1946) figured part of an egg mass of *Clathrus turtonis* with capsules of about the same size and shape as those described here, covered with accurately arranged particles of fine sand.

Natica is popularly known as the necklace snail owing to the shape of its egg mass which has the appearance of a collar stiffened with sand (fig. 196E). Projections from the surface indicate egg cases arranged in horizontal rows, usually one above the other. The collars, which have the shape of the old-fashioned Eton style, may be rigid or flexible. They are less than 3 mm thick and of varying height and length. When newly formed each has a smooth, viscid texture due to a thin surface layer of mucus, which, however, is lost after a day or so. Beneath is clear jelly (j, fig. 196F) in which the egg cases (cp) are embedded and throughout the thickness of the jelly are particles of sand, fine or coarse, and sometimes bits of shell (snd). The nature of these particles depends upon the substratum in which the collar is formed. The size of the egg cases varies according to the species; when large they form prominent bulges from the surface of the collar, while with the smaller cases its walls are smooth. Each has a thin limiting membrane which surrounds the albuminous fluid containing the eggs. In *N. pallida* the individual cases are large, measuring 2·5–3·0 mm across and there are only 40–50 of them in an egg mass (Thorson, 1935). They each contain a single egg and the embryo hatches in the creeping stage. In contrast *N. poliana* (Lebour, 1937) produces egg cases which are only 0·24 mm across and not visible to the naked eye, and the egg collar has smooth and flexible walls; it measures about 25 mm long and 7·8 mm high—half the size of the collar of N. *pallida*. The single egg in each case develops in 3 weeks to the veliger stage and is then freed. The spawn of N. *catena* is the best known and was first described by Jeffreys (1867) who stated that the embryonic stage lasts at least 2 months. The largest collar recorded is 45 mm high with a basal length of 160 mm (Lebour, 1936); others may be half this size. The egg cases are large (1·425 by 1·925 mm), arranged in rows one above the other and bulge prominently from the rigid walls. Each contains 2–4 eggs and numerous food eggs (nu); Ankel (1930b) mentioned 50 food eggs and Thorson (1946) an average of 62. The food eggs cleave in an atypical way and serve as food for the developing embryos (Ankel, 1930b); this is the only known naticid which has them. In other prosobranchs in which the embryos are similarly fed there is typically no pelagic phase in development, and Thorson (1946) described the suppression of this phase in N. *catena* from Danish waters. He suggested that the larva dredged from sandy bottoms near Plymouth and provisionally assigned to this species by Lebour (1936), belongs to N. *montagui,* though no undoubted adult of this species has been identified there recently. However, he did not exclude the possibility of N. *catena* having a pelagic phase in the warmer waters of the English Channel (Thorson, 1950).

The only detailed description of the formation of the egg collar of a naticid was given by Giglioli (1955) whose observations were made on the intertidal species *Polinices triseriata* and *P. heros*. Their distribution extends from the Gulf of St Lawrence to N Carolina. There is no reason to believe that the formation of the collar in other naticids differs from this, except perhaps in unimportant details. The initial phase of collar formation begins on the surface of the sand when the tide is nearing low water. The snail lies on its side with the foot fully extended longitudinally and folded to make a median longitudinal furrow. From the anterior end of the furrow pours mucus, which is dispersed into long festoons by the water movements (Wheatley, 1947); it is produced by gland cells on the foot, mainly the goblet cells of the propodium (Giglioli, 1955). With the beginning of the flood tide the snail burrows into the sand to a depth of 5–10 cm leaving the gelatinous festoons on the surface. It rests on its side with the apex of the shell directed obliquely upwards, and then begins to move slowly in a clockwise direction along a circular route, forcing its way through the sand. The large propodium (pp, fig. 296D) is characteristically folded back over the shell covering the head, and between propodium and shell (sh) the egg ribbon emerges (er), pressed between the two. As the ribbon loses contact with the foot the pressure between shell and sand continues to mould it, determining its breadth and distributing the egg cases evenly. At the same time, sand particles are rolled into the intercapsular jelly. In some naticids the upper and lower margins of the collar are free from capsules, which are pressed away from these borders by the moulding action of the foot. When all the eggs have been spawned the snail moves spirally around the collar smoothing the foot over the outer and inner walls and progressing in a clockwise direction. At the same time a thin pellicle of mucus is added which binds the collar until the matrix jelly has hardened. Each time the snail moves under the collar when it is travelling along its spiral course, the spawn is lifted and tilted towards the surface. As soon as it reaches the surface it is left. Giglioli (1955) was of the opinion that a certain amount of the intercapsular jelly comes from the propodium, but he also expressed the view that it may come from the expanded glandular chamber of the lower oviduct. He has found that in *P. triseriata* the capsular membrane and the albumen appear to control the ionic balance between the pregastrular embryo and the sea water, with which the early embryo is not isotonic (Giglioli, 1949). Later changes in the capsular jelly, conditioned by the developing embryo, progressively diminish the ability of the collar to withstand desiccation. It becomes brittle and crumbles under the flood tide releasing the larvae. The embryos of deep sea naticids hatch by boring through the capsular walls (Thorson, 1935). [See p. 670.]

In one family of prosobranchs the female guards the egg capsules until the young escape. This is the Calyptraeidae, and the reproductive behaviour of its genera is associated with their sedentary mode of life. The egg capsules, characteristically thin-walled, are concealed beneath the anterior part of the shell, so that they lie along the course of the inhalant flow of water which, in these sedentary, ciliary-feeding prosobranchs, is strong. In *Capulus ungaricus* a single spawn mass in the shape of a thin-walled, somewhat sausage-shaped sac, is held in a fold of the propodium (Ankel, 1936a). It contains several hundred eggs. A single mass figured by Thorson (1946) has at least 5,000 eggs. The young escape as echinospira larvae through a fissure along the under surface of the cocoon and the parent then discards the embryonic membranes. Should the egg mass be abandoned earlier the larvae do not live long. The eggs of a single spawning of *Calyptraea chinensis* are contained in a number of capsules which are attached to the substratum, usually a stone. They are covered by the part of the body in front of the foot (Lebour, 1937). Each is a very soft-walled bag, triangular in outline and measuring about 3 mm in length, and all are fixed in a bunch by their narrow ends. A capsule contains only 12–24 eggs, concentrated towards the broad end, and the young escape as miniatures of the adult with an embryonic shell of $1\frac{1}{2}$ whorls. This shell usually shows clearly in the young adult (fig. 43D). Egg capsules of *Crepidula fornicata* are similar (fig. 204), though somewhat more balloon-shaped and stalked and are united by their stalls to form a bunch of up to 70 or more. The bunch is fastened to the substratum on which the parent lives, either a stone or the shell of the underlying limpet (fig. 204B), and is hidden from view. If a chain of individuals be separated during the breeding season it is not unusual to find a female holding the stalks of a number of capsules by the anterior tip of the foot (fig. 204A). It may be that the foot collects them as they leave the oviduct and later fixes the whole bunch. A single capsule measures 2–4 mm across at the broad end and contains 250–300 eggs which float in the common mass of albumen. The young escape as veligers and remain some time in the plankton. Coe (1942) found that in capsules of *C. onyx* half the embryos frequently disintegrated and were used as food by the survivors and Thorson (1940a) stated that in some capsules of *C. walshi* from the Persian Gulf there was a suggestion that embryonic cannibalism had occurred, for some embryos were much larger than others and these appeared to be disintegrating. This is not a normal occurrence in the genus *Crepidula* and can only have been accidental. In capsules in which embryos are not separated from one another but share a supply of albumen it seems likely that a healthy individual will automatically devour disintegrating tissues with the albumen which is used as food. Lebour (1934a) similarly found that from egg capsules of *Rissoa membranacea* which have 40–60 eggs apiece, a smaller number of embryos developed and suggested that the others were used as food eggs. This may have been an abnormality due to the fact that the eggs were laid in captivity.

The egg masses of *Turritella communis* (fig. 204B) are not unlike those of *Crepidula* in that each is a cluster of spherical or ovoid capsules held together by the intertwining of their stalks. However, in *Turritella* some hundreds of capsules form a single cluster (Lebour, 1933c). A capsule measures at first about 1 mm across, but stretches as the eggs develop so that its walls become thinner. It contains 6–20 eggs which develop to the veliger stage in a week or 10 days and hatch.

Some carnivorous prosobranchs and some ectoparasites secure their egg capsules to the animals on which they live. The spawn of *Odostomia scalaris* (fig. 202C), which is attached to

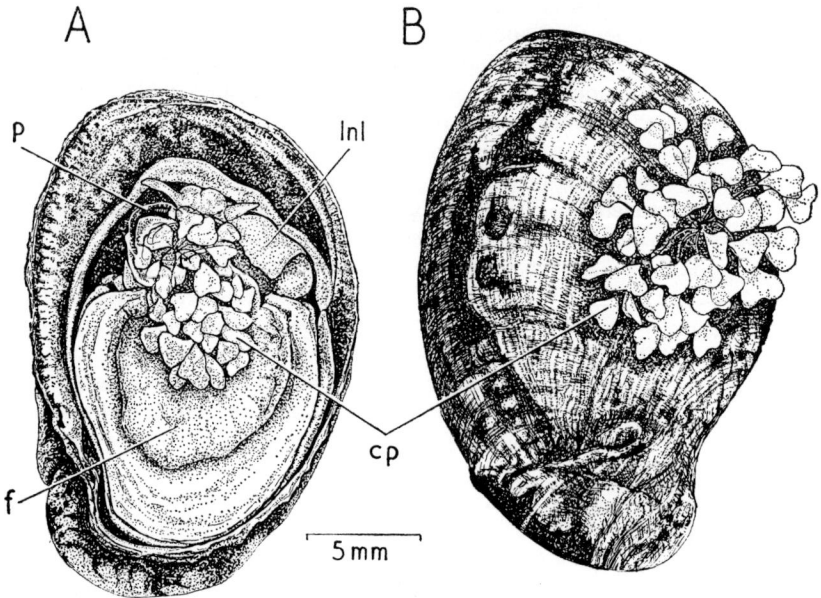

FIG. 204.—*Crepidula fornicata:* A, female, from below, with egg capsules held under the foot; B, empty shell from above, with attached egg capsules.
cp, egg capsule; f, foot; lnl, left neck lobe; p, vestigial penis, unusually large for a female.

the shell of *Mytilus,* has already been described, and *O. eulimoides,* an ectoparasite of *Chlamys opercularis* and *Pecten maximus,* deposits a similar type of spawn on the ears or valves of these lamellibranchs (fig. 202A). *Pelseneeria stylifera* lives on the surface of sea-urchins and its egg capsules may also be found there. They are not unlike those of the calyptraeids in that each is cushion-shaped, approximately triangular in outline, colourless and transparent and with a short stalk for attachment. However, they are deposited singly and not grouped together in bunches. Each measures about 1·2 × 1·1 mm and contains 60–400 eggs. The capsules are usually fixed around the bases of the spines, which afford protection, and they are apparently untouched by pedicellariae. *Pelseneeria* is recorded from Plymouth on *Psammechinus miliaris* and *Echinus esculentus,* and Lebour (1932b) described the capsules on *Psammechinus.* Ten to a dozen were found on one individual, all on the aboral surface. The young escape as veligers which probably remain long in the plankton.

Other prosobranchs which browse on the soft tissues of colonial animals spread their egg masses on the surface of these or bite holes from the tissues and sink them within. The only British species which feeds on alcyonarians is the poached egg shell, *Simnia patula* (fig. 288), and its spawn is laid as a continuous layer on *Alcyonium digitatum.* A collection of capsules appears as a circular mass an inch or more in diameter. The capsules are pressed against one another so that the outline of each is polygonal rather than circular. A single one measures about 3·5 mm across and contains several hundred eggs all of which appear to develop and the young escape as veliger larvae (Lebour, 1932a). These remain long in the plankton and only gradually assume their diagnostic features—the velum of 4 long narrow lobes and the reticulate markings of the shell (fig. 241). The 2 species of *Cerithiopsis* which live on sponges produce lens-shaped capsules like those of rissoids and place them singly in holes made in the sponge. The capsules of *C. tubercularis* are laid in *Hymeniacidon* and each contains a large number of minute eggs which measure only 0·06 mm in diameter (Lebour, 1933b). *C. barleei*, living on *Suberites domuncula*, embeds its nests of eggs in this sponge so that the top of each is hardly visible from the surface. One specimen kept in captivity (Lebour, 1933b) laid 20 capsules which were distributed at intervals of 5 mm or more apart and each contained about 200 eggs. In both species the young escape as veligers which remain in the plankton for some time.

The test of compound ascidians is used as the spawning ground of *Velutina velutina*, *Lamellaria perspicua* and *Trivia monacha*. These species sink their capsules into holes. Those of *L. perspicua* (fig. 205) are laid in the tissues of *Leptoclinum*, *Polyclinum* and probably other ascidians. Each is pot-shaped, with a rounded base and tapers slightly towards the circular opening which is filled with a plug. It measures 2–3 mm high. Only the plug surrounded by a low rim of capsule wall (cr) is exposed at the surface of the ascidian. The wall of the capsule is thin and is divided into approximately equal halves by a suture which runs down its length and can be seen across the surface of the plug (sup); it reflects the bilobed structure of the pallial oviduct (Fretter, 1946b). Externally the wall has a fibrillar appearance, the fibrillae running in a circular direction. There is also a thin inner layer to the wall (ilc) which completely surrounds the albumen (al), and so is continuous beneath the substance of the plug. The plug is composed of two different layers both of which are distinct from the wall of the capsule. The outer opercular layer (olp) is the harder and its surface markings suggest that it is made up of concentric layers of material; the concentric markings are bisected by the suture. The thickness of the inner layer (ilp) varies considerably from capsule to capsule. It forms a cementing substance which appears to diminish as the capsule ages and at the same time there is some reduction in the diameter of the overlying operculum. Both these events seem to be preparatory to the opening of the capsule by the loss of the plug. Through the transparent plug can be seen the contents of the capsule—many unshelled eggs (em) in an albuminous fluid. According to Ankel (1935) the number of eggs varies from 1,000–3,000 and they are arranged in a continuous layer, one egg thick, around the wall. Giard (1875) and Pelseneer (1911) believed that some of these are food eggs devoured by the developing embryos, but Ankel (1935) could find no evidence of this. The plug is freed from the opening of the capsule when the echinospira larvae are ready to escape. *Lamellaria*, unlike *Trivia*, possesses no ventral pedal gland for the final moulding and deposition of the egg case. With the aid of the radula it bites small round holes in the compound ascidian and the foot apparently receives the capsules from the oviduct and embeds each vertically in a hole so that they are scattered only a few millimetres apart. The test of the ascidian thickens around the capsule and secures it more firmly and may form a protecting rim around the plug. *Velutina velutina* lays similar capsules in the test of *Styela* (Diehl, 1956) and only the transparent plug is visible on the surface; the young escape as echinospira larvae.

The egg capsules of *Trivia monacha* (fig. 206) may be found in the tissues of *Polyclinum*, *Diplosoma* and *Botryllus*. Each projects as a funnel from the surface of the test. The capsule is an erect vase-shaped structure, circular in transverse section, rounded at the base and, above, the constricted neck broadens to a tall funnel. The plug (pl) at the base of the funnel blocks the entrance and closes off the sac in which the eggs float in an albuminous fluid (eec). The capsule is about 5 mm high, the funnel making up about half of this, and the diameter in the broadest region of the egg sac is about 2·5 mm. The wall has a fibrillar texture and shows 2

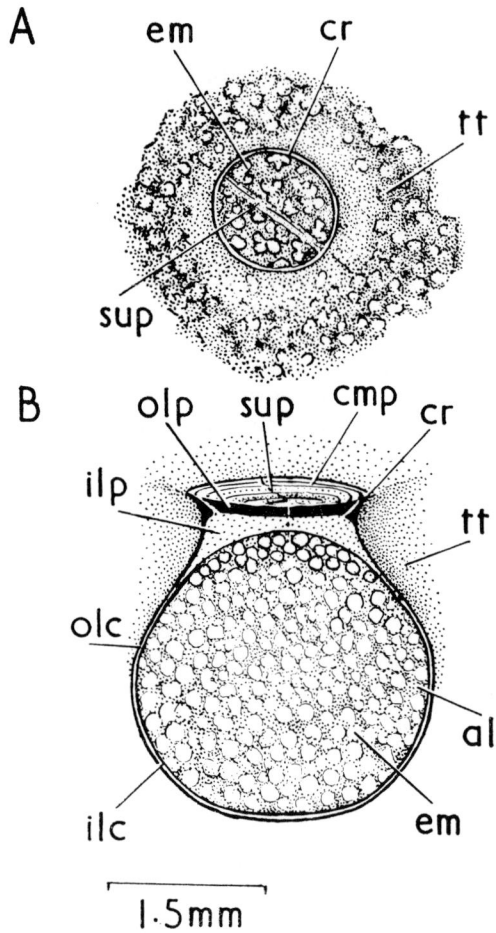

FIG. 205.—*Lamellaria perspicua:* A, surface view of egg capsule buried in tunicate test; B, vertical section of tunicate colony showing embedded capsule.

al, albumen; cmp, concentric markings on plug; cr, rim of egg capsule; em, embryo; ilc, inner layer of capsule wall; ilp, inner layer of plug; olc, outer layer of capsule wall; olp, outer layer of plug; sup, suture on plug; tt, test of tunicate.

longitudinal lines of thickening which divide the capsule into two equal halves, and a suture can be traced across the plug. Female individuals have a ventral pedal gland (fig. 68) which appears as a pit in the mid-ventral region of the sole (p. 116). In anaesthetized animals a small papilla (vp) surrounded by a deep groove is protruded from the pit and the size and shape of the papilla suggest that it is concerned with the moulding of the funnel of the egg capsule. After the capsule has been placed in the hole bitten from the test of the ascidian the gland drives it into position, at the same time gripping the pliable portion of the capsule wall and fashioning it into its final form. Its action is lubricated by secretion from glands in the epithelium and in the connective tissue below (gvp).

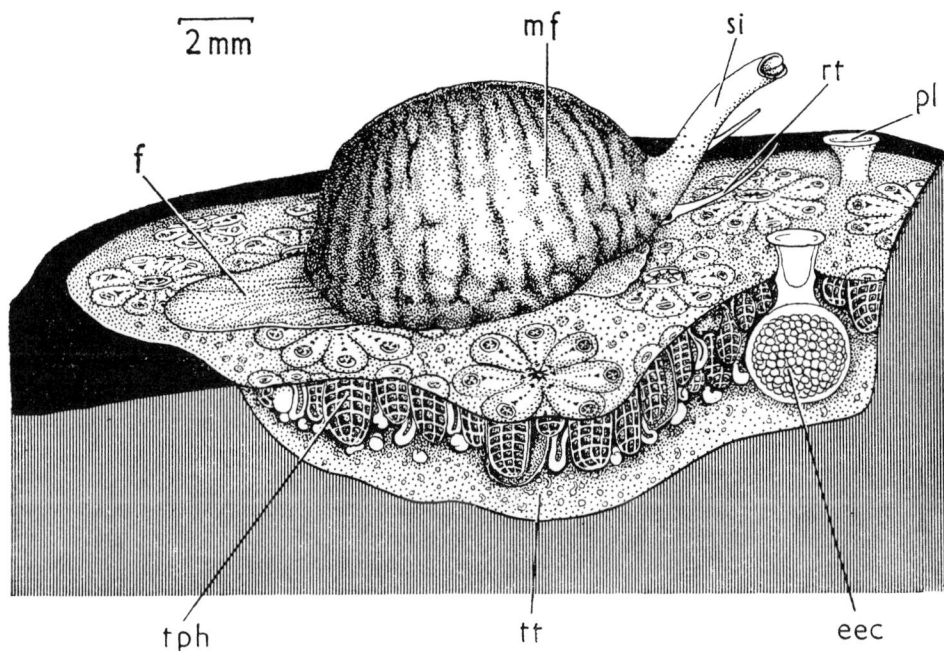

FIG. 206.—*Trivia monacha:* this figure represents a cowrie crawling over the surface of a colony of the tunicate *Botryllus schlosseri* which has been cut vertically in the foreground to show the zooids and an egg capsule embedded in it. In the right background the neck of a second capsule is visible.

   eec, eggs in egg capsule; f, foot; mf, mantle fold covering shell; pl, plug; rt, right tentacle; si, inhalant siphon; tph, pharynx of zooid of tunicate; tt, test of tunicate.

   Three species of *Ianthina* are stranded from time to time along the W and SW coasts of the British Isles, carried northwards by winds and currents. These prosobranchs are holopelagic and usually occur in shoals in warmer waters. One species, *I. janthina* (fig. 291), is viviparous and retains the eggs in the oviduct where they develop to veligers which are liberated. The others, *I. exigua* (fig. 196G) and *I. pallida,* produce egg capsules which, however, are not freed to the plankton, but are grouped along the under surface of the float which is made of air bubbles coated with mucus, and is attached to the propodium. The float is present in all individuals, males and females. According to Simroth (1895) the veliger larvae also have a long mucous strand attached to the foot, with a small swelling at the end which contains air bubbles; this is the first flotation device the animal possesses. Laursen (1953) stated that in the oviparous species the mesopodium, which is short and broad, has 2 funnel-shaped depressions lying side by side into which glands open. The mucus for the formation of the float comes from the right funnel, and from the left pass the egg capsules, which are then attached to the float. He found that the oviduct does not open into the mantle cavity (from whence according to Simroth (1895) the eggs pass to an ovipositor for deposition), but penetrates the mantle and passes obliquely down into the left side of the foot to this funnel, where it opens. Thiele (1897) also described the sole of the foot as grooved and glandular, providing the mucus for the float and helping in the manufacture of the capsules. However,

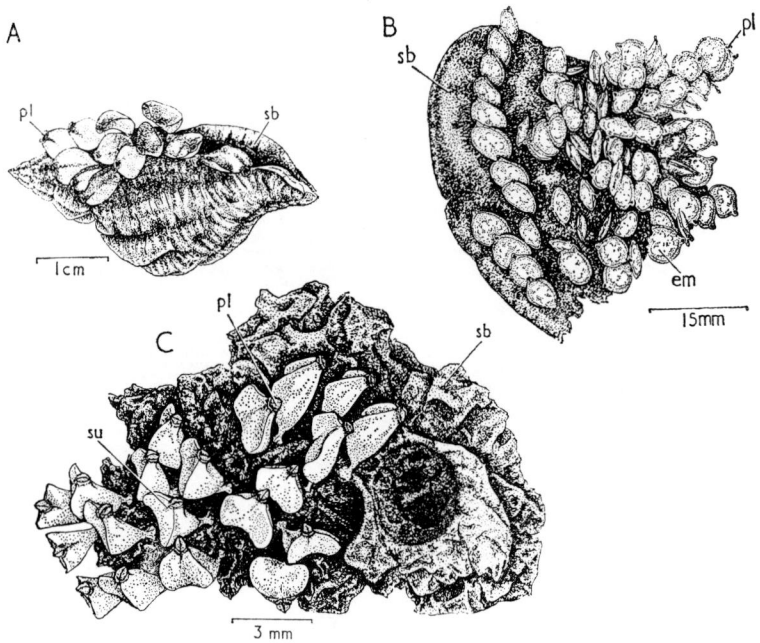

FIG. 207.—Egg capsules of stenoglossans. A, *Urosalpinx cinerea* on shell of same species; B, *Nassarius reticulatus* on *Ensis* shell; C, *Ocenebra erinacea* on stone.
em, embryo; pl, plug; sb, substratum; su, suture.

he mentioned no opening of the oviduct on the sole, where, indeed, it is wrongly placed [p. 529].

The egg capsules of *Ianthina* (fig. 196G) are pear-shaped with a stalk-like projection at one end for attachment to the float; this end is the first to emerge from the foot according to Laursen (1953). At the opposite end small spines project from the surface, scattered or in rows. In *I. exigua* they extend over two-thirds the length of the capsule and in *I. pallida* they are scattered over half the length; the longest are distally placed. In any one egg-raft capsules of all ages may be found, with embryos in stages of development up to the veliger, and finally empty capsules. The number of eggs in a capsule and the number of capsules attached to the float vary with the size of the individual and with the species. In *I. exigua* the capsules, each about 2 mm long, are arranged in rows and number about 250 in a float 5 cm long. The average number of eggs per capsule is 175, and none of these appears to be a food egg. The embryos feed on the albuminous fluid which surrounds them. *I. pallida* has a longer float, about 7 cm, and larger capsules which range from 4–7 mm in length. They are more closely packed along the float and the number of eggs in each averages 5,500 (Laursen, 1953).

The most conspicuous egg masses of prosobranchs and the most familiar to the older naturalists are those of certain rachiglossans such as *Buccinum undatum* whose spawn is frequently cast upon the shore. The capsules are comparatively large and the walls tough, taking long to weather away after the young have escaped. Moreover, some species are gregarious at the time of spawning and the spawn of several females unites to give rise to a massive array of capsules which can hardly pass unnoticed. At spawning time male and female *Nucella lapillus* congregate in rock crevices in the lower half of the intertidal zone. The animals in such a collection cease feeding for the while and, of the females, some will be copulating and others laying vase-shaped capsules, securely fastened to the rock close to one another. Spawning takes place throughout the year, chiefly in winter and spring. A whelk takes more than an hour to produce 1 capsule and 10 may be laid at intervals over a period of 24 hours (Pelseneer, 1935). The number laid by any one individual ranges from 6–31, the average being 15 (Pelseneer, 1935). The spawn of the whelk *Buccinum undatum* has a very different appearance for it is in the form of ball-shaped clusters of several hundreds of capsules piled irregularly on one another, though with spaces between (fig. 208). The masses are sometimes referred to as 'sea wash balls' for not only have they the appearance of a sponge, but are said to be used by sailors as such. One whelk may produce up to 2,000 capsules, fastening those at the base of the pile to rock or a shell (though they frequently come free), but often several individuals combine to form a single cluster which may contain up to 15,000 capsules (Dons, 1913). These may be of different sizes, the smaller ones coming from smaller whelks. A capsule has the shape of a concavo-convex lens with a diameter of about 12 mm, and is fastened to its neighbours by peripheral projections from the wall. The convex surface faces away from the centre of the cluster, and the plug, which closes the capsule until the young escape, is located near the margin of the concave face, away from the points of attachment. The young have escaped from most of the clusters cast up by the tide, but sometimes younger capsules containing embryos are attached to the emptied ones. A newly laid capsule contains several hundred eggs: Portmann (1925) stated 50–2,000 or more, and 3,200 have been obtained from the largest capsule collected around Plymouth. From such capsules only 10–30 young emerge, for the rest of the eggs fail to develop. They retain their original shape whilst the successful embryos develop and then these eat the uncleaved eggs, swallowing them whole. Each embryo devours 100 or so food eggs, and when these are finished it may turn cannibal and eat unhealthy individuals. The embryonic phase lasts about 2 months and when the young creep away they measure about 1 mm in length. The spawn of the closely related *Neptunea antiqua* is similar, though the capsules are in smaller clusters of about a dozen to a 100, and from each only 2–4 embryos hatch (Jeffreys, 1867). They are large, with a length of 6–8 mm, and it is estimated that each has devoured 2,500 eggs. Even this number is small as compared with the supply of food eggs in the northern whelk *Volutopsius norwegicus*. It lays solitary capsules which are about 2·5 mm in diameter, subhemispherical and attached by the flattened base. Thorson (1940*b*) has estimated that each of the 2–4 embryos which hatch as miniatures of the adult had 20–40 times as many food eggs as an embryo of *Neptunea*, that is 50,000–100,000. There is thus considerable intra- and inter-specific variation in the amount of food available for a single embryo in the family Buccinidae, which affects the length of time the embryos stay in the capsule and their size and stage of development when they emerge. This is particularly well seen in *Colus islandicus* where the size of the embryo varies from 3·5–8·5 mm in length according to the number developing in the capsule.

There is doubt as to whether all the eggs in the capsule of the dog whelk *Nucella lapillus* undergo early cleavage stages, Portmann (1925) alleging that division stops in the food eggs after a small sphere of cells has been formed and Staiger (1951) stating that they stop development before cleavage begins. These eggs form a central mass of yolk on which the embryos feed. A capsule may have several hundred eggs, even up to a thousand, and about 15–30 undergo normal development, occasionally fewer. [See chapter 29]. At an early veliger

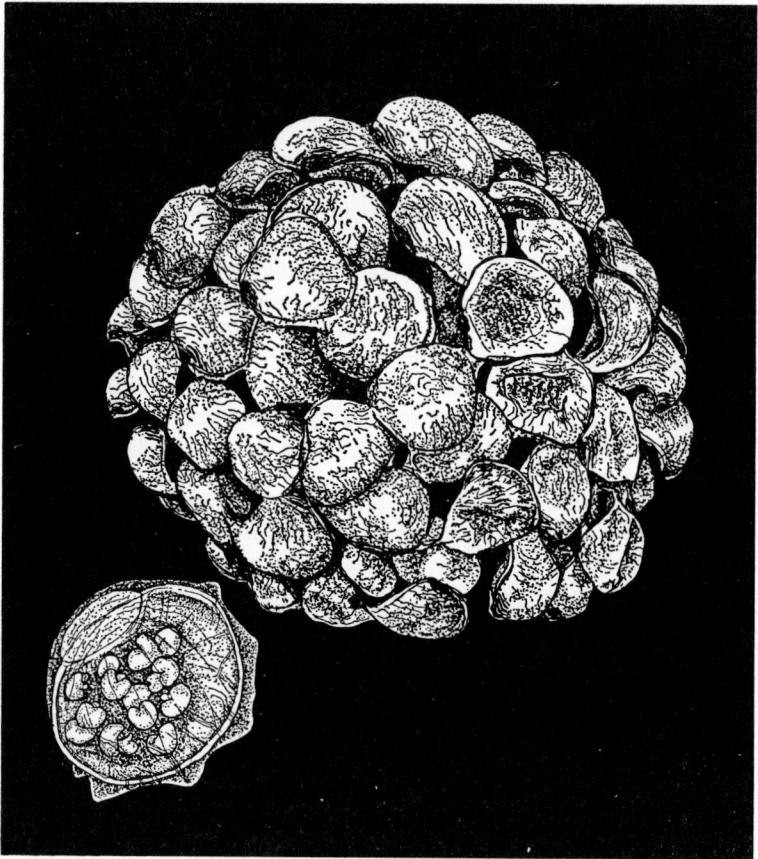

FIG. 208.—*Buccinum undatum*: egg capsules agglomerated into spawn mass, with left, below, a single capsule containing young whelks. In the single capsule note (above) the plug crossed by a suture joining right and left halves; note also (round its lower edge) the peripheral flange by which it is attached to neighbouring capsules.

stage (p. 426) these attach themselves by the sucking lips to the column of food (fig. 234A), and in contrast to the embryos of *Buccinum* development is arrested until the food is eaten. At a later stage unhealthy embryos may also be devoured. The size of the embryos varies: Risbec (1937) observed one capsule 5 mm tall with only 2 embryos one of which attained 2·5 mm. The vase-shaped capsule (p. 305) is fitted with an apical plug of mucus which is loosened and lost at the time the embryos hatch. Ankel (1937b) has shown that the plugs of fresh capsules are softened to a pulp by a mixture of embryo dog whelks in sea water, though not by crushed salivary glands of the adult, and concluded that they are normally dissolved by an enzyme liberated by the embryos. In a related purpurid of the Persian Gulf, *Thais hippocastanea*, the egg capsules have a similar appearance except that the top is covered by a thin, transparent membrane through which the contents can be seen (Thorson, 1940a). The food eggs seem to be eaten whole by the 4 or 5 embryos which develop, and it would appear that these escape from the capsule by rasping a single large hole in its wall. In emptied capsules the apical membrane is intact and the wall perforated; Thorson (1940a) has suggested that the embryos are too large to pass through the apex. This unusual means of escape has been recorded by Pope (1910–1911) for the oyster drill *Urosalpinx cinerea*. He watched young drills cut small circular holes in the egg case and then push their way through. However, he believed that this method of escape occurs only when embryos near the apex of the capsule are not developed sufficiently to emerge and they obstruct the normal passage.

The capsules of *Urosalpinx* (fig. 207A) and of the tingle *Ocenebra erinacea* (fig. 207C) are constructed in a similar way to that of *Nucella*, and are of about the same size, though details in the shape vary. Each is vase-shaped with a narrow stalk which expands to a broad basal attachment to the substratum, and distally a circular opening which is filled with a plug (pl). Unlike the capsule of *Nucella* the walls are flattened or concave on one side and convex on the other. The curvature is more pronounced in *Ocenebra* in which the middle of the convex surface may be produced into a slight keel so that in transverse section the capsule is approximately triangular. There are 2 longitudinal sutures visible on the surface of the smooth yellow walls running up the centre of the concave and convex faces and continuous over the plug. In *Urosalpinx* the exposed surface of the plug is flat and flush with the rim of the capsule. It shows concentric markings which are bisected by the suture so that it is similar to the plug which closes the capsule of *Lamellaria perspicua*. Moreover, as in *Lamellaria*, there is a second inner layer to the plug forming a complete seal over the opening and actually separating the outer horny layer from the rim of the capsule (Hancock, 1956). In *Ocenebra* the plug projects beyond the rim of the capsule to a conspicuous keel along the line of the suture and it is composed of only one type of secretion which is mucoid. The detailed structure of the wall shows all the characteristics which have been described for the capsule of *Nucella*: an outer fibrous layer, with circular fibres externally and longitudinal ones internally, over-lying a homogeneous semitransparent layer, which is thicker in *Urosalpinx* than *Ocenebra*, and then a thin inner skin which surrounds the contents of the capsule. The fibrous part of the wall and the underlying homogeneous layer are protein, probably conchiolin, and between the fibres is a mucoid substance. Each of these 2 secretions forms a layer of the plug of *Urosalpinx*, the outer layer being similar to the homogeneous skin of the capsule wall. [See p. 665.]

A capsule of *Urosalpinx* contains up to 35 eggs which float in an albuminous fluid, though occasional capsules contain much larger numbers (Hancock, 1957). An average-sized cap-sule of *Ocenebra* also contains about 35 eggs. Lebour (1937) who quoted the number as 12–20 was probably dealing with the smaller capsules, whereas in large ones, 9–13 mm high, Hancock (1957) found a range of 52–167 eggs, and Lamy (1928) up to 196. These larger numbers are exceptional. During the breeding season a female *Ocenebra* deposits 30–40 capsules. In any one a few eggs may be very small, and give rise to retarded veligers (Risbec, 1937) which are devoured by the normal embryos. However, there is no regular practice of embryonic cannibalism either in *Urosalpinx* or *Ocenebra*. Risbec (1937) stated that there are often clumps of ciliated cells in the capsules of *Ocenebra*, and suggested that they are fragments of the velum which have been cast off and not absorbed as in *Nucella*. The embryos of *Urosalpinx* develop to the crawling stage in about 2 months at 15·9°C and those of *Ocenebra* in about 3 months at 10–19°C (Hancock, 1957); they escape through the apical opening of the capsule. During their embryonic development the plug which closes the capsule gradually loosens and is lost by the time the young are ready to emerge. Hancock (1956) suggested that in *Urosalpinx* the embryos liberate a substance, probably an enzyme, which is responsible for this.

The spawn of *Nassarius reticulatus* (fig. 207B) consists of capsules which are slightly convex on one surface, flattened on the other and are fastened to any hard object in the vicinity; some have been found attached to an egg collar of *Natica*. The capsules are frequently laid evenly spaced in rows and as in *Urosalpinx* and *Ocenebra* several individuals may lay alongside one another. Each is about 5 mm high and 4 mm across at the broadest point; distally it tapers to a very small round opening closed by a plug, and proximally it has rather a broad basal attachment to the substratum. The walls have a fibrillar structure, but are thinner and more transparent than in the capsules of other rachiglossans. The lateral margins of the flattened sides extend to form a flange and this is continuous over the distal end, forming a rim to the plug on one side. A capsule contains from 50–2,000 eggs which after about a month are liberated as veligers; Ankel (1929) suggested that they secrete a substance which dissolves the plug. The egg capsules of *N. incrassatus* and *N. pygmaeus* are similar, though smaller and with fewer eggs.

The differences in shape of the egg capsules of the various species of Rachiglossa are not due to any fundamental dissimilarities in the structure of the female genital ducts, but to specific differences in the shape of the ventral pedal gland which moulds the capsules passed to it in an unfinished state from the oviduct. The capsule is passed along a temporary groove on the foot to the pedal gland which grips it so that the plug is directed to the inner end. A limited amount of secretion from the gland acts as a lubricant and a hardening mixture, but it has never been suggested that this adds to the substance of the capsule. When the capsule of *Nassarius reticulatus* leaves the genital duct it is biconvex, with 2 longitudinal lines of thickening extending from the plug to the thick basal plate; its walls are flexible. Within the pedal gland the flanges are shaped from the longitudinal thickenings and the basal disc is constricted from the capsule, which now assumes its final flattened shape in accordance with that of the gland. The capsule is retained for some minutes whilst it is kneaded by the pedal musculature, the wall smoothed and hardened and the disc pressed against the substratum, to which it adheres. In *Buccinum undatum* the convex face of each egg case is, at this stage, stamped with a reticulate pattern. The fashioning by the pedal gland takes 15–20 min in *Thais haemastoma* (Franc, 1941b) and then the whelk lifts the mesopodium and leaves the capsule firmly attached.

In *Ocenebra* the 2 longitudinal sutures, which result from the bilobed structure of the genital duct, are accurately placed one in the middle of the concave surface and the other in the middle of the convex, and in every capsule this is the same. The orientation of the capsules in the ventral pedal gland of individuals of any one species must always be remarkably similar.

A second type of egg case is found in the Stenoglossa (fig. 209). It is lens-shaped with a flattened base circular or oval in outline, by which it is attached, and a convex upper surface. The periphery is edged with a narrow brim and in the middle of the convex face is the round or oval exit hole, covered by a membrane and surrounded by a thick edge from which 2 sutures diverge. These run to the base of the capsule and divide the wall into 2 equal halves. The wall is semitransparent so that the contents are visible. Such capsules are laid by *Trophon* spp. (Lebour, 1936; Thorson, 1946) and contain less than a dozen rather large eggs which hatch as miniatures of the adult. They also occur in all the British turrids of which the spawn has been described, but contain a large number of smaller eggs all of which develop to free veliger larvae. According to Lebour (1934b) the size of the capsules of the turrids in the vicinity of Plymouth varies from a diameter of about 1·6 mm in *Mangelia nebula*, in which there are about 60 eggs per capsule, to 3·4 mm in diameter in *Philbertia gracilis* (fig. 209) which

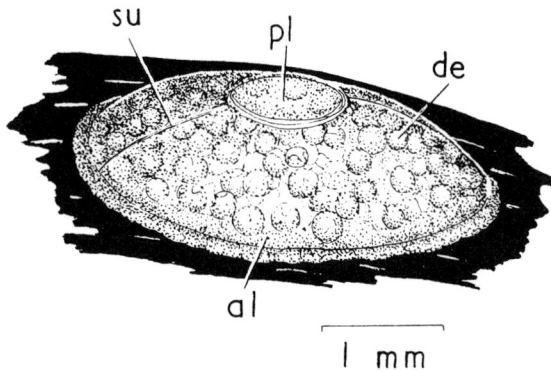

FIG. 209.—*Philbertia gracilis:* egg capsule on stone. The surface bears reticulate markings not shown in the figure.
al, albumen; de, developing egg; pl, plug; su, suture.

has 40–80 eggs (Lebour, 1933*d*). However, there is probably considerable variation within any one species for Jeffreys (1867) gave the diameter of the capsule of *P. linearis* as 5 mm and its contents as 200–300 eggs, whereas Lebour (1934*b*) gave the corresponding measurement as ranging from 1·5–2·0 mm and the number of eggs as 60–80. *P. purpurea* lays an even larger number of eggs, 350–400, in a single capsule (Franc, 1950).

Like other rachiglossans, *Trophon* has a ventral pedal gland in females only, which presumably moulds the egg capsules. Of the turrids, however, *Mangelia brachystoma* (Robinson, 1955) is reported to have no such gland, although the remarks of Ankel (1936*a*) would lead one to suppose that one is present in *Philbertia gracilis*. The exposed surface of the capsule of *P. gracilis* has a coarse reticulate marking, presumably due to the moulding action of the foot, and this may vary slightly from one group of capsules to another showing characteristics of the individual which produced them (Ankel, 1936*a*).

There is only one British member of the family Fasciolariidae, *Troschelia berniciensis,* and this is rare. Thorson (1940*b*) has described the egg capsules as ovate, with a greater diameter of 16 mm and a lesser of 11 mm; a longitudinal ridge around the wall gives it a walnut shape. It contains only one embryo, which is presumably nourished by food eggs like the embryos of other species of this family (Hyman, 1923).

CHAPTER 16

DEVELOPMENT

THE molluscs along with the platyhelminths and annelids have eggs which cleave spirally. Conklin's work (1897) on the development of the egg of the prosobranch *Crepidula* may be regarded as the foundation of our knowledge of spiral cleavage in molluscs. The size of the egg in prosobranchs varies considerably, for in some species there is little yolk and in others much. In *Viviparus* the diameter of the egg is only about 18 μm, in *Littorina littorea* 130 μm, and this is far exceeded in the Stenoglossa, for the corresponding measurement of the egg of *Trophon muricatus* is 480 μm. The quantity of yolk affects the difference in size between megameres and micromeres, which is very marked in the Stenoglossa. Typically the third cleavage is dexiotropic, the fourth laeotropic and subsequent cleavage planes alternate. This results in the formation of a dextral snail. Occasionally sinistral individuals or races occur, and there are some species (e.g. *Triphora perversa*) which are consistently sinistral. Sinistrality may be the result of the reversal of the third and subsequent cleavages, for such a change in cleavage pattern (Crampton, 1894) is known to give the sinistral individuals of certain pulmonates (*Ancylus, Physa, Planorbis*).

Cell lineage has been studied for a number of molluscs. Besides *Crepidula* these include the prosobranchs *Patella caerulea* (Patten, 1886a; Wilson, 1904), *Theodoxus fluviatilis* (Blochmann, 1882), *Trochus* (= *Gibbula*) *magus* (Robert, 1902) and *Littorina littoralis* (Delsman, 1914). The results show general agreement as to the origin of the germ layers. The first three quartettes of micromeres give rise to all the ectoderm: the first forms the pretrochal ectoderm which will be the ectoderm of the head, the second and third the rest. These micromeres may also form ectomesoderm. It is from 2*d* that much of the dorsal and ventral ectoderm of the body is derived, including the shell gland and the ectoderm of the foot. Endoderm arises from 4*A*, 4*B*, 4*C* and also the fourth quartette of micromeres 4*a*, 4*b*, 4*c*. Erlanger (1892), Conklin (1897), Robert (1902) and Wilson (1904), working with *Bithynia, Crepidula, Gibbula magus* and *Patella caerulea* respectively, described the division of the megamere of quadrant *D* as giving 4*D*, which also forms endoderm, and 4*d* which is exclusively responsible for the teloblastic mesoderm as it is in annelids. However, Wilson (1904) expressed some doubt as to the identification of 4*D* and 4*d* in the embryos of *Patella*. The time of origin of 4*d* in molluscs varies. In *Gibbula* it is at the 64-cell stage, in agreement with annelids, whilst in *Crepidula* (and also the nudibranch *Fiona* (Casteel, 1904)) it appears immediately after a short rest period which follows the 24-cell stage. Heath (1899) has accurately traced the origin of the mesoblast in *Ischnochiton* at the 72-cell stage. He was unable to decide whether any of the descendants of 4*d* form endoderm, nor could Robert (1902) determine this for *Gibbula*, but this occurs in other prosobranchs and supports the endodermal nature of the teloblastic mesoderm, which is shown by the destiny of the sister cell 4*D* (*see* Table 12). Three divisions of the endomesoderm cell are concerned with the formation of endoderm in *Crepidula* and *Fiona*. Smith (1935), Crofts (1937) and Creek (1951) suggested that in *Patella vulgata, Haliotis* and *Pomatias* the megamere of the quadrant *D* produces mesoderm and that all the endoderm is derived

TABLE 12

The origin of endomesoderm in:

A. *Patella coerulea*, after Patten (1886a);

B. *P. coerulea*, after Wilson (1904). This diagram shows only the probable state, for Wilson (p. 206) expresses some doubt as to the identification of 4D and 4d;

C. *Crepidula*, after Conklin (1897);

D. *P. vulgata*, after Smith (1935); *Haliotis*, after Crofts (1937); *Pomatias elegans*, after Creek (1951).

endomesoderm cells, their
origin not stated

A. 4 megameres give endoderm

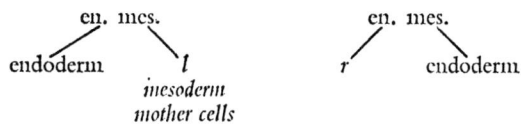

B.

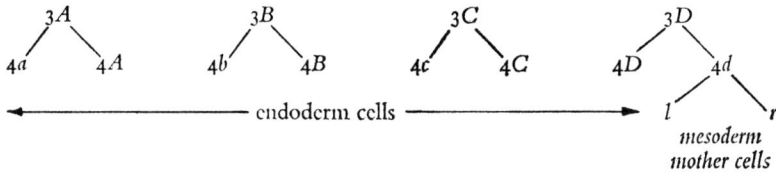

C.

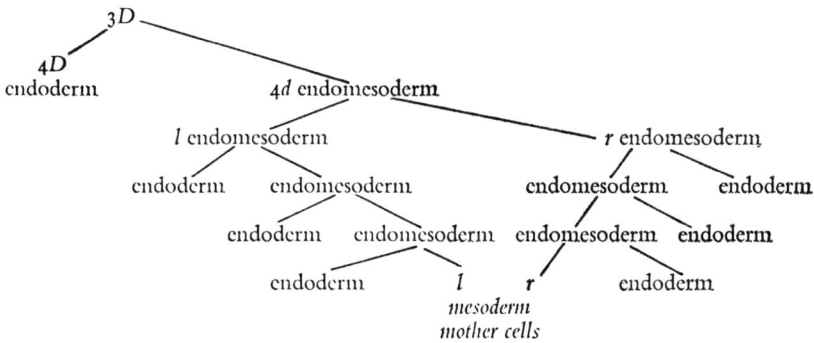

D.

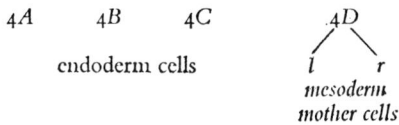

from the other three megameres, though these authors have not followed the full details of cell lineage.

The type of blastula varies with the size of egg. In species with little yolk there is a moderate (*Patella caerulea, Pomatias*) or even wide cleavage cavity (*Littorina littoralis, Viviparus, Bithynia*), but before gastrulation begins, or at its onset, the animal and vegetative poles may flatten, narrowing the blastocoel. The flattening is pronounced in *Littorina, Pomatias* and *Bithynia,* and also occurs in *Viviparus* (Dautert, 1929) where, however, the blastocoel remains wider. In eggs with considerable yolk and consequently large megameres the resulting blastula is a more or less solid sphere (*Crepidula, Nassarius, Fusus, Ocenebra, Urosalpinx, Nucella*). A large quantity of the yolk is held by 4D, and in *Nucella* and *Nassarius* 4D has a volume as large, or even larger, than the combined volume of the other megameres.

The gastrula is formed in different ways according to the size of the megameres. In *Viviparus* all blastomeres are of nearly equal size (fig. 210A), the micromeres being as big as the megameres, which invaginate into the segmentation cavity so that a wide archenteron (an) is formed. The blastopore (bl) remains as an inlet for the albumen, which is absorbed by cells of the endoderm at an early stage (abe) and compensates for lack of yolk. At the time the gastrula is formed no mesoderm has developed. In species with a small to moderate amount of yolk gastrulation is brought about by epibolic growth of the micromeres over the mega-meres, together with some movement of these larger cells into the segmentation cavity; in contrast to *Viviparus* the primordial mesoderm cells appear during this process. Epiboly and invagination are of equal importance in the production of the gastrula of *Pomatias* and *Littorina,* but in the slightly more yolky eggs of *Patella vulgata, Haliotis* and *Gibbula magus* gastrulation is effected mainly by epibolic growth. In *Pomatias* the flattening of the gastrula at the 64-cell stage brings the 4 megameres to the region of the vegetative pole, and at the same time one of the megameres moves into the blastocoel, which is being gradually obliterated. This is presumably 4D behaving differently from the other 3 megameres; Creek (1951) con-cluded that it gives rise to mesoderm only. Gastrulation is continued by the invagination of the megameres into the concavity formed by the epibolic downgrowth of the micromeres. These border a wide circular opening to the archenteron, the blastopore. In *Haliotis* and *Patella* no flattening of the blastula occurs and gastrulation begins by the inward growth of the megameres 4A, 4B, 4C, 4D which reduce the segmentation cavity and for a short time remain in connexion with the surface by long necks. At the same time the ectodermal cap of micromeres extends over them, spreading downwards to reach the vegetative pole. The region where the megameres are exposed is the blastopore, and the archenteron appears only after division of the invaginated cells. Of the megameres, that of quadrant D is the first to divide. According to Smith (1935) and Crofts (1937) it gives, during the gastrulation of the embryo of *Patella vulgata* and *Haliotis,* two equal-sized cells which are spherical or sub-ovoid in shape. They are considered to be the teloblastic mesoderm cells situated near the blasto-pore (pm, fig. 216) and they contrast with the elongated appearance of the other 3 undivided megameres (en).

In *Crepidula* (fig. 210B) and the Stenoglossa (fig. 211) gastrulation is by epiboly and after the formation of the fourth quartette there is an interval before the yolk-laden megameres divide. Meanwhile the micromeres multiply and spread over them as a thin layer (ect). In *Crepidula* the nuclei of the megameres (en) become very large and lie in the protoplasmic portion of the cell, away from the mass of yolk and near the surface just in advance of the edge of the micromeres, and in this position they move towards the vegetative pole. In both *Crepidula* and *Fusus* (Bobretzky, 1877) small cells are divided from the megameres chiefly in

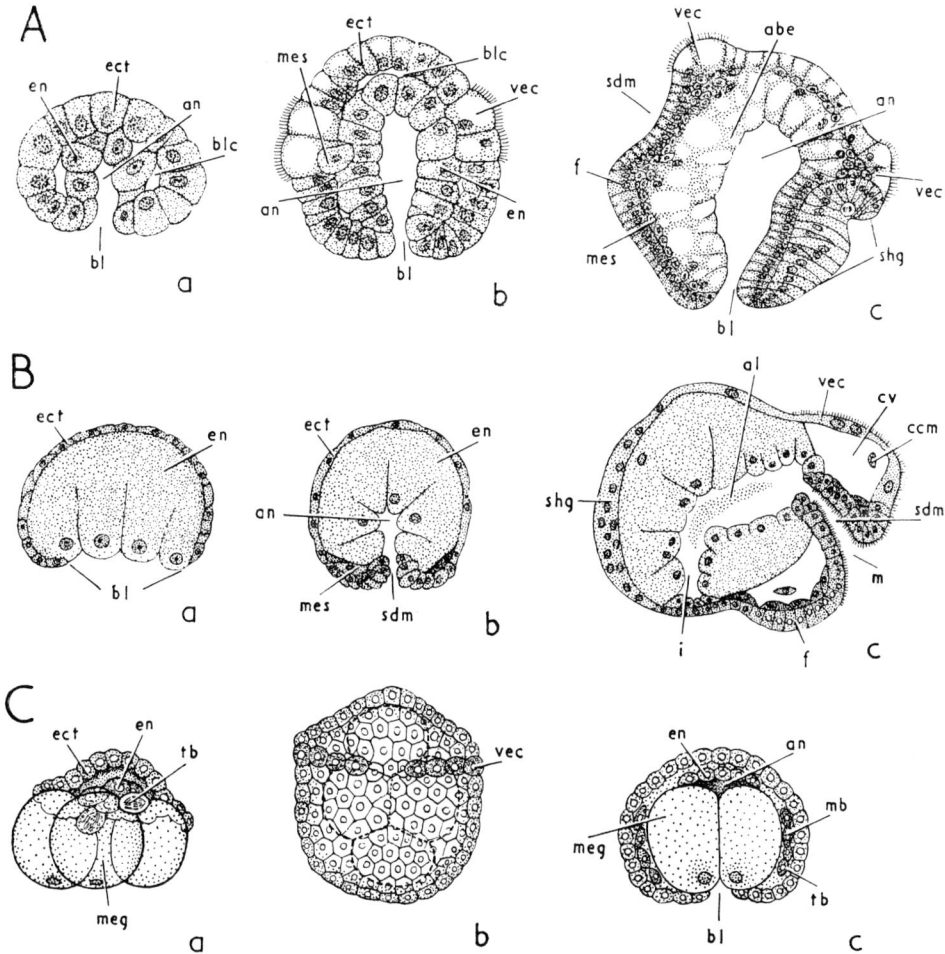

FIG. 210.—A, *Viviparus viviparus: a*, vertical section of gastrula (about 180 cells) which is formed by invagination of the endomeres. The gastrula is radially symmetrical. *b*, sagittal section of early trochophore (about 400 cells); the prototroch (velum) comprises 2 rows of ciliated cells; the ventral surface of the embryo is to the left and here the ectoderm posterior to the velum gives rise to the mesoderm, which is thus formed some time after gastrulation. *c*, sagittal section of later embryo similarly orientated. From the ventral surface protrudes the rudiment of the foot and anterior to this is the oesophageal invagination; the prevelar area is displaced dorsally; mesoderm now separates ectoderm from endoderm except mid-dorsally; the dorsal ectoderm forms the shell gland and the underlying endoderm cells are closely applied to it and lower than elsewhere; the larger endoderm cells are concerned with the absorption of albumen; the blastopore remains open and becomes the anus. After Dautert.

B, *Crepidula fornicata: a*, vertical section of embryo during gastrulation, which is by epiboly; the ectoderm forms a thin layer of cells which is surrounding the yolk-laden megameres. *b*, vertical section of gastrula; small mesoderm cells which have separated from 4*d* now lie on either side of the blastopore; this becomes the mouth. *c*, oblique longitudinal section of embryo at early veliger stage; at the posterior end the section lies to the right of the mid-line; the lumen of the gut contains ingested albumen; the ciliated cells of the velar area rotate the embryo in the albuminous fluid; with the growth of the foot, the expansion of the shell gland and the enlargement of the whole region posterior to the velum, the mouth comes to lie further and further forward and the stomodaeum now runs posteriorly from the mouth to the mesenteron. After Conklin.

the region of the blastopore, and they form the ventral wall of the archenteron which is clothed dorsally by the megameres. The distribution of the yolk in the embryos of *Nassarius, Urosalpinx* and *Nucella* is such that one megamere is far larger than the other three. As the micromeres grow over the megameres to approach the vegetative pole the 3 smaller mega-meres divide, and the resulting endoderm cells shift around the large yolky cell towards this pole to line the greater part of the archenteron (fig. 211). This, however, is a relatively inconspicuous cavity and, with the teloblastic mesoderm cells formed during gastrulation, forms part of a germ disc which lies over the surface of the megameres.

The movement of endoderm cells during the development of the archenteron in *Theodoxus* (Blochmann, 1882) is in the reverse direction (fig. 210C). Small endoderm cells (en) are cut off from the megameres (meg), which are small compared with those of the Stenoglossa, during the early stages of gastrulation, whilst the micromeres (ect) are enclosing them by epibolic growth. These endoderm cells migrate through the blastocoel towards the animal pole to form a continuous layer beneath the ectoderm and so become the dorsal wall of the primitive gut.

The primary or teloblastic mesoderm is at first represented by two cells, the teloblasts, lying in the blastocoel near the blastopore, one on either side of the archenteron, and these divide to give off new cells anteriorly. In typical cases two short mesodermal bands, bilaterally arranged, are formed by the activity of the teloblasts and the sub-division of the cells formed by them. They are at first dorsal in position. In *Crepidula* they are formed after the 65-cell stage and eventually each band comprises only 8 or 9 cells (fig. 213A, mb). In *Haliotis* there is a total of 10 mesoderm cells at the time the trochophore larva is freed (fig. 217A) and more are formed later (rmc, fig. 217C). During the pretorsional larval stage of *Haliotis* and *Patella* the anterior ends of the bands are broken up into irregular, spindle-shaped cells which will develop into muscle cells of the head and foot, and also into vascular tissue in *Haliotis*. Others surround the dorsal and lateral parts of the ectodermal foregut. The remainder of the bands will give rise to the velar retractor muscle (developed from the right one alone and respon-sible for the first or rapid stage of torsion), the columellar muscle (developed from the left band alone), and the walls of the kidney and pericardium from which the gonad will later arise. In *Crepidula* torsion is complete before the veliger is freed and it is differential growth which leads to the mesenteron being looped into a figure-of-eight. No larval retractor muscle was found by Conklin (1897) and he stated that all signs of teloblasts and mesodermal bands have disappeared as such before the beginning of the torsion of the gut, and before the early differentiation of the nervous system. However, when the veliger larva hatches there is a well developed larval retractor muscle of unknown origin (lr, fig. 228A) which in this prosobranch

---

C, *Theodoxus fluviatilis: a,* early gastrula from the left; the ectoderm of this side has been removed to reveal the small endoderm cells budded from the megameres and one of the 2 teloblastic mesoderm cells; gastrulation is by epiboly. *b,* gastrula seen from above with the future anterior end towards the top; the 4 megameres are seen by transparency. *c,* vertical section of gastrula passing through the blastopore; the small endoderm cells have shifted beneath the ectoderm of the animal pole to form a continuous layer which, with the megameres, encloses the archenteron; the megameres form the rudiment of the digestive gland; the mesodermal bands are laterally placed. After Blochmann.

abe, absorbing cell of endoderm; al, albumen in mesenteron; an, archenteron; bl, blastopore; blc, blasto-coel; ccm, cerebral commissure; cv, cephalic vesicle; ect, ectoderm; en, endoderm; f, foot; i, intestine; m, mouth; mb, mesodermal band; meg, megamere; mes, mesoderm; sdm, stomodaeal invagination; shg, shell gland; tb, origin of mesodermal band; vec, velar cells.

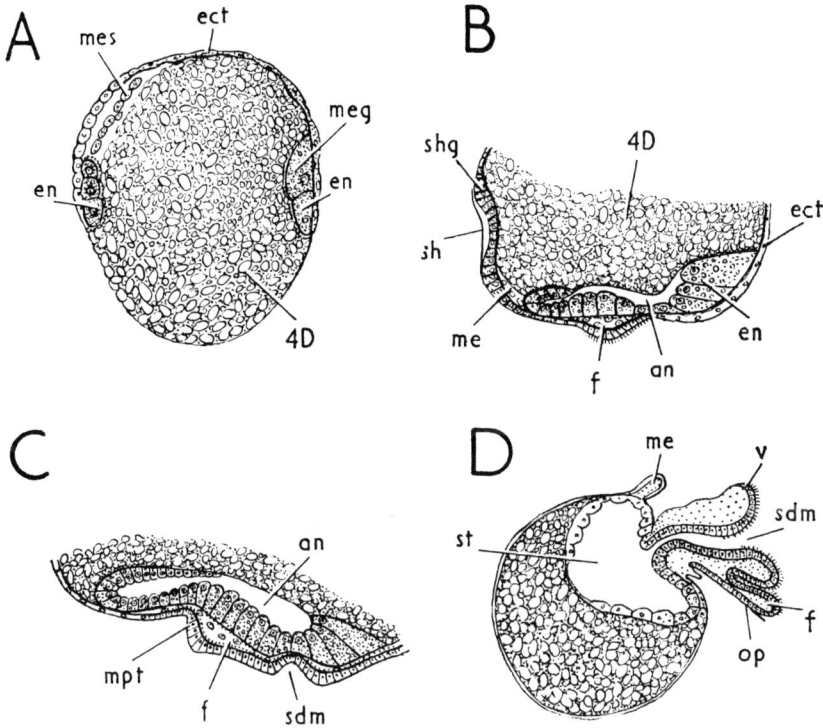

FIG. 211.—*Nassa mutabilis:* A, vertical section of early gastrula; most of the section is occupied by megamere 4D; the micromeres are spreading over the megameres and at the same time endoderm cells are dividing from 4A, 4B, 4C, 4a, 4b and 4c; the mesoderm is derived from 4d.

B, the vegetative part of a section through an embryo in which the micromeres cover the vegetative pole; the small endoderm cells have also shifted to this region and, with the large megameres, line a cavity, the archenteron; 4D forms a central core of yolk; rudiments of the foot and shell have appeared.

C, part of a section through an embryo to show the stomodaeal invagination.

D, longitudinal section through a much later embryo to show the relationship between the gut and the yolk. After Bobretzky.

an, archenteron; ect, ectoderm; en, endoderm; f, foot; me, mantle edge; meg, megameres; mes, mesoderm; mpt, metapodium; op, operculum; sdm, stomodaeum; sh, shell; shg, shell gland; st, stomach; v, velum.

is apparently late to develop and is not concerned with torsion. Unfortunately we have no knowledge of the development of tissues from the teloblastic mesoderm in *Crepidula,* for details of the internal development of later embryos have not been worked out. In fact later developmental stages have been followed in adequate detail only in *Viviparus* (Erlanger, 1891b; Tönniges, 1896; Drummond, 1903; Andersen, 1924a, 1924b), *Patella* (Patten, 1886a; Smith, 1935; Crofts, 1955), *Pomatias* (Creek, 1951) and *Haliotis* (Crofts, 1937, 1955).

The terrestrial prosobranch *Pomatias elegans* completes development to the crawling stage within the egg capsule. The egg (fig. 197) which measures 0·14 mm in diameter, has little yolk and at an early stage the embryo absorbs albumen through special ectoderm cells in the swollen cephalic region (cv, fig. 224A). The embryo is characterized by the early development of abundant teloblastic mesoderm which, by the gastrular stage, forms a compact layer

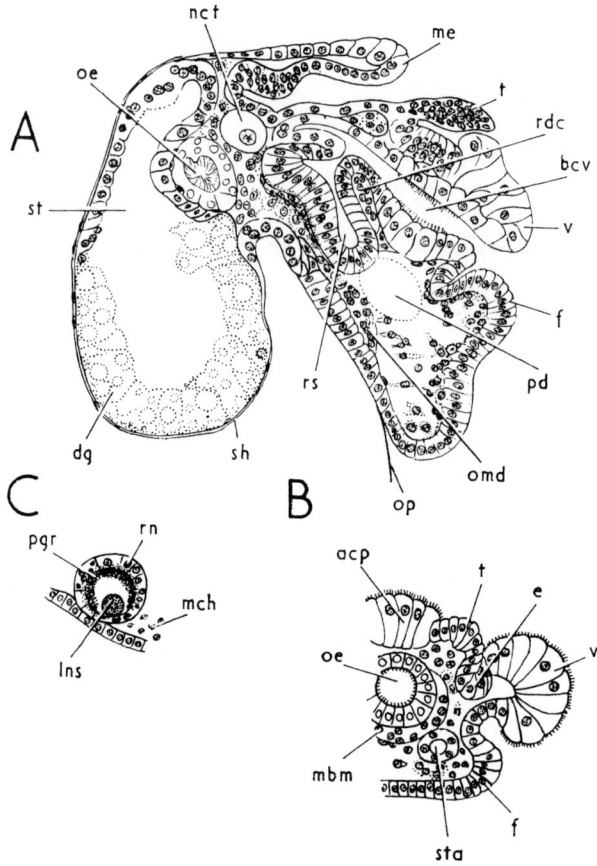

FIG. 212.—*Littorina littoralis:* A, longitudinal section of embryo after torsion; B, transverse section through head-foot of younger embryo; C, section of developing eye, showing closed vesicle. After Delsman.

acp, apical cell plate; bcv, buccal cavity; dg, digestive gland; e, eye; f, foot; lns, lens; mbm, muscles of buccal mass; mch, mesenchyme; me, mantle edge; nct, nephrocyst; oe, oesophagus; omd, developing opercular branch of columellar muscle; op, operculum; pd, projection of developing pedal ganglia on to sagittal plane; pgr, pigment granule; rdc, radular cartilage; rn, retinal cell; rs, radular sac; sh, shell; st, stomach; sta, statocyst; t, tentacle; v, velum.

between ectoderm and endoderm, and is thickest on the right and left sides. By the early veliger stage these lateral thickenings, corresponding to the mesodermal bands of other species, are no longer distinguishable owing to the increase of mesoderm dorsally and ventrally. A considerable quantity has also developed in the foot and visceral mass. However, none of the cells shows specific differentiation—in fact the differentiation of muscle is delayed until torsion is complete. With the loss of the free larval stage both contractile tissue and blood are late to develop.

In addition to the mesoderm which is bilateral and teloblastic in origin, cells may pass into the blastocoel from the ectoderm after the appearance of the teloblasts. In *Crepidula* their

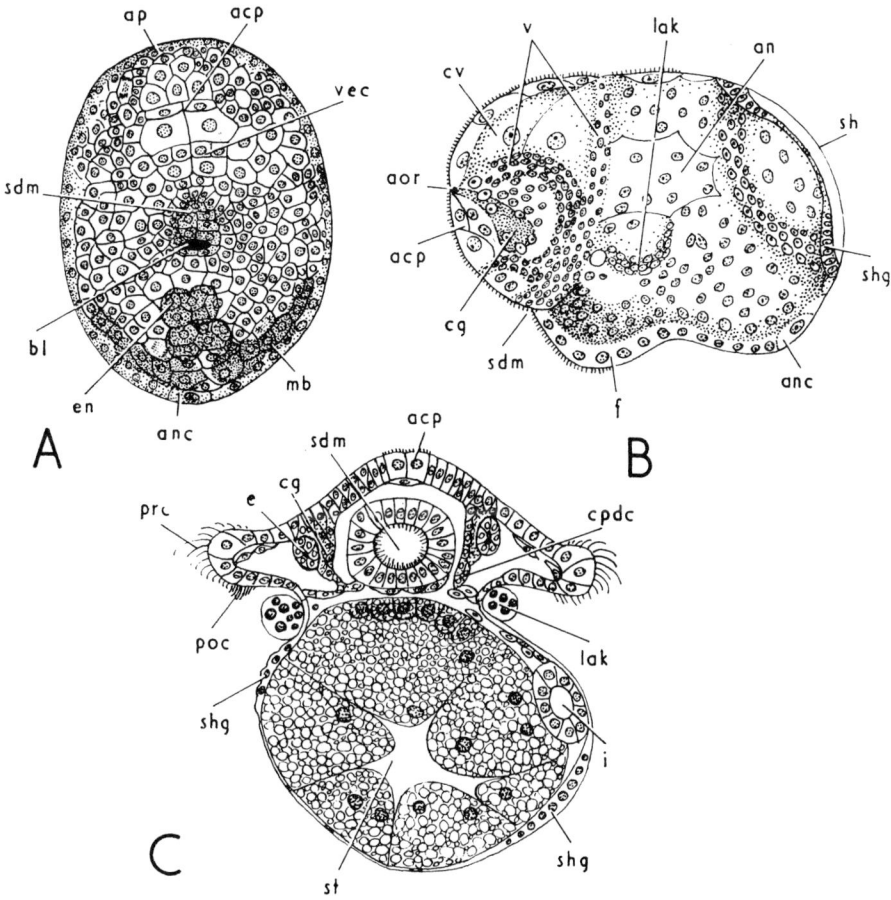

FIG. 213.—*Crepidula fornicata*: A, embryo showing closure of the blastopore; B, side view of older embryo; C, horizontal section of still older embryo at a ventral level. After Conklin.

acp, apical cell plate—the end of the guide line lies at the centre of a group of 7 cells which will give rise to this structure; an, archenteron; anc, anal cells; aor, apical organ; ap, animal pole now migrated so as to be visible in ventral view; bl, blastopore; cg, cerebral ganglion; cpdc, cerebropedal connective; cv, cephalic vesicle; e, eye; en, enteroblasts; f, foot; i, intestine; lak, larval kidney; mb, left mesodermal band; poc, postoral cilia of velum; prc, preoral cilia of velum; sdm, stomodaeal depression beginning, in C, stomodaeum; sh, shell; shg, shell gland; st, stomach; v, velum; vec, velar cells, arranged in 2 transverse rows.

origin has been traced to the descendants of the second quartette of ectomeres, 2a, 2b, 2c. They are scattered cells, chiefly concerned in the formation of muscle fibres in the foot, and also in the head vesicle and velum, which are embryonic or larval organs. This ectomesoderm is often referred to as larval mesoderm or mesenchyme, though adult tissues are allegedly derived from it. However, it appears to be neither constant in origin nor has it been found in all the embryos of prosobranchs which have been studied. In *Patella caerulea* (Patten, 1886a), *P. vulgata* (Smith, 1935), *Haliotis* (Crofts, 1937), *Pomatias* (Creek, 1951) and *Theodoxus*

(Blochmann, 1882) all mesoderm appears to come from the mesodermal bands, and Robert (1902) stated that if ectomesoderm is formed in trochids its development must be very late for he was unable to trace it. However, it is developed in *Littorina littoralis* and originates from the cells $3a^{2111}$, $3a^{2211}$ and $3b^{2111}$, $3b^{2211}$ (as in *Fiona, Physa* and *Planorbis*), and Delsman (1914) observed the ectomesoderm joining with the cells of the endomesoderm so closely that in this species the two cannot be differentiated later. Indeed he found it impossible to decide whether the heart and kidney arise from ectomesoderm or endomesoderm, or a combination of both. In the trochophore stage of polychaetes larval mesenchyme is maximal in species which have little yolk and minimal in larvae with a rich food supply (Meyer, 1901), but no such correlation can be made in prosobranchs.

The aberrant freshwater prosobranch *Viviparus,* in which the young complete development in the oviduct, appears to stand apart from all other molluscs in the mode of origin of the mesoderm. The small yolkless eggs develop rapidly to a trochophore stage so that the archenteron can take up the albuminous fluid (fig. 210A). Only at this stage does mesoderm formation begin. Erlanger (1891a and b; 1894) was the first to study the embryology and he stated that a median bilobed pouch which pushes into the blastocoel from the ventral surface of the intestine is the origin of the mesoderm. The pouch separates from the gut, loses its cavity, and gives rise to two irregular mesodermal bands which extend forwards at the sides of the gut. Anterior cells of each band form a larval kidney, a compact posterior mass cavitates to form a vesicle which will become part of the pericardial sac, and the cells of the mid-region scatter in the blastocoel. Such enterocoelic mesoderm formation has not been described for any other mollusc and seems an incongruity in an embryo developing from an egg which cleaves spirally.

Erlanger's results were soon challenged by Tönniges (1896) who stated that all adult mesoderm in *Viviparus* arises as an ectodermal proliferation; later, he repeated this investigation and obtained the same result (Otto & Tönniges, 1906). Dautert (1929) agreed that only ectomesoderm is developed and traced its origin to cells of the second and third quartettes. He found that it appears at the ventral surface of the embryo (mes, fig. 210A, c), beginning in the region of the prototroch and spreading to the blastopore. The mesoderm spreads laterally between ectoderm and endoderm, but not mid-dorsally, where the two layers remain in contact. No cavity is formed and the mesoderm gradually disperses. However, a slightly later and less detailed investigation by Fernando (1931) supported the enterocoelic method of mesoderm formation. In interpreting his results Dautert (1929) suggested that endomesoderm is normally formed in gastropods during the early stages of development, and that these are telescoped in *Viviparus* so that its development is suppressed. Consequently only ectomesoderm, which is of later appearance, is found in this genus. The only other account claiming that all mesoderm is derived from the ectoderm is given by Sarasin (1882) for *Bithynia,* but this was conclusively disproved by Erlanger (1892), who described a teloblastic origin.

The spiral cleavage of the eggs of platyhelminths, annelids and molluscs typically results in the production of ectomesoderm from the second and third quartettes of micromeres, and its failure to develop in certain prosobranchs and some other gastropods (*Aplysia, Siphonaria, Limax*) is undoubtedly secondary. This mesoderm seems to have appeared early in evolution and is present in the Radiata; in the Deuterostomia it does not develop. Its loss in prosobranchs seems to have no evolutionary significance since it apparently fails to develop in some archaeogastropods with a primitive life history and yet is present in mesogastropods in which the larval stages are abbreviated or suppressed. Conklin (1897) emphasized the radial

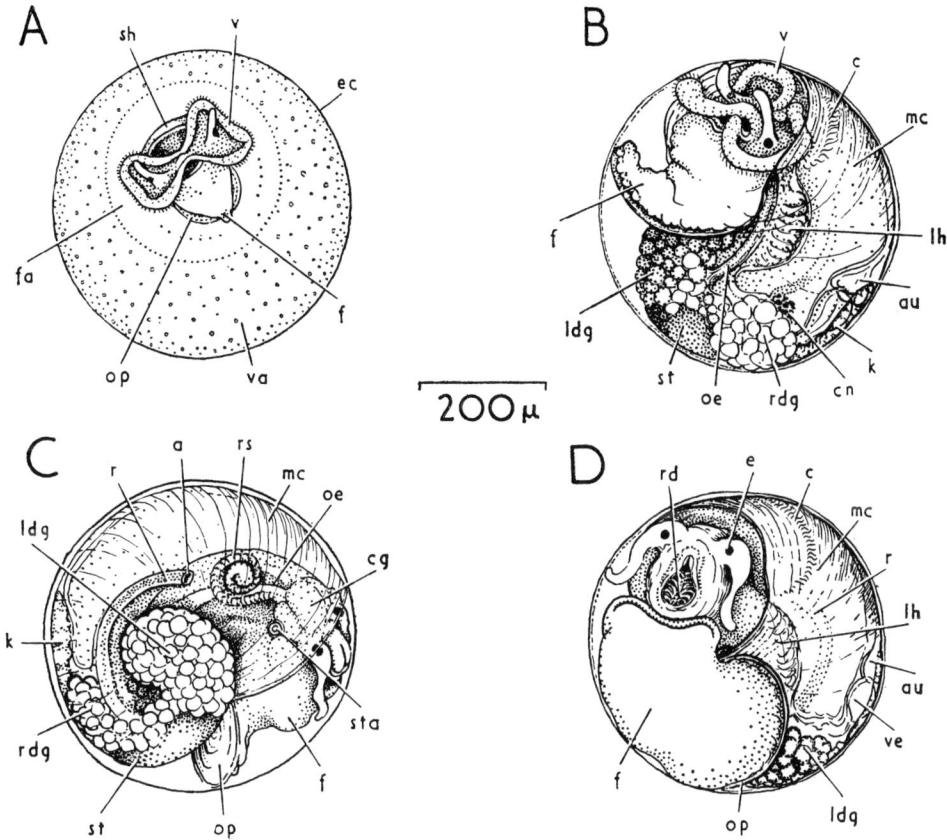

FIG. 214.—*Littorina saxatilis:* embryos from brood pouch. A, veliger stage; B, late veliger; C, just before hatching; D, hatching by rasping through egg membrane.

a, anus; au, auricle; c, cilia on mantle skirt; cg, cerebral ganglion; cn, connective tissue loaded with calcareous granules; e, eye; ec, egg covering; f, foot; fa, fluid albumen; k, kidney; ldg, left lobe of digestive gland; lh, larval heart; mc, mantle cavity; oe, oesophagus; op, operculum; r, rectum; rd, radula; rdg, right lobe of digestive gland; rs, radular sac; sh, shell; st, stomach; sta, statocyst; v, velum with slight ciliation; va, viscous albumen; ve, ventricle.

origin of the total mesoderm in *Crepidula,* for ectomesoderm arises in the three quadrants A, B and C, and endomesoderm in quadrant D. He suggested that this radial origin may be a primitive character which is lost in other gastropods.

With the possible exception of *Viviparus* it is from endomesoderm derived from teloblasts in the vicinity of the blastopore that the adult musculature, the renopericardial complex and haemal tissue are produced, though some ectomesoderm may contribute to their formation. The paired mesoderm bands which are developed are associated with a slight elongation of the embryo. In *Patella* and *Haliotis* they unite beneath the primitive rectum (fig, 217C) where, during the early part of torsion, solid rudiments of the kidneys and pericardium appear; the cavities of these organs are not present until the late veliger stage when torsion is complete.

No pouches are developed in the mesodermal bands, as happens in annelids, nor is there any sign of segmentation of the mesoderm.

Remane (1950) suggested that the mesoderm of molluscs corresponds with that which forms the larval segments of polychaetes and arthropods, and that the later or adult segments of these groups are not represented in the phylum. If this is so then molluscs have either lost, during the course of evolution, the metamerism which this larval mesoderm displays, or the mesoderm was never at any time segmented. The adult anatomy of *Neopilina* may suggest the former view, though nothing more can be said until its development is known.

During gastrulation the blastopore is formed at the vegetative pole of the embryo so that the primary axis of the ovum—a straight line passing through animal and vegetative poles—is retained. As the trochophore stage develops a new axis is established which is the principal one of the larva and adult: in all bilaterally symmetrical animals this is different from the primary one. In the typical trochophore of annelids and molluscs this change in direction of the axis is indicated by the apparent migration of the blastopore along the ventral surface towards the prototroch; the movement may be accompanied by its gradual closure from behind forwards. In both phyla the displacement is due to rapid multiplication of cells in the posterodorsal region, which will become the posterior extremity of the larva, and can be directly attributed to the activity of the derivatives of 2*d* and the underlying cells, the derivatives of 4*d*. In *Crepidula,* Conklin (1897) has described the region of greatest activity as just ventral to the future shell gland. More or less regular rows of cells radiate from this region and are particularly well marked on the ventral surface. Cells derived from 2*d* and 4*d* are also associated with the formation of the trunk in annelids where the continued elongation of this region is accompanied by metameric segmentation. In gastropods, however, this growth results in a ventral curvature of the trunk, so that mouth and anus remain at no great distance from one another. It is also associated with the formation of the shell on the posterodorsal surface, and the growth of the foot ventral to the stomodaeum.

The flow of ectoderm cells from the area of rapid multiplication results in the bending of the egg axis. In the trochophore larva of *Polygordius* it is the vegetative pole, marked by the blastopore, which appears to move forward along the ventral surface, and Smith (1935) has described a similar reorientation of the axis in the trochophore of *Patella vulgata*. After the formation of the blastopore in *Patella* two cells dorsal to it form the anal tuft or telotroch, and a flattening and division of cells on the dorsal surface, between this region and the prototroch, drives the blastopore ventrally and then forwards. The cells between the anal tuft and blastopore also multiply and spread anteriorly, continuing the forward thrust. Such a change in position of the blastopore is brought about in a similar way in *Haliotis* (Crofts, 1937) and *Gibbula magus* (Robert, 1902), and in these three diotocardians there is at the same time a reorientation of the endoderm cells and the formation of mesoderm. There is probably some shifting forward of the animal pole with the hypertrophy of the dorsal area, but it can only be slight in these species for they hatch at an early stage as trochophore larvae; these show no forward movement of the apical region, and retain throughout the cell migrations the circumapical symmetry of the prototroch which is essential to their movements through the water. There are other gastropods in which the main direction of movement of the proliferated cells in the dorsal region is anterior and displaces the animal pole. These have no free trochophore stage and their embryos have a considerable amount of yolk. In them, the vegetative pole may remain more or less fixed and the change of axis be due to a shifting of the animal pole. Thus in *Crepidula* the uppermost part of the vertical axis of the gastrula is

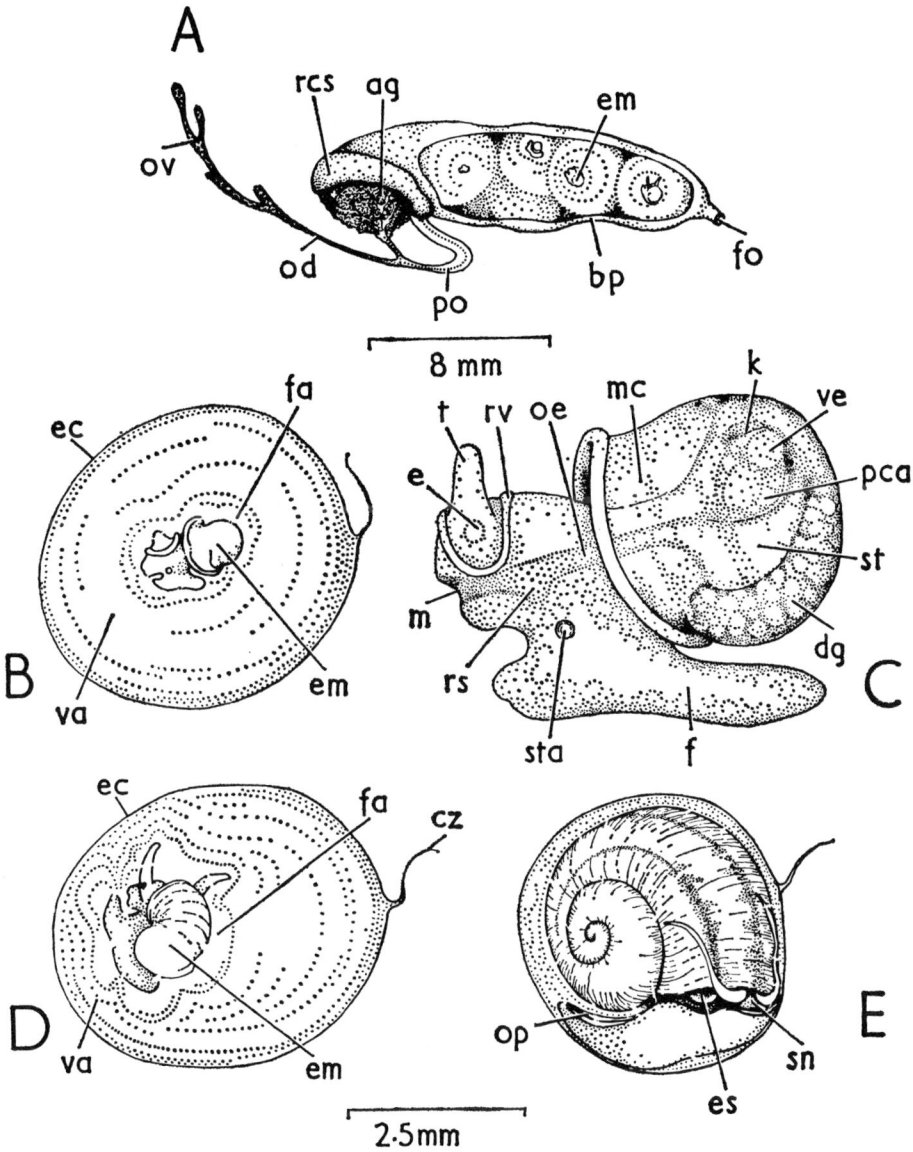

FIG. 215.—*Viviparus viviparus:* A, diagram of female reproductive system with the brood pouch opened; B, C, D and E, 4 developmental stages, B a young veliger, E just before hatching. In C the embryo has been dissected out of its coverings.

ag, albumen gland; bp, brood pouch; cz, chalaza; dg, digestive gland; e, eye; ec, egg covering; em, embryo; es, exhalant siphon; f, foot; fa, fluid albumen; fo, female opening; k, kidney; m, mouth; mc, mantle cavity; od, ovarian duct; oe, oesophagus; op, operculum; ov, ovary; pca, pericardial cavity; po, pallial oviduct; rcs, receptaculum seminis; rs, radular sac; rv, remains of velum; sn, snout; st, stomach; sta, statocyst; t, tentacle; va, viscous albumen; ve, ventricle.

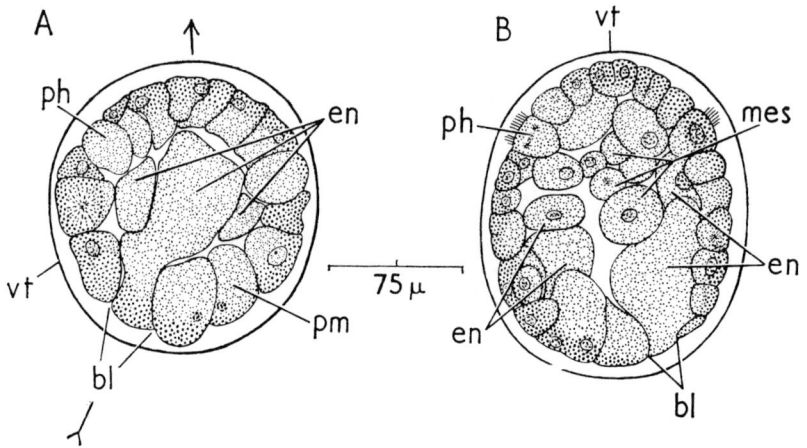

FIG. 216.—*Haliotis tuberculata*: A, vertical section of gastrula, 7 hrs after fertilization. Arrow marks embryonic axis passing through animal and vegetative poles. B, vertical section of early trochophore, 11 hrs after fertilization. After Crofts.

bl, blastopore region; en, endoderm cells; mes, mesoderm cells; ph, prototroch; pm, primitive mesoderm cells; vt, vitelline membrane.

bent forwards at an angle of 90° and the position of the blastopore itself is hardly changed, nor are the endoderm cells influenced by the shifting of the ectoderm.

The end result of these cell movements is the same and the blastopore comes to lie mid-ventrally, immediately beneath the cells of the prototroch or preoral velum. With one exception the mouth is formed in the region of the blastopore. In *Viviparus* the hypertrophy of the dorsal ectoderm causes some forward displacement of the animal pole, but the blastopore remains in its original position and finally becomes the anus.

Only a few gastropods have a trochophore larva and these include the diotocardians *Haliotis, Patella, Patina, Acmaea, Gibbula, Monodonta* and *Tricolia*. In other prosobranchs a corresponding embryonic stage can be recognized. The larva is top-shaped with the proto-troch at its greatest diameter. This consists of one row of cells in *Haliotis* (fig. 217A), *Acmaea, Gibbula* and *Monodonta*, two in *Patella* (fig. 226A), and in the corresponding embryonic stage it consists of two rows of cells in *Viviparus,* three in *Bithynia* and even more in *Crepidula* (fig. 213C). In the centre of the pretrochal area a tuft of apical cilia may develop from an ectodermal thickening, the apical plate cells. Unlike the cilia of the prototroch they are not motile; it has been suggested that they have a sensory function and may help to preserve equilibrium in the actively moving larva. (In a later stage in the development of *Crepidula* an apical sense organ arises (p. 424) and becomes connected with the cerebral ganglia, but it is not associated with a bunch of large cilia, for over the apical and many surrounding cells the cilia are short and fine.) From the pretrochal ectoderm on either side of the apical plate cells a sense plate (cephalic plate) develops (acp, figs. 212B, 213). These later give rise to the cerebral ganglia by a proliferation of cells inwards from the surface, the eyes, which are formed in connexion with the ganglia as separate involutions, and the tentacles.

The blastopore may close for a brief period after its anterior migration. Before its closure it is sunk into a depression, the stomodaeal invagination, in *Crepidula,* though the depression is formed after closure in *Haliotis* and the trochids. In *Patella* and *Littorina littoralis* (Delsman, 1914) the blastopore remains open and is carried inwards to communicate with the larval stomach. The narrowest part of the stomodaeal invagination will later give rise to the oesophagus, which is therefore ectodermal in origin. However, in the development of *Crepidula adunca*, Moritz (1939) stated that it is the anterior end of the oesophagus which marks the junction of ectoderm and endoderm. In the late trochophore stage there may be a thickening in the posterior wall of the buccal region of the stomodaeum which foreshadows the radular sac. In contrast to the trochophore of annelids the archenteron is a blind sac with, perhaps, an oral opening, though when the trochophore of *Haliotis* is freed even this is not formed: ano-pedal flexure has begun before the proctodaeum is developed and is complete before the stomodaeum opens to the gut. The position of the future anus may be marked externally by anal cells. This is a single large ectodermal cell near the telotroch in *Patella vulgata*, but in some other genera a number of anal cells (anc, fig. 213) appear in or near a position once occupied by the posterior margin of the blastopore (*Haliotis*, *Littorina*, *Crepidula*, *Calyptraea*, *Nassarius*, *Nucella*). Finally the hind gut connects with the ectoderm between the anlagen of the shell gland and foot and here the ectoderm forms a shallow proctodaeal invagination. The anal opening may be late to form and in the embryo of *Crepidula* this is after torsion has occurred.

In the trochophore of annelids paired larval or head kidneys are formed. These are protonephridia with solenocytes. In prosobranch gastropods no free trochophore or veliger has been found to possess a larval kidney though such structures do occur in the embryonic veliger stage. Only in two genera, *Viviparus* and *Bithynia*, do these resemble the protonephridia of the annelid trochophore (p. 424). [See p. 633.]

The development of *Haliotis tuberculata* (though not the cell lineage) has been carefully studied by Crofts (1937, 1955) and as she is the first author to describe the details of torsion her account will be followed here. The main events may be summarized thus:

1. Pelagic phase:
  Trochophore freed from egg membranes: 8–13 hrs after fertilization (fig. 217A).
  Early veliger larva prior to torsion: 9–27 hrs after fertilization (fig. 217B, C).
  First 90° torsion of veliger: begins 29–35 hrs after fertilization and takes 3–6 hrs (fig. 219C).

2. Transition from pelagic to benthic life: not less than 3 days after fertilization.

3. Benthic phase:
  Benthic veliger stage established: not less than 3 days after fertilization (fig. 220A).
  Second 90° torsion of veliger: begins about 5 and ends about 12 days after fertilization; velum disappears during this phase (figs. 220B, 221A).
  Final phase of metamorphosis: begins 12 days after fertilization and ends about 2 months after fertilization (fig. 221B).

The first two divisions of the fertilized egg give four blastomeres which then divide by a process of spiral cleavage similar to that found in platyhelminths and annelids. The third cleavage, which is spiral, gives four large megameres with much yolk, and four micromeres. The megameres bud off successive quartettes of micromeres towards the animal pole and a spherical blastula is formed with a very reduced blastocoel. Gastrulation is by epiboly, and

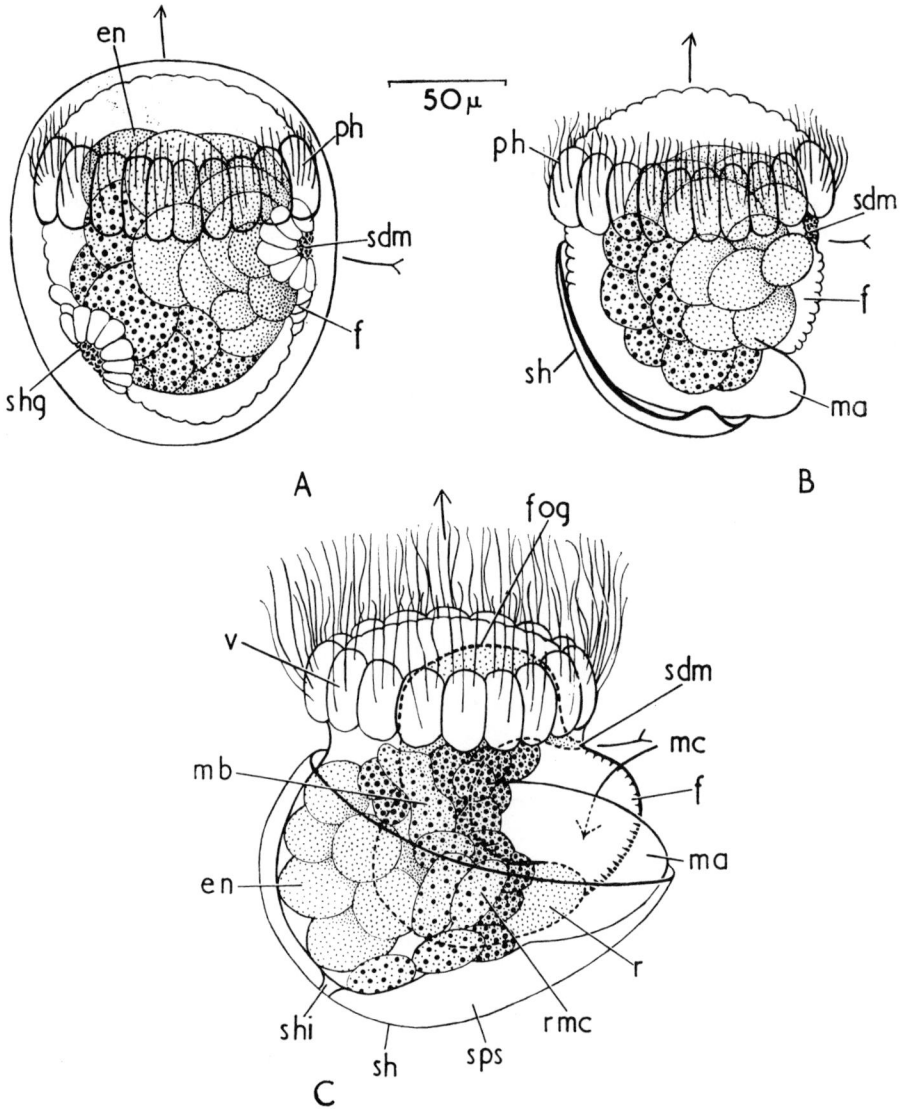

FIG. 217.—*Haliotis tuberculata:* A, trochophore larva immediately before hatching, from right side; B, early veliger, 16 hrs after fertilization from right side; C, early veliger, 19 hrs after fertilization, from right side. Endoderm shaded with small dots, mesoderm with large and small dots. Arrows mark embryonic axis passing through original position of animal and vegetative pole. After Crofts.

en, endoderm; f, foot; fog, fore gut; ma, mantle skirt; mb, mesodermal band; mc, mantle cavity; ph, prototroch; r, rectum; rmc, rudiment of muscle cell of right mesodermal band; sdm, stomodaeal invagination in region of closed blastopore; sh, shell; shg, shell gland; shi, attachment of shell to integument; sps, space between visceral mass and shell; v, velum.

the megameres 4A, 4B, 4C and 4D are pressed into the segmentation cavity and obliterate it. 4A, 4B and 4C are irregular in shape and elongated in the polar direction, whilst 4D, which is nearer the blastopore, divides equally to form two almost spherical cells (pm, fig. 216A). Crofts suggested that, as in *Patella vulgata* (Smith, 1935), these two cells give rise to mesoderm, whereas the endoderm is derived from the other three megameres (en). The invaginated cells are so large that there is no archenteric cavity in the early gastrula, and this appears only later when these cells have divided. During gastrulation the embryo elongates in the direction of the axis through the apical plate, and the prototroch (ph, fig. 216A, B) forms as a ring of ciliated cells encircling the apical area. The cilia rotate the embryo within the vitelline membrane from about 8 hours after fertilization. The blastopore (bl), at first posterior and centred on the axis of the apical field, moves towards the prototroch along the ventral surface, leaving a temporary groove along this path. The migration of the blastopore appears to be due to a multiplication of mesoderm cells, which have now become dorsolaterally placed, and, at the same time, an increase in ectoderm cells along the dorsal surface, and is accompanied by the gradual shifting of the axis of the endoderm and mesoderm.

The trochophore larva (fig. 217A) is freed from the vitelline membrane 8–13 hrs from the time of fertilization. It is then about 0·13 mm long and differs superficially from the trocho- phore of *Patella* (fig. 226A) in the absence of apical cilia and telotroch, and in the fact that the prototroch comprises a single and not a double row of ciliated cells. The rudiments of the molluscan characters have scarcely begun to appear at this stage, although a shallow depression on the dorsal surface of the body, slightly towards the right, indicates the developing shell gland (shg, fig. 217A), and immediately posterior to the ventral lip of the blastopore is a median swelling, the rudiment of the foot (f). At about this stage the blastopore is closed and the stomodaeal pit (sdm) formed near its point of closure. The mesoderm now comprises 10 cells which are orientated in the post-trochal region. The trochophore is positively phototactic and swims near the surface of the water with the prototroch directed upwards, the lashing of the cilia causing a rotatory movement. This larval stage is of short duration.

The transition to the veliger is accompanied by little modification of the swimming organ, for the velum retains the simple character of the preoral ciliated prototroch—a circle of ciliated cells constricted from the rest of the body, and comprising at first 16 cells, with an increase in the older veliger to 24. As the larva grows the cilia become more numerous and longer. The whole pretrochal area is involved in the formation of the velum.

The early veliger, the larval stage preceding torsion, is characterized externally by the development of the shell (sh, fig. 217B, C) and mantle (ma) and an increase in the size of the foot (f). The cells of the shell gland, which in the trochophore line a shallow depression, multiply rapidly, and this, together with the enlargement of the primitive mid-gut and mesodermal bands, causes the eversion of the gland at about 14 hrs after fertilization. A delicate shell of watch-glass shape is secreted; it is of transparent conchiolin and without calcareous spicules. At about 18 hrs the shell gland has spread over most of the dorsal region of the body and the shell has increased around its margin to form a deep saucer. Its expansion is most rapid over the posterior end of the body, where it begins to spread on to the ventral surface. The shell is attached to the body at its periphery, and rapid multiplication of cells near its right ventral edge produces a thickening which is the mantle fold or skirt (ma). The fold increases in size and spreads over the foot (f), and the cavity which it encloses is the mantle cavity (mc). In *Haliotis* this cavity is not at first in the mid-ventral line as in *Patella* (Smith, 1935) and *Trochus* (Robert, 1902), for the mantle skirt is developed on the right side of

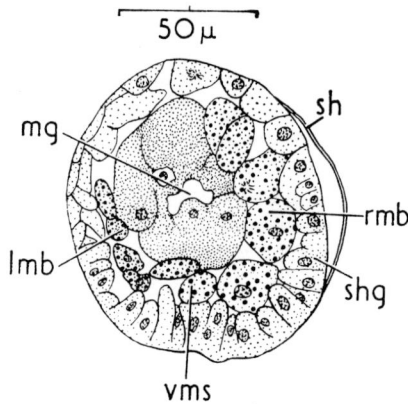

FIG. 218.—*Haliotis tuberculata:* transverse section of early veliger (17 hrs) to show the larger mesodermal band
on the right side; the section has passed through the shell gland and shell on this side. After Crofts.
   lmb, left mesodermal band; mg, cavity of mid-gut; rmb, right mesodermal band; sh, shell; shg, shell gland;
vms, ventral union of right and left mesodermal bands.

the body and then spreads to the left over the pedal rudiment. However, by 29 hrs—just
before torsion begins—the deepest part of the cavity is mid-ventral. The development of the
mantle skirt is accompanied by corresponding changes in the shell (sh) which, early on the
second day, has a wide ventral opening and is constricted dorsally between the enlarging
visceral mass and the velum. At this stage the veliger presents an all but symmetrical external
appearance and has a nautiloid exogastric shell. The opening of the mantle cavity is a broad
transverse slit curving under the foot, with the mouth directly anterior to it. The visceral mass
contracts away from the inner surface of the shell (sps, fig. 217C), giving a space which may
contain gas; in other veligers, this has been thought to have a hydrostatic function.
      Before torsion begins, the digestive tube, as in all gastropod larvae, is curved into a U-
shape, so that the rectum (r), instead of being directed posteriorly as in chitons, is directed
ventrally and anteriorly. It is separated from the stomodaeum by the foot (f). This dorsoventral
flexure (=ano-pedal flexure) is, like the shape of the shell, another consequence of the
differential growth of the visceral hump. In *Haliotis* the flexure begins at about 19 hrs after
fertilization, that is before the proctodaeum has been formed, and is completed before the
stomodaeal invagination has established connexion with the endoderm cells. At 27 hrs these
cells have formed the rudiments of the larval stomach, the digestive gland and intestine. The
digestive gland occupies the whole of the left and most of the dorsal part of the visceral dome
and has a single opening into the left wall of the stomach. The small larval stomach is lined by
cubical cells which have extended anteriorly to meet the stomodaeum. The rudiment of the
radular sac does not appear until after the first phase of torsion.
      During dorsoventral flexure of the digestive tube the mesoderm cells give rise to a right and
a left mesodermal band (mb). At 17 hrs after fertilization the right band is distinctly the larger.
Five large mesoderm cells on the right side elongate at about 22 hrs after fertilization to form
spindle-shaped muscle cells (rmc), and as these grow the visceral hump is gradually displaced
to the left so that even before torsion the veliger is asymmetrical. Posteriorly the muscle cells
converge and are attached to a projection of the mantle which is joined to the apex of the

shell on the right side (lra, fig. 219). There is a similar attachment of mantle and shell though, as yet, with no related muscle on the left side. Anteriorly the muscle cells diverge: three pass forwards along the right side of the larva and have attachments to the mantle (1, 2), the velum (2, 3) and the stomodaeum (3); the remaining two twist dorsally over the gut to the left side of the foot (5, 6), velum (4, 5, 6) and stomodaeum (4), and one of these (4, 5) divides after the first half of torsion, thus giving three muscle cells related to each side of the anterior region of the body. This asymmetrically placed larval retractor muscle of the head and left side of the foot is the cause of the beginning of torsion ontogenetically.

As expressed by Garstang (1928), the posteroventral position of the mantle cavity and the exogastric shell of the early veliger are disadvantageous since the larva cannot withdraw wholly into the shell. When torsion is complete the mantle cavity lies over the head with the opening directed anteriorly: it has been displaced 180° relative to the head and foot. In such a position the head and then the foot can be withdrawn to its shelter, and in the majority of prosobranchs an operculum on the metapodium closes the cavity. In Haliotis an operculum is secreted by unicellular glands of the ectoderm of the metapodium just before torsion begins, and it is still present in the post-larva of 44 days old; it is never calcified and later disappears.

At the stage which precedes torsion the body consists essentially of two halves, an anterior cephalopedal mass and a posterior visceral hump, the two connected by a narrower 'neck'. As the process of torsion, which is about to be described, involves the relative movement of these two halves of the body on each other it is necessary to explain that in dealing with the changes which are involved all references will be made in relation to the dorsal and ventral sides of the adult body. These adult axes coincide with the larval axes so far as the cephalo-pedal mass is concerned, but the larval axes of the visceral hump, though coincident with these before torsion, are altered by that process. It may be, as Crofts' (1955) description of Haliotis suggests, that the first part of the actual rotatory movement appears to be one of the head-foot on the visceral hump, but as the twisting takes place when the larva is planktonic, and not related to any substratum, and as the orientation of the cephalopedal mass is that on which adult anatomy is based, we have chosen to use that orientation as the standard to which the changes in development will be related.

Torsion takes place in two phases. The first 90° of rotation occurs as soon as the larval muscle cells have any contractile power, 29–35 hrs after fertilization, and is completed within 3–6 hrs. To understand this phase it is necessary to consider the relationships of the right larval retractor muscle. This runs from an attachment on the right side of the shell (lra, fig. 219A) forwards along the right side of the visceral hump to the head and foot. Most of the fibres curve dorsally and some pass across to the left side of the body so that the muscle forms a fan of fibres originating on the shell and inserted on head and foot. As soon as these fibres are sufficiently differentiated their contraction will tend to straighten this curve. When this happens the visceral hump and head-foot will be twisted one on the other in such a way that the head-foot is turned through 90° so that its ventral side faces the left side of the visceral hump, that is, it moves to the left. As a result, the mantle cavity lies on the right side of the cephalopedal axis (mc, fig. 219C) and it is possible for the velum and foot to retract partly within the endogastric shell. The asymmetrical right larval retractor muscle is now straightened and lies dorsal to the primitive gut.

The second phase of torsion takes place slowly during the transition from planktonic to benthic life. When the larva is about 40 hrs old it rests on the bottom between intervals of swimming; the intervals shorten and after the third day it becomes entirely benthic. The larva attempts to creep, though with little success at first, for the foot is unable to get a suitable

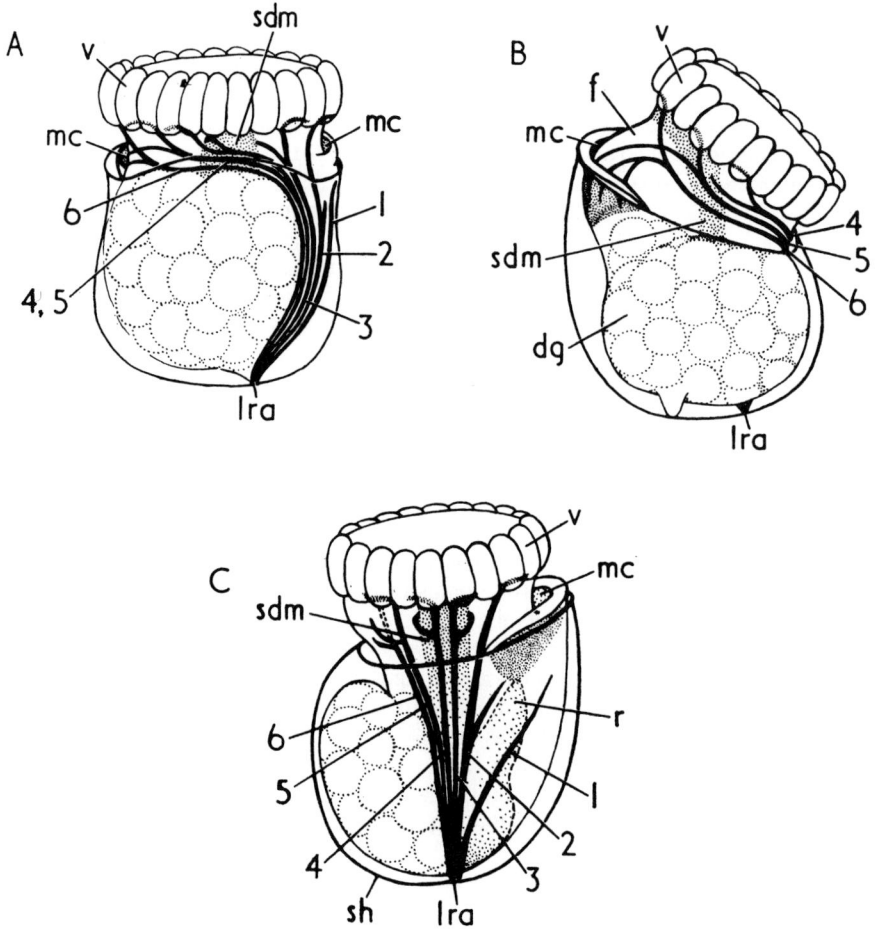

FIG. 219.—*Haliotis tuberculata*: veliger larva to show the course of the larval retractor muscle cells before torsion, A and B, and after 90° torsion, C. A, dorsal view; B, from the left; C, dorsal view with respect to foot (see text). After Crofts.

dg, digestive gland; f, foot; lra, attachment of larval retractor muscle to shell; mc, mantle cavity (stippled); r, rectum; sdm, stomodaeum; sh, shell; v, velum.

1–6 cells of larval retractor muscle; cells 4 and 5 are incompletely separated before torsion. The muscle cells originate on the shell and are inserted:

1. to outer region of mantle on right side.

2. one process to mantle near deepest part of mantle cavity, and a second under ventral right side of velum.

3. one process to epithelial cells under velum on dorsal right side and a second to right side of stomodaeum.

4. one process to epithelial cells under velum on dorsal left side, and a second to left side of stomodaeum. The proximal part of this muscle cell is incompletely divided from 5 before torsion.

5. one process to left side of velum and a second to left posterior region of foot.

6. one process to left side of velum and two processes to left region of foot.

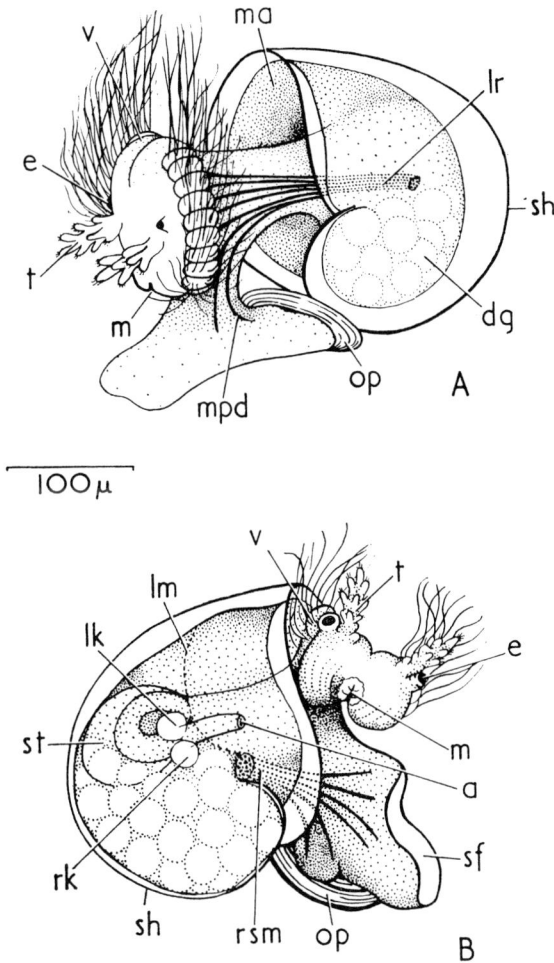

FIG. 220.—*Haliotis tuberculata*: A, benthic larva 4½ days old. The first phase of torsion is complete and the mantle cavity is on the right side of the cephalopedal mass. B, benthic larva, 6 days old, at the beginning of the second phase of torsion. After Crofts.

a, anus; dg, digestive gland; e, eye; lk, left (dorsal) kidney; lm, posterior limit of mantle cavity; lr, larval retractor muscle; m, mouth; ma, mantle; mpd, fibres of larval retractor extending into foot; op, operculum; rk, right (ventral) kidney; rsm, right shell muscle; sf, sole of foot; sh, shell; st, stomach; t, tentacle; v, velum which is lost mid-dorsally and mid-ventrally.

hold owing to the difficulty of balancing the weighty visceral hump and to the small size of the plantar surface. Between attempts the larva will lie with the left side of the shell, on which the margin is shorter, on the ground. By the time the musculature and the sole of the foot are well developed, and the creeping habit established, the larva is about 6 days old (fig. 220B) and the second 90° of torsion has begun. It is not completed until about a week later when the early post-veliger stage is reached (fig. 221A). This second phase of torsion is brought about by

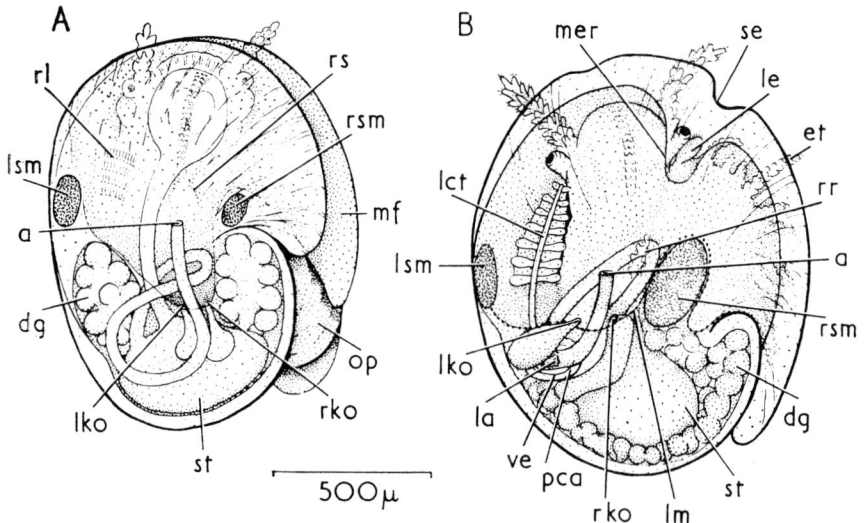

FIG. 221.—*Haliotis tuberculata*: A, post-larva 12 days old. The velum has been recently lost. Note the extensive mantle fold on the right side and the dextral coiling of the shell. B, post-larva 44 days old. The pallial cavity is already displaced somewhat to the left by hypertrophy of the right shell muscle. After Crofts.

a, anus; dg, digestive gland; et, epipodial tentacle; la, left auricle; lct, left ctenidium; le, larval process of epipodium; lko, left kidney opening; lm, posterior limit of mantle cavity; lsm, left shell muscle developed from larval retractor muscle; mer, mantle emargination; mf, mantle fold; op, operculum; pca, pericardium; rko, right kidney opening; rl, rudiment of left ctenidium; rr, rudiment of right ctenidium; rs, radular sac; rsm, right shell muscle; se, shell emargination; st, stomach; ve, ventricle.

differential growth, for which the development and migration of the columellar muscle are mainly responsible.

The origin of the larval retractor muscle migrates gradually from its posterodorsal attachment towards the anterior left side, and so comes progressively nearer the left margin of the shell (lr, fig. 220A). After the creeping stage has been reached additional muscle fibres extend into the foot, and it becomes mainly a left pedal retractor by the time the velum is lost (lsm, fig. 221A). Until then (about 9–10 days after fertilization) it is also the sole head retractor. A small retractor of the post-torsional right side of the foot develops from the pretorsional left mesodermal band and becomes obvious during the third day of development, though it is not functional until the fifth day. This muscle will become the columellar or right shell muscle of the adult. When it is first developed it runs from the rudiment of the columella to the foot, and retracts the cephalopedal mass and operculum into the shell (rsm, fig. 220B). It has, at this time, the same arrangement as the columellar muscle of gastropods with typical dextrally coiled shells. During the later veliger stages of *Haliotis,* however, the columella fails to develop further and the shell attachment of the muscle migrates anteriorly towards the centre of the last whorl (rsm, fig. 221A, B). As it does so it gains in size and functional significance and the pallial cavity expands over the dorsal aspect of the head-foot from the right side. By the time the velum is lost—by the cells being gradually nipped off—the topographical left larval retractor and the right retractor or columellar muscle are about equal in size (fig. 221A). Torsion is now complete and the shell is fully endogastric, but the anus (a) is still on the right side of the mid-dorsal pallial cavity. From the external viewpoint the larva at this stage presents a comparatively symmetrical appearance. The essential difference in organization from the moment at which torsion began is the altered disposition of the head-foot and visceral mass which is the consequence of that process.

The later asymmetry of the body which gives the characteristics peculiar to the adult, is brought about by the hypertrophy of the columellar muscle displacing the mantle cavity to the left. The muscle becomes a stout central pillar (rsm, fig. 221B) which with the pedal muscles can contract the whole body against the rock surface. In doing this it may help in producing the flattened shell into which the animal can no longer retract completely.

The dextral coiling of the shell can be followed from the third day of development, when the pallial fold is on the right side of the body and before the second phase of torsion has begun. A more rapid growth of the shell on the right side causes the creeping larva to bear towards the left, and the tilting of the shell is emphasized by the unequal distribution of weight in the visceral mass due to the presence of the digestive gland and velar retractor muscle on the left with little but the mantle cavity to counterbalance them on the right. During the end of the second and throughout the third week of development (after torsion is complete) the right mantle fold secretes shell so rapidly as to roll the left margin inwards. The expanded right side of the shell now becomes attached to and wraps itself around the apex (fig. 221B) so as to produce the exaggerated development of the right side which is characteristic of the adult (fig. 84).

In the larva which is undergoing the second phase of torsion the left ctenidium and osphradium are represented by a longitudinal ridge of thickened, ciliated epithelium on the mantle skirt, separated from the rectum by epithelium which in the late veliger stage will form the left hypobranchial gland. These rudiments of the pallial complex are distinguishable only during the last quarter of the 180° torsion (rl, fig. 221A). Along the small part of the ridge which will become the osphradium the cilia are short, but the branchial part is more thickly covered with longer cilia, and in specimens 23 days old the rudiments of the first gill lamellae are formed.

There is only room enough on the right side of the rectum for the right renal opening until the post-veliger is about a month old (rko, fig. 221A). At 6 weeks, however, the mantle cavity between the columellar muscle and the rectum has expanded and the rudiment of the right ctenidium and osphradium is present (rr, fig. 221B); gill lamellae develop about a fortnight later. This ctenidium and the right hypobranchial gland are always the smaller. This post-ponement of the development of the organs of the post-torsional right side of the mantle cavity to the period of metamorphosis foreshadows the conditions in the more advanced Diotocardia and the Monotocardia in which these organs are lost. [See p. 731.]

It is assumed (Crofts, 1937) that until the right ctenidium is developed the flow of water through the pallial cavity of the metamorphosing *Haliotis* is similar to that of the gastropod with a single ctenidium, that is, it enters by the left side of the cavity and leaves above the anus on the right. At the time the right ctenidium is developing the first shell hole is formed. A small slit appears in the mantle margin (mer, fig. 221B) and in the overlying shell (se) on the right side of the head, and later the edges of the slit fuse. Peripheral shell growth completes the first shell hole when the ormer is about 2 months old and both gills are present. The pallial flow of water now assumes the adult condition, entering on both right and left sides and passing vertically to leave the mantle cavity through the hole. As the ormer grows and the pallial cavity moves forwards there is an alternate splitting and closing of the margin of the mantle to form a series of holes, one in front of the other; those formed first and no longer of

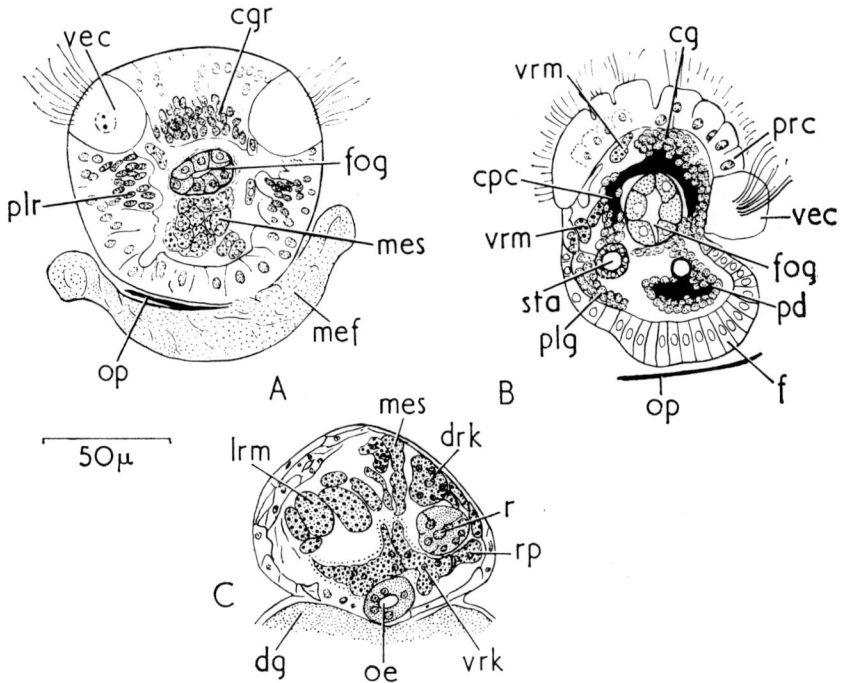

FIG. 222.—*Haliotis tuberculata*: A, transverse section through cephalopedal mass of veliger immediately before torsion begins; B, transverse section through cephalopedal mass of veliger after the first phase of torsion; C, transverse section of veliger after first phase of torsion to show renopericardial rudiments. After Crofts.

cg, cerebral ganglion; cgr, rudiment of cerebral ganglion; cpc, cerebropleural connective; dg, digestive gland; drk, dorsal rudiment of kidney; f, foot; fog, fore gut; lrm, larval retractor muscle; mef, mantle fold; mes, mesenchyme cells; oe, oesophagus; op, operculum; pd, pedal ganglion; plg, pleural ganglion; plr, rudiment of pleural ganglion; prc, prevelar cell; r, rectum; rp, rudiment of pericardium; sta, statocyst; vec, cells of velum; vrk, ventral rudiment of kidney; vrm, velar retractor muscle.

use are filled with shell (fig. 84). Garstang (1928) suggested that the holes are formed as a consequence of the fouling of the pallial stream of water in the region of the anus and kidney apertures (p. 89).

Before torsion begins the right and left mesodermal bands unite beneath the primitive rectum and it is the cells of this region which later form the kidneys and the pericardium. However, at this stage in the development of *Haliotis* and in the development of *Patella* (Smith, 1935), only the solid renal rudiments can be recognized, and during the first phase of torsion they comprise, together with the proctodaeum, the only indications of the pallial complex. In veligers showing 90° torsion the presumptive kidneys, still solid masses of cells, lie dorsal and ventral to the rectum (drk, vrk, fig. 222C), on either side of which they are linked by mesoderm which will form the pericardium (rp). It will be understood that at this stage the gut has also undergone 90° torsion and that its topographically dorsal side is morphologically right and its topographically ventral side is morphologically left (p. 401). When torsion is complete the kidneys are in their definitive position, right and left of the rectum, but it is not until the late veliger, when the reduction of the velum begins, that a cavity appears in each.

The two kidneys are then almost equal in size and are lined by cubical cells which are ciliated. The presumptive pericardial cells now form a delicate wall surrounding the rectum and enclosing the pericardial cavity (pca, fig. 221B). This communicates with each kidney by a renopericardial canal. Ingrowing cells of the pericardial wall become arranged around the rectum to form the ventricle (ve) and from this rudiment of the heart cells arise to form the left auricle (la). Later a smaller right auricle is similarly developed. In the early post-veliger the heart begins to pulsate.

The openings of the kidneys to the mantle cavity are not established until 12 days after fertilization (lko, rko, fig. 221A), although on the third day simple ectodermal invaginations from which they will arise are formed in the deepest part of the cavity, the post-torsional left one dorsal and the post-torsional right one ventral to the anus. The openings are not ciliated.

In specimens $3\frac{1}{2}$ weeks old the right kidney is the larger and its epithelium is thrown into a few folds which foreshadow the complex folding of the adult. In the adult this kidney, by far the larger of the two, communicates with the gonad, and its duct is a urinogenital duct.

At the end of the pelagic phase of torsion the stomodaeal invagination has developed into the foregut which communicates with the oesophagus. This contrasts with the development of *Littorina littoralis* (Delsman, 1914) in which the blastopore does not close, the stomodaeal invagination extends to the stomach and the oesophagus is ectodermal. The digestive gland now opens into the stomach by a single left duct. The anterior diverticula of this gland are ventral so that there is room in the dorsal region of the shell for the retracted head. The larva does not feed until early benthic life and then it is a filter feeder: until the velum disappears particles in suspension are collected from a current which is maintained by the velar cilia and those surrounding the mouth. At the end of the second week there has developed a cylindrical snout with jaws and the radular ribbon supported by two cartilages. The larva now collects loose fragments such as diatoms and Foraminifera from the stones on which it creeps. The gut is ciliated throughout its course. The stomach and oesophagus have enlarged and displaced the digestive gland considerably; the intestine has elongated and is thrown into loops on the dorsal part of the visceral hump. A new diverticulum of the digestive gland with a separate opening into the stomach does not appear until the close of metamorphosis; it comes to lie on the right side of the stomach and crop and, in the adult, curves around the left of the columellar muscle. The caecum of the stomach develops in the young ormer of about 2·5 mm long.

The differentiation of the nervous system begins before torsion, 27–36 hrs after fertilization. Mitosis in the ectoderm of the prevelar area gives rise to rounded nerve cells which aggregate at the base of the ectoderm and form the rudiments of the cerebral ganglia (cgr, fig. 222A). The delamination of the cells is not complete until after the first half of torsion, when they form a strap-shaped band around the dorsal and dorsolateral part of the foregut (cg, fig. 222B). The rudiments of the pedal ganglia arise slightly later, though still in the pretorsional stage, and are first seen as anterolateral thickenings of the pedal ectoderm which later sink into the foot (pd). Before the pelagic half of torsion is complete the formation of the pleural ganglia begins: a collection of cells sinks from the ectoderm on either side of the head (plr, fig. 222A), dorsal to the rudiments of the pedal ganglia, though they are not completely delaminated before the end of the second day of development (plg, fig. 222B). The pleural ganglia then lie close to the pedals and form a pleuropedal nerve mass. The anterior nerve ring is completed at the same time by the migration of nerve cells and fibres to form the cerebropedal and cerebropleural (cpc) connectives; the limits of the ganglia are not precise since (as in chitons) the nerve cells spread into the cords.

Ectoderm of the body wall in the deepest part of the pallial cavity gives rise to the pleurovisceral nerve cords at the end of the second day of development. They are delaminated from the ectoderm of the twisted 'neck' region soon after it has been involved in the first half of torsion. These cords have the appearance of short outgrowths from the pleural ganglia. The one associated with the right ganglion is short and on the right side of the oesophagus; it will form the supra-oesophageal connective. The other, associated with the left pleural ganglion, is longer, ventral to the oesophagus and pointing to the right. This sub-oesophageal connective is the only part of the nervous system to show the influence of the 90° torsion: the ectoderm from which it is formed was on the left side of the body before torsion began, but at the time of delamination it is ventral in position. In older veligers the visceral processes consist of nerve cells surrounding nerve fibres.

In the early post-veliger the supra-oesophageal pleurovisceral cord is displaced somewhat to the left by the enlarging columellar muscle. It passes dorsal to the gut to join the single branchial ganglion which is formed from nerve cells in the mantle skirt on the seventh or eighth day of development. This ganglion innervates the left ctenidial and osphradial rudiment. The pallial cavity has now completed 180° torsion and the proximal halves of the pleurovisceral cords show the streptoneurous condition; the visceral ganglion develops later when the typical figure-of-eight of the visceral loop is completed. The branchial ganglion of the post-torsional right side is not obvious until 44 days after fertilization and it does not join the sub-oesophageal connective until the ormer is 2 months old (rg, fig. 223). At this time the pleurovisceral cords have posterior prolongations beneath the epithelium of the floor of the pallial cavity and these unite to complete the visceral loop and form the single visceral ganglion (vg). This posterior part of the visceral loop is dorsal to the oesophagus. Dialyneury (p. 43) on both right and left is established later. The branchial ganglia of the adult are on short off-shoots of the visceral loop and not, like true parietal ganglia, on its course.

The buccal ganglia (bg) can be seen between the oesophagus and the radular sac before the velum disappears. They arise as outgrowths of the labial processes of the cerebral ganglia; in this they agree with *Acanthochitona* (Hammersten & Runnström, 1926) and differ from *Patella* (Smith, 1935). In the adult the buccal ganglia are hardly distinguishable from their commissure and connectives since all have a peripheral layer of nerve cells.

The sense organs of the veliger comprise the tentacles, eyes and statocysts and in the early post-veliger the first epipodial tentacle appears. In larvae which have undergone 90° torsion a pair of invaginations of the ectoderm near the junction of the laterodorsal part of the foot and the head are the developing statocysts. The two cup-shaped depressions close off from the surface and sink into the foot by the third day (sta, fig. 222B). The statocysts then lie close to the pleuropedal ganglia though they are innervated by nerves from the cerebral ganglia. The tentacles arise from the lateral part of the prevelar area and can be recognized in larvae two and a half days old. They soon elongate and point ventrally, and papillae develop over the surface, the longer ones having tufts of immobile cilia which are sensory. An eye develops at the base of each tentacle on the outer side. It first appears as a group of pigmented epithelial cells which become raised on a small projection, the optic tubercle. The pigment cells sink and form a shallow retinal cup. The retinal cells secrete cuticular outgrowths to initiate the rods and the crystalline lens, and by the end of the second week the lens blocks the opening to the cup and projects from the tip of the optic stalk. The cup is never closed. Both eyes and tentacles are innervated by the cerebral ganglia by nerves bearing small ganglia (on, tn, fig. 223).

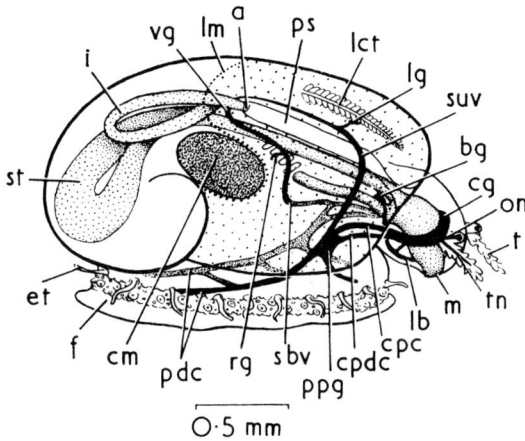

FIG. 223.—*Haliotis tuberculata:* diagram to show the nervous system in a post-veliger 2 months old, after metamorphosis is complete. Mantle cavity stippled. After Crofts.

a, anus; bg, buccal ganglion; cg, cerebral ganglion; cm, right shell muscle; cpc, cerebropleural connective; cpdc, cerebropedal connective; et, epipodial tentacle; f, foot; i, intestine; lm, posterior limit of mantle cavity; lb, labial commissure; lct, left ctenidium; lg, left branchial ganglion; m, mouth; on, optic nerve; pdc, pedal cord; ppg, pleuropedal ganglion mass; ps, pallial slit; rg, right branchial ganglion adjacent to rudiment of right ctenidium; sbv, sub-oesophageal pleurovisceral connective; st, stomach; suv, supra-oesophageal pleurovisceral connective; t, tentacle; tn, tentacular nerve; vg, visceral ganglion.

The first epipodial tentacle arises on the right side of the foot on a process which supports the operculum. Additional tentacles (et) appear during the third week of development on dorsal enlargements on both sides of the foot. They develop papillae with sensory processes similar to those of the cephalic tentacles.

Crofts (1955) has shown that the mechanism of torsion in the other primitive gastropods *Patella vulgata, Patina pellucida* and *Calliostoma zizyphinum* (the *Trochus conuloides* of Robert (1902)) is essentially similar to that of *Haliotis tuberculata.* Mesoderm cells give only a right retractor muscle before torsion begins, so that the pretorsional veliger is asymmetrical. This muscle, which goes to the velum, stomodaeum and left side of the foot, is responsible for the first 90° of torsion. The extrinsic musculature of the right side of the foot develops from the pretorsional left mesodermal band after the first phase of torsion and assists in the second phase; it contributes to the single columellar muscle of *Calliostoma* and the large right shell muscle of *Haliotis.* In *Haliotis, Calliostoma* and *Patella* the position of the shell attachment of the pretorsional right larval retractor muscle alters while the second half of torsion is taking place: it migrates to the left and forwards. By the time the velum has disappeared and torsion is complete it is equal in size to the definitive right shell muscle and the two occupy corresponding positions left and right of the main axis. In *Calliostoma* the larval retractor is reduced in the post-larval stage and then lost, as presumably happens in other prosobranchs with a single shell muscle. In *Haliotis* it persists as a small muscle attached to the shell on the definitive left side, and it occurs in a similar way in many diotocardian and in some monotocardians with broad limpet-like shells (p. 134). In *Patella* and *Patina* both left and right shell muscles probably unite to form the horseshoe-shaped shell muscle of the adult. The specialized characteristics of the four genera investigated by Crofts (1955) develop after torsion is complete and the velum has been lost for some days.

There are differences in the early development of these genera which may be correlated with the amount of yolk in the egg, and the stage at which the embryos are freed from their membranes. In *Patella* the eggs are somewhat smaller than in *Haliotis,* having less yolk, and development has gone on for about 60 hrs before torsion begins; that is about twice as long as in *Haliotis*. The pretorsional larva is therefore further advanced: statocysts have developed and the rudiments of the pedal and cerebral ganglia are already delaminated. The pelagic phase of torsion takes 10–15 hrs and is brought about by the contraction of the 6 cells of the larval retractor muscle which have a distribution similar to those in *Haliotis*. These cells, however, differ in being remarkably slender and without yolk. The second phase of torsion, which is due to differential growth, is rapid and takes about 30 hrs. The greater speed may be attributed to the fact that the larvae are older and their tissues further differentiated than in *Haliotis*.

The eggs of *Calliostoma* are heavily yolked, and in contrast to the other two genera the veliger stage is completed before hatching, and the young emerge from the spawn in the crawling stage, when the velum is lost and cephalic tentacles with eyes, and epipodial tentacles are developed. In the pretorsional veliger the 6 cells comprising the larval retractor muscle are shorter and stouter and their attachment to the shell is in the middle region of the right side instead of at the posterior end—a displacement due to the larger amount of yolk in the visceral mass. The first phase of torsion involves a rotation of rather more than 90° owing to this position and takes about 4 hrs. From the onset of torsion development is speeded up as compared with *Haliotis* and the second phase is completed in about 30 hrs.

Crofts' admirable account (1955) of the mechanism of torsion in the diotocardians enables us to assess the importance of some of the hypotheses which have been formulated concerning this process. Torsion occurs in the development of all prosobranchs either in the pelagic larvae or in the embryo, and Garstang (1928) believed that it arose as a larval mutation which persisted owing to its survival value. It seems of immediate advantage to the larva since the rotation of the mantle cavity to an anterior position enables the head and foot to be withdrawn into it for protection. In fossils there are no animals showing an intermediate stage of the 180° rotation: Garstang (1928) envisaged its rapid completion, and up to the time when his theory was made known this appeared probable, for Boutan (1899) had claimed that it occurred in a few minutes in both *Haliotis* and *Acmaea* and Robert (1902) had given the time as 6–8 hrs in different trochids. Boutan (1886, 1919) suggested that torsion was due to antagonism between the growth of the foot and the shell during development: if both were equally well developed at the same time they interfered with one another and the pressure of the one against the other brought about the rotation of the head-foot on the visceral mass. Boutan (1886, 1919), Naef (1913, 1926) and Garstang (1928) all suggested that torsion involves a certain amount of twisting by muscular contraction. Garstang (1928) was the first to suggest that two muscles were concerned and that the origin of torsion is based on a mutation which affected their symmetry and produced a cephalic retractor with a posterior attachment and a left pedal retractor with a more anterior attachment, the two functioning more or less at right angles to one another. However, Smith's account (1935) of the development of *Patella vulgata,* Crofts' (1937) of *Haliotis tuberculata* and Ramamoorthi's (1955) of *Melania crenulata* show that at the time of the onset of torsion there is only one retractor muscle which is asymmetrically placed with its posterior attachment on the right side of the shell apex (Ira, fig. 219). It runs forwards to the right and left sides of the head and the left side of the foot (Crofts, 1955). It is this muscle which is the main mechanical cause for the beginning of torsion. Crofts' studies on *Patella* (1955) do not support Smith's view that the muscle was originally an

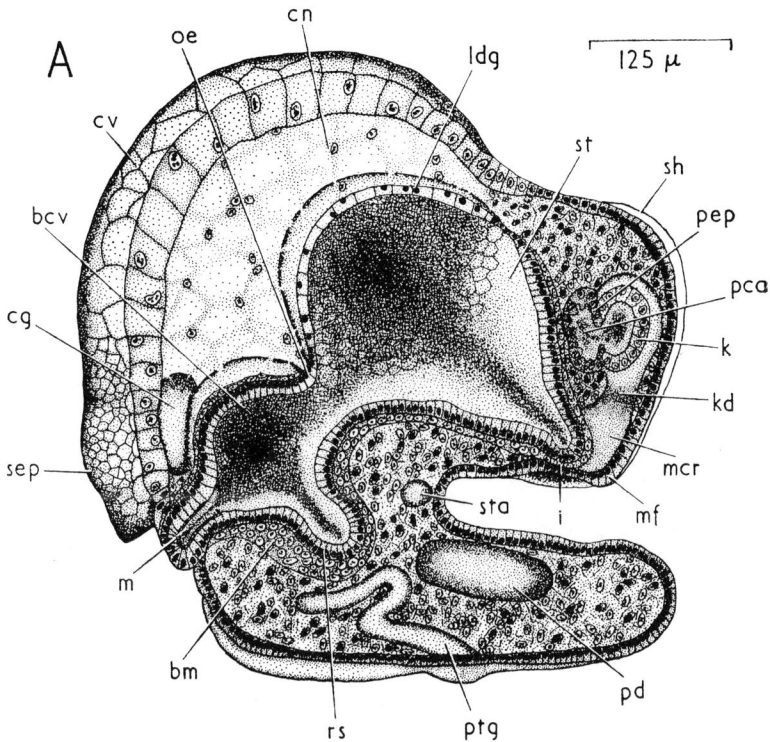

FIG. 224.—*Pomatias elegans:* A, reconstruction of right half of embryo at the mid-veliger stage, before torsion; B, reconstruction of right half of embryo at the late veliger stage, torsion almost complete. After Creek.

a, anus; apg, anterior pedal gland; bcv, buccal cavity; bg, buccal ganglion; bm, buccal mass; ccm, cerebral commissure; cg, cerebral ganglion; cm, columellar muscle; cn, connective tissue; cv, cephalic vesicle; dfc, dorsal food channel of oesophagus; f, foot; h, heart; i, intestine; k, kidney; kd, kidney duct; ko, kidney opening; ldg, left lobe of digestive gland; m, mouth; mc, mantle cavity; mcr, right invagination of mantle cavity; mf, mantle fold; oe, oesophagus; oea, oesophageal opening; op, operculum; pca, pericardial cavity; pcm, pedal commissure; pd, pedal ganglion; pep, pericardial epithelium; plg, pleural ganglion; ptg, posterior pedal gland; rdg, right lobe of digestive gland; rs, radular sac; sbg, sub-oesophageal ganglion; scm, spiral caecum of stomach; sep, sense plate; sg, salivary gland; sh, shell; sog, supra-oesophageal ganglion; st, stomach; sta, statocyst.

unpaired dorsal one, nor that the complete 180° rotation occurs in the pelagic phase of the larva. The components of the larval retractor relate to the velum and the left side of the pedal rudiment and so function as Garstang suggested for the two separate retractors, at right angles to one another. Pelseneer (1911), Naef (1913, 1926) and Smith (1935) suggested that the columellar muscle of the adult is derived from the velar retractor. This is now disproved (at least for certain diotocardians) since the velar retractor, if it persists, lies on the post-torsional left side. In the development of the Diotocardia there is a pair of retractor muscles and because of the asymmetry of the primitive digestive gland (which is directed towards the pretorsional left side of the visceral mass) the retractor of that side, the future columellar muscle, is delayed in development until the first half of torsion is complete and creeping

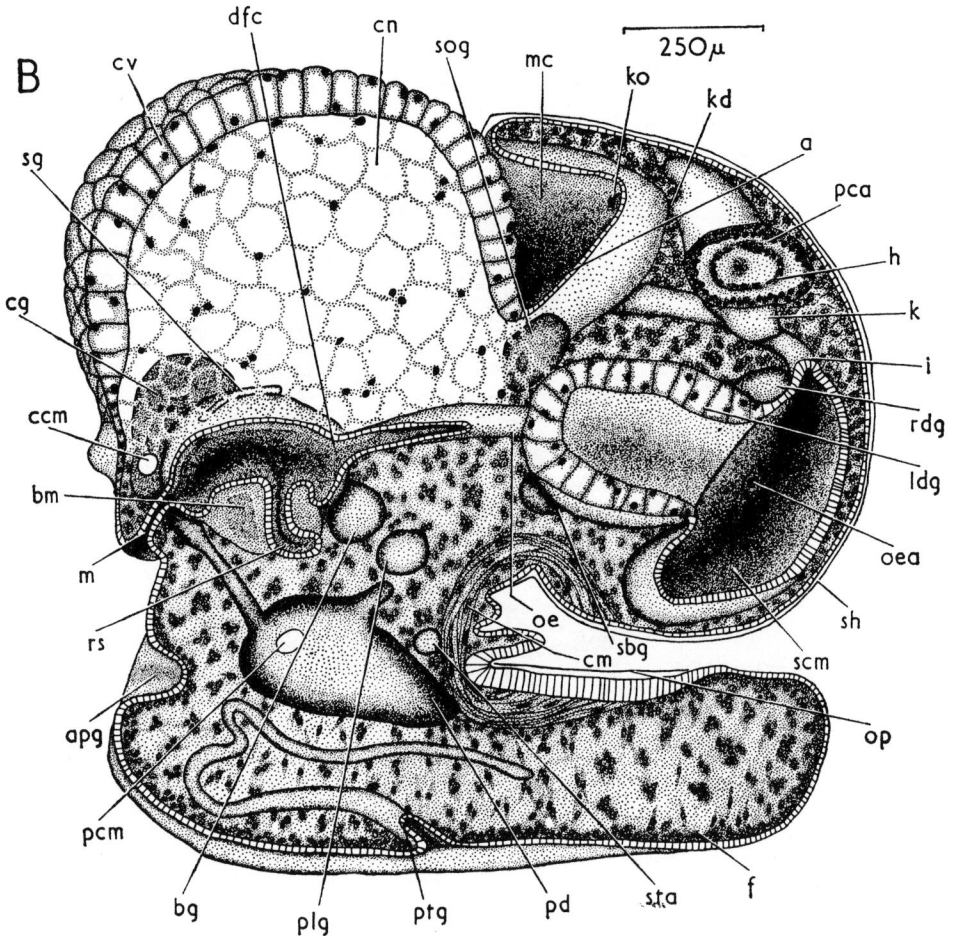

begins. Garstang's hypothesis may therefore be modified in the light of Crofts' investigations to suggest that the postulated mutation caused a separation in the time of development of the right and left larval retractor muscles rather than a differentiation in the strength each exerts.

The first and rapid phase of torsion occurs when the body is of small size and the muscle is just beginning to differentiate, so that the reorientation of the larval tissues is easily accomplished by its contraction because of the undeveloped histological state of the rest of the body of the larva, and because it is pelagic and the rotation is unhampered by drag on a substratum. The benthic phase of torsion is slow and brought about by differential growth. It occurs during the early stages of organogeny when the pedal sole is developing and the larva or embryo creeps about its environment. Naef (1913) concluded from the results of the

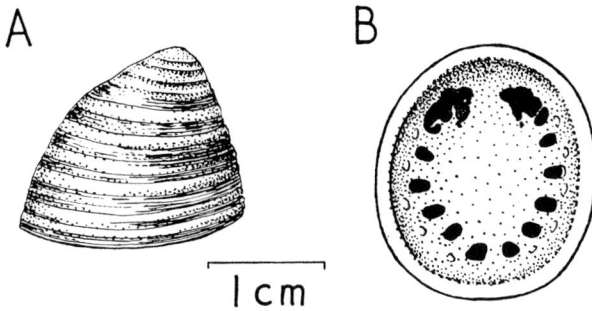

FIG. 225.—*Archaeophiala antiquissima*: A, from the right; B, from below, showing muscle scars (black) and shadow scars. After Knight.

investigations on *Acmaea* (Boutan, 1899) and *Viviparus* (Drummond, 1903; Erlanger, 1891*b*) that torsion brought about by differential growth must be a secondary modification found only in the less primitive gastropods. He stands alone in considering the ancestral gastropod as a free-swimming nautiloid form with the head-foot connected to the visceral mass by a narrow waist, the flexibility of which permitted an easy twist to bring the mantle cavity to an anterior position when the gastropod became benthic. He assumed that torsion occurred rapidly in post-larval life to meet the needs of the benthic adult. Neither palaeontology nor phylogeny supports these views. Naef would regard the veliger larva as a phylogenetic reminiscence of the ancestral free-swimming gastropod and not a simple modification of the trochophore larva in which the prototroch is elaborated to meet the demands of a longer pelagic period.

In some specialized monotocardians no larval muscles are developed before torsion is accomplished and the whole is brought about slowly by differential growth. Perhaps the most outstanding example of this is *Pomatias elegans* (Creek, 1951), a terrestrial member of the Littorinacea, in which development takes about 3 months. A free larva is suppressed and the embryonic velum is swollen into a cephalic mass which retains the cilia in some areas and can rotate the embryo in its ample supply of albumen. Torsion begins $3\frac{1}{2}$ weeks after the first cleavage and proceeds by differential growth taking $31\frac{1}{2}$ days. No musculature is developed in the body until later than this and none ever enters the transformed velum. Fig. 224A shows that in the mid-veliger stage before torsion begins the nervous, alimentary and renoperi-cardial systems are further developed than in the pretorsional stages of *Haliotis*, yet the mesoderm which will give rise to the musculature is undifferentiated, and not until torsion is almost complete do the first pedal muscles appear.

The other prosobranchs in which torsion is known to be brought about only by differential growth are *Littorina littoralis* (Delsman, 1914), *Crepidula* (Conklin, 1897; Moritz, 1939), *Ocenebra aciculata* (Franc, 1940) and *Viviparus* (= *Paludina*) (Drummond, 1903). In *Littorina* and *Ocenebra* development is completed within the egg capsule, and in *Viviparus* the young are retained in the oviduct until they are miniatures of the adult. Fig. 212A shows a sagittal section of an embryo of *Littorina littoralis* in the veliger stage after torsion and the cells of the distal part of the columellar muscle (omd) have not yet differentiated, though in the proximal region (not shown in the figure), they have begun to elongate.

Throughout the rocks of the Cambrian age there are fossils of the most primitive of all molluscs, the Monoplacophora, with low, cup-like shells and 5–8 pairs of adductor muscle scars symmetrically arranged (fig. 225). Nowhere is there space between the scars for a pallial cavity of the gastropod type, so presumably they had a pallial groove as in *Neopilina* (Lemche, 1957). These fossils, which may be closely related to the gastropods (Knight & Yochelson, 1958), suggest that the soft parts were bilaterally arranged and that the animal had not undergone torsion. Knight, the distinguished palaeontologist, has said (1952) that a prerequisite for the initiation of torsion would be the reduction in the number of these shell muscles to a single pair. This might occur with the development of a high, narrow shell which would cause the pairs of muscles to be crowded together, and as a consequence, perhaps, the elimination of all but a single pair, though Knight & Yochelson (1958) no longer find support for this in the fossil record. Or it might be brought about by the crowding together of the muscles on each side as suggested by the Devonian genus *Cyrtonella* (Knight, 1947). In the Lower Cambrian genus *Helcionella* are species with a low cup-shaped shell and, concurrent with these, others with high narrow shells; within this range this kind of modification of the muscles might be expected, but unfortunately the muscle scars of this genus are unknown and the effects of the decreased breadth of shell cannot therefore be checked. These animals have been classified as gastropods by Rasetti (1957) and Knight & Yochelson (1958) though they are excluded from that class by others, agreeing with Garstang (1928) who suggested that torsion may be due to a mutation. Knight & Yochelson (1958) believed that torsion occurred just as suddenly phylogenetically as it does today in the development of some modern gastropods and that this spectacular rotation of the visceral hump on the head-foot was initiated somewhere before the bellerophonts. They regarded the bellerophonts as diotocardians with a high degree of bilateral symmetry. The shell is typically coiled in a close spiral, the coiling being plane rather than helicoid, but the most important characteristic for the present consideration is that these animals possessed two symmetrical retractor muscles inserted one at each end of the columella, which in them runs transversely from right to left. In this position they would serve as retractors of the head and foot. The shell also has an emargination, as in the later fossil pleurotomarians. In the Recent genus *Pleurotomaria*, as described by Bouvier & Fischer (1902) there is, however, only a single columellar muscle.

The bellerophonts are contemporary with the earliest known pleurotomarians which date back to the Cambrian, over 400 million years ago, and were varied and abundant throughout succeeding Palaeozoic time. Knight suggested that the bellerophonts had paired ctenidia, osphradia, hypobranchial glands, auricles and kidneys, as do the Recent members of the families Pleurotomariidae, Haliotidae and Scissurellidae all of which were included by him in the superfamily Pleurotomariacea (= the families of Thiele's Zeugobranchia with spirally coiled shells when adult). Bellerophonts differ from pleurotomariaceans in being bilaterally symmetrical.

If we assume with Knight & Yochelson (1958) that the bellerophonts were gastropods then torsion would have been achieved by means of an embryonic course similar to that exhibited by *Haliotis,* involving asymmetrical development of the larval musculature. Once torsion had occurred, however, a re-modelling of the animal's structure produced an adult with a bilaterally symmetrical body. There is therefore no connexion between larval asymmetry (the cause of torsion) and the adult asymmetry and loss of organs characteristic of so many prosobranchs. Adult asymmetry is a separate phenomenon to be associated with the adoption of the helicoid rather than the plane coiling of the visceral hump: wherever that occurs, asymmetry to a greater or less degree is found; whenever it is absent, either primitively as in Bellerophontacea, or secondarily as in Fissurellidae and Patellacea, bilateral symmetry is characteristic.

From the preceding account it will be seen that torsion is not a phenomenon which lends itself to a gradual step-by-step development, and yet the oldest theories concerning its origin assume that it came about by stages in the adult. These ideas often originated from the fact that early workers failed to regard torsion and the lateral asymmetry of the gastropod body as two distinct phenomena brought about by different causes. Lang (1891) held that the twisting of the visceral mass on the head-foot was stimulated by the top-heavy, conical shell carried by the adult, and Plate (1895) attempted to explain both torsion and the asymmetrical coiling of the shell as due to asymmetry of the lobes of the digestive gland. However, such theories are only of historical interest.

CHAPTER 17

LARVAL FORMS

T HE British prosobranch gastropods exhibit great variety in the spawn masses they produce and in the provision made for their embryos, and in many species this is related to the state of development of the young at the time of hatching. Within their investments the eggs may complete their development to the crawling stage, although in the majority of species the young hatch as pelagic larvae, and, after a period of time varying with the species, become benthic and attain the habitat of the adult. Only in certain diotocardians are the eggs freed singly to the plankton, unprovided with membranes except those secreted by the ovum itself (*Haliotis tuberculata, Patella* spp., *Patina pellucida, Acmaea virginea, Gibbula magus, G. cineraria, G. umbilicalis, Monodonta lineata, Tricolia pullus*). These species have a free-swimming trochophore, whereas in all other gastropods the embryo becomes a larva when the trochophore stage has been passed through and the coiled shell is formed. The free trochophore stage (fig. 226A) is brief and the larva soon develops (a) a shell, which comes to occupy the entire dorsal surface behind the prototroch, (b) an enlargement of the prototrochal girdle which forms the velum, and (c) ventrally, behind the mouth, a protuberance which is the developing foot (fig. 226B). These are the principal external changes in the transformation to the veliger, the characteristic larva of the Mollusca. The velum is the feeding and swimming organ of the larva, fulfilling only larval needs, and in the diotocardians it remains comparatively inconspicuous (fig. 217C), unlobed and projecting only slightly from the preoral surface, whereas in the monotocardians, with typically a longer pelagic life, it is enlarged to form lateral lobes which are beset around the edge with a thick covering of cilia (fig. 227). By the time the velum of the mesogastropod has grown to any size torsion has taken place and it can be withdrawn and folded into the anterior mantle cavity, the opening to which is then closed by the operculum (op) of the retracted foot. It is not until the end of larval life that the velum gradually disappears and is finally lost. As in other groups with pelagic larvae the final settlement is probably susceptible of a certain amount of control by the larva which allows it to select an appropriate substratum, but little work on factors affecting larval settlement has been done with molluscs. It is known, however, that the settling of larvae of those species of *Odostomia* which parasitize the polychaet *Pomatoceros* is accelerated by the presence of the worms and delayed by their absence. [See p. 673.]

In *Patella* the gelatinous layer which surrounds the egg when it is spawned soon disappears, although in top shells which shed their eggs singly (*Gibbula, Monodonta*) it protects them during their embryonic stage and the trochophore larva can be seen revolving within the covering. The embryo of *Patella* hatches 24 hrs after fertilization and the trochophore, measuring 0·18 mm across, has a tuft of apical cilia (at, fig. 226A), and a prototroch (apt, ppt) of 2 rows of ciliated cells surrounds the larva at its greatest diameter. The prototrochal cilia beat in a clockwise direction and rotate the top-shaped larva through the water. Short cilia cover the area between the prototroch and apical tuft. On either side of the apical plate there is a prominent patch of stiff cilia which have undoubtedly (Smith, 1935) some sensory function

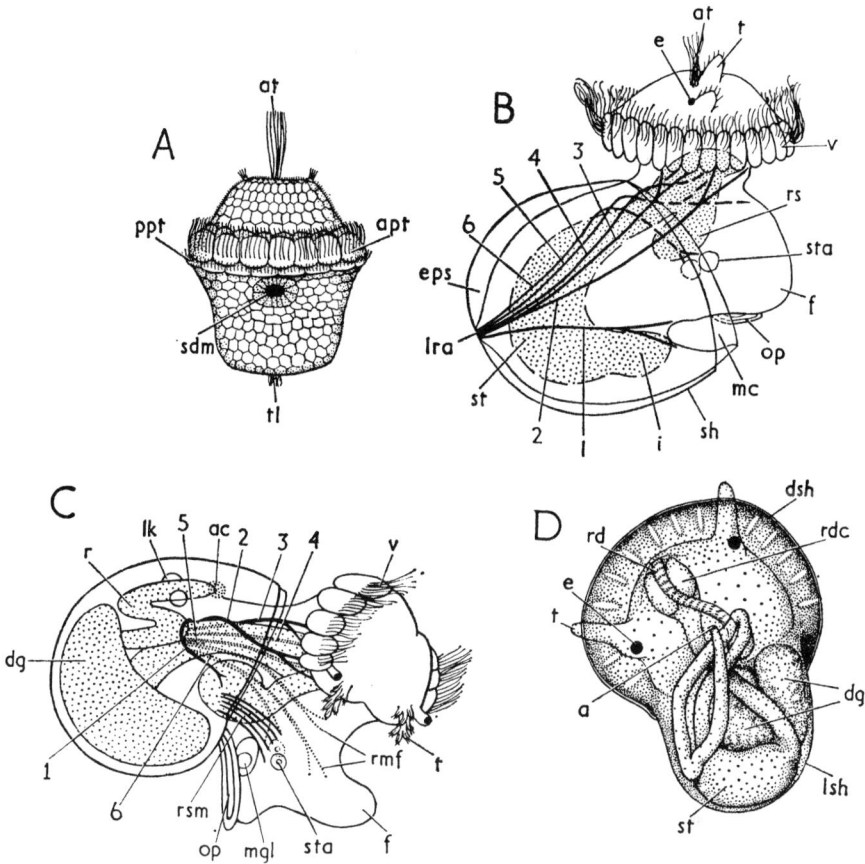

FIG. 226.—*Patella vulgata:* A, early trochophore, after Smith; B, pretorsional veliger (about 70 hrs after fertilization), after Crofts, showing the asymmetrical larval retractor muscle responsible for the first 90° of torsion. C,
post-torsional veliger (about 96 hrs after fertilization), after Crofts, showing dorsal position of rectum and
larval retractor muscle, and right shell muscle (= columellar muscle). D, dorsal view of larva during metamorphosis, after Smith, showing the definitive shell appearing as an outgrowth from the mouth of the larval shell.
The latter will ultimately be lost.

a, anus; ac, anal cell; apt, anterior cilia of prototroch; at, apical tuft; dg, digestive gland; dsh, adult shell; e,
eye; eps, extra-pallial space; f, foot; i, intestine; lk, rudiment of left kidney; lra, attachment of larval retractor
muscle to shell; lsh, larval shell; mc, mantle cavity; mgl, mucous gland; op, operculum; ppt, posterior cilia of
prototroch; r, rectum; rd, radula; rdc, radular cartilage; rmf, pedal retractors; rs, radular sac; rsm, right shell
muscle; sdm, stomodaeum; sh, shell; st, stomach; sta, statocyst; t, tentacle; tl, telotroch; v, velum; 1–6, cells of
larval retractor muscle.

(similar tufts are also present in the larva of *Acmaea* (Boutan, 1899)) and 2 cells bearing similar
cilia form the telotroch (tl). Two days after fertilization changes have been effected by which
the larva is transformed into a pretorsional veliger: the shell and foot (sh, f, fig. 226B) are
present and the anterior row of ciliated cells of the prototroch (v) has developed longer and
more numerous cilia, whilst the posterior row has been lost. Torsion begins while the larva is
free-swimming. The first 90°, brought about by contraction of the 6 cells of the larval retractor

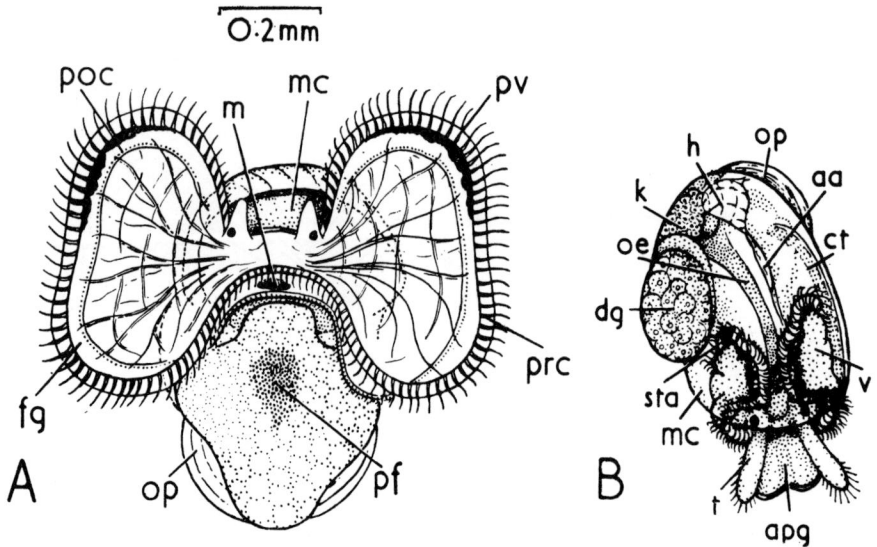

FIG. 227.—*Littorina littorea:* veligers at swimming-crawling stage. A, in ventral view, swimming; B, in dorsal view, crawling. The anterior border of the velum bears a characteristic band of dark pigment visible through the colourless shell when the animal crawls. Older larvae have also pigment on the sole of the foot. The digestive gland is yellow-green.

aa, anterior aorta; apg, anterior pedal gland; ct, ctenidium; dg, digestive gland; fg, food groove; h, heart; k, kidney; m, mouth; mc, mantle cavity; oe, oesophagus; op, operculum; pf, pigment on foot; poc, postoral cilia of velum; prc, preoral cilia of velum; pv, pigment on velum; sta, statocyst; t, tentacle; v, velum.

muscle, is accomplished in 10–15 hrs, and the second phase, due to differential growth, takes about 30 hrs (Crofts, 1955). The larva both swims and crawls during the second phase and, although the velum is retained, there is a break in the girdle of ciliated cells mid-dorsally and mid-ventrally (v, fig. 226C). Larvae in which torsion is complete are $3\frac{1}{2}$–4 days old and their pelagic life has ended. The velum does not finally disappear until the third week when the snails are actively crawling (Smith, 1935), and at about the same time the operculum is lost. The shell of the veliger is coiled dextrally, though with hardly more than one whorl, and at the time of metamorphosis, when in the crawling stage, a secondary symmetry is developed in both shell and animal (lsh, dsh, fig. 226D). The new shell which is added at the peristome of the larval one is on a new axis inclined downwards and to the left of the original larval axis. At the end of metamorphosis a shell plate is secreted across the base of the larval shell which is then lost (pch, fig. 260).

The embryos of diotocardians which develop from eggs laid in gelatinous masses or ribbons (*Diodora apertura, Margarites helicinus, Gibbula tumida, Calliostoma zizyphinum, C. papillosum, Cantharidus striatus, C. exasperatus, Skenea serpuloides*) pass through both trochophore and veliger stage, and the young creep from the gelatinous coverings as miniatures of the adult, with sometimes a vestige of the velum. In *Calliostoma zizyphinum* the veliger rotates in its albuminous covering (which was secreted by the egg whilst still in the ovary) and gradually devours it, and as the muscles of the foot develop it creeps around the wall of its spherical cell. Torsion is complete by the end of $3\frac{1}{2}$–4 days of development, rotation through 180° taking a total of about 36 hrs, which is slightly more rapid than in *Patella* and considerably more rapid than in *Haliotis* (Crofts, 1955). From the very onset of torsion development is speeded up in *Calliostoma* as compared with these two zeugobranchs. This is partly owing to the larger amount of yolk, which displaces the larval retractor muscle and so causes rather more than 90° rotation during the first phase of torsion, and also because of the retention of the larva within the egg membranes. By the fifth day of development (when the embryo may hatch) the first epipodial tentacles are present, whereas in *Haliotis* they appear on the tenth day. In *Cantharidus striatus* Robert (1902) stated that the larvae have epipodial tentacles at 124 hrs and are hatched in the crawling stage at this time, but the free larvae of *Gibbula magus* reach the same stage of development at 150 hrs. It would thus appear that the rate of development in those members of the Diotocardia in which there is no free larval stage is more rapid.

No diotocardian is known in which the young hatch as veligers, yet all monotocardians with an indirect development emerge from the egg mass in this form, and when torsion is complete. The pelagic larval phase of the monotocardian is considerably longer than that of the diotocardian and may extend to 2 months, as in the British species of *Nassarius*, *N. reticulatus* and *N. incrassatus*, for many species have exploited this period of their life history, so that it becomes of greater importance in distribution. About a third of the British mesogastropods hatch in the crawling stage and these emerge from egg masses which are usually deposited in the vicinity of the adults. They include the minute specialized proso-branchs of rock pools—*Rissoella diaphana*, *R. opalina*, *Omalogyra*, *Skeneopsis*, *Cingulopsis*; some small rissoaceans—*Cingula cingillus*, *C. semicostata*, *Barleeia rubra*; the ovigerous species of *Ianthina* and also *Calyptraea chinensis*, *Littorina littoralis* and *Lacuna pallidula*. Amongst the Stenoglossa the undoubted occurrence of a larval stage is restricted to a small number of species.

Each lobe of the velum of the monotocardian gastropod (fig. 227A) consists of an upper and lower epithelium with an underlying nerve net, muscle fibres which are components of the larval retractor muscle, and, in certain species, pigment cells (pv), which may be a mark of identification of the species. In the thickness of the lobe are large blood spaces. The ciliation is confined to a marginal zone (fig. 228). The edge is thickened and bears long cilia on the upper surface (prc) which are set in transverse rows, and on the under surface there is a sulcus covered by short cilia (fg), with its inner border formed by a somewhat prominent ridge with cilia of intermediate length (poc). Owing to the transparency of the tissues this inner ciliated band is always visible on the upper surface of the velum (fig. 227A); ventrally it becomes postoral and forms the posterior lip of the mouth (m), whilst the upper ciliated band is preoral. The two are homologous with the prototroch and metatroch of a trochophore. When the velum is expanded the cilia are constantly beating and a current in the sulcus wafts food—nanoplankton—into the mouth. Along the outer (preoral) ciliated band, which, with the muscles, makes the velum a powerful swimming organ, the metachronal waves of the cilia travel in a clockwise direction (fig. 229A) with their effective beat passing at right angles to this, that is into the food groove (Knight-Jones, 1954). The cilia are long and the metachronal waves particularly conspicuous. Within the sulcus the metachronal waves, at least in some genera (*Nassarius*), move in the opposite direction to the effective beat of the cilia. The natural position of the larva when swimming in the plankton is with the shell below and the foot and velum above, but the velum, especially when long lobes are present, may alter its position in many ways, the lobes flapping slowly, incompletely covering the shell or remaining outspread. The withdrawal of the velum into the mantle cavity is rapid and accompanied by a slowing

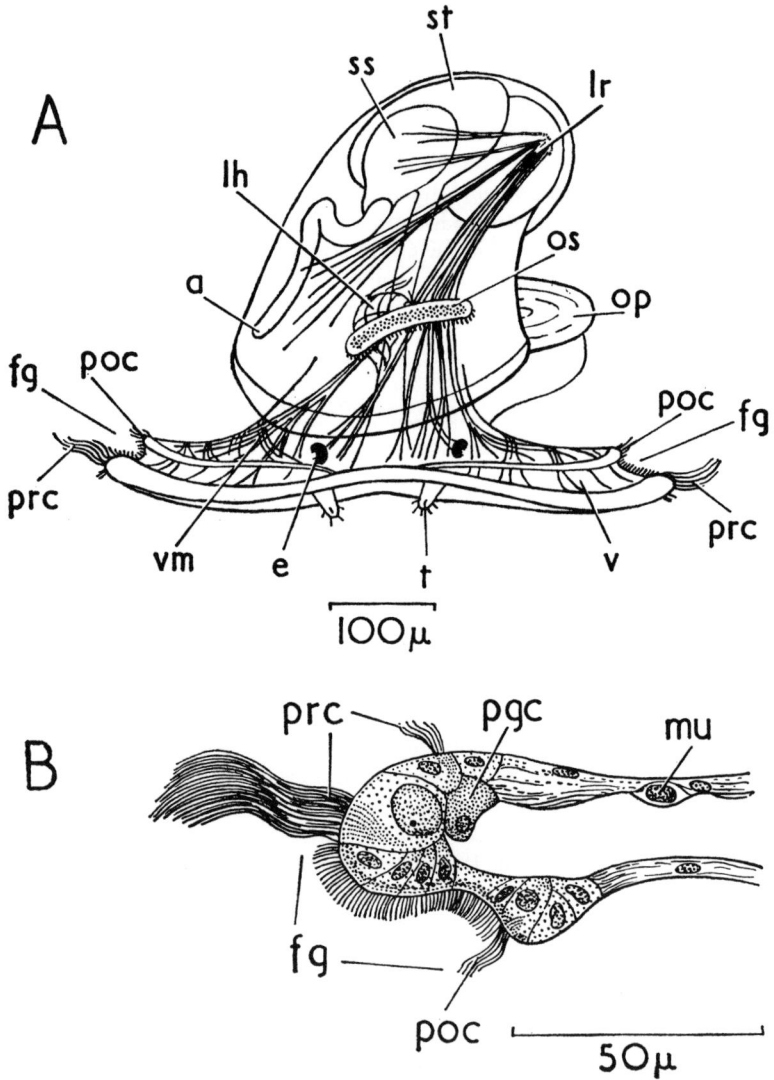

FIG. 228.—*Crepidula fornicata:* A, free-swimming veliger seen from above; B, T.S. edge of velum, after Werner. a, anus; e, eye; fg, food groove; lh, larval heart; lr, larval retractor muscle; mu, muscle cell; op, operculum; os, osphradium; pgc, pigment cell; poc, postoral cilia of velum; prc, preoral cilia of velum; ss, style sac region of stomach; st, stomach; t, tentacle; v, velum; vm, velar muscle.

down and cessation of ciliary beat. With this the larva sinks slowly through the water. Food particles collected by the velum are transferred to the stomach by the ciliated lining of the oesophagus and in the stomach they are kept in rotary movement by ciliary action. From time to time contractions of the wall of the stomach may be seen, but no particulate material

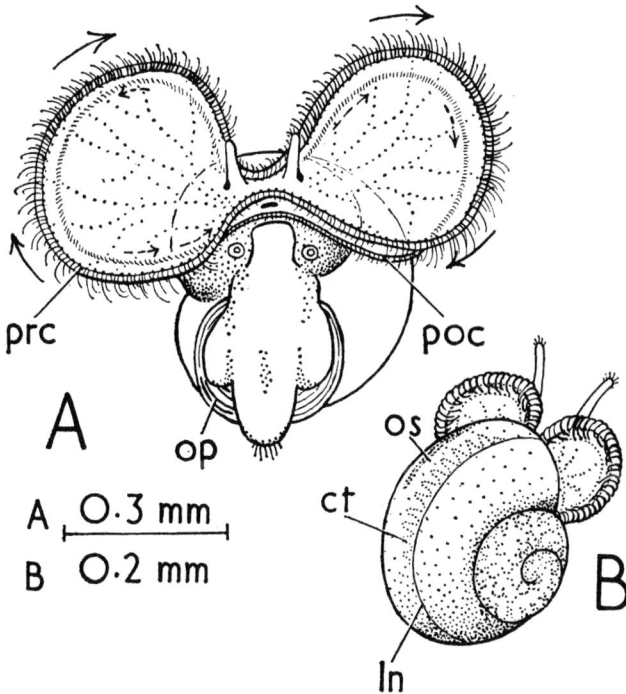

FIG. 229.—*Rissoa parva*: A, veliger, ventral view, swimming. Arrows indicate direction of movement of metachronal waves of velar cilia. B, veliger, dorsal view, creeping.
    ct, ctenidium; ln, line along periphery of shell characteristic of late larvae of this species; op, operculum; os, osphradium; poc, postoral cilia of velum; prc, preoral cilia of velum.

appears to pass to the ducts of the digestive gland. In the embryo this gland plays an important part in the uptake and digestion of the albumen, which can be seen in large vacuoles distending its cells. [See p. 671.]

Associated with the hypertrophy of the velum and its important vascular supply is the development of a larval heart which is formed and starts to function before hatching (lh, figs. 233, 237). It arises as a vesicular differentiation of the ectoderm. Franc (1943) has shown that in *Pisania maculosa* (fam. Buccinidae) it first appears between the anus and the right wall of the developing foot. When torsion occurs it is brought forwards to a dorsal position on the floor of the mantle cavity and appears as a blood sinus with contractile walls which lies obliquely above the oesophagus (oe) at the base of the velum (v), and the network of powerful muscle fibres in its walls can be seen through the overlying tissues. These muscles are usually held to be of mesenchymal origin, though Franc (1943) preferred to interpret them as differentiations of the original ectodermal layer. The heart has an anterior opening connected with blood spaces in the velum and foot, and a posterior opening to blood spaces in the visceral sac, and Werner (1955) has shown that in *Crepidula* there is an anterior and posterior valve. The fluid passes through it in an anteroposterior direction, and then from the visceral mass the flow is anterior to the foot and velum. The blood spaces in the velum make it an important respiratory organ and its continuous movement assists circulation. Before the larva hatches the rate of beat of the larval heart is comparatively low: in *Crepidula fornicata* it is about 24–35/min (Werner, 1955) though about twice as fast after hatching; when the velum is withdrawn, however, the beat is reduced or even stopped. The larval heart is present in species with no free larval stage and Franc (1940) has shown that in the embryo of the stenoglossan *Ocenebra aciculata* its rate of beat is similar to that of *Crepidula* before hatching. In all species which have been observed the heart persists for some time after the true heart has been formed, though there seems to be no relationship between the beat of the two, and it finally becomes incorporated in the anterior aorta.

   The foot (fig. 230) in the early veligers of gastropods is a small ventral projection behind the mouth, which narrows posteriorly, and from an early stage is provided with an operculum (op). It gradually lengthens and ultimately grows forwards into a flexible process which in older larvae may be seen licking the food groove of the velum in the vicinity of the mouth as though keeping the tract clear. Musculature in the foot is differentiated at an early stage in species which have a free larva and the late embryo may be seen creeping around the wall of the capsule some days before it escapes as a young snail. At the base of the foot are large, paired statocysts (sta) and dorsal to its origin, on each side of the neck, are in some species the so-called larval kidneys (lak, fig. 233). These are typically ectodermal projections of one or more cells in which, it has been held, excretory products accumulate during the embryonic phase. They are particularly conspicuous in the Rachiglossa and have also been described (Pelseneer, 1911) in the embryos of a number of Taenioglossa (*Littorina* (fig. 194A), *Lacuna, Rissoa, Natica, Lamellaria, Calyptraea, Crepidula*). In *Crepidula* (Conklin, 1897) each consists of several ectodermal cells which swell up and gradually lose their nuclei and boundaries, and the protoplasm becomes vacuolated. Within the vacuoles small granules accumulate. Gradually the whole mass becomes constricted at its base by the ingrowth of surrounding ectodermal cells, and ultimately the kidney is pinched off. It has gone by the time the larva hatches (Werner, 1955). The larval kidney of the Stenoglossa has been described in some detail by Portmann (1930) in *Buccinum undatum* and Franc (1940) in *Ocenebra aciculata*. In *Buccinum* (fig. 235B) the kidneys (lak) are at their maximal size when the larva is full of food eggs, and they are obvious owing to their yellow-green colour. Each consists of a single cell which projects from the surface of the body and also spreads far beneath the ectoderm, and into which pass two types of amoebocyte concerned with excretion. One type, filled with crystals of waste, is shed from the larval kidney, the other disintegrates within the renal cell and the waste which it contains is either emitted from the surface of the kidney or passes into the amoebocytes and leaves with them. In this way unwanted material is shed into the egg capsules in *Buccinum,* and the larval kidney is not, as suggested by Pelseneer (1911) for this and other species, an organ of accumulation which entraps the toxic waste during the long intracapsular life. However, in *Ocenebra,* Franc (1940) had no evidence that amoebocytes are concerned and described each kidney as a single cell (with a stalk cell added later) which is stuffed full of granules of many sizes, and often with yolk, and which elongates transversely so that the two kidneys form an almost complete collar around the head of the embryo. When other larval organs diminish the kidneys are reduced in size. They become detached later and may be found in the capsule after the larva has left. Franc suggested that the yolk granules are digested in the renal cell, which would account for its regression in size. He also found (1941a) yolk granules in the larval kidneys of *Thais haemastoma,* and these facts, together with the observations that the larval kidneys get smaller in the late embryo of this and other species, led him to the conclusion that the so-called excretory bodies may be deposits of nutritive

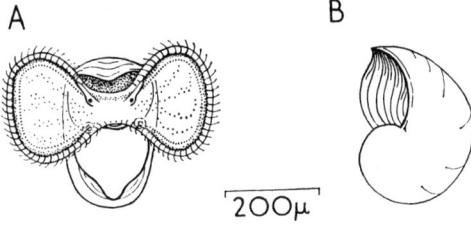

A

B

200μ

0.25mm

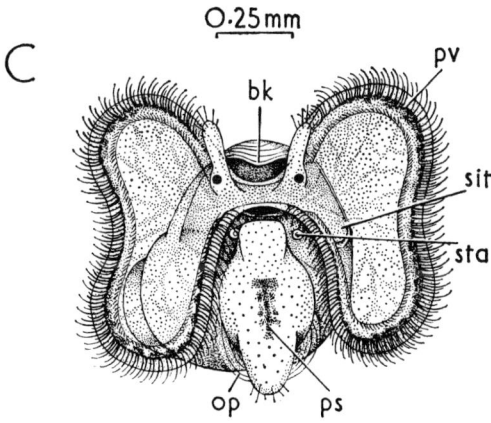

C

pv

bk

sit

sta

op   ps

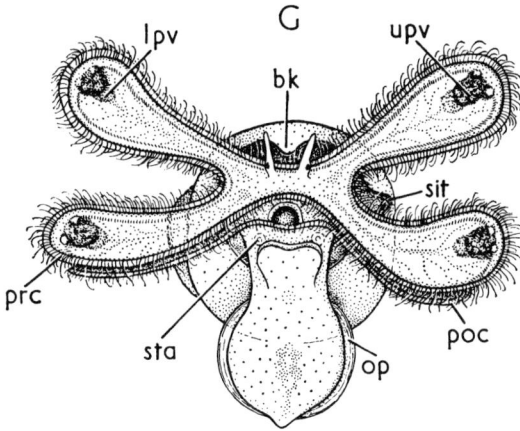

G

lpv   upv

bk

sit

prc

poc

sta   op

0.5mm

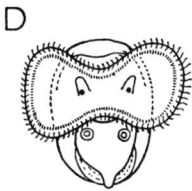

D

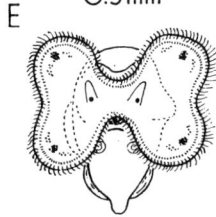

E

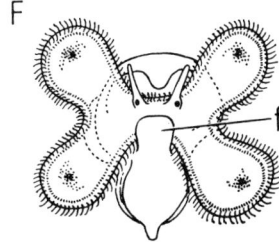

F

f

materials. Moreover, he found that in *Pisania maculosa,* a rachiglossan of warmer seas which ingests numbers of food eggs in the embryonic stage, the larval kidneys are not loaded with obvious excreta. Waste leaves the body by cells which escape from the posterior part of the sole of the foot, the free edge of the mantle and the intertentacular zone of the velar area.

Unicellular structures, one on either side of the oesophagus in the veliger stage of *Littorina littoralis,* were referred to by Delsman (1914) as nephrocysts (nct, fig. 212). They are similar to the unicellular larval kidneys or nephrocysts of the nudibranch *Fiona* (Casteel, 1904) and the tectibranch *Aplysia* (Saunders & Poole, 1910), and lie between the ectoderm and mesoderm, not projecting from the surface of the body. However, Delsman did not trace the origin of these cells in *Littorina,* nor their fate. They do not apparently amass any conspicuous accumulations like the corresponding cells in *Aplysia* which may contain brightly coloured lipoid droplets.

In the freshwater pectinibranch *Viviparus viviparus* there are undoubted larval kidneys in approximately the same position. They are apparently mesodermal in origin, and according to Erlanger (1894) are developed from the anterior end of the mesodermal band on each side. Each is a V-shaped tube with the apex directed forwards; the internal or upper limb is a solenocyte and opens to the lower limb which leads to the exterior. A second freshwater form, *Bithynia,* is the only other pectinibranch known to have similar kidneys in embryonic development, though they occur in pulmonate gastropods, and in *Physa* (Wierzejski, 1905) their origin is similar to that in *Viviparus.* It is tempting to homologize this type of kidney with the protonephridium of a trochophore, but that is an ectomesodermal organ (Meyer, 1901). Its occurrence in freshwater species may mean that it is concerned with osmoregulation. [See p. 633.]

In the early development of the embryonic veligers of some prosobranchs an apical sense organ has been described. Patten (1886a), Wilson (1904) and Smith (1935) figured a prominent one in *Patella* (at, fig. 226A) and Crofts (1937) mentioned the occurrence of a group of small cilia on 2–3 cells of the apical plate of *Haliotis,* which soon disappear. It was first mentioned in some detail by Conklin (1897) in *Crepidula* (acp, fig. 213), and consists of 4 apical cells, initially slightly indented on their outer surface, with a few cells which they have proliferated inwards towards the cavity of the head vesicle (acp), and these meet a strand of tissue growing out from each cerebral ganglion (cg). The organ is therefore V-shaped with the apex of 4 cells on the surface covered by fine cilia; there is no bunch of long cilia as in many trochophore larvae. The organ arises from the point at which the polar bodies were extruded. Later the out-growths from the cerebral ganglia meet one another and form the cerebral commissure, and the connexion with the apical organ and the organ itself is lost. In the newly hatched veliger of *Crepidula* Werner (1955) described the remains of the organ as consisting of large, sub-epidermal, vacuolated cells between the bases of the tentacles, and the epidermis which

---

FIG. 230.—*Nassarius:* veliger larvae. A, B, C, *N. reticulatus:* A, just after hatching; B, its shell; C, about 40 days old. D, E, F, G, *N. incrassatus:* D, 2 weeks; E, 3 weeks; F, about 13 weeks after hatching; G, swimming-crawling stage. The 2 species differ in the longer velar lobes, each with 2 black apical pigment patches, of *N. incrassatus* contrasted with the shorter lobes, each with continuous, reddish-brown, marginal pigment band of *N. reticulatus.*

bk, beak of shell; f, foot; lpv, lower pigment patch of velum; op, operculum; poc, postoral cilia of velum; prc, preoral cilia of velum; ps, pigment on sole of foot; pv, pigment on velum; sit, siphonal tube; sta, statocyst; upv, upper pigment patch of velum.

oe ols m   f

mc

i

dg   v

col

st   r   op

lh   mst

280 μ

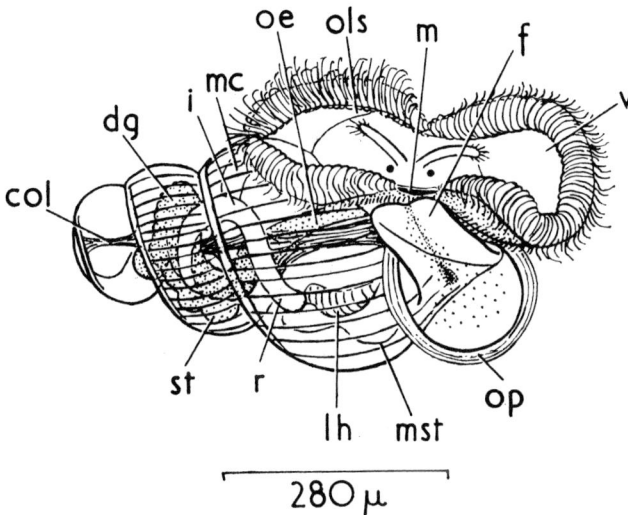

FIG. 231.—*Epitonium* sp.: free-swimming veliger. Note that the coiling of the shell is so tight as to bring the dorsal surface of the body, where the larval heart lies, to the underside of the body whorl as seen in the figure. The larval shell is characterized by ribs on the younger whorls and by the colour: initial whorl brown, subsequent whorls bluish and columella violet.

col, columella; dg, digestive gland; f, foot; i, intestine; lh, larval heart; m, mouth; mc, mantle cavity; mst, mantle skirt; oe, oesophagus; ols, outer lip of shell; op, operculum; r, rectum; st, stomach; v, velum.

covers them has short cilia, though no cilia occur elsewhere on the head except those associated with the velum. In *Trochus* Robert (1902) has described an apical invagination of cells ($1q^{111}$ and $1q^{112}$) which is developed at the 97-cell stage and has disappeared about the 145-cell stage. It may represent a transitory sense organ. However, the invaginated area is not ciliated, though Robert stated that sometimes in older embryos, especially of *Trochus striatus* (= *Cantharidus striatus*), there are some very short cilia in this position. A similar apical invagination was described by Blochmann (1882) for *Neritina*, but Conklin (1897) was of the opinion that the embryos in which it was described were abnormal. Pelseneer (1911) figured the apical organ in veliger embryos of *Lacuna pallidula*, *Littorina saxatilis* and *L. littoralis* (aor, fig. 194A) as a protuberance and did not mention the histological structure. However in *Littorina littoralis* Delsman (1914) described a group of large vacuolated cells covered by short cilia between the bases of the developing tentacles and referred to it as the apical cell plate (acp, fig. 212B). A tentacle, eye and cerebral ganglion arise from cells on either side between this plate and the velum, though there appears to be no connexion between apical cells and cerebral ganglia as in *Crepidula*. Pelseneer (1911) stated that in the rachiglossans *Nassarius*, *Nucella* and *Buccinum* there is a sense organ at the junction of the right and left halves of the cerebral commissure, which forms a projection from the surface of the head, and he figured it in older embryos of *Buccinum* as a swelling on the commissure well away from the surface tissues. The function of this organ is obscure, and old authors, like McMurrich (1886), regarded it as a rudimentary structure.

The velum in embryos which are never freed is typically bilobed, and the lobes are small (fig. 233). The preoral girdle of cilia (prc) rotates the embryo in the albuminous fluid in which it

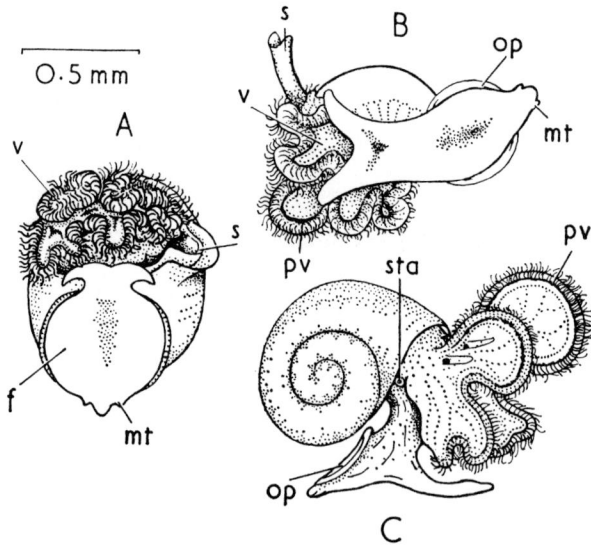

FIG. 232.—*Nassarius reticulatus:* veliger larvae at swimming-crawling stage, about seven weeks after hatching.
The foot is acquiring the characteristic shape of the adult.
    f, foot; mt, metapodial tentacle; op, operculum; pv, pigment on velum; s, siphon; sta, statocyst; v, velum.

is embedded and which is used as food, though the locomotory power of the velum is weak
compared with that of free larvae, and its cilia and musculature correspondingly less developed. Franc (1940) has suggested that in *Ocenebra aciculata* the movement of the veliger
mixes enzymes which the embryo produces with the surrounding albumen, so that it
becomes less viscous and is conveyed more easily by cilia to the mouth. In this species
development is completed in 46 days, but it is not until the 45th day that the velum is
completely absorbed and all trace has gone. In *Fusus,* Portmann (1955) stated that the fluid is
ingested by swallowing movements of the stomodaeum. A thickening is developed in the
dorsal wall of the stomodaeum by means of which a rhythmical closure is effected. It
disappears when the albumen is exhausted and only then does further differentiation of the
stomodaeum occur and the radular sac appear. The albumen is taken up by the epithelial
cells of the albumen sac, a development of the mid-gut in the region from which the stomach
will later differentiate: it is a large and conspicuous part of the gut, for the differentiation of the
stomach and digestive gland is delayed. This larval structure eventually becomes reduced in
size and disappears, though even when the young escape from the capsule it occupies the
apical region of the shell. In yet another respect the embryonic development of this stenoglossan is specialized, for the 4 megameres, filled with yolk granules, persist to a late stage.
After giving rise to the endoderm they help to line the mesenteron until, with the extension of
the true lining of the gut, they are gradually excluded from the lumen and degenerate.
    The development of the veliger embryos of the dog whelk (*Nucella lapillus*) is unusual since
it is arrested at an early stage, during the period of uptake of all the available food eggs in the
capsule. At this time (fig. 234) the cephalic region, foot and visceral mass are differentiated
and the secretion of the shell (sh) has begun. [See p. 576.] The cephalic region is composed of

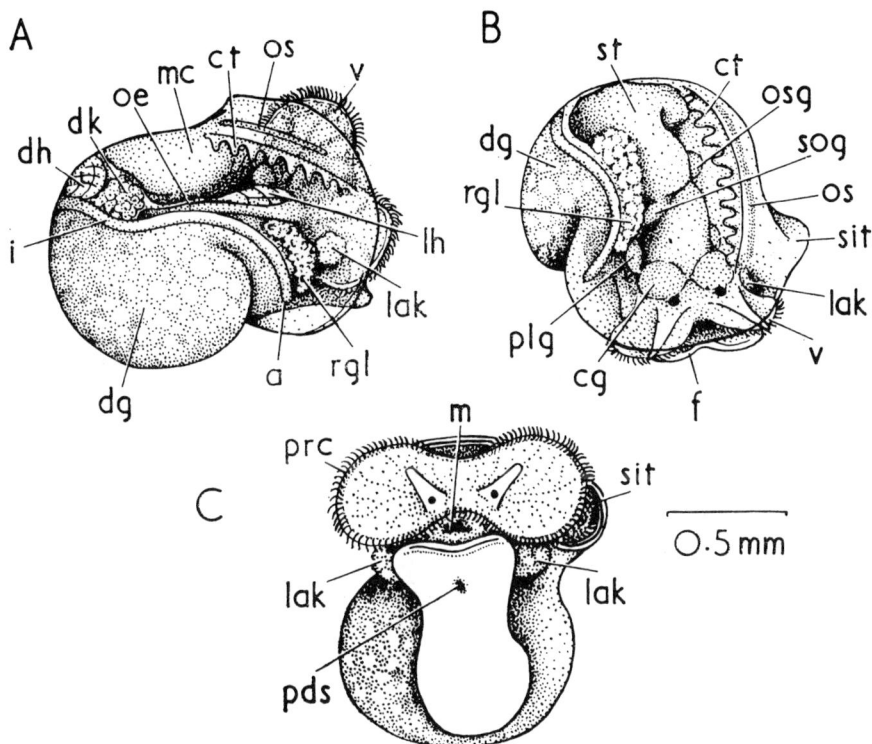

FIG. 233.—*Nucella lapillus:* late embryos from egg capsule; A, in lateral, B, in dorsal and C, in ventral view. Note special larval heart and kidneys, reduced velum and large rectal gland. The digestive gland is swollen with ingested food.

a, anus; cg, cerebral ganglion; ct, ctenidium; dg, digestive gland; dh, developing heart; dk, developing kidney; f, foot; i, intestine; lak, larval kidney; lh, larval heart; m, mouth; mc, mantle cavity; oe, oesophagus; os, osphradium; osg, osphradial ganglion; pds, accessory boring organ; plg, pleural ganglion; prc, preoral cilia of velum; rgl, rectal gland; sit, siphonal tube; sog, supra-oesophageal ganglion; st, stomach; v, velum.

large cells and has no rudiment of eye or tentacle; it is reminiscent of the cephalic mass of the veliger stage of *Pomatias elegans,* though much smaller. It is bordered by ciliated cells (prc) which represent the preoral band of the velum, though the cilia are not thickly set, and its tissues are transparent like those of the foot, so that the course of the gut and the statocysts are visible. The velum (v) has neither food-collecting groove nor postoral band of cilia, for the food eggs are sucked into the gut and the embryos remain attached to the food, which forms a central core in the egg capsule (fe), until all has gone. The lips of the large circular mouth of the embryo, surrounded dorsally by long cilia, are applied to the food, and the stomodaeum, also strongly ciliated (sdm), pumps it into the stomach where it is stored. If the embryos are detached they rotate slowly through the albuminous fluid, but soon attach themselves again. In the second phase of development (fig. 233) the embryonic or larval heart (lh) and kidneys (lak) are formed, and later the true kidney and heart (dk, dh). The differentiation of the liver (dg) is delayed and occurs only when the first coil of the shell is complete and some larval

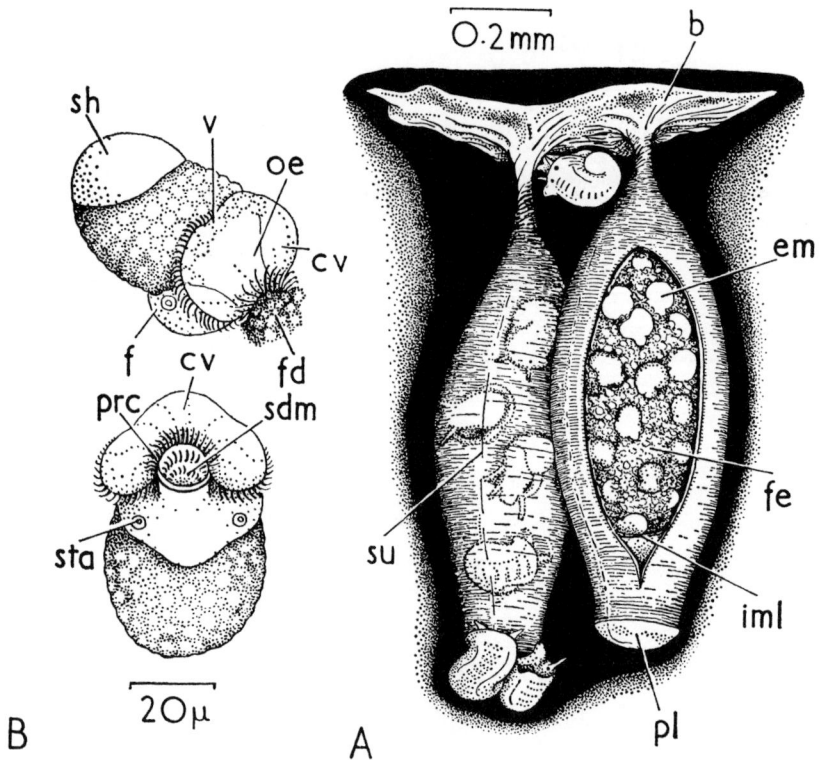

FIG. 234.—*Nucella lapillus:* A, 2 egg capsules attached to substratum. The right one has been opened to show
    young embryos (em) ingesting food eggs (fe); the left one contains a few larvae ready to hatch, and two others
    are emerging after loss of the plug (pl). B, young embryos in lateral and ventral view showing gut gorged with
    food eggs and poor development of other structures. The upper embryo is sucking food.

   b, base of capsule; cv, cephalic vesicle; em, embryo within capsule; f, foot; fd, food; fe, food eggs within
    capsule; iml, inner mucous layer of capsule; oe, oesophagus; pl, plug; prc, preoral cilia of velum; sdm,
    stomodaeum; sh, shell; sta, statocyst; su, suture; v, velum.

characters are disappearing. As the tentacles and eyes develop the velar lobes become more
pronounced and are constricted from the head, though compared with those of related
Stenoglossa, such as *Nassarius,* they are inconspicuous in size and ciliation. The embryos of
Buccinum undatum (fig. 235) also depend on food eggs for their nourishment: these are stored
in the mid-gut, which is greatly distended, and digested and absorbed by a special vesicular
enlargement of the intestine which later disappears. These eggs retain their original spherical
shape and no cleavage occurs, and development proceeds while the embryo feeds, though
the development of the digestive gland is delayed (Portmann, 1925).

   In the terrestrial snail, *Pomatias elegans* (Creek, 1951), the egg is small and surrounded by a
large amount of albumen which is used by the embryo as food. Intracapsular development
continues until the animal is a miniature adult and this takes about 3 months. The embryonic
veliger stage has the cephalic region highly modified into a swollen area (cv, fig. 224) which

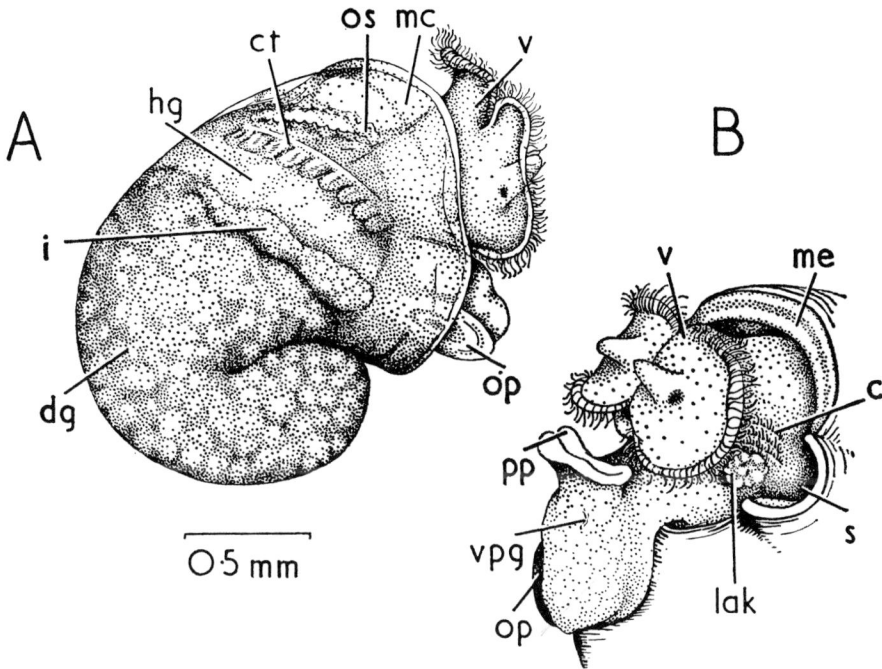

FIG. 235.—*Buccinum undatum:* embryos from egg capsule. A, from right; B, details of head and foot from left. Note the conspicuous patch of cilia (c) on the left side of the head. In other respects these larvae resemble those of *Nucella lapillus.*

c, cilia; ct, ctenidium; dg, digestive gland; hg, hypobranchial gland; i, intestine; lak, larval kidney; mc, mantle cavity; me, mantle edge; op, operculum; os, osphradium; pp, propodium; s, siphon; v, velum; vpg, ventral pedal gland.

has some short cilia (not shown in fig.), and rotates the embryo in the albumen. No velum is developed. In the early veliger stage, 15 days after the first cleavage, the cephalic region covers three-quarters of the external surface. It has undoubtedly an absorptive function, and its precocity in development is shared with the digestive gland, which assumes quite early its role as a storage organ, accumulating albuminous fluid. The cephalic mass is at its maximal development during the mid and late veliger stages, and is then steadily reduced and disappears by the time of hatching, when the supply of albumen is exhausted and the volume of the embryo, as compared with the first cleavage stage, has increased fifteen-fold. Embryos of *Pomatias* have neither larval heart nor larval kidneys.

Cephalic vesicles which are smaller though superficially similar to the cephalic mass of *Pomatias,* are developed not only in the early veliger stage of *Nucella* (cv, fig. 234), *Fusus* and *Buccinum,* but also in some *Crepidula* spp. (Conklin, 1897). In these genera the vesicle is reduced and lost as the velum develops, and there is no evidence that it has an absorptive role. However, in *Crepidula adunca* of the N Californian coast, there is no free larval stage and the head vesicle is larger and the velum much smaller than in species with an indirect development. Delsman (1914) suggested that hypertrophy of the cephalic region is an adaptation to provide a greater surface area for the absorption of oxygen, and that it is absent in *Littorina littoralis,* the embryology of which he studied, because it develops in the littoral surf zone where the water is supersaturated with oxygen and supplementary respiratory organs are not required. It is doubtful whether this argument could be applied to the early stages of development of *Crepidula* species with a free veliger stage or to the stenoglossans, though it may hold for the embryos of the terrestrial pulmonates *Helix pomatia* (Fol, 1880) and *Agriolimax* (Carrick, 1938). However, Moritz (1939) suggested that in *Crepidula adunca* the vesicle is the larval organ of respiration since its thin membrane may easily allow the exchange of gases between the haemocoel and the exterior.

In all planktonic veligers of Monotocardia the velum is at first bilobed, and as the larva grows it increases in size to meet the increased needs of food and buoyancy (fig. 230). In the majority of species it remains as a single lobe on either side of the head, the two meeting ventrally in the region of the mouth and mid-dorsally behind the tentacles, where in most species the girdle of cilia is interrupted (fig. 227). In larvae with a short pelagic life—*Hydrobia ulvae, Turritella communis*—the shape of the velum does not appear to change. In others its proportions, and sometimes its pigmentation, alter rapidly during the early pelagic phase. One of the abundant veligers of our coasts is that of the common winkle, *Littorina littorea* (fig. 227), which is easily recognized by the conspicuous dark purple or black pigment around the border of the velum (pv) and, in older larvae, by a similar colour on the sole of the foot (pf). In the earliest planktonic stage the lateral lobes of the velum are scarcely constricted from one another and the girdle of cilia is continuous around the head. Gradually, as the velum grows, the lobes elongate and each narrows at the base becoming more independent of the head and capable of extensive swimming movements, and, at the same time, the bands of cilia are interrupted mid-dorsally. In this species the planktonic life may be little more than a fortnight, though Thorson (1946) stated that in The Sound the larvae may not settle for a month or more. In the three genera *Bittium, Cerithiopsis* and *Triphora* (fig. 236C) the velar lobes (v), as they enlarge and are constricted at the base, retain the rounded shape, but grow unequally. The inequality of the lobes is least pronounced in *Bittium reticulatum* in which the right, when fully grown, is somewhat larger than the left; the right is distinctly larger in *Cerithiopsis,* whereas in *Triphora perversa,* in accordance with the reversal of the arrangement of organs in the mantle cavity, the left is the larger.

Larvae which remain long in the plankton and are of considerable size before they settle have the most elaborate vela. In the turrid *Mangelia nebula* (fig. 236D), in which the shell attains a length of 1 mm and has $3\frac{1}{2}$ whorls before metamorphosis, the velar lobes are broad and elongated. Each has a slight mid-lateral constriction tending to divide it into dorsal and ventral parts and the breadth at the constriction, the narrowest part, far exceeds the height of the shell. The swimming of the larva is by the undulating movements of the velum and the metachronal beating of the cilia, though the cilia are not particularly long and seem of less importance. In the later stages of development the velum is held over the shell so that the animal is covered and swims completely hidden by it. The velum is conspicuous not only for its size but also for its ornamentation of brilliant orange spots (pv) which vary in number and arrangement and increase with age, tending to form a border as well as being scattered irregularly over the surface. In the related turrid *Philbertia linearis* (fig. 236A, B), also with a long larval life, the shell has $4\frac{1}{2}$ whorls at metamorphosis and a length of about 1 mm. The velum is colourless, at first bilobed, then longer and slightly indented at the sides to form 4 blunt lobes. These elongate whilst the breadth of the velum increases only slightly, and eventually the lobes when deflected posteriorly may project some way beyond the apex of the shell. The velum has thus an extensive margin for its area. The marginal cilia are long and, perhaps more

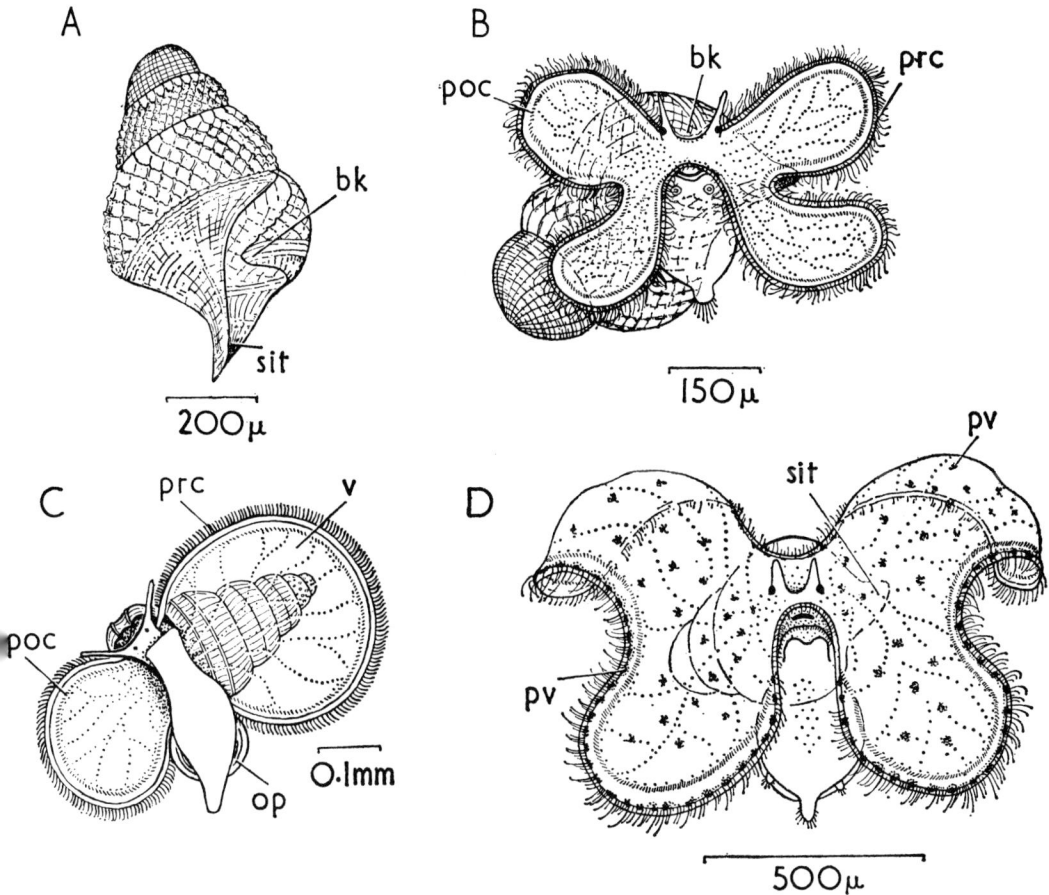

FIG. 236.—A and B, *Philbertia linearis*, shell and veliger in ventral view; C, *Triphora perversa*, veliger in ventral
view; D, *Mangelia nebula*, veliger in ventral view.
   bk, beak; op, operculum; poc, postoral cilia of velum; prc, preoral cilia of velum; pv, pigment of velum; sit,
siphonal tube; v, velum.

than the muscles of the velum, are responsible for swimming. *Aporrhais pespelicani* reaches a
larger size (1·25 mm) before metamorphosis and its velum, at first bilobed, soon changes to 4
and then 6 lobes which are deep.
   The dissimilar appearance of the larva of the two turrids shows that the form of the fully
developed velum in larvae with a long pelagic phase is of no taxonomic value. This is also
illustrated by the larvae of the two British species of *Nassarius* which occur in the lower parts
of the intertidal zone, *N. reticulatus* extending to a depth of a few fathoms, and *N. incrassatus*
to about 46 fathoms. In the adult the latter is about half the size of the former and the newly
hatched larva is also considerably smaller. The following measurements are given by Lebour
(1931a) for the early larval stages of these species:

N. reticulatus	Breadth of body whorl of shell	0·28–0·30 mm
(figs. 230A, B, C; 232)	Breadth of velum	0·29–0·32 mm
N. incrassatus	Breadth of body whorl of shell	0·18–0·20 mm
(fig. 230D, E, F, G)	Breadth of velum	0·20–0·24 mm

Both larvae are at first similar in appearance except that in *N. reticulatus* the velum has a continuous reddish brown border just inside the margin, whereas in *N. incrassatus* there is no pigmentation for the first few days. The shell is smooth, unsculptured, very transparent and consists of one whorl. The outer lip is drawn out slightly at the centre to form an incipient beak (bk), and the hollow on either side supports a lobe of the velum. As the beak grows it is bent inwards and the hollows deepen as the velum enlarges. The larvae grow quickly, but have a long free-swimming period, 2 months in *N. reticulatus* and probably longer in *N. incrassatus*. After about 3 weeks the velum becomes 4-lobed and the animal in both species looks like a butterfly with outspread wings. By this time pigment is beginning to concentrate to form large spots in each corner of the velum of *N. incrassatus* (lpv, upv), one associated with the upper ciliary band, one with the lower, and these velar lobes now grow out to an enormous length. When the shell is approximately 0·8 mm long the lobes are at their maximal development and each is twice as long as the shell and often longer. The velum of *N. reticulatus* is never so large and remains only slightly 4-lobed. This species attains only about two-thirds the size of *N. incrassatus* at metamorphosis, although the adult and the first larval stage are much larger. The long velar lobes of *N. incrassatus* are sometimes out-spread with little motion except the metachronal beating of the cilia, and the shell hangs downwards from them; at other times they move like the wings of a pteropod and propel the animal along. Towards the end of larval life the larvae can both crawl and swim (fig. 232); this stage immediately preceding metamorphosis has been called the veliconcha by Werner (1939). The shell mouth has no longer the projecting beak, but the margin is simple with a slightly crenated edge, and the siphonal tube of the shell is fully formed and contains the mantle siphon (s). The shell of *N. reticulatus* then measures about 0·8 mm long and that of *N. incrassatus* 1·2 mm.

The larvae of some mesogastropods have a comparatively long life in the surface waters and, as well as a large velum, an elaborate shell, which helps to keep them afloat It consists of an inner layer of conchiolin reinforced with calcareous matter (is, fig. 237A, B) which is closely applied to the mantle and, except around the mouth, is separated from a much larger layer of conchiolin by a fluid-filled space. The fluid appears initially to be extra-pallial like that which occurs in all molluscs between the living tissues and shell, and which at the edge of the mantle lobes of bivalves, in particular, also fills the space between the periostracum and calcareous layers of the shell. The outer layer of the shell of these specialized larvae has been regarded hitherto as an accessory shell, the scaphoconch, and the inner layer as the true shell. The name echinospira has been given to them and they are found in 7 British proso-branchs (Lebour, 1935a, 1937): 3 in the family Lamellariidae—*Velutina velutina*, *Lamellaria perspicua* (figs. 237, 238) and *L. latens*; 3 in the family Eratoidae—*Trivia monacha* (fig. 239), *T. arctica* and *Erato voluta*—and one in the family Capulidae—*Capulus ungaricus*. The shell has the typical double appearance in the embryo and it grows rapidly during larval life. The outer layer (sca, fig. 237), which is thin, colourless and glass-like in transparency, coils in one plane in *Lamellaria* and has the sides flat with 2 peripheral keels (ok, fig. 238); the inner calcareous layer (is, fig. 237) grows less rapidly and, in contrast to the outer, forms a helicoid

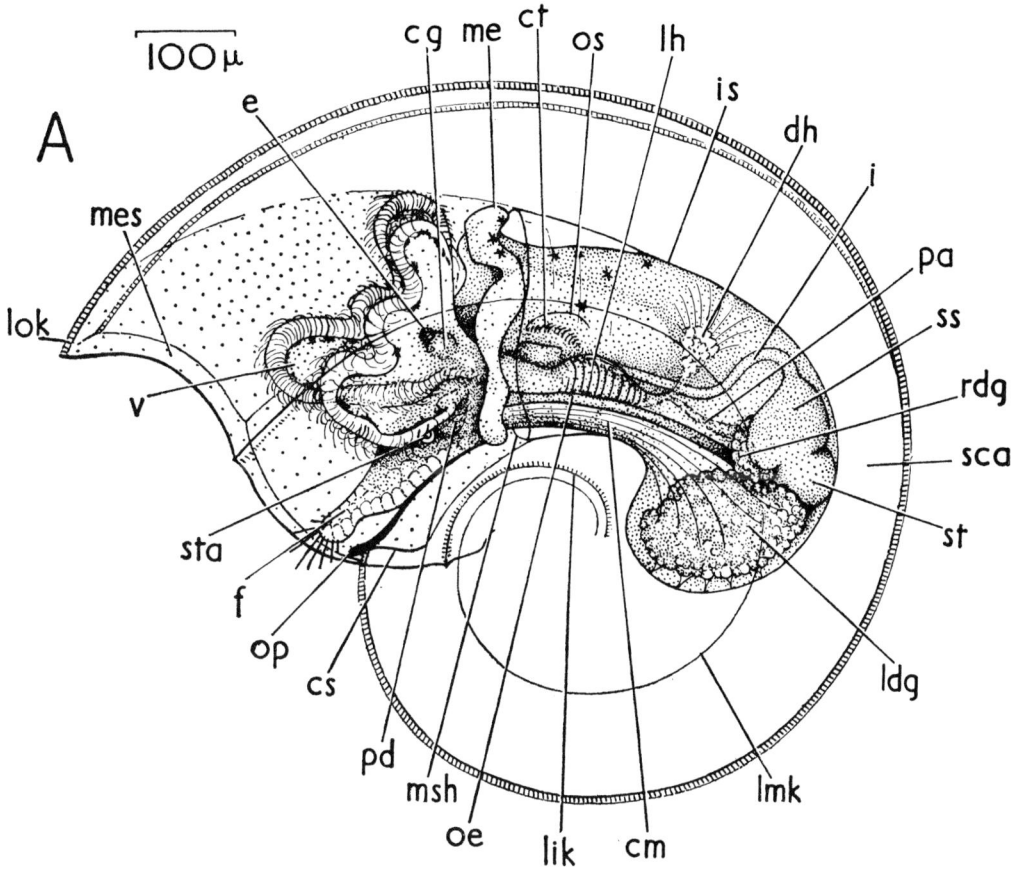

FIG. 237.—*Lamellaria perspicua*: echinospira larva. A, from the left, partially retracted into shell; B, from the right, somewhat less than fully expanded. Chromatophores of a tawny colour occur on the mantle and velum.

cg, cerebral ganglion; cm, columellar muscle; cs, chiselled surface of shell; ct, ctenidium; dh, definitive heart developing; e, eye; f, foot; i, intestine; is, inner part of echinospira shell; l, point where lip rests on main part of echinospira shell; ldg, left lobe of digestive gland; lh, larval heart; lik, left inner keel of shell; lmk, left middle keel of shell; lok, left outer keel of shell; me, mantle edge; mes, mouth of outer part of echinospira shell or scaphoconch; msh, mouth of inner part of echinospira shell; oe, oesophagus; op, operculum; os, osphradium; pa, posterior aorta; pd, pedal ganglion; rdg, right lobe of digestive gland; rik, right inner keel of shell; rmk, right middle keel of shell; sca, space between scaphoconch and inner part of shell; ss, style sac region of stomach; st, stomach; sta, statocyst; v, velum.

spiral. Each has not more than $1\frac{1}{2}$ whorls. In *Velutina* the shell is similarly coiled though not carinated, whereas in *Trivia* (fig. 239) and *Erato* (fig. 240) both periostracal and calcareous parts are helicoid, and the surface is smooth. In *Velutina* and *Capulus* the outer layer is almost gelatinous in its softness. The velum (v) is well developed in all species: it is bilobed in *Capulus*, *Velutina*, *Erato* and *Trivia monacha*, 4-lobed in *T. arctica* and 6-lobed in *Lamellaria*.

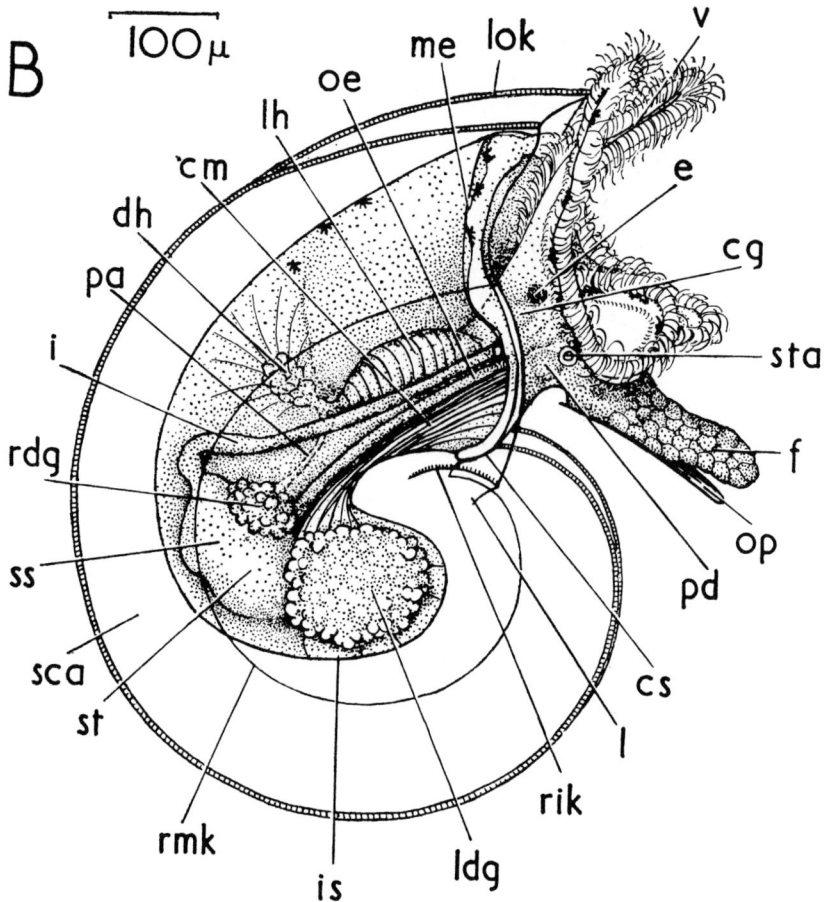

Simnia patula (fig. 241), being the only British cypraeacean, is without an echinospira larva. Its egg capsules are deposited in a single layer over the surface of *Alcyonium* and each contains numerous eggs. These are pink at first, but as development proceeds become brown owing to the development of the shell. In the newly hatched larva the shell has $1\frac{1}{2}$ whorls and is sculptured with irregular granules. At the outer lip a reticulate marking appears and a characteristic tooth and as the shell increases the reticulate pattern persists (fig. 241). The larva is supported in the surface waters by the conspicuous velum of 4 long, narrow lobes and when the shell is about 0·6 mm long, from apex to shell siphon, the velum measures 3·0 mm across. Larval life is probably long (Lebour, 1932a). At the early crawling stage the velum is lost and the mantle begins to grow over the shell; the beak of the shell disappears and the reticulate marking is replaced by straight ribs. The adult shell gradually envelops the larval one.

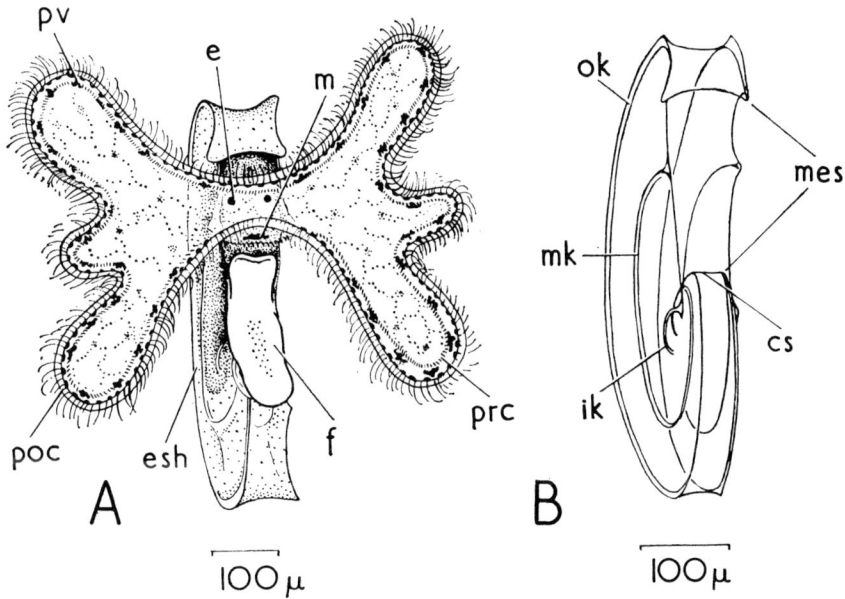

FIG. 238.—*Lamellaria perspicua*: A, echinospira larva extended, seen from in front; B, its scaphoconch.
cs, chiselled surface of shell; e, eye; esh, outer part of echinospira shell or scaphoconch; f, foot; ik, inner keel; m, mouth; mes, mouth of outer part of echinospira shell or scaphoconch; mk, middle keel; ok, outer keel; poc, postoral cilia of velum; prc, preoral cilia of velum; pv, pigment on velum.

The earliest observations on echinospira larvae concern *Lamellaria*. Giard (1875) studied the embryology of *Lamellaria perspicua* and noticed two apparent shells, and his work was confirmed and elaborated by Pelseneer (1911). Simroth (1895) gave a summary, with illustrations, of the echinospira larvae from the Plankton Expedition, and later (1911) described the northern forms including *Lamellaria*. None of these workers, however, observed the later larval stages, and it was not until Lebour (1935a) successfully reared the larvae that the external development from the egg to metamorphosis was observed. Likewise she extended the preliminary observations of Pelseneer (1926b) on the larva of *Trivia* (Lebour, 1931b), and also gave an account of other British echinospira larvae (Lebour, 1937). Her observations on *Lamellaria perspicua* inspired Garstang (1951), and his amplification of her work is incorporated in his poem 'Echinospira's Double Shell'. This gives a most accurate account of the larva, though it still holds that there are 2 shells, the entrance to which cannot be closed by the operculum.

The idea that the echinospira larva has 2 shells, the relationship between which has always been problematic, seems a much more artificial explanation than the suggestion that the 2 shells are merely the periostracal and calcareous parts of the gastropod shell separated from one another to a greater extent than usual. Although *Simnia* has no echinospira shell a pointer to its nature is perhaps given in the revealing statement by Lebour (1932a): 'It was interesting to find that when these embryos were fixed in Bouin-Duboscq a film separated off from the shell, presumably the periostracum, forming a covering of exactly the shape of the

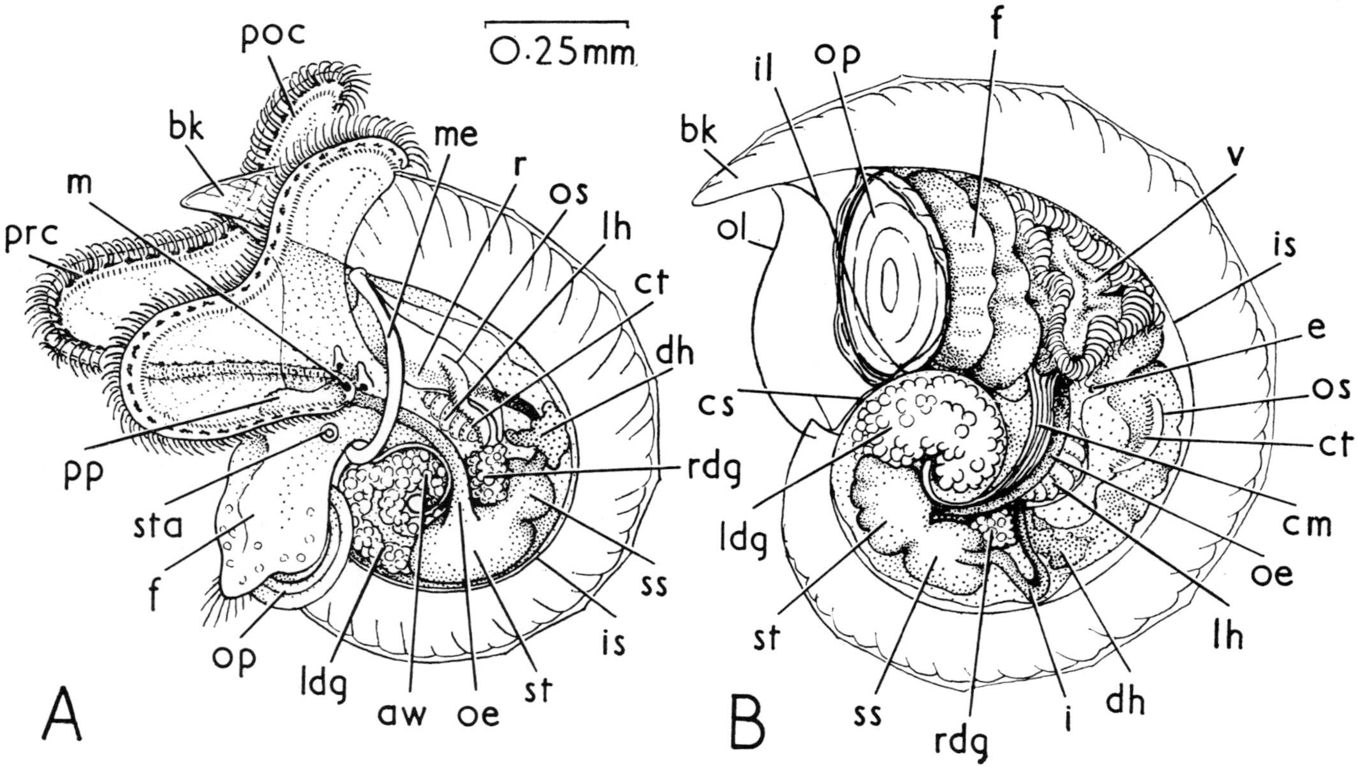

FIG. 239.—*Trivia monacha*: echinospira larva from the left. A, extended; B, fully retracted. Note the spiral coiling of the scaphoconch, its beak and the crumpling of the opercular edge on retraction. The larva is more heavily pigmented than that of *Lamellaria*, with brownish black gut and brown spots on velar edge only.

aw, apical whorl of shell; bk, beak; cm, columellar muscle; cs, chiselled surface of shell; ct, ctenidium; dh, definitive heart; e, eye; f, foot; i, intestine; il, inner lip of scaphoconch; is, inner part of echinospira shell; ldg, left lobe of digestive gland; lh, larval heart; m, mouth; me, mantle edge; oe, oesophagus; ol, outer lip of scaphoconch; op, operculum; os, osphradium; poc, postoral cilia of velum; pp, propodium; prc, preoral cilia of velum; r, rectum; rdg, right lobe of digestive gland; ss, style sac region of stomach; st, stomach; sta, statocyst; v, velum.

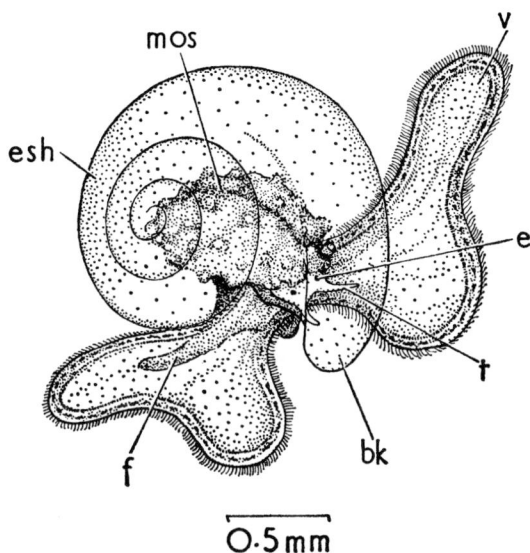

FIG. 240.—*Erato voluta:* echinospira larva during metamorphosis, after Lebour. The mantle covers the true shell and the operculum has been lost, but the echinospira shell and the velum persist.

bk, beak; e, eye; esh, echinospira shell; f, foot; mos, mantle fused over shell; t, tentacle; v, velum.

accessory shell of *Trivia*.' In the growth of the shell of any gastropod the formation of a layer of conchiolin, which is known as the periostracum, precedes the formation of a layer of conchiolin impregnated with calcareous salts. The areas which produce them at the mantle edge form two concentric circles separated by an indifferent epithelium. In the embryo the shell gland first appears as a pit which everts and spreads over the visceral mass. It secretes a layer of conchiolin which represents the outer layer of the echinospira shell and, as the visceral mass grows, the gland cells which produce this come to form a ring at the mantle edge. The second ring of gland cells of the adult, more posteriorly placed and secreting calcareous salts in addition to conchiolin, is also derived from the shell gland, and, in the echinospira, its cells produce the inner calcified layer of the shell after the outer layer has grown away from the visceral mass as noted by Pelseneer (1911) in the embryos of *Lamellaria perspicua*. In veligers in which the protoconch is not calcified the activity of these cells is delayed. As the shell of the echinospira larva grows they add to the lip of the inner layer which embraces the visceral hump, and the outer ring of glands, which, from the start, is considerably more active, adds to the larger outer layer. The echinospira veliger differs from all others in that the growing edges of the 2 layers of the shell diverge markedly as the larva grows; this divergence exaggerates the degree of separation of the periostracal and the inner layer of the shell so that their union to form a single shell, as in other gastropods, never can occur. The distance between the growing edges represents the indifferent mantle epithelium which separates the secreting areas. The mantle edge of all echinospira larvae (me, figs. 237A, B; 239) is thickened and very mobile. When the larva is extended to swim and feed it surrounds the opening of the periostracal part of the shell and adds secretion to it; the radiating striae mark these periods of secretory activity. When the veliger withdraws, the mantle edge is folded

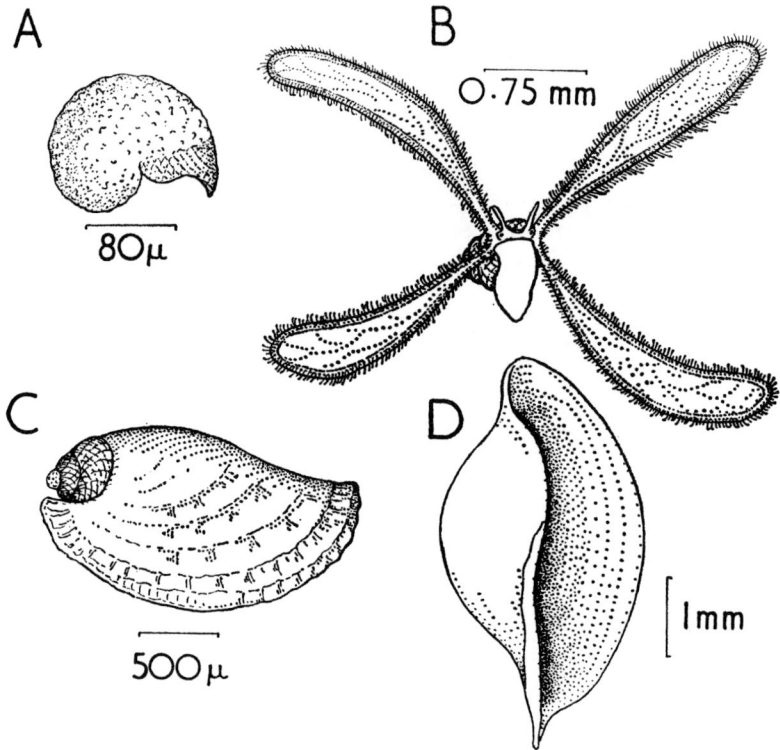

FIG. 241.—*Simnia patula:* A, shell of larva 3 days old. The granular surface of the protoconch contrasts with the reticulate pattern appearing on the outer lip. B, larva with 4-lobed velum which is colourless. C, shell of crawling stage (veliconcha) from shelly gravel, showing granular surface of protoconch, reticulate pattern of larval whorls and beginning of adult shell with ribs. D, shell showing adult characters. After Lebour.

back on itself (fig. 239B), the velar lobes fold dorsally over the head, and the operculum (op) (absent in the adult) closes the shell, fitting against the opening of the inner layer, which is small as compared with the mouth formed by the periostracum. The larva, as it grows, is hampered by the restricted space into which it must withdraw, and from the moment that the mouth of the shell starts to overgrow the first whorl—which would then project into it and constrict it still further—the larva sets about enlarging the space by chiselling away the periostracal layer of this whorl (cs, figs. 237, 238B, 239B) with the ventral edge of the operculum. This has the effect of opening the space between the 2 layers of shell to the external medium so that it becomes filled with sea water, and of giving the larva slight freedom of movement within this space. The operculum of *Trivia* (op, fig. 239) becomes considerably larger than the opening which it closes, so that as the larva withdraws into the shell the rim of the operculum is bent forwards to fit the opening, and, ventrally, folds into a pointed V-shaped gouge. When the larva emerges this gouge comes against the periostracum and slices it away horizontally (cs, fig. 239B); the chiselled surface is then smoothed by mucus from the mantle (Garstang, 1951).

All this happens some time before the larva settles, but is a preliminary to the loss of the periostracal layer. *Lamellaria perspicua* metamorphoses when the echinospira shell is about 2·25 mm across; at this time the mantle is no longer extended to the edge of the larval shell, but has gradually inserted itself between the periostracal and calcareous layers. These separate from one another. The outer layer is cast off, the operculum lost and the mantle spreads over the inner layer of the shell. From this moment onwards the further secretion of the periostracal layer will be associated directly with the calcareous material as it is in the adult prosobranch. The outer shell of echinospira larvae is therefore seen as a larval development of periostracum, a larval organ facilitating pelagic life and shed at metamorphosis. During the metamorphosis of *Velutina* the outer layer of the shell is similarly cast off (Lebour, 1935a), and for *Trivia* Lebour (personal communication) tells us that 'further investigation has shown that my statement concerning the absorption of the echinospira shell by the mantle is wrong, for although the covering is extremely fragile it can be shed whole by the late larva as a very thin helicoid shell'. *Erato voluta* (fig. 240) differs in that the mantle (mos) envelops the true shell while the echinospira shell (esh) still covers it and this is not cast off for some time; moreover the velum (v) persists even longer than the shell.

In all other embryos of monotocardians the shell, or protoconch, is a layer of conchiolin which is closely applied to the mantle and, at some time, may be calcified. It first appears as a shallow cap which deepens as the shell gland spreads over the visceral mass, and in species with free larvae it is already spirally twisted with about $1\frac{1}{2}$ whorls at the time of hatching. Its surface may be smooth or ornamented with spiral striae, pits or tubercles. The shell which is added to the protoconch during larval life may be similar, though frequently the sculpture is more complex, foreshadowing the patterning of the adult, and the changes in the shell are synchronous with the change in mode of life. During adult life the soft underlying tissues move forward from the embryonic and larval coils of the shell, which may then be lost. In species in which the adult shell is a tall spire and there is a fairly long pelagic stage, the shell of the larva at metamorphosis may have several whorls, up to 8 or 9 in *Triphora perversa* according to Fischer (1887), $5\frac{1}{2}$ in *Philbertia teres* (Lebour, 1934b), 5 in *Balcis alba* (Lebour, 1935b), $4\frac{1}{2}$ in *Cerithiopsis tubercularis*, *C. barleei* and *Philbertia gracilis* (Lebour, 1933b, 1933d, 1934b), $3\frac{1}{2}$ in *Mangelia nebula* (Lebour, 1934b). In all these the surface has features of the adult shell. In *Balcis alba* it is smooth and colourless and the protoconch grades into the teleoconch without abrupt change; in the adult the tip of the spire (the embryonic and larval whorls) is nearly always broken off, and is filled with calcareous secretion from the underlying mantle. In specimens of *Triphora perversa*, which have 6 whorls at metamorphosis, the embryonic shell is tuberculate (fig. 236C), the following whorls are sculptured with longitudinal striae and are keeled, whilst the sixth is tuberculated as in the adult. The embryonic shell of *Philbertia gracilis* is sculptured with irregular dots and flecks and the following whorls are keeled and have oblique striations. This shell is unlike that of related species, for the remaining turrids may be divided into 2 groups, those with smooth apices representing the embryonic shell (*Haedropleura*, *Lora*, *Mangelia*) and those with reticulate or elaborately sculptured apices (*Philbertia*, fig. 236A). The embryonic and larval whorls of turrids show plainly in the adult though the sculpture may be worn. All these are shells of planktotrophic larvae and prosobranchs of this habit agree in having a protoconch which is smaller, usually shows some sculpture and is narrowly twisted. By contrast, the protoconch of prosobranchs without a larval stage is larger, smoother and inflated to an almost spherical shape. The pelagic larva experiences a relatively abrupt change of life at metamorphosis; the non-pelagic does not. This is normally reflected in an abrupt change from protoconch to teleoconch in the one case

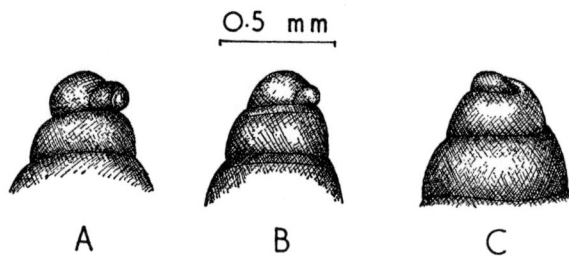

FIG. 242.—Shell apices of pyramidellids: A, *Odostomia unidentata*; B, *O. plicata*; C, *O. eulimoides*. A and B show heterostrophy.

and by a gradual transition in the other. Inspection of the apical region of a shell therefore permits one to say whether a pelagic or non-pelagic development has occurred. This is the shell apex theory first suggested by the work of Dall (1924).

There is one feature developed in larval shells of unrelated species: a large outgrowth from the upper lip which is bent over the mouth of the shell so that a velar lobe is supported in the concavity on either side (bk, figs. 230, 236B, 239, 240). A similar curved beak has already been described in *Nassarius* and is pronounced in the larval shells of *Bittium, Triphora, Cerithiopsis* and all turrids (fig. 236); it disappears at the end of larval life, and has not been described (or does not occur) in species in which the veliger stage is passed through in the egg capsule. The siphonal canal (sit), characteristic of the shell of a number of monotocardians, is developed in the shell of the veliger, either free or embryonic (fig. 233).

In species with a short pelagic phase the shell is less developed. In *Turritella communis*, which has a tall spire in the adult, there are only $2\frac{1}{2}$ whorls at metamorphosis, the last developing the adult pattern of spiral ribbing. *Rissoa membranacea* has only 2 whorls as compared with 4 in *R. parva* and $3\frac{1}{2}$ in *R. sarsi*, both with a longer pelagic phase. When the free-swimming stage is suppressed the crawling stage may start with only $1\frac{1}{2}$–2 whorls as in *Barleeia rubra, Cingula cingillus* and *Colus islandicus* (Lebour, 1937), and they are heavy-looking as compared with the coils formed during a free veliger stage. In *Colus* the embryos are nourished by food eggs and the size of the crawling stage varies with the nourishment available in the capsule. Thorson (1935) described one capsule from which only one embryo emerged and it was 8·5 mm long, with a shell of $3\frac{1}{2}$ whorls, a little more than half a whorl being spirally sculptured as in the adult. The embryos of other Stenoglossa (*Ocenebra* and *Nucella*) which are similarly nourished emerge from the capsule with up to $2\frac{1}{2}$ coils to the shell.

In some shells the protoconch and teleoconch are remarkably dissimilar owing to a change in the direction of coiling, or to the fact that whereas the initial coils are contiguous the later ones unroll more or less completely and continue their course in a much looser spiral (*Vermetus*) or in a nearly straight line (*Caecum*, fig. 41B); or the adult shell expands to a patelliform shape (fig. 226D). The planktonic larvae of a number of animals traditionally regarded as monotocardians have a sinistral shell though the adult is dextral. They belong to the Pyramidellidae, and the sinistrality of the larval shell is one of their opisthobranch characters. This shell is colourless and smooth. The adult shell is dextrally coiled and carries the minute protoconch (fig. 242A, B) horizontally across its apex (*Odostomia plicata, O. turrita, O. unidentata, O. acuta, O. conoidea, Turbonilla crenata, T. elegantissima, Eulimella macandrei, E. laevis, E. nitidissima*). According to Thorson (1946) these have a free larval stage. There are

TABLE 13

Occurrence of Prosobranch Larvae in Inshore Plankton at Plymouth, 1940–45 (after Lebour, 1947)

Species	Jan.	Feb.	Mar.	Apr.	May	Jun.	Jul.	Aug.	Sep.	Oct.	Nov.	Dec.
						Month						
Patella vulgata	⊠	×	×	×					×	×	x̲	⊠
Patina pellucida		×						×	×		×	×
Littorina littorea egg capsules	x̲	⊠	⊠	x̲	x̲	x̲	×	x̲	x̲	x̲	x̲	x̲
Littorina littorea larvae	×	×		×	×	×	×	×	×	×	×	×
Littorina neritoides egg capsules	×	×	×	×						×		×
Rissoid larvae	×	×	×	x̲	x̲	x̲	x̲	x̲	x̲	x̲	x̲	x̲
Trivia arctica	×	×	×	×	×							
Trivia monacha				×	×	×	×	×	×			
Lamellaria perspicua	×	×	×	×	×	×	x̲	x̲	×	×	×	×
Simnia patula						×	×	×	x̲			
Natica catena			×	×	×	x̲	x̲	×	×	×		
?Bittium reticulatum							×	×				
Triphora perversa							×	×	×	×		
Cerithiopsis tubercularis	×					×	×	×	×	×		
Cerithiopsis barleei								×				
Balcis sp.									×			
Caecum sp.	×						×	×	×	×	x̲	×
Nassarius reticulatus			x̲	x̲	x̲	x̲	×	×	×	×		
Nassarius incrassatus			×		x̲	x̲	x̲	x̲	×	×		
?Haedropleura septangularis					×	×	x̲	×	×			
Mangelia nebula							×	×	×	×	×	
Philbertia gracilis					×	×	x̲	x̲	×			
Philbertia linearis						x̲	⊠	x̲	x̲	x̲		

× = Recorded. x̲ = Common. ⊠ = Abundant.

other members of the pyramidellids in which all the shell whorls have a normal dextral appearance and there is no indication of any aberrant larval coiling either externally (Thorson, 1946), or when the apex of the shell is broken and examined internally (*Chrysallida decussata, C. indistincta, C. obtusa, C. spiralis, Menestho divisa, M. obliqua, M. clavula, Odostomia scalaris* and *O. eulimoides* (fig. 242C)). It seems probable that in these species the pelagic phase is greatly reduced or absent. The adult shell of *Caecum imperforatum* (fig. 41B) is a slightly arched tube with a round mouth at the broader end and a rounded septum posteriorly, and there is no trace of coiling. The larva, however, has a spiral planorbiform shell of about $2\frac{1}{2}$ whorls which in the adult is separated by a calcareous septum and cast away. The Chinaman's hat shell (*Calyptraea chinensis*) is of contrasting shape: it is patelliform and retains at the apex the embryonic shell of $1\frac{1}{2}$ whorls (fig. 43D). There is no free larva in this species.

Veligers of a few species are easily recognizable in the plankton not only on account of specific characteristics of the velum or the coiling and sculpture of the shell, but more particularly by its colour. The larval shell with the most distinctive colour is that of *Epitonium* sp. (fig. 231), which is a purplish blue with a purplish red umbilicus and columella. Closer

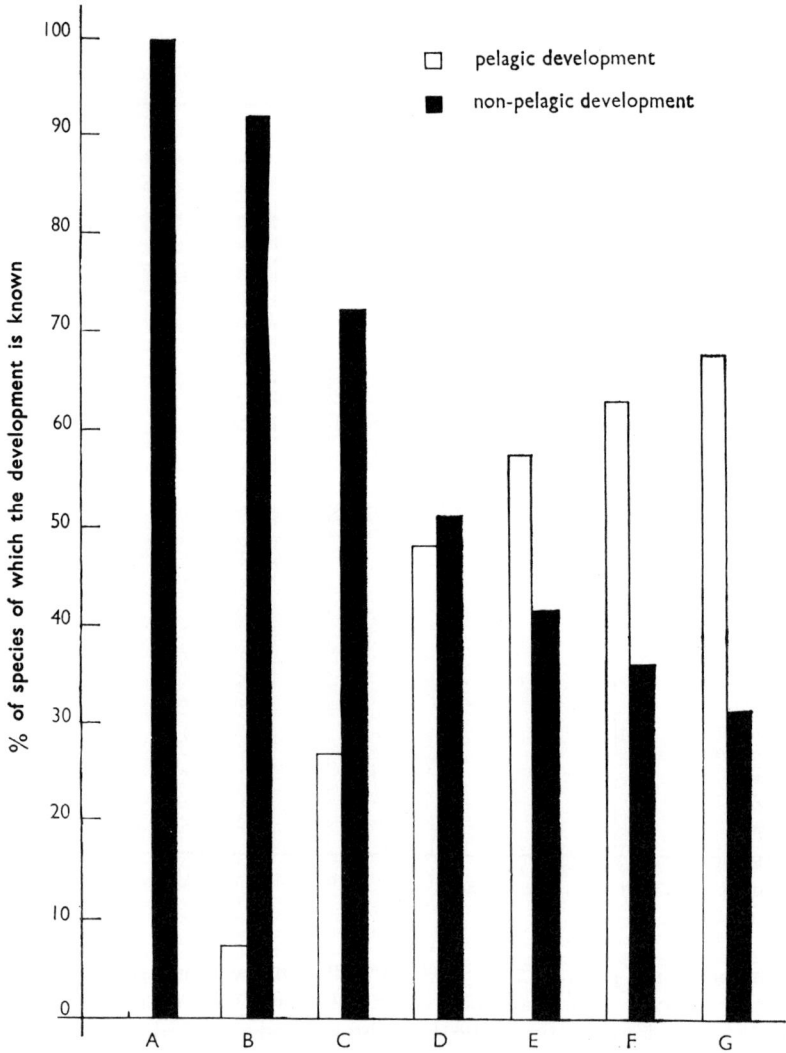

		p. — 0%	n.p. 100%
A	E Greenland	p. — 0%	n.p. 100%
B	N & E Iceland	7·5	92·5
C	W & S Iceland	27·3	72·7
D	Faroes, Shetland, Orkney	48·7	51·3
E	S Norway, W Sweden, Denmark	58	42
F	S England, Channel Is.	63·5	36·5
G	Canary Is.	68	32

FIG. 243.—Graphs giving the percentage of prosobranchs (for which the development is known) with free-swimming larvae (white columns), and those in which the larval stage is suppressed (black columns).
 A, E Greenland; B, N and E Iceland; C, W and S Iceland; D, Faroes, Shetland, Orkney; E, S Norway, W Sweden, Denmark; F, S England, Channel Is.; G, Canary Is.; n.p., non-pelagic larval development; p., pelagic larval development. After Thorson.

examination shows that the first whorl is colourless and undecorated, the others tinted and with several fine longitudinal ribs like the adult shell. Horn colour is a more usual tint for the larval shell, as in *Alvania punctura, Cerithiopsis* and *Triphora,* and the larvae of *Philbertia gracilis* are easily identified since the heavily sculptured shell is a dark brown.

These various characteristics of the larval shell of gastropods have been used for taxonomic purposes—in fact Iredale (1911) suggested that no genus should be permitted to contain species with more than one type of apex. Clearly this produces an impossible situation, for within a single genus like *Cingula* or *Natica,* some species have no pelagic phase whilst others have, and this fact alone may give rise to differences in the shell apex. However, with a certain caution, the larval shell may be used in the identification of adult specimens, though diagnostic markings are often worn or obliterated. Its greatest value is undoubtedly in the recognition of the planktonic larva.

Our knowledge of the molluscan larvae of the English Channel comes from the work of Lebour. This mainly concerns the larvae of gastropods, for they are the most easily recognizable in the plankton, whereas the larvae of many species of bivalves are so much alike that it is difficult or impossible to determine the individual species. During the period 1940–45 records were kept of the larval forms in the inshore plankton of Plymouth (Lebour, 1947) and the results for the prosobranch gastropods which were identified are shown in Table 13; unidentified gastropod larvae were present in every month. The rissoids are amongst the commonest, occurring throughout the year, often in large numbers (the actual species of these larvae were not recorded). Their abundance may be partly explicable in the light of the observations made by Smidt (1938) that in *Rissoa membranacea* at least one-eighth of the eggs laid complete their development, a much greater proportion than in other prosobranch species; this rissoid, however, has a short pelagic stage.

Lebour's studies of the eggs and larvae of the prosobranchs of Bermuda (Lebour, 1945) show that the majority of molluscs in these warmer seas have free-swimming larvae, and Thorson (1950), using her investigations, has calculated that it concerns 85% of the species; this is the maximum known for any marine area, even tropical seas. Thorson (1950) gave 68% for the Canary Islands. In the high Arctic seas (E Greenland) not a single prosobranch is known to have a pelagic phase in its development; in fact 95% of all marine species of bottom invertebrates living there have a direct development; there are, however, pelagic larvae of polychaetes, echinoderms and lamellibranchs (Thorson, 1936). The Antarctic plankton has no molluscan larvae (Simroth, 1911). Thorson (1950) stated that in order to survive in the high Arctic areas a planktotrophic larva must complete its development, from the time of hatching to metamorphosis, at a temperature below 4·5°C and within 1–1½ months, which includes the very short period of phytoplankton production. The prosobranchs more than any other invertebrates are sensitive to environmental changes, and it is therefore not surprising to find that the percentage of species in the N Atlantic with a pelagic phase varies consistently with latitude (fig. 243).

CHAPTER 18

THE ROCKY SHORE: LIMPETS

T HE sublittoral part of the rocky shore is thought to have been the ancestral home of the prosobranch molluscs, and the most primitive prosobranchs living today retain a similar habitat. They are rock clingers living near ELWST, or deeper, and avoiding well lit areas; gathering their food from the surface of the substratum in the vicinity of their home; spawning into the sea, and after a planktonic larval period settling on the rocks to take up their final niche. The evolution of the ability to endure variations in temperature, light, humidity and salinity allowed the prosobranchs to escape from the sublittoral areas and colonize the beach. Here their distribution is governed by the extent to which their emancipation has proceeded. Food preferences and the ability to endure periods of restricted feeding, together with their response to gravity, are other factors governing their distribution on the shore, which, for some species, lies between very narrow limits. Certain pre-adaptations, anatomical and behaviourial, must have speeded this evolution. These concern especially the shell, into which the snail retreats and virtually shuts itself off from the external environment, and the foot, which maintains balance as the animal creeps, though if waves are strong the foot will hold the animal motionless against the rocks. Prosobranchs colonizing the upper parts of the shore near high tide level are still dependent on the tide, for in no littoral species can reproduction be fulfilled without the snails being able to discharge eggs or young to the sea, or the fixed spawn being immersed periodically. Even in the viviparous *Littorina saxatilis* the eggs do not develop normally, and the young do not apparently leave the brood pouch, unless the female is at times covered with water. The young snails are more sensitive to the changing conditions than the adults: they more or less avoid them, and are found either at lower levels on the beach or they hide.

The intertidal level at which a species lives is no true indication of its tolerance of exposure, for on an irregular shore the S and SW faces of the rocks will receive maximal illumination and heat, whereas adjacent N and NW faces and overhangs, sheltered from the sun, provide shade and lower temperature. A few species of prosobranchs can withstand the force of the summer sun on dry exposed rocks, and even during abnormally hot periods this does not appear to be a lethal factor. Indeed *Monodonta* and *Littorina littorea* are recorded as living on boulders where the temperature may be 40°C which is exceptionally high for the British Isles. The temperature of their body tissues during the period of exposure may remain a degree or so lower than that of the air (Deshpande, 1957), due to the fact that water in the pallial cavity is slow to heat and loss by evaporation through the shell mouth, though slight, will tend to lower the temperature. However, contrary results have been obtained by Southward (1958) who has shown that the body temperature of limpets exposed to air under varied weather conditions is higher than might be expected from the sea and air temperatures. Up to 20°C there is accurate correspondence between the temperature of the limpet and that of an inanimate object; between 20° and 30°, however, the animal remains a little cooler than its surroundings. Results of a similar experiment with 'a top shell' showed the animal tissue had a

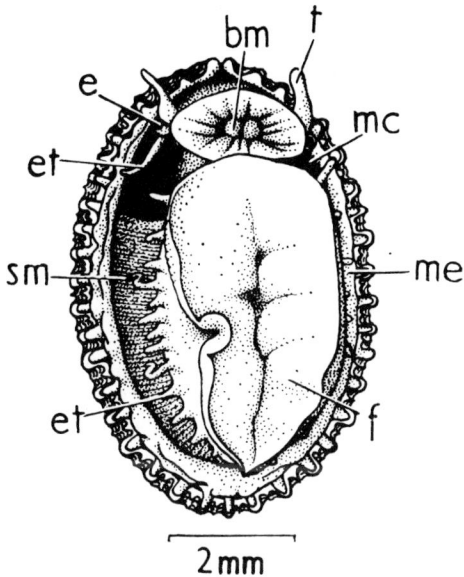

FIG. 244.—*Emarginula conica*: animal in ventral view.
   bm, buccal mass; e, eye; et, epipodial tentacle; f, foot; mc, nuchal cavity; me, mantle edge; sm, shell muscle; t, tentacle.

higher temperature over the range 16·3–30·0° which was covered in 90 min. The rising tide brings a sudden fall in temperature and this may help to stimulate the animals to activity. Evans (1948) studied the effect of temperature, apart from desiccation, on 11 common species of British littoral molluscs during the summer—*Patella vulgata, P. intermedia, P. aspera, Gibbula cineraria, G. umbilicalis, Monodonta lineata, Littorina littorea, L. saxatilis, L. neritoides, L. littoralis, Nucella lapillus*. The animals were immersed in sea water with a temperature increment of 1°C/5 min. At 30° most species showed some signs of distress and active movement usually ceased. The temperature of heat coma of each species was registered and the thermal death point. *G. cineraria,* which occurs at very low levels, has the lowest thermal death point at 36·2°C and irritability is no longer recorded at 34–35°C (heat coma). In *L. neritoides,* which lives well above HWST, death occurs at 46·3° and heat coma at 38°; this littorinid, most tolerant of heat, has the most southerly distribution, extending into the Black Sea and N Africa. The degree of heat tolerance of each species appears to be related to the temperature range which it experiences in the precise niche of the shore which it inhabits. *N. lapillus* has a rather similar vertical range to *L. littorea,* but is more restricted to shade near its upper limit and correlated with this is its much lower lethal temperature of 40° as compared with 46° for *L. littorea*. The thermal death points of the 11 species which were studied are high in comparison with the normal temperatures of the environment, and it can only be concluded that temperature alone can have no direct limiting effect on their vertical distribution. For instance, in its natural environment, *Nucella* rarely needs to withstand air temperatures above 25°. Gowanloch & Hayes (1926) working at Halifax, Nova Scotia, found that *L. littorea* and *L. saxatilis* from different levels on the shore have correspondingly different death temperatures, but Evans recorded no significant differences with animals from Cardigan Bay. Both these winkles and also *L. littoralis* occur in N Russia and workers there have found *L. saxatilis* the most tolerant of low temperatures surviving more than 27 hrs in air at −9·4° (Gurjanova, Sachs & Uschakov, 1930).

When conditions are adverse snails tend to remain inert, and consequently the activity of species which inhabit the shore varies with the tide, which imposes a rhythm on behaviour. Activity is influenced also by the alternation of day and night. As the tide retreats the prosobranchs on rocks become inert before the surface dries, many having returned to a home or protective crevice where the period of ebb tide will be spent. However, during darkness, if the air be still and the habitat moist, there is considerable activity. Many individuals in rock pools, both those which are more or less permanent inhabitants and the casual winkle or topshell, will continue to move about and feed at LW even in bright sunlight, and carnivores such as the dog whelk may continue to suck the flesh of their prey in some sheltered gully. However, for the majority, especially the herbivores and scavengers, there is a rhythm of locomotory activity and feeding. It has been suggested that this may be an endogenous rhythm rather than a direct induction by environmental factors (Brown, Fingerman, Sandeen & Webb, 1953) and it may persist when external conditions remain constant. Bohn (1904) described a fortnightly rhythm of activity in *L. saxatilis*, coinciding with the spring tides, and stated that when snails are brought into uniform conditions in the laboratory they become active at 15-day intervals corresponding with the spring tides. Stephens, Sandeen & Webb (1953) have shown that only about 2% of a population of the mud snail *Nassa obsoleta* may be active at LW as against 54% at HW, and individuals collected from localities with different tidal times—not only from between tide-marks, but also from below—show activity patterns correlated with their local conditions. The expression of the tidal pattern of activity of these animals persisted in the laboratory for approximately 36 hrs at 22° and for considerably longer at 12°, and it appeared to be revived by 24–48 hrs stay at 12° after the animals had become apparently arhythmic at the higher temperature.

Rhythmical activity will be reflected in oxygen consumption. Gompel (1937) found that oxygen consumption in *Patella vulgata* is maximal at high tide and minimal at low, and Sandeen, Stephens & Brown (1954) described a persistent diurnal and tidal rhythm of consumption in *Littorina littorea* and *Urosalpinx cinerea*. The diurnal rhythm in *Urosalpinx* is one with maximum oxygen consumption between 04.30 and 06.30 hrs and between 19.30 and 21.30 hrs with a lesser maximum at 14.30, and minimal consumption between 00.30 and 01.30, between 11.30 and 13.30 and at 16.30 hrs. The persistent tidal rhythm involves minimal rates of oxygen consumption about 5 hrs after low tide and maximal rates 2–3 hrs before low tide. Two such rhythms of different frequencies have been demonstrated in other invertebrates including the lamellibranchs *Crassostrea virginica* (Brown, 1954) and *Venus mercenaria* (Bennett, 1954). The resultant pattern of such rhythms is of semilunar frequency and shows peaks coincident with spring tides; these may be part of the mechanism regulating spawning. When *Littorina irrorata* is exposed after a long period of submergence oxygen consumption increases (Newcombe, Miller & Chappel, 1936) and is recorded as 0·31 ml/gm/hr at 32°C. Subsequent submergence at the same temperature brings about a reduction to 0·179 ml/gm/hr. These results partly reflect differences in the availability of oxygen in air as compared with water. The consumption varies with the temperature; it is maximal for *L. irrorata* at 35° and decreases rapidly at lower temperatures, and below 10° it is difficult to obtain reliable results.

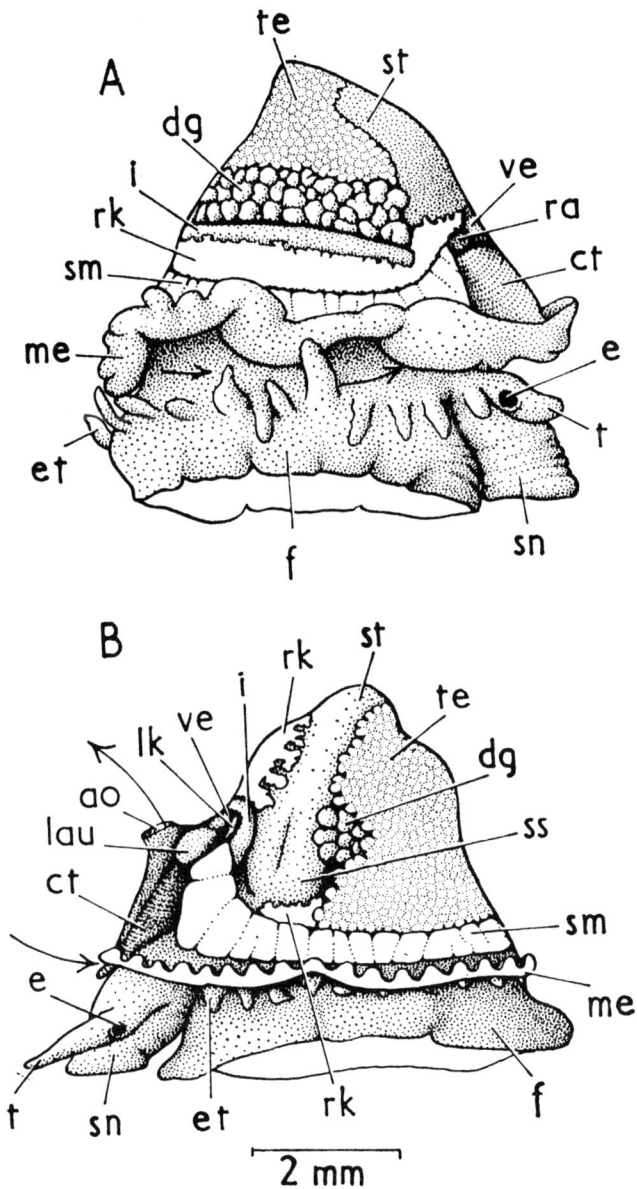

FIG. 245.—*Emarginula conica:* animal removed from shell and seen A, from the right; B, from the left side. Arrows indicate ciliary and water currents.

ao, apical opening of mantle cavity; ct, ctenidium; dg, digestive gland; e, eye; et, epipodial tentacle; f, foot; i, intestine; lau, left auricle; lk, left kidney; me, mantle edge; ra, right auricle; rk, right kidney; sm, shell muscle; sn, snout; ss, style sac region of stomach; st, stomach; t, tentacle; te, testis; ve, ventricle.

For easy discussion the prosobranchs which live on rocky shores have been dealt with here under the three general headings 'limpets', 'whelks and tingles' and 'other prosobranchs'. These will be found immediately below and in Chapters 19 and 20 respectively.

Like the word 'worm' the word 'limpet' is not an accurate zoological term, but is merely the name of a shape, the shape of the simple, conical shell which is found in those gastropods that live habitually in one place, clinging to the substratum and often withstanding the wash of waves or the current of a torrential stream. It is often assumed that the ancestral gastropod had a simple cap-shaped shell like that shown by recent limpets, but in no living limpet is the shape of the shell primary: in every instance it has been secondarily derived from a helicoid spiral shell which may still be seen in the young stages (fig. 226D). As the animals grow, however, the body whorl of the shell grows disproportionately large and provides accommodation for all the soft parts, whilst the original spire is usually wholly worn away. The mouth of the shell becomes round or oval and the sole of the foot acquires more or less the same shape. The operculum is lost, because the animal never normally lets go of the substratum to which it clings, and the columellar muscle usually becomes horseshoe-shaped (fig. 42B), probably formed by the union of originally separate right and left shell muscles.

Limpets may be found in all groups of gastropods. *Ancylus,* the freshwater limpet, and *Siphonaria,* found on tropical beaches, are both pulmonates, but most limpets are prosobranchs and, in particular, belong amongst the more lowly members of that group such as the Zeugobranchia and Patellacea (=Docoglossa). A few monotocardians have also adopted this rock-clinging mode of life like the Chinaman's hat shell (*Calyptraea chinensis*) and the Hungarian cap shell (*Capulus ungaricus*).

The Zeugobranchia are perhaps the least successful of these three groups of prosobranch limpets. They are an archaic group dating back to late Cambrian, differing from other prosobranchs in a number of important ways. The two other groups are perhaps much younger, the Patellacea dating certainly from the Trias (Wenz, 1938) and perhaps from the Silurian (Knight, 1952) and the Calyptraeacea from the lower Cretaceous. In both these groups, too, the calculations of Schilder (1947) suggest that a flowering into many species has occurred only within geologically very recent times.

Of these animals the calyptraeaceans are protandrous hermaphrodites and some approach to this condition is exhibited by the docoglossan limpets. With a few exceptions these are the only undoubted prosobranchs which show this—the other hermaphrodites being perhaps more truly opisthobranchs or having opisthobranch affinities—and the association of the hermaphrodite condition with the limpet mode of life may be significant.

The Zeugobranchia are a small group of prosobranchs which are not richly represented in British waters since these lie in too cold an area of the world. Most zeugobranchs prefer warmer conditions, although the restriction of the family Scissurellidae to cold water areas shows that it is not beyond their power to adapt themselves in this way. Of 18 genera listed in Thiele's *Handbook* (1929–35) only 5 are to be found in British seas, even when this phrase is interpreted in the broadest sense to include the Channel Islands. These animals are the ormer, *Haliotis tuberculata* L., which is exclusively Sarnian, three species of the genus *Emarginula,* the slit limpets *E. reticulata* Sowerby, *E. conica* Lamarck, and *E. crassa* Sowerby, the keyhole limpet *Diodora apertura* (Montagu), *Puncturella noachina* (L.), and *Scissurella crispata* Fleming.

With the exception of *Haliotis* and *Scissurella* these animals appear to lead essentially the same kind of life—that of the rock-clinger spending long periods in one and the same spot from which it makes periodic excursions for feeding. This may be verified by direct

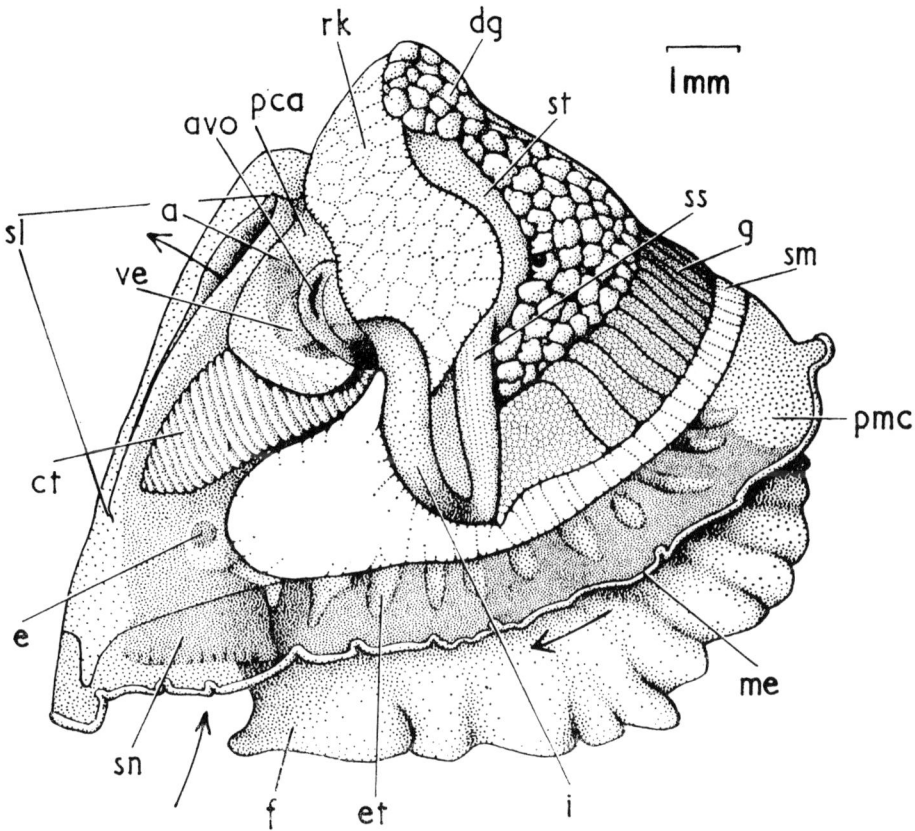

FIG. 246.—*Puncturella noachina:* animal removed from shell and seen from the left. Arrows indicate ciliary and water currents.

a, anus, seen through mantle skirt; avo, auriculo-ventricular opening; ct, ctenidium, seen through mantle skirt; dg, digestive gland; e, eye, seen through mantle skirt; et, epipodial tentacle, seen through mantle skirt; f, foot; g, gonad; i, intestine; me, mantle edge; pca, pericardial cavity; pmc, posterior part of mantle cavity; rk, right kidney; sl, extent of slit in mantle skirt; sm, shell muscle; sn, snout, seen through mantle skirt; ss, style sac region of stomach; st, stomach; ve, ventricle.

observation of *Emarginula reticulata* and *Diodora apertura,* which live between tidemarks, but is only an inference in the case of those other members of the group which live in deeper waters. For this type of life they are adapted by the limpet-like shape, the conical shell and the foot which gives them considerable powers of adhesion, though never equal to those of the docoglossan limpets. As in these the shell muscle has a horseshoe-shaped attachment to the shell and is cup-shaped, broken anteriorly, roofed over by the shell and fringed sideways by the mantle, with the head occupying the gap in the horseshoe at the front. In *Haliotis* and *Scissurella* there are two shell muscles right and left, and the hypertrophy of the right one in *Haliotis* gives it a median attachment to the shell (rsm, fig. 84) and displaces the mantle cavity to the left. In many limpet-like animals the mantle carries sensory processes, but in the

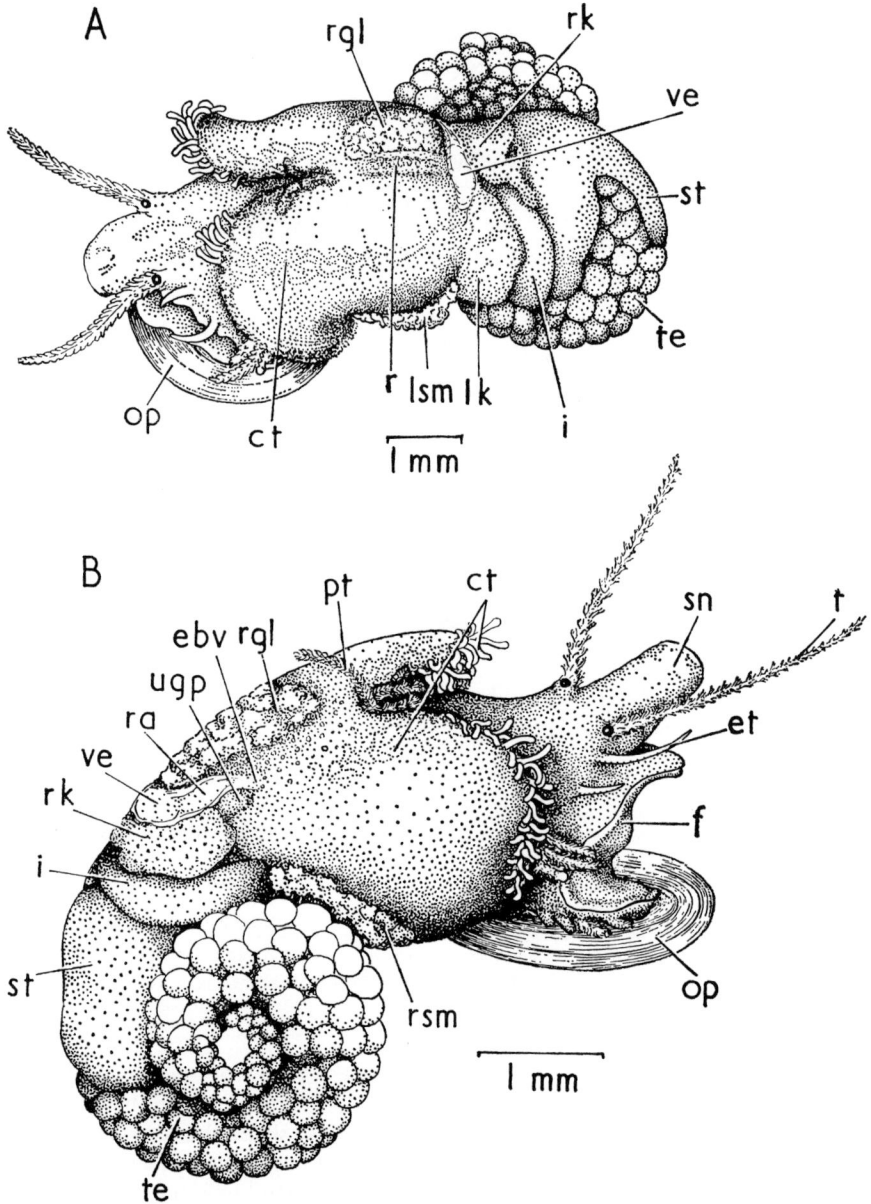

FIG. 247.—*Scissurella crispata*: animal removed from shell and seen A, from the dorsal side; B, from the right. ct, ctenidium with filiform filaments; ebv, efferent branchial vessel; et, epipodial tentacle; f, foot; i, intestine; lk, left kidney (papillary sac); lsm, left shell muscle; op, operculum; pt, pallial tentacle; r, rectum, seen through mantle skirt; ra, right auricle; rgl, rectal gland; rk, right kidney; rsm, right shell muscle; sn, snout; st, stomach; t, tentacle; te, testis; ugp, opening of right kidney (urinogenital papilla); ve, ventricle.

zeugobranchs this tendency is exaggerated and the edge of the mantle skirt thickened and made warty with processes of sensory importance. These papillae are least numerous in *Puncturella*. Augmenting this sensory equipment is the epipodium, which varies in nature from genus to genus. It is at its most elaborate in *Haliotis*, where it forms a collar which projects rather high up the side of the foot and is beset with numerous sensory bosses, small tentacles and, especially dorsally, long tentacles. In *Diodora*, *Emarginula* (figs. 244, 245) and *Puncturella* (fig. 246) the epipodium is less developed and takes the form of a circlet of separate tentacles running round the sides of the foot, and in *Scissurella* (figs. 247, 248) these are reduced in number.

Each epipodial tentacle is, presumably, generally sensitive to tactile stimuli since it contains a prominent nerve related to the pedal ganglia, but in addition it bears a special group of sensory cells lying at its base on the ventral side (eso, fig. 79). The precise function of these organs is not known, but they must add considerably to the total sensitivity of the margin of the body of the limpets. An epipodial tentacle lies behind the right tentacle in *Emarginula*, *Diodora* and *Puncturella*, where it is bifid. This has, because of its position, been assumed to act as a penis (Odhner, 1932) but is almost certainly not so since it occurs in both sexes.

During early life the anterior edge of the mantle skirt in all the Zeugobranchia shows a cleft which is repeated as a slit in the overlying shell. Under this lie the anus, excretory and reproductive apertures. The slit remains as such in the pleurotomariids, scissurellids (sls, fig. 248) and Emarginulinae (fig. 33), its inner end being closed by secretion of shell material as the animal grows and leaving a track over the older whorls known as the slit band (slb, fig. 248A). In the genus *Haliotis* (fig. 84) the slit is at regular intervals closed near the edge of the shell by deposition of shelly material, so that its upper end remains open for a period during which further marginal growth may form a new slit. This may be repeated a few times so that the slit is represented by a series of holes (hs) under one of which the anus is placed. Eventually the older holes get closed by growth of shell. In *Puncturella* and in *Diodora* the young animal has a slit mantle skirt and shell, but as it gets older the mantle edges fuse across the slit basally leaving only the upper part as an opening. Where the pallial fusion has occurred secretion of shell substance goes on with the result that the slit becomes converted into a hole lying on the anterior face of the shell. It remains thus in *Puncturella* (fig. 246) but differential growth in *Diodora* converts it into an apical aperture (fig. 33).

Whatever its final form, this hole first appears as a bay at the edge of the mantle skirt. The edge of the skirt often protrudes through a slit or hole as a kind of short exhalant siphon, and bears tentacles, as in *Puncturella*, where there are six, *Scissurella* (figs. 247, 248) and *Haliotis*, where three small ones pass through three of the separate holes (usually the oldest and two others according to Crofts (1929), though they may vary their position).

The head, foot, mantle and mantle cavity of some of the Zeugobranchia are bilaterally symmetrical (*Emarginula*, *Diodora* (fig. 249), *Puncturella*). Even where this is not true (*Haliotis* (fig. 221B) and less so, *Scissurella*), a double set of pallial organs occurs, with those on the left larger than those on the right. The gills (fig. 249) are bipectinate and are slung over the greater part of their length by efferent membranes (efm) passing from the floor of the mantle cavity to the ctenidial axis and by afferent membranes connected to the roof. The anus (a) lies in the central compartment of the mantle cavity flanked by two apertures, an excretory one on the left, a urinogenital one on the right (rko). The roof of the cavity between these and the gill axis forms right and left hypobranchial glands, so that at first sight these limpets look very much as one supposes the original symmetrical gastropod may have looked. They may be used to suggest this type of animal indeed, but it must be realized that they are far

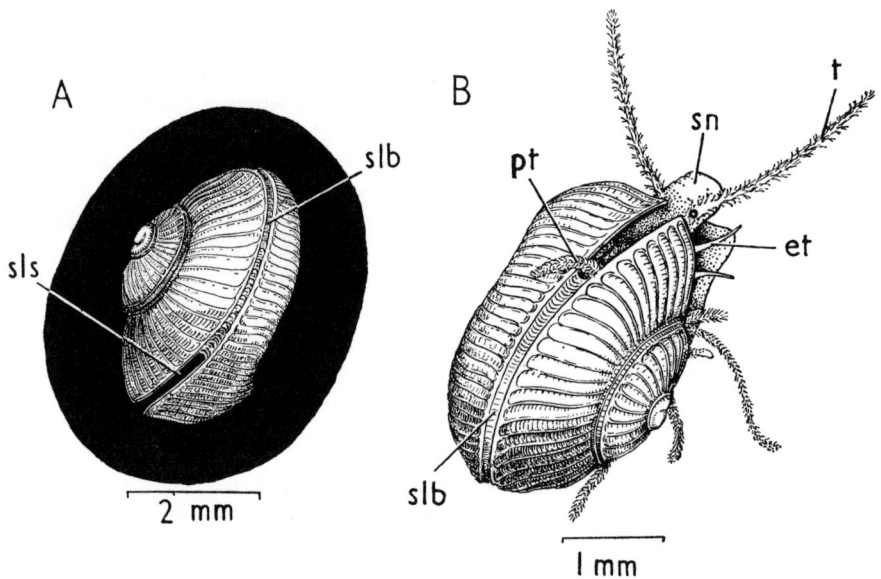

FIG. 248.—*Scissurella crispata:* A, shell; B, animal, alive, in dorsal view.
et, epipodial tentacle; pt, pallial tentacle; slb, slit band; sls, slit in shell; sn, snout; t, tentacle.

from being primitive and are highly specialized for their rock-clinging mode of life. So far as their external features are concerned this is reflected in the slit or apical hole which they possess in the mantle: this permits water to enter the mantle cavity anterolaterally, to wash over the ctenidial leaflets and to escape from the mantle cavity dorsally, where the out-thrust mantle edge acts as a siphon and directs the water current away from the animal's head.

This is the solution to the sanitary problems of the gastropod mantle cavity found by the zeugobranchs, but, whilst it is good in its way, the animal probably loses as much as it has gained by its adoption. The introduction of an apical hole into the shell of a keyhole limpet certainly allows the escape of the exhalant water current from the mantle cavity, but also means that the shell is no longer a waterproof cover, and this may seriously affect the mollusc's ability to live in an intertidal habitat. It certainly seems to imply that it is impossible for any zeugobranch limpet to live in the same kind of exposed situation as would be occupied by the common limpet, and all zeugobranchs are to be found at low water level or below—that is, in situations where the tendency to desiccation is least. [See p. 576.]

In correlation with the presence of slit or apical hole the intestine is short and not concerned with the elaboration of faecal pellets (p. 220); the stomach has lost its caecum and sorting areas (p. 214). A siphon runs along the intestine (p. 223), which passes through the ventricle. The left kidney is minute and has a renopericardial canal as Crofts (1929) recorded in *Haliotis*. The right is responsible for most of the excretory activity and also acts as a conduit for the genital products. In *Diodora nubecula* von Medem (1945) stated that the spermatozoa are shed in packets surrounded by testicular epithelium and that these enter the female gonad in which the eggs are fertilized: this is certainly not true of other zeugobranchs in which sperm are emitted in the usual way and fertilization is external. Apart from these features the general organization is what would be expected in a primitive prosobranch.

The mode of life of *Haliotis tuberculata* has been described by Stephenson (1924) and Crofts (1929). The animals live on rocky shores, for the most part, where there is little sand or other sediment which might clog the respiratory apparatus. They usually live at LWST though they may be found to a depth of 5–6 fathoms, or at higher levels in tidal pools of such size as not to heat up too rapidly. They avoid light and therefore spend daylight hours hidden under a rock or in a crevice, attached by their powerful foot and coming out only at nightfall. They tend to hide in any appropriate situation rather than have a definite 'home'. The shell, the epipodium and the sides of the foot (which cannot be drawn under its shelter) are protectively coloured with blotches and streaks of brown, green or red, and mimic the normal encrusting growth of the rocks to which the animal clings. Their attachment to the rock is extremely tenacious and the pull on the shell effected by the shell muscles will tear that from the body rather than detach the body from the rock. When they do move they move rapidly, Stephenson recording a speed of 5–6 yards per minute, though Crofts cautioned that this distance would not be actually traversed in such a short time.

The food of *Haliotis* is largely seaweed, particularly the more delicate red weeds like *Delesseria* and *Griffithsia*, but the animal is in part a detrital feeder and in the course of its browsing takes detritus and pieces of almost all encrusting organisms into its gut. The animals appear to be susceptible to lack of oxygen, to brackishness and stillness of the water and, except when minute, are difficult to keep in aquaria for these reasons. They seem to have a rather extended breeding season. Fertilization is external and a free-swimming larva is produced.

Of the three species of *Emarginula* occurring in this country only *E. reticulata* is at all likely to be found by the ordinary shore collector, *E. crassa* and *E. conica*, which can occur occasionally in that habitat on W British coasts, being usually dredged. *E. reticulata* may be collected from the underside of stones not normally higher than LWST and extending thence downwards to considerable depths. All species are particularly fond of stones with rough pitted surfaces into the crannies of which they withdraw and out of which it is often difficult to prise them. Unlike *Haliotis* they do not object to sediment and are often found on stones covered with a thin deposit of fine mud. This is stirred up as the limpet creeps and particles caught between foot and mantle edge are carried forward to the mantle cavity and leave by the pallial siphon (ao, fig. 245B). Similar rejection currents along the side of the foot occur in *Puncturella*. Their mode of life appears to be much more like that of a patellid than that of *Haliotis* and they seem to make excursions for feeding purposes and return to a 'home'. Their food is sometimes said to be algae (Eigenbrodt, 1941), though they are often to be found in positions where the search for these would involve them in lengthy forays. Their food certainly includes considerable quantities of sponge, but it is likely that they ingest much mixed detrital material as they feed: diatom frustules are often to be seen in their stomach contents. Their reproduction is not known but is likely to be as in *Diodora*.

Diodora apertura is not uncommon in the same kind of situations as provide a suitable habitat for *Emarginula reticulata*, but it is more restricted and does not occur on the E coasts of England or Scotland. Littoral specimens are usually found only at LWST on pitted rocks with a covering of slimy mud and they are smaller than specimens from deeper waters. The animal is sluggish, like *Emarginula*, and appears to live in the same way and feed on similar food. It is characterized by an elaboration of the mantle edge (figs. 79, 249) which is exposed as the limpet creeps, but is covered by the shell when it is gripping the rock, though not

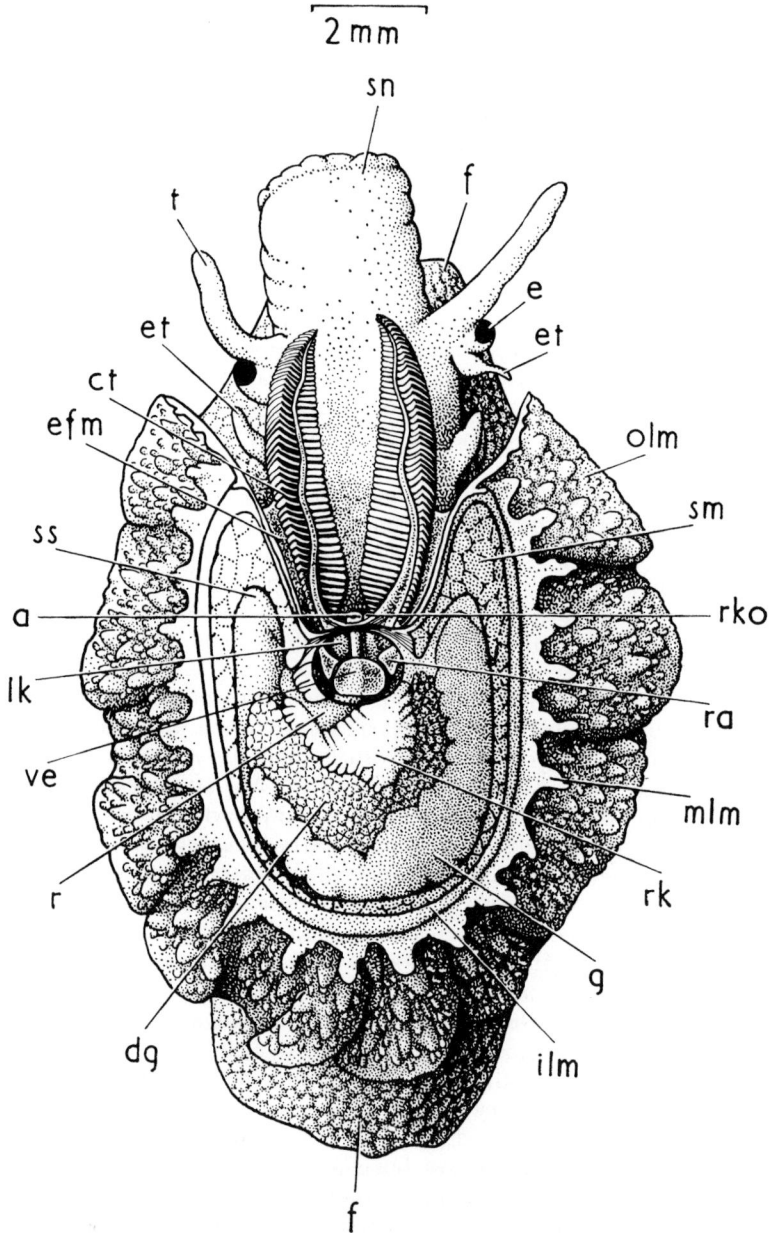

FIG. 249.—*Diodora apertura:* animal removed from shell and seen in dorsal view. The mantle skirt over the main pallial cavity bearing the hypobranchial glands has been removed and the pericardial cavity opened.

a, anus; ct, ctenidium; dg, digestive gland; e, eye; efm, efferent membrane of ctenidium; et, epipodial tentacle; f, foot; g, gonad; ilm, inner lobe of mantle edge; lk, left kidney; mlm, median lobe of mantle edge; olm, outer lobe of mantle edge; r, rectum; ra, right auricle; rk, right kidney; rko, opening of right kidney (urinogenital papilla); sm, shell muscle; sn, snout; ss, style sac region of stomach; t, tentacle; ve, ventricle.

completely so, for the tentacles of the middle pallial layer (mlm) lie at the ends of the ridges radiating from the apex of the shell and so can never be completely protected (me, fig. 43A). These zeugobranchs are intolerant of brackish water, *Diodora apertura* dying out in the estuary of the Rance when the salinity falls to 21‰ (Fischer, 1931) and this seems to be a general characteristic of all the zeugobranchs. Fischer also noted that it can withstand prolonged starvation. The limpet appears to have a breeding season extending over a considerable period, December to May (Pelseneer, 1935), the eggs (140 μm diameter) being yellow in colour and adhering to the substratum after extrusion from the mantle cavity. They hatch in the crawling stage, possessing a shell with $1\frac{1}{2}$ whorls and neither hole nor slit, and having lost the velum. Later a slit appears at the edge and is converted to a hole which gradually moves to an apical position. The mouth of the shell expands and the spire becomes minute and in older stages is worn off so as to leave a conical shell with an apical hole.

Puncturella noachina is not a littoral animal but may be dredged on rocky or clay bottoms particularly off the northern parts of Britain. It is, like *Scissurella*, a cold water animal and occurs widespread in the Arctic, Antarctic and cold temperate Atlantic and Pacific Oceans. In lower latitudes it finds the coldness it appreciates at greater depths and whereas it may be found at a depth of 20 m off E Greenland it occurs at 1100 m off Portugal (Thorson, 1944). Little is known about its biology, but there is no reason to suppose that it is significantly different in its way of life from other zeugobranchs. Specimens have been seen to eat diatoms and detritus which they rasp from the surface of stones as they creep.

Scissurella crispata (figs. 247, 248) is the least well known of this group of British proso-branchs, being recorded only off the Shetland Islands, on shelly and stony ground off the Orkneys, NW Scotland, Antrim and W Ireland, though it occurs widely over arctic seas and at a greater depth further south. One living specimen—the animal drawn in figs. 247 and 248—was dredged in 105–110 f at 47° 35′ N, 7° 13′ W in shell gravel, and in this locality the species may be moderately abundant, for a small sample of the dredging yielded three empty shells and two others with dead animals. The living specimen was found by Dr Daphne Atkins who kindly allowed us to examine it. It was remarkably active, creeping with speed over the substratum and rasping the surface for food particles as it went. In this respect it resembled the related *Incisura* (*Scissurella*) *lytteltonensis* described by Bourne (1910) from seaweed in rock pools at Lyttelton Harbour, New Zealand, one specimen of which was observed to crawl for a distance of nearly half an inch in the space of a quarter of an hour. The animal has a shell coiled in a spiral like that of a pleurotomariid with a prominent slit and slit band. The snout (sn) is large and through the outer tissues can be seen the stout jaws and posteriorly the salivary glands. The body tissues appear a greyish white against the pearl-white shell which is semi-transparent and of a delicate texture. When the animal is active a number of tentacles, pallial and pedal (et, pt), are extended, and the longer ones are waved slowly through the water. Two or more pallial tentacles project from the slit of the shell. They, and also shorter ones, arise from the edge of the mantle skirt and their surface is tuberculate, as is also this edge. The epipodial tentacles may be similar, though two conspicuous ones behind each eye are short and stout, pointed and with the edge entire. The specimen, which was sectioned, was a male with no accessory genital structures. The reproductive habits of *Scissurella* are unknown and likely to remain so until more extensive local collecting can be undertaken.

The limpets which are classified in the Patellacea comprise one of the most successful groups of gastropods. [For recent work see chapter 31.] Some species colonize the most exposed parts of the seashore with a degree of freedom not reached by any other mollusc. They include *Patella, Patina, Acmaea* and also *Lepeta* and *Propilidium,* the least abundant in

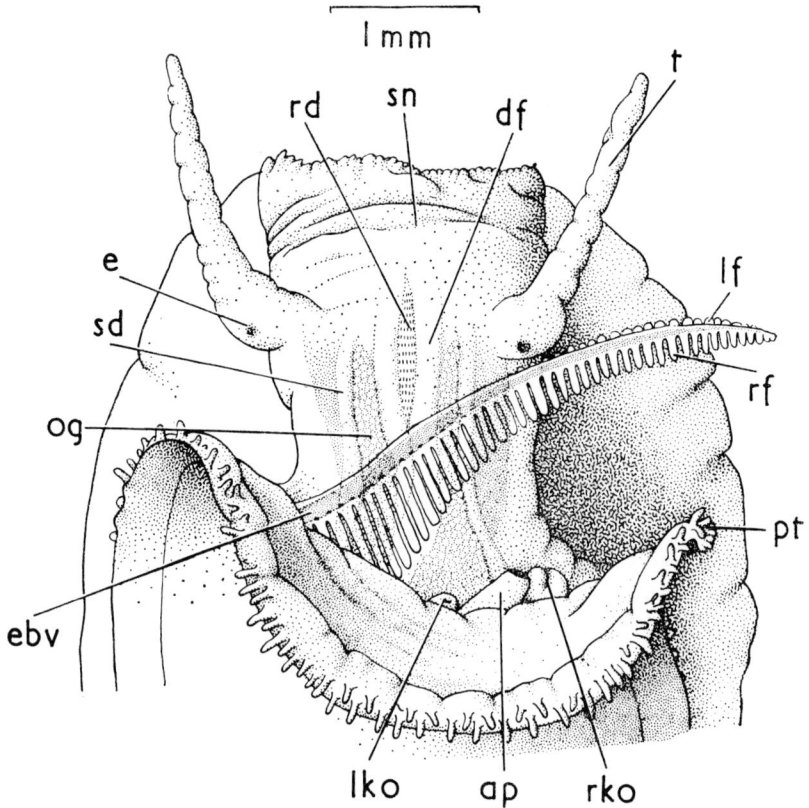

FIG. 250.—*Acmaea tessulata:* the animal has been removed from the shell and the roof of the nuchal cavity folded back to show the contents; in dorsal view.

ap, anal papilla; df, dorsal fold of oesophagus, seen by transparency; e, eye; ebv, efferent branchial vessel; lf, ctenidial leaflet on left side of axis; lko, opening of left kidney; og, anterior end of oesophageal gland, seen by transparency; pt, pallial tentacle; rd, radula, seen by transparency; rf, ctenidial leaflet on right side of axis; rko, opening of right kidney; sd, left salivary duct, seen by transparency; sn, snout; t, tentacle.

our seas. In these genera the overgrowth of the mantle edge (which is concurrent with the development of the cap-shaped shell of the limpet) encloses a pallial groove which encircles the foot and is continuous with the pallial cavity, or nuchal cavity, anteriorly. The groove is not restricted by the development of an epipodium as in zeugobranchs, and the anterior mantle cavity is smaller than in those forms. The mantle margin is typically fringed with tentacles which may compensate for the lack of epipodial tentacles. The direction of the water current through the nuchal cavity and pallial grooves differs from genus to genus. Only *Acmaea* (figs. 250, 251, 252) has a ctenidium. It is a left one (ct), which is elongated and held horizontally across the nuchal cavity, so that the tip can sometimes be seen projecting from the right anterior margin of the shell as the animal creeps. When the limpet comes to rest the muscles of the ctenidial axis contract and withdraw the gill into the cavity; it may be reduced to about half of its original length. The respiratory current is drawn into the nuchal cavity on

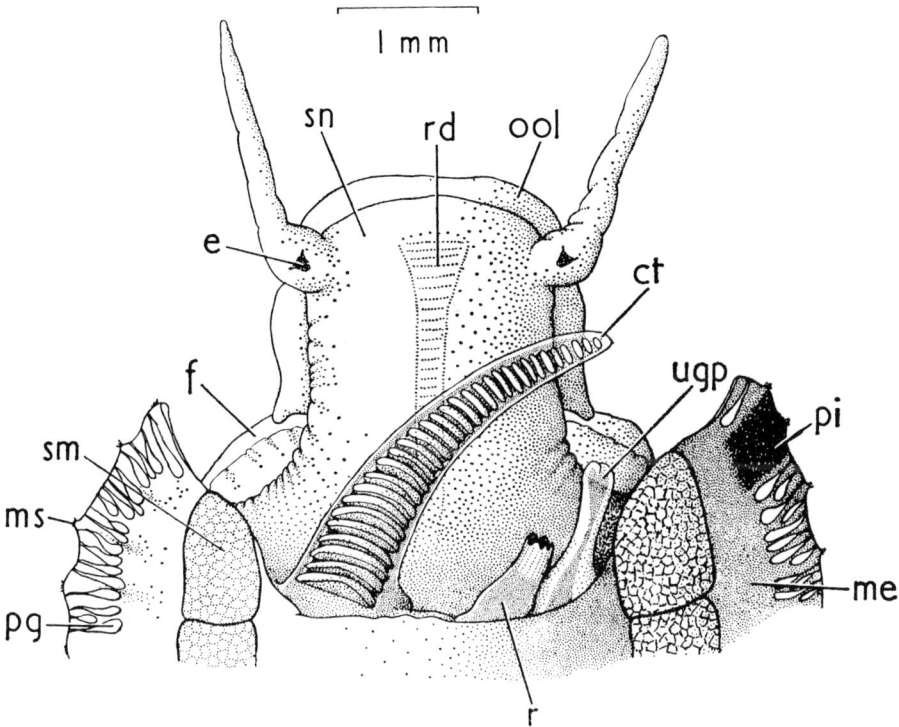

FIG. 251.—*Acmaea virginea:* the animal has been removed from the shell and the roof of the nuchal cavity cut
away to show the contents; in dorsal view.
    ct, ctenidium; e, eye; f, foot; me, mantle edge; ms, sensory organ on mantle edge; ool, outer lobe of outer
lip; pg, repugnatorial pallial gland; pi, pigment patch in edge of mantle skirt; r, rectum; rd, radula, seen by
transparency; sm, shell muscle; sn, snout; ugp, urinogenital papilla (opening of right kidney).

the left side and passes between the ctenidial filaments towards the right and posteriorly, and
the cilia of the epithelium lining the floor of the cavity lead particulate matter in the same
direction. Along each pallial groove is a powerful ciliary current beating posteriorly (instead of
anteriorly as in *Emarginula* and *Puncturella*) which carries material from the nuchal groove to
the mid-line posteriorly where the two currents meet (fig. 73). All sediment, faeces, renal and
perhaps reproductive products, leave the limpet by this route and are expelled from under
the posterior edge of the shell.
    The genera *Patella* and *Patina* have pallial gills (pag, figs. 50, 253, 258, 259B) set on the outer
edge of the pallial groove. They hang down from its roof and encircle the head-foot in *Patella*,
but in *Patina* the ring is incomplete anteriorly. In *Patina* the exhalant current from the mantle
cavity conveying away all material leaves the body from the right side of the nuchal cavity.
The inhalant current is drawn into the pallial groove in all regions where gills occur, by the
cilia on the gills. Sediment entering with the current is entrapped in mucus from a glandular
region around the foot, and passes anteriorly along the pallial groove with the water flow, and
through the nuchal cavity to leave on the right. *Patella*, with a complete circlet of pallial gills,

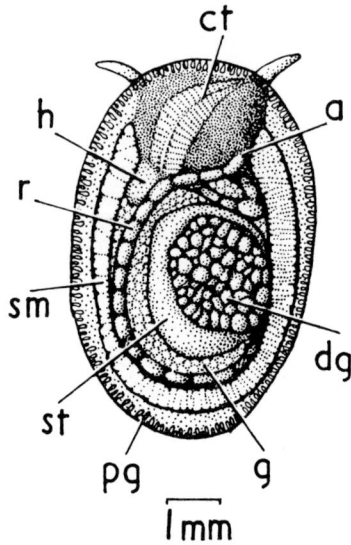

FIG. 252.—*Acmaea virginea:* animal removed from shell, dorsal view.
a, anus, seen through mantle skirt; ct, ctenidium, seen through mantle skirt; dg, digestive gland; g, gonad; h, heart in pericardial cavity; pg, repugnatorial pallial glands; r, rectum, with faecal pellets; sm, shell muscle; st, stomach.

has yet another arrangement of currents in the mantle cavity and of disposal of material from the body (figs. 48C, 259B). A gentle inhalant current is drawn in all around the margin of the mantle by the cilia on the gills, and a weak current leaves the mantle cavity ventral to this inhalant flow. Particulate material is directed from the body by cilia on the sides of the foot and on the roof of the nuchal cavity, which conduct it to a position midway along the right pallial groove. Here it accumulates and is expelled from time to time, not by ciliary action, but by sharp contractions of the shell muscle. Limpets removed from rocks often leave an accumulation of faecal pellets in this situation which have been collected from the nuchal cavity and await expulsion.The forceful expulsive movement compensates for the lack of a glandular streak around the foot of *Patella* except in young individuals, and may be a decisive factor in enabling this limpet to live in situations where there is considerable sediment. The ciliary currents in this process are strongly developed only in limpets less than 10 mm in length, in which the transverse, left to right current through the nuchal cavity is marked. *Lepeta* and *Propilidium* (fig. 254) have no gills. A flow of water enters the nuchal cavity and is directed posteriorly by cilia on the walls of the pallial groove to leave the body posteriorly.

Limpets of the genus *Patella* are amongst the most ubiquitous animals on the rocky shore and the most abundant representative of the Patellacea in the British Isles. The stout, conical shell, concealing the soft tissues of the body and ribbed from apex to margin, is a familiar sight on rocks and boulders to which the limpet clings with proverbial strength [see p. 583]. Its grip is increased by nearby vibrations or when it is touched. Otherwise, when water covers the animal, whether it be stationary or on the move, the edge of the shell is raised slightly and there is a gentle, respiratory flow of water to the circlet of gills (fig. 48C). Empty shells of limpets are common amongst the jetsam on the beach, on cliff ledges where gulls have been

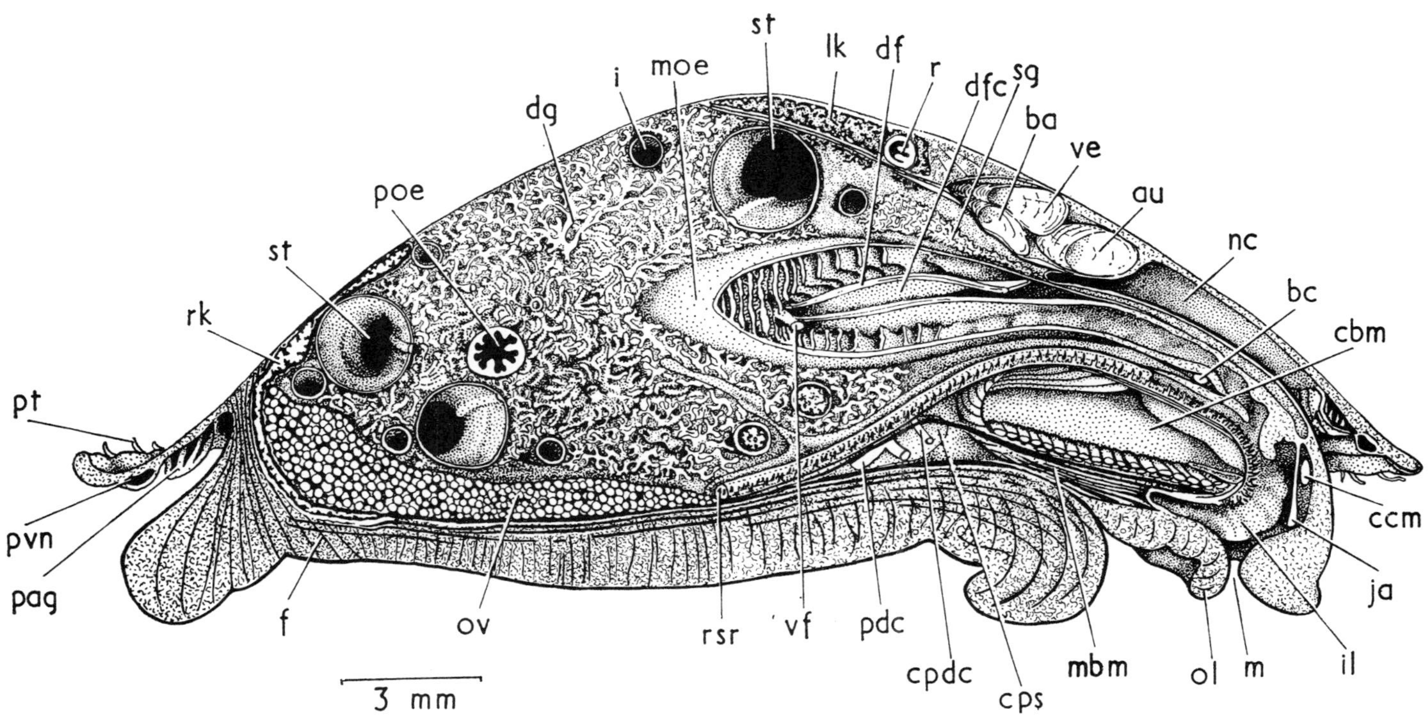

FIG. 253.—*Patella vulgata*: sagittal half seen from the cut surface.

au, auricle; ba, bulb of aorta; bc, buccal commissure; cbm, cartilage of buccal mass; ccm, cerebral commissure; cpdc, cerebropedal connective; cps, cephalopedal blood sinus; df, dorsal fold of oesophagus; dfc, dorsal food channel; dg, digestive gland; f, foot; i, intestine, which is cut (but not labelled) 7 other times; il, inner lip; ja, jaw; lk, left kidney; m, mouth; mbm, muscles of buccal mass; moe, mid-oesophagus; nc, nuchal cavity; ol, outer lip; ov, ovary; pag, pallial gill; pdc, pedal cord; poe, posterior oesophagus; pt, pallial tentacle; pvn, pallial vein; r, rectum; rk, right kidney; rsr, radular sac in radular sinus; sg, salivary gland; st, stomach also cut (but unlabelled) once more; ve, ventricle; vf, ventral fold of oesophagus.

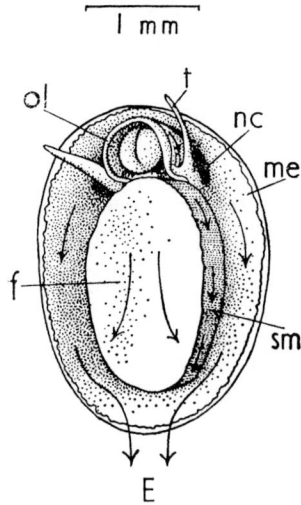

FIG. 254.—*Propilidium exiguum:* animal in ventral view. Arrows show direction of water currents.
E, exhalant current; f, foot; me, mantle edge; nc, nuchal cavity; ol, outer lip; sm, shell muscle; t, tentacle.

feeding, and in the kitchen middens of olden times when, more than now, this proso-branch was used as food. The commonest species is *Patella vulgata*. It may be found on both rough and smooth rock as well as on pebbles which are not subjected to too much movement, and it is frequent on all angles of slope. It occurs in conditions ranging from considerable exposure to those offering most shelter from the surf, and extends from high on the beach to MLWS, where the population density falls off rapidly. The limpet does not particularly favour sheltered regions with a thick growth of fucoids, and in areas covered with a felt of smaller weeds, such as *Enteromorpha, Porphyra* and *Bangia,* it clusters in clearings amongst the algae. The level which these limpets reach on the upper shore is dependent upon the exposure and aspect of the locality: in an extremely well splashed, shaded situation Evans (1947) found them to a height of 8 ft above EHWS. The presence of algal food does not in itself lead to the colonization of the higher levels (Lewis, 1954a), though this is the factor emphasized by Das & Seshappa (1947). In particular situations the seasonal variation in sunshine causes a downward migration of limpets in spring and summer and a return upward movement in autumn and winter, though the upward movement is not confined to individuals which originally occupied the high levels (Lewis, 1954a). *P. vulgata* invades estuaries where there is sufficient rock or stone on which it may live, and in such muddy habitats, with abundant silt and detritus, the growth rate is rapid (Fischer-Piette, 1948). Pollution also increases the size of the shell and the rate of growth. The limpet can endure low salinities: in the Rance, Fischer (1931) found that it dies out only when the salinity is reduced to 3–1‰.

There are two other species of *Patella* on British shores, *P. intermedia* and *P. aspera*. The former is less tolerant of sheltered conditions than *P. vulgata* and is usually found in exposed places (Southward & Orton, 1954) and only along the S and SW shores of England. It may be mixed with *P. vulgata* on bare rock, but does not follow it into sheltered or polluted waters, nor on to boulders or shingle. Its lower limit is vague: it is not found below MLWS and in most places dies out between MLWN and MLWS. It extends up the shore to MHWN (Evans, 1947), and this upper limit may be raised by splash to EHWS, or at this level it may be found in Lithothamnion-encrusted pools. P. aspera is characteristic of exposed reefs on the lower part of the shore where fucoids are poor or absent and Corallina is the chief alga (Eslick, 1940; Evans, 1947); here it particularly favours shallow pools or runnels of water, and it also frequents overhanging rocks and the banks of gullies. This species is common down to the lowest level of the intertidal area and appears to extend into the sublittoral. On very exposed, wave-beaten surfaces it may extend up the shore to MTL, though on more sheltered reefs it is found only to MLWN. In the Lithothamnion pools it is often more abundant than P. vulgata and P. intermedia, even above high water.

Along some parts of the S and SW coast of England the three species of Patella can be found in quantity within an area of a few square metres, and with a little experience they can be readily identified in the field (Table 14). If they are removed from the rock and viewed ventrally for comparison it will be seen that the colour of the foot is greyish-green in P. vulgata and P. intermedia, though it is darker in the latter species; in P. aspera it is light in colour, orange to creamy pink or cream. The marginal tentacles of the mantle also differ. They are transparent in P. vulgata and coloured like the background, a feature by which this species is readily distinguished; a cream colour in P. aspera, and they are most conspicuous in P. intermedia in which they are opaque cream or white like the tentacles of Sagartia. If the limpets are removed from their shells certain differences in the inner surface of the shell (including the head scar) are noticeable. In P. vulgata there is a nacreous lining which may show a green or blue iridescence, and in older cells a yellowish cast; around the margin of young shells there may be red, orange, yellow, brown or perhaps green rays. The head scar, of variable size, may be a silvery nacreous area, or an opaque white or grey. In P. aspera the inner surface of the shell has a porcellanous appearance and may be white, and the head scar, in contrast, orange or creamy yellow; marginal rays are never conspicuous. These rays, however, are a distinctive feature of P. intermedia since they are dark in colour. The whole inside of the shell is usually dark in this species and the head scar is dusky cream with perhaps an orange tinge. The species may also be separated on their breeding season which has been recorded by Pelseneer (1911), Orton (1946) and Fischer-Piette (1946) (see Table 14).

Over a hundred years ago it was realized that the British species of Patella could be distinguished by the characteristics of the radula (Woodward, 1851–56; Cooke, 1917). If the radulae are removed and the length of each measured against that of the shell, the range of the means of the ratio (length of radula/length of shell) separates P. aspera (1·05–1·15) from the other two species (1·51–2·10) (p. 166), and this difference is supported by differences in the pluricuspid teeth of the radula (fig. 255). These teeth have each three cusps with characteristic size and shape associated with the species, and there is an incipient fourth cusp in P. vulgata and P. aspera. The appearance of the three cusps in P. aspera is distinct from that of the other two species, which are much more closely related in this character, and are also more variable (Evans, 1953). In P. aspera the innermost cusp is the smallest, and if the other two are of unequal size it is the outer which is larger. In P. vulgata and P. intermedia the three cusps are typically unequal, the innermost is the smallest and the median largest, though there are variants in each species in which the middle and outer cusps approach one another in size. In both species the median cusp has unequal sides, the inner being shorter than the outer. In P. vulgata the cusps are sharply pointed, whereas in P. intermedia they taper to blunt tips. The shape of the pluricuspid is so variable that on this basis Evans (1953), dealing mainly

TABLE 14

Feature	Patella vulgata	Patella intermedia	Patella aspera
Colour of foot . .	Grey-green	Dark grey-green	Cream to orange
Marginal tentacles .	Transparent, no white	Opaque cream or white	Cream
Inner surface of shell .	Green or blue nacre, yellower in old shells	Dark	Porcellanous, white
Marginal rays . .	Red-brown at margin only	Dark, conspicuous	Not conspicuous
Head scar . . .	Silvery, opaque white	Dusky cream	Cream to orange
$\dfrac{\text{Length of radula}}{\text{Length of shell}}$.	Range of means 1·51–1·75	1·60–2·10	1·05–1·15
$\dfrac{\text{Length of radula}}{\sqrt[3]{\text{Shell volume}}}$.	3·2–4·8 Mean 3·90	4·1–5·7 Mean 4·71	2·5–3·5 Mean 2·90
Pluricuspid tooth .	See	fig. 255,	p. 463
Microstructure of shell fibres	Stout and thick	Intermediate	Long and narrow
Relation to weed (see p. 464)	On *Fucus*	On *Himanthalia*	On *Himanthalia*
Urceolaria (see p. 464) .	Without zooxanthellae	With zooxanthellae	With zooxanthellae
Breeding period .	Winter; max. Oct.– Dec. Rests spring, early summer	Summer Rests Jan. (perhaps)	Summer Rests Jan.
Posterior shell outline	rounded	angulated	angulated

with the limpets along the S coast, subdivided each species into groups. However, these will not be dealt with here and only the most frequent types of pluricuspid are shown in fig. 255.

Some authors describe differences in the shape and sculpturing of the outer surface of the shell of the three species. Jeffreys (1865) stated that in *P. aspera* the shell is depressed, with fine sharp ribs and the apex near the anterior end, and that in *P. intermedia* it is smaller, flatter and more oval than in *P. vulgata,* and has fewer ribs. However, Fischer-Piette (1935*b*) and later workers noted the inconstancy of these characters, which appear to give little reliable help in the separation of the species. Some variations are undoubtedly phenotypic. Thus specimens of *P. vulgata* living in estuaries and under thick growths of weed have a thinner, more regular shell with finer external markings than individuals in exposed positions.

FIG. 255.—Outlines of the pluricuspid radular teeth of limpets, after Evans. A, *Patella vulgata*; B, *Patella intermedia*; C, *Patella aspera.*

The microstructure of the shell of the limpet shows a fibrous texture in sections viewed under low magnification. The fibres run through the thickness of the shell, perpendicular to the internal and external surfaces. They are stout and thick in *P. vulgata*, long and narrow in *P. aspera* and intermediate in *P. intermedia* (Lhoste, 1944, 1946).

These three species of *Patella* also occur on the coast of Brittany where they are easily distinguishable (Fischer-Piette, 1935b), but further S, along the Basque coast, a number of intermediate types are found, making identification more difficult, and here some features of the Mediterranean species *P. caerulea* are also displayed. Fischer-Piette (1948) held that *P. vulgata* and *P. aspera* maintain their identity in this area as they do in Brittany, and that all transitional types are forms of *P. intermedia*. From his survey he concluded that the limpets may have arisen from a polymorphic species, 'Patella depressa-caerulea' which gave rise to *vulgata* and *aspera*, and from which *intermedia* emerges as a distinct species to the N and *caerulea* to the S. He made a far less thorough investigation of the limpets along part of our Channel coast, and found the three species unmistakable in the W, but their distinction obscured on travelling E. On the Isle of Wight he found the characteristics of the limpets approaching those of the Basque coast, even to the occasional appearance of features of the Mediterranean form, *caerulea*. A more detailed investigation of the region was made by Evans (1953) who confirmed that transitional types of shell and pluricuspid tooth do occur with increasing frequency from the W to the Isle of Wight. E of this island neither *intermedia* nor *aspera* is found, which agrees with their distribution on the channel coast of France (Fischer-Piette, 1941b). However, Evans (1953) found *vulgata* at least as variable as *intermedia* on the S coast of England and suggested that *aspera*, *intermedia* and also *caerulea* are more closely related to *vulgata* than to each other. Indeed he found that the *caerulea* facies is not displayed by *intermedia* but by *vulgata*. So he discarded Fischer-Piette's hypothesis of a *depressa-caerulea* root-stock and tentatively substituted a *vulgata*-like type. From this he envisaged the following stages in the emergence of the species: at Bognor reef, E of the Isle of Wight, the *vulgata* population on wave-beaten rocks, but not on the sheltered pier, shows some characters of *aspera* and *intermedia*, though neither of these is present as distinct species. The population on the Isle of Wight shows, according to Evans (1953), a further stage in evolution since *aspera* is isolating itself as a true species; it is only much further W (beyond Torquay) that undoubted *intermedia* form a significant fraction of the limpets; *caerulea* has never separated as a species in the British Isles although some of its features are indicated in the *vulgata* stock.

Limpets are undoubtedly very variable, for along the eastern part of the S coast where only *P. vulgata* persists this species shows variations which are different from those on the Isle of Wight, and not in the direction of *aspera, intermedia* or *caerulea* (Evans, 1953). The occurrence of only *vulgata* in this area is curious. Presumably ecological conditions favour only this species in both the adult and larval stage, and they may also be responsible for the variations which are different from those further W. On the Dutch coast *P. vulgata* is again the only species (Lucas, 1954), though all three species occurring in the Channel are occasionally cast up on weeds. Limpets on *Fucus* are invariably *P. vulgata,* all those adhering to *Himanthalia* are either *P. intermedia* or *P. aspera,* neither of which persist. Despite these observations no correlation between weed and any stage in the life history of the limpet has been recorded on the Channel coasts from which these presumably came.

Parasites and symbionts often show specific preferences, and it is interesting to find two physiological races of the peritrich protozoon, *Urceolaria patellae* (Brouardel, 1948) related to the species of *Patella*. These ciliates are epizoic and may even be ectoparasites (Hyman, 1940). Two strains occur, one with zooxanthellae. The strain without the symbiotic cells settles for preference on the gills of *P. vulgata:* the other avoids this species and prefers *P. aspera* and *P. intermedia*. The differentiation of this species suggests that speciation of *Patella* is no recent event. It would be interesting to see how *Urceolaria patellae* behaves in regions like the Isle of Wight and Basque coast.

Limpets move about when the tide is in and the sea not too rough, and browse on detritus and algal growth; under thick fucoid covering, and in other sheltered localities when the weather is damp individuals may be found on the move during the day, after the tide has fallen. Numbers browse over damp rocks at night. It is generally accepted that each medium or large individual returns after feeding to a so-called home, that is, a definite position on the rock where it remains stationary. The broad flat foot fixes the animal to its home and the elliptical edges of the conical shell fit the particular patch of rock so accurately that each ridge is covered by an exactly fitting indentation of the shell. This means that every time the limpet settles its orientation is the same and its shell, as it grows at the edge, fits into the irregularities of the home. Consequently when the shell is pulled on to the exposed rock at the ebb tide on a sunny day, little or no evaporation can take place under its edge and the soft tissues are sealed off from the external environment. The limpet remains motionless, the head raised from the rock and withdrawn into the nuchal cavity, the tentacles relaxed and the foot spread over the substratum, though not to grip it tightly. At any near vibration the shell muscles contract vigorously, clamping the animal to the rock. The full strength of pull of the pedal muscles has been estimated as 3.5 kg/cm^2 (Fischer, 1948): it is their force (together with the fact that the conical shell offers little resistance) which secures the animal against the action of the waves in the most exposed situations. On soft rock the surface of the home is worn away by the limpet and a scar is made. These scars are often abandoned by the individuals which formed them, for there is usually some movement of limpets from one place to another; this is more frequent on smooth rock, especially if damp (Orton, 1929b), than on uneven surfaces and on rocks thickly covered with *Balanus* where the number of homing places is limited. Sometimes, on smooth rocks, limpets will return to rest anywhere within a specific area of a few square inches. Jones (1948) observed the movements of a number of limpets (*P. vulgata*) on smooth, flat, limestone rock which was bare of fucoids and *Balanus* and dry at low tide: many of them had home scars suggesting that they had been there for some time. He found that each week some of the 182 individuals had moved to a new site, and after 6 months only 9 were occupying their original area. There was no obvious common direction of migration and the distance moved was usually only a few yards, though one individual moved 30 yards down the shore.

Although homing is usually regarded as part of normal limpet behaviour it appears to be absent in some species of *Acmaea* (Villee & Groody, 1940) though Hewatt (1940) recorded examples of it in a species in which Villee & Groody failed to substantiate its occurrence. The mechanism by which homing occurs is unknown. The capacity to home seems to vary with the circumstances in which the limpets live—nature of rock surface, vegetation and the like—and with age, young limpets being more given to vagrancy than adults. This perhaps explains the difference between the results of Villee & Groody and those of Hewatt. Morgan (1894) showed that the percentage of limpets successfully returning to their home fell off greatly when the distance over which they were experimentally moved exceeded their normal feeding range. This and other experiments suggest that the animals rely either on the trails which they leave as they go out from their home (Piéron, 1909; Hewatt, 1940), or on a knowledge and memory of their territory. The sense-data which the animals use are probably olfactory and are appreciated predominantly by the marginal pallial tentacles (Davis, 1895). Similar statements may be made about *Acmaea, Diodora* and the pulmonate limpet *Siphonaria* (Willcox, 1950b).

When *Patella* feeds it moves systematically around its home rasping with the radula anything which it happens to meet. Consequently there is considerable variation in diet. The head and tentacles protrude from the margin of the shell and the anterior part of the body swings slowly from side to side in a pendulum action, directing the radular teeth across the substratum. Where the surface permits, as on rocks with a layer of sediment, these feeding tracks are often obvious and it can be seen that the limpet frequently follows the outward track back home again, or it may return by a different route. The whole journey will cover two feet or so, sometimes following a zigzag path. Small limpets occasionally home on the shells of older ones and, if removed and placed a short distance away, will return. Individuals removed several feet from their scars do not appear to make their way home again. Individuals living in the *Balanus* zone collect diatoms, silt and debris, others living amongst *Enteromorpha, Ascophyllum* and *Fucus* fill the gut with algae. The larger algae can be utilized as food since the limpet has enzymes capable of digesting laminarin (Dr V. C. Barry, quoted by Jones, 1948), and fucoidin and of de-esterifying their carbohydrate sulphates (Dodgson & Spencer, 1954). Fischer-Piette (1948) found that specimens of *P. vulgata* living under the cover of fucoids in sheltered localities at St Malo, Dinard, have a more rapid growth rate, a higher ratio between the volume of the animal and that of the shell, and attain a larger size and earlier sexual maturity than individuals on exposed rocks with barnacles. He did not believe that the quieter waters can be the factor which is directly responsible for these differences, since waves and currents aid growth (Hatton, 1938; Fischer-Piette, 1948), and although pollution and silting are favourable for nutrition, they do not affect the onset of sexual maturity. He regarded *Fucus* as the important factor, especially as a food rather than a shelter. He recognized the fucoid mantle as a mechanical barrier to the settlement of the larvae on the underlying rocks, thus accounting for the difference in the density of adults under *Fucus* and on bare rock with *Balanus*. Large concentrations of limpets were found in some localities between belts of *Fucus* and bare rocks with barnacles, the size of the individuals being intermediate between those of these two zones. Fischer-Piette likened these concentrations to those of *Nucella* which may be found bordering a mussel bed, and stated that where the limpets are most abundant the *Fucus* is destroyed by their attack.

Jones (1948), who studied the limpet population in localities at Port St Mary, Isle of Man, found that individuals living on rock which is bare of both fucoids and *Balanus* are still larger than those amongst *Fucus*. The population density is highest on rocks with barnacles, where it may attain 240/sq m, and lowest on bare rock: high density is correlated with low mean size and low density with high mean size. Samples from under beds of thick fucoids have a low density and a medium average size, whilst those among thin *Fucus serratus* have a fairly high density and a fairly high mean size, resembling the population described by Fischer-Piette. Jones concluded that although the presence of fucoids favours the growth of *Patella*, the determining factor is the amount of feeding space over which the limpet can browse without competing with others. This space is perhaps restricted by the stems of algae in thick fucoid zones where the limpets do not seem to make the best use of the available food (Jones, 1948), and on rocks covered with barnacles it is limited by interspecific competition (Hatton, 1938; Evans, 1947). There must, of course, be a critical size at which population density begins to affect growth rate, though nothing is known of this. Das & Seshappa (1947) found that on the rocky shore at Cullercoats, Northumberland, the population density decreases slowly from LWNT to HWST, and that generally the larger limpets predominate at higher levels, even though they are covered by the tide for a shorter time and so their feeding time is less. Again, this may suggest that differences in the mean size between limpets from the lower and upper shore can be explained by the feeding space available, though the possibility of an upshore migration of the larger limpets to areas where competition is less cannot be neglected. Orton (1929b) held that in some high-water situations adults occur where younger and smaller individuals, less than 25 mm, could not exist.

If a limpet (*Patella vulgata*), motionless on dry rock, be splashed with sea water, the front edge of the shell is lifted and the head and tentacles extended (Arnold, 1957). With repetition of this treatment it moves away. However, if fresh water trickles down the shell and comes in contact with the mantle edge and pallial tentacles, the limpet withdraws and clamps the shell on to the substratum. These responses are effected without the entry of water to the mantle cavity where the osphradia are situated: the perception of salinity is by receptors in the mantle fringe and cephalic tentacles. Arnold has shown that where limpets are similarly stimulated by sea water of reduced salinity the magnitude of the response is approximately correlated with the salinity. This is especially obvious in limpets living high on the shore which are much less sensitive to reduced salinity than those from low tidal level. They are stimulated to creep by half-normal sea water or less whereas the others fail to creep when the salinity is 80% normal or less. Similarly, limpets from a high level are more readily stimulated by the splash of normal sea water than those from a low level. There appears, indeed, to be an increasing tolerance of low salinities from the lower to the upper limit of distribution of the species on the shore, and also a more immediate and larger response to splash by the high tide individuals which are exposed for longer periods. It may be that the degree of sensitivity to salinity limits the position of the individual limpet on the shore; or the same potentialities may be possessed by all individuals but they become acclimatized to a particular level on the beach.

It has been estimated that during the first year of life an area of about 75 sq cm/cc of limpet is required to provide sufficient food for the maintenance of an individual. Moore (1938c) calculated this from regions in a thick felt composed of *Enteromorpha* sp., *Porphyra umbilicalis* and other algae, kept clear of new growth by browsing limpets which homed there. Limpets devour the spores of algae as they settle on the rock and before they have a chance of establishing themselves as sporelings, and will destroy, though slowly, a matting of fine weeds such as *Enteromorpha*. These green algae (which often appear first on a cleared area) prevent the settling of barnacles, and may provide protection for the fucoids to grow until they are relatively immune from the attack of limpets. The established plants seem to withstand the depredations of the limpets for some time, and Jones (1948) held that the regression of a belt of fucoids is slow, for the mollusc does not make a determined attack on it, but browses along the edge and eats the alga when it comes against it. However, he did not encounter anything approaching the population density described by Fischer-Piette in the immediate proximity of belts of fucoids.

The population of *Patella vulgata* on the shore is apparently recruited from spat settling with maximal abundance on rocks near LWNT (Jones, 1948), and in pools and moist crevices at higher levels (Orton, 1929b). The planktonic larva settles when it is not more than 0·2 mm long (Smith, 1935). The mortality of such minute forms is likely to be higher on bare rock which dries as the tide ebbs than on rock covered with *Balanus* and, more especially, with scattered plants of *Fucus*, which remain comparatively moist. Jones (1948) found the highest density of settlement in shallow pools, where it may reach a number of 150–250/sq m, and from the pools the spat moves out, at a size of 3 mm or more, to populate surrounding areas. These observations are contrary to those of Hatton (1938) who held that the newly settled spat will not remain in pools, and described a denser settling on exposed surfaces than on sheltered ones. From experiments he deduced that the alternation of immersion and exposure seems to be necessary for the very young individuals. The rate of growth of the young limpet varies considerably. Russell (1909) estimated that limpets in certain established populations in Scotland attain a length of about 29 mm in the first year, during which time they reach sexual maturity, and probable sizes at the end of the second and subsequent years are 38 mm, 44 mm, 48 mm, 53 mm. Shells over 50 mm may be considered more than 5 years old. This growth rate is slow as compared with that of the first limpets which settled on concrete piles of a new wharf at Plymouth during very favourable weather (Orton, 1928b). They measured 26–35 mm in length at the end of the first year and 53 mm at the end of the second; this second measurement approaches that given by Fischer-Piette for the largest individuals at Dinard. He found that longevity is greatest where growth is slowest and *vice versa*: limpets living in the most normal habitat, on rock with cirripedes, though not under the influence of strong currents, lived up to 15 years, and those under *Fucus* only about 3 years (1941a). [See chapter 31.]

The shells of limpets which live at a high level on the shore are, at the adult stage, higher than those of individuals near low water level or in rock pools (Russell, 1907). Russell suggested that the difference is due to exposure. Orton (1929b; 1932) correlated these differences with differences in the degree of exposure to desiccation. The limpets near low water remain fairly damp during the short time they are left by the tide, whereas those at high water are uncovered for the greater part of the day and their surroundings will dry up seriously, especially on a south-facing slope, unless the rock is covered by weed. Orton suggested that the tall spire with a relatively narrow base is produced by the limpet which holds close to the rock to prevent the desiccation of its tissues: the continued downpull of the muscles during the hours of exposure pulls in the mantle skirt which is responsible for new growth at the margin of the shell. Consequently a smaller peripheral increment of growth will be made by these individuals than by others living lower on the shore, or in rock pools, in which the shell muscles are relaxed. Some limpets living at high levels show a more gradual steepening of the shell than others, and this may indicate that they started life lower on the shore and ascended to the higher and more exposed levels as the thickening shell made it possible (Moore, 1934). Very occasionally a shell is encircled by a ledge at a certain distance from the apex (fig. 259A), which is due to an abrupt alteration in the angle of growth and indicates a sudden change from a dry to a damp habitat (Moore, 1934). Flat shells are comparatively thin, but very tall ones are thick especially in the region of the apex where they may attain a thickness of about 1 cm. This thickening gives protection against exposure.

In the Pacific our common limpet, *Patella*, is not represented and its position on the shore is occupied by species of the related genus *Acmaea*. These limpets occur in great numbers in a variety of situations, approaching the size of *Patella* and clinging to the rock with no less tenacity. They retain a single bipectinate gill. *Acmaea scutum* and *A. cassis* (= *pelta*) occur throughout the intertidal range on the Oregon coast (Shotwell, 1950) and, contrary to *Patella*, the larger individuals by volume are more abundant on the lower parts of the shore, and no relationship between height of shell and position on the shore can be established. Shotwell, using the findings of Abé (1931), suggested that the smaller limpets have a relatively greater mantle cavity, and therefore retain a larger amount of water when uncovered by the tide and that this protection against desiccation enables them to survive high on the beach. Another species, *A. limatula*, also with a wide vertical distribution along the coast of Southern California has no correlated difference in size. However, individuals at higher levels have thicker and heavier shells and a relatively larger extra-pallial space separating the shell from the body tissues than those on the lower part of the shore. This space is filled with fluid which may act as a reservoir against water loss by evaporation during the hours of exposure (Segal, 1956).

On British shores the genus *Acmaea* is represented by two species; neither is abundant and both occur only from near low water to a depth of a few fathoms. The two differ anatomically in several points though these do not appear to be correlated with differences in habitat. *A. virginea*, the pink-rayed limpet, is to be found on all rocky coasts, while *A. tessulata*, the tortoiseshell limpet, has a more restricted range which is N of the Humber and N Wales, also in N Ireland. This species also occurs on the Atlantic Coast of America and ranges from the Arctic Seas to Long Island Sound, New York. It is the larger and may attain more than an inch in length, though this is rare. Both feed on diatoms and encrusting algal growth, and spend long periods on the same position on the rock, leaving it only for feeding excursions. They move slowly, Willcox (1905a) recording a maximal speed of 3 in/min for *A. tessulata*. She also stated that in autumn it retires below the low water mark. Lilly *et al.* (1953) have shown that *A. virginea*, whilst not much affected by current strength, has a preference for the tops of boulders, probably to avoid silt.

The blue-rayed limpet, *Patina pellucida*, sometimes known as peacock's feathers, is associated with species of *Laminaria* and other brown weeds near low water; the younger stages may also be found on rocks and stones favouring situations where the movement of the water is neither too great nor too little (Ebling *et al.*, 1948). It is much smaller in size than the common limpet and the approximately conical shell, which is horn coloured or brown, has conspicuous blue rays radiating backwards from the summit; in certain lights these are an iridescent green. There are two ecotypes of the species each with characteristics of habitat and structure, the latter affecting mainly the shell (figs. 33, 256). One variety, *pellucida*, which lives on the fronds of *Laminaria* has a smooth, elongated oval and low shell, with the summit placed near the anterior end. It is normally devoid of either epiphytic or epizoic growth, so that the rays, 2–8 in number, are not concealed. The shell is translucent and the soft underlying tissues of the body are pigmented and visible by transparency. The second variety, *laevis*, lives in caves in the holdfasts of *Laminaria* and, in contrast, has a rough and

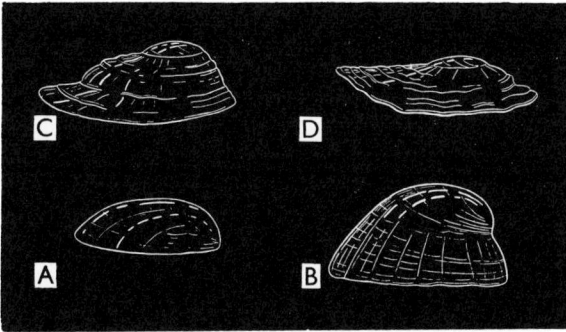

FIG. 256.—*Patina pellucida:* profiles of 4 types of shell, × about 3. A, young *pellucida* from frond of *Laminaria;* B, old *pellucida* from frond of *Laminaria;* C, D, 2 shapes of *laevis* from *Laminaria* holdfast.

usually high shell which is approximately circular in outline with a central summit, though the proportions are more variable than in the variety *pellucida.* The shell is pale brown or greenish brown in colour and opaque, with 2–46 blue rays, and there are also red-brown rays which alternate with them; the whole surface, however, may be concealed with growths of various kinds. The soft tissues of the body of this variety are not pigmented. For many years these two ecotypes were considered to be different species, but a study of the life history proved otherwise (Graham & Fretter, 1947). The limpets breed maximally in winter and spring, and numbers of planktonic larvae are settling in May as spat about 2 mm long; they are all alike with the characteristics of *pellucida.* These characters persist into the adult stage in all individuals except those which migrate into the holdfasts of *Laminaria.* The limpets on the fronds feed on the substance of the weed or on diatoms and similar minute material which settles there. As they grow they retain the thin shell characteristic of the newly metamor- phosed animal. Because of exposure to light the soft tissues become pigmented and with exposure to wave action the shell develops into a rather low structure which reduces the risk of it being swept away. *Patina* which are carried by water currents or migrate into the space inside the holdfast of *Laminaria* excavate caves directly under the stipe or sometimes on the outside of the lowest part of the stipe. They grow a thick stony, calcareous shell which fits into the more limited space available and is high because it is sheltered from the waves. The prominent red lines which develop may be due to the difference in diet. The number of limpets which ultimately end in the holdfasts is greater than the number which fell there by chance as spat and those that migrate to this habitat show an abrupt change in shell structure which marks the time of their migration. The oldest part of the shell has the proportions, colouring and texture of the thin variety *pellucida,* the more recently formed part the characteristics of the variety *laevis.* So far as the soft parts of the animals are concerned there are only trifling differences between the two. For instance, in both, the general plan of the gut coils is the same though their exact disposition varies so as to give in the one a lay-out elongated in one direction and in the other in a direction at right angles to that. The radula has apparently the same arrangement in both thick- and thin-shelled varieties and bears a close similarity to that of *Patella* but there is a greater degree of wear in those animals in the holdfast. They feed on the tissues of the stipe which they undercut so that their diet is mainly undiluted *Laminaria.*

The young which settle as spat in May grow to a length of about 5 mm in the following autumn and about 10 mm after a year of sedentary life, becoming sexually mature at 5 mm length. The majority die after a settled life of 12 months and nearly all those that live longer belong to the variety *laevis,* and have the more sheltered life. These figures concerning length of life are comparable with those for *Patella vulgata* (Fischer-Piette, 1948) growing at a maximal rate, e.g. under *Fucus.*

The movements of the blue-rayed limpet on the fronds of *Laminaria* are associated with the life cycle of the weed. *Laminaria digitata* has a long fruiting season, with maxima in spring and autumn. After the autumn fruiting the distal half of the frond gradually disintegrates and breaks away, and a certain amount of breakdown occurs after the spring fruiting. During the autumn there is a movement of limpets down the frond giving a concentration on its basal half. The migration takes place about the time when the animals are becoming sexually mature and is of very great importance as a means of securing a base on which they may survive the winter. The stimulus for migration may be chemical changes in the plant tissues, for these are eaten by the limpet. In spring it seems that a certain amount of the spat-fall will be cast away with the disintegrating frond, though a much greater amount will be lost from the settlement on rocks and other unsuitable substrata where prolonged life is not possible. *Laminaria cloustoni* fruits during winter and the old growth is carried up on the new frond which grows at its base until the spring, when it is cast off as 'Mayweed'. The limpets migrate down the frond in late autumn but movement seems to stop when the animals reach the basal part of the old frond. Limpets must frequently be cast off with the remains of the old frond though this will not happen until their breeding period is over and they are then probably exhausted and moribund.

The mesogastropods which have adopted a limpet-like form are members of the Calyptraeacea—*Capulus, Calyptraea* and *Crepidula.* [See also chapter 32.] They have a cap- or slipper-shaped shell and only in the adult stage of *Calyptraea* is there no suggestion of a dextral coil. In *Capulus* (fig. 27) the shell is not unlike a jester's cap in shape and internally can be seen the scar of attachment of the horseshoe-shaped shell muscle. In the other two genera the shell muscle is attached to a shelf-like projection, which in the low conical shell of *Calyptraea* projects obliquely across the posterior wall, near the apex, and in *Crepidula* forms a transverse partition (sl, fig. 42A) across the posterior half of the base, set only a short distance in from the edge. Thus when the shell of *Crepidula* is turned over it has the appearance of a rounded slipper with the shelf forming the toe-cap. These 2 genera of limpets have complete or effective loss of movement. They are found in comparatively clear water or in silty habitats, though since they are fixed to stones, rock or shells, they live above the bottom deposits. They filter the sea water for their food and, as in lamellibranchs, the necessity for providing feeding currents and extensive food-collecting surfaces has been met by a modification of the gill. The mantle cavity is long and deep and with a breadth anteriorly corresponding to that of the shell, and the mono-pectinate gill with elongated filaments hangs diagonally from its roof and is many times longer than that of the more typical snail (fig. 58). It maintains the strong flow of water which traverses the cavity from left to right. The radula is no longer used for rasping as in the diotocardian limpets. It draws into the gut the mucous masses laden with the particles which have been collected without selection save for size (p. 97). These limpets lay egg capsules which are incubated by the female (p. 373). They are retained beneath the anterior part of the shell, either held by the propodium or fixed to the substratum, and so lie in the course of the current of water maintained by the ctenidium. Observations in

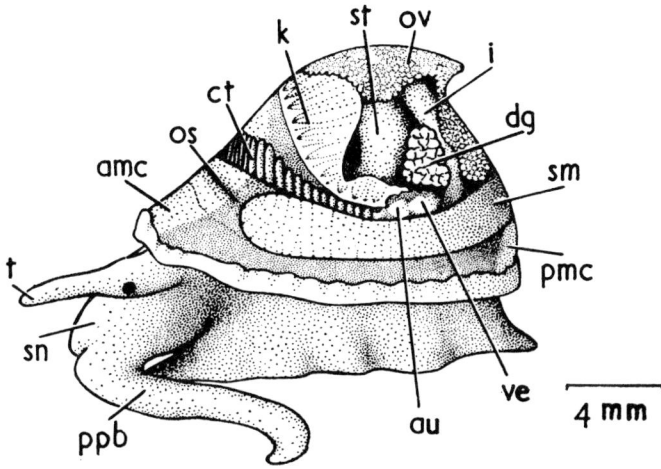

FIG. 257.—*Capulus ungaricus:* animal removed from shell, from the left.
    amc, anterior part of mantle cavity; au, auricle; ct, ctenidium seen through mantle skirt; dg, digestive
gland; i, intestine; k, kidney; os, osphradium; ov, ovary; ppb, proboscis; pmc, posterior mantle cavity; sm, shell
muscle; sn, snout; st, stomach; t, tentacle; ve, ventricle.

Essex (Chipperfield, 1952) suggest that these limpets usually spawn twice a year, some
time between April and October, and that there may be a period of concerted spawning
at, or immediately after, neap tides.

The protandrous hermaphrodite *Calyptraea chinensis* is dredged on stony ground in
shallow water around the SW coast, and in the sheltered and perhaps silty lower reaches of
rivers. It occurs with less frequency intertidally, where it may be collected from the undersides of stones or shells. It grips these objects with the broad sucker-like foot and remains in
the same position for long periods of time, though it will move about over limited distances,
especially during the breeding season when pairs come together. At Plymouth copulation
occurs all the year, chiefly October to April, and the females deposit egg capsules between
April and September (Wyatt, 1957). The habits of *Crepidula* are better known, for this genus
has the peculiar habit of living in chains of up to a dozen or more, each clinging to the shell of
the one beneath, and the sex and age of the individuals vary with their position in the chain
(fig. 189). The sexual biology of the species is discussed on pp. 343, 657. Only males, the last to
settle, may leave the chain and move elsewhere. In older individuals the foot is used only to
keep the limpet fixed and loses its power of creeping so that a chain once dislodged will be
cast about by the sea. *Crepidula fornicata* was introduced to the British Isles from the Atlantic
coast of America; the earliest record of its occurrence, though based only on collected shells,
was, according to McMillan (1938–39), at Liverpool in 1872, where it may have been introduced with the American clams (*Venus mercenaria*) which were laid down in the neighbourhood at that time; but here it is now extinct—if ever a breeding population were established.
Consignments of American oysters to Brightlingsea on the Essex coast in 1890 introduced
the limpets to that area, where now it forms masses inches deep over the bottom of sheltered
creeks in which oysters were abundant. *Crepidula* and oysters compete in the same habitat
and the shell of living oysters is often selected by young slipper limpets as a surface for

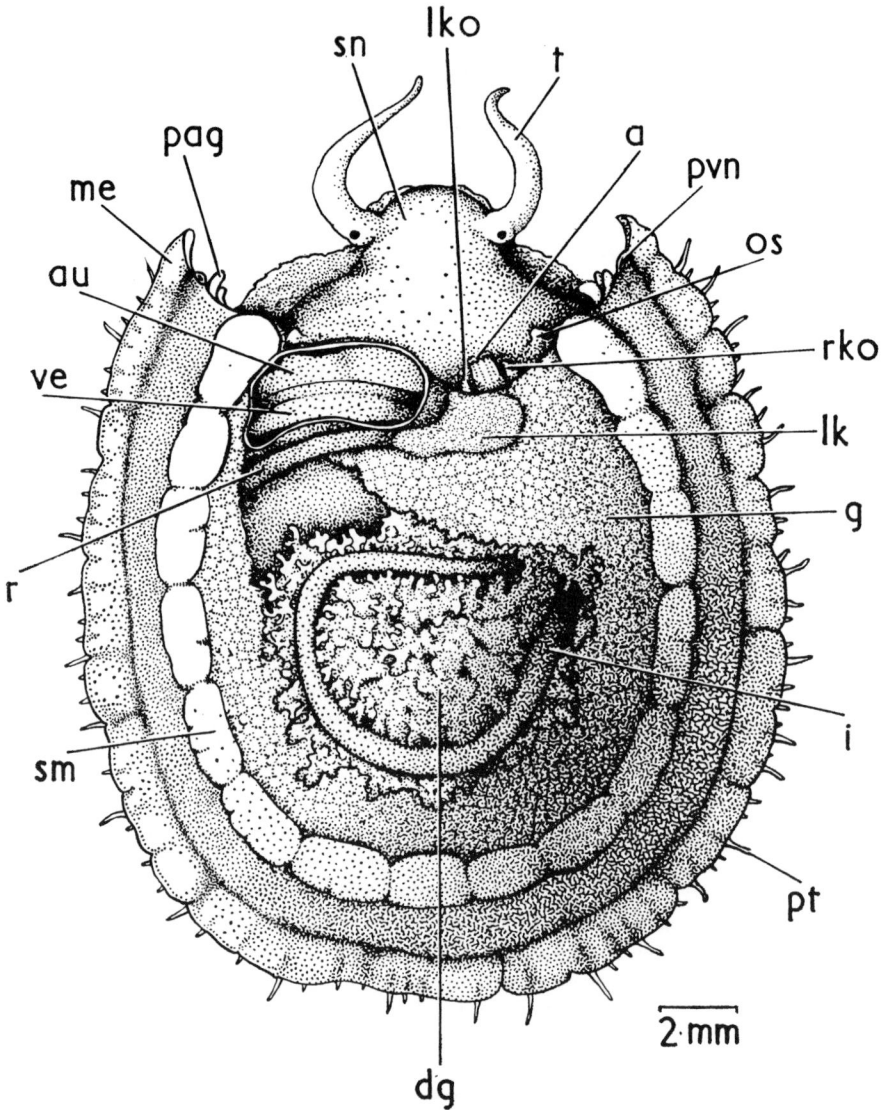

FIG. 258.—*Patella vulgata:* the animal has been removed from the shell and the roof of the nuchal and pericardial cavities cut away to show the contents; dorsal view.

a, anus; au, auricle; dg, digestive gland; g, gonad; i, intestine; lk, left kidney; lko, left kidney opening; me, mantle edge; os, osphradium; pag, pallial gill; pt, pallial tentacle; pvn, pallial vein; r, rectum; rko, right kidney opening (urinogenital); sm, shell muscle; sn, snout; t, tentacle; ve, ventricle.

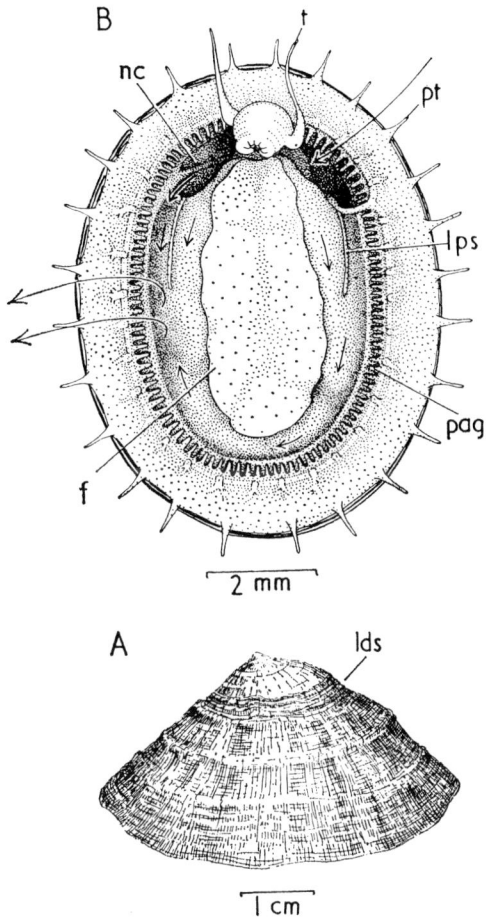

FIG. 259.—*Patella vulgata*: A, shell of animal collected in cave at high water mark, Langland Bay, Glamorgan, to show ledging; B, small limpet in ventral view. Arrows show direction of ciliary and water currents.

f, foot; lds, ledge on shell; lps, lateral pallial streak; nc, nuchal cavity; pag, pallial gill; pt, pallial tentacle; t, tentacle.

settlement; indeed, the limpet thrives and breeds most effectively on grounds which produce oysters of finest quality, though it is more resistant to extreme cold and low salinities. Consequently its rapid spread along the coasts of Britain has been a matter both of general zoological interest and of practical concern to those whose oyster beds are threatened. The combined strengths of the feeding current set up by thousands of individuals on the shore and the accumulation of faeces and pseudofaeces cause the deposition of mud on the oyster beds so that oyster spat will not settle. To reclaim grounds thus lost to cultivation necessitates the removal of up to 20 tons of *Crepidula* per acre, or even more (Cole & Hancock, 1956). Once such ground is cleared of the largest colonies it is not less than 10 years before a climax density is again reached. The species has spread from Essex N to the Scottish border and

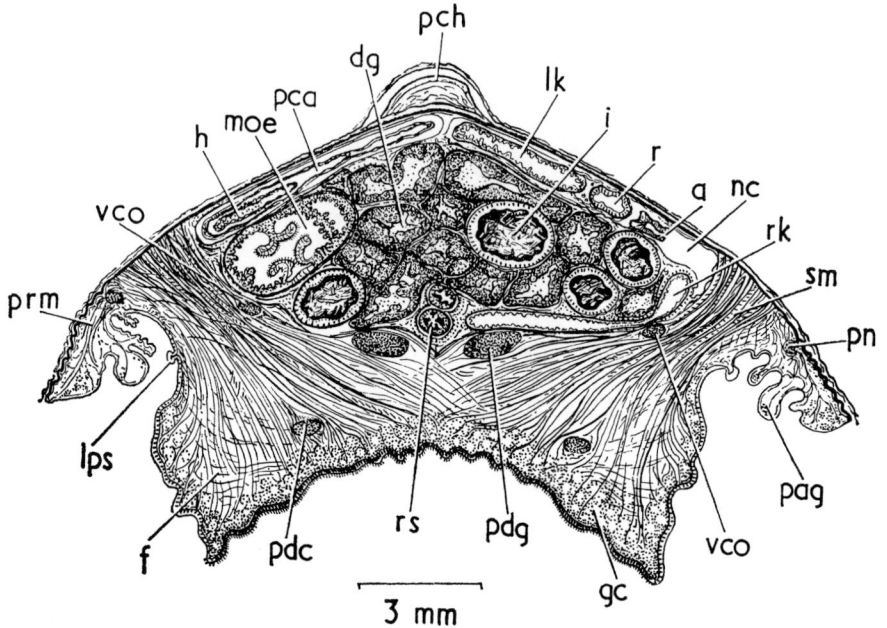

FIG. 260.—*Patella vulgata*: transverse section of young animal.
a, anus; dg, digestive gland; f, foot; gc, gland cells of sole gland; h, heart; i, intestine; lk, left kidney; lps, lateral pallial streak; moe, mid-oesophagus; nc, nuchal cavity; pag, pallial gill; pca, pericardial cavity; pch, proto-conch; pdc, pedal ganglion cord; pdg, pedal ganglion; pn, pallial nerve; prm, retractor muscle of mantle edge; r, rectum; rk, right kidney; rs, radular sac in radular sinus; sm, shell muscle; vco, visceral connective.

along the S coast as far as Land's End (Cole, 1952). It has also been found in Milford Haven, Pembrokeshire (Cole & Baird, 1953). The rate of spread along the Cornish coast is compara-tively slow: perhaps conditions are not so favourable for it there. However, the dread that it might become common can be deduced from the fact that in the late forties when it first appeared in the Helford River five shillings was paid for each *Crepidula* collected, though later the rate was reduced to a penny (Cole & Hancock, 1956). Between the years 1949 and 1953 the number of limpets taken on the Helford River oyster grounds increased about eighty-fold. The spread of *Crepidula* from place to place is speeded by its rapid rate of increase, for under favourable conditions three broods of larvae are produced each year, and especially by its easy transport on vessels, floating wreckage and seaweed: it is supposed that in this way it crossed the North Sea about 20 years ago and subsequently spread along the shores of Holland, Germany and Denmark. In 1949 a small number was found in two places on the French coast which may have been conveyed there with installations for the Normandy invasion.

Occasionally fixed solitary female limpets are found, with no indication that they have ever formed the lower member of a chain, and sometimes these individuals are guarding embryos. Orton (1950a) suggested that they may be self-fertilizing and that sperm retained from the male phase is carried over to the female phase. If this does occur it will be of extreme importance in the distribution of the species.

Capulus ungaricus (figs. 257, 60) has limited powers of movement and is often found attached to the shells of lamellibranchs which are dredged from comparatively shallow water. It also occurs on gravel ground attached to stones, rock or valves of shells, and very occasionally frequents the lower parts of the littoral zone. In this limpet the lower lip is prolonged to form a proboscis (ppb) which has a dorsal groove leading posteriorly to the mouth. The proboscis rests on the flat upper surface of the anterior prolongation of the foot, and food particles from the mantle cavity are led there and then licked up by the proboscis and conveyed to the mouth. The proboscis is capable of considerable extension and individuals which live on shells of living lamellibranchs may use it to reach the food in the food grooves of the gill lamellae, or on the recurrent ciliated path along the edge of the mantle of the host. The relationship with lamellibranchs is facultative, for when removed from the shell the limpet will settle elsewhere and collect adequate food by its own ciliary feeding mechanism. *Capulus* has been found associated with *Pecten, Chlamys, Modiolus, Monia* and *Astarte* and probably occurs on other bivalves. It has also been collected from the ciliary feeding gastropod *Turritella communis*. It occupies a characteristic position at the edge of a valve of the lamellibranch or the mouth of the gastropod shell, away from the exhalant current. The anterior margin of its shell overlaps the valve edge which at this point may have a semilunar gap where growth has been delayed owing to the presence of the limpet. There is a possibility that it may rasp this edge with its radula (Sharman, 1956). Limpets in captivity have remained in this settled position up to three months (Sharman, 1956), but still retain the ability to creep elsewhere. The point on the shell where the limpet was sitting is marked by a scar, which may be just a cleaner patch, or a circular area within which the sculpture has gone. More than one individual may be attached to a shell or stone, and sometimes the limpets are associated in pairs, a smaller one resting on a larger. The smaller individual is orientated in the same direction as the larger, and is usually about midway between the apex and the anterior edge. *Capulus* is a protandrous hermaphrodite (p. 346) and this association may occur only during the breeding season. There is no evidence that solitary individuals are self-fertilizing.

Sometimes *Crepidula* settles on the edge of a scallop shell, and although it has no proboscis to tap the food collected by the bivalve it may benefit from the feeding current. This trend towards the establishment of an association with other animals is followed further by *Thyca*, a capulid of warmer seas, which is completely dependent on echinoderms. It is attached throughout life to its host and, by means of a long proboscis, penetrates the tissues and sucks the coelomic fluid (Koehler & Vaney, 1912).

CHAPTER 19

THE ROCKY SHORE: WHELKS AND TINGLES

T HOSE prosobranchs which are grouped in the suborder Stenoglossa did not appear before the Upper Cretaceous period and more than 40% of known species are Recent. They have departed furthest from the ancestral, rock-clinging gastropod, especially in their feeding habits, for all are either flesh or carrion feeders and attack their prey by means of a proboscis (fig. 269). The most advanced forms, the cones, which are not represented in the British fauna, are well known for inflicting poisonous wounds by which they immobilize their prey. They attack actively moving creatures, whereas the more primitive carnivores choose sedentary prey—bivalves, barnacles and tube worms—and may get at the flesh by the laborious fashion of boring a hole in the shell to make an entrance for the proboscis. As the name of the suborder indicates, the radula is no longer a broad ribbon with a large number of teeth in each row. Indeed it shows the greatest reduction in tooth number of all prosobranchs, and there is a corresponding reduction in the size of the buccal mass. In the superfamilies Muricacea and Buccinacea (fig. 104E) each radular row has typically 3 teeth, a median and 2 laterals, but in the majority of Toxoglossa (fig. 104F) both median and laterals are absent and 2 marginals represent a row. Moreover, the radular membrane is usually lost, so that the teeth are separate from one another; only one at a time is brought into action in the cones, which show the peak of radular specialization. The alimentary canal of the stenoglossans is short, compatible with the carnivorous diet, and its complexities are associated with the development of the proboscis. This has involved the forward migration of the oesophagus through the nerve ring, which comes to occupy a more posterior position than in other prosobranchs, and the salivary ducts no longer pass through it. The glandular tissue, which has been stripped from the midoesophagus so that that is a simple tube, forms the gland of Leiblin in the Rachiglossa and perhaps the poison gland of toxoglossans. It is essentially such specializations of the gut that separate these higher prosobranchs from the mesogastropods (figs. 112, 115); the rest of their anatomy calls for little comment. The plan of the reproductive ducts in species which have been investigated resembles that of *Littorina*, though the histology, especially in the female, is more elaborate. This elaboration (fig. 163) is associated with the complexity of the wall of the egg capsules which the duct secretes, and with the development of a special sperm-absorbing gland (ig) from part of the receptaculum seminis (rcs).

The Stenoglossa are almost all marine and avoid brackish water, though *Nucella*, *Buccinum*, *Urosalpinx* and *Nassarius reticulatus* can tolerate a moderate amount. The shell, which is always external, is siphonate and the pallial siphon (in some species very long) can be used as a distance receptor. Working in conjunction with the osphradium (os, fig. 268) it may test the environment into which the animal moves, and can be seen examining the substratum and possible food at some distance from the body. By means of the siphon a flow of clean water is passed through the mantle cavity while the mollusc is feeding on putrefying flesh or is all but buried in mud or sand.

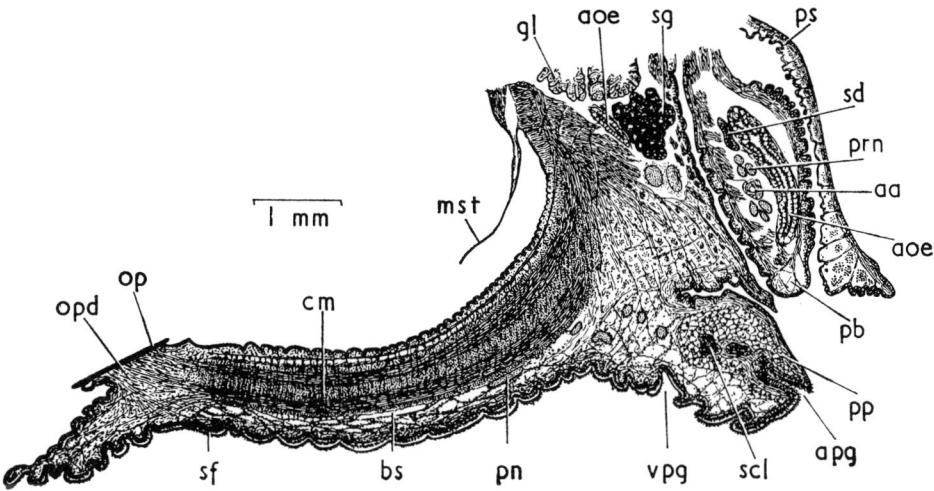

FIG. 261.—*Nassarius reticulatus:* sagittal section of head and foot of a young female.
aa, anterior aorta in proboscis; aoe, anterior oesophagus cut twice, once in proboscis, once in body; apg, opening of duct of anterior pedal gland; bs, blood space in foot; cm, columellar muscle, which is highly organized in this genus; gl, gland of Leiblein; mst, posterior part of mantle skirt; op, operculum; opd, operculigerous disc; pb, proboscis; pn, pedal nerve; pp, propodium; prn, proboscis nerve; ps, proboscis sheath; scl, sagittal canal of anterior pedal gland; sd, salivary duct; sf, sole of foot; sg, salivary gland; vpg, opening of ventral pedal gland.

The most primitive stenoglossans, the muricid whelks, are small and on the whole slow-moving. They are predatory and bore the shells of other molluscs to get at their flesh, and all have an accessory boring organ (= pedal sucker of Fretter (1946a) and accessory proboscis of Carriker (1943)) which helps in boring (Carriker, 1959; and see p. 621). All, except species of *Trophon,* are intertidal, so that their habits are, on the whole, well known, more especially as some species ravage young oysters and have been the concern of oyster farmers for many years. Even before this muricids were familiar shells, for the Tyrian purple of the ancients (secretion from the hypobranchial gland) was obtained from certain species of *Murex* and *Thais.* These species are believed to be *Murex brandaris, M. trunculus* and *Thais haemastoma*—none of them British—though British species of related genera secrete the fluid to some extent. There is no free larval stage in the life history so that dispersal is essentially limited to the snail's ability to crawl about.

Nucella lapillus (fig. 38) is the common carnivorous prosobranch of most rocky shores, occurring in large numbers within the balanoid zone. [For recent work see chapter 35.] The upper limit of the species at most places is between EHWN and MHWN, but it may be found in sheltered crevices up to MHWS. It is not tolerant of heavy surf unless shelter is available and is not normally found under a thick covering of weed, perhaps because of the lack of the barnacles which it eats, though it may seek shelter under weed which is near its feeding grounds. The lower level of abundance lies between MLWN and MLWS where food and sheltered crevices are available, and it occurs in the sublittoral zone down to a depth of 10 fathoms (Moore, 1936). The whelks are frequently found clustering in rock crevices in the intertidal area either for shelter or for breeding. Fischer-Piette (1935a) suggested that they are driven to shelter by extreme cold or the risk of drying up, and it is known that in the Gullmar Fjord, Sweden, they go below low water for the winter, presumably to avoid the ice (Gislén, 1930). The northern limit of distribution of the species is near the 0°C winter isotherm of oceanic water (Moore, 1936) which suggests the presence of ice as a limiting factor. The southern limit is close along the 19°C summer isotherm of oceanic water, though the intertidal temperatures would be considerably higher; the upper lethal temperature has been estimated as about 35°C (Gowanloch, 1927). Butler (1954) stated that in the related American species *Thais f. floridana* spawning is initiated at a water temperature of 20°C and ceases at temperatures exceeding 30°C.

There is some doubt as to the tolerance of low salinities, for whereas Agersborg (1929) recorded *N. lapillus* in freshwater pools along the edge of high water in N Norway and suggested that there it remains out of sea water most of the time, Pelseneer (1935, p. 323) stated that it can survive 9·5 days in fresh water, and Fischer (1931) found that it dies out only at the low salinity of 10‰ in the Rance, though the eggs are killed at low salinities tolerated by the adults. Other records suggest far less tolerance. It does not penetrate beyond the extreme mouths of the Tamar and Yealm estuaries at Plymouth (Moore, 1938*b*), nor beyond the mouth of the Tees, which is highly polluted (Alexander, Southgate & Bassindale, 1935): earlier records (Fischer, 1928) show it in the estuary of the Rance only down to a salinity of 22·8‰, and around the island of Tromö, S Norway (Økland, 1933), it does not occur when the summer salinities fall below 20–25‰ even though *Mytilus* and *Balanus* are in abundance. It seems that its distribution must be regulated by a number of factors and not by salinity alone.

The diet which the adult seems to prefer is barnacles, *Balanus balanoides* or *Chthamalus stellatus,* and it eats various kinds of molluscs, the favourite being *Mytilus edulis.* The whelks are commonly found on mussel beds, sometimes trapped in byssus threads which are anchored around them whilst they are inactive during an interval between periods of feeding, or whilst the shell of another mussel is being bored. This process is slow and may take 2 days. Although the whelk characteristically bores the shells of molluscs to get at the soft parts (p. 245), it opens barnacles by forcing the valves apart and Dubois (1909) claimed that hypobranchial secretion anaesthetizes the prey (p. 120). Fischer-Piette (1935*a*) has shown that the rate of destruction by the dog whelk may be sufficient to change the balance of life on a stretch of shore. He studied an area covered by barnacles with *Nucella* feeding on them, and on which *Mytilus* settled as the barnacles were cleared. When all available barnacles had been destroyed the prosobranch started on the mussels, though its efforts to eat them seemed clumsy at first, and empty shells were sometimes bored from either inside or out. Later, with greater efficiency, the whelk forced the shells of the smaller mussels apart without boring them. So many mussels were killed that barnacles were able to recolonize the rocks and the *Nucella* returned to their original diet, though the barnacles were not attacked until they were about 6 months old and had attained a fair size. The dog whelk may also destroy oyster spat. In the up-river nursery grounds of oysters in Essex giant *Nucella* of about twice the normal height may be collected (Cole, 1956). The young whelks may have a different diet. They emerge from capsules laid in the vicinity of the adults and Colton (1916) suggested that they feed on young *Mytilus* which are presumably near the empty capsules. However, Moore (1938*b*) held that their feeding grounds are lower on the shore. Here he has found numbers of young *Nucella*—perhaps carried by wave action—living on the underside of stones among the tubicolous polychaete *Spirorbis borealis*. The young whelks which he kept in captivity ate these worms and not small barnacles, although they tended to change to this diet when they had attained a height of about 8–10 mm. Perhaps at this size they normally migrate to the

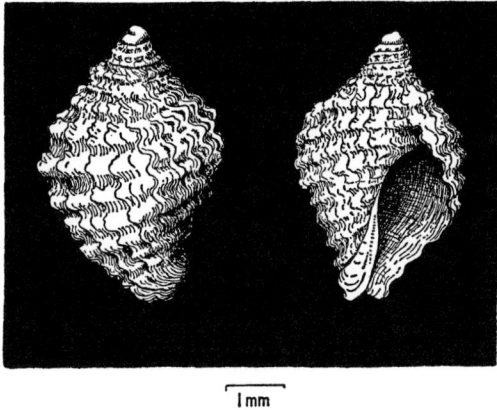

1 mm

FIG. 262.—*Nucella lapillus:* shells of young dog whelks, the right one in apertural view, the left in abapertural view.

barnacle zone where they are frequent in sheltered crevices. Adults and young may frequently be found, nevertheless, on shores devoid of *Spirorbis* e.g. Rhossili Bay, S Wales; perhaps in these circumstances other worms are eaten since at Rhossili vast colonies of *Sabellaria* occur.

Mature (fig. 38) and immature (fig. 262) shells of *N. lapillus* differ considerably. The shell which is still growing, is characterized by the outer lip being thin and sharp. At the onset of maturity growth of the lip ceases and this region of the shell thickens, sometimes up to 5 mm. The outer lip becomes rounded and develops a series of teeth along its inner edge which restrict the opening. Sometimes another tooth develops on the inner lip. Thickened shells of sexually mature individuals have been recorded as small as 13 mm high (Moore, 1936) though 45 mm is the size of a large specimen from the shore. However, a thin type as large as 63 mm in height and others approximating to this size have been found in deeper water: it is suggested that in the sublittoral habitat growth is not inhibited at sexual maturity, though an examination of the sex organs of these individuals has not been made. In a study of the growth rate of animals from the littoral zone Moore (1938a) found that in the first year the shells attained 10–15 mm in height, grew another 11 mm in the second year and less in the third when the whelks attained sexual maturity. He found no differences in the rate for the two sexes. The shell of this whelk is initially imbricated (Labbé, 1926) and in the intertidal zone small individuals in their first year are often found with the longitudinal striae forming a flounce-like ornamentation (fig. 262). In larger snails subjected to wave action this characteristic is lost. However, animals in deeper water (which have been regarded as the variety *imbricata*) sometimes retain it in the adult, and Colton (1922) found this type of shell characteristic of sheltered mud flats in Mount Desert Island, Maine.

The colour of the shell of *Nucella* is extremely variable. It may be a uniform white, or yellow which may shade into orange, and some shells are banded with yellow on white or on black or brown. Other shells may be brown shading into black and this is the colour which shows the strongest tendency to banding. Finally shells may be mauve, grading occasionally into pink, and the mauve may be overlaid with bands of black or brown. The breadth of the bands varies from individual to individual but they are constant in position throughout the life of the animal. *Thais lamellosa* fills a similar ecological niche on the Pacific coasts of North America and is even more variable in colour and form than *Nucella*. Colton (1916) stated that in the Mount Desert region the abundance of coloured *Nucella* is greater with increasing wave exposure, and also shows a slight increase in places of extreme shelter. He found more white shells in the *Balanus* zone than in the *Mytilus* zone. However, Colton (1922) concluded that variation in colour and shell sculpture is due to hereditary factors, and that the proportion of any one colour variety found in a single station is determined by natural selection, more light snails being found on light environments than on dark ones. Moore (1936) has held that the colour types brown-black and mauve-pink are unquestionably dependent on the abundance of *Mytilus* in the diet, and that the pigments are derived from the prey. Moreover he suggested that this diet delays sexual maturity and so decreases the proportion of mature shells in the population and consequently increases the shell height at which sexual maturity sets in; it produces a fatter shell, a more open spiral and hence a wider aperture. White shells are associated with a *Balanus*-fed community. Moore transferred individuals with coloured shells to a diet of *Chthamalus, Balanus* and *Mytilus* respectively. Those on mussels showed no change in pigmentation either in the brown or the mauve types, whereas the new shell growth which appeared on the barnacle-fed individuals had a reduced pigmentation, though the speed of the reduction varied considerably. When these animals were returned to their normal diet of *Mytilus* the pigmentation reappeared. However, shells from a *Balanus*-fed population showed no pigmentation in their newly formed shell after a period of 6 months on a diet of *Mytilus,* suggesting that it requires long periods of feeding on mussels before sufficient pigment is accumulated in the animal to appear in the shell. The banding of a shell is produced by localization of pigment in certain regions of the mantle edge (fig. 38), and occasionally these bands are very narrow. Moore suggested that in extreme cases the pigment-susceptible areas in the mantle edge may be narrowed to extinction, resulting in an animal which, though feeding on *Mytilus,* has a white shell. He considered the relationship of yellow shells to environmental factors as more doubtful (Moore, 1936), though he found this type most abundant on coasts with a moderate degree of wave exposure and practically absent in extreme shelter and extreme exposure. In some localities the true pigmentation of the shell is masked by boring filamentous algae which give it a greenish brown colour.

Agersborg (1929) basing his conclusions on a study of *Nucella lapillus* in N Norway suggested that variations in the shell of this species, concerning size, thickness, spire and surface texture, were due to an interaction of environmental and genetic factors, and more recently Staiger (1950a, b, 1954, 1955) has followed up this problem. He showed that at Roscoff there exist two races of *N. lapillus,* one with 18 pairs of chromosomes, the other with 13. Because of the chromosome numbers found in other stenoglossan species it seems reasonable to suggest that the race with 13 pairs is an evolutionary novelty which would require a certain degree of isolation to arise. The two races cross and segregation gives animals with all possible numbers of chromosomes from 13–18 pairs. Investigation shows that the 13 race predominates in exposed areas, the 18 race in sheltered habitats and intermediates occur in places with moderate exposure. The thickness and length of the shell are also related to the chromosomal composition of the individual: the thickness is greater in a mixed population than in a pure one (13 or 18 chromosome pairs). Heterogeneity, however, increases the versatility of the population in tolerating a variety of ecological conditions and for this reason it is not possible to explain the successful occupation of a micro-ecological niche in terms of one factor alone. For example, the mean body-size of a heterogeneous

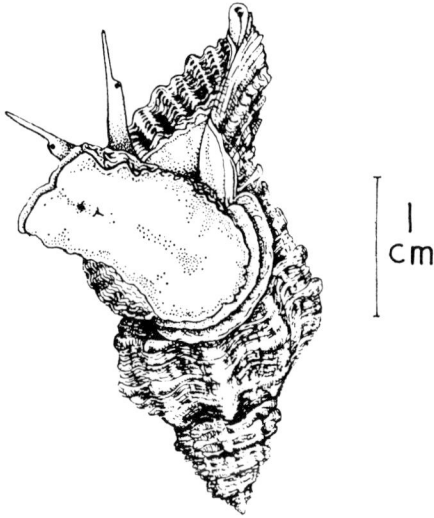

FIG. 263.—*Ocenebra erinacea:* the rough whelk-tingle seen from below as it creeps on a glass plate. The apex of the shell is below and its siphonal canal points upwards. The siphon projects from its tip. The animal is a female and on the sole of the foot may be seen 2 apertures: the anterior is that of the pit in which the accessory boring organ lies, the posterior is that of the ventral pedal gland.

population increases with exposure: this is due to a balance between the increased vigour of the mixed stock and its lesser adaptation to exposure and the lesser vigour of the 13-chromosome race coupled with its greater adaptation to exposure. No correlation between chromosome number and colour of shell has yet been made.

Some egg capsules of *Nucella* (fig. 234A) are mauve or brownish in colour instead of the usual yellow, and Moore (1936) also attributed this colouring to the pigment derived from a diet of *Mytilus*. However, the pigment may come from secretion of the hypobranchial gland (p. 125) which can easily contaminate the capsules as they are laid, and, in extreme cases, even affect the colour of the yolk. The odour of this same secretion has been claimed to be the cause of sexual aggregation (p. 120).

Ocenebra erinacea, the European sting winkle (fig. 263), may be found with *Nucella* near low water on some rocky shores. It is, however, predominantly a sublittoral form occurring down to about 50 fathoms, and its invasion of the littoral zone is dependent upon temperature: in some parts of the SW, where recently its numbers have increased, it may be found at mid-tide in sheltered crannies especially in the warmer months, though not in exposed places like *Nucella*. It is now distributed along the W and SW coasts and is scarce in the SE where, previously, it occurred in greater abundance. During the unusually cold winter of 1928–29 *Ocenebra* suffered virtual extinction in the region of the River Blackwater, Essex, whereas the closely related American oyster drill *Urosalpinx* increased in abundance (Orton & Lewis, 1931). *Ocenebra* is now reappearing in the area (Mistakidis & Hancock, 1955). The food selected is commonly lamellibranchs, *Tapes, Cardium, Venus,* which are associated with more sheltered habitats than the prey of the dog whelk, for in its sublittoral habitats *Ocenebra* frequents silty grounds with rock, stones and perhaps clinker. On the shore the animal may be found eating barnacles, small tubicolous worms, mussels and anomiid bivalves and is frequently associated with oyster beds, especially when other food is less abundant. In the Helford Estuary (Cornwall) it is a significant predator on the beds of cultivated oysters and here the American drill *Urosalpinx,* a more destructive oyster pest which is common on the Essex coast, does not occur. Both eat oyster spat as well as boring the adults, and in the months of May and June which are their breeding time, they are hand picked from the oyster beds. It would appear that specimens of *Ocenebra,* when removed from their natural habitat, retain a preference for their accustomed food although a greater variety may be offered. Thus Orton (1929*a*) found that 264 specimens from Plymouth (where no oysters are found) attacked only 10 of the oysters which were provided with other food, while 30 smaller individuals from West Mersea, Essex, ate 45. There is evidence that the winkle sometimes attempts to eat *Crepidula* which is a recent intruder in its habitat, though the shells are often perforated at unsuitable points such as the shelf or the edge overlying the mantle and foot (Orton, 1950*b*). The size of the adult varies. At Plymouth it ranges from less than 20 to 40 mm in length, whilst further W, at Falmouth, there is an increase of about a third of this. Hancock (1957) has suggested that diet may influence the size though probably more than one factor is concerned. Females are larger than males and may be at least twice as numerous on the shore.

Both prosobranch gastropods which have been introduced into this country from America are important oyster pests and were shipped over with oysters. *Urosalpinx cinerea* seems to have been introduced some years after *Crepidula* and was first recognized here in 1928 (Orton & Winckworth, 1928; Orton, 1930) when it was realized that it had been confused previously with the English sting-winkle (for it resembles *Ocenebra* in size, shape and habits); specimens were then discovered which had been collected in 1920. So far as is known, *Urosalpinx* is confined to Essex and Kent, where it has shown itself to be a danger to the shell-fish industry. Indeed, less than 10 years after it was first recognized there its numbers had so increased that one shilling per 1,000 drills was paid to dredgermen on the River Roach, Essex (Cole, 1942). In 1936–37 the sum paid out amounted to £9 5s., representing a total of 185,000 drills. In the River Crouch, Essex, the drill is now present on grounds up to 14 miles from the mouth and concentrations of up to 10,000 per acre have been recorded. Here its density has increased considerably during the past 15 years (Mistakidis, 1951; Hancock, 1959). It feeds in the same way as *Ocenebra* by drilling an entrance through the shells of other molluscs to reach the flesh, which is devoured. It chooses a variety of hosts including *Ostrea edulis* and its spat, *Mytilus edulis, Cardium, Paphia, Crepidula;* it also relishes barnacles, and frequently cannibalism has been observed. Observations on the Essex oyster beds show that the drills feed preferentially on oyster spat, and each destroys about 40 spat during the feeding season. Oysters larger than spat are not seriously damaged when spat is available.

Unlike *Crepidula,* this pest has not spread rapidly from the two centres where it was introduced, which are believed to be Brightlingsea and West Mersea, Essex (Cole, 1942), and there are no records of it in the oyster beds of Holland, Belgium and France (Carriker, 1955). This may be due to the fact that *Urosalpinx* lacks a free-swimming larval stage and its distribution depends entirely on its slow rate of creeping and on possible transportation by human or other agencies. Moreover it lacks the gripping power which in the limpet has accounted for successful transference over long distances on floating wood and weed. In the United States of America the drill is present from Maine to Florida on the E coast and in Washington and California on the W, and the steady decline in oyster production from 231 million pounds in 1910 to 77 million pounds in 1950 has been attributed to it (Glancy, 1953). It reaches a greater size in England than in its natural habitat on the Atlantic Coast of the States and appears to be more prolific on the British oyster beds (Cole, 1942).

The occurrence of *Urosalpinx* in the intertidal region of the shore in SE England varies with the season. Although an area between LWST and 12 m or more below is perhaps the most frequented zone, large numbers of the mollusc migrate on shore during spring and early summer, travelling over fairly firm mud where there are shells, living mussels, *Littorina* and oysters, and avoiding soft ground devoid of hard objects. Many migrate to near mid-tide level. A single individual can move 2 m a day, and although there may be a certain amount of horizontal migration the main one is vertical (Hancock, 1959) and is associated with spawning. Some egg capsules are deposited on the shore, fixed to stones, oysters and other objects, but most are deposited just below LWST. Little food is taken while the drills breed in May and June, but as the peak of the spawning period passes feeding begins and continues until the end of October. Feeding is also dependent on temperature: Hanks (1957) has shown that it does not occur below 5°C, is intermittent below 10°, rises steadily to a maximum at 25° and ceases again at 30°.

The young begin to emerge from the capsules early in July after an incubation period of about 8 weeks. Since each female may deposit 25 capsules in a single laying (Cole, 1942) and about a dozen young emerge from each, there is a considerable annual increase in numbers. Moreover, both males and females may live to an age of 13–14 years. The young drills attack oyster spat which has recently settled and do considerable damage. In autumn there is a reverse migration of both young and old followed by winter hibernation beneath the surface of the mud. Hibernation generally occurs when the temperature falls below 7°C and the drills remain quiescent until the temperature again reaches 9–10°C when the snails become active and feeding begins, together with the spawning migration (Hancock, 1959).

Several methods have been tried to control the pest: the most successful in the United States is claimed to be the trapping of the drills during the breeding season. A wire bag about 12 x 15 in in size is partly filled with small bait oysters and shell to give added weight and a holdfast for the victims. The bags are attached to lines about 8 ft apart on ground infested with drills and at regular intervals of time the mollusc and its spawn are cleared away and the bait renewed. In Britain this method appears to have little success and a more effective control has been repeated dredging for adults and spawn during the spring and summer, and also hand picking them from the shore (Cole, 1956), though a cheaper and simpler method is the use of curved roofing tiles which have been exposed on the shore over a winter to remove newness, and are then placed in rows near LWST. The drills congregate on the curved under-surface and are destroyed about once a fortnight (Hancock, 1959).

The largest of the British prosobranch gastropods are some of the whelks which fall into the family Buccinidae. Occasional specimens of *Neptunea antiqua* attain a shell height of 8 in. Their size and palatable flesh have made this group of molluscs well known, and also the fact that the shell is stout and slow to disintegrate after the whelk is dead. Only one species, *Buccinum undatum* (fig. 264), the common whelk, is found in all parts of the British seas and on every kind of ground from near low water to the greatest depth. Its local abundance has made it in times past an important item of food. In the Whitstable area in the middle of the last century the whelk fishery yielded £12,000 a year, for those days a considerable sum. Important areas for the fishery now lie between Grimsby and Southampton, though this is probably on account of the large markets in SE England. The mollusc was once regarded as a greater delicacy than now. In Roman kitchen middens in this country whelk shells have been found mixed with oyster shells and Jeffreys (1867) recorded that in 1504 the enthronement

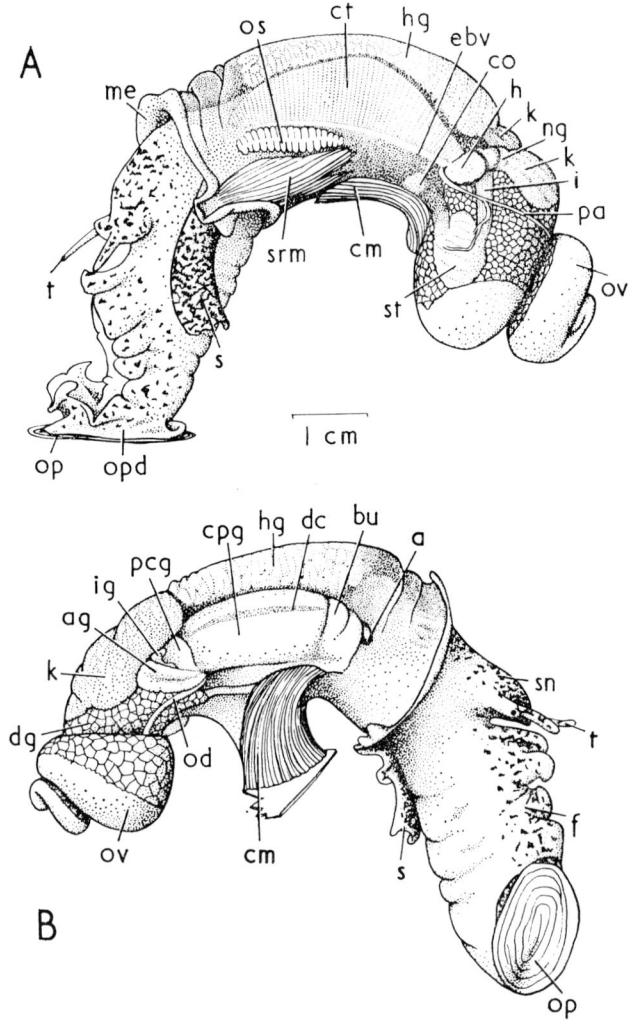

FIG. 264.—*Buccinum undatum:* the animal has been removed from the shell and is seen A, from the left; B, from the right.

a, anus; ag, albumen gland; bu, bursa copulatrix; cm, columellar muscle; co, oesophageal caecum, seen by transparency; cpg, capsule gland; ct, ctenidum; dc, dorsal channel along capsule gland; dg, digestive gland; ebv, efferent branchial vessel; f, foot; h, heart; hg, hypobranchial gland; i, intestine; ig, ingesting gland; k, kidney; me, mantle edge; ng, nephridial gland; od, ovarian duct; op, operculum; opd, operculigerous disc; os, osphradium; ov, ovary; pa, posterior aorta; pcg, posterior lobe of capsule gland; si, siphon; sn, snout; srm, siphonal retractor muscle; st, stomach; t, tentacle.

feast of the Archbishop of Canterbury, William Wareham, included 8,000 whelks costing 5 shillings per 1,000. *Neptunea antiqua,* the red whelk, is also marketable and may be included in catches with the common whelk. It is a sublittoral form found in shallow waters, but does not occur in S and SW England nor S Wales. Whelks which are marketed from the Northumbrian and Scottish coasts may also include *Colus gracilis* though this is accidental as, for some reason, *Colus* is not used as food. Murray & Hjort (1912) stated that *Neptunea antiqua* and *Colus gracilis* occur in the North Sea everywhere from Denmark to the Scottish coast, sometimes in great numbers, and in one haul from a depth of 96 m 130 *Neptunea* and 375 *Colus* were taken. *Buccinum* is also found here down to 100 m, though never in such abundance. Little is known of other species of *Buccinum,* except for their rather occasional occurrence in dredgings. On the whole all are characteristic of soft substrata, and they can plough through the surface layers of sand or mud and retain a clean current of water through the mantle cavity since the pallial siphon is long and kept clear of the bottom (si, fig. 268). At the breeding period the females seek some hard object on which to deposit the egg capsules. They are hemispherical or semi-ovoid, attached by the broad base and only in *Buccinum undatum* and *Neptunea antiqua* are they heaped on one another to form an agglomerate mass (fig. 208). Those of other whelks are conspicuous owing to their size: a capsule of *Volutopsius norwegicus* measures 3 cm across the base. As far as is known no species has a free larval stage.

Buccinum may be left uncovered by the tide and on rocky shores it will then retreat into its shell and behave like other univalves, clinging to the rock until the tide returns. Sometimes on muddy and sandy shores a low spring tide leaves numbers of large whelks exposed. This may be seen where the tidal range is great, as at Oxwich Bay, S Wales, and more dramatically in the Bay of Fundy on the N Atlantic coast of Canada. In these circumstances, instead of withdrawing into the shelter of the shell and sealing it with the operculum, or seeking some nearby stone on which to cling, the whelks continue to creep about, sometimes moving on to drier ground further from the sea. Eventually, as a result of enfeeblement due to desiccation, many lie senseless on the shore, and some are eaten. Gowanloch (1927) concluded that the deep sea population of *Buccinum* forms a reserve from which a small temporary intertidal one is drawn. These molluscs fail to become successful shore-living forms because they lack the protective behaviour which ensures survival. Unlike *Littorina littorea* they show no tendency to migrate to their normal level on the shore when displaced to a position above their usual habitat, and they may die from desiccation within a metre of shelter. There is some evidence that the whelk moves from deeper to shallower water at spawning time, which is winter in most localities, and many females may come to the same place to spawn, forming separate smaller or combining to form larger heaps of eggs. [See p. 664.]

The common whelk, and probably other members of the Buccinacea, are carnivores with a tendency to scavenging and will attack animals which are moribund, and fresh corpses. It is uncertain whether they eat putrefying flesh. The living animals which are attacked are primarily worms. *Buccinum* is also known to eat a variety of bivalves and reach the flesh not by the slow boring of the shell, but by taking advantage of its greater size and strength. It will open the shells of living cockles and scallops by settling on or near the upper valve and then, with a sudden quick movement, insert the edge of its own shell between the open valves, preventing their closure. The proboscis is inserted and the soft tissues attacked. Dakin (1912) found that the proboscis first destroyed the adductor muscle of a *Pecten maximus* and so disabled the closing mechanism of the valves. *Mytilus, Ostrea* and *Mya* are also eaten; the last is an easy prey since its soft tissues are never totally concealed by the shell. Although the whelk occurs on oyster beds Hancock (1957) found that in the laboratory oysters are only occasionally eaten and cockles are preferred. The maximum feeding rate was recorded in April when 2 cockles a week sufficed: the opening and eating of each may take less than an hour. Although on the shore more food may be required, the actual time spent in feeding must be remarkably little. Unlike *Ocenebra* and *Urosalpinx* the common whelk continues to feed all the winter in the laboratory, provided that the water temperature does not fall below 5°C.

The whelk's habit of eating moribund animals, or those entrapped, and of sensing them from a distance, is better known, and is exploited by the fisherman. Whelk pots, made of withy and twine woven on an iron frame, are baited with freshly dead crabs or fish, and crabs strung in a bunch and sunk in the sublittoral zone will also attract whelks from a considerable radius. The whelk will penetrate the carapace of the crab with its radula. On one occasion (fig. 269) a moribund holothurian was seen to be attacked from the other side of a partition, the proboscis of the whelk being directed through a crevice on to the prey. It then reached a length of 18 cm, just over twice the length of the shell. It is stated that in Denmark the whelk may be a menace to the plaice fisheries (Petersen, 1911) for it attacks fish entangled in nets, bores through the skin and devours the tissues. A dozen or more whelks may be attached to the same fish. The animals occur in such numbers that in one area $3,845\frac{1}{2}$ bushels of *Buccinum* mixed with *Nassarius reticulatus* were trapped between 5 April and 8 November. *Buccinum*, in its turn, is preyed on by fish, and is used as bait in fishing. Jeffreys (1867) found 30 to 40 shells in the stomach of a single cod and occasional ones may be found in the stomach of rays and dogfish.

The empty shells of whelks, as well as those which are inhabited, are favourite homes of other animals. Murray & Hjort (1912) found that every one of the 375 *Colus gracilis* in a haul from the North Sea had the sea anemone *Hormathia digitata* on its shell, and several *Neptunea antiqua,* the larger whelk, had two large actinians of a different species, *Tealia felina* and *Metridium senile.* The anemones are thus carried on to feeding grounds which may provide no holdfast for them. Other anemones, *Calliactis parasitica* and *Adamsia palliata,* live only on shells inhabited by hermit crabs. The older stages of *Eupagurus bernhardus* and *E. prideauxi* leave the shells of trochids, periwinkles and dog whelks, which they inhabit when young, and are dependent on whelk shells when fully grown. The shell then becomes the home of the numerous commensals associated with the pagurids and is protected.

There is considerable variation in the shell of *Buccinum undatum* and *Neptunea antiqua:* Jeffreys (1867) listed 6 varieties of the former each associated with a different type of habitat. They are marked by differences in the size of the whole shell, in the body whorl and the spire, and in the thickness and ribbing. Small forms deprived of decoration, which are found in brackish water, constitute one of the 6, and the unusually large specimens of the Dogger Bank which have thin shells with long spires are another. Besides these variations monstrosities are sometimes found: the shell may be sinistral; the whorls keeled; the spire may be extended to a considerable length and the whorls flattened; the body whorl may be compressed and elongated to give the shape of a *Voluta;* the operculum may be patelliform or there may be two or three.

Perhaps the most agile of all British prosobranchs are the species of *Nassarius.* Two of these, *N. reticulatus* (figs. 265, 38) and *N. incrassatus,* are common and may be collected near low water: the former also lives in brackish water, when the shell grows thicker (Berner, 1942). The third species, *N. pygmaeus,* is sublittoral and rare. They are all scavengers living on what can be found of dead or decaying flesh, and consequently favour silty places where organic remains accumulate. *N. incrassatus* congregates under stones and in crevices of rocky shores; sometimes a dozen or more may be found together. *N. reticulatus* (fig. 266), the larger intertidal form, favours a greater abundance of sand and silt. When the tide is out it buries

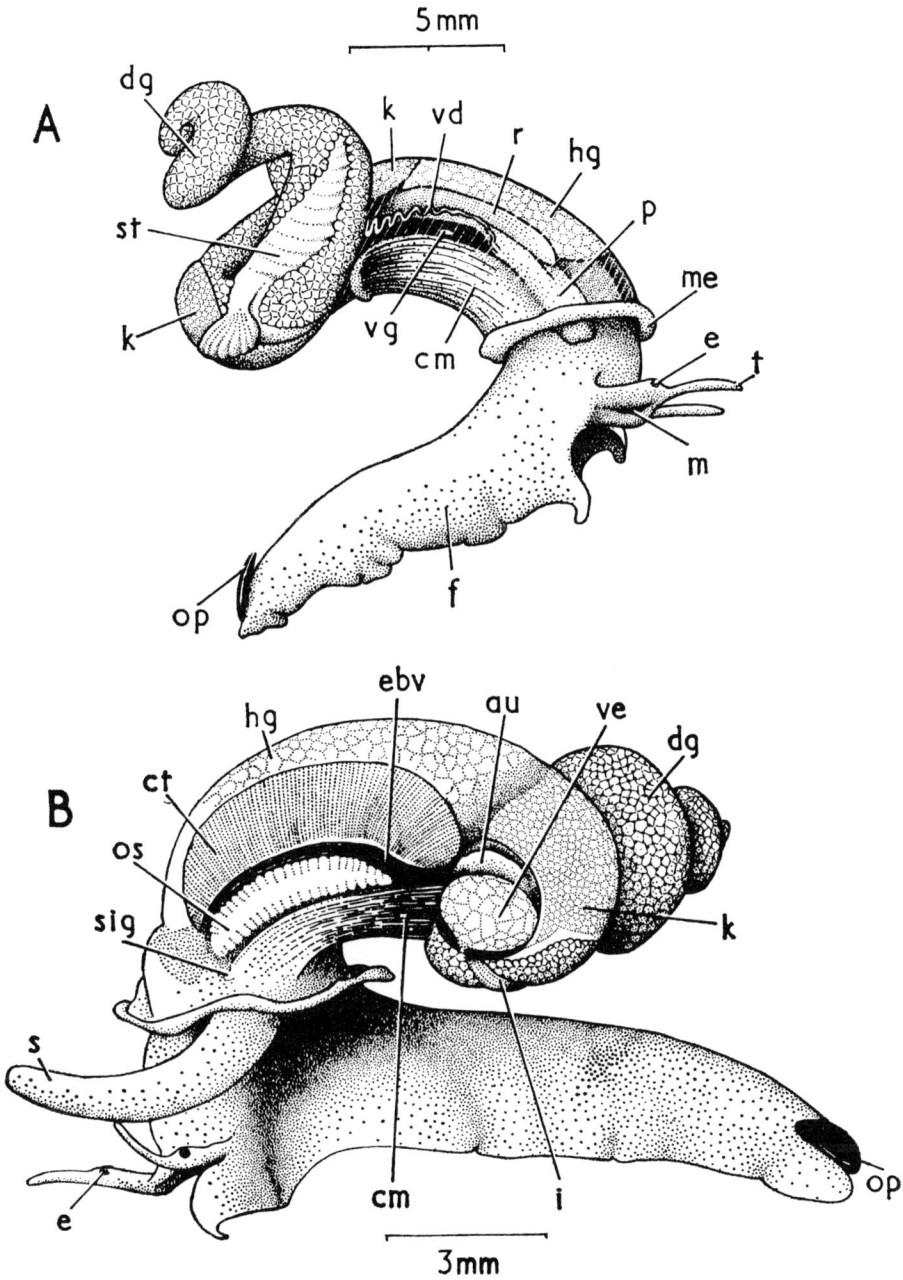

FIG. 265.—*Nassarius reticulatus:* the animal has been removed from the shell and is seen A, from the right; B, from the left.

au, auricle; cm, columellar muscle; ct, ctenidium; dg, digestive gland; e, eye; ebv, efferent branchial vessel; f, foot; hg, hypobranchial gland; i, intestine; k, kidney; m, mouth (opening of proboscis pouch); me, mantle edge; op, operculum; os, osphradium; p, penis within mantle cavity; r, rectum; si, siphon; sig, siphonal ganglion; st, stomach; t, tentacle; vd, vas deferens; ve, ventricle; vg, visceral ganglion.

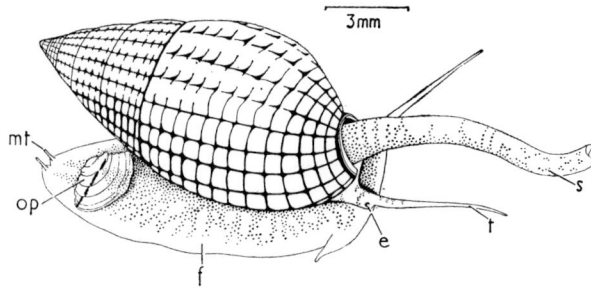

FIG. 266.—*Nassarius reticulatus:* animal creeping.
e, eye; f, foot; mt, metapodial tentacle; op, operculum; s, siphon; t, tentacle.

itself just beneath the surface in a slanting position, or shelters beneath stones and few individuals are active, though some may continue to feed. As the water covers it again the mollusc starts to move and glides with remarkable rapidity over the surface of the soft substratum, or ploughs through the surface layers, often hidden from view except for the long pallial siphon (fig. 266). A study of the behaviour of other species in this genus shows a rhythm of activity caused by the tide. The American species *Nassa obsoleta* has maximal activity at high tide and at low tide relatively few animals are moving about. Snails collected from localities with different tidal times show activity patterns correlated with their local tides (Stephens, Sandeen & Webb, 1953). Ohba (1952) has shown that in *Nassarius festivus* the quiescence during low tide may be broken by the presence of food, to which they are chemically attracted. This whelk is, as a rule, more active at night than by day and this diurnal rhythm appears to mask the tidal activity at night. The activity of the animals begins to increase at dusk and is maximal about an hour later (Ohba, 1954); it then decreases, though about 40% of the animals remain active until dawn. Light appears to be the factor which regulates the rhythmic activity of this species.

The siphon is relatively longer than that of other Stenoglossa and may be extended to a length exceeding that of the shell; sometimes only the tip is visible when the animal is scavenging deeper in the mud or sand. The mollusc moves forwards with the tube directed backwards, clearing the dorsal surface of the shell, and through it can be seen to pass the strong inhalant water flow. Sometimes, however, the siphon is fully extended and waved through the water as if to examine the vicinity, and when food is approached may be directed around it before the proboscis is protruded. Near the base of the siphon is a ganglion (sig, fig. 267) from which nerves pass towards the tip, and distal to the ganglion a flange of tissue, projecting into the lumen, may act as a valve in regulating the inhalant water current. In the absence of stimulation from food the animals are not affected by water currents, but when stimulated they exhibit a positive rheotaxy and move upstream to the food, which is found over the last 21 cms by a phobic mechanism (Henschel, 1932). They react positively to a great variety of chemical substances e.g. 1% glucose, 0·5% sucrose, 0·1-1·0% starch, 0·1% glycocol, 0·025% glycogen, and to skatol; a mixture of these reinforces the stimulating effect.

The foot of *Nassarius* is large, broad in front and with angulated corners which are recurved, deeply grooved and richly innervated; it narrows posteriorly and has a median notch, on either side of which is a short metapodial tentacle. It is exceedingly mobile not only in swift gliding movements over rock and as a ploughshare through the sand, but also in manipulating fragments of food, pushing them towards the surface and often wrapping itself around whatever is eaten. The mouth (fig. 132) is at the end of the long eversible proboscis which can be extended to a length over half as long again as the shell. The end of the proboscis has a sucker action so that the whelk can cling to the food while the radula rasps it, and, in this way, food at some distance may be eaten. Both N. incrassatus and N. reticulatus are found in lobster pots and whelk pots, attracted by the carrion there, for scent undoubtedly plays a great part in directing them to food. A dead fish or bivalve which they relish will bring whelks from some distance, as will fish caught in nets. They differ from Buccinum in devouring putrefying flesh with avidity. According to Starmühlner (1956) N. incrassatus eats mainly decaying sponges.

Nassarius reticulatus is preyed upon by starfish, though sometimes it does not fall an easy catch, for it responds to their attack with a quick evading reaction. If the posterior end of the foot is touched there is a rapid response. The shell is moved forwards to the right or left and twisted on the head-foot so that the apex is directed anteriorly and the mouth faces upwards. Then, without pause and with equal speed, the shell is flung across the head-foot as the columellar muscle gives a sudden contraction, raises the foot off the ground and swings the body in the opposite direction to that of the shell. The whelk thus scampers to and fro, for the leaping movement may be repeated several times in an attempt to evade the predator. The ability to do this is undoubtedly related to the organization of the dorsal pedal muscula-ture (fig. 261) which is similarly elaborate in N. incrassatus. Hoffmann (1930) suggested that when the starfish touches the mollusc the stimulus is received by special cells on the dorsal surface of the foot, not, perhaps, in the metapodial tentacles, for the removal of these is without effect. The stimulant may be a protein secreted by special glands in the skin of the starfish, though the same reaction may be induced by a great variety of stimuli.

Nassarius is less timid than most prosobranchs and so its activities can be observed more easily—even when depositing spawn the female is not easily distracted by intruders. The spawning behaviour is probably similar to that of other stenoglossans. The siphon is first directed over the hard surface, shell, stone or rock, on which the egg cases will be deposited, as though examining it, and the surface may then be cleaned by the radula. Soon the whelk settles with the right side of the foot raised from the ground so as to form a groove which links the genital aperture with the ventral pedal gland. When an egg capsule passes from the oviduct it is manipulated along the groove by the musculature of the foot and directed into the gland, where it may remain for about 5 min. The sole of the foot is then pressed against the substratum while the capsule is moulded and the visceral mass may be rocked from side to side whilst the base of the egg case is secured. Eventually the anterior part of the foot is drawn up and back, as the muscles of the pedal gland relax, and the capsule is left. The next one is deposited in a similar way nearby, so that several come to be arranged in equidistant rows (fig. 207B). Frequently, as in other Stenoglossa, more than one female will use the same spawning ground. The eggs hatch as veligers which have a rather long pelagic life, whereas in the larger whelks of the superfamily Buccinacea the larval stage is suppressed.

A general obscurity surrounds the British members of the Toxoglossa: many species are listed, but all, except perhaps for one or two species of Mangelia, are said to be scarce or rare. The food of many species is not known, but is probably polychaetes. The structure of the gut of Mangelia (Robinson, 1955) shows that these animals are undoubtedly carnivores of a highly specialized kind (fig. 103). Consequently their distribution will be restricted to the proximity of their prey and when that is known for each species these molluscs may be secured in greater abundance. The prey of Mangelia species is undoubtedly a soft-bodied creature which will be

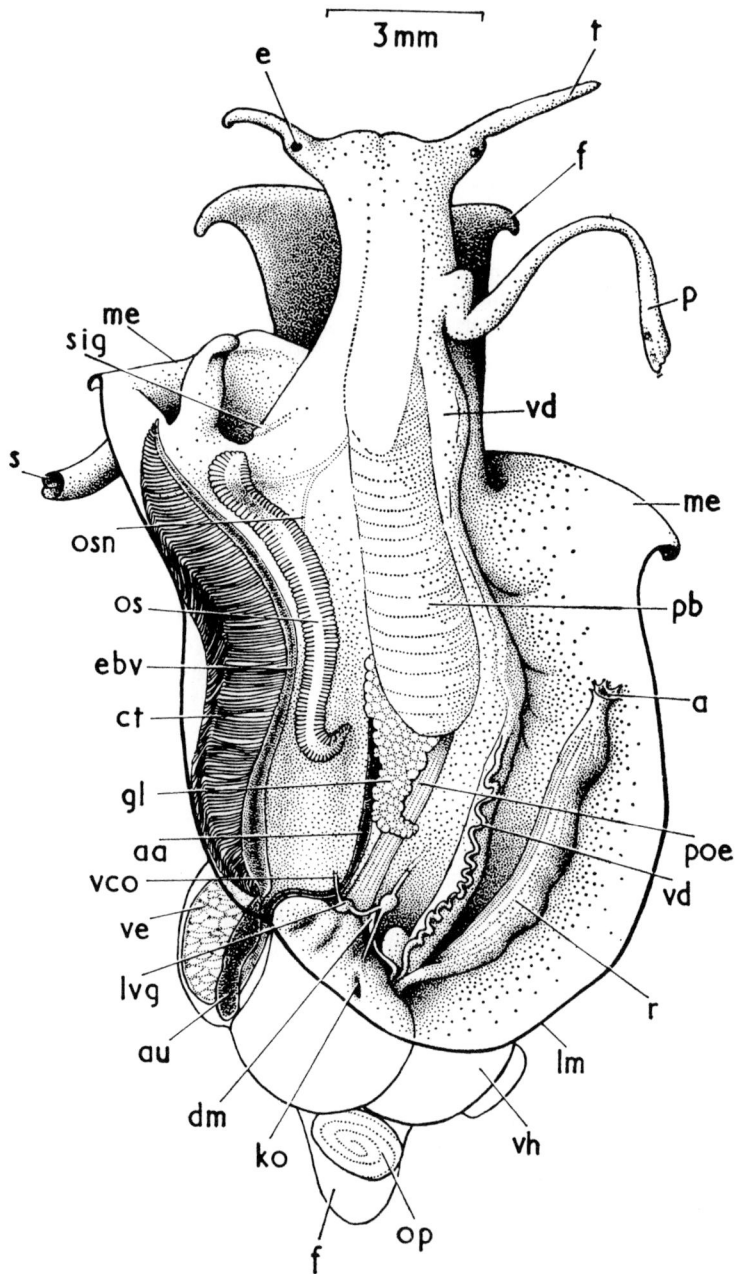

FIG. 267.—*Nassarius reticulatus:* the animal has been removed from the shell and the mantle skirt cut longitudinally along the mid-dorsal line. In addition to the contents of the mantle cavity some structures are seen by transparency.

a, anus; aa, anterior aorta; au, auricle; ct, ctenidium; dm, diverticulum of male duct opening to mantle cavity; e, eye; ebv, efferent branchial vessel; f, foot; gl, gland of Leiblein; ko, opening of kidney to mantle cavity; lm, cut edge of mantle skirt; lvg, left visceral ganglion; me, mantle edge; op, operculum; os, osphradium; osn, osphradial nerve; p, penis; pb, proboscis; poe, posterior oesophagus; r, rectum; s, siphon; sig, siphonal ganglion; t, tentacle; vco, left part of visceral loop; vd, vas deferens; ve, ventricle; vh, visceral hump.

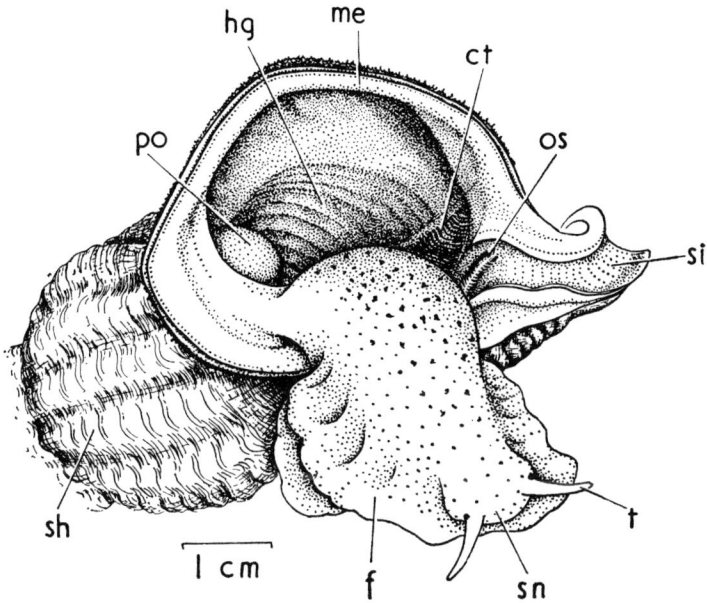

FIG. 268.—*Buccinum undatum:* animal extended from its shell and showing contents of mantle cavity. ct, ctenidium; f, foot; hg, hypobranchial gland; me, mantle edge; os, osphradium set across base of siphon; po, pallial oviduct; sh, shell; si, siphon; sn, snout; t, tentacle.

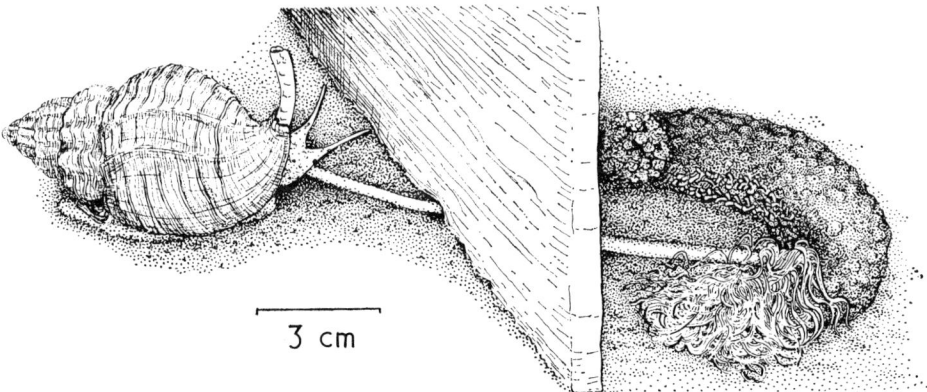

FIG. 269.—*Buccinum undatum:* animal feeding on moribund holothurian. The predator and its prey are separated by the wooden partition between 2 aquarium tanks. The prey must have been sensed by smell and the whelk has extended its proboscis underneath the partition in order to reach its food.

pierced by the point of a radular tooth and injected with secretion from the poison gland. The secretion is directed into the buccal cavity by the contraction of the muscular bulb at the blind end of the gland (pg, fig. 103) and fills the cavity of the tooth. After stabbing the prey is presumably manipulated by the proboscis and ingested whole. This is similar to the attacking action of many cones. [For recent work see chapter 25 and chapter 26.]

Most of the British toxoglossans are found on soft ground, mud or sand with perhaps some gravel, though some species of *Philbertia* favour stony bottoms. In these situations there also occur errant annelids which may be their prey, and may be hunted by the mollusc burrowing through the surface layers of the substratum. These toxoglossans show adaptations for this. The shell (fig. 29B) is the shape of an elongated spindle offering little resistance if dragged through the soil. The head-foot is separated from the visceral mass by a moderately long neck, which allows considerable play of the visceral mass as the animal creeps forwards and encounters obstacles. The mantle cavity, which is wide and depressed, may be closed to the exterior by the thickened edge of the mantle which forms a collar leaving only the inhalant and exhalant passageways exposed. There is an extensible inhalant siphon in *Mangelia* with a valve at its base which regulates the inhalant water current (fig. 34). The exhalant siphon is a small gutter formed by the mantle where the free edge of the shell meets the body whorl, and is bounded on either side by a prominent ridge of ciliated epithelium. Specimens of *Mangelia powisiana* collected from the muddy sand at White Patch, Plymouth, where it occurs with large numbers of the annelids *Lumbriconereis latreilli* and *Melinna palmata*, on which it perhaps feeds, have been seen to burrow with astounding rapidity, and Jeffreys (1867) stated that he observed *M. nebula* burrowing in the sand at Oxwich, near Swansea, where tubicolous polychaetes also abound, at the retreat of the tide. It is probably for this reason that specimens of our toxoglossans are so rarely seen. The egg capsules of a number of species have been described, for they are fixed on stones or shells, and also some of the veliger larvae. As far as is known all species have free veligers.

CHAPTER 20

THE ROCKY SHORE: OTHER PROSOBRANCHS

T HE family Trochidae includes the animals popularly known as top shells which favour
shores with broken reefs and undisturbed boulder beaches and are intolerant of
excessive exposure, since they have little power of adhesion. [For recent work see
chapter 32.] *Gibbula cineraria, G. umbilicalis* and *Monodonta lineata* are common intertidal trochids and a fourth, *Calliostoma zizyphinum,* occurs at MLWS and below. *G. cineraria*
(figs. 271, 272) is the only one which is found on all coasts of the British Isles, and *Monodonta*
has the most restricted occurrence (McMillan, 1944). It may be found in the S and W between
Poole in Dorset and the Mersey, and in the S and W of Ireland (Southward & Crisp, 1954).
Even within these areas its distribution is patchy. On sheltered coasts it is typical of the
region between high and low neap tide, though there are areas (e.g. R. Yealm at Noss, S
Devon) where it occurs abundantly to a higher level and in reduced salinity. Outside this
range it cannot survive, but the reason is unknown. *Monodonta* is conspicuous owing to the
large size of its robust shell, the exposed mother-of-pearl on the apex, the tooth (tcl, fig. 273)
on the columella (from which it is named) and its occurrence on bare rocks and stones which
are sheltered from the sea though subjected to the full force of the sun. Only when winter
temperatures are very low does it move away from exposed surfaces. Smaller individuals are
not readily visible, for they tend to live apart, hidden in pits and crevices of the rock. They
settle from the plankton at lower levels on the beach and need the protection of crevices as
they migrate upwards to the adult position.

Gibbula umbilicalis is a more familiar species for it is widespread and rather more tolerant
of wave action than *Monodonta* (Southward & Orton, 1954); otherwise it favours similar
grounds and, with *G. cineraria,* is often associated with fucoids and, in exposed regions, may
be restricted to pools and the undersides of stones and boulders which may be bare of weed.
It is absent from the E coast of Scotland and England, and in the NW is less abundant than
G. cineraria and somewhat patchy in its distribution. Its upper limit on the shore is slightly
lower than that of *Monodonta* and lies between EHWN and MHWN, though this may be
raised by splash. Its numbers begin to dwindle below MLWN, which is about the upper limit of
G. cineraria, and below MLWS the species may be absent; in contrast to *Monodonta* it is
sometimes dredged. Young animals of the two species of *Gibbula* are often difficult to find for
they are concealed in damp crevices, under stones and in rock pools along the lower part of
the beach.

The shells of *G. umbilicalis* and *G. cineraria* are found together and on occasions seem
difficult to distinguish (fig. 274). The shell of *G. umbilicalis* may be more depressed and never
pyramidal in form as is sometimes that of *G. cineraria.* Moreover the spiral ridges on the
whorls are fewer (sharper in young individuals), though this feature is not seen in worn
specimens in which the mother-of-pearl at the apex is exposed. The wear here may be so
great that an upper opening to the canal in the centre of the columella is produced. The
longitudinal lines of colour are more conspicuous in *G. umbilicalis*: they are red or purple,

broader and fewer in number, sometimes zigzag and never interrupted by the spiral ridges to give a speckled appearance. The patterning on the shell of *G. cineraria* is so fine that the surface has an ashy hue, and the narrow, closely set and oblique streaks of dark purple or brown may be interrupted by the spiral ridges and hardly separately distinguishable. The umbilicus of this shell is rather small, and in both species it may be obliquely funnel-shaped.

Gibbula cineraria is the abundant top shell on the E coast and extends further up the shore where there is more moisture at low tide than in the W, perhaps through lack of competition from *G. umbilicalis*. It is also sublittoral, though beyond a depth of about 15 f the numbers are small. Its vertical distribution overlaps that of *Calliostoma zizyphinum*, which occurs deeper and is rarely found above MLWS, tolerating only 10% exposure in the Plymouth vicinity (Colman, 1933), whereas *G. cineraria* can tolerate up to 30%. Both species favour a sheltered habitat and it is of interest to find that they are recorded from pools at the level of the topmost barnacles on the exposed rocky shores of Caithness (Lewis, 1954b). Ebling et al. (1948) showed that in situations where the current was strong specimens of *G. cineraria* were washed off weed to which they were clinging.

Calliostoma zizyphinum (figs. 29D, 71) is one of the most conspicuous diotocardians owing to its size and the colour of both shell and body tissues, which may be flecked with purple, crimson, reddish brown and yellow. Even the smallest individuals recently hatched from the egg ribbon are pigmented. Perhaps as in other primitive prosobranchs these pigments may be porphyrins (p. 130) of dietary origin. The glandular equipment of the foot is more pronounced than in the other trochids for besides the anterior pedal gland which discharges between propodium and mesopodium and consists of subepithelial mucous cells and PAS-negative cells, glands in and under the epithelium, with similar staining properties, open to the surface of the sole. Over the metapodium, posterior to the operculum, is a conspicuous V-shaped area grooved transversely and with abundant mucous cells, less pronounced in other genera in which the glands on the sole are well developed. Subepithelial glands diminish towards the posterior end of the sole. They are typically mucous glands and their secretion enables the snail to maintain a purchase on the rock. Ankel (1947) has observed that if *Gibbula cineraria* is disturbed when creeping the sole of the foot immediately produces a copious supply of mucus which helps the snail to increase its grip (p. 136).

Another top shell, the large species *Gibbula magus*, is occasionally collected on the shore and, indeed, is recorded as locally common on the rocks of Cardigan Bay (Flattely & Walton, 1922) and has been found in numbers on the coast of Connemara, Ireland, where *Lithothamnion calcareum* forms beaches between rocky headlands. Here up to 40/sq m have been seen at LWST browsing over the calcareous alga. However, its more frequent habitat is on muddy sand below LW.

In contrast to these species, both in appearance and habits, is the smallest and most primitive British trochid, *Margarites helicinus* (fig. 275). Its shell, thin and semi-transparent, is orange or reddish brown in colour and has a blue-green or purple lustre. It is primitive in that the whorls dip to the sutures and are not angulated, the umbilicus is wide and the spire low. Moreover it lacks surface ornamentation except for faint spiral striae. This species is locally abundant in areas N of the Humber on the E and the Bristol Channel on the W, and also in N Ireland, occurring gregariously amongst weed and under stones and boulders—never on exposed upper surfaces. The animal is very active and bold, and as it creeps the epipodial tentacles, of which there are 6 pairs, can be seen waving in the water (et). At its base each has a pigmented papilla with the appearance of an eye (eso). The tentacles, cephalic (t) and epipodial, are contractile, and their surface is annulated and setose. As in other trochids these

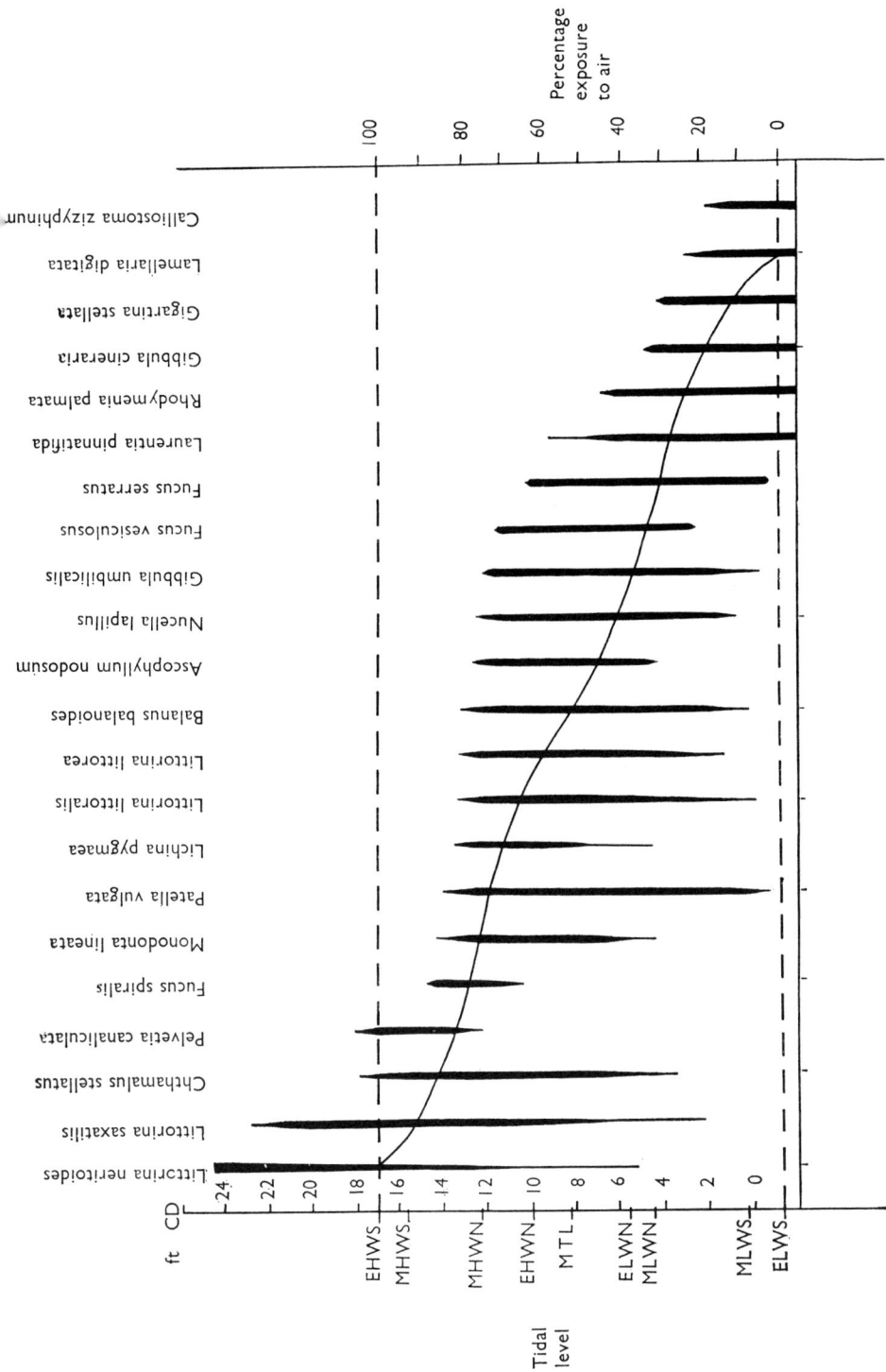

FIG. 270.—Diagram relating the vertical range of common littoral animals and plants of rocky shores to height on the shore and degree of exposure to air, based on Evans' data for Church Reef, Wembury (1947) (*Calliostoma zizyphinum* based on Colman (1933)). The distribution of each organism is given by the extent of the black line under its name. Tidal levels are given on the vertical axis at the left as are heights, in feet above chart datum. The percentage exposure to air is given along the vertical axis to the right. The superimposed curve represents the percentage exposure to air at different tidal levels.

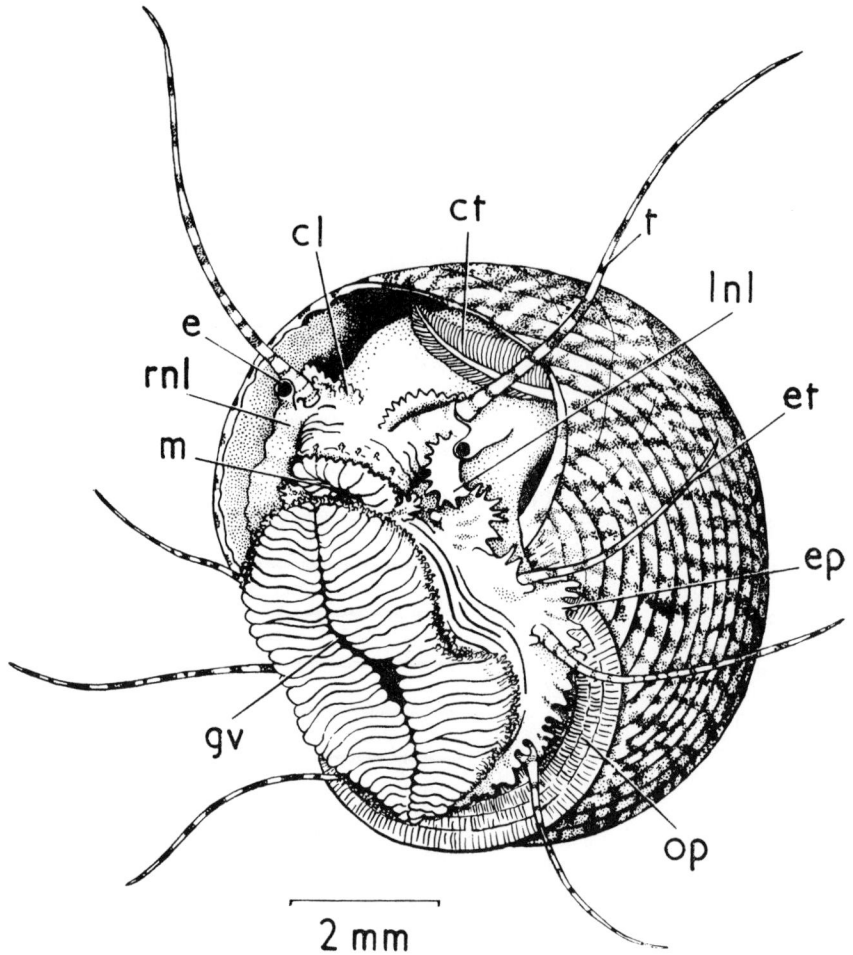

FIG. 271.—*Gibbula cineraria*: living animal.
    cl, cephalic lappet; ct, ctenidium; e, eye; ep, epipodium; et, epipodial tentacle; gv, groove along sole of foot; lnl, left neck lobe; m, mouth; op, operculum; rnl, right neck lobe; t, tentacle.

epipodial outgrowths, innervated by pedal nerves, presumably increase the tactile efficiency of the foot. Although in shell form and as the possessor of 6 pairs of epipodial tentacles, *Margarites* is primitive, yet the spawn masses found with the adults are similar in their make-up to those of mesogastropods, and a free larval stage is suppressed. Evidence suggests that this prosobranch lives little more than 1 year.

    As might be expected in such lowly prosobranchs as the trochids epipodial outgrowths are a conspicuous feature, 4 pairs occurring typically in *Calliostoma* and 3 in other species (et, fig. 271); in addition there is a neck lobe behind each cephalic tentacle. The margins of the lobes may be entire (rnl, lnl, fig. 275) or fringed (lnl, fig. 271). Each is ciliated dorsally and covered by a

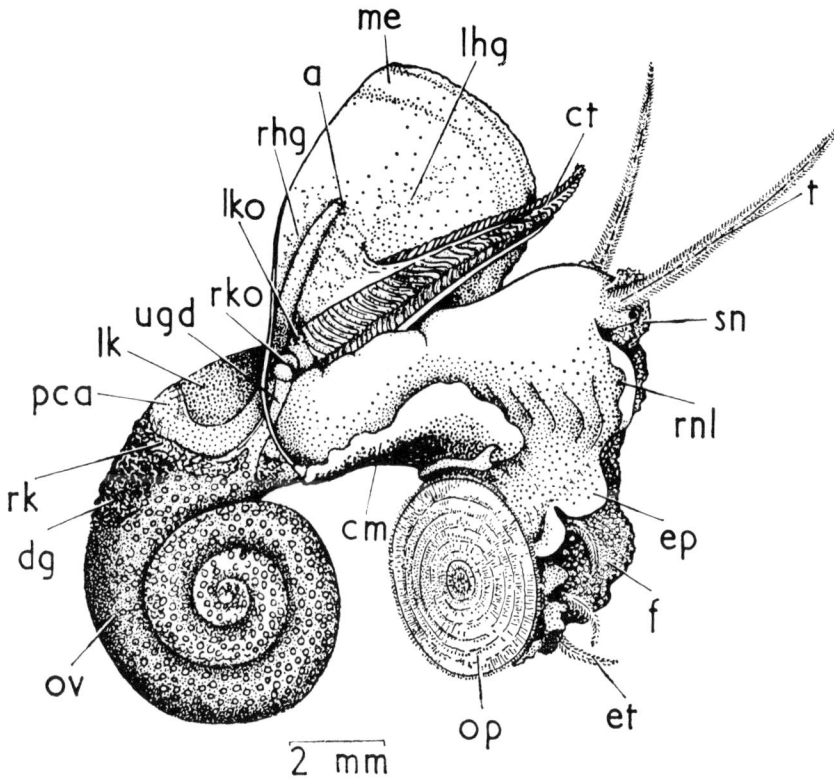

FIG. 272.—*Gibbula cineraria:* animal removed from shell and with the mantle cavity opened by a cut along the right side of the mantle skirt; seen from the right.

a, anus; cm, columellar muscle; ct, ctenidium; dg, digestive gland; ep, epipodium; et, epipodial tentacle; f, foot; lhg, left hypobranchial gland; lk, left kidney or papillary sac; lko, opening of left kidney to mantle cavity; me, mantle edge; op, operculum; ov, ovary; pca, pericardial cavity; rhg, right hypobranchial gland; rk, right kidney; rko, opening of right kidney to mantle cavity; rnl, right neck lobe; sn, snout; t, tentacle; ugd, neck of right kidney which acts as a urinogenital duct.

thin cuticle ventrally. On the left lobe the cilia beat in an inhalant direction and in the reverse direction on the right lobe. When the animal is submerged each forms a half siphon along which water passes into or out of the mantle cavity. Other lobes, innervated by nerves supplying the snout, lie one at the base of each cephalic tentacle on its median side. They are not developed in *Margarites,* are rudimentary in *Calliostoma* but large and fringed in *Monodonta* and *Gibbula* (cl). A further sensory process arises from the base of the right eye stalk in most topshells, and the sides of the foot are beset with numerous papillae covered by immobile cilia which give the surface a wrinkled appearance. These are absent in *Margarites.*

Tricolia pullus (fig. 276A) is the only British member of the Turbinidae, a family well represented in warmer seas and closely related to the trochids. It is small, the shell measuring up to 7 mm in height. It occurs in rock pools near LWST creeping rapidly over the weed with a shuffling motion, the snout being swung from side to side as the animal feeds; a longitudinal

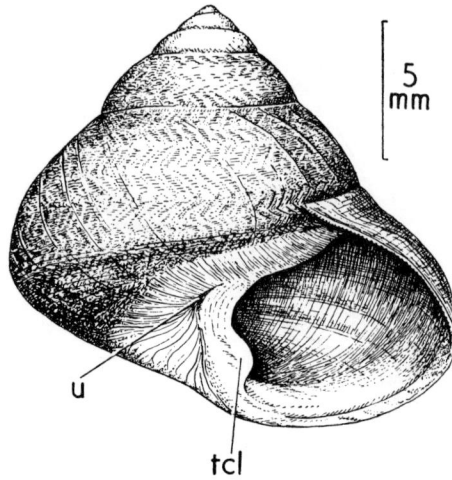

FIG. 273.—*Monodonta lineata:* shell in apertural view.
tcl, tooth on columella; u, umbilicus.

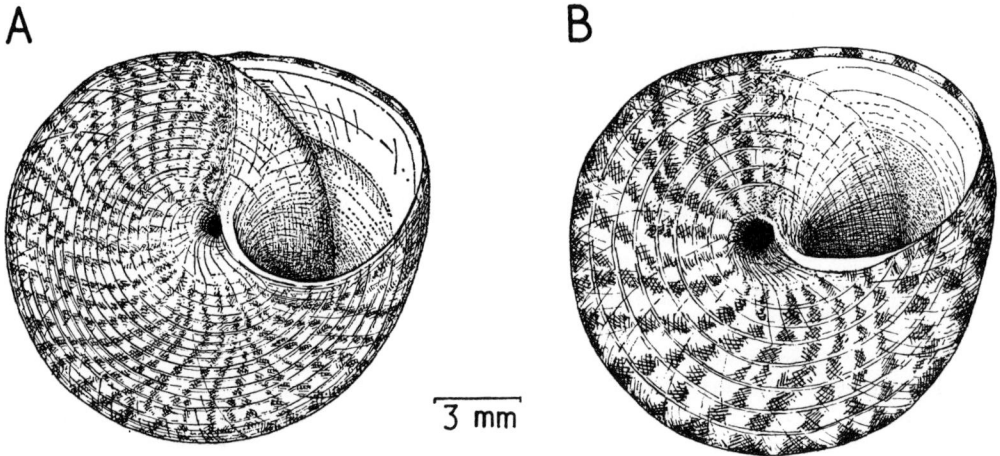

FIG. 274.—Shells of *Gibbula:* A, *G. cineraria;* B, *G. umbilicalis.*

groove down the middle of the foot marks the division between the two halves which are
advanced alternately. The red weed *Chondrus crispus* is particularly favoured, streaks on the
shell and body of the snail matching its colour; these may be due to the pigment of the food
as Ino (1949) has shown for *Turbo cornutus.* The red or purple streaks on the yellowish shell
are frequently zigzag and contrast with the emerald green mantle and green markings on the
rest of the body. This has also pink or purple lines on a background of pale yellow, whilst the

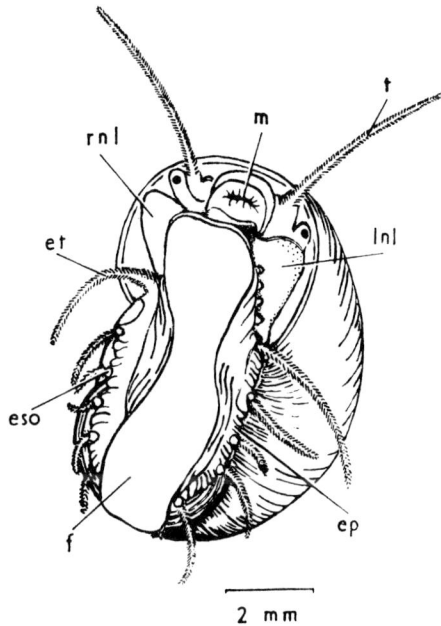

FIG. 275.—*Margarites helicinus*: living animal creeping on surface film of water, seen in ventral view. ep, epipodium; eso, epipodial sense organ; et, epipodial tentacle; f, foot; lnl, left neck lobe; m, mouth; rnl, right neck lobe; t, tentacle.

head is reddish brown. The tentacles, both cephalic and epipodial, are setose and extensile and wave through the water as the animal creeps. The animals resemble trochids in general anatomy, though differing in one obvious external feature, the operculum of the shell. In a trochid the operculum is thin, horny and multispiral: in a turbinid it is calcareous and thick with a convex outer surface. Its colour in *Tricolia* is white (fig. 39).

Another family, the Littorinidae, includes 4 periwinkles, *Littorina neritoides, L. saxatilis, L. littoralis* and *L. littorea* which may be collected on almost any rocky or stony shore, zoned from top to bottom in that order, but overlapping considerably. [For recent work see chapter 30.] *L. littoralis* may usually be separated from the three other species very readily on account of the extremely depressed spire of the shell and because of its close association with *Fucus vesiculosus* or *Ascophyllum nodosum*, but it is often a difficult matter to distinguish between specimens of the other three species, particularly when, because of age differences, they are all of similar size. For these reasons a number of critical points (mainly in their external appearance) which help in identification are summarized in Table 15.

The 2 periwinkles *Littorina neritoides* (fig. 277) and *L. saxatilis* may be collected from above EHWS especially on exposed coasts where there is considerable wave splash; the former may then be over 20 ft above the reach of the tide, and the latter about half this distance, though a height of 60 ft was recorded by Hunter (1953a) for *L. saxatilis* in the Garvelloch Islands, Argyllshire. *L. neritoides* is typically the higher, though in very sheltered places where the upper limit of both species is lowered so that they extend only as high as MHWS, the distance

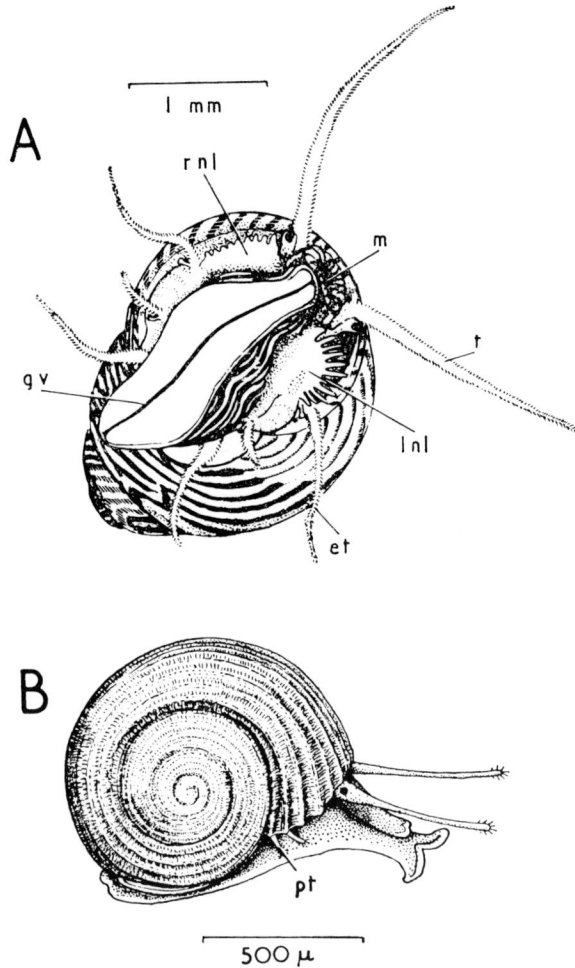

FIG. 276.—A, *Tricolia pullus,* living animal in ventral view. B, *Circulus striatus,* living animal from the right.
et, epipodial tentacle; gv, groove along mid-line of sole of foot; lnl, left neck lobe; m, mouth; pt, pallial
tentacle; rnl, right neck lobe; t, tentacle.

of a foot may separate them. *L. neritoides* favours steep, rough rocks which are very exposed, even where splash is strong, but requires cracks or crevices in which to shelter; its position on the shore is probably explicable in terms of gravity, light and the presence or absence of water (Fraenkel, 1927a). This species may also live submerged in high rock pools and water-filled pits in the rock. Increased wave action not only raises its upper limit of distribution, but also lowers its lower limit and may bring it below MLWN. In very sheltered places the vertical distribution is, in contrast, narrow, and the snail may not be found below MHWN. *L. saxatilis* favours a less exposed habitat and is reduced in numbers where exposure is high. It collects in crevices of irregular and broken rock, especially where there are plenty of pools and is

TABLE 15

Diagnostic Characters of Littorina *spp. with the Most Useful Italicized*

	Littorina neritoides	Littorina saxatilis	Littorina littoralis	Littorina littorea
Spire	High	High	*Very low*	High
Columella	*Dark*	Light–dark	Light–dark	*White*
Junction of outer lip with body whorl	Acute	*Rectangular*	Acute	Acute
Periostracum . . .	*Lip to aperture*	None visible	None visible	None visible
Black stripes on tentacles .	Longitudinal	Longitudinal	Longitudinal	*Transverse*
Breeding	Oviparous	*Viviparous*	Oviparous	Oviparous

abundant among *Pelvetia* and other weeds. Unlike *L. neritoides* it is also found on sheltered, level, rock faces. The lower limit of the species is variable: specimens may be collected at MLWS though usually the numbers dwindle at ELWN. Occasionally it is dredged for it tolerates permanent immersion. Thorson (1944) stated that living specimens are known in E Greenland from depths of 57 m and in W Greenland from 94 m. The occurrence of this species on a rocky shore in the British Isles is more regular than that of *L. neritoides* and it is plentiful on stony and silty beaches and invades estuaries where it is found with the edible winkle. At Rum Bay, Plymouth Sound, Moore (1940) found up to 3,000/sq m in a stony gully considerably overgrown with fucoids, and on the intertidal mud flats at St John's Lake, the Tamar Estuary, up to 1,100/sq m have been recorded (Spooner & Moore, 1940), the highest figures coinciding with the occurrence of *Zostera*. The snail may be found attached to weed, stone or any solid holdfast on the flats between high water and mid-tide. It often abounds in shallow brackish waters and Howes (1939) recorded it swarming on *Ruppia* in a saline Essex creek. Its rapid colonization of such habitats may be associated with its viviparity.

The high position of these 2 periwinkles on the shore indicates that they can withstand a considerable amount of exposure to air and fresh water which reaches them as rain. *Littorina neritoides* is the more resistant. It can survive absence of moisture for at least 5 months (Patanè, 1933) and regains activity within a few minutes when replaced in sea water. Colgan (1910) found that it is unharmed after exposure to dry air for 42 days, and it will survive when the water content of the body is reduced to 66% (Fischer, 1948). The gill in this species, and in *L. saxatilis*, has a reduced number of leaflets (ct, fig. 277C) as compared with the other periwinkles and the leaflets are small. Respiration is remarkably efficient as the winkles can breathe in water when its oxygen content is as low as 0·24 ml/l and they use 99·4% of the available oxygen before dying (Fischer, Duval & Raffy, 1933). The mantle cavity contains water even after long exposure to dry air, and as this moisture is reduced the operculum withdraws further into the shell. When quiescent on dry rock the metabolic rate of the winkles is low. It has been calculated that the oxygen consumption of *L. neritoides* in dry air is 5–6 times lower than in water (Fischer, Duval & Raffy, 1933). Both species can withstand immersion in fresh

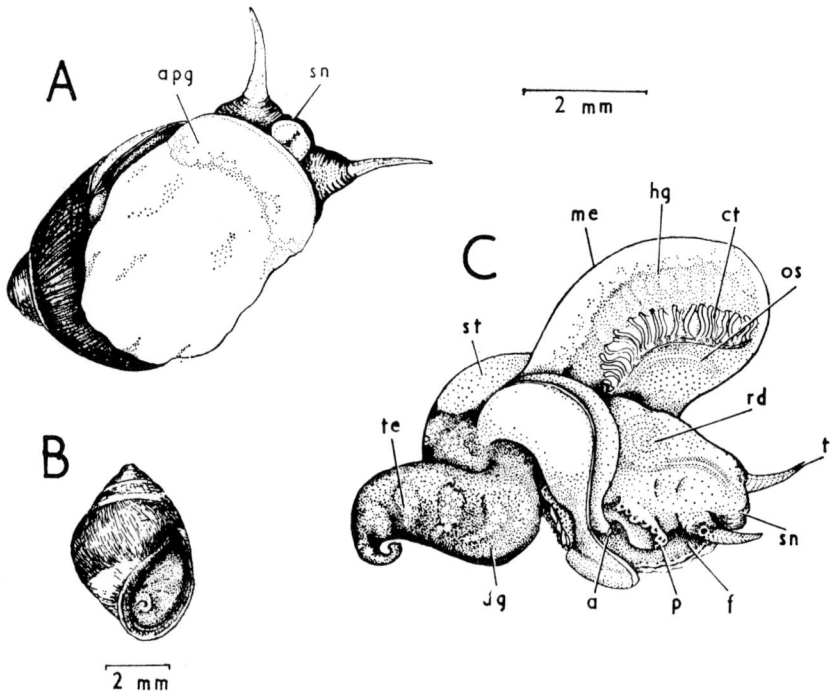

FIG. 277.—*Littorina neritoides*: A, living animal in ventral view; B, animal withdrawn into shell; C, animal removed from shell with the mantle cavity opened by a median dorsal cut, seen from the right.

a, anus; apg, anterior pedal gland seen by transparency; ct, ctenidium with reduced number of leaflets; dg, digestive gland; f, foot; hg, hypobranchial gland; me, mantle edge, cut; os, osphradium; p, penis with glands; rd, radula in radular sac seen through floor of mantle cavity; sn, snout; st, stomach; t, tentacle; te, testis.

water for a week, and *L. neritoides* for 11 days. In light rain they are often very active and will creep about the rocks and feed. Some of the high rock pools in which they live have a high salinity, especially in summer, and Colgan (1910) found that both species survive in sea water of treble salinity for about a week. Lysaght (1941) suggested that *L. neritoides* avoids highly saline water, for when the salinity increases at low tide the snail crawls out of the pits which it inhabits on the Plymouth Breakwater, and returns when the water has been renewed by the tide.

The food of these winkles is lichens, minute algae and diatoms which are gathered from the substratum. *Littorina saxatilis* browses over *Pelvetia* and *Ascophyllum*, sometimes rasping the surface and eating decaying tissues, but usually collecting diatoms. The animals move about and feed when their habitat is moist, and if it is calm they are as active in slight rain as when covered by the tide. On warm days under heavy sea mist the majority of individuals are on the move when the tide is out, and it is at such a time that they may be seen to copulate. Hot dry weather may limit the feeding period for those living above high tide, and they appear to be able to go for weeks without food. Snails of all sizes, except the very young of *L. neritoides*, are found together in sheltered places and the smaller ones may be clustered in empty barnacle shells or rock crevices with the adults. These small snails are able to survive emersion for a much shorter period than the larger ones. The choice of a dry and apparently adverse habitat by the adult may be correlated with the need for shelter from the force of the waves, and, except for spawning which appears to be associated with certain spring tides, the presence or absence of water seems to be of secondary importance since they have become better able to withstand desiccation. [See p. 685.]

Females of *Littorina saxatilis* are distinguished from males by the round aperture and the larger size of the fully grown individual. The brood pouch contains embryos in all months of the year, but the numbers may be small during the summer and are highest in spring. A fluctuation like this is pronounced only on certain shores—indeed Linke (1933) working on animals from the North Sea found the brood pouch full all the year. Berry (1956) studied a population at Whitstable, Kent, where there is a decline in reproductive activity between June and mid-August, when the penis and vesicula seminalis become reduced in the male and the ovary in the female. He found that snails living high on the beach grow faster than those at lower levels since they have a longer period of activity when uncovered by the tide: they attain a larger size and produce more eggs. The varieties described below appear to differ in breeding times.

Lysaght (1941) studied the growth rate of *Littorina neritoides* in a population on the Plymouth Breakwater where conditions for shelter and continuous feeding are very favourable and lead to the production of large individuals—one with a shell height of 10·4 mm was found. Many of the larger snails live more or less permanently in water in pits on the top of the Breakwater, and the smaller ones are most abundant on the exposed southern slopes. The growth rate appears to decrease with age and is very low in snails of 6 mm or more in height, from which it follows that the larger individuals must be many years old. There is evidence that females of *L. neritoides* grow faster than males for they are significantly more numerous amongst the larger individuals, and this difference in growth rate between the sexes appears to hold for *L. saxatilis* too (Pelseneer, 1926a). The oldest specimens of *L. saxatilis* were suggested by Thorson (1944) to have lived for at least 6 years.

The colour of the shell and of the exposed parts of the body of *Littorina saxatilis* is very variable. [See p. 573.] The former is frequently yellow, brown, orange or red and sometimes purple, black or pure white, and the body tissues various shades of white, yellow, brown or flesh-colour streaked with deep purple. It is unknown to what extent these colours are linked with food or other environmental factors and to what extent they are genetic. Varieties of the species have been based on them and on differences in shape, size and sculpture of the shell and on habitat. One, *saxatilis* of Johnston, is normally very small, with a shell height of not more than 4 mm, though it may grow larger than this in favourable circumstances. The largest specimens which we have seen were 9 mm high and were collected by Professor L. A. Harvey on wave-beaten rocks on Lundy Island; others approaching this size have been obtained by G. M. Spooner, Esq., on the Eddystone. This variety prefers rock crevices and empty barnacle shells throughout the barnacle zone and above, extending to a lower level on the beach than other varieties. In addition to its small size the shell is recognizable by its squat shape (due to the fact that the spire is small in relation to the body whorl), and the presence of spiral brown lines on a grey to white background. The lines vary in breadth and number, sometimes simple, parallel to and alternating with the spiral striae and usually with a broader one below the periphery, sometimes interrupted, sometimes cutting across the striae and interacting to give a reticulate pattern which appears to be associated with the most exposed habitats. The junction of outer lip and body whorl is rectangular; the spiral striae are not pronounced.

A second variety, *patula* of Jeffreys, is found high up on rocky shores at about the upper limit of barnacles, extending above and below and not avoiding a slight amount of weed cover. It is of the normal size associated with this species. Its shells vary in colour and are often black, grey, yellowish-grey or yellow, without colour patterning when adult though frequently clearer and paler in the neighbourhood of the aperture and darker on the spire. Some young ones may show occasional brown streaks. Coarse spiral striae are of regular occurrence. The variety may be further recognized by its moderately elongated shape, the sudden increase in size of the penultimate whorl and by the way in which the mouth is a little constricted by the thickening of the outer lip. The variety *tenebrosa* of Montagu has, in most cases, a smooth shell of about 6 whorls with a prominent spire. The colour is dark, usually with a brownish or orange tinge, and there is frequently a mottling of irregular, angulated streaks of light colour, though this is not as conspicuous as in *saxatilis*. The shell is thin, the outer lip never thick-edged and it meets the body whorl practically at a right angle; the outer lip is slightly dilated and there is rarely any trace of umbilicus. The largest variety, *L. saxatilis rudis* of Donovan, which may measure up to 28 mm in height and which the careless eye may easily mistake for *L. littorea,* has a very prominent spire of ventricose whorls (fig. 29A), with the apical angle less than in the other three varieties. The colour is yellow, orange or darker, or it may be white with dark spiral bands (sometimes then known as *nigrolineata*). The surface is smooth or very weakly striated below the periphery of the body whorl. The mouth is distinctly small for such a big shell and this is due almost entirely to the thickening of the peristome where inner and outer lips meet. Here, too, the mouth is turned out slightly to the left. There is almost always a dimple marking the position of the umbilicus and the parietal lip in this region is thick. The shell may be easily distinguished from that of *L. littorea* by the swollen whorls, the dark coloured columellar region and the straightness of the inner lip.

It is well to add that these varieties grade easily into one another and intermediates are to be found which cannot be satisfactorily placed in any one of them.

The edible winkle *Littorina littorea* (fig. 8), one of the most abundant of littoral gastropods, ranges from HWNT to LWEST on various types of shore. Its upper limit of distribution appears to be raised by splash (Moore, 1940; Evans, 1947), and Waterston & Taylor (1906) described it high above the sea on St Kilda. It may occur lower on stones and gullies than on a stretch of shore covered by fucoids. This is the largest of our British periwinkles and occasional specimens may attain a height of 38 mm. The shell is commonly soot-coloured, or it may be reddish orange, brown and, rarely, white. Fulvous and brown shells may be zoned with narrow rings of red or grey. The body has usually a dark hue with closely set black lines of pigment on a yellowish ground, except for the sole of the foot which is yellowish. The tentacles are annulated or streaked across with black, and appear flattened when contracted. Their pigmentation is a useful feature in distinguishing between small individuals of this species and those of the other British species of *Littorina* for in the latter there are two darker longitudinal stripes towards the tip of the tentacle, one on the upper and the other on the lower surface. On occasions the dark pigmentation of *L. littorea* is lacking. The winkle lives on rocks and amongst small stones, on gravel and on wooden structures, even on soft mud or sand (particularly if sheltered or poorly drained), but only if stones, boulders or tufts of weed occur to provide a firm base amidst these soft surroundings. It avoids a shifting substratum such as shingle, and is infrequent on chalk and limestone rock, and shores exposed to the full force of storms, for it is less tolerant of surf than *L. saxatilis* (Evans, 1947). On sheltered rocky coasts with clear water it is found at the same level as *L. littoralis,* but unlike this species, which is less resistant to drought and fresh water, away from dense algal covering. It can tolerate water of a salinity as low as 10‰ (Fischer, 1948) and often penetrates estuaries which are polluted. Indeed it can survive for some days in water without oxygen, lying meanwhile with the operculum closing the shell (Thamdrup, 1935). In the Blackwater Estuary, Essex, winkles have been dredged in considerable numbers in certain channels below low water (Wright, 1936), and there are records of specimens being dredged in other localities around our coasts to a depth of about 35 fathoms; it seems probable that some individuals may be submerged all their life. Until about 25 years ago there was an important periwinkle fishery at Maldon in the Blackwater Estuary and fisheries occurred elsewhere. The official statistics show that in 1928 22,290 cwts of periwinkles were collected in England and Wales with an average value of 11s. per cwt. Since that date the fishery has declined and is now negligible.

Littorina littorea tends to aggregate in damp or sheltered places at low tide and clusters may be seen resting in shallow pools and depressions in the rock, filling the angle between a vertical and horizontal slope, or heaped around stones and small clumps of algae in sandy and muddy areas. Estimations of population density around the Plymouth area show that in some localities on a rocky shore concentrations of over 950/sq m may be encountered during a low-tide period when the winkles are inactive, whereas not a single individual may occur on adjacent rocks. These clusters appear more or less permanent, for a favourable site is usually filled, many of the same animals returning repeatedly to it after feeding forays, though sometimes they join other groups. If the surface is damp the winkle clings to it with its foot, its head withdrawn into the mantle cavity. On bare rocks and boulders groups of periwinkles may appear completely dry on a hot day when the tide is out. Most individuals are then orientated with the head and the lip of the shell uppermost, and the apex directed downwards (Haseman, 1911). The mollusc is not clinging to the dry rock with the foot, but the lip of the shell is stuck to it by a film of mucus which becomes hard and brittle (Wilson, 1929). This secretion, presumably from the anterior pedal gland, is produced as the substratum dries, and as it hardens the foot is withdrawn. Its strength is sufficient to hold the snail even on a vertical surface, though on gusty days individuals may be blown off. Occasional periwinkles orientate themselves with the top of the shell upwards, but they invariably topple over. The same habit is found in L. neritoides and L. saxatilis, in the former of which it is facilitated by the presence of a pliable flap of periostracum along the outer lip of the shell (pms, fig. 281). If an individual be picked off the dry rock and the operculum touched it will withdraw further into the shell and expel water from the mantle cavity, for the winkle is not completely retracted when it lies motionless and the mantle cavity retains a supply of water. Thamdrup (1935) has calculated that the oxygen consumption of the edible winkle in air is only 26% of the consumption in water. It can live in dry air attached by its mucous support for at least 3 weeks and when splashed with water again it soon secures a holdfast with its foot and begins to creep.

The periwinkle gathers micro-organisms and detritus from the surface on which it lives, including the young sporelings of weeds and the young stages of other attached organisms, and so it must contribute to the control of settlement. It also eats decaying plant and animal tissue. As it feeds the radula erodes a soft rock surface. North (1954) studying Littorina planaxis in S California found that a snail feeding in a rock pool refills the gut 4–8 times daily, the food taking $2\frac{1}{2}$–6 hrs to pass through the body; the erosion of sandstone is estimated as deepening a tidal pool 1 cm every 16 years. L. littorea will feed at high tide on calm days, and at low tide if the substratum is not too dry and the humidity high (Thamdrup, 1935). Gowanloch & Hayes (1926) recorded the rate of creeping as 0·8–2·4 m/hr (mean 1·3 m) and from this Thamdrup (1935) calculated that during an ebb tide 4–5 m might be covered, but since the winkle often pauses to feed the longest distance travelled may be only 1·5–2·0 m. Where winkles live on a soft surface the passage of the foot makes a well defined track $\frac{3}{4}$–1 cm broad and with low edges. At Whitstable, where the winkles settle on pebbles scattered over flat sand, Newell (1958a) described U-shaped tracks radiating from the stones, each indicating a period of activity and a return to the site of rest. Those which are visible on the wet sand as the tide recedes during the day are mostly orientated with respect to the sun, indicating that the winkle first crawls towards the light and then, after a time, reverses its direction of movement (Newell, 1958b). Newell stated that on this shore the animals are active as the tide ebbs and as the incoming tide reaches them, but that they remain stationary at most other times. Owing to the turbidity of the water he was unable to follow their movements after the tide had covered them. Haseman (1911) working on a different type of shore observed that when the periwinkle is well submerged it crawls at random.

Gowanloch & Hayes (1926) carried out migration experiments with L. littorea in the harbour at Halifax, N.S. Their conclusions support a belief that displaced individuals migrate back to the intertidal level from which they were removed, thus presumably seeking the zone for which they are most adapted, for Hayes (1927a) has shown that winkles from different levels of the beach are characterized by different levels of phototactic behaviour. Gowanloch & Hayes (1926) found that in winter, when temperatures are low, the periwinkles show a decreased activity and are unable to cling to the substratum, so that they get washed to lower levels where they are just covered at low tide. They stay there until the spring when, according to the authors, it is negative geotaxis, positive phototaxis and response to heat stimuli, with perhaps other factors, which cause them to regain the higher level. In extreme cold in this country one finds that winkles become inactive, are unable to cling to the shore, and are washed down by the tide.

This littorinid was introduced into the United States of America from the British Isles and was recorded on the coast of Maine in 1868. By 1880 it was one of the most abundant snails along the Massachusetts coast (Morse, 1880) and has been credited with driving the indigenous species L. palliata from the shores of New England. Its introduction into Canada was earlier for there are records of it in New Brunswick in 1855.

The growth rate of the edible winkle has been studied by Moore (1937a). He found that specimens from Drake's Island, Plymouth, collected from stones from MT to between MLWN and MLWS, have a shell height of about 14 mm at 6 months old; a year later when the winkle is sexually mature this is increased to 17·4 mm. At $2\frac{1}{2}$ years the average height of the shell is 22·4 mm and at $4\frac{1}{2}$ years 27·3 mm; the number of individuals exceeding this height is few, though shells 36 mm high are found in this locality. In the different habitat of Trevol (=Treval of O.S. maps), where winkles cluster on the mud flat in St John's Lake, the growth rate is similar, though it is less in the Yealm where the ground is stony and the winkles cluster on the few fucoids present. Hayes (1929) found a progressive decrease in the growth rate of the adults with increasing height above low water at St Andrews, N.B., and Moore (1940) also observed that the largest are rare or absent at the top levels, though medium-sized individuals occur throughout the local vertical range of the species. Females grow faster than males and this explains their preponderance in the larger sizes. Up to a height of 25 mm the sexes may be equal in numbers, but in taller shells the proportion of females rises rapidly to three-quarters of the population and females made up 76·9% of a sampling from Trevol with a mean shell height of 29·8 mm (Moore, 1937a). Whether they live longer than males is not known; the oldest record for an individual kept in captivity is nearly 20 years (Woodward, 1913).

In his survey of the population at Trevol Moore (1937a) sieved samples down to a depth of about 5 cm to find the smallest winkles. In this way he procured at MT individuals with a shell height of 0·5 mm which must have settled very recently. Between mid-June, when this method of sampling began, and mid-July the shell height increased to 5 mm. During the period immediately after settlement the mortality rate is undoubtedly very high, and from the size distribution curves of the Trevol population Moore (1937a) calculated that the loss of the recently settled individuals is at least 94% in July and August. For the rest of the first year the percentage mortalities are estimated at 66%, and in later years 57%. For the other two localities which were studied at the same time, Moore stated that sieving out the minute forms is difficult, as the ground is stony, and consequently the proportion of first year shells to older ones collected is too low up to a height of about 1 cm. Moore's work implies that the population of winkles on the shore is reinforced by young ones settling there. He did not study the movements of the young after settlement to find whether there was any suggestion of migration to different levels, or whether those which fortuitously settle on a favourable area merely remain there and others die.

Smith & Newell (1955) studied a population of edible winkles on a stretch of flat stony beach which covers the upper half of the middle shore at Whitstable. Here they found that adults with a shell height of 13–16 mm, which were regarded as second year individuals, make up the bulk of the population. Few winkles survive beyond this age. The specimens were hand picked and search was made for the smallest. However, only two individuals of a shell height of 2·5 mm, and nothing smaller, were seen out of several thousand collected, and the population never contained more than 4% with a shell height less than 5 mm. From this it is argued that the stock of periwinkles on the shore is renewed by the migration on to the beach of young which have been living a benthic life off-shore, or that these individuals may be cast on the shore by waves. They considered that during the first year the winkles move over the stony beach, establishing themselves in the adult zoning and thereafter their distribution pattern remains more or less constant.

Littorina littoralis [this account may include references to *L. mariae*; see chapter 30] is readily distinguished from other British periwinkles by the remarkably depressed spire of the shell (for which reason it is known as the flat periwinkle) and also for the variability in shell colour. This winkle, which occurs both on the E coast of N America and the W coast of Europe, is referred by Dautzenberg & Fischer (1914) to a subspecies of *Littorina obtusata*. Winckworth (1932), in his list of British marine Mollusca, followed Forbes & Hanley (1850) in giving it the status of a species and so separating it from *L. obtusata*, which has a subarctic distribution. Jeffreys (1865), unfortunately, referred to the British as distinct from the continental form as *L. obtusata* and, indeed, of the two it has the more obtuse shell. In both species there are varieties in form as well as colour, though the form varieties of *L. littoralis* show many transitions and are often difficult to distinguish, particularly in the young stages. Linke (1934a) and Barkman (1955) have suggested that they are phenotypic variations. Form and colour vary independently. The colour is not known to be influenced by light, salinity or water movements, nor is it certain that these varieties are correlated with particular species of algae (Linke, 1934b); transitional forms are comparatively rare. Dautzenberg & Fischer (1914) recognized 12 colour varieties. In 5 of these the shell has a uniform colour: *citrina* (yellow), *aurantia* (red-orange), *rubens* (vermilion), *fusca* (blackish) and *olivacea* (olive green). The other 7 varieties have the body whorl either banded in a contrasting colour, marked with fine zigzag or reticulate lines, or banding and reticulate pattern may be combined: *zonata* (single band of light colour), *alternata* (2 yellow bands), *inversicolor* (shell light with 2 dark bands), *rhabdota* (shell light with numerous reddish brown bands), *ziczac* (shell light with numerous narrow dark zigzag lines), *reticulata* (shell light with a criss-cross of darker lines), *inversicolor-reticulata* (a combination of these two varieties). The shells of *citrina* are some times a bright green owing to the presence of unicellular algae, and algae may camouflage the colouring of other varieties. Milk-white shells, the variety *albescens,* also occur.

In his extensive work on this species Barkman (1955) concluded that most of the colour varieties are genetically distinct, though he suggested that *reticulata* is the juvenile form of *fusca,* and he also pointed out that young specimens of *inversicolor* have the appearance of *citrina* or *albescens.* Bakker (1959) suggested that young *olivacea* cannot be distinguished from young *citrina.* In NW Scotland Barkman found 9 of the 12 colour varieties originally described by Dautzenberg & Fischer (1914), only *rubens, alternata* and *inversicolor-reticulata* being absent; *fusca* is by far the most abundant. The variety *albescens* is also there.

This winkle is typically littoral and is especially associated with the fronds of the large brown algae *Fucus* and *Ascophyllum,* which offer protection against surf, desiccation and rain, as well as providing food and a spawning place. The presence of algae determines the upper limit of the species. Van Dongen (1956) has shown that the winkles are attracted by the scent of the Fucaceae from a distance of 1 m. The attraction is strongest to *F. vesiculosus* and least to *F. serratus,* and the two varieties which he worked on, *citrina* and *olivacea,* showed no marked difference in reaction. At close quarters both scent and taste play a part in the selection of weed. The snail is also found in smaller numbers on *Laminaria, Pelvetia, Chondrus* and a few other algae, though it prefers members of the Fucaceae as food; this has been confirmed by Bakker (1959). When the tide is out the winkle shelters amongst the weeds and if the humidity is high it will feed, rasping the tissues and anything which may have settled there. It favours somewhat sheltered, rocky coasts where the water is clear, avoiding turbid waters, and is intolerant of heavy surf. The lower limit of the vertical distribution of both *Littorina littoralis* and *L. littorea* is influenced by surf and shifts to higher levels on surf-beaten shores. However, 3 varieties of *L. littoralis*—*citrina, inversicolor* and *ziczac*—do not appear to thrive where surf is slight (Barkman, 1955).

Together with the other species of British periwinkles *Littorina littoralis* displays negative geotaxis. This is equally strong under and out of water when the humidity is high. The reaction remains constant even after long periods of immersion (370 hrs) in a confined space, and observations on the variety *olivacea* indicate no difference in this behaviour in individuals collected from different levels. This variety shows a stronger negative geotaxis than var. *citrina,* which lives at a lower level (Barkman, 1955), and its negative phototactic responses are more pronounced; it usually burrows into the tangle of algal fronds whereas the latter variety lives mainly on the surface of the weed. Reactions to gravity and light are eliminated by desiccation, which causes the snail to withdraw into its shell. Under laboratory conditions it can withstand 6 days emersion (Colgan, 1910).

Tidal movements influence the behaviour of the snail, though this has been clearly displayed only on a stretch of shore from which algal growth had been removed. Wubben observed (quoted by Barkman, 1955) that as the tide begins to rise the animals leave the water and climb the shore until desiccation stops them. As the rising tide moistens them they cling to the stones and then when covered by water climb again, but stop before the tide is full. On the falling tide they move downward, though this movement is much slower than the upward one, and they finally reach a level where the stones are dry, and come to rest. A trickle of water draining over the snails may induce them to move downwards, following the water, and it is this reaction, together with the combined influences of gravity, desiccation and light, which may account for movements synchronous with the tides.

Female individuals of *Littorina littoralis* are larger than males and slightly more numerous (Pelseneer, 1926a). Some hours after copulation, which may last 10–85 min (Linke, 1934a), spawn masses are deposited on the fronds of fucoids. Apparently no spawning takes place on other algae. The young hatch as miniatures of the adult and colonize the same habitat, though they get washed from the fucoid fronds and may be collected from more sheltered habitats nearby. Colman (1940) recorded numbers of juveniles of this species along with juveniles of *L. littorea* and *L. saxatilis* on *Lichina pygmaea*. This lichen grows as small, closely set tufts 2–3 cm in height and its finely branched fronds offer considerable protection to small littoral animals. Other young periwinkles may be found amongst the dense weed growth of rock pools. The young are less tolerant of salinity changes than the adult and, according to Hertling (1928), the eggs cease to develop below 25‰ total salinity. Adults have been recorded in water of a salinity as low as 10‰ (Fischer, 1931).

The chink shells, species of *Lacuna,* are classified with *Littorina* as the most primitive marine mesogastropods. Their popular name is derived from the shape of the prominent umbilicus, which is an elongated slit lying alongside the inner lip and continuing the spiral coiling of the columella. In their general anatomy they resemble *Littorina* (fig. 278). However, they are restricted to the lower level of the shore, occurring near low water and below, and no species can endure long periods of desiccation. They may be found in slightly brackish water: in the Baltic Jessen (1918) recorded *L. vincta* and *L. pallidula* at a salinity of 12–13‰. They are herbivorous, collecting food from the surface of weeds and from rocks and stones. *Lacuna vincta* is the most abundant species, living on brown, green and red weeds and depositing its spawn on the fronds in spring and early summer. As it creeps with an apparently unsteady and awkward gait, the 2 metapodial tentacles (mt), characteristic of the genus, can be seen projecting beyond the operculum. During this, secretion is poured out from a vast anterior pedal gland which, at least in *L. pallidula,* extends into the haemocoel below the buccal mass, and from a well developed sole gland lying in two parts, one in each lateral half of the foot. The position of the sole gland can be associated with the bipedal stepping locomotion. *L. pallidula* sometimes occurs with *Littorina littoralis* which it resembles in shell shape, for the spire is flat and the body whorl expands outwards. The spawn of these 2 species is so alike that there is often confusion over the identification (p. 359 and 691) and in both the larval stage is suppressed. This species is sexually dimorphic, and the male, 5–6 mm in shell length, may be found clinging to the shell of the female which is twice the size (fig. 195); they are annuals. The male dies after a period of copulation and the female lives only a month or two longer (Gallien & Larambergue, 1936, 1938).

Among the most numerous prosobranchs of some rocky shores are the rissoids, though they may easily pass unnoticed because of their small size, and of the fact that they live in obscure situations. The shell, which measures only a few millimetres in height, is produced into a conical spire, and there is considerable variation in its sculpture. The surface may be smooth or ribbed (fig. 39), striated spirally (fig. 43C) or cancellated. In *Alvania* and *Rissoa* there is a labial rib. There are 24 species of the family Rissoidae listed for the British Isles, all originally so classified on the bases of shell characters, radula and such soft parts as are seen when the animal creeps about. Characteristic of many members of the family are the tentacles developed from the opercular lobe of the foot (mt, figs. 279, 280), and one or two tentacles at the edge of the mantle skirt, which project over the edge of the shell when the animal is extended. There may be only one pallial tentacle (pt, fig. 67), though two, one on each side (pt,

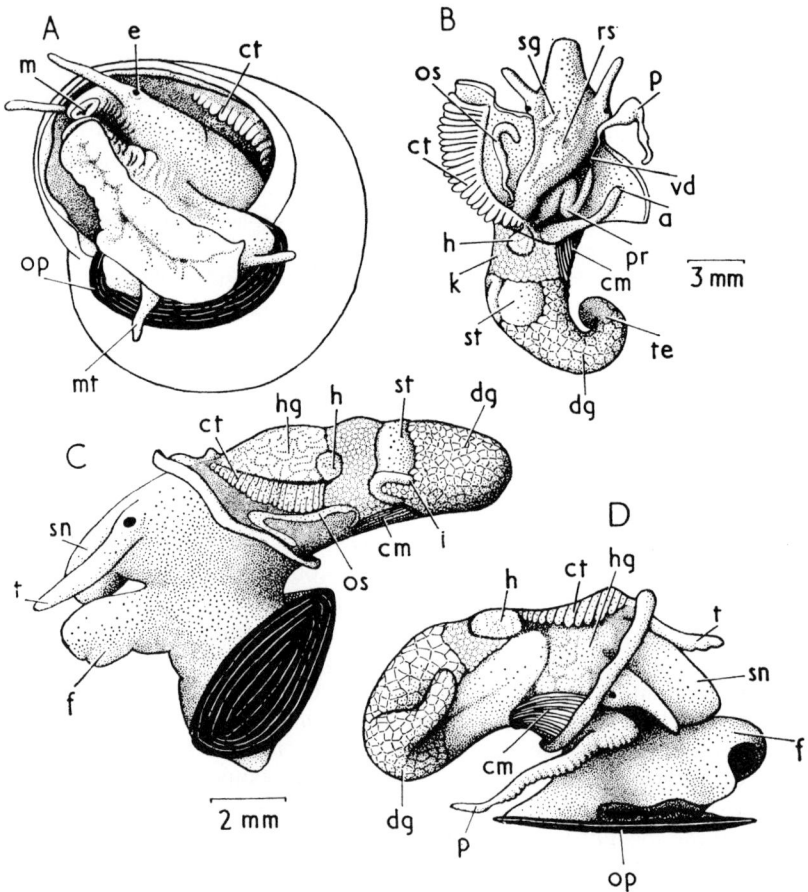

FIG. 278.—*Lacuna pallidula*: A, living animal in ventral view; B, male animal removed from shell with the mantle cavity opened by a mid-dorsal cut, in dorsal view; C, animal removed from shell, from the left; D, male animal removed from shell, from the right.

a, anus; cm, columellar muscle; ct, ctenidium; dg, digestive gland; e, eye; f, foot; h, heart in pericardial cavity; hg, hypobranchial gland; i, intestine; k, kidney; m, mouth; mt, metapodial tentacle; op, operculum; os, osphradium; p, penis; pr, prostate gland; rs, radular sac seen through dorsal wall of head; sg, salivary gland seen through dorsal wall of head; sn, snout; st, stomach; t, tentacle; te, testis; vd, vas deferens.

fig. 280), occur in some species of *Alvania* (*A. cancellata, A. carinata, A. abyssicola, A. punctura*). In other members of the family neither pallial nor metapodial tentacles are developed (*Cingula vitrea, C. proxima*). A silty habitat is favoured, though this may be on exposed, wave-beaten shores where there are rocks with deep sheltered crevices in which the fauna lives protected from the rapid changes of the external environment. Other species find protection among the dense tufts of smaller weeds such as *Ceramium, Gelidium, Porphyra, Nitophyllum* or the corallines of rock pools and damp crevices. The numbers are greater on such weeds as gather silt.

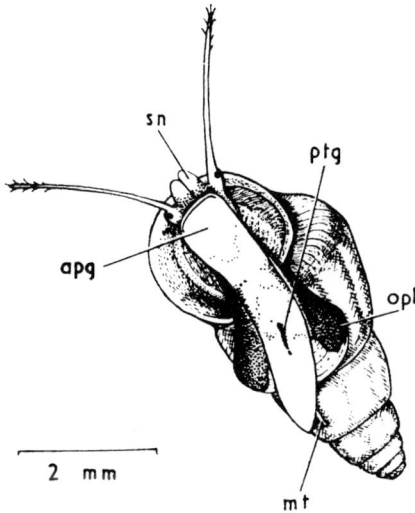

FIG. 279.—*Rissoa parva:* living animal seen from below.
apg, anterior pedal gland; mt, metapodial tentacle; opl, operculigerous lobe; ptg, opening of posterior pedal gland to sole of foot; sn, bifid snout.

In all species which have been studied there is a large posterior pedal mucous gland (ptg, figs. 279, 280) as well as an anterior pedal gland, and, except in *Barleeia rubra* (fig. 281B), the secreting tissue of this gland spreads into the haemocoel of the head. In *Barleeia* it is confined to the foot, in the epithelium of which there are also numerous mucus-secreting cells. The secretion from the posterior pedal gland hardens in sea water to form tough threads which anchor the rissoid in position. Or a thread may be spun from the opening of the gland to lower the animal through the water from the surface film along which it commonly creeps or swims in an inverted position. A posterior pedal gland with similar function is developed in other small prosobranchs which are intertidal and live alongside the rissoids, and are related to them in structure and mode of life. These include *Cingulopsis fulgida* (fig, 70), *Skeneopsis planorbis* (ptg, fig. 282), *Omalogyra atomus* (fig. 176A) and *Rissoella* spp. (ptg, figs. 198A, 283).

The rissoids gather diatoms, detritus and fragments of algae for their food, The snout penetrates interstices of sand or silt and explores the surface of weeds. Fragments are drawn into the gut by the radula perhaps aided by jaws. A characteristic rissoid radula is shown in fig. 284. It is typically taenioglossan, bearing many transverse rows of sharply cusped teeth, 7 to a row. The cusps of the rachidian and lateral teeth vary in number, but the marginals are almost identical in all the British species which have been studied. Secretion from a pair of salivary glands (sg, fig. 67) lubricates the action of the radula and cements together the food particles collected by it. The oesophagus and intestine are simple conducting tubes and what complexity there is in the gut is encountered in the stomach (st). Here a crystalline style is present with the accompanying gastric shield, ciliated sorting areas and typhlosoles. The style sac (ss), as described by Graham (1939) in *Rissoa parva*, lies at the anterior end of the stomach, with the intestine running along one side between it and the oesophagus; in this species the sac communicates with the intestine for about half its length. The lay-out of the alimentary canal is remarkably uniform throughout the group.

Other systems of the body show a generally uniform structure: the organs of the mantle cavity, the nervous system and, to a large extent, the reproductive system (chs. 13, 14). The arrangement of the organs in the pallial cavity is similar to that of a typical prosobranch, though the gill is short and has few filaments in some species—in *Cingula cingillus*, a species which lives high on the shore, there are as few as 6, 10–15 have been recorded in *C. semicostata* and an intermediate number in *C. semistriata*. The number increases in *Rissoa parva* and *R. lilacina* to about 20, and *Barleeia rubra* may have even more. The nervous system shows a high degree of concentration (p. 280).

Less than half the species of rissoids recorded for the British Isles occur above LWST, and of these only a few are common. Least is known about the genus *Alvania*, the members of which are typically infralittoral and live on varying types of substratum. Of the 9 species, 5 are rare or even scarce, and we have no knowledge of the reproduction of 6; for the rest free veliger larvae have been described. The life history is known for a single species, *Alvania punctura* (fig. 280), which occurs amongst *Laminaria* in rather sheltered situations on all coasts, though the egg capsules may be attached to weeds higher on the shore.

Species of the genus *Rissoa* are associated with weeds near the infralittoral fringe and below, and the egg capsules are usually deposited on the fronds. The young hatch as veligers. Individuals may be seen creeping over submerged weeds which are lit by the brightest sun, Several species favour *Zostera*, especially *R. membranacea*, and the abundance of this species would appear to be correlated with the abundance of the weed. It collects its food from the surface of the plant, deposits the egg capsules there and when the young settle from the plankton they congregate on the weed. The growth rate of the recently metamorphosed individual is rapid and it may be fully grown in about a month (Smidt, 1938). The total life span is only a year or a little over. In Britain the species was abundant before *Zostera* was destroyed by disease, but from about 1938 the numbers have dwindled and now it is rare. A similar decline in population is quoted for Danish waters (Thorson, 1946), Other species of *Rissoa* with a lesser dependence on *Zostera* appear to have suffered little on account of its destruction. [With the return of *Zostera* numbers have recently increased.]

A most prolific species is *Rissoa parva*, which is known not only for its abundance, variability in sculpture and colour of shell, but also for its remarkable agility and liveliness. Both the type, which has a sculptured shell with strong and slightly curved longitudinal ribs, and the variety *interrupta*, which is ribless, occur together (fig. 39). Some shells are fully ribbed, some half ribbed and some have only a trace of ribs on one or two larger whorls. Presumably there is a genetic explanation of these polymorphs, but in view of the difficulty of rearing successive generations of animals with pelagic larvae, none has yet been demonstrated. Jeffreys stated in 1867 that the variety was more common in the N; the type is now everywhere less common and in many places relatively scarce. The species is easy to recognize on account of (1) the dark purplish-brown markings at the junction of the head and body on each side; (2) a similarly coloured opercular lobe of the foot (opl, fig. 279); (3) the lateral constrictions of the foot which divide it into a smaller anterior portion and a posterior one of about twice the length; and (4) the brown comma-shaped marking on the body whorl of the shell which separates this species from all other rissoids (fig. 39). Females are more numerous than males, and according to Pelseneer (1926a) comprise about 63% of the population. This is also true of *R. lilacina*.

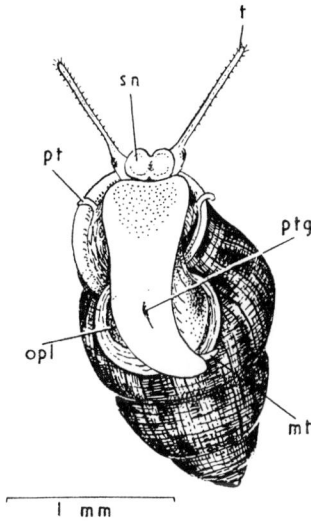

FIG. 280.—*Alvania punctura:* living animal seen from below.
mt, metapodial tentacle; opl, operculigerous lobe; pt, one of 2 pallial tentacles; ptg, opening of posterior pedal gland; sn, snout; t, tentacle.

Some counts have been made by Miss M. A. Perry to illustrate the abundance of *Rissoa parva* in the intertidal zone of Whitsand Bay, Cornwall, which is an exposed stretch of shore and we are indebted to her for permission to quote these here. In a shallow coralline pool with no depth greater than 6 in and a surface area of 3 sq ft, 3,766 individuals were collected on one occasion in September. They were mainly on the weeds, though some were in the silt at the bottom. Of this number only 13 possessed shells with ribs, the remainder being ribless. Among the smooth variety the shells of 58 were very heavily calcified, and some of these, as well as shells of other individuals, were encrusted with the stony alga *Melobesia.* In summer and early autumn large numbers of young individuals are found with the adults; although the species may breed all the year round the peak breeding time is spring and summer when individuals may hatch from the egg, pass through the pelagic larval stage and attain maturity. During the winter months many individuals are washed from the weed and perish, or migrate to join the infralittoral population. A single square foot of rock covered with weed (mainly *Chondrus* but with an admixture of coralline algae and other red weeds) was found to yield 7,114 specimens in September. In early February only 457 individuals were taken from a similar area. Of the September individuals all but 17 were the smooth variety and of the February ones all but 7. A study of similar populations at Cullercoats, Northumberland, showed a similar reduction in winter. The numbers at all times of the year were smaller than at Whitsand Bay and in the areas studied every individual had a ribless shell. A similar seasonal fluctuation in numbers has been described for *R. membranacea* and *R. inconspicua.* Petersen (1918) found that they are annuals, attaining maximal numbers during late summer in the Limfjord and he recorded 100,000 individuals/sq m with a weight of 100 gm.

Barleeia rubra may be found with *Rissoa parva* on weeds near low water and in rock pools along the shores of SW England and Ireland. It is not gregarious, is less active than *R. parva* and its shell and body colouring are distinctive. The shell, without sculpture and with whorls rather swollen, is remarkably strong, dark red, claret, yellowish-brown or tawny in colour. Sometimes, however, it may be lighter with a broad band of reddish-brown encircling each whorl, or this band may be divided into two (fig. 281B). Another variety is white. The body is yellowish-white, usually with smoke coloured lines, but the opercular lobe of the foot is a dark purplish-brown (opl) and bears a thick, dark crimson operculum. This is one of the three British rissoids known to have no free larva. Egg capsules (which are fixed to weed) contain only one egg and a considerable supply of albumen which is used by the embryo as food.

Three of the 6 species of the genus *Cingula* are gregarious intertidal forms (fig. 285) favouring a silty habitat. The others live in deeper water and at least 2 (*proxima* and *vitrea*) on a muddy substratum, though they are rare and little is known about them. One of the most abundant species—the one which lives highest on the shore—is *C. cingillus*. It is found in silty crevices of rock, even those not covered by neap tides, and under stones where silt and diatoms collect. The solid, conical shell, which is buff and has chocolate or reddish-brown bands, makes it one of the most easily recognizable members of the crevice fauna. The animals extend down the shore to MTL and below and sometimes live in the silt at the bottom of rock pools, but always avoid strongly lit ones. On exposed rocks deep crevices, which are always moist, provide an almost constant environment and here clusters of individuals of all ages may be found together, for the egg capsules are laid in the crevices and there is no free larva. Each capsule contains a single egg. *C. semicostata* has a similar life history and may share the same habitat towards MT, though it is more frequent lower on the shore, often under stones and small boulders where, in sheltered gullies, colonies of over 50 individuals may be found closely aggregated.

Both *Cingula cingillus* and *C. semicostata* are able to live in sea water of reduced salinity. *C. semicostata* will spawn in water of a salinity of about 18‰ (Rasmussen, 1951) which is little more than half that of other waters in which it lives. Rasmussen (1951) found that given a choice of weed or empty mollusc shells, the animal chooses the latter on which to deposit the capsules, for it prefers a hard surface; spawn may be found on the stones, boulders or sides of crevices where it lives. The capsules of this species and of *C. cingillus* have a particularly thick and tough wall, and no distinct exit as in the species of *Rissoa*.

A variety of *Cingula cingillus*, var. *rupestris*, is distinguished by its cream coloured or milk-white shell, which is bandless. It is found in the deepest crevices of rock such as the Dartmouth slate reefs along the coast of Cornwall. It seems to stay in these to avoid light, perhaps because of its lack of pigment. It is gregarious but is probably never found along with the pigmented animals of the same species.

The third intertidal species of *Cingula*, *C. semistriata*, is found under stones near those which shelter *C. semicostata*, and in crevices, but it also lives on weeds. The egg capsules are deposited on fronds, and sometimes other surfaces, and the young escape as veligers.

Classified near the Rissoidae is a number of small marine prosobranchs the systematic position of which is uncertain. They resemble rissoids in certain structural features (though in some cases the resemblances are few), live in a similar habitat and select a similar type of food. All are of small size and by comparison with the rissoids show (in some respects) a simplicity of structure and (in others) a high degree of specialization which may be associated with their habitat and their smallness. They are so different among themselves that they have been ascribed to a number of families. In the majority the shell has no resemblance to that of a rissoid, for it is roughly discoidal in shape, with an extremely short and rounded spire, and a wide umbilicus, as in *Circulus striatus* (fig. 276B), *Tornus* spp. and *Skeneopsis planorbis* (fig.

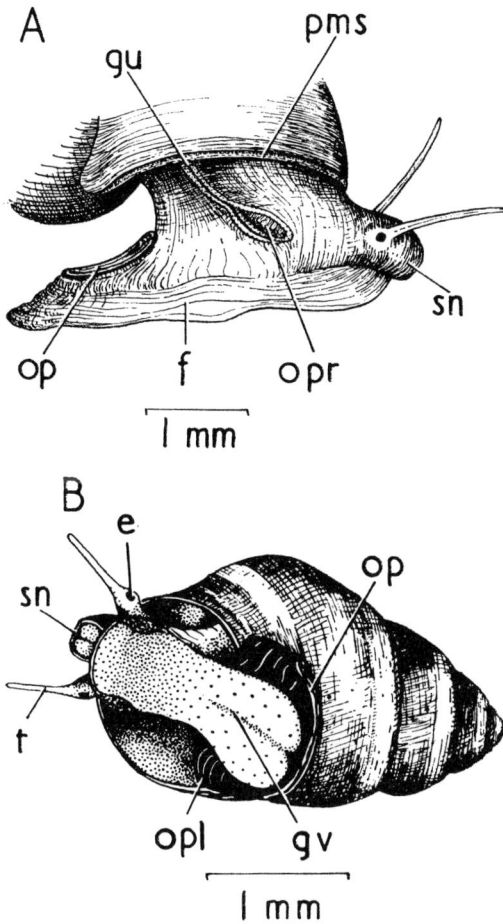

FIG. 281.—A, *Littorina neritoides*, female from the right. B, *Barleeia rubra*, living animal in ventral view.
e, eye; f, foot; gu, groove from female aperture to ovipositor; gv, groove on sole of foot; op, operculum; opl, operculigerous lobe; opr, ovipositor; pms, periostracal margin of shell; sn, snout; t, tentacle.

282), or it is coiled in a plane spiral with the umbilicus so widely open that the interior of the spire is exposed as in *Omalogyra atomus* (fig. 198B) and *Ammonicera rota*. This second type of shell is concave on both sides and has a bilateral symmetry about the sagittal plane of the animal. A conoidal shell is characteristic only of *Cingulopsis fulgida* and *Rissoella* sp. (figs. 198A, 283).

Some of these prosobranchs are very rare. The wide distribution of *Tornus subcarinatus* given by Jeffreys is based on the occurrence of dead shells only, and since 1898, when Woodward described the anatomy of this species from 3 individuals obtained from Guernsey, no living specimens have been recorded. The only British record of *Circulus striatus* is shells from Bundoran (Jeffreys, 1865). In both these genera the detailed structure of the

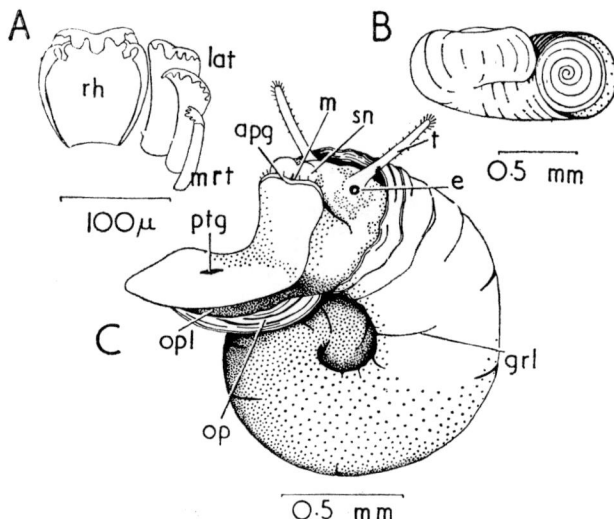

FIG. 282.—*Skeneopsis planorbis:* A, half row of radular teeth; B, shell, with operculum; in apertural view; C, living animal emerging from shell.

apg, opening of anterior pedal gland; e, eye; grl, growth line; lat, lateral tooth; m, mouth; mrt, marginal tooth; op, operculum; opl, operculigerous lobe; ptg, opening of posterior pedal gland; rh, rachidian tooth; sn, snout; t, tentacle.

radular teeth is similar, and approximates most closely to that of the rissoids. The two also agree in having a pair of pallial tentacles which project from the right posterior corner of the mantle skirt (pt, fig. 276B). *Circulus* is more closely related to the rissoids than *Tornus,* especially in the structure of the alimentary canal, the kidney and the male reproductive system; the female system shows certain specializations (Fretter, 1956). In some parts of its anatomy *Tornus* is remarkably unlike any rissoid for it has oesophageal pouches and the male has neither penis nor accessory glands; in the nervous system there is less concentration of the ganglia.

The other genera under consideration, *Cingulopsis, Skeneopsis, Omalogyra, Ammonicera* and *Rissoella,* are associated with the finer sea weeds from MTL and lower, and they often occur in rock pools. Only *Ammonicera* is scarce. These gastropods have a large posterior pedal gland and their ability to produce vast quantities of mucus appears to be associated with their habitat. Associated with their small size is perhaps the reduction or loss of the ctenidium and the position of the kidney in the mantle skirt. *Skeneopsis* is larger than the others and has the typical arrangement of organs in the mantle cavity (fig. 176B). The number of ctenidial filaments is reduced to 9, the kidney lies near the posterior end of the cavity and the hypobranchial gland extends far back along the right side. In *Cingulopsis* the ctenidium is reduced to 3 or 4 small filaments, in *Rissoella* it is represented only by a patch of ciliated cells and in *Omalogyra* there is not even this vestige. In these four the kidney has grown forward into the mantle skirt bringing with it a rich vascular supply and increasing the chances for the oxygenation of the blood. In the absence of a ctenidium the animals depend entirely on pallial respiration, and the stream of water through the mantle cavity is maintained by strips

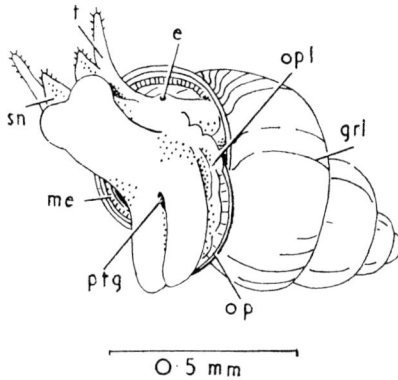

FIG. 283.—*Rissoella diaphana:* animal from below.
    e, eye; grl, growth line; me, mantle edge; op, operculum; opl, operculigerous lobe; ptg, opening of posterior pedal gland; sn, snout; t, tentacle.

of ciliated epithelium which pass forward from the anus to the mantle's edge. These cause a strong exhalant stream and carry away the faecal pellets; there is a compensating inhalant flow. Associated with these ciliated strips are secreting cells which represent the hypo-branchial gland. In the vicinity of the anus a group of relatively enormous gland cells opens into the mantle cavity. They can be seen through the shell and in *Rissoella* are associated with dark brown or black pigment granules: in *R. diaphana* only one such grouping of glands occurs, but in *R. opalina* (fig. 198A) there are 3, for 2 others open elsewhere into the mantle cavity.

The food of *Cingulopsis, Skeneopsis* and *Rissoella* is diatoms, algal filaments and some detritus, and since they live in damp situations they may continue to feed when the tide is out. *Omalogyra* sucks the sap from algal cells which are punctured by the stiletto-like rachidian teeth of the radula. The teeth are reduced to 3 in each row. *Ulva* is a favourite food, and as the mollusc creeps over a thallus the head moves like a pendulum leaving behind a zigzag feeding trail (fig. 198B). The stomach and cells of the digestive gland may be coloured green by the sap of the weed. In the stomach of these prosobranchs there is no style as in rissoids, though in *Cingulopsis, Rissoella* and *Skeneopsis* there is a gastric shield. Only in *Cingulopsis* are there oesophageal pouches. The food of *Omalogyra* needs no mechanical treatment prior to digestion and the stomach (st, fig. 176A) has a lining of digestive cells, similar to those of the digestive gland. This is reminiscent of the embryonic condition in which liver cells cover the wall of the stomach, and are constricted only later. Unlike that of some larger intertidal gastropods which are herbivores and inactive at low tide, the intestine (i) is extremely short and does not enter the coils of the visceral mass, whilst the anus (a) opens well within the mantle cavity. These differences may be correlated with the fact that these small prosobranchs live in rock pools or near low water and so are able to maintain a more or less constant flow of water through the mantle cavity to carry away the faeces; a short and histologically simple intestine suffices them for it is not concerned with the elaboration of faecal matter. A short intestine is also characteristic of rissoids—for the same reason—though in them the anus is far forward.

Cingulopsis, Skeneopsis, Omalogyra and *Rissoella* show a maximum of numbers during the summer months when there is ample food; owing to their rapid growth and reproduction

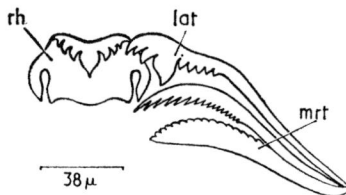

FIG. 284.—*Rissoa parva:* half radular row.
  lat, lateral tooth; mrt, marginal tooth; rh, rachidian tooth.

several generations of each species may co-exist. In the winter the population is at a minimum and has a high percentage of immature individuals. The rapid increase in numbers during a favourable period is reflected in the hypertrophy and complexity of the reproductive organs to provide for the protection and feeding of the embryos but in no two genera—even the hermaphrodite forms *Omalogyra* and *Rissoella*—are these alike. Relatively large egg capsules are produced, each fixed to the weed on which the animals live and containing 1–3 heavily yolked eggs, surrounded by albumen, and a thick protective wall. In contrast to the rissoids there is no free veliger larva. The young hatch from the capsule in about 2 weeks. Since they are annuals and lay large, yolky eggs, considerable expenditure of energy is involved in their reproductive activities.

Omalogyra atomus (fig. 176A), one of the most minute of British molluscs, as suggested by its trivial name, has a shell which measures only about 1 mm at its broadest diameter. There are many points in its anatomy in which it differs from other prosobranchs and some of these are specializations associated with its smallness. Yet it has a remarkable array of characters in common with the most primitive opisthobranchs, the pyramidellids, which superficially resemble prosobranchs and have been classified with them until recently. These characters are in the gut, the reproductive system and the organs of the mantle cavity.

Another small prosobranch which has adopted a mode of life similar to that of some rissoids is *Bittium reticulatum,* which belongs to a group higher in the evolutionary scale, the Cerithiidae. It lives under boulders and on shelving rock faces in the lower part of the beach and below, with algae and in both sunny and shady positions, feeding on diatoms (especially *Licmophora, Grammatophora* and *Gomphonema* according to Starmühlner, 1956), filamentous algae, diatoms and detritus. The fixed egg coils (fig. 196C) are unlike those of any other British operculate. Though not common in Britain it is gregarious, and several may be found under one stone. It appears to be of that numerous group of prosobranchs the numbers of which have decreased in the last century; indeed Marshall (1910–12) writes 'the sea shore at Falmouth consists largely of this species'. It may still be encountered in fair numbers in suitable situations in SW Ireland, e.g. Lough Ine.

A variety of carnivorous gastropods feed on semi-sedentary or sessile prey. The rocky shore with its abundance of sedentary and sessile animals is the habitat of a number of them of which the eratoids, muricaceans and pyramidellids are the best known. They attack forms which have a protective outer shell—bivalves, barnacles and other prosobranchs—and usually bore the calcareous covering to get at the soft tissues. Their feeding habits and ecology are dealt with elsewhere (p. 230). Others are ectoparasites and suck fluids. The majority of these are pyramidellids which settle near the host, accurately shoot the proboscis

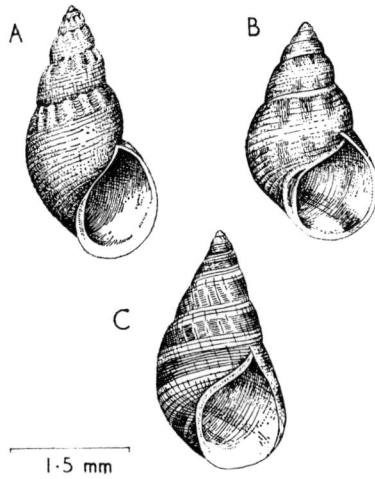

FIG. 285.—Shells of *Cingula* in apertural view: A, *C. semicostata*; B, *C. semistriata*; C, *C. cingillus*.

in the vicinity of the food, attach it by the terminal sucker and pierce the tissues with a stylet, the modified jaws; at the slightest disturbance feeding stops and the proboscis is instantly withdrawn. Each species has its food preference, which may be a particular bivalve, sedentary polychaete, echinoderm or coelenterate. These minute gastropods feed when the tide covers them, for they attack the prey when it is active. The larger animals, *Pelseneeria stylifera* and the eulimids, suck fluids and ingest loose cells of echinoderms. Their proboscis is stout and armed with neither radula nor jaws and, in contrast to the pyramidellids, it may become so firmly embedded in the host that it, rather than the foot, holds the animal attached. The third grouping of carnivorous gastropods feeds on sponges, coelenterates and ascidians. They may be seen moving over the surface of the colony, feeding as they go, and their egg capsules are attached to the surface or deposited in holes made in the tissues. Some select the deeper and softer parts of the body of the host for feeding, whilst others rasp the surface in a rather haphazard fashion.

In addition to the zeugobranchs *Diodora* and *Emarginula* which have an unspecialized method of feeding on sponges, *Cerithiopsis tubercularis*, *C. barleei*, *C. jeffreysi* and *Triphora perversa* have also adopted this diet, though not exclusively, for they also feed on diatoms and detritus. The larger specimens of these species attain a height of 8 mm. *Triphora* is readily distinguishable since it is sinistrally coiled, the only truly sinistral species of prosobranch in the British Isles. The organs of the mantle cavity and the nervous system are mirror images of those of the dextrally coiled *Cerithiopsis*. When these prosobranchs are examined alive it can be seen that the foot produces an abundant secretion which helps the mollusc to secure a holdfast as well as providing a viscid climbing rope by which it can lower itself through the water. The sole is truncated in front and tapers to a blunt point posteriorly, and there is frequently a transverse groove dividing it into an anterior half containing the anterior pedal gland and a triangular posterior half with the opening of the posterior pedal gland at its anterior border; this gland spreads into the haemocoel of the head. There are also glands opening to the surface of the sole lying in

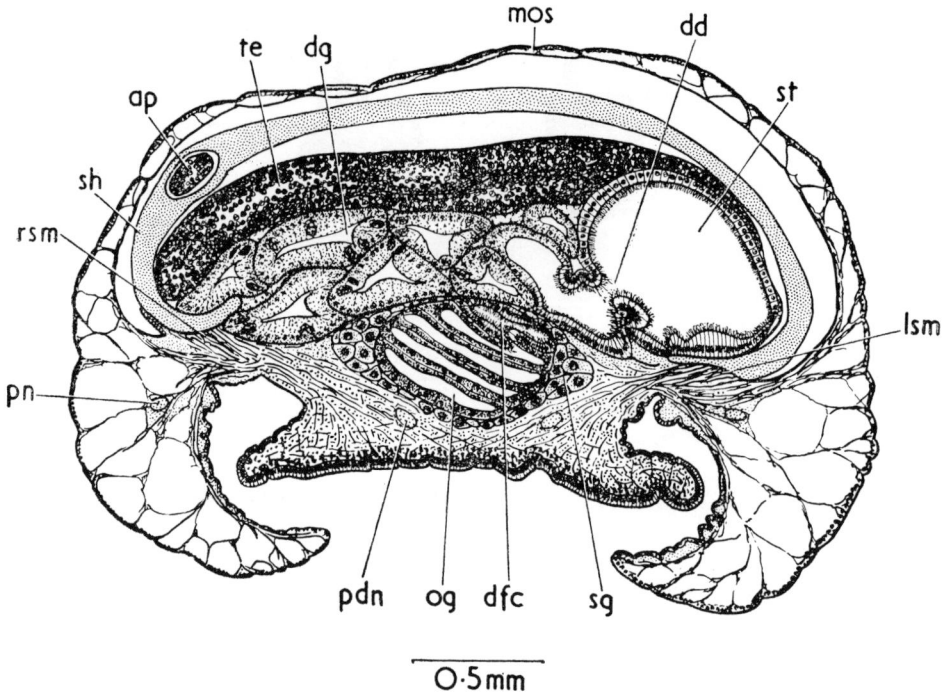

FIG. 286.—*Lamellaria perspicua:* transverse section through middle region of body. The spaces between shell, visceral hump and mantle are artificial.

ap, apex of visceral hump; dd, opening of duct of digestive gland; dfc, dorsal food channel in mid-oesophagus; dg, digestive gland; lsm, left shell muscle; mos, mantle lying over shell; og, oesophageal gland; pdn, pedal nerve; pn, pallial nerve; rsm, right shell muscle; sg, salivary gland; sh, shell; st, stomach; te, testis.

the epithelium and below. These carnivores have adopted a different method of feeding on sponges. They have a long acrembolic proboscis (Fretter, 1951b) which is narrow enough to pass through an osculum and reach the softer tissues, which are loosened by the jaws and raked into the buccal cavity by the numerous fine radular teeth. The spicular cortex which protects the sponge externally is avoided. *Triphora perversa* is comparatively rare intertidally. Its egg capsules have been seen by Pelseneer (1926b), fixed to a dead shell of *Glycymeris;* they have not been described from this country though they are of such a size that they could easily be overlooked. The larvae, however, are common in the Plymouth plankton, both inshore and offshore, and the adults are dredged on rock and gravel grounds. *Cerithiopsis tubercularis* is occasionally common on the SW and W coasts, clustered on the surface of *Hymeniacidon sanguinea,* in the tissue of which the egg capsules are deposited. It is also dredged, generally on sponges. Another species, *C. barleei,* is similarly associated with *Suberites domuncula,* feeding and depositing egg capsules on the sponge.

With the exception of *Velutina plicatilis,* the British members of the families Lamellariidae and Eratoidae feed on tunicates. *Velutina velutina* favours the large solitary forms *Phallusia* and *Styela* while *Lamellaria, Erato* and *Trivia* eat compound ascidians. In members of these families the mantle is reflected over the shell to a varying extent. In *Velutina* (fig. 127) the pallial edge is thickened by large connective tissue cells and blocks the entrance to the mantle cavity, except for the inhalant and exhalant passages, and there is only a slight reflection over the shell. In contrast, the shell of *Lamellaria* (sh, fig. 286) is permanently enclosed in the mantle (mos) from the late larval stage and the exposed pallial surface is roughened by warty tubercles in the adult and the edge thickened as in *Velutina*. In these members of the Lamellariidae there is no pallial siphon. The mantle in the eratoids and cypraeids (figs. 206, 288) spreads over the shell when the animal is active, often covering it completely, whilst it is produced into a long siphon (si) anteriorly.

Only *Lamellaria* and *Trivia* are intertidal and may be found with their food under boulders or on the slopes of gullies. They feed when the tide covers them. *Lamellaria* browses on *Leptoclinum* and *Polyclinum*, swallowing the test (which it cannot digest) to get at the zooids. Peach (quoted by Jeffreys, 1867) observed that *Lamellaria* migrates to shallower water to spawn, depositing the capsules in the test of the compound ascidian on which it feeds. According to Jeffreys spawning occurs from February to March. He recognized only one species of this genus, *L. perspicua,* and described it as dimorphic: the male, smaller, with the shell almost flat and the female with a convex, boat-shaped shell. Colour differences also exist between these 2 forms, which are now recognized as 2 different species (McMillan, 1939). The smaller *L. latens* has a thinner mantle and is light sandy-brown, yellowish or white and is flecked with black and sometimes yellow. *L. perspicua* is even more varied in colour and may be purplish or lilac-grey flecked with white spots, yellow mottled with red and white, or lemon with clear spots.

Trivia is one of the most conspicuously coloured of British prosobranchs on account of the pigmentation of its tissues, especially the mantle which covers the shell. In the young (fig. 287) the mantle is studded with conical projections, often branched, and is blotched with brown pigment; its colour is less pronounced than in the adult (Pelseneer, 1932). The adult mantle is yellow or orange-brown and the dark blotches run together to form conspicuous transverse lines (fig. 206). The foot is also pigmented with yellow and streaked with lighter or darker longitudinal lines. *Trivia* is the only British cowrie, formerly classified in the family Cypraeidae which abound in tropical seas. The warm water cowries attain a larger size and have a thicker and often deeply pigmented shell. In *Trivia* the shell is a pale flesh colour or is whitish and *T. monacha* is distinguished from *T. arctica* by 3 purplish-brown spots across the body whorl, though these are sometimes faint (Lebour, 1933a). The shell has no periostracum and the surface is always glossy. It is smooth in young specimens, even up to the time the adult shape has been acquired, when the body whorl hides the spire (fig. 31). In older specimens there is a sculpture of 20–25 fine thread-like ribs which cross the shell and extend to the mouth; some of these anastomose or are shorter and placed between others of full length. These ribs are added by the mantle when flexed over the shell.

When *Trivia* (fig. 206) is about to feed the siphon shows some activity and its tip sweeps over the surface of the ascidian. The animal frequently shows a preference for *Diplosoma listerianum* var. *gelatinosum,* in which the test is gelatinous and without spicules, and the zooids, which are in clusters, have a meagre layer of test separating them. The tip of the proboscis is glandular and moves smoothly over the tunicate colony. The animal attacks any part of the test, biting pieces from it with the jaws to expose the zooids, which are swallowed whole (Fretter, 1951a). The test is not digested and can be traced into the faeces. The zooids are black and their passage one by one along the dorsal channel of the oesophagus,

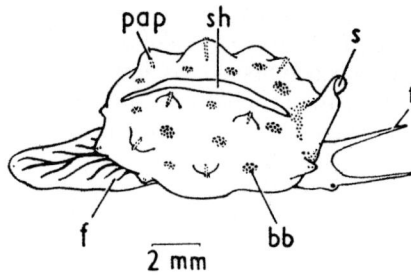

FIG. 287.—*Trivia* sp.: young animal, alive.
bb, blotches of pigment on mantle skirt; f, foot; pap, papilla on mantle skirt; s, siphon; sh, shell; t, tentacle.

embraced by the longitudinal folds can be followed owing to the transparency of the tissues of the mollusc. In *Erato* (fig. 35) the inhalant siphon is used more obviously in the search for food. The animal moves slowly over the surface of *Botryllus* or *Botrylloides* lubricating its passage with secretion from the anterior and posterior pedal glands, and the long inhalant siphon is used as a sensory organ to lead the animal to a group of zooids. It will bend so that its tip momentarily covers the mouth of each zooid in turn, as though to test the inhalant water current (Fretter, 1951a). Chemoreceptors may also be concerned with the detection of food, though these perhaps lie in the osphradium on to which the siphon directs the water. It is the fully opened mouth of a feeding zooid which the mollusc seeks, and immediately one is selected the siphon is erected and the blunt introvert is plunged like a piston into it, its thickness stretching the opening. The mollusc settles for 20 min or more with the proboscis hidden in the zooid. It may be seen moving among the viscera, the jaws biting the tissues and the radula dragging them into the mouth. The test is avoided. Two zooids may be eaten in rapid succession. The pleurembolic proboscis is longer than that of *Trivia* and the tip is not glandular.

   Velutina velutina may be collected from *Styela coriacea* and, like many of those proso-branchs which live on sponges and tunicates, resembles it in colour and in surface texture. Secretion from anterior and posterior pedal glands and from the thickened edge of the mantle lubricates its passage over the test into which it bites holes with its jaws and then extends the proboscis to feed on the soft tissues. The epithelium of the proboscis has abundant mucous glands. A second species, *V. plicatilis*, is associated with *Tubularia* and other hydroids in the coralline zone of NE England and Scotland. It shares this liking for coelenterates with *Simnia patula*, which eats the flesh of *Alcyonium digitatum* and *Eunicella verrucosa*, and has also been found among tufts of *Tubularia indivisa*. Unlike most other carnivores *Simnia* has no introvert (fig. 288). There is a short snout which expands distally around the mouth. This expansion covers the contracted polyps, which are bitten off by a pair of strong jaws aided by the median and lateral teeth of the radula. These teeth have short blunt cusps. The marginals, 3 in number in each half row, are long and each tooth is deeply divided longitudinally into as many as 40 approximately equal parts, each shaped like a fork with 3–4 prongs. They brush up the tissues with a sweeping motion and direct them into the dorsal food channel of the buccal cavity. Thus they function like the simpler and more numerous teeth in the rhipidoglossan radula—indeed the feeding mechanism of *Simnia* is very similar to that of the sponge feeder *Diodora*.

7 mm

FIG. 288.—*Simnia patula:* 2 animals creeping over the surface of a colony of *Alcyonium digitatum,* on which they feed. The lower animal shows the blotched mantle spreading over the shell, the head and siphon (left) and the foot (right); the upper animal shows the anterior view of its foot, tentacles and siphon nearly withdrawn, head with short snout expanding distally around mouth.

The eulimids are ectoparasites and the least modified of a series of prosobranchs parasitic on echinoderms. They are free-living and specialized in respect of their method of feeding and the organs associated with this. Of those listed in the British fauna only 2, *Balcis alba* (fig. 139) and *B. devians* (fig. 289), are widely distributed and may be collected in numbers in their characteristic localities. *B. devians,* in which the shell measures 3 mm in length, is found in the same dredgings as *Antedon bifida* (Fretter, 1955a), roaming freely or attached to the base of a pinnule from which it sucks food (fig. 138) and one specimen has been found attached to its disc in the anterior interradius (Dimelow, 1959). It is not easy to disturb when feeding and is slow to withdraw the proboscis from the host. Its body is flecked with crimson and yellow like the tissues of the feather-star, though the colour is lost when the mollusc is starved. *B. alba,* which is often 18 mm long, is dredged in fairly shallow water and its frequent occurrence with *Spatangus purpureus* at Plymouth suggests that this echinoid may be its host, though the gastropod has never been seen feeding. It is probable that when the hosts of the various eulimids recorded in the British fauna lists are known, and their true ecological niches explored, they will appear in numbers greatly exceeding the solitary specimens which are now occasionally discovered.

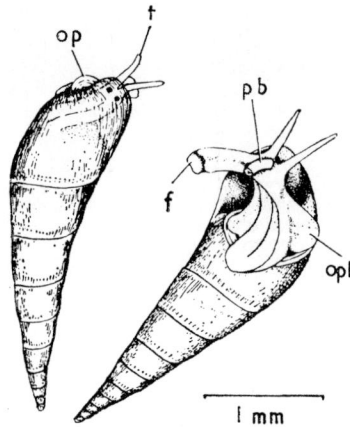

FIG. 289.—*Balcis devians:* two living animals, the right in ventral view, the left in dorsal.
f, foot; op, operculum; opl, operculigerous lobe; pb, proboscis partially protruded; t, tentacle.

Pelseneeria stylifera, also parasitic on echinoderms, is recorded from Plymouth on *Psammechinus miliaris* and *Echinus esculentus* and Jeffreys (1867) recorded it on *Echinus* from many other British localities at a depth ranging from 30–80 ft. Since then, however, it has become scarce. He observed the parasite creeping among the spines on the upper surface of the urchin, where he found as many as 40 clusters of spawn. Each globular egg sac is attached by a stalk to the test and contains 60–80 eggs (Lamy, 1928; Lebour, 1932*b*) which develop into typical veliger larvae. The protoconch and larval whorls persist as the characteristic styliform apex of the adult shell which has 3–4 tumid and rapidly enlarging whorls. The pseudopallium, rudimentary in the eulimids, where it lies towards the distal end of the proboscis, is larger in *Pelseneeria,* where it arises near the base of the proboscis and is reflected over the shell as the animal feeds; in more advanced parasites, which become embedded in the tissues of their host, it completely envelops the shell (*Gasterosiphon*) and even replaces it as a covering to the viscera (*Entocolax*). The proboscis of *Pelseneeria* and the method of feeding are similar to those of *Balcis. Pelseneeria* is said to be hermaphrodite (Koehler & Vaney, 1908), in contrast to *B. alba* in which the sexes are separate, and also, perhaps, *B. devians,* though no males of this species have been found.

Until recently surprisingly little was known about the pyramidellids and even now more than half of the 39 species mentioned by Jeffreys (1867) are described only as shells. However, the field naturalist is now alert to the possibility of finding a relationship between one of these ectoparasites and perhaps an annelid, echinoderm or molluscan host. Not only the minute size of these gastropods, but also the specialized habitat of each species, account for the obscurity which has surrounded the group. On the SW coast of England 2 species of *Odostomia* may be collected from the calcareous tubes of *Pomatoceros triqueter,* often hidden in crevices for protection. These are *O. unidentata,* the shell of which has a bluish tinge and straight-sided spire (fig. 134), and *O. lukisi* with an ivory white shell which appears very solid for its size. Sometimes *O. plicata* is in the same habitat and in contrast is yellowish in shell and soft parts. The feeding of the first 2 species is easily observed, for the animals are in no way timid (p. 241). The mollusc approaches the entrance to the calcareous tube of the host

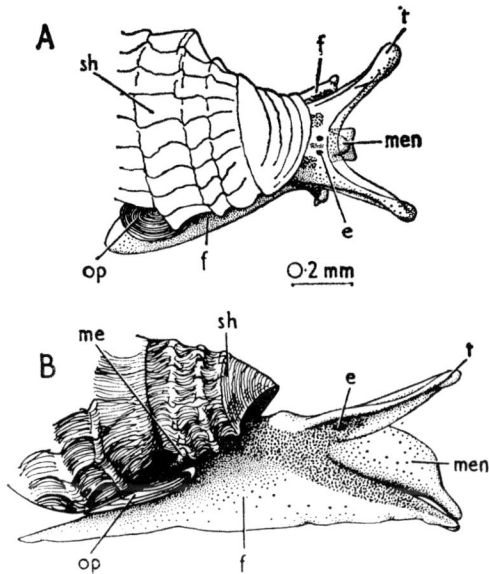

FIG. 290.—*Turbonilla fenestrata:* A, animal in dorsal view; B, seen from the right; only part of the shell is drawn.
e, eye; f, foot; me, mantle edge; men, mentum; op, operculum; sh, shell; t, tentacle.

and may wait there for some time before the worm emerges to feed. It will attach its proboscis to a tentacle, pierce and then suck the worm for a minute or so at a time, unless disturbed. When feeding is over the pyramidellid may retreat to a more sheltered position and rest, with the foot and head partially withdrawn and the concavity of the tentacles directed forwards and downwards as though no longer interested in its surroundings (t, fig. 290B). *O. scalaris* settles at the edge of the shell of *Mytilus edulis* (fig. 137) and pierces the pallial tissues when the mussel is feeding, to suck blood from the pallial vein. The mussel seems undisturbed even though 2–3 parasites are feeding at one time. In rough conditions and between meals this species may be found sheltering amongst the byssus threads.

Similarly *Chrysallida spiralis* may lurk in any sheltered cranny between the irregular sandy tubes of *Sabellaria*. It feeds on this polychaete, though not by sucking the tentacles, which, unlike those of *Pomatoceros*, are moved about to pick up food particles. When the proboscis of *Chrysallida* is rolled out from the head its tip is gently directed on to the median face of a tentacle along which it slides towards the mouth and gradually disappears out of sight, hidden in the gut of the polychaete. The region which it punctures and sucks is unknown.

In contrast to these species *Turbonilla elegantissima* burrows to seek its food (fig. 136). It is found amongst silt and sand which spreads between boulders and ledges and provides suitable niches for the sedentary polychaetes *Audouinia tentaculata* and *Amphitrite gracilis*. The tentacles of these polychaetes extend between the particles for a considerable distance and are sought by the mollusc, which pierces them and sucks blood; it will also attack the gills of *Amphitrite*. The proboscis, stouter than that of *Odostomia*, is pushed through the silt to reach the worm, accurately sensing its direction. It emerges from a subterminal opening on the dorsal surface of the mentum which gives it a firm support. While the mollusc feeds buried in the silt, particles mixed with hypobranchial secretion can be seen leaving the exhalant channel of the deep and narrow mantle cavity; the wall of this channel is extended beyond the shell as a spoon-shaped siphon. At other times the parasite will wander some distance from its host to a more superficial position on the shore, the remarkably short foot slowly dragging behind it the elongated, pointed shell, which is occasionally just raised from the ground.

One pyramidellid, *Turbonilla jeffreysi,* has been found in association with the hydroid *Halecium,* and in a short terminal region of the oesophagus every cell is filled with unexploded nematocysts, though they have not been traced elsewhere. It seems likely that these must be voided later, for there is no indication that they are used in defence or stored.

CHAPTER 21

THE PROSOBRANCHS OF OTHER MARINE HABITATS

O CCASIONAL pelagic prosobranchs may be taken from the waters of our W and N coasts. They belong to two genera, *Ianthina* and *Carinaria,* and are characteristic of tropical and subtropical seas. *Ianthina* is a surface form dependent on wind rather than water movements and the mass strandings on our coasts may be correlated with prolonged periods of westerly winds. Wilson & Wilson (1956) have summarized the occurrence of *I. janthina* on the N coasts of Cornwall and Devon in August 1954 when the strandings were probably the largest for 50 years. The siphonophore *Velella* drifts with this carnivorous prosobranch and is its food, and perhaps the source of its purple colour. Both are gregarious, drifting together apparently by chance.

Ianthina (fig. 291) floats at the surface with the foot (pp, msp) uppermost, by means of a bubble raft (mfl) which spreads back over the mesopodium and rests on the surface film (ws). No operculum is present in the adult. The coiled visceral mass and the mantle cavity are enclosed in a delicate violet-blue shell (sh). This is deeper in colour on the underside, which is directed uppermost in the water, and light below. The body of the animal is heavily pigmented and has a deep purple tinge, and the bifid tentacles (t) are black. The right epipodium of the foot (ep) has a more delicate appearance and expands over the shell to form a transparent membrane which, on occasions, shows rhythmical undulations. The head is large, the pretentacular region being extended to form a broad proboscis (pb) with a large terminal mouth. Behind the tentacles is the stout mobile neck which may swing the head

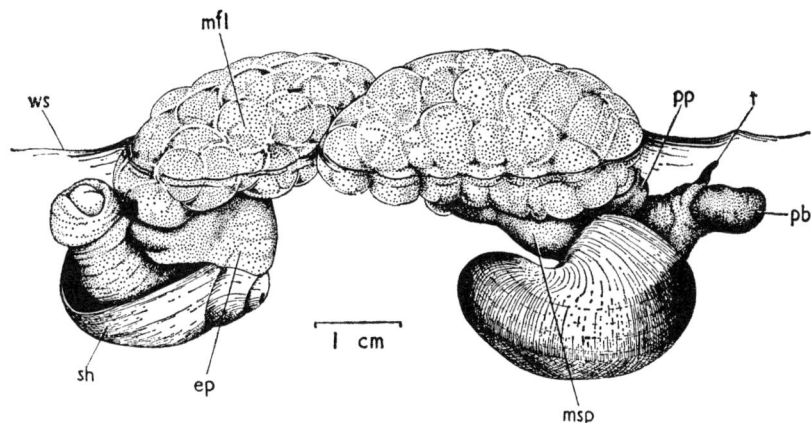

FIG. 291.—*Ianthina janthina:* animals in natural position at surface of water.
ep, epipodium; mfl, mucous float; msp, mesopodium; pb, proboscis; pp, propodium; sh, shell; t, tentacle; ws, surface of water.

rapidly over an arc of more than 180°. The animal has no visible eyes, so that other senses must be used to recognize the food which it hunts probably by the well developed olfactory sense in the tentacles placed lateroventrally on the head. The outer branch of the tentacle was thought by Bouvier (1886) to be an eye stalk; there is, however, no evidence for this and it appears to be part of the tentacle. Minute eyes are present (Thiele, 1928) each with a lens, lying ventral to the cerebral ganglia under the muscles of the body wall posterior to the base of the tentacle.

Mr Peter David (Hardy, 1956) has observed *Ianthina* feeding on *Velella* and gradually clearing the tentacles and blastostyles from the underside of the float until only the horny skeleton remained. The food is attacked by the radula [p. 621], which has in each row a large number of similar teeth (laterals), and jaws protect the buccal cavity of the mollusc against its own teeth (Simroth, 1895). The proboscis is moderately distensible and the radula can be protruded some distance from the mouth. The range of food is unknown, but Laursen (1953) found remains of siphonophores in the gut and also of *Halobates* and copepods. Bouvier (1886) gave *Porpita* as the food of some species and Risbec (1953) and Laursen (1953) both have evidence that cannibalism is practised. Thus it would appear that the carnivore seizes anything that comes within its reach. Unfortunately the specimens stranded around our coasts refuse to feed even when offered *Velella,* and the gut is empty when they are found: perhaps this is not surprising in view of the altered conditions of salinity and temperature in which they are then living. The mantle cavity of *Ianthina* contains a large monopectinate gill and a hypobranchial gland which produces a purple secretion. This secretion escapes from the cavity and gradually diffuses through the water, and in some species may tint the float. It has been suggested that the fluid anaesthetizes the prey (p. 120) and although there is little evidence for this David (Hardy, 1956) has observed that *Velella* which are being attacked by the mollusc appear lifeless, do not shed their tentacles as they normally do when they die, nor contract them as much as usual when placed in formalin.

If the float of *Ianthina* be removed the animal slowly sinks and may retreat into the shell, though turbulence in the water keeps it near the surface. A new float is formed only if the anterior third of the foot (the part somewhat loosely called the propodium in the literature) comes into contact with the surface film. Air is then trapped by the anterior tip reaching above the surface film like a hand spread out to grasp. The foot closes over the bubble of air and brings it beneath the surface, passing it back into the funnel. This is a transverse depression lying across the middle of the foot into which open many glands (Thiele, 1897). Mucus from these, and from the anterior pedal mucous gland, forms a skin around the bubble, which is then freed for this to harden in contact with the water. The float appears to be attached to the posterior end of the sole, the surface of which is deeply pigmented and carries numerous longitudinal glandular ridges. Successive bubbles are pressed against one another and cemented together by further secretion as the propodium glides over them, and eventually mutual compression causes them to assume a polyhedral shape. During the early stages of this process the movements of the foot with its large epipodial lobes help to keep the animal at the surface. The fully formed float is strong and difficult to crush or puncture so that it successfully maintains the animal's position in rough seas, and must, in general, be an important orientating organ in view of the fact that neither statocysts nor effective eyes occur (Thiele, 1929–35). David (quoted by Wilson & Wilson, 1956) observed that many small *Ianthina* feeding on *Velella* had no floats and suggested that they may be a hindrance during feeding (when attachment to the siphonophore will keep the mollusc at the surface) and so are abandoned.

Three species of *Ianthina*, *I. janthina*, *I. exigua* and *I. pallida*, are occasional visitors to our shores. Two of these, *exigua* and *pallida*, attach egg capsules to a median longitudinal band of mucus on the underside of the float (Fraenkel, 1927b) which may have a length of 5 cm in *I. exigua* or 7 cm in *I. pallida*, the number of egg capsules being approximately 50/cm length in *I. exigua* and 60–80/cm in *I. pallida*. Each capsule in *exigua* contains 17 eggs on the average: in *pallida* the corresponding figure is 5,500 (Laursen, 1953). *I. janthina* is the most common visitor to this country and is viviparous (p. 347), the young being born as veligers with an operculum.

These prosobranchs are protandrous hermaphrodites and the reproductive processes are unusual in that during the male phase there is no penis, and spermatozeugmata are responsible for the transportation of sperm to the female and that fertilization seems to occur in the ovary (p. 347). In other internal features the ianthinids approach the scalids, possessing tubular salivary glands, no trace of oesophageal glands, a ptenoglossan radula and a tendency to produce a purple hypobranchial secretion. They also differ from other prosobranchs in that the spermatozeugmata and packets of larvae leave the mantle cavity on the left (inhalant) side (Wilson & Wilson, 1956). To escape in this way against the respiratory current they must be expelled by muscular contraction and perhaps directed to this side by the movements of the propodium observed by these authors. The normal escape route on the right is blocked by the epipodium.

In contrast to *Ianthina* only isolated individuals of the two heteropods *Carinaria mediterranea* and *C. lamarcki* are found off Britain, and these do not occur at the surface. Indeed one specimen of *C. lamarcki* came from a depth of 1,000 m W of the Outer Hebrides (Hardy, 1956), and others from 250 m (information from Dr J. H. Fraser, who kindly lent the specimens). It is uncertain as to whether this is their normal depths at such latitudes. One suggestion is that they are carried with water of high salinity which passes outwards from the Mediterranean though the Straits of Gibraltar, beneath the influx of Atlantic water which replaces it, and travels northwards at a low level. It is surprising that these planktonic forms appear to maintain themselves in such changing conditions.

Like other heteropods *Carinaria mediterranea* (fig. 292) actively pursues its prey and the form of the body is considerably modified. It swims in an inverted position with the foot (pp, mpt) uppermost. The head-foot is large, elongated and somewhat laterally compressed and is transparent. The visceral mass is small and is covered by a patelliform shell which protects only the viscera and organs of the pallial complex, for neither head nor foot can be withdrawn into its shelter. The shell is glossy and characterized by a serrated dorsal keel (mk) marked with radial lines due to the extension of growth lines into it; at the apex is the persistent helicoid larval shell (pch). It is so fragile that whole specimens are come by only with difficulty and Woodward (1913) mentioned that as much as £100 has been paid by collectors for a single *Carinaria* shell. The enlargement of the head-foot is due to a post-tentacular elongation of the head as well as a smaller terminal enlargement to form a snout. There is a terminal mouth surrounded by a band of circular muscles which allows the protrusion of the large odontophore. The prey is caught and pulled into the gut by the very powerful lateral and marginal teeth of the radula. The head bears a pair of retractile tentacles (t) and at the base of each is an eye (e) surrounded by a capsule and capable of being turned in all directions by its musculature. The eye can be protruded from the capsule or withdrawn for protection. The foot lies ventral to the visceral mass, where it expands into a large muscular fin, and also tapers posteriorly. Along the posteroventral edge of the fin in both sexes is a sucker (pp). It is said to be used by the copulating partners to secure a firm hold of one another and is aided in this by glands lying over its surface (Gegenbaur, 1855). It is confined to males in *Pterotrachea*

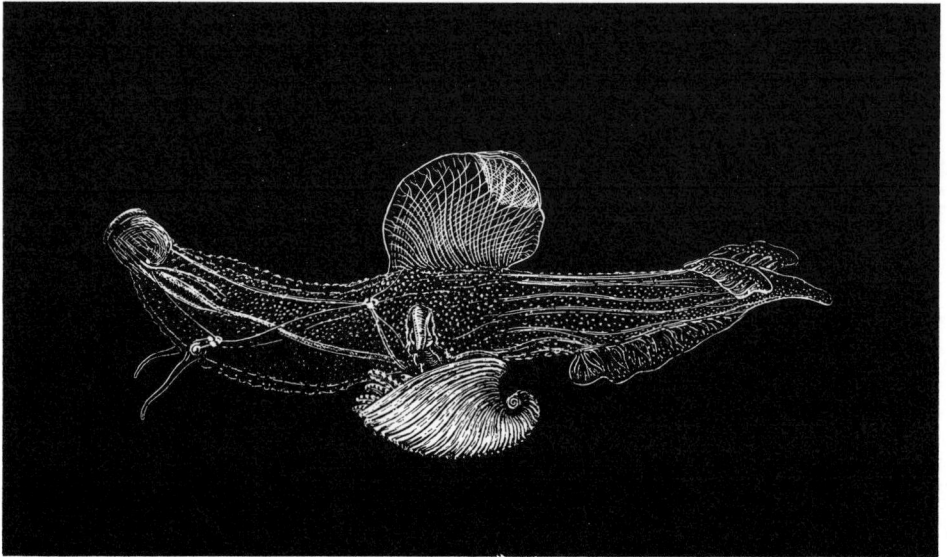

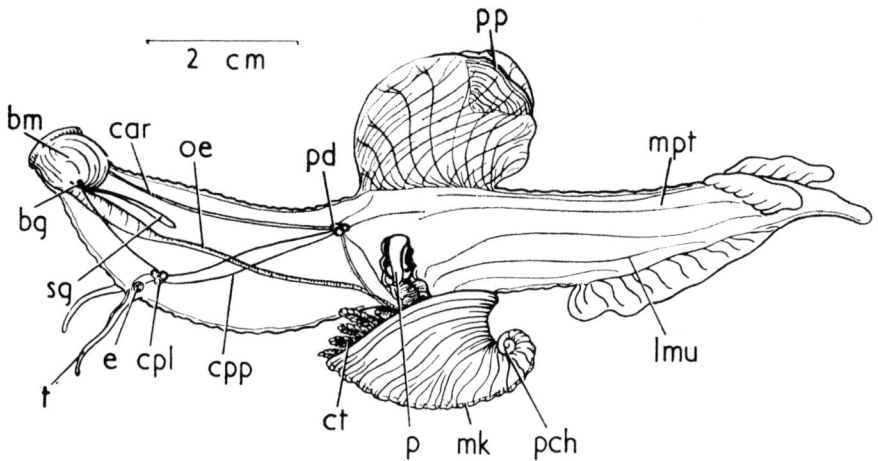

FIG. 292.—*Carinaria mediterranea:* male drawn in swimming position with the morphologically ventral surface uppermost. The body wall is sufficiently transparent to allow most of the contents of the head-foot to be seen except the visceral loop, which is too delicate. The shell obscures the organs in the visceral mass which are shown in fig. 293. For clarity the columellar muscle and the muscles of the head, except those of the buccal mass, have been omitted.

bg, buccal ganglion; bm, buccal mass; car, cephalic artery; cpl, cerebropleural ganglion; cpp, cerebro-pleuropedal connective; ct, ctenidium; e, eye; lmu, longitudinal muscle; mk, median keel on shell; mpt, metapodium; oe, oesophagus; p, penis; pch, protoconch; pd, pedal ganglion; pp, propodium, forming sucker on mesopodium; sg, salivary gland; t, tentacle.

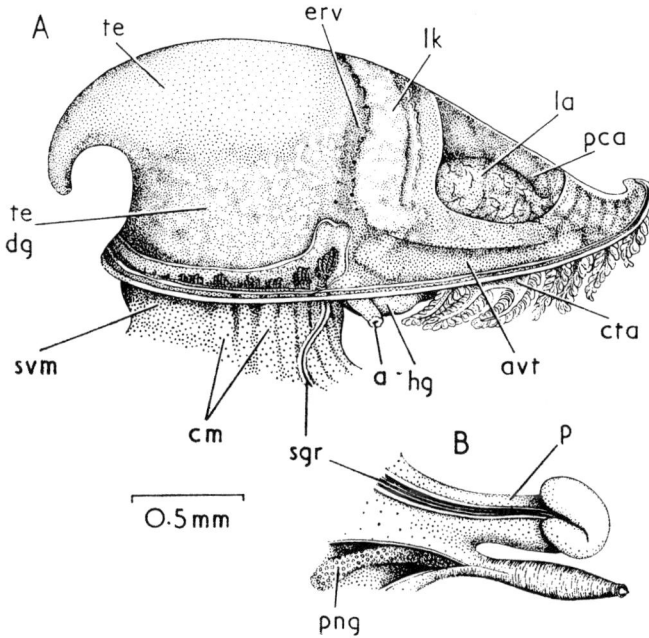

FIG. 293.—*Carinaria mediterranea:* A, visceral mass of male, from the right; the shell has been removed; B, details of penis, the appendage to which has been opened at its base.

a, anus; avt, afferent branchial vessel partly formed from transverse pallial vein; cm, columellar muscle; cta, ctenidial axis; dg, digestive gland; erv, efferent renal vein; hg, hypobranchial gland; la, auricle; lk, left kidney; p, penis; pca, pericardial cavity; png, gland on penial appendage; sgr, seminal groove; svm, stalk of visceral mass; te, testis.

where Krasucki (1911) has described a ganglion associated with its musculature. Franc (1949) has shown that in *Firoloides desmaresti* the fin is formed by the mesopodium and its sucker by the propodium, whilst the metapodium bears an operculum, later lost. The outer layer of the body wall is composed of epithelium overlying a transparent connective tissue of a very specialized nature and is papillated. The muscles are deeper and are grouped in bundles running for the most part in a longitudinal or oblique direction, to form a sheath which encloses the spacious haemocoel, some being attached to the shell to form the horseshoe-shaped shell muscle (cm, fig. 293). The papillae are formed of chondroid tissue which is also found in other parts of the body, particularly (in *Pterotrachea*) the snout, which it helps to shape and make rigid for feeding (Krasucki, 1911). In the connective tissue are stellate cells surrounded by a clear jelly, which they secrete, and this is the main substance of the outer layer of the body wall. Embedded in it are other types of cells (Simroth, 1896–1907) and plates of a special reflecting tissue.

The ctenidium (ct, cta, figs. 292, 293A) projects from beneath the anterior part of the shell and is unusual in its structure and position. The axis, which is pectinate, extends across the opening of the shell and the branches, which are graded in length from the base of the axis to the tip, bear bipectinate filaments. The gill therefore resembles that of the diotocardians more than that of monotocardians. The mantle cavity is of negligible size, for the wall to which the gill is attached has grown forwards and downwards (bringing with it the heart) until it is almost level with the mouth of the cavity. The heart (la, fig. 293A), like the gill, is median and anterior and is of considerable size, in keeping with the haemocoel. The organs normally associated with the right side of the mantle cavity of a prosobranch—rectum (a), genital duct, renal opening and hypobranchial land (hg)—are clustered in a very restricted area. This appears to be due to the hypertrophy of the left side of the mantle skirt and the ctenidium. The osphradium, however, retains its original position on the left edge of the mantle skirt and appears as a long narrow strip of ciliated and glandular cells.

The sexes of heteropods are separate and males of *Carinaria* are distinguished by the penis (p, figs. 292, 293B), which is split into dorsal and ventral halves and lies on the right side of the body, midway between the base of the fin and the free edge of the shell. An open seminal groove (sgr, fig. 293) passes from the male pore in the mantle cavity to the penis and continues along the median wall of the dorsal and stouter of the two branches, the true penis. Here the groove is broad and deep and ends distally at the swollen tip. The penis appendage bears at its tip the opening of a prostate gland (png) which is a blind tube extending into the haemocoel. No observations have been made on the function of this appendage. The secretion which it discharges may be prostatic, or it may be ejected in the vicinity of the female opening to secure the position of the penis during copulation.

Through the transparent body wall can be seen the long oesophageal region of the gut (oe, fig. 292), the strap-like salivary glands which open to the buccal cavity (sg), and much of the nervous system. The oesophagus is a straight tube without glandular outpouchings running back to the left side of the visceral mass, and the anterior part is swollen to form a crop; it is about half the length of the body. The animals are rapacious feeders capable of pursuing rapidly moving prey. Woodward (1913) stated that he took 6 small fish from the gut of a single *Carinaria,* each nearly as long as the mollusc itself. They were presumably packed into the oesophagus, where digestion occurs (Hirsch, 1915). *Pterotrachea* (found in the Mediterranean and N Atlantic) will seize living Heteronereis by the middle of the body so that the two halves fold together and the prey is swallowed whole. This is a rapid process and a worm 2·7 cm long may be swallowed by a *Pterotrachea* 6·6 cm long in 10–20 minutes. The whole body of the heteropod is flung into strong contractions, bending back and forth during the swallowing. The mollusc can also break pieces from larger masses of food which cannot be swallowed whole. The mid-gut, into which the digestive gland opens, is short and hidden in the visceral mass, and can be little but a passage-way between oesophagus, digestive gland and the short intestine.

The nervous system of *Carinaria* has a most unusual plan, for with the post-tentacular elongation of the head the pedal ganglia (pd) are far removed from the cerebropleurals (cpl) and lie near the anterior end of the fin. Consequently the cerebropedal and pleuropedal connectives are elongated and have become united on each side (cpp). The visceral loop is normal but extremely long, since the ganglia on it lie in the stalk of the visceral mass. The nerves are very slender and cross in the neighbourhood of the ganglia. Another feature which imparts a strange appearance to the system is the occurrence of pedovisceral connectives linking the supra-oesophageal ganglion to the left pedal and the sub-oesophageal ganglion to the right (Brüel, 1915, 1924a, b). This anomalous connexion may have a functional basis in making the correlation of pedal and visceral activities more effective.

A minor consequence of the elongation of the nerve ring which removes the pedal ganglia from the cerebrals is the fact that the nerve to the statocyst, which is very large, does not pass through the pedal ganglion on its way from the cerebral ganglion to the sense organ.

The eggs of *Carinaria* are laid in strings. Fertilization is internal and each egg is embedded in albumen and then provided with a shell. They escape from the female aperture in a single row close to one another (Gegenbaur, 1855) and surrounded by a protective secretion. Veligers with a bilobed velum develop in 3 days. The albumen and surrounding shell in *Pterotrachea* are secreted by glands attached separately to the oviduct but communicating with one another. The eggs passing down from the ovary, according to Krasucki (1911), enter the albumen gland and receive a layer of albumen, pass thence to the shell gland, where they are enclosed in a shell, and then return to the oviduct down which they pass to the mantle cavity.

Two prosobranchs which are dredged from soft substrata and do not occur intertidally are *Aporrhais* and *Turritella*. They are specialized for this habitat in different ways. *Turritella communis* is a sedentary filter feeder usually living just beneath the surface of muddy and sandy gravel, where its position can be detected by two depressions marking the inhalant and exhalant openings to the mantle cavity; the tip of the exhalant siphon is often visible. A veil of tentacles around the inhalant opening prevents the ingress of large particles and excessive silt. *Aporrhais pespelicani* lives at the surface of sandy mud, muddy sand or sand, or will construct itself a burrow just beneath the surface (see fig. 295C) with inhalant and exhalant channels further apart than those of *Turritella*. Yonge (1937) concluded that it remains there for a day or so at a time whilst the long proboscis searches around for food. This may be morsels of plants, though at the greater depths at which this genus is found (70–100 f) it must feed on other things, probably dead organic matter. Both prosobranchs have a shell with a long spire which may be dragged behind as they creep. In *Turritella* the shell (fig. 40A) is light yet progress appears laborious, for the foot is small. The shell of *Aporrhais* (figs. 32B, 294) is thick and heavy, and unique amongst British prosobranchs in the enormous expansion of the outer lip (ol), developed only in fully formed shells. The lip is produced into a broad flap with three angulated processes, also a spur running parallel with the spire and fused with it along the greater part of its length, and finally a terminal process (tol) which shields the inhalant pallial siphon and has its tip bent downwards. The mantle (me) spreads towards the edge of these extensions and the effective beat of its cilia is away from the mantle cavity. The opening to this cavity is constricted and the operculum (op) which closes it is a narrow, though stout, transverse strip with a surprisingly small area of attachment to the foot. When the animal is creeping it can be seen that the free blade-like part of the operculum projects upwards and to the left, away from the expanded lip of the shell.

Aporrhais can move over the surface in spasmodic jerks, displaying remarkable agility. When it does this the sole of the foot is placed firmly on the substratum, the shell and head lifted high off the ground by the upper part of the foot narrowing to form a stalk on which these pivot, and in a single swing forwards the mouth of the shell and the head are brought in front of the foot. The shell then sinks to the ground and with the aid of the enlarged outer lip maintains its position while the foot is lifted clear of the surface and brought forwards to the level of the snout. This completes a single step, and after a rest the sole is firmly pressed on the ground once more and the cycle recommences. The completion of one cycle takes about 10 secs, with a similar or longer rest period separating successive steps (Weber, 1925). Obviously this method of movement is correlated with the shape and weight of the shell. When the direction of movement is changed the head and visceral mass are rotated on the foot and may be pivoted round an obtuse angle left or right. If the shell be overturned the animal rights itself by extending the head and foot (in the usual manner for a prosobranch) stretching the foot posteriorly and bringing its tip to the substratum, always on the columellar side of the shell away from the expanded lip. Here it gains a firm purchase and will press the

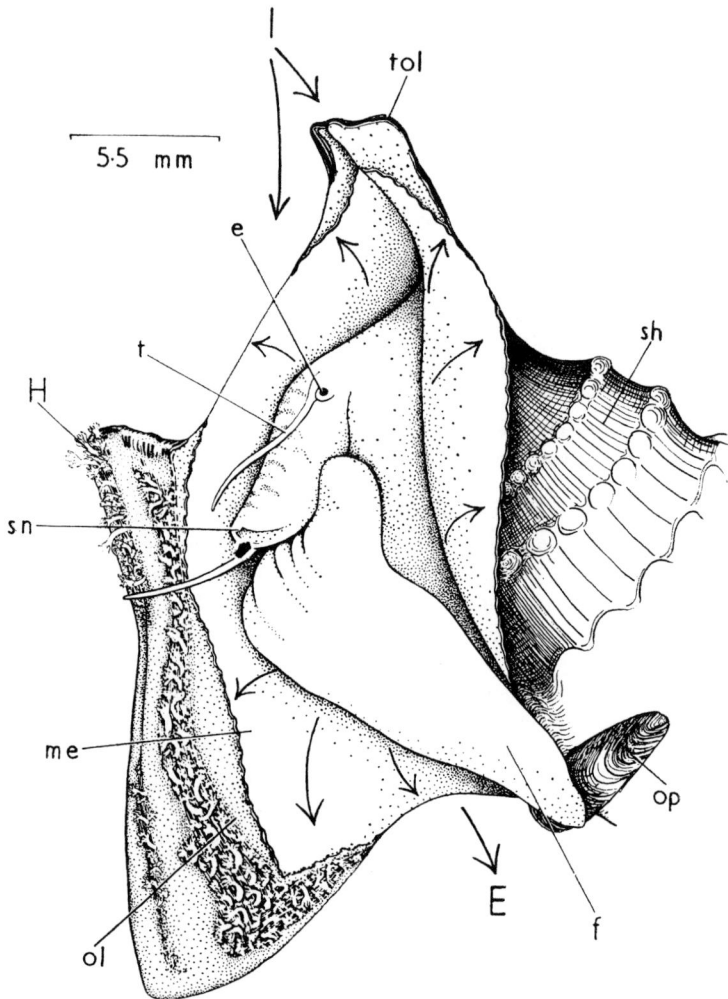

FIG. 294.—*Aporrhais perspelicani:* animal partly extended. Arrows on mantle edge show direction of ciliary rejection currents.

E, exhalant area; e, eye; f, foot; H, *Hydractinia* growing on outer lip; I, inhalant area; me, mantle edge; ol, outer lip showing thickening due to age; op, operculum; sh, shell; sn, snout; t, tentacle; tol, terminal projection of outer lip over inhalant area.

blade-like expansion of the operculum into the soil. Using this as a pivot the body is swung over.

The shell of *Aporrhais* thickens considerably with age and the sculpture of nodules and tubercles is worn down, partly with burrowing. The under surface of the expanded lip is also smoothed for this is dragged along the ground and covered to some extent by the mantle. It is perhaps surprising to find unworn grooves on this under surface covered by epizoic growth.

Fig. 294 shows a colony of *Hydractinia* (H) near the edge of the thickened lip (ol) and adjacent to the area covered by the mantle (me), where it will benefit from the water currents created by the snail. Epizoic growths are not uncommon on the upper surface of the shell, many of them being obvious to the naked eye and sometimes they are so thickly clustered that the shell is recognizable only from its outline. Out of a haul of 60 *Aporrhais pespelicani* from Kames Bay, Isle of Cumbrae, Barnes & Bagenal (1952) found 30 thickly covered with *Balanus crenatus* and only 6 quite free. In other areas a higher percentage of shells is free from epifauna. The habitat in which the mollusc lives is poor in sites on which sessile animals needing a firm holdfast may settle and their colonization of the shell suggests that the prosobranch spends less time beneath the surface than was once thought (Yonge, 1937). The firmness of the substratum and the abundance of food, at or beneath the surface, may determine the extent to which burrowing occurs. It has been suggested that the consistent absence of sedentary organisms on young *Aporrhais,* in which the outer lip has not yet expanded, and their presence on the older shells in the same locality, may be due to a change in habitat (Barnes & Bagenal, 1952). Young, unworn shells of prosobranchs, however, are usually free of epizoic growth whatever their habitat. It would certainly seem that the breadth of the expanded lip would make burrowing more difficult and provide resistance against sinking. A second species *A. serresiana,* occurring on soft mud in deep water off the Shetland Isles, has an even more expanded lip with longer digitations and the shell is lighter and the animal more delicate. It tends to sink but then pushes forwards rather than burrowing downwards. According to Yonge (1937) *A. pespelicani* will actively burrow unless placed on soft mud. It moves obliquely downwards, the proboscis frequently pushing aside the muddy gravel or sand and making a way for the foot to move forward. The worn terminal process is the first part of the shell to be pressed into the substratum and then the expanded lip cuts its way through as the columellar muscle pulls somewhat jerkily on the shell. A succession of movements, often separated from one another by a considerable pause, brings the animal beneath the surface, where it takes up a horizontal position and its presence is indicated only by the displacement of the ground. Soon, however, at the edge of the mound which covers it the tip of the long snout appears as it makes an inhalant channel. An exhalant one is similarly constructed when the head is moved round under the expanded lip of the shell and the proboscis is pushed upwards (fig. 295C). A strong pallial flow of water is now set up and maintained by the lateral cilia on the filaments of the large ctenidium. The walls of the inhalant and exhalant channels are strengthened by the proboscis which moulds them and applies mucus from unicellar glands over its surface.

Secondary sexual differences in *Aporrhais pespelicani* concern not only the penis and the elaboration of the pallial genital ducts (p. 330) but also certain conspicuous ciliated areas leading from the exhalant region of the mantle cavity. In the male (fig. 171A) such an area (ca) lies between the anus (a) and base of the penis (p), and in the female (figs. 171B, 295B) a tract (ctr) leads down the side of the foot (f) to the anterior pedal mucous gland (apg, fig. 295A). It is probable that the eggs, which measure only 0·24 mm across and are surrounded by albumen and a tough outer covering, make their way along this gutter to be attached to sand grains or pieces of debris by secretion from the pedal gland; there is no ventral pedal gland which could effect this. They are deposited singly or 2 or 3 together (Lebour, 1937).

Turritella communis is gregarious and a hundred or so animals may be collected from a small area in a favourable locality. They occur at a depth ranging from 3–100 fathoms where, undisturbed by the tide, they may feed continuously, perhaps remaining buried in the same position for long periods of time. The many-whorled shells (fig. 40A) are referred to as 'screws'

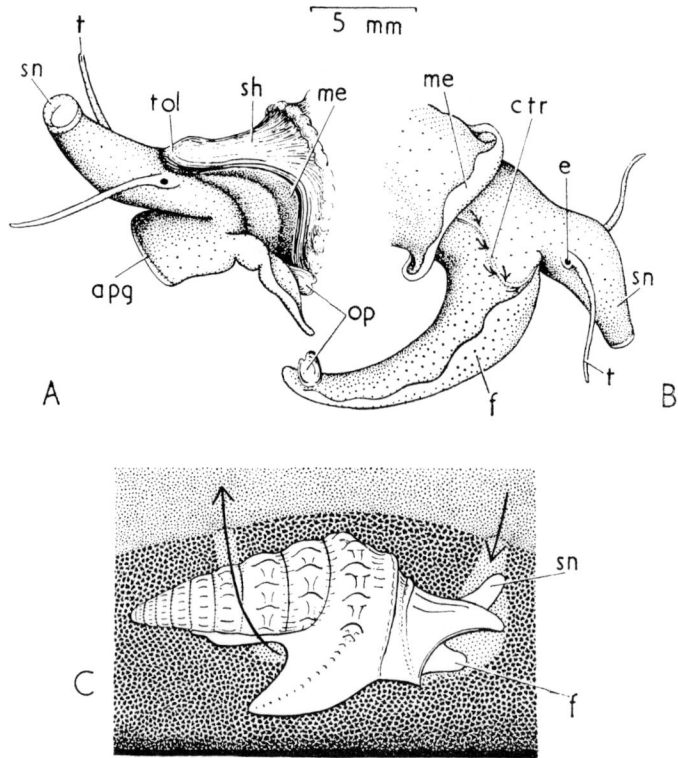

FIG. 295.—*Aporrhais pespelicani:* A, animal partly extended from shell showing extended snout; B, right side of
anterior part of body of female showing ciliary tract from mantle cavity to anterior pedal gland; C, diagram of
animal burrowed into substratum; arrows indicate direction of water and ciliary currents. C, after Yonge.
        apg, anterior pedal mucous gland; ctr, ciliated tract; e, eye; f, foot; me, mantle edge; op, operculum; sh,
shell; sn, snout; t, tentacle; tol, terminal expansion of outer lip of shell over inhalant area.

by old English naturalists. Their proportions reflect the shape of the mantle cavity which is
long and tapering and is subdivided into inhalant and exhalant compartments by the large
ctenidium. In a frontal view of the animal the openings of these two compartments are
readily distinguishable, for the wide inhalant one is hung with a curtain of pinnate tentacles
(pt, fig. 64). The ctenidial filaments, like those of other filter feeders, are narrow and elongated.
They are laterally flattened and each is attached to the ctenidial axis by its base and fixed to
the adjacent mantle skirt on the right of the axis by the basal third of its dorsal surface. In this
way their long axes lie roughly parallel with the roof of the mantle cavity and their tips curve
over to rest on the floor near the deep food-collecting gutter (fg, fig. 57 and see p. 99).
Particles which enter this gutter from the gill and from the ciliary currents on the floor of the
mantle cavity are entangled in mucus. This is secreted by the hypobranchial gland (hg), by a
strip of cells (the so-called endostyle) along the ctenidial axis and by the gill filaments and the
walls of the groove. The food-string can be seen rotating forwards, following its course along
the right side of the head, to collect in a spoon-shaped projection at the end of the groove

which forms a platter from which the mollusc feeds. At intervals, one stroke of the radula pulls in a considerable length of rope from the platter and the groove.

The exhalant siphon (es), projecting a short distance from the shell, is a physiologically closed tube formed by two folds of pallial tissue. The force of the exhalant current is strengthened by the cilia over the rectum and the open pallial genital duct, so that the ovoid faecal pellets of compacted particles (fig. 126D) are shot clear of the animal's head. Their expulsion is reinforced by a slight but sudden retraction of the head-foot into the mantle cavity.

In captivity animals will burrow close to one another. The resultant disturbance in the water, due to the combined feeding currents, appears considerable, yet there is surprisingly little disturbance of the mud. The process of burrowing is rather slow, with minimal displacement of the substratum. The foot pushes diagonally downwards in a series of jerky movements until the spire disappears and a mound marks the position of the larger whorls. The foot then moves the mud away from the entrance to the mantle cavity, pushing it to the right in front of the head where a small heap accumulates, and secretion from the pedal gland consolidates the particles (Yonge, 1946). The head and foot are now partially withdrawn and the operculum brought forward so that its spinous edge acts as a subsidiary filter; the entrance to the mantle cavity is enlarged, and the mollusc lies apparently motionless. The exhalant stream of water displaces the mud to the right of the head so that a channel leading from the siphon is established.

The gregarious habit of Turritella and the delicate adjustment of organs associated with the mantle to guard against the rigours of living in a muddy situation may be correlated with the lack of a penis in the male and the open pallial duct in both sexes.

Lebour (1933c) suggested that the veligers of this prosobranch stay only a short time in the plankton and settle to the crawling stage when the shell has only about $2-2\frac{1}{4}$ whorls and the larvae are 2–3 weeks old. The larva has no special characteristics foreshadowing the adult structure and resembles many other prosobranch veligers. It would appear that the ciliary method of feeding characteristic of the adult is not adopted until the mollusc is older, and that the radula alone is responsible for collecting the food in the younger stages. Graham (1938) suggested that even in the adult the radula may be used at times for collecting food from the substratum and Lebour (1933c) has observed them browsing on the algal growth on the sides of plunger jars. Young individuals may be found living with the adults (Cullercoats) or may live apart from them (Plymouth, Millport). At the last place they have been collected with rissoids at 5 fathoms in Kames Bay.

Species of Natica, sometimes called necklace or sand-collar snails on account of their spawn masses (fig. 196E, F), are our only mesogastropods restricted to large sandy bays, where their empty shells provide one of the few types of home available for small hermit crabs. They are carnivores preying on lamellibranchs, which are hunted in the sand. In the British Isles there are 5 species, though 2, N. pallida and N. fusca are stated to be rare. These and N. montagui are dredged and do not occur on the shore; indeed N. pallida has been recorded elsewhere at a depth of 2,300 m. Probably these so-called rare animals are more abundant than is realized. In the United States and Canada the family is well represented in numbers and species, some growing to a much larger size than any naticid does here (as much as 4 in across); in some areas the snails are a serious threat to the well-being of the clam industry.

N. catena, the larger of the two British species, specimens of which may be picked up near LWST, measures 1·5 in across the shell and N. alderi is about half this size. They may be seen

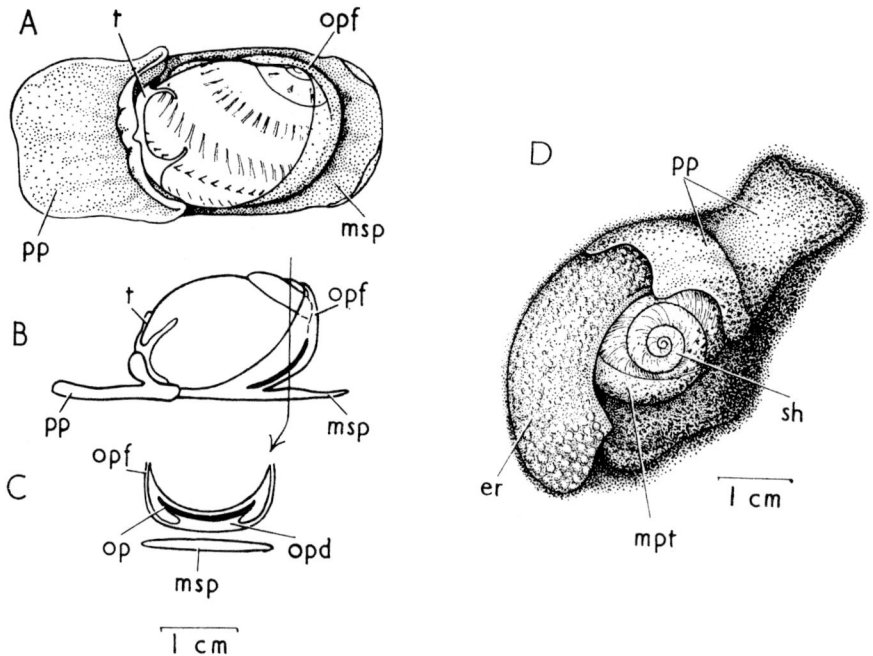

FIG. 296.—*Natica catena:* A, animal creeping, seen from above to show disposition of parts of foot; B, diagram of the same seen in left lateral view; C, ventral half of a vertical section at plane indicated by arrow; D, spawning beneath surface of sand.

er, egg ribbon; msp, mesopodium; mpt, metapodium; op, operculum; opd, operculigerous disc; opf, opercular fold of metapodium; pp, propodium; sh, shell; t, tentacle.

moving, sometimes with remarkable rapidity, just beneath the surface of the sand and the course of each is marked by a trail which is left behind the heap of sand covering the shell. The prey would appear to be sensed at some distance for the mollusc may be seen in pursuit of a bivalve which is a foot or so away. Moreover, it has been found that areas recently colonized by *Venus* spp. are soon discovered by the carnivores which invade the area in numbers (A. D. Ansell, personal communication). In 1934 an abundant population of *N. alderi* occurred on the Hunterston Sands, Firth of Clyde, which survived until 1936, when it had become greatly reduced. It was apparently feeding on the cockles (*Cardium edule*) in the sand and was itself eaten by overwintering swans.

The foot of *Natica* is exceptionally broad and large for the size of the shell (fig. 296). Not only is it used as a plough in moving through the sand, but an extension of the propodium (pp) is reflected over the shell anteriorly, covering the entrance to the mantle cavity, and an extension of the opercular lobe of the metapodium (opf, figs. 296, 297) covers the posterior part of the shell and conceals the operculum (op). In this way the shell is all but hidden. The cephalic tentacles (t) are widely separated, flattened and joined to one another by a transverse fold which, with the tentacles, is pressed against the shell by the propodium as the animal pushes forward. The eyes are covered, for they are small, each at the base of a tentacle and sunk beneath the surface tissues; in some naticids eyes are lost. The mantle cavity, so

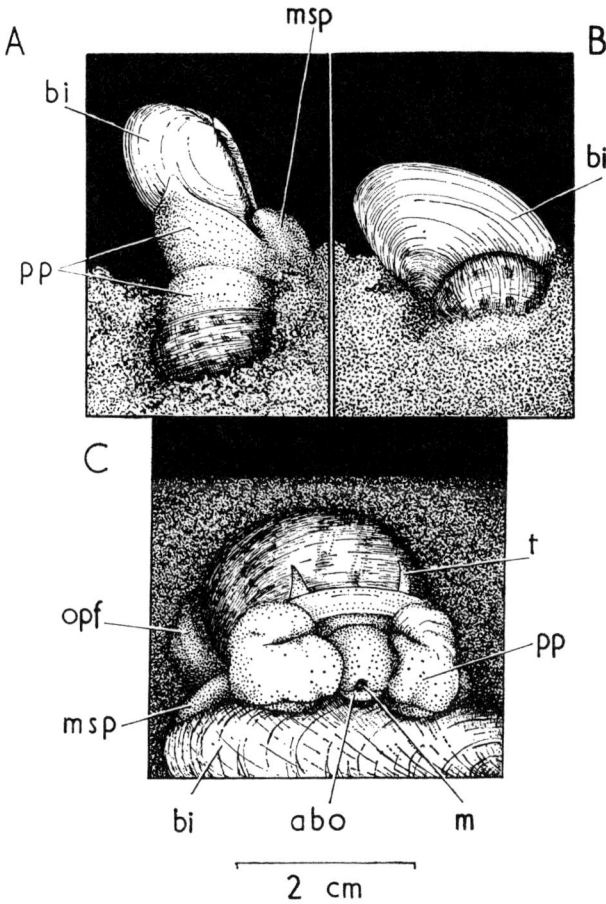

FIG. 297.—*Natica alderi:* stages in the manipulation of a bivalve prior to boring. A, the bivalve is gripped and raised preparatory to burrowing; B, the prosobranch is dragging the bivalve into the sand, having turned it so that it enters hinge first; C, the prosobranch has assumed the boring position, the proboscis is beginning to extend and is supported by the inturned corners of the propodium; the bivalve now lies with the hinge away from the prosobranch. After Ziegelmeier.

abo, accessory boring organ; bi, bivalve; m, mouth with radula; msp, mesopodium; opf, opercular fold; pp, propodium; t, tentacle.

effectively isolated from the environment, is large and has a well developed gill (ct, fig. 172A). Delicate movements of the propodium regulate the pallial stream of water and keep from it all but the smallest particles which are stirred up by the animal as it passes through the sand. *N. catena* and *N. alderi* withdraw rapidly into the shell when disturbed and the opening is then closed by the operculum. In contrast species of the common American genus *Polinices* need considerable stimulation before withdrawing, and as the foot contracts and is folded back into the mantle cavity, fountains of water are ejected from pores around its edge (Ziegelmeier, 1958) for the foot has a water vascular system (Schiemenz, 1884, 1887) which is used to enlarge it. No such water pores have been found in British species. *P. josephinus* swims by undulating movements of the foot, mainly the propodium. The propodial expansion and the ability to swim is a convergent resemblance with the lower tectibranchs, the Bullidae; indeed the word *Natica* is supposed to be derived from the natatory habit of the mollusc (Jeffreys, 1867).

Naticids attack their bivalve prey beneath the surface of the sand, enveloping the shell with the propodium. The prey remains passive while a circular hole is drilled through the shell (p. 234) and then the proboscis, accurately fitted to the hole, ingests the contents. These are dragged in by the radula and pieces cut off by the powerful jaws. According to Ziegelmeier (1958), if a naticid is creeping in the sand and crosses the point where a bivalve is buried, it starts to dig at once in search of the prey. The sensitivity appears to be located in the propodium. Once the bivalve has been found it is gripped by the foot, covered with slime and the shell examined for damage to see if it can be eaten without boring. (Naticids are said to eat freshly killed flesh.) The naticid then converts its grip on the shell to one which will allow it to burrow through the sand (fig. 297A), which it does dragging the prey behind it (fig. 297B). Beneath the surface the grip is rearranged so that the bivalve lies with the shell apex at the anterior end of the propodium—the position for boring. The middle of the propodial flap now allows the proboscis to emerge (fig. 297C), and this is held securely in contact with the bivalve by folding the right and left halves of the propodium towards the mid-line. [See chapter 34.]

Another prosobranch which lives in a somewhat similar habitat—muddy sand—and which may be encountered intertidally is the wentletrap shell, *Clathrus clathrus*. It is occasionally found in the SW of England but is rare in the N. Three other species of the same genus, which, one supposes, will probably prove to have the same habitat and habits as *C. clathrus,* but which are much more seldom seen, are *C. clathratulus, C. trevelyanus* and *C. turtonis. C. clathrus* is more commonly met on the shore during spring and summer, and this suggests that there may then be an onshore migration for spawning purposes. The spawn (fig. 203) consists of a series of small pyramidal egg cases encrusted with sand particles (Vestergaard, 1935; Clench & Turner, 1950) which is laid either on mud or muddy sand or on clean sand. In muddy environments the spawn is said to be laid on or near green weeds. From the capsules there emerge free-swimming veliger larvae which develop a smooth protoconch. Later the shell (fig. 141) develops a series of varices which give extra strength since its successive whorls are only just in contact one with the other, and may perhaps prevent boring by naticids by not allowing these predators to grip the shell correctly (Ankel, 1938b).

The mode of life of these animals is largely unknown, but such observations as have been made indicate that they attack anemones. Their internal anatomy (figs 99, 100), partly described by Thiele (1928), and observations by Ankel (1936a) fit with those that we ourselves have made in suggesting that the animals are carnivores, and probably specialized carnivores. The facts upon which this conclusion rests are given on p. 250. Ankel (1936a, 1938b) suggested that *Clathrus* feeds on sea anemones and Thorson (1958) observed the American scalid *Opalia crenimarginata* feeding on the anemone *Acanthopleura xanthogrammica* by pushing the proboscis, which can elongate to a considerable length, into the tissue of the host and 'sucking on its host for hours or even days'. [For more recent observations see chapter 26.]

CHAPTER 22

THE PROSOBRANCHS OF BRACKISH AND FRESH WATERS AND OF THE LAND

A NUMBER of prosobranch gastropods, all of which are herbivores or detritus feeders, have spread inland from their ancestral littoral home and colonized both fresh water and the land. Useful summaries of the habitats and distribution of these in the British Isles will be found in Boycott (1934, 1936) and in the census of the Conchological Society (1951). The 10 British species which live in fresh water retain the ctenidium, which indicates that they have reached this habitat by way of estuaries where there are a few other species capable of withstanding the fluctuations of salinity encountered there. In addition a number of marine forms penetrate the lower reaches of estuaries (e.g. *Patella vulgata, Gibbula umbilicalis, Littorina saxatilis, L. littorea, Buccinum undatum, Nassarius reticulatus*). These animals presumably retain isotonicity and their tissues can tolerate transitory low salt concentration, but their lack of osmotic independence means that they have not the ability to enter fresh water. However, one species, *Potamopyrgus jenkinsi*, appears to have overcome this barrier within recent times, for, though it was confined to brackish water until the end of the nineteenth century, it has subsequently spread with remarkable rapidity to both hard and soft fresh water masses, large and small. *Potamopyrgus* belongs to the Hydrobiidae, the same family as 3 more of our freshwater snails, 2 species of *Bithynia* (*B. tentaculata* and *B. leachi*) and *Bythinella scholtzi*, and also 4 brackish water species—*Hydrobia ulvae, H. ventrosa, Assiminea grayana* and *Pseudamnicola confusa*. Five of the 6 remaining freshwater species are lower mesogastropods belonging to 2 families, the Valvatidae of the northern hemisphere and the Viviparidae which, except for S America (where ampullariids occupy corresponding niches), have a world-wide distribution. *Theodoxus fluviatilis*, the last species of our inland waters, is an advanced diotocardian occurring also in Europe and N Africa. Though in other groups of the animal kingdom freshwater animals have often given rise to terrestrial forms, in the prosobranchs it is from a littorinid stock that our 2 terrestrial species have evolved. The marine members of this family are typically shore-dwellers and some exhibit strong terrestrial tendencies.

Like the majority of animals which live in fresh water the prosobranchs have modified their reproductive methods. The need to suppress the free larval stage and to enclose eggs in capsules has required the development of copulatory organs and since the more primitive freshwater forms belong to groups in which these are not normally found they have been evolved from parts of the body not used for this purpose elsewhere in the prosobranchs (Neritidae, Viviparidae). The penis of *Bithynia* may also be anomalous as its innervation is unorthodox (p. 318). Additional modifications of reproductive behaviour are found in viviparity and parthenogenesis. Three of the species are viviparous (*Viviparus viviparus, V. contectus, Potamopyrgus jenkinsi*) and retain the young in the pallial oviduct until they are small snails, although viviparity occurs in only 2 of all the British marine species (*Littorina saxatilis, Ianthina janthina*). *P. jenkinsi* is also parthenogenetic. Freshwater forms produce few eggs compared with their marine relatives (and all are characterized by the relatively small size of the gonad). The eggs are heavily yolked and protected by a thick capsule which is securely fastened to weed or some other substratum.

Some pre-adaptation for freshwater and terrestrial conditions is found in the operculate snails of the intertidal zone of the rocky shore which are subjected to periods of desiccation and rain, and a considerable range of temperature. The external shell, which protects the larger part of the surface against the osmotic inflow of water, is responsible for their resistance to these conditions. Moreover the method of reproduction of species with no free larval stage ensures that the young never leave the part of the shore occupied by the parent, though sheltered crannies afford protection until the snails are large and their shells thick. The salt concentration in fresh water may be 500 times less than in the sea and this presents the most formidable barrier to the majority of molluscs. All which have overcome it (lamellibranchs and gastropods) have an external shell characterized by a well developed periostracum which guards against erosion, and they are able to maintain a considerable osmotic gradient between their body tissues and the external medium. For some it has been shown that absorbing cells in the external surface and in the kidney tubule take up salts from dilute solutions (Krogh, 1939), which is presumably the deciding factor in the colonization of inland waters.

An examination of the literature reveals a vast number of mutually contradictory accounts of the morphology and biology of the freshwater prosobranchs. This seems to be partly traceable to the fact that these animals are much more sensitive to variation in their environment than are their marine relatives and also to the fact that freshwater environments differ more among themselves than do their marine counterparts. For these reasons it is desirable to consider the locality in which a particular piece of work was carried out as an integral part of its results.

A few operculates can live in soft waters where the concentration of calcium salts is low. The most resistant of the British species in this respect is *Valvata piscinalis,* which is the only prosobranch of Loch Lomond (Hunter, 1953b; 1957). The calcium content of the water is 2·3–3·3 mg/l, yet *Valvata* has considerable stores of calcium carbonate in the body apart from its shell. These are found in cells of the digestive gland (Cleland, 1954) and in large connective tissue cells which penetrate all spaces amongst the viscera. This special connective tissue, which is also present in other freshwater prosobranchs, was described by Bourne (1908) in *Theodoxus fluviatilis* though he did not observe the calcareous concretions within it. They can be seen readily in living tissue, and in most freshwater species the refringent spherules are visible through the skin.

A richer fauna of prosobranchs is found in inland waters which are hard. These tend to have more plant growth and consequently more places for shelter, particularly against currents, and more humus which provides food. To what extent the degree of hardness is a precise factor in determining distribution is not known, nor easy to find out, since each habitat has a complexity of ecological factors which are difficult or impossible to study singly. In the Thames between Oxford and Reading 8 of the 10 species of prosobranchs occur and only the 2 rare ones, *Bythinella scholtzi* and *Valvata macrostoma,* are absent. In the case of *Bythinella* this may be due to the chance of its introduction in the N, rather than an unsuitability of the habitat. *V. macrostoma* is recorded only in marshes and drainage ditches with a good fauna in the counties of Hampshire and Sussex and in E Anglia.

The quiet and brackish waters of estuaries, salt marshes and saline pools are favoured by the 2 species of *Hydrobia.* [See chapter 33.] The phenomenal abundance of one of these,

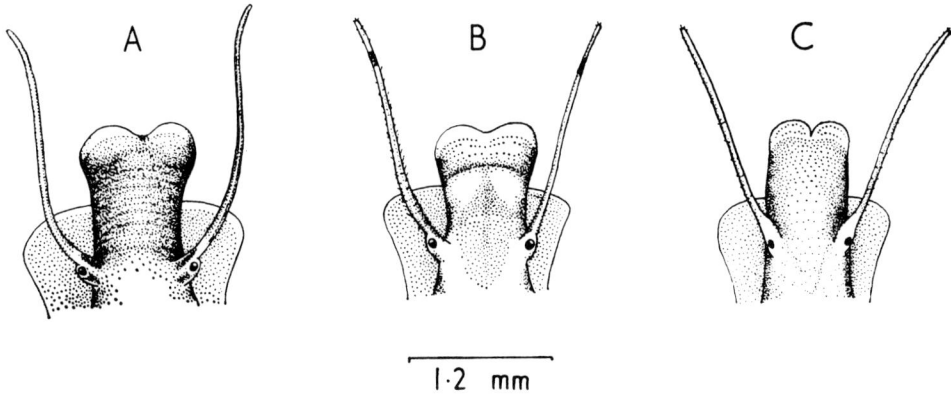

1·2 mm

FIG. 298.—Dorsal view of head of A, *Potamopyrgus jenkinsi;* B, *Hydrobia ulvae:* C, *H. ventrosa* to show pattern of pigmentation. In heavily pigmented animals the pattern may be more obscure.

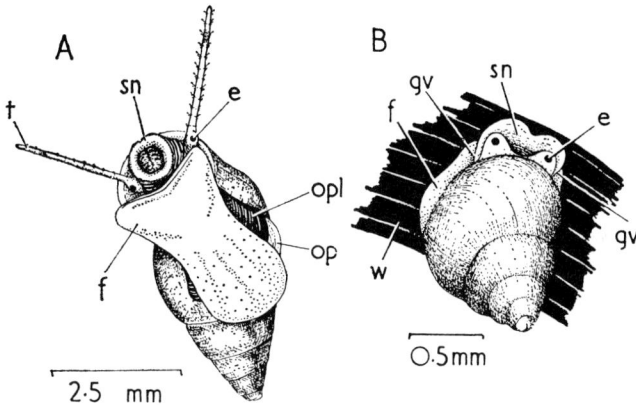

2·5 mm

0·5mm

FIG. 299.—A, *Hydrobia ulvae:* swimming and feeding on surface film. B, *Assiminea grayana:* male crawling, seen from above.
    e, eye; f, foot; gv, groove from mantle cavity to edge of foot; op, operculum; opl, opercular lobe embracing anterior and lateral margins of operculum; sn, snout; t, tentacle with characteristic pigment patch; w, weed.

H. ulvae (figs. 298B, 299A), has attracted the attention of many workers so that more is known about it than the second, *H. ventrosa* (fig. 298C). In some places the distribution of the two overlaps. The association of *H. ulvae* with the sea lettuce *Ulva lactuca* accounts for its specific name, though this link is by no means as definite as was once thought and the mollusc will browse over the surface of other weeds such as *Enteromorpha,* and occurs in myriads on apparently bare mud. The shell is conical with an elongated blunt spire, the body whorl making up about half the height, which may be as much as 10 mm, and the breadth 5 mm (though such a size is exceptional). It is rather thick and smooth and may be yellowish, brown, jet-black, reddish or even white in colour. Nicol (1936) recognized a difference in the distribution of the colour varieties in the lochs of North Uist. Here the strand form has a smooth, brown shell with very shallow sutures and is often 9 mm in height. Small numbers of this form may also be found in the lochs, where the typical individual has a rough eroded shell with deep sutures, and is dark reddish-brown; the erosion of the shell suggests that these individuals are parasitized (Rothschild, 1941b). Lambert (1930) found red the dominant colour in ditches on the marshes of the Thames estuary and attributed it to rust accumulating on the surface, which may also be green with algal growth. In females the spire of the shell is longer than in males and the whorls more tumid (Quick, 1920). The shell of *H. ventrosa* is smaller (8·4 × 3·2 mm), smooth, glossy and thin, with 6 rounded or ventricose whorls and the mouth auricular. It is horn coloured but appears dark with the pigmentation of the underlying tissues.

The different habitats of these 2 members of the family Hydrobiidae and of a third, *Potamopyrgus jenkinsi* (distinguished from each other by the pigmentation of the head (fig. 298) (Seifert, 1935)), may be illustrated by reference to the neighbourhood of the estuary of the River Adur at Shoreham, Sussex, where they were described by Ellis as occupying quite distinct areas and not occurring together (Ellis, 1932). In parts of the Ladywell Stream, not far from its junction with the estuary, and in dikes associated with it, *P. jenkinsi* was found living in water of a salinity of 0·88‰ or less, and usually associated with a fauna of freshwater insects. In a lagoon cut off from the estuary, but into which sea water percolated at high tide *H. ventrosa* was plentiful as a member of a marine fauna; the salinity of the water was 24·9‰. *H. ulvae* occurred in shallow pools on the saltings which are reached only by high spring tides and at other times may dry. After rain the salinity of the pools may fall to 13‰ and in hot weather, when a maximum temperature of 30°C has been recorded, it may rise as high as 34·7‰. *H. ulvae* flourishes in these extremes of salinity and temperature, sharing the habitat with a number of euryhaline crustaceans. Ellis also found it in a gravel pit on the river bank living amongst sea blite with the small, rare *Truncatella subcylindrica*. The salinity here at high spring tides was 24·8‰. During neap tides the floor of the basin dried and terrestrial arthropods invaded the area and were active whilst the molluscs were dormant.

Somewhat different salinity tolerances of these hydrobiids have been recorded in other localities in the British Isles and under experimental conditions suggesting that a combination of features regulates their distribution and not salinity alone. Although McMillan (1948b) found that under artificial conditions *H. ventrosa,* collected from Larne Lough, Co. Antrim, requires a minimum salinity of 24·5‰, which is similar to that in which it was living in the Adur estuary, yet this species occurs on peaty sand at Loch Obisary, North Uist, where the salinity is recorded as 10‰. Indeed Nicol (1936) found its salinity tolerance similar to that of *H. ulvae* (10–34‰). Johansen's records (1918) of the molluscs of Randers Fjord showed that *ulvae* can endure a salinity as low as 1‰, whereas the lowest record for *ventrosa* is 5‰. Jessen (1918) recorded the former species in fresh water in the Baltic, and Frömming (1956), in his study of the biology of mid-European snails, gave the salinity range for *H. ventrosa* as 10·9–2‰, and for *H. ulvae* water of full salinity to a salinity only as low as 10‰. *H. ulvae* is usually thought of as the more marine of the 2 species since it is less rarely cut off from the direct reach of the sea, and is indeed common on open beaches, where it may protect itself against the tide and desiccation by burrowing. *H. ventrosa,* on the other hand, lives in quiet lagoons not in direct communication with the sea and never where the salinity approaches that of normal sea water. In favourable places large numbers occur.

McMillan (1948a) has suggested that in the British Isles distinct biological races of *H. ulvae* can be distinguished, which differ in salinity tolerance, and perhaps the same holds for

H. ventrosa. Her experimental results showed that a collection of *H. ulvae* from Burton Marsh, Cheshire, behaved normally in 5% sea water (0·17‰), whereas another from 9 miles away was inactivated by 22% sea water (7·7‰), and a third from Holywood, Co. Down, died in water of this salinity. These differences in tolerance may be correlated with conditions in the habitat of each group of animals: only in Burton Marsh are the animals normally exposed to such low salinities; in the Holywood ditch they are subjected to inflows of normal sea water.

Thamdrup (1935), in his study of certain species of the Wattenfauna of Denmark, related oxygen consumption and temperature with habitat. The optimum temperature is highest for *Corophium, Pygospio* and *H. ulvae,* which are active at the surface of relatively dry, ebb tide areas—indeed this temperature is higher than the highest water temperature recorded for the area. Oxygen consumption rises abruptly with temperature and for *Hydrobia* it is calculated as 35 mg/kg/hr at 2°C and 490 mg/kg/hr at 20°C. Ellis (1925), working on the same species of *Hydrobia,* found that its acclimatization to fresh water is improved by warmth.

Hydrobia ulvae favours a rather firm mud or muddy sand and is also found in abundance on gravel and fragmented 'shillet' (clay slate) with silt (Spooner & Moore, 1940); in fact its distribution is almost independent of the nature of the ground. It dominates salt marshes, where large numbers congregate in the mud around plants or in other damp places, and is recorded on the shore of Plymouth Sound, though its presence in such marine habitats is linked with the local occurrence of fresh water. Its usual level on intertidal mud flats is near mid-tide, or on the upper third of the beach, and it is one of the dominant species of the *Macoma balthica* community or its *Scrobicularia* variation (Linke, 1939). In mud flats of the Cardiff area, however, the greatest frequency of the snail is towards low water, for the mud is too soft in the upper reaches of the shore (Rees, 1940). The areas it inhabits may be flooded only at spring tides, and so dry at neaps that the mud fissures, but *Hydrobia* can withstand long periods of exposure and survives in an apparently semi-desiccated state (Quick, 1920). In pools and damp hollows on the saltings conditions for growth are often more favourable than on mud flats and the snail attains a greater average size (Rothschild & Rothschild, 1939). *H. ulvae* is very sensitive to oxygen lack and appears to avoid putrid mud where *H. ventrosa* remains alive and apparently unharmed (Lambert, 1930), nor does it, like this species, appear to favour continuous immersion. *H. ventrosa,* perhaps owing to its smaller size, occurs on fine soft mud and often prefers weed to any other habitat. Nicol (1936) found it on all types of substratum except pure shell sand in the brackish water lochs of North Uist.

The intertidal area of St John's Lake in the Tamar Estuary, Plymouth, is only one of many localities around our coasts where a dense covering of *H. ulvae* spreads over acres of mud and fig. 300 shows the abundance of the species in a section across this Lake. The population density, which is based on numbers per sq m at 13 stations (indicated by vertical lines), is shown diagrammatically above the traverse. This survey was completed before the *Zostera* plague, when the ground between +1·0 m and −0·25 m was covered by a fairly continuous growth of *Zostera hornemanniana.* Here the density of *Hydrobia* appeared to increase, but the snail was also conspicuously abundant over stretches of mud devoid of such vegetation, and its numbers have not suffered significantly with the decline of the weed.

The behaviour of *Hydrobia ulvae* on exposed mud flats devoid of macroscopic vegetation shows how accurately it is adjusted to the changing conditions which prevail there. The snail creeps over the surface which is damp or covered by quiet water, but when the mud is stirred and the water made turbulent by the incoming tide, or if the surface dries, it burrows, sometimes to a depth of 1 cm. It emerges when favourable conditions return and creeps

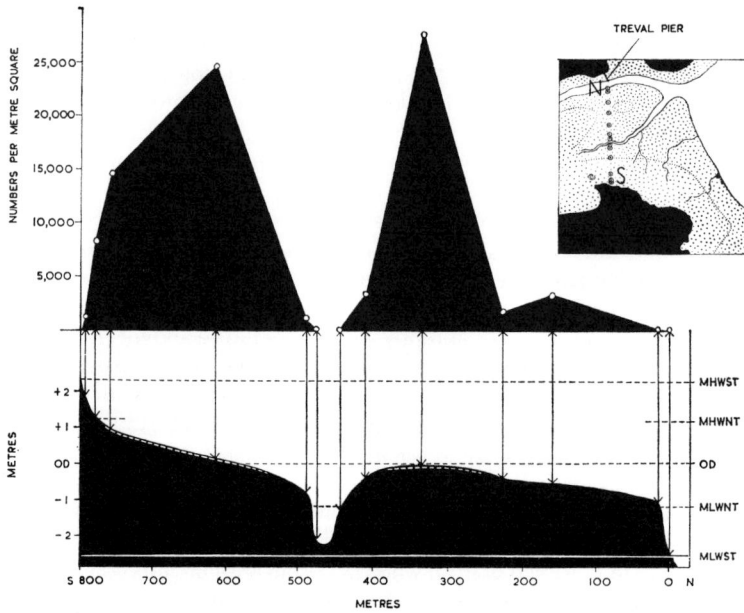

FIG. 300.—Graphs showing distribution of *Hydrobia ulvae* in St John's Lake, Tamar estuary, Cornwall. The lower half of the figure is a section along the line NS indicated in the map inset. Stations on this are marked by circles on the map; permanent water below MLW is white, land black and the dotted line on the stippled area is approximately MTL. The position of these stations is marked on the figure by arrow-headed lines and their distance from the channel at Treval pier is given in metres on the abscissa; ordinates on the left of the lower half of the figure give heights in metres from ordnance datum (OD); on the right tidal levels; ordinates for the upper half of the figure give number of snails per m². The broken line over the lower figure indicates occurrence of *Zostera;* white circles on the upper figure position of stations.

about, grazing on surface diatoms, blue-green algae or fine detritus; the long snout is swung from side to side and the movement of the radula and jaws can be seen through the surface tissues. As in other rissoaceans the tentacles have non-vibratile and vibratile cilia which sweep particles across them. In *H. ulvae* (figs. 298B, 299A) there is a hypertrophy of the ciliation of the left cephalic tentacle in the form of oblique lateral rows and this, together with the way in which the tentacle is used, suggests that it is of considerable sensory importance. The operculum (op) is not readily visible, for, when the animal creeps, its anterior and lateral edges are overhung by the opercular lobe of the foot (opl) as a protection against the ingress of silt between it and the underlying metapodium. The rapidity with which it creeps and the rhythmical movement of the snout (sn) are characteristic of rissoaceans, and a further similarity is the possession of a pallial tentacle which protrudes from the shell on the right side. Tracks are left over mud similar to those of *Littorina littorea,* but only 1·5–2 mm broad. They are often irregular and twisting, measuring 15 cm or so in length, though some may be twice as long (Linke, 1939). A trail of mucus (from the anterior pedal gland and a well developed sole gland) is laid along the path, and the closely-set, intertwining slime bands of countless snails form a network over the surface of the flat. Here, also, the spindle-shaped faecal pellets, which escape when the animal is active, accumulate in a layer sometimes

1 mm thick. On the Jadebusenwatt, Wilhelmshaven, where fantastic numbers of these *Hydrobia* occur, Linke (1939) described aggregations of pellets 15 cm deep. In drained areas, especially of soft mud, conspicuous pits, which are regularly spaced and often only a few millimetres apart, mark the positions of snails which are buried. Each lies in a more or less horizontal position with the mouth of the shell facing upwards. Snails which do not burrow when the surface dries (typically young ones recently settled on the shore) get air entrapped in the mouth of the shell, and the surface of the shell may dry, so that when water covers them they float and are lifted on the tide. On marshes they are soon caught up on the herbage, but on mud flats they may be dispersed over considerable distances. In stormy weather adults and young are scattered by the tide and they may be washed together in great masses. At the base of a dike wall on the Jadebusenwatt Linke (1939) described an accumulation of living snails 20 m long, 2 m broad and on an average 2 cm thick, though in places 20 cm thick, and he estimated the number as 55,000,000. This phenomenal population on mud flats and marshes provides food for fish and birds, and Linke (1939) also recorded the opisthobranch *Retusa truncatula* as an important predator.

The egg capsules of *Hydrobia ulvae* are fixed to the shells of other members of the community (fig. 196A), a habit which helps to ensure their survival, and occasional ones may be found elsewhere, as on the shell of a *Mytilus*. They are camouflaged and protected by particles of detritus and sand grains adhering to the capsule wall. Each contains 3–25 eggs, according to locality (p. 368). On the Jadebusenwatt, where breeding occurs between May and June (before and after this egg laying is rare), 92% of the snails have egg capsules attached to their shells in the former month. The range in numbers on each shell is 1–22 and a single capsule has 5–18 eggs. Linke (1939) calculated that a female will lay 300 eggs during a breeding period, which is few compared with the thousands which may be laid by a single specimen of *Littorina littorea*. Veligers emerge from the capsule in about 3 weeks, though undoubtedly the rate of development varies with the temperature, and Rothschild (1941a) stated that in captivity the larvae are freed 10 days after the eggs are fertilized. The pelagic phase may last a month or more (Linke, 1939), yet some authors have suggested that it is very short (Lebour, 1937; Thorson, 1946) and Smidt (1944) concluded that in certain places it may be suppressed. After the velum is lost the small snails with shells about 1 mm in length get washed about by the tide and dispersed. Smidt (1951) working on a population at Franø, Denmark, calculated that 10% of the eggs survive to the young snail stage, which is a high survival rate compared with that of lamellibranchs in the same habitat. The success of the species is partly due to its rapid multiplication and means of dispersal, for it will colonize newly available areas with remarkable speed. Such colonization has been recorded for the mud flats of the River Tamar and other estuaries in the Plymouth vicinity (Spooner & Moore, 1940), where during a long period of drought early in 1938 higher salinities than usual were registered upstream, and, in the same year, the species extended its range into these areas.

Rothschild (1941b) studied the growth rate of a population of *Hydrobia ulvae* inhabiting an isolated pool in the saltings of the River Tamar where conditions for growth are favourable. Here the main spawning period is autumn, though on nearby grounds it extends into the spring, and spring and summer are quoted as the breeding time in other parts of the British Isles. Rothschild found that by February the largest of the snails from the previous autumn spawning measured 1·25–1·75 mm long. Maximal growth occurs in April–July when this autumn spat group overtakes the previous year groups and all merge to form a population in which the measurements for length of shell show a peak of about 5·75 mm. By November, when early individuals of the next spat fall are appearing on the bottom, a further growth of barely 0·25 mm has been made, and over the rest of the life span an increase of not more than 1 mm may occur. From the age of 17 months the snails begin to die off. In captivity they are known to live over 5 years (Quick, 1924).

Assiminea grayana (fig. 299B), which is another brackish water species, has in comparison a very limited distribution. It is locally abundant on the E coast of England between Kent and the Humber and occurs at levels which are under water only at spring tides. Here it may be found in large numbers on grass and sedge stems and in the mud at their base. It seems to dislike immersion and in captivity crawls out of water to regions of high humidity, feeding on decaying herbage. It is the only British representative of the Assimineidae, a family closely related to the rissoids and resembling them in organization. However there are certain obvious differences. The tentacles are extremely short and broad, suggesting that each is represented only by the base which bears the eye; they are similar to the fully contracted tentacles of *Bythinella* (t, fig. 304C). The snout, unusually broad and bilobed, moves over the substratum as the animal creeps, and the foot is broad, short and abundantly supplied with glands of which the largest is the anterior gland provided with a lengthy sagittal duct; there are neither pallial nor metapodial tentacles. On both right and left sides a groove leaves the mantle cavity and passes forwards and downwards towards the sole of the foot (gv, fig. 299B), a feature which is also found in *Truncatella* (fig. 307A). There is no ctenidium, but 2 exhalant ciliated ridges, one on the mantle skirt and the other on the dorsal body wall, maintain water movements through the mantle cavity, the mantle epithelium acting as respiratory surface; on occasions, a bubble of air may be seen in the cavity. The intestine is long and elaborately coiled. The male is much smaller than the female and the penis is relatively enormous and arises from the mid-dorsal surface of the head; the prostate on the vas deferens is hyper-trophied and, in a mature individual, is the largest gland in the body. The organization of the female duct resembles that of *H. ulvae* except that the pallial section is narrow, markedly muscular and runs through the centre of the glands which open into it; the muscles are concentrated in a circular layer beneath the ciliated epithelium lining the duct. The duct ends on a long papilla which is equally muscular and is probably to be related to the deposition of egg capsules in a semi-terrestrial habitat. The bursa copulatrix at the upper end of the pallial duct is extremely large and like many other organs is surrounded by the kidney sac which acts as a body cavity. The kidney, indeed, is a vast thin-walled space pushing amongst the viscera and resembling in this respect the kidney of the hydrobiids which live in fresh and brackish water. In this species, however, it is even larger, and, as in *Bithynia*, sends an extension forwards alongside the rectum and pallial genital duct. The kidney opens to the posterior end of the mantle cavity and only its central part adjacent to the opening is lined by excretory tissue. Unlike the freshwater hydrobiids *Assiminea* appears to have no special connective tissue in which calcareous matter is stored.

Assiminea spawns in late spring and summer. At the time of copulation the male mounts the shell of the female and inserts the penis into the oviduct whilst the female creeps about. The egg capsules, each with a single egg, may be deposited singly in the mud, dropping, as the animal crawls, from the right groove which leaves the mantle cavity in the vicinity of the anus and genital opening (Sander, 1950, 1952). Alternatively, a conglomerate mass of capsules, faecal pellets (which pass along the same groove) and mud may be packed together by the foot, so that up to 80 capsules may be aggregated. The wall of the capsule is tough and thick for the secretion which forms it is well compacted by the muscular oviduct.

Truncatella subcylindrica is a small hydrobiid snail, sometimes known as a 'looping snail' because of its method of locomotion, which recalls that of a looper caterpillar (fig. 307). It is restricted in the British Isles to a special type of habitat on the S coast of England, but is reasonably abundant where it does occur: it lives in muddy places at the level of high tide, where it may be only occasionally wet by sea water, and is commonly associated with the plants *Suaeda maritima* and *Halimione portulacoides* (Ellis, 1932). In these circumstances it is more exposed to air than water, but retains a small ctenidium in the mantle cavity and can withstand continuous immersion for a considerable period. When moving in air over the surface of the muddy ground on which it lives, it extends the snout, which is very extensible, and grips the substratum with its tip (fig. 307A); it then pulls the foot up to grasp the ground just behind the snout (fig. 307B), dragging the shell in its rear, releases the snout and starts the process once again. Sometimes the foot slides along the surface of the ground as it is drawn forwards, sometimes it is lifted clear. Related to this method of movement are the expanded tip of the snout and the small, rounded sole of the foot. According to Pilsbry (1948) the animals are said to creep in the ordinary way when placed in water, but this was not the case with those specimens which we have seen. The animals appear to feed on detritus of vegetable origin, and on small unicellular plants in which their habitat is rich, and their alimentary tract is similar to that of other rissoaceans, with a simple oesophagus, a crystalline style and a short intestine. The nervous system is concentrated with cerebral, pleural and parietal ganglia in contact on each side, though the visceral ganglia are in the usual posterior position. The eyes are large, crescentic and lie displaced towards the mid-line at the base of the tentacles (e, fig. 307C). The kidney is not large as in *Assiminea* and possesses a small nephridial gland. The viscera lie in a reticulate connective tissue.

The most unusual features of *Truncatella subcylindrica*, however, relate to the shell and the reproductive apparatus. The former, in the mature animal, is short and consists of about 3 whorls with almost parallel sides (fig. 306B), because of the fact that at maturity the snail breaks off the apical part of the spire having previously sealed the opening with a calcareous plate, an activity from which the generic name is derived. In young animals, before this truncation, the shell has the form of a tall cone, as in most hydrobiids, though more slender (fig. 306A). The sexes are separate and the male has a long penis with a well developed prostate on the vas deferens. In the female the genital duct (fig. 306C) is more complex: the ovarian duct (od) is connected to the pericardial cavity by a long gonopericardial duct (gpd), distal to which are a receptaculum seminis (rcs) in the form of a finger-shaped tube, and a pouch, presumably the bursa copulatrix (bcp). It then enters the capsule gland (cp) in which albumen and mucus-secreting parts may be distinguished, running through these as a ciliated ventral channel. Unexpectedly, however, the neck of the bursa copulatrix, the functional significance of which we have not established, connects with the left kidney (lk), an arrangement also found in *Circulus striatus* (Fretter, 1956; see fig. 180A), and there is also a duct linking the bursa copulatrix directly to the base of the receptaculum seminis. If the bursa is indeed used for the reception of seminal fluid this tube would then permit the sperm to migrate easily to the receptaculum. The eggs are presumably enclosed in capsules and deposited amongst the plants and stones on which the animals live, perhaps escaping from the mantle cavity by the ciliated groove which runs between the foot and the base of the proboscis (gv, fig. 307A) as in *Assiminea*. The larval stage is suppressed but no other observations on the reproduction of this species appear to have been made.

Potamopyrgus jenkinsi (Smith), rissoid-like in appearance and measuring about 5 mm in shell length, has been a popular species for study amongst field naturalists. Though Frömming (1956) suggested that it may have been introduced into this country as early as 1859, it was first described in 1889 from specimens taken from brackish water at Plumstead, in the marshes alongside the Thames (Smith, 1889). It was recorded in fresh water in 1893 when Daniel took it from a canal at Dudley. Its sudden appearance from an unknown source, rapid invasion of fresh water and subsequent abundance there aroused considerable interest. Small size and abundance favoured its rapid distribution for it may be easily caught up on the bills or feet of birds and transported elsewhere; for instance, Coates (1922) recorded the mollusc adhering to the outside and entangled in the lamellae of the bill of a scaup duck shot at Perth. In about 30 years *Potamopyrgus* spread through the greater part of England and Wales by active migration and passive transport, though it is still unrecorded from consider-able areas in Scotland (W and N) where the first record was 1906 (Hunter & Warwick, 1957). There is a suggestion that its relatively slow spread in Scotland as compared with England may be due to the lack of a canal system which is so important in the distribution of a freshwater fauna (Boycott, 1936). The snail spread from the British Isles to other countries in NW Europe where the first records show it in brackish water (Bondesen & Kaiser, 1949; Hubendick, 1950) and later ones reveal its spread to fresh water, though in Sweden the species had not been observed in inland localities up to 1947 (Hubendick, 1947).

Taylor (1900) and Boycott (1917) were the first to observe the lack of males in this species, and later (1919) Boycott stated from experimental evidence that it is parthenogenetic; this was later confirmed by Quick (1920). With the exception of *Campeloma rufum* (Mattox, 1938) and members of the family Melaniidae (Jacob, 1957) it is the only undoubted parthenogenetic prosobranch which is known and like these species it is viviparous (Robson, 1920, 1923)—in the pallial oviduct there may be a brood of about 35–40 (bp, em, fig. 301). These features, together with the fact that this is one of the prosobranchs which may reproduce throughout the year, contribute to its abundance and rapid spread. According to Frömming (1956) the young are full-grown in 4–5 months and few individuals live longer than 7 months. In some areas it disappears as rapidly as it came and the reason for this is unknown, though such behaviour is commonly found in species recently introduced into a new habitat.

The shell, typically yellowish, is often so encrusted as to appear black, while other indi-viduals in the vicinity remain clean; the deposit may obliterate the sutures and become con-centrated in wart-like blotches. Warwick (1953) suggested that it is due to bacterial activity, which may be high in the mud where *Potamopyrgus* spends part of its time. The animal may be found on stones and weed, or buried in the superficial layers of mud and sediment of brackish or freshwaters where there is some current, and there may be numbers of these snails in the merest trickle; it is rarer in standing water. In favourable places it occurs in enormous numbers, blackening the weed and mud over which it crawls. It is known to provide food for carp (Dean, 1904) and large numbers are reported from the stomachs of trout (Whitehead, 1935). It can live and reproduce in salinities up to 17‰, but according to Adam (1942) none of the young hatching at a salinity above 18‰ attains maturity. Perhaps the occurrence of this species in North Uist (Nicol, 1936) in a salinity as high as 23‰, and in Randers Fjord in a salinity of 20‰ (Johansen, 1918), implies that factors other than salinity may be involved, or perhaps we are dealing with different races. It seems to be able to tolerate water with little calcium, for Boycott (1936) recorded it from Lough Leane at Killarney, along with 2 other prosobranchs—*Valvata piscinalis* and *V. cristata*—where the calcium content is only 7–11 mg/l. Moreover the density of the population in pits at Whitecross on Windermere was estimated at about 10,000/sq m in 1937 (Macan, 1950) and here, too, the water is soft.

There has been much discussion as to the origin of this prosobranch. There are suggestions that it is a mutant developed from *Hydrobia ventrosa* (Steusloff, 1927), though Boettger (1951) considered that its shell characters are identical with those of *Potamopyrgus badia* Gould

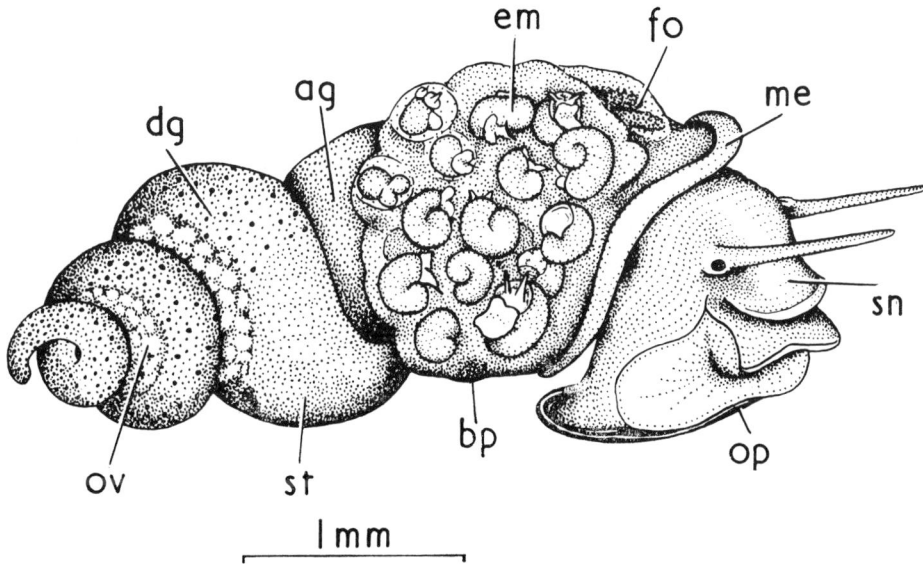

FIG. 301.—*Potamopyrgus jenkinsi:* animal removed from shell and seen from the right.
ag, albumen gland; bp, brood pouch; dg, digestive gland; em, embryo in brood pouch; fo, female aperture; me, mantle edge; op, operculum; ov, ovary; sn, snout; st, stomach.

from S Island, New Zealand. Because of the theory that the species has been introduced, records of specimens from deposits of known age have become of great interest. Those from recent deposits in the British Isles have been summarized by Kennard (1941) though his conclusion that the species was living in England in Roman times has yet to be confirmed (Bondesen & Kaiser, 1949; Warwick, 1955). Uncertainty as to the origin of *P. jenkinsi* still remains.

The cytology of the parthenogenesis has been studied by Rhein (1935) and Sanderson (1939) who found that there is only one maturation division in oogenesis and that it is non-reductional. Sanderson found the number of chromosomes in the British individuals (36–44) twice that of the continental individuals (20–22) described by Rhein. She therefore suggested that there are 2 races, a diploid one living on the continent and a tetraploid one in the British Isles which, as Peacock (Sanderson, 1940) has pointed out, may prove to be more hardy and so a more successful form. However, Rhein's diploid race is described from only one locality, so further evidence is required for the confirmation of this suggestion. Undoubtedly the species is genetically unstable, and perhaps this may be associated with its tendency to die out in habitats for which it was initially adapted. It is represented in Britain by a number of strains which differ from one another phenotypically and genotypically (Warwick, 1952), and a single population may consist of one, two or even three strains living side by side. Moreover, each of the three may show substrains. The commonest has a shell like that pictured by Ellis (1926), which is slender and the sutures not deep; the mantle pigment is rather pale. A second, also abundant, and found in brackish and inland water, has a less slender shell with the whorls more convex and the sutures correspondingly deeper; the mantle surface is deeply pigmented. It grows more slowly, does not attain such a large size and appears to be less prolific.

The third, occurring apparently only at St Bride's Bay, Pembrokeshire, is somewhat smaller and has very pale or patchy mantle pigmentation. In any of these shells there may be a keel represented by a row of bristles, or a slight ridge, or ridge and bristles may co-exist (all these are now regarded as var. *carinata* Marshall, 1889, and the last is sometimes distinguished as var. *aculeata* Overton, 1905). Starting at the third whorl this keel follows the coiling of the whorls near the periphery; rarely, accessory keels above and below this may occur. The keel is produced by a small blunt lobe in the mantle edge. Its occurrence in some individuals of a population has led to much speculation as to its origin, and since the snail breeds readily in captivity considerable work has been carried out in this connexion. However the results seem contradictory. Early workers (Robson, 1926; Boycott, 1929; Seifert, 1935) regarded the keel as due to the action of environmental factors such as salt concentration. Warwick (1944, 1952) was the first to suggest that genetical factors might be involved, acting in combination with algal metabolites in the water. Boettger (1949), on the other hand, considered it to be the result (in optimal conditions) of such environmental factors as food, oxygen, temperature and pH. In favour of this is the report of the occurrence of keeled and aculeate varieties of *Bithynia tentaculata* (Steusloff, 1939) and *Valvata piscinalis* (Haas, 1938) suggesting that shell sculpture and environmental factors are interlinked in other freshwater prosobranchs.

The keeled form was found before the smooth-shelled form both in the British Isles (Marshall, 1889) and in Belgium (Adam, 1942). It is interesting to note that *P. badia,* from which *P. jenkinsi* may have arisen (Boettger, 1951), has a keeled shell. Moreover Bondesen & Kaiser (1949) stated that all older records of *P. jenkinsi* mentioned the keel as typical of the shell and suggested that this was lost when the species invaded fresh water. Recent records from the British Isles and Europe show that the keeled shell is more common in brackish or polluted waters than in clean fresh water.

Recently a single male specimen of *Potamopyrgus jenkinsi* var. *carinata* Marshall has been recorded in the Thames at Sonning, Berkshire (Patil, 1958), though intensive search had previously been made for this sex, and subsequent search has also failed. Males must occur very infrequently, but the discovery of a single individual of this sex suggests that the species had originally separate sexes, and that parthenogenesis, evolved in comparatively recent times, has been accompanied by the gradual reduction in the number of males, which are becoming extinct. The general plan of the reproductive system of the male agrees with that of *Hydrobia ulvae* and *H. ventrosa,* and the testis, though small, shows active spermatogenesis; this is in accordance with the sporadically occurring males of species of *Melanoides—tuberculatus* and *lineatus*—in which the reproductive system is normal and similar to that of the males of related functionally bisexual species (Jacob, 1958). Krull (1935) and Thorson (1946) stated that *P. jenkinsi* may be hermaphrodite, though only Krull gave evidence for this, describing both eggs and sperm in the gonad and suggesting that self-fertilization takes place. The single male specimen which has been described is of average size for the variety *carinata,* has a well developed penis, a large prostate gland and shows no trace of oogenesis in the gonad. There is thus no suggestion that the species is hermaphrodite, and examination of small individuals to investigate whether sex change occurs failed to support this idea. [See chapter 33.]

Correlated with the recent appearance and rapid spread of *Potamopyrgus jenkinsi* is, perhaps, its apparent freedom from parasitic trematodes. *Hydrobia ventrosa* and *H. ulvae* are often heavily infected, and Robson (1923) has found 90% of the former in which the gonad is entirely destroyed. Yet when the 3 hydrobiids are living together *P. jenkinsi* is immune.

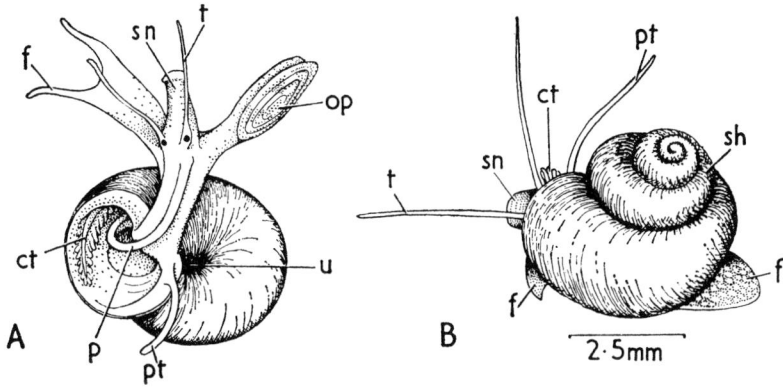

FIG. 302.—*Valvata piscinalis*: A, animal extended from shell; B, animal crawling, seen from the left.
ct, anterior tip of ctenidium projecting from mantle cavity; f, foot; op, operculum; p, penis; pt, pallial tentacle; sh, shell; sn, snout; t, tentacle; u, umbilicus.

Bithynia tentaculata is the largest of the British hydrobiids, for the shell may attain a height of 15 mm and a breadth of 6 mm; the second species, *B. leachi,* is about half this size. The shell is conical with a rounded base and has 5 or 6 convex whorls in the larger species and 4 or 5 in the smaller; the peristome is continuous; the operculum impregnated with calcium salts and concentrically ringed. *B. leachi* is distinguished by its more tumid whorls, a deeper suture, a sharply pointed apex to the short spire and a small umbilicus (this is scarcely discernible in *B. tentaculata*). Both species are calciphile and where their distribution coincides they may be found in the same habitat. However, *B. leachi* is more local and less abundant, especially in the N and W, and since it was introduced in Scotland it has become established only in limited localities. It favours water rich in plant growth, being more sensitive to oxygen lack than *B. tentaculata,* and avoids running water. Schäfer (1953a) showed that *B. tentaculata* is not affected by organic pollution and lives in the quieter stretches of rivers, quoting as maximum density in the River Lahn (W Germany) 81 animals in 250 cm^2. Both species may spread into slightly brackish water, and Johansen (1918) found them in salinities averaging 6–7‰ in Randers Fjord. He stated that they prefer a depth of 70–180 cm, but they are not uncommonly found to depths of 9–10 m.

In spring and summer *Bithynia* may be seen creeping over the surface of weed or stones and depositing egg capsules. Food is collected by the long snout, cleft in front, which incessantly searches the substratum when the animal is active. Intensive cleansing of an area will be observed before the female spawns for she carefully prepares the surface where the capsules are to be laid. They are fastened by the foot in tightly packed rows which are accurately arranged (fig. 199). An extension of the mantle edge on the right side forms a short siphon which directs the exhalant water current from the pallial cavity away from the head, carrying with it the ovoid faecal pellets, and it is from this region that the eggs are received by the foot.

An examination of the faecal pellets suggests that only the contents of ruptured plant cells are digested. Throughout the winter when there is little plant growth and water currents are stronger, the snails retreat to the surface layer of mud and go to deeper water. They feed on

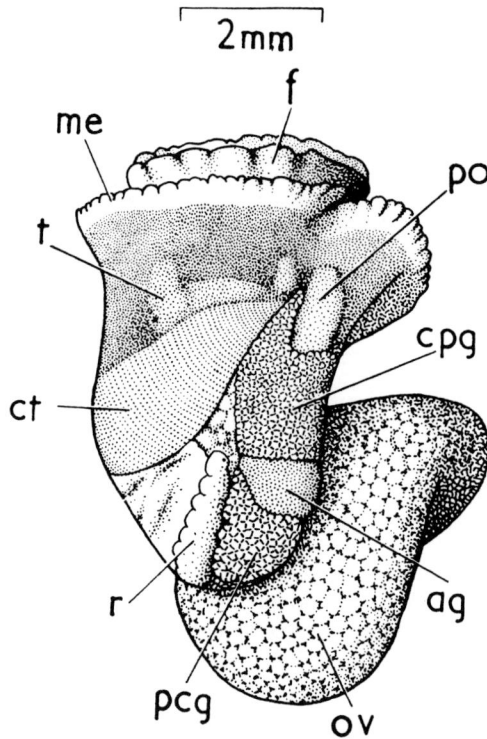

FIG. 303.—*Theodoxus fluviatilis:* animal removed from the shell and seen in dorsal view. The head has been fully
retracted into the mantle cavity. Some organs are seen by transparency.
    ag, albumen gland; cpg, capsule gland; ct, ctenidium; f, foot; me, mantle edge; ov, ovary; pcg, posterior
lobe of capsule gland; po, pallial oviduct; r, rectum; t, tentacle.

detritus, decaying weed and carrion. Particles stirred up from the bottom and entering the
mantle cavity may also be collected and eaten by *B. tentaculata* (and presumably by *B. leachi*),
though the importance of this method of feeding as compared with the selection of food by
the radula is not known. Starmühlner (1952) and Lilly (1953) were unable to prove that the
mollusc did more than concentrate the detritus in the ciliated gutter which traverses the floor
of the mantle cavity (fg, fig. 55). Lilly saw the concentrations of particles at the exhalant end of
the groove removed by the foot, or wiped on to some object which the animal brushed
against. However, Schäfer (1952) observed *B. tentaculata* eating them, and Hunter (1957)
stated that in certain localities the greater part of the food of this species is obtained in this
way. Werner (1953) suggested that this is only a supplementary method of feeding and
Frömming (1956) agreed with this. Perhaps the frequency with which this method of feeding is
used depends on the abundance of other foods and the nature of the suspended particles.
The gill of *Bithynia* is large (ct), its lamellae triangular, each with a broad base and the tip
hanging over the ciliated gutter. The gutter is not so well developed as in *Viviparus* where the
ctenidial filaments are narrow and elongated as in other prosobranchs which filter the pallial
flow of water for their food.

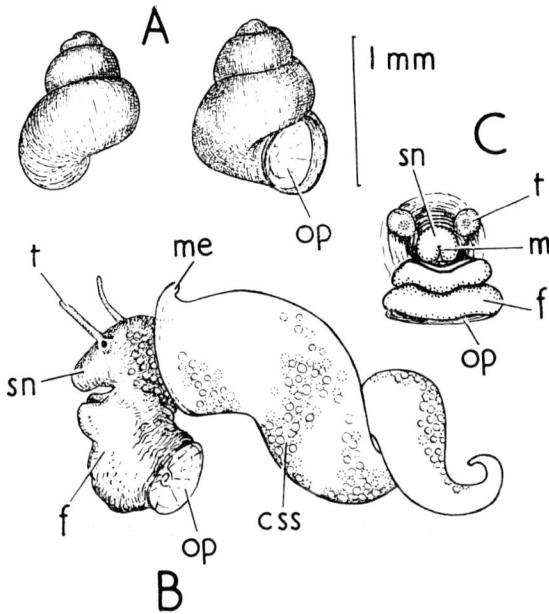

FIG. 304.—*Bythinella scholtzi*: A, shell in apertural (right) and abapertural (left) views; B, animal removed from the shell, seen from the left; C, animal seen from in front, to show shape of contracted tentacles.
css, calcareous spherules; f, foot; m, mouth; me, mantle edge; op, operculum; sn, snout; t, tentacle.

During winter floods snails living in shallower water may be drifted above the normal water course and left amongst the tangled herbage when the floods subside. They lie with the operculum closing the shell, though not tightly, and with the gut empty, yet activity is regained as soon as water covers them. They can survive in air of high humidity for several weeks, apparently unharmed.

Lilly (1953) studied the growth rate of a population of *B. tentaculata* at Wickford, Essex, and found that individuals which have a shell height of 6 mm or less in January have developed from the previous year's spawn, and that larger snails have survived a second or third winter. The growth rate is fairly constant throughout the year. By March some animals have shells 9 mm high and in its first spring the snail is mature, males before females. This growth rate is more rapid than that described by Hubendick (1948) for a population at Lake Malaren, Sweden, where the winter is more extreme. The average adult size of the snails in the two places is similar, the shell attaining a height of 10 mm. Hubendick found that growth rings on the operculum are unreliable for calculating age, since extra rings are unpredictably interpolated between the winter ones.

Bythinella scholtzi (fig. 304), the smallest British hydrobiid, which attains at most 3 mm in height, has a blackish shell due to an earthy deposit (Ellis, 1926), the cleaned surface being brown with a greenish tinge. The shell and soft parts present the general hydrobiid facies, but the former exhibits ventricose whorls and a blunt apex, and the latter shows an extraordinary contractility of the tentacles (t) and an excessive deposition of calcareous granules in the connective tissue (css). *Bythinella* was introduced from America and first recorded here in

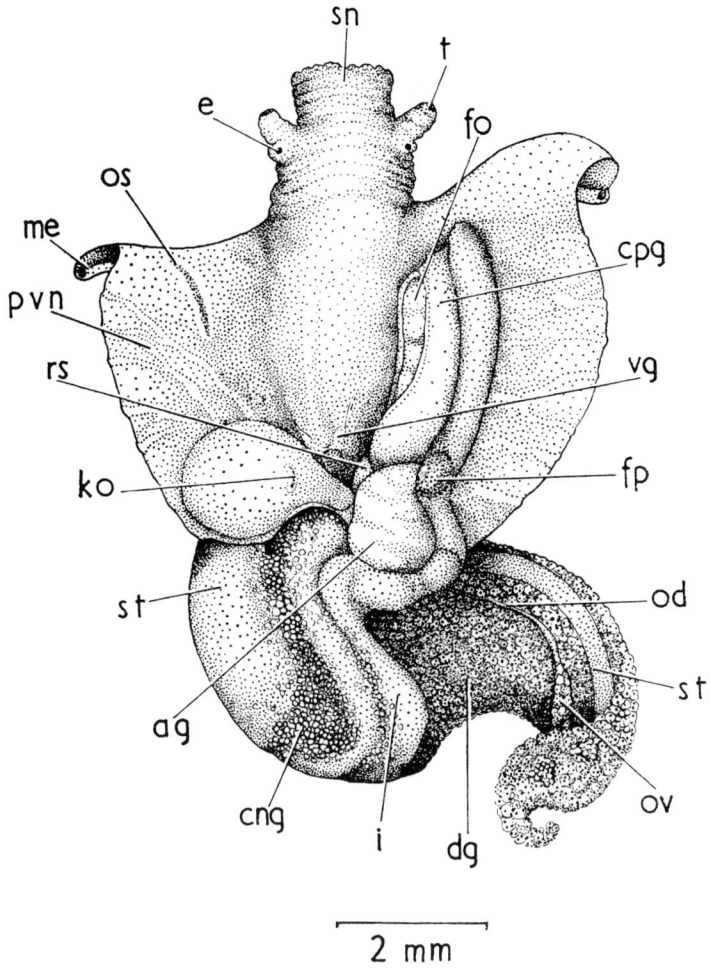

FIG. 305.—*Pomatias elegans:* female removed from shell. The mantle cavity has been opened by a median cut. Some organs are seen by transparency.

ag, albumen gland; cng, concretion gland; cpg, capsule gland; dg, digestive gland; e, eye; fo, female aperture; fp, faecal pellet in rectum; i, intestine; ko, kidney opening; me, mantle edge; od, ovarian duct; os, osphradium; ov, ovary; pvn, pallial vessel; rs, receptaculum seminis; sn, snout; st, stomach; t, tentacle; vg, visceral ganglion.

1900 from canals in Lancashire and Cheshire (Jackson & Taylor, 1904) and later in a timber dock at Grangemouth, Stirlingshire (Waterston, 1934) (where it no longer exists). In two of these places, Cheshire and Grangemouth, it was associated with the reed meadow grass *Glyceria,* of which it favoured the decaying leaves and stems. The snail lives well in captivity, and breeds. The egg capsules, deposited singly in late spring or early summer, have each a single egg which is large for the size of the snail. The capsule is attached by a flat base, circular in outline, and a keel projects across the convex upper surface (Jackson & Taylor, 1904). The young snail devours the albumen in which it is embedded and comes to fill the entire space within the tough outer wall before it escapes.

Viviparus is the largest of our freshwater prosobranchs, indeed of any in Europe, and lives longer. Jeffreys (1862) mentioned a specimen of *V. contectus* measuring as much as 2 in high and 1·75 in broad, and *V. viviparus* may be of similar dimensions—but is often slightly smaller. The shell of *V. contectus* is glossy and thin, the sutures deep and the whorls swollen, the apex pointed and the umbilicus distinct and deep; the dark green periostracum is rather thick and the brown spiral bands are seldom conspicuous (Watson, 1955). In contrast, the shell of *V. viviparus* is not very glossy, is thicker and whiter, the sutures not so deep, the apex blunt and the umbilicus scarcely visible; the periostracum is thinner and the spiral brown bands are often more conspicuous. The operculum of this second species is thicker and less nearly circular than that of *V. contectus*. Both live in hard water with slight to moderate current and may inhabit brackish areas, though they are very sensitive to any increase in salinity above 3‰. They are widely distributed in England (and throughout the greater part of Europe) and are often found together, for their habitats seem to be identical. However, *V. viviparus* has a wider distribution. It will live in reservoirs and has been recorded from standing water with a considerable accumulation of humous acids (pH 4·8).

Spoel (1958), working in Holland, found that *V. contectus* commonly reaches an age of 7 years, sometimes even 10, and *V. viviparus* an age of 6 years and sometimes 11; females live about half a year longer than males. He calculated this from growth rings on the shell and operculum, and stated that the darker winter rings are obvious and that 2–3 lying close together indicate a winter season during which there have been periods of activity and feeding. However Hubendick (1948) held that, as in *Bithynia,* the operculum is unreliable for calculating ages. A vast amount of similar work has been done in estimating the age of *Viviparus* from growth rings on the shell (Hazay, 1881; Goldfuss, 1900; Franz, 1938), but it seems unwise to adopt these as a criterion of age since they can be apparently produced at times other than winter. Spoel found that in both species of *Viviparus* the growth in height of the shell in spring exceeds that of the width, but in summer the width increases more rapidly; there are no differences in the height-width relationship in the two sexes, but Kessel (1933) said that females of both species are distinguishable on account of their more swollen whorls.

Males of *Viviparus* are less numerous than females and may be distinguished by the right cephalic tentacle which is modified to form a penis (p, fig. 175) and appears short, blunt and deformed. Behind each tentacle is a neck lobe. The right, which is much longer than the left (is), is used as a siphon (es) which directs the exhalant flow of water and the faeces away from the body of the snail. In the possession of this siphon and also the food-collecting gutter in the mantle cavity, *Viviparus* resembles *Bithynia.* Both features may be associated with a habitat where there is considerable silt.

Viviparus indeed, appears to thrive in water where there is considerable particulate matter in suspension. It will be found crawling on stones, creeping over mud, or sometimes lying half-buried in the soft bottom partially retracted into the shell which has the aperture directed upwards. Food may be collected from the surface of stones or weed by the radula or it may be filtered from the pallial water current. As the snail creeps over the mud the anterior edge of the broad foot pushes just beneath the surface, stirring up clouds of particles over the head; behind it the snail leaves a sunken tract where the particles are compressed. Suspended matter is drawn into the mantle cavity with the inhalant flow of water. The ctenidial filaments are long and narrow, resembling those of the ciliary feeder *Crepidula* (ct, fig. 331). They are muscular and each can be seen moving independently of its neighbours. The tips of the filaments hang over a conspicuous food-collecting groove (fg) which runs diagonally across the floor of the mantle cavity from its deepest part to beneath the right cephalic tentacle. Cilia on adjacent strips of body wall direct mucus and particles into the groove. Sediment is caught on the gill filaments and carried to their tips, from where it falls off immediately or is passed forwards from filament to filament before ultimately reaching the groove. Here a food-mucus string is formed, carried forwards and rolled into a ball (fg, fig. 56). At intervals the animal turns its head and eats the accumulation (Cook, 1949). The radular teeth show little signs of wear and are probably used for picking up particulate matter rather than rasping hard surfaces. The radula is short, only a fifth the height of the shell, whereas in *Littorina* it is $2\frac{1}{4}$ times the shell height. When *Viviparus* remains half buried in the mud it may be feeding on suspended particles in the water, although appearing inert. No pallial tentacles guard the mantle cavity against excessive silt as in *Turritella communis,* but the entrance is partly obstructed by the head and foot.

Embryos which have not hatched by the end of summer are carried over the winter months in the brood pouch (bp, fig. 331) and the young are liberated in spring. This appears to be due to rising temperature. Jeffreys (1862) mentioned that Millet counted as many as 82 embryos of different sizes from a single *V. viviparus.* A few are freed at a time but they are not all of the same age. Each embryo is embedded in albumen which is used as food and then surrounded by a tough shell drawn out into a stalk. Sometimes the young are born in this cocoon, which then breaks. Embryos removed prematurely from the parent will complete their development in water, and it can be seen that when they are ready to hatch they break through the cocoon using the operculum (Heywood, personal communication). Newly emerged snails may differ from the older ones in some external features: in *contectus* the shell has spiral rows of recurved bristles (sbs, fig. 56) which overlie the positions later occupied by the coloured bands, and hanging from the edge of the mantle skirt on the right side there are 3 short tentacles (pt) (Cook, 1949, where the species is mistakenly called *viviparus*); both these features are inconspicuous in *viviparus* (Heywood, personal communication). The young of *contectus* have a shell with a sharply pointed spire and those of *viviparus* are flatly coiled (Spoel, 1958).

Hubendick (1948) studying a population of *V. viviparus* at Lake Malaren, Sweden, found that 33% of the total growth of an individual occurs before its first winter; this includes intra-uterine growth, which accounts for 14%. Between the first and second winter another 21% is added and a similar amount between the second and third.

Two of the 3 species of *Valvata, piscinalis* and *cristata,* have the widest distribution in Britain of all the freshwater prosobranchs, for they occur in both hard and soft water. Ankel (1936a) stated that they live in brackish water with a salinity of about 2‰ in the North and Baltic Seas. The family to which they belong is of ancient origin, having been found in freshwater beds of the Oolitic period (Forbes & Hanley, 1850). In respect of the organization of the mantle cavity (fig. 343), many features of which may be associated with a muddy habitat, and also the complexity of the hermaphrodite reproductive system (p. 350), they stand apart from all other mesogastropods. One of their most diagnostic features is the bipectinate gill (ct, figs. 302, 343) which may project from the anterior edge of the shell when the animal is active. In some older drawings it is represented as projecting for nearly its entire length, which for a healthy snail appears to be an exaggeration. Frömming (1956) stated that it projects when distended by blood. There appears to be no correlation between the extent to which the gill is protruded and the degree of oxygenation of the water (Heywood, personal communication). The shell is primitive in having a loose winding of the whorls (most pronounced in *cristata* and least pronounced in *piscinalis*), an umbilicus and circular peristome (fig. 302). In *V. piscinalis* it is top-shaped with 5 or 6 rapidly increasing whorls which are rounded and have a conspicuous suture line. It may attain a breadth of 7 mm and the height exceeds the breadth only slightly. In *V. macrostoma* the shell is depressed with the spire only slightly raised, the umbilicus broadly open and the suture deep. The shell of *V. piscinalis* shows considerable variation in some localities and may be confused with this species, while that of *V. cristata* is markedly different for it has 5 whorls coiled in a disc-like fashion, the suture deep and the umbilicus very wide. The measurements given by Ellis (1926) for *cristata* are height 1·25 mm, breadth 2–4 mm, and for *macrostoma,* height 2 mm, breadth 3·5–4 mm.

When *Valvata* is active 3 long, extensile tentacles are seen to protrude from the body. Two are cephalic (t, fig. 302) and the third arises at the junction of the mantle skirt and body wall, in the position of an exhalant siphon. This pallial tentacle (pt) is extended upwards or laterally away from the body. Behind the right cephalic tentacle is the penis (p) which, in its resting condition, is directed posteriorly against the body wall so that its tip projects into the deepest part of the mantle cavity, for it is relatively longer than the penis of any other British prosobranch. The long narrow snout (sn) is wrinkled transversely and expands around the mouth. It appears to be continually in search of food, moving about between the 2 acute angles of the foot. The front edge of the foot (f) is deeply cleft and has a conspicuous double margin marking the opening of the pedal mucous gland. All exposed surfaces of the body have ciliary currents which cleanse them from silt. The cilia beat posteriorly over the foot, from left to right across the snout, across the cephalic tentacles from median to lateral edge and from base to tip on the pallial tentacle.

The gill extends only across the anterior third of the mantle cavity (ct, fig. 343) where it is attached to the roof by a short efferent membrane (efm) and to the rectum by an afferent membrane (afm). Between these attachments and the overlying pallial skirt a small pocket of mantle cavity is separated from the rest and into it the kidney opens posteriorly. The ctenidial axis and the filaments, which lack skeletal rods, are very mobile. The gill is moved freely within the limits of its attachments and its powerful ciliation can be seen as it projects from the shell. A strong current is directed forward along both afferent and efferent axes, carrying particles to the tip. In contrast to other Monotocardia there is no rejection current (A) associated with the inhalant current to the left of the head (Yonge, 1947) so that all particulate matter is swept across the floor by a strong current which beats towards the exhalant region on the right. Particles dropping from the tip of the gill are caught in this current or in a similar one on the head. The strength of the exhalant flow is increased and directed away from the body by the extremely extensile pallial tentacle, which may also have a tactile function. The anterior part of the mantle skirt is glandular and its secretion entangles particles and also the faeces, which are loosely compacted. The anus is on a papilla so that the pellets are swept clear of the body in the exhalant stream. The floor of the posterior part of the mantle cavity and the mantle above the ctenidium lack cilia.

During the autumn and winter months the snail may be found burrowing in the surface layer of mud and feeding there, yet the mantle cavity is kept clear of silt by the strong currents which sweep across its anterior parts and along the upturned pallial tentacle. There is no device for collecting and eating particles which enter the cavity. The snail grazes on detritus, diatoms, desmids and may rasp filamentous algae. In spring and early summer *Valvata* creeps on to plants (often entering into shallower water) and deposits egg capsules (fig. 200), or these may be fastened to any other solid object, even to the shells of bivalves. The young snails spend only a short period on the weed before they make their way down to the mud and may creep about on the surface for a time before eventually burying themselves. Cleland (1954), studying a population of *V. piscinalis* in the River Colne, Middlesex, found that all were beneath the surface by October. The older animals die off at this time and during the winter the next generation grows slowly. Growth increases in spring and the maximum size is reached by June.

Theodoxus fluviatilis is the only British representative of the Neritacea, a group of the Diotocardia which is of interest not only anatomically, since its members display a combination of primitive and advanced characters, but also ecologically, since they have exploited a variety of habitats. Such exploitation is partly attributable to their resistance to changing salinities, and to their method of reproduction, in respect of which they exhibit their most advanced characteristics. Three genera may be taken to illustrate the varieties of habitat in which members of the Neritidae are found: *Nerita,* a genus of tropical seas, occurs on rock in the littoral zone and some species ascend rivers and frequent brackish water and marshes; *Theodoxus,* the European representative, lives in brackish and fresh water with a fairly high calcium content; and finally *Neritodryas,* which occurs in the E Indies, lives on bushes and high up in trees, sometimes remote from water. *Theodoxus* extends from the Baltic to the Mediterranean. It is found on the Dutch, Danish and Baltic coasts, and in a salinity as high as 13‰ in Randers Fjord (Johansen, 1918). In Finnish brackish waters (salinity 2–6‰) it is the most abundant species in the littoral zone, living amongst *Fucus vesiculosus,* though it is not found in the inland lakes where the water is soft and fresh (Segerstråle, 1949). The size of the snail and of its egg capsules is reduced in saline water. Johansen has shown that in a salinity of 0·2–0·5‰ the shell length averages 9·5 mm, whereas between 7 and 12‰ it averages only 7·4 mm. In saline waters, too, the shell is lighter in colour.

The distribution of *Theodoxus* in the British Isles appears to depend upon the conjunction of a certain degree of hardness and of movement in the water. Thus it occurs in rivers, lakes and canals wherever these conditions are satisfied, favouring the wave line and shallower parts of lakes—not below 7–8 m—and the more rapid reaches of rivers. It lives on stones, sunken wood and aquatic plants, moving slowly over the surface as it gathers its food of microscopic organisms and detritus and it will rasp the surface tissues of plants. It is sluggish in habit and avoids strong light. As the snail creeps the shell, semi-globular in shape, is scarcely raised from the ground, so that the large head and broad snout are not exposed and only the pointed tips of the tentacles are visible. The shell is strongly built, rarely exceeding 6 mm in height and 11 mm in length, and the spire is short and oblique. Its colour may be yellow, brown or (more rarely) black and it is flecked with white, pink or purple or may have darker bands. The columella forms a broad septum behind the semilunar aperture into which fits the partly calcified operculum, orange or yellow in colour. There are 3 convex whorls but their internal parts and the associated columella are absorbed so that the visceral mass is hardly spirally coiled. The mantle cavity is broad and moderately deep with the single bipectinate gill directed across it (ct, fig. 303). The ctenidium is attached to the posterior wall on the left side (see fig. 52) and is free anteriorly, giving considerable freedom of movement. There is no hypobranchial gland comparable to that of other prosobranchs. The epithelium of the mantle skirt has scattered gland cells and in the thickness of its posterior part are glandular tubules which open by a common duct into the mantle cavity. Since this opening is between the 'organe creux' of Lenssen (1899), a possible vestige of the right gill (ho, fig. 52), and the anus, the gland is taken to represent the right hypobranchial gland (p. 121).

Males are distinguished from females by the penis, which is median to the base of the right tentacle and is stout proximally and tapering distally. Pairs may be seen copulating most freely

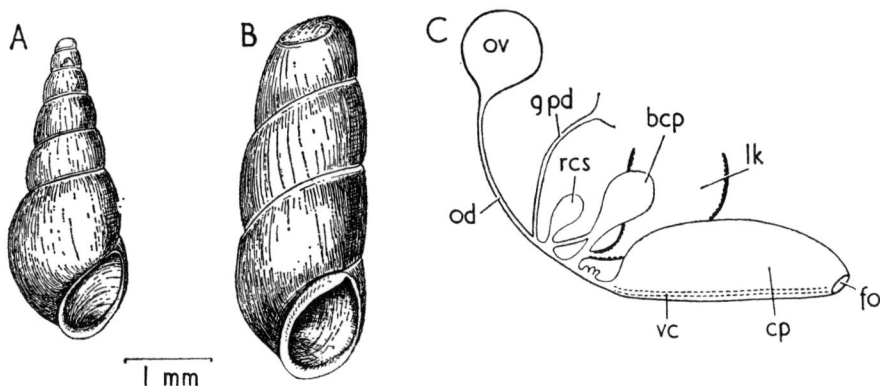

FIG. 306.—*Truncatella subcylindrica*: A, shell of young animal in apertural view; B, shell of mature animal in apertural view; C, diagram of female reproductive system.

bcp, bursa copulatrix; cp, capsule gland; fo, female opening; gpd, gonopericardial duct; lk, left kidney; od, ovarian duct; ov, ovary; rcs, receptaculum seminis; vc, ventral channel.

during late spring and early summer. The egg capsules, which have their walls reinforced by grit from the crystal sac (csa, fig. 187), are securely attached to a firm substratum by the foot (fig. 201), and are most common then too, but a few may be found at other times of the year (Berg, 1938). They may be fastened to the shell of an individual of the same species and have been found on the shell of *Viviparus*. From each capsule a single snail hatches after an embryonic life of 3 weeks or more. According to Becker (1949) 2 generations may be passed through in 12 months provided that the temperature of the water does not fall below 12°C. The animals live for 2 years, a few surviving into a third.

There are only 2 terrestrial prosobranchs which are indigenous in Britain, *Pomatias elegans* and *Acicula fusca*. Two other land operculates, natives of S Europe, were discovered on a wall at Kearsney in Kent in 1918 by Mr H. C. Huggins. These were species of *Hartmannia*, *septemspiralis* (Razoumovsky) and *patula* (Draparnaud), and had probably been deliberately imported (Huggins, 1919). *Pomatias* and *Acicula* are members of the Littorinacea, a group represented in most diverse environments. *Pomatias elegans* spends much of its time burrowing in the upper 4 or 5 in of soil during cold or dry weather and appears on the surface when the atmosphere is warm and moist; normally 95% R.H. is necessary for activity (Kilian, 1951). This is reminiscent of the behaviour of the related winkle *Littorina saxatilis*, which will creep actively over the rock at low tide when a warm sea mist covers it, but otherwise lies motionless or even hidden in rock crevices when the tide retreats. *Pomatias* requires a soil of loose texture to burrow in and this is provided in calcareous soils by the flocculating action of calcium carbonate on the clay particles. Indeed the records of its occurrence show that the species is associated with soils having high calcium content: Kilian (1951) found it numerous when the calcium content is 21% and over, and absent at 6% or under. It is not recorded from inland calcareous sand, but occurs on marine sandhills in Devon (Longstaff, 1910) and Cornwall (Boycott, 1934), habitats well suited for burrowing, and, with their proximity to the sea, rich in fragments of calcareous shells. The snail is found in woods, alongside hedges and sometimes on open grassland where a thin layer of comparatively alkaline soil (pH 7·5–7·9 covers calcareous rock (Creek, 1951). In favourable localities it is abundant. It has a daily rhythm of activity which is maximal during the early morning and minimal around midday; it becomes more active in the evening and will creep about all night. In winter, when the soil temperature drops, it hibernates and can survive a temperature of $-6°C$; the heart beat falls considerably during hibernation and may be only 2–4/min, as against 53/min at 25-30°C (Kilian, 1951). The snail awakes from hibernation and becomes active again when the soil temperature rises to 10–12°C.

Acicula fusca has a shell only about 2 mm high. It is also found on calcareous soils amongst dead leaves and wood in damp, shady places (especially favouring beech woods), but prefers a more acid soil than Pomatias, so that although both species may occur in the same area they are rarely found together. Its habitat is so damp that it is almost aquatic for so small an animal, and certain specializations in its structure may be related to size as much as habitat.

The terrestrial environment requires adaptations in respiratory structures and in repro-duction and also a considerable capacity for osmoregulation. Both Pomatias and Acicula (Creek, 1951, 1953) have lost the ctenidium, and the heart and kidney lie in the thickness of the mantle skirt, which is therefore highly vascularized and is the respiratory organ (a similar modification is found in some minute marine prosobranchs, p. 516). The mantle skirt is neither thickened anteriorly nor partially fused with the head as in pulmonates. The kidney opens (ko, fig. 305) into the posterior end of the mantle cavity as in other littorinaceans: this is unusual for a terrestrial gastropod in which there is no exhalant water current to carry away urine. In adult specimens of Pomatias the kidney concentrates large quantities of uric acid and this is also found in a special gland (cng), the concretion gland (Quast, 1924a, b; Kilian, 1951); in Acicula there are also large accumulations of crystals in the kidney tissue. These facts suggest that it is an accumulation kidney, and that the water which it excretes is almost pure. This could be used to keep the mantle cavity moist, and its surface is, indeed, always moist or even wet. In Acicula cilia on the head and hypobranchial gland set up currents in the water which probably facilitate gaseous exchange over the respiratory surface; the mantle cavity is narrow and deep and contains a large bubble of air which may be partially extruded as the animal creeps; this, like 'Notatmung' in fish, is probably a respiratory device. In both these terrestrial prosobranchs the osphradium and hypobranchial gland are retained.

The reproductive habits and processes are known only in the case of Pomatias elegans. In warm moist weather during spring and summer copulation may be frequently observed and during this period the females spawn. Each egg is laid in a large spherical capsule, about 2 mm in diameter (fig. 197), which is discharged into the mantle cavity from the swollen pallial oviduct. The opening of the duct occupies the greater part of its ventral wall. The capsule is forced out by muscular movements and passed down the right side of the head to the foot. The sole of the foot is cleft by a deep groove into right and left halves, which are moved alternately in walking. A posterior pedal gland (composed of a number of tubules which penetrate the deeper tissues of the foot) opens into the groove, but it is uncertain whether this provides secretion for the final moulding of the egg capsule; it is present in both sexes. There seems no justification for Thiele's suggestion (1929–35) that it acts as a lung. The foot covers the capsule with a coating of soil before depositing it beneath the surface. The capsule wall is permeable to water and salts which pass in from the surrounding soil and there is also some factor present in natural soil water which is necessary for development. The egg of this terrestrial prosobranch is thus a typical non-cleidoic egg and may be provided with water in excess of the requirements of the embryo as a guard against temporary drought. Embryonic development is slow and takes 3 months, and as in some other members of the Littorinacea, Littorina littoralis and L. saxatilis, the larval stage is suppressed and the embryo develops into a miniature of the adult before it leaves the capsule. In Pomatias, however, the veliger stage is

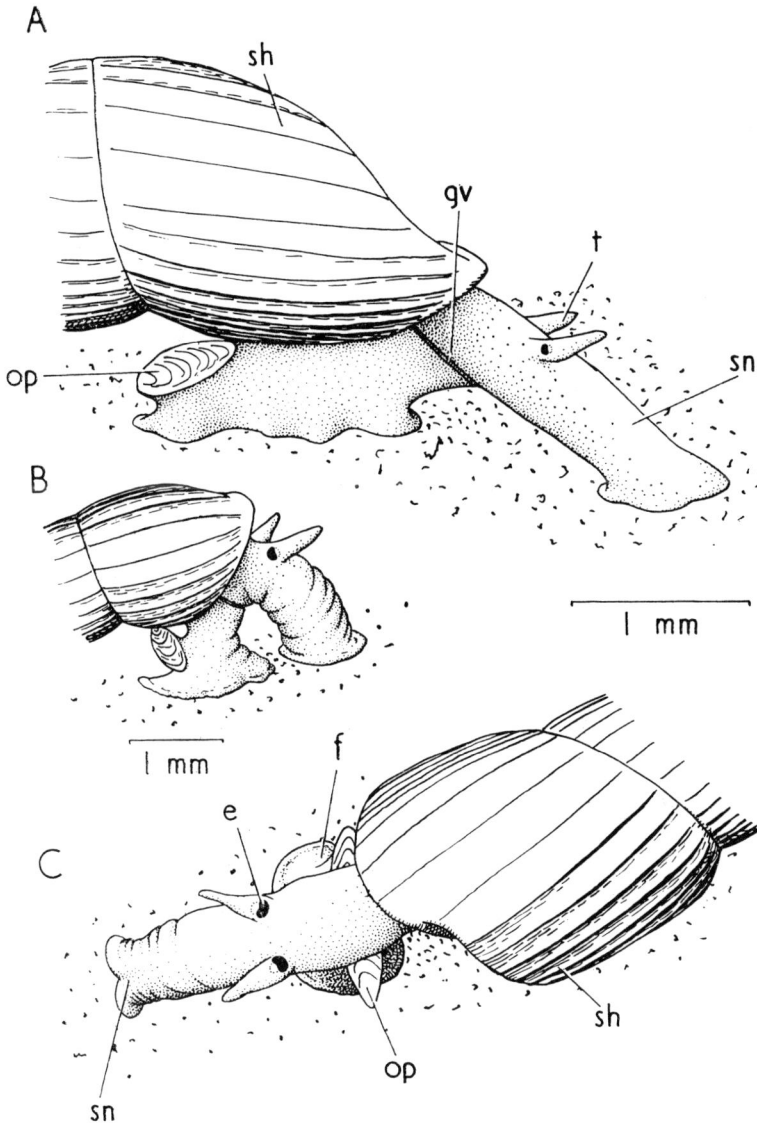

FIG. 307.—*Truncatella subcylindrica* to show method of locomotion: A, proboscis extended and gripping substratum; B, proboscis contracted and foot brought forward to its base; C, intermediate position in dorsal view.

e, eye; f, foot; gv, groove from mantle cavity; op, operculum; sh, shell; sn, snout; t, tentacle.

considerably modified, especially by the large cephalic vesicle which develops in the velar region (cv, fig. 224). A vesicle superficially similar is present in the embryo of other terrestrial gastropods and a variety of functions has been attributed to it. It has been considered as an adaptation to provide a greater surface area for the absorption of oxygen in *Helix pomatia* (Delsman, 1914); in *Agriolimax* (Carrick, 1938) it is vascular and contractile, functioning as a larval heart. In *Pomatias elegans* Creek (1951) has shown that the vesicle has an absorptive function and its cells become filled with albumen which is gradually used by the embryo.

Both these terrestrial prosobranchs feed on decaying vegetable matter, and *Pomatias* is known to digest the cellulose in its food, though it is uncertain as to whether this is done by the mollusc's own enzymes or by the bacteria which are present in the gut (Kilian, 1951). *Acicula* takes in minute particles and may select fungal hyphae from among decaying plants. Its gut, as in other small prosobranchs, is simplified. The glandular part of the oesophagus contains only mucous cells, the secretion from which may counteract the acidity of the decaying food. The stomach has neither gastric shield nor style, though there is a caecum homologous with that of winkles. In *Pomatias* the stomach is elaborated as in *Littorina* and in addition a style is present and consequently there are no oesophageal glands.

A peculiar tissue surrounds the intestine and borders the nephridium of *Pomatias elegans* and is conspicuous on account of the snow-white, rounded granules it contains. Claparède (1858) described it as the 'glande à concrétions'. Its cells contain glycogen, pigment and urates, and between them may be phagocytes with round or polymorphic nuclei. There are also bacteria which, according to Meyer (1925), are similar to those in the intestine. The concretion gland varies in size according to the season: it is maximal during autumn and decreases during hibernation (Kilian, 1951). It is thus affected by the food supply and increases with increased food. The gland would appear to be a temporary storage organ which can be called upon in times of reduced protein metabolism, and together with the nephridium it is responsible for excretion. There is a rich blood supply by which materials may enter or leave its tissues. Meyer (1925) suggested that the bacteria may aid in the transformative processes and so benefit the mollusc: they may produce enzymes which during periods of starvation transform waste material into substances of metabolic value. The snail is infected with these bacteria during its early life. None is present in the egg capsule though they occur on the outer surface, and the newly hatched individuals are said to eat this shell and so infect themselves.

CHAPTER 23

THE PROSOBRANCH SHELL

T HE molluscan shell is a composite structure made of an organic matrix containing a protein rich in the amino acids glycine and alanine mixed with some polysaccharide (Currey, 1980), and loaded with calcium carbonate, sometimes in the crystalline form calcite, more commonly aragonite. The matrix was originally regarded as merely providing centres of crystallization for the calcium salts, but recent work (Wainright et al., 1976; Kramptitz et al., 1983) suggests that its role is greater, and that it may control many of the properties of the shell and its crystalline components.

With the advent of scanning electron microscopy the nature and arrangement of the crystalline structures embedded within the matrix have become both more easily and more completely known than before. As a consequence shell structure is now commonly defined in terms of crystalline micromorphology instead of its simple mineralogical make-up, as was once done (MacClintock, 1967; J. D. Taylor et al., 1969, 1973; Grégoire, 1972; Carter, 1979). Molluscan shell micromorphology is often more elaborate than that of calcified structures in other groups of invertebrates and must result from highly organized and coordinated activities (about which next to nothing is known) in the cells of the mantle responsible for creating the shell. Study of patterns of crystallization has allowed some statements to be made about putative molluscan fossils, though frequently these only suggest possible relationships rather than pronounce certainties (Runnegar et al., 1975).

Of the patterns of crystalline calcium carbonate occurring in molluscan shells that known as nacre (fig. 308) is usually taken as the most primitive. It occurs in gastropod, bivalve and cephalopod shells and may well have been inherited from a previous monoplacophoran ancestor, since it forms the inner layer of a Neopilina shell (Schmidt, 1959) and has also been found in fossil monoplacophoran shells (Erban et al., 1968). Nacre consists of a matrix densely packed (to 95% capacity) with layers of flat aragonitic plates more or less parallel to the shell surface; each plate is thin, only $0.3–0.5\ \mu m$ thick, and the edges of the plates in different layers do not coincide, so tending to prevent a crack spreading far across the shell, though the shell is liable to collapse suddenly if sufficiently stressed. J. D. Taylor (1973) has suggested that the primitive molluscan shell consisted of an inner layer of nacre, clad externally by a layer of calcite in the form of polygonal columns up to several millimetres long, depending on the size of the shell, though only $10–200\ \mu m$ in diameter. Such a shell would be strong, largely because the nacreous component has been shown (Currey, 1980) to resist most deforming forces, particularly tensional ones rather than compressional, better than other crystalline arrangements found in shells, and, indeed, in most other objects made of calcium carbonate. From some such arrangement, the like of which is still found in some primitive gastropods, the other patterns have presumably evolved, the precise form of crystal occurring in the shell matrix perhaps depending on such factors as their speed of growth, and the concentration of salt in the mother liquor when the crystals were forming. The temperature of the environment is also important: Lowenstam (1954) found a correlation between lower temperatures

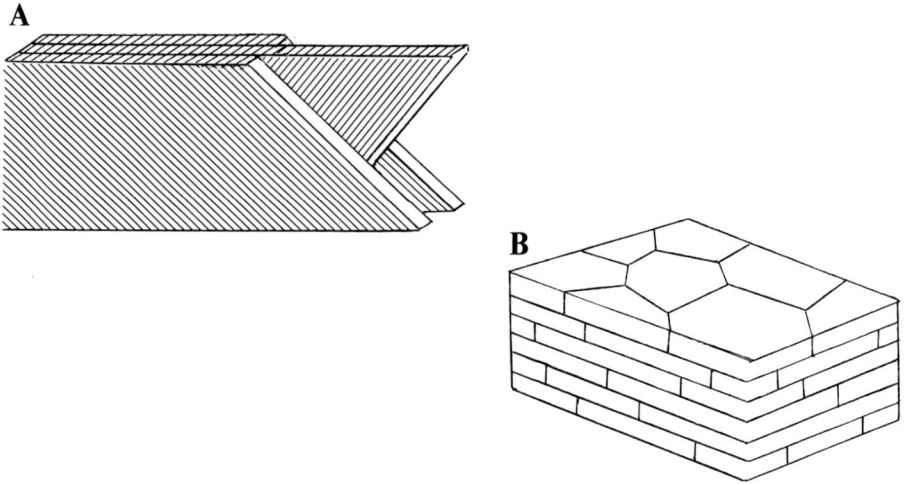

FIG. 308.—Diagrams to show A, crossed lamellar shell pattern; B, nacreous shell pattern; the shell surfaces lie approximately parallel to the top of each diagram. A shows part of three lamellae, each a stack of rectangular plate-like crystals, approximately at right angles to its neighbours. The thickness of a lamella varies from 10–15 μm and that of each crystal from 1·0–1·5 μm. B shows a stack of flat crystals each 0·3–0·5 μm thick.

and greater calcitic content and between higher ones and more aragonite; a similar finding has also been recorded by Taylor & Reid (1990) with respect to littorinids. They relate it to the lesser solubility of calcite than of aragonite at lower temperatures.

Of these patterns the crossed lamellar (fig. 308) is by far the most common in prosobranchs: in this the shell is composed of a series of layers each made of stacks of rectangular plate-like crystals lying obliquely to the main axes of the layers. The layers themselves are constructed so that the narrow ends of the component crystals ususally lie parallel to the surface of the shell and those of each layer lie approximately at right angles to those in neighbouring layers. The whole organization of the crossed lamellar pattern, both in its make-up and in its mechanical properties, resembles plywood. Though it is not as strong as nacre its criss-cross layering is more effective in stopping the spread of cracks (Jackson et al., 1988) such as would be started by predators, and this property may well underlie its frequent occurrence, particularly in larger shells. Patellogastropods, however (Lindberg, 1988), possess an outer layer of foliate structures, flat, lath-shaped components, arranged in superimposed layers more or less parallel to the outer surface of the shell; this type of structure, though found in some monoplacophorans and bivalves, is not otherwise present in gastropods.

The strength of a shell varies with the crystalline pattern which it exhibits, but it seems that prosobranch shells have, on the whole, a microstructure which imparts good resistance to all the different kinds of stress to which they may be exposed, rather than outstanding resistance to any one particular hazard (Currey, 1988). Considering that the shell of an intertidal prosobranch must be strong enough to support its owner's weight in air, resist being flung against boulders by waves or crushed beneath them, offer protection against the attack of predators whether these attempt to crush it, like fish, or bore it, as a muricid would do, this seems a sensible strategy. As shown by Currey & Hughes (1982), moreover, mechanical strength of shell is often part of the adaptation of a species to the detailed requirements of its environment. These workers measured the force necessary to break the shells of *Nucella lapillus* ranging from 17 mm to 22 mm in height and living in exposed, sheltered and very sheltered habitats in North Wales. Their results gave average values for the three habitats of 0·24 kN, 0·40 kN, and 0·58 kN respectively. Crabs, the main predators of these animals, are most abundant in the sheltered habitats; whelks that live in these seem, therefore, to have developed the shells most capable of resisting their attacks, despite one's feeling that exposure might demand the strongest shells.

Largely because of the attempts of palaeontologists to deduce from fossil material as much information as possible about the way of life of extinct forms, some important ideas relating shell shapes and apertural shape to different modes of life have been put forward recently (Linsley, 1977, 1978a, 1978b; McNair et al., 1981). The shell of a prosobranch gastropod is most efficiently carried by the animal when (1) it is balanced symmetrically over the mid line of the cephalopedal mass; (2) in particular when its long axis corresponds to the long axis of the head-foot; and (3) when the aperture lies in such a way that it allows the pallial water stream to be easily maintained, and that its plane is more or less parallel to the substratum and sufficiently above it not to introduce too much particulate matter into the mantle cavity. In the most primitive mobile prosobranchs the shell is not held in a way that satisfies these conditions, since it lies horizontally nearly at right angles to the head-foot, with the apex to the animal's right. Linsley (1978a, 1978b) has indicated adjustments, some already described by Naef (1913), by which the optimum position of the shell might be reached from such a starting point. One involves an upward tilting of the shell apex so that the whole shell lies more vertically; a second is a process which he called regulatory detorsion, in which the shell axis is rotated clockwise on the head-foot, so decreasing the angle between its long axis and that of the head-foot. The two processes together bring the shell and the cephalopedal mass more or less parallel and the snail employs that combination of the two which gives the lowest centre of gravity and at the same time makes the plane of the aperture more or less parallel to the substratum. This is, as it must be, a compromise position, not interfering with respiration, allowing rapid clamping of the aperture to the substratum should danger threaten, whilst at the same time reducing drag and so giving a faster rate of creeping than would otherwise be possible. Such a position of the apertural plane is most easily attainable when the outer lip lies tangentially to the periphery of the last whorl. Shells with this property are the high angle shells of A. M. Davies (1939) and this is the probable, but at the same time unspecified, advantage which he claimed for them.

McNair et al. (1981) have interpreted apertural shapes in terms of the need to produce an opening through which the head-foot can pass, and to provide an effective, close-fitting attachment to the substratum when the animal is withdrawn. Round apertures (length/width = 0·75–1·12), as seen for example in littorinid, rissoid, and many trochid shells, let pass a broad foot which adapts for life in rough, hard, intertidal habitats, and provide a good clamping surface for protection against predators, waves and desiccation. Long apertures (length/width = 2·5 or more), often also narrow, as in cypraeid and cone shells, provide an appropriate opening for a long narrow foot which adapts for rapid and agile movement. They do not allow effective clamping to the substratum but do not, on the other hand, give predators easy access to the inside of the shell. Since the respiratory current enters the mantle cavity at the base of the aperture and leaves at the apical end, an elongate aperture is advantageous in separating the two streams of water. Though these ideas may have some general validity, Vermeij (1981) regarded apertural shape as more easily correlated with protection against predators than with habitat or movement, and there are clear examples of round-mouthed shells (such as naticids) from habitats where clamping cannot occur, though for other reasons the foot has to be broad.

In addition to apertural shape other features of shell architecture may be related to the animal's mode of life. Apertural spines certainly help to keep epifaunal snails on the surface of soft substrata, as in muricids and some strombids. Their value in other circumstances may be more as a defence against predation by some fish (Palmer, 1979): they increase the effective diameter of the shell and so make ingestion by the predator less easy, and they also make more robust the points where pressure is applied by the fish as it attempts to break the shell. A particularly clear example of the advantageous use of a spine has been described by Perry (1985): the muricid *Acanthina spirata* has a spine at the base of the outer lip which is uses to open barnacles; if experimentally deprived of this, and so forced to bore through the barnacle shell, it takes up to five times longer to get at the food. Labial varices and the internal thickenings which often accompany them strengthen the shell against the attack of such predators as crabs and fish which nip or bite pieces off the outer lip to reach the body of their prey (Bertness & Cunningham, 1981).

Palmer (1980) has questioned the assumption that some of the changes in shell form and position described by Linsley (1977, 1978a, 1978b) relate primarily to facilitating locomotion, and has suggested that the features are better explained as helping to meet the total constraints of the habitat in which a snail lives. There are, indeed, studies, mainly on burrowing prosobranchs, which suggest that many features of a prosobranch shell which are superficially merely decorative, do play a part in locomotor activity. Epifaunal species living on soft substrata, such as naticids and *Bullia*, often move by cilia, which provide faster rates of movement than muscular waves (S. Miller, 1974a; Palmer, 1980) provided that the foot has a large sole area. This, in turn, requires for its passage the large round apertures seen in naticids or cassid shells. When these animals, which are rather large, burrow, the smoothness of the shell compensates to some extent for the difficulty of pulling such a large object into the substratum, and the large aperture permits the passage of the powerful muscles required; naticids, indeed, do burrow rapidly despite their large and globular shells (Vermeij & Zipser, 1986). It would seem likely that long, slender shells, such as are found in *Turritella* and many turrids, should burrow more easily and rapidly despite the fact that they enter the substratum broad, basal, end first. The small size of their aperture and foot, however, makes this, instead, a slow process. Signor (1982, 1983) has shown how the details of their shell sculpture make the process of burrowing more efficient than the weakness of their musculature might suggest. In many turrid shells each whorl has a flattened subsutural shelf with its surface more or less at right angles to the long axis of the shell, and the costae have an escarpment-like profile, with a steep adapical end, and taper towards the base of the shell, thus offering little resistance to forward movement through the substratum. At a later stage in the digging cycle, the shell becomes immobilized at the end of the end of the first step whilst the foot probes forward to gain anchorage for the second step. At this stage both the subsutural shelves and the steep ends of the costae hold the shell firm against any reaction from the probing foot. Shells of similar shape belonging to the cerithiid *Argyropeza* have tubercles like those of turrids which are presumed by Kohn (1986) to act in a similar way. The fact that they lie in rows set obliquely across the whorl and that the rows on each whorl are offset in relation to those on its neighbours in the spire ensures that the tubercles provide, overall, a nearly continuous surface preventing slip.

It may well be true that every feature in the ornament of a shell has selective value in relation to some part of the animal's activity and it is our ignorance which prevents our appreciation of this. Thus Palmer (1977) has shown the value of varices to the muricid *Ceratostoma foliatum*, which has a shell about 50 mm high with three flat varices on each whorl. This animal lives on vertical surfaces from which it is liable to be dislodged. When the muricid *Nucella lamellosa* and the cymatiid *Fusitriton oregonensis*, of similar size and shape but lacking varices, were allowed to fall through water, they invariably landed aperture upwards and so open to attack by predators. In similar experiments, however, *C. foliatum* fell more frequently aperture downwards and thus more safely. The varix by the aperture was of the greatest importance in ensuring this. There are other differences in shell shape, however, which so far elude explanation. Thus amongst the Hydrobiidae, *Hydrobia ulvae*, which lives on open coasts in high salinity, has the whorls of the shell with nearly straight-sided profiles, whereas in *H. ventrosa* and *H. neglecta*, which favour quieter, enclosed, lagoon-like habitats, often with reduced salinity, the whorls are tumid. Populations of *ulvae* living in the Salts Hole, Holkham, Norfolk, which is land-locked and lagoon-like (O. D. Hunt, 1971) appear to respond to this by acquiring the swollen whorls of lagoon species (Barnes, 1988), even although the salinity to which they are exposed corresponds to that of the open sea to which the Hole is linked by underground springs.

In some prosobranch species there is a marked gradient of size between individuals occupying the upper and lower limits of their intertidal range. As noted by Vermeij (1972) species that are typically found at higher intertidal levels show a gradient of decreasing size down the shore; species characteristic of low levels show the reversed gradient, becoming smaller at higher levels. The apparent explanation of this seems to vary from species to species: Sutherland (1970) showed that the gradient in the acmaeid limpet *Collisella scabra* (largest at higher levels) arises from decreased competition allowing increased growth towards the top of the shore, whereas Paine (1969) demonstrated that in the trochid *Tegula funebralis* the larger snails migrate down the beach so as to escape from the competition in the populations at higher levels, and thus come to enjoy greater concentration of food. Despite being preyed upon at this level by the starfish *Pisaster* their reproductive rate increased. Bertness (1977) explained the gradients in size exhibited by four intertidal thaidid species, which are largest at lowest tide levels, as a correlation with similar size gradients in the species of barnacle which form their staple food. Palmer (1983), who examined the food and the growth rates of three of the four species studied by Bertness, carried the idea of the prey-predator size link a stage further, and found that each species of thaidid grows best on a particular size of barnacle and that the part of the shore which each occupies is precisely where the size of barnacle giving maximal growth also occurs, an arrangement allowing optimal sharing of resources by the predators. Prey-predator relationships (which depend in part on size) are therefore also important in defining the particular habitat niche which a species may occupy. Vermeij (1972) offered different ideas as to the factors underlying the gradients in size exhibited by littoral prosobranchs, laying emphasis on the importance of juvenile stages, which were, he maintained, living in that zone of the beach where their mortality would be minimal, an argument criticized by Bertness (1977).

The greatest and most varied volume of work on prosobranch shell characteristics, both in relation to their own properties and to their role in the animal's life, refers to the genera *Littorina* and *Nucella*; this may be summarized since many of the points dealt with are likely to have much wider applicability. The shells of these two genera, like those of other proso-branchs, are subject to variation in several respects, particularly in shape, size, thickness, and

colour, variation most easily correlated with such things as exposure and predation, major
selective forces in all littoral gastropods. Since *Nucella lapillus* and some *Littorina* species,
especially those in the *saxatilis* complex, lack a free larval stage they may be bred in the
laboratory without too much difficulty (Warwick, 1982) and so offer the possibility of genetical work not easily carried out with many other prosibranchs. Much useful information of
Littorina has been summarized by Raffaelli (1982) and Janson (1986), and on *Nucella* by
Crothers (1985) and Day & Bayne (1988).

Variations in size may be considerable beween populations of *Nucella lapillus*, from mean
shell height of 17·0 mm (Crothers, 1983a) to a mean of 47·6 mm (Crothers, 1974). Mean size is,
on the whole, inversely correlated with the degree of exposure of the shore and the level on it
at which the animals live, the largest shells being found on sheltered shores (Kitching, 1977;
Cambridge & Kitching, 1982) and sublittorally (H. B. Moore, 1936). Though there is a variation
in the prominence of the spiral ridges of the shell, variation in ornament mainly affects the
degree of development of lines crossing the whorls. In most animals these are simple growth
lines, but in the variety *imbricata* they are raised to form arched scales where they cross spiral
ridges. Imbricate form has usually been regarded as a response to quiet environmental
conditions such as obtain sublittorally. Largen (1971) has, however, produced evidence
suggesting that it is more likely to be under genetic control; imbricate *Nucella* from
Whitstable and smooth ones from Cornwall raised in identical conditions in an aquarium
both bred true.

Variation in shell shape is a more complex matter and appears to involve an interplay of
genetic and environmental factors. The most variable features are the relative heights of the
aperture and spire and the shell breadth, all of which are susceptible of modification to give a
shell adapted to resist the main dangers of the shore—damage by exposure, by loss of grip,
by crab or bird predation, and by desiccation. Kitching et al. (1966) showed that whelks
(*Nucella*) in which the apertural area was great (allowing the passage of a large foot) and the
spire small (offering less surface to the water), the two features together giving a broad, squat
shell, resisted wave wash and other currents better than those in which the aperture was
small and the spire high (Kitching, 1977; Cambridge & Kitching, 1982; see also Etter, 1988 and
p. 717). Where conditions are quiet selection pressures towards this shape are reduced and
the aperture becomes relatively smaller and the spire longer. Intermediate degrees of exposure give rise to intermediate shapes of shell. Crothers (1983a, 1985) has summarized many
observations from different European and American localities which demonstrate the close
linkage between degree of exposure and shell shape in *Nucella lapillus*. He had earlier (1977,
1980) shown that a large size in this species was attained by more rapid juvenile growth rather
than by an extended growth stage and had therefore suggested that this was under genetic
control.

There is no reason to suppose that *Littorina* differs significantly from *Nucella* in this respect,
so that the numerous observations that winkles from exposed shores have relatively larger
apertures than those from sheltered ones are most sensibly explained as adaptations against
dislodgement by waves and currents. The observations refer primarily to species in the
saxatilis complex: to *saxatilis* itself (Newkirk & Doyle, 1975; Heller, 1976; P. G. Moore, 1977;
Raffaelli, 1978b; J. E. Smith, 1981; Janson, 1982a; Atkinson & Newbury, 1984; B. Johannesson,
1986; Grahame & Mill 1989), to *arcana* (Grahame & Mill, 1989; Dytham et al., 1990), to
nigrolineata (Sacchi, 1975; Heller, 1976; Raffaelli, 1976; Naylor & Begon, 1982), to *neglecta
scotia* (S. M. Smith, 1979), and also to *L. obtusata* and *L. mariae* (Reimchen, 1974). None seems
to have been made on *L. littorea*, which tends to avoid the most exposed shores, nor on

Melarhaphe neritoides which, though frequently living in very exposed conditions, may find the need for a relatively small aperture, so reducing water loss during periods when it is not wetted, more important than that for adhesion, which is satisfied by a retreat into crevices. Janson (1982a, 1982b) has not only described these differences between exposed and sheltered populations of *L. saxatilis*, with intermediate characteristics in sites of intermediate exposure, but has demonstrated clearly by transfer experiments that they are largely genetically based, and this been supported by the work of Boulding (1990). It is also indicated by the observation made by J. E. Smith (1981) that unborn young of *saxatilis* may vary in the shape of their shell.

A similar relationship between apertural area and degree of exposure has also been shown (O'Loughlin & Aldrich, 1987a) to occur in the trochid *Calliostoma zizyphinum* although it is not known whether this is also genetically based.

Though the shells of *Littorina saxatilis* may resemble those of other prosobranch species in the general relationship between shape and exposure, they are more variable than most and exhibit differences amongst themselves even in one and the same locality (Hart & Begon, 1982). Grahame & Mill (1989) have added another factor apparently involved in the determination of shape, especially the width, of the shell. Their investigations showed that shells of *L. saxatilis* were more globose if they came from a beach on which *L. arcana* also occurred, and less globose if from one on which *L. arcana* was absent, suggesting a character displacement where the species occur sympatrically. This condition parallels the displacement of characters described by Fenchel (1975a, 1975b) for hydrobiid species when they are living sympatrically, though in the littorinids neither mechanism nor possible benefits are obvious.

A wide aperture may permit sufficiently strong adhesion to prevent dislocation of a prosobranch by waves but it eases the work of a crab breaking open a shell by presenting more space for the insertion of a chela, whilst the reduced spire offers little space for the retreat of the mollusc. The taller shells with narrow apertures characteristic of quieter shores give more protection against crab predation by offering less space for a chela to grip the apertural edge and enabling the mollusc to retreat further up the spire. Crabs are more abundant in the quieter sites where this shape is the norm so that there seems to be a nice balance between the need to withstand exposed conditions and the need to withstand predation, perhaps tilted a little to the mollusc's advantage since the crabs are less frequent where the shell must be squat and more abundant where it can afford to be narrow.

Should a prosobranch, despite its larger aperture and foot, lose its grip on the substratum on an exposed shore, a thick shell protects against damage due to being thrown about by waves or crushed by boulders. Shell thickness has also proved to be a further factor protecting *Littorina* against predation by crabs (Elner & Raffaelli, 1980). Since crabs are more abundant on sheltered than on exposed shores one might expect shell thickness to increase on both types of shore, though for different reasons. Analysis is complicated by the facts that thickness of shell is directly linked to age, larger animals having thicker shells by virtue of their age alone; alternatively it may, as a result of selection pressure, be a genetic characteristic of a given population: two shells of equal age and size from different populations may therefore not have equally thick shells. Hylleberg & Christensen (1977) have, indeed, demonstrated that *L. littorea* has a relatively thicker shell on exposed shores and interpreted this as giving protection against mechanical damage, and Naylor & Begon (1982) have pointed out a correlation in *L. nigrolineata* shells, presumably of genetic origin, between large size and increased thickness on the one hand, and, on the other, sites where mechanical strength is important, whether this is to protect against crushing by boulders or by crab predation.

B. Johannesson (1986) has indicated how, in *L. saxatilis*, a compromise may be reached between the contrasted needs of a small aperture as protection against crab predation and a larger one to give adequate tenacity. He pointed out that the aperture of a shell may have, in effect, two sizes in one and the same individual. There is an outer aperture of which the area is delimited by the edge of the outer lip, and an inner apertural area, the dimensions of which are determined by the cross-section of the throat a little in from the peristome. In sites with high exposure to wave action the outer apertural area of *L. saxatilis* shells is large, and where crab predation is the greater danger the inner area is small: where both exposure and predation are high risks the shell shows a large outer aperture with the lip thickening rapidly in the throat to give a small inner one. Crabs, Johannesson (1986) showed in his experiments, always selected shells with a large outer aperture when offered mixed populations.

Naylor & Begon (1982) found that winkles living in crevices, where they were safe both from storms and crabs, had relatively thin shells. Those living on exposed cliffs described by Heller (1976) and by Raffaelli (1976) had also thin shells, probably for the same reasons, but perhaps also, as Elner & Raffaelli (1980) have suggested, as an energy-saving measure, an idea accepted by Janson (1982a) to explain the thinner shells of her morph E of *L. saxatilis* from exposed shores. Exceptions to this general trend occur. Thus Raffaelli (1978a) found that *L. saxatilis* living in salt marshes showed little sign of crab predation yet had thick shells: the explanation offered later (Raffaelli, 1982) may be that the juveniles are preyed upon by juvenile crabs (Hughes & Roberts, 1980a) to counter which selection has acted to thicken shells, but those which Raffaelli examined were naturally only those which had escaped predation. Thickness, important in early life, was carried into later stages when it might have been dispensed with.

O'Loughlin & Aldrich (1987a) found no difference in the thickness of the shells of *Calliostoma zizyphinum* from exposed or sheltered habitats.

Sculpture of the shell of littornid species is restricted to the development of spiral ridges and growth lines with, in some species, small tuberosities where they cross. Amongst local species the spiral ridges are most marked in *Littorina littorea, nigrolineata* and *saxatilis*, the growth lines in *Melarhaphe neritoides*; in all the development of the ridges is most marked in young shells and lessens with age, especially in *L. littorea*. There are only a few indications that variation in sculpture in this genus can be linked with differences in habitat. Fischer-Piette *et al.* (1961) and Fischer-Piette & Gaillard (1961) have, however, demonstrated such a link in *L. saxatilis*, where smooth shells. usually of yellow colour, were found to be characteristic of sheltered sites, and ridged shells, not necessarily of any particular colour, of exposed ones. These findings are the opposite of those of Struhsaker (1968) on *L. picta* in Hawaii. She recorded sculptured shells in sheltered sites and smooth ones (offering, one presumes, less surface currents to act on) on exposed shores, relationships which are the same as those which Wigham (1975) found in *Rissoa parva*. Struhsaker (1968) demonstrated that shell sculpture in *L. picta* was genetically controlled, ridged and smooth-shelled animals both breeding true. She assumed that settlement was random but that selection operated on populations to produce different results on different shores. Though sculpture was the most obvious feature in which ridged and smooth shells of *picta* differed it was not the only one, since size and growth rate were also affected: it may, indeed, be that they, or something else less obvious, were the critical adaptation. This might explain the differing results of Struhsaker (1968) and Fischer-Piette *et al.* (1961), but it would also suggest that sculpture is not of fundamental importance in adapting for different habitats.

More definite statements may be made about the inheritance of colour in the shells of both *Nucella* and *Littorina*. The early explanations offered for the coloration of *Nucella* shells attributed it to environmental factors such as food and exposure. H. B. Moore (1936) suggested that brown shell colour correlated with a diet of mussels, white or grey with one of barnacles, and that yellow colour characterized shells from exposed areas, though the causal link was not obvious here. More recent views minimize environmental factors and replace them with genetical ones. R. J. Berry & Crothers (1974) found some positive correlation of colour with exposure but none with food, and it is a common observation that many dog whelks associated with mussels show no trace of brown pigment. The change in pigmentation paralleling change in diet on which Moore based his conclusion has been shown to accompany any marked shock such as the removal from shore to laboratory to which his animals had been subjected. Some banding patterns of *Nucella* shells are known to be under genetic control (Castle & Emery, 1981), like those of the related *N. lamellosa* (Spight, 1976) and *N. emarginata* (Palmer, 1984, 1985), but the conditions which lead to their expression and determine the details of the pattern remain unknown; in view of the great range of possible patterns exhibited even in one locality by a species known for interbreeding, these may be complex. As banding patterns do not confer any obvious advantage in themselves, except perhaps disruption of shape, they may be pleiotropic effetcs of some gene or group of genes which does (R. J. Berry, 1983). As pointed out by Spight (1976) for the related *Nucella lamellosa* (though this does not apply to *N. lapillus*) banding patterns may be weathered out of existence and all older animals then look alike, making their selective importance doubtful.

O'Loughlin & Aldrich (1987b) have shown that the white morphs of the trochid *Calliostoma zizyphinum* often exhibit a greater number of signs of previous damage by crabs than do coloured ones, presumably due to the conspicuousness of the white shell. Since they abound in certain circumstances their conspicuousness must be counterbalanced by some other factors. These, O'Loughlin & Aldrich (1987b) proposed, are higher rates of movement, protection against overheating when emersed, and, since *C. zizyphinum* has direct development leading to isolation of local populations, a greater dispersal rate.

Reimchen (1974) has shown experimentally that the inheritance of colour in *Littorina obtusata* and *L. mariae* is under genetic control; Warwick (1982) has also demonstrated genetic control in some colour morphs of *L. saxatilis*. It is likely that with further work this could be extended to a completely general statement, though the hazards of breeding species with larval stages make proof difficult. All the British littorinid species show a range of basic colours and sometimes also a range of pattern; the range is least in *L. littorea* and *Melarhaphe neritoides*, greatest in *L. saxatilis*, *L. mariae*, and *L. obtusata*. Populations therefore tend to be polymorphic; the factors maintaining the polymorphism vary, but seem to be primarily associated with predation (Pettitt, 1973; Atkinson & Warwick, 1983). Thus Heller (1975b) found that in *L. nigrolineata* red shell colour is common where the snails live on red sandstone shore, so gaining protection against such visual predators as fish and birds. This is particularly marked on sheltered shores where barnacles tend to be absent; as they become more abundant with increasing exposure so red morphs tend to become fewer and winkles with pale shells more frequent. Heller also found that on sheltered shores with a thick covering of weeds yellow-shelled winkles were common on the underside of fucoid fronds, an unexpected finding but explained by the fact that, under water, these shells, especially if their shape were interrupted by black spiral lines, as is common in *nigrolineata*, became nearly invisible (to the human eye) against the fucoid background, which transmits only yellow light. Reimchen (1979) found the yellow morph *citrina* of *L. mariae* similarly in sheltered areas under *Fucus serratus* fronds. On exposed shores the morph *dark reticulata* occurs and this is limited to the dark stems of the weed. Juveniles with a white apex to the shell (common in one form of the species) were restricted to fronds bearing white serpulid tubes, and Reimchen (1981) was able to show that they were less frequently taken by predators than yellow or brown snails of similar size.

Giesel (1970), working with the limpet *Acmaea digitalis* in Oregon, found that animals with light-coloured shells always lived amongst growths of the equally light-coloured barnacle *Pollicipes*, where their colour made them inconspicuous. Limpets living on rock faces above the barnacle zone had dark shells which matched the colour of the substratum there. Giesel believed that shell colour was genetically controlled since repaired areas always matched the original shell in colour. In both situations young limpets (less than 4 mm shell length) showed both light and dark shells: those without cryptic shell colour were quickly removed either by predators or by migration to places where they no longer stood out against the background. If a migratory phase is not part of the life history of every individual, a situation which by itself would lead as many young limpets into places where their shell stood out against the background as it led to places where they were cryptic, then it must be supposed that the urge to migrate affects only those that find themselves on the 'wrong' background. This would seem to imply that the limpets have some visual sensitivity both to their own colour and to that of the substratum on which they find themselves.

Atkinson & Warwick (1983) have described colour morphs in *L. saxatilis* and *L. arcana* and found that both species exhibited a comparable range, with most shells brown or fawn in colour. Even where the two species lived sympatrically, however, it was noticed that over large areas some morphs were limited to one species. Thus in south-east Scotland shells with orange banding were exclusively *arcana*, whilst of nine samples from various parts of the British Isles in which a morph characterized by dark colour in the spiral grooves of the shell was present, eight consisted of specimens of *arcana* only. Since these are genetic characters it is plain that the two species involved do not normally hybridize. From the fact that most shells in both species were similar in colour it was deduced that this is primarily a cryptic coloration produced and maintained by selection, presumably on the part of visual predators. Not all colour effects can, however, be explained in this way: most *arcana* shells are lighter in colour than most *saxatilis* shells, a difference which, these authors suggested, may be linked with a difference in the need for temperature regulation because of the preference of *arcana* for sites on the better insolated vertical surfaces which *saxatilis* avoids, or perhaps with the control of water loss (Grahame et al., 1990). Dytham et al. (1990), working at Robin Hood's Bay, also found differences in colour pattern between *saxatilis* and *arcana*, banded shells being much less frequent in the latter species.

Torelli (1978), studying the isolated Mediterranean population of *L. saxatilis* in the Venice Lagoon, found its colour variability to be nearly the same as that of Atlantic populations. The most highly coloured morphs were absent, however, along with those whose shells bore spiral coloured bands, though reasons for this are not obvious. *Reticulata*, as in the Atlantic, was the common morph of more exposed sites and unicoloured ones of more sheltered places.

Colour provides one component of any cryptic mechanism, and patterning of the colour may improve its value or provide some other device protecting against predators. Broad bands of contrasting colour running across the shell may be disruptive of shape and so deceive predators, as the narrow dark spiral grooves of *L. nigrolineata* and some morphs of *saxatilis* are said to do (Heller, 1975b). Reticulate patterns probably act to some extent in the same way, suggesting a variegated substratum to a predator where a uniform patch might arouse interest. The change with growth in the pattern of *L. neglecta* shells from banded to speckled reported by Ellis (1984) may have this effect. Other relationships of colour morphs are not necessarily explained as cryptic, though this is certainly part of their role. D. A. S. Smith (1976) recorded the distribution on various weeds of different colour morphs of *L. obtusata* from British shores. The morph *olivacea* (shell uniform green to brown) is the most frequent on most shores, particularly the more sheltered ones, but tends to be relatively commoner on southern and western coasts than elsewhere. In these areas it is replaced by *citrina* (uniform yellow to orange), and *reticulata* (a uniform green, brown or black shell always overlaid by a reticulate pattern), both of which, but essentially *reticulata*, prefer exposed situations. The other morphs are rare and formed less than 2% of Smith's collection of nearly 18,000 shells. The morph *citrina* is particularly associated with *Fucus spiralis*, on which three other morphs, *inversicolor* (lightish green with two dark spiral bands and, sometimes, a reticulate pattern on the last whorl), *alternata* (brown with two yellow spiral bands on the last whorl), and *fusca* (uniform brown-black) are also found. The morph *olivacea* is usually on *Fucus vesiculosus* and *Ascophyllum nodosum*, and *reticulata* is found with *A. nodosum* and *F. serratus*. These morphs are therefore linked to certain tidal levels.

Sacchi has also shown that different colour morphs of *L. obtusata* must differ in a number of physiological activities since they have different susceptibilities to predators (1961a), have different distributions on exposed and sheltered shores (1961b), have different behaviour in relation to light (1961a, 1963), and differ in their resistance to desiccation (1963).

The fact that growth of the molluscan shell exhibits a periodicity has been appreciated for a long time and is clearly apparent in such gastropods as possess costae or varices or show the interruptions in growth known as winter rings. That these features were only the long term representatives of a family of rhythms determined by cycles which may be as short as those of the tides, became obvious with the work of Barker (1964) on four marine bivalves, and later, the simultaneous demonstration by House & Farrow (1968) and Pannella & MacClintock (1968) of a daily rhythm in the secretion of the shell of *Cardium* and *Mercenaria* respectively. The rhythm is revealed in sections of shells by a series of alternating darker and lighter bands. These correspond to areas richer or poorer in calcareous matter relative to organic base and reflect times when the shell was open for pumping or shut during emersion, when acid accumulation may well have affected the calcareous component (Lutz & Rhoads, 1977). Most of the work on rhythmical shell secretion has been done on bivalves, but Ekaratne & Crisp (1982, 1984) have shown that in such intertidal prosobranchs as *Patella*, *Littorina*, and *Nucella*, secretion of the calcareous matter of the teleoconch, and presumably also of the periostracum, occurs with a tidal rhythm, each high water coinciding with the laying down of a microscopic layer of shell. Though a variety of factors has been suggested as interfering with the rate of secretion, these workers found that it was primarily dependent on temperature, though it fell in *Patella vulgata* in summer, perhaps because of the active spermatogenesis which occurs in that species at that time.

It has been generally assumed that while some, ususally the more primitive, prosobranchs are capable of continuous growth throughout their lifetime, others grow to a size limit, perhaps genetically and environmentally determined, and then cease further elongation of the helicocone. The work of Ekaratne & Crisp (1984), however, suggests that growth is not totally halted at this time, only drastically reduced. In many prosobranchs this event is marked by a thickening of the outer lip of the apertures—continued growth, indeed, but in a changed direction—and the appearance of internal teeth. These features are commonly taken to indicate maturity, though animals without them are often mature in the sense that they can breed, and some individuals apparently fail to develop them altogether. Recent work, however, suggests that the formation of teeth is linked to any event which stops spiral growth, whether that be temporary or permanent. Thus Feare (1970a) noted that all specimens of *Nucella* on the north-east coast of England stopped growing each winter and all developed internal teeth at that time, so presenting multiple sets when old. Similarly Bryan (1969) noted that dog whelks affected by chemicals used to disperse oil spills developed internal teeth, often underneath an external mark on the shell, over the period when other growth was arrested. Cowell & Crothers (1970) also recorded formation of internal teeth when growth in *Nucella* was stopped by starvation, and it appears that any disturbance such as handling (Hughes, 1980), or filing the shell (Williamson & Kendall, 1981) may stop the growth of a prosobranch shell, though this is not accompanied by tooth or varix development unless these are normal features of the species.

The extra openings on the shell of *Haliotis*, like those on the shells of fissurelloideans, have usually been assumed to be exhalant, allowing the respiratory current to escape from the mantle cavity far from the head. Voltzow (1983) showed that in *H. kamtschatkana* the main inhalant stream enters the mantle cavity left of the mid line over the head, but also by the first opening. More posterior openings are exclusively exhalant. Murdock & Vogel (1978) have shown that geometry of the shell of keyhole and slit limpets facing into a stream of water is such that water enters the mantle cavity anteriorly and leaves posteriorly by the slit or apical hole because of pressure differences, without the animal expending energy in creating a current. According to the authors' calculations the energy thus saved is trivial but they argue that the flow of water so induced may reduce the amount of oxygen which has to be extracted from a given volume and so allow a more rapid gas exchange. It has long been assumed that the presence of secondary openings in these shells was a contribution towards improving the sanitation of the mantle cavity: this finding may provide another reason for its occurrence, additional or alternative.

The shell of many prosobranchs is clearly divisible into two parts, each often with different ornament; the older of the two, occupying the apical part of the spire, is the protoconch, the younger the teleoconch, produced after the settlement of a larva or the hatching of a juvenile. The protoconch is itself often divisible into two parts. The older is an embryonic shell, sometimes known as protoconch I (Ockelmann, 1965; Thiriot-Quiévreux, 1972), secreted before the hatching of larvae, where these occur, and during the early stages of development if they are absent, and derived from the shell gland at the apex of the visceral mass. The second part, frequently called the larval shell, is secreted by the mantle edge during a free larval stage, if one exists, or during the suppressed larval phase which, in many forms, is passed within the egg capsule. Because the use of the word 'larval' in this connection is confusing the neutral term protoconch II is perhaps more proper for this part of the shell. The embryonic shell (protoconch I) typically shows no, or only minimal ornament, whilst the larval shell (protoconch II) may present elaborate sculpture, usually lacking, however, in freshwater species. In prosobranchs at the archaeogastropod level in which a free larval stage, if present at all, is virtually non-feeding and very brief, no larval shell is produced and the embryonic shell and teleoconch abut. In more advanced forms with a well developed veliger stage in their life history, this often lasts long enough, either free-living or suppressed within the capsule, but particularly in the former case, for a larval shell of considerable size and with its own characteristic ornament, as for example in turrids, to be interposed between the embryonic shell and the teleoconch. Examination of

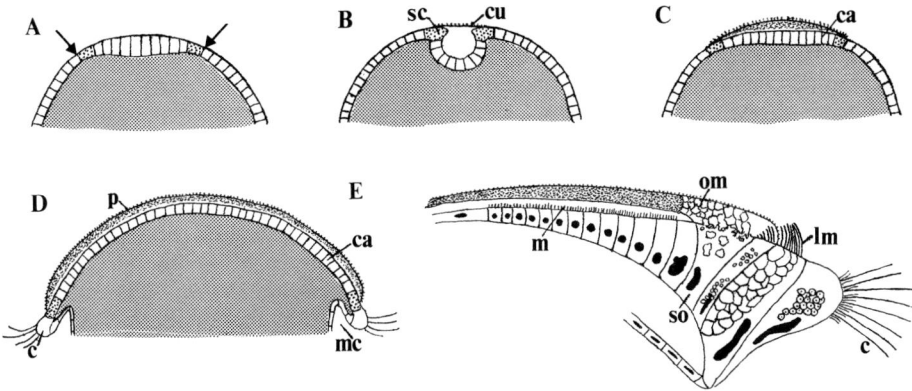

FIG. 309.—Diagrams of parts of dorsoventral sections through the apical region of the visceral mass of a gastropod embryo to show A, early differentiation of the shell field, the extent of which is indicated by arrows; B, its invagination to form the shell gland; C, the eversion and spread of the gland over the surface of the visceral mass; D, a later stage at the time of formation of the mantle cavity; E, a sagittal section of the mantle edge of early veliger to show certain details of shell formation in *Crepidula fornicata*. (E, after Maeda-Martinez, 1985).

c, cilia at edge of mantle fold; ca, calcium-secreting cells; cu, cuticle, future periostracum; lm, enlarged microvilli; m, microvilli of calcium-secreting cells; mc, mantle cavity; om, organic matrix; p, periostracum; sc, cells secreting cuticle; so, cells secreting organic matrix.

a prosobranch shell may thus reveal the general course of the life history, especially in distinguishing between that type which is lecithotrophic (with a free-swimming but non-feeding larval stage) or direct (with no free larval stage, the young hatching as juveniles or 'crawlaways') and that type which is planktotrophic, in which a free, feeding larval stage occurs (Thorson, 1950).

Recent work by Eyster (1983, 1985), Eyster & Morse (1984), and Maeda-Martinez (1985) has shown that the above concept of the shell and the formation of the embryonic shell (proto-conch I) in relation to that of the larval shell (protoconch II) needs modification. The first shell (fig. 309) arises from an ectodermal area at the apex of the embryonic visceral mass called the shell field by Kniprath (1981). The centre of the field invaginates to form a pit, the shell gland, at first minute but increasing in diameter later. The shell itself first appears as a membrane of conchiolin stretched over the mouth of the invagination, secreted by the cells lying at its lips. With growth of the embryo these cells migrate down the sides of the visceral mass, always forming a ring round it, and as they do so they increase the area covered by conchiolin, though this remains uncalcified. At a slightly later stage the shell gland evaginates and secretes calcareous material on to the pre-existing conchiolin membrane, with which its cells make contact. This process gradually extends over the whole of the visceral mass and mantle skirt, with the production of a conchiolin layer, the periostracum, at its leading edge and the addition of underlying calcareous material over the area central to that, an arrangement which persists throughout the life of the mollusc. The distinction between shell secreted by the shell gland and that produced at the mantle edge, which permeated earlier literature, therefore loses its validity: the shell is a single entity, its parts distinct only in topographical and temporal respects.

CROSS REFERENCES

CHAPTER 24

THE FOOT AND LOCOMOTION

T HOUGH some prosobranchs may swim—indeed the genus *Natica* owes its name to its supposed ability to do so—the typical mode of movement of all, and the sole mode of most, is creeping. Whatever the method used, however, it is inevitably the foot which is the locomotor organ since it is the only part of the body sufficiently emergent from the shell and not preoccupied with other functions to be available for this activity. The foot is not a purely locomotor organ since it has also the functions of lubricating movement, of gripping the substratum, of producing the operculum which closes the aperture when the snail retracts within its shell, and sometimes of other functions not relating to movement.

The gastropod foot is traditionally regarded as divisible into three parts, the propodium, mesopodium, and metapodium, with the condition found in the genus *Atlanta* usually quoted as showing these parts most clearly. Since this animal is a heteropod, an advanced prosobranch considerably modified for a pelagic existence, and since the supposed triple division is not obvious in the most primitive groups of prosobranchs, this may be accepted as a convenient way of describing the organizational pattern of the foot in a majority of prosobranchs but little related to the ancestral condition. Though the matter is complicated by the fact that many of the most primitive living prosobranchs are limpets and so have a foot modified at least by loss of the operculum in the adult state, it appears that the foot was, in the early stages of prosobranch evolution no more than a ventral muscular mass used for creeping and adhesion, and bearing an operculum posterodorsally. Secretions aiding locomotion and adhesion came from glands opening to a groove running round the foot sole near its edge, best developed anteriorly and progressively less well posteriorly. Other glands, not directly concerned with locomotion, opened on the lateral walls of the foot. With growth in efficiency the glands opening to the anterior part of the groove increased in number because of the need to lubricate the new substratum which the snail was reaching, more posteriorly placed glands gradually decreasing. As a consequence of this the groove deepened into an anterior depression and ultimately formed a deep gland, the anterior pedal gland. The anterior lip of the opening of this gland (the original anterior edge of the foot) now appeared as a somewhat separate lobe to which the name propodium was given; at the same time the posterior end of the foot (the part of the dorsal surface bearing the operculum) became more independent from the sole on which the snail moved and was distinguished as the metapodium, the sole as the mesopodium. The distinctness of the three parts is closely dependent on the animal's mode of life and is most pronounced where their separate development and differential functioning are most advantageous, as, for example, in naticids and strombids, where each part has its own special role in locomotion.

If it is accepted that molluscs originally stemmed from a group akin to platyhelminthes then their ancestral locomotor method was probably that of flatworms, ciliary creeping helped at times by local muscular activity, an adequate method for creatures that were probably not greater than a few millimetres long (Runnegar, 1985). The two effectors, cilia and muscles, are still regular inter-related parts of prosobranch creeping, though with increasing size muscular effort has become the more prominent partner in most species. Ciliary activity is either the sole means of movement or is aided by an inconspicuous amount of muscular action in many small aquatic prosobranchs moving over smooth surfaces such as hydrobiids (Elves, 1961), skeneopsids (Gersch, 1934), small rissooideans (Fretter, 1948), and *Velutina velutina* (S. L. Miller, 1974b), and in a group of larger ones which live on soft, mobile, substrata not giving adequate purchase for muscular locomotor waves. This group includes the naticids, the tonnids and cassids, and a variety of neogastropods, especially in the families Nassariidae, Olividae, and Conidae (S. L. Miller, 1974b). Ciliary movement is not usual in terrestrial prosobranchs, though it does occur in *Acicula fusca* (Creek, 1953), but this small animal lives in the film of water covering leaf litter in damp woods and is, in effect, aquatic. Most of these snails are also able, when circumstances permit or require, to use muscular movement. The most obvious adaptations for ciliary creeping are a long foot, a great increase in the area of the sole, and often (as in naticids) but not necessarily (as in cones) an increased area of the aperture of the shell allowing its passage. Though cilia are probably the least powerful of the locomotor mechanisms employed by prosobranchs they are, indeed, more energy-effective than muscles (Denny, 1980), and give a faster rate of creeping (S. L. Miller, 1974a): in prosobranchs with a shell less than 15 mm high ciliary movement is 2–3 times faster than muscular, and in cones the mean speed of movement by cilia (1·56 mm.s^{-1}) is thrice that achieved by muscles (0·43 mm.s^{-1}) (Palmer, 1980).

In most prosobranchs locomotion is effected by waves of muscular contraction which may be seen passing along the sole of the foot. Different patterns of wave have been long known (Vlès, 1907; Parker, 1911), and their correlations with habitat and activity described recently by S. L. Miller (1974a, 1974b). A descriptive approach was the only one used, however, until their basis in terms of muscular and other activities was analysed in the experiments of Lissman (1945, 1946) and Jones & Trueman (1970). Summaries of this have been given by Jones (1975) and Trueman (1983).

The two main patterns of wave, direct and retrograde, though apparently similar apart from their direction of travel, differ fundamentally in respect of the events of which they are the visible sign (fig. 310). Direct waves move forward from the rear of the foot, are waves of compression marking an area relatively more contracted than the intervening stationary attached parts: retrograde waves move back from the anterior pedal edge, and mark areas more elongated than the stationary parts. In the former the parts of the foot are pulled forward, in the latter they are pushed forward from behind. Waves are of low amplitude: Jones & Trueman (1970) measured a gap of only 50–350 μm in *Patella* between the sole and the substratum at points occupied by waves. Denny (1981), however, has suggested that there is no lifting of the sole from the substratum at points occupied by waves but that these mark areas at which the sole is moving over the substratum, whereas elsewhere it is firmly attached to it.

Between substratum and sole lies a layer of mucus which must have different properties where there is a wave (lifting/movement) and where there is none (adhesion), liquefied at the former, solidified at the latter. The change from the one state to the other, Denny (1981) showed in the slug *Ariolimax columbianus*, occurred in 0·151s, a period which, though short, might set up an upper limit to the speed with which a snail could move. The existence or not of a space between foot sole and substratum, and what might fill it if it exists, has not yet been confirmed or disproved. In *Pomatias elegans*, which moves by alternate steps of each half of the foot, a mechanism which merely represents the extreme of a ditaxic wave with a

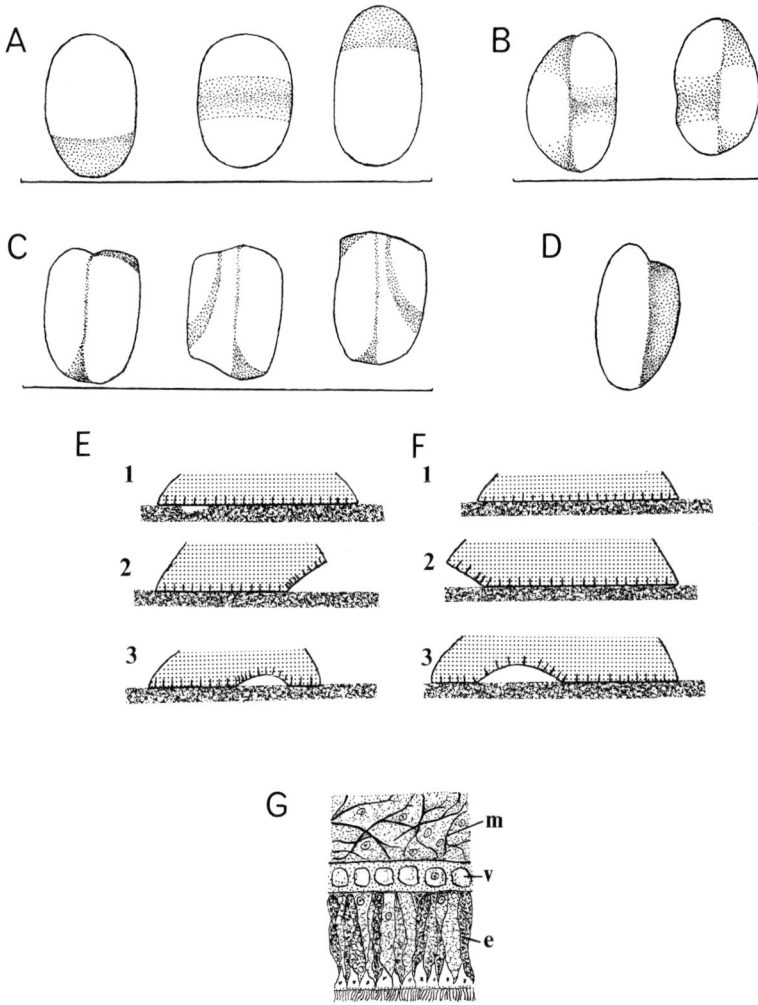

FIG. 310.—A–D, diagrams to show the passage of locomotor waves along the sole of the foot. Stippled areas mark the position of waves. The snail is moving towards the top of the page, and right hand figures represent later stages in its movement. A, direct monotaxic wave (see below E 1–3); B, direct ditaxic wave; C, direct oblique ditaxic wave; E, ditaxic pattern of *Pomatias elegans*, in which wave length equals the length of the sole. E–F, diagrams showing the initiation and propagation of locomotor waves along the sole of the foot as seen in vertical section. The direction of movement of the snail in each case is from right to left. Short vertical lines mark equal distances along the resting foot in contact with the substratum (E1, F1). These move apart when elongation occurs and become closer where the sole is contracted. E2, E3, direct wave; E2, posterior tip of foot raised by contraction of dorsoventral muscles (not shown) and shortened; E3, posterior tip reattached and wave of contraction starting to move forwards. F2, F3, retrograde wave; F2, anterior tip raised and extended forward; F3, anterior tip reattached and wave of elongation starting to move backwards. G, part of a vertical section through the sole of the foot of the patellogastropod *Lepeta caeca* to show the space layer beneath the epidermis.

e, epidermis; m, muscles of foot passing through connective tissue; v, vesicles of space layer.

wavelength equal to the foot length, each half of the foot is undoubtedly lifted off the ground. If lifting is thus a characteristic of the locomotion of one snail, even if it is one with an extreme wavelength, it may also happen in others in which the wavelength is small. *Pomatias*, however, is terrestrial and so less affected by the need to retain a grip on the substratum than aquatic forms and that need is satisfied by the large area of sole still attached. A further indication possibly supporting raising of the sole is the universal presence of dorsoventral muscles.

The foot of prosobranchs is an extremely muscular organ and, unlike most muscular organs used in locomotion, functions in most species without obvious endoskeletal, exoskeletal, or hydrostatic support. In the whelk *Bullia*, however, an animal which is an active crawler and speedy burrower, the foot exhibits an advanced state in that a large blood sinus occupies its central region and antagonistic muscles in the surrounding walls are largely responsible for the changes in shape necessary for crawling and burrowing (Trueman & Brown, 1987b). In typical prosobranchs the foot is formed almost entirely of a rather dense and complexly interwoven series of muscle bundles or layers working against one another and on local fluid-filled spaces. Its shape is maintained by the interplay of these amongst themselves and by the internal blood pressure (Jones, 1988). The foot can therefore be described as a muscular hydrostat (Kier & Smith, 1985; Kier, 1988). In most parts of the foot in most prosobranchs the muscle fibres are smooth, but striated fibres are known in *Scissurella* (Fretter & Graham, 1962), patellids (Frescura & Hodgson, 1990) and in *Bullia*, more particularly in its propodial region (Trueman & Brown, 1987b). The reasons for their presence in two animals, one adapted primarily for adhesion, the other for agile motion, cannot be identical.

The statement that the foot is devoid of endoskeletal support misrepresents its histological structure since recent work (Plesch, 1977; Voltzow, 1985, 1988, 1990; Frescura & Hodgson, 1990; Trueman & Hodgson, 1990) has shown that collagen is an important component of the foot, often forming a kind of secondary supporting network interwoven with that of the muscles. This probably acts in helping to maintain pedal shape and in providing attachments for the fibres of the muscular network in local contractions (Trueman, *in litt.*). Since the shape of the collagen framework can be kept without expenditure of energy its presence reduced the metabolic cost of preserving the foot shape and of locomotion.

One section of the pedal musculature is extrinsic, formed from the dorsoventral fibres of the columellar muscle or, in limpets, of the shell muscle; another section, lying in what is sometimes called the tarsos (Voltzow, 1988, 1990), is intrinsic, comprising an elaborate web of longitudinal, transverse, vertical and oblique fibres. The former section is mainly responsible for maintaining and altering the relationship of head-foot and shell, the latter for locomotion. Fibres of the columellar or shell muscle originate on the shell and run mainly to the more central parts of the sole and to the operculum. In patellogastropods the former group tend to lie in right and left halves, not crossing in the mid line (Voltzow, 1988; Frescura & Hodgson, 1990), but they do so markedly in other prosobranchs except in the most anterior parts of the foot. The majority of the fibres are dorsoventral in direction and their contraction brings about the powerful withdrawal of the body in the shell and the clamping of the shell to the substratum where that is possible. For long no antagonists to this action were known and it was assumed that extrusion of the body from the shell must be brought about by blood pressure. Brown & Trueman (1982) and Trueman & Brown (1985) have, however, found muscles traversing and encircling the columellar muscle of *Bullia* and the shell muscle of *Haliotis* (but doubtless of general occurrence) which on contraction elongate the dorsoventral muscle and so push the body out of the shell in the one case and raise the visceral mass away from the substratum in the other. This action will be helped by the presence of collagen fibres amongst the muscles.

The formation of a locomotor wave on the sole of the foot involves the muscles of the tarsos immediately under the epithelium of the sole, bounded dorsally by a layer of transverse muscles. From this layer muscles, largely oblique, cross to insertions on the epithelium or on the underlying connective tissue; contraction of these fibres raises the sole from the substratum (if that indeed happens) or may be part of the mechanism bringing about liquefaction of mucus (if that does not), and would certainly pull the sole backwards or forwards as a wave passes. In *Patella* only dorsoventral fibres occur here, running from the sole epithelium to the sheet of transverse fibres. Trueman (1983, 1984) assumed that the muscles act to deform small haemocoelic spaces among the fibres so that the area occupied by a wave is elongated, the transverse diameter of the foot being kept constant by contraction of the transverse fibres, whilst other local contractions prevent escape of blood from the wave area. This mechanism therefore causes elongation and later shortening of the foot in the absence of the longitudinal muscles that might have been expected to occur. The layer of vesicles of cellular origin found by Grenon & Walker (1982) under the sole epithelium of patellid limpets, and probably of general distribution amongst patellogastropods, may have some part to play in the propogation of locomotor waves, but this, as shown by Trueman (1984), can effectively be achieved without their involvement.

Waves may be monotaxic, one set occupying the whole breadth of the sole, or ditaxic, when two sets are present, each occupying half the sole and out of phase with one another. The former is probably the primitive pattern in gastropods (S. L. Miller, 1974b; Voltzow, 1988, 1990). Amongst prosobranchs there is little or no correlation between wave pattern and systematic position, but ditaxic waves are of more frequent occurrence than monotaxic and retrograde than direct. The advantage of ditaxic over monotaxic waves is likely to be found in the greater flexibility that it gives to movement, allowing more rapid turning, for example, by changing the rate of wave travel on one half compared with the other. Retrograde waves, which are also found in chitons, are for that reason supposed to be primitive, and are known (S. L. Miller, 1974b) to give greater manoeuvrability than direct waves, which, according to S. L. Miller (1974a) represent an advanced type, are correlated with a limited shape and size of foot, and so with limited and particular patterns of mobility. They are found in only seven prosobranch families: Haliotidae, Trochidae, Pomatiasidae, Truncatellidae, Calyptraeidae, Naticidae, and Muricidae; some members of this grouping may be regarded as having limited and specialized locomotion, but this is not obvious in others.

In addition to direct and retrograde waves a variety of other patterns has been described. Diagonal waves are ditaxic, start in the mid line of the foot anteriorly if retrograde, posteriorly if direct, and move to its posterolateral and anterolateral corners respectively. Though the wave fronts travel obliquely each point on them is translated anteriorly. Other patterns are found, such as the waves called composite by Olmsted (1917), which cross the sole from one side to the other in cypraeids, the directions sometimes opposite at the two ends of the foot, allowing rapid turning, a manoeuvre of obvious advantage to such carnivores. In other prosobranchs, often those with tall, narrow shells (though the reason for this association is not obvious), waves are totally irregular. Composite waves are compression waves and so presumably derived from an ancestral direct pattern, irregular waves are areas of elongation and must come from some retrograde source.

The foot is the organ by which a snail grips the substratum, an activity of importance to the many prosobranchs living where wave action, or predators, might dislodge them. Creeping involves more than the pedal muscles: between foot and substratum is a layer of mucus, most of which comes from the anterior pedal gland, but some from goblet cells on the sole. This is spread across the sole by cilia and as shown by Denny (1980, 1981) is liquefied where the sole is affected by waves, solidified and so ensuring grip in the areas between waves. This is at least part, and probably most of the mechanism underlying tenacity. The traditional view of how such an animal as a patellid limpet adheres to the substratum has always been that the shell muscle pulls up the central part of the sole of the foot, creating an area of reduced pressure, sealed at the edges by secretion from pedal glands. Such an arrangement, however, cannot keep a limpet attached to a rock surface on a beach if the dislodging pull on it exceeds 1·033 kg.cm^{-2} of foot sole (Grenon & Walker, 1981) and this is well below the forces that can be withstood—up to 5·18 kg.cm^{-2} by *Patella cochlear* (Branch & Marsh, 1978). This high tenacity appears to be due to the familiar strong adhesion between two closely applied surfaces separated by a thin film of viscous material. Branch & Marsh (1978), using *Patella* species, found tenacity was enhanced when the area of attachment of the shell muscle was large, the secretion of mucus reduced, and the pedal haemocoelic spaces were small, producing an inflexible type of foot. Adaptations for high tenacity like these greatly affect the mode of life. Thus *Patella granularis, P. oculus* and *P. granatina* live high on South African shores where the need for high tenacity is reduced, the last two, indeed, in sheltered places. This allows the foot to be mobile, encouraging extensive foraging, which results in rapid growth and high reproductive capacity. *P. cochlear, P. argenvillei,* and *P. longicosta,* on the other hand, live at low levels, the first two in exposed situations where high tenacity is essential if the limpets are not to be swept away. Mobility is thus restricted, feeding is confined to a limited area which can be explored with minimal movement and minimal exposure to the threat of dislodgement, and which, because of its restricted size, is defended by *cochlear* and *longicosta* against poachers; growth and reproductive capacity are both reduced.

The ability of a limpet to withstand a dislodging force varies with the direction of that force, a vertical one being withstood more easily than a horizontal one; it also depends upon whether the animal is stationary or moving, the latter allowing less firm attachment since only a fraction of the sole is likely to be adhering to the substratum. Further complications arise from the fact that dislodging currents may possibly strike the shell from different angles and though a limpet may orientate itself in relation to the main direction of water flow when at home, changing winds and currents may bring about rapidly changing direction of water flow. Denny et al. (1985) pointed out that currents produced by the breaking of waves and during their backflow were often accelerating currents, and whereas increased size of sedentary animals gave increased grip in relation to constant currents, the force exerted by accelerating currents grew more rapidly than the growth in tenacity with increasing size of limpet. There is therefore an upper limit to size for those species of limpet that live intertidally. In other prosobranchs such as *Thais* biological rather than physical mechanisms seem to be responsible for setting the size to which the animals grow. Etter (1988) has investigated the adaptations of *Nucella lapillus* to life in exposed and sheltered situations on New England shores and shown that whelks from exposed shores have a larger foot-sole area than those from sheltered shore [as had been shown long previously by Kitching et al. (1966)]. This seems to be a direct response of the foot to environmental conditions since the area of the foot in newly hatched young is identical whatever kind of locality they inhabit. The foot, however, grows faster in exposed than in sheltered sites. Transplantation experiments showed that foot size increased when whelks were transferred from sheltered to exposed sites but little change occurred when the transfer was in the opposite direction.

Mucus is important in locomotion and adhesion, but its benefits are gained at some cost. Where cilia are prominent in locomotion much power is required for their beating because of its high viscosity, and large amounts, demanding much water, are needed to form the trail over which a snail moves. Denny (1980) calculated that the slug *Ariolimax columbianus*, which creeps over a mucus layer 10 μm deep, uses 9·5–13·3% of its total energy needs for this activity alone, and Kofoed (1975b) that 9% of the carbon assimilated by *Hydrobia ventrosa* is secreted in the mucus produced. M. S. Davies et al. (1990) measured the cost of mucus production, largely for locomotion, as 23% of the energy budget of *Patella vulgata*.

Many prosobranchs can increase their speed of movement, primarily to escape from predators rather than improve their chances of catching prey, and many have also special escape reactions. A doubling of speed (S. L. Miller, 1974a) may be obtained whilst retaining the original length of the locomotor waves provided that their rate of travel over the sole is also doubled. The same increase in speed may also be obtained by doubling the wavelength. The former method calls for an increased rate of muscle contraction which, if it is not actually beyond the capability of the pedal muscles, would certainly soon exhaust them; the latter method has the advantage of not increasing contraction rate above normal and is, indeed, that commonly found.

Accelerated locomotor behaviour and accelerated righting reactions form the bases of the escape reactions brought about by attacks by starfishes and carnivorous gastropods and described by many observers: a useful review has been given by Ansell (1969).

In most species the defensive reaction is evoked only when the predator makes contact with the snail; *Lacuna marmorata*, however, reacts to water-borne chemicals according to Fishlyn & Phillips (1980) and shows tentacle waving and shell rotation before a starfish has actually touched it. Contact is a signal for it to loosen its hold on weed and fall off. Most escape reactions, but not all, involve the speed-up movement described above and variously called galloping (Feder, 1963) or running (Margolin, 1964), or the accelerated stepping of strombids described by Gonor (1966); this is not, however, normally part of the reactions of fissurellid and some patellid limpets (Ansell, 1969). Before many prosobranchs take to flight other responses are commonly observed: in fissurellids and *Haliotis* the cephalic and epipodial tentacles are extended and waved about, the mantle edge protruded and often extended dorsally over the edge of the shell. The shell muscle is narrowed and elongated by contraction of the circular and transverse fibres (Trueman & Brown, 1985) so that the shell is lifted up from the substratum in a process aptly described as mushrooming by Bullock (1953), and it may be rotated on the head-foot through an angle of up to 180° (Parsons & Macmillan, 1979). A similar raising of the shell with extension of cephalic and pallial tentacles, especially in the neighbourhood of the part touched by the predator, is observed in patellids and acmaeids and may well have been inherited from a monoplacophoran ancestry, as it is exhibited by *Vema* (Lowenstam, 1978). Mushrooming is followed by an escape movement in *Helcion* (Clarke, 1958) and *Acmaea* species (Bullock, 1953; Margolin, 1964) but not in *Patella*. The significance of the waving of tentacles is presumably to locate the threatened attack; that of mantle exposure and shell raising is probably to expose surfaces covered with mucus to which, unlike the shell, tube feet cannot adhere. This is well exemplified by the escape reaction of *Lunatia catena* described by Thorson in his Christmas card of 1953 and illustrated by Fretter & Graham (1962, 1981), a reaction which is also present in *L. alderi* (Ansell, 1969) and probably all naticids. In these animals the propodium is large and contains many channels connected to the external sea water by pores lying along its anterior edge; these can take up water and inflate the propodium considerably so that it covers the animal's head and the anterior face of the shell. In addition there are folds arising from the metapodial region of the foot which can be drawn over the shell from behind. If the animal is touched by a starfish both events take place so that the whole shell is covered and its aperture concealed, mucus is secreted copiously and the snail crawls rapidly away.

More exaggerated responses have been described in *Hinia* (Weber, 1924) and *Calliostoma* (Ansell, 1969). In *C. zizyphinum* attacked by *Asterias rubens* the first responses are elongation of tentacles and mushrooming. These are followed by contraction of the columellar muscle pulling the shell down to lie on one side, the foot becoming free. The foot is then extended and twisted to gain a new grip on the substratum, the columellar muscle contracted, pulling the animal to a new situation. This process may be repeated several times. The reaction of *Hinia* is similar. If the posterior tip of the foot is touched by a tube foot the shell is swung through 180° on the head-foot so that the aperture faces backwards and upwards; the foot is then loosened from the substratum and by contraction of the columellar muscle the animal is jerked off the substratum to land, with normal orientation of the body once more, some way from the starting point. The cycle may then be repeated. This reaction is a special one evoked only by the touch of a starfish: other stimuli are treated as tactile and induce, at most, withdrawal. This is, indeed, the reaction of most prosobranchs without a repertory of special tricks and it often places them at the mercy of the predator unless some other behavioural pattern intervenes. Hadlock (1980), for example, found that *Littorina littorea* living in rock pools reacted to the presence of crushed *Littorina* tissue in the water by crawling into crevices, under stones, or into algal clumps, moving at a rate four times the normal, so that all had hidden themselves in ten minutes or less. The most probable source of winkle tissue in the water would be a snail crushed by a crab, a process which Hadlock found in the laboratory took a mean time of just under the ten minutes which it took winkles to hide. A comparable response has been described in *Ilyanassa obsoleta* by M. Crisp (1969): when it detects water from injured whelks it reverses its usual taxes and crawls away. Water from healthy *I. obsoleta* is attractive, leading to an aggregation of snails.

In two species of prosobranch a more aggressive type of behavioural response to predators has been described: the columbellid neogastropod *Alia carinata* has been said (Fishlyn & Phillips, 1980) to attack the starfish *Leptasterias hexactis* with its proboscis (presumably using the radula). This behaviour has also been described by Pratt (1974) in *Crepidula fornicata* when attacked by *Urosalpinx cinerea*. Fishlyn & Phillips (1980) found that *Notoacmaea paleacea* showed none of the usual reactions to starfish, yet was rarely eaten by them: they suggested that it either had a chemical defence of its own or used a chemical (perhaps a flavonoid) derived from its plant food to make it distasteful or to camouflage its presence, a device akin to the defence of eolid nudibranchs by means of second-hand nematocysts.

As already indicated the foot is glandular, the most conspicuous glands being the anterior pedal gland and those of the sole, which together lubricate locomotion and provide for adhesion. In addition to these, however, other glands are present, usually subserving some special locomotor device. One such is the posterior pedal gland; this discharges by a duct opening to the mid line of the posterior part of the sole and is of common occurrence in rissoids and some related small prosobranchs. The secretion is moved by cilia along a groove leading to the posterior tip of the foot and forms a thread along which the animal can travel, particularly from site to site directly through the water rather than over the substratum. A further area of specialized glandular activity is the so-called funnel of janthinids: this is a depressed region in the anterior part of the sole to which many mucous cells discharge. The secretion, as described by Fraenkel (1927b), is used to entangle a bubble of air as the mollusc floats, upside down, at the surface of the sea. The bubble, its mucous coat hardened by contact with air, is moved out of the funnel to the posterior half of the sole, which bears several longitudinal grooves, and joined to others to form the float which sustains the animal near the surface and without which it sinks.

The histochemistry of the glandular equipment of the prosobranch foot has received considerable attention in recent years with the work of Vovelle and his colleagues (Vovelle, 1967, 1969a, 1969b; Vovelle et al., 1977; Vovelle & Grasset, 1979, 1982; Grasset & Vovelle, 1982) on the secretion of the operculum, and with that of Delhaye (1974a, 1974b), Simkiss & Wilbur (1977), Grenon & Walker (1978), Bensalem & Chétail (1982), and Shirbhate & Cook (1987) dealing more particularly with the glands responsible for lubricating locomotion and maintaining a grip on the substratum. All agree in demonstrating the complexity of the total assemblage apparently necessary for these last two functions. Hermans (1983) has suggested that the molluscan foot, like organs in many other phyla used for temporary adhesion, depends upon a 'duo-gland' system, one set of secretions primarily concerned with setting up adhesion to a surface, a second set with abolishing it.

Grenon & Walker (1978) examined the two limpets Patella vulgata and Acmaea tessulata; the former lives high on the beach where it requires high tenacity and resistance to drying, the latter at the lower level where emersion is more restricted and its consequences less. In Patella nine different kinds of gland cell were found, six of which also occurred in Acmaea and were the only ones found in that animal. As in many limpets (Fretter, 1988, 1989) there is a groove round the margin of the sole of the foot, rich in gland cells secreting proteinaceous material. This is swept across the sole by cilia and added to by acid sulphated and non-sulphated mucopolysaccharides from numerous goblet cells in the sole epithelium, and by secretion from four different types of gland cell lying deep in the connective tissue of the foot, but discharging to the sole. Three of these secrete sulphated and non-sulphated mucopoly-saccharides, the fourth a protein, providing thus a mixture of substances which, according to Denny & Gosline (1980) and Hermans (1983), provides the properties of adhesion and lubri-cation necessary for locomotion. These six types of gland occur in both genera but Patella possesses a further three types not found in Acmaea, which may be related to its ability to live at higher levels on the shore. Two of these, which secrete protein and mucoprotein respect-ively, are found on the side walls of the foot and were regarded by Grenon & Walker as antidesiccant, whilst the third produces acid mucopolysaccharide and is confined to a glandular streak found only in young animals, which helps to consolidate the particulate material led out of the nuchal cavity for rejection. In older animals this is more effectively voided by pulling the shell towards the substratum by contraction of the shell muscle.

Shirbhate & Cook (1987) have made a similar survey of pedal glands and their properties in the terrestial prosobranch Pomatias elegans, the freshwater snail Bithynia tentaculata, and the marine Littorina littorea. Their general findings are similar to those of Grenon & Walker (1978) but the main interest of their work lies in the comparison it allows of the glandular equipment of animals living in different habitats. Not surprisingly there is closer agreement between the two aquatic forms and limpets, and evidence of greater specialization in the terrestrial Pomatias, though, because of its special requirements, this animal should not be taken as typical of terrestrial gastropods. In L. littorea Shirbhate & Cook (1987) found two types of cell in the anterior pedal gland; both produced mucoprotein, and three produced acid and neutral mucopolysaccharides and further mucoprotein, thus forming the same lubricating and adhesive mixture as in limpets, though from only five types of gland cell. The situation seems simpler in *Bithynia* as regards the number of cell types involved—one in the anterior pedal gland and two on the sole of the foot, one of which is restricted to a band across the mid region. Their secretions, however, produce the same mixture of mucoprotein (from the anterior pedal gland) and mucopolysaccharides (from the sole gland) as in the other animals investigated.

More interest attaches to Shirbhate & Cook's findings in *Pomatias* since the locomotion of this terrestrial snail is unusual. The foot is modified for a pattern of movement in which right and left halves step alternately, and the glandular equipment for locomotion is restricted to a median longitudinal groove which separates the two halves. The anterior pedal gland discharges to the anterior end of the groove a complex secretion from four different cell types, containing protein, calcium, and a variety of mucopolysaccharides. To this further protein and mucopolysaccharide secretions are added from two cell types opening directly to the median groove, which distributes the whole mixture along the length of the foot. Like *Patella*, *Pomatias* is exposed at times to desiccating conditions and, although it usually escapes from these by burrowing, it possesses a series of glands on the side of the foot which, like those of a limpet, may be antidesiccant.

The pedal glands of *Pomatias* have also been described by Delhaye (1974*b*) and by Bensalem & Chétail (1982), whose findings on their histochemistry are broadly the same as those of Shirbhate & Cook (1987), but which include activities not studied by these authors. *Pomatias* is known (Kilian, 1951) to be strongly calciphile, and to occur only in areas where calcium is present, derived either from calcareous rocks or from sea spray. Bensalem & Chétail (1982) have suggested that calcium ions are trapped in the mucous layer lying over the epithelium of the sole and are then absorbed to maintain the very high internal salt concentration to which the tissues are accustomed; this, at about 385 milliosmoles, is almost twice that of other land gastropods, and is heavily dependent on its calcium content (20·2 mM $CaCl_2$ per litre of blood). The high internal salt concentration, they argued, allowed the snails to obtain water from the soil even when its water content was low. The amount of calcium absorbed in this way is apparently more than sufficient to maintain this level; some of the excess may be stored in cells lying within the connective tissue of the foot, but, according to Bensalem & Chétail (1982), it is regularly excreted through a pair of long, coiled, thin-walled tubular glands which lie in the pedal haemocoel and discharge to the median groove. The organization of these structures Bensalem & Chétail compared to that of the salt excreting glands of marine turtles and birds. The glands had been previously described by Delhaye (1974*b*) who thought that they were organs for the uptake of soil water, but it is difficult to understand how this might take place, both from the point of view of the mechanism and their siting. The problem requires further work.

The mechanism by which a prosobranch burrows has been elucidated by Trueman and his co-workers and the facts summarized in his review of 1983. As in other locomotor activities the foot is the principal agent by which burrowing is achieved, but it must work in relation to the shell, and some features of that structure are important in making burrowing more effective. These are discussed on p. 568.

Entry to a substratum is made, usually at a low angle to the surface, by elongation of the foot, distended with blood. When sufficiently far inserted the terminal part of the foot becomes swollen with blood to form what Trueman called a terminal anchor, holding that part of the foot firmly in the substratum. The pedal branches of the columellar muscle then contract with the result that the shell and it contents are pulled close to the terminal anchor, which then collapses in preparation for the next step in the burrowing process. At this stage the shell acts as a penetration anchor, holdings its position in the substratum whilst the foot elongates and presses deeper into the substratum, at the end of which process a further terminal anchor is formed and the digging cycle is repeated. Occasionally, as in *Bullia* (Trueman & Brown, 1976), and more commonly in burrowing bivalves, a jet of water is expelled from the mantle cavity as the foot elongates, so loosening the substratum and making penetration easier.

It seems, at first sight, unlikely that the large size of foot shown, for example, by a naticid, which is an adaptation for rapid motion over soft substrata, might also accompany efficiency in burrowing, yet this is implicit in suggestions put forward by Signor (1982, 1983) and by Vermeij & Zipser (1986). Burrowing is a process calling for about ten times the energy used in crawling (Trueman & Brown, 1976) so that the smoothness of a naticid shell must facilitate its passage through sand. The large size of the shell, however, despite its smoothness, makes entry into the substratum difficult (Trueman & Brown, 1989) and calls for expenditure of much energy. To some extent, however, this is compensated for by advantages once entry has been achieved: the large size of the aperture and foot allow the development of the powerful muscles required to pull such a large shell through the substratum, and the shell itself forms a large penetration anchor. Nevertheless figures recently obtained by Trueman & Brown (1989) for digging rates in three species of the nassariid whelk *Bullia* suggest that despite these advantages the possession of a large shell does make burrowing an energetically more expensive activity than in snails with smaller ones.

The metapodium or operculigerous disk of the prosobranch foot secretes in most species an operculum by which the aperture of the shell is closed when a snail withdraws. In most the operculum arises from the secretion of gland cells opening to a groove running across the anterodorsal surface of the operculigerous disk. Their secretion is normally produced at the right end of the groove either in greater volume, or at a greater rate, than elsewhere so that over a given unit of time a wedgeshaped piece of secretion is formed. The repeated production of units of this shape gives rise to a spiral structure which, as it grows, is rotated (in animals with dextral shells) in a clockwise direction over the surface of the operculigerous disk to produce a spiral operculum coiling anticlockwise. The details of how this rotation is achieved over the underlying epithelium without losing adhesion are not known nor are those explaining the same ability of shell muscles to move their attachment to the shell with growth.

In a majority of prosobranchs investigated the operculum is a continuum, but in a few (e.g. *Buccinum, Neptunea, Viviparus*) the operculum is put together by the apposition of separate pieces (Grasset & Vovelle, 1982; Vovelle & Grasset, 1982). This may represent a more derived method but does not relate to systematic position, and the fundamental secretory processes are similar in all. In species in which the groove is broad the spiral has only a few turns, giving a paucispiral operculum: where the groove is narrow the operculum is multispiral, with a large number of narrow whorls. E. Kessel (1942) suggested that the paucispiral type was primitive and the multispiral derived, though the latter occurs more frequently in more primitive groups; there is, however, little evidence to decide which is the evolutionary direction.

The common type of operculum is described as horny, and consists of an outer layer and an inner one, originally called the varnish: both are proteinaceous, though different in their component amino acids (S. Hunt, 1976). Grasset & Vovelle (1982) and Vovelle & Grasset (1982) have shown that in its chemical composition, in the fact that its protein component is tanned, and in the nature of the enzymes responsible for this the operculum matches almost precisely the periostracum of the shell as described by S. Hunt (1971) in *Buccinum*. This near

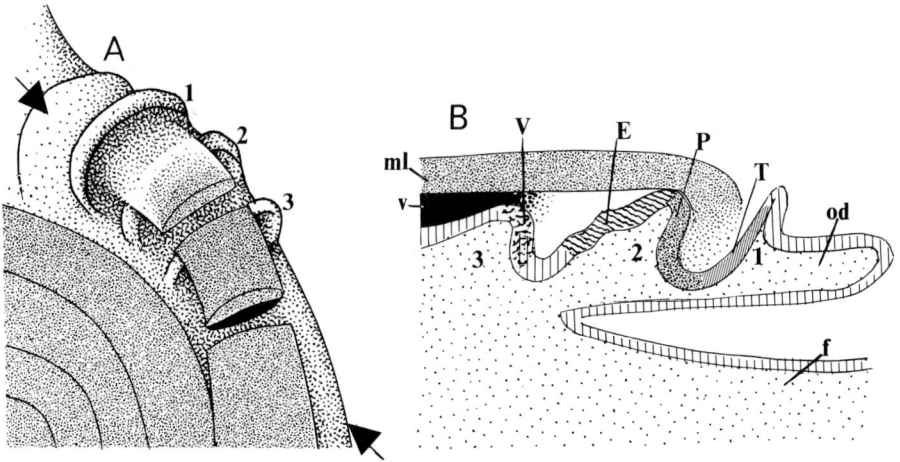

FIG. 311.—A, diagrammatic representation of the area producing opercular material in a prosobranch with a multispiral operculum. The most recently secreted part of the spiral has been cut vertically to show the layering of the main layer (stippled) and the varnish layer (black). B, longitudinal section through the area along the plane indicated by the arrows in A. 1, 2, 3, are the folds involved in the production of opercular constituents. (Based on Vovelle, 1967).

E, area of epithelium producing polyphenol oxidase; f, foot; ml, main layer (tanned protein); od, edge of operculigerous disk; P, area of epithelium producing protein of the main layer; T, area of epithelium producing polyphenol; polyphenol and polyphenol oxidase interact to tan the protein; V, area of epithelium producing the varnish; v, varnish (black).

identity has led Chétail & Krampitz (1982) to conclude that the gastropod operculum is, indeed, a 'periostracum'. This suggestion should not, however, be taken as leading back to the old belief that the operculum might represent a second valve of the shell and the columellar muscle an adductor. Biochemical similarity more probably reflects a limited repertoire of biochemical ability on the snail's part.

In *Gibbula magus* (Vovelle, 1967), which has a horny, multispiral operculum, the secretory field on the operculigerous disk presents three folds separated by two grooves, from the more anterior of which the outer layer arise (fig. 311). It is formed from tanned protein: the protein is secreted from cells on the posterior wall of the anterior groove, and is tanned by polyphenol from cells on its anterior wall by means of enzymes from cells on the middle fold. All these substances are mixed in the grooves to give rise to a sheet which becomes the initial part of the operculum or is attached to the edge of a pre-existing part. Vovelle's account differs from those of such earlier workers as Houssay (1884) and Kessel (1942) in showing that the inner so-called varnish layer is added to the underside of the outer one as it moves over the most posterior fold by cells situated on the posterior wall of the posterior groove, and not, as had earlier been believed, from glands on the dorsal surface of the foot at the level of the posterior end of the operculum. Shirbhate & Cook (1987) agree with this siting of these cells in their account of *Littorina*, but describe them as subephithelial in position. Apart from such details the formation of a horny operculum seems to be similar throughout the prosobranch series.

The formation of calcareous opercula has been followed in *Pomatias elegans* (Vovelle *et al.*, 1977; Vovelle & Grasset, 1979), the turbinid *Astralium* (Vovelle, 1969a), in *Tricolia* (Vovelle,

1969b), and in neritids (Vovelle *et al.*, 1977). These do not form a homogeneous group since in *Pomatias* the layer of tanned protein, which presumably corresponds to the outer layer of a horny operculum, lies internal to the calcareous part, whereas in neritids the calcareous layer is either mainly (*Nerita* species) or wholly (*Theodoxus* species) internal to the protein layer. To achieve these different dispositions of the layers the fold secreting the calcium has to lie either anterior (*Pomatias*) or posterior (neritids) to the source of the conchiolin layer. A further difference is that the calcareous matter in *Pomatias* is aragonite, in *Nerita* calcite, but this difference may well relate more to the nature of the organic matrix in which it is deposited than to its position. Fundamentally the production of a calcareous operculum is as in *Gibbula*, the protein base being secreted on the posterior wall of the anterior furrow of the opercular groove, the tanning agents at its base. The calcareous matter is held in cells lying in connective tissue and escapes through the overlying epithelium.

Bithynia and *Viviparus* differ from the other prosobranchs mentioned here in having an operculum of the concentric, not the spiral type, in which growth occurs by the addition of marginal rings. Shirbhate & Cook (1987) have shown that in *B. tentaculata* this is due to the presence of cells secreting opercular materials all around its attachment to the operculiger-ous disk. This presumably represents an extension of the opercular groove from a primitive anterodorsal position, though it is not obvious what selective advantage a concentric pattern has over a spiral one: a possible link with apertural shape does not seem to hold.

In many archaeogastropods (vetigastropods) the dorsal surface of the foot behind and under the operculigerous disk is crossed by a number of transverse ridges and grooves so that it presents a corrugated surface, except in the mid line where there lies a longitudal cleft. Normally the right and left halves of this area are kept folded upwards along the cleft so that the ridging is not seen. The use of this structured area has long been unknown; recently, however, H. D. Jones (1984) has described how in *Calliostoma zizyphinum* it appears to be used to keep the shell clean, and this may be its function wherever it is found.

CROSS REFERENCES

CHAPTER 25

THE ALIMENTARY SYSTEM

R ECENT work on the structure, function, and probable evolution of the alimentary canal of molluscs has been reviewed by Salvini-Plawén (1981, 1988a) and has led to a revision of ideas concerning certain of its parts.

One of these is the nature of the primitive gastropod radula. It has for long been assumed to be the type known as rhipidoglossate, in which each transverse row contains a large, sometimes a very large, number of teeth. These are differentiated into a central or rachidian tooth which is flanked on each side by a small number of stout teeth known as laterals, beyond which lies a large number of more delicate marginals. From this beginning it has been supposed that two other principal types of prosobranch radula, the taenioglossate and the rachiglossate, have been evolved by a process known as oligomerization. This involves a narrowing of the radular ribbon by loss of the more laterally placed teeth to form first the taenioglossate type, with only seven teeth per transverse row, and, after a second reduction, the rachiglossate, which has usually three teeth per row, but may have only one. The different steps of the oligomerization process accompany changes of some magnitude in the use to which the radula is put.

Though this evolutionary scheme suggests a smooth transition from a broad ancestral to a narrow derived type of radula there are difficulties in accepting it whole-heartedly. Some of these are of long standing, others have arisen more recently from new knowledge. All lead to the suspicion that the rhipidoglossate radula has been assumed to be primitive largely because the molluscs which possess it were assumed to be primitive on other grounds.

Since it was first emphasized by Macdonald (1869), and again recently by Golikov & Starobogatov (1975), it has been recognized that radular teeth may be either campylodont*, with a posteriorly curving shaft arising from a basal plate sitting on the radular membrane, or orthodont, with a straight shaft arising from the radular membrane directly without a basal plate. Campylodont teeth occur in rhipidoglossate and taenioglossate radulae, but those of the rachiglossate type are orthodont. The transition from taenioglossate to rachiglossate, therefore, is not just a simple reduction in the number of teeth per row, which is easy to envisage in relation to the accompanying evolution of a narrow proboscis, but involves simultaneously a change in the type of tooth, which is less easy.

The idea that the primitive prosobranch radula had a very large number of teeth in each row conflicts with what has been found in Neopilina, a member of a group of molluscs which seem probable ancestors of prosobranchs. Here the radula has only eleven teeth per row, three rather slender median teeth flanked by four stouter ones on each side, the third of which has a comb-like cusp (McLean, 1979; Wingstrand, 1985). This arrangement is similar in its general pattern to that of polyplacophorans, which have seventeen teeth per row, to that of patellogastropods, with up to thirteen teeth per row, and is not unlike that of scaphopods and aplacophorans—in other words it resembles the radulae of a wide scatter of primitive molluscan groups.

*As this derives from the Greek word καμπύλος , curved, it should be spelt with a -y- rather than an -i-.

The docoglossate radula of patellogastropods cannot, as pointed out by Ankel (1938) and emphasized by Golikov & Starobogatov (1975), following Troschel (1863), be directly related either in structure or in mode of action to that of other prosobranchs, only to that of chitons, scaphopods, and *Neopilina*, and even then not wholly. As in these groups the radular teeth scrape the substratum longitudinally, and neither erect nor swing laterally at the odonto-phoral tip (the bending plane) on outward movement, nor does the radula fold longitudinally on retraction, a condition called stereoglossate by Salvini-Plawén & Haszprunar (1987). The radula of other prosobranchs, called the flexoglossate type by these authors, works in an altogether more complex way. At rest on the odontophore the radular ribbon is folded longitudinally to fit into a groove; on protrusion it flattens, its teeth erect and swing laterally as they are pulled outwards over the bending plane, whilst on retraction they carry out a median swing to their resting position. During this latter movement they exert an obliquely transverse rather than a longitudinal scraping action on the substratum. The forces involved in this are less strong than those produced in the action of a docoglossate radula, the teeth of which are stengthened to resist them by addition of mineral salts, absent from rhipidoglossate teeth.

It therefore seems probable that the earliest prosobranchs had a radula with some ten to twenty teeth in each row, much as patellogastropods still have and, as a corollary, that the much greater number of teeth per row in the rhipidoglossate type has arisen as a modifi-cation of that, in adaptation for a particular and different method of feeding. Since typically it has about eleven stout teeth centrally, bounded on each side by a large and indefinite number of marginal ones, it is tempting to suppose that it was formed from an earlier type with about the same number of teeth in each row as in monoplacophorans. The rhipido-glossan gastropods can therefore be regarded as having added a secondary radula, rep-resented by the increased number of marginal teeth, to the primary ancestral one persisting in its more central parts. This new arrangement permits a great broadening of the area covered by the radula during feeding and may well have allowed the exploitation of an otherwise relatively untouched or less efficiently used food resource. The need to fold away a broad radula within the narrow confines of the buccal cavity necessitated the formation of the flexoglossate radula. How broadening was achieved is not known, though a clue might be obtained if the development were studied. It may have been simply by the addition of new groups of odontoblasts to those already present in the secreting area of the radular sac; perhaps the multicuspid marginal teeth of *Neopilina* and of some hydrothermal vent limpets (McLean, 1981; Hickman, 1984; Fretter, 1989) suggest a way in which subdivision of existing teeth might arise if it affected their formative cells permanently instead of only briefly.

Though it seems likely that the polyodont radula of rhipidoglossan prosobranchs is a secondary formation it is not possible to be sure about the nature of its precursor—whether or not it was docoglossate, that is, with longitudinal action, without articulations between mineralized teeth, and stereoglossate, as in most modern patellogastropods. The widespread occurrence in primitive groups of radulae with approximately that assemblage of characters suggests that it may have been. McLean (1990), however, has shown that the radula of one family of recently discovered patellogastropod limpets from hot vents, the Neolepetopsidae, differs from the typical docoglossate pattern in some significant features. Though there are only eleven teeth per row, the radula does show some slight degree of longitudinal folding and some articulation between neighbouring teeth, features which also occur in, and charac-terize, rhipidoglossate radulae. In addition the neolepetopsid radula shows no mineralization of its teeth. The neolepetopsid radula may therefore represent a condition akin to that of

more primitive classes and near a type from which, by differential emphasis on some characters and the invention of others, both typical docoglossate and rhipidoglossate radulae might
have evolved.

Sirenko & Minichev (1975) have described the development of the radula in a group of
polyplacophorans in which, they stated, it first appears during the pelagic larval phase. At this
stage and at first settlement the radula consists of a series of rows each containing only a
single tooth which later divides to form three teeth set in a transverse row. The central one of
these ultimately disappears, but it is soon replaced by a new single tooth which, in its turn
subdivides into three giving a transverse row containing in total five teeth. The two original
persisting teeth, the outermost of the five, undergo further divisions to produce the polyodont row of the adult. These divisions must be due to the separation into distinct groups of
the formative cells of the radular sac which secrete the teeth.

The development of the radula in a further series of chitons has also been followed by
Eernisse & Kerth (1988) whose findings differ from those of Sirenko & Minichev (1975) in that
no radular structures appeared in the species which they studied until the larvae had
metamorphosed. Further, the initial monoseriate radula described by the earlier workers was
never observed nor, apart from one instance, when a median tooth divided to give a
definitive rachidian flanked on each side by a first lateral, was any of the fragmentation of
teeth which Sirenko & Minichev (1975) had also described. Instead, Eernisse & Kerth (1988)
found that the radula first appeared in a bilaterally symmetrical pattern of paired teeth and
only later did its median members develop. Increase in breadth was achieved, not by
fragmentation, but by the intercalation of extra series of teeth between those already
present.

Nothing like the picture of radular development given by Sirenko & Minichev (1975) has
been observed in gastropods. Kerth (1979, 1983) has described the course of radular development in some pulmonates: here he found that the first teeth to appear in most of the genera
which he studied were the first laterals, followed a little later by the second laterals, and then
by the third. Only at this stage did a rachidian tooth develop to fill the gap between the right
and left sets of laterals. Furthermore, the rachidian tooth, when it did appear, was double and
it was only at a still later stage that the two rudiments came together and fused, giving a single
median tooth uniting the right and left halves of each radular row into a superficially single
whole. Further broadening of the gastropod radula was achieved by the addition of new
marginal series of teeth, not by intercalation as in chitons.

As pointed out by Warén (1990) the intention of such work was to compare the gastropod
radula with that of other classes; incidentally it may have also attempted to settle the old
controversy as to whether the molluscan radula is fundamentally a paired structure of right
and left halves (distichous) only secondarily united, as proposed by Nierstrasz (1905) and
Boettger (1955b), or whether it is a median structure (monostichous) as has been most
recently maintained by Salvini-Plawén (1978, 1985, 1988a) who, on the basis of his work on
the two molluscan groups Solenogastres and Caudofoveata, regards the distichous condition as derived from a previous monostichous one. If the distichous condition is indeed the
starting point of gastropod radular evolution the gastropod rachidian tooth must then be
regarded as either a fusion of two teeth (perhaps the original innermost laterals) as suggested
by the observations of Kerth (1979), or as a new structure.

Warén (1990) has, however, himself investigated the developmental stages of the radula
of a number of trochoidean species, extending earlier work on similar lines on atlantids
(Richter, 1961), *Tricolia* (Robertson, 1985b), cones (Nybakken & Perron, 1988) and some

mesogastropods (Fretter & Montgomery, 1968). Most of these workers agree that the earliest stages in all groups share an unspecialized radula, which at a slightly later stage rather abruptly develops a series of specializations adapting each species for the particular use and diet of the adult. Warén (1990) points out that while this is compatible with the idea that radular structure reflects radular function, since all these early stages feed on the same bacterial-algal film, it also has the advantage of not committing the animal to a particular adult diet, leaving open the possibility of the adoption of a new one. Two trochid subfamilies, however, Calliostomatinae and Solariellinae, have, even at this early stage, already acquired specializations committing them to a particular diet and so have narrowed their evolutionary potential. During the lifetime of a prosobranch the radular gland at the inner end of the radular sac is continually adding new teeth to the radular ribbon at a rate which compensates for their destruction at the tip of the odontophore. This also correlates with the length of the radular sac: where tooth replacement rate is high, movement of teeth along the sac is rapid, but since their formation is completed as they migrate and time is necessary for this, the sac must be long. If the replacement rate is low and movement slow a short sac allows adequate time. Fujioka (1985) found that secretion of radular teeth in *Thais bronni* and *T. clavigera* was affected by temperature, being stopped by persistent temperatures below 10°C. From his observations he calculated that the radula was entirely replaced 2-2·5 times each year, and as the animals lived for about five years, 10-13 times during the lifetime of a whelk.

For a discussion of the toxoglossan radula see p. 614.

It has been customary for some workers to divide the initial part of the prosobranch gut (between the jaws and the openings of the oesophagus and radular sac) into an anterior section, the buccal cavity, and a posterior section, the pharynx. It is not particularly appropriate to subdivide this region since the two parts are so closely interlinked and interactive as to form a single functional unit, nor is there anything to differentiate them in development. Extension of the name buccal cavity to cover the whole of this space seems proper since the entire mass within which it lies is most frequently called the buccal mass or bulb.

The roof of the buccal cavity is marked by two longitudinal ciliated folds with a ciliated channel between them, all three structures running back into the oesophagus and extending along the whole length of its glandular mid section. The floor of the oesophagus immediately posterior to the buccal region also bears two ciliated folds but these commonly rapidly converge and the channel between them narrows and ultimately disappears, though the fold itself may persist much longer, often with a bifid tip representing its twofold origin. The detailed arrangements vary from group to group (see Salvini-Plawén, 1988a). The basic structure of the oesophagus therefore seems to comprise a dorsal and a ventral channel (Fretter, 1990) along which food passes, each bounded laterally by longitudinal folds, with the dorsal channel typically of greater importance than the ventral, which is frequently lost (patellogastropods and most caenogastropods). The side walls of the oesophagus between the channels are glandular and expanded laterally into pouches, an arrangement found, or hinted at, in several other molluscan groups. In vetigastropods the surface of the lateral pouches is increased by the presence of finger-shaped papillae: in patellogastropods and many caenogastropods the same end is achieved by the development of transverse septa; this seems the more primitive arrangement. The connection between glandular pouches and the main oesophageal channel is commonly extensive, but may, as in neritids, be limited to a short stretch. It is probable that the true junction between buccal cavity and oesophagus is represented by an oblique plane running from the dorsal lip of the radular sac forwards and dorsally to the anterior end of the dorsal folds on what appears to be the buccal roof. If this delimitation be accepted then the salivary glands (commonly one pair, but two in many patellogastropods), which discharge immediately lateral to the anterior ends of the dorsal folds, are to be interpreted as the anteriormost sections of the same glandular tracts as give rise to buccal (or oesophageal) pouches where these occur (e.g. *Littorina*), and still more posteriorly to the oesophageal glands. The tubular or accessory salivary glands of neogastropods and some other genera, which open elsewhere, may be a separate development. Separation of the buccal pouches and oesophageal glands, which are more or less continuous in archaeogastropods, is due to the constriction of the gut introduced by the change from a hypoathroid to an epiathroid nervous system, a change which gives rise to a narrow circumoesophageal nerve ring.

In *Neopilina* Wingstrand (1985) has shown that on each side the mid oesophagus expands into a very large glandular sac that insinuates itself between muscles and viscera to give an apparently unconnected series of pouches. These were originally taken to be kidney sacs (Lemche & Wingstrand, 1959) and used to support the idea that *Neopilina* showed metamerism. It is now clear that the apparent metamerism is imposed on the glands by their relationship to the regularly arranged musculature. A similar swelling of mid oesophageal glands was described by Fretter *et al.* (1981) in *Neomphalus* and has been found in other deep-sea limpets by Fretter (1988, 1989) and by Haszprunar (1988b). In these animals the walls of the pouches lack conspicuous folds and, though still glandular, appear less so (perhaps because of their greater surface area) than in more familiar prosobranchs. A similar tendency to inflate oesophageal glands, though these still retain glandular septa, was noted by Fretter (1984a) in the neritoidean limpet *Phenacolepas*. The significance of the expansion is not obvious: Wingstrand (1985) suggested that it might form an internal hydrostatic skeleton helping in the extrusion of the odontophore. In some other deep-sea prosobranchs the oesophageal gland also assumes unusual shapes: in the cocculiniform genus *Addisonia* (which is unusual in several respects) it takes the form of a series of separate tubules connected through a common aperture to the oesophagus (Haszprunar, 1987c), whilst in the genus *Osteopelta* it has the form of two large tubes each opening to the oesophagus by a wide aperture at the level of the posterior end of the radular sac (Haszprunar, 1988d).

Oesophageal glands are often found to be reduced or absent in prosobranchs in which a crystalline style is present. Examples of families with a crystalline style and in which the oesophageal glands are lost are calyptraeids and capulids (Graham, 1939), turritellids (Graham, 1939), vermetids (J.E. Morton, 1965), some other cerithioidean families (Houbrick, 1988) and many families in the Truncatelloidea (= Rissooidea) (Ponder, 1988b). There are other families, however, in which style and oesophageal glands co-exist as in the Cerithioidea (Houbrick, 1988) and the neogastropod *Ilyanassa obsoleta* (Jenner, 1956). The matter is complicated, however, as there is sometimes no distinction made in the literature between a true crystalline style (with absorbed enzymes) and a protostyle (J. E. Morton, 1952), which is only the initial stage in the formation of a faecal rod and contains no enzymes. The presence of a true style cannot be deduced from the presence of a style sac unless that is separate from the intestine since in many species the style is a temporary structure present only at certain stages of the feeding cycle. In those animals in which a true, crystalline, style has been shown to be present the oesophageal gland region, if not wholly simplified, is often slightly dilated. Little or no investigation has been made of the nature of the cells in this dilated region and of what their function may be: they might represent a vestigial gland, perhaps still functioning at a low level, or an early stage in the formation of a crop. As suggested by Houbrick (1988) there may well be a continuous series of structural and functional stages between a normal glandular mid oesophagus and a simple conducting tube represented by different species.

The original contention of Yonge (1930) that crystalline styles and extracellular proteases cannot occur together has proved incompatible with the results of subsequent research, starting with that of J. E. Morton (1956) on the bivalve *Lasaea*.

One of the outstanding characteristics of neogastropods is the organization of the anterior part of their alimentary canal, which shows an arrangement superficially quite unlike that of other prosobranchs. No intermediate forms exist to bridge the gap between lower gastropods and neogastropods and to suggest how their anatomy arose, and no help is apparently forthcoming from a study of development since Fretter (1972) found that in *Hinia* the larval gut is rapidly replaced at metamorphosis by a preformed adult one in which all the features of the neogastropod are already present and have developed directly.

In muricids, which amongst living forms present a state not too far from the original neogastropod arrangement the buccal cavity leads into an elongated tube marked by two prominent ventrolateral folds, representing the dorsal folds of the buccal region of other prosobranchs, which have migrated to a more ventral position than usual. This tube runs as far as the nerve ring, immediately anterior to which it expands into a conical structure, the valve of Leiblein, within which lies an incompletely funnel-shaped projection bearing an apical fringe of long cilia. This acts as a valve preventing forward movement of the contents of the more posterior gut. Behind this structure the gut contracts to penetrate the nerve ring and then dilates since the folds (now anatomically as well as morphologically dorsal because of the effects of torsion) become extremely glandular and increase greatly in size. They run back to, and then enter, a duct leading from a mass of glandular tissue lying otherwise free in the haemocoel, the gland of Leiblein. Behind this point the oesophagus runs straight to the stomach.

The interpretation of the morphological nature of this section of the alimentary canal rests upon the work of Amaudrut (1898), Graham (1941), and Ponder (1973). Amaudrut (1898) first proposed that the gland of Leiblein was the homologue of the oesophageal glands of other prosobranchs which had been stripped off the oesophagus from before backwards to form a separate gland, and this view was supported by Graham (1941) who found a scar marking the line of the original separation, and showed that it followed the same twisting course due to torsion as did the oesophageal glands. He suggested that the elongated portion of the gut in front of the valve of Leiblein was formed from the anterior part of the oesophagus, but Ponder (1973) preferred to interpret it as an elongation of the posterior part of the buccal cavity—an area, as suggested above, easily attributed to either. Ponder also described the fold within the valve of Leiblein as derived from the oesophageal valve, a structure commonly found at the junction of buccal cavity and oesophagus in less highly modified prosobranchs, though against this view is the fact that the valvular flap is a dorsal structure, whereas the oesophageal valve is ventral. The functional reason for these changes may be in part the evolution of a proboscis, which calls for the great elongation of some anterior portion of the gut. Since the odontophore remains in the neighbourhood of the mouth the section of gut undergoing elongation necessarily lies posterior to that. The valve of Leiblein prevents forward flow and possible dilution of the enzymes which originate in the gland of Leiblein, the secretory activities of which Martoja (1963) has shown in nassariids have a rhythm parallel to that of feeding. It may, indeed, be the advantages derived from the concentration of enzymes in a limited area of the alimentary canal that were responsible for the conversion of the primitive oesophageal gland, adding its secretions along the length of the oesophagus, into the gland of Leiblein with its duct. A somewhat comparable narrowing of the connection between gland and gut is found in neritoideans, though here situated at the anterior end of the gland. The salivary glands have also been affected by development of the proboscis and their duct have elongated, passing over the nerve ring, not through it, as is more usual in mesogastropods.

Some variations of this muricoideam condition are found amongst the neogastropods that are not toxoglossans (Graham, 1941). Among other things these affect the precise point at which torsion occurs: for example, within the valve of Leiblein in muricids, behind it in buccinids and nassariids. The degree of glandular development in the dorsal folds is consider-able in muricids where they form a structure so elaborate that it was called the 'glande framboisée' by Amaudrut (1898), but it is little or none in buccinids and nassariids. The size and complexity of the valve of Leiblein is considerable in muricids, but it is reduced or absent in buccinids and nassariids. It lies behind the proboscis in these three families but within it in cancellariids. The tract of scar tissue marking the line of separation of the gland from the main food channel of the oesophagus is clear in muricids and buccinids, but it is absent in nassariids. These differences suggest that different families followed slightly different evolutionary path-ways as they departed from an original type in which the general anatomical pattern was already in course of establishment.

Neogastropods include not only the rachiglossans (muricoideans) but also the toxoglossans (conoideans = cones, turrids, and terebrids), all with a proboscis and with the anterior end of the gut highly modified, but in apparently different ways from that of rachiglossans. Indeed as our knowledge of this group has recently been enlarged by the work of Marsh (1971), Sheridan et al. (1973), B. A. Miller (1975, 1979), Shimek (1983), J. D. Taylor (1985, 1990), J. A. Miller (1989), Kantor & Sysoev (1989, 1990), Kantor (1990b), and J. D. Taylor & Miller (1990) it has become plain that the toxoglossans present wider variation in anterior gut anatomy than was suspected. Little of this new knowledge has yet been incorporated into our understanding of the evolutionary relationships of toxoglossans but schemes have been proposed by Sheridan et al. (1973), Kantor (1990b), and J. D. Taylor (1990).

In its most advanced state the conoidean gut is characterized by a poison apparatus consisting of a muscular terminal bulb connected to the point where oesophagus and buccal cavity join by a long, convoluted, duct-like gland, by great reduction of the odontophore, by a radula in which the teeth are hollow and come to lie singly, free from the radular mem-brane, and by the absence of a valve of Leiblein and any apparent gland of Leiblein, the oesophagus running as a simple tube from buccal cavity to stomach. Prosobranch structure, no matter how apparently highly modified, can, nevertheless, usually be analysed so as to show its derivation from some simpler state. In this expectation the idea has persisted that the toxoglossan anatomy should be traceable, preferably to that of other neogastropods, but at least to that of some lower prosobranchs. A possible evolutionary pathway from the latter was suggested by Amaudrut (1898) who supposed that the toxoglossan poison gland was homologous with the mesogastropod oesophageal glands, but, unlike the rachiglossan gland of Leiblein, had separated from the oesophagus from behind forwards and without leaving any scar—at least none has so far been found. Such a course, however, would imply that rachiglossans and toxoglossans were not as closely linked as they are commonly held to be and might not deserve inclusion in a single group Neogastropoda. Amaudrut (1898), Ponder (1973), Sheridan et al. (1973), J. D. Taylor et al. (1980), Shimek & Kohn (1981), and J. D. Taylor & Morris (1988) have all sought to place the origin of the neogastropod stock somewhere within the mesogastropods, though at different levels therein.

An alternative scheme to that of Amaudrut for the evolution of the venom gland has been put forward by Ponder (1970a, 1973) (Fig. 312). This has as its starting point the conditions found in muricoideans, already shown by Amaudrut (1898) and Graham (1941) to be easily derived from a mesogastropod origin. This scheme makes the poison gland the homologue of the gland of Leiblein (as has always seemed the most likely conclusion) but also involves the 'glande framboisée' (the glandular dorsal folds lying between the valve and the gland of Leiblein), which is a less obvious homology. Though no intermediate forms fill the gap between the organization of the lower prosobranch gut and that of rachiglossans, Ponder (1970a, 1973) has described a number which may be interpreted as intermediate between the rachiglossan and the toxoglossan so that the latter is a derived condition evolved from a state similar to that seen in primitive neogastropods. Analogous steps are seen within the muricoideans which give clues as to how this occurred.

The first step in the transformation of the fore-gut of a rachiglossan into that typical of toxoglossans is the fusion of the tips of the dorsal folds so as to separate a dorsal tube, with which the duct of the gland of Leiblein connects, from a ventral one which leads to the posterior oesophagus and the stomach. This separation affects first the most posterior part of the mid oesophagus, but in different species can be seen to extend more and more anteriorly until it almost reaches the valve of Leiblein, as in the volutid *Alcithoe* (Ponder, 1970b), in which a considerable glandular mass lies along the dorsal surface of the oesophagus. The large part of its secretion comes from the dorsal folds, and the gland of Leiblein produces a less conspicuous contribution.

In another series of neogastropods, represented by some marginellids which have reached more or less the same stage of organization as the volutids, a further important change has taken place in that the separation of the channel formed by the fusion of the dorsal folds has extended anteriorly, ventral to the valve of Leiblein. In the forms which Ponder investigated (1970a) the extension opens to the floor of the gut just anterior to the valve, brought ventrally by torsion. This process of separation many now be presumed to have continued to move forward until the connection with the gut lies at the level of the buccal cavity as in some marginellids (Graham, 1966), and in turrids and cones.

The importance of the gland of Leiblein as a source of secretion in this sequence becomes gradually less, a trend obvious even in the rachiglossans where it is reduced to a small caecum in the family Galeodidae (Haller, 1888), and has been lost in *Melongena* (Vanstone, 1894). On the other hand, the activity of the dorsal folds increases and the original gland becomes more muscular. In the most advanced forms the so-called venom 'gland' is in fact only a muscular pump, and its 'duct' is the source of the secretion. It is one of the merits of Ponder's suggestion that it offers an adequate explanation of such an apparently anomalous arrangement. It also offers confirmation of the original proposal of Amaudrut, made nearly a century earlier (1898), that while the gland of Leiblein was formed by stripping off the glands of the mid oesophagus from in front backwards, the poison gland arose by a stripping off from behind forwards: where Amaudrut went wrong, however, was in supposing that it was the same glandular tissue that was finally functional. The formation of this part of the conoidean fore-gut thus seems to result from an extraordinary and, indeed, superficially improbable sequence of events involving two successive separations of glandular tissue from the main channel of the gut, one in one direction, the other in the opposite. Though it is possible in this way to present a plausible explanation of the anatomical changes leading to the establishment of the conoidean fore-gut, there are difficulties in accepting such a devious evolution wholly; furthermore it commits us to a particular pattern of evolution within the neogastropods.

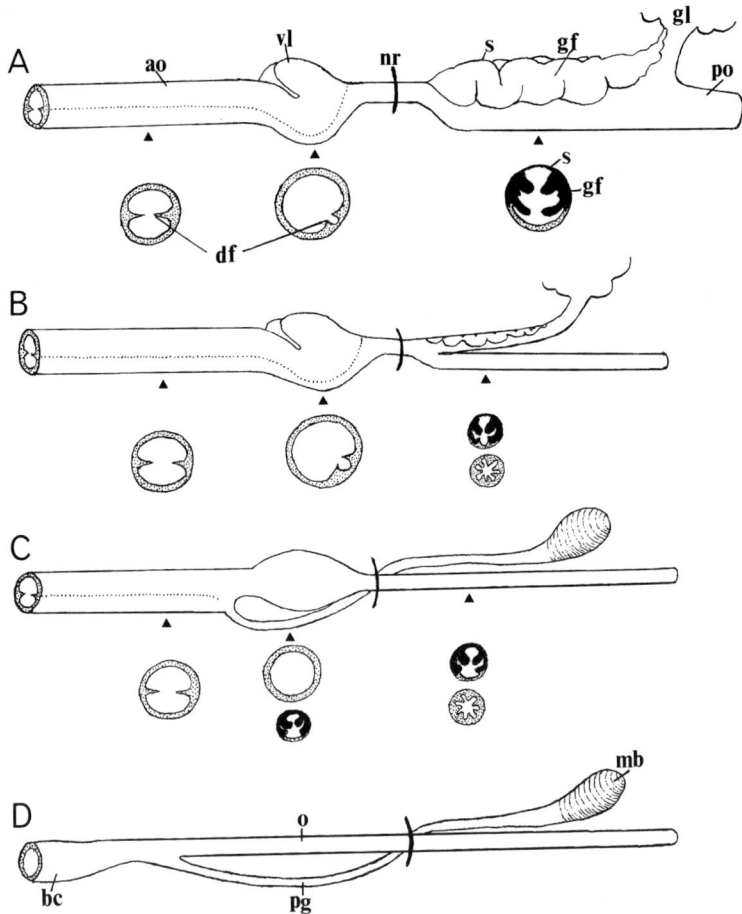

FIG. 312.—A series of diagrams to illustrate the possible derivation of the toxoglossan venom gland from the 'glande framboisée' of a rachiglossan. Anterior is to the left. A, typical muricid rachiglossan such as *Nucella*; B, *Alcippe* (volutid); C, *Marginella* (marginellid); D, a toxoglossan. The level of the transverse sections is indicated by the arrowheads. (In part after Ponder, 1970a).

ao, anterior oesophagus; bc, buccal cavity; df, dorsal fold; gf, glandular fold ('glande framboisée'); gl, opening of gland of Leiblein; mb, muscular bulb; nr, position of nerve ring; o, oesophagus; po, posterior oesophagus; pg, venom gland; s, scar marking the position of stripped off glandular tissue; vl, valve of Leiblein.

The scheme demands the prior presence of the gland of Leiblein and the neighbouring glandular folds of the oesophagus (though perhaps not of a valve of Leiblein) in the stock from which the toxoglossans arose, an organizational pattern assumed to satisfy functional needs introduced by the presence of a proboscis. Ponder's scheme therefore suggests that a proboscis would have been present in the ancestral conoidean; if so, it would have been pleurembolic since that is found in most neogastropods apart from conoideans. That type is also present in the mesogastropod tonnoideans, whose relationship with neogastropods is suggested by several lines of evidence (Healy, 1983, 1986; Haszprunar, 1985a; J. D. Taylor &

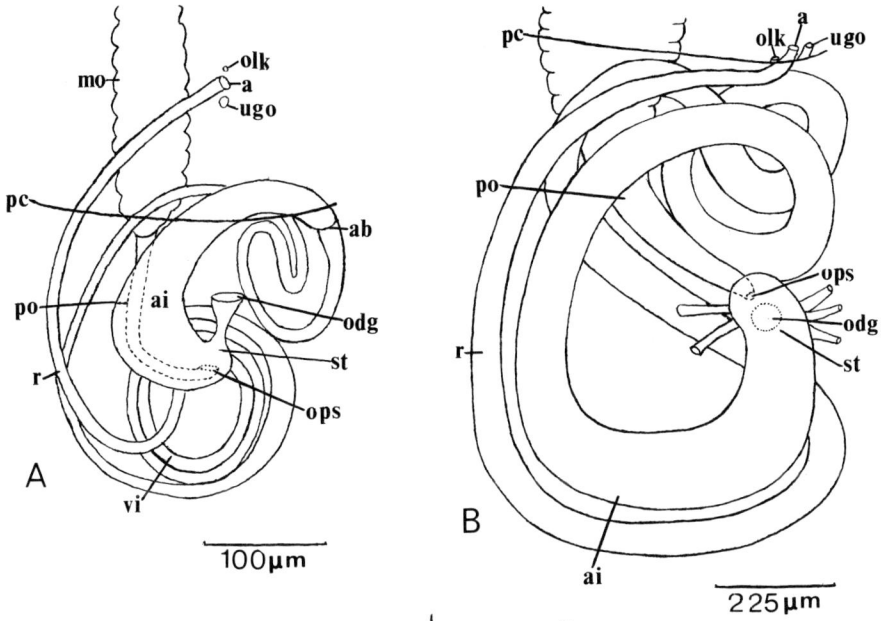

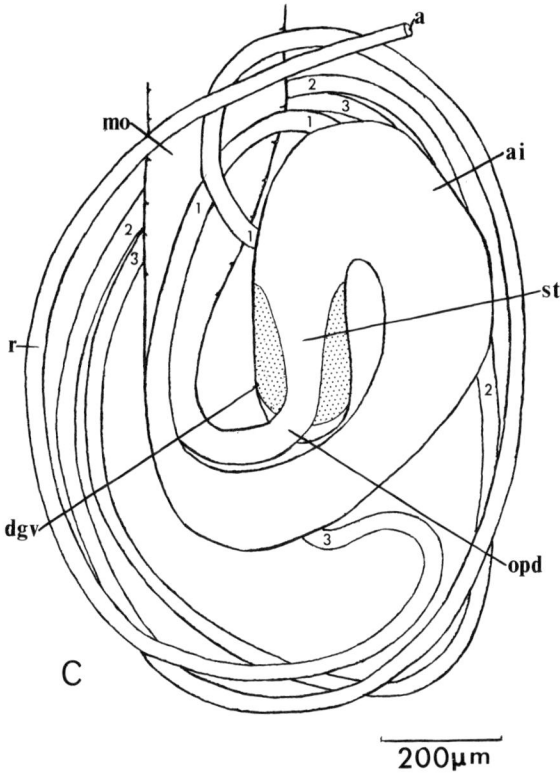

Morris, 1988). In these mesogastropods, however, it is not accompanied by either a valve or gland of Leiblein, and though some have poison or acid glands these are part of the normal salivary glands and not homologous with the accessory glands of neogastropods. Such anatomical arrangements are compatible with the idea that neogastropods are a mono-phyletic group as claimed by Ponder (1973) and J. D. Taylor & Morris (1988), in which first a pleurembolic proboscis, then a gland of Leiblein evolved, giving a starting point for conoidean evolution. The fact that conoideans lack a valve of Leiblein suggests that their evolution and that of other neogastropods diverged at that point or that the valve has been lost. If the conoidean proboscis is derived directly from an acrembolic one, as accepted by Sheridan et al. (1973) and Kantor (1990b), neogastropods would appear diphyletic, with rachiglossans and toxoglossans following different pathways; the evolution of the poison gland might then have to be explained differently.

It cannot be assumed that the changes described by Ponder (1970a) occurred in isolation: the organization of the conoidean gut involves not just formation of a poison apparatus but also of a proboscis which is unlike that of the rachiglossans in that the accompanying elongation of the gut is commonly produced by increased length of the oral tube, that part of the gut between the mouth and the odontophore and buccal cavity, not by increased length posterior to the odontophore. J. D. Taylor (1985) and J. D. Taylor & Miller (1989) have described species of the turrid Turricula in which the odontophore does lie close to the mouth and the oral tube is short. Comparative information on turrid anatomy is so slight that it is impossible to be certain whether this is a stage intermediate between the rachiglossan and toxoglossan states or represents a special condition. (See also p. 617).

The prosobranch stomach commonly exhibits, in the form which Salvini-Plawén (1988a) calls the style sac stomach, a topographically posterior, though morphologically anterior, more or less globular section to which the oesophagus opens ventrally, anterior to which lies a tubular style sac leading to the intestine, or perhaps itself an intestinal region integrated into the stomach. Ducts from the digestive gland open to the globular part near the oesophageal opening and also close to the beginning of the style sac. In many prosobranchs part of the globular section is lined with cuticle; this area is extensive in such forms as vetigastropods but is often reduced in those at the caenogastropod level. Part of the surface which is not cuticularized is elaborately ciliated, ridged and grooved, and forms a sorting area. A prominent ridge with a groove alongside emerges from one digestive gland duct and runs into the style sac, where a second fold arises to border the other side of the groove; all three structures run into and along the intestine, often for a considerable distance. In some archaeogastropods, most markedly in vetigastropods, the globular part of the stomach is drawn out into a spirally coiled caecum, sometimes small, but prominent with several turns in pleurotomarians (Fretter, 1964, 1966) and trochoideans (Graham, 1949).

The style sac stomach has apparently been modified in two groups of prosobranchs, those that have adopted a carnivorous diet and those that have become limpet-like, though some

FIG. 313.—Diagrams of mid and hind gut in dorsal view. A, Lepeta caeca; B, Acmaea tessulata; C, Propilidium ancyloide. 1, 2, 3, indicate continuity of intestinal coils.

a, anus; ab, abrupt ventral bend of mid intestine; ai, anterior intestine; dgv, opening of digestive gland ducts to lateral and ventral walls of stomach; mo, mid oesophagus; olk, opening of left kidney; opd, opening of posterior oesophagus to dorsal wall of stomach; ops, opening of posterior oesophagus to ventral wall of stomach; pc, posterior limit of nuchal cavity; po, posterior oesophagus; r, rectum; st (in A & B), stomach, (in C) ciliated tract forming roof of stomach; ugo, urinogenital opening; vi, most ventral coil of the intestine.

features of more typical gastric structure can often be found. In the carnivorous neogastro-pods style sac and globular regions are usually recognizable, often with a small cuticularized area in the latter (*Trophon truncatus* and *Colus gracilis*: E. H. Smith, 1967b; *Hinia* species: Graham, 1949; E. H. Smith, 1967b) enlarged into a gastric shield in the few species which have developed a crystalline style (*Ilyanassa obsoleta*: Jenner, 1956; *Cyclope neritea*: Morton, 1969). A sorting area has been described by E. H. Smith (1967b) in *Colus gracilis*, *Neptunea antiqua*, and *Hinia incrassata*, but in advanced carnivores such as muricids (Wu, 1965), turrids (E. H. Smith, 1967b), marginellids (Ponder, 1970a), and mitrids (Ponder, 1972) none of these features persists and the stomach is merely a sac with folded walls, sometimes elaborated. (*Alcithoe*: Ponder, 1970b), where oesophagus, digestive gland ducts, and intestine all converge. Similar modifications also affect naticoidean carnivores.

The second collection of prosobranchs in which modification appears evident are those which are limpet-like. Here, however, caution is called for, since these animals fall into at least five groups, not necessarily closely related: (1) the vetigastropod limpets; (2) the neritoidean limpets; (3) the patellogastropods; (4) other 'archaeogastropod' limpets; (5) the mesogastro-pod limpets. In the mesogastropod group, exemplified by calyptraeids and capulids (Graham, 1939, 1954), and in the fissurelloidean (Graham, 1939) and neritoidean (Fretter, 1984a) limpets the stomach is clearly recognizable as of the style sac type, with cuticle, sorting area and caecum, though this is usually simple. The stomach of the fourth group, as exemplified by *Neomphalus* (Fretter et al., 1981), other vent genera (Fretter, 1988, 1989) and cocculiniform limpets (Haszprunar, 1988b, 1988c, 1988d, 1988e), shows reduced versions, sometimes only indications, of the features of this type and could be interpreted either as a much reduced style sac stomach, or as an ancestral form from which the style sac type could have evolved. It cannot be argued very forcibly that the adoption of the limpet facies must lead to the simplified stomach that this group exhibits since in vetigastropod, neritoidean, and meso-gastropod limpets the stomach has not been significantly affected by such a change. That fact leaves the idea that the fourth group may exhibit a primitive condition the more attractive.

Like many other features of their organization the stomach of patellogastropods cannot be easily related to any pattern found in other gastropod groups. Though details of its internal structure are little known, where they are (Bush, 1988) the stomach shows no obvious differentiation into two regions, has no, or only minimal sorting areas, little or no cuticular lining, and is without a caecal appendage. It does receive ducts from the digestive gland and there are typhlosoles, but these are the sole features shared with the style sac type, and, indeed, are only those which designate a part of the alimentary tract as stomach throughout the phylum.

The patellogastropod stomach is not clearly differentiated and its boundaries hard to determine, especially that with the intestine (Fig. 313). In *Patella*, *Lepeta*, *Propilidium*, and *Helcion*, this part of the gut is somewhat greater in diameter than what is clearly intestine, and has two folds along it; in other genera, such as *Lottia* and, particularly, *Erginus* (Lindberg, 1988), the difference in diameter is much greater. A similar dilatation of a region of the gut posterior to the digestive gland ducts occurs in the related neolepetopsid patellogastropods (Fretter, 1990), and in the aberrant genus *Addisonia* (Haszprunar, 1987c) though in this genus dila-tation affects a length separated from the stomach by a region of normal dimensions. This dilated part is regarded as the stomach by some workers. It may not be unreasonable to suppose that it represents a part of the intestine which in other prosobranch groups has been elaborated into the style sac at the same time as the other features were being formed.

That this region of the patellogastropod gut and the style sac of other prosobranchs are homologous is further suggested by the fact that compaction of a protostyle occurs in both. The formation of the style sac may have provided a more effective way of doing this; it certainly proved to be a prerequisite for the conversion of the protostyle into a crystalline style, a feat not apparently achieved by any patellogastropod though widespread through-out other molluscan groups from monoplacophorans onwards—an amazing independent production of a nearly identical and complex apparatus in no less than five, perhaps six, groups: monoplacophorans according to Lemche & Wingstrand (1959) and Wingstrand (1985), though not yet proved by detection of enzymes; neritoideans (Seshaiya, 1934); meso-gastropods (Yonge, 1954; Fretter & Graham, 1962); neogastropods (Jenner, 1956; Morton, 1960); thecosomatous pteropods (Yonge, 1926; Howells, 1936), though perhaps not here functional (Lalli & Gilmer, 1989), and the lamellibranch bivalves.

The relationship between the patellogastropod stomach and the style sac stomach is not clear. If its anatomy has been brought about by reduction from a style sac type the change is much more extreme than in any other group. It cannot be regarded as similar to that of monoplacophorans as described by Lemche & Wingstrand (1959) but is perhaps more akin to the *Hanleya* type of chiton stomach described by Plate (1901). In that case it may represent an archaic type from which the style sac stomach may have come. The resemblances between the style sac stomach as seen in vetigastropods for example and in many bivalves—even to some extent in cephalopods—suggest that it may have been an ancestral conchiferan pattern. If this near identity of structure with all its wealth of details of organization and functioning, as believed by Haszprunar (1988a, 1988c), it is an outstanding example of parallel development.

In the immediate proximity of the anus of some prosobranchs, mainly trochids and muricids, there lies a glandular caecum of the rectum, the anal or rectal gland. Its structure was investigated by Fretter (1946; see p. 223) who concluded that in *Nucella lapillus* it was a secondary source of excretory material, voided by the anus. This structure together with superficially similar organs in trochids, has been re-examined by E. B. Andrews (1992). The trochid gland proves to be a simple evagination lined by the same type of epithelium, though richer in mucous cells, as lines the rest of the rectum. Its function is apparently to add further lubricant to the gut at a point where faeces are shed.

The anal gland of *Nucella* is more complex, larger, elaborately branched, its cells with numerous granules and melanin particles, and its lumen filled with a population of bacteria living on cell debris. The cells are ciliated and their surface forms a concave cup inhabited by many bacteria, as in some other molluscs with extracellular bacteria (Southward, 1986). The gland has reserves of lipid and glycogen. Andrews (1992) suggests that the gland, placed in a vascular sinus around the rectum, receives blood from the kidney but still containing some metabolites. These are advantageously removed before the blood reaches the ctenidium by its cells which absorb them, convert them to substances deposited in residual bodies and discharge these to the lumen, where they form the food utilized by the bacteria.

CROSS REFERENCES

Radular adaptations:	605–18 passim, 682–3
Use of proboscis in neogastropods:	612–23 passim, 718
Radular evolution in conoideans:	615–17
Development of fore-gut in *Lacuna* and *Hinia*:	675–8

CHAPTER 26

THE FEEDING OF PROSOBRANCHS

I N nearly all gastropods the radula, the original food-gathering organ of molluscs, persists with the same role. It is lost only in a small number of species, primarily those which are true parasites [see, for example, Lützen (1972), Gooding & Lützen (1973), Lützen & Nielsen (1975)], or, like some eulimids (Warén, 1983) and pyramidellids (Fretter & Graham, 1949) are ectoparasitic and take liquid food, and those which, like some turrids and terebrids, ingest prey whole by suction. Much of prosobranch radiation relates to changes in radular form and function, and during the nineteenth century this provided, in the hands of Gray (1853), Troschel (1863) and P. Fischer (1880), the accepted basis for the clasification of prosobranchs.

By 1938, when Ankel published his comparative account, the general mode of functioning of the different radular types had been established. The arrival of the scanning electron microscope, however, radically altered the study of radular structure and activity by giving for the first time a clear picrture of the finer organization of the teeth, and allowed it to be related to their use; it is now standard practice to present scanning electron micrographs in describing radulae. On this basis much information about tooth structure and function in archaeogastropods has been given by Hickman (1980, 1984) and Hickman & Morris (1985), and on toxoglossan teeth by Marsh (1977), Mills (1977), James (1980), Shimek & Kohn (1981) and Kantor & Sysoev (1990). Many of the recent findings on archaeogastropods relate to devices for strengthening teeth against the stresses to which they are exposed, either as individual teeth or as units in the functioning of the radula as a whole.

A curved campylodont (see p. 592) radular tooth acts after extrusion of the odontophore by being drawn forward over its under surface towards the odontophoral tip and the mouth, and at the same time across the substratum on which the animal is feeding. The rachidian (central) teeth move with the morphologically posterior (but in this situation anteriorly facing) concave side leading, along a straight line at right angles to their breadth. In radulae other than docoglossate all other teeth follow a curved, oblique path so that stresses acting on them, though generally similar, are continually changing. The leading face of every tooth is therefore exposed to a stretching (tensile) force, the opposite face to a compression, and the base which attaches it to the radular membrane to a bending force. Teeth in taenioglossate and rhipidoglossate radulae are strengthened against these stresses by alterations in their shape: the base is expanded and a buttressing thickening appears along the convex side. The leading concave face is less modified as this might well interfere with its action in collecting food. In patellogastropods, with a docoglossate radula, the strength of the individual teeth appears to be increased chemically rather than morphologically by the tanning of its protein component and by the incorporation into its substance of iron and silicon. These are added to each tooth rudiment, as it migrates forward towards the dorsal surface of the odonto-phore, from the cells lining the roof of the radular sac. The substances are distributed within the matrix of chitin so as to strengthen the posterior wall of the tooth (Runham et al., 1969), which forms the leading edge when it is drawn across the substratum. In tooth cusps the iron is in the form of crystals of goethite, in tooth bases in a microcrystalline form resembling

FIG. 314.—The central part of a few rows of the radula of the hydrothermal vent limpet *Nodopelta subnoda* (Archaeogastropoda), showing interlocking at the base of the rachidian and first lateral teeth; other articulations are present but not visible in this SEM photograph. × 650.
　　　arrowhead, rachidian tooth.

goethite (St. Pierre *et al.*, 1986). S. J. Hawkins *et al.* (1989) give the illuminating comparisons of the hardness of rhipidoglossate and taenioglosssate teeth (2·0–2·5 on Moh's scale) with the hardness of human fingernail (2·0) or the soft mineral gypsum, and of the hardness of docoglossate teeth (4·0–5·0) with human teeth (5·0) or Derbyshire spar or Blue John.

Perhaps of even greater significance in ensuring efficient functioning of a radula than the straightforward strengthening of the component teeth are the adaptations which interlock neighbouring teeth in a row or in successive rows (fig. 314). Without such devices the teeth form a coherent rasping surface only by virtue of their common attachment to the under-lying, flexible radular membrane. By the development of processes on one tooth which articulate with neighbouring teeth, either those alongside or those in more anterior and posterior rows, or both, the whole active part of the structure can be transformed into a functional unit: stresses operating upon teeth in contact with the substratum are spread to others and dissipated, and the movements of teeth are coordinated as they curve from one position to another at the tip of the odontophore and fold away on its dorsal surface on withdrawal.

Steneck & Watling (1982) have correlated the different types of radula found in plant-eating prosobranchs with the different types of algae which they eat. They pointed out that amongst herbivorous prosobranchs the preferred foods are microalgae, filamentous forms, and those with an expanded thallus; tougher plants are less attacked. A rhipidoglossate radula with its very large number of marginal teeth covers a wide belt of substratum during feeding but may not have enough power to do more than collect loose or weakly attached particles. This is supported by their observation that rhipidoglossans take primarily diatoms, cyanobacteria, together with filamentous and some foliose algal species. Observations by Hickman & Morris (1985) suggest, however, that a rhipidoglossate radula is capable not only of this brushing action, which is carried out by the more laterally placed marginal teeth, but also of a more powerful scratching of the substratum carried out by the more medially placed marginal teeth, which have sharper and more markedly denticulate cusps.

A reduction in number allows each tooth in a row to be stronger and to have increased excavating or cutting ability. Though it grazes a narrower belt of the substratum a taenio-glossate radula achieves this end. If, however, its teeth become secondarily multicuspid then the kind of sweeping ability possessed by the rhipidoglossate type may be partly regained, whilst some of the greater power is retained. Steneck & Watling (1982) gave some examples of how the number of contact points between radula and substratum seems to affect a snail's grazing ability, and so its choice of food. Thus the radula of *Littorina littorea* makes, on average, twenty contacts per row with the substratum, whereas that of *L. obtusata* has seventeen: this may not seem much of a difference and is perhaps not the only one, but it may explain why *L. littorea* feeds mainly, according to these workers, on diatoms, *Ulva*, and *Enteromorpha*, which it much prefers to *Fucus*, the chosen diet of *L. obtusata*. The food preferences of three species of *Littorina*, *littorea*, *obtusata*, and *mariae*, have also been investi-gated by Watson & Norton (1985, 1987). Their findings agree in general with those of Steneck & Watling and provide further details. According to their observations *L. obtusata* and *L. mariae* both showed a preference for fucoids, the latter with a narrower distribution on the shore also showing a narrower diet, mainly *Fucus serratus*. Some preference related to the parts of the plant eaten: thus *L. obtusata* preferred the reproductive areas of the thallus to the vegetative parts and took young rather than fully grown *Ulva*. These authors found both species taking fucoid tissues rather than their epiphytic growths as suggested by Reimchen (1974) and Lubchenco (1982); *mariae* ate more of them than did *obtusata*, and according to G. A. Williams (1990) is largely dependent upon them. This difference may relate to differ-ences in the shape of the cusps of the lateral teeth which are more pointed in *obtusata*, but broader and flatter in *mariae*, Sacchi et al. (1981) suggested that *L. littorea* eats fucoids wherever these are available and ingests microalgae only in their absence: it is microphagous only under compulsion.

As a further example of radular structure affecting feeding Steneck & Watling (1982) quoted the relationship between *Fissurella angulata* and *Acmaea jamaicensis* which live sympatrically on coralline algae. They may do so because they are not competing for food, the former, with a rhipidoglossate radula, taking epiphytic microscopic and filamentous algae, the latter, with a docoglossate one, eating the corallines themselves. Petraitis (1989), in a study of the interrelationship of a taenioglossan (*Littorina littorea*, taking microscopic and filamentous algae) and a docoglossan (*Notoacmaea tessulata*, taking firmly attached algae) has shown that the presence of the two species may be mutually beneficial; in particular, winkles, by removing filamentous weeds, encourage the growth of the algae which form the food of the limpets, leading to their improved growth and survival.

The docoglossate radula of patellogastropods is the only type capable of dealing with the toughest weeds: it does so because it operates on a different principle from both the rhipido-glossate and the taenioglossate. The attachment of each tooth to the radular membrane is extensive and secure, several rows of teeth applied to the substratum simultaneously, and all teeth act in the same longitudinal direction as the movement of the odontophore, not obliquely to that as in other types (Ankel, 1938).

The function of this radular type is made more efficient by two structures peculiar to patellogastropods (figs 315, 316): one is the single, powerful, arched dorsal jaw, the second is a structure known as the licker (Davis & Fleure, 1903). The latter lies on the anterodorsal wall of the sublingual pouch just posterior to the anterior tip of the radular membrane. It is a thinkening of the cuticular subradular membrane, and its surface is diversified differently in different species, presumably in relation to their type of food or the type of substratum from which it is gathered. The licker may be dilated by blood passed from odontophoral sinuses and has retractor muscles by whose contraction it can be partly withdrawn. When a limpet feeds (fig. 315) the dilated licker is everted and applied to the substratum just before the radular teeth and just posterior to the point on which they are set. At the same time the edge of the jaw is applied to the substratum level with the dorsal lip of the mouth; the two structures delimit the area over which the radula will work. As the teeth move over this their raspings may adhere to them, or may separate and pile up against the jaw, or be left on the substratum; the licker, moving along with but behind the teeth, collects anything left on the substratum as well as what may have accumulated behind the jaw and pulls it through the mouth on retraction. Though apposition of licker and jaw invites the idea of a biting action, this does seem to have evolved. S. J. Hawkins & Hartnoll (1983a) came to similar conclusions as Steneck & Watling (1982) about the grazing efficiencies of different radular types. Raffaelli (1985), too, after a study of grazing prosobranchs in New Zealand, found that groups of animals with similar diets were usually groups with similar radulae. He argued, however, that the situation is by no means as circumscribed as these findings suggested. Prosobranch radulae are versatile tools (Markel, 1966) and may be used in different ways in different circumstances; microalgae are often an important incidental uptake by a radula capable of taking tougher food, or the sole uptake in its absence. It must be remembered that the radula of a small prosobranch is a weaker implement than the radula of a similar pattern of a larger snail: the taenioglossate radula of a winkle can take fucoids, but the taenioglossate radula of a small rissoid can take only epiphytic diatoms. The relationship between the power of a radula and the size of its owner has been well shown by Vahl's work (1971) on *Helcion pellucidum*. The larvae settle widely over a beach, but until they have grown to a shell length of four millimetres their radula is too weak to allow them to eat the tissues of laminarians, the food of the adult limpets. Until they reach that size they can eat only softer weeds: at that size they secret a mucous float and migrate to the laminarians on which the rest of their life will be spent. A similar change in the power of the radula with consequent change in the type of food eaten has been shown by Ndifon (1979) in *Bulinus globosus* and by Thomas et al. (1985) in *Biomphalaria glabrata*. The latter workers were able to show that the odontophore increases in size relative to other structures as the animal grows and so in relative power and in the area of substratum over which it works.

Rissoids (Fretter & Manly, 1979), hydrobiids (R. C. Newell, 1965), and many other small prosobranchs depend upon microalgae and bacteria epiphytic on weeds or growing on sand

FIG. 315.—Left: *Helcion pellucidum*, anterior end in ventral view. The mouth is open and the odontophore partly protruded, showing the anterior part of the radula anteriorly and the swollen, triangular licker posteriorly. The surface of the licker shows a few transverse grooves. The bulk of the mouth is occupied by the horseshoe-shaped jaw around the tip of which the inner lips are visible laterally. Zonation of the outer lip is visible, as is the circumpedal groove on the foot.

Top right: *Patella vulgata*, left side of everted licker. The outer surface is marked by deep transverse grooves separated by plate-like ridges, largest dorsally, decreasing ventrally; the edge of each ridge is thickened and turned dorsally. The structure is a formation of the subradular membrane and the anterior edge of the radular membrane is visible, overlying it dorsally.

Bottom right: *Acmaea tessulata*, right side of everted licker, anterior to the right, dorsal above. The surface is somewhat irregularly covered with scales. The anterior tip of the radular membrane and a few teeth are visible above on the left and some muscle fibres displaced in preparation on the right. All bars = 100 μm.

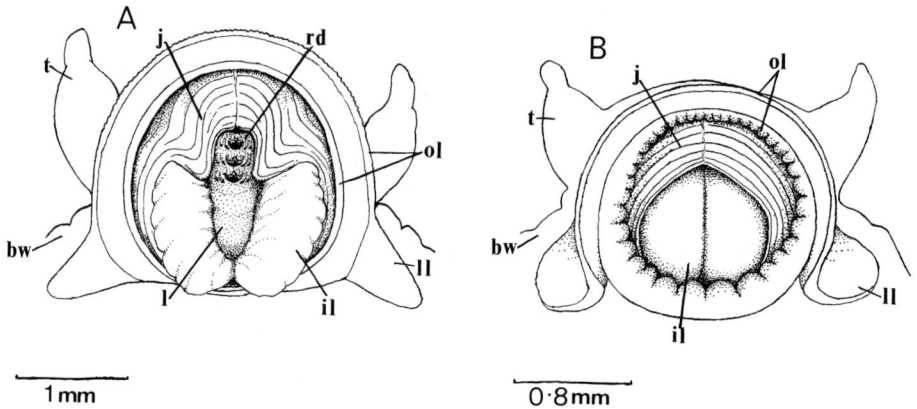

FIG. 316.—Oral view of the head of A, *Lepeta caeca*, showing licker at odontophoral tip; B, *Iothia fulva*, with mouth open and inner lips approximated.
    bw, body wall; il, inner lip; j, jaw; l, licker; ll, labial lobe; ol, outer lip; rd, radula; t, tentacle.

grains. Jensen & Siegismund (1980) have shown that the diatoms are more important elements of the diet of the hydrobiids than the bacteria; Skoog (1978) demonstrated that this also applies to *Theodoxus fluviatilis* and Whitlatch & Obretski (1980) that it is true for some cerithiids, so that it may be valid for many small animals with this habit. In the deep sea, however, where diatoms are absent, such prosobranchs as the limpets associated with hydrothermal vents rely largely on bacterial mats for food (Jannasch & Wirsen, 1981). Cocculiniform deep sea limpets, however, live where bacterial populations are not found and feed on such unlikely objects as submerged wood, cephalopod beaks, crab carapaces, elasmobranch egg cases, whale skulls, and other such carrion (see, for example, Haszprunar, 1988d). Hydrobiids (Lopez & Kofoed, 1980) have strict limits to the size of particle which they can ingest, perhaps correlated with their own size and with an unusual method of feeding which they sometimes use and which allows them to utilize food on particles too big to be swallowed. They take sand grains into the buccal cavity, rasp them clear of adherent food and then reject them, a method of feeding which Lopez & Kofoed (1980) called epipsammic browsing. Their observations explain the earlier ones of Fenchel et al. (1975) who found a consistently higher diatom content in the gut of hydrobiids than in the sediments from which they were feeding.

It is unlikely that the feeding of any prosobranch is totally random, though material of little or no nutritive value is commonly ingested incidentally. In the initial stages of a feeding foray there may be some degree of randomness in their movement, or it may even be totally random, but the actual finding of food is almost certainly a response to the detection of edible material. This is well illustrated by the observations of Hughes (1986) on *Galeodea echinophora*, which moves, apparently randomly, over sand until it detects a buried echinoderm, when it stops and attacks. The primary distance receptor for the information on which this behaviour is based is the chemosensory osphradium (Haszprunar, 1985a) bathed in the water which enters the mantle cavity. In more primitive prosobranchs the water is what comes, rather randomly, from the space in front of the animal, but when, as in many tonnoideans and neogastropods, the mantle edge forms a siphon, the snail is able to test water reaching it from any selected direction by turning its siphon or its shell to face it.

This ability is obviously more important to a carnivore whose food, unlike that of a herbivore, is usually less abundant, more mobile, and sometimes capable of taking avoiding action (Menge, 1974; Black, 1978). Correlated differences affect the degree of development of the osphradium in herbivores and carnivores. J. D. Taylor & Miller (1989) have shown that in most herbivorous prosobranchs the osphradial surface is relatively small, since it is only that of a ridge running parallel to the gill axis, as in *Littorina*. In some cerithiids (Houbrick, 1974, 1978, 1985), mainly feeding on detritus and microalgae, the surface is enlarged since the axis is extended on each side into a series of leaflets; in many carnivorous prosobranchs this development is more pronounced and each leaflet is often grooved. There seems to be a positive correlation between the size and complexity of osphradial structure [and perhaps also of its histology (Haszprunar, 1985a)] and the nature of the prey which an animal seeks. Thus the osphradium of *Nucella*, which takes attached and easily located prey such as barnacles, is distinctly less elaborate than that of naticids or nassariids which have to hunt for mobile prey, and the osphradium of *Conus striatus*, which catches fish, is the largest and most complex of all so far described. In agreement with this Lubchenco (1987), for example, found that herbivores in an aquarium seemed to show little or no ability to sense their food from a distance. In the field, however, Lubchenco (1978), Steneck & Watling (1982), Watson & Norton (1985, 1987), and G. A. Williams (1990) have all shown how littorinids have very marked food preferences, based (in *L. littorea*: Imrie *et al.*, 1989) on the perception of chemical signals. Much more precise information is available for carnivorous or carrion-feeding prosobranchs, particularly the mud snail *Ilyanassa obsoleta* (M. Crisp, 1969). This animal everts its proboscis readily when appropriate chemicals stimulate its osphradium (Carr, 1967a) and so has provided a ready test for its distance chemosensitivity to a variety of substances (Carr, 1967b). Most of these have proved to be protein or peptide in nature (Gurlin & Carr, 1971; Carr *et al.*, 1974). These substances provide the first clue to a predator of the near presence of prey; secondary information for its precise location comes from the anterior edge of the foot and the lips (Brock, 1936; Pearce & Thorson, 1967; M. Crisp, 1971; Nielsen, 1975; B. A. Miller, 1975, 1979) but as these bear only contact receptors they are valuable only in the final stages of the search.

The ability to detect signals emanating from potential food is, as would be expected, innate: for example, Rittschof *et al.* (1983) showed that newly hatched oyster drills (*Urosalpinx cinerea*), obviously without any experience of prey detection, responded to water in which barnacles (a common food) were living by migrating towards its source and were still able to recognize it after a 200-fold dilution. Odours may also have a pheromonal function, carrying further information to the hunting predator: Pratt (1976) found that specimens of *U. cinerea* that had been starved for a month chose to migrate towards neutral water rather than towards water from starved tingles, but moved towards water from well fed animals rather than neutral water.

The development of a carnivorous habit has occurred in several groups of prosobranchs, from those regarded as primitive to those counted as the most advanced, and has been so successful that gastropods are the most abundant invertebrate predators on some beaches. In its more highly developed forms the carnivorous habit involves a more complex series of behavioural patterns than does algal grazing. Nevertheless the habit probably evolved from the generalized browsing of primitive gastropods which must often have included some animal material; by selection of appropriate feeding areas this may have become the main component of the diet. Food taken in this way can consist only of attached organisms such as sponges, coelenterates, bryozoans, barnacles and tunicates, and these are indeed what the least specialized gastropod carnivores take. Some species of *Calliostoma* (Savini-Plawén, 1972a; Perron, 1975; Perron & Turner, 1978) and *Velutina* (Ankel, 1936) eat hydroids, whilst *Trivia* (Fretter, 1951a) rasps and swallows bits of tunicate test, though it cannot digest these, to reach the polyps, which it can. The development of a proboscis opens up new feeding opportunities: it allows entry to crevices where more rewarding prey lurks, gives direct access to the inner organs of such things as tunicate polyps (*Erato*: Fretter, 1951a), and allows such predators as *Buccinum undatum* (Nielsen, 1975) to feed through a gap as small as 1·3 mm which has been forced between the valves of a bivalve shell.

The evolution of a proboscis of gradually increasing length has permitted many proso-branchs to adopt an ectoparasitic mode of life. A short one allows feeding on the surface epithelium of a host, like the eulimid *Pelseneeria* on echinoids (Ankel, 1938); with increased length penetration of the body cavity of a host becomes possible, as in many other eulimids associated with echinoderms (Lützen & Nielsen, 1975; Warén, 1983) and in the blood-sucking pyramidellids (Fretter & Graham, 1949). A blood-sucking habit has also been demonstrated in some neogastropods. One such group is the cancellariids, which possess a long, narrow proboscis containing a peculiar radula in which the teeth resemble blades of grass with serrated edges at their tips. The feeding habits of these animals have long been a puzzle, but O'Sullivan et al. (1987) have recently described how one species, *Cancellaria cooperi*, uses them to puncture the skin of electric rays and suck blood. Other species may feed in the same way. Bouchet (1989b) has described a marginellid which draws blood from sleeping fish, though its radula is less specialized.

For such things to be possible a proboscis must be narrow, and for efficient working the odontophore must be protrusible through the mouth at its tip when it is extended. There may well have been an evolutionary link between the development of a proboscis and the oligomerization of the radula through the prosobranch series. Even where, as in pteno-glossans, the radula remains (or has become) wide, it takes the form of two halves which collapse on one another until the buccal cavity is everted during feeding; only then do the halves spread. A proboscis is not found in rhipidoglossans: their wide radula and odontophore may well have proved incompatible with such a structure.

The head of the typical prosobranch ends anteriorly in a snout projecting in front of the cephalic tentacles for a short distance and carrying the mouth at its tip. It can be shortened slightly by contraction of cephalic branches of the columellar muscle. A proboscis is pro-duced by great elongation of this snout, accompanied by the development of the ability to withdraw the proboscis into the body when not used for feeding. Since the muscles originally responsible for such retraction of the snout as was possible insert on the lips of the mouth it seems likely that the primitive proboscis would then be invaginated, with the mouth coming to lie at the innermost end of an introvert: this is the type of proboscis known as acrembolic. The opening on the surface of a snail with this type of proboscis when it is not feeding is that of the introvert, commonly called the rhynchostome: when the proboscis is everted the introvert is abolished and the mouth and buccal cavity lie at its tip.

An apparently more effective type of proboscis, since it involves shorter lengths of retrac-tor muscle, is the pleurembolic: here the main bundles of the retractor muscles no longer insert on the lips of the mouth (though some others still persist there) but about halfway along the length of the proboscis. The result of their contraction is that it is this point, not the tip, which comes to lie at the deepest part of the introverted area, and the distal half of the

THE FEEDING OF PROSOBRANCHS

proboscis with the true mouth at its tips lies, shortened but not invaginated, within the introvert, proboscis sac or rhynchodaeum, as it is commonly known. In the non-feeding animal the apparent mouth is still the rhynchostome; when the proboscis is used in feeding, however, the mouth and the buccal cavity come to lie at its tip as in the acrembolic type, allowing the radula to be used in the same way as in animals without a proboscis. In both types of proboscis the anterior part of the gut, usually the oesophagus, shares in the elongation of this part of the body; it is therefore flung into coils when the proboscis is retracted and straightens when it is extended.

Some confusion exists over the nomenclature of the parts involved here, especially in relation to the anatomy of the toxoglossan proboscis. Thus the term rhynchodaeum is used as the name of the space in which a retracted pleurembolic proboscis lies by E. H. Smith (1967a), Sheridan et al. (1973), and J. A. Miller (1989). Kantor & Sysoev (1989), however, call this space (appropriately enough) the rhynchocoel, and use rhynchodaeum for its outer lining, attached to the body wall, a name for which the etymology of the word does not adapt it.

In addition to the acrembolic and pleurembolic types a number of other patterns of proboscis structure have recently been described, and further information about the familiar patterns has been presented by J. D Taylor et al. (1983). Day (1969), Houbrick & Fretter (1969), and Hughes & Hughes (1981) have described the pleurembolic proboscis found in the meso-gastropod tonnoideans. In these the buccal cavity and the odontophore lie at the summit of the extended proboscis when the animal is feeding and the radula can thus be protruded through the mouth and used to rasp the prey, which is either an echinoderm or another gastropod.

Discussions of the type of proboscis found in neogastropods have been given by Ponder (1973), and for toxoglossans by E. H. Smith (1967a), Sheridan et al. (1973), B. A. Miller (1970, 1975, 1979), Kantor (1987, 1988, 1990b), Kanot & Sysoev (1989, 1990), J. A. Miller (1989), and J. D. Taylor & Miller (1990).

New information allows better understanding of the relation between proboscis structure and feeding methods than was previously possible. Thus the length of the extended pro-boscis may be correlated with the food which the mollusc eats and the manner in which it does so: the proboscis is short in such animals as olivids and volutids (Ponder, 1973), which hold their prey in the foot, and in naticids, which hold it similarly whilst it is being drilled. A long proboscis is found in those which explore crevices or the depth of soft substrata for prey or carrion, such as buccinids, nassariids, and fasciolariids. When great length is called for the proboscis sac or rhynchodaeum, in which the proboscis is coiled and stored on retraction, may have to be dilated to accommodate it.

The situation which Kantor (1985) described in species of the buccinid Volutopsius as a new type of proboscis is only slightly different from a normal pleurembolic proboscis, and enables the mollusc to reach the deeper parts of the body of its echinoderm prey. In these animals, however, the walls of the buccal cavity are everted at the tip of the proboscis so that the mouth disappears, and the visible external entrance to the gut becomes the opening from the buccal cavity to the oesophagus. The odontophore and radula then lie at the tip of the extruded buccal wall so that food can be rasped and pulled into the body on the return of the buccal region to its normal disposition. This arrangement, except for the fact that it occurs at the tip of an otherwise normal pleurembolic proboscis, offers a precise parallel to that shown by janthinids (see p. 621 and fig. 322) during feeding (Graham, 1965), and which has also been shown to occur in the pulmonate Testacella maugei (Crampton, 1975).

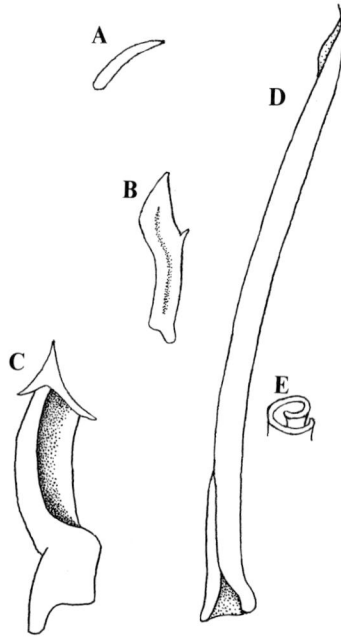

FIG. 317.—Diagrams of a series of marginal teeth from genera within the family Turridae suggesting their evolution. (Largely after Shimek & Kohn, 1981). Teeth are oriented so that their tips are uppermost. A, *Calliclava albolaqueata*, slicing tooth; B, *Hindsiclava militaris*, slicing tooth with longitudinal groove; C, *Imiclava unimaculata*, slicing–stabbing tooth with increased development of the folds bordering the longitudinal groove (which is now pronounced), with sharp apex, and barbs for anchoring tooth in prey; D, *Ophiodermella inermis*, simple unbarbed hyperdermic tooth with complete enclosure of central canal, its elongation increasing the amount of venom which will be delivered; E, transverse section of shaft of D. All teeth except that shown in A are detachable from the radular membrane. A, B, C, × 200; D, × 90.

Recent work on the proboscis of toxoglossans has revealed a surprising range of structure, and occasional doubt as to the interpretation of the parts. Comparisons are not eased by the variable nomenclature used by different workers, a situation which J. A. Miller (1989) has attempted to improve, nor by the fact that few species have been seen feeding, so that ascribed function is largely deduced. The evolution of the toxoglossan proboscis has presumably occurred in parallel with that of their remarkable radula, which may therefore be summarized first.

The most extreme modification of radular structure amongst carnivorous prosobranchs is undoubtedly that found in the neogastropod conoideans (that is mainly the turrids, cones and terebrids) (fig. 317). The biology and detailed anatomy of this large prosobranch group, which probably contains several thousand species, is very inadequately known, but as our knowledge of it increases, largely due to work by W. F. Ponder, J. D. Taylor, and Y. Kantor, it proves to be much more diverse than was originally supposed. See also Barinaga (1990).

Most members of the group possess the type of radula known as toxoglossate, with which some species are able to capture such active prey as fish, though most are content with such slower animals as worms, enteropneusts, or mollusc (see list of food given by J. A. Miller, 1989).

The ultimate refinements of the toxoglossate radula are associated with, and dependent upon, the presence of a poison gland, a derivative of the oesophageal glands of other prosobranchs (see p. 598). Evolutionary schemes for this type of radula have been proposed by Shimek & Kohn (1981), J. D. Taylor & Morris (1988), and Kantor (1990b). These authors assume that it arose from an ancestral taenioglossate radula rather than the rachiglossate one characteristic of other neogastropod groups. Details of tooth structure have also been given by Marsh (1977), Mills (1977), James (1980), Starobogatov (1990), J. D. Taylor & Miller (1990), and Kantor & Taylor (1991), and the development of the hollow teeth of cones described by Nybakken & Perron (1988). In the most primitive toxoglossans (family Turridae, subfamily Pseudomelatominae) the radula is set on a well formed odontophore (Kantor, 1988) and has five teeth per row, a large rachidian with two lateral teeth on each side, all attached to the radular membrane and providing powerful cutting blades. Since this radular pattern is limited to this subfamily it suggests that it represents an evolutionary line distinct from other toxoglossans (Kantor, 1990b). In the subfamilies Clavinae (also called Drillinae) and Turriculinae the central rachidian tooth has been lost, and with further evolutionary advance the inner lateral teeth also disappear. A further stage is represented by animals in which the remaining teeth, the outer laterals, are lost alternately on the left and right is successive rows, giving a radula in which each row contains only a single tooth. The most advanced arrangement is found only in mangeliine turrids, the common toxoglossans of the North East Atlantic, and in the tropical cone shells and some terebrids. In these the attachment of each tooth to the radular membrane is reduced to a thread-like ligament which breaks when the tooth is about to be used, allowing it to lie free in the buccal cavity.

The anatomy of the buccal cavity is also highly modified by the formation of a diverticulum, which we call the odontophoral sac (fig. 318), which opens posteriorly from the main part of the cavity. In this sac, in the primitive turrids described by Kantor (1988) and Kantor & Sysoev (1990), lies a well formed odontophore, and to it open the salivary gland ducts and the radular sac. In more advanced forms such as the turrid *Haedropleura septangularis* (Sheridan *et al.* (1973) the odontophore is much reduced though still supported by cartilages, and a similar arrangement is present in the terebrid *Hastula bacillus* (J. D. Taylor & Miller, 1990). In the most advanced forms the odontophore becomes vestigial, though it may still be recognized as such in the form of a pad of connective tissue (*Conus flavidus*: Marsh, 1977). Whatever its nature, however, the mature teeth from the radular sac come to lie over its surface, attached to it by the radular membrane or by the ligaments which represent that, and are stored there until used in feeding.

Accompanying these evolutionary steps is another series, affecting the structure of the teeth. Teeth in primitive toxoglossan genera (Shimek & Kohn, 1981; Kantor & Sysoev, 1989, 1990; Kantor, 1990b), have on the whole the same general shape as those of a taenioglossate radula stripped of its marginal teeth, and lie on the odontophore, forming what Kantor (1990b) calls the 'whole radula'. They are used to cut up the prey after ingestion and whilst it lies within the buccal cavity. In the more advanced toxoglossan genera the teeth which represent the reduced radula have become rolled to form scrolls with a sharply pointed tip and barbed edges, and are used in a different manner. The teeth often also become very elongate: Kohn *et al.* (1960) illustrated one 10·9 mm long from *Conus striatus*, a species which may have a shell length of up to 120–125 mm. Prior to use a tooth is detached from the odontophore and passed into the main buccal cavity. At the same time, presumably, secretion from the poison gland is also released to the cavity and fills the central space of the tooth, while some clings to its surface. When the animal is ready to feed the charged tooth is

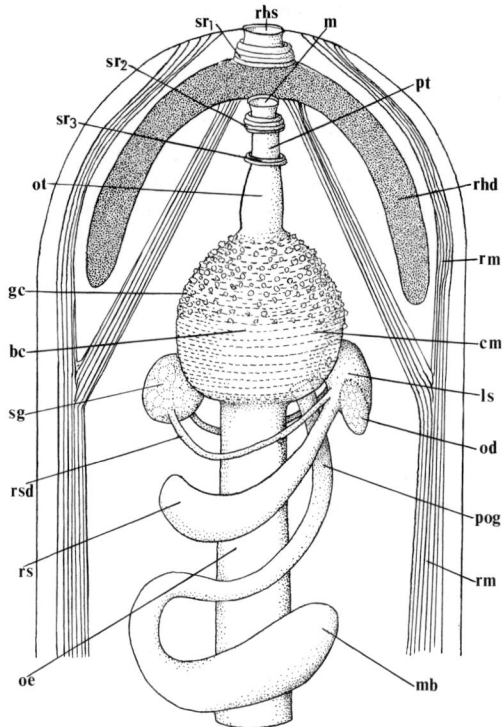

FIG. 318.—Diagram of the proboscis and anterior end of the gut of the turrid *Haedropleura septangularis*. (Based on Sheridan *et al.*, 1973).

bc, buccal cavity; cm, circular muscles; gc, protein gland cells; ls, 'ligament sac' (odontophoral sac); m, mouth; mb, muscular bulb of poison gland; od, odontophore in odontophoral sac opening to buccal cavity; oe, oesophagus; ot, oral tube; pog, poison gland opening to buccal cavity; pt, part of oral tube occupied by radular tooth during attack on prey; rhd, rhynchodaeum; rhs, rhynchostome; rm, retractor muscles of proboscis; rs, radular sac opening to odontophoral sac; rsd, right salivary duct opening to odontophoral sac; the opening of the left duct can be seen by transparency; sg, fused salivary glands; sr₁, sphincter closing rhynchostome; sr₂, oral sphincter; sr₃ sphincter closing pt from rest of oral tube.

manipulated forwards until it is held in a pocket at the tip of the proboscis, gripped by the lips of the mouth and prevented from slipping back by a sphincter muscle at the base of the pocket. The proboscis, thus armed, is then shot out rapidly, the prey stabbed and injected with the poison. This contains a complex mixture of neuropeptides, the composition varying from species to species to some extent in relation to the type of prey attacked. Each peptide apparently affects a different ion channel or receptor target in the neuromuscular system of the prey and brings about rapid immobilization (Olivera *et al.*, 1990), permitting its leisurely ingestion whole.

The comparative anatomy of the anterior part of the conoidean gut in advanced forms has not always been clearly understood. The radular sac proper and the odontophoral sac have often been treated as limbs of the radular sac, its 'long arm' and 'short arm' respectively, meeting at right angles; some authors have termed the latter the 'ligament sac'. Examination of what occurs in these parts, however, clearly demonstrates their true nature. Radular teeth are secreted at the inner end of the 'long arm' as flexible structures; they migrate towards the point where it joins the 'short arm' (or 'ligament sac'), becoming hardened during their passage by the activity of the cells on the wall opposite that carrying the radular ribbon. These events are therefore precisely parallel to those occurring during the movement of teeth along the radular sac of typical prosobranchs, and indicate that the 'long arm' is the homologue of that structure. At the mouth of this true radular sac the teeth enter the 'short arm' and lie over the pad of tissue representing the odontophore (Marsh, 1977), as they do in more typical prosobranchs. The 'short arm' (or 'ligament sac') is thus revealed as part of the buccal cavity secondarily separated from the main chamber; it is what is called the odontophoral sac above.

The anatomical arrangements just described for conoideans are surprisingly closely duplicated in some other neogastropod groups. Some marginellids, for example, posses an apparatus comprising a muscular bulb connected by a long, coiled, glandular tube to the buccal cavity (Graham, 1966; Ponder, 1970a), so nearly identical in appearance with the poison gland of toxoglossans as to suggest that it probably has a similar function. The absence of a valve and a gland of Leiblein makes it likely, too, that it could have arisen only by a similar process (see p. 598). Further resemblances also appear in some of these snails in that in species in which an odontophore is developed (fig. 319), it lies not in the main bucccal cavity but in a separate sac opening posteriorly from it. The presumably more primitive species possess a well developed muscular odontophore, whereas the more advanced ones have none, though still having an odontophoral sac. How the secretion of the poison gland (if that is what it is) is used to affect the prey is not known; in species with an odontophore the prey may well be drawn into the buccal cavity by the everted odontophore, whilst in those without, it must be manipulated by the lips of the snout. In both cases envenomation would occur in the buccal cavity.

In mitrid and the closely related costellariid neogastropods (Ponder, 1972), which feed primarily on sipunculans (J. D. Taylor, 1989), the salivary glands open at the tip of a protrusible and retractile structure known as an epiproboscis. This may be used to deposit saliva on the prey (Maes, 1971), especially on areas where toxins (if these are present) might penetrate. In the genus Thala, if the observations of Maes & Raeihle (1975) are substantiated, this can apparently take place without rupture of the skin, inducing rapid immobility and allowing easy ingestion. The salivary glands of some other neogastropods, most notoriously those of the red whelk, Neptunea antiqua, contain a poison (Fänge, 1957, 1958) (tetramine) and have been proved responsible for a number of cases of poisoning among human beings (Fleming, 1971).

In 1967 E. H. Smith (1967a) described two types of proboscis from toxoglossans which he called intraembolic and polyembolic (fig. 320). Their occurrance and possible evolution have been discussed by Sheridan et al. (1973), J. A. Miller (1989), Kantor (1990b), and Kantor & Sysoev (1990). The former type, found in simple or modified form in many conoideans, has appeared to be the more primitive—it is the basic conoidean form according to J. A. Miller (1989)—and, according to Sheridan et al. (1973) and Kantor (1990b) may be directly derived from an acrembolic type, since the retractor muscles are inserted at the lips of the mouth. It differs from that type, however, in three ways: (1) introversion of the proboscis is never total and its distal part remains uninverted though withdrawn into the rhynchodaeum, as in the pleurembolic type; (2) the outer wall of the rhynchodaeum never wholly straightens on protrusion of the proboscis, as it does in the pleurembolic type where it contributes to the

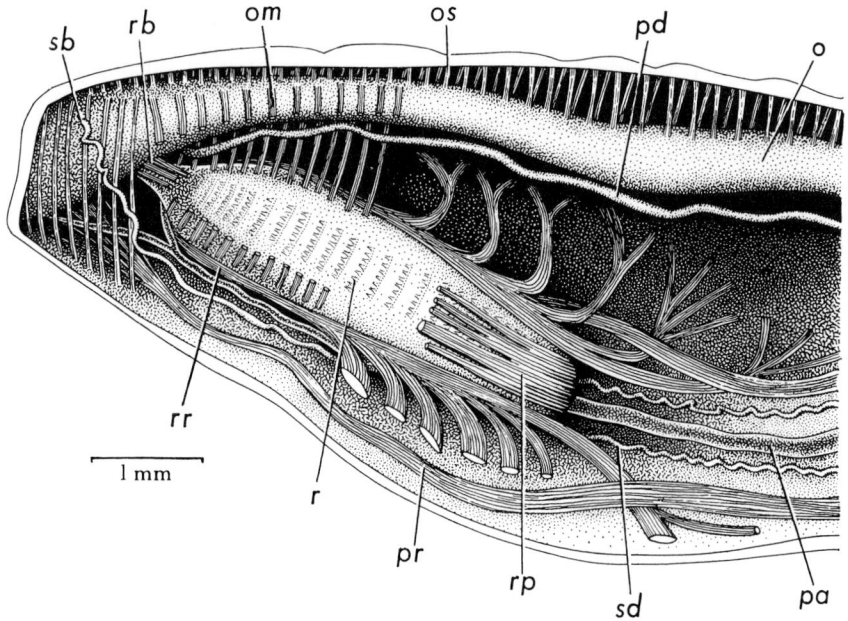

FIG. 319.—*Marginella* sp. Dissection of the anterior part of the proboscis, showing the odontophore with the radula withdrawn into the odontophoral sac.

o, oesophagus; os, suspensory muscle of oesophagus; pa, proboscis artery; pd, duct of poison gland; pr, retractor muscle of proboscis; r, radula on odontophore; rb, opening of odontophoral sac to buccal cavity; rp, radular protractor muscle; rr, radular retractor muscle; sb, opening of salivary duct to buccal cavity; sd, salivary duct.

length, but remains as an external support round the proboscis base; (3) the retractor muscles originate in the foot, not as part of the columellar muscle. In most conoideans which possess this type of proboscis the odontophore and radula remain at the base of the extended proboscis, lying in a buccal region connected to the mouth by a long oral tube, and so cannot be extruded to rasp prey. In forms such as the pseudomelatomine turrids, in which the radular teeth are not modified, the prey must therefore be captured by the proboscis, passed down the oral tube to the buccal cavity and there torn by the radular teeth moving in typical prosobranch fashion over the odontophore after its eversion from the odontophoral sac (Maes, 1971; Kantor, 1987, 1990b; Kantor & Sysoev, 1990). Envenomation may occur as the prey passes along the oral tube or even as it is captured, which would make manipulation easier. This does occur in other genera such as the turriculine and mangeliine turrids and the cones, in which radular teeth are modified and detachable: they are moved to the lips of the proboscis to immobilize prey but their structure prevents their use in tearing it up: this implies that prey is swallowed whole. Digestion probably begins in the buccal cavity with enzymes from the salivary glands or the digestive gland which, in the absence of a valve of Leiblein, are free to move forwards. J. D. Taylor (1985), however, has described two species of turriculine turrid in which the radula, which has retained unmodified teeth, can be brought to the mouth by eversion of a shortened oral tube and used there to rasp prey.

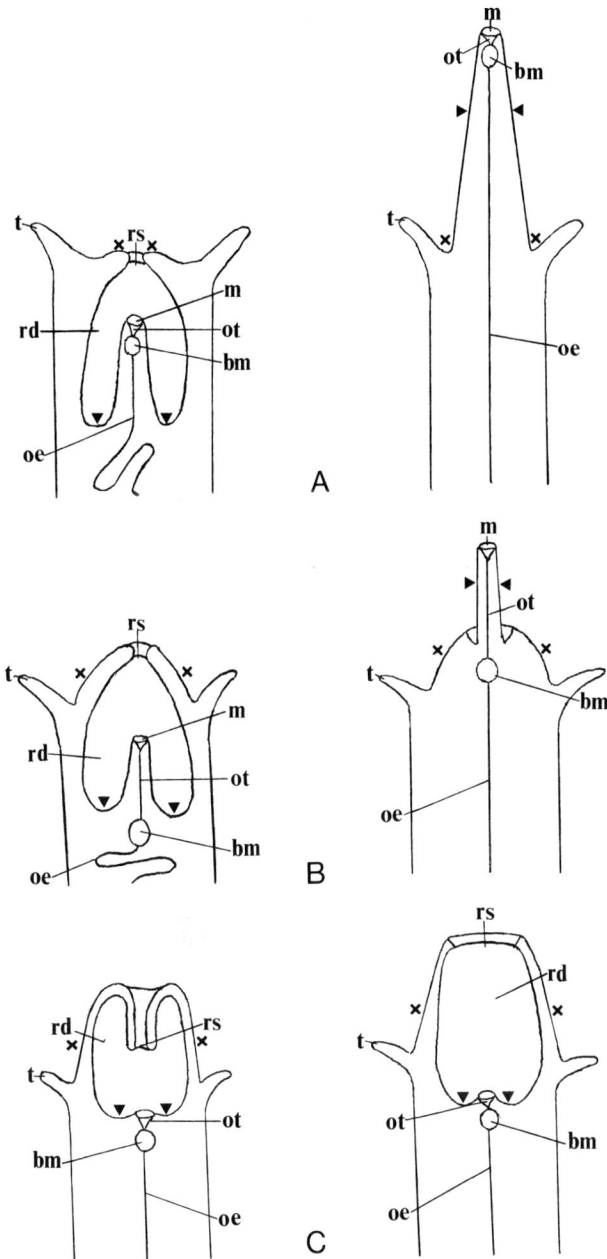

FIG. 320.—Diagrams showing the basic structure of A, a pleurembolic proboscis; B, an intraembolic proboscis; C, a polyembolic proboscis. In each set the left diagram shows the proboscis retracted and the right the proboscis extended. Arrowheads and crosses mark corresponding anatomical points.

bm, buccal mass; m, mouth; oe, oesophagus; ot, oral tube; rd, rhynchodaeum; rs, rhynchostome; t, cephalic tentacle.

A more complex version of the intraembolic proboscis has been described (E. H. Smith, 1967a; Sheridan et al., 1973) in some species of Mangelia. Here a fold arises between the proboscis in the centre of the rhynchodaeum and the rhynchodaeal wall, probably stabilizing the proboscis during food capture. This fold may be either a new structure arising from the rhynchodaeal floor or a mere reduplication of the proboscis base: it cannot, probably, be ascertained how, or from what, it has evolved since all the structures (proboscis, fold, rhynchodaeal wall) are morphologically part of a pretentacular snout, with indentical structure and innervation.

The second type of proboscis described by E. H. Smith (1967a) is the polyembolic, which is related to a different method of feeding and may have had an intraembolic origin. The principal change in the evolution of this type seems to have been the gradual decrease in importance of the original primary proboscis and of the fold around it, so that the mouth comes to lie on the flat base of the rhynchodaeum. The prehensile part of the polyembolic proboscis is now formed from the rhynchodaeal and body walls, which have elongated to form a secondary proboscis, also called a pseudoproboscis by Rudman (1969) and a labial tube by B. A. Miller (1979) and J. A. Miller (1989), opening at the rhynchostome. Two variants of this type may be mentioned. In one, shown by the turrid Philbertia purpurea, which has detachable radular teeth and a poison gland, feeding probably involves manipulation of a tooth laden with poison to the mouth. It may be held here to stab prey in the rhynchodaeum, or possibly passed through the mouth, gripped by the rhynchostomial lips and used there to stab prey. In either case the prey is probably digested within the rhynchodaeum by regurgitated enzymes, since the radula cannot shred it. The second arrangement, perhaps derived from the first, is found in the turrids P. linearis and the genus Cenodagreutes, and accompanies a loss of poison gland and radula. Here prey, presumably rather slow-moving, is manipulated by the rhynchodaeal walls and held for digestion in the rhynchodaeum. In both types, with the need for mobility and the provision of a roomy rhynchodaeum, the rhynchodaeal walls elongate and, just as the primary proboscis is withdrawn when not used in feeding, so is the secondary proboscis, the distal parts of which are folded into the rhynchodaeum

A parallel evolution in proboscis structure (fig. 321) has been described by B. A. Miller (1970, 1975, 1979) and extended by J. D. Taylor (1990) and J. D. Taylor & Miller (1990) in the family Terebridae, leading from forms such as Hastula inconstans, with an intraembolic proboscis, poison gland and radula to Terebra gouldi, with a secondary rhynchodaeal proboscis and neither radula nor poison gland. In addition, however, some terebrid species [Terebra affinis (B. A. Miller, 1970); Hastula bacillus (J. D. Taylor & Miller, 1990)] have developed a structure not encountered, so far as is known, in other conoideans, an accessory proboscis. This arises as a tentaculiform process from the inner side of the rhynchodaeal wall and can be extended through the rhynchostome. Its structure is that of a muscular hydrostat (Kier & Smith, 1985) and it is mobile and probably chemosensory (J. D. Taylor, 1990; J. D. Taylor & Miller, 1990). Its function is therefore most likely to probe the substratum in search of the small polychaetes which these animals eat.

The effects of these variations on conoidean classification, possibly serious, have not yet been examined.

Retraction of an extended probiscis is achieved by the contraction of well developed retractor muscles which are morphologically slips of the columellar muscle, though they often lose their original attachment to the shell for a more anterior one on the body wall or foot. The method of extrusion of the proboscis is less obvious, though it has always been assumed to be by means of blood pressure. This has been demonstrated by Trueman &

Brown (1987a) in *Bullia*, and there is little reason to suppose that the mechanism they describe is not also found in other groups. They compare the proboscis hydrodynamically to the body of an annelid, in particular to an aseptate region, and show that elongation is due to the simultaneous contraction of the circular muscles of its wall and relaxation of the longitudinal. In the absence of septa, however, such activity could bring about elongation both forwards and backwards, the latter until the pressure within the proboscis haemocoel equalled that in the cephalic sinus. To some extent contraction of the circular musclees at the base of the proboscis closes the passage from proboscis to head and so functionally replaces a septum, but the pressure within the proboscis haemocoel responsible for its extension can never exceed that in the cephalic sinus without bringing about some backward elongation and so undoing the desired effect.

Kier & Smith (1985) and Kier (1989) have shown that in muscular hydrostats (extensible organs such as tentacles which act by antagonistic muscles and have neither hard nor fluid skeletons) the rate of extension is related to the resting length–width ratio. On the assumption that this rule also applies to organs with a hydrostatic skeleton Trueman & Brown (1987a) calculated that in *Bullia*, with a ratio equal to 8:1, the proboscis should extend with some rapidity. In *Bullia* and other nassariids which do not seek active prey this speed is obviously adequate: in cones and turrids, however, whose success as predators is more dependent upon rapid extrusion of the proboscis, speed may have to be increased by changing the ratio or by some other device. In the absence of definite measurements it is not clear whether the modified proboscis structures described by E. H. Smith (1967a) would produce this effect or not.

More methods of attack on living prey than simple grazing, but less specialized than those of conoideans are found in some prosobranchs. The development of powerful jaws is one such: a radula working against these allows biting, though the jaws themselves only occasionally act as scissors. Species of *Janthina* (Graham, 1965) and of *Epitonium* (Perron, 1978) respectively bite pieces off the pelagic coelenterates *Physalia*, *Porpita* and *Velella*, and off various anemones. These animals possess the type of radula known as ptenoglossate which is characterized by the possession of numerous similar, backwardly pointing, fang-like teeth in each row; it is carried on a deeply cleft odontophore, an arrangement which allows the pieces bitten off to be grasped and drawn into the buccal cavity. To achieve this the entire buccal wall of *Janthina* is everted during feeding so that the odontophore and the radula which it bears project from the anterior end of the head (fig. 322). In *Epitonium* species salivary ducts open on cuticularized projections near the tip of the everted proboscis in such a way as to suggest that the saliva contains some narcotizing substance facilitating the attack on the prey. Though all *Epitonium* species seem to exploit coelenterate prey (Robertson, 1963, 1970, 1983) each seems to have its own special method, some gnawing the column wall (Perron, 1978), others attacking the tentacles (Sisson, 1986), still others boring through the column to reach the coelenteron (Peterson & Black, 1986).

A further device which gives the prosobranch carnivore access to prey which is protected by a calcareous shell is the ability to bore, a skill limited to capulids, naticids, and to the muricid neogastropods (Kabat, 1990). The mechanism of boring in muricid prosobranchs has been very thoroughly explored in the oyster drill, *Urosalpinx cinerea*, by Carriker and his co-workers, and he has summarized the conclusions reached (Carriker, 1981).

The two parts of the body involved are the radula and the accessory boring organ (ABO), which is lodged in a pit on the sole of the foot in muricids and on the ventral tip of the proboscis in naticids, which also bore. Most of the rasping in *Urosalpinx* is done by the

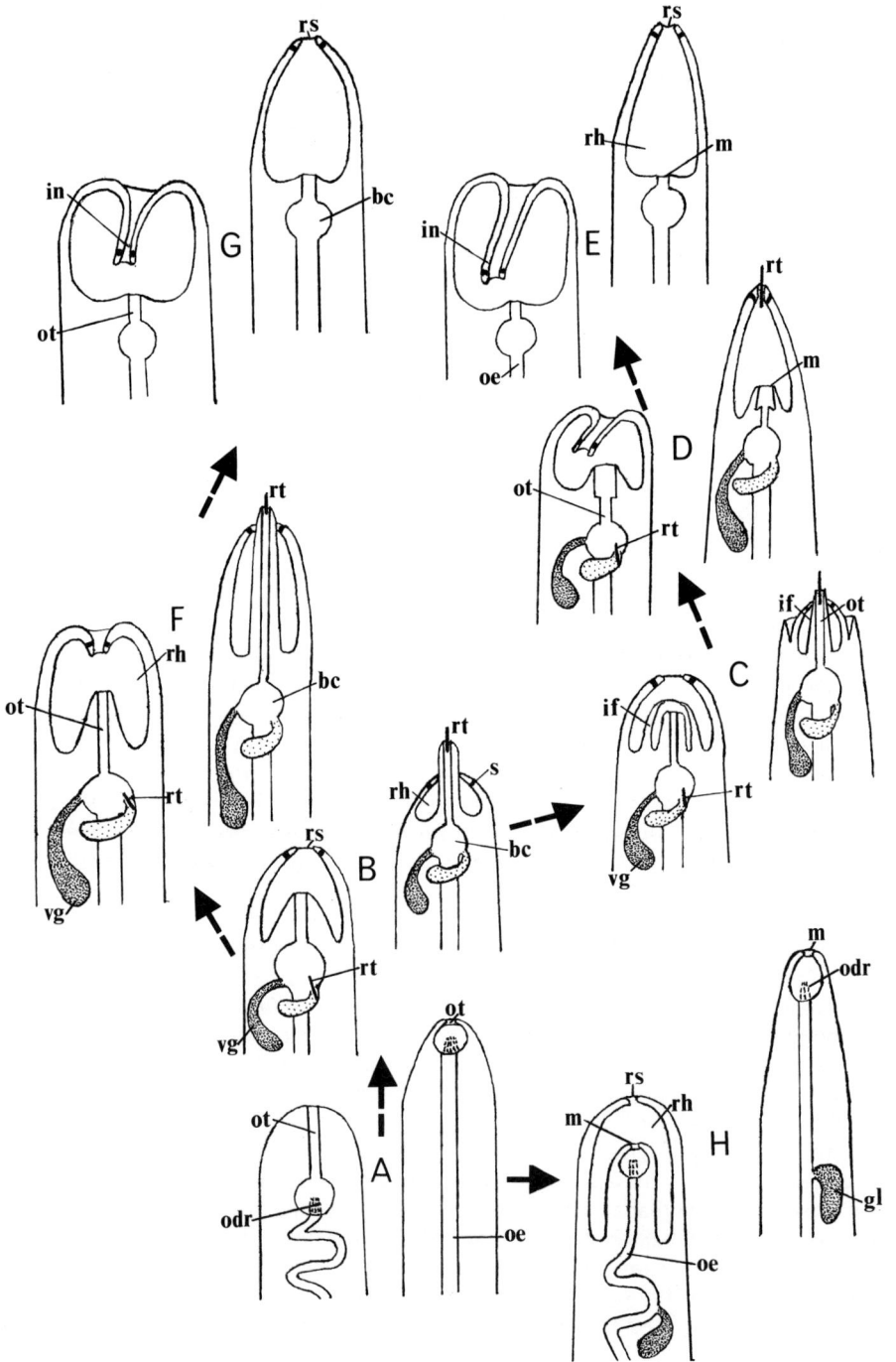

rachidian teeth of the radula, their marks on the shells of the prey running in many directions due to rotation of the odontophore. Each stroke of the radula occupies about one second, and strokes are usually grouped in minute–long bursts (Carriker, 1977). Ultimately rasping stops, and the proboscis withdrawn from the borehole; this is then wiped with the tip of the propodium, the purpose of which is uncertain. The next phase involves the ABO, which is everted from the pit on the sole of the foot and swollen with blood until it is more or less equal to the proboscis in size. It is inserted into the borehole, remains there for 30–40 minutes, and is then withdrawn. This cycle of radular activity, propodial wiping, and use of the ABO is repeated until the shell is perforated.

The chemical activity of the ABO has been largely deciphered by Carriker et al. (1963, 1978), Carriker & Williams (1978), and Carriker (1978), and further information on the histology and functioning of the ABO of Nucella lapillus has been given by Chétail et al. (1968). The attack on the shell is initiated by secretion of a proteolytic enzyme to dissolve the periostracum; further secretion, probably involving hydrochloric acid and a chelating agent, then softens the calcareous material, while other enzymes attack the conchiolin. Carbonic anhydrase may also be involved (Webb & Saleuddin 1977). Mucus prevents dispersion of the enzymes. The radular teeth of Urosalpinx are, some, harder, others, softer than the shell of the prey, but the weakness of some is not a serious disadvantage since they are largely clearing up softened shell debris. The secretion of the ABO is only mildly acid (pH 3·8–4·0) and small in volume (a few μl) so that in combination with the dense texture of molluscan shell only a slow rate of penetration can be achieved, about 0·5 mm depth of shell being removed in a day: a total of 72 hours may be required to bore through an oyster shell.

Carriker left it undecided as to whether the accessory salivary glands of Urosalpinx play any part in boring, though opening as they do at the tip of the proboscis they might seem likely to be involved. The work of Hemingway (1978) also left it uncertain what their role in feeding, if any, might be. He used what he called the 'salivary-accessory salivary gland complex' in experiments with the thaidid Acanthina and found that its extracts had some paralysing effect on the prey. It is possible, nevertheless, that secretion from the ordinary salivary glands was the effective source of this, since Fänge (1957, 1958) had found toxic substances in them in the buccinid Neptunea, which lacks accessory glands. Andrews et al. (1991), however, have found clear evidence in the accessory salivary glands of Nucella of physiologically active substances with effects on the heart and anterior byssus retractor muscle of Mytilus comparable to those of choline esters; others were found in the secretion of

FIG. 321.—Diagrams illustrating the possible derivation and evolution of the proboscis in conoidean neogastro-pods, primitive stages at the base of the diagram, advanced ones at the top. Each stage is represented by two diagrams, the left showing the proboscis in the retracted state, the right in the extended. The gland of Leiblein and the venom gland are heavily stippled, the radular sac lightly. The position of a radular tooth in more advanced genera when prey is being attacked is indicated in diagrams B, C, D, and F. A, ancestral meso-gastropod with acrembolic proboscis; B, Haedropleura; C, Mangelia, with an intraembolic proboscis; D, Philbertia; E, Cenodagreutes, which has lost poison gland and radula and has a polyembolic proboscis. B–E form a series in the family Turridae. A similar evolution is shown for the family Terebridae, represented by F, Hastula; and G, Terebra. H shows the proposed evolution of the pleurembolic proboscis of rachiglossans. [Based on Sheridan et al. (1973); Kantor (1990b); J. D. Taylor (1990)].

bc, buccal cavity; gl, gland of Leiblein; if, inner proboscis fold; in, invaginated tip of polyembolic proboscis; odr, odontophore; rh, rhynchodaeum; rs, rhynchostome; rt, radular tooth; s, rhynchostomial sphincter; vg, venom gland.

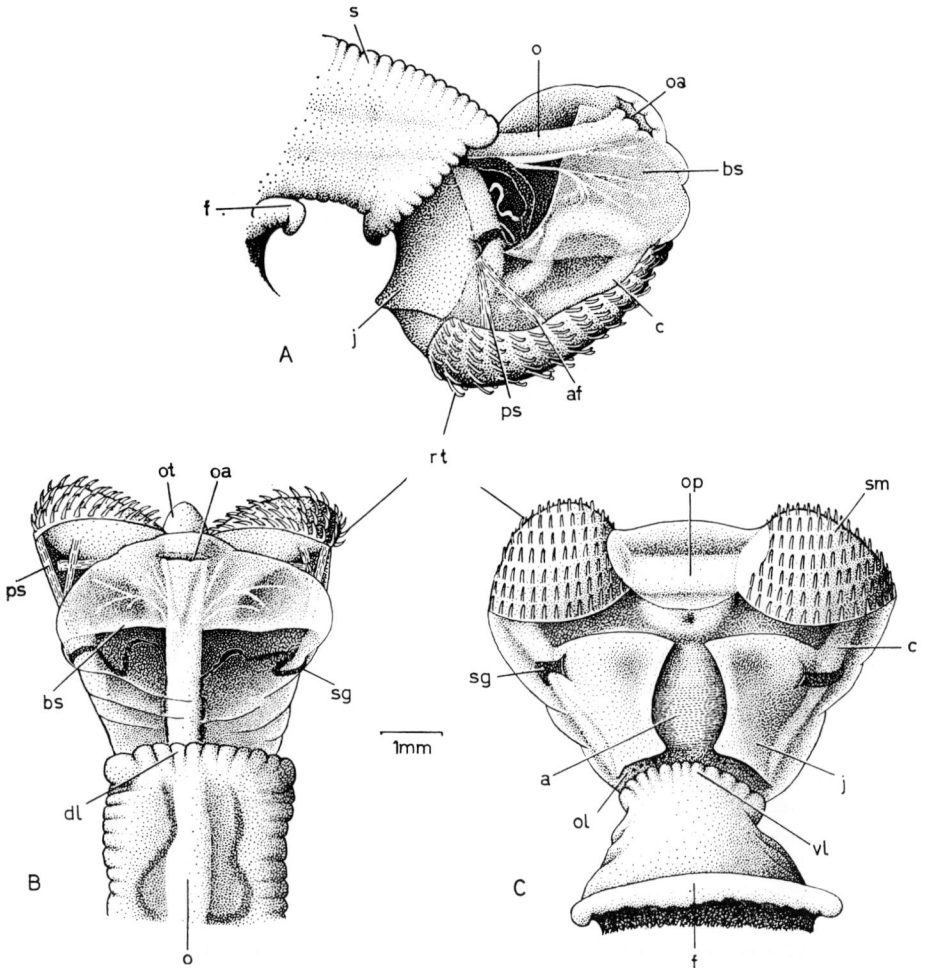

FIG. 322.—*Janthina janthina*. The snout of an animal in the feeding position, with extroverted buccal cavity. A, from the right; B, in dorsal view; C, in ventral view. From Graham (1965) with the permission of the Malacological Society of London.

a, anterior transverse muscle; af, accessory flexor muscle of the cartilage; bs, basal septum; c, odontophoral cartilage; dl, dorsal lip; f, anterior edge of foot; j, jaw; o, oesophagus with underlying retractor muscle spreading into basal septum; oa, opening of oesophagus on dorsal surface of extrovert; ol, opening of labial gland; op, outer posterior transverse muscle; ot, anterior tip of odontophore; ps, protractor muscle of the subradular membrane; rt, radular tooth; s, snout; sg, salivary gland; sm, subradular membrane; vl, ventral lip.

the hypobranchial gland, though it is not clear how, in view of its situation, they could be involved in boring or feeding. The accessory salivary glands of *Nucella* also contain a cement-like substance (Andrews, 1991) which may have some importance in attaching the proboscis to the prey.

The role of the accessory salivary glands of neogastropods remains problematic, and attempts to understand it are not eased by the apparent randomness of their occurrence. The glands appear to be generally distributed in muricids, to be not uncommon in volutids (Woodward, 1900; Pace, 1902) and olivids (Marcus & Marcus, 1968), and occasional in marginellids (Graham, 1966, Ponder, 1970a), turrids (Ponder, 1973) and some other families. They were first recorded in eighteen species of cones by Marsh (1971) and have been investigated by Schultz (1983). It was initially believed that in muricids they were related to boring, or killing prey or both, and Marsh (1971) linked them with the vermivorous habit shown by seventeen of her species, though the eighteenth attacked fish. In view of the recent findings of Andrews et al. (1991) it may well be that their secretion plays some part in the subjugation of prey, whatever the manner of its capture. In cones, however, the presence of a venom gland suggests that this role for accessory salivary glands is unnecessary and (at least on present knowledge) most manage without them.

Though fundamentally similar in their biochemistry the boring methods of naticids and muricids differ in some respects (J. D. Taylor et al., 1983). Naticids (see the review by Kabat, 1990) have to wrap their prey, often an unattached bivalve or gastropod, in their foot in order to hold it firmly enough for boring to proceed, correlated with which the ABO is situated on the tip of the proboscis. As a consequence they tend to take prey within certain size limits—small enough to be held firmly, but large enough to provide adequate reward for the considerable amount of energy expended in drilling (Kitchell et al., 1981; Ansell, 1982b, 1982c). In species which drill through the central part of a valve the boreholes tend to be sited wherever the proboscis happens to lie when the prey is held in the position giving the most secure grip, and so in a pattern charactistic for each prey species, determined by its shape and size (Berg & Porter, 1974; Negus, 1975; Bayliss, 1986). Many species of naticid, however, bore their prey at the edge of its valves, and the borehole then coincides with the thinner edge of the shell (Ansell, 1960; Ansell & Morton, 1985, 1987) (see p. 712). Naticids also tend to select prey with thin rather than thick shells (Wiltse, 1980a) or those with incompletely closing shells (Schneider, 1981). Muricids do not hold prey in the foot and so are free of these constraints. Indeed they prefer prey such as oysters, barnacles, and mussels, which are firmly attached to the substratum, and most take the largest animals available (Connell, 1970; Menge, 1974; Barnett, 1979). They may, however, perch on these and drill in such a fashion as to suggest that they are seeking a position directly over the most nutritious organs of the prey. J. D Taylor et al. (1983) pointed out that when muricids attack the exposed parts of burrowing animals (the only available parts) a spurious appearance of choice of boring site arises. Barnett (1979) found that Nucella, if forced to bore barnacles, always bores in the neighbourhood of, or at, the margin of the valves, their weakest point, though this may have resulted from failure to gain access by a thrust of the proboscis, and the animal was already poised in that position: once again a spurious appearance of choice is produced. Since the nutritious parts of a bivalve (such as the gonad) usually lie under the thickest shell, and since ingestion is done by a mobile and flexible proboscis, the naticid choice of thin areas to bore seems the more energetically economical method.

Kitchell et al. (1981) have proposed that naticids take prey in such a way as to make their energy gain optimal. This is achieved by a behaviour pattern involving the likelihood of a successful attack, mainly based on the predator/prey size ratio, the availability of different prey species, and the energetic cost of dealing with them, together with the likely food value. The assumptions made in this model have been claimed to be largely corroborated by investigations based on the presence of boreholes in assemblages of fossil shells by Kitchell *et al.* (1986) and by Hoffman & Martinell (1984). These seem, however, to be less securely based when the diversity of naticid feeding methods, especially those not involving boring, is taken into consideration. The same factors are likely to be involved in the feeding strategies of other carnivorous prosobranchs.

A further group of prosobranchs with the abillity to attack prey with a calcareous skeleton is the cassids (Tonnoidea), which eat mainly echinoderms. The secretion of sulphuric acid by these animals to achieve this is now well documented (Fänge & Lidman, 1976; Hughes & Hughes, 1981). It is produced in what are sometimes called proboscis glands (Day, 1969; Hughes, 1986), closely associated with, and probably only a specialized part of the salivary glands. They are covered with a basketwork of muscle fibres which is responsible for the rather forceful ejection of their secretion. This has a pH of 0·13 in *Galeodea echinophora* (Hughes, 1986), of 1·1 in *Argobuccinum argus* (Day, 1969), and is mainly a sulphuric acid solution, perhaps also with a chelating agent, and sometimes with a neurotoxic substance (Houbrick & Fretter, 1969 dealing with *Cymatium nicobaricum* and *Bursa granularis*). In those which attack echinoderms a hole is cut in the test, less commonly on the periproct or peristome region, by means of the acid, the sole role of the radula being to remove the calcium sulphate which is produced. In *Galeodea* the hole may be up to 9 mm in diameter, depending on the size of the predator and is rather rapidly made at a rate of 0·1 mm per minute, a figure which reflects the relatively open nature of the echinoderm skeleton and the high acidity of the salivary secretion. It is made much more rapidly than the borehole of muricids and compares with the rate of drilling of molluscan shells by species of *Octopus*, which achieve entry in one to two hours (Nixon & Macconachie, 1988). Though many species of *Cymatium* and *Bursa* eat ascidians and bivalves (Laxton, 1971) these two species attack gastropods and worms respectively, biting the former near the heart by thrusting the proboscis into the mantle cavity; the prey is almost at once immobilized by the poison and is rapidly ingested.

The collection of food by means of ciliary mechanisms on the ctenidia was originally supposed to be limited to mesogastropods (Yonge, 1938), since the arrangement of currents in the mantle cavity of prosobranchs with paired ctenidia would not permit the transport of the material collected to the mouth (Yonge, 1947). Ciliary feeding is now known, however, among diotocardian prosobranchs. The sand-living trochid *Umbonium vestiarium* (Fretter, 1975), the neomphaloidean *Neomphalus* (Fretter *et al.*, 1981), and some other hot vent limpets, the Lepetodriloidea (Fretter, 1988), all appear to gather at least some of their food in this way. The long-held belief that this mode of feeding is not compatible with two gills is, nevertheless, probably still valid, since all these animals have lost the right ctenidium. In *Umbonium*, as in most trochids and in lepetodriloideans, much of the ctenidium is pectinibranch and the arrangements in the mantle cavity are therefore largely comparable to those in the ciliary-feeding monotocardian. *Neomphalus*, however, differs in that the whole length of its elongated ctenidium is bipectinate. The disposition of the filaments has changed so that both sets extend to the right across the mantle cavity, and the material which they collect is deposited in a food groove on that side. The course of the groove is unusual in that it passes to the mouth dorsal to the base of the right tentacle instead of ventral to it as in other ciliary-feeding prosobranchs. The details of the ciliation on the gill filaments of ciliary-feeding diotocardians are in general the same as in monotocardians. In lepetodriloideans, however, (Fretter, 1988), the bands of frontal and lateral cilia expand at the tip of each filament to form prominent pads, and some terminal cilia become extremely long. These seem to be adaptations for the transport of food particles towards the mouth, but as no living animals have been examined their precise function must remain speculative.

Werner's work (1959) showed that the food-collecting mechanism of ciliary-feeding proso-branchs resembles that found in Amphioxus and tunicates—the capture of particles in a continuous, moving sheet of mucus, which in those animals is rolled up in the mid dorsal line of the pharynx and carried to the oesophagus; in prosobranchs it is rolled up at the free edge of the ctenidium and then travels across the floor of the mantle cavity to the mouth. The mechanism thus differs in a basic way from that of bivalves, in which Owen (1974) has shown that food particles are trapped in the pinnately branched laterofrontal cirri of the gill fila-ments in *Mytilus*. The evidence available from the anatomy of the ctenidium and its appear-ance in scanning electron micrographs suggests that lepetodriloideans differ from other gastropods in this respect collecting their food like bivalves, on cilia (Fretter, 1988).

The mechanism by which the ciliary-feeding freshwater prosobranch *Bithynia* collects food differs in another way from that used by other genera (Schäfer, 1952). It takes the form of a mucous net secreted by the hypobranchial gland and slung between that structure and the floor of the mantle cavity on the left, the gill not being involved except in so far as it creates the current of water to be strained. At intervals the net breaks and it is carried by cilia to the right of the mantle cavity and then forward to the mouth. The prosobranch *Capulus ungaricus* is normally able to collect food from whatever animal it is associated with by means of its proboscis but can supplement this with food collected on its gill.

Thorson (1965) found a dwarf form of what he presumed to be this animal living in association with *Turritella*, usually clinging to the shell near the aperture. These animals lacked a proboscis and so must have been dependent on their ciliary feeding. Their situation, however, placed them where they would be able to intercept the current of water entering the mantle cavity of their host, iteself a ciliary feeder.

Branch (1975) has show how *Patella* species control the algal growths on a shore so as to maintain a flora advantageous for their feeding, and Lubchenco (1978) has demonstrated the same ability in *Littorina littorea* living in high intertidal pools. According to Steneck (1982) the limpet *Collisella tessulata* and the coralline alga *Clathromorpha circumscripta* have a relation-ship which benefits both. The density of *Collisella* on the shore is greatest on *Clathromorpha*, a site which minimizes the time the limpet spends looking for food, and so its exposure to predators. The removal of an epidermis about six cell layers deep (overlying a meristematic layer), which is what the limpet eats, prevents the coralline from being smothered by epiphytic growths. A nice balance is normally obtained since the feeding rate of the limpet and the replacement rate of the algal epidermis are equal.

To some extent carnivores are capable of the same control of the population of prey in the areas in which they live,but they are more likely to have drastic effects upon the structure of their community (Wiltse, 1980a). The balance between supply and demand is delicate and easily disrupted by any change in trophic relationships. Though there is an inbuilt system by which a balance is restored it is not necessarily the same community in terms of specific make-up which is reproduced: it is best, therefore, for an animal, especially a carnivore, to be conservative in its feeding habits. *Nucella lapillus* is well known for its addictive feeding habits, tending to take one form of prey even to exhaustion of the stocks (Wood, 1968; Murdoch, 1969; Hughes & Dunkin, 1984b) and even a herbivore such as *Biomphalaria glabrata* takes time to adjust to new food (J. D. Thomas, 1982). Such a pattern is not universal, however: L. West (1988) found great variations in the feeding of *Thais melones* on the Pacific coast of Panama, some snails having a well diversified diet whilst others (even those living alongside) had an extremely limited one. Diet had some effect on growth rate at certain sizes, generalists then growing faster than snails with a restricted diet.

Carnivores, living at higher trophic levels than herbivores, are more affected by competition for food, both intraspecific and interspecific. Selective removal of bivalves by naticids, as discussed by Wiltse (1980a), especially of non-dominant forms, may alter the diversity of the fauna by removing them altogether, leaving an already dominant species as the exclusive form. Wiltse (1980a) investigated *Polinices duplicatus* feeding on *Mya arenaria* of 15–45 mm shell length, the preferred size of one of its favourite foods, in Barnstaple Harbor, Massachusetts. Her calculations showed that each naticid consumed 0·27 *Mya* per day; the density of *P. duplicatus* was 0·53 per sq. metre so that 52 specimens of *Mya* in each sq. metre were removed each year. Since the total population of all bivalves at this site is maximally 84·7 per sq. metre, pressure on food supply is very high, and with the unpredictable variations in spatfall characteristic of bivalves, likely to lead to occasional crises in the population of predatory gastropods, and to changes in the specific composition of the bivalve population.

In tropical areas the number of competing species of carnivores is much greater than in higher latitudes (Pianka, 1966; Kohn, 1971). Competition there, however, may be reduced by the facts that there is a greater number of trophic niches, and that each species of carnivore tends to confine itself to one or only a few species of prey (J. D. Taylor et al., 1980). Other ways of sharing resources also operate: as an example Levings & Garrity (1983) described how two sympatric species of *Nerita*, *scabriscosta* and *funiculata*, successfully co-exist on a shore in Panama whilst constrained by fish predation and the need to avoid overheating and desiccation. *N. funiculata* lives in crevices in the mid littoral and forages briefly at ebbing and flowing tides, never moving more than 50 cm from a crevice. *N. scabricosta* on the other hand lives in crevices above high water mark and has a different feeding pattern, following the retreating tide and feeding as it goes, returning to its crevice as the water rises again. The two species are therefore hardly in competition with one another.

The following table gives the common foods of British prosobranchs about which positive statements can be made, based on reports in the literature or on personal observations. It may, however, be assumed with reasonable probability that related species, especially when not sympatric, have comparable diets: if a definite, statement can be made about one species of *Rissoa*, or *Alvania*, or *Lunatia*, it is likely (but not more than that) that the same statement, if made in sufficiently general terms, will be applicable to other species in the genus though the details may well differ: a large *Lunatia* such as *catena* will take larger bivalves than a small one such as *alderi*. The term 'detritus' is used to cover the scrapings from rock surfaces eaten by many molluscs, the similar layer covering soft bottoms, and the material which gathers entangled in the attachments of many seaweeds. It includes a number of ingredients, some of which, like sand grains and sponge spicules, have themselves no nutritive value but may provide a substratum for bacterial growth. It also contains free diatoms and bacteria, fragments of algal tissue, less commonly of animal remains, and, on hard surfaces, sporelings and other small developmental stages of both plants and animals. It therefore provides a relatively rich diet which may be ingested even by small or minute prosobranchs.

Scisssurella crispata: detritus, growths on sand grains.
Haliotis tuberculata: algae and their epiphytes (*Delesseria* spp. mainly; *Griffithsia*, *Chondrus*, young *Corallina* and *Lithothamnion* (Crofts, 1929; Vahl, 1971); diatoms (Hayashi, 1980).
Emarginula fissura: sponges, algae?, detritus? (Graham, 1939).

Diodora graeca: sponges (*Halichondria*, *Hymeniacidon*), detritus? (Graham. 1939).

Acmaea virginea: encrusting algae, detritus.

Collisella tessulata: red algae (*Ralfsia*, *Cruoriella*, *Lithothamnion*), detritus (Ankel, 1936; Rasmussen, 1973).

Helcion pellucidum: *Laminaria digitata*, *L. hyperborea*, *Alaria esculenta* (not *L. saccharina*, *Saccorhiza polyscheides*), some *Fucus serratus*, *Himanthalia elongata*; not on laminarians until shell is 4 mm long (Vahl, 1971).

Patella vulgata: fucoid sporelings; edges of fronds of *Fucus*, and *Ascophyllum*; detritus (Fretter & Graham, 1976).

Margarites helicinus: detritus, microalgae (Fretter, 1955).

Solariella amabilis: predator, prey unknown (Nordsieck, 1968); suspected of microphagy (Fretter & Graham, 1977).

Gibbula cineraria: epiphytes, microalgae, same macrophytes, detritus (Fretter & Graham, 1977).

Gibbula umbilicalis: fucaceans (Vahl, 1971).

Monodonta lineata: cyanobacteria, developmental stages of macrophytes (Fretter & Graham, 1977).

Calliostoma zizyphinum, *C. granulosum*: coelenterates (Salvini-Plawén, 1972a; Lowry et al., 1974; Perron, 1975) to what extent this is deliberate or incidental to the ingestion of vegetable food, which is certainly taken, is not known.

Calliostoma occidentale: the alcyonacean *Gersemia rubiformis*, hydroids (Perron & Turner, 1978).

Tricolia pullus: red weeds and their epiphytes (*Lomentaria*, *Laurencia*, *Gigartina*, *Plumaria*, *Chondrus*, *Nitophyllum*, *Ceramium*) (Vahl, 1971; Fretter & Manly, 1977a).

Theodoxus fluviatilis: predominantly diatoms (Skoog, 1978).

Viviparus viviparus, *V. contectus*: detritus; plants (Frömming, 1956); ciliary feeding (Cook, 1949).

Pomatias elegans: vegetable matter, dead leaves (Kilian, 1951; Frömming, 1954).

Acicula fusca: decaying leaves, fungi (Creek, 1953).

Hyrobia ulvae, *H. ventrosa*, *H. neglecta*: all species take mainly diatoms (Jensen & Siegismund, 1980) but also take sand grains into the buccal cavity, rasp off attached diatoms, bacteria and then eject the sand grains (Lopez & Kofoed, 1980). *H. ulvae* cannot swallow particles over 200 μm in size, *H. ventrosa* and *neglecta* only those up to 120 μm. *H. ventrosa* feeds to a considerable extent on its own recycled faecal pellets (Lopez & Levinton, 1978). *H. ulvae* also eats delicate weeds and, when floating, planktonic organisms caught in a mucous float (R. C. Newell, 1962). *H. ventrosa* eats primarily diatoms, fungi, gram-negative bacteria (Hylleberg, 1976); *H. neglecta* takes organic detritus preferentially (Hylleberg, 1976).

Potamopyrgus jenkinsi: diatoms, bacteria rasped from sand grains or plants. Takes particles up to 180 μm (Lopez & Kofoed, 1980).

Marstoniopsis scholtzi: diatoms, decaying plants, detritus (Dussart, 1977; Fretter & Graham, 1978a).

Truncatella subcylindrica: microalgae, vegetable detritus (Fretter & Graham, 1978a).

Bithynia tentaculata: diatoms. filamentous algae, epiphytes, detritus (Schäfer, 1952; Lilly, 1953); ciliary feeding (Schäfer, 1952; Tsikhon-Likanina, 1961a, 1961b).

Assiminea grayana: fresh and decaying weeds, diatoms, detritus (Ankel, 1936).

Cingula trifasciata: detritus (Fretter & Graham, 1978b).

Onoba semicostata; detritus (Fretter & Graham, 1978b).

Manzonia crassa: corallines (Pelseneer, 1935; Vahl, 1971); ?detritus (Fretter & Graham, 1978b).

Manzonia jeffreysi: foraminifera (?).

Alvania punctura: detritus (Fretter & Graham, 1978b).

Alvania carinata: red algae (Pelseneer, 1935; Vahl, 1971); detritus (Fretter & Graham, 1978b).

Alvania semistriata: corallines (Pelseneer, 1935); detritus (Fretter & Graham, 1978b).

Rissoa parva: diatoms and epiphytes of mainly red weeds (*Lomentaria, Plumaria, Callithamnion, Ceramium, Corallina, Gigartina, Rhodymenia*), also *Fucus, Ulva, Enteromorpha*: detritus (Pelseneer, 1935; Fretter & Graham, 1978b).

Rissoa guerini: *Codium* (Vahl, 1971).

Pusillina inconspicua: detritus, weed epiphytes (Fretter & Graham, 1978b).

Rissostomia membranacea: *Zostera* (Ankel, 1936); other weeds and their epiphytes (Rehfeldt, 1968; Fretter & Graham, 1978b).

Barleeia unifasciata: detritus, epiphytes of small red weeds (Fretter & Graham, 1978b; Southgate, 1982).

Cingulopsis fulgida: detritus (Fretter & Graham, 1978b).

Rissoella diaphana, R. opalina: diatoms, fine weeds, algal filaments, detritus (Ankel, 1936; Fretter, 1948).

Omalogyra atomus: diatoms, algae (mainly fine reds), algal filaments, detritus (Ankel, 1936; Fretter, 1948); *Ulva* (Vahl, 1971).

Ammonicera rota: this animal pierces algal cells with the radula and sucks out the contents (Franc, 1948).

Skeneopsis planorbis: diatoms, algal filaments, fine red weeds, detritus (Ankel, 1936; Fretter, 1948).

Caecum glabrum: diatoms (Götze, 1938).

Tornus subcarinatus : algae (Pelseneer, 1935); more probably the diatoms and bacteria on sand grains (Fretter & Graham, 1978b).

Lacuna vincta: laminarians, fucoids (*Fucus vesiculosus* and *F. serratus*), some red weeds (D. A. S. Smith, 1973; Fretter & Manly, 1977a).

Lacuna parva: fucoids, the red weeds *Ceramium, Gigartina* (Pelseneer, 1935), *Phyllophora, Delesseria* (Ocklemann & Nielsen, 1981).

Lacuna pallidula: laminarians, *Fucus serratus* (D. A. S. Smith, 1973).

Littorina littorea: *Ulva, Enteromorpha*, fucoid sporelings, detritus (Nicotri, 1977; Watson & Norton, 1985).

Littorina obtusata: *Pelvetia, Fucus vesiculosus* (especially the reproductive receptacles), *F. serratus, Ascophyllum nodosum*, young *Ulva* (Daguzan, 1976b, Watson & Norton, 1987).

Littorinia mariae: *Fucus vesiculosus, F. serratus, Ascophyllum nodosum* and their epiphytes (Watson & Norton, 1987); the epiphytes more or less exclusively (G. A. Williams, 1990).

Littorina nigrolineata: macrophytic fucaceans (Sacchi *et al.*, 1977, 1981).

Littorina neglecta: unicellular algae (Fretter & Graham, 1980).

Littorina saxatilis, L. arcana: macrophytic fucaceans, their epiphytes, detritus, unicellular algae (Raffaelli, 1976; Daguzan, 1976a; Sacchi *et al.*, 1977, 1981).

Melarhaphe neritoides: black lichens (Daguzan, 1976a); detritus (Fretter & Graham, 1980).

Turritella communis: ciliary feeding (Graham, 1938).

Bittium reticulatum: diatoms (Starmühlner, 1956); detritus (Fretter & Graham, 1981).

Aporrhais pespelecani: vegetable fragments, detritus (Yonge, 1937).

Trichotropis borealis: detritus (Fretter & Graham, 1981).

Capulus ungaricus: ciliary feeding (Yonge, 1938; Thorson, 1965); detritus. pseudofaeces of bivalves (Sharman, 1956).

Crepidula fornicata: ciliary feeding (Orton, 1912).

Calyptraea chinensis: ciliary feeding (Werner, 1953).

Lamellaria perspicua, L. latens: the ascidians *Botryllus, Leptoclinum, Polyclinum* (Fretter & Graham, 1981).

Velutina plicatilis: *Tubularia indivisa*, compound ascidians (Ankel, 1936).

Velutina velutina: the ascidians *Phallusia, Styela* (Ankel, 1936; Diehl, 1956).

Erato voluta: compound ascidians (*Diplosoma, Botryllus, Botrylloides*) (Fretter, 1951a).

Trivia monacha, T. arctica: compound ascidians (*Diplosoma, Botryllus, Botrylloides, Polyclinum*) (Fretter, 1951a).

Simnia patula: polyps of *Eunicella, Alcyonium, Tubularia* (Fretter, 1951a).

Lunatia alderi, L. catena: bivalves, especially tellinoideans (Ansell, 1960).

Charonia lampas: starfish (*Echinaster*) (mussels, squid in aquarium) (Amouroux, 1974); holothurians (Percharde, 1972; Thomasin, 1976).

Carinaria lamarcki: jelly-fish, salps, worms (Franc, 1949; Thiriot-Quiévreux, 1973).

Cerithiopsis tubercularis: *Halichondria, Hymeniacidon* (Fretter, 1951b; Fretter & Manly, 1977a).

Cerithiopsis barleei: *Suberites* (Lebour, 1933).

Triphora adversa: *Halichondria, Hymeniacidon* (Fretter, 1951b).

Epitonium clathrus: *Anemonia sulcata* (Fretter & Graham, 1982).

Janthina janthina, J. pallida, J. exigua: *Porpita, Velella, Physalia* (Hardy, 1956; Bayer, 1963).

Eulima glabra, E. bilineata: ectoparasitic on ophiuroids (Warén, 1983).

Melanella alba: ectoparasitic on holothurians (*Neopentadactyla mixta*) (Cabioch et al., 1978).

Polygireulima sinuosa, P. monterosatoi: ectoparasitic on echinoids (Warén, 1983).

Vitreolina philippii: ectoparasitic on crinoids, holothurians, ophiuroids (Warén, 1983).

Pelseneeria stylifera: ectoparasitic on echinoids (*Echinus, Psammechinus, Strongylocentrotus, Paracentrotus*) (Ankel, 1936).

Nucella lapillus: barnacles, mussels, other molluscs (Moore, 1936; Largen, 1967a; Crothers, 1985).

Urosalpinx cinerea: oysters and other bivalves (Carriker, 1955; Hancock, 1960).

Ocenebra erinacea: bivalves (*Cardium, Venus, Venerupis*), barnacles (Orton, 1929; Hancock, 1960).

Volutopsius norwegicus: echinoderms, mainly ophiuroids (Kantor, 1985, 1990a).

Neptunea antiqua: bivalves, annelids (Pearce & Thorson, 1967; J. D. Taylor, 1978).

Buccinum undatum: polychaetes (mainly *Lanice*), bivalves (mainly *Cardium*), crustaceans (Nielsen, 1975; J. D. Taylor, 1978).

Hinia reticulata, H. incrassata: carrion [Fretter & Graham, n. d. (1985)].

Oenopota turricula: small polychaetes (Eliason, 1920).

Oenopota trevelliana: annelids (*Prionospio, Spiophanes, Ammotrypane*) (Pearce, 1966).

Mangelia brachystoma: spionid worms (Taylor, quoted by J. A. Miller, 1989).

Chrysallida obtusa: oysters, other bivalves (Cole & Hancock, 1955; Rasmussen, 1973).

Partulida spiralis: *Pomatoceros* (Ankel, 1959); *Sabellaria* (Fretter & Graham, 1949).

Evalea diaphana: *Phascolion* (Ankel, 1959; Kristensen, 1970).

Liostomia clavula: *Pennatula* (Maas, 1965).

Brachystomia rissoides: mainly *Mytilus* but also occasionally many other molluscs (F. Ankel & Christensen, 1963; Rasmussen, 1973).

Brachystomia eulimoides: *Pecten, Chlamys* (Jeffreys, 1867); *Turritella* (Ankel, 1959).

Brachystomia lukisi: *Pomatoceros*, serpulids (Fretter & Graham, 1949).
Odostomia plicata: *Pomatoceros* (Ankel, 1959).
Odostomia turrita: *Homarus* (Sneli, 1972).
Odostomia unidentata: *Pomatoceros*, serpulids (Fretter & Graham, 1949).
Odostomia conoidea: *Astropecten* (Nordsieck, 1972).
Odostomia umbilicaris: bivalves (?) (Jeffreys, 1867).
Turbonilla lactea: cirratulids, terebellids (Fretter, 1951c).
Turbonilla jeffreysi: hydroids (*Halecium, Antennularia, Hydrallmania*) (Fretter & Graham, 1949).

   The suggestion, made originally by Fretter & Graham (1949), that each species of pyramidellid is normally restricted to one host has been disproved, W. E. Ankel (1959) and Boss & Merrill (1965) having shown that a given species of pyramidellid will feed on different (though often related) animals. It still remains true if one knows, for example, that *Odostomia unidentata* commonly, though not exclusively, feeds on *Pomatoceros*, the most reliable and easy way of finding the animal is to search amongst aggregations of such worms.

CROSS REFERENCES

CHAPTER 27

THE EXCRETORY AND VASCULAR SYSTEMS

O F all the systems in the body of the prosobranch two, the excretory and the closely linked vascular system, have aroused less obvious interest amongst malacologists than most, perhaps because they have proved more difficult to investigate, especially experimentally. So far as the excretory system is concerned there was little investigation after the early work of Perrier (1889) and Cuénot (1899) (much of which was, for its date, surprisingly interested in function) until the field was reopened by such workers as Turchini (1923), Delaunay (1931), and, more recently, with an approach partly based on experiment, partly on new information provided by electron microscopy, by Potts, Delhaye, Andrews and Little. A similar history is true of the circulatory system: such early work as that of Milne-Edwards (1846) remained the main basis of knowledge in this field until that of Skramlik (1941) and the modern experimental work of Krijgsman and Divaris, of a group of Manchester workers led by Trueman, and of Mangum and her co-workers, though much of this was on pulmonates.

The nomenclature of excretory organs in molluscs is in disarray (Martin, 1983). Goodrich (1946) made a clear distinction between two types of organ which develop in animals and may become involved in excretory activity. The first of these is the nephridium which grows inwards from the epidermis and so is of ectodermal origin; in their primitive form (proto-nephridium) the nephridia form branching systems of blind tubules, with flame cells at the inner ends of the branches, and are both excretory and, where necessary, osmoregulatory. Protonephridia are found in platyhelminthes and annelids and occur in the developmental stages of some molluscs, mostly euthyneurans (Meisenheimer, 1899). Amongst prosobranchs they have been described in *Bithynia tentaculata* (Erlanger, 1891) and *Viviparus* (Erlanger, 1894). Some doubt about the homology of these molluscan organs with those found in lower phyla rests on the fact that they appear to arise from mesoderm rather than ectoderm (Fretter & Graham, 1962; Salvini-Plawén, 1988b). Their occurrence in prosobranchs, however, may be wider than is currently accepted in view of the work of Ruthensteiner & Schaefer (1991). Structures known as 'larval kidneys', in the form of a pair of small papillae, are well known in the veliger larvae of many prosobranchs (Portmann, 1930; Fioroni, 1966). Each comprises only a few cells and lies on the head-foot, just posterior to the attachment of a velar lobe, projecting into the mantle cavity, and is laden with presumed excretory granules. In *Hinia reticulata* Ruthensteiner & Schaefer (1991) have found that these cells lie in such close proximity to the openings of what appear to be true, though microscopic, protoneph-ridia as to suggest that the 'larval kidneys' may be no more than hypertrophied cells properly belonging to their lips. The protonephridial part of this structure is so small that it might easily have been overlooked in some species, and have become vestigial or lost in others. Should closer investigation of 'larval kidneys' make this appear more likely, then some representation of protonephridia would occur in a much wider spread of prosobranchs than is at present accepted. That both *Bithynia* and *Viviparus* are freshwater animals, in which osmoregulation is important, may explain why more typical protonephridia are obvious there.

A more advanced type of nephridium, the metanephridium, opens by a nephrostome to some body cavity, often the coelom, a device presumably speeding up excretory processes and allowing the passage of particulate material. Metanephridia occur in many annelids.

The second type of organ described by Goodrich is the coelomoduct which, as its name implies, is a coelomic, and therefore a mesodermal, outgrowth, of which the primary function is the passage of sex cells to the exterior. In some animals (capitellid polychaetes for example) the two sets of ducts co-exist, at least during the breeding season, and are clearly distinct. A certain duplication of function is, however, obvious is such a situation and there is a pronounced tendency for the one set of ducts to persist whilst the other either becomes temporarily or permanently fused with them forming structures known as nephromixia or mixonephridia, or it disappears altogether. In molluscs, as in vertebrates, it is the nephridia which go, leaving coelomoducts to act as excretory, osmoregulatory, and genital ducts.

It is therefore unsatisfactory, when the words nephridium and coelomoduct have been thus strictly defined as structures with distinct development and function, that many writers use the word nephridium for the molluscan excretory organ or even write such sentences as '[their] excretory provision typically consists of a pair of metanephridia, which are true coelomoducts'. In such confused circumstances the only sensible usage is to employ functional names rather than those based on real or supposed comparative embryology and anatomy, and to speak of the organs as kidneys or, if preferred, Hoffman's (1937) clumsier term emunctoria, and as genital ducts.

One anomaly remains: in many prosobranchs part of the left kidney becomes specialized for the control of the chemical composition of the blood. This structure is invariably known as the nephridial gland, a name which, despite its inappropriateness, appears likely to remain in use until a more suitable one is found.

That the kidneys of prosobranchs are derived from coelomoducts is obvious from their development. A single embryonic rudiment represents the mesoderm from which three sets of organs develop, the pericardial cavity and heart, the gonad, and the kidneys. The last retain their connection with the pericardial cavity by links which are known as renopericardial canals, and the gonad connects with the right kidney, either directly (patellogastropods) or, more commonly, by an extension of its wall forming a gonadial duct. In many primitive living prosobranchs, and perhaps also in the earliest, the pericardial rudiment comes to enclose the rectum and when the heart forms by immigration of cells from the pericardial walls these group themselves round that structure so that the rectum appears to penetrate the ventricle. Like the heart of the adult monoplacophoran *Neopilina* (Lemche & Wingstrand, 1959) the pericardial rudiment of prosobranchs (Erlanger, 1892; Drummond, 1903) is double and, like other organs in this part of the visceral mass, shows a bilateral organization. The original description of *Neopilina* suggested that the kidneys drained coelomic pouches: these are known (Wingstrand, 1985) to be parts of oesophageal glands, and the precise arrangement of the kidneys remains doubtful, particularly whether they open independently to the exterior or whether those on each side link with one another to form an elongated saccular organ comparable to the kidney of polyplacophorans, with a single posterior opening. In the earliest prosobranchs the pericardial cavity, flanked on either side by a kidney, lay in the most anterior part of the visceral hump, the two kidneys opening to the innermost part of the mantle cavity on either side of the anus. Since in gastropods, unlike other molluscs, the gonad connects with the right kidney the right aperture was also the genital pore. Study of the development of this area shows that the terminal part of each kidney duct, as well as that of the rectum, is formed from a shallow invagination of the wall of the mantle cavity (fig. 323).

A

C

E

B

D

F

G

H

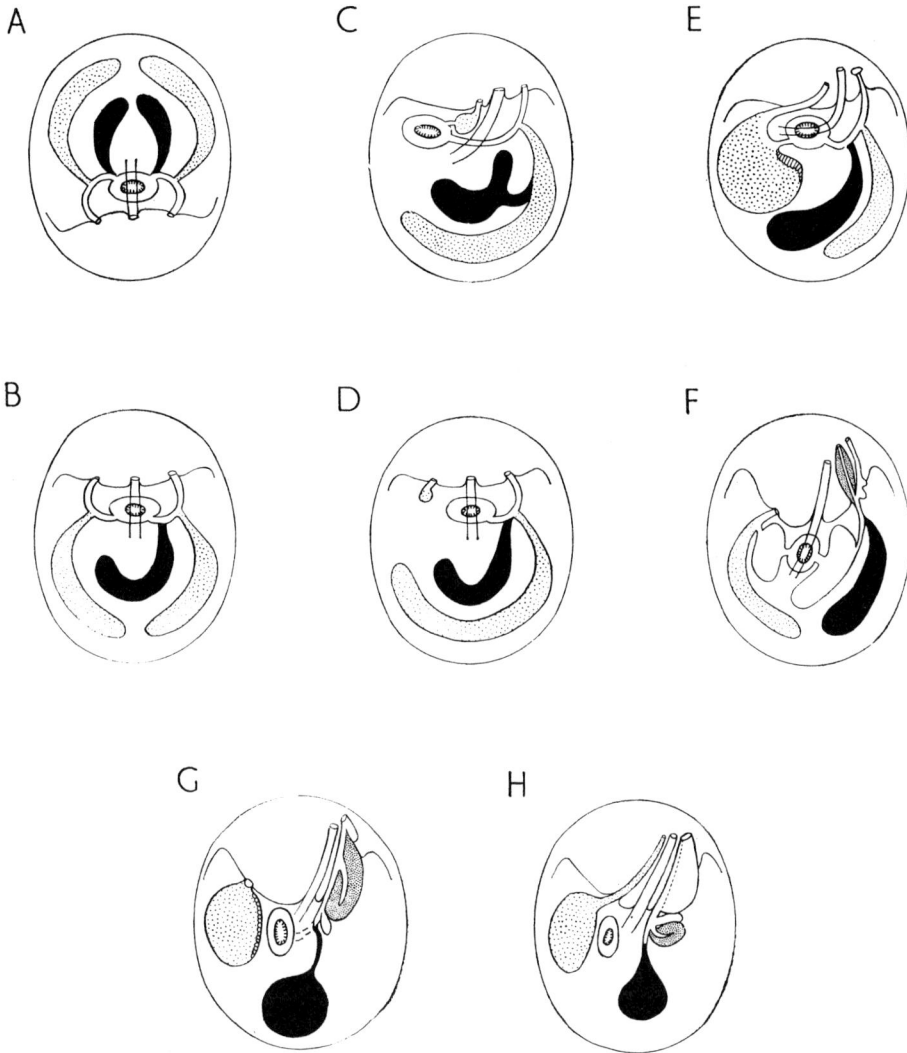

FIG. 323.—Diagrams to show the inter-relationships of rectum, gonad, pericardial cavity, kidneys and mantle cavity in a series of molluscs. The left kidney is always shown to the left of the rectum and pericardial cavity, the right kidney and its connexions to the right, so that the disposition of the parts does not necessarily match their anatomical arrangement. A, progastropod, with the mantle cavity directed posteriorly; B, ancestral prosobranch, post-torsional, with the mantle cavity facing forwards; C, Patellacea; D, Diodora; E, Trochacea; F, Theodoxus female; G, monotocardian female; H, Viviparus female. In the diagrams the pericardial cavity and the ducts from it to the mantle cavity are unshaded, kidneys are lightly stippled, the gonad is black; glandular enlargements on the urinogenital or genital ducts are heavily stippled and the ventricle is shown in the pericardial cavity. Note the occurrence of a nephridial gland on the pericardial wall of the left kidney in Trochacea (E) and Monotocardia (G), the occasional persistence of a gonopericardial duct in Monotocardia (G), and the enlargement of the renopericardial canals in Theodoxus (F) (and other Neritacea) to produce a body cavity.

The two systems, vascular and excretory, are intimately linked not only developmentally but also functionally. As in many groups of animals urine is produced as a filtrate of the blood, passed through the wall of some part of the vascular system, normally by blood pressure, into some part of the excretory system. The filter is sufficiently delicate to prevent the escape of cells and of such large molecules as proteins from the blood, but the filtrate, the primary urine, still usually contains substances of metabolic or other value. Its composition is modified by their resorption, and usually also by the direct addition of excretory material as the fluid traverses the excretory organs, so that the composition of the urine discharged is quite unlike that of the original filtrate. The kidneys, therefore, are not so much organs originating urine as organs controlling its ultimate composition in relation to the activities and current needs of the animal and to the habitat. In this respect the gastropod kidney is no different from that of most other groups of animals. In one respect, however, it differs profoundly since the filtration site, where the primary urine is produced, is the heart, as first surmised by Picken (1937). The duality of function, pumping and filtration, which results from this arrangement, has produced a complexity of structure and function in the heart which might not have been necessary had the two activities been carried out by separate organs. The gastropod arrangement appears to have been the original molluscan one since it is found also in polyplacophorans (Økland, 1980) and in some bivalves (Odhner, 1912; Jennings, 1984). In the course of the evolution of the latter group, however, pumping and filtration have become secondarily separated in many species, pumping being the sole activity of the heart and filtration confined to glands placed remotely on the pericardial wall. The same separation has occurred in cephalopods in which filtration occurs in the appendages of the branchial hearts, 'the true auricles' (Martin & Aldrich, 1970), leaving the heart as a purely circulatory pump, and there is an approach to the same end in some gastropods, but to a much more limited extent.

Though the two kidneys may well have been similar in size and position so long as the ancestral prosobranch retained bilateral symmetry, or something near to that, there is no reason to suppose that they remained so with the advent of helical coiling, the left, like many other visceral organs perhaps being larger than the right in such an early stage of prosobranch evolution. It is probable, too, that the two organs were similar in structure and function, as they are in all other molluscan classes. At this stage it would appear that each kidney received its major supply of blood from vessels draining the head-foot in one direction and the rest of the visceral mass in the other. After passing through a network of vessels on the renal walls the blood passed to the mantle skirt, from which it travelled, perhaps by channels through the hypobranchial glands, to an afferent branchial vessel situated in the axis of each ctenidium. Each kidney was therefore part of a renal portal system; nitrogenous waste was removed from the blood before it was sent to the gills for oxygenation and thence to the heart. Since the removal of waste involves the expenditure of energy each kidney was presumably also supplied with oxygenated blood by renal arteries arising from the posterior aorta.

The conditions just described, however, seem to have been representative of a relatively brief phase in prosobranch history, and in the further evolution of archaeogastropod prosobranchs right and left kidneys have diverged in many respects. In prosobranchs at this level (fig. 324) both kidneys persist in the adult though they differ in position, in size, in structure, and in function. The right invariably retains the original position in the base of the visceral mass and the original relationships with the vascular system. However, unlike most of the organs of the right side of the pallial complex, which become reduced in size or disappear, it enlarges, forming what is often the largest cavity in the body (*Haliotis, Diodora, Patella*),

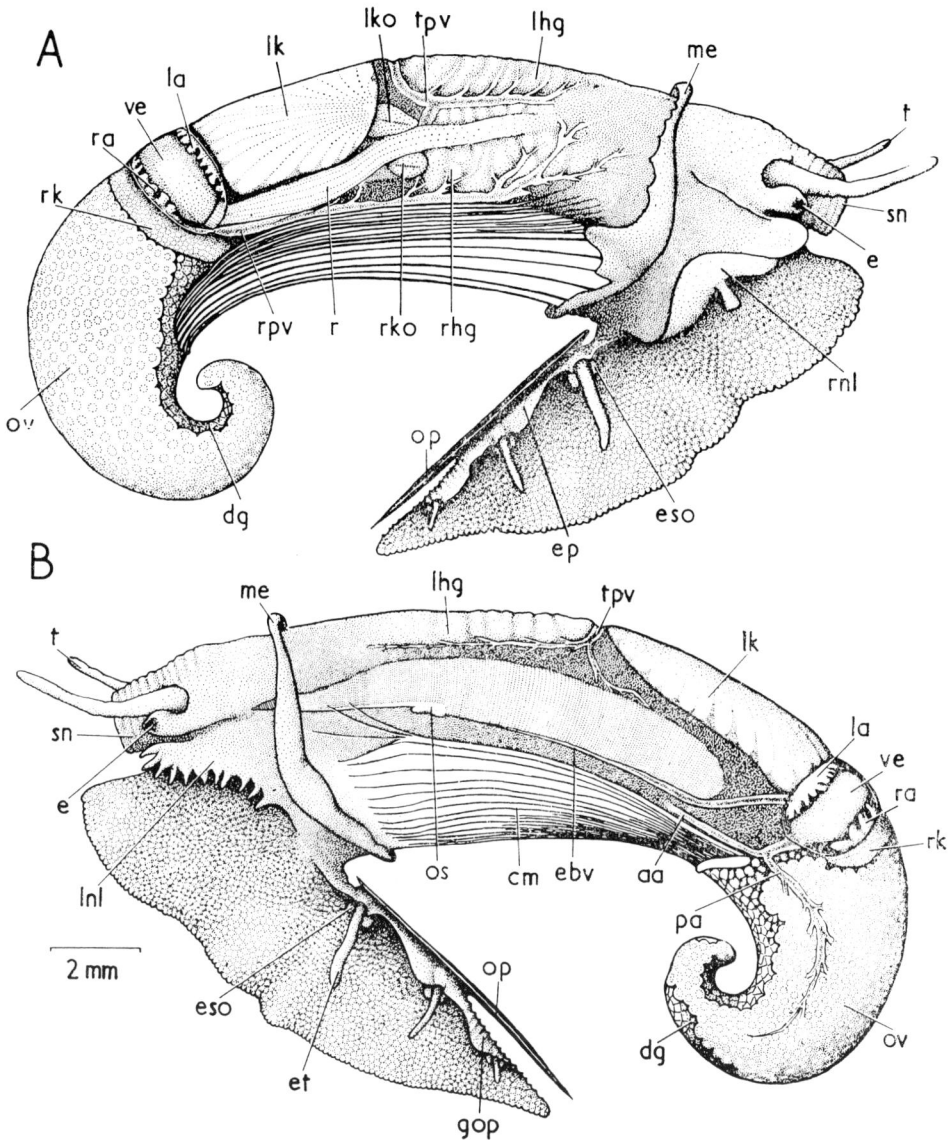

FIG. 324.—*Monodonta lineata*. Animal removed from shell and seen A, from the right; B, from the left.

aa, anterior aorta; cm, columellar muscle; dg, digestive gland; e, eye; ebv, efferent branchial vessel; ep, epipodium; eso, epipodial sense organ; et, epipodial tentacle; gop, glandular area below operculum; la, left auricle; lhg, left hypobranchial gland; lk, left kidney (papillary sac); lko, left kidney opening; lnl, left neck lobe; me, mantle edge; op, operculum; os, osphradium; ov, ovary; pa, posterior aorta; r, rectum; ra, right auricle; rhg, right hypobranchial gland; rk, right kidney; rko, opening of right kidney (urinogenital); rnl, right neck lobe; rpv, right pallial vein; sn, snout; t, tentacle; tpv, transverse pallial vein; ve, ventricle around rectum.

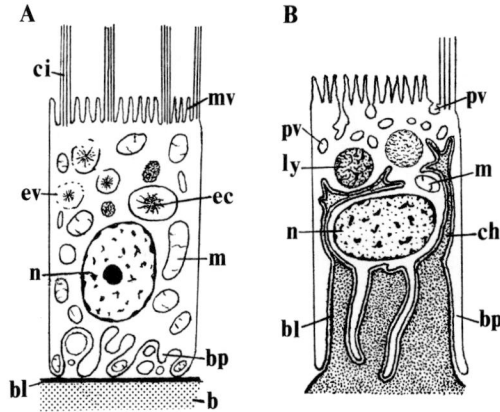

FIG. 325.—Diagrammatic representations of vertical sections of epithelial cells of the right and left kidneys of trochids. A, excretory cell of right kidney; B, cell of the papillary sac. (After E. B. Andrews, 1981).

b, blood; bl, basal lamina; bp, basal process; ch, blood-filled channel penetrating cell body; ci, cilium; ec, excretory concretion; ev, excretory vacuole; ly, lysosome; m, mitochondrion; mv, microvillus; n, nucleus; pv, pinocytotic vesicle.

wrapping round the viscera and sending lobes amongst them so that the relationship between kidney and viscera becomes close. It often extends far to the left side as if to compensate for the fact that the left kidney has lost most of its ability to deal with nitrogenous excretion. Its walls bear low folds overlying the network of blood vessels which they bear, though sometimes the folds are sufficiently prominent to give the organ a slightly spongy texture. On these folds are concentrated the cells responsible for excretion. In trochoideans the right kidney is still capacious, but a mass of tubular extensions from its main cavity, which abuts the digestive gland, gives it a more solid appearance. These form the main site of excretory activity, though all the cells in the renal epithelium share in this to some extent. The trochoidean kidney also differs in that the blood from the head-foot does not pass through it, though that from the visceral mass still does (Nisbet, 1953). In prosobranchs at the archaeogastropod level, except patellogastropods, there is a region close to the external aperture of this kidney which is nearly or wholly devoid of excretory activity and acts primarily to convey the urine and, in the breeding season, the sex cells to the mantle cavity; in some species (e.g. lepetodriloidean limpets) the part carrying gametes may be nearly completely separate from that carrying urine (Fretter, 1988).

Right kidney cells (fig. 325) (Delhaye, 1976; E. B. Andrews, 1981, 1985) appear all to belong to one type, though the details of their organization vary to some extent in relation to stages in their cycle of activity. They are columnar and their most obvious feature is a series of vacuoles containing concretions which, when fully formed, are discharged to the lumen. Delhaye (1976) has shown that these contain purines and a variety of other substances such as melanin, chromolipids, and lipofuscins, which are responsible for the colours, probably of dietary origin, which the archaeogastropod right kidney often shows. The base of each cell is usually elaborately branched and mitochondria are frequently in the branches, suggesting the active removal of solutes from the underlying blood; pinocytotic uptake of particulate matter has also been observed in *Monodonta* (Delhaye, 1976; E. B. Andrews, 1981). The presence of apical microvilli, between the bases of which vesicles discharge to the lumen, helps to confirm the idea that these cells extract material from the blood, and either transport it directly to the lumen of the kidney or incorporate it into the spherules which they elaborate and which are ultimately shed in blebs of cytoplasm nipped off from the surface of the cell. During the early phases of the excretory cycle the cells often bear some cilia amongst their microvilli, but as the cycle proceeds the cilia become less obvious in sections and are perhaps lost. In patellids, however, the ciliary covering is dense and permanent (E. B. Andrews, 1985). The cells of the right kidney in *Patella* differ from those in vetigastropods in further details, being more heavily pigmented, having less clearly crystalline concretions, and in a greater tendency for cells at the same physiological stage to be grouped within the epithelium.

The left kidney of the diotocardian prosobranchs is a much more variable organ than the right, from which it differs in almost all respects, making it unique amongst molluscs. It lies frequently in the mantle skirt (*Haliotis*, trochoideans, *Neomphalus*, lepetodriloideans, *Addisonia*) but in the posterior wall of the mantle cavity in fissurelloideans and patello-gastropods and rarely extends amongst the viscera; it is always less capacious than the right and in patellogastropods becomes small and in fissurelloideans minute; its histology is unlike that of the right kidney, and there is often a specially elaborated area forming what is known as the nephridial gland; finally, the relationship of the kidney to the vascular system and, in particular, to the left auricle of the heart is both special and complex.

In vetigastropods (apart from slit and keyhole limpets) the left kidney has the form of a papillary sac so called because the wall bears a large number of digitiform papillae projecting into the lumen. These are richly vascularized and often contain crystals, first noted by Perrier (1889) and then believed to be protein reserves, but shown by E. B. Andrews (1985) to be polymerized haemocyanin in crystalline form. The kidney lies on the left side of the body in the mantle skirt, its posterior wall adjacent to the anterior wall of the pericardial cavity and its left side close to the efferent vessel from the left ctenidium. Along the parts contiguous with the pericardium and efferent branchial vessel in trochoideans (but not in pleurotomarioideans) there lies the nephridial gland, distinguishable by the fact that tubular extensions of the lumen of the kidney open between the papillae, giving that whole area a spongy texture.

Away from the nephridial gland the walls of the papillary sac are covered by cells of one type the ultrastructural features of which (Delhaye, 1976; E. B Andrews, 1985) show clearly that they are concerned with the transport of material: they are cuboidal, their outer surface covered with microvilli and some cilia, much of their volume occupied by a large nucleus and devoid of concretions, their apical cytoplasm rich in vesicles of which some open to the lumen and which may interlink to form an endocytotic canal system similar to that described in cells of the digestive gland by Owen (1972). The base of each cell is drawn out into many long and delicate processes which extend into the underlying blood spaces. Blebs of cytoplasm are nipped off from the cell surfaces into the lumen of the kidney.

The cells lining the nephridial gland are, on the whole, similar but have an even more extended and folded basal surface in contact with blood and this part of the cell is rich in mitochondria. The inner and outer surfaces of the gland itself are highly folded so that the total area exposed to blood and to the fluid in the kidney becomes very large indeed, providing not only an extensive site for exchange of materials but also a place where blood may be stored. The nephridial gland also contains many muscle fibres the activity of which speeds the circulation of blood through the gland. This circulation is correlated with cardiac activity (see p. 653)

In two other groups of archaeogastropods the left kidney shows differences from that of trochoideans and pleurotomarioideans: these are the two groups of limpets, the patello-gastropods and the fissurelloideans. In both the kidney lies in the posterior wall of the nuchal cavity and so in the anterior face of the visceral mass, the ancestral position of kidneys. This may be a persistent primitive character but, since it is shared by two groups not otherwise closely related, it is more probably secondary, due to shortening of the mantle skirt as the animals assumed the limpet form. In both groups, too, it is small, so small indeed is fissurelloi-deans as to suggest that it is a vestigial organ; in neither group does it possess papillae, though their loss may be a direct consequence of small size, nor does it have a conspicuously distinct nephridial gland. Despite its small size—in *Emarginula fissura* only about 300 μm diameter (E. B. Andrews, 1985)—the kidney is a fully functional organ, connected to the pericardial cavity by a well developed renopericardial canal, and its cells resembling those of the trochid papillary sac. The histology of the patellogastropod left kidney differs from both the trochoi-dean and fissurelloidean pattern, but resembles them in others: the cells are richly ciliated, possess few basal processes and microvilli, lack concretions, and rest on a thick basal lamina over constricted blood spaces. The part of the kidney wall which abuts the pericardial cavity is thick and spongy, with tubules, and is therefore very reminiscent of a nephridial gland.

In addition to the ultrastructure of its cells the left kidney of diotocardians also differs from the right in its relationships with the vascular system. In *Haliotis* (Milne-Edwards, 1846; Crofts, 1929), in contrast to the right kidney, it receives no venous blood directly, either from the haemocoelic spaces of the head-foot or from those of the visceral hump, but may do so indirectly by way of a link with the afferent of the left ctenidium; its sole other vascular connection is with the efferent of the same organ. The left kidney is richly supplied with blood so that these vessels are not functionless, but difficulties appear in attempting to assess their function. If the vessel from the branchial afferent feeds blood into the kidney it brings blood which has passed through the right kidney but which is deoxygenated; if, in conjunction with this, it is assumed that the vessel to the branchial efferent is a straightforward outlet for blood from the kidney, it would mean that that organ received no supply of oxygen except what it might pick up directly from the water in the mantle cavity, which has already passed over the gill surface. As a further complication deoxygenated blood from the kidney would then be added to the oxygenated blood in the branchial efferent. To avoid this situation there appears to have arisen an important change in that the link with the ctenidial efferent has become the source of oxygenated blood as well as its drain, an idea first supported by Crofts' (1929) observation of a possible tidal flow in the vessel. That this is the most significant vascular connection of the left kidney is stressed by the fact that the link with the afferent branchial is lost in *Emarginula*. The same vascular connections exist in *Monodonta* as in *Haliotis*, but the amount of deoxygenated blood entering the kidney from the afferent branchial is minimal, most being oxygenated blood from the link with the efferent branchial. In trochids (and also in monotocardians) this vessel relates mainly to the nephridial gland, for which reason it is commonly called the nephridial gland vein. In *Monodonta*, as demonstrated by E. B. Andrews (1985), the meeting point of the nephridial gland vein and the branchial efferent lies where both connect with the auricle—indeed some blood spaces in the gland open directly to that heart chamber—and the anatomical arrangements of the vessels, together with changing pressures as the heart beats, bring about alternating inflow and outflow of the blood in it. An even more direct connection between auricle and nephridial gland is shown in peltospiroi-dean limpets from hydrothermal vents (Fretter, 1989): here (fig. 326) the nephridial gland extends as an outpouching from the kidney between the pericardial wall on one side and the

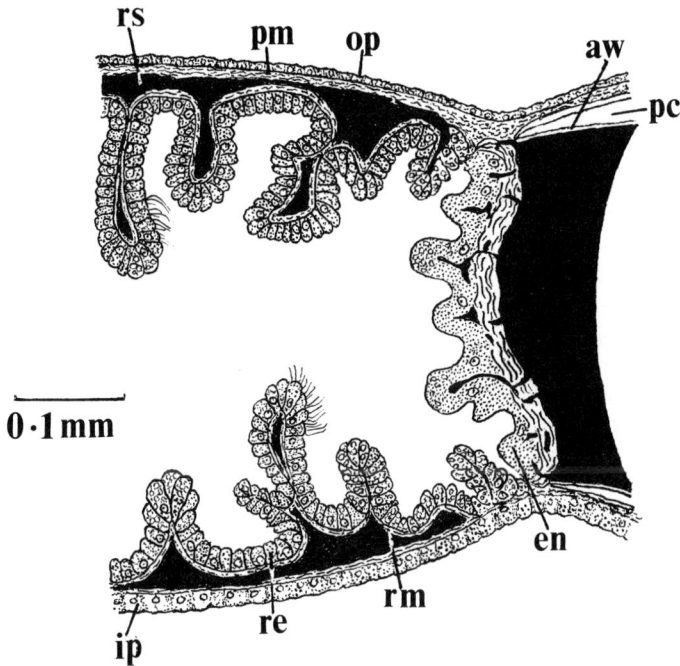

FIG. 326.—*Rhynchopelta concentrica*. Diagrammatic section of part of the mantle skirt showing the posterior part of the kidney and the anterior wall of the auricle, where renal, pericardial and auricular walls are fused; perforations in the network of muscles allow passage of blood between the auricle and blood spaces in the nephridial gland. The blood flow must be controlled by the heart beat: at auricular systole the perforations will be closed and will open at diastole, when blood from spaces in the nephridial gland will be sucked into the auricle. Blood is shown in black.

aw, auricular wall; en, epithelium of nephridial gland; ip, inner pallial epithelium; op, outer pallial epithelium; pc, pericardial cavity; pm, pallial muscles; re, renal epithelium; rm, renal muscles; rs, renal sinus (deoxygenated blood).

auricular wall on the other, to both of which it is fused. There are perforations, however, in the fused kidney-auricular wall which allow blood to flow between the auricle and the vascular spaces in the kidney. These are therefore flushed with blood at each contraction of the heart and drained at each relaxation.

That the two kidneys of archaeogastropods have different physiological roles has been clear from the early researches of Perrier (1889) and Cuénot (1899, 1914) and has been confirmed since by the work of Harrison (1962) and Delhaye (1976). From this it is clear that the right kidney is the primary excretory organ of the archaeogastropods, probably mainly responsible for the elimination of nitrogenous waste. This it may excrete directly as ammonia, or, if purine, elaborated into concretions (Delhaye, 1976; E. B. Andrews, 1981). It is also involved in the excretion of such substances as acid dyes (Harrison, 1962), and can remove particulate material from the body (Delhaye, 1976). Since ammonium ions are known to interfere with the mechanism of resorption, that activity cannot be pronounced in this kidney, though it is known to occur slowly. As much of the nitrogenous excretory material is from metabolic activity in the digestive gland the close connection between gland and kidney, together with the situation of the latter, interrupting the blood flow to the respiratory organs and heart, are likely to facilitate the process of excretion. The large capacity of the kidney allows the storage of urine prior to discharge, which gives time for a slow resorptive process to act, and may be a matter of importance to archaeogastropods like patellids which live at high levels on the shore.

The left kidney, on the other hand, seems less involved in nitrogenous excretion; it has not lost all excretory importance, however, but exhibits more pinocytosis and deals with larger molecules, voiding such substances as basic dyes (Harrison, 1962). It acts more as a resorptive organ, withdrawing such soluble materials as glucose from the urine in its lumen (Martin et al., 1954) by more efficient mechanisms than those operating in the right kidney, still functional because of the absence of ammonium ions in the blood with which it is supplied. Its relatively small capacity, great surface area, and the pumping of blood in and out of its vascular spaces (all factors which favour this resorptive activity) allow rapid treatment of small volumes of urine (E. B. Andrews, 1985). In addition, this kidney, in particular the nephridial gland, is rich in amoebocytes which phagocytose particulate matter in the blood and then escape through its walls.

The situation of the left kidney in the mantle skirt, its unusual vascular connections, and its resorptive role have all been linked by the work of E. B. Andrews (1985). The vascular connections of the two kidneys (fig. 327) are such that the blood reaching each differs in several important points: that to the right, coming from the viscera and head-foot, is deoxygenated, has a lower pH, and is rich in metabolites ready for excretion, some of which antagonize resorption; that to the left, coming from the ctenidium, is oxygenated, has a higher pH, and has a low content of metabolites since it has already passed through the right kidney, circumstances which permit resorptive mechanisms to operate. All these features have allowed the two organs to become specialized for different functions. The right kidney retains the original situation of a gastropod kidney and functions in the traditional manner: the left is a new type of organ, operating in a new way. The conditions necessary to achieve this are (1) its migration into the mantle skirt; (2) the provision of a supply of oxygenated blood; and (3) the loss of contact with the blood laden with certain metabolites. The first may have occurred as the left side of the body enlarged with the assumption of helical coiling; the second is an innovation dependent upon the formation of a new vascular link with the efferent ctenidial vessel; the third is a straightforward loss. The successful functioning of the arrangement depends on the reduced possibility of venous blood reaching the kidney from the afferent branchial and, once a through flow of blood has been stopped, upon a tidal flow in and out of the kidney from the efferent ctenidial vessel. This, E. B. Andrews (1985) suggested, is determined in Monodonta, and in other species (see p. 653), by the detailed anatomical arrangements of this vessel and the nephridial gland vein and their close connections with the left auricle, to which some of the vascular spaces in the kidney discharge directly, and the beating of which affects and controls the blood flow.

Since right and left kidneys are both supplied by way of the renopericardial canals with the same filtrate of blood, the primary urine, as is found in the pericardial cavity, it would not seem an efficient arrangement if the one kidney could only add nitrogenous waste to it, letting some valuable solutes escape unresorbed, whilst the other had no excretory importance. The lack of the power to excrete nitrogenous material in the left kidney might perhaps be thought of minor significance since the blood supplied to it has already passed through the right kidney and the ctenidium, and may therefore be assumed to have lost most of its load of

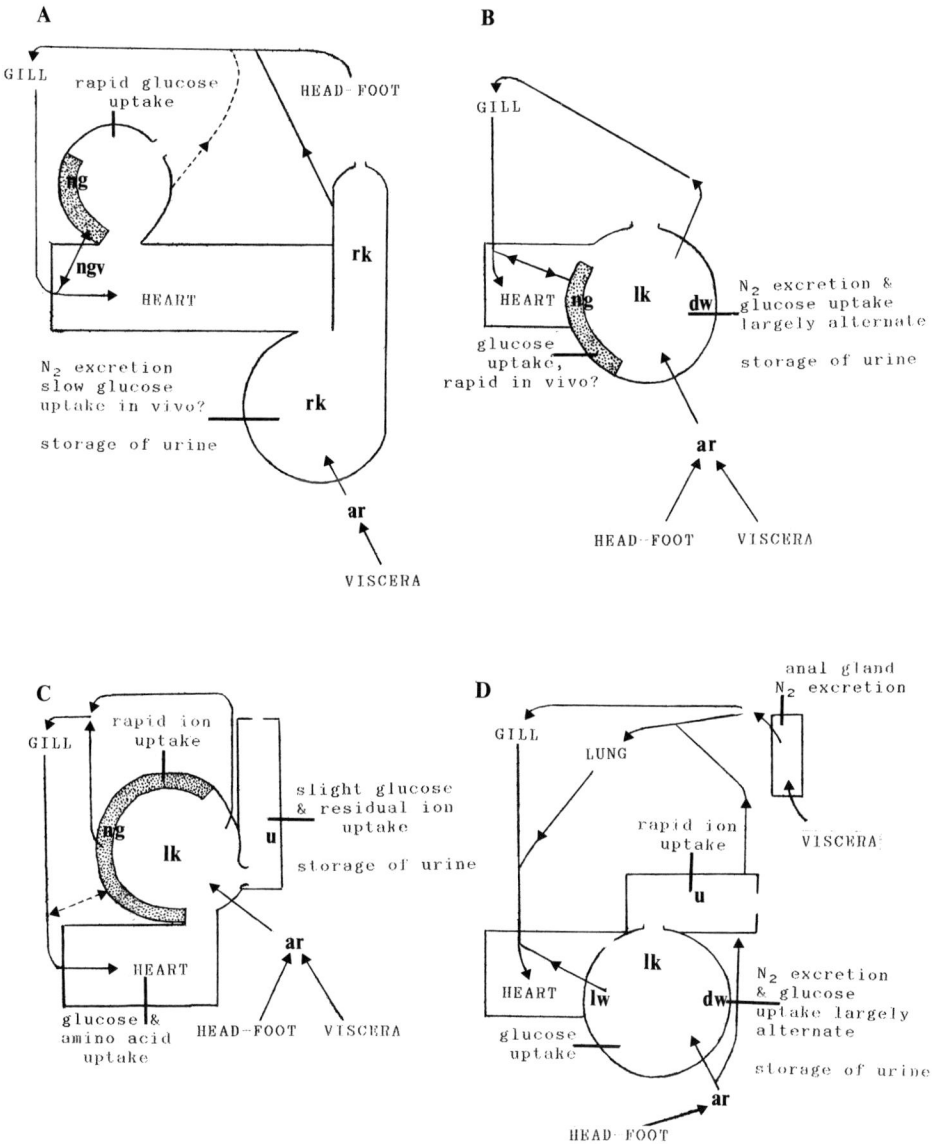

FIG. 327.—Diagrammatic representation of the excretory system and associated vessels, also of the connections between the kidneys, pericardial cavity and mantle cavity, in A, *Monodonta* (marine archaeogastropod); B, *Littorina* (marine caenogastropod); C, *Viviparus* (freshwater architaenioglossan); and D, *Pomacea* (amphibious architaenioglossan). Arrows indicate the direction of blood flow. Dashed line represents vessel open only when the snail retreats into its shell. In *Pomacea* the kidney receives blood only from the head-foot: its resorptive capacity is reduced since some resorption is undertaken by the ureter; resorption from the visceral blood occurs in the anal gland. (After E. B. Andrews & Taylor, 1990).

av, afferent renal vessel; dw, dorsal wall of left kidney; lk, left kidney; lw, left wall of left kidney; ng, nephridial gland; ngv, nephridial gland vein; rk, right kidney; u, ureter.

excretory material. The lack of resorption in the right kidney could, however, constitute a serious drain on the animal's resources. Several factors appear to prevent such a situation arising. Since both kidneys share a common origin some resorptive and some excretory potential might be expected to persist in both, as has, indeed, been known since the earliest work on kidney function. In addition, however, as has been shown by Harrison (1962), the right renopericardial canal of *Haliotis rufescens* is narrower than the left; Crofts (1929) failed to find the right canal in *H. tuberculata*, and for a while some doubt existed about that in *Patella*, suggesting that it is narrow throughout the archaeogastropod series, though Fretter (1988) found it unusually broad in some deep-sea limpets. In addition it tends to be long so that resistance to the movement of liquid through it will be greater. In these circumstances the flow of primary urine will be slower and less to the right than to the left kidney. Since the capacity of the right kidney is great it appears likely that the urine will accumulate there and possibly stay for a long period, whereas the left can treat only a small volume but has a rapid turnover. In the right kidney the lower capability for resorption is compensated for by the long stay of its contents, whereas in the left the combination of small volume and large surface allows resorption as well as such excretion as occurs there.

Prosobranchs other than archaeogastropods (those known as caenogastropods or monotocardians) are distinguished by the great reduction or total loss of the right organs of the pallial complex. Despite the fact that throughout the archaeogastropod series the right kidney is enlarged, often markedly so, and is the main organ of excretion, when monotocardians are examined it is found to have apparently abruptly vanished as an excretory organ, as have also the right ctenidium, auricle and osphradium (fig. 327).

In the developmental stages of monotocardians, however, as in archaeogastropods, the rudiments of two kidneys appear. The right rudiment fails to develop as an excretory organ but as it still receives the developing gonadial duct it comes to constitute part of the adult genital tract, the original connection of the rudiment to the pericardial cavity persisting in the females of some species as a gonopericardial canal, whilst the original connection of the rudiment to the mantle cavity becomes the genital pore. Thus whereas the part of the archaeogastropod right kidney which is concerned with excretion has aborted, the part which acted as conduit and outlet for gametes has persisted. The rudiment of the left kidney, on the other hand, develops to form the definitive excretory organ, linked to the pericardial cavity by a renopericardial canal and opening to the mantle cavity. Though the organ coincides in these details and usually in the possession of a nephridial gland with the left kidney of archaeogastropods, with which it is clearly homologous, it does not resemble it in others, most of which depend on the fact that, as the sole excretory organ of the caenogastropod, it has to undertake all the functions carried out by the two kidneys of the archaeogastropod. It differs further in not being a papillary sac, in its relations with the vascular system, and in not lying in the mantle skirt but in the base of the visceral mass, except in viviparids and terrestrial species, where it retains its pallial position.

The left kidney lies alongside the pericardial cavity but is more median in position; along its left wall, abutting the pericardial cavity, is the nephridial gland from which a blood vessel runs to join the efferent ctenidial vein at a point close to, or at, its entry to the auricle. The relationships of the nephridial gland, the nephridial vessel, the efferent branchial vein, and the auricle are strictly comparable to those found in archaeogastropods, and there exists the same tidal flow of oxygenated blood between the efferent branchial and the vessels of the nephridial gland, its timing dictated by the heart beat, as in these animals (E. B. Andrews & Taylor, 1988).

Away from the region of the nephridial gland the wall of the kidney (the dorsal wall) presents a series of folds invaded by blood vessels which branch from an afferent renal vein formed by fusion of veins coming from sinuses draining the head-foot and the visceral mass. This part of the kidney thus receives blood which is deoxygenated and rich in metabolites and so comparable to the blood received by the right kidney of an archaeogastropod. In theory, there is the possibility of some mixture of the two types of blood entering the kidney, but the areas supplied by the two vessels are well separated and it is unlikely that in normal circumstances mixing occurs to a significant extent, if at all, except when the snail is retracted within its shell. In effect, therefore, the single left kidney of the caenogastropod is a compound organ with two distinct parts, one, the nephridial gland area, functionally equivalent to the left kidney of the archaeogastropods, and the other, the dorsal wall, to their right.

This equivalence extends, though never wholly, to the ultrastructure of the component cells (fig. 328). In the nephridial gland where, typically, numerous tubules project from the main surface of the kidney into underlying blood spaces, most cells resemble those in the corresponding situation in archaeogastropods; there are also many interspersed ciliated cells maintaining currents over the surface (Delhaye, 1974c, 1975; E. B. Andrews, 1981). In addition to these E. B. Andrews (1981) described a 'mucoid' cell which she regarded as absent from archaeogastropods. Since then however, further study (Andrews in litt.) has persuaded her that these cells are homologous with the excretory cells on the dorsal folds of the kidney. The cells in the nephridial gland differ from them, nevertheless, in being devoid of uric acid or other purines since these have been removed from the blood before it reaches the gland.

The dorsal wall of the kidney, its surface increased by folds vascularized with deoxygenated blood, is the main site of nitrogenous excretion. Two types of cell occur there, one bearing cilia and microvilli, and the other with microvilli only. The former has the obvious function of circulating fluid over the surface of the epithelium, but it is also involved in the transport of material from blood to lumen as well as in the opposite direction (Delhaye, 1974c, 1975), activities for which the microvillous covering and the basal processes showing evidence of pinocytosis are important adaptations. In addition the apical cytoplasm contains many vesicles, some opening to the surface. The second type of cell is that responsible for excretion of nitrogenous waste taken up from underlying blood spaces, gathered into vacuoles, and passed to the lumen of the kidney in blebs of cytoplasm. In the typical marine monotocardian the excretory material in the vacuoles is in the form of small granules (Delhaye, 1974c, 1975; E. B. Andrews, 1981) and does not become aggregated into large concretions as in archaeogastropods. The difference may relate to water supply since aggregation does happen in snails living in places where they are exposed to possible dehydration, like the salt marsh snail Assiminea grayana (Little & Andrews, 1977), or which have to aestivate during a dry season, like some pilids (E. B. Andrews, 1976a) and cyclophorids (E. B. Andrews & Little, 1972). In this part of the left kidney one of the two types of cell, the ciliated cell, is identical with that found as the sole type in the main part of the papillary sac of an archaeogastropod; the second, though presenting some resemblance to the excretory cell of the right kidney, shows features, such as delicate and elongated basal processes, which suggest that they are papillary sac cells altered in response to their changed biochemical environment to carry out a new function.

The kidney which has been described up to this point is that which would be found in such marine monotocardians as littorinids or rissoids. Various modifications of this pattern are found, some of which seem to be related to diet, others either to the passage from a marine to

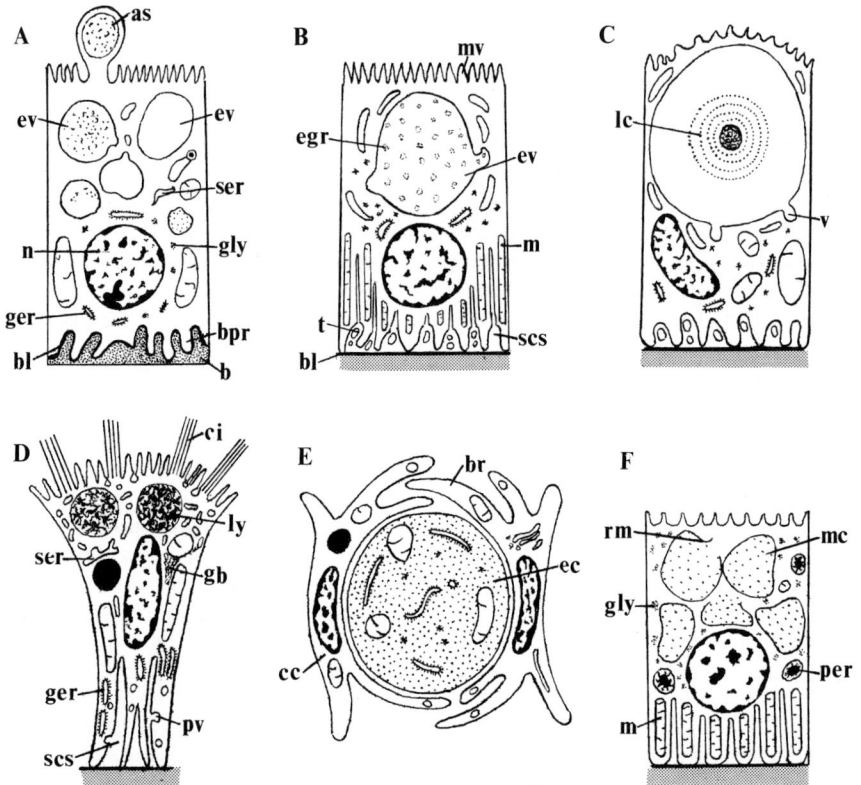

FIG. 328.—Diagrams of vertical sections (except E) of the types of cell in the kidney of some monotocardians. A, excretory cell of *Littorina littorea* (full length of basal processes not shown); B, excretory cell of *Assiminea*; C–F, cells of ampullariid or cyclophorid; C, excretory cell; D, typical pigmented ciliated cell; E, horizontal section of an excretory cell embraced by two ciliated cells; F, mucoid cell of nephridial gland.

as, apocrine secretion; b, blood; bl, basal lamina; bpr, basal process; br, branch of ciliated cell interdigitating with those of neighbouring cell; cc, ciliated cell; ci, cilium; ec, excretory cell; eg, excretory granule; ev, excretory vacuole (numerous in A, single in B); g, golgi body; ger, granular endoplasmic reticulum; gly, glycogen; lc, layered concretion; ly, lysosome; m, mitochondrion; mc, mucoid secretion; mv, microvilli; n, nucleus; per, peroxisome; pv, pinocytotic vesicle; rm, ruptured membrane; scs, subcellular space containing blood filtrate without haemocyanin (this varies in size with the availability of water and its salinity); ser, smooth endoplasmic reticulum; t, cut tips of basal processes; v, vacuole feeding excretory vacuole. (After E. B. Andrews, 1981).

a brackish or freshwater environment, or to the adoption of a terrestrial mode of life. These changes are in part anatomical, but in so far as they accord with the idea put forward by Needham (1935) that the end product of nitrogen metabolism relates to habitat, they are also in part biochemical.

A change in organization perhaps related to a carnivorous diet is found in the mesogastropod naticids (fig. 329), eratoids, and cypraeids, and throughout the neogastropods, all these animals being marine. In the mesogastropods the folds on the dorsal wall occur in two regions, in one of which they are identical with those of a littorinid; in the other the folds differ

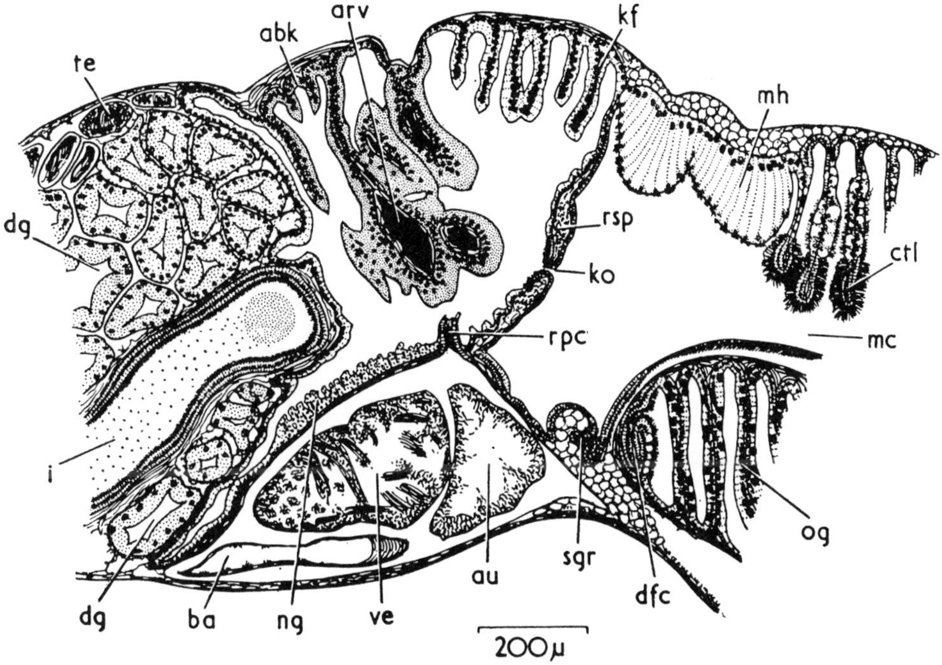

FIG. 329.—*Lunatia alderi*. Longitudinal section through the upper part of the mantle cavity and the base of the visceral hump to show the kidney sac in relation to surrounding organs.

abk, resorbing epithelium of kidney; arv, afferent renal vessel; au, auricle; ba, aortic bulb; ctl, ctenidial leaflet; dfc, dorsal food channel of oesophagus; dg, digestive gland; i, intestine; kf, folded surface of kidney; ko, kidney opening; mc, mantle cavity; mh, mucous cells of hypobranchial gland; ng, nephridial gland; og, oesophageal gland; rpc, renopericardial canal; rsp, renal sphincter muscle; sgr, seminal groove; te, testis; ve, ventricle.

in their cellular organization and presumably, therefore, in their function, though what that may be is not known. In the neogastropods, in addition to the folds which occur regularly in all caenogastropods, there has developed a second series which alternate with them. In *Nucella* (fig. 330) the former receive blood from a branch of the afferent renal vein running over the ventral wall of the kidney sac, and the latter from a branch running over its dorsal wall. Both sets therefore receive the same kind of blood and both are similar histologically: this modification thus seems to be simply an increase in the surface of the kidney which is available for voiding the increased nitrogenous excretory material which might be expected in such carnivorous animals. In *Buccinum*, however (E. B. Andrews, 1988), the blood supply to the two sets of folds is not identical, the primary set receiving blood from the visceral mass and the secondary set blood from the head-foot. The latter may well be both richer in oxygen and poorer in excretory products and this may correlate with histological differences observable in the two sets.

Modifications of kidney structure are more profound in those prosobranchs which live in brackish and fresh water and on land. The former must prevent their tissues being flooded by osmotic uptake of water and, whilst eliminating that, ensure that salts are not lost with it. The

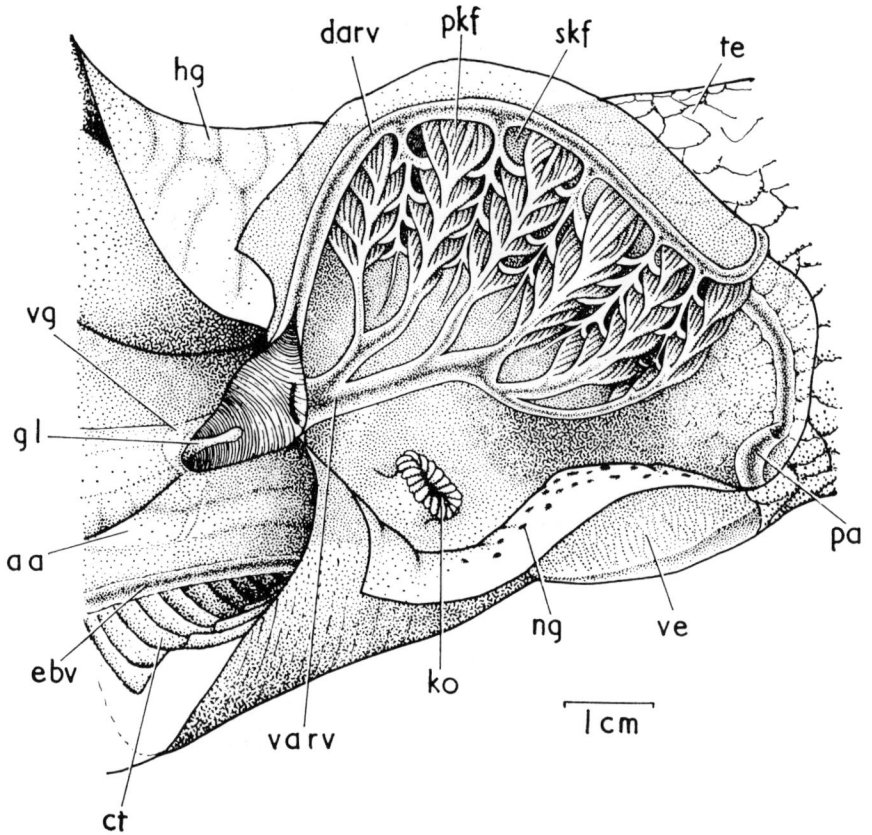

FIG. 330.—*Nucella lapillus*. Kidney sac opened to show double series of folds on the wall.

aa, anterior aorta; ct, ctenidium; darv, dorsal branch of afferent renal vein; ebv, efferent branchial vessel; gl, posterior tip of gland of Leiblein; hg, hypobranchial gland; ko, kidney opening; ng, nephridial gland; pa, posterior aorta; pkf, primary kidney folds; skf, secondary kidney folds; te, testis; varv, ventral branch of afferent renal vein; ve, ventricle; vg, visceral ganglion.

two activities, as emphasized by Little (1985), are counter-effective: the more rapidly excess water is voided the more likely is resorption from it to be incomplete. Terrestrial species, on the other hand, must conserve water, especially if they live where aestivation over a dry season may be necessary.

The most thoroughly investigated freshwater prosobranchs from the point of view of excretory activity are the viviparids and pilids, of which the former are confined to fresh water whereas the latter are amphibious. In *Viviparus* (E. B. Andrews, 1979, 1981) the kidney (figs 327, 331) is a rather small structure confined to the mantle skirt and connected to a pore lying alongside the anus by a long ureter, probably derived from the epithelium lining the mantle cavity as it is in the pilid *Marisa* (Demian & Yousif, 1973). Though its total volume is considerable its walls are so spongy that the lumen is much reduced. The pericardial cavity, lying in the visceral mass, is greatly enlarged and seems to form a storage chamber for fluid about to

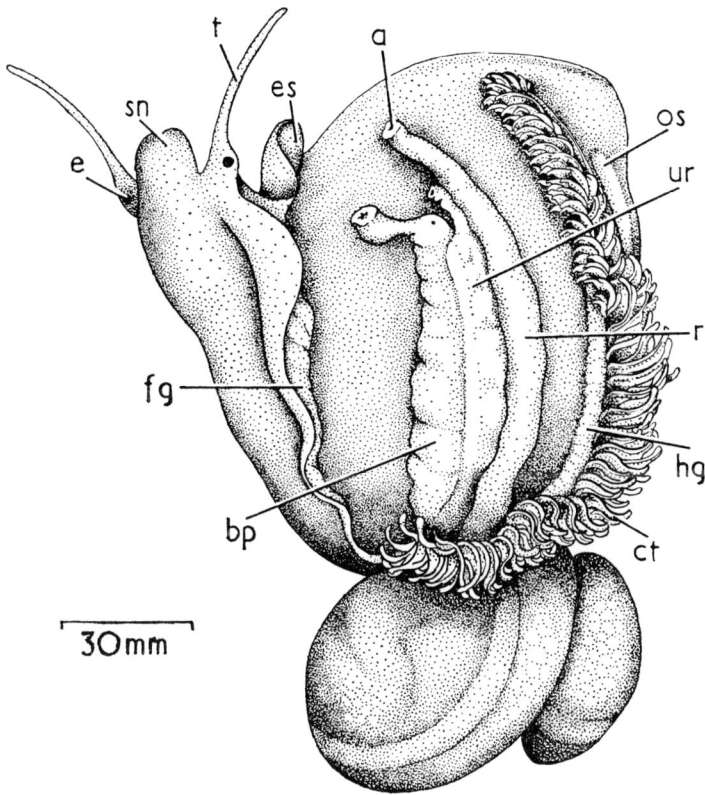

FIG. 331.—*Viviparus viviparus*. Female dissected to show contents of the mantle cavity. The animal has been removed from the shell and the mantle skirt cut along the left and turned to the right.

a, anus; bp, brood pouch; ct, ctenidium; e, eye; es, exhalant siphon; fg, food groove; hg, hypobranchial gland; os, osphradium; r, rectum; sn, snout; t, tentacle; ur, ureter.

be passed to the kidney in quantities the small size of which is dictated by the restricted cavity of that organ. In the kidney it is rapidly circulated, not only by the ciliated cells in its epithelium, though these are few in comparison with marine species, but to a greater extent by muscular activity of the walls (Little, 1965b). Although both areas characteristic of the caenogastropod kidney are recognizable, the folds of the dorsal wall have been lost and the kidney consists almost entirely of what must be regarded as an enlarged nephridial gland.

The form in which nitrogenous waste is excreted in viviparids has been explored by Spitzer (1937), Little (1981) and P. M. Taylor & Andrews (1991). There appears to be little uric acid in the urine, though there is what may be a long-term store in underlying connective tissue (E. B. Andrews, 1981) and in the related *Sinotaia* there is equally little urea (Horne & Barnes, 1970). The situation is therefore most easily interpreted on the supposition that what nitrogen is not dumped as purine in connective tissue, is lost primarily as ammoniacal nitrogen, which may escape from many places other than the kidney. The digestive gland here, as in many other freshwater prosobranchs, is also involved in nitrogenous excretion. The excretion of ammoniacal nitrogen, so frequent in aquatic invertebrates, is not available to terrestrial ones. Of such prosobranchs the cyclophorids produce excretory spherules containing much uric acid. In *Pomatias elegans*, however (Martoja, 1975), uric acid is absent: this is presumably to be linked with the presence of the concretion gland (Kilian, 1951) which gathers and stores uric acid, whilst excretory activity has developed in the cells of the pericardial wall.

Some activity in the kidney of *Viviparus* produces spherules in the tubules of the nephridial gland which Delhaye (1974c) took to be excretory. E. B. Andrews (1979), however, believed that they were part of what is clearly the most important function of this kidney, the resorption of ions from the primary urine whilst eliminating water. P. M. Taylor & Andrews (1991), using X-ray microanalysis, have shown that they contain virtually no nitrogenous waste. Since they contain some materials found in similar spherules in other species they may be produced in homologous cells. An indication of the size of this task is given by the fact that the final production of urine is up to twenty times greater than in a marine prosobranch (Little, 1985). Since, nevertheless, most of the water of the primary urine is, perhaps rather surprisingly, resorbed (Potts, 1975), the fact that 94% of its ions are captured (Burton, 1983) indicates that the mechanism of resorption is highly efficient. Not all resorption, and apparently none of organic solutes, is carried out in the kidney: resorption of glucose (Little, 1979) and of amino acids (P. M. Taylor & Andrews, 1987) is effected by cells in the ventricular walls from the primary urine in the pericardial cavity, and there is further removal of ions from the urine through the walls of the ureter (Little, 1965a).

In cyclophorids, which are terrestrial and may aestivate, E. B. Andrews & Little (1972) have described how the hypobranchial gland has become markedly elaborated. It is divisible into two histologically distinct areas, anterior and posterior, and its surface becomes folded to form a series of flask-shaped pockets opening narrowly to the surface. The gland is partly secretory and partly excretory, the latter function presumably linked to the animals' aestivating habit.

The heart of the prosobranch consists in lower forms of a ventricle flanked by left and right auricles, the latter often the smaller, in higher forms of a ventricle and one auricle, the left. It has usually been accepted that it works on the constant volume principle proposed independently by Ramsay (1952) and Krijgsman & Divaris (1955) to explain, primarily, how the auricle is filled in a monotocardian heart, on the supposition that the driving force of the ventricular contraction is exhausted at the end of the systemic and branchial circulation. It supposes that the total volume of the pericardial space and contained heart is constant so that when the ventricle contracts pressure is lowered in the pericardial cavity and blood is sucked into the auricle. It rests on the assumptions that auricular and ventricular volumes are equal and that the pericardial walls are rigid or, since that is clearly not the case, that change in shape in one place is compensated by change in shape elsewhere. Though the formation of urine in molluscs by filtration through the heart wall into the pericardial cavity and its subsequent escape to the kidney by way of the renopericardial canal had been previously proposed by Picken (1937), the consequences of these activities for the constant volume theory were not considered, largely because it was propounded in relation to pulmonates, where filtration through the heart wall was believed not to occur. The work of Civil & Thompson (1972) showed that, at least in *Helix*, the pericardium did constitute an essential part of the cardiac mechanism, and the constant volume principle has been claimed to operate in certain bivalves (Florey & Cahill, 1977; H. D. Jones, 1983), though the auricle is not necessarily involved in the production of primary urine in these molluscs (Jennings, 1984; E. B. Andrews, 1988). Sommerville (1973) has given convincing reasons that the constant volume theory is only approximately true in *Helix*. The most detailed investigation of this point, however, is that of E. B. Andrews & Taylor (1988) described below.

The wall of the heart exhibits three layers: on the pericardial side it is covered with (1) the epicardium, a cubical or nearly squamous epithelium resting on (2) a basal lamina, internal to which is (3) a layer of muscle cells from which irregularly arranged trabeculae cross the lumen. The muscle layer is never complete so that some blood comes into direct contact with the basal lamina of the epicardium.

Separation of blood for filtration and blood for circulation is achieved primarily in the two auricles of an archaeogastropod or the single auricle of a caenogastropod, and their increasing structural complexity as one moves from the archaeogastropod grade of organization to the caenogastropod allows a progressively more complete separation. Some porosity also occurs in the ventricle of some species, but its contribution to the filtrates is normally insignificant.

Two devices permit the movement of fluid from the lumen of the heart to the pericardial cavity. The first filtrating device is the podocyte, a specialized epicardial cell similar to those found in the renal corpuscles of the vertebrate kidney and first found in prosobranchs by E. B. Andrews & Little (1971). In this type of cell (fig. 332) the cytoplasm extends into a number of stout branches, about 14 μm long in *Gibbula cineraria* (E. B. Andrews, 1976b), from which arise numerous finer processes (pedicels), about 500 nm long and 100 nm in diameter, often running under the main body of the podocyte. They are separated by gaps only a few nanometres in width, about 33 nm in *Gibbula cineraria*, which are reduced still further by projections (diaphragms) from the side of the pedicels. The podocytes and their pedicels lie over the basal lamina and that, in turn, is in contact with the blood in the underlying heart chamber. Driven, or sucked, by the difference in pressure between the heart and pericardial cavities blood is forced through the basal lamina, which acts as a fine primary filter, and then between the diaphragms and pedicels, which form a less fine secondary one, into a space, the subcellular space, between the pedicels and the main bodies of the podocytes. The second device which allows fluid to pass through the heart wall is a series of channels, about 300 nm in diameter in *Gibbula cineraria*, which penetrate the cell bodies of the podocytes (and sometimes of other epicardial cells as well). The primary urine thus passes two filters on its journey from the cavity of the heart, and finally enters the pericardial cavity by way of the channels through the epicardial cells.

Though it is unlikely to correspond with the evolutionary pathway actually followed, E. B. Andrews (1976b, 1981, 1985, 1988) has charted a series of stages which suggest what this may have been. In the archaeogastropods *Gibbula cineraria* (E. B. Andrews, 1976b) and *Patella vulgata* (Økland, 1982) the epicardium of the auricles is composed entirely of podocytes, and as this is also true of chitons (Økland, 1980) it may well be the original condition found throughout the phylum. Even at this level of organization, however, some indication of separation of filtratory and pumping activity is obvious (E. B. Andrews, 1981, 1985) with the formation of special filtration chambers, as seen in *Haliotis* and *Emarginula*. In the latter podocytes are largely limited to a bay in each auricle which can be isolated from the main auricular chamber by muscles; in the former similar chambers are visible on each auricle. A similar localization of podocytes to form filtration chambers occurs in the auricle of caenogastropods, but with a more elaborate organization. Here small groups of podocytes occur at intervals in local areas of the epicardium, the cells lying between being simple squamous epithelial cells. The processes and pedicels of the podocytes spread under these cells and over the underlying basal lamina, which is in contact with the blood in the auricle.

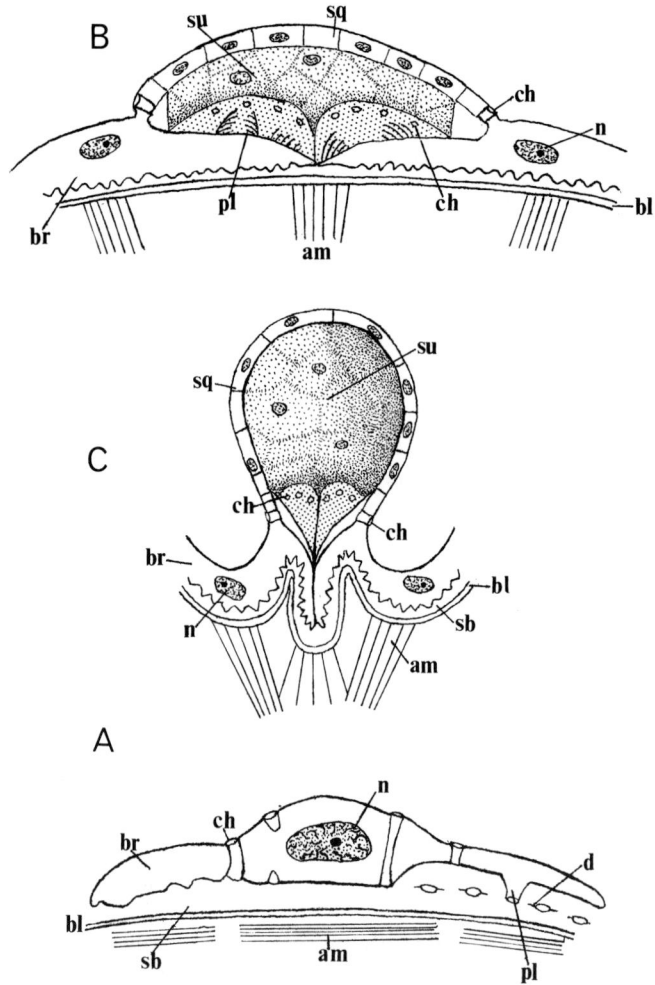

FIG. 332.—Diagrammatic sections showing auricular epicardial structure in A, *Gibbula cineraria*, B and C, *Viviparus*; in B the auricle is relaxed, in C, contracted. In each diagram the pericardial cavity is towards the top and the auricular cavity towards the base of the figure. Cut surfaces of auricular walls are not shaded; the internal surface of the filtration pouch is stippled. (Based on E. B. Andrews, 1976, 1979).

am, auricular muscles; bl, basal lamina; br, branch of podocyte; ch, channel through podocyte; d, diaphragm (shortened for clarity); n, nucleus; pl, pedicel; sb, subcellular space; sq, squamous cells forming roof of filter chamber; su, filter chamber formed by expansion of subcellular space.

There is thus formed an enlargement of the subcellular space which is floored by podocyte processes and roofed by the epithelial cells, which have become raised from the basal lamina: this is the filtration pouch. Into this space blood is filtered from the auricle to escape to the pericardial cavity by way of the channels in the bodies of the podocytes. Pouches such as this may be relatively numerous but become fewer and more localized in higher prosobranchs;

they also vary in number in relation to habitat. The arrangement of the auricular musculature becomes modified as the evolution of filtration chambers proceeds, usually in such a way as to produce a partial isolation of the central part of the auricle from the more peripheral part underlying the filtration chambers: blood in the former is passed at auricular systole straight to the ventricle, that in the latter is at least in part forced into the pouches, and some of that filtered through their walls into the pericardial cavity.

The most complex examples so far described (E. B. Andrews & Taylor 1988) of the separation of the two kinds of blood and, incidentally, revealing the incorporation of the nephridial gland into the process, as if it were part of the heart, is shown by *Littorina littorea* (fig. 333).

The heart of this animal has one auricle, anterolaterally placed, and one ventricle, dorso-median in position, from which the aortae emerge by way of an aortic bulb. The ventricle is a relatively simple chamber which shortens and narrows on contraction, expelling all its contents to the bulb, valves preventing any reflux to the auricle. The auricle, by contrast, is more elaborate: its base connects with a space which is the meeting place of the efferent vessel from the ctenidium and another from the nephridial gland region of the kidney, previously regarded as an efferent vessel supplementing the more important one which takes blood from the kidney to the ctenidium. The auricle is subdivided into two parts, a narrow, more tubular, dorsal one, separated by a nearly complete sheet of longitudinal muscles from a more spherical and more capacious ventral part, the wall of which includes the only site of podocytes in the heart. The dorsal part connects anteriorly only with the nephridial gland vessel, and posteriorly only with the ventricle. The ventral chamber has no link with the auriculoventricular opening, being separated from that by the longitudinal muscles: anteriorly it can connect, but because of the detailed anatomical arrangements only at different stages of the cardiac cycle, either with the ctenidial efferent or with the nephridial vessel.

The events of the cardiac cycle as described by E. B. Andrews & Taylor (1988) are as follows. At systole the ventricle pushes blood into the aortae; the simultaneous expansion of the auricle sucks blood from the ctenidial efferent into the ventral auricular chamber with its filtration area, and from the nephridial vessel into the dorsal part, the latter process aided by muscular activity within the nephridial gland. A variety of anatomical devices, mainly the sheet of longitudinal muscle, ensures that the two streams do not mix or that admixture is minimal. At auricular systole increasing pressure opens the auriculoventricular valve and the contents of the dorsal auricular chamber are passed to the ventricle. At the same time rising pressure in the ventral chamber brings about filtration of the primary urine and the transfer of the rest of the contents round the side of the contracted sheet of longitudinal muscle to the nephridial vessel and its branches. The course of the blood through the heart is thus from the ctenidial vessel to the ventral part of the auricle, thence to the nephridial vessel and blood spaces of the nephridial gland, from which it later returns to the dorsal auricular chamber and so to the ventricle. The heart is thus in effect functionally three-chambered, and there is a tidal flow of postbranchial blood in and out of the nephridial vessel.

It follows from this analysis of cardiac action that the auricle must be capable of holding more blood than is discharged from the ventricle at each heart beat: at the beginning of auricular systole it holds a volume equal to the ventricular discharge in its dorsal chamber alone, and it has a further quantity in its ventral chamber. This latter volume is probably slightly greater than the former since it will be reduced by filtration of primary urine to the pericardial cavity (calculated as 0·7% of cardiac output), and by some possible resorption in the nephridial gland and storage in its blood spaces. At any rate the equivalence of auricular and ventricular volumes necessary for strict application of the constant volume mechanism

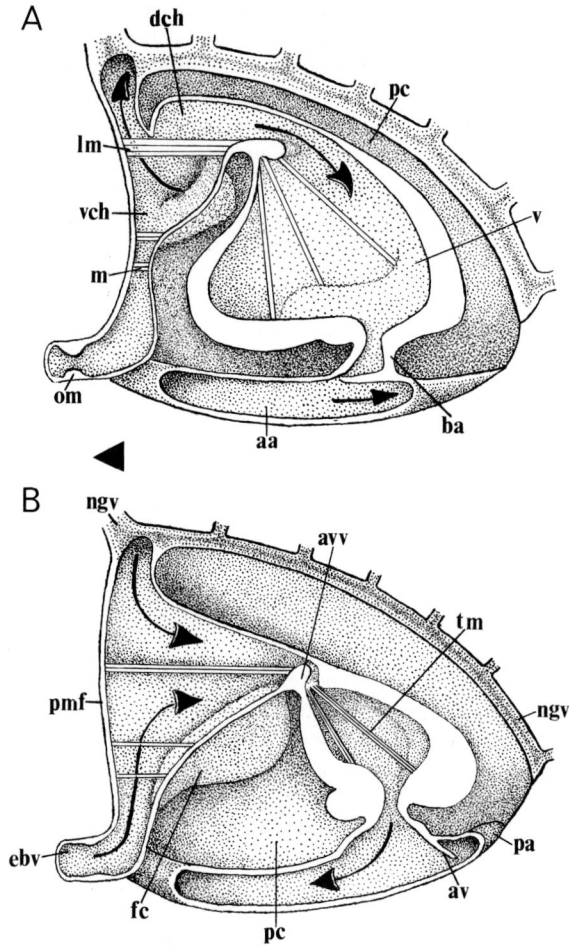

FIG. 333.—Diagrammatic representation of the right sagittal half of the heart of *Littorina littorea*. Cut surfaces are white. Arrows indicate the direction of blood flow, the arrowhead the anterior end. A, heart at auricular systole; B, heart at ventricular systole. (Based on E. B. Andrews & Taylor, 1988).

aa, anterior aorta; av, aortic valve; avv, auriculo-ventricular valve; ba, bulbus aortae; dch, dorsal chamber of auricle; ebv, efferent branchial vein; fc, filter chamber; lm, longitudinal muscles forming floor of dorsal auricular chamber; m, longitudinal muscles preventing reflux to efferent branchial vein; ngv, nephridial gland vein; om, occlusor muscle; pa, posterior aorta; pc, pericardial cavity; pmf, fused anterior wall of pericardium and posterior wall of mantle cavity; tm, trabecular muscles of ventricle; v, ventricle; vch, ventral chamber of auricle.

is clearly not achieved, even if it is assumed that escape of primary urine to the kidney along the renopericardial canal exactly balances what is gained by filtration from the auricle. The pericardial cavity, however, must still be accepted as an integral and important part, though not the whole, of the mechanism which controls the filling of the chambers of the heart. Though demonstrated only in *Littorina littorea* so far, the regular occurrence in prosobranchs of a nephridial gland vessel connected to the efferent ctenidial at the base of the auricle suggests that this may be the general mode of functioning of the heart.

Though the kidney is the organ most clearly concerned in the control of the internal environment of those prosobranchs living in brackish or fresh water and on land, the heart shows some modification too, since it is the source of the urine. By comparison with that of marine species the auricular surface of freshwater forms carries many more filtration pouches. In *Viviparus*, for example (E. B. Andrews, 1976b), and the same appears to be true of *Bithynia* (E. B. Andrews, 1988), filtration pouches cover the entire auricle with the result that the rate of production of primary urine can be very high: Little (1965b) gave a maximum filtration rate in *Viviparus* of 0·91 μl.g^{-1}.min^{-1}, a rate which demands a high resorption rate if material is not to be washed out of the body with the urine. A converse situation is encountered in terrestrial prosobranchs, especially those which face periods of aestivation, though the small number of prosobranch species which have succeeded in this habitat suggests that the problems of water conservation have proved more difficult to solve than that of resorbing substances from a copious urine flow. Conservation of water is most easily achieved by a restricted production of primary urine. Filtration pouches are few in number and the filtering area of each podocyte is reduced in terrestrial cyclophorids, for example (Andrews & Little, 1982), though the channels through the epicardial cells may persist. In the terrestrial littorinoidean *Pomatias elegans* and the high littoral *Melarhaphe neritoides* podocytes are lost, as they are also in the salt marsh snail *Assiminea grayana* (E. B. Andrews, 1988). As a consequence the rate at which urine is filtered through the heart wall is reduced: the figure given by E. B. Andrews & Little (1972) is 0·39 μl.g^{-1}.min^{-1}.

Recent investigation of circulatory function has been greatly facilitated by the arrival of sufficiently small and sensitive pressure transducers to permit measurements within various parts of the vascular system. Even so the small size of prosobranchs and the difficulty of access to many of their organs has limited study of their vascular functioning to a small number of appropriate species: in a recent review of this topic H. D. Jones (1983) mentioned only four.

Bourne & Redmond (1977) measured the pressures in *Haliotis corrugata* responsible for circulating the blood. At ventricular systole blood is expelled from the heart by a pressure of 8 cm of water above ambient, in aliquots of 100–180 μl at each heart beat, and a total flow of 2·1–3·6 ml.min^{-1}. The pressure falls to 4 cm as the blood approaches the ctenidia after passing through the various blood spaces of the body, and when it has traversed the gills drops still further to 1·9 cm in the efferent ctenidial vein, reaching the auricles with a residual pressure of 1 cm. In *Patella vulgata* H. D. Jones (1970) and H. D. Jones & Trueman (1970) recorded a ventricular systolic pressure of 5 cm of water, which dropped to 2 cm at auricular diastole. In *Littorina littorea* Kamel (1979) reported that ventricular systole produced a pressure of about 4 cm of water, a stroke volume of only 2·5 μl, and a total flow of 65 μl.min^{-1}, small, as might be expected in an animal of that size. A more recent investigation of haemodynamics in the same species (E. B. Andrews & Taylor, 1988) gave a ventricular systolic pressure of 3·0 $\pm$ 0·54 cm of water which had dropped to 0·7–1·3 cm in the afferent branchial vessel and to only slightly less, 0·7–1·1 cm, in the efferent branchial vein. Mean diastolic pressure in the auricle was 0·7 cm and systolic pressure 1·59 $\pm$ 0·27 cm. It is a little surprising that the figures given by Bourne & Redmond (1977) and E. B. Andrews & Taylor (1988) both suggest that after passing through the equivalent of three capillary beds—systemic, renal, ctenidial—there is still some degree of pressure left in the efferent branchial vessel; however, the still lesser pressure in the expanding auricle is an important factor in causing the blood to complete its course round the body, and some vessels have enough elastic and muscular tissue in their walls to produce a pulse.

In several gastropods the circulatory system possesses a structure allowing escape of blood if, because of sudden contraction of parts of the body, pressure within the system were to rise to levels which might damage the heart. This was first demonstrated when Lever & Bekius (1965) found a haemal pore in the pulmonate *Lymnaea stagnalis*; since then Fretter (1982) has found similar pores in some species of *Littorina*, though only an area of weakness in *L. littorea*, and Depledge & Phillips (1986) found evidence of blood loss, perhaps through the gill, when the neogastropod *Hemifusus tuba* was provoked into sudden withdrawal.

There are so many sites in the body of a prosobranch at which some exchange of respiratory gases with the environment is possible that it is difficult to imagine an animal getting into respiratory difficulties provided that it can expose the head-foot or mantle cavity to an environment rich enough in oxygen. Since respiratory exchange requires a moist external surface sufficiently delicate to allow diffusion of gases most of these sites are within the mantle cavity: where the ctenidia (or ctenidium) lie, there are found the secondary gills of patellogastropods, and there the mantle skirt itself is richly vascularized. All these structures drain blood directly to the auricle(s) so that the heart is a systemic heart, supplying oxygenated blood to all parts of the body. Though the blood system is technically what is called 'open', the spaces to which the arteries lead and which connect, in turn, with venous sinuses are, as noted by H. D. Jones (1983), not markedly different in dimensions from the capillaries of animals with 'closed' systems. The lack of cellular wall, however, probably implies the absence of control of circulation and permeability exercised by true capillaries.

Pelseneer (1935) measured the ctenidial respiratory surface of a series of prosobranchs and found figures ($cm^2.g^{-1}$) within the range 7·1 (*Nucella lapillus*) to 8·2 (*Gibbula cineraria*), suggesting a rather high degree of constancy, whether the animal had an aspidobranch or a pectinibranch ctenidium. Even his figure, 9·36, for *Patella vulgata*, which lacks a ctenidium but has pallial gills, is not far outside this range. The figures were quoted by Yonge (1947) who explained the very different values calculated for bivalves as due to the fact that bivalve ctenidia are of greater importance as food-collecting organs than as sites of respiratory exchange, for which ample surface has been provided in the enlarged mantle folds.

Brown et al. (1989) have recalculated figures for gill size in relation to body size in a series of new measurements, mainly of prosobranchs. Their results largely support the Pelseneer-Yonge idea that whatever type of gill may be present there is, on the whole, a reasonably limited range in the area of respiratory surface per unit of body mass. Their figures are lower than the older ones, ranging from 3–7 $cm^2.g^{-1}$, but confirm that the ratios are comparable whether there are two bipectinate gills (fissurellids), one bipectinate gill (trochids), or one monopectinate gill as in caenogastropods, compensation in gill area obviously occurring.

CROSS REFERENCES

CHAPTER 28

ENDOCRINES

I N the hands of a group of French workers calyptraeoideans and, to a lesser extent, patellogastropods have provided almost all available information on the endocrine control of sex differentiation in prosobranchs and also the bulk of our total knowledge of prosobranch endocrinology. The emphasis of the French work on these two groups summarized by Choquet (1971) for patellogastropods, by Lubet *et al.* (1973) for calyptraeoideans and, in general, by Martoja (1972), Joosse (1979), and Joosse & Geraerts (1983)] was perhaps stimulated by the fact that the animals are protandrous hermaphrodites which undergo sex change. They are thus good material for analysis of the influences controlling the development of the primary and secondary characters of both sexes. Fortuitously and fortunately their organs proved to be rather readily cultured *in vitro*. Whether the control of the target organ is direct from a neurosecretory cell or by way of an intermediate endocrine organ, the source of the stimulus is apparently invariably within the central nervous system.

All reproductive activity in calyptraeoideans appears to depend upon the presence of a neurohormone produced in the cerebral ganglia, since removal of these structures at any time in the life history brings it to an end and their reimplantation restores it (Lubet & Streiff, 1969; Lubet & Silberzahn, 1971; Lubet *et al.*, 1973). *In vitro* cultivation of a gonad which is in the male phase has likewise no success unless the medium includes cerebral ganglia from an active male; blood from a limpet in the male phase is equally effective in maintaining the gonad in function, implying that the control is humoral. In the absence of the androgenic hormone, however, though further spermatogenesis stops, oocytes begin to appear (Streiff, 1966) suggesting that the gonad is basically female. Development of the oocytes, however, halts before the start of vitellogenesis and that process begins only in the presence of cerebral ganglia from limpets in the female phase.

As young males mature ripe sperm come to fill the vas deferens and as this happens the sperm groove and the penis develop, not under the influence of the gonad as might perhaps be expected, but under that of a neurohormone formed in the pedal ganglia, released to the blood and, in a way not understood, concentrated and stored in blood spaces at the base of the right tentacle (S. Le Gall, 1974, 1978, 1981). This hormone is also essential for the persistance of these structures throughout the male phase. At the end of this stage, when the animal is changing sex, the male organs regress rather rapidly and are replaced by those of the female, a process which takes 60 days at Helgoland (Pandian, 1969). For this to happen two processes appear to complement one another: the gradual disappearance of the androgenic substances from the cerebral and pedal ganglia, and the production of a third hormone in the pleural ganglia bringing about their dedifferentiation (Streiff, 1966; Lubet & Streiff, 1969; S. Le Gall & Streiff, 1975). In *Crepidula fornicata* both penis and sperm groove disappear completely but in *Calyptraea chinensis* a vestigial penis persists behind the right tentacle of the female. Though the male system has been lost the competence of the area behind the tentacle to form a penis has not disappeared and the vestigial penis of a female *C. chinensis* grows large, and the penis area of C. *fornicata* regenerates a penis if either is cultivated *in vitro* with the eyestalk of an active male (Streiff, 1966). The vestigial penis of a post-breeding male *Littorina littorea* may also be redeveloped by cultivating it with the right eyestalk of a functional male (Le Breton, 1969, 1970). With the loss of the endocrines protecting the male system and in the presence of feminizing ones the female system becomes established and, on the whole, once formed is, unlike that of the male, fixed and does not require to be maintained by hormonal influences, though the production of ripe female gametes does.

Associated with the habit of living in stacks, sexuality in *Crepidula fornicata* is interactive, with the stack members influencing one another. P. le Gall (1980) has shown that there is two-directional transmission of sex-determining substances along the length of a stack. All the animals in a stack make contact with their neighbours, above and below, by means of the edge of the mantle skirt. The males at the apex of the stack produce a feminizing hormone in the tentacles at the pallial edge which is picked up, in an unknown way, by the mantle edge of the next lower limpet and so is passed down the stack towards the base. At the same time the females at the base of the stack produce a masculinizing substance at their mantle edge which is picked up by the pallial tentacles of the next higher animal and passed in this manner up the chain; the whole mechanism tends to stabilize the sexual organization of the stack. Since these substances originate in the pallial edge they are presumably unlike those of nervous origin responsible for the primary determination of sex as distinct from its maintenance once established, but perhaps distinct from these too.

The general patten of hormonal control of sex change in *Patella* is similar to that of calyptraeoideans though simpler since there are no secondary sexual organs to modify. A few differences have been noted (Choquet, 1971). Oocytes never develop spontaneously in a cultured gonad: they do so only if female cerebral ganglia are also present in the medium; similarly male cells require cerebral ganglia or blood from an active male. A further hormone produced in the tentacles of *Patella* antagonizes this cerebral one, giving rise, during the male phase, to resting periods when no sperm are formed, and, at its end, to the total stoppage of spermatogenesis. The cerebral androgenic hormone does not disappear in patelloideans when the limpet changes sex as it does in calyptraeoideans, but it does change function, promoting the multiplication of the gametes within the ovary. To some extent the development of the unisexual gonad in *Viviparus viviparus* (Griffond, 1969, 1975) is similar, both male and female cells arising only in the presence of the appropriate hormone.

It seems probable that at least some of these hormones have a corresponding role to play in the sexual activities of gonochoristic prosobranchs. For example, Streiff & Le Breton (1970) and Streiff *et al.* (1970) showed that the pleural ganglia of *Littorina littorea* produce a hormone which is effective in causing penial regression in *Crepidula*, and S. Le Gall (1981) that the pedal ganglia of male *Littorina* and of *Buccinum undatum* are as potent in stimulating the formation of a penis in *Crepidula* as those of *Crepidula* itself. Ram (1977) has implicated the nervous system, in particular the parietal ganglia (= supra- and suboesophageal), in stimulating the production of egg capsules in two species of the whelk *Busycon*. This cannot, however, be the sole factor involved in this activity since all the capsules which were observed in his experiments were empty; apparently the timing of egg release and that of capsule formation are controlled by different mechanisms and must normally be coordinated.

In addition to various parts of the nervous system some other structures in the prosobranch body seem possible endocrine organs. The most likely of these is the juxtaganglionar organ, first described in opisthobranchs by Martoja (1965a, 1965b) but also known in a number of prosobranchs (*Diodora mamillata* and *Monodonta turbinata*: Martoja, 1965c;

Haliotis: Joosse, 1972; *Patella vulgata*: Choquet & Lemaire, 1969; *Gibbula umbilicalis*: Herbert, 1982) though not, so far, in any caenogastropod. This organ may have some control of reproductive activity as it has in the opisthobranchs. Though separate from the nervous system it is likely that it represents cells included within ganglia in other groups of gastropods (Highnam & Hill, 1977). Reproduction is the most obvious prosobranch function which is under endocrine control, but Boquest *et al.* (1971) and Davidson *et al.* (1971) have reported the occurrence of insulin-like hormones in intestinal cells of *Buccinum undatum* and this has been suggested (Bush, 1988) as a possible function of one type of gland cell in the intestinal epithelium of *Patella vulgata*. Like vertebrate insulin this promotes the deposition of glycogen.

CROSS REFERENCES

CHAPTER 29

REPRODUCTIVE ACTIVITIES

THE different patterns of life history exhibited by invertebrates have been described and summarized by Grahame & Branch (1985) whose account included much information about prosobranchs. These molluscs have adopted three developmental strategies, oviparity [sometimes called ovulparity (Guillette, 1991) if the female gametes are discharged unfertilized], semiovoviviparity, and ovoviviparity. The first of these predominates: the eggs are released and the parent pays little or no further attention to them. In some species, however, such as *Capulus ungaricus* and *Crepidula fornicata*, the female broods them under the foot until larvae escape. Less directly protective behaviour, but still offering some degree of care, is shown by numerous species (e.g. *Hydrobia ulvae, Neptunea despecta*) which attach egg capsules to their shell; this habit is also found in the trochid *Clanculus* in which single eggs are laid in rows in the spiral grooves of the shell and covered with a thin layer of mucus. Most frequently eggs are fastened to a substratum, grouped into spawn masses or contained in capsules. The different strategies and the number of eggs produced are related to the mortality to which larvae or young snails are exposed, a greater reproductive effort being associated with indirect development.

Oviparity associated with deposition of spawn on a substratum may result in direct development, when the young hatch as juveniles or 'crawlaways'. In species of archaeogastropods which broadcast fertilized eggs a larval stage is always present, at first in the form of a trochophore which is in most respects like that of other groups with spiral cleavage. Later this larva assumes some of the features of a veliger larva, though this should, perhaps, be distinguished as a 'protoveliger' since it lacks some characters of the veliger of higher level prosobranch groups. In most prosobranch species, however, the trochophore stage is passed within the spawn mass or capsule and, if a larval stage is present it is a true veliger which hatches.

Larvae may be lecithotrophic, having sufficient food reserve to tide them over a (usually brief) larval life without feeding, or they may be planktotrophic, when they feed. Because food for the embryo is least in archaeogastropods these have the shortest embryonic life and produce the smallest larvae or young snails. This is particularly clear in trochoideans and applies whether they undergo direct or indirect development. The eggs of *Monodonta lineata*, which are shed singly to the plankton, have an overall diameter of about 200 μm, rising to 500 μm after the swelling of the jelly coat. The trochophore-veliger stage hatches after 30 h (at 15–20°C) and starts to settle in 4–5 days, the shell then measuring 1+ mm across (Desai, 1966). This larva is virtually, or perhaps wholly lecithotrophic. By contrast *Calliostoma zizyphinum* has direct development (Robert, 1902). The eggs measure 260–300 μm on laying, each surrounded by a jelly coat; trochophore and veliger stages are passed within the spawn mass and young snails emerge in 7–8 days with shells measuring 300–350 μm across. Hatching of the snails in *Jujubinus exasperatus* occurs only 5–6 days after fertilization (Robert, 1902).

On the whole patellogastropods take a longer time to become juvenile snails: this is in part correlated with the presence of a prolonged swimming-benthic stage during which feeding occurs and organogeny is completed. *Acmaea tessulata* hatches as a trochophore larva in about 4 days, the egg being only 140 μm in diameter. The larvae are planktonic for 3–4 days, 19 h as trochophores and 56 h as veligers; feeding starts at about 60 h. Although metamorphosis begins at about 15 days it is not completed until 42 days (M. M. Kessel, 1964). In *Patella vulgata* the planktonic eggs, 160 μm in diameter, rising to 230 μm with the jelly coat, hatch as trochophores in about 24 h (F. G. W. Smith, 1935) and torsion is complete in 4–5 days (Dodd, 1957). Spat settle 10–14 days after fertilization, when they have a transparent protoconch produced by the shell gland, measuring about 0·2 mm long and surrounded anteriorly by a flange of teleoconch. Some argument has centred on whether the protoconch shows signs of coiling, but the majority of workers have accepted that some twist is present, clearly shown in *Patella coerulea* (Warén, 1988). The flange completely encircles the aperture of the protoconch 4 weeks after fertilization (Dodd, 1957) and only then does the young snail emerge from the wet cracks in which the spat settled (Bowman, 1981) at a length of 0·5–1·0 mm. Protoconch and teleoconch are not coaxial and the young shell shows a spiral twist as well as the coiling of the protoconch, though this has nothing to do with the dextral coiling of the typical prosobranch since it is sinistral. Eleven weeks after fertilization the viscera have retracted from the protoconch which is then lost and replaced by a septum laid across its base.

In oviparous caenogastropods and neritoideans, which have a more elaborate glandular genital duct, the embryos are provided with albumen as well as yolk, and there may be food eggs; the proportions of these foods vary between species. Most caenogastropods and neritoideans have a planktotrophic larva; the planktonic stage is prolonged as compared with more primitive forms and transformation into the adult is relatively abrupt. Two examples illustrate the difference from archaeogastropods. The eggs of *Lacuna vincta*, 94–125 μm in diameter, with yolk and albumen as food, develop in 2–3 weeks to free-swimming veligers with a shell 200 μm in breadth and height. There is a long planktonic life (Thorson, 1946) during which the shell grows to 2·5 whorls with a diameter of about 500 μm and a height of 2–3 mm. At this stage the larva normally settles on algae and metamorphoses rapidly, ingesting the velum (Fretter, 1972). In the neogastropod *Hinia reticulata* the eggs are larger, 160 μm in diameter, and development slower (1–2 months). The planktonic veliger stage lasts for 2–3 months, at the end of which it has a shell of three whorls, the last with a diameter of 700–800 μm. At this stage it undergoes a rapid metamorphosis (Fretter, 1972).

Few monotocardians have lecithotrophic veligers: they have been reported in species of *Melanella* (Warén, 1983) and in *Hydrobia ulvae* at Plymouth (Lebour, 1937; Pilkington, 1971) though the same species is described as planktotrophic in the Dovey estuary (Fish & Fish, 1977) and elsewhere (A. Anderson, 1971; Barnes, 1988).

Direct development has enabled a number of species to colonize fresh water (avoiding the effects of currents on larvae), polar marine conditions (avoiding the reduced availability of planktonic food), and the land, where free larvae cannot survive. Direct development in monotocardians depends upon the provision of an increased food supply for the embryo— yolk, albumen or food eggs. Two neogastropods illustrate alternative nutritional strategies. The eggs of *Busycon carica* measure 1,700 μm in diameter (Fioroni, 1966) as a result of having a considerable amount of yolk which, together with a limited amount of albumen, is sufficient for the embryo to complete development within the egg capsule. In contrast the embryos of *Nucella lapillus* are provided with food eggs and develop from eggs only 170–188 μm in diameter; each may devour 15 eggs and some albumen. This condition must be the more primitive, the former the more derived.

Some families of monotocardians are exclusively freshwater (pilids, ampullariids, viviparids, bithyniids) whilst others include many freshwater species (thiarids, hydrobiids), and some only a few (neritids, cerithiids). Reproduction in non-marine habitats typically necessitates internal fertilization, a cleidoic egg, and suppression of a larval stage. Fewer eggs are produced, often only one per capsule, which is enlarged for the increased supply of food.

Most freshwater gastropods living in the temperate zone are semelparous (L. C. Cole, 1954), the overlap of parents and juveniles being minimal since the parents die sooner or later after a single reproductive season. Others are iteroparous, the snails undergoing more than one reproductive season, with the result that parents and offspring overlap and to that extent may compete for resources. However, depending on the conditions of the environment, one type of life cycle may transform into another. For example, in Europe *Bithynia tentaculata* breeds at the age of 1 year, again at 2 years, and possibly at 3 years, when they die, whereas those experiencing the extreme winters of Eastern Canada are semelparous (Pinel-Alloul & Magnin, 1971). Similarly the freshwater *Theodoxus fluviatilis* normally lives for two years, in the second of which it spawns and dies; in warmer habitats (over 12°C) it may show two generations per year (Becker, 1949). A mock semelparity is exhibited by *Nassarius pauperatus* in habitats where conditions are degenerating, since more eggs are laid if food is short, resulting in more abundant larvae for dispersal to other habitats, and a weakening of the adults leading to their death (McKillup & Butler, 1979).

The majority of marine prosobranchs are perennial and iteroparous. In adapting to a freshwater habitat they have replaced iteroparity with semelparity, probably as an adaptation to the reduced resource availability and the need to produce eggs with enough food to allow more fully developed juveniles at hatching (Calow, 1978). This is achieved to a certain extent by producing fewer eggs, though there must be a limit to this to ensure survival of the species. Any theory which relates semelparity simply to resources, however, cannot be the whole story: size must also be involved since a number of minute marine prosobranchs with abundant resources are semelparous (e.g. *Rissoella* spp., *Rissoa parva*, *Skeneopsis planorbis*). According to Underwood (1979) small molluscs find themselves in the dilemma of either not being able to produce sufficient (and large enough) eggs to perpetuate the species through direct development, or of producing enough larvae to survive the losses which larval existence imposes.

In terrestrial forms the albumen supplies both food and water. Some cyclophorid prosobranchs have the ability to live in places where they must tide over a dry season, though their reproductive activity is limited to wet periods. The best known example of terrestrial development is that described for *Pomatias elegans* by Creek (1951). The spherical capsule, secreted by the pallial oviduct, is 2 mm in diameter and contains a single egg, 120 μm in diameter, surrounded by albumen and a gelatinous capsule wall, both of which are used as food during three months of intracapsular development. Absorption of soil water is necessary for development: it contributes oxygen and also dilutes the albumen, which is absorbed by a large cephalic vesicle in the velar area. The vesicle disappears near the time of hatching and the remaining albumen is then ingested. A modification of the velar area with functional similarity is found in cyclophorid prosobranchs (Kasinathan, 1975) and in pulmonates (Régondaud, 1964).

Semiovoviviparity, in which young are liberated as veligers from egg capsules retained within the mother, is known in relatively few monotocardians. The embryos are retained in the pallial oviduct, the mantle cavity or in a pouch formed as an ectodermal invagination passing deeply into the head-foot at the site of the ovipositor. Retention in the pallial oviduct occurs in *Polygireulima* (Warén, 1983) and *Janthina janthina* (Wilson & Wilson, 1956), in the mantle cavity of *Littorina scabra* (Struhsaker, 1966; Rosewater; 1980), and in a head-foot brood pouch in a number of cerithioideans, such as *Melanoides tortulosa* (Seshaiya, 1940) and populations of *Planaxis* from Queensland, Australia, though not apparently in others (Houbrick, 1987). For further examples see Fretter (1984b).

In some monotocardians protection of the embryo is enhanced by ovoviviparity, fertilized eggs provided with albumen being retained within a brood pouch until they hatch as young snails. The pouch may be a modified part of the oviduct (*Littorina saxatilis*, viviparids, *Potamopyrgus jenkinsi*), or a head-foot pouch as in the vermetid *Pyxipoma* (J. E. Morton, 1951) and a number of cerithioideans (Houbrick, 1987). There is no proof that any ovoviviparous prosobranch produces a histotrophe (V. Y. Berger, 1975) for the nutrition of the embryo, hence true viviparity does not occur. However in *Viviparus viviparus* it has been shown (Alyakrinskaya, 1969) that oxygenation of fluid in the brood pouch is augmented by increased vascularization of its walls and this, together with the presence of haemocyanin in the albumen, facilitates the respiration of the developing young. Moreover Stadnichenko (1970) has shown that in *V. contectus* there is a difference in the protein content of the blood of males and females and between that of virgin and pregnant females, suggesting the presence of a histotrophe.

Potamopyrgus jenkinsi [properly *P. antipodarum* according to Ponder (1988a)], living in fresh or slightly brackish water, is the only British prosobranch which reproduces partheno-genetically. Although males have been recorded, the first in 1958 (Patil, 1958), there is still uncertainty as to the part they play in the reproductive activity of the community. More recently Wallace (1985) found males constituted up to 20% of some European populations and 9% of Australian ones (Wallace, 1978, 1979), whereas in some New Zealand populations the percentage was as high as fifty. Although these high numbers suggest that sexual repro-duction might well occur in many populations, though possibly only at low frequency (Ponder, 1988a), this has yet to be proven. Hermaphrodites are also found in small numbers, the individuals having a penis and a brood pouch; there is the possibility that this may relate to imposex.

The ways in which the eggs of prosobranchs are released and deposited are more varied than in other groups of molluscs. Primitively, as in *Patella, Haliotis* and *Monodonta*, unencap-sulated eggs are broadcast into the sea: even when egg capsules were invented this habit sometimes persisted, as in *Littorina littorea* and *Melarhaphe neritoides*. All these hatch as larvae. On the other hand unencapsulated eggs of diotocardians may be attached to the substratum. They may form a thin layer adherent to the substratum and to one another by a mucous coat produced by the ovum, as in *Acmaea virginea* (M. M. Kessel, 1964) and *Diodora graeca*; in the former the young hatch as trochophores whereas in the latter larval life is suppressed. Alternatively, the eggs may be embedded in a mucous cord produced primarily by the enlarged urinogenital papilla of the female, augmented by hypobranchial gland secretion; the spawn ribbon is attached by the foot to weed or stones along its length in *Jujubinus exasperatus*, but only at intervals in *Calliostoma zizyphinum*.

Monotocardian spawn masses liberated from the pallial oviduct are, with few exceptions, fixed to a firm substratum selected by the parent; though in some cases the reason for the selection is obvious, in others it appears obscure. The oviparous littorinid *Littorina arcana* and the ovoviviparous *L. saxatilis* are commonly sympatric, but during September and October the breeding population of *arcana* migrates downshore to damper zones to deposit egg masses in sheltered places where conditions are more favourable for the young snails (Ellis, 1985). A downshore spawning migration has also been reported by R. J. Berry & Crothers (1968) in *Nucella lapillus*. On the contrary *Epitonium turtonis*, living on muddy sand 5–70 m deep, migrates onshore to L.W.S.T. where silt particles are less abundant, and deposits a string of pyramidal capsules, each covered in sand and larger mud particles (Vestergaard, 1935). Similarly an upshore spawning migration is reported in *Urosalpinx cinerea* (Hancock, 1959) and, in *Nucella*, several females migrate to the same place to deposit capsules in damp, shady areas.

Many monotocardians attach gelatinous egg masses to the weed on which they feed and on to which the young will hatch: this applies to *Littorina obtusata, L. mariae* (Goodwin, 1975; Goodwin & Fish, 1977), *Lacuna pallidula* (Grahame, 1977) and *L. parva* (Ockelmann & Nielsen, 1981). The clear, lenticular capsules of rissoids are also deposited on weed, attached by their flattened bases (e.g. *Rissoa parva, R. guerini, Rissostomia membranacea*), and the veliger larvae, after a short planktonic life, settle in the same habitat as the adults.

Snails living on soft substrata must take advantage of any available firm surface for deposition: for *Hinia reticulata* this is usually weed, *Zostera* if available, and harder surfaces are neglected. Barnett *et al.* (1980) offered these snails five species of brown and five species of red algae, three species of bivalve shells, and sandstone pebbles as possible deposition sites: only four species of red alga were used, the favourite being *Gigartina stellata*, perhaps because it offered the most reliable surface for secure attachment. Other neogastropods requiring a more solid surface use any available stones or the shells of their congeners, or construct a holdfast from a few empty capsules (Kohn, 1961). In the first case several individuals may have to share a single attachment site. In *Buccinum undatum* the resultant assemblage of capsules may be very large: one photographed with G. Thorson was 45 cm across and 20 cm high [see fig. 337 in Fretter & Graham, n.d. (1985)]. Female *Neptunea despecta* (Pearce & Thorson, 1967) and *Liomesus ovum* deposit capsules on the columellar side of the aperture of their own shell: in the latter the highest recorded number of capsules attached there is 34, but it may be greater in the former species. Alternatively the capsules may remain unattached on laying and be protected against predators by a camouflaging covering of soil particles either adhering to a surface layer of mucus or embedded in it. The spawn may consist of single capsules (*Pomatias*: Creek, 1951; *Pomatiopsis*: Dundee, 1957) to whose surface earth or mud clings, or many capsules may form a conglomerate mass covered with faecal pellets and mud compacted by the foot (*Assiminea*: Sander, 1951, 1952). In epitoniids sand particles are partly embedded in the wall of capsules connected to one another to form a short chain, but the most complex spawn of this type is that of naticids. Here a mass of mucous secretion, mainly of propodial origin, is passed into the mantle cavity at times when egg capsules are escaping from the genital aperture and the two become mixed together. A pallial fold directs sand grains into the mass as it is forced out of the mantle cavity between the body and the mantle skirt, an action which imparts the familiar collar shape to it. The result is a smooth mass of secretion with a thin covering of mucus, soon lost, in which are embedded many capsules, with sand grains in the intervening mucous matrix.

Before deposition of the egg capsules the surface of the substratum may be tested by the siphon (*Hinia*), or the propodial surface of the foot (*Melongena*: Bingham & Albertson, 1973), moulded by the radula to form a nest (in sponges: *Cerithiopsis*; in tunicates: *Lamellaria, Trivia, Austromitra*), or by the foot (*Pila globosa*: Bahl, 1928); in others the anterior pedal gland deposits a thick layer of mucus upon which successive capsules are deposited (*Pomacea canaliculata*: E. B. Andrews, 1964). During deposition the foot regulates the placing of the spawn whether it is a continuous string of capsules in irregular, or regular, tightly packed coils (*Bittium*), or pressed together in rows parallel to the edges of the leaf to which it is attached (*Bithynia*: Lilly, 1953).

The structure of the tough wall of the egg capsule of neogastropods has been given further study since the work of W. E. Ankel (1937) and Fretter (1941). Capsules may be vase-shaped, as in muricids (*Nucella, Ocenebra, Urosalpinx*), or hemispherical, as in buccinids (*Buccinum, Neptunea*), with a basal stalk or flange for attachment and a plugged opening on the opposite wall. They house the developing embryos for several weeks or even months. Perron (1981) has shown a positive correlation in *Conus* species between capsule strength and length of intracapsular development. All workers agree that the wall, manufactured in the pallial oviduct is multilayered. The outermost part [= outer and inner fibrous layers of Hancock (1956) and C. J. Bayne (1968); =L1 of Tamarin & Carriker (1967) and of Sullivan & Maugel (1984); =L1 + L2 of L. E. Hawkins & Hutchinson (1988)] consists of protein fibres embedded in a mucopolysaccharide matrix, more superficial fibres running mainly in a circular direction, the deeper ones longitudinally. This fibrous part is followed internally by a homogeneous layer lacking well defined structural elements, though some subdivision into three sections can be detected. It is said to determine the shape of the capsule and is commonly the thickest part of the wall, constituting about two thirds of its thickness in *Urosalpinx* (Hancock, 1956; Tamarin & Carriker, 1967), about 80% in *Ilyanassa obsoleta* (Sullivan & Maugel, 1984); in *Ocenebra erinacea* and *Nucella lapillus* (fig. 334) the homogeneous layer is thin (Hawkins & Hutchinson, 1988). The innermost two layers are thin and have few or no structural elements; C. J. Bayne (1968) found their structure and that of the plug apparently identical in *Nucella*. The chemical constitution of the four layers and of the plug was studied by Hawkins & Hutchinson (1988) in *Ocenebra*: each gave positive tests for amino acid groups and carbohydrates; the homogeneous layer also contained tryptophan, and the innermost layer mucopolysaccharides. In *Buccinum undatum* S. Hunt (1971) has shown that the protein has a chemical composition resembling keratin rather than the conchiolin suggested by earlier workers, and this may be true for other species.

The capsule of all these neogastropods is manufactured in the pallial oviduct, then passed to the ventral pedal gland in which it is moulded and by which it is attached to a substratum (fig. 334). According to C. J. Bayne (1968) the gland adds a thin coat of mucopolysaccharide in *Nucella*: this is probably an incidental contaminant lubricating movement in and out of the gland. Tamarin & Carriker (1967), however, suggested that the ventral pedal gland was responsible for much more than the shaping of the capsule and its deposition and attachment to the substratum, which were the only activities previously attributed to it. They proposed that its secretions brought about a polymerization of the protein fibres produced in the capsule gland, so imparting to them their characteristic resistance to exposure. Sullivan & Maugel (1984) have described a fibrous layer on the outer surface of the capsule of *Ilyanassa obsoleta* which they believe is added by this gland since it is not present on capsules as they leave the oviduct. The few records of the time a capsule remains within the gland show considerable variation. They are 2 min for *Olivella verreauxii* (Marcus & Marcus, 1959), 45 min for *Melongena corona* (Bingham & Albertson, 1973), and 5–12 hours for *Colus stimpsoni* (West, 1979).

Earlier descriptions of prosobranch spermatogenesis and sperm microanatomy given by Franzén (1955) have been much extended by Healy (summary, 1988) who has had the great advantage of being able to study the subject with the aid of electron microscopy. His results,

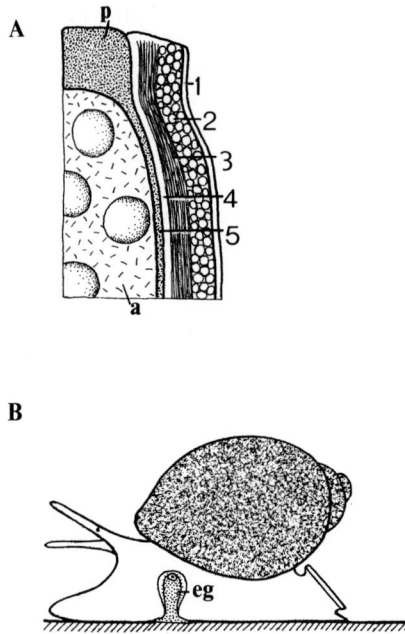

FIG. 334.—*Nucella lapillus*. A, composition of wall of egg capsule (after C. J. Bayne, 1968). B, deposition of capsule by ventral pedal gland.

a, albumen in which eggs are embedded; eg, egg capsule; p, plug; 1, mucous layer; 2, circular fibres; 3, longitudinal fibres; 4, homogeneous layer; 5, inner layer continuous with plug.

in addition to adding detail to our knowledge, have proved to offer a further base on which prosobranch relationships may be assessed. In archaeogastropods other than neritimorphs the testis produces only one type of spermatozoan, euspermatozoa, which are eupyrene and are capable of fertilizing ova; in neritimorphs and caenogastropods, including architaenioglossans, two types of sperm are present, euspermatozoa and a second type, paraspermatozoa, which are to a greater or lesser extent in different species deficient in their chromosome content and so incapable of fertilizing ova. This does not, however, mean that they are without a role in the reproductive activities of the animal, though what that may be is often as obscure as the reason which causes their development to differ from that of the euspermatozoa. In a few groups paraspermatozoa play a clear role in fertilization since large numbers of euspermatozoa become attached to them, are perhaps to some extent nurtured by them, and are transported by them from male to female. These compound structures, known as spermatozeugmata, have been described in epitoniids and janthinids (W. E. Ankel, 1926), and in cerithiopsids (Fretter & Graham, 1962). All these animals are aphallic and the role of the spermatozeugmata is thus reasonably obvious. Though littorinids have a penis and copulate W. E. Ankel (1930a) described in males what he regarded as probable spermatozeugmata [an identification accepted by Buckland-Nicks & Chia (1977)], formed by an association between a food-laden paraspermatozoan and a number of euspermatozoa. The spermatozeugmata in littorinids break up only after they have reached the bursa copulatrix of the female. The reason why these events occur in littorinids but not in other copulating species is not clear: it may be that the nutritive role of the paraspermatozoan is emphasized.

Sperm are also transferred from male to female prosobranchs by way of spermatophores—non-motile packets of sperm encased in secretion from the prostate gland, where spermatophores are manufactured one by one. At least in *Cerithium muscarum* (Houbrick, 1973) each contains both euspermatozoa and paraspermatozoa. Spermatophores have been found in members of the Neritidae (W. E. Ankel, 1936; E. A. Andrews, 1937; C. J. Bayne, 1968), in heteropods (Tesch, 1949), in a single British pyramidellid, *Chrysallida obtusa* (Höisaeter, 1965), all of which have a penis, as well as in cerithioideans (Houston, 1985) with no penis and open genital ducts. They appear to have developed sporadically throughout the prosobranchs. Indeed in the Triphoridae both spermatozeugmata and spermatophores occur, though presumably not in the same species (Robertson, 1989). The behaviour of *Cerithium muscarum* during spermatophore transfer has been described by Houbrick (1973); in this species pairing lasts from one to three hours.

There are various categories of cell providing for the needs of developing ova or embryos: follicle cells of somatic origin (Cowden, 1976), derived from ovarian supporting tissue; nurse cells, originating from the germinal epithelium; and food eggs, also of germinal origin but, unlike the other two types, liberated with the viable ova and in some species undergoing some early stages of development. Follicle cells occur in the ovary of pulmonates (Raven, 1961), chitons, and cephalopods (E. Anderson, 1969; Cowden, 1976) but not in prosobranchs. They practically or completely surround the ovum, may transfer material to it, and produce a secondary egg membrane. Nurse cells become connected to oocytes by confluent cytoplasmic bridges (E. Anderson, 1974; Huebner & Anderson, 1976). They are either engulfed by the oocyte or destroyed after transferring to it material which they have elaborated. These cells occur in *Loligo*. There is confusion in the literature over the use of the terms 'food egg' and 'nurse cell' and the two are often wrongly used interchangeably. Cannibalism occurs when one developing embryo attacks another which has become vulnerable because it is late in developing or has been damaged.

In the egg capsules of some prosobranchs in which the ova have a common supply of albumen, but only in those, embryos complete development to the crawlaway stage by feeding on their companions, either food eggs or early embryos. The occurrence of this phenomenon is erratic even within a genus. It is common in neogastropods and also occurs in the unrelated mesogastropods *Rissostomia membranacea, Lunatia catena, Planaxis sulcatus* [according to Thorson (1940a) in a population in the Persian Gulf, though denied by Houbrick (1987) for populations in Queensland], and in *Theodoxus fluviatilis*. The only mesogastropod family in which it has been recorded in more than one species is the Calyptraeidae (Hoagland, 1986), these being *Calyptraea novaezelandiae* (Pilkington, 1974), *Crepidula adunca* (Coe, 1949), *C. dilatata* (Gallardo, 1977b), *C. fecunda, C. maculosa* (Hoagland & Coe, 1982), *C. (=Maoricrypta) monoxyla* (Pilkington, 1974; Nelson & Morton, 1979) and *C. philippiana* (Gallardo, 1977a). Rehfeldt (1968) described the use of food eggs, or possibly of cannibalism, in a single population of *Rissostomia membranacea* from the narrow part of the Roskildefjord (Denmark). Here, as elsewhere in the fjord, the embryos of this species normally display no cannibalism and all eggs in a capsule develop and hatch as free veligers; in this case the number of eggs in a capsule ranges from 40–100 and their diameter from 90–100 μm. In the population in which food eggs were present the number was less (18–75) and they were larger (120–125 μm diameter). The two types of development may indicate that we are dealing with sibling species (Munksgaard, 1990); this may also apply to *Planaxis sulcatus*. Hoagland & Robertson (1988) and Bouchet (1989a), in reviews of poecilogony (the occurrence of two modes of development in one species), all conclude that it is of very limited occurrence in prosobanchs, though found in other gastropod groups, and that its apparent presence is probably due to failure to recognize two separate species. *Lunatia catena* is the only known naticid with food eggs (W. E. Ankel, 1930b; Thorson, 1946). Each capsule in the spawn collar has 50–180 eggs (mean number 86) and except for about two (range 2–19) all are food eggs which cleave in an atypical way, become adherent to the inner wall of the capsule, are disintegrated by the action of the velar cilia of the developing embryos, and are ingested. *Theodoxus*, in contrast to the other species whose development is dependent on these eggs, lives in rivers, though it is also found in brackish environments. The size of the capsules and the number of food eggs vary with the salinity of the habitat. Bondesen (1940), working in Denmark, found that in fresh water the hemispherical capsule (1·0–1·2 mm across) contained 140–150 eggs, in brackish water the capsule was not larger than 0·8 mm in diameter and contained 50–60 eggs. In both habitats only one snail develops to a crawlaway and comes to fill the capsule before hatching: it is larger in fresh water (shell length 0·75–0·9 mm), smaller in brackish water (shell length 0·65–0·75 mm) reflecting, no doubt, the number of food eggs to which it has had access.

In neogastropods the number of food eggs available to the developing snail is commonly greater: this permits a more prolonged development (Portmann & Sandmeier, 1965) and larger crawlaways. However, in one species, *Thais rustica*, the young hatch as free veligers after using the meagre ration of food eggs provided in the capsule: here the number of eggs per capsule is estimated at 1,070, of which 400 undergo normal development (D'Asaro, 1970). In contrast, an exceptionally high number of food eggs is enclosed in the large, egg-shaped capsule (33 mm in diameter, 20 mm high) of the buccinid *Volutopsius norwegicus*: over 50×10^3 has been estimated by Thorson (1940b) and these are used by only 2–4 developing snails, whose shell on hatching has 2–3 whorls and is 20 mm in height and 11 mm broad. The size and precocity of the emerging snail are directly related to the food available and even when only one or two emerge may vary considerably. The five thousand eggs in a capsule of *Neptunea antiqua*, which is about one third smaller than that of *V. norwegicus*, though the eggs are larger (300 μm in diameter), give rise to one or two snails, hatching with a shell of 2·0–2·25 whorls, but the shell length may vary from 6·0–12·7 mm (Pearce & Thorson, 1967). In this species intracapsular life has been estimated at 6 months in nature. A greater contrast has been observed in the size of snail emerging from a single capsule in *Colus islandicus*, for although these may number only two or three, resulting from 6,500 eggs (200 μm in diameter), as many as sixteen may hatch from another capsule with comparable food supply. The range in shell length for the crawlaways is 3·5–8·5 mm. Larger size at hatching followed by faster growth reduces predation risks (Rivest, 1983).

Several theories have been put forward to account for the occurrence of food eggs: failure to fertilize (Staiger, 1951), fertilization without karyogamy, which may be associated with the possession of unbalanced chromosome sets as in *Murex trunculus* and *Buccinum undatum*. Other workers (Portmann, 1927, 1931) have suggested that food eggs are sterile because they have been fertilized by abnormal, oligopyrene spermatozoa (paraspermatozoa), but such sperm have not yet been found in *Lunatia catena*. The abnormality may, however, be related rather to the ova (Burger & Thornton, 1935). Rivest (1983) pointed out that the ratio of food eggs to viable eggs within prosobranch capsules suggests that the former are genetically determined and that their distribution to capsules is random. This would explain why some embryos have more food and grow to a larger size. There must, however, be some mechanism to ensure that at least one viable egg enters each capsule, otherwise many would be aborted, a point which needs further investigation.

Reasons for the occurrence of food eggs in prosobranchs remain obscure. To take an extreme example: in the species *Volutopsius norwegicus* it is believed that as each egg, the diameter of which is 200 μm, develops to hatching, it has used fifty thousand other eggs as food during the process (Fioroni, 1966, table 17). Had this supply of food been incorporated as yolk into a single ovum its diameter would have been over 7 mm. Whilst a number of animal groups, including the cephalopods, have evolved methods of dealing with such yolky eggs, this may well have proved beyond the power of a gastropod, and food eggs offer an alternative to yolk with which their developmental mechanism can cope. Large eggs, moreover, would lower the rate of development (Spight, 1975b).

Embryos feeding on food eggs may break them up before devouring them, in which case their development is continuous: this has been described in *Thais canaliculata* by Spight (1977). In other species the food eggs are swallowed whole during a pronounced ingestive phase, when development is temporarily delayed. This ingestive phase occurs in the late preveliger stage in *Nucella lapillus* and lasted about a week in a population studied by Pechenik et al. (1984). At that time only the primordia of ectodermal organs (apical, pedal, and anal plates, cerebral and pedal ganglia, statocysts) are present (Stöckmann-Bosbach, 1988). A dense circumoral field of cilia attaches the embryo to the central column of food eggs and these are passed into the archenteron, the cilia assisting in the swallowing as in *N. crassilabrum* (Gallardo, 1979). During ingestion the differentiation of the mid gut and head-foot is delayed, as well as the digestion of the food eggs. The ingestion of food eggs in *Buccinum undatum* takes place when the first phase of torsion has occurred and all the essential organ rudiments established except for the pericardial-renal complex and the organs of the mantle cavity (Giese, 1978). Development proceeds while the embryo feeds though that of the digestive gland and the completion of torsion are delayed (Portmann, 1925; Giese, 1978). The food eggs are taken into a special, temporary enlargement of the intestine which does not form in *Nucella*. Perhaps the most voracious ingestion of food eggs (in this case unsegmented) was described by Franc (1943) in *Pisania maculosa*, where 80–90 may fill the archenteron to such an extent as to lead eventually to its rupture and the death of the embryo.

The capsule wall of *Nucella lapillus* and related species has been shown by Pechenik (1982, 1983) to be readily permeable to water and inorganic ions, but only slightly to small organic molecules. The capsule is normally turgid since the osmotic concentration of its contents (25–30 mOsm.1^{-1} above ambient) increases its volume and allows space for some movement of the embryos as they develop. In dilute sea water salts, which account for 80% of the internal osmotic concentration, escape through the capsule wall and the internal and external osmotic concentrations approach equality. Pechenik (1982) concluded that the value of the capsule wall in this respect was not to prevent osmotic change but to control the rate at which it took place, so preventing osmotic shock. Whilst this may be true, it would seem to be of value only if the exposure to dilute sea water is of short duration, since Feare (1970a) showed that successful development is reduced to 27% in capsules deposited in places where they are regularly washed by fresh water at low tides. All embryos might be killed in capsules regularly exposed to drying at these times. Pechenik et al. (1984) assumed that the intracapsular fluid (albumen) plays a minor nutritive role in the first few days of the four-month developmental period when its main importance lies in giving support to the embryos by virtue of its viscosity. Ingestion of intracapsular fluid may occur at the post-gastrular stage when the stomodaeum already communicates with the archenteron, though much of the latter is occupied by cell 4D, in which yolk is stored (Stöckmann-Bosbach, 1988). Scheidegger & Fioroni (1983) have shown that in *Nucella* some fraction of the inorganic components of the fluid is incorporated into the large vacuolated cells of the embryonic surface, and this may also occur in *Buccinum undatum* (Giese, 1978). In *Nucella*, when food eggs have been ingested whole and the embryos have enlarged in consequence, the average capsule contains only 1·1 μl of intracapsular fluid per embryo.

Prosobranch embryos hatch from their protective membranes as trochophore or veliger larvae, or as miniatures of the adult. Mechanisms causing hatching have been reviewed by Davis (1968). The capsule membranes deteriorate with time and their disintegration is nicely adjusted to the stage of development at which the embryos are capable of hatching and surviving. Osmotic swelling may accompany deterioration. The planktonic egg capsules of *Littorina littorea* and *Melarhaphe neritoides* gradually swell and after about a week, when the veligers are ready to escape, the membranes burst with the rising pressure (Linke, 1933, 1935). Hertling (1928) described a similar osmotic mechanism in the benthic egg mass of *Lacuna vincta*, which lasts for three weeks: in this species he suggested that the velar cilia help the embryos to escape from their weakened encasements, though cilia can provide little force. In *Assiminea grayana*, which lives near the upper limits of salt marshes, the rate of osmotic swelling of the egg mass, and therefore of the hatching of the veligers, is inhibited by high salinity, as induced by drying (Sander & Sibrecht, 1967); hatching occurs only after long immersion by spring tides or freshwater flooding. A mechanical device providing escape for larvae has been recorded in *Lunatia alderi* by Ziegelmeier (1961). The deterioration of the cement of the egg collar is such that it falls apart when the veligers are ready to hatch: as the contained sand grains are freed they break open the capsule walls and liberate the larvae.

When embryos develop to miniatures of the adult the radula may be used to let them bite their way to freedom. This is the means of escape of young *Lacuna pallidula* and *Littorina obtusata* (Hertling, 1928); it may also occur in *Urosalpinx cinerea*. As in other stenoglossans the eggs of *Urosalpinx* lie in a communal capsule with tough walls, fixed at the base, and with a plugged exit hole. The cementing substance securing the plug gradually dissolves with time (Hancock, 1956) and hatching is by way of the escape hole. Pope (1910–11), however, stated that young drills will eat holes in the wall of the capsule if the area of the plug is obstructed by other embryos not ready to escape, though normally the first drill pushes out the plug and others follow. Enzymes are probably responsible for loosening the plug (Hancock, 1956) and this also applies to *Nucella* (W. E. Ankel, 1937; Kostitzine, 1940), several Mediterranean prosobranchs such as *Pisania* (Franc, 1943), and cones (Kohn, 1961). The plug itself softens and disappears, or it may be eaten.

In one species, *Viviparus viviparus*, in which the young develop in capsules in the terminal part of the oviduct, the capsule wall is broken by pressure from the edge of the operculum during movement of the young, a mechanism occasionally also used by hatching larvae of the neogastropod *Columbella rustica* (Franc, 1943).

The typical monotocardian larva is a veliger which is planktotrophic, feeding on nano-plankton and microplanktonic cells, especially dinoflagellates, and to a lesser extent diatoms. On hatching from benthic or planktonic capsules feeding veligers rise towards the water surface where they aggregate in areas of high light intensity, though avoiding phytoplankton blooms. The density and species of phytoplankton may affect their activity (Mapstone, 1970; Fretter & Shale, 1973). This is true even of the larvae of species whose adults live at depths of

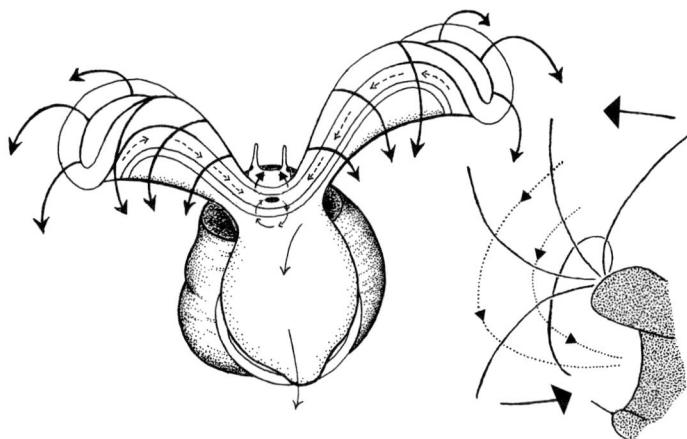

FIG. 335.—Left, swimming veliger larva of *Rissoa parva*, foot and mouth uppermost. Arrows show the effective beat of the preoral cilia (thick arrows), the cilia of the food groove (broken arrows), and those causing rejection currents (thin arrows). Mid dorsally, where the mantle cavity is exposed, the effective beat of the preoral cilia is the reverse of elsewhere. Right, diagram of a small portion of the food groove to show a prevelar compound cilium in a range of positions of its effective beat, and the position of the postoral ciliary band. (After Strathmann & Leise, 1979). Dotted lines indicate the course of two particles sucked into the food groove (see text). Cilia of the groove omitted. Larger arrow heads indicate the direction of beat of the cilium, smaller ones the direction of movement of the food particles.

2,000–4,000 m, as shown by Bouchet (1976) for *Benthonella tenella*. Our knowledge of larvae in the life history of prosobranchs living at great depths is comparatively recent. Rex & Warén (1982) studied the larval shell of such mesogastropods and neogastropods living in the western North Atlantic. From this they deduced the occurrence of lecithotrophy and planktotrophy and found that the latter increased in species living at greater depths. They also found that most planktotrophic larvae were from carnivorous adults and related this to the advantages of dispersal in an environment where food for carnivores is patchily distributed as compared with that of detritus feeders.

Veligers spend days or weeks in the plankton, according to the species. They feed almost continuously when the concentration of food is low, rising through the water, with the velum fully extended and the foot uppermost, to gather organic and inorganic particles, then withdrawing the velum and sinking, which necessarily interrupts the feeding process. Food is collected at the velar edge which increases in length as the animal grows, so meeting increased demands. At the velar edge is a preoral band of long, compound cilia under nervous control, which borders a groove leading to the mouth (fig. 335). The groove has short cilia beating oralwards, and is limited postorally by another ciliary band, with cilia shorter than the preoral and not compound. The preoral cilia, responsible for both swimming and feeding, display a metachronism which is diaplectic (at right angles to the direction of the effective beat) and laeoplectic (each cilium beating after its right hand neighbour) (Knight-Jones, 1954). Strathmann & Leise (1979) filmed the capture of particles by veligers of *Ilyanassa obsoleta*, the opisthobranch *Tritonia diomedea*, and the bivalve *Crassostrea gigas*. In all three the preoral cilia beat metachronally into the food groove, as also do the postoral. Many particles enter the groove in the water current created by the preoral cilia, which in *Ilyanassa* beat 2–3 times faster than the postoral, without contacting them. It was also shown that in this species the preoral cilia have the greatest length and that their beat has the highest angular velocity (70 radians.s^{-1}), the two combining to give a clearance rate of 17 × 10^3 $\mu m^3.s^{-1}$. μm of velar edge, as against 4,900 for *Tritonia* and 3,600 for *Crassostrea*. The particles are probably not sieved out by the combined action of the pre- and postoral cilia, since the spacing of these does not set a minimal capture size. The postoral cilia may set up and maintain a fluid pressure barrier in the food groove which retains the trapped particles, and they may help in sucking particles into the groove if the recovery stroke of the preoral cilia is inadequate for this.

If a particle enters the food groove of a prosobranch veliger it is automatically passed to the mouth, but it may or may not be eaten. The unwanted food either circles the vicinity of the mouth for a limited period before being rejected along the length of the foot (Fretter, 1967) or it arrives at the edge of the oral ciliary field and is immediately taken away by the pedal rejection currents. Such rejection is not related to the type of food, organic or inorganic. It occurs in the still waters of an aquarium with high concentration of digestible algal cells when a larva, previously starved, will fill the stomach in a few moments and then stop feeding, after which it sinks and remains at a lower level by reduced movements of the preoral cilia while digestion of the food is under way (Fretter & Montgomery, 1969). Acceptance or rejection of food at the oral area is undoubtedly governed by the state of the digestive processes. If the velum collects particles of no food value feeding is nevertheless continuous and a few particles are rejected, but they pass rapidly through the gut and digestive processes are not initiated. However, the presence of even a single digestible cell in the stomach is enough to trigger off a series of events characterizing the digestive process. Digestible cells are subjected to both mechanical and enzymatic treatments: vigorous rotation of algae against the gastric shield either weakens the cell wall (e.g. *Cricosphaera carterae*, *Exuviella baltica*) so that digestive secretions can penetrate it, or may fragment it (e.g. *Monochrysis lutheri*, *Isochrysis galbana*). This is accompanied by rhythmic pulsations of the digestive diverticula which become filled with the contents of the stomach, so that food is brought into direct contact with the digestive cells, then emptied to the stomach again along with secretions and waste from the digestive epithelium. Contents of plant cells, but not the cell walls, are taken up by the ingesting cells which become pigmented by their presence.

Rates of larval growth must be mainly a function of the quantity and quality of the food ingested and assimilated. However at least some larvae are capable of taking up dissolved organic matter (DOM), especially amino acids, from the micromolar concentrations present in sea water. It has been shown that veligers of *Crepidula fornicata* deprived of particulate food and exposed to eleven free amino acids and several vitamins do not show any increase is shell length after five days. Their weight and protein content, however, were maintained, whereas controls lost both. Concentrations of amino acids of 25–150 μM were equally effective (Lord & Pechenik, 1984). Thus in natural conditions free amino acids may be of value if particulate food is scarce.

As a veliger grows the size of the shell increases by addition of successive strips of flexible organic matrix at the mantle edge (Fretter & Pilkington, 1971), and at the same time there is an increase in diameter of the shell corresponding to the increased volume of the viscera. The organic matrix which allows this increase in diameter also provides sites for the deposition of calcareous concretions which, at metamorphosis, apparently initiate the formation of crystals.

In the English Channel prosobranch veligers of all ages are found at all depths, the highest concentration of the youngest ones being towards the surface. As the population ages more larvae spend more time at greater depths, and if the environment is favourable, they meta-morphose (Fretter & Shale, 1973). Since larvae are at the mercy of currents many will be brought by chance to the appropriate rock crevice (Fretter & Manly, 1977a, 1977b), weed, soft substratum, or (for parasites) host. However, the differing horizontal movements of layers of the water mass may sweep some far from the area from which they originated, the (possibly restricted) habitat of the adult. In such circumstances larvae of *Lacuna vincta* competent to metamorphose continue to feed and grow, and may settle if suitable conditions ultimately become available. Indeed this mechanism on a large scale, perhaps due to delay in the onset of competence (Dobberteen & Pechenik, 1987) may have played an important part in establishing the wide distribution of *Cymatium parthenopeum* and *Bursa* (Scheltema, 1966, 1972) on both sides of the Atlantic. Their larvae, called teleplanic by Scheltema, have a long development and, though competent to metamorphose, may delay settlement for extended periods (Scheltema, 1986).

Transport of such larvae by oceanic currents has been most intensively studied in the tropical Atlantic (Robertson, 1964; Scheltema, 1971a, 1971b; Laursen, 1981), and to a lesser extent in the Indo-Pacific (Scheltema & Williams, 1983). Scheltema (1971b) estimated the duration of pelagic development for certain prosobranchs whose larvae were collected from the open waters of the North Atlantic. He concluded that for some—*Charonia variegata*, *Cymatium nicobaricum, Tonna galea, T. maculosa, Phalium granulatum*—this was over three months. The velocity of the tropical North Atlantic surface currents shows that transoceanic dispersal of these species is possible within that period. Anatomical similarity of the species occurring on both sides of the Atlantic bears a direct relationship to the abundance of veligers in the North Atlantic gyre, supporting the hypothesis that their teleplanic larvae are important in maintaining genetic continuity. Indeed for species whose veligers are amongst the most abundant—*Cymatium parthenopeum, Charonia variegata, Tonna galea*—no geographical subspecies are known.

The success of the veligers of benthic marine prosobranchs depends on metamorphosis on a favourable substratum, and, if left to chance, would, indeed, be remote. Factors relating to this success include the possibility of delaying metamorphosis, which is quoted by a number of authors (e.g. Pechenik, 1986) and must be of general occurrence, and a rapid response to a stimulus received from an appropriate environment when it is encountered. Most observations have been made on intertidal species. Veligers swept over the shore are likely to be briefly trapped, allowing a thigmotactic and probably chemical response to initiate acceptance or rejection of a substratum, the criteria for which may be a complex of factors. Scheltema's work (1961) on the mud snail *Ilyanassa obsoleta*, a deposit feeder, demon-strated that in the laboratory veligers may metamorphose seventeen days after hatching from the egg capsule if presented with a suitable environment, but in the absence of this, planktonic life might be extended by twenty days after the creeping-swimming stage is attained, when the larva is competent to metamorphose, though metamorphosis cannot be indefinitely delayed. A substratum is favourable for *I. obsoleta* not just as a direct function of the physical characteristics of the sediment but also as a response to its biological and chemical properties. The factor inducing metamorphosis appears to be a water-soluble substance that escapes to the superjacent water.

The benthic life of a number of marine prosobranchs is confined to certain algae on which the veligers settled. Kiseleva (1967) has shown that metamorphosis of the veligers of *Rissoa splendida* and *Bittium reticulatum* is induced by chemical stimuli from *Cystoseira*, a discrimi-nation by the larvae essential for the survival of the species. Such substances may be either attractants or repellents. Chemicals produced by at least one species of alga, *Ralfsia*, have been known to kill veligers (Conover & Sieburth, 1965). Not only the species of alga but the epiphytic diatoms associated with it may be the attractant. A film of micro-organisms, including diatoms, on the surface of natural and artificial substrata is known to be of importance in contributing to the initiation of metamorphosis in *Littorina picta* (Struhsaker & Costlow, 1968), *Rissoa splendida*, and *Bittium reticulatum* (Kiseleva, 1967). In the case of *Cerithiopsis tubercularis* the response of the larva is to the sponge on which the adults feed (Fretter & Manly, 1977a) rather than to the alga associated with the sponge. It may be assumed that chemotactic and thigmotactic responses initiate the settlement of many kinds of veligers on certain algae and help to maintain aggregations of the adults in suitable habitats. Many larvae settle in the vicinity of adults, suggesting that their exudates, such as mucous trails, may be the attractant (Wells & Buckley, 1972).

The speed of metamorphic changes essential for survival is well illustrated by reference to *Melarhaphe neritoides* (Fretter & Manly, 1977b), in which the site of settlement of the larva must provide conditions that tide the prosobranch over a period of transition from planktonic to semi-terrestrial life. A study of this species at Wembury, S. Devon, showed that at the base of the vertical seaward face of a rock exposed to tidal surge and composed of quartzite traversed by numerous quartz veins, young snails occurred in crannies and pits formed by the loss of crystals from the quartz. Adults lived at higher levels on the same rock. Veligers cast on to this area may be hurled into crystal pits, and, if conditions are favourable, they adhere to their wall by an instantaneous exudation of secretion from the foot, simultaneous with the metamorphic changes. No swimming-crawling stages were evident, and if such do occur in *M. neritoides* they would be sucked away by the surge of the water. The fact that, with few exceptions, recently metamorphosed snails are confined to the base of the rock, though physical conditions higher on its surface appear equally favourable, indicates that metamorphosis is a response to some specific factor operating there, which may be a biotic attractant. The favoured pits are those with walls covered with algae which the snail rasps within an hour from settlement. The algae include a coccoid blue-green, a red (unidentified), and the conchocelis stage of *Porphyra* and *Bangia*. As the snail feeds the digestive gland enlarges and this, together with some uptake of water, leads to the growth in shell diameter allowed by the flexibility of the organic matrix; this secures the snail more firmly in the crystal pit before the shell calcifies.

The secretion from the foot which anchors a veliger to its favoured settlement site at metamorphosis comes from a large gland of subepithelial cells lying ventral to the pedal ganglia and spreading back to open near the posterior tip of the mesopodium. It is developed in the late veliger. Observations on *Lacuna vincta*, *Rissoa parva*, *Cerithiopsis tubercularis*, *Aporrhais pespelecani*, *Hinia incrassata*, and *H. reticulata* show that after metamorphosis the gland is lost.

When a monotocardian planktonic veliger hatches its mantle cavity lies dorsally and is deep enough to accommodate on retraction both the velum and the foot with its operculum, which may close it. In an active larva a flow of water is maintained through the cavity, passing, as in the adult, from left to right. This flow is brought about by strips of ciliated cells bordering each side of the osphradial ganglion, and by a ciliated area on the underlying body wall. The osphradium and its accompanying ciliated strips may come to occupy a third of the breadth of the mantle skirt on the inhalant side (Fretter, 1972). This pallial water current is maintained while the larva swims and also when it is partially withdrawn into the shell. It stops suddenly if adverse conditions cause a rapid withdrawal of velum and foot, but on their stepwise

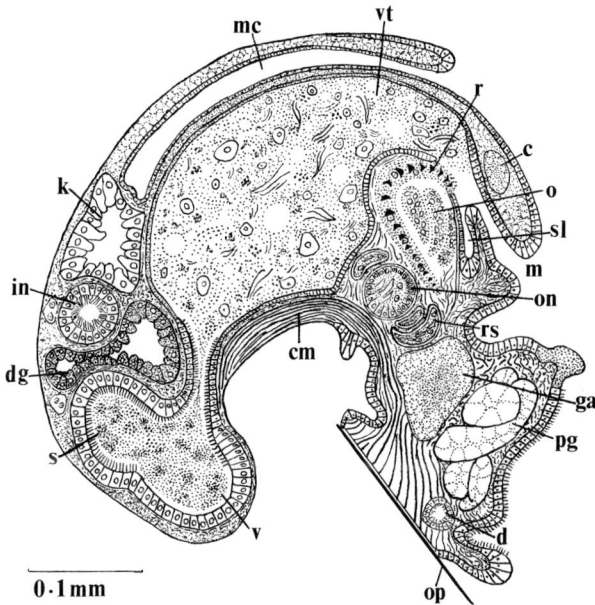

FIG. 336.—*Lacuna vincta*. Parasagittal section of metamorphosing larva to show ingested velum. (After Abro, 1969).

c, cerebral commissure; cm, columellar muscle; d, duct of posterior pedal gland; dg, digestive gland; ga, pedal ganglion; in, intestine; k, kidney; m, mouth; mc, mantle cavity; o, odontophoral cartilage; on, odonto-blast nest; op, operculum; pg, posterior pedal gland; r, radula; rs, coil of radular sac; s, style sac region of stomach; sl, future sublingual pouch; v, ventral (digestive) region of stomach; vt, velar tissue, distending oesophagus.

extension starts again rhythmically, as if each sample of water was cautiously tested. The osphradium is similar in the early veligers of lower mesogastropods (e.g *Lacuna*, *Rissoa*), of the higher ones (e.g. *Lamellaria*, *Trivia*), and of neogastropods (e.g. *Hinia*) regardless of any differences in their final structure.

The gill is first seen as a longitudinal fold of the inner epithelium of the mantle skirt to the right of the osphradium, covered by columnar ciliated epithelium and related to blood channels within the mantle skirt. The gill has not been seen to differentiate further until after metamorphosis in *Lacuna vincta*, *Littorina littorea*, *Rissoa* species and *Aporrhais*, but in the higher mesogastropods *Lunatia alderi*, *Lamellaria perspicua*, *Trivia* species, and in the neogastropods *Hinia incrassata* and *H. reticulata* the fold has subdivided to form a few gill leaflets by the late veliger stage.

While the monotocardian veliger collects food with the velum the odontophore is developing beneath the larval gut, though it in no way deforms the ciliated larval foregut, which remains unchanged until the time of metamorphosis, when the radula first becomes functional. Changes which bring this about have been studied in *Lacuna vincta* (Fretter, 1972); some are shown in figs 336, 337. By the time that the veliger is competent to settle a long radular sac already lies in the haemocoel immediately posterior to the pedal commissure, curving anteriorly from its opening to the gut, from which it has been formed as a ventral

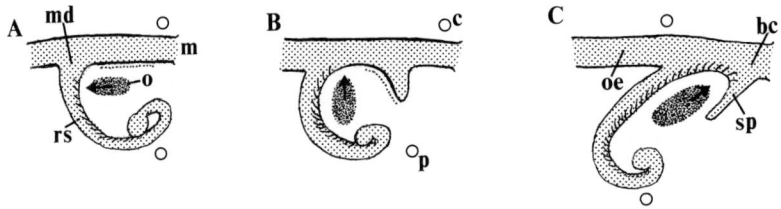

FIG. 337.—*Lacuna vincta*. Diagrams to show stages in the development of the anterior gut and rotation of the odontophore. A, early veliger stage; B, mid veliger; C, late veliger. Anterior is to the right. The dotted line marks the area of the floor of the larval gut which will form the roof of the sublingual pouch.

    bc, buccal cavity; c, cerebral commissure; m, mouth; md, mouth of diverticulum; o, odontophoral cartilage (arrow marks definitive anterior end); oe, oesophagus; p, pedal commissure; rs, radular sac; sp, sublingual pouch.

diverticulum. The sac is coiled in a vertical planispiral round the bulbous odontoblast nest, and is filled with rows of teeth secreted by the odontoblasts, the oldest row at its opening. A proliferation of cells around the ventral wall of the larval gut, posterior to the pedal commissure and anterior to the opening of the radular sac, differentiates to form paired odontophoral cartilages and the protractor, retractor, and tensor muscles of the odontophore and radula. These are orientated at this stage so that the future anterior end is topographically posterior and the future dorsal side is ventral.

    The end of larval life is abrupt. The larva, now benthic, displays sporadic contractions of the velar retractor muscles which are then severed from their attachments to the head, and right and left velar lobes are simultaneously ingested, distending the mid oesophagus and stomach (fig. 336). The mid oesophagus, its lumen as yet unrestricted by the development of glands, is expanded to four times its previous diameter, bringing about a simultaneous reorientation of the odontophore and radular sac. These structures, pushed by the bulky velar tissue, pass anteriorly through the nerve ring, the posteriorly placed future anterior tips of the cartilages are rotated upwards, and their future posterior tips directed downwards. As a final part of the reorientation the odontophore is rotated outwards through the mouth of the diverticulum and comes to lie on the floor of the buccal cavity, having undergone a total movement through about 180°. Underneath it lies a sublingual pouch the walls of which have been formed from that part of the floor of the original larval gut which lay over the developing odontophore (fig. 337). The odontophore is repeatedly protruded towards the mouth and then withdrawn. The radula spreads over the odontophore as though pulled by its reorientation and rhythmic action, its final spread accelerated by an increase in the rate of production of tooth rows by the odontoblasts, though this may, more probably, reflect their response to changing tension. The salivary ducts now open into the buccal cavity, the glands and ducts having differentiated earlier. After the digestion of the velar tissues the radula starts to gather food.

    The development of the odontophore of *Lacuna* illustrates the ability of the larva to feed and grow unhampered by the simultaneous development of the feeding apparatus essential for the young snail after metamorphosis. A study of the veliger of *Hinia incrassata* and *H. reticulata* indicates that this also applies to the development of their more complex gut and pleurembolic proboscis. In the young larva of *H. incrassata*, about two weeks old, a ventral diverticulum grows out from the gut posterior to the pedal commissure. Soon after its appearance cells are proliferated from its epithelium and underlying tissues, especially

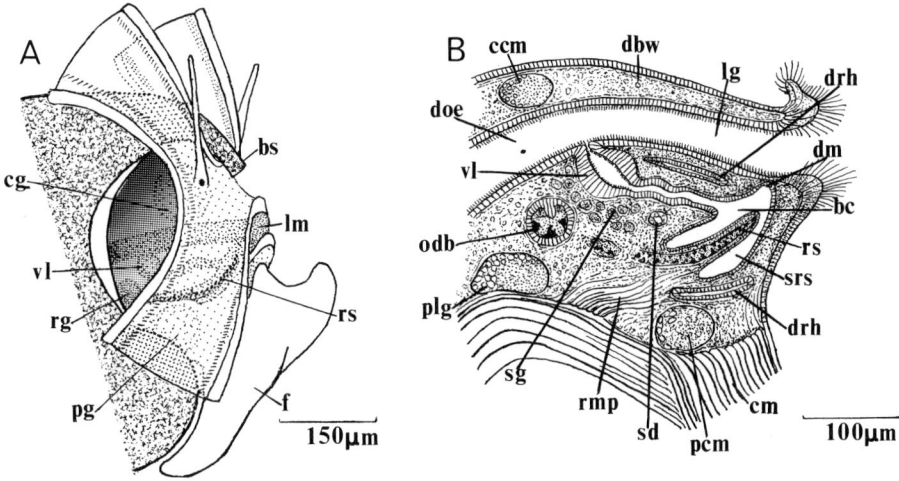

FIG. 338.—*Hinia incrassata*. A, right view of head of veliger. The velar lobes have been cut off towards their base. Some internal structures are seen by transparency (rhynchodaeum omitted). B, sagittal section of head of veliger at the same developmental stage as A. Note that the development of the foregut and its final situation are anterior to the nerve ring.

bc, definitive buccal cavity; bs, beak of shell; ccm, cerebral commissure; cg, cerebral ganglion; cm, fibres of columellar muscle; dbw, dorsal body wall; dm, site of definitive mouth; doe, definitive oesophagus; drh, developing rhynchodaeum (proboscis sac); f, foot; lg, larval foregut; lm, larval mouth, future rhynchostome; odb, odontoblasts; pcm, pedal commissure; pg, pedal ganglion; plg, pleural ganglion; rg, radular gland; rmp, retractor muscles of proboscis; rs, radular sac; sd, salivary duct; sg, tubules of salivary gland; srs, subradular sac (later sublingual pouch); vl, valve of Leiblein.

laterally, and there is some proliferation of cells from the gut wall near the origin of the diverticulum. These cells are concerned with the formation of the proboscis, the salivary glands and their ducts, and the muscles and cartilages of the odontophore (fig. 338). The diverticulum lengthens, passes anteriorly and curves back ventrally at an acute angle, to become V-shaped. In a three-week old larva it can be seen that the ventral limb is the developing radular sac which leads posteriorly to a spherical odontoblast nest situated beneath the origin of the diverticulum from the larval gut. Rows of teeth fill the sac, which is later bordered by rod-shaped cartilages and associated muscles. The opening of the radular sac is at the angle of the diverticulum, the dorsal limb of which will become the initial part of the adult oesophagus, and the area into which the teeth project will become the most anterior part of the buccal cavity of the adult. The dorsal limb elongates, giving the undulatory course capable of running through the extended proboscis of the metamorphosed snail. The epithelium of the diverticulum at its origin from the larval gut thickens to form the valve of Leiblein, which marks the junction of the anterior and mid oesophagus.

As the veliger grows differential growth brings the anterior oesophagus of the veliger, with the underlying diverticulum and radular sac, anterior to the nerve ring, and the valve of Leiblein to the adult position. The salivary glands, which have developed alongside the definitive anterior oesophagus, and their ducts, pass forward to the buccal cavity, but do not open until after metamorphosis; unlike those of mesogastropods they do not pass through the nerve ring.

All these components of the definitive adult foregut become enclosed in a cylinder of dense tissue which ultimately forms the muscles of the proboscis and the epithelium of the proboscis sac or rhynchodaeum which appears as a cavity in the anterior part of the dense tissue. On first appearing this is an internal space, not connected to any other cavity. Later, when the velum is cast off at metamorphosis (it is not ingested as in *Lacuna*) the wall separating the cavity from the ventral wall of the initial part of the larval gut breaks down, as also does that between the cavity and the developing buccal region. At the same time that part of the larval gut lying between the opening of the rhynchodaeum anteriorly and that to the valve of Leiblein posteriorly starts to occlude and ultimately disappears. The course of the gut after these events is from the larval mouth, which has become the rhynchostome, to the 'true' mouth leading to the buccal cavity and finally to the valve of Leiblein, where it rejoins the posterior part of the larval oesophagus. As this part of the gut forms a pleurembolic proboscis when development is complete, the outer and inner walls of the rhynchodaeum telescope outwards on protrusion and the rhynchostome disappears.

This development raises a few points to be noticed. In *Lacuna* the larval gut becomes broadly the adult gut, the only addition to it being the odontophore, the radular sac, and perhaps also that small part of the oesophagus called the anterior oesophagus, all of which have developed from a ventral diverticulum ready for the change in feeding at metamorphosis. This pattern is primitive. In *Hinia* the complexity of the anterior part of the gut, especially the formation of a proboscis, combined with the need for it to be immediately functional at metamorphosis, have required a greater length of ventral diverticulum to be involved in its production, with the result that it replaces, not just adds to, a stretch of larval oesophagus. These developmental patterns suggest that a separation of buccal cavity and anterior oesophagus is morphologically unsound: the two form a developmental unit, though they are functionally distinct. Despite the fact that in many species the valve of Leiblein shows the twist due to torsion and signs of the separation of oesophageal glands, its development is of no help in confirming that this is how the glands were formed.

The opening of the gut to the rhynchodaeum in neogastropods is usually taken to be the true mouth, with the rhynchostome representing an essentially secondary structure, a mere fold in the retracted pleurembolic proboscis. The development of *Hinia*, however, suggests otherwise, since the larval mouth, which corresponds to the adult mouth in archaeogastropods and mesogastropods, becomes the rhynchostome, and the gut opens secondarily to the rhynchodaeum, which could, therefore, be regarded as an expansion of that part of the gut between buccal cavity and mouth. The development of this area is so unusual and so little known, however, that the homologies are best taken as doubtful.

That the presence of a proboscis is the main factor complicating the development of the foregut is supported by the fact that the acrembolic proboscis of *Natica* and *Lamellaria* also develops beneath the larval gut (Abro, 1969), the anterior end of which is occluded and lost after metamorphosis. Since the proboscis of *Lamellaria* is small its development in no way interferes with the ingestion of the velum, which stretches the mid oesophagus before the oesophageal glands develop. In general it appears that the velum of a prosobranch veliger is commonly ingested at the end of larval life; only where a long and therefore bulky proboscis is being formed is the velum cast off, perhaps because its ingestion would interfere with the proper development of that organ. When ingestion occurs right and left halves of the velum may be engulfed simultaneously as in *Lacuna*, but Pilkington (1970) has shown that in an atlantid heteropod occurring off New Zealand, one side of the velum is ingested and then the other.

Metamorphosis in *Lacuna* and *Hinia* involves several changes in the organization of the body in adaptation to a new mode of life; it does not, however, involve serious alterations in the shape of the shell, which continues to elongate in much the same way as it did in the larva. This is not true of those prosobranchs which are limpet-like in adult life. The limpet shell may be attained in two different ways, one of which is found in the vetigastropods and patellogastropods, the other, involving more complex changes, in the calyptraeoideans. In the first group the limpet shape is reached by differential growth of the lips of the primary shell aperture, the initial coils of the larval shell often remaining visible at the apex. In calyptraeoideans, however, as described by Werner (1955) and Fretter (1972), a secondary aperture is produced, partly from the post-torsional right side of the original larval shell, and partly from a new structure secreted by an everted fold of the mantle, the two fusing to give an apparently continuous structure. The initial coils of the shell in these animals are covered over by the secondary growth and become invisible.

Metamorphosis, however, does not leave the calyptraeoidean body unaffected. At that moment the muscles running into the velar lobes break at their point of entry, presumably at a preformed plane of breakage, and in *Crucibulum*, *Crepipatella*, and probably *Crepidula*, though not in *Calyptraea*, which has no free larval stage, the discarded velar lobes are rapidly sucked into the gut and digested. At the same time the branches of the columellar muscle which run to the operculum separate from that structure, which falls off. The foot is thus converted into the more suctorial type characteristic of a limpet and subsequent action of the muscle pulls the shell to the substratum rather than the body into the shell. Similar changes occur in the vetigastropod and patellogastropod limpets though in some from hydrothermal vents (Fretter, 1988, 1990) and in the neritoideans *Septaria* and *Phenacolepas* (Fretter, 1984a) the operculum may be retained as a vestigial structure, sometimes secondarily enclosed in an internal space.

Even in species in which velar tissue is cast off at metamorphosis the blood which has been used to expand the lobes during larval life is returned to the body prior to their loss by contraction of the velar muscles. Along with the blood some velar material may also return: in *Mangelia nebula*, for example, pigment cells, originally confined to the velum, are withdrawn with the blood and settle posterior to the cephalic tentacles, where they are the first indications of the ultimate adult pigmentation (Fretter, 1972).

CROSS REFERENCES

CHAPTER 30

PERIWINKLES AND CHINK SHELLS: LITTORINIDAE

T HE taxonomy of littorinid species found in the British Isles has had a curious history, and cannot even now be regarded as settled to the agreement of all. Linnaeus (1758) recognized three species of British periwinkle, which he placed in the genus *Turbo*, *littorea*, *neritoides*, and *obtusata*, and a fourth, *littoralis*, which he placed in the genus *Nerita*. Near the end of the eighteenth century Olivi (1792) added *saxatalis* to the genus *Turbo* (the specimens on which the species was based came from Venice and this has remained the only Mediterranean locality for the species) and Maton (1797) had recognized *Turbo rudis* in south-west England. By the middle of the nineteenth century many further species had been named, some of which, however, were rapidly perceived by most malacologists as synonymous with, or at most varieties of, others. A number, some of which have proved permanent (or possibly only more long-lasting) may be mentioned: *tenebrosa* Montagu, 1803, *jugosa* Montagu, 1803, *fabalis* Turton, 1825, *nigrolineata* Gray, 1839, *neglecta* Bean, 1844 and *patula* Thorpe, 1844. By 1850 or thereabouts, therefore, it seemed that at least ten good species of periwinkle were to be met with on the shores of the British Isles. This proved, however, to be only a transient view of the situation and when the genus *Littorina* was dealt with by Jeffreys (1865) the number was reduced to four, *littorea*, *neritoides*, *rudis*, and *obtusata*. He regarded *saxatilis* Olivi as a synonym of *neritoides* and treated most of the others as varieties of *rudis*, though sometimes without expressly saying so; *fabalis* became a variety of *obtusata* and later (1869) he added *aestuarii* as another variety of the same species. The influence of Jeffreys's work was so great that this arrangement persisted largely unchanged for the next hundred years.

During this long period these four species of periwinkle were widely used to introduce students to two of the basic ideas of littoral biology, that species are zoned, and that their reproductive methods are adapted to the conditions in which they live: *Littorina saxatilis*, at the top of the shore, was ovoviviparous, *L. obtusata*, below it, laid eggs in jelly masses which underwent direct development and hatched as juvenile snails, whilst *L. littorea*, the lowest, had planktonic egg capsules from which veliger larvae emerged. Nothing definite was known about the breeding of *neritoides*, but as it lived higher on the shore than *saxatilis*, and as *saxatilis* was ovoviviparous it was assumed that it had to be ovoviviparous too. The first sign, indeed, that the apparent beauty of this series of adaptations was artificial was the simultaneous discovery by Lebour (1935) and Linke (1935) that *neritoides*, the highest species on the shore, far from being ovoviviparous, was like *littorea*, the lowest species, and had pelagic eggs and free veliger larvae.

At a later date the taxonomic simplicity of Jeffreys's arrangement also started to crumble, though two of the four species, *littorea* and *neritoides*, have persisted unchallenged to the present day. In 1966, however, Sacchi & Rastelli showed that *obtusata* could be split into *obtusata* proper and a species which they called *mariae*, though S. M. Smith (1982) believes that *mariae* is in fact *fabalis* Turton under another name. Whatever the name the two species are now generally accepted as valid; Reimchen (1981) has suggested on the basis of differences in colour morphs and periostracal patterns between larger and smaller animals that *mariae* may comprise two sibling species. Moyse *et al.* (1982) have concluded on the basis of structure and enzyme content that Jeffreys's variety *aestuarii*, sometimes regarded as an independent species, is a variety of *obtusata* and not related to *mariae*.

A more profound alteration of accepted ideas was initiated by Heller (1975a) who split the species *saxatilis* back into some of the original constituents out of which Jeffreys had assembled it—*nigrolineata*, *neglecta*, *rudis*, and *patula*. A little later Ellis (1978) separated, on the grounds of its oviparity, a species *arcana* from Heller's *rudis* grouping, but agreed (1979) with Raffaelli's argument (1976) that *patula* could not be separated from *rudis*. On the grounds of strict priority the name *rudis* Maton, 1797 should be replaced by *saxatilis* Olivi, 1792 if the two species are regarded as the same, a position originally taken by perhaps a majority of malacologists, though rejected by J. E. Smith (1981). Their unity may, perhaps, now be regarded as proved by Warwick's breeding of fertile offspring from a *saxatilis-rudis* cross (Atkinson & Newbury, 1984) and is accepted by Reid (1989) in his study of the Littorinidae. Of a series of interspecific crosses attempted by Warwick *et al.* (1990) this was the only wholly successful one; all others failed except for that between male *saxatilis* and female *arcana*. The true specific status of *L. arcana*, which has sometimes been regarded as merely a reproductive variant of *L. saxatilis* (= *rudis*), seems also to have been demonstrated by Ward & Warwick (1980), Ward & Janson (1985) and Mill & Grahame (1988) on the basis of enzyme studies, and by Atkinson & Warwick (1983) on the basis of its polychromaticism. Though it has been argued (Ellis, 1979) that *nigrolineata* was involved, the discovery of *arcana* finally clears up a puzzle which arose as far back as 1947 when Seshappa described 'oviparity in *Littorina saxatilis*' from the Northumberland coast, an observation to the significance of which all contemporary malacologists remained blind. Heller's collection of species as modified by Ellis is frequently referred to as the *saxatilis* species-complex.

From the position which obtained intil 1966, in which four species represented the genus *Littorina* in the fauna of the British Isles, we have now returned almost full circle to that of the mid nineteenth century with no fewer than eight species of winkle accepted as probably valid (*littorea, neritoides, obtusata, mariae, neglecta, nigrolineata, saxatilis, arcana*); a possible ninth (*tenebrosa*) has been shown by Janson & Ward (1985) to be better regarded as a form of *saxatilis*. Even this list may not represent the truth as the status of some is still uncertain; in particular, B. & K. Johannesson (1990) and K. & B. Johannesson (1990) argue that *neglecta*, a distinctly variable taxon, may be no more than a morph of *saxatilis*, though Warwick *et al.* (1990) could not successfully cross the two. One unfortunate consequence of the changing names and status of littorinid taxa, especially the more recent ones, is that the precise specific make-up of the populations studied by many workers is left uncertain and the applicability of their conclusions unreliable.

Though the local littorinids are currently acceptable as placed in eight species their inclusion in the single genus *Littorina* is not. The species *neritoides* stands apart from the others in respect of several characters [longitudinal black lines along the tentacles; closed vas deferens; absence of mamilliform glands on the penis; radula; allozyme content (Ward, 1990); kidney structure (E. B. Andrews, personal communication)]. It has been accepted by several authors, most recently by Reid (1989), based on a cladistic analysis of littorinid structure, as properly placed in the genus *Melarhaphe*. Reid also places *littorea* in the subgenus *Littorina* (with pelagic larvae) and the others in the subgenus *Neritrema* (with direct development).

Littorinids are herbivores, eating algae and their epiphytes, actively avoiding sessile animals (Frey, 1986), but inevitably incidentally ingesting some detrital material. Their radula is a typical taenioglossate one, with only moderate ability to rasp, which might suggest that they are likely to rely on softer and smaller types of algae than on tougher ones. Where winkles live on rocks without a covering of macrophytes they must, as pointed out by Sacchi et al. (1977) for L. saxatilis and L. nigrolineata, but also undoubtedly true for all species, be dependent on unicellular algae (mainly diatoms), newly settled sporelings, and what is known as 'detritus', material containing much fragmented organic remains. This must also apply, but for a different reason, to such small animals as L. neglecta which live where macrophytes are usually absent, but are probably too weak to attack them if they were.

Melarhaphe neritoides lives high above the level at which the common intertidal seaweeds are found, often on rock which seems totally incapable of providing any food. To some extent, however, the place of weeds is taken by lichens, and Daguzan (1976a) has stated that neritoides subsists primarily on black lichens; but there are certainly places where the snails occur but not the lichen and they must then be dependent on the detritus which is abundant in nearly all sites.

When winkles live in association with weeds these are used as food. Though the impression left by different studies is that winkles are fundamentally opportunistic generalists eating whatever plant food is available in their environment, at least some species appear to have marked preferences for some weeds and avoid others. This has been most clearly shown for L. littorea and L. obtusata. Lubchenco (1978), Steneck & Watling (1982), and Watson & Norton (1985) have all agreed that the edible winkle has a strong liking for soft ephemeral weeds such as Ulva, Enteromorpha, Cladophora, and Ectocarpus, particularly when these are young. Fucoids are in general much less frequently taken and some, such as Ascophyllum nodosum, are refused altogether, as are corallines, even when the animals are hungry, according to Watson & Norton (1985). Sacchi et al. (1981), however, reported that L. littorea living amongst fucoids eats them willingly and pointed out that the microphagy ascribed to littorinid species is often not their preferred mode of feeding but is imposed on them by the absence of any large plants in the environment. L. littorea living near the top of a beach where none of its favourite food is growing is similarly forced (Watson & Norton, 1985) to depend on whatever weeds are washed up to the strand line; as these are often decaying this may allow the snails to deal with food which, alive, was too tough for them to eat and which, because of some decomposition of the tissues may also provide more easily assimilable substances. The same preferences of L. littorea have also been demonstrated by Imrie et al. (1989) in experiments using plant extracts. This result strongly suggests that the winkles use chemosensory and probably also gustatory cues in deciding what plants to use as food. In a later paper Imrie et al. (1990) have suggested that L. littorea tends to a conservative diet, showing the same ingestive conditioning as Nucella lapillus.

L. obtusata was shown by van Dongen (1956) to be strongly attracted to fucoids, especially Fucus vesiculosus, on which it is commonly found, and Bakker (1959) confirmed that this was its preferred food. L. mariae is similarly related to F. serratus at a lower tidal level. In laboratory trials (Watson & Norton, 1987) both species preferred fucoids to other weeds. L. obtusata exhibited a preference for the reproductive receptacles rather than vegetative tissues and also took a number of young stages, sometimes choosing these rather than the fully grown plant (Ulva), sometimes the other way round (Fucus serratus). The radula of obtusata, it has been suggested by Steneck & Watling (1982), is probably stronger than that of littorea, forming a coarser but more powerful tool for abrading the tougher weeds on which it feeds. Watson & Norton (1987) reported differences between the radular teeth of *obtusata* and *mariae*, mainly affecting the number of cusps on the dominant lateral teeth. These differences suggest that *obtusata* is better able than *mariae* to gouge weed tissue and that *mariae* is to that extent more dependent on the epiphytic flora, an idea confirmed by G. A. Williams (1990).

L. *obtusata* has limited food preferences and does not appear to affect the local abundance of weeds significantly, nor interfere with their natural succession: this is not always the case with L. *littorea*. Lubchenco (1978, 1982, 1983) has shown that the composition of the flora of high level rock pools, predominantly composed of ephemeral algae, is largely maintained by the grazing activities of the winkles (L. *littorea*) which live there, preventing their natural replacement by perennial seaweeds, less preferred by the snails as food, which takes place on the shore outside the pools. A similar but at the moment less definite role has been attributed to L. *mariae* whose feeding and abundance are closely linked to *Fucus serratus*. G. A. Williams (1990) has suggested that by the removal of the epiphytes the winkle contributes to the maintenance of populations of this weed. Since the activity of L. *littorea* keeps the flora of the pools in a state advantageous to the winkles, and since that of L. *mariae* does the same, they present a parallel to the pattern of shore management carried out by patellid limpets (Branch, 1975) with the same result.

Intertidal winkles such as L. *littorea* (R. C. Newell & Branch, 1980) show a rhythm in their feeding, starting to graze on re-immersion after a long tide period, and at the same time discharging excretory matter from the kidney (Daguzan, 1970). The rate of feeding, measured by the rate of odontophoral activity (R. C. Newell *et al.*, 1971), depends upon temperature, but has also been said to be linked to the length of the preceding period of immersion; as a consequence those winkles that live high on the shore feed more rapidly than do those at lower levels, thus compensating for the fact that their feeding period is shorter. L. *saxatilis*, a detritivore, differs from L. *littorea*, according to A. J. Berry (1961), in that it feeds when emersed and not when under water, perhaps an indication of its growing independence of water.

The rhythmical feeding imposed upon winkles by tidal cycles affects the activity of the cells of the digestive gland. This has been described most clearly by Boghen & Farley (1974) in *Littorina saxatilis*. Merdsoy & Farley (1973) also demonstrated a cycle of digestive gland activity in L. *littorea* but this was not strictly tied to the tidal cycle, a fact perhaps reflecting its lower level on the shore than *saxatilis*. In L. *saxatilis* most of the digestive cells in the tubules of the gland are, at any given moment, in the same phase of a cycle which starts with their ingestion of food, either finely particulate or in solution, into food vacuoles and its transfer to heterophagosomes (Owen, 1972), vacuoles lying at a deeper level in the cell. Lysosomes fuse with these to form heterolysosomes which are the sites of digestion and lie at a still deeper level. After assimilation of the digested food the heterolysosomes are transformed into residual bodies, composed of indigestible material, which move to the surface of the cell and are voided to the lumen of the tubule.

Survival of intertidal periods requires that a littoral animal has the ability to withstand degrees of desiccation, of heat and cold, and, possibly, of anoxia. It has usually been assumed that these abilities, which become progressively more essential higher up the shore since they relate more or less to the length of the period of emersion, are possessed to a corresponding degree by the animals living at different levels. K. Johannesson (1989) and later McMahon (1990) have argued, however, that instead of this gradient the high tide mark indicates a critical level separating two areas which call for different adaptations from their inhabitants.

The lower area is the eulittoral zone, characterized by regularly alternating and not over-lengthy periods of emersion and immersion, without extremes of temperature or dryness and therefore, on the whole, a not unkindly zone; the upper area, the eulittoral fringe, is a much more severe habitat, with extended periods of emersion and accompanying heat, or cold, and lack of water. These differences have developed different morphological and physio-logical adaptations in their molluscan inhabitants, described by McMahon (1990). Species living in the eulittoral fringe, especially in the warmer parts of the world, have smaller shells which are less heavily pigmented so that their reflectance is increased and reduces heating; eulittoral species are larger and darker. Eulittoral fringe species have a higher tolerance of heat, and their greater tendency to stay withdrawn in their shells when emersion lasts for a long time reduces, or even avoids, water loss; eulittoral animals are less tolerant of overheat-ing but can cool themselves by evaporation of water since its loss can be made good at the next flowing tide.

Zonation of littoral species implies adaptation to varying degrees of stress, especially in the ability to resist desiccation when emersed and varying salinity when under water. *Melarhaphe neritoides* is certainly able to resist some dilution of salinity since it can be seen actively browsing whilst rain falls, but how much of the body is actually exposed to the water and how saline the rain has become after falling on a spray-swept shore, is doubtful, and no figures are available. More information relates to *saxatilis* (Sandison, 1966, 1967; A. J. Berry & Hunt, 1980; Sundell, 1985). Specimens of this species from exposed Swedish shores showed a mean survival time in fresh water of 16.8 ± 0.2 days whilst those from sheltered shores were more affected and lasted 13.8 ± 0.2 days. Both types continued normal activity until the salinity was reduced to 16‰. Much more work has been carried out on *L. littorea* which, living lower on the beach might show greater sensitivity to reduced salinity. The position is complicated by the dependence of survival in low salinities on temperature: thus Todd (1964) found that *littorea* lived for 17 days in water of 9.6‰ at 5°C but died at 14 days when the temperature was 15°C, and A. J. Berry & Hunt (1980) showed that adults and juveniles of *saxatilis* are affected badly by salinities less than 15‰ but that adults, though not juveniles, would survive the low salinity provided that the temperature was 0–5°C.

Arnold (1972) working with animals from Scotland and eastern Canada concluded that the winkles became inactive at salinities below 20–25‰, whereas Gowanloch & Hayes (1926), also working in eastern Canada, had them living indefinitely at salinities of 15‰, and Rosenberg & Rosenberg (1973) described specimens of *littorea* as 'rather active' at a salinity of 10‰ and thriving at salinities between 11.5 and 30‰. Arnold's animals, however, were kept at 20–25°C whereas those of the Rosenbergs were at 12–13°C. When salinities are liable to be lowered by winter rain or snow, therefore, the accompanying fall in temperature may moderate their impact. From a study of *L. littorea* collected at sites in Denmark and Sweden with naturally different salinities Rosenberg & Rosenberg (1973) concluded that local popu-lations show adaptation to the salinity to which they are accustomed, an adaptation prob-ably developed after settlement rather than genetically determined. Just as there seem to be populations with different salinity tolerances so populations of *littorea* respond differently to exposure to heat, showing adaptations to local conditions. R. G. Evans (1948) reported heat coma in winkles from Cardigan Bay at 39°C and heat death at 46°C, whereas Sandison's (1967) corresponding figures for those living in the Firth of Forth were 32°C and 42°C.

When exposed to diluted sea water winkles remain isosmotic with the external medium to 50% sea water; below that they regulate their volume. This appears to be achieved mainly passively by withdrawal from contact (Todd, 1964; Berger *et al.* (1978). If they are prevented from doing this they continue to be isosmotic (Mayes, 1960, 1962; Avens & Sleigh, 1965; Hoyaux et al., 1976) though swelling is offset by a degree of active uptake of ions. Some control of body volume is also exercized through mobilization of amino acid reserves (Berger et al., 1978)—alanine has been observed by Wieser (1980)—but this method of osmotic control seems to be much less important than in bivalves.

Survival of intertidal periods requires the ability to withstand desiccation and, possibly, periods of anoxia. Both relate to the amount of water which may be retained within the mantle cavity (Boyle et al., 1979) and to the topography of the aperture and operculum (Atkinson & Newbury, 1984; Lowell, 1984). Houlihan et al. (1981) measured the amount of water retained in the mantle cavity of emersed L. littorea and, like Innes & Houlihan (1985), found that in their experimental animals aerobic respiration continued throughout the entire period of emersion, though Patanè (1946, 1955) had already suggested that the long periods of inactivity which some littorinids can endure must involve some anaerobic metabolism.

If the feeding excursions of intertidal prosobranchs were totally random the regularity of their zonation would be rapidly lost unless it is assumed that all individuals migrating more than a certain distance up or down the shore are removed by predation or other means. It is, therefore, to be expected that the animals have behaviour patterns which confine them within certain limits and allow (or compel) them to return to them if they exceed them. Definite, though temporary, change in the level at which some species live, however, has been reported as an accompaniment either of breeding or of changing temperatures.

Breeding migrations have been described in Melarhaphe neritoides, but to what extent they are of general occurrence is not known. Lysaght (1941) stated that spawning takes place only when the females are submerged, a constraint easily understood since the egg capsules are pelagic. The distribution of the animals on a shore is obviously patchy, sometimes on a small scale, though the factors determining patchiness are not known; it may be that it reflects the limitation of the snails to places which, even when high, are always wet by spring tides, times with which their spawning is known to be correlated. On some shores, however, it may be necessary for females living at the highest levels to make some downward migration. Ellis (1985) has recorded a regular breeding migration in the oviparous species L. arcana. These animals normally live alongside the ovoviparous species saxatilis at a high level on the shore; those of reproductive age, however, migrate downshore in autumn and deposit their egg masses in damper situations than would otherwise be available. A further advantage conferred by this migration is that it separates the breeding populations of arcana and saxatilis, an important point in a genus known for extended breeding periods and promiscuous reproductive habits.

Most British littorinid species are hardy enough to withstand the summer-winter fluctuations of temperature they normally encounter without undertaking seasonal migrations. The behaviour of L. littorea, however, seems to involve, at least in certain areas, a migration downshore as temperatures fall in autumn and an upshore one as they rise in spring. Gowanloch & Hayes (1926) and Lambert & Farley (1968) reported that this was true for winkles in Nova Scotia, as did Rasmussen (1973) for those living in the Isefjord in Denmark. Gendron (1977), however, found only limited migration at Wood's Hole. I. C. Williams & Ellis (1975) recorded a downward movement of edible winkles in N. E. England in autumn (and assumed a return one in spring) whilst Underwood (1973) reported a winter concentration of winkles in low pools on a beach in Plymouth Sound and an upward dispersal in spring. H. B. Moore (1937), however, also working in S.W. England, found no such migration or concentration. The behaviour of littorea, it would seem, is thus very much a function of local winter temperatures and, perhaps, local habitat, migration taking place when temperatures are low enough. By contrast no temperature-related migration occurs in species of the *saxatilis* complex: these, even in the severe conditions of northern Russia (Matveeva, 1955), overwinter in crevices.

A number of workers have investigated the behaviour of winkles and other prosobranchs which have moved up or down a beach from the level at which they normally live, experiments which might reveal the mechanism by which they keep their normal stations. In general all then seek to return to that level, using a series of responses to light, gravity (F. Evans, 1961; Frank, 1965; Gendron, 1977; Doering & Phillips, 1983) or, on occasion, wave action (Gendron, 1977), presumably the same behaviour patterns as allowed those that settled after a pelagic larval phase to find this level in the first place (Desai, 1966; Fretter & Manly, 1977b). These stimuli cannot, however, be the sole ones effective in controlling behaviour since other movements are carried out (in relation to feeding, for example) which do not necessarily fit that pattern. If the stimuli are responsible for confining animals to definite limits, too, one might expect to find weakening, extinction, or even reversal of responses as limits are approached or transgressed. One of the earliest analyses of littorinid behaviour, that of Fraenkel (1927a) for *Melarhaphe neritoides*, is also one of the clearest illustrations of how a species finds and keeps its normal level on the beach. *M. neritoides*, crawling under water on a horizontal surface such as the floor of a rock pool or crevice, is negatively geotactic and negatively phototactic; when it is crawling upside down over the roof of a crevice, the phototactic response changes and becomes positive: such behaviour thus leads the snails into and then out of water-filled crevices. When, however, the animals are creeping in air the reversal of the phototaxis does not take place, so producing a piece of behaviour which leads them to stay within the crevice, as many, indeed, are found to do. The negative geotaxis leads *neritoides* to creep high up a shore from the lower level at which the larvae settle (Fretter & Manly, 1977b). Since they do not keep on climbing indefinitely, however, and since they emerge from crevices to feed when not immersed, it is clear that other factors are involved in controlling the rhythm of their daily life. Petpiroon & Morgan (1983) have shown that in *L. nigrolineata* this daily rhythm seems to be partly a response to immersion and emersion, both of which cause increased activity, and partly to an endogenous rhythm.

So long as winkles are living within the limits of the zone in which they are normally found their movements are either apparently random (*L. littorea*: Petraitis, 1982), though perhaps more truly related to something unnoticed or unnoticeable by the observer, or they carry out the type of underwater feeding excursion which was described by G. E. Newell (1958a, 1958b) or something similar. In this a winkle moves initially towards the sun, feeding as it goes, but after some time its reaction to light reverses and it moves back to the neighbourhood of its starting point or to that point precisely. Since tide time and sun position are both variable from day to day this arrangement means that the animal tends to sample different parts of its environment at different times.

Most Atlantic littorinids have planktonic eggs and larvae (Bandel, 1974). The collection of species found in N. W. Europe is unusual in that two (*L. littorea, M. neritoides*) show this typical life history, whereas others (*L. obtusata, mariae, arcana, nigrolineata*) lay egg masses attached to a substratum from which juveniles hatch, and still others (*saxatilis, neglecta*) are ovoviviparous. It is to be assumed that these differing reproductive methods are the result of selection and are efficient in the particular ecological situation in which each species lives.

Species of littorinid illustrate more clearly than any other group of prosobranchs the modification of the pallial oviduct in relation to the type of spawn produced (Ellis, 1979;

Fretter, 1980). In each species an ovarian duct leads from the ovary to an albumen gland, anterior to which is a membrane gland, but the ultimate section of the pallial oviduct leading to the genital aperture may be modified in one of three ways. It may be (1) a jelly gland producing a viscous fluid which hardens on exposure and in which a number of egg capsules are embedded before deposition (*L. arcana, L. obtusata, L. mariae, L. nigrolineata*). (2) It may exhibit two components, an initial glandular part and a relatively thin-walled duct leading thence to the genital aperture; the glandular section, which spreads laterally and dorsally over the proximal part of the oviduct, receives incomplete capsules from the membrane gland, embeds them in a fluid less viscous than the albumen, and passes them to the duct, the walls of which produce the elements of their outer wall (*L. littorea, Melarhaphe neritoides*). (3) In ovoviviparous species it forms a brood pouch divided into compartments (*L. saxatilis, L. neglecta*). The path of the channel followed by eggs through the pallial duct is not straight (A. J. Berry & Chew, 1973; Reid, 1989) but thrown into loops ('spirals' of Reid, 1989) so lengthening the pathway, an arrangement which, according to Reid, permits a greater egg production. The eggs are passed to an ovipositor which either liberates them to the plankton or attaches them to a substratum. The ovipositor of *Melarhaphe neritoides* as figured by Fretter & Graham (1962) represents that of an anaesthetized animal: in mature, unanaesthetized individuals the walls are higher than those figured and its form and dimensions correspond to those of the biconvex egg capsule.

A free larval stage is distributive and permits colonization of distant sites with a speed and freedom denied to species lacking one. Populations of species with a larval phase may have been recruited from many sources, of those without one wholly, or almost so, from animals already living at that site. Variation of characteristics brought about by genetic drift and by selection of features of local adaptive value is therefore likely to be less in the former group and greater in the latter (Shuto, 1974). Indeed Scheltema (1977) has found that the long-lived larvae of *Cymatium parthenopeum* and *Tonna galea* can drift across the Atlantic and maintain indistinguishable populations in Africa and America. E. M. Berger (1972, 1973) found that there was a three- to fourfold greater genetic heterogeneity (measured in terms of the esterases which the animals possessed) in populations of *L. saxatilis* and *L. obtusata* (no larval stage) than in those of sympatric *littorea* (with a free larva). Populations of *L. littorea* on the coasts of New England (Vermeij, 1982a, 1982b) exposed to spreading populations of *Carcinus* failed to adapt to the increasing predation by thickening of the shell since any local change towards this was swamped by larvae from areas not affected by the crab. Sympatric populations of *Nucella* (no free larva), not so diluted, reacted by development of a thicker shell. Similar indications of the effect of a larval stage were found by Gooch et al. (1972) and Snyder & Gooch (1973) who showed that populations of *L. saxatilis* were more variable than those of *Ilyanassa obsoleta* (with a larva) living alongside. Gaines et al. (1974) found that populations of the mangrove periwinkle, *L. angulifera*, which has no larval stage, could differ significantly when as little as 300 m apart.

Much debate, summarized by Raffaelli (1982), has arisen as to whether the differing life histories of *Littorina* species are better explained by demographic theory or by selection to meet local conditions. Most workers in this field (Hughes & Roberts, 1980a, 1980b, 1981; Roberts & Hughes, 1980; Hart & Begon, 1982) concluded that selection has been the most potent factor in controlling the mode of reproduction employed. Almost certainly this must have arisen in response to many pressures and need not fall neatly into the *r* and K selection scheme as described, for example, by Stearns (1976). Thus a small animal such as *Melarhaphe neritoides*, living in a habitat where feeding may be difficult, may not acquire sufficient reserves to produce eggs rich enough in food to allow complete development: there may then be no escape from a self-supporting larva as part of the life history. Hughes & Roberts (1980b) found that growth rates and the reproductive activity of this species were extremely variable and in some years no growth occurred. Even were it possible in theory for a snail living at this level to lay large eggs producing sufficient juveniles to maintain the population, the environmental harshness to which spawn would be exposed would compel development to take place in some kindlier ambience.

At other levels on the shore it seems that pressures exerted by the dangers of desiccation and crab predation are the significant ones determining reproductive modes. Hart & Begon (1982) suggested that differences in the size and number of young produced by a population of L. saxatilis living on a cliff face and another on a boulder shore reflected the dangers from crab predation and moving boulders in the latter situation. To avoid such dangers snails expend energy on fast growth and strong shells, postponing reproduction. Hughes & Roberts (1980a) noted similar differences but explained the larger eggs and embryos of a cliff population as protection against the more probable danger of desiccation in such a place.

The size to which adult winkles may grow, and therefore the size of the population which a given locality can support, is at least in part dependent upon the number of refuges against storm damage and desiccation which are available. Emson & Faller-Fritsch (1976) and Raffaelli & Hughes (1978) have all shown that populations of L. saxatilis are affected. The first two workers found that after they had experimentally increased the number of crevices on a boulder beach the population size rose without any accompanying decrease in growth rates, showing that the resources of the habitat were not being fully used, whilst the latter two found a correlation between the size of crevice available at a given level and the size of winkle living there. This factor is presumably of little or no importance to L. obtusata and L. mariae which find shelter in the very large number of 'crevices' available amongst clumps of weed.

The reproductive activities of littorinid species have been much studied, and whilst some findings seem to have some general applicability within a species, the observations which have been made on other points differ so much as to suggest that local conditions are important and may modify behaviour considerably.

The sex of a winkle is usually easily determined by the presence of a penis in males, the features of which exhibit specific differences (Heller, 1975a), and of an ovipositor in a corresponding position in females. In L. littorea, however, the penis is shed at the end of each breeding season (Grahame, 1969) and the ovipositor regresses, both reappearing at the beginning of the next. Reduction in penis size has also been reported by A. J. Berry (1961) in specimens of L. saxatilis from Whitstable, but was said not to occur in those from southern Brittany examined by Daguzan (1976a), nor has it been found elsewhere or in any other species. Males are usually believed to mature earlier than females (L. littorea: E. E. Williams, 1964a; Fish, 1972; L. obtusata: Daguzan, 1976b, Goodwin & Fish, 1977), though Daguzan (1976b) stated that females were mature at a smaller size than males in L. littorea. Males have also been claimed to be more numerous than females in the younger age classes, but this varies, and the sex ratio appears to fluctuate about 1:1 in most species. In L. saxatilis Daguzan (1976a) showed that all large animals are female, and suggested that it resulted from a faster rate of growth; but large size also indicates longer life, so males may die earlier.

Many species seem to be capable of breeding throughout the year with a peak in late spring, but the length and timing of the breeding period are extremely dependent on climatic conditions and may be shortened in the early months of the year by winter cold and interrupted or ended in summer by exhaustion. Differences in breeding times reported by different workers may therefore only reflect the conditions in the different localities studied during the usually rather short period over which observations were made. J. E. Smith & Newell (1955) recorded continuous breeding in *L. littorea* at Whitstable and Goodwin & Fish (1977) and Goodwin (1978) reported the same for *L. obtusata* in Cardigan Bay, though with a winter reduction. *L. nigrolineata* has been said both to be capable of breeding throughout the year (Hughes, 1980) and to breed only during winter-spring (Faller-Fritsch, 1975). The same is true of *Melarhaphe neritoides* which was said by Lysaght (1941) to spawn from September to April at Plymouth, was stated to breed January to June in North Wales by Hughes & Roberts (1980b), but to have no interruption in its breeding in Brittany (Daguzan, 1976a). Fretter & Manly (1977b) reported some larval settlement, and therefore presumably breeding, through-out the year in a population in south Devon. *L. saxatilis* is capable of breeding throughout the year (A. J. Berry, 1961; Daguzan 1976a) but more commonly appears to have its breeding restricted to spring (E. E. Williams, 1964a; Bergerard, 1971; Rasmussen, 1973).

The number of eggs laid by a winkle relates to its reproductive strategy, being greater in those species with planktotrophic larvae. Since both the production of spawn masses and of larvae make great demands on female winkles it is not surprising that the breeding period tends to consist of a series of spells of laying separated by recuperative pauses. In *L. littorea* Daguzan (1976b) recognized ten periods of activity, shown by J. E. Smith & Newell (1955), Fish (1979) and Alfierakis & Berry (1980) to possess a lunar periodicity; in each period a female produced about 500 capsules each of which might contain up to five eggs. Grahame (1973) calculated that a female with a shell 27 mm high (and therefore at about the height of her reproductive powers) produced 109×10^3 eggs in a year, a number which might well require more than the 5,000 capsules which Daguzan had presumed she produced, if all the eggs were in fact to be laid. Hughes & Roberts (1980b) found extremely variable but much lower figures to be characteristic of *Melarhaphe neritoides*, as might be predicted for a smaller animal inhabiting a harsher environment. The mean egg production of a female of this species was 300 eggs, and the highest figure which they recorded (3,866 eggs from a female with a shell 7 mm high) is a long way short of Grahame's figures of *littorea*. As in that species capsules were produced cyclically, in bursts preceding or coincident with spring tides.

The number of eggs produced in those species which deposit spawn masses (*obtusata*, *mariae*, *nigrolineata*, *arcana*) is almost certainly less than in *L. littorea*, but though it is known that a single mass may contain 100–200 eggs it is not known how many masses a female may lay in the course of a year; the same is true for *L. neglecta* and *L. saxatilis*, which are ovoviviparous.

Juveniles of ovoviviparous species and of such oviparous species as produce spawn masses escape from spawn or brood pouch into the habitat in which their parents live, and there is little evidence that they undertake any migratory movements, or need to do so. The settling larvae of *L. littorea* and *M. neritoides*, however, have to find their proper level on the shore. Most investigations of settlement of *littorea* suggest that it is spread widely over the intertidal area, and since adult edible winkles have a correspondingly wide distribution it is not necessary to postulate a migration. E. E. Williams (1964a), for example, recorded densities of over 4×10^3 animals per square metre with shells less than 5 mm high at all levels of a beach at Aberystwyth. J. E. Smith & Newell (1955), on the other hand, investigating the behaviour of *L. littorea* at Whitstable, found at most 4% of the total population on the beach with shells of that height, and as this would not sustain the population concluded that at least on that shore settlement was sublittoral and that the young winkles migrated up the beach to their final level. Similar findings by Lambert & Farley (1968), who recorded a gradient from 2% of the population at high shore levels being one-year old winkles to 80% at low levels, and by Gardner & Thomas (1987), who found 8·4% of the population they studied in the Bay of Fundy had shells less than 5 mm high, support the idea of sublittoral settlement followed by upshore migration.

These findings agree with what has been observed in work with *Melarhaphe neritoides*. The larvae of this species settle at a level lower on the shore than they will occupy as adults, in empty barnacle shells or in cracks between barnacles (Lysaght, 1941; Daguzan, 1976a), or in pits in quartz rocks produced by loss of crystals (Fretter & Manly, 1977b), the precise points of settlement probably due to attraction to minute algae growing there, which provide the metamorphosed animals with their first food. Attachment to the substratum is achieved by means of special glands on the foot, later lost, and by radial inflation of the shell, which helps to wedge the animal into the crevice. Within a day the snails develop a negative geotaxis which prompts them to start the upward migration which ultimately leads them to their characteristic level on the shore. Since settlement occurs at M.H.W.N. and their final is H.W.S.T. or above, this involves much time and effort on the part of an animal with an initial shell height of about 0·4 mm.

The genus *Lacuna* (chink shells) has commonly been regarded as related to littorinids but sufficiently distinct in several ways to deserve separation from them in a family Lacunidae. Recent opinion (Ponder, 1976; Arnaud & Bandel, 1978; Reid, 1989), however, has favoured their union, though *Lacuna* species are placed in a separate subfamily.

Lacuna is a northern genus and the British Isles are near the southern edge of the range of some species, the animals always being more abundant in their northern parts. *L. vincta*, for example, is rare in France but in north-east England there may be 300 per square metre (D. A. S. Smith, 1973) and comparable numbers occur in Denmark (Hagerman, 1966), whilst Rusanova (1963) recorded as many as 1,170 on a single plant of *Rhodymenia palmata* in the White Sea and M. L. Thomas & Page (1983) 1,570 snails per square metre in eastern Canada.

Though chink shells such as *Lacuna vincta* are amongst the common prosobranchs of rocky shores rich in weeds the small size of the animals makes them unfavourable subjects for experimental work despite the great numbers available. These are especially marked in spring when an onshore migration brings sublittoral populations to intertidal levels for breeding. On the other hand their ecology has been investigated by several workers. This applies to the three species *pallidula*, *parva*, and especially *vincta*; *crassior* is sufficiently rare for such work to be impossible. All chink shells are herbivores and the snails are often voracious. If they are present in great numbers they may more or less totally destroy the local vegetation: Fralick *et al.* (1974) described how a bed of laminarians was reduced to a collection of stipes and holdfasts by the attack of a horde of *L. vincta*, which eats only the frond, the rest being too tough and too rich in polyphenols (Johnson & Mann, 1986). They recorded populations of 227 snails per plant, each snail eating 0·326 cm^2 of frond per day. Though not uncommon on laminarians *Lacuna* species are more usually associated with *Fucus serratus*, sometimes *F. vesiculosus*, and it is the distribution of these algae on the shore rather than tidal height that determines their intertidal occurrence (M. L. Thomas & Page, 1983). The same algae as the adult eat are also chosen for the attachment of their spawn. Chink shells also frequent red weeds and Ockelmann & Nielsen (1981) described *L. parva* in Denmark as particularly common on *Delesseria* and *Phyllophora*, using these both as food and as substratum for their spawn.

D. A. S. Smith (1973) has shown how nicely the annual cycles of *L. pallidula* and weed are matched. *Fucus serratus* sporelings become established over the winter and grow to form a thick cover from February to May, after which their growth decreases. The spring growth period of the weed coincides with the breeding period of *Lacuna pallidula* and spawn is laid on the growing weed. As this species has no free larval stage the juveniles hatch on to a rich pasture, their feeding largely responsible for the check in growth of *F. serratus* seen in early summer. By June all the parent snails have died, so that the animals are strict annuals, but the juveniles persist, growing rapidly, increasing their shell height by 0·91 mm per month until October, then more slowly (0·27 mm each month). They probably copulate in mid winter (Gallien & Larambergue, 1936, 1938; D. A. S. Smith, 1973), breed in spring, and then die. Males never grow more than half the size of females, the result of a slower growth rate throughout their lifetime. The entire life cycle thus takes place on, and is dependent upon, *F. serratus*.

D. A. S. Smith (1973) estimated that each female on the shore on which he worked in north-east England (Whitburn, Tyne and Wear) produced over the breeding period an average of 700 eggs, laid in about 11 spawn masses containing 30–150 eggs apiece: Grahame (1977), working on an aquarium population, suggested a figure nearly twice that. The result of such figures and of the abundance of adults is that the post-hatching population density is high, over 2,000 m^{-2} (D. A. S. Smith, 1973). The spawn mass is oval or circular in outline and has sometimes been confused with that of *Littorina obtusata* but may be distinguished (Goodwin, 1979) by its elevated centre, where the eggs lie two layers deep, its firm attachment to the substratum even when old and weathered, the eggs extending to the outer margin, and the capsules in which they are enclosed pressed tightly against one another so as to be angulated in profile.

Like *L. pallidula*, *L. parva* has direct development (Ockelmann & Nielsen, 1981). In the Øresund, where they studied the snails, the annual cycle is like that of *pallidula*, spawn being deposited March–May, hatching taking place May–July. The juveniles grow rather rapidly until November, then more slowly over the winter, reaching the normal adult size of about 4 mm shell height by February. Copulation occurs in mid winter and the animals have all died by July–August after breeding. Female fecundity is comparable with that of *pallidula*: Ockelmann & Nielsen (1981) recorded an average number of eggs per female as 3,343 laid in 428 spawn masses, which therefore contain fewer eggs than those of *pallidula* but otherwise resemble them. They also pointed out that Danish animals of this species differ from those living further south in their food preferences and in some shell details: some caution should therefore be exercized in supposing that breeding is identical throughout the range of the species.

Lacuna vincta differs from both *L. pallidula* and *L. parva* in having a free-swimming larval stage in the life history. According to Underwood (1979) this may be because individual *vincta*, not being as large as individual *pallidula*, cannot amass enough yolk to produce an adequate number of lecithotrophic eggs to maintain the species. When that was written the development of *parva*, the smallest species, was unknown, but since it has proved to be direct the size argument fails, and there is no very obvious reason for the distinction between *vincta* and other species. Correlated with its free larval stage, however, goes the fact that it is probably the most abundant species. The life history is similar to that of the other species though the spawning period usually lasts longer, February–June in Britain and until October in the Isefjord (Rasmussen, 1973), hatching occurring about one month after laying. The spawn is characteristic, a cylinder coiled in a circle or a spiral with the ends just overlapping, the whole about 3 mm in diameter and fastened to weed, usually *Fucus serratus*, but also to *Laminaria* and to some reds. The larval stage lasts for 2–3 months, the main settlement occurring late May and June in the south (Fretter & Manly, 1977a), September in the north

(D. A. S. Smith, 1973). If no suitable settlement site is available when the larvae first become competent to metamorphose, larval life can be considerably extended until one is found (Fretter & Shale, 1973) by which time the larvae have become very large. The animals are prolific, as would be expected of a species with self-supporting larvae. D. A. S. Smith (1973) recorded about 1,000–1,500 eggs in each spawn mass, each egg measuring 94–125 μm in diameter compared with 265–290 μm for those of L. pallidula and L. parva. He calculated a minimum total of eggs laid of about 21,000, but the calculation was based on only part of the breeding season and Grahame (1977) suggested that a more realistic average total, based on his laboratory observations, was 53,500 eggs per female per season. M. L. Thomas & Page (1983), working in the Bay of Fundy, found a much lower reproductive effort, recording the production of only two egg masses, each with about 1,150 eggs, per female per month.

CROSS REFERENCES

CHAPTER 31

PATELLOGASTROPOD LIMPETS

T HE ecology of *Patella* species has been much studied in recent years as the animals are abundant, accessible on the shore and easily indentifiable as individuals. Some further investigation of structure has allowed more reliable identification of the species without removing animals from the substratum. This is based upon the shape of the posterior part of the shell. In *P. vulgata* this is in general smoothly rounded, with all the radiating ridges more or less equally developed. In *P. aspera* and *P. depressa* the backward growth of more prominent right and left posterolateral ridges and the straight edge of the shell between them give this area an angulated, somewhat squarish outline. In addition, the shell apex is usually central or nearly so in *vulgata*, the anterior slope steeper than the posterior, but it is distinctly anterior in *aspera* and even more so in *depressa*.

As a result of work by Bowman (1981) identification of newly-settled limpets may also be made. Once again *P. vulgata* stands apart from the other two species in that the sagittal plane of the shell is occupied by ridges, whereas in the other species coloured rays, placed between ridges, occupy that position. The other distinguishing features of *vulgata* (central apex, rounded outline) are also already present (Fretter & Graham, 1976). Five radial ridges lie on the right half of the shell, four on the left, eleven in all; their course is often slightly skewed clockwise. The pallial tentacles are markedly extensile and, like those of the adult, lack white pigment.

The young of *P. aspera* and *P. depressa* may be distinguished from one another by the shell outline and the detailed pattern of the coloured rays. In *aspera* the anterior half of the shell is a little narrowed, whereas in *depressa* it is markedly so. The lateral colour rays in *aspera* are greatly widened as compared with the more anterior and posterior ones, producing a prominent wedge of colour on each side running straight from apex to margin; those of *depressa* are not so prominent and are skewed, those on the anterior half forwards, those on the posterior half backwards. The number of ridges on the young shell is also different in the two species, eight in *aspera*, ten in *depressa*. White pigment occurs in the tentacles of both species.

The fusion of specific characters described by R. G. Evans (1953) as occurring at the Isle of Wight, which he ascribed to hybridization, has been confirmed by Gaffney (1980) who, however, offered a different explanation of its nature. Using gel electrophoresis he investigated enzymes of the three *Patella* species in animals collected between Plymouth and the Isle of Wight. The three were easily distinguishable and maintained their enzymatic distinctness throughout the area; even though their superficial appearances merged at its eastern end there was no indication of any hybridization. Gaffney suggested that the variation in external appearance reflected the fact that, since the Isle of Wight forms the eastern limit of their occurrence along the Channel coast, *P. aspera* and *P. depressa* must there be living in conditions not optimal for their needs. In what respects the habitat is deficient, however, is still not clear, nor why the deficiencies should affect the animals in this particular way.

P. vulgata, unlike the other species, occurs all round the British Isles wherever it finds a suitable substratum. Its limits abroad are northern Norway and south-eastern Portugal. These are set (Bowman & Lewis, 1977) in the north by the ability of newly-settled young to survive cold, and in the south by their ability to withstand the heat and dryness of the substratum. South of northern Spain even adult limpets survive only in shady and damp situations.

The feeding behaviour of limpets appears to be very labile: it varies markedly from place to place and its relationship to external factors must still be regarded as occasionally enigmatic. It is probably determined by a number of interacting factors, and any explanation based on only one or two is unlikely to be universally satisfactory. In addition to external influences internal ones must also be considered, since Funke (1968) showed that in the laboratory *Patella* exhibits an endogenous rhythm which he could relate both to the light-dark cycle and to the immersion-emersion cycle to which the animals were exposed.

Recent work on foraging patterns (S. J. Hawkins & Hartnoll, 1983b; Little, 1989; Little *et al.*, 1990), however, has begun to show some interlinking of the pattern, the tidal cycle, the weather, and the nature of the shore, though probably no worker would claim that he had a complete explanation of the behaviour that he studied. One important factor in feeding behaviour seems to be whether the surface over which the limpet moves is horizontal or vertical. Cook *et al.* (1969) and Hartnoll & Wright (1977) recorded that *P. vulgata* living on horizontal surfaces foraged only when they were under water. S. J. Hawkins & Hartnoll (1982), however, found that a population of *vulgata* living on a vertical surface in Port Erin Bay, Isle of Man, foraged, like those on horizontal surfaces, when covered by water during daytime, but also when they were uncovered at night. As they also fed emersed by day, provided that they were in shade or the humidity was high, this suggests that the presence of water is relevant only when they could not otherwise forage without danger of desiccation. This idea is supported by the finding that a population in Alderney (Cook *et al.*, 1969) and another at Roscoff (Funke, 1968), where desiccation rates may be high by day, fed only at night but when emersed. Complicating observations made at these sites were that the Manx limpets never fed at night when immersed, and the French ones never fed, either by day or night, when immersed.

In another study of limpet feeding (Little & Stirling, 1985; Little *et al.*, 1988, 1990), this time at Lough Ine (Hyne), south-west Ireland, still other activity patterns were recorded. Feeding here was primarily restricted to low water or to the period immediately before emersion, but was also linked to the level of the shore at which the limpets lived, to the day-night rhythm, and to the spring-neap tidal cycle. In general, however, limpets living high on the shore were active only at nocturnal low waters whereas those living low on the shore foraged then, or just before emersion, but mainly at daytime low tides. These behaviour patterns were explained on the supposition that they allowed high level limpets to avoid desiccation and low level ones to avoid predation by crabs, which are most active at night. These may well be factors involved in the timing of foraging patterns of limpets but it seems unlikely that they explain why Alderney limpets are apparently active at the very times that crabs are, though perhaps successfully evading them by feeding when not covered by water. Though smaller limpets, on account of their greater mobility (N. S. Jones, 1948), are the most probable victims of crab predation, larger ones may also be taken, especially when circumstances favour the crabs (Little *et al.*, 1990), as when limpet adhesion is reduced. At Lough Ine these workers noted that limpets living at lower levels were apt to forage less in calm weather, when crabs were active, and more in rough conditions, which, they suggested, reduced crab movement; all limpet foraging was reduced by heavy rain. Light-dark, low-high water, and perhaps other cycles, together with their interaction (early morning and late evening low water spring tides in the Isle of Man, midnight and midday in the Channel Islands), are probably all involved in setting the behavioural patterns of a given population of limpets.

Little (1989) has suggested that the physiological state of the algal food of patellid limpets may be an important factor in the control of feeding times. At the end of a day's photosynthetic activity algae are rich in photosynthates and so good food: when competition is less and predation low, it therefore profits a limpet to postpone feeding until then. If competition is high, however, Little suggested that it is more profitable to arrive early on the foraging ground in order to secure a share of the available food, even if that is less rich. Calow (1979) has suggested that at least in some instances the mucous trail left by a snail acts as a trap for some organisms used as food by the snail, and Connor & Quinn (1984) have shown that the mucus, in fact, stimulates their growth. A further study of the properties of the pedal mucus of patellids as a medium available for the growth of organisms is given by Davies et al. (1992).

The homing of limpets after foraging has been re-investigated by Cook et al. (1969) and in great detail by Funke (1964, 1968). Homing is not a universal feature of P. vulgata (J. R. Lewis, 1954) though its absence in the population which he observed was probably related to life on a smooth rock surface. The mechanism by which a limpet returns to its home commonly involves, as in many molluscs, the chemoreceptive retracing of a mucous trail laid by the animal on its outward journey. Sometimes other routes are followed (Little, in litt.) but what then guides the animal is doubtful. The precise area regarded as home is recognized by the mucus deposited there over the various periods that it has been occupied by the limpet. Patella species seem to fall into two categories, aggressive and non-aggressive, the former defending the home area and expelling trespassers (Funke, 1968; Branch, 1975).

The energy budgets of patellid species have been studied by a number of workers, sometimes in relation to the cost of the production of mucus, sometimes in relation to their growth and reproduction.

The production of mucus by prosobranchs as a lubricant and adhesive in locomotion (see p. 585) is an energetically expensive activity. An energy budget for Patella vulgata prepared by Wright & Hartnoll (1981) calculated that the secretion of mucus, mainly for this purpose, accounted for 4% of all the energy expended. A revised budget for the same species (M. S. Davies et al., 1990), suggests that this figure considerably underestimated the cost of this activity. The figure proposed by these workers is 23%, a level which compares with that of 23–29% for Haliotis tuberculata (Peck et al., 1987). It may be sufficiently high to limit the extent of foraging excursions, especially over certain types of substratum where locomotion is difficult.

Breeding, population dynamics, and energy budgets have been studied for Patella vulgata by Choquet (1966), Blackmore (1969a) Lewis & Bowman (1975), Bowman & Lewis (1977), Thompson (1980), Bowman (1981), Wright & Hartnoll (1981), and Workman (1983); for P. aspera by Thompson (1979), and Bowman (1981); and for P. depressa by Bowman (1981). From the findings of these workers it is clear that recruitment and growth fluctuate from year to year and from place to place to such an extent that it is hardly possible to make statements that have more than a local validity. A few points appear to be generally accepted: the gonad is largely inactive in the early part of the year, ripening of gametes occurring in summer and gonad weight becoming greatest in autumn; spawning occurs in autumn, perhaps also in winter, initiated by rough weather; larval life is brief, settlement occurring not more than a fortnight after fertilization. Bowman (1981) has pointed out that the traditional statements about patellid breeding seasons are not universally valid for the British Isles. P. vulgata is a winter breeder only in southern England: in the north of Scotland it breeds in August and in north-east England in September. Similarly *P. aspera* breeds June–September in north-east England but on the warmer western British and Irish coasts this may continue into November. *P. depressa* may exhibit a prolonged breeding period from spring to autumn, but this may be split into two parts by a summer pause.

Settling larvae show a preference for wet places; those that settle elsewhere are probably rapidly killed as they have little protection against desiccation. Those of *P. vulgata* (Bowman, 1981) have a shell length of 0·2 mm at this stage and hide in pools or wet cracks bearing growths of red algae or small mussels; they do not emerge from such places until the shell is 0·5–1·0 mm long, that is at about five weeks after fertilization. If they have settled higher on the beach, where emergence might offer greater risk of drying, they may stay in the crack for up to six months, by which time the shell may be 2–3 mm long. *P. aspera* and *P. depressa* both settle on wet areas or on *Lithothamnion* at a low level on the shore, measuring about 1 mm in shell length, and stay in these situations until they have reached a length of about 5 mm.

After settlement the course of events is largely dependent on circumstances. Bowman & Lewis (1977) found spat very sensitive in frost, so that should the rough weather which prompted spawning be followed by frost mortality among the spat was high; recruitment to a population thus varied unpredictably from year to year. Thompson (1980), on the other hand, noticed little annual fluctuation in recruitment; he was, however, working in south-west Ireland, whereas Bowman & Lewis were recording in north-east England, where frosts are common. This sensitivity to cold is most marked in spat since adults were little affected by the cold winter of 1963 (D. J. Crisp, 1964) and Ekaratne & Crisp (1984) found adult limpets continuing to grow over winter when temperatures fell to $-6°C$, and stopped only by still more severe weather. Little (*in litt.*) has observed loss of adhesion after exposure of limpets to $-13°C$, with many falling off rocks and so becoming easy prey to crabs or birds. Those that survived gradually recovered, but this was a slow process and where the protection of a weed cover was absent mortality was as high as 90%.

The rate of growth after settlement is markedly dependent on a variety of factors, in particular level on the shore and nature of the substratum. On the whole limpets at a given level thrive better on what appears to be a bare rock than on barnacle-covered or densely weed-covered surfaces, where their food is less, because of competition or exclusion, and where feeding excursions are energetically more expensive because of difficulty of movement (S. J. Hawkins & Hartnoll, 1982). Thompson (1980) gave the following sizes for *P. vulgata* living with barnacles: at 1 yr, 3·6 mm shell length; at 2 yr, 7–12 mm; at 3 yr, 11–15 mm; limpets on bare rock, however, measured at 1 yr, 11–18 mm; at 2 yr, 20–30 mm; at 3 yr, 24–34 mm, figures not very different from those recorded at comparable sites by Bowman & Lewis (1977) at Robin Hood's Bay, and by Wright & Hartnoll (1981) in the Isle of Man. Little *et al.* (1988) also concluded that the limpets which they studied found travel over barnacle-covered surfaces difficult. Those living at a high level on the shore tended to direct their feeding excursions to a still higher level, an action which was interpreted as avoidance of the barnacle zone below; limpets living at a low level travelled horizontally to feed, so avoiding the barnacles at a higher level. Only limpets at a mid tide level, living unavoidably alongside barnacles, were forced to travel over them and feed with them: as a consequence they made shorter excursions, moved more slowly, fed and grew less.

Though fucoids may slow growth they do offer some compensating advantages: juvenile mortality is less amongst limpets living there, the young presumably gaining some protection from the weed cover, and growth is less seasonal (Choquet, 1968). Patellid limpets are relatively long-lived prosobranchs: in the population studied by Wright & Hartnoll (1981) nearly half the animals were 5–6 years old, the oldest of these calculated to be 17 years of age. J. R. Lewis & Bowman (1975) found that limpets at higher shore levels lived longer than those at lower ones.

With good growth goes good reproduction. A limpet must have plentiful food reserves to get successfully through the period of gametogenesis which occurs in autumn (Choquet, 1967, 1970). Blackmore (1969b) showed that in July, with gametogenesis just beginning, the carbohydrate level of tissues other than the gonad of Patella vulgata was five times greater than when it finished; similarly lipid stores were maximal in August and fell as the gonad grew. Workman (1983) calculated that the fraction of energy production used in gamete formation rose as animals aged, being 11–15% when they first matured (as three-year olds) and rising to 65–96% at ten years and older. Wright & Hartnoll (1981) obtained somewhat similar figures, calculating that limpets used 50% of their annual production for reproduction, rising to 90% in older animals. Males always make a greater reproductive effort than females and since both P. vulgata and P. aspera are predominantly protandrous hermaphrodites this slightly compensates for the increasing cost of reproduction with age. Up to two years of age all the limpets examined by Wright & Hartnoll (1981) had a neuter gonad; all three-year olds, 92·5% of four-year olds and 58% of older limpets were male. Workman (1983) found that 80% of older limpets at Robin Hood's Bay were female. Though most animals thus change sex as they get older a considerable percentage seem genetically determined as males unable to change. The endocrine control of sex change is discussed on p. 658.

The littoral habitats of Patella species have been described by Ebling et al. (1962) and by Fretter & Graham (1976). Evidence from a variety of sources leads to the conclusion that P. vulgata flourishes best on shores that are moderately exposed, being replaced by P. aspera on the most exposed ones. On moderately exposed shores vulgata is better off when neither too high, where desiccation is maximal and food less, nor too low, where it seems to suffer competition from aspera.

In an intertidal situation the ability to withstand desiccation and possible loss of respiratory and feeding opportunity are important adaptations, varying in significance with the level on the shore at which the animal lives. These have been investigated in Patella species by P. S. Davies (1966, 1967, 1969, 1970) and by Houlihan et al. (1981). Patella species are useful animals for this purpose since vulgata appears able to live at both high and low levels, whereas aspera is confined to the latter. P. vulgata living at high levels is obviously at the greatest risk of both lethal high temperatures and of desiccation. P. S. Davies (1970) found that in summer, if exposed to the sun, especially if they were small or living on dark rock, they might warm to 35°C, whereas in shade their temperature equalled that of the air. They all survived water loss of 60% (P. S. Davies, 1969) though this must be near a critical level, since none survived a loss of 65%. P. vulgata living low on the shore are not exposed to the same length of insolation and their temperature never rose so high. They are more sensitive to desiccation and were all killed when water loss exceeded 55–60%. P. aspera, living at the same level, were still more sensitive, and could not survive a water loss greater than 40%

Accompanying these varying degrees of sensitivity are differences in respiratory rates. These are already known to vary with tidal conditions, being maximal at high water, when the animals are immersed, and less at low water, when they are exposed to air (Gompel, 1937; H. D. Jones, 1968). In winter (P. S. Davies, 1966) there is no difference in the respiratory rates of comparable limpets, vulgata or aspera, high or low. In spring and summer, however, whereas the rate of vulgata high on the beach remains unchanged, there is a marked rise in that of low

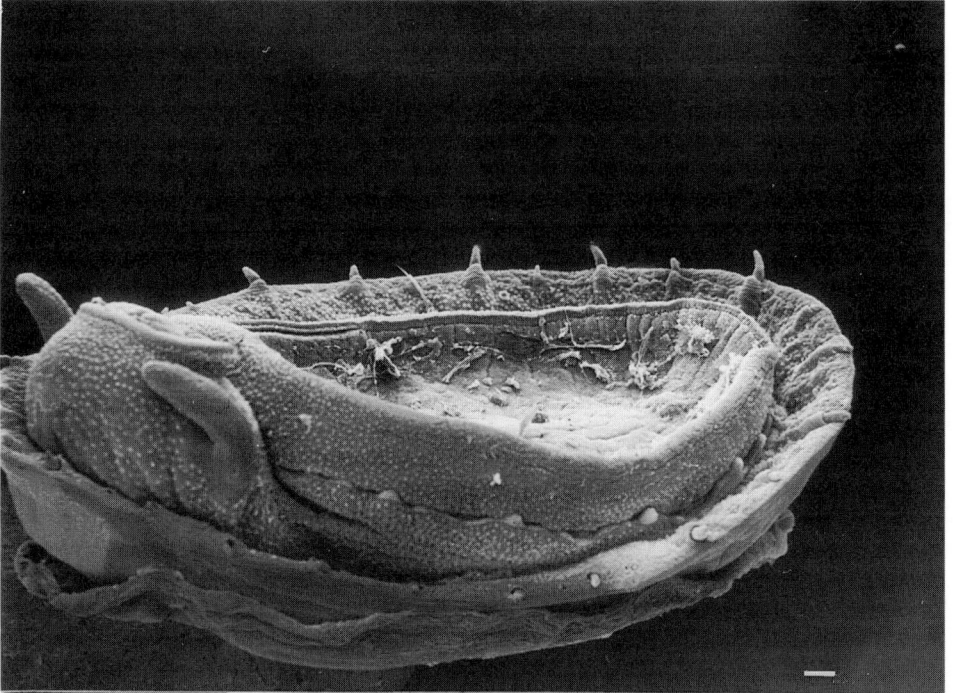

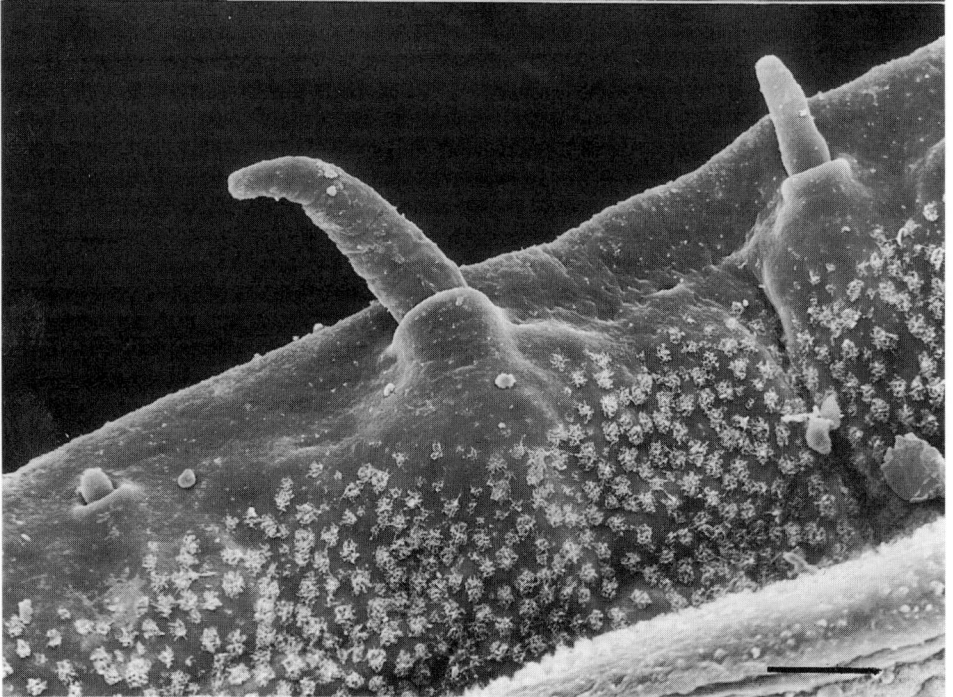

level *vulgata* and *aspera*. According to Davies the lack of change in high level *vulgata* is an acquired ability which is lost if an individual is transplanted from a high to a low level. The inability of *aspera* to acquire this ability is presumably one obvious factor excluding that species from the higher parts of a shore.

It has been claimed that differences in the shape of the shell of limpets living high and low on the shore correlate with differences in the amount of water which is retained in the mantle cavity at low tide, and therefore, perhaps, with ability to withstand desiccation. Houlihan *et al.* (1981) measured the amount retained by *P. vulgata* but found little link between that and the amount of emersion which the animals experienced. On the other hand Lowell (1984), working with an acmaeid (*Collisella pelta*) rather than a patellid limpet, described water loss as dependent upon the length of the shell edge (the only link between water store and environment) rather than upon the volume of the mantle cavity. In *Patella* tall-shelled limpets characteristic of higher parts of the shore, which were supposed to be adapted for survival there by their greater pallial volume, do also have a shorter shell perimeter for a given volume and may therefore be adapted by a reduced rate of water loss rather than by a greater volume of reserve.

The highest densities of population of *P. vulgata* coincide with moderate exposure, an approximately mid tide shore level, and a relatively bare substratum. As has been shown by N. S. Jones (1948) and Workman (1983), by Branch (1975) for the southern African *P. cochlear*, and by Frank (1965) for *Acmaea digitalis*, high densities give rise to a lower growth rate and a lower mean size, and the resultant smaller animals produce fewer eggs. This might in the long run destroy the population were it not that recruitment occurs by settlement of larvae from less densely populated areas. High density, however, has been shown to be advantageous to *P. cochlear* in that it prevents the settlement of algae and of the larvae of other animals on the encrusting corallines which form its preferred food. Thus by accepting that they must compete for a limited amount of food the limpets are able to keep the shore in a condition appropriate for their continued survival. *Mutatis mutandis*, *P. vulgata* manages shores in a similar fashion.

The streamlined profile of limpet shells is also of importance in increasing their tolerance of water movement, and this is undoubtedly one factor in determining the different shape of limpets high and low on a beach. Warburton (1976) investigated the ability of *Helcion pellucidum* to resist currents tending to wash it off a substratum, a need which their position on laminarian fronds may have increased, and which may be related to the presence throughout their lifetime of the lateral glandular streak on their foot (fig. 339). He found that animals of 6·0–13·5 mm shell length (the latter figure the common maximal size) could resist constant currents up to 0·9–1·3 m.s^{-1}. Younger, smaller animals were even more resistant, and all were extremely sensitive to the direction of the flow, aligning themselves to face any current faster than 0·5 m.s^{-1}. To some extent these findings are not in agreement with those of Vahl (1972) who investigated the orientation of *Helcion pellucidum* on the weed *Laminaria*

FIG. 339.—*Helcion pellucidum*. Above: foot uppermost, to show outer part of pallial fold with retractile tentacles, the circumpedal groove, and (on the animal's right) the lateral glandular streak with its dorsal lip bearing contractile tentacles; at a few points the tips of some strongly contracted gills can be seen in the pallial groove. The open mouth of an optic vesicle is visible dorsally near the base of the right cephalic tentacle. Below: ventral surface of pallial edge showing two extended tentacles, one slightly extended one, and the mouth of two pits into which tentacles have been withdrawn. Bars = 100 μm.

hyperborea. Limpets living on the basal third of a frond predominantly orientated themselves so as to face the free tip of the weed, those on the distal third were placed so that most faced the attachment of the weed to the substratum, whilst those on the central third were more variable, though usually lying across the main axis of the frond. These results may be reconciled with those of Warburton only if one supposes that current direction is different at the stipe end of the frond and at its distal tip. It may also be true that currents amongst kelp fronds are not by any means constant either in rate or direction. The ability of limpets to resist accelerating as distinct from constant currents may set a limit to the kind of habitat which they can occupy and to the size to which they can grow (Denny et al., 1985). As an indication of the severity of the physical conditions which such animals as limpets may have to survive, Denny (1984) calculated that the currents set up (often in the backwash of breaking waves) could, on occasion, reach speeds of 20 m.s^{-1}, with accelerations up to 500 m.s^{-1}, a velocity stated by Vogel (1981) to be equivalent to a wind speed of 1,000 miles per hour. An important point in this connection is therefore the tenacity with which a limpet can grip the substratum: this depends on the glandular secretions and the size of the foot, and is dealt with on p. 583.

Details of the structure of the pallial gills of *Patella vulgata* have been given by Nuwayhid et al. (1978). Each gill is a triangular leaflet attached to the underside of the pallial skirt by its base, and they are arranged round the mantle edge in a primary set of larger lamellae and a secondary set of smaller ones. Their area is increased by corrugations of the surface. Blood flows from the visceral and pedal haemcoels into a vessel lying along the median edge of each lamella, trickles through small spaces to another vessel along the lateral margin, and then passes to efferent vessels in the pallial edge. The water current, driven by scattered tufts of cilia which are especially abundant along the gill edges, travels over the lamellae in the opposite direction, so maintaining a counter-current system. Along the distal half of the lateral margin of each gill plate runs a deep and densely ciliated groove: this is responsible for driving particles caught on the gill surface to its tip, where they drop off. Though clearly adapted for aquatic respiration the gills are reported by McMahon (1988) to function well in aerial respiration when a limpet is emersed, since in these circumstances limpets do not accumulate the end products of anaerobic metabolism (Brinkhoff et al., 1983).

The single ctenidium found in the mantle cavity of acmaeid limpets presents some unusual features and differs from that of most prosobranchs in a number of ways. It arises from the posterior wall of the mantle cavity on the left side and curves to the right; it is free from the mantle skirt and is unsupported by membranes along its whole length. The gill is twisted so that the axis lies horizontally, the efferent side to the left, with the result that the lamellae lie in a dorsal and ventral series; it is also unusual in that the axis often appears to have the same vascular organization as the lamellae—irregular blood channels rather than distinct vessels to which the lamellae drain. These, however, connect with definite afferent and efferent vessels running along opposite sides of the axis, a nerve close to the latter. The arrangement of the gill has the effect of maintaining a counter-current system of blood and water move-ment, and it also helps to accommodate the gill within the restricted space of the nuchal cavity, aided by the fact that the lamellae are short and without internal skeleton. Savini-Plawén (1980) regarded these features, in some of which acmaeids resemble chitons, as primitive, and rejected the starting point for ctenidial evolution within the gastropods pro-posed by Yonge (1947), of which supporting membranes and gill skeletons were prominent features. Some, however, may well be adaptations to fit the gill within a small cavity. One consequence of their lack of skeletal supports is that the lamellae collapse on one another if the mantle cavity is drained of water, and so the ctenidium loses its value as a respiratory organ. The roof of the mantle cavity, however, stays damp in these conditions and, as it is richly vascularized, remains functional as a site of gaseous exchange. This explains the finding of Kingston (1968) that the ctenidium is the main respiratory organ of an immersed acmaeid, but the mantle is that of an emersed one.

CROSS REFERENCES

CHAPTER 32

TOPSHELLS AND SLIPPER LIMPETS: TROCHIDAE AND CALYPTRAEIDAE

S INCE some species of trochid are amongst the commonest prosobranchs of intertidal regions it is not surprising that their ecology, their population structure, and some aspects of their physiology have been recently investigated in some detail. This refers in particular to *Monodonta lineata*, a conveniently large and conspicuous trochid which is somewhat erratically distributed around the British Isles (Southward & Crisp, 1954; D. J. Crisp & Southward, 1958; Hawthorne, 1965). In these latitudes it is very sensitive to cold and whole populations were destroyed in the cold winter of 1963 (D. J. Crisp, 1964). E. E. Williams (1965) studied a population of *M. lineata* on a shore near Aberystwyth, and Williams & Kendall (1981), Garwood & Kendall (1985), Kendall (1987), and Kendall *et al.* (1987) have made a prolonged and detailed study of a rich population at Aberaeron, some way further south, as well as at other sites. In addition Desai (1966) made observations in North Wales and Underwood (1972a, 1973) in Plymouth Sound, whilst the reproduction of a southern African species, *M. australis*, has been studied by Lasiak (1987). In general all these investigators have reported similar findings, though some differences in details are to be expected and were found.

In *M. lineata* gametogenesis starts in autumn in animals which have a shell width of at least 11·0–11·5 mm and are about two years old. This is, however, a slow process and spawning does not occur until the following midsummer or autumn (Desai, 1966; Garwood & Kendall, 1985; Lasiak, 1987), when eggs measuring about 150 μm in diameter are laid, a few at a time, each in a jelly coat of ovarian origin which swells in sea water to give a total egg diameter of 500 μm (Desai, 1966). In the laboratory larvae started to feed 10–15 hours after egg laying and they settled in 4–6 days when their shell was 1–2 mm high. Multiple spawnings must take place since Kendall *et al.* (1987) found three cohorts in the population at Aberaeron which they studied. Newly settled juveniles grow rapidly to a shell width of 7–8 mm before the onset of winter stops growth, an event marked on the shell by a winter ring (Williams & Kendall, 1981). At one year old juveniles have a shell width of about 10 mm and one of about 15 mm at their second winter (E. E. Williams, 1965); thereafter growth becomes slow. *M. lineata* is capable of long life: Williamson & Kendall (1981) reported that animals over nine years old were found at all the sites that they examined and that some snails had an estimated age of fifteen years. Kendall *et al.* (1987) found snails up to and over seven years old.

Comparable figures are available for *Gibbula umbilicalis* which has often been studied alongside *Monodonta lineata* (E. E. Williams, 1964b; Garwood & Kendall, 1985; Kendall *et al.*, 1987). This species, too, is sensitive to cold, and the population which Williams studied near Aberystwyth was destroyed in 1963. The animals first spawn when two years old (Garwood & Kendall, 1985) after a prolonged gametogenesis lasting about a year. Eggs were laid in Williams' (1964b) population in May–June, whereas Kendall *et al.* (1987) recorded mid August. Several spawning periods, however, give rise to distinct cohorts in the population. Larval life, as in trochids generally, is brief and growth after settlement rapid until cold brings about a winter stoppage at a shell diameter of 3·0–3·5 mm. Growth resumes in May and one-year old animals have a shell breadth of 5–7 mm, those aged two years about 13 mm, after which growth, as in *Monodonta*, proceeds very slowly to a maximum of about 17 mm.

In the British Isles both *Monodonta lineata* and *Gibbula umbilicalis* are living close to the northern limits of their geographical range and might therefore be expected to show effects of this: one such, indeed, is the marked sensitivity of both species to cold. Kendall (1987) has suggested that such a distributional limit might be set either by the animals' failure to reproduce with sufficient success to maintain populations, or, if they can do this, by failure to survive after settlement. The former state would be revealed by an irregular age structure of the population with some year classes absent and a predominance of older, larger animals, the latter by their absence and a predominance of juveniles. These features were looked for in populations of *G. umbilicalis* by Kendall & Lewis (1986) and in those of *M. lineata* by Kendall (1987). Several separate populations of *G. umbilicalis* showed identical irregularities in their recruitment of juveniles, a pattern which might have been brought about by climatic factors, which would in this case be responsible for limiting further spread of the species. On the other hand Kendall investigated seventeen populations of *M. lineata* between the north coast of Brittany and that of Mid Wales without finding any obvious consequences of their marginal siting, suggesting that these were more kindly habitats than might have been expected. Were the British Isles less marginal, however, the distribution of *Monodonta* might well be less scattered than it is.

The worse aspects of winter may be partly tempered for *M. lineata* by its habit, noted by E. E. Williams (1965) and Desai (1966), of migrating downshore in autumn and upwards with the return of spring and summer. This, and a circadian vertical migration, are brought about by interacting phototactic and geotactic responses, noted by Desai (1966) and Micallef (1968), and investigated in the related Mediterranean species *M. turbinata* by Chelazzi & Focardi (1982). When the latter animals are kept in an aquarium exposed to a light/dark cycle simulating that in the field, they migrate up in the dark and down in the light, taking up the same respective positions as they adopt in constant darkness or light. The phototactic response remains constant, but as it is inoperative in the dark, a negative geotaxis brings the animals upwards; by day the phototactic response is the more powerful and causes downward movement. Chelazzi & Focardi (1982), however, considered that the geotaxis reverses its sign in light and reinforces the phototaxis; similarly Desai (1966) assumed that the geotaxis was positive as temperature fell in autumn, but reversed with rising temperature to produce the upshore movement of spring.

Underwood (1972b) had earlier explored the behaviour of *M. lineata* in a tidal aquarium in an attempt to explain its zonation on the shore along with that of *Calliostoma zizyphinum*, *Gibbula cineraria* and *G. umbilicalis*. *M. lineata* was the only one whose position in the tidal tank corresponded even approximately with its natural one, the other animals following the water level and showing no consistent behaviour pattern.

Some further distinctions amongst intertidal trochids are their respiratory adaptations relative to the level at which they normally live. Micallef (1966), Bannister et al. (1966), Micallef & Bannister (1967), and Houlihan & Innes (1982) have shown that in littoral trochids there is a direct relationship between the oxygen consumption in air, and an inverse relationship between the oxygen consumption in water, and the height at which each species lives on the beach, thus suggesting increasing adaptation for keeping active even when emersed without recourse to anaerobiosis. To some extent this depends upon the ability of the snails to retain sufficient water in the mantle cavity, if only to prevent collapse of the gill leaflets. Houlihan *et al.* (1981) showed that the oxygen consumption of *Gibbula cineraria* in air varies with the volume of fluid in the mantle cavity which, even when the animal is attached, still has contact with the atmosphere. Similarly Houlihan & Innes (1982) found that the Mediterranean *Monodonta turbinata* and *M. articulata* showed a significantly depressed oxygen consumption in air if the water in the mantle cavity was lost.

By comparison with *Monodonta* the equally large trochid *Calliostoma* has not been studied to the same extent, probably because local populations of its species are less dense. Some recent interest has been shown in the shell shape and colour of *C. zizyphinum* in relation to habitat. Thus Seed (1979) found that a morph (or morphs) with coloured shells occurred only on open Irish Sea coasts of County Down, whereas a white morph, var. *lyonsii* (occasionally regarded as a separate species) dominated populations living in sheltered areas within Strangford Lough. The investigation has been taken further by O'Loughlin & Aldrich (1987a, 1987b) who demonstrated that the shell of this trochid showed similar trends to that of *Nucella lapillus*, having a greater aspical angle and so a wider aperture, giving passage, presumably, to a more powerful foot where the animals occupied an exposed habitat, and a lower angle and narrower aperture where they were sheltered. Roberts & Kell (1987) have shown that the white morphs often exhibit a greater number of signs of previous damage by crabs than do coloured ones, presumably due to the conspicuousness of the white shell. Since they abound in certain localities their conspicuousness must be balanced by some other factor(s). These, it is proposed, are higher rates of movement, protection against overheating when emersed, and, since *Calliostoma zizyphinum* has direct development leading to isolation of local populations with possible overcrowding, the faster movement may lead to a greater dispersal rate and consequent reduction in population density and competition.

Salvini-Plawén (1972a) suggested that at least some species of *Calliostoma* eat coelenterates, and this has been verified for three species, *C. annulatum*, *C. variegatum*, and *C. ligatum*, by Perron (1975) and by Perron & Turner (1978) for *C. occidentale*. Feeding on coelenterates has not been proven unequivocally for *C. granulatum* or *C. zizyphinum*, though such material as hydroids sessile on weeds may well form part of their diet, whether deliberately searched for or only incidentally ingested.

Crepidula fornicata, the slipper limpet, was accidentally introduced to Britain with oysters imported from North America in 1887–1890. The first permanent British locality was Essex, from where it has spread over the years both northwards and westwards. Northward migration has led to its presence being recorded as far north as Northumberland but does not seem to have given rise to permanent populations as its migration southwards and westwards has done; there it has established itself at various places along the south coast of England. Occasional animals have been found in the Irish Sea and Kerry, but no permanent population has apparently developed in Ireland.

Though the veliger larval stage of the slipper limpet has successfully extended the range of the species over these considerable distances it does not appear to have made any such successful entry to continental Europe from English sources. According to Polk (1962a, 1962b) present populations on continental North Sea and Atlantic coasts seem to stem from a separate introduction of American oysters to the Netherlands in 1929. From this the limpets spread rapidly as far north as Denmark, which was reached in 1935, and they have also been found on the Scandinavian shores of the Skagerrak and Kattegat. Their extension westwards has been slower and they did not reach Calvados in France until 1955, though they are now present in the northern part of the Bay of Biscay. Presumably the course of currents in the North Sea and Channel explains why northward travel on the British east coast has been a failure when northward extensions on eastern North Sea coasts has not, and why westward movement along the English coast of the Channel has been more rapid than on the French. It may well be that casual transfer by human means has sometimes been involved. Parenzan (1979) has recorded the occurrence of slipper limpets in Sicily and the Adriatic Sea: this must represent still another introduction from elsewhere.

The successful spread of *Crepidula fornicata* depends upon the fact that it has a vigorous veliger larva (Werner, 1955) since the adults are immobilized in chains (stacks) and a juvenile has only a brief period of activity before it joins an already existing stack as a young male, or becomes the founding matriarch of a new one. Sex at this stage of the life history is labile and determined by the environment in which the metamorphosing larva finds itself, though it appears to be biased towards maleness as the larvae actively seek chains on which to settle and are able to postpone metamorphosis until they meet one. Fretter & Shale (1973) found that the larvae of many prosobanchs can delay settlement until an appropriate substratum is encountered, but as they carry on feeding during this period of search they continue to grow and are recognizable by their unusually large size. Pechenik (1986) has found that this ability is also present in *C. fornicata* and noted that 5–20% of the larvae of that species which he found in N. W. Atlantic plankton in June 1974 fell into that category.

Hoagland (1978, 1986) has investigated relationships between the presence and absence of stacking and the type of life history in a number of calyptraeid species. *C. fornicata* is an example of one pattern in which the adult animals live in chains and there is a free larval stage, the sex of which on settling is determined by its environment. *C. convexa* and *Calyptraea chinensis* do not form stacks nor do they have a free larval stage, hatching as juveniles which all enter a male phase. Sex change from male to female is delayed in the former group and the male phase may last for some years, maintained by pheromones from the associated females, whereas it takes place early in the latter group; males are found with a female only during the breeding season and move about freely at other times. The presence of a distributive planktonic larval stage in the life cycle of *C. fornicata* ensures that the source of the animals which comprise a stack is likely to be varied, with the result that outbreeding is probably the norm for this species and all those with a similar life history. In species without a larval stage females tend to be extremely immobile though males wander. Some outbreeding is thereby achieved but its degree is probably less, as, too, is likely to be rate of spread of the species. In accord with this Hoagland (1978) noted a greater variability in a number of features in species with a larval stage than in those without, though her studies of New England populations of *Crepidula fornicata* (Hoagland, 1985) showed that the larval stage was there a unifying factor, keeping populations similar; interestingly, a population from near Portsmouth, England, was not significantly different from those in New England. In species with a larval stage (Hoagland, 1986) the eggs in the capsules are small (136–212 μm in diameter): in those in which juveniles are hatched they may still be small (150–260 μm diameter) but they are then enclosed with food eggs or may cannibalize their siblings, or they may be large (300–440 μm), in which case they are rich in yolk.

According to Lubet & Le Gall (1972) new stacks of *Crepidula fornicata* tend to have 2–4 members to begin with—the founder plus 1–3 males which she has attracted. In such stacks the females grow faster than the males. The rates reverse, however, 2–3 years later so that size differences vanish. Recruitment of new males, however, maintains the appearance of graduated size.

Calyptraeoideans are ciliary feeders (Orton, 1912; Werner, 1953, 1959; Nelson & Morton, 1979) with the gill and mantle cavity highly modified for this function. Bulnheim (1970) has shown that *Crepidula*, like many bivalves, does not pump water through the mantle cavity continuously but with a cyclical rhythm. This exhibits a short term periodicity of 10–15 minutes, with a superimposed longer cycle of about two hours. The flow through the mantle cavity at 15°C showed a maximum of about 4 ml.min^{-1}, falling in the short term to about 0·5 ml.min^{-1} and in the longer cycle to 0·3 ml.min^{-1}. R. C. Newell & Kofoed (1977) have measured the rates at which the alga *Phaeodactylum* is filtered by *C. fornicata* and shown that, like most activities of cold-blooded animals, it is affected by temperature, being maximal at 15°C in animals acclimated to 10°C. At 15°C it was calculated that a hypothetical standard limpet of 160 mg dry weight (i.e. of about 30 mm shell length) would clear about 250 ml of water with a concentration of *Phaeodactylum* equal to 4×10^5.ml^{-1} in one hour. Flow rate and filtration rate agree here, but they need not do so, since respiratory exchange is necessarily a more continuous process than feeding. Nelson & Morton (1979) have shown that in the calyptraeoidean *Maoricrypta monoxyla* this rhythm in feeding produces a corresponding rhythm in the activity of the digestive gland similar to that demonstrated by Merdsoy & Farley (1973) in *Littorina littorea*, in *Rissoa parva* by Wigham (1976), in *Ilyanassa obsoleta* (Curtis, 1980), and in many bivalves. The gland, as suggested by Owen (1972), is probably capable of dealing with a steady flow of food but reacts to cyclical supplies externally imposed by tides, or to reduced food making ingestion unprofitable, by cyclical functioning (Robinson et al., 1981).

CROSS REFERENCES

CHAPTER 33

MUD SNAILS: HYDROBIIDAE

W HILST most recent work on prosobranch biology has been concentrated on such easily accessible and abundant animals as limpets, winkles and dog whelks, some lesser, though still considerable effort has been put into the study of animals of other groups, especially if they may be found in reasonable numbers without difficulty. These include, in particular, a few top shells (trochids), some mud snails (hydrobiids), and necklace shells (naticids). Most of the other prosobranchs which occur around the British Isles are not sufficiently easily available to permit investigation of their activities except to the modest extent allowed by their chance capture.

The family Hydrobiidae contains many species of small snails typically associated with the borderline between marine and freshwater habitats, some becoming purely freshwater whilst others tend towards a more terrestrial mode of life. All are small but often so numerous that their biomass is large. They are quantitatively the most important members of the macrofauna of mud-flats in N. W. Europe (Muus, 1967) with densities sometimes up to three hundred thousand per square metre (Smidt, 1951). Walters & Wharfe (1980) recorded up to ninety thousand specimens of *Hydrobia ulvae* per square metre in the estuary of the Medway, Kent. Indeed at some population densities well below the maximum recorded *H. ulvae* is adversely affected both in its growth rate and in its mortality rate (Morrisey, 1987). At a density of 0.3 snails per cm^2 growth over a 23-week period averaged 5 mg, but at 0.6–0.7 animals per cm^2 it fell to 0.5 mg, and a mortality of 15% in a population of 100, increased to 50% when a population of 600 was kept in the same area. How these changes are brought about is not known. The mechanism does not seem to be comparable to that producing a similar effect in *Hydrobia ventrosa* (Levinton, 1979) which become disturbed as population density rises, move about continuously and so feed less; neither movement nor feeding are changed in *H. ulvae*, though in *H. totteni* [properly *H. truncata* according to Davis *et al.* (1989)] both are affected when the substratum on which they feed is ploughed over by such large animals as *Ilyanassa obsoleta* (Levinton *et al.*, 1985). Control of size in *Hydrobia* species has been ascribed to a variety of environmental factors: the degree of crowding, the extent of parasitization (Rothschild, 1941), though the effect of this has been minimized by recent work, and the relative richness in food of different substrata (Chatfield, 1972; Fish & Fish, 1974), perhaps correlated with particle size (Cammen, 1982; Morrisey, 1990). The last author has extended the list further by suggesting that water movement is also involved and has emphasized the obvious complexity of the situation.

All hydrobiids are microphagous and some species are frequently found together so as to suggest that they may be in direct competition for food. Much of the recent work on this group has reference to their choice of habitat, their feeding, and their sharing of resources whenever they happen to live together.

Hydrobiids are usually described as detritus feeders. This term has, however, been shown by Lopez & Kofoed (1980) to cover two different ways of eating the detritus used as food: it may be what these workers term deposit feeding, in which particles picked up from the substratum (partially decomposed organic matter, bacteria, diatoms, protists, inorganic particles) are taken into the gut, swallowed, the digestible matter removed in stomach and digestive gland, and the rest voided. Or it may be what they described as epipsammic browsing: in this particles of the substratum such as sand grains are taken into the buccal cavity and cleaned of adherent bacteria and diatoms by radular action; the food is then swallowed but the particles themselves are spat out. In both methods there is a limit to the size of particle which can be swallowed or which can be ingested for cleaning, and it is differences in these limits, and also in the size of particle which allows fastest feeding, which are part of the means by which sympatric species of hydrobiid avoid competition. Thus in the experiments of Lopez & Kofoed (1980) *Hydrobia ventrosa* fed most rapidly on particles <40 μm in size, *H. ulvae* on those <80 μm, *H. neglecta* on those between 40 and 80 μm, and *Potamopyrgus jenkinsi* [recently shown by Ponder (1988a) to be properly called *P. antipodarum*] on those 80 to 160 μm in size. The sharing of resources in one and the same habitat is further facilitated by the changes observed in sympatric species by Fenchel (1975a, 1975b) who noted that in populations in which two species of *Hydrobia* occurred together the modal size of one species tends to become smaller and that of the other larger than when they live apart: *H. ventrosa* becomes smaller than usual when found with *ulvae* or *neglecta*, and these become larger, though *neglecta* diminishes in size where it is found with *ulvae*. These displacements in size thus accentuate the differences in the size of food particles taken. Incidentally Fenchel also found a displacement of the breeding seasons of the two sympatric species, the one breeding earlier, the other later than usual, so reducing the chances of interbreeding. The mechanisms which might bring about such adjustments remain unknown, and whether they are indeed to be attributed to interspecific competition has been questioned by a number of subsequent workers (Levinton, 1982; Cherrill & James, 1987; Barnes, 1988; Morrisey, 1990) (see below).

For most species of hydrobiid diatoms seem to be the most important constituent of the diet (Muus, 1967; Jensen & Siegismund, 1980), whether they are taken by epipsammic browsing or picked singly (Fenchel & Kofoed, 1976). The partially decomposed matter of the detritus, on the other hand, seems to be the most important ingredient of the food of *neglecta* (Hylleberg, 1976). Enzymes allowing the digestion of different fractions of the detritus to varying extents were found by this worker to characterize the different species of hydrobiid, so that even when they were eating the same food they were nevertheless exploiting different parts of it, and so reducing competition.

If particle size is important in the feeding of mud snails it might be expected that they would exhibit some ability to select appropriate areas of sediment. Barnes & Greenwood (1978) and Barnes (1979) have, indeed, shown that *Hydrobia ulvae* has a preference for a substratum of particles 90–179 μm in size and can discriminate between sands differing in mud content by as little as 3% of their volume. These distinctions, however, appeared to be based on assessment of particle size only, without reference to potential food value, since all their attempts to make a sediment more attractive by the addition of organic material failed. These results are in agreement with those of R. C. Newell (1965) and Fenchel et al. (1976). On the other hand Forbes & Lopez (1986), working with the American species *H. truncata*, described observations that suggest that areas of substratum rich in food are detectable by this species: snails were found to decrease their rate of crawling over patches of substratum rich in chlorophyll-a so that they lingered there. Since the rate of feeding did not alter their slower movement over such patches meant that their food intake increased. When the snails were over areas composed largely of newly egested material the feeding rate was reduced but the crawling rate did not change, so that food intake decreased. If, however, such patches were experimentally enriched with chlorophyll-a the decrease in the rate of feeding was abolished. This ability to detect areas rich in food must be additional to any enabling the snails to detect areas of appropriate particle size, since F. E. Wells (1978) found that *H. truncata* (quoted as *totteni*) exhibited a preference for a substratum of particles 125–250 μm in size over other sizes, though all had been cleaned so that presence or absence of food did not enter into the choice. Though *H. truncata* reacts thus to areas of freshly voided faeces it presumably does not do so once they have been broken down by bacterial or other means. It may be like *H. ventrosa*, which was shown by Lopez & Levinton (1978) to be able to exist within a system formed largely of its own faecal material, browsing on the bacteria which have broken it down.

In addition to feeding preferences, those relating to salinity and other features of the habitat are also important in keeping species apart (Muus, 1967; Cherrill & James, 1985; Falniowski, 1987). The ranges of salinity which each species can tolerate frequently overlap, but *H. ulvae* is most common in higher salinities, often in nearly fully marine habitats, and *Potamopyrgus jenkinsi* in water that is either fresh or of a salinity not greater than about 20‰, with *H. ventrosa* and *H. neglecta* preferring intermediate values: the former does not endure salinities less than 1‰ (Barnes, 1987), the latter less than 10‰ though both tolerate those up to full strength sea water (Muus, 1967; Hylleberg, 1975; Bishop, 1976; Cherrill & James, 1985; Barnes, 1989). What has been described as a salinity-dependent separation of species, though with some overlap, has been shown in the Kysing Fjord estuary in Denmark by Siegismund & Hylleberg (1987). Here *H. ulvae* lives in open sea water at the mouth of the estuary; further away from the sea is an area dominated by *H. neglecta*, then one by *H. ventrosa*, with *Potamopyrgus jenkinsi* concentrated at the inner end where salinity is lowest.

Other conditions than salinity must be satisfied before a habitat is suitable. Thus *H. ulvae* is almost always found in open sites with marked movement of the water, commonly tidal; *P. jenkinsi* also requires some movement in the water, whereas *H. neglecta* and, in particular, *H. ventrosa* prefer sheltered, rather enclosed, lagoonal sites with little movement (Muus, 1967; Barnes, 1987). Barnes (1989 and *in litt.*) has expressed doubts about salinity having much or any significance in determining the distribution of hydrobiid species. He points out that in Denmark, where a decrease in salinity and an increase in the lagoon-like features of the habitat coincide, the species do distribute themselves along the salinity gradients, but that elsewhere, where these two gradients do not coincide, there is no such relationship with salinity: other features are presumably more critical.

Blandford & Little (1983) have shown that *H. ulvae* and *P. jenkinsi* are sensitive to changing salinity in the water in which they are crawling, detected, perhaps, more by its osmotic effect than directly. The former snails are the more sensitive, detecting and reacting by changing direction to salinity differences of 18–26%; *P. jenkinsi* responds only if the difference is 53–55%, a crude sensitivity which can contribute little to the control of its distribution.

In 1962 R. C. Newell described a behavioural pattern exhibited by *H. ulvae* on a beach at Whitstable, Kent, which combined a feeding foray with a mechanism maintaining the snail at its correct level on the shore. As the tide ebbed, and for so long as the substratum remained damp, the animals moved about, feeding as they went; as the substratum dried they sought dampness by burrowing, climbing to the surface again as, or even before, the flowing tide reached them. They then became attached to the surface film and were floated back and forth until they settled, more or less at their starting level. Their position in the surface film was maintained by a mucous float which was ultimately eaten with any particles which had been trapped in it. The climbing activity of these snails had been previously noted by Linke (1939) and interpreted by him as a device leading to floating for dispersal, or, by Whitlatch & Obretski (1980), for the avoidance of competition; the former idea was adopted by Holme (1949) and, for *H. ventrosa*, by Levinton (1979). A. Anderson (1971), however, studying *H. ulvae* in the Ythan estuary, Aberdeenshire, found many snails climbing and believed that this was primarily so that they could reach a position from which they might become pelagic and exploit food in the water surface. Though there is no doubt that hydrobiids can climb and float, it does not seem that the cycle of activities described by R. C. Newell (1962) can be regarded as a regular feature of the behaviour of *H. ulvae*. Vader (1964) and Little & Nix (1976) found the number of snails which float at a given tide to be usually not more than 1% of the total population, and that there was no anticipation of the return of the tide. Siegismund & Hylleberg (1987) have also discounted floating as a major distributive device by comparison with movement brought about by tidal bottom currents.

Barnes (1981a, 1981b , 1986) has reinvestigated the behaviour of *Hydrobia ulvae* and has concluded that climbing is part of the normal foraging activity of the animal, since a signifi-cantly larger number of snails climb objects which bear food than those which are clean, but that of those that do climb very few are floated off, and that this is not its object. He has also shown that more animals are active in the dark than in the light and when they are under water than when emersed. These environmental factors seem to control the behaviour of *H. ulvae*, though experiments revealed an endogenous rhythm of activity, too weak, apparently, to influence the response of snails to external conditions.

The species of *Hydrobia* differ amongst themselves in their reproductive methods, though all lay eggs in capsules, commonly attached in *ulvae* to their own shells or those of their neighbours, to the substratum in *ventrosa* and *neglecta* (Fish & Fish, 1981). *H.ventrosa* and *H. neglecta*, apparently without exception, exhibit direct development, hatching as young snails. Correlated with this their eggs are large (125–185 μm in diameter) and there is usually only a single egg in each capsule, whilst suppression of a free larval stage is appropriate for the quiet and enclosed habitats which they frequent. *H. ulvae*, on the other hand, shows more variability in its reproduction: its eggs are smaller (70–90 μm in diameter, therefore about a fifth to half the volume of those of *ventrosa* and *neglecta*), are more numerous, varying at Plymouth from 3–7 per capsule (Lebour, 1937), up to 25 (Pilkington, 1971), and to a maximum of 40, with a mode of 7–8, in the Ythan, Aberdeenshire (A. Anderson, 1971), or to 50 on Norfolk coasts, with a mode of 21–22 (Barnes, 1988). There have been different accounts of their later history. According to Lebour (1937) and Pilkington (1971) *H. ulvae* hatches at a shell length of 360–400 μm and settles after a non-feeding and non-planktonic larval stage of only two days' duration. Fish & fish (1977), however, who worked with animals from the Dovey estuary in Mid Wales, found that *H. ulvae* there hatched at a size of 140–150 μm and had a subsequent planktotrophic existence lasting four weeks before settlement, and that they differed in other respects from those described by Pilkington (1971). A. Anderson (1971) obtained similar results in his study of *H. ulvae* from the Ythan estuary.

Calculations based on the figures for egg numbers and egg and capsule sizes for *H. ventrosa* and *H. neglecta* given by Fish & Fish (1981) and for *H. ulvae* by Lebour (1937) show that the albumen per egg available for food in *ventrosa* and *neglecta* before hatching is about one tenth of that available in *ulvae*. Even when the greater numbers of eggs per capsule in *ulvae* noted by Pilkington (1971), twenty-five, and Barnes (1988), up to fifty, are taken into calcu-lation, this still remains true, though the disparity is much reduced. These differences in albumen supply suggest that development to hatching juveniles (*ventrosa*, *neglecta*) is more efficiently achieved by the provision of yolk than albumen, but that development to settling veligers (ulvae) is easier by albumen supplemented by planktotrophy, though at the cost of many larval lives lost through predation. The differences in egg size and number in hydrobiid species are those common to many prosobranchs and relate to the relative risks attached to different reproductive strategies (see, for example, p. 660). Barnes (1990) has, however, suggested that the differing types of life history and reproductive adaptations to environment in *H. ulvae* are due to the fact that it is historically an iteroparous species, though it is normally rendered effectively semelparous by the high mortality which it suffers in most of the habitats which it occupies. Selection acting on this has apparently driven it to a certain extent towards a true semelparity.

H. ventrosa, H. neglecta and *H. ulvae* [in the type of life history described by Fish & Fish (1977)] show an increase in volume from zygote to hatching of between four and ten times. Growth of the last species, if its life history is as described by Pilkington (1971), however, would be an order of magnitude higher.

The combination of numerous small eggs, little intracapsular growth, and planktotrophic larvae, common in many prosobranchs, characterizes apparently the majority of populations of *H. ulvae*, whereas at Plymouth a developmental mode more akin to that of *H. ventrosa* and *H. neglecta* appears to operate, suggesting that *H. ulvae*, like *Rissostomia membranacea* (Rehfeldt, 1968; but see p. 667), can alter its reproductive strategy to suit its environmental conditions. Recent work by Barnes (1988), however, seems to contradict this idea. He studied a population of *H. ulvae* living in the Salts Hole at Holkham, Norfolk, a salt water pond without direct connection with the sea but linked to it by underground springs (O. D. Hunt, 1971). In its isolation and freedom from wave and tidal action it therefore resembles the usual habitats of *H. ventrosa* and *H neglecta*, and the shells of *ulvae* living in it do exhibit some convergence with the shells of those species in their size, the increased tumidity of their whorls, and narrower spire. But although the environmental factors operating in the Salts Hole have seemingly acted to produce a displacement of the shell features of *ulvae* towards those of *ventrosa* and *neglecta*, they have not operated to drive its reproduction in the same direction. This remains that characteristic of most populations of *ulvae*, with long-lived veliger larvae, though the number of eggs produced per annum or lifetime differs greatly from that in populations living on open coasts (Barnes, 1990).

It is for such reasons that Barnes (1988) was led to question the extent to which the changed characteristics of the species of *Hydrobia* in mixed populations described by Fenchel (1975b) are truly interspecific reactions rather than simply environmental, a point of view shared with several other workers on hydrobiids. It also reinforces the suggestion that Plymouth *ulvae* (or at least some of them) must represent an altogether separate race, or even species.

CROSS REFERENCES

CHAPTER 34

NECKLACE SHELLS: NATICIDAE

THE biology and ecology of naticid prosobranchs have, like those of muricids, with which they share the ability to bore shells, been much explored in recent years. Since naticids form part of the simpler ecosystem of sandy shores their study has perhaps shown more clearly than that of the rocky shore muricids what effects predators have on the fauna of which they form part. The prey of naticids is other molluscs, but since they live where gastropods are not as frequent as bivalves the latter are their usual prey, together with smaller members of their own species. A. J. Berry (1982, 1984), however, recorded a population of *Natica maculosa* on a Malaysian shore which subsisted entirely on a diet of the trochid *Umbonium vestiarium* even though bivalves were also abundant, and Hughes (1985) found *N. unifasciata*, in Panama, taking frequent gastropods of several species. Indeed, Kabat (1990), who has extensively reviewed naticid feeding on a worldwide basis, has recorded the taking of about two thirds as many species of gastropods as of bivalves, though one suspects that their biomass would be relatively still less.

Feeding invariably takes place when the naticid has burrowed into the sand; even if prey is captured on the surface it is wrapped in mucus and carried on the posterior part of the foot until its captor can burrow. It is then moved to the anterior part of the foot and held there for consumption.

Selection of prey seems to depend upon several factors, but of these it is generally agreed [Edwards & Huebner, 1977 (*Polinices duplicatus*); Griffiths, 1981 (*Natica tecta*); Wiltse, 1980b and Kitchell *et al.*, 1981 (*P. duplicatus*); A. J. Berry, 1982 and Broom, 1982 (*Natica maculosa*); Bayliss, 1986 (*P. duplicatus* and *Lunatia alderi*)] that the size of the prey is one of the most important of these, if not the decisive one. This relates to more than simple appreciation of the fact that larger predators can cope with larger prey. One essential for a successful attack which involves drilling, is that the prey be held firmly, a condition which calls for a rather precise topographical relationship between predator and prey. Whilst the surface of the former is more or less constant that of the prey varies from species to species. This may mean that a firm hold may be obtained on one species only when it falls into a particular size range but on another at a different size. It also implies that according to what species is being bored the borehole is likely to be sited at different points on the shell of the prey. It used to be believed that boreholes lay over the most nutritious organs of the body of the prey (Pelseneer, 1925) and even that the naticid learned where to bore to reach them (Verlaine, 1936). However, it is now generally accepted (Ansell, 1960; Berg & Porter, 1974; Negus, 1975; Kitchell *et al.*, 1981; J. D. Taylor *et al.*, 1983; Bayliss, 1986) that the siting of the borehole relates primarily to the morphology of the shell of the prey, in particular its shape, its ornament and its thickness, rather than to what lies immediately under the part bored.

Ansell (1960) found that *Lunatia alderi* (cited as *Natica*) bored *Venus striatula* near the ventral margin of the shell, a site which he believed was chosen because the shell was thinner there than elsewhere. Bayliss (1986), however, offered the bivalve *Spisula* of varying size to *L. alderi* and found that neither size by itself nor the thickness of shell that went with increasing size was the criterion underlying choice of prey; this seemed to rest primarily on the topology of its shell in relation to the gripping and boring apparatus of the predator. This need produces a degree of regularity in the siting of boreholes, as shown by Berg & Porter (1974), who observed that whilst the boreholes made by three naticids, *Polinices duplicatus*, *Lunatia heros* and *L. triseriata*, all lay at the same dorsoventral level on the shells of *Mya arenaria* given to them as prey, each species bored at a different point along the anteroposterior axis of the bivalve according to where it obtained the surest hold.

J. D. Taylor *et al.* (1980) supposed that naticid feeding evolution started with the probing of potential prey by the proboscis. Later the process was elaborated and improved by the holding of the prey by the foot whilst it was attacked, firstly, presumably, only mechanically, but later chemically as well. The attack was then directly on the surface of a valve or whorl (side-boring), and this is still the method normally used by all the British naticid species.

Though the drilling of shells may still be the most common method of feeding used by naticids recent work has shown that their feeding is more diverse in its methods than was at one time supposed, and that Ansell's idea of the value of edge-boring was more prescient than seemed at the time. Different species have different methods of attacking prey so as to minimize handling time and maximize nutritional gain. A summary and discussion are given by Kabat (1990). Perhaps the most frequent of these less traditional methods is edge-boring, the attacking of prey by boring, or, if possible, by thrust of proboscis between the valves of a bivalve near or along the mid ventral line, as suggested by Ansell (1960). Further evidence of the frequency of this method has been given by J. D. Taylor (1970), Vermeij (1980), and by Ansell & Morton (1985, 1987). The main advantage offered by this mode of attack is the speed of entry permitted by the thinness of the shell and the avoidance of shell ridges, spines, or the other ornaments by which bivalves defend themselves against predators. Hughes (1985) has suggested that gastropod prey may similarly be more easily overcome by an attack on the edge of the operculum. Despite this, Ansell & Morton (1985) describe some naticid species as exclusively side-borers, getting their food the hard way.

A further method by which some naticids (*Polinices tumidus*: Vermeij, 1980; Ansell & Morton, 1987; and possibly other species) kill their prey is a development of the way in which prey is transported wrapped in mucus in a depression of the foot to a site where the naticid can burrow. If the period during which the prey is held is prolonged the bivalve may be suffocated, its valves gape, and it can be eaten without the expense of boring. Failure to take into account the possible diversity of feeding methods may, however, invalidate the assessment of the degree of predation in collections of shells, Recent or fossil, made on the basis of counts of boreholes.

Despite the apparent care which observations suggest naticids exercize in the selection of prey it seems that their attacks are not by any means invariably successful. Kitchell *et al.* (1986) have analysed encounters between the naticid *Polinices duplicatus* and the gastropod *Terebra dislocata*. Out of 34 attacks by the naticid there were 27 which did not lead to feeding and only seven which did. The success rate would probably have been higher with bivalve than gastropod prey, escaping being not uncommon in gastropods but much less so in bivalves. Most molluscs remain relatively or wholly passive whilst their shell is being bored and react only when penetration, which is achieved by a thrust of the proboscis (Carriker & van Zandt, 1972), is complete.

To the normal relationship of predator to prey size must be added a further factor—the need for the predator to obtain sufficient energy from its food to sustain growth and reproduction. This need fluctuates, for, as with many other prosobranchs, activity is much affected by temperature (Macé & Ansell, 1982) and may cease in winter. Edwards & Huebner (1977) recorded no feeding by *Polinices duplicatus* for four months in winter. As naticids grow their increasing need for food can be satisfied by taking larger and larger prey since in normal circumstances these are available. Thus Wiltse (1980b) found that *Polinices duplicatus* fed on the bivalve *Gemma gemma* almost exclusively until its shell reached a height of 8 mm, taking bigger and bigger specimens as it grew. At still larger sizes, however, it practically ceased attacking this prey and changed to other, larger, species. Ansell (1982a, 1982b, 1982c) and Macé & Ansell (1982) have studied this area and some accompanying physiological changes, working with *Lunatia alderi*, and *L. catena* (both cited as *Polinices*) so as to make it one of the most carefully investigated fields of physiological ecology of prosobranchs. In the wild the annual cycle of each species exhibits a growth phase followed by a reproductive phase during which growth is greatly reduced. *L. alderi*, in the presence of ample food, grows at a rate which is determined by temperature and by the species of prey available, some of which, such as *Tellina tenuis*, promote rapid growth, whereas others, such as *Spisula*, *Donax*, and *Venus*, allow only a more moderate rate. As the naticids grow their rising needs are met by their ability to take larger prey but, as boring is a slow affair, there is a limit to the number of attacks which can be successfully completed over a given period of time, and so to the amount which they are able to eat . At the peak of its reproductive phase *L. alderi* takes about 0·5 *Tellina tenuis* per day, at other times a third of this; *L. catena* can rarely consume more than one specimen of *T. tenuis* per day at summer temperatures and only about one per week at winter ones. When this limit is reached growth is restricted and the reproductive rate may be lowered. The same results follow if food intake is decreased by the absence of large prey, compelling the animals to eat small ones, though this state may be partially alleviated by increased searching. A. J. Berry (1982) recorded a similar effect on *Natica maculosa*, which ate only the gastropod *Umbonium*. Its food consumption when young and growing was 1·5 *Umbonium* per day, falling to 1·27 when adult; as a proportion of body weight, however, its consumption dropped steadily, a feature which Berry attributed to the lack of sufficient numbers of large prey, forcing it to take smaller ones to an inadequate extent.

Like all carnivores naticids have an important regulatory or directive effect on the comminuty of which they are a part, more particularly as they switch from one species of prey to another more readily than do muricids. Wiltse (1980a) has studied this on a shore in Massachusetts where *Polinices duplicatus* was abundant. Exclusion of the predator enriched the fauna, not only the molluscan component as might be expected, but also the non-molluscan one, which she assumed had been disturbed by the movement of the naticids. The nature of the bivalve population of the shore was largely controlled (for both good and bad) by the selective feeding of *P. duplicatus*: if it preferred the rarer species its feeding consolidated and increased the dominance of the already dominant bivalve and the fauna was impoverished in the number of species it contained, though not necessarily in biomass; if it selected the dominant species the fauna was diversified overall, but because of its choice of prey of a certain size, whole year classes of its selected food, especially if they were numerically small, could be eliminated, sowing the seeds of future change. The predation pressure exerted on the bivalve population was high: Wiltse (1980a) calculated that *P. duplicatus*, with a density of $0·53.m^{-2}$ and an average feeding rate of 0·27 *Mya* per snail per day (Edwards & Huebner, 1977), removed 52 $Mya.m^{-2}.year^{-1}$, on a shore where the total population of molluscs with a shell over 3 mm varied from 4·5 to $84·7.m^{-2}$.

In 1884 and again in 1887 Schiemenz reported that the foot of naticids was penetrated by a series of channels connected to the external medium by pores along the anterior edge of the propodium. The channels could thus be filled with sea water and, together with an influx of blood to the pedal spaces, inflate the whole propodium and make it turgid, so that it acted as a plough when the snails moved through the sand. Ziegelmeier (1958) confirmed this for the Mediterranean species *Polinices josephinus*, having seen that when he handled the animals streams of water were ejected from the pores. With the loss of water the propodium was greatly reduced in size, became flaccid, and rendered useless for burrowing. Russell-Hunter & Russell-Hunter (1968) and Russell-Hunter *et al.* (1968) examined the system in two American naticids, *P. duplicatus* and *Lunatia heros*, and both they and Trueman (1968), who worked with *P. josephinus*, quantified the changes in volume which occurred. Both sets of workers discovered that the uptake of water was so great that when it was complete the animals had become too big to withdraw into the shell. Trueman (1968), like Schiementz (1884) before him, found that the volume of water expelled from the foot of *P. josephinus* was 2–3 times the shell volume, and Russell-Hunter & Russell-Hunter (1968) showed that the weight of a fully expanded naticid was 3·5 times that of a contracted one. Some of this change (5–7%) was due to expansion of the mantle cavity and a still smaller amount (2%) to entry of water into a space between mantle and shell which appears when the propodium is extended, but the remainder entered the water spaces of the foot. Only when all these spaces are emptied, a process which, as it is achieved by muscular contraction, is rapid (2·5–4·0 s), can the naticid withdraw completely. Expansion, which Russell-Hunter & Russell-Hunter (1968) believed was due to the elastic recoil of the pedal tissues, is a much slower process, taking up to eight minutes to complete. Normally, naticids do not have to eject water—it is an emergency measure to permit withdrawal into the shell—and as even those found intertidally live in a habitat and eat at levels where they are rarely emersed, they remain in the expanded state for days, or even weeks, the water pores closed by contraction of local muscles. The existence of this system of water spaces has not yet been demonstrated in any British naticid but it presumably exists: Russell-Hunter & Russell-Hunter (1968) failed to find it in *L. alderi* but explained this on the fact that these animals are much smaller than the other species which they examined and their techniques could not detect it.

Trueman (1968) also investigated the process of burrowing in naticids, a group of animals which burrow at a speed greater than that of most burrowing gastropods and equal to that of many bivalves. The process resembles that of most soft-bodied burrowing invertebrates in that the main body of the animal is held fixed in position by what Trueman called a penetration anchor, whilst a part of the body is extended into the substratum and dilated to form a terminal anchor. The penetration anchor is then abolished and the body pulled towards the terminal anchor. This sequence of events, the digging cycle, is then repeated until the animal has entered the substratum. In naticids the penetration anchor is formed by the shell and mesopodium, which is dilated by blood, whilst the terminal anchor is formed by expansion of the propodium after it has pushed its way into the substratum. There is an interplay between the two parts of the foot, since contraction of the dorsoventral muscles of the propodium, which brings about its elongation, tends to displace blood to the mesopodium and maintain its swollen state as part of the penetration anchor; as the columellar muscle later contracts to pull the animal towards the expanded propodium so do the dorsoventral muscles of the mesopodium, thus tending to move blood to the propodium and maintain the terminal anchor. No blood appears to be displaced into the visceral mass during this cycle, perhaps because internal pressure there is kept high by contraction of local muscles, nor does any water appear to be discharged from the propodial water pores to loosen the substratum into which the foot is being pushed, as happens in some bivalves.

CROSS REFERENCES

CHAPTER 35

THE DOG WHELK, *NUCELLA LAPILLUS*

F OR the excellent reasons summarized by Crothers (1985) *Nucella lapillus* has become, and remains, a favourite animal for ecological investigations and, to a lesser extent, for the study of some aspects of its physiology.

Dog whelks have no free larval stage in their life cycle and hatch as juvenile snails. They are also inactive as adults and though they wander throughout their immediate surroundings, particularly at night when immersed and males more than females, they never go far. Connell (1961) gave 100 mm as the average movement of a whelk during one tidal cycle, Morgan (1972b) 123 mm per day over barnacles, 329 mm per day over the easier going of a cockle bed, and Castle & Emery (1981) stated that adults do not move more than 30 m in a lifetime. The related *Nucella emarginata* moves less than 5 m in a year (Palmer, 1984). This is a mode of life permitted by their reliance on abundantly available, attached prey, largely barnacles and mussels. As a consequence each local population is isolated and there is little, perhaps no, interchange of genes between them. These are the very conditions which lead to increasing variability between local populations and may well explain much of that notice- able in this species, affecting in particular shell shape and colour. R. J. Berry & Crothers (1968) found that variability in a given population was greatest among young animals and became less in older ones, the more extreme variants being lost by one means or another. Most were destroyed before breeding age was reached and the population stabilized more or less around its previous norm. But though selection may thus minimize change it does not necessarily abolish it.

Some variability of dog whelk shells is known to be genetically based and Staiger's (1957) demonstration that at Roscoff *Nucella lapillus* shows two distinct races, one, more frequent in general occurrence and commoner on exposed shores, with 13 pairs of chromosomes, the other, more characteristic of sheltered situations, with 18 pairs, seemed to offer the beginnings of an understanding of this, especially as there was also some relationship between chromosome number and shell thickness. This genome pattern has proved to be localized: Mayr (1963) did not find it in north-east North America nor did Hoxmark (1970, 1971) in Norway nor Staiger himself (1957) in the channel east of Roscoff. Bantock & Cockayne (1975), however, have recorded it from the British shores of the western Channel and Bantock (quoted by Crothers, 1985) has found the two races in Pembrokeshire and the Firth of Clyde.

The distribution of dog whelks on the shore is centred around mid-tide level (about 10–75% emersion). Like other prosobranchs there it is affected by the need to survive desiccation and to maintain its respiration when emersed. Much of these needs is satisfied by behavioural adaptations, the whelks retiring at low water under stones and into crevices, unless humidity is high, when they remain active. In these circumstances two interdependent things prove important: the shape of the shell and the amount of water retained in the mantle cavity. The wide aperture of *Nucella* shells from exposed shores (Kitching *et al.*, 1966), with its greater peristomial length (Lowell, 1984), allows greater possibility of evaporation from the mantle cavity than does the narrow one with shorter peristomial length of animals from sheltered shores, and their shorter spire permits less water storage than the longer one of sheltered animals. Balancing these, however, is the fact that desiccation rates are likely to be lower on exposed shores and greater on sheltered ones. That evaporation of mantle cavity water does occur in intertidal dog whelks is suggested by the observations of Boyle *et al.* (1979). In laboratory experiments the initial amount of water in the mantle cavity (39% of the total water content) fell after 6 h drying at 20°C to 33%, partly, it was true, by simple drainage, but also, as indicated by raised internal solute concentration, by evaporation. *Nucella* on the whole showed similar changes when out of water, those at higher levels, exposed to longer emersion, exhibiting a greater reduction in the volume and greater increase in salt concentration. Coombs (1973a), working at Brighton (her results may not be universally valid), had linked level on the shore with ability to withstand desiccation with age: her younger animals (shell height <16 mm, up to 1 year old) were more resistant to water loss, 50% dying if it exceeded 50–55%, whereas older ones (shell height 25 mm or more, 3–4 years old) were more sensitive, a loss of 35–40% being sufficient to kill 50%. Those animals retaining most water show a consistently higher oxygen consumption than those retaining less (Houlihan *et al.*, 1981) but all can apparently maintain normal aerobic respiration during the intertidal period (Innes & Houlihan, 1985).

A further risk to which intertidal animals are exposed is lowered salinity, though this is usually of limited duration. *Nucella lapillus* can withstand this but is not common in areas such as estuaries where dilution of normal marine salinity is more permanent. Stickle & Bayne (1982) investigated the reaction of *Nucella* to lowered salinities brought about both abruptly and by slower acclimation. Snails over 20 mm shell height showed little difference in their tolerance of reduced salinity, however it was brought about and at whatever temperature within the range 5–20°C the experiment was carried out. The lowest salinity tolerated fell between 14·2 and 16·2‰. Smaller animals appeared more tolerant, surviving at salinities as low as 12·7‰. That this was a period of stress for all, however, is indicated by the fact that their oxygen consumption was maximal at normal marine salinities and was reduced as salinity fell, and that a similar reduction occurred in their rate of nitrogen excretion.

Nucella lapillus is probably the most familiar local prosobranch carnivore and it is well known that it may feed by boring the shells of other molluscs. The method of boring is similar to that of *Urosalpinx* (p. 621). The food is commonly barnacles and mussels, but Morgan (1972a) has recorded them attacking cockles and Largen (1967a) has suggested that the animals are more opportunistic than is generally believed and eat whatever kind of mollusc is available and which they can overcome. This idea, however, conflicts with the fact that they are animals of habit, or what Wood (1968) called ingestive conditioning and Morgan (1972b) associative learning, become used to a particular type of prey and are reluctant to switch to another, even if other prey abounds, until their usual stocks are exhausted. They then show themselves inept at tackling strange food and need practice to become skilled in dealing with it. Dog whelks are able to bore the shells of prey but if they can enter its body by a simple thrust of the proboscis this method is preferred as energetically more economical—it may take up to three days to enter a mussel by boring. The method chosen often reflects the predator's strength: older whelks can enter barnacles by direct parting of the valves, whereas young ones, not strong enough to do this must bore.

A period of more detailed investigation of how *Nucella* feeds was initiated by the work of Connell (1961) who showed that the amount of food taken was greatest in autumn, when it averaged 1·1 barnacles per day, and was less at other times of the year. Indeed, as indicated by Feare (1970a) and Stickle *et al.* (1985), feeding stops altogether when the temperature drops to winter levels. The latter workers have also shown that it is affected by extremes of salinity and is greater in animals exposed at low water than in those permanently immersed. A different estimate of the daily food ration required, this time in terms of mussel flesh, was given by B. L. Bayne & Scullard (1978b) as 2·2 mg for a whelk with a shell height of 23·3 mm. To a certain extent these figures are apparently determined by the way in which *Nucella* feeds, Connell (1961) reporting that the number of barnacles opened was independent of the whelk's size, as was also, B. L. Bayne & Scullard (1978b) noted, the time spent in boring and eating a mussel. Moran (1985), however, working with another muricid, *Morula marginalba*, reported that the size of both predator and prey affected feeding rates.

B. L. Bayne & Scullard (1978b), Stickle *et al.* (1985), and Hughes and his co-workers (Hughes & Elner, 1979; Hughes & Dunkin, 1984a, 1984b; Dunkin & Hughes, 1984; Hughes & Drewett, 1985; Burrows & Hughes, 1988) have described the feeding behaviour of *Nucella lapillus* and how it varies with the type of prey, with age, and with environmental conditions. The pattern comprises a feeding period followed by a resting phase, apparently required for digestion; the relative lengths of these probably depend on the nutritive value of the food. After a dog whelk has spent 1–3 tidal periods eating barnacles the post-feeding period lasts for 2–4 tidal periods, but after a meal of mussels this extends to 7–8 tides. With seasonal changes in the amount of food and the amount of feeding required this means that the life of a dog whelk is divided amongst its various activities in a different way in summer and winter. Thus Connell (1961) reported that in summer *Nucella* spends 60% of its time feeding, but only 13% in winter. This difference is also reflected in the rate of respiration: according to B. L. Bayne & Scullard (1978a) the oxygen consumption of comparable animals recently fed is 82 μl per hour in summer and 61 μl in winter, the latter figure probably reflecting long periods when no feeding took place. Burrows & Hughes (1989) have also shown that though a majority of dog whelks on a given shore tend to show similar foraging patterns, these vary from place to place depending on the type of shore, its exposure, and on local weather conditions. Dog whelks on sheltered shores foraged less in sunny and warm weather, favouring periods when stress from these factors was less; animals from exposed shores were, as might be expected, most affected by roughness of water and fed mainly in calm periods even when these were sunny.

As with other carnivorous animals the size of predator in relation to size of prey is also a factor of importance in the feeding behaviour of *Nucella*. Garton (1986) compared the efficiency of feeding as measured by the energy available for growth and reproduction in the related *Thais haemastoma* when smaller and larger whelks were offered smaller and larger oysters as food in standard conditions of temperature and salinity. Small whelks fared better on small oysters than on large ones, though larger whelks seemed to flourish equally well on both smaller and larger oysters. It is clearly important, for smaller predators in particular, to use the prey which can be most easily managed, an observation which has also been made on the basis of study of naticids (Edwards & Huebner, 1977; Griffiths, 1981; A. J. Berry, 1982; Broom, 1982; Bayliss, 1986). Increase in predator size demands more frequent feeding, B. L. Bayne & Scullard (1978b) noticing a shortening of the resting phase in the feeding cycle of *Nucella* as the animals become larger.

It might be supposed that when a barnacle or mussel is attacked by a dog whelk its reaction would be to close the shell and await, fatalistically, the result of the attack. This may well be true of barnacles, but does not appear to be the reaction of the population of mussels that form a bed. Wayne (1980) and Petraitis (1987) have made observations showing that a dog whelk attacking a mussel in a mussel bed may experience a counter-attack from the mussels. According to Petraitis (1987) a dog whelk in these circumstances is likely to have byssus threads attached to its shell by the three or four mussels nearest to the one attacked. The threads are tightened, with the result that the whelk is pulled off its potential prey and turned over, so that its foot, despite its efforts, cannot make contact with the substratum and it is effectively trapped. This reaction seems to be aimed directly at *Nucella* since shells of *Littorina littorea* (an animal which poses no threat to mussels) were much less frequently bound in this way.

This type of mussel behaviour has its effects on the foraging of dog whelks: they avoid mussels in beds (95% of the dog whelks in the area surveyed by Petraitis (1987) were outside the mussel beds); they prefer solitary mussels and smaller to larger ones (Hughes & Dunkin, 1984b), even if the food gained is less; and they bore near the posterior adductor muscle, well away from the foot and byssus (Hughes & Dunkin, 1984b). These workers claimed that this position gave the highest energy gain: Petraitis (1987) pointed out that it is also the safest.

Feare (1970b) has added much to our knowledge of the reproduction of *Nucella lapillus* from a study of a population at Robin Hood's Bay, North Yorkshire. Here breeding occurs April to May, hatching taking place five months later. Development depends upon environmental temperatures, being about a month faster in the southern parts of the British Isles, but lasting as long as seven in the cold of the White Sea, where the severe conditions also reduce the number of capsules laid and the number of eggs per capsule, the total reproductive effort being about a fifth of that of animals in most Atlantic sites (Matveeva, 1955).

The number of eggs actually escaping from the ovary of a female dog whelk is of little significance since most (up to 94%) do not develop into embryos but are used as food by those that do. Spight (1975a) combined the average number of 22 young emerging from each capsule calculated by Feare (1970b) with the average number of capsules which Hughes (1972) calculated were laid by a female each breeding season, to arrive at a total production of 1,030 hatching young per female per year. The two figures may not be strictly comparable since Feare's calculations referred to North Sea animals and those of Hughes to animals from the colder waters of Nova Scotia and temperature effects on the breeding of *Nucella* are well known. The figure, however, compares with that for the Eastern Pacific *N. emarginata* (Spight, 1975a), which also has food eggs, though it is considerably higher than that calculated by H. A. Cole (1942) for the related *Urosalpinx cinerea* from Essex waters and than the range (153–424) that may be calculated from the figures given by Carriker (1955) for animals of the same species from New Jersey. The figures for *Nucella lapillus* and *N. emarginata* are also of the same order as those for *Thais canaliculata* and *N. lamellosa* (Spight, 1975a) which have direct development, but yolky eggs and no food eggs, a fact which provokes doubts as to what strategic benefits food eggs confer. In both groups, however, the number of eggs is lower, often markedly, and their size greater than in species of neogastropod in which planktotrophic veligers occur. Whatever device is used total reproductive efforts may well be equivalent.

Growth rates of young *Nucella lapillus* appear to be roughly comparable throughout the range of the species. Juveniles measure about 10 mm in shell height at 1 year at Robin Hood's Bay (Feare, 1970b), 10–15 at Plymouth (H. B. Moore, 1938a), 15–16 at Black Rock, Brighton (Coombs, 1973b); the corresponding figures for two-year olds are 15, 12–26, 26+ mm. Maturity is calculated to be reached at an age of 2·5 years. Since growth then stops and year classes can no longer be distinguished it is doubtful how much longer the whelks live, perhaps another three years. The proportion of the original population reaching such an age is inevitably very small: Feare (1970b) assessed mortality at 90% in the first year, with half of the survivors dying in the second and a further quarter of those left in the third, leaving about 1·25% of the original population surviving. These figures relate well to those for other muricids (Spight *et al.*, 1974) in which 90–99% of juveniles die during the first year of life, though the survivors may last for another two decades.

The sexes are separate in *Nucella lapillus*, as in most neogastropods, and males are easily recognized by the penis. In 1970 Blaber noticed the occurrence of a rudimentary penis in some spent females from Plymouth, an observation which had previously been made on the related *Urosalpinx cinerea* in North America by Griffiths & Castagna (1962) and was also made a little later on *Ocenebra erinacea* in France (Poli *et al.* 1971). At almost the same time B. S. Smith (1971) described a similar imposition of male characters on female specimens of *Ilyanassa obsoleta* in Long Island Sound, a phenomenon to which he gave the name imposex and which he linked with some chemical arising from local marinas. Since that time the prevalence of imposex—most easily recorded and its intensity assessed by the degree of penis development in females relative to that of males [= degree of imposex of Bryan *et al.* (1986), or relative penis size (RPS) of Gibbs *et al.* (1987)]—has increased markedly and it is now of common occurrence along the French Atlantic coast (Féral, 1980), and throughout the British Isles (Bryan *et al.*, 1986, 1988; Bailey & Davies, 1989). This may be illustrated by the conversion of Blaber's figures (percentage of females showing a penis) to the degree of imposex used by Bryan *et al.* (1986) In a population of *Nucella lapillus* living in front of the Plymouth Laboratory the degree of imposex in 1970 was under 5%, and in a population at Rum Bay less than 0·1%: the 1985 levels found by Bryan *et al.* (1986) at the same two sites were 67% and 48%.

Imposex has been shown to be due, in general, to the use of anti-fouling paints and, in particular, to their main biotoxic ingredient tributyltin (TBT) (B. S. Smith, 1981; Féral & Le Gal, 1982; Bryan *et al.*, 1986), though some other organotin compounds have similar but usually lesser effects (Bryan *et al.*, 1988). The amount of TBT in the environment has grown rapidly, parallel to the increasing popularity of fish farming and of sailing as a leisure pursuit and is reflected in the high incidence of imposex in areas where these activities are common; aquaculture has been responsible for some TBT contamination but to a lesser extent than boating (Bailey & Davies, 1989). Though its use as an anti-fouling paint for small boats and in fish farms is now banned in this country the effects of TBT on populations of *Nucella* may be long-lasting because of the particular way in which it affects the mollusc.

The impact of imposex on *Nucella lapillus* has proved to be more profound than on other gastropods which show it, and has led to the near or complete destruction of populations in which it became marked (Bryan *et al.*, 1987). Gibbs *et al.* (1987) described (fig. 340) the stages of changing anatomy, primarily in the development of the imposed vas deferens, associated with different levels of environmental TBT; these form a series known as the vas deferens sequence (VDS). The seriousness of the effect of TBT poisoning on *Nucella* is due to the fact that in affected females the imposed penis and vas deferens grow in such a way as to block the pallial part of the oviduct (Gibbs & Bryan, 1986; Gibbs *et al.*, 1987). Though ova may still be produced and enclosed in capsules the blockage means that these cannot be laid and the females are effectively sterilized.

More recent investigations (Gibbs *et al.*, 1988) have shown that the effect of tributyltin is not confined to the growth of male organs so as to obstruct the female tract, but that it can, if applied early enough in the life history and at sufficient strength, cause a complete sex reversal, presumably by acting on the endocrine control of the reproductive system. Gibbs

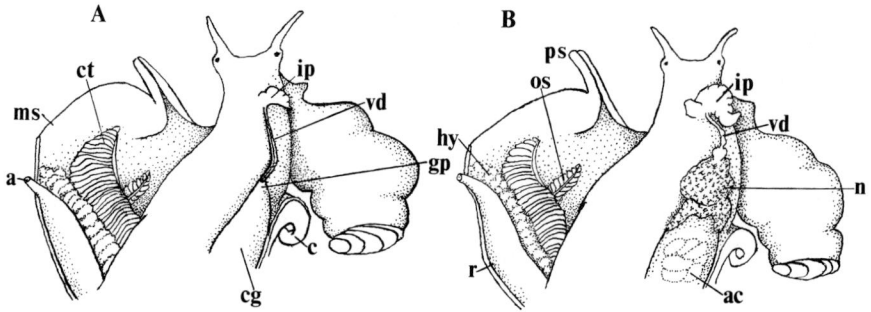

FIG. 340.—*Nucella lapillus*, female, the mantle skirt cut and reflected to expose the anterior end of the pallial genital duct and related structures. A, an intermediate stage of imposex in which a penis is present and a closed vas deferens leads from the female opening to the penis. B, late imposex stage in which proliferation of tissue has blocked the female opening and forms tumour-like nodules; the latter may also occur in males. (After Gibbs & Bryan, 1986, 1987).

a, anus; ac, aborted egg capsules in capsule gland; c, columellar muscle; cg, capsule gland; ct, ctenidium; gp, genital papilla; hy, hypobranchial gland; ip, imposed penis; ms, mantle skirt; os, osphradium; ps, siphon; r, rectum; vd, vas deferens.

et al. (1988) have assayed the amounts of tributyltin responsible for varying degrees of imposex: a concentration in water of less than 1 ng Sn.l^{-1} ($=2 \cdot 5$ ng TBT.l^{-1}) initiates signs of imposex; a concentration of 1–2 ng Sn.l^{-1} ($=2 \cdot 5$–$5 \cdot 0$ ng TBT.l^{-1}) may sterilize a female, and one of 10 ng Sn.l^{-1} ($=25$ ng TBT.l^{-1}) causes a sex reversal, spermatogenesis completely replacing oogenesis. *Nucella* is most sensitive to exposure to TBT during the first two years of its life: if the concentration of TBT is moderate at that time then all females are sterilized, if high enough the whole population is masculinized; in either case it will come to an end, and since the arrest of reproduction requires only a low concentration of TBT any immigrant juveniles are likely to suffer this chemical castration. Since dog whelks have direct development and are also so inactive the replacement of a destroyed population is likely to be a very slow process.

Gibbs *et al.* (1990) have also made investigations into the effects of TBT on *Ocenebra erinacea*, the rough tingle. In general these are similar to those produced in *Nucella* in that they lead to masculinization of the body in females and a more intense expression of masculine characters in males. Since the organization of the normal male system in *Ocenebra* is different from that in *Nucella* the changes imposed on female *Ocenebra* are different from those imposed on female *Nucella*. In particular the pallial oviduct of *Ocenebra* splits along its length: this parallels the condition in males in which the duct develops as an open groove, later closed by fusion of its lips except in the deepest part of the mantle cavity, where it remains permanently open. In the development of females, presumed to be like that of males except that closure of the duct is complete, it would seem that TBT prevents closure. The other changes due to imposex in *Ocenebra* include destruction of the female genital papilla and malformation of the bursa copulatrix, all leading to reduced reproductive activity and ultimate sterility.

The sensitivity of this species to TBT, according to Gibbs *et al.* (1990), seems to be a little less than that of *Nucella*: closure of the pallial duct appears to be stopped by a concentration of TBT in the water in which development is occurring of $1 \cdot 0$ ng Sn.l^{-1} ($=2 \cdot 5$ ng TBT.l^{-1}).

CROSS REFERENCES

CHAPTER 36

THE RELATIONSHIPS AND CLASSIFICATION OF PROSOBRANCHS

T HE identification of different kinds of animals is the historical approach of mankind to their classification, the recognition of groups of similar organisms: it may be necessary as a purely utilitarian measure, as (to take a molluscan example) in separating the whelk *Buccinum undatum*, which is safe to eat, from the red whelk *Neptunea antiqua*, which is safe to eat only after removal of the toxic salivary glands. Classification takes matters a stage further, with a variety of ends in view, but mainly as a device for making identification easier or more certain, or, at a less practical level, as the equivalent of a genealogical tree, suggesting relationships and evolutionary pathways. It is doubtful if the two goals are wholly compatible, and the second may not be realizable, since it attempts to cover the events of a time span of 500–600 million years, most of which is represented only by a fragmentary fossil record.

Whatever the objective, classification depends upon the use of the similarities and differences in organization which organisms show, primarily of those in their anatomy. Over the years the choice of organ system or systems whose variation forms the main basis on which a classification rests, has ranged widely, and almost every system in the body of a prosobranch has been used by some malacologist as the base on which his scheme rests: for a historical summary see Cox (1960a). The choice is, indeed, crucial, since such differences and similarities as the system exhibits must be as free as possible (ideally totally free) from the effects of convergence, a process which leads to close resemblance, or even indistinguishable identity, of organs used for similar activities in unrelated stocks: many parts of proposed classifications have crumbled because they rested on such a false foundation.

At present interest focuses mainly on the evolution which has occurred within the gastropods: the origin of the class from a stem related to monoplacophorans and, less closely, to polyplacophorans, and the origin of the phylum Mollusca itself from some group of invertebrates at approximately the platyhelminth level of organization, are both more or less generally accepted. Gastropods share many features with other molluscan groups, but are unique in having undergone torsion; this is, therefore, in the language of cladistics, a synapomorphy (an advanced or derived character marking all members of a taxonomic group). Torsion is so unusual a process as to make it improbable that it would have occurred in two unrelated stocks, so making the gastropods a monophyletic group, all its members derivable from one ancestral stem. Despite this, the peculiarities of patellogastropod organization and some other features in prosobranchs have led some (Horny, 1963; Shilenko, 1977; Wingstrand, 1985; Bandel, 1988) to wonder (almost yearningly) whether a torsion process distinct from that of other prosobranchs is an idea wholly impossible to contemplate: it would certainly offer the easiest solution of some problems.

Since the publication of Thiele's 'Handbuch der systematische Weichtierkunde' (1929–1931) most malacologists have used the classification of gastropods therein, which divided the class into the subclasses Prosobranchia, Opisthobranchia, and Pulmonata, with the first of these subdivided into three orders, Archaeogastropoda, Mesogastropoda, and Stenoglossa.

The last was later renamed Neogastropoda by Wenz (1938) to avoid the change of taxobase implicit in Thiele's name. Two modifications of this scheme have been proposed: Cox (1960a) suggested that the characters linking mesogastropods and neogastropods were so numerous and so close that their separation in different orders was unjustified, and he joined them into a single group which he called Caenogastropoda. For similar reasons Boettger (1955a) preferred to unite the opisthobranchs and pulmonates in a single subclass called by the long established name Euthyneura. Recently Haszprunar (1985b) has followed the same course but has called the united group Pentaganglionata. To most malacologists these changes reflect real relationships.

Extended knowledge of the anatomy of familiar animals, of that of animals known but unfamiliar, plus that of many new to science, the results of investigation of such topics as sperm morphology, together with the use of novel taxobases, have all combined to produce a degree of dissatisfaction with the Thiele-Wenz classification, and, based mainly on work by Kosuge (1966), Golikov & Starobogatov (1975), Salvini-Plawén (1980), Haszprunar (1985a, 1985b, 1985c, 1988a, 1988b, 1988c), Salvini-Plawén & Haszprunar (1987), and Lindberg (1986, 1988), there has been an almost complete recasting of prosobranch classification, and the following pages reflect most of their ideas. In this revision interest has centred primarily on a limited number of topics: the validity of an archaeogastropod taxon, on what animals may be included in it and their relationships, on the nature of groups lying at the archaeogastropod-caenogastropod boundary, and to what prosobranch groups the opisthobranch-pulmonate line may perhaps be related.

It has long been customary to suppose that amongst living prosobranchs the organization of those included in the superfamily Zeugobranchia by Thiele (= Pleurotomarioidea of Swainson, 1820), together with those he placed in the superfamily Trochoidea, is the most primitive. The work of Golikov & Starobogatov (1975), Salvini-Plawén (1980), Haszprunar (1988c), and Lindberg (1988) has shown that this is probably not true in the absolute sense, yet from the point of view of looking for a group which might be ancestral to the higher prosobranchs, it is justifiable, since it is only from some stock possessing many of the characters which they exhibit that any of the higher prosobranchs could be derived. The Zeugobranchia, as conceived by Thiele, also contained a group of limpets, the keyhole and slit limpets, now often placed in a separate superfamily Fissurelloidea, so similar in organization to pleurotomarioideans and trochoideans, despite the modifications imposed by their limpet facies, that Salvini-Plawén (1980) united all three superfamilies into a single group Vetigastropoda (the 'ancient gastropods'), thus giving substance to and formalizing an impression already widespread amongst students of archaeogastropods, and readily acceptable as a natural grouping.

The main features of vetigastropod anatomy may be enumerated. (1) The adult shell retains its protoconch and is helically coiled except in the patelliform fissurelloideans, though even here helical coiling is visible in the protoconch; (2) except in some trochoidean families the shell contains nacre. (3) The mantle cavity is deep and (4) typically contains a double set of pallial organs (gills, osphradia, hypobranchial glands) though in trochoideans the right gill and osphradium (but not the right hypobranchial gland) are lost. (5) From the shell paired dorso-ventral muscles pass to the head-foot; in fissurelloideans they extend posteriorly and unite to give a horseshoe-shaped shell muscle, not, however, divided into blocks as in patellogastropods. (6) The ctenidia (including at least part of the single gill of trochoideans) are bipectinate and their axes attached by membranes for some part of their length to the body and mantle skirt; (7) the axes and lamellae have internal skeletal supports which prevent their collapse and that of the blood vessels they contain; (8) the lamellae also have invaginated sensory pits (bursicles) (Szal, 1971; Haszprunar, 1987a). (9) The cephalic tentacles have a fringe of papillae each bearing a complex terminal sensory structure (M. Crisp, 1981); (10) lateral to each tentacle is an eye stalk carrying at its tip an open eye cup, its mouth plugged by a cuticular lens. (11) The sides of the foot are papillated and bear a fold, the epipodium, with tentacles, each provided with sensory papillae like those on the cephalic tentacles, and often with other sense organs at their base. Except in limpet forms the posterodorsal surface bears a multi-spiral operculum and, in addition, a glandular area which is used for cleaning the shell (H. D. Jones, 1984) and is folded away when not in use. (12) The mantle edge shows three parallel folds; (13) any tentacles which it carries are not retractile into pits. (14) The heart has two auricles even in trochoideans with only one ctenidium, the left receiving blood from the left kidney as well as from the left gill. (15) The lips of the mouth may be fringed, but are not otherwise elaborated apart from a mid ventral split between right and left halves. (16) In the buccal cavity jaws are paired, but are neither large nor heavily cuticularized. (17) The radula is rhipidoglossate and (18) the salivary glands are small. (19) There is no licker. (20) The secretory area of the oesophageal glands, which open widely to the oesophagus, is increased by papillae. (21) The stomach has a cuticularized area, the gastric shield, a spiral caecum (much reduced or vestigial in limpet forms), ciliated and ridged sorting areas, and is linked to the intestine by a style sac, probably of intestinal origin; (22) the rectum runs through the ventricle. (23) There are two kidneys of different structure and function: the right lies in the visceral mass and expands among the viscera, the left (the papillary sac) lies partly in the mantle skirt and, except in fissurelloideans in which it is minute, its wall is papillated. (24) The sperm are of one sort; gametes escape through a short gonadial duct to the right kidney in which there is often a special tract leading them to the mantle cavity; they are broadcast and, except in rare cases (e.g. *Bathymargarites*) fertilized externally. A lecithotrophic larval stage soon settles. (25) The nervous system is hypoathroid, with anterior cerebral ganglia, separate labial ganglia united by a commissure passing ventral to the gut, and (26) a long visceral loop commensurate with the size of the mantle cavity.

Three other groups of prosobranchs have traditionally been placed in the archaeogastro-pods along with the vetigastropods: the patelloidean limpets, cocculiniform limpets, and the neritoideans, the first and last well known, the deep-sea cocculiniform limpets little known until relatively recently (see the reviews by Haszprunar, 1988b, 1988c) and represented by only two species in the British fauna. Before discussing the possible interrelationships of the groups it is convenient to summarize their main features.

The patelloidean limpets have long been known to stand apart from other gastropod groups in many ways. (1) The adult shell is invariably limpet-shaped; it loses its protoconch at an early stage either by wear or by shedding, the point of detachment being sealed over by secretion of shelly material. (2) The shell commonly contains nacre. (3) The mantle cavity over the head (the nuchal cavity) is shallow in an antero-posterior direction and (4) does not contain a double set of pallial organs, though it may have paired osphradia and peculiar structures of unknown function called wart organs. It never has more than one ctenidium and hypobranchial glands are always absent. (5) From the shell paired dorsoventral muscles pass to the head-foot, the right and left muscles having extended their origin on the shell posteriorly and united to form, as in vetigastropod limpets, a horseshoe-shaped shell muscle, but, unlike these, it is here formed of blocks of muscle between which blood vessels pass. (6) In the genera which possess a ctenidium it is bipectinate, and it is the left which is present, its axis arising at the inner end of the nuchal cavity and curving to the right to fit within the shallow confines of that space; it is not attached to the mantle skirt or body by membranes; (7) the axis and lamellae have no skeletal rods, and (8) lack bursicles, so resembling those of lower classes in the phylum. (9) The head bears a pair of smooth tentacles; (10) on the base of each is an eye formed of an open pit with no cuticular lens. (11) The sides of the foot are smooth and lack an epipodium, though a glandular streak present in patellids has sometimes been regarded as representing such; an operculum and the special posterior area of the vetigastropod foot are both absent, but the margin of the sole is marked by a glandular groove not present in vetigastropods, and an operculum is present in the larva. (12) The mantle edge is formed of only a single fold and bears respiratory outgrowths on its underside, elaborated into pallial gills in a variety of species; (13) its marginal tentacles retract into pits (fig. 339). (14) The heart has a single auricle receiving blood from the ctenidium (if one is present), the pallial edge, and left kidney; the blood lacks haemocyanin. (15) The lips of the mouth are often elaborated, cuticularized, and used in feeding; (16) a single horseshoe-shaped, powerful jaw lies in the roof of the buccal cavity, its inner side backed by a pair of well developed inner lips; (17) the radula is docoglossate, and (18) salivary glands, sometimes two pairs present, are well developed, with long ducts. (19) At the tip of the odontophore, ventral to the most anterior part of the radula, lies a cuticularized structure known as the licker, used in feeding. (20) The secreting area of the oesophageal gland, which connects with the oesophagus along its whole length, is increased by the formation of septa, whilst (21) the stomach shows no gastric shield, spiral caecum, or obvious sorting areas; an expanded initial part of the intestine may represent the style sac; the rest of the intestine is long and much coiled. (22) The rectum does not pass through the ventricle. (23) There are two kidneys, the right a capacious sac, the left reduced in size, both without papillae. (24) The sperm are of one type, like but not identical with those of vetigastropods and like the ova, escape directly, not by way of a gonadial duct, to the right kidney, which has no special tract for their transport to the nuchal cavity. The eggs are broadcast, typically fertilized externally, and give rise to short-lived lecithotrophic larvae. (25) The nervous system is hypoathroid, with long pedal cords, labial ganglia separate from the cerebrals and linked by a commissure ventral to the gut; statocysts lie lateral to the pedal ganglia. (26) The visceral loop is short, and often little more than a transverse band between the pleural ganglia; from it arise the osphradial nerves and these still cross (fig. 341).

Cocculiniform limpets, though reminiscent in some ways of patellogastropod limpets, show features in some respects primitive, in others suggestive of closer relationships with vetigastropods, and have several characteristics peculiar to themselves, related to their deep-sea habitat and mode of life. They are largely carrion eaters, feeding variously on such unlikely objects as submerged wood, cephalopod beaks, crab carapaces, fish skeletons, elasmobranch egg cases, or whale skulls.

As in patellogastropods the nuchal cavity over the head is shallow and its contents are modified, especially on the right side, where the ctenidium, osphradium, and hypobranchial gland are all lost. Respiratory activity, marked in the nuchal roof as in all archaeogastropods, is supplemented by a variable range of structures on the left side and in the left pallial groove, though their main function may be as much the creation of a water current for other purposes as for gas exchange. Some of these structures show some similarities with vetigastropod gills in the presence of sensory pits and skeletal rods, perhaps, however, independently evolved. As in patellogastropods, too, the heart has only one auricle and the left kidney is reduced in size, whereas the right commonly expands. As might be expected from the varied diet the cocculiniform gut shows several different patterns. All, however, possess a radula which is

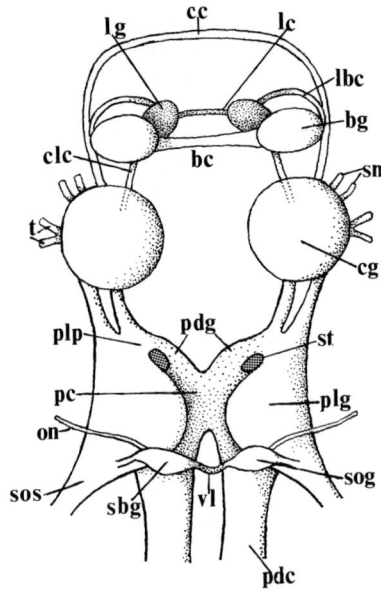

FIG. 341.—Diagram of the central nervous system in dorsal view. This figure adequately represents the condition both in *Acmaea tessulata* and in *Lepeta caeca*.

bc, buccal commissure; bg, buccal ganglion; cc, cerebral commissure; cg, cerebral ganglion; clc, cerebro-labial connective; lbc, labiobuccal connective; lc, labial commissure; lg, labial ganglion; on, osphradial nerve; pc, pedal commissure; pdc, pedal cord; pdg, pedal ganglion; plg, pleural ganglion; plp, pleuropedal connective; sbg, suboesophageal ganglion; sn, nerves to snout; sog, supraoesophageal ganglion; sos, nerve to shell muscle and osphradium; st, statocyst; t, tentacular nerve, bifid at base; vl, visceral loop.

basically of a rhipidoglossate type, and lack the voluminous salivary glands and, usually, the single strong jaw of the patellogastropods. In most the stomach has a gastric shield and thus approaches the vetigastropod more closely than the patellogastropod type. In the more primitive families the rectum passes through the pericardial cavity, though not the ventricle. Correlated with their habitat, their sedentary mode of life, and the dispersed nature of their population groups, the animals are hermaphrodite (except in the family Choristellidae) and, commonly, males have adapted the right cephalic tentacle for copulation, though a separate penis is developed in some.

The third group not included in the Vetigastropoda is the Neritoidea (= Neritopsina Cox & Knight, 1960, and Neritimorpha Golikov & Starobogatov, 1975) which differ profoundly in several ways from all other archaeogastropods and show approaches to caenogastropods. Such shared characters as they possess suggest a closer link with vetigastropods than with patellogastropods. Their main features are: (1) the shell is helicoid though low-spired, and often loses the internal septa separating whorl from whorl, a device economizing calcium and creating greater internal space. It retains its protoconch but (2) has no nacre. (3) The mantle cavity is deep, as in helicoid vetigastropods, but (4) contains only a left, bipectinate ctenidium, osphradium (lost in terrestrial forms), and hypobranchial gland; those on the right may be vestigial but still recognizable. (5) From the shell paired but unequal dorsoventral muscles pass to the head-foot. (6) The ctenidium arises from the base of the mantle cavity: its axis is attached near its base to the body supported by membranes (Bourne, 1908; Fretter, 1965). (7) Its lamellae have neither internal skeleton nor (8) bursicles. (9) The cephalic tentacles may be fringed and have (10) alongside them, stalked eyes, which are closed vesicles with an internal cuticular lens. (11) The sides of the foot are simple and there is no glandular streak and no posterior ridged glandular area; a reduced epipodium is occasionally present, marked in deep-sea species. The operculum is paucispiral. (12) Despite the vestigial nature of the right ctenidium the heart still has two auricles, in the more primitive forms, though the right may be vestigal (most terrestrial forms). (13) The lips of the mouth may be slightly fringed; (14) no jaws are present in the buccal cavity but a transverse dorsal fold may correspond to the inner lips associated with jaws in patellogastropods. (15) The radula is rhipidoglossate. (16) There are no salivary glands, nor (17) is there a licker. (18) The oesophageal glands have internal septa but are stripped off the main oesophageal channel and communicate with it only at their anterior end. (19) The stomach has a large cuticularized area, ciliated sorting area, but no, or only a vestigial, spiral caecum, and is linked to a lengthy intestine by a style sac; (20) primitively the rectum penetrates the ventricle. (21) There is only one kidney, the left, but this is not a papillary sac and is not sited within the mantle skirt. (22) Neritoideans produce both euspermatozoa and paraspermatozoa, the former showing some resemblance to those of caenogastropods; gametes escape through a duct partly formed from the rudiment of the right kidney, which has lost all excretory activity. The female system has become complex in relation to the reception and storage of sperm and to the provision of glands for the secretion of egg capsules, so is superficially like that of many caenogastropods; the glands, however, are probably derived from the right kidney rather than from the closure of a glandular pallial groove. Correlated with these activities males, except in the terrestrial helicinids and hydrocenids have a penis, a modified cephalic lappet, and make spermatophores transmitted in copulation, whilst females are frequently diaulic, retaining the original opening of the right kidney as a vaginal aperture and acquiring a new and separate one for egg laying. Many neritoideans are freshwater or terrestrial, in which case larval stages are suppressed; marine forms have a planktotrophic larval stage. (23) The nervous system is hypoathroid; the pleural ganglia give rise to a long, sometimes incomplete, visceral loop, whilst the pedal ganglia extend into pedal cords with many transverse connections. The cerebral commissure is long; there are separate labial ganglia typically with a commissure ventral to the gut.

Two questions now arise: what are the relationships of these groups to one another, and are they sufficiently closely linked to merit inclusion in a single higher level taxon Archaeogastropoda? Traditionally the more primitive members of the vetigastropod group, mainly those such as the pleurotomarioideans with a slit or hole in their shell, have been regarded as the most primitive living prosobranchs. This belief, with its accompanying emphasis on shell coiling, shell slit, and double set of pallial organs, may well be related to the fact that coiled shells with slits are also found in the fossil bellerophonts, a possible ancestral group that has hovered sometimes within the confines of the gastropod class (as it still does for some palaeontologists), sometimes outside it (as it still does for others). In bellerophonts, however, the coiling is plane, not helicoid as in vetigastropods, the shell is nearly always strictly bilaterally symmetrical, and there is no proof that the animals have undergone torsion and therefore that they are gastropods: indeed there are difficulties in accepting that they are.

Since the animals in the molluscan classes other than Gastropoda are bilaterally symmetrical, progastropod snails were also presumably bilaterally symmetrical. Prosobranch gastropods, however, are marked by torsion through 180°, a process which, however brought about, can hardly be imagined as occurring without causing some degree of asymmetry in body and shell. In agreement with this all gastropods are asymmetrical in the possession of only one gonad, and the archaeogastropod groups contain animals in all of which some degree of asymmetry is visible internally, though their external appearance may show none. External bilateral symmetry is to be associated with the rather static, bottom-clinging mode of life adopted by both fissurelloidean vetigastropod and patellogastropod limpets. An original coiled condition is suggested by its perpetuation in the fissurelloidean protoconch; this is not so obvious in the patellogastropod protoconch but has been seen clearly by Warén (1988, p. 680, figs 12, 13) in *Patella coerulea* and (Warén *in litt.*) in *Acmaea*, so confirming many earlier statements (Jeffreys, 1865; Davis & Fleure, 1903; Dodd, 1957; Bowman, 1981; Wingstrand, 1985). In *Erginus*, a patellogastropod which is aberrant in several respects (Lindberg, 1988), the protoconch is symmetrical, but this is presumably a modification of an ancestral coiled condition. Ancestral coiling is compatible with the original presence of a double set of pallial organs as, indeed, is found in vetigastropods, though often modified; greater changes in organization have occurred in patellogastropods, linked perhaps to a shortening of the nuchal region as the animals assumed the limpet form: no more than one ctenidium (and that of different structure) is ever present; paired osphradia may occur, however, since they lie on the head-foot and are unaffected by torsion.

As pointed out by Runnegar (1985) amongst others, the progastropod and the earliest gastropod molluscs were predominantly small, with shells a few millimetres high at most, though occasional fossils suggest dimensions more comparable with those of modern snails of moderate size. Over the ages gastropods have increased in size, sometimes relatively enormously. On the assumption that gastropods are a monophyletic class, as their shared torsion certainly suggests, evolution from the minute to the large has followed two main pathways which must have diverged at a very early stage as animals adopted different modes of life and so came to possess different morphologies. One of these lines has proved to have much more evolutionary potential than the other, and from points along it all higher prosobranchs must have arisen. At the archaeogastropod level it is exemplified by the vetigastropods, though these exhibit some derived characters (such as bursicles) which must have arisen after separation from the main ascending line. Many of the changes involved in the evolution of the molluscan classes, especially the gastropods, from such small beginnings have been pointed out and discussed recently by Haszprunar (1992).

Possibly the most important feature of vetigastropod organization which has led to their success and to that of more advanced prosobranchs, has been the helicoid shell which, balanced over the head-foot, adapts them for efficient locomotion, allows them to accommodate within its shelter both the visceral mass and, when necessary, the head-foot, and so frees them from a relatively static mode of life. Vetigastropods have retained the ancestral double set of pallial organs and the size of these has increased in relation to larger body size and a more active life. This is achieved by elongation of the mantle cavity and the strengthening of the ctenidia by the development of supporting skeletal structures within their axis and lamellae, as has independently, and for similar reasons, occurred in bivalves. Exposure of the head during locomotion has developed a high degree of cephalic sensitivity rather than the all-round pallial sensitivity of a patellogastropod, an arrangement not possible with a helicoid shell, though partially compensated for by the presence of circumpedal epipodial organs. In caenogastropods, which inherited this mode of life from a vetigastropod-like ancestry, exposure of the head has also encouraged cephalization, the development of a brain with the change from a hypoathroid to an epiathroid nervous system, and ultimately, even in animals with large, globose shells, permits the contact necessary for internal fertilization, so allowing invasion of freshwater and terrestrial habitats. An advanced feature which developed in trochoideans within the vetigastropod stock was loss of the right components of the pallial complex, except for the right kidney and the right hypobranchial gland, the former retained as an active excretory organ, the latter as an accessory reproductive structure. This loss has, in whole or in part, also occurred in, or been inherited by, caenogastropods, which, in addition, lose the right kidney as a functional excretory organ. It has also occurred in patellogastropods, though the functional reasons underlying the loss were probably not identical throughout (see below).

The second pathway has produced only limpets, and to that extent appears to be evolutionarily a dead end, not having given rise to any advanced groups: limpets can, apparently, produce nothing but more animals that are, fundamentally, limpets, however successful they may be. At the archaeogastropod level this line is represented primarily by the patellogastropod limpets, regarded by some modern malacologists (Golikov & Starobogatov, 1975; Salvini-Plawén, 1980; Haszprunar, 1987b, 1988b, 1988c; Lindberg, 1988) as the most primitive of living groups of prosobranchs and as exhibiting, in their limpet shape, the original prosobranch body form, however modified their internal anatomy may be. The features leading to this conclusion are primarily the microstructure of the shell, the fine anatomy of the gill, the docoglossate and stereoglossate radula, the pit-like, lensless eyes, the position of the statocysts, and the lack of a gonadial duct and of any special genital tract through the right kidney. Many of their features are highly adaptive and one can never be certain that all apparently primitive characters are truly so. It is probable that the difference between right and left kidneys is a basic prosobranch feature since it occurs in both patellogastropods and vetigastropods, brought about in the first place by a torsion-induced asymmetry, followed by a modification in function due to a special relationship established with the efferent ctenidial vessel in the case of the left kidney, and the absence of that plus an association with the viscera in the case of the right (E. B. Andrews, 1985). The ctenidium, where present, is clearly more primitive than that of a vetigastropod, but resembles that of chitons (Yonge, 1939) and caudofoveates (Salvini-Plawén, 1985) in its ciliation, its lack of axial and lamellar skeletal supports, and the pattern of the blood spaces. Its presence within the group is limited to the acmaeids and, within them, to the left side of the body, thus presenting a superficial resemblance to trochoidean vetigastropods.

The monobranchiate condition of trochids and, by extension, of acmaeids, has traditionally been attributed to the reduction in area of the right side of the mantle cavity and to the weight of the overlying shell in a dextrally coiling animal, and interpreted as giving improved ventilation within the cavity. Salvini-Plawén (1980) has suggested that a comparable absence of right pallial organs in caenogastropods may be due to the introduction of a prolonged planktotrophic larval stage into the life history, at the end of which the anlagen of the right organs have lost their competence to develop further. However valid this may be for caenogastropods it cannot explain the condition in archaeogastropods since, with the exception of neritoideans (J. B. Lewis, 1960), they have no such larval stage, and it therefore casts doubts on its value as an explanation of loss in caenogastropods. Crofts (1937) showed that in *Haliotis* development of the pallial organs occurred in two stages: in the first (larval) stage the organs of the post-torsional left appear, but those of the right do not do so until the larva has settled and at a time when little or no helical coiling has yet occurred. She suggested that if development of pallial organs was halted at the time of settlement then an adult stage like that of an acmaeid (or trochid) would result. Lindberg (1988) has also proposed that some aspects of the organization of patellogastropods are most readily explained as due to paedo-morphosis, perhaps to progenesis, which is the adoption by a descendant animal, as its adult condition, of an ancestral juvenile state, brought about by precocious sexual maturity. This, using Crofts' observations, would explain the presence of only a left gill in acmaeids. The simplified structure of the patellogastropod stomach as compared with that of vetigastro-pods, which, since it is directly comparable with that of bivalves, and even with that of cephalopods, may be taken as the ancestral conchiferan type, may well be due to a similar process, an idea supported by the remark of Crofts (1937) (still the only account of the development of internal organs of an archaeogastropod) that the appearance of the spiral caecum, one of the major features of the vetigastropod stomach, is very late in development.

A similar process may be responsible for the shape of the shell in patellogastropods, as the idea that the original gastropod was limpet-like in shape and symmetry is not wholly accept-able. There are too many signs of coiling amongst the lowest prosobranchs for this to be so, and some degree of asymmetry was an almost certain consequence of torsion. In patello-gastropods, presumably the earlier evolutionary line to become distinct, if asymmetry was not too great, stability was obtained by regaining the ancestral external symmetry with increased adaptation for limpet life. In vetigastropods exploitation of an incipiently helicoid shell balanced over the head-foot, gave equilibrium during locomotion and led ultimately to their escape to a freer mode of life.

How far may the four groups Patellogastropoda, Cocculiniformia, Vetigastropoda, and Neritimorpha be properly linked with one another in a single order Archaeogastropoda, as has been customary? This question may be posed in another way: is Archaeogastropoda the name of an order of genuinely related animals or merely, in the form archaeogastropod, a word conveying something about their level of organization? It is clear from what has been said above that one could regard the differences shown by the groups as sufficient to deny their close, though not their ultimate relationship: patellogastropods certainly do not relate closely to the other groups and cannot be regarded as ancestral to any higher forms. The same is largely true of neritimorphs, and several workers (Morton & Yonge, 1964; Fretter, 1965; Healy, 1988) are willing to place them in a taxon separate from both archaeogastropods and caenogastropods and of equal rank. Lindberg's (1988) view of the systematic position of the patellogastropods is similar.

Neritimorph organization is in part reminiscent of, or identical with, that of archaeogastro-pods (for example, radula, pedal cords, kidney structure) but stands apart in others, particu-larly in the elaboration of their reproductive system which has involved the destruction of the right kidney as an excretory organ, but has formed the base on which their invasion of freshwater and terrestrial habitats rests. Healy (1988), summarizing work on their sperm morphology, has stated that this is so different from that of both archaeogastropods and caenogastropods as to suggest that their classification should be with neither, but as originally proposed by Morton & Yonge (1964), in an independent order.

The two remaining groups, Cocculiniformia and Vetisgastropoda, were originally placed together when Salvini-Plawén (1980) established the latter group. Haszprunar's recent work (1988b), however, separates them, though the rest of the original vetigastropods remains a cohesive grouping of molluscs.

We are therefore left with a situation in which the traditional archaeogastropod collection of prosobranchs has been shown to comprise four distinct groups, two of which (patello-gastropod and neritimorphs) have been stated by their most recent students to be so unlike the others that they should certainly be classified separately. The choices open to systematists are (1) to accept that these archaeogastropod groups are sufficiently devoid of linking, shared characters as to deserve separation, each in its own taxon and all of equal taxonomic rank, a step which reduces the term archaeogastropod to a description of certain levels of organization. This, however truthfully it may reflect actual evolutionary history, is described as taxonomic inflation by Hickman (1988). (2) To find sufficient common fundamental features to justify the belief that the four groups form a natural group of related animals, though the point of common origin must, in that case, be very remote in time: this is the traditional view. (3) To allow a term Archaeogastropoda to be used as the name of a paraphyletic taxon (a group of animals with an ultimate common ancestor, but not including all its descendants, useful only as indicating an evolutionary stage). Choices (1) and (2) are extreme, neither, for different reasons, perhaps acceptable, but both logical, differing only in the point at which one draws a reference or starting line; choice (3) is a compromise, but has the apparent advantage of retaining a familiar taxonomic scheme. None properly reflects true phylogeny as currently perceived.

The transition from an archaeogastropod to a caenogastropod level of organization appears to have been attempted, with partial success, several times in the course of prosobranch evolution, giving rise to a number of groups with different mixtures of old-fashioned and newer features. One such is the Neritimorpha, others are the architaenioglossans and neomphalids (Fretter et al., 1981). (Until recently the valvatids would have been added to this list, but Rath (1988) has shown that they are not properly placed at the archaeogastropod-caenogastropod boundary.) The classification of such groups is sensitively dependent on the choice of taxo-base: if, for example, the type of osphradial cell (Haszprunar, 1985a) or of nervous system be used, then architaenioglossans, with a hypoathroid arrangement, which Haszprunar (1988a) claims is diagnostic, are archaeogastropods, though they must then be placed in a different group from all the others mentioned above; if radular pattern, shell microstructure, or renopericardial and reproductive arrangements be chosen then they are caenogastropods, as accepted by Ponder & Warén (1988). The truth seems to be that many primitive proso-branchs experimented with a variety of ways of improving efficiency, or their degree of adaptation to new habitats, changing those parts of their organization which brought this about, but leaving the rest as it was unless it was positively disadvantageous so to do. This process leads to mosaic evolution and produces an ill-defined area between undoubted lowly forms and equally undoubted advanced ones. According to one's choice of criterion the boundary drifts raggedly from one position to another. Classificatory schemes as at present organized cannot cope with such a situation without the danger of becoming either over-elaborate or procrustean (Hickman, 1988).

Most of the families included in Thiele's Mesogastropoda form a relatively cohesive grouping, though they are marked by the many detailed specializations in relation to differing modes of life which form the main substance of previous sections of this book. They present many features reminiscent of trochoidean gastropods—a single, left ctenidium, now however monopectinate, with its axis fused to the mantle skirt along its whole length and breadth (so not needing supporting membranes) and with the osphradium as a consequence appearing to arise independently (fig. 342). The heart has only one auricle which retains a direct or nearly direct connection with the nephridial gland in the (left) kidney. The right kidney is lost as an excretory organ, but its opening to the mantle cavity persists as a genital one, and its connection with the pericardial cavity may be retained in females as a gonopericardial canal. From the genital aperture a groove, probably originally open, though closed in more advanced forms, runs along the right side of the mantle cavity, probably not leading to a penis in males,

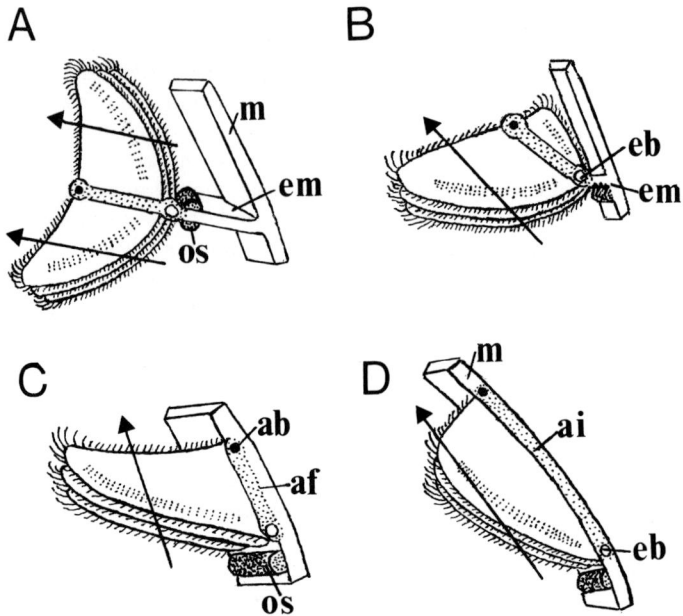

FIG. 342.—Diagrams of thick transverse sections of ctenidium, osphradium, and associated mantle skirt to show the change from the bipectinate ctenidium of the archaeogastropod to the monopectinate ctenidium of the caenogastropod, along with concurrent change in the position of the osphradium. A, archaeogastropod with pronounced efferent membrane and osphradium on the ctenidial axis; B, hypothetical intermediate stage showing reduction of efferent membrane and median gill lamellae; C, hypothetical later intermediate stage after loss of efferent membrane and median gill lamellae and fusion of axis with the mantle skirt; D, caenogastropod condition with axis incorporated in mantle skirt and osphradium separate from but parallel to the axis. Arrows indicate direction of water current; ctenidial axis lightly stippled.

ab, afferent branchial vessel; af, ctenidial axis fused with mantle skirt; ai, axis incorporated in mantle skirt; eb, efferent branchial vessel; em, efferent membrane; m, mantle skirt; os, osphradium.

though this organ is usual in higher forms. In the absence of a copulatory organ sperm transfer may have been by simple approximation of snails and use of the respiratory water current through the mantle cavity. Both open groove and aphallism occur (or recur) sporadically throughout the caenogastropods but are characteristic of the cerithioideans, a superfamily generally accepted as the most primitive of Recent mesogastropods. As in neritimorphs the eyes are situated on stalks fused to the cephalic tentacles, and are closed vesicles with an internal lens; the cephalic tentacles are often smooth and the epipodium, if not lost, is represented by only a few tentacles on the lobe of the foot bearing the paucispiral operculum. Internally the formation of a narrow nerve ring as part of an epiathroid nervous system has had the effect of constricting the gut posterior to the buccal mass; this has resulted in the once continuous series of oesophageal glands being split into pouches anterior to the nerve ring and glands behind it. The salivary glands are acinous and have also come to lie behind the nerve ring, their ducts passing through it. The stomach has the same features as that of vetigastropods though the caecum is reduced or vestigial; the intestine is short. In two of the basal mesogastropod groups, Cerithioidea and Rissooidea, as well as in the more advanced

Calyptraeoidea, a crystalline style is formed in the stomach, in which case the oesophageal glands are much reduced or absent; a style, however, cannot be a primitive character of mesogastropods as it has appeared in several other groups of molluscs. The female genital duct becomes glandular since the eggs are fertilized internally and laid in protective cases from which veliger larvae emerge; sometimes these have a long planktotrophic life before settling. The presence of egg capsules has allowed certain mesogastropod superfamilies, in particular cerithioideans and rissooideans, to spread with outstanding success into freshwater habitats, when larvae are suppressed and the young hatch as juveniles. Attempts to find groupings of mesogastropod superfamilies, as tried by Fretter & Graham (1962), have largely failed, and were probably based on convergent characters rather than on synapomorphies.

The more primitive mesogastropod superfamilies contain animals which are microphagous or herbivorous: the more advanced ones (Naticoidea, Cypraeoidea, Heteropoda, Tonnoidea) are carnivorous and have developed an elongated proboscis to facilitate prey capture, as well as other, less obvious, adaptations. This type of feeding is also found in most neogastropods and has, indeed, determined many of the organizational changes which they have undergone. The origin of neogastropods is still the subject of argument, and both one from a primitive group (Ponder, 1973) and one from an advanced mesogastropod one (J. D. Taylor & Morris, 1988; Healy, 1988) have been suggested, a situation which probably arises from the difficulties of distinguishing similarities due to convergence from those due to kinship.

In most of the families in the group Mesogastropoda of the Thiele-Wenz classification the radula is taenioglossate. In several, however, it is ptenoglossate, with many similar fang-like teeth, or it does not conform to the taenioglossate pattern in other ways, usually having more than seven teeth in each row. These animals also show other differences from typical taenioglossans in their structure. Their apartness was first emphasized by Habe & Kosuge (1966) who proposed their separation from other mesogastropods and their union in an order which they called Heterogastropoda, members of which shared some features with opisthobranchs (Robertson, 1974). Though it is now generally accepted that some of these animals should be separated in classification from the rest of the prosobranchs, it is now also believed that the original heterogastropod group is polyphyletic; residual argument centres on how its break-up and its heterogeneity are best expressed in classification. Closer examination of the organization of the snails placed here (Kosuge, 1966; Climo, 1975; Robertson, 1974, 1983, 1985a; Healy, 1982; Haszprunar, 1985a, 1985b, 1985d, 1985e; Bieler, 1988), stimulated by the heterogastropod idea, has demonstrated that the group contains at least two unrelated collections of animals, any common features being due to convergence. The interrelationships of the groups to one another, of each to other prosobranch groupings, and of the superfamilies within each, are still provisional; current groupings may prove to be polyphyletic, and future work may well change their taxonomic position.

A first group contains a number of Recent families akin in several ways and, on the basis of the pattern of cells in the osphradia (Haszprunar, 1985b), generally agreed to be caenogastropods, though their allocation to superfamilies is uncertain. These are the Triphoridae, the confusingly similarly named Triforidae, the Cerithiopsidae, Epitoniidae, Janthinidae, Aclididae, and Eulimidae. The first of these, Triphoridae, contains animals nearly all with sinistral shells [though in the British fauna one of its members *Metaxia* (or *Cerithiopsis*) *metaxa* is dextral], whereas in all others the animals have dextral shells. Triphorids (*Triphora*), triforids (*Cerithiella, Eumetula*) and cerithiopsids share such characters as a long acrembolic proboscis and a radula with claw-like teeth (though with different numbers) which together allow them to feed on the tissues of the sponges in association with which they live; their salivary glands, the right and left of different size, are tubular, not acinous, their oesophageal structure anomalous though perhaps still derivable from a taenioglossan pattern; the sexes are separate, males are aphallic and spermatozeugmata, each fashioned from an enlarged paraspermatozoon, carry euspermatozoa to the female duct. The considerable resemblance between the families suggests their inclusion in one superfamily, Triphoroidea.

Epitoniids and janthinids, though possessing shells of widely different appearances, are sufficiently alike in anatomy (though some resemblance may be merely adaptive convergence) to suggest that they too fall into a single superfamily, Epitonioidea or Janthinoidea. Both families contain animals attacking coelenterates with a strictly ptenoglossate radula and powerful jaws, epitoniids with a very long acrembolic proboscis, janthinids with eversible buccal walls; as in triphoroideans the salivary glands are tubular, and in epitoniids their ducts end in stylets, perhaps suggesting that the saliva may be injected to sedate the prey; in addition to these salivary glands an accessory pair, also tubular, is present in both families, and the oesophagus is not glandular. In both families the hypobranchial gland produces a purple secretion to which a sedative or toxic action has been attributed. Males, as in tri-phoroideans, are aphallic and sperm transfer is by spermatozeugmata. The family Aclididae, the organization of whose members is still inadequately known, is included in this grouping by Ponder & Warén (1988): it was previously placed along with the family Eulimidae in a superfamily Eulimoidea which, after its removal, contains the single family Eulimidae. Eulimids differ from the members of the other superfamilies in not having paraspermatozoa and cannot therefore produce spermatozeugmata: males have a penis. They are ectopara-sites of echinoderms and many have lost the radula and feed suctorially; in such as possess a radula it is small and ptenoglossate, used as a filter to strain coelomocytes from the body fluids of their prey (Warén, *in litt.*). Oesophageal glands are always absent. The pedal glands have hypertrophied, their secretion perhaps helping to maintain the snail on its host, and many have developed a protective pseudopallial fold at the base of the acrembolic probos-cis, which is kept more or less permanently everted in the host. The fold may completely cover the snail and even provide a brood chamber for the development of eggs.

All three superfamilies are clearly related and were placed by Haszprunar (1985b), largely on the basis of the fine structure of their osphradium, in a group which he called Heteroglossa (to distinguish it from Taenioglossa), later (1988a) Ctenoglossa, a subdivision of the Apogastropoda (his name for those prosobranchs not archaeogastropods); Ponder & Warén (1988) placed them in a suborder Ptenoglossa of the order Neotaenioglossa, the latter roughly equivalent to Thiele's Mesogastropoda. All three authors therefore treat the animals as part of what may be regarded as mainstream prosobranch evolution, unlike those falling into the second collection of the original heterogastropods.

When we turn to these any conclusions that may be reached rest on much less secure grounds, partly because many of the animals are both so little known in their anatomy, and that is often so unusual, that their relationships are difficult to assess; and partly because many are small, so that it is difficult to be certain that some of their features are not attributable to that fact alone. The families of the original heterogastropods left for inclusion in this group after removal of those dealt with above, are the Omalogyridae, Rissoellidae, and Pyramidellidae amongst British forms, together with the Mathildidae (Climo, 1975; Haszprunar, 1985e) and Architectonicidae (Robertson, 1974; Haszprunar, 1985d; Bieler, 1988).

The animals are unusual in many different ways, but share enough characters to suggest that they form a natural, but not a closely knit, assemblage. These basic features include the common but not universal occurrence of a heterostrophic protoconch; generally a loss of the ctenidium and its replacement as a device for ventilating the mantle cavity (whose surfaces therefore become the principal areas for gas exchange) by ciliated ridges, one on the floor and one on the apparent roof of the cavity. These may be derivatives of the ciliated bands originally placed alongside an osphradium, as still seen in the larval stages of more typical prosobranchs (Fretter, 1972), moved to their new positions by expansion of the mantle skirt. If a gill is present, as in architectonicids, it does not conform to the typical ctenidial type, but is foliobranch, with short and thick projections from the main axis. Respiratory exchange may also be helped by the situation of the richly vascularized kidney lying in the mantle skirt. Salivary glands are tubular and oesophageal glands lost. A pigmented organ of uncertain function, perhaps a modified hypobranchial gland, perhaps some kind of persistent larval kidney (Robertson, 1974, 1985a), is also of common occurrence. A similar structure is also present in epitoniid and janthinid ptenoglossans: if it is homologous it may perhaps indicate a link with these groups. The snails are hermaphrodite and the penis is invaginable. In pyramidellids the eyes approach the mid line between the tentacles and sink inwards; indications of similar change may be seen in some rissoellids. The nervous system is euthyneurous. In the egg capsules of architectonicids and pyramidellids the individual eggs are interlinked by chalazae, though these are not found in omalogyrids or rissoellids; in these snails, however, the capsule contains one, or at most two eggs, and chalazae would not, therefore, be expected.

In many ways the animals show links with such lower opisthobranch families as the Acteonidae (Fretter & Graham, 1954) and Ringiculidae (Fretter, 1960). Pyramidellids, specialized for a predatory (or ectoparasitic) mode of life, may on balance, be better placed with opisthobranchs than with any section of the prosobranchs (Fretter & Graham, 1949; Fretter et al., 1986; Haszprunar, 1988c; Healy, 1988). If we remove them the other families in this area fall between undoubted prosobranchs on one side and equally undoubted opisthobranchs on the other, a situation comparable to what was described in the similar region lying between archaeogastropods and mesogastropods. Here, as there, the animals are best interpreted as forms which have failed to make the full journey from prosobranch to opisthobranch but represent important and suggestive staging posts on the way followed by those others that completed it successfully.

In his admirable discussion of gastropod classification Cox (1960a) wondered 'if in striving after a phylogenetic classification we are attempting the impossible', a feeling also occurring to Hickman (1988). Such a classification, even if satisfactorily imagined, may prove inexpressible in the current language of taxonomy. Classifications are of necessity twodimensional since they are normally given on a printed page; yet organisms evolved in a fourdimensional space-time continuum. Something more complex than a printed table may be required to present their relationships and until that is achieved we may be able to show only an inadequate, partial, and perhaps erroneous aspect of the truth.

Within recent years the animals included in this second grouping of heterogastropods were placed by Golikov & Starobogatov (1975) in a superorder Pyramidellomorpha, but this included those in the first heterogastropod group as well, and it is clear that the two should be separated. Haszprunar (1985b) suggested the name Allogastropoda for the second group, and it has also been used by Salvini-Plawén & Haszprunar (1987); they did not include in it, however, either the Rissoellidae or the Omalogyridae. These, together with the valvatids, were united, along with the allogastropods, in a larger group Triganglionata (a name referring to the three ganglia on the visceral loop, supraoesophageal, suboesophageal, and visceral), which was contrasted with a group containing the opisthobranchs

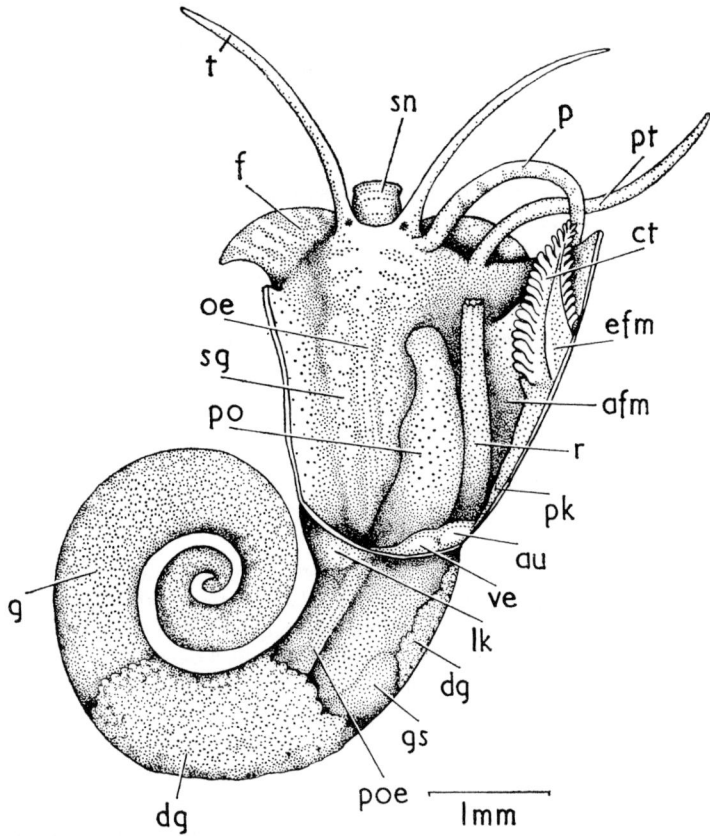

FIG. 343.—*Valvata piscinalis*. Animal removed from the shell, mantle cavity opened along the left side and the mantle skirt folded to the right. Some organs are seen by transparency.

afm, afferent membrane of gill; au, auricle; ct, gill; dg, digestive gland; efm, efferent membrane of gill; f, foot; g, gonad; gs, gastric shield; lk, left kidney; oe, oesophagus; p, penis; pk, pallial extension of kidney; po, pallial oviduct; poe, posterior oesophagus; pt, pallial tentacle; r, rectum; sg, salivary gland; sn, snout; t, tentacle; ve, ventricle.

and pulmonates (so roughly equivalent to Euthyneura) to which was given the name Penta-ganglionata (with five ganglia on the visceral loop, the three of the triganglionates plus the two parietal ganglia characteristic of euthyneurans). This classification therefore emphasizes the opisthobranch links of these animals, though it formally excludes even the pyramidellids from that group.

In the phyletic scheme presented later by Haszprunar (1988a, 1988c) the Allogastropoda has been extended to include the families Rissoellidae and Omalogyridae, still includes pyramidellids, but now excludes valvatids. Allogastropods were regarded as apogastropods (non-archaeogastropod prosobranchs), so retaining their prosobranch character. It must be emphasized, however, that these terms were used as the names of organizational grades, not necessarily of phylogenetic clades. In other systematic schemes (Salvini-Plawén & Haszprunar, 1987; Ponder & Warén 1988) the animals are placed in an order Heterostropha (a name referring to the apparently reversed twist of the larval shell) of the subclass Hetero-branchia, a taxon of equal standing with Prosobranchia. This arrangement separates them from prosobranchs and expresses opisthobranch affinities openly. In the present state of knowledge, and in the light of taxonomic judgements arrived at in the course of its current rapid expansion, it is inevitable that some such conceptual differences should arise.

Some recent alterations in classificatory schemes are connected with studies of the family Valvatidae. These snails have long been regarded (see p. 558) as a group of rather unusual prosobranchs lying somewhere at the base of the mesogastropods. They differ from these in having a bipectinate gill (fig. 343) placed far forward on the mantle skirt, in being hermaphro-dite (and sometimes viviparous), in lacking a free larval stage, in having the kidney in the mantle skirt, and in having neither buccal cartilages nor oesophageal glands. Some of these characters can be plausibly explained as adaptations for the life in fresh waters to which all valvatids are limited (kidney, hermaphroditism, direct development) and others (gill) as part of the mix of archaeogastropod and caenogastropod features which prosobranchs at the archaeogastropod-caenogastropod boundary so often show. Recently, however, Haszprunar (1985a) and Rath (1988) have brought forward evidence which shows con-clusively that valvatids cannot be properly be placed here. Rath (1988) has pointed out that despite its superficial similarity the valvatid gill is not a true ctenidium but a new develop-ment, and that other features of their organization [a buccal mass devoid of supporting cartilages, a non-glandular oesophagus, a kidney in the mantle skirt, the absence of para-spermatozoa, and the presence of chalazae (Cleland, 1954) linking the eggs in an egg capsule] not only separate valvatids from mesogastropods but positively associate them with the Heterostropha as understood by Ponder & Warén (1988) (= Allogastropoda of Haszprunar, 1985b). To make their taxonomic position more precise, however, Haszprunar (1988a) has pointed out that valvatids lack the pallial ciliated ridges of other animals in that group, as well as their heterostrophic shell, and should therefore be classified separately: he placed them in a suborder for which he revived the name Ectobranchia Fischer, 1884, a name based on the fact that the valvatid gill, because of its forward situation rather than the degree of oxygenation of the ambient water as has been claimed, is usually largely exposed in active animals.

One of the peculiarities of the family Valvatidae is that all its known members are restric-ted to fresh water, a fact that fits ill with their supposedly primitive evolutionary position. Recently, however, Ponder (1990) has described animals placed in a family Orbitestellidae which, he showed, share certain characters with valvatids, and which he placed with them in the Heterobranchia. Unlike valvatids, however, orbitestellids are marine and therefore prob-ably closer to the position occupied by basal heterobranchs. The relationships of the two families to one another and how their organization helps in understanding heterobranch origins are discussed by Ponder (1991).

REFERENCES USED IN INTRODUCTION AND PART I

ABBOTT, R. T. 1954. 'American Seashells.' New York, van Nostrand.

ABE, N. 1931. Ecological observations in *Acmaea dorsuosa* Gould. *Sci. Rep. Tohoku Imp. Univ.* (4), **6**, 403–427.

ADAM, W. 1942. Notes sur les gastéropodes XI. Sur la répartition et le biologie de *Hydrobia jenkinsi* Smith en Belgique. *Bull. Mus. R. Hist. nat. Belg.,* **18**, 1–18.

ADANSON, M. 1757. 'Histoire naturelle du Sénégal (Coquillages). Avec le relation abrégée d'un voyage fait en ce pays pendant les années 1749–53'. Paris, C. J. B. Bauche.

AGERSBORG, H. P. K. 1929. Factors in the evolution of the prosobranchiate mollusc, *Thais lapillus. Nautilus,* **43**, 45–49.

ALBRECHT, P. G. 1921. Chemical study of several marine mollusks of the Pacific coast. *J. biol. Chem.,* **45**, 395–405.

ALBRECHT, P. G. 1923. Chemical study of several marine mollusks of the Pacific coast. The liver. *J. biol. Chem.,* **57**, 789–794.

ALEXANDER, W. B., B. A. SOUTHGATE & R. BASSINDALE. 1935. Survey of the River Tees. Pt. 2. The estuary—chemical and biological. *Tech. Pap. Wat. Pollut. Res., Lond.,* **5**, 1–171.

ALPERS, F. 1932. Zur Biologie des *Conus mediterraneus* Brug. *Jena. Z. Naturw.,* **67**, 346–363.

AMAUDRUT, A. 1898. La partie antérieure du tube digestif et la torsion chez les mollusques gastéropodes. *Ann. Sci. nat. Zool.* (7), **8**, 1–291.

ANCEY, G. F. 1906. Observations sur les mollusques gastropodes sénestres de l'époque actuelle. *Bull. sci. Fr. Belg.,* **40**, 187–205.

ANDERSEN, K. 1924a. Entwicklungsgeschichtliche Untersuchungen an *Paludina vivipara.* i Teil. Die Formengestaltung der Sumpfschnecke (*Paludina vivipara*) während der Larvenzeit. *Morph. Jb.,* **53**, 211–258.

ANDERSEN, K. 1924b. Entwicklungsgeschichtliche Untersuchungen an *Paludina vivipara.* ii Teil. Die Entwicklung des Nervensystems bei *Paludina vivpara,* zugleich eine kritische Studie über die Torsion und Chiastoneurie der Gastropoden. *Morph. Jb.,* **54**, 157–204.

ANDREWS, E. A. 1935. The egg capsules of certain Neritidae. *J. Morph.,* **57**, 31–59.

ANKEL, W. E. 1924. Spermatozoen-Dimorphismus und Befruchtung bei *Bythinia tentaculata* L. und *Viviparus viviparus* L. *Senckenbergiana,* **6**, 1–12.

ANKEL, W. E. 1925. Zur Befruchtungsfrage bei *Viviparus viviparus* L. nebst Bemerkungen über die erste Reifungsteilung des Eies. *Senckenbergiana,* **7**, 37–54.

ANKEL, W. E. 1926. Spermiozeugmenbildung durch atypische (apyrene) und typische Spermien bei *Scala* und *Janthina. Verh. dtsch. zool. Ges., Zool. Anz.* Suppl. **2**, 193–202.

ANKEL, W. E. 1929. Über die Bildung der Eikapsel bei Nassa-Arten. *Verh. dtsch. zool. Ges., Zool. Anz.* Suppl. **4**, 219–230.

ANKEL, W. E. 1930a. Die atypische Spermatogenese von *Janthina* (Prosobranchia, Ptenoglossa). *Z. Zellforsch.,* **11**, 491–608.

ANKEL, W. E. 1930b. Nähreierbildung bei *Natica catena* (da Costa). *Zool Anz.,* **89**, 129–135.

ANKEL, W. E. 1930c. Über das Vorkommen und die Bedeutung zwittriger Geschlechtzellen bei Prosobranchiern. *Biol. Zbl.,* **50**, 513–531.

ANKEL, W. E. 1933. Untersuchungen über Keimzellbildung und Befruchtung bei *Bythinia tentaculata* L. 11. Gibt es in der Spermatogenese von *Bythinia tentaculata* eine Polymegalie? *Z. Zellforsch.,* **17**, 160–198.

ANKEL, W. E. 1935. Das Gelege von *Lamellaria perspicua. Z. Morph. Ökol. Tiere,* **30**, 635–647.

ANKEL, W. E. 1936a. Prosobranchia. In GRIMPE, G. & E. WAGLER: *Die Tierwelt der Nord-und Ostee,* **IX**b1. Leipzig, Akademische Verlagsgesellschaft.

ANKEL, W. E. 1936b. Die Fresspuren von *Helcion* und *Littorina* und die Funktion der Radula. *Verh. dtsch. zool. Ges., Zool. Anz.,* Suppl. **9,** 174–182.

ANKEL, W. E. 1937a. Wie bohrt *Natica? Biol. Zbl.,* **57,** 75–82.

ANKEL, W. E. 1937b. Der feinere Bau des Kokons der Purpurschnecke *Nucella lapillus* (L.) und seine Bedeutung für das Laichleben. *Verh. dtsch. Zool. Ges., zool. Anz.,* Suppl. **10,** 77–86.

ANKEL, W. E. 1937c. Wie frisst *Littorina?* 1. Radula-Bewegung und Fresspuren. *Senckenbergiana,* **19,** 317–333.

ANKEL, W. E. 1938a. Erwerb und Aufnahme der Nahrung bei den Gastropoden. *Verh. dtsch. Zool. Ges., Zool. Anz.* Suppl. **11,** 223–295.

ANKEL, W. E. 1938b. Beobachtungen an Prosobranchiern der schwedischen Westküste. *Ark. Zool.,* **30A** (9), 1–27.

ANKEL, W. E. 1947. Über einen 'Anklebe'-Reflex von *Gibbula. Arch. Molluskenk.,* **76,** 167–168.

ANKEL, W. E. 1948. Die Nahrungsaufnahme der Pyramidelliden. *Verh. dtsch. zool. Ges., Zool. Anz.* Suppl. **13,** 478–484.

ARNOLD, D. C. 1957. The response of the limpet, *Patella vulgata* L., to waters of different salinities. *J. mar. biol. Ass. U.K.,* **36,** 121–128.

AUBIN, P. A. 1892. The limpet's power of adhesion. *Nature, Lond.,* **45,** 464–465.

BACCI, G. 1947. L'inversione del sesso ed il ciclo stagionale dell gonade in *Patella coerulea* L. *Pubbl. Staz. zool. Napoli,* **21,** 183–217.

BACCI, G. 1951. L'ermafroditismo di *Calyptraea chinensis* L. e di alni Calyptraeidae. *Pubbl. Staz. zool. Napoli,* **23,** 66–90.

BÄCKER, R. 1903. Die Auge einiger Gastropoden. *Arb. zool. Inst. Univ. Wien,* **14,** 259–290.

BAKKER, K. 1959. Feeding habits and zonation in some intertidal snails. *Arch. néerl. Zool.,* **13,** 230–257.

BALL, E. G. & B. MEYERHOF. 1940. The occurrence of cytochrome and other haemochromogens in certain marine forms. *Biol. Bull., Wood's Hole,* **77,** 321.

BARKMAN, J. J. 1955. On the distribution and ecology of *Littorina obtusata* (L.) and its subspecific units. *Arch. néerl. Zool.,* **11,** 22–86.

BARNES, H. & T. B. BAGENAL. 1952. The habits and habitats of *Aporrhais pes-pelicani* (L.). *Proc. malac. Soc. Lond.,* **29,** 101–105.

BATAILLON, C. 1921. Spermies couplées et hétérochromosome dans la lignée typique d'une Turritelle. *C. R. Soc. Biol., Paris,* **84,** 219–222.

BECK, K. 1912. Anatomie der deutschen *Buliminus*-Arten. *Jena. Z. Naturw.,* **48,** 187–262.

BECKER, K. 1949. Untersuchungen über das Farbmuster und das Wachstum der Mollusken-schale. *Biol. Zbl.,* **68,** 263–288.

BEEDHAM, G. E. 1958a. Observations on the mantle of the Lamellibranchia. *Quart. J. micr. Sci.,* **99,** 181–197.

BEEDHAM, G. E. 1958b. Observations on the non-calcareous component of the shell of the Lamellibranchia. *Quart. J. micr. Sci.,* **99,** 341–357.

BELDING, D. L. 1930. The soft shelled clams of Massachusetts. *Commonwealth of Mass. mar. Fish. Ser.,* **1.**

BENNETT, M. F. 1954. The rhythmic activity of the quahog, *Venus mercenaria,* and its modification by light. *Biol. Bull., Wood's Hole,* **107,** 174–191.

BERG, K. 1938. Studies on the bottom animals of Esrom Lake. *K. danske vidensk. Selsk.,* Sect. Sci. (9), **8,** 1–255.

BERNARD, F. 1888. Recherches anatomiques sur la *Valvata piscinalis* C. R. Acad. Sci., Paris **107,** 191–194.

BERNARD, F. 1890. Recherches sur les organes palléaux des gastéropodes prosobranches. *Ann. Sci. nat. Zool.* (7), **9,** 89–404.

BERNER, L. 1942. La croissance de la coquille chez les gastéropodes. *Bull. Inst. océanogr. Monaco,* **816,** 1–16.

BERRY, A. J. 1956. 'Some factors affecting the distribution of *Littorina saxatilis* (Olivi).' Ph.D. Thesis. University of London.

BEVELANDER, G. & P. BENZER. 1948. Calcification in marine molluscs. *Biol. Bull., Wood's Hole,* **94,** 176–183.

BIEDERMANN, N. W. 1905. Studien zur vergleichenden Physiologie der peristaltischen Bewegungen. 11. Die lokomotorischen Wellen der Schneckensohle. *Pflüg. Arch. ges. Physiol.,* **107,** 1–56.

BLOCH, I. 1896. Die embryonale Entwicklung der Radula von *Paludina vivipara.* Jena. *Z. Naturw.,* **30,** 350–392.

BLOCHMANN, F. 1882. Über die Entwicklung der *Neritina fluviatilis* Mül. *Z. wiss. Zool.,* **36,** 125–174.

BOBRETZKY, N. 1877. Studien über die embryonale Entwicklung der Gastropoden. *Arch. mikr. Anat.,* **13,** 75–169.

BOETTGER, C. R. 1930. Die Lage der Bohrstelle beim Angriff der Raubschnecken aus der Familie Naticidae. *Z. wiss. Zool.,* **136,** 453–463.

BOETTGER, C. R. 1949. Hinweise zur Frage der Kielbildung auf der Schale der Wasserschnecke *Potamopyrgus crystallinus jenkinsi* (E. A. Smith). *Arch. Molluskenk.,* **22,** 63–72.

BOETTGER, C. R. 1951. Die Herkunst und Verwandtschaftsbeziehungen der Wasserschnecke *Potamopyrgus jenkinsi* (Smith), nebst einer Angabe über ihr Aufreten im Mediterrangebiet. *Arch. Molluskenk.,* **80,** 57–84.

BOETTGER, C. R. 1954. Die Systematik der euthyneuren Schnecken. *Verh. dtsch. zool. Ges., Zool. Anz.* Suppl. **18,** 253–280.

BØGGILD, O. B. 1930. The shell structure of the mollusks. *K. danske vidensk. Selsk.* (9), **2,** 235–325.

BOHN, G. 1904. Périodicité vitale des animaux soumis aux oscillations du niveau des hautes mers. *C. R. Acad. Sci., Paris,* **139,** 610–611.

BONDESEN, P. 1940. Preliminary investigations into the development of *Neritina fluviatilis* L. in brackish and fresh water. *Vidensk. Medd. dansk naturh. Foren. Kbh.,* **104,** 283–318.

BONDESEN, P. & W. KAISER. 1949. *Hydrobia (Potamopyrgus) jenkinsi* Smith in Denmark illustrated by its ecology. *Oikos,* **1,** 252–281.

BONSE, H. 1935. Ein Beitrag zum Problem der Schneckenbewegung. *Zool. Jb. (Zool. Physiol.),* **54,** 349–384.

BOSS, K. J. 1971. Critical estimate of the number of Recent Mollusca. *Occ. Pap. mollusks Harv.,* **3,** 81–135.

BOUCHILLOUX, S. & J. ROCHE. 1955. Contribution à l'étude biochimique da la pourpre des *Murex. Bull. Inst. océanogr. Monaco,* **1054,** 1–23.

BOURNE, A. 1894. On certain points in the development and anatomy of some earthworms. *Quart. J. micr. Sci.,* **36,** 11–33.

BOURNE, G. C. 1908. Contribution to the morphology of the group Neritacea of aspidobranch gastropods. Part I. The Neritidae. *Proc. zool. Soc. Lond.,* 810–887.

BOURNE, G. C. 1910. On the anatomy and systematic position of *Incisura (Scissurella) lytteltonensis. Quart. J. micr. Sci.,* **55,** 1–47.

BOUTAN, L. 1886. Recherches sur l'anatomie et le développement de la Fissurelle. *Arch. Zool. exp. gén.* (2), **3,** 1–173.

BOUTAN, L. 1892. Sur le développement de l'Haliotide, et sur l'utilité du Scaphandre dans les recherches zoologiques. *C. R. Ass. franc. Ad. Sci.,* **2,** 522–525.

BOUTAN, L. 1899. La cause principale de l'asymétrie des mollusques gastéropodes. *Arch. Zool. exp. gén.* (3), **7,** 203–342.

BOUTAN, L. 1900. Gastéropodes. La Patelle commune. In *Zoologie Descriptive,* **2,** edited by L. Boutan. Paris, O. Doin.

BOUTAN, L. 1919. Considérations nouvelles sur les affinités réciproques des mollusques gastéropodes. *Act. Soc. linn. Bordeaux,* **71,** 5–116.

BOUVIER, E. L. 1886. Contributions à l'étude des protobranches pténoglosses. *Bull. Soc. malac. Fr.,* **3,** 77–130.

BOUVIER, E. L. 1887. Système nerveux, morphologie générale et classification des gastéropodes proso-branches. *Ann. Sci. nat. Zool.* (7), **3,** 1–510.

BOUVIER, E. L. & H. FISCHER. 1902. L'organisation et les affinités des gastéropods primitifs d'après l'étude anatomique du *Pleurotomaria beyrichi. J. Conchyliol.,* **50,** 117–272.

BOYCOTT, A. E. 1917. Where is the male of *Paludestrina jenkinsi? J. Conch., Lond.,* **15,** 216.

BOYCOTT, A. E. 1919. Parthenogenesis in *Paludestrina jenkinsi. J. Conch., Lond.,* **16,** 54.

BOYCOTT, A. E. 1929. The inheritance of ornamention in var. *aculeata* of *Hydrobia jenkinsi* Smith. *Proc. malac. Soc. Lond.,* **18,** 230–234.

BOYCOTT, A. E. 1934. The habitats of land Mollusca in Britain. *J. Ecol.,* **22,** 1–38.

BOYCOTT, A. E. 1936. The habitats of the fresh-water Mollusca in Britain. *J. Anim. Ecol.,* **5,** 116–186.

BOYCOTT, A. E., C. DIVER, S. L. GARSTANG & F. M. TURNER. 1930. The inheritance of sinistrality in *Limnaea peregra* (Mollusca, Pulmonta). *Phil. Trans. R. Soc. B,* **219,** 51–131.

BOZLER, E. 1930 . Untersuchungen zur Physiologie der Tonusmuskeln. *Z. vergl. Physiol.,* **12,** 579–602.

BREGENZER, A. 1916. Anatomie und Histologie von *Bythinella dunkeri. Zool. Jb. (Anat. Ont.),* **39,** 237–292.

BROUARDEL, J. 1948. Etude du mode d'infestation des Patelles par l'*Urceolaria patellae* (Cuénot). Influence de l'espèce de Patelle. *Bull. Lab. marit. Dinard,* **30,** 1–6.

BROCK, F. 1933. Analyse des Beute- und Verdauungsfeldes der Wellhornschnecke *Buccinum undatum* L. *Verh. dtsch. zool. Ges., Zool. Anz.* Suppl. **6,** 243–250.

BROCK, F. 1936. Suche, Aufnahme und enzymatische Spaltung der Nahrung durch die Well-hornschnecke *Buccinum undatum* L. *Zoologica, Stuttgart,* **34** (92), 1–136.

BROWN, C. H. 1952. Some structural proteins of *Mytilus edulis. Quart. J. micr. Sci.,* **43,** 487–503.

BROWN, F. A., Jr., M. FINGERMAN, M. I. SANDEEN & H. M. WEBB. 1953. Persistent diurnal and tidal rhythms of colour change in the fiddler crab *Uca pugnax. J. exp. Zool.,* **123,** 29–60.

BROWN, F. A., Jr. 1954. Persistent activity rhythms in the oyster. *Amer. J. Physiol.,* **178,** 510–514.

BRÜEL, L. 1915. Über das Nervensystem der Heteropoden. I. *Pterotrachea. Zool. Anz.,* **45,** 530–548.

BRÜEL, L. 1924a. Über das Nervensystem der Heteropoden. II. Das Nervensystem von *Carinaria* und seine Herleitung von den Prosobranchiern. *Zool. Anz.,* **59,** 113–127.

BRÜEL, L. 1924b. Über das Nervensystem der Heteropoden. II. Das Nervensystem von *Carinaria* und seine Herleitung von den Prosobranchiern (Schluss). *Zool. Anz.,* **59,** 190–199.

BUDDENBROCK, W. von. 1915. Die Statocyste von *Pecten,* ihre Histologie und Physiologie. *Zool. Jb. (Zool. Physiol.),* **35,** 301–356.

BURGER, J. W. & C. S. THORNTON. 1935. A correlation between the food eggs of *Fasciolaria tulipa* and the apyrene spermatozoa of prosobranch molluscs. *Biol. Bull., Wood's Hole,* **68,** 253–257.

BUTLER, P. A. 1954. The southern oyster drill. *Proc. nat. Shellfish Ass.,* **1953,** 67–75.

CARLSON, A. J. 1905a. The physiology of locomotion in gasteropods. *Biol. Bull., Wood's Hole,* **8,** 85–92.

CARLSON, A. J. 1905b. Comparative physiology of the invertebrate heart. The innervation of the invertebrate heart. *Biol. Bull., Wood's Hole,* **8,** 123–167.

CARRICK, R. 1938. The life-history and development of *Agriolimax agrestis* L., the grey field slug. *Trans. R. Soc. Edinb.,* **59,** 563–597.

CARRIERE, J. 1889. Über Molluskenaugen. *Arch. mikr. Anat.,* **33,** 378–402.

CARRIKER, M. R. 1943. On the structure and function of the proboscis in the common oyster drill *Urosalpinx cinerea* Say. *J. Morph.,* **73,** 441–506.

CARRIKER, M. R. 1946. Observations on the functioning of the alimentary system of the snail *Lymnaea stagnalis appressa* Say. *Biol. Bull., Wood's Hole,* **91,** 88–111.

CARRIKER, M. R. 1951. Observations on the penetration of tightly closing bivalves by *Busycon* and other predators. *Ecology,* **32,** 73–83.

CARRIKER, M. R. 1955. Critical review of biology and control of oyster drills *Urosalpinx* and *Eupleura*. *Spec. sci. Rep. U.S. Dept. Inst. Fish.,* **148,** 1–150.

CARRIKER, M. R. 1959. Comparative functional morphology of the drilling mechanism in *Urosalpinx* and *Eupleura* (muricid gastropods). *Proc. XVth int. Cong. Zool. Lond.,* 373–376.

CASTEEL, D. B. 1904. The cell-lineage and early larval development of *Fiona marina,* a nudibranch mollusk. *Proc. nat. Acad. Sci. Philad.,* **56,** 325–405.

CATE, J. TEN. 1922. Quelques observations sur la locomotion de l'escargot des vignes. *Arch. néerl. Physiol.,* **7,** 103–111.

CATE, J. TEN. 1923. Quelques recherches sur la locomotion des limaces. *Arch. néerl. Physiol.,* **8,** 377–393.

CAULLERY, M. & P. PELSENEER. 1910. Sur la ponte et le développement du vignot (*Littorina littorea*). *Bull. sci. Fr. Belg.,* **44,** 357–360.

CHARIN, N. 1926. Über die Nahrung des Embryo von *Paludina vivipara. Bull. Soc. Nat. Voronèje,* **I,** 60–66.

CHASTER, G. W. 1896. *Adeorbis unisulcatus,* new species, from the Irish coast. *J. Conch., Lond.,* **8,** 373.

CHIPPERFIELD, P. N. J. 1952. The breeding of *Crepidula fornicata* (L.) in the river Blackwater, Essex. *J. mar. biol. Ass. U.K.,* **30,** 49–71.

CHUMLEY, J. 1918. 'The Fauna of the Clyde Sea Area.' Glasgow, Glasgow University Press.

CLAPAREDE, E. 1858. Beitrag zur Anatomie des *Cyclostoma elegans. Arch. Anat. Physiol., Lpz.,* **2,** 1–34.

CLARK, W. C. 1958. Notes on the mantle cavities of some trochid and turbinid Gastropoda. *Proc. malac. Soc. Lond.,* **33,** 57–64.

CLARKE, A. H. & R. J. MENZIES. 1959. *Neopilina* (*Vema*) *ewingi,* a second living species of the paleozoic class Monoplacophora. *Science,* **129,** 1026–1027.

CLELAND, D. M. 1954. A study of the habits of *Valvata piscinalis* (Müller), and the structure and function of the alimentary canal and reproductive system. *Proc. malac. Soc. Lond.,* **30,** 167–203.

CLENCH, W. J. 1947. The genera *Purpura* and *Thais* in the Western Atlantic. *Johnsonia,* **2,** 61–91.

CLENCH, W. J. & R. D. TURNER. 1948. The genus *Truncatella* in the Western Atlantic. *Johnsonia,* **2,** 149–164.

CLENCH, W. J. & R. D. TURNER. 1950. The genera *Sthenorytis, Cirsotrema, Acirsa, Opalia* and *Amaea* in the Western Atlantic. *Johnsonia,* **2,** 221–246.

CLIMO, F. M. 1975. The anatomy of *Gegania valkyrie* Powell (Mollusca: Heterogastropoda: Mathildidae) with notes on other heterogastropods. *Trans. R. Soc. N.Z.,* **5,** 275–288.

COATES, H. 1922. Exhibit. *J. Conch., Lond.,* **16,** 319.

COE, W. R. 1938a. Conditions influencing change of sex in mollusks of the genus *Crepidula. J. exp. Zool.,* **77,** 401–424.

COE, W. R. 1938b. Influence of the association on the gastropods having protandric consecutive sexuality. *Biol. Bull., Wood's Hole,* **75,** 274–285.

COE, W. R. 1942. The reproductive organs of the prosobranch mollusk *Crepidula onyx* and their transformation during the change from male to female phase. *J. Morph.,* **70,** 501–512.

COE, W. R. 1944. Sexual differentiation in mollusks. II. Gastropods, amphineurans, scaphopods and cephalopods. *Quart. Rev. Biol.,* **19,** 85–97.

COE, W. R. 1948. Nutrition and sexuality in protandric gastropods of the genus *Crepidula. Biol. Bull., Wood's Hole,* **94,** 158–160.

COE, W. R. 1953. Influences of association, isolation and nutrition on the sexuality of snails of the genus *Crepidula. J. exp. Zool.,* **122,** 5–19.

COLE, H. A. 1942. The American whelk tingle, *Urosalpinx cinerea* (Say), on British oyster beds, *J. mar. biol. Ass. U.K.,* **25,** 477–501.

COLE, H. A. 1952. The American slipper limpet on Cornish oyster beds. *Fish Invest., Lond.,* ser. 2, **17, 7,** 1–13.

COLE, H. A. 1956. Benthos and shellfish of commerce. In 'Sea Fisheries, their Investigations in the United Kingdom.' (M. Graham ed.). London, Macmillan.

COLE, H. A. & R. H. BAIRD. 1953. The American slipper limpet (*Crepidula fornicata*) in Milford Haven. *Nature, Lond.,* **172,** 687.

COLE, H. A. & D. A. HANCOCK. 1955. *Odostomia* as a pest of oysters and mussels. *J. mar. biol. Ass. U.K.,* **34,** 25–31.

COLE, H. A. & D. A. HANCOCK. 1956. Progress in oyster research in Britain 1949–54, with special reference to the control of pests and diseases. *Rapp. Cons. Explor. Mer.,* **140,** 24–29.

COLGAN, N. 1910. Notes on the adaptability of certain littoral Mollusca. *Irish Nat.,* **19,** 127–133.

COLMAN, J. 1933. The nature of the intertidal zonation of plants and animals. *J. mar. biol. Ass. U.K.,* **18,** 435–476.

COLMAN, J. 1940. On the faunas inhabiting intertidal seaweeds. *J. mar. biol. Ass. U.K.,* **24,** 129–183.

COLTON, H. S. 1908. How *Fulgar* and *Sycotypus* eat oysters, mussels and clams. *Proc. nat. Acad. Sci., Philad.,* **60,** 3–10.

COLTON, H. S. 1916. On some varieties of *Thais lapillus* in the Mount Desert region, a study of individual ecology. *Proc. nat. Acad. Sci., Philad.,* **68,** 440–451.

COLTON, H. S. 1922. Variation in the dog whelk *Thais* (*Purpura auct.*) *lapillus*. *Ecology,* **3,** 146–157.

COMFORT, A. 1951. The pigmentation of molluscan shells. *Biol. Rev.,* **26,** 285–301.

CONCHOLOGICAL SOCIETY OF GREAT BRITAIN AND IRELAND. 1951. Census of the distribution of British non-marine Mollusca. *J. Conch., Lond.,* **23,** 171–244.

CONKLIN, E. G. 1897. The embryology of *Crepidula*. *J. Morph.,* **13,** 1–226.

COOK, P. M. 1949. A ciliary feeding mechanism in *Viviparus viviparus* (L.). *Proc. malac. Soc. Lond.,* **27,** 265–271.

COOKE, A. H. 1895. Molluscs. In 'The Cambridge Natural History', **3,** (S. F. Harmer & A. E. Shipley eds). London, Macmillan.

COOKE, A. H. 1917. *Patella vulgata,* Linnæus, and its so-called variety, *Patella depressa,* Pennant. *Proc. malac. Soc. Lond.,* **12,** 135–137.

COOKE, A. H. 1920. Evolution in the molluscan radula. *J. Conch., Lond.,* **16,** 145–150.

COPELAND, M. 1919. Locomotion in two species of the gastropod genus *Alectrion* with observations on the behaviour of pedal cilia. *Biol. Bull., Wood's Hole,* **37,** 126–138.

COPELAND, M. 1922. Ciliary and muscular locomotion in the gastropod genus *Polinices*. *Biol. Bull., Wood's Hole,* **42,** 132–142.

CORNET, R. & I. MARCHE-MARCHAD. 1951. Inventaire de la faune marine de Roscoff. Mollusques. *Trav. Sta. biol. Roscoff,* suppl. **5,** 1–81.

COX, L. R. 1955. Observations on gastropod descriptive terminology. *Proc. malac. Soc. Lond.,* **31,** 190–202.

COX, L. R. 1960a. Thoughts on the classification of the Gastropoda. *Proc. malac. Soc. Lond.,* **33,** 239–261.

COX, L. R. 1960b. Gastropoda. General characteristics of Gastropoda. In 'Treatise on Invertebrate Paleontology' (R. C. Moore ed.). I, Mollusca **1,** 84–169. Lawrence, Kansas, Geological Society of America and University of Kansas Press.

CRAMPTON, H. E. 1894. Reversal of cleavage in a sinistral gasteropod. *Ann. N.Y. Acad. Sci.,* **8,** 167–170.

CREEK, G. A. 1951. The reproductive system and embryology of the snail *Pomatias elegans* (Müller). *Proc. zool. Soc. Lond.,* **121,** 599–640.

CREEK, G. A. 1953. The morphology of *Acme fusca* (Montagu) with special reference to the genital sytsem. *Proc. malac. Soc. Lond.,* **29,** 228–240.

CROFTS, D. R. 1929. *Haliotis. L.M.B.C. Memoir,* **29.** Liverpool, University Press of Liverpool.

CROFTS, D. R. 1937. The development of *Haliotis tuberculata,* with special reference to the organo-genesis during torsion. *Phil. Trans. R. Soc. B,* **208,** 219–268.

CROFTS, D. R. 1955. Muscle morphogenesis in primitive gastropods and its relation to torsion. *Proc. zool. Soc. Lond.,* **125,** 711–750.

CROSSE, H. & P. FISCHER. 1882. Note complémentaire sur la résorption des parois internes du test chez les *Olivella*. *J. Conchyliol.,* **30,** 177.

CUVIER, G. L. C. F. D. 1798. 'Tableau Elémentaire de l'Histoire Naturelle des Animaux'. Paris, Baudonin.

DAKIN, W. J. 1912. *Buccinum. L.M.B.C. Memoir,* **20.** London, Williams & Norgate.

DALL, W. H. 1889. Reports on the results of dredging, under the supervision of Alexander Agassiz, in the Gulf of Mexico (1877–78) and in the Caribbean Sea (1879–80), by the U.S. Coast Survey Steamer 'Blake', Lieut.-Commander C. D. Sigsbee, U.S.N., and Commander J. R. Bartlett, U. S. N., commanding. XXIX. Report on the Mollusca. Part II. Gastropoda and Scaphopoda. *Bull. Mus. comp. Zool. Harvard,* **18,** 1–492.

DALL, W. H. 1924. The value of the nuclear characters in the classification of marine gastropods. *J. Wash. Acad. Sci.,* **14,** 177–180.

DANIEL, A. T. 1893. *Hydrobia jenkinsi* Smith in an inland locality. *J. Conch., Lond.,* **7,** 325.

DAS, S. M. & G. SESHAPPA. 1947. A contribution to the biology of *Patella*: on population distribution and sex-proportions in *Patella vulgata* Linnaeus at Cullercoats, England. *Proc. zool. Soc. Lond.,* **117,** 653–662.

DAUTERT, E. 1929. Die Bildung der Keimblätter bei *Paludina. Zool. Jb. (Anat. Ont.),* **50,** 433–496.

DAUTZENBERG, P. & H. FISCHER. 1914. Etude sur le *Littorina obtusata* et ses variations. *J. Conchyliol.,* **62,** 87–128.

DAVIES, A. M. 1939. Some palaeontological problems. *Proc. malac. Soc. Lond.,* **23,** 336–344.

DAVIS, J. R. A. 1895. The habits of limpets. *Nature, Lond.,* **51,** 511–512.

DAVIS, J. R. A. & H. J. FLEURE. 1903. *Patella. LMBC. Memoir,* **10.** Liverpool, University Press of Liverpool.

DEAN, J. D. 1904. Fish and their relation to *Paludestrina jenkinsi. J. Conch., Lond.,* **II,** 15.

DEAN, J. D. 1936. Conchological cabinets of the last century. *J. Conch., Lond.,* **20,** 225–252.

DELAGE, Y. 1887. Sur une fonction nouvelle des otocystes comme organes d'orientation locomotrice. *Arch. Zool. exp. gén.* (2), **5,** 1–26.

DELSMAN, H. C. 1914. Entwicklungsgeschichte von *Littorina obtusata. Tijdschr. ned. dierk. Ver.,* **13,** 170–340.

DESHPANDE R. D. 1957. 'Observations on the anatomy and ecology of British trochids'. Ph.D. Thesis, University of Reading.

DIEHL, M. 1956. Die Raubschnecke *Velutina velutina* das Feind and Bruteinmieter der Ascidie *Styela coriacea. Kieler Meeresforsch.,* **12,** 180–185.

DIMELOW, E. J. 1959. 'Some aspects of the biology of *Antedon bifida* (Pennant).' Ph.D. Thesis, University of Reading.

DODD, J. M. 1956. Studies on the biology of limpets. II. Breeding of *Patella vulgata* L. in Britain. III. Hermaphroditism in the three British species of *Patella. J. mar. biol. Ass. U.K.,* **35,** 149–176 and 327–340.

DODD, J. M. 1957. Artificial fertilisation, larval development and metamorphosis in *Patella vulgata* L. and *Patella coerulea* L. *Pubbl. Staz. zool. Napoli,* **29,** 172–186.

DODGSON, K. S. & B. SPENCER. 1954. Studies on sulphatases. 7. A preliminary account of the glycosulphatase of *Littorina littorea. Biochem. J.,* **57,** 310–315.

DONGEN, A. VAN. 1956. The preference of *Littorina obtusata* for Fucaceae. *Arch. néerl. Zool.,* **11,** 373–386.

DONS, C. 1913. Zoologisker notiser II. Om egglaegningen hos enkelte Buccinider. *Tromsø Mus. Aarsh.,* **35 & 36,** 11–22.

DRUMMOND, I. M. 1903. Notes on the development of *Paludina vivipara*, with special reference to the urinogenital organs and theories of gasteropod torsion. *Quart. J. micr. Sci.,* **46,** 97–143.

DUBOIS, R. 1902a. Sur le mécanisme intime de la formation de la pourpre chez *Murex brandaris. C.R. Soc. Biol., Paris,* **54,** 82–83.

DUBOIS, R. 1902b. Sur le mécanisme intime de la formation de la pourpre. *C.R. Acad. Sci., Paris,* **134,** 245–246.

DUBOIS, R. 1903a. Sur la formation de la pourpre de *Purpura lapillus*. *C.R. Acad. Sci., Paris,* **136,** 117–118.

DUBOIS, R. 1903b. Sur le vénin de la glande à pourpre de *Purpura lapillus*. *C.R. Soc. Biol., Paris,* **55,** 81.

DUBOIS, R. 1909. Recherches sur la pourpre et sur quelques autres pigments animaux. *Arch. Zool. exp. gén.,* **42,** 471–590.

DUGES, A. 1829. Observations sur la structure et la formation de l'opercule chez les mollusques. *Ann. Sci. nat. Zool.* (I), **18,** 113–133.

EBLING, F. J. 1945. Formation and nature of the opercular chaetae of *Sabellaria alveolata*. *Quart. J. micr. Sci.,* **85,** 153–176.

EBLING, F. J., J. A. KITCHING, R. D. PURCHON & R. BASSINDALE, 1948. The ecology of the Lough Ine rapids with special reference to water currents. 2. The fauna of the *Saccorhiza* canopy. *J. Anim. Ecol.,* **17,** 223–244.

EIGENBRODT, H. 1941. Untersuchungen über die Funktion der Radula einiger Schnecken. *Z. Morph. Ökol. Tiere,* **37,** 735–791.

ELLIOT, G. F. S., M. LAURIE & J. B. MURDOCH. 1901. 'Fauna, Flora and Geology of the Clyde-Area.' Glasgow, Local Committee for the meeting of the British Association.

ELLIS, A. E. 1925. Experimental acclimatization in *Sabanaea ulvae* (Pennant) to freshwater. *Ann. Mag. nat. Hist.* (9), **15,** 496–497.

ELLIS, A. E. 1926. 'British Snails. A Guide to the Non-Marine Gastropoda.' Oxford, Clarendon Press.

ELLIS, A. E. 1932. The habitats of Hydrobiidae in the Adur estuary. *Proc. malac. Soc. Lond.,* **20,** 11–18.

ERLANGER, R. VON. 1891a. Zur Entwickelung von *Paludina vivipara*. *Morph. Jb.,* **17,** 337–379 and 636–680.

ERLANGER, R. VON. 1891b. Zur Entwickelung von *Paludina vivipara*. Vorläufige Mittheilung. *Zool. Anz.,* **14,** 280–283.

ERLANGER, R. VON. 1892. On the paired nephridia of prosobranchs, the homologies of the only remaining nephridium of most prosobranchs, and the relation of the nephridia to the gonad and genital duct. *Quart. J. micr. Sci.,* **33,** 587–623.

ERLANGER, R. VON. 1894. Zum Bildung des Mesoderms bei der *Paludina vivipara*. *Morph. Jb.,* **22,** 113–118.

ERSPAMER, V. 1947. Ricerche chimiche e farmacologiche sugli estratti di ghiandola ipobranchiale di *Murex* (*Truncularia*) *trunculus* (L.), *Murex* (*Bolimes*) *brandaris* (L.) e *Tritonalia erinacea* (L.). *Pubbl. Staz. zool. Napoli,* **20,** 91–101.

ERSPAMER, V. 1952. Wirksame Stoff der hinteren Speicheldrüsen der Octopoden und der Hypo-branchialdrüse der Purpurschnecken. *Arzneimittelforsch.,* **2,** 253–258.

ESLICK, A. 1940. An ecological study of *Patella* at Port St Mary, Isle of Man. *Proc. Linn. Soc. Lond.,* Sess. **152,** 45–58.

EVANS, R. G. 1947. The intertidal ecology of selected localities in the Plymouth neighbourhood. *J. mar. biol. Ass. U.K.,* **27,** 173–218.

EVANS, R. G. 1948. The lethal temperatures of some British littoral molluscs. *J. Anim. Ecol.,* **17,** 165–173.

EVANS, R. G. 1953. Studies on the biology of British limpets—the genus *Patella* on the south coast of England. *Proc. zool. Soc. Lond.,* **123,** 357–376.

FÄNGE, R. 1957. An acetylcholine-like salivary poison in the marine gastropod *Neptunea antiqua*. *Nature, Lond.,* **180,** 196–197.

FÄNGE, R. 1958. Paper chromatography and biological extracts of the salivary gland of *Neptunea antiqua* (Gastropoda). *Acta zool. Stockh.,* **39,** 39–46.

FÄNGE, R. & A. MATTISSON. 1958. Studies on the physiology of the radula-muscle of *Buccinum undatum*. *Acta zool. Stockh.,* **39,** 53–64.

FERNANDO, W. 1931. The origin of the mesoderm in the gastropod *Viviparus* (=*Paludina*). *Proc. R. Soc.* B, **107,** 381–390.

FISCHER, E. 1928. Recherches de bionomie et d'océanographie littorales sur La Rance et le littoral de la Manche. *Ann. Inst. océanogr. Monaco,* **5,** 201–429.

FISCHER, E. 1931. Sur la pénétration des diverses espèces marines sessiles dans les estuaires et sa limitation par l'eau douce. *Ann. Inst. océanogr. Monaco,* **10,** 213–243.

FISCHER, H. 1892. Recherches sur la morphologie du foie des gastéropodes. *Bull. sci. Fr. Belg.,* **24,** 260–346.

FISCHER, P. 1865. Note sur les moeurs du *Murex erinaceus. J. Conchyliol.,* **13,** 5–8.

FISCHER, P. 1887. 'Manuel de Conchyliologie.' Paris, Savy.

FISCHER, P.-H. 1922. Sur les gastropodes perceurs. *J. Conchyliol.,* **67,** 3–56.

FISCHER, P.-H. 1925. Sur la rôle de la glande purpurigène des *Murex* et des Pourpres. *C. R. Acad. Sci., Paris,* **180,** 1369–1371.

FISCHER, P.-H. 1940a. Structure et évolution de l'épithelium de l'opercule chez *Purpura lapillus* L. *Bull. Soc. zool. Fr.,* **65,** 199–204.

FISCHER, P.-H. 1940b. Observations sur la ponte de quelques Muricidés. *Bull Soc. zool. Fr.,* **65,** 205–211.

FISCHER, P.-H. 1948. Données sur la résistance et le vitalité des mollusques. *J. Conchyliol.,* **88,** 100–140.

FISCHER, P.-H., M. DUVAL & A. RAFFY. 1933. Etudes sur les échanges respiratoires des Littorines. *Arch. Zool. exp. gén.,* **74,** 627–634.

FISCHER-PIETTE, E. 1935a. Histoire d'une moulière. Observations sur une phase de déséquilibre faunique. *Bull. biol.,* **69,** 154–180.

FISCHER-PIETTE, E. 1935b. Systématique et biogéographie—les Patelles d'Europe et d'Afrique du Nord. *J. Conchyliol.,* **79,** 5–66.

FISCHER-PIETTE, E. 1941a. Croissance, taille maximum, et longévité possible de quelques animaux intercotideaux en fonction du milieu. *Ann. Inst. océanogr. Monaco,* **21,** 1–28.

FISCHER-PIETTE, E. 1941b. Observations biométriques sur Patelles de la Manche. *J. Conchyliol.,* **84,** 300–306.

FISCHER-PIETTE, E. 1946. Review of Biology of *Patella* in Great Britain, par J. Orton, *Nature, Lond.,* **158,** 173, 3 Août 1946. *J. Conchyliol.,* **87,** 83–84.

FISCHER-PIETTE, E. 1948. Sur les éléments de prospérité des Patelles et sur leur spécificité. *J. Conchyliol.,* **88,** 45–96.

FLATTELY, F. W. & C. L. WALTON. 1922. 'The Biology of the Sea-shore.' London, Sidgwick & Jackson.

FLEISCHMANN, A. 1932. Vergleichende Betrachtungen über das Schalenwachstum der Weichtiere (Mollusca). II. Deckel (Operculum) und Haus (Concha) der Schnecken (Gastropoden). *Z. Morph Ökol. Tiere,* **25,** 549–622.

FLEURE, H. J. 1903. Notes on the relations of the kidney in *Haliotis tuberculata* etc. *Quart. J. micr. Sci.,* **46,** 77–97.

FOL, H. 1880. Sur le développement de gastropodes pulmonés. *Arch. Zool. exp. gén.* (1), **8,** 103–232.

FORBES, E. & S. HANLEY. 1849–53. 'A History of British Mollusca, and their Shells.' 4 vols. Vol. **2** (1849), **3** (1850), **4** (1852, 1853). London, van Voorst.

FRAENKEL, G. 1927a. Beiträge zur Geotaxis und Phototaxis von *Littorina. Z. vergl. Physiol.,* **5,** 585–597.

FRAENKEL, G. 1927b. Biologische Beobachtungen an *Ianthina. Z. Morph. Ökol. Tiere,* **7,** 597–608.

FRAISSE, P. 1881. Über Molluskenaugen mit embryonalem Typus. *Z. wiss. Zool.,* **35,** 461–477.

FRANC, A. 1940. Recherches sur le développement d'*Ocinebra aciculata,* Lamarck (mollusque gastéropode). *Bull. biol.,* **74,** 327–345.

FRANC, A. 1941a. Sur les reins larvaires de certains mollusques prosobranches. *C. R. Soc. Biol., Paris,* **135,** 1487–1489.

FRANC, A. 1941b. Sur la formation des oothèques des mollusques prosobranches. *C. R. Soc. Biol., Paris,* **135,** 1609–1611.

FRANC A. 1943. 'Etudes sur le développement de quelques prosobranches méditerranéens.' Thèse, Université d'Alger.

FRANC, A. 1949. Notes histologiques sur la métamorphose de *Firoloida desmaresti* Lesueur (mollusque hétéropode). *Bull. Soc. zool. Fr.,* **74,** 141–146.

FRANC, A. 1950. Ponte et larves planctoniques de *Philbertia purpurea* (Montagu). *Bull. Lab. marit. Dinard,* **33,** 23–25.

FRANC, A. 1952. Notes éthologiques et anatomiques sur *Tritonalia (Ocinebrina) aciculata* (Lk.). (Mollusque prosobranche). *Bull. Lab. marit. Dinard,* **36,** 31–34.

FRANK, E. J. 1914. Beiträge zur Anatomie der Trochiden. *Jena. Z. Naturw.,* **51,** 377–486.

FRANZ, V. 1938. Die europäische Flussdeckelschnecke (*Viviparus fasciatus*) im starkem Strom des Njemen. *Arch. Molluskenk.,* **70,** 9–30.

FRANZÉN, A. 1955. Comparative morphological investigations into the spermiogenesis among Mollusca. *Zool. Bidr. Uppsala,* **30,** 399–456.

FRANZÉN, A. 1956. On spermiogenesis, morphology of the spermatozoon, and biology of fertilization among invertebrates. *Zool. Bidr. Uppsala,* **31,** 356–482.

FREDERICQ, H. 1939. Action des nerfs du coeur d'*Aplysia limacina.* Analyse au moyen de la caféine. *Arch. int. Physiol.,* **49,** 299–304.

FRETTER, V. 1941. The genital ducts of some British stenoglossan prosobranchs. *J. mar. biol. Ass. U.K.,* **25,** 173–211.

FRETTER, V. 1946a. The pedal sucker and anal gland of some British Stenoglossa. *Proc. malac. Soc. Lond.,* **27,** 126–130.

FRETTER, V. 1946b. The genital ducts of *Theodoxus, Lamellaria* and *Trivia,* and a discussion on their evolution in the prosobranchs. *J. mar. biol. Ass. U.K.,* **26,** 312–351.

FRETTER, V. 1948. The structure and life history of some minute prosobranchs of rock pools: *Skeneopsis planorbis* (Fabricius), *Omalogyra atomus* (Philippi), *Rissoella diaphana* (Alder) and *Rissoella opalina* (Jeffreys). *J. mar. biol. Ass. U.K.,* **27,** 597–632.

FRETTER, V. 1951a. Some observations on the British cypraeids. *Proc. malac. Soc. Lond.,* **29,** 14–20.

FRETTER, V. 1951b. Observations on the life history and functional morphology of *Cerithiopsis tubercularis* (Montagu) and *Triphora perversa* (L.). *J. mar. biol. Ass. U.K.,* **29,** 567–586.

FRETTER, V. 1951c. *Trubonilla elegantissima* (Montagu), a parasitic opisthobranch. *J. mar. biol. Ass. U.K.,* **30,** 37–47.

FRETTER, V. 1952. Experiments with P^{32} and I^{131} on species of *Helix, Arion* and *Agriolimax. Quart. J. micr. Sci.,* **93,** 133–146.

FRETTER, V. 1953. The transference of sperm from male to female prosobranch, with reference, also, to the pyramidellids. *Proc. Linn. Soc. Lond.,* Sess. **164,** 1951–52, 217–224.

FRETTER, V. 1955a. Observations on *Balcis devians* (Monterosato) and *Balcis alba* (Da Costa). *Proc. malac. Soc. Lond.,* **31,** 137–144.

FRETTER, V. 1955b. Some observations on *Tricolia pullus* (L.) and *Margarites helicinus* (Fabricius). *Proc. malac. Soc. Lond.,* **31,** 159–162.

FRETTER, V. 1956. The anatomy of the prosobranch *Circulus striatus* (Philippi) and a review of its systematic position. *Proc. zool. Soc. Lond.,* **126,** 369–381.

FRETTER, V. & A. GRAHAM. 1949. The structure and mode of life of the Pyramidellidae, parasitic opisthobranchs. *J. mar. biol. Ass. U.K.,* **28,** 493–532.

FRETTER, V. & A. M. PATIL. 1958. A revision of the systematic position of the prosobranch gastropod *Cingulopsis* (= *Cingula*) *fulgida* (J. Adams). *Proc. malac. Soc. Lond.,* **33,** 114–126.

FRIZA, F. 1932. Zur Kenntnis des Conchiolins der Muschelschalen. *Biochem. Z.,* **246,** 29–37.

FRÖMMING, E. 1956. 'Biologie der mitteleuropäischen Süsswasserschnecken.' Berlin, Duncker & Humblot.

GABE, M. 1951a. Données histologiques sur la neurosécrétion chez les Pterotracheidae (hétéropodes). *Rev. canad. Biol.,* **10,** 391–410.

GABE, M. 1951b. Données histologiques sur les organes du complexe palléal chez la Fissurelle. *Bull. Lab. marit. Dinard,* **35,** 1–14.

GABE, M. 1953a. Particularités morphologiques des cellules neurosécrétrices chez quelques proso-branches monotocardes. *C. R. Acad. Sci., Paris,* **236,** 333–335.

GABE, M. 1953b. Particularités histologiques des cellules neurosécrétrices chez quelques gastéropodes opisthobranches. C. R. Acad. Sci., Paris, **236,** 2161–2163.

GABE, M. 1954. La neurosécrétion chez les invertébrés. Ann. biol., **30,** 5–62.

GABE, M. & M. PRENANT. 1949. Contribution à l'étude cytologique et histochimique du tube digestif des polyplacophores. Arch. Biol., Paris, **60,** 39–77.

GALLIEN, L. & M. DE LARAMBERGUE. 1936. Cycle et dimorphisme sexuel chez Lacuna pallidula da Costa (Littorinidae). C. R. Acad. Sci., Paris, **203,** 409–412.

GALLIEN, L. & M. DE LARAMBERGUE. 1938. Biologie et sexualité de Lacuna pallidula da Costa (Littorinidae). Trav. Sta. zool. Wimereux, **13,** 293–306.

GALTSOFF, P. S., H. F. PRYTHERCH & J. B. ENGLE. 1937. Natural history and methods of controlling the common oyster drills (Urosalpinx cinerea Say and Eupleura caudata Say). Cir. U.S. Bur. Fish., **25,** 1–24.

GARNAULT, P. 1887. Recherches anatomiques et histologiques sur le Cyclostoma elegans. Acta Soc. linn. Bordeaux, **41,** 11–158.

GARSTANG, W. 1928. The origin and evolution of larval forms. Presidential Address, Section D. Brit. Ass. Rep., Glasgow, 77–98.

GARSTANG, W. 1951. 'Larval Forms'. Oxford, Blackwell.

GEDDES, P. 1879. On the mechanism of the odontophore in certain Mollusca. Trans. zool. Soc. Lond., **10,** 485–491.

GEGENBAUR, C. 1855. 'Untersuchungen über Pteropoden und Heteropoden.' Leipzig, Engelmann.

GEORGE, J. C. & C. JURA. 1958. A histochemical study of the capsule fluid of the egg of a land snail Succinea putris L. Proc. Acad. Sci. Amst., **61,** C, 598–603.

GERMAIN, L. 1930. Mollusques terrestres et fluviatiles. Faune de France, **22,** 1–444.

GERSCH, M. 1934. Zur experimenteller Veränderung der Richtung der Wellenbewegung auf der Kriechsole von Schnecken und zur Rückwartsbewegung von Schnecken. Biol. Zbl., **54,** 511–518.

GERSCH, M. 1936. Der Genitalapparat und die Sexualbiologie der Nordseetrochiden. Z. Morph. Ökol. Tiere, **31,** 106–150.

GERSCH, M. 1959. Neurohormone bei wirbellosen Tieren. Verh. dtsch. zool. Ges., Zool. Anz. Suppl. **23,** 40–76.

GHISELIN, M. T. 1966. The adaptive significance of gastropod torsion. Evolution **20,** 337–348.

GIARD, A. 1875. Sur l'embryologie du Lamellaria perspicua. C. R. Acad. Sci., Paris, **80,** 736–739.

GIBSON, R. J. H. 1887. Anatomy and physiology of Patella vulgata. Part I, Anatomy. Trans. R. Soc. Edinb., **32,** 601–638.

GIESE, M. 1915. Der Genitalapparat von Calyptraea sinensis Linn., Crepidula unguiformis Lam. und Capulus hungaricus Lam. Z. wiss. Zool., **114,** 169–231.

GIGLIOLI, M. E. C. 1949. Some observations on the biology of the whelk Polynices heros Say (1822) and Polynices triseriata Say (1826), at Belliveau Cove, Nova Scotia. MS Rep. Fish. Res. Bd. Canada, **398,** 1–140.

GIGLIOLI, M. E. C. 1955. The egg masses of the Naticidae (Gastropoda). J. Fish. Res. Bd. Canada, **12,** 287–327.

GISLÉN, T. 1930. Epibioses of the Gullmar Fjord I. A study in marine sociology. Skriftser. Kristinebergs zoologiska Station 1877–1927, **3,** 1–123.

GLANCY, J. B. 1953. Oyster production and oyster drill control. Conv. Pap. nat. Shellf. Ass., New Orleans.

GLASER, O. C. 1906. Über den Kannibalismus bei Fasciolaria tulipa (var. distans) und deren larve Exkretionsorgane. Z. wiss. Zool., **80,** 80–121.

GOLDFUSS, O. 1900. 'Die Binnenmollusken Mitteldeutschlands.' Leipzig, Engelmann.

GOLIKOV, A. & STAROBOGATOV, Y. T. 1975. Systematics of prosobranch gastropods. Malacologia, **15,** 185–232.

GOMPEL, M. M. 1937. Recherches sur la consommation d'oxygène de quelques animaux aquatiques littoraux. C.R. Acad. Sci., Paris, **205,** 816–818.

GOSTAN, G. 1958. Correlation entre la croissance d'un prosobranche (Rissoa parva da Costa) et le développement des organes internes. C.R. Acad. Sci., Paris, **247**, 2193–2195.

GÖTZE, E. 1938. Bau und Leben von Caecum glabrum (Montagu). Zool. Jb. (Syst.), **71**, 55–122.

GOULD, H. N. 1919. Studies on sex in the hermaphrodite mollusc Crepidula plana. III. Transference of the male-producing stimulus through sea-water. J. exp. Zool., **29**, 113–120.

GOULD, H. N. 1947. Conditions affecting the development of the male phase in Crepidula plana. Biol. Bull., Wood's Hole, **93**, 194.

GOULD, H. N. 1952. Studies on sex in the hermaphrodite mollusk Crepidula plana. IV. Internal and external factors influencing growth and sex development. J. exp. Zool., **119**, 93–160.

GOWANLOCH, J. N. 1927. Contributions to the study of marine gastropods. II. The intertidal life of Buccinum undatum, a study in non-adaptation. Contr. canad. Biol. Fish., N.S., **3**, 167–177.

GOWANLOCH, J. N. & F. R. HAYES, 1926. Contributions to the study of marine gastropods. I. The physical factors, behaviour and intertidal life of Littorina. Contr. canad. Biol. Fish., N.S., **3**, 133–165.

GRABAU, A. W. 1902. Characters of the gastropod shell. Amer. Nat., **36**, 917–945.

GRABAU, A. W. 1928. The significance of the so-called ornamental characters in the molluscan shell. Bull. Peking Soc. nat. Hist., **2**, 27–36.

GRAHAM, A. 1932. On the structure and function of the alimentary canal of the limpet. Trans. R. Soc. Edinb., **57**, 287–308.

GRAHAM, A. 1938. On a ciliary process of food-collecting in the gastropod Turritella communis Risso. Proc. zool. Soc. Lond., A, **108**, 453–463.

GRAHAM, A. 1939. On the structure of the alimentary canal of style-bearing prosobranchs. Proc. zool. Soc. Lond., B, **109**, 75–112.

GRAHAM, A. 1941. The oesophagus of the stenoglossan prosobranchs. Proc. R. Soc. Edinb., B, **61**, 1–23.

GRAHAM, A. 1949. The molluscan stomach. Trans. R. Soc. Edinb., **61**, 737–778.

GRAHAM, A. 1954a. The anatomy of the prosobranch Trichotropis borealis Broderip and Sowerby, and the systematic position of the Capulidae. J. mar. biol. Ass. U.K., **33**, 129–144.

GRAHAM, A. 1954b. Some observations on the reproductive tract of Ianthina janthina (L.). Proc. malac. Soc. Lond., **31**, 1–6.

GRAHAM, A. 1964. The functional anatomy of the buccal mass of the limpet (Patella vulgata). Proc. zool. Soc. Lond., **143**, 301–329.

GRAHAM, A. & V. FRETTER. 1947. The life history of Patina pellucida (L.). J. mar. biol. Ass. U.K., **26**, 590–601.

GRAY, J. E. 1850. On the operculum of the gasteropodous Mollusca and an attempt to prove that it is homologous or identical with the second valve of Conchifera. Ann. Mag. nat. Hist. (2), **5**, 476–483.

GRENACHER, H. 1886. Abhandlungen zur vergleichenden Anatomie des Auges. II. Das Auge der Heteropoden. Abh. naturf. Ges. Halle, **7**, 1–64.

GROSSU, A. V. 1956. 'Fauna republicii populare Romîne. Mollusca, 3, 2. Gastropoda Prosobranchia si Opisthobranchia.' Bucureşti, Academiei Republicii Populare Romîne.

GURJANOVA, E., I. SACHS & P. USCHAKOV. 1930. Das Littoral des Kola Fjords. III. Trav. Soc. Nat. Leningrad, **60**, 17–107.

GWATKIN, H. M. 1914. Some molluscan radulae. J. Conch., Lond., **14**, 139–148.

HAAS, F. 1938. Über potentielle Skulpturblidung bei Valvata (Cincinna) piscinalis antiqua (Sow.). Arch. Molluskenk., **70**, 41–45.

HALLER, B. 1884. Untersuchungen über marine Rhipidoglossen. I. Morph. Jb., **9**, 1–98.

HALLER, B. 1886. Untersuchungen über marine Rhipidoglossen. II. Morph. Jb., **II**, 321–430.

HALLER, B. 1888. Die Morphologie der Prosobranchier, gesammelt auf einer Erdumsegelung durch die König. italienische Korvette 'Vettor Pisani'. I. Morph. Jb., **14**, 54–169.

HALLER, B. 1894. 'Studien über docoglosse und rhipidoglosse Prosobranchier.' Leipzig, Engelmann.

HAMMERSTEN, O. D. & J. RUNNSTRÖM. 1926. Zur Embryologie von *Acanthochiton discrepans* Brown. *Zool. Jb. (Anat. Ont.)*, **47**, 261–318.

HANCOCK, D. A. 1954. The destruction of oyster spat by *Urosalpinx cinerea* (Say) on Essex oyster beds. *J. Cons. int. Explor. Mer*, **20**, 186–196.

HANCOCK, D. A. 1956. The structure of the capsule and the hatching process in *Urosalpinx cinerea* (Say). *Proc. zool. Soc. Lond.*, **127**, 565–571.

HANCOCK, D. A. 1957. 'Studies in the biology and ecology of certain marine invertebrates, with particular reference to those associated with oyster culture.' Ph.D. Thesis, University of Reading.

HANCOCK, D. A. 1959. The biology and control of the American whelk tingle *Urosalpinx cinerea* (Say) on English oyster beds. *Fish. Invest. Lond.* (2), **22**, no. 10, 1–66.

HANKO, B. 1913. Über die Regeneration des Operkulums bei *Murex brandaris. Arch. EntwMech. Org.*, **35**, 740–747.

HANKS, J. E. 1957. The rate of feeding of the common oyster drill *Urosalpinx cinerea* (Say) at constant water temperatures. *Biol. Bull., Wood's Hole*, **112**, 330–335.

HANSON, J. & J. LOWY. 1957. Structure of smooth muscles. *Nature, Lond.*, **180**, 906–909.

HANSON, J., J. T. RANDALL & S. T. BAYLEY. 1952. The microstructure of the spermatozoa of the snail *Viviparus. Exp. Cell Res.*, **3**, 65–78.

HARDY, A. C. 1956. 'The Open Sea. Its Natural History: the World of Plankton.' London, Collins.

HASEMAN, J. D. 1911. The rhythmical movements of *Littorina littorea* synchronous with ocean tides. *Biol. Bull., Wood's Hole*, **21**, 113–121.

HASZPRUNAR, G. 1985a. The fine morphology of the osphradial sense organs of the Mollusca. I. Gastropoda Prosobranchia. *Phil. Trans. R. Soc. B*, **307**, 457–496.

HASZPRUNAR, G. 1985b. The fine morphology of the osphradial sense organs of the Mollusca. II. Allogastropoda (Architectonicidae, Pyramidellidae). *Phil. Trans. R. Soc. B*, **307**, 497–505.

HASZPRUNAR, G. 1985c. The Heterobranchia—a new concept of the phylogeny of the higher Gastropoda. *Z. zool. Syst. Evolut.-forsch.*, **28**, 15–37.

HASZPRUNAR, G. 1988a. A preliminary phylogenetic classification of the streptoneurous gastropods. *Malacol. Rev.*, suppl. **4**, 7–16.

HASZPRUNAR, G. 1988b. Comparative anatomy of cocculiniform gastropods and its bearing on archaeogastropod systematics. *Malacol. Rev.*, suppl. **4**, 64–84.

HASZPRUNAR, G. 1988c. On the origin and evolution of major gastropod groups, with special reference to the Streptoneura. *J. moll. Stud.*, **54**, 367–441.

HASZPRUNAR, G. 1992. The first molluscs—small animals. *Bull. Zool.*, **59**, 1–6.

HATTON, H. 1938. Essais de bionomie explicative sur quelques espèces intercotidales d'algues et d'animaux. *Ann. Inst. océanogr. Monaco*, **17**, 241–348.

HAYES, F. R. 1927a. The negative geotropism of the perwinkle: a study in littoral ecology. *Trans. Nova Scotian Inst. Sci.*, **16**, 155–173.

HAYES, F. R. 1927b. The effect of environmental factors on the development and growth of *Littorina littorea. Trans. Nova Scotian Inst. Sci.*, **17**, 6–13.

HAYES, F. R. 1929. Contributions to the study of marine gastropods. III. Development, growth and behaviour of *Littorina. Contr. canad. Biol.*, N.S., **4**, 413–430.

HAZAY, J. 1881. Die Molluskenfauna von Budapest. Biologischer Teil. *Malak. Bl.*, N.F., **4**, 43–221.

HAZELHOFF, E. H. 1938. Über die Ausnutzung des Sauerstoffs bei verschiedenen Wassertieren. *Z. vergl. Physiol.*, **26**, 306–327.

HEATH, H. 1899. The development of *Ischnochiton. Zool. Jb. (Anat. Ont.)*, **12**, 567–656.

HENKING, H. 1894. Beiträge zur Kenntnis von *Hydrobia ulvae* Penn. und deren Brutpflege. *Ber. naturf. Ges. Freiburg i. B.*, **8**, 89–110.

HENSCHEL, J. 1932. Untersuchungen über den chemischen Sinn von *Nassa reticulata. Wiss. Meeresuntersuch. Abt. Kiel*, **21**, 131–159.

HENSEN, V. 1865. Über das Auge einiger Cephalopoden. *Z. wiss. Zool.*, **15**, 155–242.

HERRICK, J. C. 1906. Mechanism of the odontophoral apparatus in *Sycotypus canaliculatus*. *Amer. Nat.,* **40,** 707–737.

HERTLING, H. 1928. Beobachtungen und Versuche an den Eiern von *Littorina* und *Lacuna*. Bedeutung der Eihüllen. Entwicklung im naturlichen und abgeänderten Medium. *Wiss. Meeresuntersuch. Abt. Helgoland,* **17,** 1–49.

HERTLING, H. 1931. Über den Einfluss des veränderten Mediums auf die Entwicklung von *Lacuna divaricata*, besonders auf die Bildung der Schale. *Wiss. Meeresuntersuch. Abt. Helgoland,* **18,** 1–27.

HERTLING, H. & W. E. ANKEL. 1927. Bemerkungen über den Laich und die Jugendformen von *Littorina* und *Lacuna*. *Wiss. Meeresuntersuch. Abt. Helgoland,* **16,** 1–13.

HESSE, R. 1900. Untersuchungen über die Organe der Lichtempfindung bei niederen Thieren. VI. Die Augen einiger Mollusken. *Z. wiss. Zool.,* **68,** 379–477.

HESSE, R., W. C. ALLEE & K. P. SCHMIDT. 1937. 'Ecological Animal Geography.' New York, John Wiley & Sons.

HEWATT, W. G. 1940. Observations on the homing limpet, *Acmaea scabra* Gould. *Amer. midl. Nat.,* **24,** 205–208.

HILGER, C. 1885. Beiträge zur Kenntnis des Gastropodenauges. *Morph. Jb.,* **10,** 351–371.

HIRASE, S. 1932. The adaptive modifications of the gastropod *Stilifer celebensis* Kükenthal, parasitic on the starfish *Certonardea semiregularis* (Müller & Troschel). *Proc. malac. Soc. Lond.,* **20,** 73–76.

HIRSCH, G. C. 1915. Die Ernährungsbiologie fleischfressender Gastropoden (*Murex, Natica, Pterotrachea, Pleurobranchaea, Tritonium*). I Teil. Makroskopischer Bau, Nahrungsaufnahme, Verdauung, Sekretion. *Zool. Jb. (Zool. Physiol.),* **35,** 357–504.

HOFFMANN, H. 1930 Über dem Fluchtreflex bei *Nassa*. *Z. vergl. Physiol.,* **11,** 662–688.

HOFFMANN, H. 1932. Über die Radulabildung bei *Lymnaea stagnalis*. Jena. *Z. Naturw.,* **67,** 535–550.

HOFFMANN, H. 1938. Beiträge zur Kenntnis der Chitonen. 2. Zur Frage der Anheftung der Chitonen an die Unterlage. *Z. Morph. Ökol. Tiere,* **34,** 647–662.

HÖRSTADIUS-KJELLSTRÖM, G. & S. HÖRSTADIUS. 1940. Untersuchungen über die Eiweissver-dauung *in vivo* und *in vitro* bei einigen Gastropoden. *Pubbl. Staz. zool. Napoli,* **18,** 151–249.

HOUSSAY, F. 1884. Recherches sur l'opercule et les glandes du pied des gastéropodes. *Arch. Zool. exp. gén.* (2), **2,** 171–288.

HOWES, N. H. 1939. The ecology of a saline lagoon in south-east Essex. *J. Linn. Soc. (Zool.),* **40,** 383–445.

HUBENDICK, B. 1945. Phylogenie und Tiergeographie der Siphonariidae. Zur Kenntnis der Phylogenie in der Ordnung Basommatophora und des Ursprungs der Pulmonatengruppe. *Zool. Bidr. Uppsala,* **24,** 1–216.

HUBENDICK, B. 1947. Die Verbreitungsverhältnisse der limnischen Gastropoden in Südschweden. *Zool. Bidr. Uppsala,* **24,** 419–559.

HUBENDICK, B. 1948. Über den Bau und das Wachstum des Konzentrischen Operculartypus bei Gastropoden. *Ark. Zool.,* **40**A, (10), 1–28.

HUBENDICK, B. 1950. The effectiveness of passive dispersal in *Hydrobia jenkinsi*. *Zool. Bidr. Uppsala,* **28,** 493–504.

HUBENDICK, B. 1958. On the molluscan adhesive epithelium. *Ark. Zool.,* A.S., **11,** 31–36.

HUGGINS, H. C. 1919. Occurrence of *Hartmannia septemspiralis* (Razoumovsky) and *H. patula* (Drap.) in England. *J. Conch., Lond.,* **16,** 51–52.

HULBERT, G. C. E. B. & C. M. YONGE. 1937. A possible function of the osphradium in the Gastropoda. *Nature, Lond.,* **139,** 840.

HUNTER, W. R. 1953a. Notes on the Mollusca of the Garvelloch Islands. *J. Conch., Lond.,* **23,** 379–386.

HUNTER, W. R. 1953b. On migrations of *Lymnaea peregra* (Müller) on the shores of Loch Lomond. *Proc. R. Soc. Edinb., B,* **65,** 84–105.

HUNTER, W. R. 1957. Studies on freshwater snails at Loch Lomond. *Glasg. Univ. Publ. Stud. Loch Lomond,* **1,** 56–95.

HUNTER, W. R. & T. WARWICK. 1957. Records of 'Potamopyrgus jenkinsi' (Smith) in Scottish fresh
    waters over fifty years (1906–56). Proc. R. Soc. Edinb., B, **66,** 360–373.
HUXLEY, T. H. 1853. On the morphology of the cephalous Mollusca, as illustrated by the anatomy
    of certain heteropoda and Pteropoda collected during the voyage of H.M.S. 'Rattle-snake' in
    1846–50. Philos. Trans., **143,** 29–65.
HYKEŠ, O. V. 1929. Adrenalin und das Weichtierherz. Biol. Listy, **14,** 385.
HYKEŠ O. V. 1930. L'adrénaline et le coeur des mollusques. C.R. Soc. Biol., Paris, **103,** 360–363.
HYKEŠ, O. V. 1932. Adrenalwirkung am Herzen der Avertebraten. Čas. Lék. česk., 129; Ber. wiss. Biol., **22,**
    144.
HYMAN, L. H. 1940. 'The Invertebrates. I. Protozoa through Ctenophora.' New York, McGraw-Hill.
HYMAN, O. W. 1923. Spermic dimorphism in Fasciolaria tulipa. J. Morph., **37,** 307–383.
ILYIN, P. 1900. Das Gehörbläschen als Gleichgewichtsorgan bei den Pterotracheiden. Zbl. Physiol., **13,**
    691–694.
INO, T. 1949. The effect of food on growth and coloration of the topshell (Turbo cornutus Solander). J.
    mar. Res., **8,** 1–5.
IREDALE, T. 1911. On the value of the gastropod apex in classification. Proc. malac. Soc. Lond., **9,**
    319–323.
ISHIKI, H. 1936. Sex changes in Japanese slipper limpets Crepidula aculeata and Crepidula walshi. J. Sci.
    Hiroshima Univ., Ser. B, Div. 1, **3,** 91–99.
JACOB, J. 1957. Cytological studies of Melaniidae (Mollusca) with special reference to parthenogenesis
    and polyploidy. I. Oogenesis of the parthenogenetic species of 'Melanoides' (Prosobranchia-
    Gastropoda). Trans. R. Soc. Edinb., **63,** 341–352.
JACOB, J. 1958. Cytological studies of Melaniidae (Mollusca) with special reference to parthenogenesis
    and polyploidy. II. A study of meiosis in the rare males of the polyploid race of 'Melanoides
    tuberculatus' and Melanoides lineatus. Trans. R. Soc. Edinb., **63,** 433–444.
JACKSON, J. & W. F. TAYLOR. 1904. Observations on the habits and reproduction of Paludestrina taylori.
    J. Conch., Lond., **II,** 9–11.
JÄGERSTEN, G. 1937. Zur Kenntnis der Parapodialborsten bei Myzostomum. Zool. Bidr. Uppsala, **16,**
    283–299.
JEFFREYS, J.G . 1862–69. 'British Conchology.' Vols. 1–5. **1** (1862), **2** (1863), **3** (1865), **4** (1867), **5** (1869).
    London, van Voorst.
JENKINS, H. L. 1955. 'Digestive system of Littorina littorea (L.).' M.Sc. Thesis, University of Wales.
JENSEN, A. S. 1951. Do the Naticidae (Gastropoda Prosobranchia) drill by chemical or mechanical
    means? Vidensk. Medd. dansk naturh. Foren. Kbh., **113,** 251–261.
JESSEN, A. 1918. Udstroekningen af Randers Fjord i Litorinatiden. Randers Fjords Naturhistorie, **1B.**
    Copenhagen, C. A. Reitzel.
JOHANSEN, A. C. 1918. Bløddyrene i Randers Fjord. Randers Fjords Naturhistorie, **5G.** Copenhagen, C. A.
    Reitzel.
JOHANSSON, J. 1939. Anatomische Studien über die Gastropodenfamilien Rissoidae und Littorinidae.
    Zool. Bidr. Uppsala, **18,** 289–296.
JOHANSSON, J. 1942. Von diaulen Geschlechtsapparaten bei den Prosobranchiern. Ark. Zool., **34**A (12),
    1–10.
JOHANSSON, J. 1946. Von den Geschlechtsorganen bei Turritella communis nebst Bemerkungen über
    die diaulen Geschlechtsorgane der Neritaceen. Ark. Zool., **38**A (12), I–II.
JOHANSSON, J. 1947. Über den offenen Uterus bei einigen Monotocardiern ohne Kopulationsorgan.
    Zool. Bidr. Uppsala, **25,** 102–110.
JOHANSSON, J. 1948a. Über die Geschlechtsorgane der Hydrobiiden und Rissoiden und den ursprüng-
    lichen Hermaphroditismus der Prosobranchier. Ark. Zool., **40**A (15), 1–13.
JOHANSSON, J. 1948b. Über die Geschlechtsorgane von Aporrhais pespelecani nebst einigen Betrach-
    tungen über die phylogenetische Bedeutung der Cerithiacea und Architaenioglossa. Ark. Zool.,
    41A (8), 1–13.

JOHANSSON, J. 1950. On the embryology of *Viviparus* and its significance for the phylogeny of the Gastropoda. *Ark. Zool., A.S.,* **1,** 173–177.

JOHANSSON, J. 1953. On the genital organs of some mesogastropods: *Cerithium vulgatum* Brug., *Triphora perversa* (L.) and *Melanella (Eulima) intermedia* (Cantr.). *Zool. Bidr. Uppsala,* **30,** 1–23.

JOHANSSON, J. 1955. Garnault's duct and its significance for the phylogeny of the genital system of *Valvata. Zool. Bidr. Uppsala,* **30,** 457–464.

JOHANSSON, J. 1956a. Genital organs of two *Alvania* species, and a comparison with related families (Moll. Pros.). *Ark. Zool., A.S.,* **9,** 377–387.

JOHANSSON, J. 1956b. On the anatomy of *Tympanotonus fuscatus* (L.), including a survey of the open pallial oviducts of the Cerithiacea. *Atlantide Report,* no. 4, 149–166.

JONES, E. I., R. A. McCANCE & L. R. B. SHACKLETON. 1935. The role of iron and silica in the structure of the radular teeth of certain marine molluscs. *J. exp. Biol.,* **12,** 59–64.

JONES, N. S. 1948. Observations on the biology of *Patella vulgata* at Port St. Mary, Isle of Man. *Proc. Lpool. biol. Soc.,* **56,** 60–77.

JORDAN, H. J. & H. BEGEMANN. 1921. Über die Bedeutung des Darmes von *Helix pomatia. Zool. Jb. (Zool. Physiol.),* **38,** 565–582.

JORDAN, H. J. & H. J. LAM. 1918. Über die Darmdurchlässigkeit bei *Astacus fluviatilis* und *Helix pomatia. Tijdschr. nederl. dierk. Ver.,* **16,** 281–292.

JULLIEN, A. 1948. Recherches sur les fonctions de la glande hypobranchiale chez *Murex trunculus. C.R. Soc. Biol., Paris,* **142,** 102–103.

KENNARD, A. S. 1941. The geological record of *Potamopyrgus jenkinsi* in the British Isles. *Proc. malac. Soc. Lond.,* **24,** 156.

KESSEL, E. 1933. Über die Schale von *Viviparus viviparus* L. und *Viviparus fasciatus* Müll. *Z. Morph. Ökol. Tiere,* **27,** 129–198.

KESSEL, E. 1942. Über Bau and Bildung des Prosobranchier-Deckels. *Z. Morph. Ökol. Tiere,* **38,** 197–250.

KESSEL, E. 1944. Über Periostracum-Bildung. *Z. Morph. Ökol. Tiere,* **40,** 348–360.

KILIAN, F. 1951. Untersuchungen zur Biologie von *Pomatias elegans* (Müller) und ihrer 'Konkrement-drüse'. *Arch. Molluskenk.,* **80,** 1–16.

KNIGHT, J. B. 1947. Bellerophont muscle scars. *J. Paleont.,* **21,** 264–267.

KNIGHT, J. B. 1952. Primitive fossil gastropods and their bearing on gastropod classification. *Smithson. misc. Coll.,* **117,** no. 13, 1–56.

KNIGHT, J. B. & E. L. YOCHELSON. 1958. A reconsideration of the relationships of the Monoplacophora and the primitive Gastropoda. *Proc. malac. Soc. Lond.,* **33,** 37–48.

KNIGHT-JONES, E. W. 1954. Relations between metachronism and the direction of ciliary beat in Metazoa. *Quart. J. micr. Sci.,* **95,** 503–521.

KOEHLER, R. & C. VANEY. 1908. Description d'un nouveau genre de prosobranches, parasite sur certains échinides (*Pelseneeria* n.g.). *Bull. Inst. océanogr. Monaco,* **118,** 1–16.

KOEHLER, R. & C. VANEY. 1912. Nouvelles formes de gastéropodes ectoparasites. *Bull sci. Fr. Belg.,* **46,** 191–217.

KORRINGA, P. 1952. Recent advances in oyster biology. *Quart. Rev. Biol.,* **27,** 266–308; 339–365.

KOSTITZINE, J. 1934. Le cycle génital femelle de la pourpre. *C.R. Soc. Biol., Paris,* **115,** 264.

KOSTITZINE, J. 1940. Sur la ponte de la pourpre. *Bull. Soc. zool. Fr.,* **65,** 80–84.

KOSTITZINE, J. 1949. Contribution à l'étude de l'appareil reproducteur femelle de quelques mollusques prosobranches marins. *Arch. Zool. exp. gén.,* **86,** 145–167.

KOSUGE, S. 1964. On the systematic position of the family Triphoridae from their radula and operculum. *Venus,* **23,** 43–47.

KOSUGE, S. 1966. The family Triphoridae and its systematic position. *Malacologia,* **12,** 297–324.

KRASUCKI, A. 1911. Untersuchungen über Anatomie und Histologie der Heteropoden. *Bull. Acad. Sci. Cracovie, ser. B, Sci. nat.,* **5**B, 391–448 and **6**B, 449–450.

KRIJGSMAN, B. J. 1925. Die Arbeitsrhythmus der Verdauungsdrüsen bei *Helix pomatia*. *Z. vergl. Physiol.,* **2,** 264–302.

KRIJGSMAN, B. J. 1928. Die Arbeitsrhythmus der Verdauungsdrüsen bei *Helix pomatia*. II. Teil: Sekretion, Resorption und Phagocytose. *Z. vergl. Physiol.,* **8,** 187–280.

KROGH, A. 1939. 'Osmotic Regulation in Aquatic Animals.' Cambridge, Cambridge University Press.

KRULL, H. 1935. Anatomische Untersuchungen an einheimischen Prosobranchiern und Beiträge zur Phylogenie der Gastropoden. *Zool. Jb.* (*Anat. Ont.*), **60,** 399–464.

KRUMBACH, T. 1918. Napfschnecken in der Gezeitenwelle und der Brandungszone der Karstküste. *Zool. Anz.,* **49,** 96–112; 113–123.

KUSCHAKEWITSCH, S. 1910. Zur Kenntnis der sogenannten 'wurmförmigen' Spermien der Proso-branchier. *Anat. Anz.,* **37,** 318–324.

KÜTTLER, A. 1913. Die Anatomie von *Oliva peruviana* Lamarck. *Zool. Jb.,* Suppl. **13** (*Fauna chilensis,* **4**), 477–544.

LABBE, A. 1926. Contributions à l'étude de l'allélogénèse. 2e mémoire: croissance et environment. Essai d'une théorie des adaptations. *Bull. biol.,* **60,** 1–87.

LACAZE-DUTHIERS, H. DE. 1859. Système nerveux de l'Haliotide et sur la pourpre. *Ann. Sci. nat. Zool.* (4), **12,** 5–84.

LACAZE-DUTHIERS, H. DE. 1872. Otocystes ou capsules auditives des mollusques (gastéropodes). *Arch. Zool. exp. gén.,* **I,** 97–166.

LAMBERT, F. J. 1930. Animal life in the marsh ditches of the Thames estuary. *Proc. zool. soc. Lond.,* 801–808.

LAMY, E. 1928. La ponte chez les gastéropodes prosobranches. *J. Conchyliol.,* **72,** 25–196.

LANG, A. 1891. Versuch einer Erklärung der Asymmetrie der Gastropoden. *Vjschr. naturf. Ges. Zürich,* **36,** 339–371.

LANG, A. 1896. 'Text-book of comparative Anatomy,' **2**. London, Macmillan.

LANKESTER E. R. 1872. A contribution to the knowledge of haemoglobin. *Proc. R. Soc.,* **21,** 70–81.

LANKESTER, E. R. 1893. Note on the coelom and vascular system of Mollusca and Arthropoda. *Quart. J. micr. Sci.,* **34,** 427–432.

LAURSEN, D. 1953. The genus *Ianthina*. *Dana Rep.* no. **38,** 1–40.

LAWRENCE-HAMILTON, J. 1892. The limpet's strength. *Nature, Lond.,* **45,** 487.

LEBOUR, M. V. 1931a. The larval stages of *Nassarius reticulatus* and *Nassarius incrassatus*. *J. mar. biol. Ass. U.K.,* **17,** 797–818.

LEBOUR, M. V. 1931b. The larval stages of *Trivia europaea*. *J. mar. biol. Ass. U.K.,* **17,** 819–832.

LEBOUR, M. V. 1932a. The larval stages of *Simnia patula*. *J. mar. biol. Ass. U.K.,* **18,** 107–115.

LEBOUR, M. V. 1932b. The eggs and larval stages of two commensal gastropods, *Stilifer stylifer* and *Odostomia eulimoides*. *J. mar. biol. Ass. U.K.,* **18,** 117–122.

LEBOUR, M. V. 1933a. The British species of *Trivia, T. arctica* and *T. monacha*. *J. mar. biol. Ass. U.K.,* **8,** 477–484.

LEBOUR, M. V. 1933b. The life-histories of *Cerithiopsis tubercularis* (Montagu), *C. barleei* Jeffreys and *Triphora perversa* (L.). *J. mar. biol. Ass. U.K.,* **18,** 491–498.

LEBOUR, M. V. 1933c. The eggs and larvae of *Turritella communis* Lamarck and *Aporrhais pes-pelicani* (L.). *J. mar. biol. Ass. U.K.,* **18,** 499–506.

LEBOUR, M. V. 1933d. The eggs and larvae of *Philbertia gracilis* (Montagu). *J. mar. biol. Ass. U.K.,* **18,** 507–510.

LEBOUR, M. V. 1934a. Rissoid larvae as food of young herring. The eggs and larvae of Plymouth Rissoidae. *J. mar. biol. Ass. U.K.,* **19,** 523–540.

LEBOUR, M. V. 1934b. The eggs and larvae of some British Turridae. *J. mar. biol. Ass. U.K.,* **19,** 541–554.

LEBOUR, M. V. 1935a. The echinospira larvae of Plymouth. *Proc. zool. Soc. Lond.,* 163–174.

LEBOUR, M. V. 1935b. The larval stages of *Balcis alba* and *B. devians*. *J. mar. biol. Ass. U.K.* **20,** 65–70.

LEBOUR, M. V. 1935c. The breeding of *Littorina neritoides*. *J. mar. biol. Ass. U.K.,* **20,** 373–378.

LEBOUR, M. V. 1936. Notes on the eggs and larvae of some Plymouth prosobranchs. *J. mar. biol. Ass. U.K.,* **20,** 547–565.

LEBOUR, M. V. 1937. The eggs and larvae of the British prosobranchs with special reference to those living in the plankton. *J. mar. biol. Ass. U.K.,* **22,** 105–166.

LEBOUR, M. V. 1945. The eggs and larvae of some prosobranchs from Bermuda. *Proc. zool. Soc. Lond.,* **114,** 462–489.

LEBOUR, M. V. 1947. Notes on the inshore plankton of Plymouth. *J. mar. biol. Ass. U.K.,* **26,** 527–547.

LEMCHE, H. 1957. A new living deep-sea mollusc of the Cambro-Devonian class Monoplacophora. *Nature, Lond.,* **179,** 413–416.

LEMCHE, H. 1959. Protostomian relationships in the light of *Neopilina. Proc. XVth int. Cong. Zool. Lond.,* 381–389.

LEMCHE, H. & K. WINGSTRAND. 1959. The comparative anatomy of *Neopilina galatheae* Lemche, 1957 (Mollusca Monoplacophora). *Proc. XVth int. Cong. Zool. Lond.,* 378–380.

LENSSEN, J. 1899. Système digestif et système génital de la *Neritina fluviatilis. Cellule,* **16,** 179–232.

LETELLIER, A. 1889. Recherches sur la pourpre produite par le *Purpura lapillus. C.R. Acad. Sci., Paris,* **109,** 82–84.

LETELLIER, A. 1890. Recherches sur la pourpre produite par le *Purpura lapillus. Arch. Zool. exp. gén.* (2), **8,** 361–408.

LEVER, J. 1957. Some remarks on neurosecretory phenomena in *Ferrissia* (Gastropoda Pulmonata). *Proc. Acad. Sci. Amst.,* **60,** C, 510–522.

LEWIS, J. R. 1954a. Observations on a high-level population of limpets. *J. Anim. Ecol.,* **23,** 85–100.

LEWIS, J. R. 1954b. The ecology of exposed rocky shores of Caithness. *Trans. R. Soc. Edinb.,* **62,** 695–723.

LEYON, H. 1947. The anatomy of the cerebral nerves of Gastropoda. *Zool. Bidr. Uppsala,* **25,** 394–401.

LHOSTE, L.-J. 1944. Sur la microstructure interne du test des gastéropodes. *C.R. Acad. Sci., Paris,* **219,** 351–352.

LHOSTE, L.-J. 1946. Les microstructures des Patelles. *J. Conchyliol.,* **87,** 29 and 38.

LILLY, M. M. 1953. The mode of life and the structure and functioning of the reproductive ducts of *Bithynia tentaculata* (L.). *Proc. malac. Soc. Lond.,* **30,** 87–110.

LILLY, S. J., J. F. SLOANE, R. BASSINDALE, F. J. EBLING & J. A. KITCHING. 1953. The ecology of the Lough Ine rapids with special reference to water currents. IV. The sedentary fauna of sublittoral boulders. *J. Anim. Ecol.,* **22,** 87–122.

LINDBERG, D. R. 1986. Radular evolution in the Patellogastropoda. *Am. malacol. Bull.,* **4,** 115.

LINDBERG, D. R. 1988. The Patellogastropoda. *Malacol. Rev.,* suppl. **4,** 35–63.

LINKE, O. 1933. Morphologie und Physiologie des Genitalapparates der Nordseelittorinen. *Wiss. Meeresuntersuch. Abt. Helgoland,* **19,** Nr. 5, 3–52.

LINKE, O. 1934a. Beiträge zur Sexualbiologie der Littorinen. *Z. Morph. Ökol. Tiere,* **28,** 170–177.

LINKE, O. 1934b. Über die Beziehungen zwischen Keimdrüse und Soma bei Prosobranchiern. *Verh. dtsch. zool. Ges., Zool. Anz.* Suppl. **7,** 164–175.

LINKE, O. 1935a. Der Laich von *Littorina* (*Melaraphe*) *neritoides* L. *Zool. Anz.,* **112,** 57–62.

LINKE, O. 1935b Zur Morphologie und Physiologie des Genitalapparates der Süsswasser-littorinide *Cremnoconchus syhadrensis* Blanford. *Arch. Naturg.,* N.F., **4,** 72–87.

LINKE, O. 1939. Die Biota des Jadebusenwattes. *Helgoland. wiss. Meeresunters.,* **1,** 201–348.

LISSMANN, H. W. 1945. The mechanism of locomotion in gastropod molluscs. I. Kinematics. *J. exp. Biol.,* **21,** 58–69.

LISSMANN, H. W. 1946. The mechanism of locomotion in gastropod molluscs. II. Kinetics. *J. exp. Biol.,* **22,** 37–50.

LIST, T. 1902. Die Mytiliden. *Fauna und Flora des Golfes von Neapel,* **27.**

LONGSTAFF, M. J. 1910. Non-marine Mollusca found in the parish of Mortehoe, North Devon. *J. Conch., Lond.,* **13,** 15–23.

LOPPENS, K. 1922. Note sur la variabilité et l'éthologie de *Patella vulgata. Ann. Soc. zool. malac. Belg.,* **53,** 57–68.

LOPPENS, K. 1926. La perforation des coquilles de mollusques par les gastropodes et les éponges. *Ann. Soc. zool. Belg.,* **57,** 14–18.

LOWENSTAM, H. A. 1978. Recovery, behaviour and evolutionary implications of live Monoplacophora. *Nature, Lond.,* **273,** 231–232.

LUCAS, J. A. W. 1954. Het genus *Patella* in Nederland. *Basteria,* **18,** 36–40.

LYSAGHT, A. M. 1941. The biology and trematode parasites of the gastropod *Littorina neritoides* (L.) on the Plymouth Breakwater. *J. mar. biol. Ass. U.K.,* **25,** 41–67.

McLEAN, J. H. 1979. A new monoplacophoran limpet from the continental shelf off Southern California. *Contrib. Sci. nat. Hist. Mus. Los Angeles County,* **307,** 1–19.

McMILLAN, N. F. 1938–39. Early records of *Crepidula* in English waters. *Proc. malac. Soc. Lond.,* **23,** 236.

McMILLAN, N. F. 1939. The British species of *Lamellaria. J. Conch., Lond.,* **21,** 170–173.

McMILLAN, N. F. 1944. The distribution of *Mondonta* (*Trochus*) *lineata* (da Costa) in Britain. *Northw. Naturalist,* **19,** 290–292.

McMILLAN, N. F. 1948a. Possible biological races in *Hydrobia ulvae* (Penant) and their varying resistance to lowered salinity. *J. Conch., Lond.,* **23,** 14–16.

McMILLAN, N. F. 1948b. The resistance of *Hydrobia ventrosa* (Montagu) to low salinities. *J. Conch., Lond.,* **23,** 16.

McMURRICH, J. P. 1886. A contribution to the embryology of the prosobranch gasteropods. *Stud. biol. Lab. Johns Hopkins Univ.,* **3,** 403–450.

MACAN, T. T. 1950. Ecology of the freshwater Mollusca in the English Lake District. *J. Anim. Ecol.,* **19,** 124–146.

MANSOUR-BEK, J. J. 1934. Über die proteolytischen Enzyme von *Murex anguliferus* Lamk. *Z. vergl. Physiol.,* **20,** 343–369.

MARINE BIOLOGICAL ASSOCIATION. 1957. 'Plymouth Marine Fauna.' Plymouth, M.B.A.

MARSHALL, J. T. 1889. On *Hydrobiae* and *Assimineae* from the Thames valley. *J. Conch., Lond.,* **6,** 140–142.

MARSHALL, J. T. 1910–12. Additions to 'British Conchology'. *J. Conch., Lond.,* **13,** 179–190, 192–209, 223–231, 294–306, 324–338.

MATTOX, N. T. 1938. Morphology of *Campeloma rufum,* a parthenogenetic snail. *J. Morph.,* **62,** 243–261.

MEDEM, F. GRAF V. 1945.Untersuchungen über die Ei- und Spermawirkstoffe bei marinen Mollusken. *Zool. Jb.* (*Zool. Physiol.*), **61,** 1–44.

MENDEL, L. B. & H. C. BRADLEY. 1905a. Experimental studies in the physiology of the molluscs. First paper. *Amer. J. Physiol.,* **13,** 17–29.

MENDEL, L. B. & H. C. BRADLEY. 1905b. Experimental studies in the physiology of the molluscs. Second paper. *Amer. J. Physiol.,* **14,** 313–327.

MENDEL, L. B. & H. C. BRADLEY. 1906. Experimental studies in the physiology of the molluscs. Third paper. *Amer. J. Physiol.,* **17,** 167–176.

MENG, K. 1958. 5-hydroxytryptamin und Acetylcholin als Wirkungsantagonisten beim *Helix-*Herzen. *Naturwissenschaften,* **45,** 470.

MENKE, H. 1911. Physikalische und physiologische Faktoren bei der Anheftung von Schnecken der Brandungszone. *Zool. Anz.,* **37,** 19–30.

MEVES, F. 1903. Über oligopyrene und apyrene Spermien und über ihre Entwicklung nach Beobachtungen an *Paludina* und *Pygaera. Arch. mikr. Anat.,* **61,** 1–84.

MEYER, A. 1913. Das Renogenitalsystem von *Puncturella noachina* L. *Biol. Zbl.,* **33,** 564–576.

MEYER, E. 1901. Studien über den Korperbau der Anneliden. V. Das Mesoderm der Ringelwürmer. *Mitt. zool. Stat. Neapel,* **14,** 247–585.

MEYER, K. F. 1925. The bacterial symbiosis in the concretion deposits of certain operculate land molluscs of the families Cyclostomatidae and Annulariidae. *J. infect. Dis.,* **36,** 1–107.

MILNE-EDWARDS, H. 1846. Sur la circulation chez les mollusques (chez les Patelles et les Haliotides). *Ann. Sci. nat. Zool.* (3), **8**, 37–53.

MISTAKIDIS, M. N. 1951. Quantitative studies of the bottom fauna of Essex oyster grounds. *Fish. Invest. Lond.* (2), **17** no. 6, 1–47.

MISTAKIDIS, M. N. & D. A. HANCOCK. 1955. Reappearance of *Ocenebra erinacea* (L.) off the east coast of England. *Nature, Lond.*, **175**, 734.

MONTALENTI, G. 1960. Perspectives of research on sex problems in marine animals. In 'Perspectives in Marine Biology'. (A. A. Buzzati-Traverso ed.). Berkeley & Los Angeles, University of California Press.

MOORE, H. B. 1931. The systematic value of molluscan faeces. *Proc. malac. Soc. Lond.*, **19**, 281–290.

MOORE, H. B. 1932. The faecal pellets of the Trochidae. *J. mar. biol. Ass. U.K.*, **18**, 235–241.

MOORE, H. B. 1934. The relation of shell growth to environment in *Patella vulgata*. *Proc. malac. Soc. Lond.*, **21**, 217–222.

MOORE, H. B. 1936. The biology of *Purpura lapillus* I. Shell variation in relation to environment. *J. mar. biol. Ass. U.K.*, **21**, 61–89.

MOORE, H. B. 1937a. The biology of *Littorina littorea*. Part I. Growth of the shell and tissues, spawning, length of life and mortality. *J. mar. biol. Ass. U.K.*, **21**, 721–742.

MOORE, H. B. 1937b. 'Marine Fauna of the Isle of Man.' Liverpool, University Press of Liverpool.

MOORE, H. B. 1938a. The biology of *Purpura lapillus*. Part II. Growth. *J. mar. biol. Ass. U.K.*, **23**, 57–66.

MOORE, H. B. 1938b. The biology of *Purpura lapillus*. Part III. Life history and relation to environmental factors. *J. mar. biol. Ass. U.K.*, **23**, 67–74.

MOORE, H. B. 1938c. Algal production and food requirements of a limpet. *Proc. malac. Soc. Lond.*, **23**, 117–118.

MOORE, H. B. 1939. Faecal pellets in relation to marine deposits. In 'Recent Marine Sediments: a Symposium.' (P. D. Trask ed.). London, Murby for the American Association of Petroleum Geologists, Tulsa, Oklahoma.

MOORE, H. B. 1940. The biology of *Littorina littorea*. Part II. Zonation in relation to other gastropods on stony and muddy shores. *J. mar. biol. Ass. U.K.*, **24**, 227–237.

MOQUIN-TANDON, A. 1855. 'Histoire naturelle des mollusques terrestres et fluviatiles de France.' Paris, Ballière.

MORGAN, C. L. 1894. The homing of limpets. *Nature, Lond.*, **51**, 127.

MORITZ, C. E. 1939. Organogenesis in the gasteropod, *Crepidula adunca* Sowerby. *Univ. Calif. Publ. Zool.*, **43**, 217–248.

MORSE, E. S. 1880. The gradual dispersion of certain mollusks in New England. *Bull. Essex Inst.*, **12**, 3–8.

MORTON, J. E. 1955. The functional morphology of the British Ellobiidae (Gastropoda Pulmonata) with special reference to the digestive and reproductive systems. *Phil. Trans. R. Soc. B*, **239**, 89–160.

MOSELEY, H. N. 1877. On the colouring matter of marine animals. *Quart. J. micr. Sci.*, **17**, 1–23.

MURRAY, J. & J. HJORT. 1912. 'The Depths of the Ocean.' London, Macmillan.

NAEF, A. 1913. Studien zur generellen Morphologie der Mollusken. 1. Teil: Über Torsion und Asymmetrie der Gastropoden. *Ergebn. Zool.*, **3**, 73–164.

NAEF, A. 1926. Studien zur generellen Morphologie der Mollusken. 3. Teil: Die typischen Beziehungen der Weichtiere untereinander und das Verhältnis ihrer Urformen zu anderen Cölomaten. *Ergebn. Zool.*, **6**, 27–124.

NATHUSIUS-KÖNIGSBORN, W. VON. 1877. 'Untersuchungen über nicht celluläre Organismen namentlich Crustaceen-Panzer, Molluskenschalen und Eihüllen.' Berlin, Weigardt, Hempel & Parey.

NEEDHAM, J. 1938. Contributions of chemical physiology to the problem of reversability in evolution. *Biol. Rev.*, **13**, 225–251.

NEKRASSOW, A. D. 1928. Vergleichende Morphologie der Laiche von Süsswassergastropoden. *Z. Morph. Ökol. Tiere*, **13**, 1–35.

NELSON, T. C. 1918. On the origin, nature and function of the crystalline style of lamellibranchs. *J. Morph.*, **31**, 53–111.

NEWCOMBE, C. L., C. E. MILLER & D. W. CHAPPEL. 1936. Preliminary report on respiratory studies of *Littorina irrorata. Nature, Lond.,* **137,** 33.

NEWELL, G. E. 1958a. The behaviour of *Littorina littorea* (L.) under natural conditions and its relation to position on the shore. *J. mar. biol. Ass U.K.,* **37,** 229–239.

NEWELL, G. E. 1958b. An experimental analysis of the behaviour of *Littorina littorea* (L.) under natural conditions and in the laboratory. *J. mar. biol. Ass. U.K.,* **37,** 241–266.

NICOL, E. A. T. 1936. The brackish-water lochs of North Uist. *Proc. R. Soc. Edinb. B,* **66,** 169–195.

NISBET, R. H. 1953. 'The structure and function of the buccal mass in some gastropod molluscs. I. *Monodonta lineata* (da Costa).' Ph.D. Thesis, University of London.

NORTH, W. J. 1954. Size distribution, erosive activities, and gross metabolic efficiency of the marine intertidal snails, *Littorina planaxis* and *L. scutulata. Biol. Bull., Wood's Hole,* **106,** 185–197.

NOWIKOFF, M. 1912. Studien über das Knorpelgewebe von Wirbellosen. *Z. wiss. Zool.,* **103,** 661–717.

ODHNER, N. H. 1932. Zur Morphologie und Systematik der Fissurelliden. *Jena. Z. Naturw.,* **67,** 292–309.

OHBA, S. 1952. Analysis of activity rhythm in the marine gastropod, *Nassarius festivus,* inhabiting the tide pool. I. On the effect of tide and food in the daytime rhythm of activity. *Annot. zool. Jap.,* **25,** 289–297.

OHBA, S. 1954. Analysis of activity rhythm in the marine gastropod, *Nassarius festivus,* inhabiting the tide pool. II. Nocturnal activity and its artificial control by light. *Biol. J. Okayama Univ.,* **1,** 209–216.

ØKLAND, F. 1933. Litoralstudien an der Skagerrakküste Norwegens: die Verbreitung von *Purpura lapillus, Patella vulgata* und den *Littorina*-Arten in Tromö. *Zoogeographica,* **1,** 579–601.

OLMSTED, J. M. D. 1917. Notes on the locomotion of certain Bermudian mollusks. *J. exp. Zool.,* **24,** 223–236.

OLDFIELD, E. 1959. 'The embryology of *Lasaea rubra* (Montagu) and the functional morphology of *Kellia suborbicularis* (Montagu), *Montacuta ferruginosa* (Montagu) and *M. substriata* (Montagu) (Mollusca, Lamellibranchiata).' Ph.D Thesis, University of London.

ORTON, J. H. 1909. On the occurrence of protandric hermaphroditism in the mollusc *Crepidula fornicata. Proc. R. Soc. B,* **81,** 468–484.

ORTON, J. H. 1912a. An account of the natural history of the slipper limpet (*Crepidula fornicata*), with some remarks on its occurrence on the oyster grounds on the Essex coast. *J. mar. biol. Ass. U.K.,* **9,** 437–443.

ORTON, J. H. 1912b. The mode of feeding of *Crepidula,* with an account of the current-producing mechanism in the mantle cavity, and some remarks on the mode of feeding in gastropods and lamellibranchs. *J. mar. biol. Ass. U.K.,* **9,** 444–478.

ORTON, J. H. 1913a. On the breeding habits of *Echinus miliaris,* with a note on the feeding habits of *Patella vulgata. J. mar. biol. Ass. U.K.,* **10,** 254–257.

ORTON, J. H. 1913b. On ciliary mechanisms in brachiopods and some polychaetes, with a comparison of the ciliary mechanisms on the gills of molluscs, Protochordata, brachiopods and crypto-cephalous polychaetes, and an account of the endostyle of *Crepidula* and its allies. *J. mar. biol. Ass. U.K.,* **10,** 283–311.

ORTON, J. H. 1920. Sex-phenomena in the common limpet (*Patella vulgata*). *Nature, Lond.,* **104,** 373.

ORTON, J. H. 1927. The habits and economic importance of the rough whelk-tingle (*Murex erinaceus*). *Nature, Lond.,* **120,** 653–654.

ORTON, J. H. 1928a. Observations on *Patella vulgata.* Part 1. Sex-phenomena, breeding and shell-growth. *J. mar. biol. Ass. U.K.,* **15,** 851–862.

ORTON, J. H. 1928b. Observations on *Patella vulgata.* Part II. Rate of growth of shell. *J. mar. biol. Ass. U.K.,* **15,** 863–874.

ORTON, J. H. 1929a. Habitats and feeding habits of *Ocenebra erinacea. Nature, Lond.,* **124,** 370–371.

ORTON, J. H. 1929b. Observations on *Patella vulgata.* Part III. Habitat and habits. *J. mar. biol. Ass. U.K.,* **16,** 277–288.

ORTON, J. H. 1930. On the oyster drills in the Essex estuaries. *Essex Nat.,* **22,** 298.

ORTON, J. H. 1932. Studies on the relation between organisms and environment. *Proc. Lpool. biol. Soc.,* **46,** 1–16.

ORTON, J. H. 1946. Biology of *Patella* in Great Britain. *Nature, Lond.,* **158,** 173–174.

ORTON, J. H. 1950a. Recent breeding phenomena in the American slipper limpet, *Crepidula fornicata. Nature, Lond.,* **165,** 433.

ORTON, J. H. 1950b. The recent extension in the distribution of the American slipper limpet, *Crepidula fornicata,* into Lyme Bay in the English Channel. *Proc. malac. Soc. Lond.,* **28,** 168–184.

ORTON, J. H. & H. N. LEWIS. 1931. On the effect of the severe winter of 1928–29 on the oyster drills (with a record of five years' observations on sea-temperature on the oyster beds) of the Blackwater estuary. *J. mar. biol. Ass. U.K.,* **17,** 301–313.

ORTON, J. H., A. J. SOUTHWARD & J. M. DODD. 1956. Studies on the biology of limpets. Part II. The breeding of *Patella vulgata* L. in Britain. *J. mar. biol. Ass. U.K.,* **35,** 149–176.

ORTON, J. H. & R. WINCKWORTH. 1928. The occurrence of the American oyster pest *Urosalpinx cinerea* (Say) on English oyster beds. *Nature, Lond.,* **122,** 241.

OTTO, H. & C. TÖNNIGES. 1906. Untersuchungen über die Entwicklung von *Paludina vivipara. Z. wiss. Zool.,* **80,** 411–514.

OVERTON, H. 1905. Note on variety of *Paludestrina jenkinsi,* E. A. Smith. *J. Malacol.,* **12,** 15.

OWEN, G. 1956. Observations on the stomach and digestive diverticula of the Lamellibranchia. II. The Nuculidae. *Quart. J. micr. Sci.,* **97,** 541–567.

OWEN, G. 1958. Observations on the stomach and digestive gland of *Scutus breviculus* (Blainville). *Proc. malac. Soc. Lond.,* **33,** 103–114.

PARKER, G. H. 1911. The mechanism of locomotion in gastropods. *J. Morph.,* **22,** 155–170.

PATANÈ, L. 1933. Sul comportamento di *Littorina neritoides* L. mantenuta in ambiente subaero ed in altro condizioni sperimentali. *Atti Accad. naz. Lincei Rc.* (6), **17,** 961–967.

PATIL, A. M. 1958. The occurrence of a male of the prosobranch *Potamopyrgus jenkinsi* (Smith) var. *carinata* Marshall in the Thames at Sonning, Berkshire. *Ann. Mag. nat. Hist.* (13), **1,** 232–240.

PATTEN, W. 1886a. The embryology of *Patella. Arb. zool. Inst. Wien,* **6,** 149–174.

PATTEN, W. 1886b. Eyes of molluscs and arthropods. *Mitt. zool. Stat. Neapel,* **6,** 542–756.

PEILE, A. J. 1937. Some radula problems. *J. Conch., Lond.,* **20,** 292–304.

PELLEGRINI, O. 1948. Ricerche statistiche sulla sessualità di *Patella coerulea* L. *Boll. Zool.,* **15,** 115–121.

PELLEGRINI, O. 1949. Ermafroditismo proterandrico in *Calyptraea chinensis* (L.) (Gasteropoda Proso-branchiata). *Boll. Zool.,* **16,** 49–59.

PELSENEER, P. 1895. Hermaphroditism in Mollusca. *Quart. J. micr. Sci.,* **37,** 19–46.

PELSENEER, P. 1898–99. Recherches morphologiques et phylogénétiques sur les mollusques archaïques. *Mém. Sav. étr. Acad. R. Belg.,* **57,** 1–112.

PELSENEER, P. 1902. Sur l'exagération du dimorphisme sexuel chez un gastéropode marin. *J. Conchyliol.,* **50,** 41–43.

PELSENEER, P. 1906. Mollusca. In 'A Treatise on Zoology,' **5,** (E. R. Lankester ed.). London, A. & C. Black.

PELSENEER, P. 1910. Glandes pédieuses et coques ovigères des gastropodes. *Bull. sci. Fr. Belg.,* **44,** 1–9.

PELSENEER, P. 1911. Recherches sur l'embryologie des gastropodes. *Mém. Acad. R. Belg. Cl. Sci.* (2), **3,** 1–167.

PELSENEER, P. 1914. De quelques *Odostomia* et d'un Monstrilide. *Bull. sci. Fr. Belg.,* **48,** 1–8.

PELSENEER, P. 1925. Gastropodes marins carnivores *Natica* et *Purpura. Ann. Soc. zool. Belg.,* **55,** 37–39.

PELSENEER, P. 1926a. La proportion relative des sexes chez les animaux et particulièrement chez les mollusques. *Mém. Acad. R. Belg. Cl. Sci.* (2), **8,** 1–258.

PELSENEER, P. 1926b. Notes d'embryologie malacologique. *Bull. biol.,* **60,** 88–112.

PELSENEER, P. 1928. Les parasites des mollusques et les mollusques parasites. *Bull. Soc. zool. Fr.,* **53,** 158–189.

PELSENEER, P. 1932. Le métamorphose préadulte des Cypraeidae. *Bull. biol.,* **66,** 149–163.

PELSENEER, P. 1935. 'Essai d'Ethologie zoologique d'après l'Etude des Mollusques.' *Acad. R. Belg. Cl. Sci. Publ. Fondation Agathon de Potter,* **I,** 1–662.

PEREZ, C. & J. KOSTITZINE. 1930. Processus de résorption dans l'ovaire de la Turritelle. *C. R. Soc. Biol., Paris,* **104,** 1270–1272.

PERRIER, R. 1889. Recherches sur l'anatomie et l'histologie du rein des gastéropodes Prosobranchiata. *Ann. Sci. nat. Zool.* (7), **8,** 61–192.

PETERSEN, C. G. J. 1911. Some experiments on the possibility of combating the harmful animals of the fisheries, especially the whelks in the Limfjord. *Rep. Danish biol. Sta.,* **19,** 1–20.

PETERSEN, C. G. J. 1918. The sea bottom and its production of fish-food. *Rep. Danish biol. Sta.,* **25,** 1–62.

PFEIL, E. 1922. Die Statocyste von *Helix pomatia* L. *Z. wiss. Zool.,* **119,** 79–113.

PIERON, H. 1909. Contribution à la biologie de la Patelle et de la Calyptrée. *Arch. Zool. exp. gén.,* **41,** 18–24.

PIERON, H. 1933. Notes éthologiques sur les gastéropodes perceurs et leur comportement avec utilisation de méthodes statistiques. *Arch. Zool. exp. gén.,* **75,** 1–20.

PILSBRY, H. A. 1948. Land Mollusca of North America (North of Mexico). **II,** 2. *Monogr. Acad. nat. Sci. Philad.,* **3,** 521–1113.

PLATE, L. H. 1895. Bemerkungen über die Phylogenie und die Entstehung der Asymmetrie der Mollusken. *Zool. Jb. (Anat. Ont.),* **9,** 162–206.

POJETA, J. & RUNNEGAR, B. 1976. The paleontology of rostroconch mollusks and the early history of the phylum Mollusca. *Geol. Surv. Prof. Pap. (U.S.),* **968,** 1–88.

PONDER, W. F. & WAREN, A. 1988. Classification of the Caenogastropoda and Heterostropha—a list of the family-group names and higher taxa. *Malacol. Rev.,* suppl. **4,** 288–326.

POPE, T. E. B. 1910–11. The oyster drill and other predatory Mollusca. *Rep. U.S. Bur. Fish. Wash.,* (Unpublished.)

PORTMANN, A. 1925. Der Einfluss der Nähreier auf die Larvenentwickelung von *Buccinum* und *Purpura. Z. Morph. Ökol. Tiere,* **3,** 526–541.

PORTMANN, A. 1927. Die Nähreierbildung durch atypische Spermien bei *Buccinum undatum* L. *Z. Zellforsch.,* **5,** 230–243.

PORTMANN, A. 1930. Die Larvennieren von *Buccinum undatum* L. *Z. Zellforsch.,* **10,** 401–410.

PORTMANN, A. 1931a. Die Entstehung der Nähreier bei *Purpura lapillus* durch atypische Befruchtung. *Z. Zellforsch.,* **12,** 167–178.

PORTMANN, A. 1931b. Die atypische Spermatogenese bei *Buccinum undatum* L. und *Purpura lapillus* L. *Z. Zellforsch.,* **12,** 307–326.

PORTMANN, A. 1955. La métamorphose 'abritée' de *Fusus* (Gast. Prosobranches). *Rev. suisse Zool.,* **62,** fasc. suppl., 236–252.

PRASHAD, B. 1932. *Pila* (The apple snail). *Indian zool. Mem.,* **4.** Lucknow, Methodist Publishing House.

PRENANT, M. 1924. L'activité sécrétrice dans l'épithelium supérieur de la gaine radulaire chez l'escargot (*Helix pomatia* L.). *Bull. soc. Zool. Fr.,* **49,** 336–341.

PRENANT, M. 1925. Sur la permanence des odontoblastes de la radula. *Bull. Soc. zool. Fr.,* **50,** 164–167.

PRUVOT, G. 1913. Sur la structure et la formation des soies de 'Nereis'. *IXth Cong. int. Zool. Monaco,* 348–355.

PRUVOT-FOL, A. 1925. Morphogénèse des odontoblastes chez les Mollusques. *Arch. Zool. exp. gén.,* **64,** 1–7.

PRUVOT-FOL, A. 1926. Le bulbe buccal et la symétrie des mollusques. I. La radula. *Arch. Zool. exp. gén.,* **65,** 209–343.

PRUVOT-FOL, A. 1954. Le bulbe buccal et la symétrie des mollusques. II. *Arch. Zool. exp. gén.,* **91,** 235–330.

QUAST, P. 1924a. Chemische Untersuchungen des Organextrakts der Konkrementendrüse und des Nephridium von *Cyclostoma elegans* Drap. *Z. Biol.,* **80,** 211–222.

QUAST, P. 1924*b*, Der Konkrementendrüse von *Cyclostoma elegans. Z. ges. Anat. I. Z. Anat. EntwGesch.,* **72,** 169–198.

QUICK, H. E. 1920. Notes on the anatomy and reproduction of *Paludestrina (Hydrobia) stagnalis. J. Conch., Lond.,* **16,** 96–97.

QUICK, H. E. 1924. Length of life of *Paludestrina ulvae. J. Conch., Lond.,* **17,** 169.

RAMAMOORTHI, K. 1955. Studies in the embryology and development of some melaniid snails. *J. zool. Soc. India,* **7,** 25–34.

RAMMELMEYER, H. 1925. Zur Morphologie der *Puncturella noachina. Zool. Anz.,* **64,** 105–114.

RANDLES, W. B. 1905. Some observations on the anatomy and affinities of the Trochidae. *Quart. J. micr. Sci.,* **48,** 33–78.

RASETTI, F. 1957. Additional fossils from the Middle Cambrian Mt. Whyte formation of the Canadian Rocky Mountains. *J. Paleont.,* **31,** 955–972.

RASMUSSEN, E. 1944. Faunistic and biological notes on marine invertebrates. I. *Vidensk. Medd. dansk naturh. Foren. Kbh.,* **107,** 207–233.

RASMUSSEN, E. 1951. Faunistic and biological notes on marine invertebrates. II. The eggs and larvae of some Danish marine gastropods. *Vidensk. Medd. dansk naturh. Foren. Kbh.,* **113,** 201–249.

RATH, E. 1988. Organization and systematic position of the Valvatidae. *Malacol. Rev.,* suppl. **4,** 194–204.

REAUMUR, R. 1711. De la formation et de l'accroissement des coquilles des animaux tant terrestres qu'aquatiques, soit de mer soit de rivière. *Mém. Hist. Acad. Sci., Année,* **1709,** 364–400.

REES, C. B. 1940. A preliminary study of the ecology of a mud-flat. *J. mar. biol. Ass. U.K.,* **24,** 185–199.

REMANE, A. 1950. Die Entstehung der Metamerie der Wirbellosen. *Verh. dtsch. zool. Ges., Zool. Anz.* Suppl. **14,** 16–23.

RHEIN, A. 1935. Diploide Parthenogenese bei *Hydrobia jenkinsi* Smith. *Naturwissenschaften,* **23,** 100.

RIJNBERK, G. VAN. 1919. Petites contributions à la physiologie comparée. IV. Sur le mouvement de locomotion de l'escargot terrestre *Helix aspersa. Arch. néerl. Physiol.,* **3,** 539–552.

RISBEC, J. 1937. Les irrégularités et les anomalies du développement embryonnaire chez *Murex erinaceus* L. et chez *Purpura lapillus* L. *Bull. Lab. marit. Dinard,* **17,** 25–38.

RISBEC, J. 1953. Notes sur la biologie et l'anatomie de *Ianthina globosa* (Gast. Prosobranches). *Bull. Soc. zool. Fr.,* **78,** 194–201.

RISBEC, J. 1954. Observations sur les Eulimidae (Gastéropodes) de Nouvelle-Calédonie. *Bull. Mus. Hist. nat., Paris,* **26,** 109–116.

ROAF, H. E. 1906. A contribution to the study of the digestive gland in Mollusca and decapod Crustacea. *Biochem. J.,* **I,** 390–397.

ROAF, H. E. 1908. The hydrolytic enzymes of invertebrates. *Biochem. J.,* **3,** 462–472.

ROBERT, A. 1900. Gastéropodes. Le troque. In 'Zoologie Descriptive,' **2,** (L. Boutan ed.). Paris, Doin.

ROBERT, A. 1902. Recherches sur le développement des troques. *Arch. Zool. exp. gén.* (3), **10,** 269–538.

ROBERTSON, J. D. 1949. Ionic regulation in some marine invertebrates. *J. exp. Biol.,* **26,** 182–200.

ROBERTSON, R. 1957. Gastropod host of an *Odostomia. Nautilus,* **70,** 96–97.

ROBINSON, E. 1955. Observations on the toxoglossan gastropod *Mangelia brachystoma* (Philippi) and on the stenoglossan *Trophon muricatus* (Montagu). M. Sc. Thesis, University of London.

ROBSON, G. C. 1920. On the anatomy of *Paludestrina jenkinsi. Ann. Mag. nat. Hist.* (9), **5,** 425–431.

ROBSON, G. C. 1922. On the style-sac and intestine in Gastropoda and Lamellibranchia. *Proc. malac. Soc. Lond.,* **15,** 41–46.

ROBSON, G. C. 1923. Parthenogenesis in the mollusc *Paludestrina jenkinsi.* Part I. *J. exp. Biol.,* **1,** 65–77.

ROBSON, G. C. 1926. Parthenogenesis in the mollusc *Paludestrina jenkinsi.* Part II. The genetical behaviour, distribution, etc., of the keeled form ('var. *carinata'). J. exp. Biol.,* **3,** 149–159.

ROHLACK, S. 1959. Über das Vorkommen von Sexualhormonen bei der MeeresSchnecke *Littorina littorea* L. *Z. vergl. Physiol.,* **42,** 164–180.

ROSEN, B. 1932. Zur Verdauungsphysiologie der Gastropoden. *Zool. Bidr. Uppsala*, **14**, 1–67.

ROSEN, B. 1937. Vergleichende Studien über die Proteinasen von Gastropoden und dekapoden Crustaceen. *Z. vergl. Physiol.*, **24**, 602–612.

ROSEN, N. 1910. Zur Kenntnis der parasitischen Schnecken. *Acta Univ. lund.*, N.F. Afd., **2**, 1–67.

RÖSSLER, R. 1885. Die Bildung der Radula bei den cephalophoren Mollusken. *Z. wiss. Zool.*, **41**, 447–482.

ROTARIDES, M. 1934. Zum Formproblem des Schneckenfusses. *Zool. Anz.*, **108**, 165–178.

ROTHSCHILD, A. & M. ROTHSCHILD. 1939. Some observations on the growth of *Peringia ulvae*. (Penant) 1777 in the laboratory. *Novit zool.*, **41**, 240–247.

ROTHSCHILD, M. 1941a. The metacercaria of a pleurolophocerca cercaria parasitizing *Peringia ulvae* (Pennant, 1777). *Parasitology*, **33**, 439–441.

ROTHSCHILD, M. 1941b. Observations on the growth and trematode infections of *Peringia ulvae* (Pennant), 1777, in a pool in the Tamar Saltings, Plymouth. *Parasitology*, **33**, 406–415.

ROTTMANN, G. 1901. Über die Embryonalentwicklung der Radula bei den Mollusken. I. Die Entwicklung der Radula bei den Cephalopoden. *Z. wiss. Zool.*, **70**, 236–262.

RÜCKER, A. 1883. Über die Bildung der Radula bei *Helix pomatia*. *Ber. oberhess. Ges. Nat.-u. Heilk.*, **22**, 209–229.

RUDOLPHI, C. A. 1802. Fortsetzung der Beobachtungen. *Wiedemann's Arch. Zool. Zoot.*, **2**, (2), 1–67.

RUSSELL, E. S. 1907. Environmental studies on the limpet. *Proc. zool. Soc. Lond.*, 856–870.

RUSSELL, E. S. 1909. The growth of the shell of *Patella vulgata* L. *Proc. zool. Soc. Lond.*, 235–253.

SAHM, W. 1932. Bau und Wachstum des Deckels. In 'Vergleichende Betrachtungen über das Schalenwachstum der Weichtiere (Mollusca). II. Deckel (Operculum) und Haus (Concha) der Schnecken (Gastropoden).' (A. Fleischmann ed.). *Z. Morph. Ökol. Tiere*, **25**, 549–622.

SALVINI-PLAWEN, L. v. 1972. Zur Morphologie und Phylogenie der Mollusken. *Z. wiss. Zool.*, **184**, 205–394.

SALVINI-PLAWEN, L. v. 1980. A reconsideration of systematics in the Mollusca (phylogeny and higher classification). *Malacologia*, **19**, 249–278.

SALVINI-PLAWEN, L. v. 1985. Early evolution and primitive groups. In 'The Mollusca' **10** (E. R. Trueman & M. R. Clarke eds), 59–150. Orlando, Academic Press.

SALVINI-PLAWEN, L. v. & HASZPRUNAR, G. 1987. The Vetigastropoda and the systematics of streptoneurous Gastropoda (Mollusca). *J. Zool., Lond.*, **211**, 747–770.

SANDEEN, M. I., G. C. STEPHENS & F. A. BROWN, JR. 1954. Persistent daily and tidal rhythms of oxygen consumption in two species of marine snails. *Physiol. Zool.*, **27**, 350–356.

SANDER, K. 1950. Beobachtungen zur Fortpflanzung von *Assiminea grayana* Leach. *Arch. Molluskenk.*, **79**, 147–149.

SANDER, K. 1952. Beobachtungen zur Fortpflanzung von *Assiminea grayana* Leach (2). *Arch. Molluskenk.*, **81**, 133–134.

SANDERSON, A. R. 1939. The cytology of pathogenesis in the snail *Potamopyrgus jenkinsi* Smith. *Advanc. Sci. Lond.*, **1**, 46.

SANDERSON, A. R. 1940. Maturation in the parthenogenetic snail *Potamopyrgus jenkinsi* Smith and in the snail *Peringia ulvae* (Pennant). *Proc. zool. Soc. Lond.*, **110**, 11–15.

SARASIN, P. 1882. Entwickelungsgeschichte der *Bithynia tentaculata*. *Arb. zool. Inst. Würzburg*, **6**, 1–68.

SAUNDERS, A. M. C. & M. POOLE. 1910. The development of *Aplysia punctata*. *Quart. J. micr. Sci.*, **55**, 497–539.

SCHÄFER, H. 1952. Ein Beitrag zur Ernährungsbiologie von *Bithynia tentaculata* L. (Gastropoda Prosobranchia). *Zool. Anz.*, **148**, 299–303.

SCHÄFER, H. 1953a. Beobachtungen zur Ökologie von *Bithynia tentaculata*. *Arch. Molluskenk.*, **82**, 67–70.

SCHÄFER, H. 1953b. Beiträge zur Ernährungsbiologie einheimischer Süsswasserprosobranchier. *Z. Morph. Ökol. Tiere*, **41**, 247–264.

SCHARRER, B. 1935. Über das Hanströmsche Organ X bei Opisthobranchiern. *Pubbl. Staz. zool. Napoli,* **15,** 135–142.

SCHARRER, B. 1937. Über sekretorisch tätige Nervenzellen bei wirbellosen Tieren. *Naturwissenschaften,* **25,** 131–138.

SCHARRER, B. 1954. Neurosecretion in invertebrates: a survey. *Pubbl. Staz. zool. Napoli,* **24,** suppl., 38–40.

SCHEPOTIEFF, A. 1903. Untersuchungen über den feineren Bau der Borsten einiger Chätopoden und Brachiopoden. *Z. wiss. Zool.,* **74,** 656–710.

SCHEPOTIEFF, A. 1904. Untersuchungen über die Borstentaschen einiger Polychäten. *Z. wiss. Zool.,* **77,** 586–605.

SCHIEMENZ, P. 1884. Über die Wasseraufnahme bei Lamellibranchiaten und Gastropoden (einschliesslich der Pteropoden). *Mitt. zool. Stat. Neapel,* **5,** 509–543.

SCHIEMENZ, P. 1887. Über die Wasseraufnahme bei Lamellibranchiaten und Gastropoden (einschliesslich der Pteropoden). Zweiter Theil. *Mitt. zool. Stat. Neapel,* **7,** 423–472.

SCHIEMENZ, P. 1891. Wie bohrt *Natica* die Muscheln an? *Mitt. zool. Stat. Neapel,* **10,** 153–169.

SCHILDER, F. A. 1947. Die Zahl der Prosobranchier in Vergangenheit und Gegenwart. *Arch. Molluskenk.,* **76,** 37–44.

SCHITZ, V. 1920a. Sur la spermatogénèse chez *Cerithiopsis vulgata* Brug., *Turritella triplicata* Brocchi (*mediterranea* Monterosato) et *Bittium reticulatum* da Costa. *Arch. Zool. exp. gén.,* **58,** 489–520.

SCHITZ, V. 1920b. Sur la spermatogénèse chez *Murex trunculus* L., *Apporhais pespelecani* L., *Fusus* sp. et *Nassa reticulata* L. *Arch. Zool. exp. gén.,* **59,** 477–508.

SCHNABEL, H. 1903. Über die Embryonalentwicklung der Radula bei den Mollusken. II. Die Entwicklung der Radula bei der Gastropoden. *Z. wiss. Zool.,* **74,** 616–655.

SEGAL, E. 1956. Adaptive differences in water-holding capacity in an intertidal gastropod. *Ecology,* **37,** 174–178.

SEGERSTRÅLE, S. G. 1949. The brackish-water fauna of Finland. *Oikos,* **1,** 127–141.

SEIFERT, R. 1935. Bemerkungen zur Artunterscheidung der deutschen Brackwasser-Hydrobien. *Zool. Anz.,* **110,** 233–239.

SESHAPPA, G. 1947. Oviparity in *Littorina saxatilis* (Olivi). *Nature, Lond.,* **160,** 335.

SHARMAN, M. 1956. Note on *Capulus ungaricus* (L.). *J. mar. biol. Ass. U.K.,* **35,** 445–450.

SHOTWELL, J. A. 1950. Distribution of volume and relative linear measurement changes in *Acmaea,* the limpet. *Ecology,* **31,** 51–61.

SIMROTH, H. 1882. Über die Bewegung und das Bewegungsorgan des *Cyclostoma elegans* und der einheimischen Schnecken überhaupt. *Z. wiss. Zool.,* **36,** 1–67.

SIMROTH, H. 1895. Die Gastropoden der Plankton-Expedition. *Eregbn. Atlant. Planktonexped.,* **2,** 1–206.

SIMROTH, H. 1896–1907. Gastropoda Prosobranchia. In 'Klassen und Ordungen des Tierreichs,' **3,** (H. G. Bronn ed.). Leipzig, Akademische Verlagsgesellschaft.

SIMROTH, H. 1911. Die Gastropoden des nordischen Planktons. *Nordisches Plankton,* **5,** 1–36.

SMIDT, E. 1938. Notes on the reproduction and rate of growth in *Rissoa membranacea* (Adams) (Gastropoda Prosobranchiata) in the Sound. *Vidensk. Medd. dansk naturh. Foren. Kbh.,* **102,** 169–181.

SMIDT, E. L. B. 1944. Biological studies of the invertebrate fauna of the harbour of Copenhagen. *Vidensk. Medd. dansk naturh. Foren. Kbh.,* **107,** 235–316.

SMIDT, E. L. B. 1951. Animal production in the Danish Waddensee. *Medd. Komm. Havundersøg., Kbh.,* Ser. Fiskeri, **11,** no. 6, 1–151.

SMITH. E. A. 1889. Notes on British *Hydrobiae* with a description of a supposed new species. *J. Conch., Lond.,* **6,** 142–145.

SMITH, F. G. W. 1935. The development of *Patella vulgata. Phil. Trans. R. Soc. B,* **225,** 95–125.

SMITH, J. E. & G. E. NEWELL. 1955. The dynamics of the zonation of the common periwinkle (*Littorina littorea* (L.)) on a stony beach. *J. Anim. Ecol.,* **24,** 35–36.

SOLLAS, I. B. J. 1907. The molluscan radula: its chemical composition, and some points in its development. *Quart. J. micr. Sci.,* **51,** 115–136.

SOOS, L. 1936. Zur Anatomie der Ungarischen Melaniiden. *Allat. Közlem.,* **33,** 103–128.

SOUTHWARD, A. J. 1958. Note on the temperature tolerances of some intertidal animals in relation to environmental temperatures and geographical distribution. *J. mar. biol. Ass. U.K.,* **37,** 49–66.

SOUTHWARD, A. J. & D. J. CRISP. 1954. The distribution of certain intertidal animals around the Irish coast. *Proc. R. Ir. Acad.,* **57,** B, **1,** 1–29.

SOUTHWARD, A. J. & J. H. ORTON. 1954. The effects of wave action on the distribution and numbers of the commoner animals living on the Plymouth breakwater. *J. mar. biol. Ass. U.K.,* **33,** 1–19.

SPEK, J. 1921. Beiträge zur Kenntis der chemischen Zusammensetzung und Entwicklung der Radula der Gastropoden. *Z. wiss. Zool.,* **118,** 313–363.

SPENGEL, J. W. 1881. Die Geruchsorgane und das Nervensystem der Mollusken. *Z. wiss. Zool.,* **35,** 333–383.

SPILLMANN, J. 1905. Zur Anatomie und Histologie des Herzens und der Hauptarterien der Diotocardier. *Jena. Z. Naturw.,* **40,** 537–588.

SPOEL, S. VAN DER. 1958. Groei en ouderdom bij *Viviparus contectus* (Millet, 1813) en *Viviparus viviparus* (Linné, 1758). *Basteria,* **22,** 77–90.

SPOONER,. G. M. & H. B. MOORE. 1940. The ecology of the Tamar estuary. VI. An account of the macrofauna of the intertidal muds. *J. mar. biol. Ass. U.K.,* **24,** 283–330.

STAIGER, H. 1950a. Chromosomenzahlen stenoglosser Prosobranchier. *Experientia,* **6,** 54–59.

STAIGER, H. 1950b. Chromosomenzahl-Varianten bei *Purpura lapillus. Experientia,* **6,** 140–145.

STAIGER, H. 1950c. Zur Determination der Nähreier bei Prosobranchiern. *Rev. suisse Zool.,* **57,** 496–503.

STAIGER, H. 1951. Cytologische und morphologische Untersuchungen zur Determination der Nähreier bei Prosobranchiern. *Z. Zellforsch.,* **35,** 496–549.

STAIGER, H. 1954. Die Chromosomendimorphismus beim Prosobranchier *Purpura lapillus* in Beziehung zur Ökologie der Art. *Chromosoma,* **6,** 419–478.

STAIGER, H. 1955. Reziproke Translokationen in naturlichen Populationen von *Purpura lapillus* (Prosobranchia). *Chromosoma,* **7,** 181–197.

STAIGER, H. 1957. Genetical and morphological variation in *Purpura lapillus* with respect to local and regional differentiation of population groups. *Année biol.,* **33,** 251–258.

STARMÜHLNER, F. 1952. Zur Anatomie, Histologie und Biologie einheimischer Prosobranchier. *Öst. zool. Z.,* **3,** 546–590.

STARMÜHLNER, F. 1956. Zur Molluskenfauna des Felslitorals und submariner Höhlen am Capo di Sorrento (I. Teil). Ergebnisse der Österr. Tyrrhenia-Expedition 1952 Teil IV. *Öst. zool. Z.,* **6,** 147–249. II. Teil. *Öst. zool. Z.,* **6,** 631–713.

STASEK, C. R. 1972. The molluscan framework. In 'Chemical Zoology' (M. Florkin & B. T. Scheer eds), **3,** 1–44. New York, Academic Press.

STEPHENS, G. C., M. I. SANDEEN & H. M. WEBB. 1953. A persistent tidal rhythm of activity in the mud snail *Nassa obsoleta. Anat. Rec.,* **117,** 635.

STEPHENSON, T. A. 1924. Notes on *Haliotis tuberculata. J. mar. biol. Ass. U.K.,* **13,** 480–495.

STEUSLOFF, U. 1927. Die Bedeutung der *Paludestrina jenkinsi* E. A. Smith für unsere Vorstellungen über Artentstehung und Artverbreitung. *Verh. int. Ver. Limnol.,* **3,** 454–459.

STEUSLOFF, U. 1939. *Potamopyrgus crystallinus crystallinus* J. T. Marshall mit Kalkkielen auf der Schale. *Arch. Molluskenk.,* **71,** 82–86.

STINSON, R. H. 1946. Observations on the natural history of clam drills. *MS Rep. Fish. Res. Bd. Canada,* 383.

STROHL, J. 1914. Die Exkretion. In 'Handluch der vergleichenden Physiologie,' **II,** 2, (H. Winterstein ed.). Berlin, Gustav Fischer.

SUZUKI, S. 1934. On the innervation of the heart of limpets. *Sci. Rep. Tohoku Imp., Univ.* (4) (Biol.), **9,** 117–121.

SUZUKI, S. 1935. The innervation of the heart of molluscs. *Sci. Rep. Tohoku Imp. Univ.* (4) (Biol.), **10,** 15–27.

SYKES, E. R. 1903. Notes on British Eulimidae. *Proc. malac. Soc. Lond.,* **5,** 348–353.

TATTERSALL, W. M. 1920. Notes on the breeding habits and life history of the periwinkle. *Sci. Invest. Fish. Br. Ire.,* **1,** 1–11.

TAYLOR, D. W. & N. F. SOHL. 1962. An outline of gastropod classification. *Malacologia* **1,** 7–32.

TAYLOR, F. 1900. *Paludestrina jenkinsi* Smith at Droylsden, Lancashire. *J. Conch., Lond.,* **9,** 340.

TECHOW, G. 1910. Zur Kenntnis der Schalenregeneration bei den Gastropoden. *Arch. Entwick-Mech.,* **31,** 258–288.

THAMDRUP, H. M. 1935. Beiträge zur Ökologie der Wattenfauna. *Medd. Komm. Havundersøg., Kbh.,* ser. Fiskeri, **10,** no. 2, 1–125.

THIELE, J. 1897. Beiträge zur Kenntnis der Mollusken. III. Über Hautdrüsen und ihre Derivative. *Z. wiss. Zool.,* **62,** 632–670.

THIELE, J. 1928. Über ptenoglosse Schnecken. *Z. wiss. Zool.,* **132,** 73–94.

THIELE, J. 1929–35. 'Handbuch der Systematischen Weichtierkunde.' 4 parts. **1** (1929), **4** (1935). Jena, Fischer.

THIEM, H. 1917a. Beiträge zur Anatomie und Phylogenie der Docoglossen. I. Zur Anatomie von *Helcioniscus ardosiaeus* Hombron et Jaquinot unter Bezugnahme auf die Bearbeitung von Erich Schuster in den Zoolog. Jahrb., Supplement XIII, Bd. IV, 1913. *Jena. Z. Naturw.,* **54** 333–404*b.*

THIEM, H. 1917b. Beiträge zur Anatomie und Phylogenie der Docoglossen. II. Die Anatomie und Phylogenie der Monobranchen. (Akmäiden und Scurriiden nach der Sammlung Plates.) *Jena. Z. Naturw.,* **54,** 405–630.

THOMAS, I. M. 1948. The adhesion of limpets. *Aust. J. Sci.,* **II,** 28–29.

THOMPSON, D'A. W. 1942. 'On Growth and Form.' Cambridge, Cambridge University Press.

THORSON, G. 1935. Studies on the egg-capsules and development of Arctic marine prosobranchs. *Medd. Grønland,* **100,** no. 5, 1–71.

THORSON, G. 1936. The larval development, growth and metabolism of Arctic marine bottom invertebrates. *Medd. Grønland,* **100,** no. 6, 1–155.

THORSON, G. 1940a. Studies on the egg masses and larval development of Gastropoda from the Iranian Gulf. *Danish Sci. Invest. Iran,* **2,** 159–238.

THORSON, G. 1940b. Notes on the egg-capsules of some North-Atlantic prosobranchs of the genus *Troschelia, Chrysodomus, Volutopsis, Sipho* and *Trophon. Vidensk. Medd. naturh. Foren. Kbh.,* **104,** 251–265.

THORSON, G. 1944. Marine Gastropoda Prosobranchiata. *Medd. Grønland,* **121,** no. 13, 1–181.

THORSON, G. 1946. Reproduction and larval development of Danish marine bottom invertebrates. *Medd. Komm. Havundersøg., Kbh.,* ser. Plankton, **4,** 1–523.

THORSON, G. 1950. Reproductive and larval ecology of marine bottom invertebrates. *Biol. Rev.,* **25,** 1–45.

THORSON, G. 1958. Parasitism in the marine gastropod-family Scalidae. *Vidensk. Medd. naturh. Foren. Kbh.,* **119,** 55–58.

TÖNNIGES, C. 1896. Die Bildung des Mesoderms bei *Paludina vivipara. Z. wiss. Zool.,* **61,** 541–605.

TOTZAUER, R. J. 1902. Nieren- und Gonadenverhältnisse von *Haliotis. Zool. Anz.,* **25,** 487–488.

TOURAINE, J. 1952. Les glandes pédieuses des gastéropodes prosobranches monotocardes. *Bull. Soc. zool. Fr.,* **77,** 240–241.

TRINCHESE, S. 1878. Anatomia e fisiologia della *Spurilla neapolitana. Mem. Accad. Bologna* (3), **9,** 1–48.

TRUEMAN, E. R. 1949. The ligament of *Tellina tenuis. Proc. zool. Soc. Lond.,* **119,** 717–742.

TSCHACHOTIN, S. 1908. Die Statocyste der Heteropoden. *Z. wiss. zool.,* **90,** 343–422.

TULLBERG, T. 1881. Studien über den Bau und das Wachstum des Hummerpanzers und der Molluskenschalen. *K. svenska VetenskAkad. Handl.,* **19,** (3), 1–57.

TURNER, H. J. 1953. The drilling mechanism of the Naticidae. *Ecology*, **34**, 222–223.

UNDERWOOD, A. J. 1972. Spawning, larval development and settlement behaviour of *Gibbula cineraria* (Gastropoda: Prosobranchia) with a reappraisal of torsion in gastropods. *Mar. Biol.*, **17**, 541–549.

VANDEBROEK, G. 1936. Organogénèse des follicules sétigères chez *Eisenia foetida* Sav. *Mém. Mus. Hist. nat. Belg.* (2), **3**, 559–568.

VANSTONE, J. H. 1894. Some points in the anatomy of *Melongena melongena*. *J. Linn. Soc.* (Zool.), **24**, 369–373.

VERLAINE, L. 1936. L'instinct et l'intelligence chez les Mollusques. Les gastéropodes perceurs de coquilles. *Mém. Mus. Hist. nat. Belg.* (2), **3**, 387–394.

VESTERGAARD, K. 1935. Über den Laich und die Larven von *Scalaria communis* (Lam.), *Nassarius pygmaeus* (Lam.) und *Bela turricula* (Mont.). *Zool. Anz.*, **109**, 217–222.

VILLEE, C. A. & T. C. GROODY. 1940. The behaviour of limpets with reference to their homing instinct. *Amer. midl. Nat.*, **24**, 190–204.

VILLEPOIX, M. DE. 1892. Recherches sur la formation et l'accroissement de la coquille des mollusques. *J. Anat. Paris*, **28**, 461–518 and 582–674.

VLES, F. 1907. Sur les ondes pédieuses des mollusques reptateurs. *C.R. Acad. Sci.*, *Paris*, **145**, 276–278.

VOLTZOW, J. 1990. The functional morphology of the pedal musculature of the marine gastropods *Busycon contrarium* and *Haliotis kamtschatkana*. *Veliger*, **33**, 1–19.

WAELE, A. DE. 1930. Le sang d'Anodonte et la formation de la coquille. *Mém. Acad. R. Belg. Cl. Sci.* (2), **10**, no. 3, 1–52.

WALNE, P. R. 1956. The biology and distribution of the slipper limpet *Crepidula fornicata* in Essex rivers with notes on the distribution of the larger epi-benthic invertebrates. *Fish. Invest. Lond.* (2), **20**, no. 6, 1–50.

WAREN, A. 1988. *Neopilina goesi*, a new Caribbean monoplacophoran mollusk dredged in 1869. *Proc. biol. Soc. Wash.*, **101**, 676–681.

WARWICK, T. 1944. Inheritance of the keel in *Potamopyrgus jenkinsi* (Smith). *Nature, Lond.*, **154**, 798–799.

WARWICK, T. 1952. Strains in the mollusc *Potamopyrgus jenkinsi* (Smith). *Nature, Lond.*, **169**, 551–552.

WARWICK, T. 1953. The nature of shell incrustations in some aquatic molluscs. *Proc. malac. Soc. Lond.*, **30**, 71–73.

WARWICK, T. 1955. *Potamopyrgus jenkinsi* (Smith) in recent deposits in Suffolk. *Proc. malac. Soc. Lond.*, **31**, 22–25.

WATERSTON, J. & J. W. TAYLOR. 1906. Land and freshwater molluscs of St. Kilda. *Ann. Scot. nat. Hist.*, 21–24.

WATERSTON, R. 1934. Occurrence of *Amnicola taylori* (E. A. Smith) and *Bithynia leachi* (Sheppard) in Scotland. *J. Conch., Lond.*, **20**, 55–56.

WATSON, H. 1955. The names of the two common species of *Viviparus*. *Proc. malac. Soc. Lond.*, **31**, 163–174.

WEBER, H. 1925. Über arhythmische Fortbewegung bei einigen Prosobranchiern. Ein Beitrag zur Bewegungsphysiologie der Gastropoden. *Z. vergl. Physiol.*, **2**, 109–121.

WEBER, H. 1927. Der Darm von *Dolium galea* L., eine vergleichend anatomische Untersuchung unter besonderer Berücksichtigung der *Tritonium*-Arten. *Z. Morph. Ökol. Tiere*, **8**, 663–804.

WEGMANN, H. 1884. Contributions à l'histoire naturelle des Haliotides. *Arch. Zool. exp. gén.* (2), **2**, 289–378.

WEGMANN, H. 1887. Notes sur l'organisation de la *Patella vulgata* L. *Rec. zool. suisse*, **4**, 269–303.

WEISE, W. 1924. Das Nervensystem von *Calyptraea sinensis* Lin. und *Aporrhais pes-pelicani* Lam. *Z. wiss. Zool.*, **128**, 570–600.

WELSH, J. H. 1953. The action of acetylcholine antagonists on the heart of *Venus mercenaria*. *Brit. J. Pharmacol.*, **8**, 327–333.

WELSH, J. H. 1956. Neurohormones of invertebrates. I. Cardio-regulators of *Cyprina* and *Buccinum*. *J. mar. biol. Ass. U.K.*, **35**, 193–201.

WELSH, J. H. 1957. Neurohormones or transmitter agents. In 'Recent Advances in Invertebrate Physiology.' (B. T. Scherr ed.). Eugene, Oregon, University of Oregon.

WENZ, W. 1938. Gastropoda 1, 2. In 'Handbuch der Paläozoologie,' **6**, (O. H. Schindewolf ed.). Berlin, Borntraeger.

WERNER, B. 1939. Über die Entwicklung und Artunterscheidung von Muschellarven des Nordseeplanktons, unter besonderer Berücksichtigung der Schalenentwicklung. *Zool. Jb. (Anat. Ont.)*, **66**, 1–54.

WERNER, B. 1952. Ausbildungsstufen der Filtrationsmechanismen bei filtrierenden Prosobranchiern. *Verh. dtsch. zool. Ges., Zool. Anz. Suppl.* **17**, 529–546.

WERNER, B. 1953. Über den Nahrungserwerb der Calyptraeidae (Gastropoda Prosobranchia). Morphologie, Histologie und Funktion der am Nahrungserwerb beiteiligten Organe. *Helgoländ. wiss. Meeresunters.*, **4**, 260–315.

WERNER, B. 1955. Über die Anatomie, die Entwicklung und Biologie des Veligers und der Veliconcha von *Crepidula fornicata* L. (Gastropoda Prosobranchia). *Helgoländ. wiss. Meeresunters.*, **5**, 169–217.

WERNER, B. 1959. Das Prinzip des endlosen Schleimfilters beim Nahrungserwerb wirbelloser Meerestiere. *Int. Rev. ges. Hydrobiol.*, **44**, 181–216.

WESENBERG-LUND, C. 1939. 'Die Biologie der Süsswassertiere. Wirbellose Tiere.' Wien, Springer.

WESTBLAD, E. 1922. Zur Physiologie der Turbellarien. I. Die Verdauung. II. Die Exkretion. *Lunds Univ. Årssk.*, N.F. (2), **18**, 6, 9–212.

WHEATLEY, J. M. 1947. Investigations on *Polynices* and clams at Belliveau Cove, N.S. *MS Rep. Fish. Res. Bd. Canada*, 371.

WHITEHEAD, H. 1935. An ecological study of the invertebrate fauna of a chalk stream near Great Driffield, Yorkshire. *J. Anim. Ecol.*, **4**, 58–78.

WHITAKER, M. B. 1951. On the homologies of the oesophageal glands of *Theodoxus fluviatilis* (L.). *Proc. malac. Soc. Lond.*, **29**, 21–34.

WHITTAKER, V. P. & I. A. MICHAELSON. 1954. Studies in urocanylcholine. *Biol. Bull., Wood's Hole*, **107**, 134.

WIERZEJSKI, A. 1905. Embryologie von *Physa fontinalis* L. *Z. wiss. Zool.*, **83**, 502–706.

WILBUR, K. & SALEUDDIN, A. S. M. 1983. Shell formation. In 'The Mollusca', **4**, (1) (A. S. M. Saleuddin & K. M. Wilbur, eds), 236–287. New York & London, Academic Press.

WILCZYNSKI, J. C. 1955. On sex behaviour and sex determination in *Crepidula fornicata*. *Biol. Bull., Wood's Hole*, **109**, 353–354.

WILLCOX, M.A .1898. Zur Anatomie von *Acmaea fragilis*. Jena. *Z. Naturw.*, **32**, 411–456.

WILLCOX, M. A. 1905a. Biology of *Acmaea testudinalis* Müller. *Amer. Nat.*, **39**, 325–333.

WILLCOX, M. A. 1905b. Homing of *Fissurella* and *Siphonaria*. *Science*, **22**, 90.

WILLEM, V. 1892a. Contributions à l'étude physiologique des organes des sens chez les mollusques. I. La vision chez les gastropodes pulmonés. *Arch. Biol.*, **12**, 57–98.

WILLEM, V. 1892b. Contributions à l'étude physiologique des organes des sens chez les mollusques. II. Les gastropodes pulmonés perçoivent-ils les rayons ultra-violets? *Arch. Biol.*, **12**, 99–122.

WILLEM, V. 1892c. Contributions à l'étude physiologique des organes des sens chez les mollusques. III. Observations sur la vision et les organes visuels de quelques mollusques prosobranches et opisthobranches. *Arch. Biol.*, **12**, 123–149.

WILSMANN, T. 1942. Der Pharynx von *Buccinum undatum*. *Zool. Jb. (Anat. Ont.)*, **68**, 1–48.

WILSON, D. P. 1929. A habit of the common periwinkle (*Littorina littorea* Linn.). *Nature, Lond.*, **124**, 443.

WILSON, D. P. & M. A. WILSON. 1956. A contribution to the biology of *Ianthina janthina* (L.). *J. mar. biol. Ass. U.K.*, **35**, 291–305.

WILSON, E. B. 1904. On germinal localization in the egg. II. Experiments on the cleavage mosaic in *Patella*. *J. exp. Zool.*, **I**, 197–268.

WINCKWORTH, R. 1932. The British marine Mollusca. *J. Conch., Lond.,* **19,** 211–252.

WINGSTRAND, K. G, 1985. On the anatomy and relationships of Recent Monoplacophora. *Galathea Rep.,* **16,** 7–94.

WINKLER, L. R. & E. D. WAGNER. 1959. Filter paper digestion by the crystalline style in *Oncomelania. Trans. Amer. micr. Soc.,* **78,** 262–268.

WOODWARD, B. B. 1892. On the mode of growth and the structure of the shell in *Velates conoideus* Lamk. and other Neritidae. *Proc. zool. Soc. Lond.,* 528–540.

WOODWARD, B. B. 1913. 'The Life of the Mollusca.' London, Methuen.

WOODWARD, M. F. 1901a The anatomy of *Pleurotomaria beyrichii* Hilg. *Quart. J. micr. Sci.,* **44,** 215–286.

WOODWARD, M. F. 1901b. Note on the anatomy of *Voluta ancilla* (Sol.), *Neptuneopsis gilchristi* Sby., and *Volutilithes abyssicola* (Ad. and Rve.). *Proc. malac. Soc. Lond.,* **4,** 117–125.

WOODWARD, S. P. 1851–56. 'A Manual of the Mollusca.' London, Lockwood.

WOODWARD, S. P. 1875. 'A Manual of the Mollusca.' 3rd edition. London, Lockwood.

WRIGHT, F. S. 1936. Report on the Maldon (Essex) periwinkle fishery. *Fish. Invest., Lond.* (2), **14,** no. 6, 1–37.

WRIGLEY, A. 1932. Spiral sculpture and lip-denticulation of the Cymatiidae. *Proc. malac. Soc. Lond.,* **20,** 127–128.

WRIGLEY, A. 1934. Spiral sculpture and colour markings of the Cassididae. *Proc. malac. Soc. Lond.,* **21,** 111–114.

WRIGLEY, A. 1942. English Eocene *Hastula* with remarks on the coloration of the Terebridae. *Proc. malac. Soc. Lond.,* **25,** 17–24.

WRIGLEY, A. 1948. The colour patterns and sculpture of molluscan shells. *Proc. malac. Soc. Lond.,* **27,** 206–217.

WYATT, H. V. 1957. The biology and reproduction of *Calyptraea chinensis. Challenger Soc. Rep.,* **3,** 33.

YONGE, C. M. 1925a. The hydrogen ion concentration in the gut of certain lamellibranchs and gastropods. *J. mar. biol. Ass. U.K.,* **13,** 938–952.

YONGE, C. M. 1925b. The digestive diverticula in the lamellibranchs. *Trans. R. Soc. Edinb.,* **54,** 703–718.

YONGE, C. M. 1926. Structure and physiology of the organs of feeding and digestion in *Ostrea edulis. J. mar. biol. Ass. U.K.,* **14,** 295–386.

YONGE, C. M. 1930. The crystalline style of the Mollusca and a carnivorous habit cannot normally co-exist. *Nature, Lond.,* **125,** 444–445.

YONGE, C. M. 1932. Notes on feeding and digestion in *Pterocera* and *Vermetus,* with a discussion on the occurrence of the crystalline style in the Gastropoda. *Sci. Rep. Gt. Barrier Reef Exped.,* **1,** 259–281.

YONGE, C. M. 1937. The biology of *Aporrhais pes-pelicani* (L.) and *A. serresiana (Mich.). J. mar. biol. Ass. U.K.,* **21,** 687–704.

YONGE, C. M. 1938. Evolution of ciliary feeding in the Prosobranchia, with an account of feeding in *Capulus ungaricus. J. mar. biol. Ass. U.K.,* **22,** 453–468.

YONGE, C. M. 1939. The protobranchiate Mollusca: a functional interpretation of their structure and evolution. *Phil. Trans. R. Soc. B,* **230,** 79–147.

YONGE, C. M. 1946. On the habits of *Turritella communis* Risso. *J. mar. biol. Ass. U.K.,* **26,** 377–380.

YONGE, C. M. 1947. The pallial organs in the aspidobranch Gastropoda and their evolution throughout the Mollusca. *Phil. Trans. R. Soc. B,* **232,** 443–518.

ZIEGELMEIER, E. 1954. Beobachtungen über den Nahrungserwerb bei der Naticide *Lunatia nitida* Donovan (Gastropoda Prosobranchia). *Helgoländ. wiss. Meeresunters.,* **5,** 1–33.

ZIEGELMEIER, E. 1958. Zur Lokomotion bei Naticiden (Gastropoda Prosobranchiata) (Kurze Mitteilung über Schwimmbewegungen bei *Polynices josephinus* Risso). *Helgoländ. wiss. Meeresunters.,* **6,** 202–206.

ZIEGENHORN, A. & H. THIEM. 1926. Beiträge zur Systematik und Anatomie der Fissurellen. *Jena. Z. Naturw.,* **62,** 1–78.

ZUBKOV, A. A. 1934. Studies on the comparative physiology of the heart: the pace-maker of the heart of the snail (*Helix pomatia*). *Coll. Pap. Lab. comp. Physiol. Timiriasev biol. Inst.,* 52–61.

REFERENCES USED IN PART II

ABRO, A. M. 1969. 'Studies of the functional anatomy of some British monotocardian veligers, with observations on the musculature of the buccal mass.' Ph.D. thesis, University of Reading.

ALFIERAKIS, N. S. & BERRY, A. J. 1980. Rhythmic egg release in *Littorina littorea* (Mollusca: Gastropoda). *J. Zool., Lond.,* **190,** 297–307.

ALYAKRINSKAYA, I. O. 1969. Morphological adaptations to viviparity in *Viviparus viviparus* (Gastropoda, Prosobranchia). *Zool. Zh.,* **48,** 1608–1613. [In Russian.]

AMAUDRUT, A. 1898. La partie antérieure du tube digestif et la torsion chez les mollusques gastéro-podes. *Ann. Sci. nat. Zool.* (7), **8,** 1–291.

AMOUROUX, J. M. 1974. Observations sur la biolgie du mollusque *Charonia nodiferus* (Lamarck). *Vie et Milieu,* **A24,** 365–367.

ANDERSON, A. 1971. Intertidal activity, breeding and the floating habits of *Hydrobia ulvae* in the Ythan estuary. *J. mar. biol. Ass. U.K.,* **51,** 423–437.

ANDERSON, E. 1969. Oocyte-follicle cell differentiation in two species of amphineurans (Mollusca), *Mopalia muscosa* and *Chaetopleura apiculata. J. Morph.,* **129,** 89–126.

ANDERSON, E. 1974. Comparative aspects of the ultrastructure of the female gamete. *Intern. Rev. Cytol.,* suppl. **4,** 1–70.

ANDREWS, E. A. 1937. Certain reproductive organs in the Neritidae. *J. Morph.,* **60,** 191–209.

ANDREWS, E.B. 1964. The functional anatomy and histology of the reproductive system of some pilid gastropod molluscs. *Proc. malac. Soc. Lond.,* **36,** 121–140.

ANDREWS, E. B. 1976a. The ultrastructure of the heart and kidneys of the pilid gastropod mollusc *Marisa cornuarietis* with special reference to filtration throughout the Architaenioglossa. *J. Zool., Lond.,* **179,** 85–106.

ANDREWS, E. B. 1976b. The fine structure of the heart of some prosobranch and pulmonate gastro-pods in relation to filtration. *J. moll. Stud.,* **42,** 199–216.

ANDREWS, E. B. 1979. Fine structure in relation to structure in the excretory system of two species of *Viviparus. J. moll. Stud.,* **45,** 186–206.

ANDREWS, E. B. 1981. Osmoregulation and excretion in prosobranch gastropods. Part 2: structure in relation to function. *J. moll. Stud.,* **47,** 248–289.

ANDREWS, E. B. 1985. Structure and function in the excretory system of archaeogastropods and their significance in the evolution of gastropods. *Phil. Trans. R. Soc. B,* **310,** 383–406.

ANDREWS, E. B. 1988. Excretory systems in molluscs. In 'The Mollusca' **11,** 381–448 (E. R. Trueman & M. R. Clarke eds). San Diego, Academic Press.

ANDREWS, E. B. 1991. The fine structure and function of the salivary glands of the dogwhelk *Nucella lapillus* (Gastropoda: Muricidae). *J. moll. Stud.,* **57,** 111–126.

ANDREWS, E. B. 1992. The fine structure and function of the anal gland of the muricid *Nucella lapillus* (Neogstropoda) (and a comparison with that of the trochid *Gibbula cineraria*). *J. moll. Stud.,* **58,** 297–313.

ANDREWS, E. B. & LITTLE, C. 1971. Ultrafiltration in the gastropod heart. *Nature, Lond.,* **234,** 411–412.

ANDREWS, E. B. & LITTLE, C. 1972. Structure and function in the excretory system of some terrestrial prosobranch snails (Cyclophoridae). *J. Zool., Lond.,* **168,** 395–422.

ANDREWS, E. B. & LITTLE, C. 1982. Renal structure and function in relation to habitat in some cyclophorid snails from Papua New Guinea. *J. moll. Stud.,* **48,** 124–143.

ANDREWS, E. B., ELPHICK, M. R. & THORNDYKE, M. C. 1991. Pharmacologically active constituents of the accessory salivary and hypobranchial glands of *Nucella lapillus. J. moll. Stud.,* **57,** 136–138.

ANDREWS, E. B. & TAYLOR, P. M. 1988. Fine structure, mechanism of heart function and haemo-dynamics in the prosobranch gastropod mollusc *Littorina littorea* (L.). *J. comp. Physiol.,* **B148,** 247–262.

ANDREWS, E. B. & TAYLOR, P. M. 1900. Reabsorption of organic solutes in some marine and freshwater prosobranch gastropods. *J. moll. Stud.,* **56,** 147–162.

ANKEL, F. & CHRISTENSEN, A. M. 1963. Non-specificity in host selection by *Odostomia scalaris. Vidensk. Medd, naturhist. Foren.,* **125,** 321–325.

ANKEL, W. E. 1926. Spermiozeugmenbildung durch atypische (apyrene) und typische Spermie bei *Scala* und *Janthina. Verh. dtsch. zool. Ges.,* **31,** (*Zool. Anz.* suppl. 2), 193–202.

ANKEL, W. E. 1930a. Die atypische Spermatogenese von *Janthina* (Prosobranchia, Ptenoglossa). *Z. Zellforsch. mikr. Anat.,* **11,** 491–608.

ANKEL, W. E. 1930b. Nähreierbildung bei *Natica catena* (da Costa). *Zool. Anz.,* **89,** 129–135.

ANKEL, W. E. 1936. Prosobranchia. In 'Tierwelt der Nord- und Ostsee' (G. Grimpe & E. Wagler eds) **IXb,** 1–240. Leipzig, Akademische Verlagsgesellschaft.

ANKEL, W. E. 1937. Der feinere Bau des Kokons der Purpurschnecke *Nucella lapillus* (L.) und seine Bedeutung für der Laichleben. *Verh. dtsch. zool. Ges.,* **39,** (*Zool. Anz.* suppl. 10), 77–86.

ANKEL, W. E. 1938. Erwerb und Aufnahme der Nahrung bei den Gastropoden. *Verh. dtsch. zool. Ges.,* **40,** (*Zool. Anz.* suppl. 11), 223–295.

ANKEL, W. E. 1959. Beobachtungen an Pyramidellen des Gullmar-Fjordes. *Zool. Anz.,* **162,** 1–21.

ANSELL, A. D. 1960. Observations on predation of *Venus striatula* (da Costa) by *Natica alderi* (Forbes). *Proc. malac. Soc. Lond.,* **34,** 157–164.

ANSELL, A. D. 1969. Defensive adaptations to predation in the Mollusca. *Symp. mar. biol. Ass. India,* **3,** 487–512.

ANSELL, A. D. 1982a. Experimental studies of a benthic predator-prey relationship. I. Feeding, growth and egg-collar production in long-term cultures of the gastropod drill *Polinices alderi* (Forbes) feeding on the bivalve *Tellina tenuis* (da Costa). *J. exp. mar. Biol. Ecol.,* **56,** 235–255.

ANSELL, A. D. 1982b. Experimental studies of a benthic predator-prey relationship. II. Energetics of growth and reproduction and food conversion efficiencies, in long-term cultures of the gastropod drill *Polinices alderi* (Forbes) feeding on the bivalve *Tellina tenuis* (da Costa). *J. exp. mar. Biol. Ecol.,* **61,** 1–29.

ANSELL, A. D. 1982c. Experimental studies of a benthic predator-prey relationship. III. Factors affecting rate of predation and growth in juveniles of the gastropod drill *Polinices catena* (da Costa) in laboratory cultures. *Malacologia,* **22,** 367–375.

ANSELL, A. D. & MORTON, B. 1985. Aspects of naticid predation in Hong Kong with special reference to the defensive adaptations of *Bassina (Callanaitis) calophylla* (Bivalvia). In 'Proceedings of the Second International Workshop on the Malacofauna of Hong Kong and Southern China, Hong Kong, 1983' (B. Morton & D. Dudgeon eds), 635–660. Hong Kong, Hong Kong University Press.

ANSELL, A. D. & MORTON, B. 1987. Alternative predation tactics of a tropical naticid gastropod. *J. exp. mar. Biol. Ecol.,* **111,** 109–119.

ARNAUD, P. M. & BANDEL, K. 1978. Comments on six species of marine Antarctic Littorinacea (Mollusca: Gastropoda). *Tethys,* **8,** 213–230.

ARNOLD, D. C. 1972. Salinity tolerance of some common prosobranchs. *J. mar. biol. Ass. U.K.,* **52,** 475–486.

ATKINSON, W. D. & NEWBURY, S. F. 1984. The adaptations of the rough winkle, *Littorina rudis,* to desiccation and to dislodgement by wind and waves. *J. Anim. Ecol.,* **53,** 93–105.

ATKINSON, W. D. & WARWICK, T. 1983. The role of selection in the colour polymorphism of *Littorina rudis* Maton and *Littorina arcana* Hannaford-Ellis (Prosobranchia: Littorinidae). *Biol. J. Linn. Soc.,* **20,** 135–151.

AVENS, A. C. & SLEIGH, M. A. 1965. Osmotic balance in gastropod molluscs. I. Some marine and littoral gastropods. *Comp. Biochem. Physiol.,* **16,** 121–141.

BAHL, K. N. 1928. On the reproductive processes and development of *Pila globosa* (Swainson). I. Copulation and oviposition. *Mem. Indian Mus.*, **9**, 1–11.

BAILEY, S. K. & DAVIES, I. M. 1989. The effects of tributyltin on dogwhelks (*Nucella lapillus*) from Scottish coastal waters. *J. mar. biol. Ass. U.K.*, **69**, 335–354.

BAKKER, K. 1959. Feeding habits and zonation in some intertidal snails. *Arch. néerl. Zool.*, **13**, 230–257.

BANDEL, K. 1974. Studies on Littorinidae from the Atlantic. *Veliger*, **17**, 92–114.

BANDEL, K. 1988. Early ontogenetic shell and shell structure as aids to unravel gastropod phylogeny and evolution. *Malacol. Rev.*, suppl. **4**, 267–272.

BANNISTER, W. H., BANNISTER, J. V. & MICALLEF, H. 1966. A biochemical factor in the zonation of marine molluscs. *Nature, Lond.*, **211**, 747.

BANTOCK, C. R. & COCKAYNE, W. G. 1975. Chromosomal polymorphism in *Nucella lapillus*. *Heredity*, **34**, 231–245.

BARINAGA, M. 1990. Science digests the secrets of voracious killer snails. *Science*, **249**, 250–251.

BARKER, R. M. 1964. Microtextural variation in pelecypod shells. *Malacologia*, **2**, 69–86.

BARNES, R. S. K. 1979. Intrapopulation variation in *Hydrobia* sediment preferences. *Est. coastal mar. Sci.*, **9**, 231–234.

BARNES, R. S. K. 1981a. An experimental study of the pattern and significance of the climbing behaviour of *Hydrobia ulvae*. *J. mar. biol. Ass. U.K.*, **61**, 285–299.

BARNES, R. S. K. 1981b. Factors affecting climbing in the coastal gastropod *Hydrobia ulvae*. *J. mar. biol. Ass. U.K.*, **61**, 301–306.

BARNES, R. S. K. 1986. Daily activity rhythms in the intertidal gastropod *Hydrobia ulvae* (Pennant). *Est. coastal Shelf Sci.*, **22**, 325–334.

BARNES, R. S. K. 1987. The coastal lagoons of East Anglia, U.K. *J. coastal Res.*, **3**, 417–427.

BARNES, R. S. K. 1988. On reproductive strategies in adjacent lagoonal and intertidal marine populations of the gastropod *Hydrobia ulvae*. *J. mar. biol. Ass. U.K.*, **68**, 365–375.

BARNES, R. S. K. 1989. What, if anything, is a brackish-water fauna? *Trans. R. Soc. Edinb.*, **80**, 235–240.

BARNES, R. S. K. 1990. Reproductive strategies in contrasting populations of the coastal gastropod *Hydrobia ulvae*. II. Longevity and life-time egg production. *J. exp. mar. Biol. Ecol.*, **138**, 183–200.

BARNES, R. S. K. & GREENWOOD, J. G. 1978. The response of the intertidal gastropod *Hydrobia ulvae* (Pennant) to sediments of different particle size. *J. exp. mar. Biol. Ecol.*, **31**, 43–54.

BARNETT, B. E. 1979. A laboratory study of predation by the dog-whelk *Nucella lapillus* on the barnacles *Elminius modestus* and *Balanus balanoides*. *J. mar. biol. Ass. U.K.*, **59**, 299–306.

BARNETT, P. R. O., HARDY, B. L. S. & WATSON, J. 1980. Substratum selection and egg-capsule deposition in *Nassarius reticulatus* (L.). *J. exp. mar. Biol. Ecol.*, **45**, 95–103.

BAYER, F. M. 1963. Observations on pelagic mollusks associated with the siphonophores *Velella* and *Physalia*. *Bull. mar. Sci. Gulf Carib.*, **13**, 454–466.

BAYLISS, D. E. 1986. Selective feeding on bivalves by *Polinices alderi* (Forbes) (Gastropoda). *Ophelia*, **25**, 33–47.

BAYNE, B. L. & SCULLARD, C. 1978a. Rates of oxygen consumption by *Thais* (*Nucella*) *lapillus* (L.). *J. exp. mar. Biol. Ecol.*, **32**, 97–111.

BAYNE, B. L. & SCULLARD, C. 1978b Rates of feeding in *Thais* (*Nucella*) *lapillus* (L.). *J. exp. mar. Biol. Ecol.*, **32**, 113–129.

BAYNE, C. J. 1968. Histochemical studies on the egg capsules of eight gastropod molluscs. *Proc. malac. Soc. Lond.*, **38**, 199–212.

BECKER, K. 1949. Untersuchungen über das Farbmuster und das Wachstum der Molluskenschale. *Biol. Zbl.*, **68**, 263–288.

BENSALEM, M. & CHETAIL, M. 1982. Hydrocalcic metabolism and pedal glands in *Pomatias elegans* (Müller) (Mollusca, Prosobranchia). *Malacologia*, **22**, 293–303.

BERG, C. & PORTER, M. E. 1974. A comparison of predatory behaviour among the naticid gastropods *Lunatia heros*, *Lunatia triseriata* and *Polinices duplicatus*. *Biol. Bull.*, **147**, 469–470.

BERGER, E. M. 1972. The distribution of genetic variation in three species of *Littorina*. *Biol. Bull.*, **143**, 455.

BERGER, E. M. 1973. Gene-enzyme variation in three sympatric species of *Littorina*. *Biol. Bull.*, **145**, 83–90.

BERGER, V. Y. 1975. The changes of euryhalinity in ontogenesis and adaptations connected with reproduction in the White Sea mollusc *Littorina saxatilis*. *Biologya Morya, Vladivostock*, **1**, 43–50. [In Russian.]

BERGER, V. Y., KHLEBOVICH, V. V., KOVALEVA, N. M. & NATOCHIN, Y. V. 1978. The changes of ionic composition and cell volume during adaptation of molluscs (*Littorina*) to lowered salinity. *Comp. Biochem. Physiol.*, **60A**, 447–452.

BERGERARD, J. 1971. Cycle sexuel annuel de *Littorina saxatilis* (Olivi). *Haliotis*, **1**, 23–24.

BERRY, A. J. 1961. Some factors affecting the distribution of *Littorina saxatilis* (Olivi). *J. Anim. Ecol.*, **30**, 27–45.

BERRY, A. J. 1982. Predation by *Natica maculosa* Lamarck (Naticidae: Gastropoda) upon the trochacean gastropod *Umbonium vestiarium* (L.) on a Malaysian shore. *J. exp. mar. Biol. Ecol.*, **64**, 71–89.

BERRY, A. J. 1984. *Umbonium vestiarium* (L.) (Gastropoda Trochacea) as the food source for naticid gastropods and a starfish on a Malaysian sandy shore. *J. moll. Stud.*, **50**, 1–7.

BERRY, A. J. & CHEW, E. 1973. Reproductive systems and cyclic release of eggs in *Littorina melanostoma* from Malayan mangrove swamps (Mollusca: Gastropoda). *J. Zool., Lond.*, **171**, 333–344.

BERRY, A. J. & HUNT, D. C. 1980. Behaviour and tolerance of salinity and temperature in new-born *Littorina rudis* (Maton) and the range of the species in the Forth estuary. *J. moll. Stud.*, **46**, 55–65.

BERRY, R. J. 1983. Polymorphic shell banding in the dog-whelk, *Nucella lapillus* (Mollusca). *J. Zool., Lond.*, **200**, 453–470.

BERRY, R. J. & CROTHERS, J. H. 1968. Stabilizing selection in the dog-whelk (*Nucella lapillus*). *J. Zool., Lond.*, **155**, 5–17.

BERRY, R. J. & CROTHERS, J. H. 1970. Genotypic stability and physiological tolerance in the dog-whelk (*Nucella lapillus*). *J. Zool., Lond.*, **162**, 293–302.

BERRY, R. J. & CROTHERS, J. H. 1974. Visible variation in the dog-whelk *Nucella lapillus*. *J. Zool., Lond.*, **174**, 123–148.

BERTNESS, M. D. 1977. Behavioral and ecological aspects of shore-level size gradients in *Thais lamellosa* and *Thais emarginata*. *Ecology*, **58**, 86–97.

BERTNESS, M. D. & CUNNINGHAM, C. 1981. Crab shell-crushing predation and gastropod architectural defense. *J. exp. mar. Biol. Ecol.*, **50**, 213–230.

BIELER, R. 1988. Phylogenetic relationships in the gastropod family Architectonicidae, with notes on the family Mathildidae (Allogastropoda). *Malacol. Rev.*, suppl. **4**, 205–240.

BINGHAM, F. O. & ALBERTSON, H. D. 1973. Observations on the attachment of egg capsules to a substrate by *Melongena corona*. *Veliger*, **16**, 233–237.

BISHOP, M. J. 1976. *Hydrobia neglecta* Muus in the British Isles. *J. moll. Stud.*, **42**, 319–326.

BLABER, S. J. M. 1970. The existence of a penis-like outgrowth behind the right tentacle in spent females of *Nucella lapillus* (L.) *Proc. malac. Soc. Lond.*, **39**, 231–233.

BLACK, R. 1978. Tactics of whelks preying on limpets. *Mar. Biol.*, **46**, 157–162.

BLACKMORE, D. T. 1969a. Studies on *Patella vulgata* L. 1. Growth, reproduction and zonal distribution. *J. exp. mar. Biol. Ecol.*, **3**, 200–213.

BLACKMORE, D. T. 1969b. Studies on *Patella vulgata* L. 2. Seasonal variation in biochemical composition. *J. exp. mar. Biol. Ecol.*, **3**, 231–245.

BLANDFORD, P. & LITTLE, C. 1983. Salinity detection by *Hydrobia ulvae* (Pennant) and *Potamopyrgus jenkinski* (Smith) (Gastropoda: Prosobranchia). *J. exp. mar. Biol. Ecol.*, **68**, 25–38.

BOETTGER, C. R. 1955a. Die Systematik der euthyneuren Schnecke. *Verh. dtsch. zool. Ges.* 1954 (*Zool. Anz.* suppl. **18**), 253–280.

BOETTGER, C. R. 1955b. Beiträge zur Systematik der Urmollusken (Amphineura). *Verh. dtsch. zool. Ges.* 1955 (*Zool. Anz.* suppl. **19**), 223–256.

BOGHEN, A. & FARLEY, J. 1974. Phasic activity in the digestive gland cells of the intertidal prosobranch *Littorina saxatilis* (Olivi) and its relation to the tidal cycle. *Proc. malac. Soc. Lond.*, **41**, 41–56.

BONDESEN, P. 1940. Preliminary investigations into the development of *Neritina fluviatilis* L. in brackish and fresh water. *Vidensk. Medd. dansk naturh. Foren.,* **104,** 283–318.

BOQUEST, L., FALKMER, S. & MEHROTRA, B. K. 1971. Ultrastructural search for homologues of pancreatic *β*-cells in the intestinal mucosa of the mollusc *Buccinum undatum. Gen. comp. Endocrinol.,* **17,** 236–239.

BOSS, K. J. & MERRILL, A. S. 1965. Degree of host specificity in two species of *Odostomia* (Pyramidellidae: Gastropoda). *Proc. malac. Soc. Lond.,* **36,** 349–355.

BOUCHET, P. 1976. Mise en évidence d'une migration de larves véligères entre l'étage abyssal et la surface. *C. R. hebd. Acad. Sci.* sér. D, **283,** 821–824.

BOUCHET, P. 1989a. A review of poecilogony in gastropods. *J. moll. Stud.,* **55,** 67–78.

BOUCHET, P. 1989b. A marginellid gastropod parasitic on sleeping fishes. *Bull. mar. Sci.,* **45,** 76–84.

BOULDING, E. G. 1990. Are the opposing selection pressures on exposed and protected shores sufficient to maintain genetic differentiation between gastropod populations with high inter-migration rates. *Hydrobiologia,* **193,** 41–52.

BOURNE, G. B. & REDMOND, J. R. 1977. Hemodynamics in the pink abalone, *Haliotis corrugata* (Mollusca, Gastropoda). *J. exp. Zool.,* **200,** 9–22.

BOURNE, G. C. 1908. Contributions to the morphology of the group Neritacea of aspidobranch gastropods. Part 1. The Neritidae. *Proc. zool. Soc. Lond.,* 810–887.

BOWMAN, R. S. 1981. The morphology of *Patella* spp. juveniles in Britain, and some phylogenetic inferences. *J. mar. biol. Ass. U.K.,* **61,** 647–666.

BOWMAN, R. S. & LEWIS, J. R. 1977. Annual fluctuations in the recruitment of *Patella vulgata* L. *J. mar. biol. Ass. U.K.,* **57,** 793–815.

BOYLE, P. R., SILLAR, M. Y. & BRYCESON K. 1979. Water balance and the mantle cavity fluid of *Nucella lapillus* (L.) (Mollusca: Prosobranchia). *J. exp. mar. Biol. Ecol.,* **40,** 41–51.

BRANCH, G. M. 1975. Intraspecific competition in *Patella cochlear* Born. *J. Anim. Ecol.,* **44,** 263–281.

BRANCH, G. M. & MARSH, A. C. 1978. Tenacity and shell shape in six *Patella* species: adaptive features. *J. exp. mar. Biol. Ecol.,* **34,** 111–130.

BRINKHOFF, N., STOCKMANN, K. & GRIESHABER, M. 1983. Natural occurrence of anaerobiosis in molluscs from intertidal habitats. *Oecologia (Berlin),* **57,** 151–153.

BROCK, F. 1936. Suche, Aufnahme und enzymatische Spaltung der Nahrung durch die Wellhornschnecke *Buccinum undatum* L. *Zoologica, Stuttgart,* **34,** 1–136.

BROOM, M. J. 1982. Size-selection, consumption rates and growth of the gastropods *Natica maculosa* (Lamarck) and *Thais carinifera* (Lamarck) preying on the bivalve *Anadara granosa* (L.). *J. exp. mar. Biol. Ecol.,* **56,** 213–233.

BROWN, A. C. & TRUEMAN, E. R. 1982. Muscles that push snails out of their shells. *J. moll. Stud.,* **48,** 97–98.

BROWN, A. C., TRUEMAN, E. R. & STENTON-DOZEY, J. M. E. 1989. Gill size and respiratory requirement in the Mollusca, with special reference to the prosobranch Gastropoda. *S. African J. Sci.,* **85,** 126–127.

BRYAN, G. W. 1969. The effects of oil-spill removers (detergents) on the gastropod *Nucella lapillus* on a rocky shore and in the laboratory. *J. mar. biol. Ass. U.K.,* **49,** 1067–1092.

BRYAN, G. W., GIBBS, P. E., HUMMERSTONE, L. G. & BURT, G. R. 1986. The decline of the gastropod *Nucella lapillus* around south-west England: evidence for the effect of tributyltin from antifouling paints. *J. mar. biol. Ass. U.K.,* **66,** 611–640.

BRYAN, G. W., GIBBS, P. E., BURT, G. R. & HUMMERSTONE, L. G. 1987. The effects of tributyltin (TBT) accumulation on adult dog-whelks, *Nucella lapillus*: long-term field and laboratory experiments. *J. mar. biol. Ass. U.K.,* **67,** 525–544.

BRYAN, G. W., GIBBS, P. E. & BURT, G. R. 1988. A comparison of tri-n-butyltin chloride and five other organotin compounds in promoting the development of imposex in the dog-whelk, *Nucella lapillus. J. mar. biol. Ass. U.K.,* **68,** 733–744.

BUCKLAND-NICKS, J. A. & CHIA, F.-S. 1977. On the nurse cell and the spermatozeugma in *Littorina sitkana. Cell Tissue Res.,* **179,** 347–356.

BULLOCK, T. R. 1953. Predator recognition and escape responses of some intertidal gastropods in presence of starfish. *Behaviour*, **5**, 130–140.

BULNHEIM, H. P. 1970. Measurement of pumping activity in the gastropod *Crepidula fornicata*. *Experientia*, **26**, 808–809.

BURGER, J. W. & THORNTON, C. S. 1935. A correlation between the food eggs of *Fasciolaria tulipa* and the apyrene spermatozoa of prosobranch molluscs. *Biol. Bull.*, **68**, 253–257.

BURROWS, M. T. & HUGHES, R. N. 1989. Natural foraging of the dogwhelk, *Nucella lapillus* (Linnaeus); the weather and whether to feed. *J. moll. Stud.*, **55**, 285–295.

BURTON, R. F. 1983. Ion regulation and water balance. In 'The Mollusca' **5**, 291–352 (A. S. M. Saleuddin & K. M. Wilbur eds). Orlando, Florida, Academic Press.

BUSH, M. S. 1988. The ultrastructure and function of the intestine of *Patella vulgata*. *J. Zool., Lond.*, **215** 685–702.

CABIOCH, L., GRAINGER, J. N. R., KEEGAN, B. F. & KONNECKER, B. 1978. *Balcis alba* a temporary ectoparasite on *Neopentadactyla mixta* Östergren. In 'Physiology and Behaviour of marine Organisms' (D. S. McLusky & A. J. Berry eds), 237–241. Oxford, Pergamon Press.

CALOW, P. 1978. The evolution of life-cycle strategies in fresh-water gastropods. *Malacologia*, **17**, 351–364.

CALOW, P. 1979. Why some metazoan mucus secretions are more susceptible to microbial attack than others. *Am. Nat.*, **114**, 149–152.

CAMBRIDGE, P. G. & KITCHING, J. A. 1982. Shell shape in living and fossil (Norwich Crag) *Nucella lapillus* (L.) in relation to habit. *J. Conch., Lond.*, **31**, 31–38.

CAMMEN, L. M. 1982. Effect of particle size on organic content and microbial abundance within four marine sediments. *Mar. Ecol. Prog. Ser.*, **9**, 273–280.

CARR, W. E. S. 1967a. Chemoreception in the mud snail, *Nassarius obsoletus*. I. Properties of stimulating substances extracted from shrimp. *Biol. Bull.*, **133**, 90–105.

CARR, W. E. S. 1967b. Chemoreception in the mud snail, *Nassarius obsoletus*. II. Identification of stimulatory substances. *Biol. Bull.*, **133**, 106–127.

CARR, W. E. S., HALL, E. R. & GURIN, S. 1974. Chemoreception and the role of proteins: a comparative study. *Comp. Biochem. Physiol.*, **47A**, 559–566.

CARRIKER, M. R. 1955. Critical review of biology and control of oyster drills *Urosalpinx* and *Eupleura*. *Spec. sci. Rep. U.S. Dept. Int. Fish.*, **148**, 1–150.

CARRIKER, M. R. 1977. Ultrastructural evidence that gastropods swallow shell rasped during shell boring. *Biol. Bull.*, **152**, 325–326.

CARRIKER, M. R. 1978. Ultrastructural analysis of dissolution of shell of the bivalve *Mytilus edulis* by the accessory boring organ of the gastropod *Urosalpinx cinerea*. *Mar. Biol.*, **48**, 105–134.

CARRIKER, M. R. 1981. Shell penetration and feeding by naticacean and muricacean predatory gastropods: a synthesis. *Malacologia*, **20**, 403–422.

CARRIKER, M. R., SCOTT, D. B. & MARTIN, G. N. 1963. Demineralization mechanism of boring gastropods. *Publ. Am. Ass. Advanc. Sci.*, **75**, 55–89.

CARRIKER, M. R. & VAN ZANDT, D. 1972. Predatory behavior of a shell-boring muricid gastropod. In 'Behavior of Marine Animals. Current Perspectives in Research' (H. E. Winn & B. L. Olla eds) **1**, 157–244. New York, Plenum Press.

CARRIKER, M. R. & WILLIAMS, L. G. 1978. The chemical mechanism of shell penetration by predatory boring gastropods: a review and an hypothesis. *Malacologia*, **17**, 142–156.

CARRIKER, M. R., WILLIAMS, L. G. & VAN ZANDT, D. 1978. Preliminary characterization of the secretion of the accessory boring organ of the shell-penetrating muricid gastropod *Urosalpinx cinerea*. *Malacologia*, **17**, 125–142.

CARTER, J. G. 1979. Comparative shell microstructure of the Mollusca, Brachiopoda and Bryozoa. In 'Scanning Electron Microscopy' (O. Johari ed.) **2**, 439–456. Chicago, Chicago Press Corporation.

CASTLE, S. L. & EMERY, A. E. H. 1981. *Nucella lapillus*: a possible model for the study of genetic variation in natural populations. *Genetica*, **56**, 11–15.

CHATFIELD, J. E. 1972. Studies on variation and life history in the prosobranch *Hydrobia ulvae* (Pennant). *J. Conch., Lond.,* **27,** 463–473.

CHELAZZI, G. & FOCARDI, S. 1982. A laboratory study on the short-term zonal oscillations of the trochid *Monodonta turbinata* (Born) (Mollusca: Gastropoda). *J. exp. mar. Biol. Ecol.,* **65,** 263–273.

CHERRILL, A. J. & JAMES, R. 1985. The distribution and habitat preferences of four species of Hydrobiidae in East Anglia. *J. Conch., Lond.,* **32,** 123–133.

CHERRILL, A. J. & JAMES, R. 1987. Character displacement in *Hydrobia. Oecologia (Berlin),* **71,** 618–623.

CHETAIL, M., BINOT, D. & BENSALEM, M. 1968. Organe de perforation de *Purpura lapillus* (L.) (Muricidae): histochemie et histoenzymologie. *Cah. Biol. mar.,* **9,** 13–22.

CHETAIL, M. & KRAMPITZ, G. 1982. Calcium and skeletal structures in molluscs: concluding remarks. *Malacologia,* **22,** 337–339.

CHOQUET, M. 1966. Biologie de *Patella vulgata* L. dans le Boulonnais. *Cah. Biol. mar.,* **7,** 1–22.

CHOQUET, M. 1967. Gamétogenèse in vitro au cours du cycle sexuel chez *Patella vulgata* L. en phase mâle. *C. R. Acad. Sci., Paris,* **265D,** 333–335.

CHOQUET, M. 1968. Croissance et longévité de *Patella vulgata* (L.) (Gastropoda Prosobranchia) dans le Boulonnais. *Cah. Biol. mar.,* **9,** 449–468.

CHOQUET, M. 1970. Etude cytologique de la gonade de *Patella vulgata* L. au cours de changement de sexe naturel. *C. R. Acad. Sci., Paris,* **271D,** 1287–1290.

CHOQUET, M. 1971. Etude du cycle biologique et de l'inversion de sexe chez *Patella vulgata* L. (Mollusque Gastéropode Prosobranche). *Gen. comp. Endocrinol.,* **16,** 59–73.

CHOQUET, M. & LEMAIRE, J. 1969. Contribution à l'étude de la régénération tentaculaire chez *Patella vulgata* L. (Gastéropode Prosobranche). *Arch. Zool. exp. gén.,* **109,** 319–337.

CIVIL, G. W. & THOMPSON, T. E. 1972. Experiments with the isolated heart of the gastropod *Helix pomatia* in an artificial pericardium. *J. exp. Biol.,* **56,** 239–247.

CLARKE, W. C. 1958. Escape response of herbivorous gastropods when stimulated by carnivorous gastropods. *Nature, Lond.,* **181,** 137–138.

CLELAND, D. M. 1954. A study of the habits of *Valvata piscinalis* (Müller), and the structure and function of the alimentary canal and reproductive system. *Proc. malac. Soc. Lond.,* **30,** 167–203.

CLIMO, F. M. 1975. The anatomy of *Gegania valkyrie* Powell (Mollusca: Heterogastropoda: Mathildidae) with notes on other heterogastropods. *Trans. R. Soc. N.Z.,* **5,** 275–288.

COE, W. R. 1949. Divergent methods of development in morphologically similar species of prosobranch gastropod. *J. Morph.,* **84,** 383–400.

COLE, H. A. 1942. The American whelk tingle, *Urosalpinx cinerea* (Say), on British oyster beds. *J. mar. biol. Ass. U.K.,* **25,** 477–501.

COLE, H. A. & HANCOCK, D. A. 1955. *Odostomia* as a pest of oysters and mussels. *J. mar. biol. Ass. U.K.,* **34,** 25–31.

COLE, L. C. 1954. The population consequences of life history phenomena. *Quart. Rev. Biol.,* **29,** 103–137.

CONNELL, J. H. 1961. Effects of competition, predation by *Thais lapillus,* and other factors on natural populations of the barnacle *Balanus balanoides. Ecol. Monogr.,* **31,** 61–104.

CONNELL, J. H. 1970. A predator-prey system in the marine intertidal region. I. *Balanus glandula* and several predatory species of *Thais. Ecol. Monogr.,* **40,** 49–78.

CONNOR, V. M. & QUINN, J. F. 1984. Stimulation of food species growth by limpet mucus. *Science,* **225,** 843–844.

CONOVER, J. T. & SIEBURTH, J.McN. 1965. Effects of tannins excreted from Phaeophyta on planktonic animal survival in tide pools. In 'Fifth International Symposium on Seaweeds' (E. J. Young & J. L. McLachlan eds), 99–100.

COOK, A., BAMFORD, O. S., FREEMAN, J. D. B. & TEIDEMAN, D. I. 1969. A study of the homing habit of the limpet. *Anim. Behav.,* **17,** 330–339.

COOK, P. M. 1949. A ciliary feeding mechanism in *Viviparus viviparus* (L.). *Proc. malac. Soc. Lond.,* **27,** 265–271.

COOMBS, V.-A. 1973a. Desiccation and age as factors in the vertical distribution of the dog-whelk, *Nucella lapillus. J. Zool., Lond.,* **171,** 57–66.

COOMBS, V.-A. 1973b. A quantitative system of age analysis for the dog-whelk, *Nucella lapillus. J. Zool., Lond.,* **171,** 437–448.

COWDEN, R. R. 1976. Cytochemistry of oogenesis and early embryonic development. *Am. Zool.,* **16,** 363–374.

COWELL, E. B. & CROTHERS, J. H. 1970. On the occurrence of multiple rows of "teeth" in the shell of the dog-whelk *Nucella lapillus. J. mar. biol. Ass. U.K.,* **50,** 1101–1111.

COX, L. R. 1960. Thoughts on the classification of the Gastropoda. *Proc. malac. Soc. Lond.,* **33,** 239–261.

CRAMPTON, D. M. 1975. The anatomy and method of functioning of the buccal mass of *Testacella maugei* Férussac. *Proc. malac. Soc. Lond.,* **41,** 549–570.

CREEK, G. A. 1951. The reproductive system and embryology of the snail *Pomatias elegens* (Müller). *Proc. zool. Soc. Lond.,* **121,** 599–640.

CREEK, G. A. 1953. The morphology of *Acme fusca* (Montagu) with special reference to the genital system. *Proc. malac. Soc. Lond.,* **29,** 228–240.

CRISP, D. J. 1964. The effects of the severe winter of 1962–63 on marine life in Britain. *J. Anim. Ecol.,* **33,** 165–210.

CRISP, D. J. & SOUTHWARD, A. J. 1958. The distribution of intertidal organisms along the coast of the English Channel. *J. mar. biol. Ass. U.K.,* **37,** 157–208.

CRISP, M. 1969. Studies on the behavior of *Nassarius obsoletus* (Say) (Mollusca, Gastropoda). *Biol. Bull.,* **136,** 355–373.

CRISP, M. 1971. Structure and abundance of receptors of the unspecialized external epithelium of *Nassarius reticulatus* (Gatropoda, Prosobranchia). *J. mar. biol. Ass. U.K.,* **51,** 865–890.

CRISP, M. 1981. Epithelial sensory structures of trochids. *J. mar. biol. Ass. U.K.,* **61,** 95–106.

CROFTS, D. R. 1929. Haliotis. *L.M.B.C. Memoir,* **29,** 1–174.

CROFTS, D. R. 1937. The development of *Haliotis tuberculata,* with special reference to the organogenesis during torsion. *Phil. Trans. R. Soc. B,* **208,** 219–268.

CROTHERS, J. H. 1974. On variation in *Nucella lapillus* (L): shell shape in populations from the Bristol Channel. *Proc. malac. Soc. Lond.,* **41,** 157–170.

CROTHERS, J. H. 1977. Some observations on the growth of the common dog-whelk in the laboratory. *J. Conch., Lond.,* **29,** 157–162.

CROTHERS, J. H. 1980. Further observations on the growth of the common dog-whelk, *Nucella lapillus* (L.) in the laboratory. *J. moll. Stud.,* **46,** 181–185.

CROTHERS, J. H. 1983. Some observations on shell-shape variation in North American populations of *Nucella lapillus* (L.). *Biol. J. Linn. Soc.,* **19,** 237–274.

CROTHERS, J. H. 1985. Dog-whelks: an introduction to the biology of *Nucella lapillus. Field Studies,* **6,** 291–360.

CUENOT, L. 1899. L'excrétion chez les mollusques. *Arch. Biol.,* **16,** 49–96.

CUENOT, L. 1914. Les organes phagocytaires des mollusques. *Arch. Zool. exp. gén.,* **54,** 207–305.

CURREY, J. D. 1980. Mechanical properties of molluscan shell. *Soc. exp. Biol. Symp.,* **34,** 75–97.

CURREY, J. D. 1988. Shell form and strength. In 'The Mollusca' **11,** 183–210 (E. R. Trueman & M. R. Clarke eds). Orlando, Florida, Academic Press.

CURREY, J. D. & HUGHES, R. N. 1982. Strength of the dogwhelk *Nucella lapillus* and the winkle *Littorina littorea* from different habitats. *J. Anim. Ecol.,* **51,** 47–56.

CURTIS, C. A. 1980. Daily cycling of the crystalline style in the omnivorous, deposit-feeding estuarine snail, *Ilyanassa obsoleta. Mar. Biol.,* **59,** 248–250.

DAGUZAN, J. 1970. Relation entre l'excrétion de l'acide urique et le cycle de la marée chez *Littorina littorea* (L.) adulte (mollusque mésogastéropode Littorinidae). *C. R. Acad. Sci., Paris,* **270D,** 3131–3133.

DAGUZAN, J. 1976a. Contribution à l'écologie des Littorinidae (mollusques gastéropodes prosobranches). 1. *Littorina neritoides* (L.) et *Littorina saxatilis* (Olivi). *Cah. Biol. mar.,* **17,** 213–236.

DAGUZAN, J. 1976b. Contribution à l'écologie des Littorinidae (mollusques gastéropodes proso-branches). 2. *Littorina littorea* et *Littorina littoralis*. *Cah. Biol. mar.,* **17,** 275–293.

D'ASARO, C. N. 1970. Egg capsules of prosobranch mollusks from South Florida and the Bahamas and notes on spawning in the laboratory. *Bull. mar. Sci.,* **20,** 414–440.

DAVIDSON, J. K., FALKMER, S., MEHROTRA, B. K. & WILSON, S. 1971. Insulin assays and light micro-scopical studies of digestive organs in protostomian and deuterostomian species and in coelenterates. *Gen. comp. Endocrinol.,* **17,** 388–401.

DAVIES, A. M. 1939. Some palaeontological problems. *Proc. malac. Soc. Lond.,* **23,** 336–344.

DAVIES, M. S., HAWKINS, S. J. & JONES, H. D. 1990. Mucus production and physiological energetics of *Patella vulgata* L. *J. moll. Stud.,* **56,** 499–503.

DAVIES, M.S., JONES, H. D. & HAWKINS, S. J. 1992. Physical factors affecting the fate of pedal mucus produced by the common limpet *Patella vulgata*. *J. mar. biol. Ass. U.K.,* **72,** 633–643.

DAVIES, P. S. 1966. Physiological ecology of *Patella* I. The effect of body size and temperature on respiration rate. *J. mar. biol. Ass. U.K.,* **46,** 647–658.

DAVIS, P. S. 1967. Physiological ecology of *Patella* II. Effect of environmental acclimation on the metabolic rate. *J. mar. biol. Ass. U.K.,* **47,** 61–74.

DAVIES, P. S. 1969. Physiological ecology of *Patella* III. Desiccation effects. *J. mar. biol. Ass. U.K.,* **49,** 291–304.

DAVIES, P. S. 1970. Physiological ecology of *Patella* IV. Environmental and limpet body temperatures. *J. mar. biol. Ass. U.K.,* **50,** 1069–1077.

DAVIS, C. C. 1968. Mechanisms of hatching in aquatic invertebrate eggs. *Oceanogr. mar. Biol. Ann. Rev.,* **6,** 325–376.

DAVIS, G. M., McKEE, M. & LOPEZ, G. 1989. The identity of *Hydrobia truncata* (Gastropoda: Proso-branchia): comparative anatomy, molecular genetics, ecology. *Proc. Acad. nat. Sci. Philad.,* **141,** 333–359.

DAVIS, J. R. A. & FLEURE, R. J. 1903. *Patella. L.M.B.C. Memoir,* **10,** 1–76.

DAY, A. J. & BAYNE, B. L. 1988. Allozyme variation in populations of the dog-whelk *Nucella lapillus* (Prosobranchia: Muricacea) from the south west peninsula of England. *Mar. Biol.,* **99,** 93–100.

DAY, J. A. 1969. Feeding of the cymatiid gastropod *Argobuccinum argus* in relation to the structure of the proboscis gland. *Am. Zool.,* **9,** 909–916.

DELAUNAY, H. 1931. L'ecrétion azotée chez les invertébrés. *Biol. Rev.,* **6,** 265–301.

DELHAYE, W. 1974a. Histophysiologie comparée des organes excréteurs chez quelques Neritacea (Mollusca-Prosobranchia). *Arch. Biol.,* **85,** 235–262.

DELHAYE, W. 1974b. Contribution à l'étude des glandes pédieuses de *Pomatias elegans* (Mollusque, Gastéropode, Prosobranche). *Ann. Sci. nat. Zool.,* **16,** 97–110.

DELHAYE, W. 1974c. Histophysiologie comparée du rein chez les Mésogastéropodes Architaenio-glossa et Littorinoidea (Mollusca—Prosobranchia). *Arch. Biol.,* **85,** 461–507.

DELHAYE, W. 1975. Histophysiologie comparée du rein chez les Mésogastéropodes Rissoidea et Cerithioidea (Mollusca—Prosobranchia). *Arch. Biol.,* **86,** 355–373.

DELHAYE, W. 1976. Histophysiologie comparée des organes rénaux chez les Archaeogastéropodes (Mollusca—Prosobranchia). *Cah. Biol. mar.,* **17,** 305–322.

DEMIAN, E. S. & YOUSIF, F. 1973. Embryonic development and organogenesis in the snail *Marisa cornuarietis* (Mesogastropoda: Ampullariidae). I. General outline of development. *Malacologia,* **12,** 123–150.

DENNY, M. 1980. Locomotion: the cost of gastropod crawling. *Science,* **208,** 1288–1290.

DENNY, M. 1981. A quantitative model for the adhesive locomotion of the terrestrial slug *Ariolimax columbianus*. *J. exp. mar. Biol. Ecol.,* **91,** 195–218.

DENNY, M. W. 1984. Mechanical properties of pedal mucus and their consequences for gastropod structure and performance. *Am. Zool.,* **24,** 23–36.

DENNY, M. W., DANIEL, T. L. & KOEHL, M. A. R. 1985. Mechanical limits to size in wave-swept organisms. *Ecol. Monogr.,* **55,** 69–102.

DENNY, M. W. & GOSLINE, J. M. 1980. The physical properties of the pedal mucus of the terrestrial slug, *Ariolimax columbianus*. *J. exp. Biol.,* **88,** 375–393.

DEPLEDGE, M. H. & PHILLIPS, D. J. H. 1986. Circulation, respiration and fluid dynamics in the gastropod mollusc *Hemifusus tuba* (Gmelin). *J. exp. mar. Biol. Ecol.,* **95,** 1–13.

DESAI, B. N. 1966. The biology of *Monodonta lineata* (da Costa). *Proc. malac. Soc. Lond.,* **37,** 1–17.

DIEHL, M. 1956. Die Raubschnecke *Velutina velutina* das Feind und Bruteinmieter der Ascidie *Styela coriacea. Kieler Meeresunters,* **12,** 180–185.

DOBBERTEEN, R. A. & PECHENIK, J. A. 1987. Competence of larval bioenergetics of two more gastropods with widely different lengths of planktonic life *Thais haemastoma canaliculata* (Gray) and *Crepidula fornicata* (L.). *J. exp. mar. Biol. Ecol.,* **109,** 173–191.

DODD, J. M. 1957. Artificial fertilization, larval development and metamorphosis in *Patella vulgata* L. and *Patella coerulea* L. *Pubbl. Staz. zool. Napoli,* **29,** 172–186.

DOERING, P. H. & PHILLIPS, D. W. 1983. Maintenance of the shore-level size-gradient in the marine snail *Tegula funebralis* (A. Adams): importance of behavioral responses to light and sea star predators. *J. exp. mar. Biol. Ecol.,* **67,** 159–173.

DRUMMOND, M. 1903. Notes on the development of *Paludina vivipara.* with special reference to the urinogenital ducts and theories of gasteropod torsion. *Quart. J. micr. Sci.,* **46,** 97–143.

DUNDEE, D. S. 1957. Aspects of the biology of *Pomatiopsis lapidaria* (Say) (Mollusca: Gastropoda: Prosobranchia). *Misc. Publ. Mus. Univ. Mich.,* No. **100,** 1–65.

DUNKIN, S. de B. & HUGHES, R. N. 1984. Behavioural components of prey-selection by dog-whelks, *Nucella lapillus* (L.) feeding on barnacles, *Semibalanus balanoides* (L.) in the laboratory. *J. exp. mar. Biol. Ecol.,* **79,** 91–103.

DUSSART, G. J. B. 1977. The ecology of *Potamopyrgus jenkinsi* (Smith) in North West England with a note on *Marstoniopsis scholtzi* (Schmidt). *J. moll. Stud.,* **43,** 208–216.

DYTHAM, C., GRAHAME, J. & MILL, P. J. 1990. Distribution, abundance and shell morphology of *Littorina saxatilis* (Olivi) and *Littorina arcana* Hannaford Ellis at Robin Hood's Bay, North Yorkshire. *Hydrobiologia,* **193,** 233–240.

EBLING, F. J., SLOANE, J. F., KITCHING, J. A. & DAVIES, H. M. 1962. The ecology of Lough Ine XII. The distribution and characteristics of *Patella* species. *J. Anim. Ecol.,* **31,** 457–470.

EDWARDS, D. C. & HUEBNER, J. D. 1977. Feeding and growth rates of *Polinices duplicatus* preying on *Mya arenaria* at Barnstaple Harbor, Massachusetts. *Ecology,* **58,** 1218–1236.

EERNISSE, D. J. K. & KERTH, K. 1988. The initial stages of radular development in chitons (Mollusca: Polyplacophora). *Malacologia,* **28,** 95–103.

EKARATNE, S. U. K. & CRISP, D. J. 1982. Tidal micro-growth bands in intertidal gastropod shells, with an evaluation of band dating techniques. *Proc. R. Soc. B,* **214,** 305–323.

EKARATNE, S. U. K. & CRISP, D. J. 1984. Seasonal growth studies of intertidal gastropods from shell micro-growth band measurements, including a comparison with alternative methods. *J. mar. biol. Ass. U.K.,* **64,** 183–210.

ELIASON, A. 1920. Biologisch-faunistische Untersuchung aus dem Øresund. V. Polychaeta. *Acta Univ. Lund. N. F. avd 2,* **16,** (6), 1–103.

ELLIS, C. J. H. 1978. *Littorina arcana* sp. nov.: a new species of winkle (Gastropoda: Prosobranchia: Littorinidae). *J. Conch., Lond.,* **29,** 304.

ELLIS, C. J. H. 1979. Morphology of the oviparous rough winkle, *Littorina arcana* Hannaford Ellis, 1978, with notes on the taxonomy of the *L. saxatilis* species-complex (Prosobranchia: Littorinidae). *J. Conch., Lond.,* **30,** 43–56.

ELLIS, C. J. H. 1984. Ontogenetic change of shell colour patterns in *Littorina neglecta* Bean (1844). *J. Conch., Lond.,* **31,** 343–347.

ELLIS, C. J. H. 1985. The breeding migration of *Littorina arcana* Hannaford Ellis, 1978 (Prosobranchia Littorinidae). *Zool. J. Linn. Soc.,* **84,** 91–96.

ELNER, R. W. & RAFFAELLI, D. G. 1980. Interactions between two marine snails, *Littorina rudis* (Maton) and *Littorina nigrolineata* Gray, a predator, *Carcinus maenas* (L.) and a parasite, *Microphallus similis* Jagerskiold. *J. exp. mar. Biol. Ecol.*, **43**, 151–160.

ELVES, M. W. 1961. The histology of the foot of *Discus rotundatus*, and the locomotion of gastropod Mollusca. *Proc. malac. Soc. Lond.*, **34**, 346–355.

EMSON, R. H. & FALLER-FRITSCH, R. J. 1976. An experimental investigation into the effect of crevice availability on abundance and size structure in a population of *Littorina rudis* (Maton): Gastropoda: Prosobranchia. *J. exp. mar. Biol. Ecol.*, **23**, 285–297.

ERBAN, H. K., FLAJS, G. & STEHL, A. 1968. Über die Schalenstruktur von Monoplacophoren. *Akad. Wiss. Lit. Abhand. math.-naturw. Kl. Jahrgang*, **1**, 1–24.

ERLANGER, R. von. 1891. Zur Entwicklung von *Bythinia tentaculata*. *Zool. Anz.*, **14**, 385–388.

ERLANGER, R. von. 1892. On the paired nephridia of prosobranchs, the homologies of the only remaining nephridium of most prosobranchs, and the relation of the nephridia to the gonad and genital ducts. *Quart. J. micr. Sci.*, **33**, 587–623.

ERLANGER, R. von. 1894. Zum Bildung des Mesoderms bei der *Paludina vivipara*. *Morph. Jb.*, **22**, 113–118.

ETTER, R. J. 1988. Asymmetrical developmental plasticity in an intertidal snail. *Evolution*, **42**, 322–334.

EVANS, F. 1961. Responses to disturbance of the periwinkle *Littorina punctata* (Gmelin) on a shore in Ghana. *Proc. zool. Soc. Lond.*, **137**, 393–402.

EVANS, R. G. 1948. The lethal temperatures of some common British littoral molluscs. *J. Anim. Ecol.*, **17**, 165–173.

EVANS, R. G. 1953. Studies on the biology of British limpets—the genus *Patella* on the south coast of England. *Proc. zool. Soc. Lond.*, **123**, 357–376.

EYSTER, L. S. 1983. Ultrastructure of early embryonic shell formation in the opisthobranch gastropod *Aeolidia papillosa*. *Biol. Bull.*, **165**, 394–408.

EYSTER, L. S. 1985. Origin, morphology and fate of the nonmineralized shell of *Coryphella salmonacea*, an opisthobranch gastropod. *Mar. Biol.*, **85**, 67–76.

EYSTER, L. S. & MORSE, M. P. 1984. Early shell formation during molluscan embryogenesis with new studies of the surf clam *Spisula solidissima*. *Am. Zool.*, **24**, 871–882.

FALLER-FRITSCH, R. J. 1975. A note on the status of *Littorina nigrolineata*. *Littorinid Tidings*, **3**, 29.

FALNIOWSKI, A. 1987. Hydrobioidea of Poland (Prosobranchia: Gastropoda). *Folia malacologica*, **1**, 1–122.

FÄNGE, R. 1957. An acetylcholine-like salivary poison in the marine gastropod *Neptunea antiqua*. *Nature, Lond.*, **180**, 196–197.

FÄNGE R. 1958. Paper chromatography and biological extracts of the salivary gland of *Neptunea antiqua* (Gastropoda). *Acta zool., Stockh.*, **39**, 39–46.

FÄNGE, R. & LIDMAN, U. 1976. Secretion of sulfuric acid in *Cassidaria echinophora* Lamarck (Mollusca: Mesogastopoda, marine carnivorous snail). *Comp. Biochem. Physiol.*, **53A**, 101–103.

FEARE, C. J. 1970a. Aspects of the ecology of an exposed shore population of the dog-whelk *Nucella lapillus* (L.). *Oecologia (Berlin)*, **5**, 1–18.

FEARE, C. J. 1970b. The reproductive cycle of the dog whelk (*Nucella lapillus*). *Proc. malac. Soc. Lond.*, **39**, 125–137.

FEDER, H. M. 1963. Gastropod defensive responses and their effectiveness in reducing predation by starfishes. *Ecology*, **44**, 505–512.

FENCHEL, T. 1975a. Factors determining distribution patterns of mud snails (Hydrobiidae). *Oecologia (Berlin)*, **20**, 1–18.

FENCHEL, T. 1975b. Character displacement and coexistence in mud snails (Hydrobiidae). *Oecologia (Berlin)*, **20**, 19–32.

FENCHEL, T. & KOFOED, L. H. 1976. Evidence for exploitative interspecific competition in mud snails (Hydrobiidae). *Oikos*, **27**, 367–376.

FENCHEL, T., KOFOED, L. H. & LAPPALAINEN, A. 1976. Particle-size selection of two deposit feeders: the amphipod *Corophium volutator* and the prosobranch *Hydrobia ulvae*. *Mar. Biol.*, **30**, 119–128.

FERAL, C. 1980. Variations dans l'évolution du tractus génital mâle externe des femelles de trois gastéropodes prosobranches gonochoriques de stations atlantiques. *Comp. Biochem. Physiol.*, **21**, 479–491.

FERAL, C. & LE GALL, S. 1982. Induction expérimentale par un pollutant marin (le tributylétain) de l'activité neuroendocrine contrôlant la morphogenèse du pénis chez les femelles d'*Ocenebra erinacea* (Mollusque, Prosobranche gonochorique). *C. R. Acad. Sci., Paris*, **295**, 627–630.

FIORONI, P. 1966. Zur Morphologie und Embryogenese des Darmtraktes und der transitorischen Organe bei Prosobranchiern (Mollusca, Prosobranchia). *Rev. suisse Zool.*, **73**, 621–876.

FISCHER, P. 1880. 'Manuel de Conchyliologie et de Paléontologie conchyliologique ou Histoire naturelle des Mollusques vivants et fossiles.' Paris, F. Savy.

FISCHER-PIETTE, E. & GAILLARD, J.-M. 1961. Etudes sur les variations de *Littorina saxatilis*. III. Comparaison des points abrités, au long des côtes françaises et ibériques. *Bull. Soc. zool. Fr.*, **86**, 163–172.

FISCHER-PIETTE, E., GAILLARD, J.-M. & JOUIN, C. 1961. Etudes sur les variations de *Littorina saxatilis*. IV. Comparaison des points battus, au long des côtes européennes. A. Côtes ibériques. *Bull. Soc. zool. Fr.*, **86**, 320–328.

FISH, J. D. 1972. The breeding cycle and growth of open coast and estuarine populations of *Littorina littorea*. *J. mar. biol. Ass. U.K.*, **52**, 1011–1019.

FISH, J. D. 1979. The rhythmic spawning behaviour of *Littorina littorea* (L.). *J. moll. Stud.*, **45**, 172–177.

FISH, J.D. & FISH, S. 1974. The breeding cycle and growth of *Hydrobia ulvae* in the Dovey Estuary. *J. mar. biol. Ass. U.K.*, **54**, 685–697.

FISH, J. D. & FISH, S. 1977. The veliger larva of *Hydrobia ulvae* with observations on the veliger of *Littorina littorea* (Mollusca: Prosobranchia). *J. Zool., Lond.*, **182**, 495–503.

FISH, J. D. & FISH. S. 1981. The early life-cycle of *Hydrobia ventrosa* and *Hydrobia neglecta* with observations on *Potamopyrgus jenkinsi*. *J. moll. Stud.*, **47**, 89–98.

FISHLYN, D. A. & PHILLIPS, D. W. 1980. Chemical camouflaging and behavioral defenses against a predatory seastar by three species of gastropods from the surfgrass Phyllospadix community. *Biol. Bull.*, **158**, 34–48.

FLEMING, C. 1971. Case of poisoning from red whelk. *Brit. med. J.*, **3**, 520–521.

FLOREY, E. & CAHILL, M. A. 1977. Hemodynamics in lamellibranch mollusks: confirmation of constant volume mechanism of auricular and ventricular filling. Remarks on the heart as a site of filtration. *Comp. Biochem. Physiol.*, **57A**, 47–52.

FORBES, V. E. & LOPEZ, G. R. 1986. Changes in feeding and crawling rates in *Hydrobia truncata* (Prosobranchia: Hydrobiidae) in response to sedimentary chlorophyll-a and recently egested sediment. *Mar. Ecol. Progr. Ser.*, **33**, 287–294.

FRAENKEL, G. 1927a. Beiträge zur Geotaxis und Phototaxis von *Littorina*. *Z. vergl. Physiol.*, **5**, 585–597.

FRAENKEL, G. 1927b. Biologische Beobachtungen an *Ianthina*. *Z. Morph. Ökol. Tiere*, **7**, 597–608.

FRALICK, R. A., TURGEON, K. W. & MATHIESON, A. C. 1974. Destruction of kelp populations by *Lacuna vincta* (Montagu). *Nautilus*, **88**, 112–114.

FRANC, A. 1943. 'Etudes sur le développement de quelques Prosobranches méditerranéens.' Thèse, Université d'Alger.

FRANC, A. 1948. Note sur deux Homalogyridés: *H. fischeriana* et *H. atomus*. *Bull. Soc. Hist. nat. Afrique Nord*, **39**, 142–145.

FRANC, A. 1949. Hétéropodes et autres Gastropodes planctoniques de Méditerranée occidentale. *J. Conchyliol.*, **89**, 209–230.

FRANK, P. W. 1965. The biodemography of an intertidal snail population. *Ecology*, **46**, 831–844.

FRANZEN, A. 1955. Comparative morphological investigations into the spermiogenesis among Mollusca. *Zool. Bidr., Uppsala*, **30**, 399–456.

FRESCURA, M. & HODGSON, A. N. 1990. The fine structure of the shell muscle of patellid prosobranch limpets. *J. moll. Stud.*, **56**, 435–447.

FRETTER, V. 1941. The genital ducts of some British stenoglossan prosobranchs. *J. mar. biol. Ass. U.K.*, **25**, 173–211.

FRETTER, V. 1946. The pedal sucker and anal gland of some British Stenoglossa. *Proc. malac. Soc. Lond.,* **27,** 126–130.

FRETTER, V. 1948. The structure and life history of some minute prosobranchs of rock pools: *Skeneopsis planorbis* (Fabricius), *Omalogyra atomus* (Philippi), *Rissoella diaphana* (Alder) and *Rissoella opalina* (Jeffreys). *J. mar. biol. Ass. U.K.,* **27,** 597–632.

FRETTER, V. 1951a. Some observations on the British cypraeids. *Proc. malac. Soc. Lond.,* **29,** 14–20.

FRETTER, V. 1951b. Observations on the life history and functional morphology of *Cerithiopsis tubercularis* (Montagu) and *Triphora perversa* (L.). *J. mar. biol. Ass. U.K.,* **29,** 567–586.

FRETTER, V. 1951c. *Turbonilla elegantissima* (Montagu), a parasitic opisthobranch. *J. mar. biol. Ass. U.K.,* **30,** 37–47.

FRETTER, V. 1955. Some observations on *Tricolia pullus* (L.) and *Margarites helicinus* (Fabricius). *Proc. malac. Soc. Lond.,* **31,** 159–162.

FRETTER, V. 1960. Observations on the tectibranch *Ringicula buccinea* (Brocchi). *Proc. zool. Soc. Lond.,* **135,** 537–549.

FRETTER, V. 1964. Observations on the anatomy of *Mikadotrochus amabilis* Bayer. *Bull. mar. Sci. Gulf Caribb.,* **14,** 172–184.

FRETTER, V. 1965. Functional studies of the anatomy of some neritid prosobranchs. *J. Zool., Lond.,* **147,** 46–74.

FRETTER, V. 1966. Biological investigations of the deep sea. 16. Observations on the anatomy of *Perotrochus. Bull. mar. Sci.,* **16,** 603–614.

FRETTER, V. 1967. The prosobranch veliger. *Proc. malac. Soc. Lond.,* **37,** 357–366.

FRETTER, V. 1972. Metamorphic changes in the velar musculature, head and shell of some prosobranch veligers. *J. mar. biol. Ass. U.K.,* **52,** 161–177.

FRETTER, V. 1975. *Umbonium vestiarium,* a filter-feeding trochid. *J. Zool., Lond.,* **177,** 541–552.

FRETTER, V. 1980. Observations on the gross anatomy of the female genital duct of British *Littorina* species. *J. moll. Stud.,* **46,** 148–153.

FRETTER, V. 1982. An external vascular opening in *Littorina* species. *J. moll. Stud.,* **48,** 105.

FRETTER, V. 1984a. The functional anatomy of the neritacean limpet *Phenacolepas omanensis* Biggs and some comparison with *Septaria. J. moll. Stud.,* **50,** 8–18.

FRETTER, V. 1984b. Prosobranchs. In 'The Mollusca' **7,** 1–45 (A. S. Tompa, N. H. Verdonk & J. A. M. van den Bruggen eds). Orlando, Florida, Academic Press.

FRETTER, V. 1988. New archaeogastropod limpets from hydrothermal vents; superfamily Lepetodrilacea. Part 2. Anatomy. *Phil. Trans. R. Soc. B,* **319,** 33–82.

FRETTER, V. 1989. The anatomy of some new archaeogastropod limpets (superfamily Peltospiracea) from hydrothermal vents. *J. Zool., Lond.,* **218,** 123–169.

FRETTER, V. 1990. The anatomy of some new archaeogastropod limpets (order Patellogastropoda, suborder Lepetopsina) from hydrothermal vents. *J. Zool., Lond.,* **222,** 529–556.

FRETTER, V. & GRAHAM, A. 1949. The structure and mode of life of the Pyramidellidae, parasitic opisthobranchs. *J. mar. biol. Ass. U.K.,* **28,** 493–532.

FRETTER, V. & GRAHAM, A. 1954. Observations on the opisthobranch mollusc *Acteon tornatilis* (L.). *J. mar. biol. Ass. U.K.,* **33,** 565–585.

FRETTER, V. & GRAHAM, A. 1962. 'British Prosobranch Molluscs.' London, Ray Society.

FRETTER, V. & GRAHAM, A. 1976. The prosobranch molluscs of Britain and Denmark. Part 1— Pleurotomariacea, Fissurellacea and Patellacea. *J. moll. Stud.,* suppl. **1,** 1–37.

FRETTER, V. & GRAHAM, A. 1977. The prosobranch molluscs of Britain and Denmark. Part 2— Trochacea. *J. moll. Stud.,* suppl. **3,** 39–100.

FRETTER, V. & GRAHAM, A. 1978a. The prosobranch molluscs of Britain and Denmark. Part 3— Neritacea, Viviparacea, Valvatacea, terrestrial and freshwater Littorinacea and Rissoacea. *J. moll. Stud.,* suppl. **5,** 101–152.

FRETTER, V. & GRAHAM, A. 1978b. The prosobranch molluscs of Britain and Denmark. Part 4— marine Rissoacea. *J. moll. Stud.,* suppl. **6,** 153–241.

FRETTER, V. & GRAHAM, A. 1980. The prosobranch molluscs of Britain and Denmark. Part 5—marine Littorinacea. *J. moll. Stud.*, suppl. **7**, 243–284.

FRETTER, V. & GRAHAM, A. 1981. The prosobranch molluscs of Britain and Denmark. Part 6—Cerithiacea, Strombacea, Hipponicacea, Calyptraeacea, Lamellariacea, Cypraeacea, Naticacea, Tonnacea, Heteropoda. *J. moll. Stud.*, suppl. **9**, 285–362.

FRETTER, V. & GRAHAM, A. 1982. The prosobranch molluscs of Britain and Denmark. Part 7—'Heterogastropoda' (Cerithiopsacea, Triforacea, Epitoniacea, Eulimacea). *J. moll. Stud.*, suppl. **11**, 363–434.

FRETTER, V. & GRAHAM, A. 1985. The prosobranch molluscs of Britain and Denmark. Part 8—Neogastropoda. *J. moll. Stud.*, suppl. **15**, 435–556.

FRETTER, V., GRAHAM, A. & ANDREWS, E. B. 1986. The prosobranch molluscs of Britain and Denmark. Part 9—Pyramidellacea. *J. moll. Stud.*, suppl. **16**, 557–649.

FRETTER, V., GRAHAM, A. & McLEAN, J. H. 1981. The anatomy of the Galapagos Rift limpet, *Neomphalus fretterae*. *Malacologia*, **21**, 337–361.

FRETTER, V. & MANLY, R. 1977a. Algal associations of *Tricolia pullus*, *Lacuna vincta* and *Cerithiopsis tubercularis* (Gastropoda) with special reference to the settlement of their larvae. *J. mar. biol. Ass. U.K.*, **57**, 999–1017.

FRETTER, V. & MANLY, R. 1977b. The settlement and early benthic life of *Littorina neritoides* (L.) at Wembury, S. Devon. *J. moll. Stud.*, **43**, 255–262.

FRETTER, V. & MANLY, R. 1979. Observations on the biology of some sublittoral prosobranchs. *J. moll. Stud.*, **48**, 209–218.

FRETTER, V. & MONTGOMERY, M. C. 1968. The treatment of food by prosobranch veligers. *J. mar. biol. Ass. U.K.*, **48**, 499–520.

FRETTER, V. & PILKINGTON, M. C. 1971. The larval shell of some prosobranch gastropods. *J. mar. biol. Ass. U.K.*, **51**, 49–62.

FRETTER, V. & SHALE, D. 1973. Seasonal changes in population density and vertical distribution of prosobranch veligers in offshore plankton at Plymouth. *J. mar. biol. Ass. U.K.*, **53**, 471–492.

FREY, I. D. 1986. Grazing effects of *Littorina littorea* in different habitats. *Biol. Bull.*, **171**, 480.

FRÖMMING, E. 1954. 'Biologie der mitteleuropäischen Landgastropoden.' Berlin, Duncker & Humblot.

FRÖMMING, E. 1956. 'Biologie der mitteleuropäischen Süsswasserschnecken.' Berlin, Duncker & Humblot.

FUJIOKA, Y., 1985. Seasonal aberrant radular formation in *Thais bronni* (Dunker) and *T. clavigeri* (Küster) (Gastropoda: Muricidae). *J. exp. mar. Biol. Ecol.*, **90**, 43–54.

FUNKE, W. 1964. Untersuchungen zur Heimfindverhalten und zur Ortstreue von *Patella* L. (Gastropoda, Prosobranchia). *Verh. dtsch. zool. Ges. 1964 (Zool. Anz.* suppl. **29**), 411–418.

FUNKE, W. 1968. Heimfindvermögen und Ortstreue bei *Patella* L. (Gastropoda, Prosobranchia). *Oecologia (Berlin)*, **2**, 19–142.

GAFFNEY, P. M. 1980. On the number of *Patella* species in south-west England. *J. mar. biol. Ass. U.K.*, **60**, 565–574.

GAINES, M. S., CALDWELL, J. & VIVAS, A. M. 1974. Genetic variation in the mangrove periwinkle *Littorina angulifera*. *Mar. Biol.*, **27**, 327–332.

GALLARDO, C. 1977a. *Crepidula philippiana* n. sp. nueva gastrópodo Calyptraeidae de Chile con especial referencia al patrón de desarrollo. *Stud. neotropical Fauna and Environment*, **12**, 177–185.

GALLARDO, C. 1977b. Two modes of development in the morphospecies *Crepidula dilatata* (Gastropoda: Calyptraeidae) from southern Chile. *Mar. Biol.*, **39**, 241–251.

GALLARDO, C. 1979. Developmental pattern and adaptation for reproduction in *Nucella crassilabrum* and other muricacean Gastropoda. *Biol. Bull.*, **157**, 453–463.

GALLIEN, L. & LAREMBERGUE, M. de. 1936. Cycle et dimorphisme sexuel chez *Lacuna pallidula* da Costa (Littorinidae). *C. R. Acad. Sci., Paris*, **203**, 409–412.

GALLIEN, L. & LAREMBERGUE, M. de. 1938. Biologie et sexualité de *Lacuna pallidula* da Costa (Littorinidae). *Trav. Sta. zool. Wimereux*, **13**, 293–306.

GARDNER, J. P. A. & THOMAS, M. L. H. 1987. Growth and production of a *Littorina littorea* (L.) population in the Bay of Fundy. *Ophelia,* **27,** 181–195.

GARTON, D. W. 1986. Effect of prey size on the energy budget of a predatory gastropod *Thais haemastoma canaliculata* (Gray). *J. exp. mar. Biol. Ecol.,* **98,** 21–38.

GARWOOD, P. R. & KENDALL, M. A. 1985. The reproductive cycles of *Monodonta lineata* and *Gibbula umbilicalis* on the coast of mid-Wales. *J. mar. biol. Ass. U.K.,* **65,** 993–1008.

GENDRON, R. P. 1977. Habitat selection and migratory behaviour of the intertidal gastropod *Littorina littorea* (L.). *J. Anim. Ecol.,* **46,** 79–92.

GERSCH, M. 1934. Zur experimenteller Veränderung der Richtung der Wellenbewegung auf der Kriechsole von Schnecken und zur Rückwartsbewegung von Schnecken. *Biol. Zbl.,* **54,** 511–518.

GHISELIN, M. T. 1966. The adaptive significance of gastropod torsion. *Evolution,* **20,** 337–348.

GIBBS, P. E. & BRYAN, G. W. 1986. Reproductive failure in populations of the dog-whelk, *Nucella lapillus,* caused by imposex induced by tributyltin from antifouling paints. *J. mar. biol. Ass. U.K.,* **66,** 767–777.

GIBBS, P. E., BRYAN, G. W., PASCOE, P. L. & BURT, G. R. 1987. The use of the dog-whelk, *Nucella lapillus,* as indicator of tributyltin (TBT) contamination. *J. mar. biol. Ass. U.K.,* **67,** 507–523.

GIBBS, P. E., BRYAN, G. W., PASCOE, P. L. & BURT, G. R. 1990. Reproductive abnormalities in female *Ocenebra erinacea* (Gastropoda) resulting from tributyltin-induced imposex. *J. mar. biol. Ass. U.K.,* **70,** 639–656.

GIBBS, P. E., PASCOE, P. L. & BURT, G. R. 1988. Sex change in the female dog-whelk, *Nucella lapillus,* induced by tributyltin from antifouling paints. *J. mar. biol. Ass. U.K.,* **68,** 713–731.

GIESE, K. 1978. Zur Embryonalentwicklung von *Buccinum undatum* L. (Gastropoda, Prosobranchia, Stenoglossa (Neogastropoda), Buccinacea). *Zool. Jahrb. (Abt. Anat. Ont. Tiere),* **100,** 65–117.

GIESEL, J. T. 1970. On the maintenance of a shell pattern and behaviour polymorphism in *Acmaea digitalis,* a limpet. *Evolution,* **24,** 98–119.

GOLIKOV, A. & STAROBOGATOV, Y. I. 1975. Systematics of prosobranch gastropods. *Malacologia,* **15,** 185–232.

GOMPEL, M. 1937. Recherches sur la consommation d'oxygène de quelques animaux aquatiques littoraux. *C.R. Acad. Sci., Paris,* **205,** 816–819.

GONOR, J. J. 1966. Escape responses of North Borneo strombid gastropods elicited by the predatory prosobranchs *Aulica vespertilio* and *Conus marmoreus. Veliger,* **8,** 226–230.

GOOCH, J. L., SMITH, B. S. & KNUPP, D. 1972. Regional survey of gene frequencies in the mud snail *Nassarius obsoletus. Biol. Bull.,* **142,** 38–48.

GOODING, R. U. & LÜTZEN, J. 1973. Studies on parasitic gastropods from echinoderms. III. A description of *Robillardia cernica* Smith, 1889, parasitic on the sea urchin *Echinometra* Meuschen, with notes on its biology. *Biol. Skr. dan. Vid. Selsk.,* **20,** (4), 1–22.

GOODRICH, E. S. 1946. The study of nephridia and genital ducts since 1895. *Quart. J. micr. Sci.,* **86,** 113–392.

GOODWIN, B. J. 1975. 'Studies on the Biology of *Littorina obtusata* and *L. mariae* (Mollusca: Gastropoda).' Ph.D. thesis, University of Wales.

GOODWIN, B. J. 1978. The growth and breeding cycle of *Littorina obtusata* (Gastropoda: Prosobranchia) from Cardigan Bay. *J. moll. Stud.,* **44,** 231–242.

GOODWIN, B. J. 1979. The egg mass of *Littorina obtusata* and *Lacuna pallidula* (Gastropoda: Prosobranchia). *J. moll. Stud.,* **45,** 1–11.

GOODWIN, B. J. & FISH, J. D. 1977. Inter- and intraspecific variation in *Littorina obtusata* and *L. mariae* (Gastropoda: Prosobranchia). *J. moll. Stud.,* **43,** 241–254.

GÖTZE, E. 1938. Bau und Leben von *Caecum glabrum* (Montagu). *Zool. Jahrb. (Syst.),* **71,** 55–122.

GOWANLOCH, J. N. & HAYES, F. R. 1926. Contributions to the study of marine gastropods. I. The physical factors, behaviour and intertidal life of *Littorina. Contr. canad. Biol Fish. N.S.,* **3,** 133–165.

GRAHAM, A. 1938. On a ciliary process of food-collecting in the gastropod *Turritella communis* Risso. *Proc. zool. Soc. Lond.,* **A108,** 453–463.

GRAHAM, A. 1939. On the structure of the alimentary canal of style-bearing prosobranchs. *Proc. zool. Soc. Lond.* **B109**, 75–112.

GRAHAM, A. 1941. The oesophagus of the stenoglossan prosobranchs. *Proc. R. Soc. Edinb.*, **B61**, 1–23.

GRAHAM, A. 1949. The molluscan stomach. *Trans. R. Soc. Edinb.*, **61**, 737–778.

GRAHAM, A. 1954. The anatomy of the prosobranch *Trichotropis borealis* Broderip & Sowerby, and the systematic position of the Capulidae. *J. mar. biol. Ass. U.K.*, **33**, 129–144.

GRAHAM, A. 1965. The buccal mass of ianthinid prosobranchs. *Proc. malac. Soc. Lond.*, **36**, 323–338.

GRAHAM, A. 1966. The fore-gut of some marginellid and cancellariid prosobranchs. *Stud. trop. Oceanogr. Miami*, **4**, 134–151.

GRAHAME, J. 1969. Shedding of the penis in *Littorina littorea*. *Nature, Lond.*, **221**, 976.

GRAHAME, J. 1972. Breeding energetics of *Littorina littorea* (L.) (Gastropoda: Prosobranchiata). *J. Anim. Ecol.*, **42**, 391–403.

GRAHAME, J. 1977. Reproductive effort and r- and k-selection in two species of *Lacuna* (Gastropoda: Prosobranchia). *Mar. Biol.*, **40**, 217–224.

GRAHAME, J. & BRANCH, G. M. 1985. Reproductive patterns of marine invertebrates. *Oceanogr. mar. Biol. Ann. Rev.*, **23**, 373–398.

GRAHAME, J. & MILL, P. J. 1986. Relative size of the foot of two species of *Littorina* on a rocky shore in Wales. *J. Zool., Lond.*, **A208**, 229–236.

GRAHAME, J. & MILL, P. J. 1989. Shell shape variation in *Littorina saxatilis* and *L. arcana*: a case of character displacement? *J. mar. biol. Ass. U.K.*, **69**, 837–855.

GRASSET, M. & VOVELLE, J. 1982. Données histochimiques et ultrastructurales sur l'opercule de *Buccinum undatum* (L.) (Mollusca, Gastropoda). *Malacologia*, **22**, 251–255.

GRAY, J. E. 1853. On the division of ctenobranchous gasteropodous Mollusca into larger groups and families. *Proc. zool. Soc. Lond.*, **21**, 32–44.

GREGOIRE, C. 1972. Structure of the molluscan shell. In 'Chemical Zoology' (M. Florkin & B. T. Scheer eds) **7**, 45–102. New York, Academic Press.

GRENON, J.-F. & WALKER, G. 1978. The histology and histochemistry of the pedal glandular system of two limpets, *Patella vulgata* and *Acmaea tessulata* (Gastropoda: Prosobranchia). *J. mar. biol. Ass. U.K.*, **58**, 803–816.

GRENON, J.-F. & WALKER, G. 1981. The tenacity of the limpet *Patella vulgata* L.: an experimental approach. *J. exp. mar. Biol. Ecol.*, **54**, 277–308.

GRENON, J.-F. & WALKER, G. 1982. Further fine structure studies of the "space" layer which underlies the footsole epithelium of the limpet, *Patella vulgata* L. *J. moll. Stud.*, **48**, 55–63.

GRIFFITH, G. W. & CASTAGNA, M. 1962. Sexual dimorphism in oyster drills of Chincoteague Bay, Maryland-Virginia. *Chesapeake Sci.*, **3**, 215–217.

GRIFFITHS, R. J. 1981. Predation of the bivalve *Choromytilus meridionalis* (Kr.) by the gastropod *Natica* (*Tectonatica*) *tecta* Anton. *J. moll. Stud.*, **47**, 112–120.

GRIFFOND, B. 1969. Survie et évolution, en culture *in vitro*, des testicules de *Viviparus viviparus* L., Gastéropode Prosobranche à sexes séparés. *C. R. Acad. Sci., Paris*, **268D**, 963–965.

GRIFFOND, B. 1975. Analyse expérimentale en culture "in vitro" des facteurs de la gamétogenèse chez la paludine *Viviparus viviparus* (L.). *Ann. Sci. Univ. Besançon Zool.*, **12**, 73–98.

GUILLETTE, L. J. 1991. The evolution of viviparity in amniotic vertebrates: new insights, new questions. *J. Zool., Lond.*, **223**, 521–526.

GURIN, S. & CARR, W. E. 1971. Chemoreception in *Nassarius obsoletus*: the role of specific stimulatory proteins. *Science*, **174**, 293–295.

HABE, T. & KOSUGE, S. 1966. 'Shells of the World in Colour.' **2**, 1–193. Osaka, Hoikusha.

HADLOCK, R. F. 1980. Alarm response of the intertidal snail *Littorina littorea* (L.) to predation by the crab *Carcinus maenas* (L.). *Biol. Bull.*, **159**, 269–279.

HAGERMAN, L. 1966. The macro- and microfauna associated with *Fucus serratus* L., with some ecological remarks. *Ophelia*, **3**, 1–43.

HALLER, B. 1888. Die Morphologie der Prosobranchier I. *Morph. Jahrb.*, **14**, 54–169.

HANCOCK, D. A. 1956. The structure of the capsule and the hatching process in *Urosalpinx cinerea*. *Proc. zool. Soc. Lond.,* **127,** 565–571.

HANCOCK, D. A. 1959. The biology and control of the American whelk tingle *Urosalpinx cinerea* (Say) on English oyster beds. *MAFF Fish. Invest. (ser.* 2), **22,** no. **10,** 1–66.

HANCOCK, D. A. 1960. The ecology of the molluscan enemies of the edible mollusc. *Proc. malac. Soc. Lond.,* **34,** 123–143.

HARDY, A. C. 1956. 'The Open Sea. Its Natural History: the World of Plankton.' London, Collins.

HARRISON, F. M. 1962. Some excretory processes in the abalone *Haliotis rufescens*. *J. exp. Biol.,* **39,** 179–192.

HART, A. & BEGON, M. 1982. The status of general reproductive strategy theories, illustrated in winkles. *Oecologia (Berlin),* **52,** 37–42.

HARTNOLL, R. G. & WRIGHT, J. R. 1977. Foraging movements and homing in the limpet *Patella vulgata* L. *Anim. Behav.,* **25,** 806–810.

HASZPRUNAR, G. 1985a. The fine morphology of the osphradial sense organs of the Mollusca. I. Gastropoda Prosobranchia. *Phil. Trans. R. Soc. B,* **307,** 457–496.

HASZPRUNAR, G. 1985b. The fine morphology of the osphradial sense organs of the Mollusca. II. Allogastropoda (Architectonicidae, Pyramidellidae). *Phil. Trans. R. Soc. B,* **307,** 497–505.

HASZPRUNAR, G. 1985c. The Heterobranchia—a new concept of the phylogeny of the higher Gastropoda. *Z. zool. Syst. Evolut.-forsch.,* **28,** 15–37.

HASZPRUNAR, G. 1985d. Zur Anatomie und systematische Stellung der Architectonicidae (Mollusca, Allogastropoda). *Zool. Scripta,* **14,** 25–43.

HASZPRUNAR, G. 1985e. On the anatomy and systematic position of the Mathildidae (Mollusca, Allogastropoda). *Zool. Scripta,* **14,** 201–213.

HASZPRUNAR, G. 1987a. The fine structure of the ctenidial sense organs (bursicles) of Vetigastropoda (Zeugobranchia, Trochoidea) and their phylogenetic significance. *J. moll. Stud.,* **53,** 46–51.

HASZPRUNAR, G. 1987b. Anatomy and affinities of cocculinid limpets (Mollusca, Archaeogastropoda). *Zool. Scripta,* **16,** 305–324.

HASZPRUNAR, G. 1987c. The anatomy of *Addisonia* (Mollusca, Gastropoda). *Zoomorphology,* **106,** 269–278.

HASZPRUNAR, G. 1988a. A preliminary phylogenetic classification of the streptoneurous gastropods. *Malacol. Rev.,* suppl. **4,** 7–16.

HASZPRUNAR, G. 1988b. Comparative anatomy of cocculiniform gastropods and its bearing on archaeogastropod systematics. *Malacol. Rev.,* suppl. **4,** 64–84.

HASZPRUNAR, G. 1988c. On the origin and evolution of major gastropod groups, with special reference to the Streptoneura. *J. moll. Stud.,* **54,** 367–441.

HASZPRUNAR, G. 1988d. Anatomy and relationship of the bone-feeding limpets, *Cocculinella minutissima* (Smith) and *Osteopelta mirabilis* Marshall (Archaeogastropoda). *J. moll. Stud.,* **54,** 1–20.

HASZPRUNAR, G. 1988e. Anatomy and affinities of pseudococculinid limpets (Mollusca, Archaeogastropoda). *Zool. Scripta,* **17,** 161–179.

HASZPRUNAR, G. 1992. The first molluscs—small animals. *Bull. Zool.,* **59,** 1–16.

HAWKINS, L. E. & HUTCHINSON, S. 1988. Egg capsule structure and hatching mechanism in *Ocenebra erinacea* (L.) (Prosobranchia, Muricidae). *J. exp. mar. Biol. Ecol.,* **119,** 269–283.

HAWKINS, S. J. & HARTNOLL, R. G. 1982. The influence of barnacle cover on the numbers, growth and behaviour of *Patella vulgata* on a vertical pier. *J. mar. biol. Ass. U.K.,* **62,** 855–867.

HAWKINS, S. J. & HARTNOLL, R. G. 1983a. Changes in a rocky shore community: an evaluation of monitoring. *Mar. env. Res.,* **9,** 131–181.

HAWKINS, S. J. & HARTNOLL, R. G. 1983b. Grazing of intertidal algae by marine invertebrates. *Oceanogr. mar. Biol. Ann. Rev.,* **21,** 195–282.

HAWKINS, S. J., WATSON, D. C., HILL, A. S., HARDING, S. P., KYRIAKIDES, M. A., HUTCHINSON, S. & NORTON, T. A. 1989. A comparison of feeding mechanisms in microphagous herbivorous intertidal prosobranchs in relation to resource partitioning. *J. moll. Stud.,* **55,** 151–165.

HAWTHORNE, J. B. 1965. The eastern limit of *Monodonta lineata* (da Costa) in the English Channel. *J. Conch., Lond.,* **25,** 348–352.

HAYASHI, I. 1980. The reproductive biology of the ormer, *Haliotis tuberculata. J. mar. biol. Ass. U.K.,* **60,** 415–430.

HEALY, J. M. 1982. Ultrastructure of spermiogenesis of *Philippia (Psilaxis) oxytropis,* with special reference to the taxonomic position of the Archnitectonicidae (Gastropoda). *Zoomorphology,* **101,** 197–214.

HEALY, J. M. 1983. Ultrastructure of euspermatozoa of cerithiacean gastropods (Prosobranchia, Mesogastropoda). *J. Morph.,* **178,** 57–75.

HEALY, J. M. 1986. An ultrastructural study of euspermatozoa, paraspermatozoa and nurse cells of the cowrie *Cypraea errones* (Gastropoda, Prosobranchia, Cypraeidae). *J. moll. Stud.,* **52,** 125–137.

HEALY, J. M. 1988. Sperm morphology and its systematic importance in the Gastropoda. *Malacol. Rev.,* suppl. **4,** 251–266.

HELLER, J. 1975a. The taxonomy of some British *Littorina* species, with notes on their reproduction (Mollusca: Prosobranchia). *Zool. J. Linn. Soc.,* **56,** 131–151.

HELLER, J. 1975b. Visual selection of shell colour in two littoral prosobranchs. *Zool. J. Linn. Soc.,* **56,** 152–170.

HELLIER, J. 1976. The effects of exposure and predation on the shell of two British winkles. *J. Zool., Lond.,* **179,** 201–213.

HEMINGWAY, G. T. 1978. Evidence for a paralytic venom in the intertidal snail, *Acanthina spirata* (Neogastropoda: Thaisidae). *Comp. Biochem. Physiol.,* **60C,** 79–81.

HERBERT, D. G. 1982. Fine structural observations on the juxtaganglionar organ of *Gibbula umbilicalis* (da Costa). *J. moll. Stud.,* **48,** 226–228.

HERMANS, C. O. 1983. The duo-gland adhesion hypothesis. *Oceanogr. mar. Biol. Ann. Rev.,* **21,** 283–339.

HERTLING, H. 1928. Beobachtungen und Versuche an der Eiern von *Littorina* und *Lacuna.* Bedeutung der Eihüllen. Entwicklung in naturlichen und abgeänderten Medium. *Wiss. Meeresuntersuch. Abt. Helgoland,* **17,** 1–49.

HICKMAN, C. S. 1980. Gastropod radulae and the assessment of form in evolutionary paleontology. *Paleobiology,* **6,** 276–294.

HICKMAN, C. S. 1984. Implications of radular tooth-row functional integration for archaeogastropod systematics. *Malacologia,* **25,** 143–160.

HICKMAN, C. S. 1988. Archaeogastropod evolution, phylogeny and systematics: a re-evaluation. *Malacol. Rev.,* suppl. **4,** 17–34.

HICKMAN, C. S. & MORRIS, T. E. 1985. Gastropod feeding tracks as a source of data in analysis of the functional morphology of radulae. *Veliger,* **27,** 357–365.

HIGHNAM, H. & HILL, L. 1977. 'The Comparative Endocrinology of the Invertebrates.' London, Edward Arnold.

HOAGLAND, K. E. 1978. Protandry and the evolution of environmentally-mediated sex change: a study of the Mollusca. *Malacologia,* **17,** 365–391.

HOAGLAND, K. E. 1985. Genetic relationships between one British and several North American populations of *Crepidula fornicata* based on allozyme studies (Gastropoda: Calyptraeidae). *J. moll. Stud.,* **51,** 177–182.

HOAGLAND, K. E. 1986. Patterns of encapsulation and brooding in the Calyptraeidae (Prosobranchia: Mesogastropoda). *Am. malac. Bull.,* **4,** 173–183.

HOAGLAND, K. E. & COE, W. R. 1982. Larval development in *Credpidula maculosa* (Prosobranchia: Crepidulidae) from Florida. *Nautilus,* **96,** 122.

HOAGLAND, K. E. & ROBERTSON, R. 1988. An assessment of poecilogony in marine invertebrates: phenomenon or fantasy? *Biol. Bull.,* **174,** 109–125.

HOFFMAN, A. & MARTINELL, J. 1984. Prey selection by naticid gastropods in the Pliocene of Emporda (Northeast Spain). *N. Jb. Geol. Paläont, Mb.,* **7,** 393–399.

HOFFMANN, H. 1937. Über die Stammesgechichte der Weichtiere. *Verh. dtsch. zool. Ges.* **39,** (*Zool. Anz.* suppl. **10**), 33–69.

HÖISAETER, T. 1965. Spermatophores in *Chrysallida obtusa* (Brown) (Opisthobranchia, Pyramidellidae). *Sarsia,* **18,** 63–68.

HOLME, N. A. 1949. The fauna of sand and mud banks near the mouth of the Exe estuary. *J. mar. biol. Ass. U.K.,* **28,** 189–237.

HORNE, F. R. & BARNES, G. 1970. Reevaluation of urea biosynthesis in prosobranch and pulmonate snails. *Z. vergl. Physiol.,* **69,** 452–457.

HORNY, R. J. 1963. Lower Paleozoic Monoplacophora and patellid Gastropoda (Mollusca) of Bohemia. *Ustředni Ustav Geologicky, Sbornik,* **38,** 7–83.

HOUBRICK, R. S. 1973. Studies on the reproductive biology of the genus *Cerithium* (Gastropoda: Prosobranchia) in the Western Atlantic. *Bull. mar. Sci.,* **23,** 875–904.

HOUBRICK, R. S. 1974. The genus *Cerithium* in the Western Atlantic (Cerithiidae: Prosobranchia). *Johnsonia,* **5** (50), 33–84.

HOUBRICK, R. S. 1978. The family Cerithiidae in the Indo-Pacific. Part 1: the genera *Rhinoclavis, Pseudovertagus, Longicerithium* and *Clavocerithium. Monogr. mar. Moll.,* **1,** 1–130.

HOUBRICK, R. S. 1985. Genus *Clypeomorus* (Cerithiidae: Prosobranchia). *Smithsonian Contr. Zool.* No. 403, 1–131.

HOUBRICK, R. S. 1987. Anatomy, reproductive biology and phylogeny of the Planaxidae (Cerithiacea: Prosobranchia). *Smithsonian Contr. Zool.* No 445, 1–57.

HOUBRICK, R. S. 1988. Cerithioidean phylogeny. *Malacol. Rev.,* suppl. **4,** 88–128.

HOUBRICK, R. S. & FRETTER, V. 1969. Some aspects of the functional anatomy of *Cymatium* and *Bursa. Proc. malac. Soc. Lond.,* **38,** 415–430.

HOULIHAN, D. F. & INNES, A. J. 1982. Respiration in air and water of four Mediterranean trochids. *J. exp. mar. Biol. Ecol.,* **57,** 35–54.

HOULIHAN, D. F., INNES, A. J. & DEY, D. G. 1981. The influence of mantle cavity fluid on the aerial oxygen consumption of some intertidal gastropods. *J. exp. mar. Biol. Ecol.,* **49,** 57–68.

HOUSE, M. R. & FARROW, G. E. 1968. Daily growth banding in the shell of the cockle, *Cardium edule. Nature, Lond.,* **219,** 1384–1386.

HOUSSAY, F. 1884. Recherches dur l'opercule et les glandes du pied des Gastéropodes. *Arch. Zool. exp. gén.* (7), **2,** 177–208.

HOUSTON, R. S. 1985. Genital ducts of the Cerithiacea (Gastropoda: Mesogastropoda) from the Gulf of California. *J. moll. Stud.,* **51,** 183–189.

HOWELLS, H. H. 1936. The anatomy and histology of the gut of *Cymbulia peronii* (Blainville). *Proc. malac. Soc. Lond.,* **22,** 62–72.

HOXMARK, R. C. 1970. The chromosome dimorphism of *Nucella lapillus* (Prosobranchia) in relation to wave exposure. *Nytt. Mag. Zool.,* **18,** 229–238.

HOXMARK, R. C. 1971. Shell variation in *Nucella lapillus* in relation to environmental and genetic factors. *Norweg. J. Zool.,* **19,** 145–148.

HOYAUX, J., GILLES, R. & JEUNIAUX. C. 1976. Chemoregulation in molluscs of the intertidal zone. *Comp. Biochem. Physiol.,* **53A,** 361–365.

HUEBNER, E. & ANDERSON, E. 1976. Comparative spiralian oogenesis—structural aspects: an overview. *Am. Zool.,* **16,** 315–343.

HUGHES, R. N. 1972. Annual production of two Nova Scotian populations of *Nucella lapillus* (L.). *Oecologia* (*Berlin*), **8,** 356–370.

HUGHES, R. N. 1980. Population dynamics, growth and reproductive rates of *Littorina nigrolineata* Gray from a moderately sheltered locality in North Wales. *J. exp. mar. Biol. Ecol.,* **44,** 211–228.

HUGHES, R. N. 1985. Predatory behaviour of *Natica unifasciata* feeding intertidally on gastropods. *J. moll. Stud.,* **51,** 331–335.

HUGHES, R. N. 1986. Laboratory observations on the feeding behaviour, reproduction and morphology of *Galeodea echinophora* (Gastropoda: Cassidae). *Zool. J. Linn. Soc.,* **86,** 355–365.

HUGHES, R. N. & DREWETT, D. 1985. A comparison of the foraging behaviour of dogwhelks, *Nucella lapillus*, feeding on barnacles or mussels on the shore. *J. moll. Stud.,* **51,** 73–77.

HUGHES, R. N. & DUNKIN, S. de B. 1984a. Behavioural components of prey-selection by dogwhelks, *Nucella lapillus*, feeding on mussels, *Mytilus edulis* (L.) in the laboratory. *J. exp. mar. Biol. Ecol.,* **77,** 45–68.

HUGHES, R. N. & DUNKIN, S. de B. 1984b. Effect of dietary history on selection of prey and foraging behaviour among patches of prey, by the dogwhelk, *Nucella lapillus* (L.). *J. exp. mar. Biol. Ecol.,* **79,** 159–172.

HUGHES, R. N. & ELNER, R. W. 1979. Tactics of a predator, *Carcinus maenas*, and morphological responses of the prey, *Nucella lapillus*. *J. Anim. Ecol.,* **48,** 65–78.

HUGHES, R. N. & HUGHES, H. P. I. 1981. Morphological and behavioural aspects of feeding in the Cassidae (Tonnacea, Mesogastropoda). *Malacologia,* **20,** 385–402.

HUGHES, R. N. & ROBERTS, D. J. 1980a. Reproductive effort of winkles (*Littorina* spp.) with contrasted methods of reproduction. *Oecologia (Berlin),* **47,** 130–136.

HUGHES, R. N. & ROBERTS, D. J. 1980b. Growth and reproductive rates of *Littorina neritoides* (L.) in North Wales. *J. mar. biol. Ass. U.K.,* **60,** 591–599.

HUGHES, R. N. & ROBERTS, D. J. 1981. Compartive demography of *Littorina rudis, L. nigrolineata*, and *L. neritoides* on three contrasted shores in North Wales. *J. Anim. Ecol.,* **50,** 251–268.

HUMPHREYS, W. F. & LUTZEN, J. 1972. Studies on parasitic gastropods from echinoderms. I. On the structure and biology of the parasitic gastropod *Megadenus cantharelloides* n. sp. *Biol. Skr. dan. Vid. Selsk.,* **19** (1), 1–27.

HUNT, O. D. 1971. Holkham Salts Hole, an isolated salt-water pond with relict features. An account based on studies by the late C. F. A. Pantin. *J. mar. biol. Ass. U.K.,* **51,** 717–741.

HUNT, S. 1971. Comparison of three extracellular structural proteins in the gastropod mollusc *Buccinum undatum* L., the periostracum, egg capsule and operculum. *Comp. Biochem. Physiol.,* **40B,** 37–46.

HUNT, S. 1976. The gastropod operculum: a comparative study of the composition of gastropod proteins. *J. moll. Stud.,* **42,** 251–260.

HYLLEBERG, J. 1975. The effect of salinity and temperature on egestion in mud snails (Gastropoda: Hydrobiidae). A study on niche overlap. *Oecologia (Berlin),* **21,** 279–289.

HYLLEBERG, J. 1976. Resource partitioning on basis of hydrolytic enzymes in deposit-feeding mud snails (Hydrobiidae). II. Studies on niche overlap. *Oecologia (Berlin),* **23,** 115–125.

HYLLEBERG, J. & CHRISTENSEN, J. T. 1977. Phenotypic variation and fitness of periwinkles (Gastropoda: Littorinidae) in relation to exposure. *J. moll. Stud.,* **43,** 192–199.

IMRIE, D. W., HAWKINS, S. J. & McCROHAN, C. R. 1989. The olfactory-gustatory basis of food preference in the herbivorous prosobranch, *Littorina littorea* (Linnaeus). *J. moll. Stud.,* **55,** 217–225.

IMRIE, D. W., McCROHAN, C. R. & HAWKINS, S. J. 1990. Feeding behaviour in *Littorina littorea*: a study of the effects of ingestive conditioning and previous dietary history on food preference and rates of consumption. *Hydrobiologia,* **193,** 191–198.

INNES, A. J. & HOULIHAN, D. F. 1985. Aquatic and aerial oxygen consumption of cool temperate gastropods: a comparison with some Mediterranean species. *Comp. Biochem. Physiol.,* **82A,** 105–109.

JACKSON, A. P., VINCENT, J. F. V. & TURNER, R. M. 1988. The mechanical design of nacre. *Proc. R. Soc. B,* **234,** 415–440.

JAMES, M. J. 1980. Comparative morphology of radular teeth in *Conus*: observations with scanning electron microscopy. *J. moll. Stud.,* **46,** 116–128.

JANNASCH, H. W. & WIRSEN, C. O. 1981. Morphological survey of microbial mats near deep-sea thermal vents. *Appl. environ. Microbiol.,* **41,** 528–538.

JANSON, K. 1982a. Genetic and environmental effects on the growth rate of *Littorina saxatilis*. *Mar. Biol.,* **69,** 73–78.

JANSON, K. 1982b. Phenotypic differentiation in *Littorina saxatilis* Olivi (Mollusca, Prosobranchia) in a small area on the Swedish west coast. *J. moll. Stud.,* **48,** 167–173.

JANSON, K. 1986. 'Polymorphisms, Causes and Evolutionary Consequences in a marine Prosobranch Species, *Littorina saxatilis*.' Dissertation, University of Göteborg, 1–20.

JANSON, K. & WARD, R. D. 1985. The taxonomic status of *Littorina tenebrosa* Montagu as assessed by morphological and genetical analysis. *J. Conch., Lond.,* **32,** 9–15.

JEFFREYS, J. G. 1865. 'British Conchology' **3,** 1–393. London, van Voorst.

JEFFREYS, J. G. 1867. 'British Conchology' **4,** 1–486. London, van Voorst.

JEFFREYS, J. G. 1869. 'British Conchology' **5,** 1–258. London, van Voorst.

JENNER, C. E. 1956. The occurrence of a crystalline style in the mud snail *Nassarius obsoletus. Biol. Bull.,* **111,** 304.

JENNINGS, K. H. 1984. 'The organization, fine structure and function of the excretory systems of the estuarine bivalve, *Srobicularia plana* (da Costa) and the freshwater bivalve *Anodonta cygnea* (Linné) and other selected species.' Ph.D. Thesis, University of London.

JENSEN, K. T. & SIEGISMUND, H. R. 1980. The importance of diatoms and bacteria in the diet of *Hydrobia*-species. *Ophelia* suppl. **1,** 193–199.

JOHANNESSON, B. 1986. Shell morphology of *Littorina saxatilis* Olivi: the relative importance of physical factors and predation. *J. exp. mar. Biol. Ecol.,* **102,** 183–195.

JOHANNESSON, B. & JOHANNESSON, K. 1990. *Littorina neglecta* Bean, a morphological form within the variable species *Littorina saxatilis* (Olivi)? *Hydrobiologia,* **193,** 71–87.

JOHANNESSON, K. 1989. The bare zone of Swedish rocky shores: why is it there? *Oikos,* **54,** 77–86.

JOHANNESSON, K. & JOHANNESSON, B. 1990. Genetic variation within *Littorina neglecta* Bean: is *L. neglecta* a good species? *Hydrobiologia,* **193,** 89–97.

JOHNSON, C. R. & MANN, K. H. 1986. The importance of plant defence abilities to the structure of subtidal seaweed communities: the kelp *Laminaria longicruris* de la Pylaie survives grazing by the snail *Lacuna vincta* (Montagu) at high population densities. *J. exp. mar. Biol. Ecol.,* **97,** 231–267.

JONES, H. D. 1968. Some aspects of heart function in *Patella vulgata* L. *Nature, Lond.,* **217,** 1170–1172.

JONES, H. D. 1970. Hydrostatic pressures within the heart and pericardium of *Patella vulgata* L. *Comp. Biochem. Physiol.,* **34,** 263–272.

JONES, H. D. 1975. Locomotion. In 'Pulmonates' (V. Fretter & J. Peake, eds) **1,** 1–32. London & New York, Academic Press.

JONES, H. D. 1983. The circulatory systems of gastropods and bivalves. In 'The Mollusca' **5,2** (A. S. M. Saleuddin & K. M. Wilbur eds), 189–238. London, Academic Press.

JONES, H. D. 1984. Shell cleaning behaviour of *Calliostoma zizyphinum. J. moll. Stud.,* **50,** 245–247.

JONES, H. D. 1988. *In vivo* cardiac pressure and heart rate, and heart mass of *Busycon canaliculatum* (L.). *J. exp. Biol.,* **140,** 257–271.

JONES, H. D. & TRUEMAN, E. R. 1970. Locomotion of the limpet, *Patella vulgata* L. *J. exp. Biol.,* **52,** 202–216.

JONES, N. S. 1948. Observations on the biology of *Patella vulgata* at Port St. Mary, Isle of Man. *Proc. Trans. Lpool. Bio. Soc.,* **56,** 60–77.

JOOSSE, J. 1972. Endocrinology of reproduction in molluscs. *Gen. comp. Endocrinol.* suppl. **2,** 591–601.

JOOSSE, J. 1979. Endocrinology of molluscs. In 'Pathways in Malacology' (S. van der Spoel, A. C. van Bruggen & J. Lever eds), 107–137. Utrecht, Bohn, Scheltema & Holkham.

JOOSSE, J. & GERAERTS, W. P. M. 1983. Endocrinology. In 'The Mollusca' **4** (A. S. M. Saleuddin & K. M. Wilbur eds), 317–406. New York & London, Academic Press.

KABAT, A. R. 1990. Predatory ecology of naticid gastropods with a review of shell boring predation. *Malacologia,* **32,** 155–193.

KAMEL, E. G. 1979. 'The physiological effects of platyhelminth parasites on *Littorina littorea*.' Ph.D. Thesis, University of Manchester.

KANTOR, Y. I. 1985. Feeding and some features of functional morphology of the molluscs in the subfamily Volutopsinae (Gastropoda, Pectinibranchia). *Zool. Zh.,* **54,** 1640–1647. [In Russian.]

KANTOR, Y. I. 1987. Morphological analysis of the digestive system of gastropod molluscs of the order Toxoglossa (Gastropoda, Pectinibranchia). *Dok. Akad. Nauk SSSR,* **297,** 251–253. [In Russian.]

KANTOR, Y. I. 1988. On the anatomy of Pseudomelatominae (Gastropoda, Toxoglossa, Turridae) with notes on functional morphology and phylogeny of the subfamily. *Apex*, **3**, 1–19.

KANTOR, Y. I. 1990a. 'Gastropods of the subfamily Volutopsinae of the world ocean.' Moscow, Nauka. [In Russian.]

KANTOR, Y. I. 1990b. Anatomical basis for the origin and evolution of the toxoglossan mode of feeding. *Malacologia*, **32**, 3–18.

KANTOR, Y. I. & SYSOEV, A. V. 1989. The morphology of toxoglossan gastropods lacking a radula, with a description of a new species and genus of Turridae. *J. moll. Stud.*, **55**, 537–549.

KANTOR, Y. I. & SYSOEV, A. V. 1990. Special morphology and evolution of the anterior part of the alimentary system of Toxoglossa. In 'Evolutionary morphology of molluscs' (A. A. Shilenko ed.). *Arch. zool. Mus. Moscow State Univ.*, **28**, 91–134. [In Russian.]

KANTOR, Y. I. & TAYLOR, J. D. 1991. Evolution of the toxoglossan feeding mechanism: new information on the use of the radula. *J. moll. Stud.*, **57**, 129–134.

KASINATHAN, R. 1975. Some studies of five cyclophorid snails from peninsular India. *J. moll. Stud.*, **41**, 379–394.

KENDALL, M. A. 1987. The age and size structure of some northern populations of the trochid gastropod *Monodonta lineata*. *J. moll. Stud.*, **53**, 213–222.

KENDALL, M. A. & LEWIS, J. R. 1986. Temporal and spatial patterns in the recruitment of *Gibbula umbilicalis*. *Hydrobiologia*, **142**, 15–22.

KENDALL, M. A., WILLIAMSON, P. & GARWOOD, P. R. 1987. Annual variation and recruitment and population structure of *Monodonta lineata* and *Gibbula umbilicalis* populations at Aberaeron, Mid-Wales. *Est. coastal Shelf Sci.*, **24**, 499–511.

KERTH, K. 1979. Phylogenetische Aspekte der Radulamorphogenese von Gastropoden. *Malacologia*, **19**, 103–108.

KERTH, K. 1983. Radulaapparat und Radulabildung der Mollusken. II. *Zool. Jahrb. Abt. Anat. Ontog. Tiere*, **110**, 205–269.

KESSEL, E. 1942. Über Bau und Bildung des Prosobranchier-Deckels. *Z. Morph. Ökol. Tiere*, **38**, 197–250.

KESSEL, M. M. 1964. Reproduction and larval development of *Acmaea testudinalis* (Müller). *Biol. Bull.*, **127**, 294–303.

KIER, W. M. 1988. The arrangement and function of molluscan muscles. In 'The Mollusca' **11**, (E. R. Trueman & M. R. Clarke eds) 211–252. San Diego, Academic Press.

KIER, W. M. 1989. The fin musculature of cuttlefish and squid (Mollusca, Cephalopoda): morphology and mechanics. *J. Zool., Lond.*, **217**, 23–38.

KIER, W. M. & SMITH, K. K. 1985. Tongues, tentacles and trunks: the biomechanics of movement in muscular hydrostats. *Zool. J. Linn. Soc.*, **83**, 307–324.

KILIAN, E. 1951. Untersuchungen zur Biologie von *Pomatias elegans* (Müller) und ihrer 'Konkrement-drüse'. *Arch. Molluskenk.*, **80**, 1–16.

KINGSTON, R. S. 1968. Anatomical and oxygen electrode studies of respiratory surfaces and respiration in *Acmaea*. *Veliger* **11** (suppl.), 73–78.

KISELEVA, G. A. 1967. Influence of the substratum on the settlement and metamorphosis of benthic animals. In 'Benthic biocoenoses and biology of benthic organisms of the Black Sea' (V. A. Vodyanitsckii ed.), 71–84. Kiev, Akademiya Nauk SSSR. [In Russian.]

KITCHELL, J. A., BOGGS, C. H., KITCHELL, J. F. & RICE, J. A. 1981. Prey selection by naticid gastropods: experimental tests and application to the fossil record. *Paleobiology*, **7**, 533–552.

KITCHELL, J. A., BOGGS, C. H., RICE, J. A., KITCHELL, J. F., HOFFMANN, A. & MARTINELL, J. 1986. Anomalies in naticid predatory behaviour: a critique and experimental observations. *Malacologia*, **27**, 291–298.

KITCHING, J. A. 1977. Shell form and niche occupation in *Nucella lapillus* (L.) (Gastropoda). *J. exp. mar. Biol. Ecol.*, **26**, 275–287.

KITCHING, J. A., MUNTZ, L. & EBLING, F. J. 1966. The ecology of Lough Ine. XV: the ecological significance of shell and body forms in *Nucella*. *J. Anim. Ecol.*, **35**, 113–126.

KNIGHT-JONES, E. W. 1954. Relations between metachronism and the direction of ciliary beat in Metazoa. *Quart. J. micr. Sci.,* **95,** 503–521.

KNIPRATH, E. 1981. Ontogeny of the molluscan shell field: a review. *Zool. Scripta,* **10,** 61–79.

KOFOED, L. H. 1975a. The feeding biology of *Hydrobia ventrosa* (Montagu). I. The assimilation of different components of the food. *J. exp. mar. Biol. Ecol.,* **19,** 233–241.

KOFOED, L. H. 1975b. The feeding biology of *Hydrobia ventrosa* (Montagu). II. Allocation of the components of the carbon-budget and the significance of the secretion of dissolved organic material. *J. exp. mar. Biol. Ecol.,* **19,** 243–256.

KOHN, A. J. 1961. Studies on spawning behavior, egg masses and larval development in the gastropod genus *Conus.* II. Observations in the Indian Ocean during the Yale Seychelles Expedition. *Bull. Bingham Oceanogr. Collect.,* **17,** 1–51.

KOHN, A. J. 1971. Diversity, utilization of resources and adaptive radiation in shallow-water marine invertebrates of tropical oceanic islands. *Limnol. Oceanogr.,* **16,** 332–348.

KOHN, A. J. 1986. Slip-resistant silver-feet: shell form and mode of life in lower Pleistocene *Argyropeza* from Fiji. *J. Paleontol.,* **60,** 1066–1074.

KOHN, A. J., SAUNDERS, P. R. & WIENER, S. 1960. Preliminary studies on the venom of the marine snail *Conus. Ann. N. Y. Acad. Sci.,* **90,** 706–725.

KOSTITZINE, J. 1940. Sur la ponte de la pourpre. *Bull. Soc. zool. Fr.,* **65,** 80–84.

KOSUGE, S. 1966. The family Triphoridae and its systematic position. *Malacologia,* **12,** 297–324.

KRAMPITZ, G., DROESHAGEN, H., HAUSLE, J. & HOF-IRMSCHER, K. 1983. Organic matrices of mollusc shells. In 'Biomineralization and biological metal accumulation' (P. Westbroek & E. W. de Jong eds) 231–247. Dordrecht, Reidel.

KRIJGSMAN, B. J. & DIVARIS, G. A. 1955. Contractile and pace-maker mechanisms of the heart of molluscs. *Biol. Rev.,* **30,** 1–39.

KRISTENSEN, J. H. 1970. Fauna associated with the sipunculid *Phascolion strombi* (Montagu), especially the parasitic gastropod *Menestho diaphana* (Jeffreys). *Ophelia,* **7,** 257–276.

LALLI, C. M. & GILMER, R. W. 1989. 'Pelagic Snails.' Stanford, California, Stanford University Press.

LAMBERT, T. C. & FARLEY, J. 1968. The effect of parasitism by the nematode *Cryptocotyle lingua* (Creplin) on zonation and winter migration of the common periwinkle *Littorina littorea* (L.). *Can. J. Zool.,* **46,** 1139–1147.

LARGEN, M. J. 1967. The diet of the dog-whelk, *Nucella lapillus* (Gastropoda Prosobranchia). *J. Zool., Lond.,* **152,** 123–127.

LARGEN, M. J. 1971. Genetic and environmental influences upon the expression of shell sculpture in the dog-whelk (*Nucella lapillus*). *Proc. malac. Soc. Lond.,* **39,** 383–388.

LASIAK, T. 1987. The reproductive cycles of three trochid gastropods from the Transkei coast, southern Africa. *J. moll. Stud.,* **53,** 24–32.

LAURSEN, D. 1981. Taxonomy and distribution of teleplanic prosobranch larvae in the North Atlantic. *Dana Rep.,* **89,** 1–44.

LAXTON, J. H. 1971. Feeding in some Australasian Cymatiidae (Gastropoda: Prosobranchia). *Zool. J. Linn. Soc.,* **50,** 1–9.

LEBOUR, M. V. 1933. The life-histories of *Cerithiopsis tubercularis* (Montagu), *C. barleei* Jeffreys and *Triphora perversa* (L.). *J. mar. biol. Ass. U.K.,* **18,** 491–498.

LEBOUR, M. V. 1935. The breeding of *Littorina neritoides. J. mar. biol. Ass. U.K.,* **20,** 373–378.

LEBOUR, M. V. 1937. The eggs and larvae of the British prosobranchs with special reference to those living in the plankton. *J. mar. biol. Ass. U.K.,* **22,** 105–166.

LE BRETON, J. 1969. Thèse doctorale: Biologie animale. Université de Caen.

LE BRETON, J. 1970. Evolution et chute du pénis, étude de l'influence du jeûne, chez *Littorina littorea* L., mollusque, gastéropode, prosobranche. *C R. Acad. Sci., Paris,* **271D,** 534–536.

LE GALL, P. 1980. 'Etude expérimentale de l'association en chaine et de son influence sur la croissance et la sexualité chez la crépidule *Crepidula fornicata* Linné (1758) (Mollusque Mésogastéropode).' Thèse, Université de Caen.

LE GALL, S. 1974. 'Déterminisme de la morphogenèse et du cycle du tractus génital mâle externe chez *Crepidula fornicata* Phil. (Mollusque Mésogastéropode).' Thèse, Université de Caen.

LE GALL, S. 1978. Contrôle du facteur pédieux morphogénétique du pénis par les ganglions cérébropleuraux chez *Crepidula fornicata* Phil. (Mollusque hermaphrodite protandre). *C. R. Acad. Sci., Paris,* **287,** 1305–1307.

LE GALL, S. 1981. Etude expérimentale du facteur morphogénétique contrôlant la différenciation du tractus génital mâle externe chez *Crepidula fornicata* L. (Mollusque hermaphrodite protandre). *Gen. comp. Endocrinol.,* **43,** 51–62.

LE GALL, S. & STREIFF, W. 1975. Protandric hermaphroditism in prosobranch gastropods. In 'Intersexuality in the Animal Kingdom' (R. Reimboth, ed.), 170–178. Berlin, Springer.

LEMCHE, H. & WINGSTRAND, K. 1959. The anatomy of *Neopilina galatheae* Lemche, 1957 (Mollusca Tryblidiacea). *Galathea Rep.,* **3,** 9–71.

LEVER, J. & BEKIUS, R. 1965. On the presence of an external haemal pore in *Lymnaea stagnalis* L. *Experientia,* **21,** 395–396.

LEVINGS, S. C. & GARRITY, S. D. 1983. Diel and tidal movement of two neritid snails: differences in grazing patterns on a tropical rocky shore. *J. exp. mar. Biol. Ecol.,* **67,** 261–278.

LEVINTON, J. S. 1979. The effect of density upon deposit-feeding populations: movement, feeding and floating of *Hydrobia ventrosa* Montagu (Gastropoda: Prosobranchia). *Oecologia (Berlin),* **43,** 27–39.

LEVINTON, J. S. 1982. The body size-prey size hypothesis: the adequacy of body size as a vehicle for character displacement. *Ecology,* **63,** 869–872.

LEVINTON, J. S., STEWART, S. & DEWITT, T. H. 1985. Complex interaction of a deposit feeder with its resources. 2. Field and laboratory experiments on interference between *Hydrobia totteni* and *Ilyanassa obsoleta* (Gastropoda) and its possible relationship to seasonal shift in zonation on mud flats. *Mar. Biol. Prog. Ser.,* **22,** 53–58.

LEWIS, J. B. 1960. The fauna of rocky shores of Barbados, West Indies. *Can. J. Zool.,* **38,** 391–435.

LEWIS, J. R. 1954. Observations on a high-level population of limpets. *J. Anim. Ecol.,* **28,** 85–100.

LEWIS, J. R. & BOWMAN, R. S. 1975. Local habitat-induced variations in the population dynamics of *Patella vulgata* L. *J. exp. mar. Biol. Ecol.,* **17,** 165–203.

LILLY, M. M. 1953. The mode of life and the structure and functioning of the reproductive ducts of *Bithynia tentaculata* (L.). *Proc. malac. Soc. Lond.,* **30,** 87–110.

LINDBERG, D. R. 1986. Radular evolution in the Patellogastropoda. *Am. malacol. Bull.,* **4,** 115.

LINDBERG, D. R. 1988. The Patellogastropoda. *Malacol. Rev. suppl.* **4,** 35–63.

LINKE, O. 1933. Morphologie und Physiologie der Genitalapparates der Nordseelittorinen. *Wiss. Meeresunters. Abt. Helgoland,* **19** Nr **5,** 3–52.

LINKE, O. 1935. Der Laich von *Littorina (Melaraphe) neritoides* L. *Zool. Anz.,* **112,** 57–62.

LINKE, O. 1939. Die Biota des Jadebusenwattes. *Helgoland. wiss. Meeresunters.,* **1,** 201–348.

LINNAEUS, C. 1758. 'Systema Naturae' 1. Editio decima reformata. Facsimile reprint 1956. London, British Museum (Natural History).

LINSLEY, R. M. 1977. Some "laws" of gastropod shell form. *Paleobiology,* **3,** 196–206.

LINSLEY, R. M. 1978a. Shell form and the evolution of gastropods. *Am. Scient.,* **66,** 432–441.

LINSLEY, R. M. 1987b. Locomotion rates and shell form in the Gastropoda. *Malacologia,* **17,** 193–206.

LINSLEY, R. M. & KIER, W. M. 1984. The Paragastropoda: a proposal for a new class of Paleozoic Mollusca. *Malacologia,* **25,** 241–254.

LISSMANN, H. W. 1945. The mechanism of locomotion in gastropod molluscs. I. Kinematics. *J. exp. Biol.,* **21,** 58–69.

LISSMANN, H. W. 1946. The mechanism of locomotion in gastropod molluscs. II. Kinetics. *J. exp. Biol.,* **23,** 35–50.

LITTLE, C. 1965a. Osmotic and ionic regulation in the prosobranch gastropod mollusc, *Viviparus viviparus* Linn. *J. exp. Biol.,* **43,** 23–37.

LITTLE, C. 1965b. The formation of urine by the prosobranch gastropod mollusc, *Viviparus viviparus* Linn. *J. exp. Biol.,* **43,** 39–54.

LITTLE, C. 1979. Reabsorption of glucose in the renal system of *Viviparus. J. moll. Stud.,* **45,** 207–208.

LITTLE, C. 1981. Osmoregulation and excretion in prosobranchs. Part I. Physiology. *J. moll. Stud.,* **47,** 221–247.

LITTLE, C. 1985. Renal adaptations of prosobranchs to the freshwater environment. *Bull. Am. malac. Un.,* **3,** 223–231.

LITTLE, C. 1989. Factors governing patterns of foraging activity in littoral marine herbivorous molluscs. *J. moll. Stud.,* **55,** 273–284.

LITTLE, C. & ANDREWS, E. B. 1977. Some aspects of excretion and osmoregulation in assimineid snails. *J. moll. Stud.,* **43,** 263–285.

LITTLE, C. & NIX, W. 1976. The burrowing and floating behaviour of the gastropod *Hydrobia ulvae. Est. coastal mar. Sci.,* **4,** 537–544.

LITTLE, C. & STIRLING, P. 1985. Patterns of foraging activity in the limpet *Patella vulgata* L.—a preliminary study. *J. exp. mar. Biol. Ecol.,* **89,** 283–296.

LITTLE, C., WILLIAMS, G. A., MORRITT, D., PERRING, J. M. & STIRLING, P. 1988. Foraging behaviour of *Patella vulgata* L. in an Irish sea-lough. *J. exp. mar. Biol. Ecol.,* **120,** 1–21.

LITTLE, C., MORRITT, D., PATERSON, D. M., STIRLING, P. & WILLIAMS, G. A. 1990. Preliminary observations affecting foraging activity in the limpet *Patella vulgata. J. mar. biol. Ass. U.K.,* **70,** 181–195.

LOPEZ, G. R. & KOFOED, L. H. 1980. Epipsammic browsing and deposit-feeding in mud snails (Hydrobiidae). *J. mar. Res.,* **38,** 585–599.

LOPEZ, G. R. & LEVINTON, J. S. 1978. The availability of micro-organisms attached to sediment particles as food for *Hydrobia ventrosa* Montagu (Gastropoda: Prosobranchia). *Oecologia (Berlin),* **32,** 263–275.

LORD, A. & PECHENIK, J. A. 1984. Uptake and use of dissolved organic matter by larvae of *Crepidula fornicata. Am. Zool.,* **24,** 44A.

LOWELL, R. B. 1984. Desiccation of intertidal limpets: effects of shell size, fit to substratum, and shape. *J. exp. mar. Biol. Ecol.,* **77,** 197–207.

LOWENSTAM, H. A. 1954. Factors affecting the aragonite:calcite ratios in carbonate-secreting organisms. *J. Geol.,* **62,** 284–322.

LOWENSTAM, H. A. 1978. Recovery, behaviour and evolutionary implications of live Monoplacophora. *Nature, Lond.,* **273,** 231–232.

LOWRY, L. F., McELROY, A. J. & PEARSE, J. S. 1974. The distribution of six species of gastropod molluscs in a Californian kelp forest. *Biol. Bull.,* **147,** 386–396.

LUBCHENCO, J. 1978. Plant species diversity in a marine intertidal community: importance of herbivore food preference and algal competitive abilities. *Am. Nat.,* **112,** 23–39.

LUBCHENCO, J. 1982. Effects of grazers and algal competitors on fucoid colonization in tide pools. *J. Mycol.,* **18,** 511–550.

LUBCHENCO, J. 1983. *Littorina* and *Fucus*: effects of herbivores, substratum heterogeneity, and plant escapes during succession. *Ecology,* **64,** 1116–1123.

LUBET, P. & LE GALL, P. 1972. Recherches prèliminaires sur la structure des populations de *Crepidula fornicata* Philb., mollusque mésogastéropode. *Bull. Soc. zool. Fr.,* **97,** 211–222.

LUBET, P., LE GALL, P. BARBIER, J. & SILBERZAHN, N. 1973. Quoted by Lubet, P., Streiff, W. et al. (1973).

LUBET, P. & SILBERZAHN, N. 1971. Recherches sur les effets de l'ablation bilatérale des ganglions cérébroides chez la crépidule (*Crepidula fornicata,* Mollusque Gastéropode): effets somatotropes et gonadotropes. *C. R. Soc. Biol.,* **165,** 590–594.

LUBET, P. & STREIFF, W. 1969. Etudes expérimentales de l'action des ganglions nerveux sur la morpho-genèse du pénis et "activité génitale de *Crepidula fornicata* Phil. (Mollusque Gastéropode). In 'Cours et documents de biologie' **1,** 143–159. Paris, Gordon & Breach.

LUBET, P., STRIEFF, W., SILBERZAHN, N. & DROSDOWSKY, M. 1973. Endocrinologie de la différenciation sexuelle chez les mollusques prosobranches. *Bol. Zool. Biol. São Paulo N.S.,* **30,** 821–841.

LUTZ, R. A. & RHOADS, D. C. 1977. Anaerobiosis and a theory of growth line formation. *Science,* **198,** 1222–1227.

LÜTZEN, J. 1972. Studies on parasitic gastropods from echinoderms II. On *Stilifer* Broderip, with special reference to the structure of the sexual apparatus and the reproduction. *Biol. Skr. dan. Vid. Selsk.,* **19** (6), 1–18.

LÜTZEN, J. & NIELSEN, K. 1975. Contribution to the anatomy and biology of *Echineulima* n.g. (Prosobranchia: Eulimidae), parasitic on sea urchins. *Vidensk. Meddr. dansk naturh. Foren.,* **138,** 171–199.

LYSAGHT, A. M. 1941. The biology and trematode parasites of the gastropod *Littorina neritoides* (L.) on the Plymouth breakwater. *J. mar. biol. Ass. U.K.,* **25,** 41–67.

MAAS, D. 1965. Anatomische und histologische Untersuchungen am Mundapparat der Pyramidelliden. *Z. Morph. Ökol. Tiere,* **54,** 566–642.

MACCLINTOCK, C. 1967. Shell structure of patelloid and bellerophontoid gastropods (Mollusca). *Peabody Mus. nat. Hist., Yale Univ. Bull.,* **22,** 1–140.

MACDONALD, J. D. 1969. On the homologies of the dental plates and teeth of proboscidiferous Gasteropoda. *Ann. Mag. nat. Hist. (4),* **3,** 113–117.

MACÉ, A.-M. & ANSELL, A. D. 1982. Respiration and nitrogen excretion of *Polinices alderi* (Forbes) and *Polinices catena* (da Costa) (Gastropoda: Naticidae). *J. exp. mar. Biol. Ecol.,* **60,** 275–292.

MAEDA-MARTINEZ, A. N. 1985. 'Studies on the physiology of shell formation in molluscan larvae, with special reference to *Crepidula fornicata*.' Ph.D. thesis, University of Southampton.

MAES, V. O. 1971. Evolution of the toxoglossan radula and methods of envenomation. *Am. malac. Un. ann. Reps,* 1970, 69–71.

MAES, V. O. & RAEIHLE, D. 1975. Systematics and biology of *Thala floridana* (Gastropoda: Vexillidae). *Malacologia,* **15,** 43–67.

MANGUM, C. P. 1979. A note on blood and water mixing in large marine gastropods. *Comp. Biochem. Physiol.,* **63A,** 389–391.

MAPSTONE, G. M. 1970. Feeding activities of veligers of *Nassarius reticulatus* and *Crepidula fornicata* and the use of artificial foods in maintaining cultures of these larvae. *Helgol. wiss. Meeresunters.,* **20,** 565–575.

MARCUS, E. & MARCUS, E. 1959. On the reproduction of "*Olivella*". *Univ. São Paulo Fac. Filos. Cienc. Let.,* No. **232,** 189–200.

MARCUS, E. & MARCUS, E. 1968. On the prosobranchs *Ancilla dimidiata* and *Marginella fraterculus*. *Proc. malac. Soc. Lond.,* **38,** 55–69.

MARGOLIN, A. S. 1964. A running response of *Acmaea* to sea-stars. *Ecology,* **45,** 191–193.

MARKEL, K. 1966. Über funktionelle Radulatypen bei Gastropoden unter besonderer Berücksichtigung der Rhipidoglossa. *Vie Milieu,* **17A,** 1121–1138.

MARSH, H. 1971. The foregut glands of vermivorous cone shells. *Austr. J. Zool.,* **19,** 313–326.

MARSH, H. 1977. The radular apparatus of *Conus. J. moll. Stud.,* **43,** 1–11.

MARTIN, A. W. 1983. Excretion. In 'The Mollusca' **5,** (A. S. M. Saleuddin & K. M. Wilbur eds), 353–405. Orlando, Florida, Academic Press.

MARTIN, A. W. & ALDRICH, F. A. 1970. Comparison of hearts and branchial heart appendages in some cephalopods. *Can. J. Zool.,* **48,** 751–756.

MARTIN, A. W., STEWART, D. M. & HARRISON, F. M. 1954. Kidney function in the giant African snail. *J. cell. comp. Physiol.,* **44,** 345–346

MARTOJA, M. 1963. Contribution à l'étude de l'appareil digestif et de la digestion chez les Gastéropodes carnivores de la famille des Nassariidés (Prosobranches Sténoglosses). *La Cellule,* **64,** 237–334.

MARTOJA, M. 1965a. Sur l'incubation et l'existence possible d'une glande endocrine, chez *Hydromyles globulosa* Rang (*Halopsyche gaudichaudi* Keferstein), Gastéropode Gymnosome. *C. R. Acad. Sci., Paris,* **260,** 2907–2909.

MARTOJA, M. 1965b. Existence d'un organe juxta-ganglionnaire chez *Aplysia punctata* Cuv. (Gastéropode Opisthobranche). *C. R. Acad. Sci., Paris,* **260,** 4615–4617.

MARTOJA, M. 1965c. Données relatives à l'organe juxta-ganlionnaire des Prosobranches Diotocardes. *C. R. Acad. Sci., Paris,* **261,** 3195–3196.

MARTOJA, M. 1972. Endocrinology of Mollusca. In 'Chemical Zoology' (M. Florkin & B. T. Scheer eds) **7,** 349–392. New York & London, Academic Press.

MARTOJA, M. 1975. Le rein de *Pomatias* (= *Cyclostoma*) *elegans* (Gastéropode prosobranche): données structurales et analytiques. *Ann. Sci. nat. Zool.,* **17,** 535–558.

MATON, W. G. 1797. 'Observations relative chiefly to the natural history, picturesque scenery and antiquities of the western counties of England made in the years 1794 and 1796.' Salisbury, Eaton.

MATVEEVA, T. A. 1955. Ecology and the life cycles of mass species of gastropods in the Barents and White Seas. In 'Explorations of the fauna of the seas. XIII (XXI). Seasonal phenomena in the life of the White and Barents Seas.' Moscow-Leningrad, Nauk SSSR. [In Russian.]

MAYES, P. A. 1960. Physiological adaptation to environments in the Littorinidae. *J. Physiol.,* **151,** 17P–18P.

MAYES, P. A. 1962. Comparative investigations of the euryhaline character of *Littorina* and the possible relationship to intertidal zonation. *Nature, Lond.,* **195,** 1269–1270.

MAYR, E. 1963. 'Animal speciation and evolution.' Harvard University Press.

McKILLUP, S. C. & BUTLER, A. J. 1979. Modification of egg production and packaging in response to food availability by *Nassarius pauperatus. Oecologia (Berlin),* **43,** 221–231.

McLEAN, J. H. 1979. A new monoplacophoran limpet from the continental shelf off Southern California. *Contrib. Sci. nat. Hist. Mus. Los Angeles County,* **307,** 1–19.

McLEAN, J. H. 1981. The Galapagos Rift Limpet *Neomphalus*: relevance to understanding the evolution of a major Paleozoic-Mesozoic radiation. *Malacologia,* **21,** 291–336.

McLEAN, J. H. 1990. Neolepetopsidae, a new docoglossate limpet family from hydrothermal vents and its relevance to patellogastropod evolution. *J. Zool., Lond.,* **222,** 485–528.

McMAHON, R. F. 1988. Respiratory response to periodic emergence in intertidal molluscs. *Am. Zool.,* **28,** 97–114.

McMAHON, R. F. 1990. Thermal tolerance, evaporative water loss, air-water oxygen consumption and zonation of intertidal prosobranchs: a new synthesis. *Hydrobiologia,* **193,** 241–260.

McNAIR, C. G., KIER, W. M., LACROIX, P. D. & LINSLEY, R. M. 1981. The functional significance of aperture form in gastropods. *Lethaia,* **14,** 63–70.

MEISENHEIMER. J. 1899. Zur Morphologie der Urniere der Pulmonaten. *Z. wiss. Zool.,* **65,** 709–724.

MENGE, J. L. 1974. Prey selection and foraging period of the predaceous rocky intertidal snail, *Acanthina punctulata. Oecologia (Berlin),* **77,** 292–316.

MENKE, H. 1911. Physikalische und physiologische Faktoren bei der Anheftung von Schnecken der Brandungszone. *Zool. Anz.,* **37,** 19–30.

MERDSOY, B. & FARLEY, J. 1973. Phasic activity in the digestive cells of the marine prosobranch gastropod *Littorina littorea* (L.). *Proc. malac. Soc. Lond.,* **40,** 473–480.

MICALLEF, H. 1966. 'Ecology and behaviour of selected intertidal gastropods.' Ph.D. thesis, University of London.

MICALLEF, H. 1968. The activity of *Monodonta lineata* in relation to temperature as studied by means of an actograph. *J. Zool., Lond.,* **154,** 155–159.

MICALLEF, H. & BANNISTER, W. H. 1967. Aerial and aquatic oxygen consumption of *Monodonta turbinata* (Mollusca: Gastropoda). *J. Zool., Lond.,* **151,** 479–482.

MILL, P. J. & GRAHAME, J. W. 1988. Esterase variability in the gastropods *Littorina saxatilis* (Olivi) and *L. arcana* Ellis. *J. moll. Stud.,* **54,** 347–353.

MILLER, B. A. 1970. 'Studies on the biology of Indo-Pacific Terebridae.' Ph.D. dissertation, University of New Hamsphire, Durham.

MILLER, B. A. 1975. The biology of *Terebra gouldi* Deshayes, 1859, and a discussion of life history similarities among other terebrids of similar proboscis type. *Pac. Sci.,* **29,** 227–241.

MILLER, B. A. 1979. The biology of *Hastula inconstans* (Hinds, 1844) and a discussion of life history similarities among other Hastulas of similar proboscis type. *Pac. Sci.,* **33,** 289–306.

MILLER, J. A. 1989. The toxoglossan proboscis: structure and function. *J. moll. Stud.,* **55,** 167–181.

MILLER, S. L. 1974a. Adaptive design of locomotion and foot form in prosobranch gastropods. *J. exp. mar. Biol. Ecol.,* **14,** 99–156.

MILLER, S. L. 1974b. The classification, taxonomic distribution, and evolution of locomotor types of prosobranch gastropods. *J. moll. Stud.,* **41,** 233–272.

MILLS, P. M. 1977. Radular tooth structure in three species of Terebridae (Mollusca: Toxoglossa). *Veliger,* **19,** 259–265.

MILNE-EDWARDS, H. 1846. Sur la circulation chez les mollusques (chez les Patelles et les Haliotides). *Ann. Sci. nat. Zool.,* (3), **8,** 37–53.

MOORE, H. B. 1936. The biology of *Purpura lapillus.* I. Shell variation in relation to environment. *J. mar. biol. Ass. U.K.,* **21,** 61–89.

MOORE, H. B. 1937. The biology of *Littorina littorea.* Part I. Growth of the shell and tissues, spawning, length of life and mortality. *J. mar. biol. Ass. U.K.,* **21,** 721–742.

MOORE, H. B. 1938. The biology of *Purpura lapillus.* Part II. Growth. *J. mar. biol. Ass. U.K.,* **23,** 57–66.

MOORE, P. G. 1977. Additions to the littoral fauna of Rockall, with a description of *Araeolaimus penelope* sp. nov. (Nematoda: Axonolaimidae). *J. mar. biol. Ass. U.K.,* **57,** 191–200.

MORAN, M. J. 1985. Effects of prey density, prey size and predator size on rates of feeding by an intertidal predatory gastropod *Morula marginalba* Blainville (Muricidae), on several species of prey. *J. exp. mar. Biol. Ecol.,* **90,** 97–105.

MORGAN, P. R. 1972a. *Nucella lapillus* (L.) as a predator of edible cockles. *J. exp. mar. Biol. Ecol.,* **8,** 45–52.

MORGAN, P. R. 1972b. The influence of prey availability on distribution and predatory behaviour of *Nucella lapillus* (L.). *J. Anim. Ecol.,* **41,** 257–274.

MORRISEY, D. J. 1987. Effect of population density and presence of a potential competitor on the growth rate of the mud snail *Hydrobia ulvae* (Pennant). *J. exp. mar. Biol. Ecol.,* **108,** 275–295.

MORRISEY, D. J. 1990. Factors affecting individual body weight in field populations of the mudsnail *Hydrobia ulvae. J. mar. biol. Ass. U.K.,* **70,** 99–106.

MORTON, J. E. 1951. The structure and adaptations of the New Zealand Vermetidae. Parts 1–3. *Trans. R. Soc. N.Z.,* **79,** 1–51.

MORTON, J.E . 1952. The role of the crystalline style. *Proc. malac. Soc. Lond.,* **29,** 85–92.

MORTON, J. E. 1956. The tidal rhythm and action of the digestive system of the lamellibranch *Lasaea rubra. J. mar. biol. Ass. U.K.,* **35,** 563–586.

MORTON, J. E. 1960. The habits of *Cyclope neritea,* a style-bearing stenoglossan gastropod. *Proc. malac. Soc. Lond.,* **34,** 96–105.

MORTON, J.E . 1965. Form and function in the evolution of the Vermetidae. *Bull. Brit. Mus. (Nat. Hist.) Zool.,* **11,** 585–630.

MORTON, J.E. & YONGE, C. M. 1964. Classification and structure of the Mollusca. In 'Physiology of Mollusca' (K. M. Wilbur & C. M. Yonge eds) **1,** 1–58. New York & London, Academic Press.

MOYSE, J., THORPE, J. P. & AL-HAMADANI, E. 1982. The status of *Littorina aestuarii* Jeffreys, an approach using morphology and biochemical genetics. *J. Conch., Lond.,* **31,** 7–15.

MUNKSGAARD, C. 1990. Electrophoretic separation of morphologically similar species of the genus *Rissoa* (Gastropoda: Prosobranchia). *Ophelia,* **31,** 97–104.

MURDOCH, W. M. 1969. Switching in general predators: experiments on predator specifity and stability of prey populations. *Ecol. Mon.,* **39,** 335–354.

MURDOCK, G. R. & VOGEL, S. 1978. Hydrodynamic induction of water flow through a keyhole limpet (Gastropoda, Fissurellidae). *Comp. Biochem. Physiol.,* **61A,** 227–231.

MUUS, B. J. 1967. The fauna of Danish estuaries and lagoons. Distribution and ecology of dominating species in the shallow reaches of the mesohaline zone. *Meddr. Komm. Dan. Fisk. Havunders. N.S.,* **5** (1), 1–316.

NAEF, A. 1913. Studien zur generellen Morphologie der Mollusken. *Ergebn. Zool.,* **3,** 73–164.

NAYLOR, R. & BEGON, M. 1982. Variation within and between populations of *Littorina nigrolineata* Gray, on Holy Island, Anglesey. *J. Conch., Lond.,* **31,** 17–30.

NDIFON, G. T. 1979. 'Studies on the feeding biology, anatomical variations and ecology of vectors of schistosomiasis and other freshwater snails of south western Nigeria.' Ph.D. thesis, University of Ibadan, Nigeria.

NEEDHAM, J. 1935. Problems of nitrogen catabolism in invertebrates. II. Correlation between uricotelic metabolism and habitat in the phylum Mollusca. *Biochem. J.,* **29,** 238–251.

NEGUS, M. R. S. 1975. An analysis of boreholes drilled by *Natica catena* (da Costa) in the valves of *Donax vittatus* (da Costa). *Proc. malac. Soc. Lond.,* **41,** 353–356.

NELSON, L. & MORTON, J.E. 1979. Cyclical activity and epithelial renewal in the digestive gland tubules of the marine prosobranch *Maoricrypta monoxyla* (Lesson). *J. moll. Stud.,* **45,** 262–283.

NEWELL, G. E. 1958a. The behaviour of *Littorina littorea* (L.) under natural conditions and its relation to position on the shore. *J. mar. biol. Ass. U.K.,* **37,** 229–239.

NEWELL, G. E. 1958b. An experimental analysis of the behaviour of *Littorina littorea* (L.) under natural conditions and in the laboratory. *J. mar. biol. Ass. U.K.,* **37,** 241–266.

NEWELL, R. C. 1962. Behavioural aspects of the ecology of *Peringia* (=*Hydrobia*) *ulvae* (Pennant) (Gasteropoda, Prosobranchia). *Proc. zool. Soc. Lond.,* **138,** 49–75.

NEWELL, R. C. 1965. The role of detritus in the nutrition of two marine deposit feeders, the prosobranch *Hydrobia ulvae* and the bivalve *Macoma balthica. Proc. zool. Soc. Lond.,* **144,** 25–45.

NEWELL, R. C. & BRANCH, G. M. 1980. The influence of temperature on the maintenance of metabolic energy balance in marine invertebrates. *Adv. mar. Biol.,* **17,** 329–396.

NEWELL, R. C. & KOFOED, L. H. 1977. The energetics of suspension feeding in the gastropod *Crepidula fornicata* L. *J. mar. biol. Ass. U.K.,* **57,** 161–180.

NEWELL, R. C., PYE, V. & ASANULLAH, M. 1971. Factors affecting the feeding rate of the winkle *Littorina littorea. Mar. Biol.,* **9,** 138–144.

NEWKIRK, G. F. & DOYLE, R. W. 1975. Genetic analysis of shell-shape variation in *Littorina saxatilis* on an environmental cline. *Mar. Biol.,* **30,** 227–237.

NICOTRI, M. E. 1977. Grazing effects on the microflora. *Ecology,* **58,** 1020–1032.

NIELSEN, C. 1975. Observations on *Buccinum undatum* L. attacking bivalves and on prey responses, with a short review on attack methods of other prosobranchs. *Ophelia,* **13,** 87–108.

NIERSTRASZ, H. F. 1905. *Kruppomania minima* und die Radula der Solenogastren. *Zool. Jb. (Anat.),* **21,** 655–702.

NISBET, R. H. 1953. 'The structure and function of the buccal mass of some gastropod molluscs. I. *Monodonta lineata* (da Costa).' Ph.D. thesis, University of London.

NIXON, M. & MACCONACHIE, E. 1988. Drilling by *Octopus vulgaris* (Mollusca: Cephalopoda) in the Meditteranean. *J. Zool., Lond.,* **216,** 687–716.

NORDSIECK, F. 1968. 'Die europäischen Meeres-Gehauseschnecken (Prosobranchia) von Eismeer bis Kapverden und Mittelmeer.' Stuttgart, Fischer Verlag.

NORDSIECK, F. 1972. 'Die europäischen Meeresschnecken (Opisthobranchia mit Pyramidellidae: Rissoacea)'. Stuttgart, Fischer Verlag.

NUWAYHID, M. A., DAVIES, P. S. & ELDER, H. Y. 1978. Gill structure in the common limpet *Patella vulgata. J. mar. biol. Ass. U.K.,* **58,** 817–823.

NYBAKKEN, J. & PERRON, F. 1988. Ontogenetic change in the radula of *Conus magus* (Gastropoda). *Mar. Biol.,* **98,** 239–242.

OCKELMANN, K. W. 1965. Developmental types in marine bivalves and their distribution along the Atlantic coast of Europe. *Proc. 1st. Eur. malacol. Congr., 1962,* 25–35.

OCKELMANN, K. W. & NIELSEN, C. 1981. On the biology of the prosobranch *Lacuna parva* in the Øresund. *Ophelia,* **20,** 1–16.

ODHNER, N. H. 1912. Morphologische und phylogenetische Untersuchungen über die Nephridien der Lamellibranchien. *Z. wiss. Zool.,* **100,** 287–301.

ØKLAND, S. 1980. The heart structure of *Lepidopleurus asellus* (Spengler) and *Tonicella marmorea* (Fabricius). *Zoomorphology,* **96,** 1–19.

ØKLAND, S. 1982. The ultrastrucure of the heart complex in *Patella vulgata* L. (Archaeogastropoda, Prosobranchia). *J. moll. Stud.*, **48**, 331–341.

OLIVERA, B. M., RIVIER, J., CLARK, C., RAMILO, C. A., CORPUZ, G. P., ABOGADIE, F. C., MENA, E., NOODWARD, S. R., HILLYARD, D. R. & CRUZ, L. J. 1990. Diversity of *Conus* neuropeptides. *Science*, **249**, 257–263.

OLIVI, A. G. 1792. 'Zoologia adriatica.' Bassano, Venezia, G. Remondini.

OLMSTED, J. M. D. 1917. Notes on the locomotion of certain Bermudian mollusks. *J. exp. Zool.*, **24**, 223–236.

O'LOUGHLIN, E. F. M. & ALDRICH, J. C. 1987a. An analysis of shell shape variation in the painted topshell *Calliostoma zizyphinum* (L.) (Prosobranchia: Trochidae). *J. moll. Stud.*, **53**, 62–68.

O'LOUGHLIN, E. F. M. & ALDRICH, J. C. 1987b. Morphological variation in the painted topshell *Calliostoma zizyphinum* (Linnaeus) (Prosobranchia: Trochidae). *J. moll. Stud.*, **53**, 267–272.

ORTON, J. H. 1912. The mode of feeding of *Crepidula*, with an account of the current-producing mechanism in the mantle cavity, and some remarks on the mode of feeding in gastropods and lamellibranchs. *J. mar. biol. Ass. U.K.*, **9**, 444–478.

ORTON, J. H. 1929. Habitats and feeding habits of *Ocenebra erinacea*. *Nature, Lond.*, **124**, 370–371.

O'SULLIVAN, J. B., McCONNAUGHEY, R. R. & HUBER, M. E. 1987. A blood-sucking snail: the Cooper's nutmeg *Cancellaria cooperi* Gabb, parasitizes the electric ray, *Torpedo californica* Ayres. *Biol. Bull.*, **172**, 263–366.

OWEN, G. 1972. Lysosomes, peroxisomes and bivalves. *Sci. Prog. Oxford*, **60**, 299–318.

OWEN, G. 1974. Studies on the gill of *Mytilus edulis*: the eulatero-frontal cirri. *Proc. R. Soc. B*, **187**, 83–91.

PACE, S. 1902. On the anatomy and relationships of *Voluta musica* Linn.; with notes upon certain other supposed members of the Volutidae. *Proc. malac. Soc. Lond.*, **5**, 21–31.

PAINE, R. T. 1969. The *Pisaster-Tegula* interaction: prey patches, predator food preferences, and intertidal community structure. *Ecology*, **50**, 950–961.

PALMER, A. R. 1977. Function of shell sculpture in marine gastropods: hydrodynamic destabilization in *Ceratostoma foliatum*. *Science*, **197**, 1293–1295.

PALMER, A. R. 1979. Fish predation and the evolution of gastropod shell sculpture: experimental and geographic evidence. *Evolution*, **33**, 697–713.

PALMER, A. R. 1980. Locomotor rates and shell form in the Gastropoda: a re-evaluation. *Malacologia*, **19**, 289–296.

PALMER, A. R. 1983. Growth rate as a measure of food value in thaidid gastropods: assumptions and implications for prey morphology and distribution. *J. exp. mar. Biol. Ecol.*, **73**, 95–124.

PALMER, A. R. 1984. Species cohesiveness and genetic control of shell colour and form in *Thais emarginata* (Prosobranchia, Muricacea). Preliminary results. *Malacologia*, **25**, 477–491.

PALMER, A. R. 1985. Genetic basis of shell variation in *Thais emarginata* (Prosobranchia, Muricacea). I. Banding in populations from Vancouver Island. *Biol. Bull.*, **169**, 638–657.

PANDIAN, V. J. 1969. Yolk utilization in the gastropod *Crepidula fornicata*. *Mar. Biol.*, **3**, 117–121.

PANNELLA, G. & MACCLINTOCK, C. 1968. Biological and environmental rhythms reflected in molluscan shell growth. *J. Paleontol.*, **62**, 64–80,

PARENZAN, P. 1979. Genus *Crepidula* in the Mediterranean. *La Conchiglia*, **11**, (118–119), 8–9.

PARKER, G. H. 1911. The mechansim of locomotion in gastropods. *J. Morph.*, **22**, 155–170.

PARSONS, D. W. & MACMILLAN, D. L. 1979. The escape response of abalone (Mollusca, Prosobranchia, Haliotidae) to predatory gastropods. *Mar. Behav. Physiol.*, **6**, 65–82.

PATANÈ, L. 1946. Anaerobiosi in *Littorina neritoides* (L.) *Boll. Soc. Ital. Biol. sper.*, **22**, 929–930.

PATANÈ, L. 1955. Cinesi e tropismi, anidro e anaerobiosi in *Littorina neritoides* (L.). (Nota preliminare.) *Boll. Accad. Sci. nat. Gioenia sper.*, (4) **3**, 65–73.

PATIL, A. M. 1958., The occurrence of a male of the prosobranch *Potamopyrgus jenkinsi* (Smith), var. *carinata* Marshall in the Thames at Sonning, Berkshire. *Annal. Mag. nat. Hist.*, **13**, 232–240.

PEARCE, J. B. 1966. On *Lora trevelliana* (Turton) (Gastropoda: Turridae). *Ophelia*, **3**, 81–91.

PEARCE, J. B. & THORSON, G. 1967. The feeding and reproductive biology of the red whelk, *Neptunea antiqua* (L.) (Gastropoda: Prosobranchia). *Ophelia,* **4,** 277–314.

PECHENIK, J. A. 1982. Ability of some gastropod egg capsules to protect against low-salinity stress. *J. exp. mar. Biol. Ecol.,* **63,** 195–208.

PECHENIK, J. A. 1983. Egg capsules of *Nucella lapillus* (L.) protect against low-salinity stress. *J. exp. mar. Biol. Ecol.,* **71,** 165–179.

PECHENIK, J. A. 1986. Field evidence for delayed metamorphosis of larval gastropods *Crepidula plana* Say, *C. fornicata* (L.), and *Bittium alternatum* (Say). *J. exp. mar. Biol. Ecol.,* **97,** 313–319.

PECHENIK, J. A., CHANG, S. C. & LORD, A. 1984. Encapsulated development of the marine prosobranch gastropod *Nucella lapillus*. *Mar. Biol.,* **78,** 223–229.

PECK, L. S., CULLEY, M. B. & HELM, M. M. 1987. A laboratory energy budget for the ormer *Haliotis tuberculata* L. *J. exp. mar. Biol. Ecol.,* **111,** 1–22.

PELSENEER, P. 1925. Gastropodes marins carnivores *Natica* et *Purpura*. *Ann. Soc. zool. Belg.,* **55,** 37–39.

PELSENEER, P. 1935. Essai d'éthologie zoologique d'après l'étude des mollusques. *Acad. R. Belg. Cl. Sci. Publ. Fondation Agathon de Potter,* **1,** 1–662.

PERCHARDE, P. L. 1972. Observations on the gastropod *Charonia variegata*, in Trinidad and Tobago. *Nautilus,* **85,** 84–92.

PERRIER, R. 1889. Recherches sur l'anatomie et l'histologie du rein des gastéropodes Prosobranchiata. *Ann. Sci. nat. Zool.* (7), **8,** 61–192.

PERRON, F. E. 1975. Carnivorous *Calliostoma* (Prosobranchia: Trochidae) from the northeastern Pacific. *Veliger,* **18,** 52–54.

PERRON, F. E. 1978. The habitat and feeding behavior of the wentletrap *Epitonium greenlandicum*. *Malacologia,* **17,** 63–72.

PERRON, F. E. 1981. The partitioning of reproductive energy between ova and protective capsules in marine gastropods of the genus *Conus*. *Am. Nat.,* **118,** 110–118.

PERRON, F. E. & TURNER, R. D. 1978. The feeding behaviour and diet of *Calliostoma occidentale*, a coelenterate-associated prosobranch gastropod. *J. moll. Stud.,* **44,** 100–103.

PERRY, D. M. 1985. Function of the shell spine in the predaceous rocky intertidal snail *Acanthina spirata* (Prosobranchia: Muricacea). *Mar. Biol.,* **88,** 51–59.

PETERSON, C. H. & BLACK, R. 1986. Abundance patterns in infaunal sea anemones and their potential benthic prey in and outside seagrass patches on a Western Australian sand shelf. *Bull. mar. Sci.,* **38,** 498–511.

PETPIROON, S. & MORGAN, E. 1983. Observations on the tidal activity rhythm of the periwinkle *Littorina nigrolineata* (Gray). *Mar. Behav. Physiol.,* **9,** 171–192.

PETRAITIS, P. S. 1982. Occurrence of random and directional movements in the periwinkle *Littorina littorea* (L.). *J. exp. mar. Biol. Ecol.,* **59,** 207–217.

PETRAITIS, P. S. 1987. Immobilization of the predatory gastropod, *Nucella lapillus*, by its prey, *Mytilus edulis*. *Biol. Bull.,* **172,** 307–311.

PETRAITIS, P. S. 1989. Effects of the periwinkle *Littorina littorea* (L.) and of intraspecific competition on growth and survivorship of the limpet *Notoacmaea testudinalis* (Müller). *J. exp. mar. Biol. Ecol.,* **125,** 99–115.

PETTITT, C. W. 1973. An examination of the distribution of shell pattern in *Littorina saxatilis* (Olivi) with particular regard to the possibility of visual selection in this species. *Malacologia,* **14,** 339–343.

PIANKA, E. R. 1966. Latitudinal gradients in specific diversity: a review of concepts. *Am. Nat.,* **100,** 33–46.

PICKEN, L. E. R. 1937. The mechanism of urine formation in invertebrates. The excretory mechanism of certain molluscs. *J. exp. Biol.,* **14,** 20–34.

PILKINGTON, M. C. 1970. Young stages and metamorphosis in an atlantid heteropod occurring off south-eastern New Zealand. *Proc. malac. Soc. Lond.,* **39,** 117–124.

PILKINGTON, M. C. 1971. The veliger stage of *Hydrobia ulvae* (Pennant). *Proc. malac. Soc. Lond.,* **39,** 281–287.

PILKINGTON, M. C. 1974. The egg and hatching stages of some New Zealand prosobranch molluscs. *J. R. Soc. N.Z.*, **4**, 411–431.

PINEL-ALLOUL, B. & MAGNIN, E. 1971. Cycle vital et croissance de *Bithynia tentaculata* L. (Mollusca, Gastropoda, Prosobranchia) du Lac St.-Louis près de Montreal. *Can. J. Zool.*, **49**, 749–766.

PLATE, L. 1901. Die Anatomie und Phylogenie der Chitonen. Teil C. *Zool. Jb.* Suppl. **5**, 281–600.

PLESCH, B. 1977. An ultrastructural study of the musculature of the pond snail *Lymnaea stagnalis* (L.). *Cell Tiss. Res.*, **180**, 317–340.

POLI, G., SALVAT, B. & STREIFF, W. 1971. Aspect particulier de la sexualité chez *Ocenebra erinacea* (Mollusque Gastéropode Prosobranche). Note préliminaire. *Haliotis*, **1**, 29–30.

POLK, P. 1962a. Bijdrage tot de kennis der mariene fauna van de Belgische Kust. III. Waarnemingen aangaande het voorkomen, de voortplanting, de settling en de groei van *Crepidula fornicata* (L.). *Ann. Soc. roy. zool. Belg.*, **92**, 47–80.

POLK, P. 1962b. Bijdrage tot de kennis der mariene fauna van de Belgische Kust. IV. De bestrijding van de oesterplaag *Crepidula fornicata* (L.) in de spuikom te Ostende. *Biol. Jaarb.*, **30**, 37–46.

PONDER, W. F. 1970a. Some aspects of the morphology of four species of the neogastropod family Marginellidae with a discussion on the evolution of the toxoglosson poison gland. *J. malac. Soc. Aust.*, **2**, 55–81.

PONDER, W. F. 1970b. The morphology of *Alcithoe arabica* (Gastropoda: Volutidae). *Malacol. Rev.*, **3**, 127–165.

PONDER, W. F. 1972. The morphology of some mitriform gastropods with special reference to their alimentary and reproductive systems (Mollusca: Neogastropoda). *Malacologia*, **11**, 295–342.

PONDER, W. F. 1973. The origin and evolution of the Neogastropoda. *Malacologia*, **12**, 295–338.

PONDER, W. F. 1976. Three species of Littorinidae from southern Australia. *Malacol. Rev.*, **9**, 105–114.

PONDER, W. F. 1988a. *Potamopyrgus antipodarum*—a molluscan coloniser of Europe and Australia. *J. moll. Stud.*, **54**, 271–285.

PONDER, W. F. 1988b. The truncatelloidean (=rissoacean) radiation—a preliminary phylogeny. *Malacol. Rev.*, Suppl. **4**, 129–164.

PONDER, W. F. 1990. The anatomy and relationships of a marine valvatoidean (Gastropoda: Heterobranchia). *J. moll. Stud.*, **56**, 533–555.

PONDER, W. F. 1991. Marine valvatoidean gastropods—implications for early heterobranch phylogeny. *J. moll. Stud.*, **57**, 21–32.

PONDER, W. F. & WAREN, A. 1988. Classification of the Caenogastropods and Heterostropha—a list of the family-group names and higher taxa. *Malacol. Rev.*, Suppl. **4**, 288–326.

POPE, T. E. B. 1910–11. The oyster drill and other predatory Mollusca. *Rep. U.S. Bur. Fish. Wash.*, (Unpublished.)

PORTMANN, A. 1925. Der Einfluss der Nähreier auf die Larven-entwicklung von *Buccinum* und *Purpura*. *Z. Morph. Ökol. Tiere*, **3**, 526–541.

PORTMANN, A. 1927. Die Nähreierbildung durch atypische Spermien bei *Buccinum undatum* (L.). *Z. Zellforsch. mikrosk. Anat.*, **5**, 230–243.

PORTMANN, A. 1930. Die Larvalnieren von *Buccinum undatum*. *Z. Zellforsch. mikrosk. Anat.*, **10**, 401–410.

PORTMANN, A. 1931. Die Enstehung der Nähreier bei *Purpura lapillus* durch atypische Befruchtung. *Z. Zellforsch. mikrosk. Anat.*, **12**, 167–178.

PORTMANN, A. & SANDMEIER, E. 1965. Die Entwicklung von Vorderdarm, Makromeren und Enddarm unter dem Einfluss von Nähreiern bei *Buccinum, Murex* und *Nucella* (Gastropoda, Prosobranchia). *Rev. suisse Zool.*, **72**, 187–204.

POTTS, W. T. W. 1975. Excretion in gastropods. *Fortsch. Zool.*, **23**, 76–88.

PRATT, D. M. 1974. Behavioral defenses of *Crepidula fornicata* against attack by *Urosalpinx cinerea*. *Mar. Biol.*, **27**, 47–49.

PRATT, D. M. 1976. Intraspecific signalling of hunting success or failure in *Urosalpinx cinerea* Say. *J. exp. mar. Biol. Ecol.*, **21**, 7–9.

RAFFAELLI, D. G. 1976. 'The determinants of zonation patterns of *Littorina neritoides* and the *Littorina saxatilis* species-complex.' Ph.D. thesis, University of Wales.

RAFFAELLI, D. G. 1978a. The relationship between shell injuries, shell thickness and habitat characteristics of the intertidal snail *Littorina rudis* Maton. *J. moll. Stud.,* **44,** 166–170.

RAFFAELLI, D. G. 1978b. Factors affecting the population structure of *Littorina neglecta* (Bean). *J. moll. Stud.,* **44,** 223–230.

RAFFAELLI, D. G. 1979. The taxonomy of the *Littorina saxatilis* species-complex, with particular reference to the systematic status of *Littorina patula* Jeffreys. *Zool. J. Linn. Soc.,* **65,** 219–232.

RAFFAELLI, D. G. 1982. Recent ecological research on some European species of *Littorina*. *J. moll. Stud.,* **48,** 342–354.

RAFFAELLI, D. G. 1985. Functional feeding groups of some intertidal molluscs defined by gut contents analysis. *J. moll. Stud.,* **51,** 233–239.

RAFFAELLI, D. G. & HUGHES, R. N. 1978. The effect of crevice size and availability on populations of *Littorina rudis* and *Littorina neritoides*. *J. Anim. Ecol.,* **47,** 71–83.

RAM, J. L. 1977. Hormonal control of reproduction in *Busycon*: laying of egg capsules caused by nervous system extracts. *Biol. Bull.,* **152,** 221–232.

RAMSAY, J. A. 1952. 'A physiological approach to the lower animals.' Cambridge, Cambridge University Press.

RASMUSSEN, E. 1973. Systematics and ecology of the Isefjord marine fauna (Denmark). *Ophelia,* **11,** 1–495.

RATH, E. 1988. Organization and systematic position of the Valvatidae. *Malacol. Rev.,* Suppl. 4, 194–204.

RAVEN, C. P. 1961. 'Oogenesis: the storage of developmental information.' New York, Pergamon Press.

REES, W. J. 1964. A review of breathing devices in land operculate snails. *Proc. malac. Soc. Lond.,* **36,** 55–67.

REGONDAUD, J. 1964. Origine embryonnaire de la cavité pulmonaire de *Lymnaea stagnalis* L. Considérations particulières sur la morphogenèse de la commissure viscérale. *Bull. biol. Fr. Belg.,* **98,** 433–471.

REHFELDT, N. 1968. Reproductive and morphological variations in the prosobranch "*Rissoa membranacea*". *Ophelia,* **5,** 157–173.

REID, D. G. 1989. The comparative morphology, phylogeny and evolution of the gastropod family Littorinidae. *Phil. Trans. R. Soc. B,* **234,** 1–110.

REIMCHEN, T. E. 1974. 'Studies on the biology and colour polymorphism of two sibling species of marine gastropod (*Littorina*).' Ph.D. thesis, University of Liverpool.

REIMCHEN, T. E. 1979. Substratum heterogeneity, crypsis, and colour polymorphism in an intertidal snail (*Littorina mariae*). *Can. J. Zool.,* **57,** 1070–1085.

REIMCHEN, T. E. 1981. Microgeographical variation in *Littorina mariae* Sacchi and Rastelli and a taxonomic consideration. *J. Conch., Lond.,* **30,** 341–350.

REX, M. A. & WAREN, A. 1982. Planktotrophic development in deep-sea prosobranch snails from the Western North Atlantic. *Deep-Sea Res.,* **29,** 171–184.

RICHTER, G. 1961. Die Radula der Atlantiden (Heteropoda, Prosobranchia) und ihre Bedeutung für die Systematik und Evolution der Familie. *Z. Morph. Ökol. Tiere,* **50,** 163–238.

RITTSCHOF, D., WILLIAMS, L. G., BROWN, B. & CARRIKER, M. R. 1983. Chemical attraction of newly hatched oyster drills. *Biol. Bull.,* **164,** 493–505.

RIVEST, B. R. 1983. Development and the inflence of nurse egg allotment on hatching size in *Searlesia dira* (Reeve, 1846) (Prosobranchia: Buccinidae). *J. exp. mar. Biol. Ecol.,* **69,** 217–241.

ROBERT, A. 1902. Recherches sur le développement des troques. *Arch. Zool. exp. gén.,* **10,** 269–558.

ROBERTS, D. J. & HUGHES, R. N. 1980. Growth and reproductive rates of *Littorina rudis* from three contrasted shores in North Wales, U.K. *Mar. Biol.,* **58,** 47–54.

ROBERTS, D. J. & KELL, G. V. 1987. Shell colour in *Calliostoma zizyphinum* (L.) from Strangford Lough, N. Ireland. *J. moll. Stud.,* **53,** 273–283.

ROBERTSON, R. 1963. Wentle traps (Epitoniidae) feeding on sea anemones and corals. *Proc. malac. Soc. Lond.,* **35,** 51–63.

ROBERTSON, R. 1964. Dispersal and wastage of larval *Philippia krebsii* (Gastropoda: Architectonicidae) in the North Atlantic. *Proc. Acad. nat. Sci. Philad.,* **115,** 1–27.

ROBERTSON, R. 1970. Review of the predators and parasites of stony corals, with special reference to symbiotic prosobranch gastropods. *Pacific Sci.,* **24,** 43–54.

ROBERTSON, R. 1974. The biology of the Architectonicidae, gastropods combining prosobranch and opisthobranch traits. *Malacologia,* **14,** 215–220.

ROBERTSON, R. 1983. Observations on the life history of the wentletrap *Epitonium albidum* in the West Indies. *Am. malacol. Bull.,* **1,** 1–12.

ROBERTSON, R. 1985a. Four characters and the higher category systematics of gastropods. *Am. malacol. Bull.,* special edition **1,** 1–22.

ROBERTSON, R. 1985b. Archaeogastropod biology and the systematics of the genus *Tricolia* (Trochacea Tricoliidae) in the Indo-West Pacific. *Monogr. mar. Mollusca,* **3,** 1–103.

ROBERTSON, R. 1989. Spermatophores of aquatic non-stylommatophoran gastropods: a review with new data on *Heliacus* (Architectonicidae). *Malacologia,* **30,** 341–364.

ROBINSON, W. E., PENNINGTON, M. R. & LANGTON, R. W. 1981. Variability of tubule types within digestive glands of *Mercenaria mercenaria* (L.), *Ostrea edulis* L. and *Mytilus edulis* L. *J. exp. mar. Biol. Ecol.,* **54,** 265–276.

ROSENBERG, R. & ROSENBERG, R. K. 1973. Salinity tolerance in three Scandinavian populations of *Littorina littorea* (L.) (Gastropoda). *Ophelia,* **10,** 129–139.

ROSEWATER, J. 1980. Subspecies of the gastropod *Littorina scabra. Nautilus,* **94,** 158–162.

ROTHSCHILD, M. 1941. Observations on the growth and trematode infection of *Peringia ulvae* (Pennant, 1777) in a pool on the Tamar Saltings, Plymouth. *Parasitology,* **33,** 406–415.

RUDMAN, W. B. 1969. Observations on *Pervicacia tristis* (Deshayes, 1859) and a comparison with other toxoglossan gastropods. *Veliger,* **12,** 53–64.

RUNHAM, N. W., THORNTON, P. R., SHAW, D. A. & WAYTE, R. C. 1969. The mineralization and hardness of the radular teeth of the limpet *Patella vulgata* L. *Z. Zellforsch,* **99,** 608–626.

RUNNEGAR, B. 1985. Origin and early history of mollusks. *Univ. Tennessee Stud. Geol.,* **13,** 17–32.

RUNNEGAR, B., POJETA, J., MORRIS, N. J., TAYLOR, M. E., TAYLOR, J. D. & McCLUNG, G. 1975. Biology of the Hyolitha. *Lethaia,* **8,** 181–191.

RUSANOVA, M. N. 1963. A short account of the biology of some widespread species of invertebrates of the region of Cape Kartesch. In 'Data for a comprehensive study of the White Sea' (Z. G. Palenichko ed.) **2,** 53–65. Moscow-Leningrad, Akademiya Nauk, USSR. [In Russian.]

RUSSELL-HUNTER, W. D. & RUSSELL-HUNTER, M. 1968. Pedal expansion in the naticid snails I. Introduction and weighing experiments. *Biol. Bull.,* **135,** 548–562.

RUSSELL-HUNTER, W. D., RUSSELL-HUNTER, M. & APLEY, M. L. 1968. Pedal expansion in the naticid snails II. Labelling experiments using inulin. *Biol. Bull.,* **135,** 563–573.

RUTHENSTEINER, B. & SCHAEFER, K. 1991. On the protonephridia and "larval kidneys" of *Nassarius reticulatus* (L.) (Caenogastropoda). *J. moll. Stud.,* **57,** 323–329.

SACCHI, C. F. 1961a. Contribution à l'étude des rapports écologie-polychromatisme chez un Proso-branche intercotidale, *Littorina obtusata. Cah. Biol. mar.,* **2,** 271–290.

SACCHI, C. F. 1961b. Relazioni ecologia-polycromatismo nel Prosobranco intertidale *Littorina obtusata* (L.). II. Ricerche biometriche. *Boll. Zool.,* **28,** 517–528.

SACCHI, C. F. 1963. Contribution à l'étude des rapports écologie-polychromatisme chez un Proso-branche intercotidale, *Littorina obtusata.* III. Données expérimentales et diverses. *Cah. Biol. mar.,* **4,** 299–313.

SACCHI, C. F. 1975. *Littorina nigrolineata* (Gray) (Gastropoda: Prosobranchia). *Cah. Biol. mar.,* **16,** 111–120.

SACCHI, C. F. & RASTELLI, M. 1966. *Littorina mariae,* nov. sp.: les différences morphologiques et écologiques entre 'nains' et 'normaux' chez 'l'espèce' *L. obtusata* (L.) (Gastr. Prosobr.) et leur signification adaptive et évolutive. *Att. Soc. ital. Sci. nat.,* **105,** 351–369.

SACCHI, C. F., AMBROGI, A. O. & VOLTALINA, D. 1981. Recherches sur le spectre trophique de *Littorina saxatilis* (Olivi) et de *L. nigrolineata* (Gray) (Gastropoda, Prosobranchia) sur la grève de 'Roscoff. II— Cas de populations vivant au milieu d'algues macroscopiques. *Cah. Biol. mar.,* **22,** 83–88.

SACCHI, C. F., TESTARD, F. & VOLTALINA, D. 1977. Recherches sur le spectre trophique comparé de *Littorina saxatilis* (Olivi) et de *L. nigrolineata* (Gray) (Gastropoda, Prosobranchia) sur la grève de Roscoff. *Cah. Biol. mar.,* **18,** 499–505.

SALVINI-PLAWEN, L. v. 1969. Solenogastres und Caudofoveata (Mollusca Aculifera): Organisation und phylogenetische Bedeutung. *Malacologia,* **9,** 191–216.

SALVINI-PLAWEN, L. v. 1972. Cnidaria as food sources for marine invertebrates. *Cah. Biol. mar.,* **13,** 385–400.

SALVINI-PLAWEN, L. v. 1978. Antarktische und subantarktische Solenogastres (eine Monographie: 1898–1974). *Zoologica, Stuttgart,* **44,** 1–305.

SALVINI-PLAWEN, L. v. 1980. A reconsideration of systematics in the Mollusca (phylogeny and higher classification). *Malacologia,* **19,** 249–278.

SALVINI-PLAWEN, L. v. 1981. The molluscan digestive system in evolution. *Malacologia,* **21,** 371–401.

SALVINI-PLAWEN, L. v. 1985. Early evolution and primitive groups. In 'The Mollusca' **10** (E. R. Trueman & M. R. Clarke eds), 59–150. Orlando, Academic Press.

SALVINI-PLAWEN, L. v. 1988a. The structure and function of molluscan digestive sytems. In 'The Mollusca' **11,** (E. R. Trueman & M. R. Clarke eds), 301–379. Orlando, Florida, Academic Press.

SALVINI-PLAWEN, L. v. 1988b. Annelida and Mollusca—a prospectus. *Microfauna mar.,* **4,** 383–396.

SALVINI-PLAWEN, L. v. & HASZPRUNAR, G. 1987. The Vetigastropoda and the systematics of strepto-neurous Gastropoda (Mollusca). *J. Zool., Lond.,* **211,** 747–770.

SANDER, K. 1950. Beobachtungen zur Fortpflanzung von *Assiminea grayana* Leach. *Arch. Molluskenk.,* **79,** 147–149.

SANDER, K. 1952. Beobachtungen zur Fortpflanzung von *Assiminea grayana* (2). *Arch. Molluskenk.,* **81,** 133–134.

SANDER, K. & SIBRECHT, L. 1967. Das Schlupfen der Veliger-larve von *Assiminea grayana* Leach (Gastropoda Prosobranchia). *Z. Morph. Ökol. Tiere,* **60,** 141–152.

SANDISON, E. E. 1966. The oxygen consumption of some intertidal gastropods in relation to zonation. *J. Zool., Lond.,* **149,** 163–173.

SANDISON, E. E. 1967. Respiratory responses to temperature and temperature tolerance of some intertidal gastropods. *J. exp. mar. Biol. Ecol.,* **1,** 271–281.

SCHÄFER, H. 1952. Ein Beitrag zur Ernährungsbiologie von *Bithynia tentaculata* L. (Gastropoda Proso-branchia). *Zool. Anz.,* **148,** 299–303.

SCHEIDEGGER, D. P. & FIORONI, P. 1983. Die Ultrastruktur der Hautvakuolenzellen der Kopfblase bei intrakapsulären Larven von *Nucella lapillus* (Gastropoda, Stenoglossa). *Zool. Jahrb. (Anat. Ont. Tiere),* **100,** 153–166.

SCHELTEMA, R. S. 1961. Metamorphosis of the veliger larvae of *Nassarius obsoletus* (Gastropoda) in response to bottom sediment. *Biol. Bull.,* **120,** 92–109.

SCHELTEMA, R. S. 1966. Evidence for trans-Atlantic transport of gastropod larvae belonging to the genus *Cymatium. Deep-Sea Res.,* **13,** 83–95.

SCHELTEMA, R. S. 1971a. The dispersal of larvae of shoal-water benthic invertebrate species over long distances by ocean currents. In 'Fourth European Marine Biology Symposium' (D. J. Crisp ed.), 7–28. Cambridge, Cambridge University Press.

SCHELTEMA, R. S. 1971b. Larval dispersal as a means of genetic exchange between geographically separated populations of shoal-water benthic marine gastropods. *Biol. Bull.,* **140,** 284–322.

SCHELTEMA, R. S. 1972. Eastward and westward dispersal across the tropical Atlantic Ocean of larvae belonging to the genus *Bursa* (Prosobranchia, Mesogastropoda, Bursidae). *Int. Rev. ges. Hydrobiol.,* **57,** 863–873.

SCHELTEMA, R. S. 1977. Dispersal of marine invertebrate organisms: paleobiogeographic and biostrati-graphic implications. In 'Concepts and methods of biostratigraphy' (E. G. Kauffman & J. E. Hazel eds), 73–108. Stroudsburg, Pennsylvania, Dowden, Hutchinson & Ross.

SCHELTEMA, R. S. 1986. On dispersal and planktonic larvae of benthic invertebrates: an eclectic overview and summary. *Bull. mar. Sci.,* **39,** 290–322.

SCHELTEMA, R. S. & WILLIAMS, I. P. 1983. Long-distance dispersal of planktonic larvae and the biogeography and evolution of some Polynesian and western Pacific mollusks. *Bull. mar. Sci.,* **33,** 545–565.

SCHIEMENZ, P. 1884. Über die Wasseraufnahme bei Lamellibranchiaten und Gastropoden (einschliesslich der Pteropoden). *Mitt. zool. Stat. Neapel,* **5,** 509–543.

SCHIEMENZ, P. 1887. Über die Wasseraufnahme bei Lamellibranchiaten und Gastropoden (einschliesslich der Pteropoden). Zweiter Theil. *Mitt. zool. Stat. Neapel,* **7,** 423–472.

SCHMIDT, W. J. 1959. Bemerkungen zur Schalenstruktur von *Neopilina galatheae. Galathea Rep.,* **3,** 73–78.

SCHNEIDER, D. 1981. Escape response of the infaunal clam *Ensis directus* Conrad, 1843 to a predatory snail, *Polinices duplicatus* Say, 1822. *Veliger,* **24,** 371–372.

SCHULTZ, M. C. 1983. A correlated light and electron microscope study of the structure and secretory activity of the accessory salivary glands of the marine gastropods, *Conus flavidus* and *C. vexillum* (Neogastropoda, Conacea). *J. Morph.,* **176,** 89–111.

SEED, R. 1979. Distribution and shell characteristics of the painted topshell *Calliostoma zizyphinum* (L.) (Prosobranchia: Trochidae) in County Down, N. Ireland. *J. moll. Stud.,* **45,** 12–18.

SESHAIYA, R. V. 1934. A further note on the style sac of gastropods. *Rec. Ind. Mus.,* **36,** 179–183.

SESHAIYA, R. V. 1940. A free larval stage in the life history of a fluviatile gastropod. *Curr. Sci.,* **9,** 331–332.

SESHAPPA, G. 1947. Oviparity in *Littorina saxatilis* (Olivi). *Nature, Lond.,* **160,** 335.

SHARMAN, M. 1956. Note on *Capulus ungaricus* (L.). *J. mar. biol. Ass. U.K.,* **35,** 445–450.

SHAW, H. O. N. 1914. On the anatomy of *Conus tulipa,* Linn., and *Conus textile,* Linn. *Quart. J. micr. Sci.,* **60,** 1–60.

SHERIDAN, R., VAN MOL, J.-J. & BOUILLON, J. 1973. Etude morphologique du tube digestif de quelques Turridae (Mollusca-Gastropoda-Prosobranchia-Toxoglossa) de la région de Roscoff. *Cah. Biol. mar.,* **14,** 159–188.

SHILENKO, A. A. 1977. The symmetry of the Docoglossa and the problem of the origin of the order. *Bull. nat. Soc. Moscow, Biol. Ser.,* **81,** 60–65. [In Russian.]

SHIMEK, R. L. 1983. The biology of the northeastern Pacific Turridae. II. *Oenopota. J. moll. Stud.,* **49,** 146–163.

SHIMEK, R. L. & KOHN, A. J. 1981. Functional morphology and evolution of the toxoglossan radula. *Malacologia,* **20,** 423–438.

SHIRBHATE, R. & COOK, A. 1987. Pedal and opercular secretory glands of *Pomatias, Bithynia* and *Littorina. J. moll. Stud.,* **53,** 79–96.

SHUTO, T. 1974. Larval ecology of prosobranch gastropods and its bearings on biogeography and paleontology. *Lethaia,* **7,** 239–256.

SIEGISMUND, H. R. & HYLLEBERG, J. 1987. Dispersal-mediated coexistence of mud snails (Hydrobiidae) in an estuary. *Mar. Biol.,* **94,** 395–402.

SIGNOR, P. W. III. 1982. Resolution of life habits using multiple morphologic criteria: shell form and life-mode in turritelliform gastropods. *Paleobiology,* **8,** 378–388.

SIGNOR, P. W. III. 1983. Burrowing and the functional significance of ratchet sculpture in turritelliform gastropods. *Malacologia,* **23,** 313–320.

SIMKISS, K. & WILBUR, K. M. 1977. The molluscan epidermis and its secretions. *Symp. zool. Soc. Lond.,* **39,** 35–76.

SIRENKO, B. & MINICHEV, Y. 1975. Développement ontogénétique de la radula chez les Polyplacophores. *Cah. Biol. mar.,* **16,** 425–433.

SISSON, R. F. 1986. Tide pools. *Nat. Geographic,* **169,** 252–259.

SKOOG, G. 1978. Influence of natural food items on growth and egg production in brackish water populations of *Lymnaea peregra* and *Theodoxus fluviatilis* (Mollusca). *Oikos,* **31,** 340–348.

SKRAMLIK, E. v. 1941. Über den Kreislauf bei den Weichtiere. *Ergebn. Biol.,* **18,** 88–286.

SMIDT, E. L. B. 1951. Animal production in the Danish Waddensee. *Medd. Komm. Havundersøg, Kbh.,* Ser. Fiskeri no. 6, 1–151.

SMITH, B. S. 1971. Sexuality in the American mud snail, *Nassarius obsoletus* Say. *Proc. malac. Soc. Lond.,* **39,** 377–378.

SMITH, B. S. 1981. Tributyltin compounds induce male characteristics on female mud snails *Nassarius obsoletus = Ilyanassa obsoleta. J. appl. Toxicol.,* **1,** 141–144.

SMITH, D. A. S. 1973. The population biology of *Lacuna pallidula* (da Costa) and *Lacuna vincta* (Montagu) in north-east England. *J. mar. biol. Ass. U.K.,* **53,** 493–520.

SMITH, D. A. S. 1976. Disruptive selection and morph-ratio clines in the polymorphic snail *Littorina obtusata* (L.) (Gastropoda: Prosobranchia). *J. moll. Stud.,* **42,** 114–135.

SMITH, E. H. 1967a. The proboscis and oesophagus of some British turrids. *Trans. R. Soc. Edinb.,* **67,** 1–22.

SMITH, E. H. 1967b. The neogastropod stomach, with notes on the digestive diverticula and intestine. *Trans. R. Soc. Edinb.,* **67,** 23–42.

SMITH, F. G. W. 1935. The development of *Patella vulgata. Phil. Trans. R. Soc. B,* **225,** 95–125.

SMITH, J. E. 1981. The natural history and taxonomy of shell variation in the periwinkles *Littorina saxatilis* and *Littorina rudis. J. mar. biol. Ass. U.K.,* **61,** 215–241.

SMITH, J. E. & NEWELL, G. E. 1955. The dynamics of the zonation of the common periwinkle (*Littorina littorea* (L.)) on a stony beach. *J. Anim. Ecol.,* **24,** 35–56.

SMITH, S. M. 1979. *Littorina rudis* var. *scotia* and its adaptation to the extreme environment of Rockall (Mollusca: Gastropoda). *Porcupine Newsletter,* **1,** 138–139.

SMITH, S. M. 1982. A review of the genus *Littorina* in British and Atlantic waters (Gastropoda: Prosobranchia). *Malacologia,* **22,** 535–539.

SNELI, J.-A. 1972. *Odostomia turrita* found on *Homarus gammarus. Nautilus,* **86,** 23–24.

SNYDER, T. & GOOCH, J. L. 1973. Genetic differentiation in *Littorina saxatilis* (Gastropoda). *Mar. Biol.,* **22,** 177–182.

SOMMERVILLE, B. A. 1973. The circulatory physiology of *Helix pomatia.* III. The hydrostatic pressure changes in the circulatory system of living *Helix. J. exp. Biol.,* **59,** 291–303.

SOUTHGATE, T. 1982. The biology of *Barleeia unifasciata* (Gastropoda: Prosobranchia) in red algal tufts in S.W. Ireland. *J. mar. biol. Ass. U.K.,* **62,** 461–468.

SOUTHWARD, E. C. 1986. Gill symbionts in thyasirids and other bivalve molluscs. *J. mar. biol. Ass. U.K.,* **66,** 889–914.

SOUTHWARD, A. J. & CRISP, D. J. 1954. The distribution of certain intertidal animals around the Irish coast. *Proc. R. Ir. Acad.,* **57B,** 1–29.

SPIGHT, T. M. 1975a. On a snail's chance of becoming a year old. *Oikos,* **26,** 9–14.

SPIGHT, T. M. 1975b. Factors extending gastropod embryonic development and their selective cost. *Oecologia (Berlin),* **21,** 1–16.

SPIGHT, T. M. 1976. Colors and patterns of an intertidal snail, *Thais lamellosa. Res. popul. Biol.,* **17,** 176–190.

SPIGHT, T. M. 1977. Is *Thais canaliculata* (Gastropoda, Muricidae) evolving nurse eggs? *Nautilus,* **91,** 74–76.

SPIGHT, T. M., BIRKELAND, C. & LYONS, A. 1974. Life histories of large and small murexes (Prosobranchia: Muricidae). *Mar. Biol.,* **24,** 229–242.

SPITZER, J. M. 1937. Physiologische-ökologische Untersuchungen über den Exkretstoffwechsel der Mollusken. *Zool. Jb. (Zool. Physiol.),* **57,** 457–496.

STADNICHENKO, A. P. 1970. Sexual variability of blood composition in *Viviparus contectus* (Gastropoda, Prosobranchia). *Zool. Zh.,* **49,** 680–684. [In Russian.]

STAIGER, H. 1951. Cytologische und morphologische Untersuchungen zur Determination der Nähreier bei Prosobranchiern. *Z. Zellforsch. mikrosk. Anat.,* **35,** 496–549.

STAIGER, H. 1957. Genetics and morphological variation in *Purpura lapillus* with respect to local and regional differentiation of population groups. *Année biol.,* **61,** 251–258.

STAROBOGATOV, Y. I. 1990. Evolutionary changes in the radula. In 'Evolutionary morphology of molluscs' (A. A. Shilenko ed.). *Arch. zool. Mus. Moscow State Univ.,* **28**, 48–91. [In Russian.]

STEARNS, S. C. 1976. Life-history tactics: a review of the ideas. *Quart. Rev. Biol.,* **51**, 1–47.

STENECK, R. S. 1982. A limpet-coralline alga association: adaptations and defenses between a selective herbivore and its prey. *Ecology,* **63**, 507–522.

STENECK, R. S. & WATLING, L. 1982. Feeding capabilities and limitation of herbivorous molluscs: a functional group approach. *Mar. Biol.,* **68**, 299–319.

STICKLE, W. B. & BAYNE, B. L. 1982. Effects of temperature and salinity on oxygen consumption and nitrogen excretion in Thais (Nucella) lapillus (L.). *J. exp. mar. Biol. Ecol.,* **58**, 1–17.

STICKLE, W. B., MOORE, M. N. & BAYNE, B. L. 1985. Effects of temperature, salinity, and aerial exposure on predation and lysosomal stability of the dogwhelk Thais (Nucella) lapillus (L.). *J. exp. mar. Biol. Ecol.,* **93**, 235–258.

STÖCKMANN-BOSBACH, R. 1988. Early stages in encapsulated development of Nucella lapillus (Linnaeus) (Gastropoda, Muricidae). *J. moll. Stud.,* **54**, 181–196.

ST. PIERRE, T. G., MANN, S., WEBB, J., DICKSON, D. P. E., RUNHAM, N. W. & WILLIAMS, R. J. P. 1986. Iron oxide biomineralization in the radula teeth of the limpet Patella vulgata: Mossbauer spectroscopy and high resolution transmission electron microscopy studies. *Proc. R. Soc. B,* **228**, 31–42.

STRATHMANN, R. R. & LEISE, E. 1979. On feeding mechanisms and clearing rates of molluscan veligers. *Biol. Bull.,* **157**, 524–535.

STREIFF, W. 1966. Etude endocrinologique du déterminisme du cycle sexuel chez un mollusque hermaphrodite protandre Calyptraea sinensis L. I. Mise en évidence, par culture in vitro, des facteurs hormonaux conditionnant l'évolution du tractus génital mâle. *Ann. Endocrinol.,* **27**, 385–400.

STREIFF, W. & LE BRETON, J. 1970. Etude endocrinologique des facteurs régissant la morphogenése et la régression du pénis chez un mollusque prosobranche gonochorique. *C. R. Acad. Sci., Paris,* **270D**, 547–549.

STREIFF, W., LE BRETON, J. & SILBERZAHN, N. 1970. Non-spécificité des facteurs hormonaux responsables de la morphogenése et de cycle du tractus génital mâle chez les mollusques prosobranches. *Ann. Endocrinol.,* **31**, 548–556.

STRUHSAKER, J. W. 1966. Breeding, spawning, spawning periodicity and early development in the Hawaiian Littorina: L. pintada (Wood), L. picta Philippi and L. scabra (Linné). *Proc. malac. Soc. Lond.,* **37**, 137–166.

STRUHSAKER, J. W. 1968. Selection mechanisms associated with intraspecific shell variation in Littorina picta (Prosobranchia: Mesogastropoda). *Evolution,* **22**, 459–480.

STRUHSAKER, J. W. & COSTLOW, J. D. Jr. 1968. Larval development in the laboratory. *Proc. malac. Soc. Lond.,* **38**, 153–160.

SULLIVAN, C. H. & MAUGEL, T. K. 1984. Formation, organization and composition of the egg capsule of the marine gastropod, Ilyanassa obsoleta. *Biol. Bull.,* **167**, 378–389.

SUNDELL, K. 1985. Adaptability of two phenotypes of Littorina saxatilis (Olivi) to direct salinities. *J. exp. mar. Biol. Ecol.,* **92**, 115–123.

SUTHERLAND, J. P. 1970. Dynamics of high and low populations of the limpet Acmaea scabra (Gould). *Ecol. Monogr.,* **40**, 169–188.

SZAL, R. 1971. 'New' sense organ of primitive gastropods. *Nature, Lond.,* **229**, 490–492.

TAMARIN, A. & CARRIKER, M. R. 1967. The egg capsule of the muricid gastropod Urosalpinx cinerea: an integrated study of the wall by ordinary light, polarized light, and electron microscopy. *J. ultra-structure Res.,* **21**, 26–40.

TAYLOR, J. D. 1970. Feeding habits of predatory gastropods in a Tertiary (Eocene) molluscan assemblage from the Paris Basin. *Palaeontology,* **13**, 254–260.

TAYLOR, J. D. 1973. The structural evolution of the bivalve shell. *Palaeontology,* **16**, 519–534.

TAYLOR, J. D. 1978. The diet of Buccinum undatum and Neptunea antiqua (Gastropoda: Buccinidae). *J. Conch., Lond.,* **29**, 309–318.

TAYLOR, J. D. 1985. The anterior alimentary system and diet of *Turricula nelliae spurius* (Gastropoda: Turridae). *Proc. 2nd intern. Workshop Malacofauna Hong Kong S. China* (B. S. Morton & D. Dudgeon eds), 175–190. Hong Kong, Hong Kong University Press.

TAYLOR, J. D. 1989. The diet of coral-reef Mitridae (Gastropoda) from Guam: with a review of other species of the family. *J. nat. Hist.,* **23,** 261–278.

TAYLOR, J. D. 1990. The anatomy of the foregut and relationships in the Terebridae. *Malacologia,* **32,** 19–34.

TAYLOR, J. D. & MILLER, J. A. 1989. The morphology of the osphradium in relation to feeding habits in meso- and neogastropods. *J. moll. Stud.,* **55,** 227–237.

TAYLOR, J. D. & MILLER, J. A. 1990. A new type of gastropod proboscis: the foregut of *Hastula bacillus* (Gastropoda: Terebridae). *J. Zool., Lond.,* **220,** 603–617.

TAYLOR, J. D. & MORRIS, N. J. 1988. Relationships of neogastropods. *Malacol. Rev. Suppl.* **4,** 167–179.

TAYLOR, J. D. & REID, D. G. 1990. Shell microstructure and mineralogy of the Littorinidae: ecological and evolutionary significance. *Hydrobiologia,* **193,** 199–215.

TAYLOR, J. D., CLEEVELY, R. & MORRIS, N. J. 1983. Predatory gastropods and their activities in the Blackdown Greensand (Albian) of England. *Palaeontology,* **26,** 521–553.

TAYLOR, J. D., KENNEDY, W. J. & HALL, A. 1969. The shell structure and mineralogy of the Bivalvia: introduction: Nuculacea—Trigonacea. *Bull. Brit. Mus. (Nat. Hist.) Zool.,* Suppl. **3,** 1–125.

TAYLOR, J. D., KENNEDY, W. J. & HALL, A. 1973. The shell structure and mineralogy of the Bivalvia: II. Lucinacea-Clavagellacea: conclusions. *Bull. Brit. Mus. (Nat. Hist.) Zool.,* **22,** 253–294.

TAYLOR, J. D., MORRIS, N. J. & TAYLOR, C. N. 1980. Food specialization and the evolution of predatory prosobranch gastropods. *Palaeontology,* **23,** 373–409.

TAYLOR, P. M. & ANDREWS, E. B. 1987. Tissue adenosine triphosphate activities of the gill and excretory system in mesogastropod molluscs in relation to osmoregulatory capacity. *Comp. Biochem. Physiol.,* **A86,** 693–696.

TAYLOR, P. M. & ANDREWS, E. B. 1991. Non-protein nitrogen excretion in the prosobranch gastropod *Viviparus contectus. J. moll. Stud.,* **57,** 391–393.

TESCH, J. J. 1949. Heteropoda. *Dana Rep.,* **34,** 1–53.

THIELE, J. 1929–1931. 'Handbuch der systematischen Weichtierkunde' 1 (1929), 1–376; 2 (1931), 377–778. Jena, Fischer.

THIRIOT-QUIEVREUX, C. 1972. Microstructure de coquilles larvaires de prosobranches au microscope électronique à balayage. *Arch. Zool. exp. gén.,* **113,** 553–564.

THIRIOT-QUIEVREUX, C. 1973. Heteropoda. *Oceanogr. mar. Biol. ann. Rev.,* **11,** 173–217.

THOMAS, J. D. 1982. Chemical ecology of the snail hosts of schistosomiasis: snail-snail and snail-plant interactions. *Malacologia,* **22,** 81–89.

THOMAS, J. D., NWANKO, D. I. & STERRY, P. R. 1985. The feeding strategies of juvenile and adult *Biomphalaria glabrata* (Say) under simulated natural conditions and their relevance to ecological theory and snail control. *Proc. R. Soc. B,* **226,** 177–209.

THOMAS, M. L. H. & PAGE, F. H. 1983. Grazing by the gastropod, *Lacuna vincta,* on the lower intertidal area at Musquash Head, New Brunswick, Canada. *J. mar. biol. Ass. U.K.,* **63,** 725–736.

THOMASSIN, B. A. 1976. Feeding behaviour of the felt-, sponge-, and coral-feeder sea stars, mainly *Culcita schmideliana. Helgol. wiss. Meeresunters.,* **28,** 51–65.

THOMPSON, G. B. 1979. Distribution and population dynamics of the limpet *Patella aspera* in Bantry Bay. *J. exp. mar. Biol. Ecol.,* **40,** 115–135.

THOMPSON, G. B. 1980. Distribution and population dynamics of the limpet *Patella vulgata* in Bantry Bay. *J. exp. mar. Biol. Ecol.,* **45,** 173–217.

THORSON, G. 1940a. Studies on the egg masses and larval development of gastropods from the Iranian Gulf. *Danish sci. Invest., Iran,* part **2,** 159–238.

THORSON, 1940b. Notes on the egg-capsules of some North-Atlantic prosobranchs of the genus *Troschelia, Chrysodomus, Volutopsius, Sipho* and *Trophon. Vidensk. Medd. naturh. Foren. Kbh.,* **104,** 251–266.

THORSON, G. 1946. Reproduction and larval development of Danish marine bottom invertebrates. *Medd. Komm. Havundersøg. Kbh.,* ser. Plankton **4,** 1–523.

THORSON, G. 1950. Reproductive and larval ecology of marine bottom invertebrates. *Biol. Rev.,* **23,** 1–45.

THORSON, G. 1965. A neotenous dwarf-form of *Capulus ungaricus* (L.) (Gastropoda Prosobranchia) commensalistic on *Turritella commnunis* Risso. *Ophelia,* **2,** 175–210.

TODD, M. E. 1964. Osmotic balance in *Littorina littorea, Littorina littoralis* and *Littorina saxatilis* (Littorinidae). *Physiol. Zool.,* **37,** 33–44.

TORELLI, A. R. 1978. Ricerche sui rapporti tra ecologia e policromatismo in *Littorina saxatilis* (Olivi) (Gastropoda Prosobranchia) della Laguna di Venezia. *Cah. Biol. mar.,* **19,** 91–98.

TROSCHEL, F. H. 1863, 'Das Gebiss der Schnecken zur Begründung einer natürlichen Classification.' Berlin, Nicolai.

TRUEMAN, E. R. 1968. The mechanism of burrowing of some naticid gastropods in comparison with that of other molluscs. *J. exp. Biol.,* **48,** 663–678.

TRUEMAN, E. R. 1983. Locomotion. In 'The Mollusca' **4** (A.S.M. Saleuddin & K. M. Wilbur eds), 155–198. New York, Academic Press.

TRUEMAN, E. R. 1984. Retrograde locomotion in gastropods. *J. moll. Stud.,* **50,** 235–237.

TRUEMAN, E. R. & BROWN, A. C. 1976. Locomotion, pedal retraction and extrusion, and the hydraulic system of *Bullia* (Gastropoda: Nassariidae). *J. Zool., Lond.,* **178,** 365–384.

TRUEMAN, E. R. & BROWN, A. C. 1985. The mechanism of shell elevation in *Haliotis* (Mollusca: Gastropoda) and a consideration of the evolution of the hydrostatic skeleton in Mollusca. *J. Zool., Lond.,* **A205,** 585–594.

TRUEMAN, E. R. & BROWN, A. C. 1987a. Proboscis extrusion in *Bullia* (Nassariidae): a study of fluid skeletons in Gastropoda. *J. Zool., Lond.,* **211,** 505–513.

TRUEMAN, E. R. & BROWN, A. C. 1987b. Locomotory function of the pedal musculature of the nassariid whelk, *Bullia. J. moll. Stud.,* **53,** 287–288.

TRUEMAN, E. R. & BROWN, A. C. 1989. The effect of shell shape on the burrowing performance of species of *Bullia* (Gastropoda: Nassariidae). *J. moll. Stud.,* **55,** 129–131.

TRUEMAN, E. R. & HODGSON, A. N. 1990. The fine structure and function of the foot of *Nassarius kraussianus,* a gastropod moving by ciliary locomotion. *J. moll. Stud.,* **56,** 221–228.

TSIKHON-LIKANINA, E. A. 1961a. On the filtration method of feeding in *Bithynia tentaculata* (L.) and *Valvata piscinalis* (Müller) (Gastropoda, Prosobranchia). *Byull. Inst. Biol. vodokh.,* **10,** 28–30. [In Russian.]

TSIKHON-LIKANINA, E. A. 1961b. The dependence of the feeding rate and filtration in some freshwater gastropods on the concentration of food. *Byull. Inst. Biol. vodokh.,* **10,** 31–34. [In Russian.]

TURCHINI, J. 1923. Contribution á l'étude de l'histologie comparée de la cellule rénale. L'excrétion urinaire chez les mollusques. *Arch. Zool. exp. gén.,* **18,** 3–253.

UNDERWOOD, A. J. 1972a. Observations on the reproductive cycles of *Monodonta lineata, Gibbula umbilicalis* and *G. cineraria. Mar. Biol.,* **17,** 333–340.

UNDERWOOD, A. J. 1972b Tide-model analysis of the zonation of intertidal prosobranchs II. Four species of trochids (Gastropoda: Prosobranchia). *J. exp. mar. Biol. Ecol.,* **9,** 257–277.

UNDERWOOD, A. J. 1973. Studies on zonation of intertidal prosobranch molluscs in the Plymouth region. *J. Anim. Ecol.,* **42,** 353–372.

UNDERWOOD, A. J. 1979. The ecology of intertidal gastropods. *Adv. mar. Biol.,* **16,** 111–210.

VADER, W. J. M. 1964. A preliminary investigation into the reactions of the infauna of the tidal flats to tidal fluctuations in water levels. *Neth. J. Sea Res.,* **2,** 189–222.

VAHL, O. 1971. Growth and density of *Patina pellucida* (L.) (Gastropoda: Prosobranchia) on *Laminaria hyperborea* (Gunnerus) from Western Norway. *Ophelia,* **9,** 31–50.

VAHL, O. 1972. On the position of *Patina pellucida* (L.) (Gastropoda) on the frond of *Laminaria hyperborea. Ophelia,* **10,** 1–9.

VAN DONGEN, A. 1956. The preference of *Littorina obtusata* for Fucaceae. *Arch. néerl. Zool.,* **11,** 373–386.

VANSTONE, J. H. 1894. Some points in the anatomy of *Melongena melongena. J. Linn. Soc.* (Zool.), **24,** 369–373.

VERLAINE, L. 1936. L'instinct et l'intelligence chez les mollusques. Les gastéropodes perceurs de coquilles. *Mém. Mus. Hist. nat. Belg.* (2), **3,** 387–394.

VERMEIJ, G. J. 1972. Intraspecific shore-level size gradients in intertidal molluscs. *Ecology,* **53,** 693–700.

VERMEIJ, G. J. 1980. Drilling predation tactics of a tropical naticid gastropod. *Malacologia,* **19,** 329–334.

VERMEIJ, G. J. 1981. Apertural form in gastropods. *Lethaia,* **14,** 104.

VERMEIJ, G. J. 1982a. Phenotypic evolution in a poorly dispersing snail after arrival of a predator. *Nature, Lond.,* **299,** 349–350.

VERMEIJ, G. J. 1982b. Environmental change and the evolutionary history of the periwinkle (*Littorina littorea*) in North America. *Evolution,* **36,** 561–580.

VERMEIJ, G. J. & ZIPSER, E. 1986. Size of naticid foot in relation to burrowing. Burrowing performance of some tropical Pacific gastropods. *Veliger,* **29,** 200–206.

VESTERGAARD, K. 1935. Über den Laich und die Larven von *Scalaria communis* (Lam.), *Nassarius pygmaeus* (Lam.) und *Bela turricula* (Mörch). *Zool. Anz.,* **109,** 217–222.

VLES, F. 1907. Sur les ondes pédieuses des mollusques reptateurs. *C. R. Acad. Sci., Paris,* **145,** 270–278.

VOGEL, S. 1981. 'Life in moving fluids: the physical basis of flow.' Boston, Massachusetts, Willard Green.

VOLTZOW, J. 1983. Flow through and around the abalone *Haliotis kamtschatkana* Jonas, 1845. *Veliger,* **26,** 18–21.

VOLTZOW, J. 1985. 'Functional morphology and evolution of the prosobranch gastropod foot.' Ph.D. thesis, Duke Univ. N.C.

VOLTZOW, J. 1988. The organization of limpet pedal musculature and its evolutionary implications for the Gastropoda. *Malacol. Rev.,* Suppl. **4,** 271–281.

VOLTZOW, J. 1990. The functional morphology of the pedal musclature of the marine gastropods *Busycon contrarium* and *Haliotis kamtschatkana. Veliger,* **33,** 1–19.

VOVELLE, J. 1967. Sur l'opercule de *Gibbula magus* (L.), Gastéropode Prosobranche: édification, nature protéique et durcissement par tannage quinonique. *C. R. Acad. Sci., Paris,* **264D,** 141–144.

VOVELLE, J. 1969a. Complexity of the opercular materials in *Astralium rugosum* L. (Gastropoda, Prosobranchia, Turbinidae). *Proc. malac. Soc. Lond.,* **38,** 557.

VOVELLE, J. 1969b. Elaboration de la matière operculaire chez *Tricolia pullus* (L.) (Gastropoda — Prosobranchia). *Malacologia,* **9,** 203–294.

VOVELLE, J. & GRASSET, M. 1979. Approche histophysiologique et cytologique au rôlle des cellules à sphérules calciques du repli operculaire chez *Pomatias elegans* (Müller), Gastéropode Prosobranche. *Malacologia,* **18,** 557–560.

VOVELLE, J. & GRASSET, M. 1982. Etude cytologique et histo-chimique comparée de la formation de l'opercule corné chez les prosobranches. *Malacologia,* **22,** 257–263.

VOVELLE, J., GRASSET, M. & MEUNIER, F. 1977. Elaboration de l'opercule calcifié chez *Nerita plicata* Linnaeus et *Pomatias elegans* (Müller), Gastéropodes Prosobranches. *Malacologia,* **16,** 279–283.

WAINWRIGHT, S. A., BIGGS, W. D., CURREY, J. D. & GOSLINE, J. M. 1976. 'Mechanical design in organisms.' London, Arnold.

WALLACE, C. 1978. Notes on the distribution of sex and shell characters in some Australian populations of *Potamopyrgus* (Gastropoda: Hydrobiidae). *J. malac. Soc. Aust.,* **4,** 71–76.

WALLACE, C. 1979. Notes on the occurrence of males in populations of *Potamopyrgus jenkinsi. J. moll. Stud.,* **45,** 61–67.

WALLACE, C. 1985. On the distribution of the sexes of *Potamopyrgus jenkinsi* (Smith). *J. moll. Stud.,* **51,** 290–296.

WALTERS, G. J. & WHARFE, J. R. 1980. Distribution and abundance of *Hydrobia ulvae* (Pennant) in the lower Medway Estuary, Kent. *J. moll. Stud.,* **46,** 171–180.

WARBURTON, K. 1976. Shell form, behaviour, and tolerance to water movement in the limpet *Patina pellucida* (L.) (Gastropoda: Prosobranchia). *J. exp. mar. Biol. Ecol.,* **23,** 307–325.

WARD, R. D. 1990. Biochemical genetic variation in the genus *Littorina* (Prosobranchia: Mollusca). *Hydrobiologia,* **193,** 53–69.

WARD, R. D. & JANSON, K. 1985. A genetic analysis of sympatric subpopulations of the sibling species *Littorina saxatilis* (Olivi) and *Littorina arcana* Hannaford Ellis. *J. moll. Stud.,* **51,** 86–94.

WARD, R. D. & WARWICK, T. 1980. Genetic differentiation in the molluscan species *Littorina rudis* and *Littorina arcana* (Prosobranchia: Littorinidae). *Biol. J. Linn. Soc.,* **14,** 417–428.

WAREN, A. 1983. A generic revision of the family Eulimidae (Gastropoda, Prosobranchia). *J. moll. Stud.,* suppl. **13,** 1–95.

WAREN, A. 1988. *Neopilina goesi,* a new Caribbean monoplacophoran mollusk dredged in 1869. *Proc. biol. Soc. Wash.,* **101,** 676–681.

WAREN, A. 1990. Ontogenetic changes in the trochoidean (Archaeogastropoda) radula, with some phylogenetic interpretations. *Zool. Scripta,* **19,** 179–187.

WARWICK, T. 1982. A method of maintaining and breeding members of the *Littorina saxatilis* (Olivi) species complex. *J. moll. Stud.,* **48,** 368–370.

WARWICK, T., KNIGHT, A. J. & WARD, R. D. 1990. Hybridisation in the *Littorina saxatilis* species complex (Prosobranchia: Mollusca). *Hydrobiologia,* **193,** 109–116.

WATSON, D. C. & NORTON, T. A. 1985. Dietary preferences of the common periwinkle *Littorina littorea* (L.). *J. exp. mar. Biol. Ecol.,* **88,** 193–211.

WATSON, D. C. & NORTON, T. A. 1987. The habitat and feeding preferences of *Littorina obtusata* (L.) and *L. mariae* Sacchi & Rastelli. *J. exp. mar. Biol. Ecol.,* **112,** 61–72.

WAYNE, T. A. 1980. Antipredator behavior of the mussel, *Mytilus edulis. Am. Zool.,* **20,** 789.

WEBB, R. S. & SALEUDDIN, A. S. M. 1977. Role of enzymes in the mechanism of shell penetration by the muricid gastropod, *Thais lapillus* (L.). *Can. J. Zool.,* **55,** 1846–1857.

WEBER, H. 1924. Ein Umdreh- und Fluchtreflex bei *Nassa mutabilis. Zool. Anz.,* **60,** 261–269.

WELLS, F. E. 1978. The relationship between environmental variables and the density of the mud snail *Hydrobia totteni* in a Nova Scotia salt marsh. *J. moll. Stud.,* **44,** 120–129.

WELLS, M. J. & BUCKLEY, S. K. L. 1972. Snails and trails. *Animal Behav.,* **20,** 345–355.

WENZ, W. 1938. Gastropoda, Teil 1: allgemeiner Teil und Prosobranchia. In 'Handbuch der Paläo-zoologie' (O. H. Schindewolf ed.), **6,** 1–1639. Berlin, Borntraeger.

WERNER, B. 1953. Über den Nahrungserwerb der Calyptraeidae (Gastropoda Prosobranchia). Morphologie, Histologie und Funktion der am Nahrungserwerb beiteiligten Organe. *Helgol. wiss. Meeresunters.,* **4,** 260–315.

WERNER, B. 1955. Über die Anatomie, die Entwicklung und Biologie des Veligers und der Veliconcha von *Crepidula fornicata* L. (Gastropoda Prosobranchia). *Helgol. wiss. Meeresunters.,* **5,** 169–217.

WERNER, B. 1959. Das Prinzip des endlosen Schleimfilter beim Nahrungserwerb wirbelloser Meerestiere. *Int. Rev. ges. Hydrobiol.,* **44,** 181–216.

WEST, D. L. 1979. Reproductive biology of *Colus stimpsoni* (Prosobranchia: Buccinidae). III. Female genital system. *Veliger,* **21,** 432–438.

WEST, L. 1988. Prey selection by the tropical snail *Thais melones:* a study of interindividual variation. *Ecology,* **69,** 1839–1854.

WHITLATCH, R. B. & OBRETSKI, S. 1980. Feeding selectivity and coexistence in two deposit-feeding gastropods. *Mar. Biol.,* **58.** 219–225.

WIESER, W. 1980. Metabolic end products in three species of marine gastropods. *J. mar. biol. Ass. U.K.,* **60,** 175–180.

WIGHAM, G. D. 1975. Environmental influences upon the expression of shell form in *Rissoa parva* (da Costa) [Gastropoda: Prosobranchia]. *J. mar. biol. Ass. U.K.,* **55,** 425–438.

WIGHAM, G. D. 1976. Feeding and digestion in the marine prosobranch *Rissoa parva* (da Costa). *J. moll. Stud.,* **42,** 74–94.

WILLIAMS, E. E. 1964a. The growth and distribution of *Littorina littorea* (L.) on a rocky shore in Wales. *J. Anim. Ecol.,* **33,** 413–432.

WILLIAMS, E. E. 1964b. The growth and distribution of *Gibbula umbilicalis* (da Costa) on a rocky shore in Wales. *J. Anim. Ecol.,* **33,** 433–442.

WILLIAMS, E. E. 1965. The growth and distribution of *Monodonta lineata* (da Costa) on a rocky shore in Wales. *Field Studies,* **2,** 189–198.

WILLIAMS, G. A. 1990. *Littorina mariae* — a factor structuring low shore communities? *Hydrobiologia,* **193,** 139–146.

WILLIAMS, I. C. & ELLIS, C. 1975. Movements of the common periwinkle, *Littorina littorea* (L.), on the Yorkshire coast in winter and the influence of infection with larval Digenea. *J. exp. mar. Biol. Ecol.,* **17,** 47–58.

WILLIAMSON, P. & KENDALL, M. A. 1981. Population age structure and growth of the trochid *Monodonta lineata* determined from shell rings. *J. mar. biol. Ass. U.K.,* **61,** 1011–1026.

WILSON, D. P. & WILSON, A. 1956. A contribution to the biology of *Ianthina janthina*. *J. mar. biol. Ass. U.K.,* **35,** 291–305.

WILTSE, W. I. 1980a. Effects of *Polinices duplicatus* (Gastropoda: Naticidae) on infaunal community structure at Barnstaple Harbor, Massachusetts, U.S.A. *Mar. Biol.,* **56,** 301–310.

WILTSE, W. I. 1980b. Predation by juvenile *Polinices duplicatus* (Say) on *Gemma gemma* (Totten). *J. exp. mar. Biol. Ecol.,* **42,** 187–199.

WINGSTRAND, K. G. 1985. On the anatomy and relationships of Recent Monoplacophora. *Galathea Rep.,* **16,** 7–94.

WOOD, L. 1968. Physiological and ecological aspects of prey selection by the marine gastropod *Urosalpinx cinerea* (Prosobranchia: Muricidae). *Malacologia,* **6,** 267–320.

WOODWARD, M. F. 1899. On the anatomy of *Adeorbis subcarinatus* Montagu. *Proc. malac. Soc. Lond.,* **3,** 140–146.

WOODWARD, M. F. 1900. Note on the anatomy of *Voluta ancilla* (Sol.), *Neptuneopsis gilchristi,* Sby., and *Volutilithes abyssicola* (Ad. & Rve.) *Proc. malac. Soc. Lond.,* **4,** 117–125.

WOODWARD, M. F. 1901. The anatomy of *Pleurotomaria beyrichii* Hilg. *Quart. J. micr. Sci.,* **44,** 215–268.

WORKMAN, C. 1983. Comparisons of energy partitioning in contrasting age-structured populations of the limpet *Patella vulgata*. *J. exp. mar. Biol. Ecol.,* **68,** 81–103.

WRIGHT, J. R. & HARTNOLL, R. G. 1981. An energy budget for a population of the limpet *Patella vulgata*. *J. mar. biol. Ass. U.K.,* **61,** 627–646.

WU, S.-K. 1965. Comparative functional studies of the digestive system of the muricid gastropods *Drupa ricina* and *Morula granulata*. *Malacologia,* **3,** 211–233.

YONGE, C. M. 1926. Ciliary feeding mechanisms in the thecosomatous pteropods. *J. Linn. Soc.,* **36,** 417–429.

YONGE, C. M. 1930. The crystalline style of the Mollusca and a carnivorous habit cannot normally co-exist. *Nature, Lond.,* **125,** 444–445.

YONGE, C. M. 1937. The biology of *Aporrhais pes-pelecani* (L.) and *A. serresiana* (Mich.). *J. mar. biol. Ass. U.K.,* **21,** 687–704.

YONGE, C. M. 1938. Evolution of ciliary feeding in the Prosobranchia, with an account of ciliary feeding in *Capulus ungaricus*. *J. mar. biol. Ass. U.K.,* **22,** 453–468.

YONGE, C. M. 1939. On the mantle cavity and its contained organs in the Loricata (Placophora). *Quart. J. micr. Sci.,* **81,** 367–390.

YONGE, C. M. 1947. The pallial organs in the aspidobranch Gastropoda and their evolution throughout the Mollusca. *Phil. Trans. R. Soc. B,* **232,** 443–518.

YONGE, C. M. 1954. Feeding mechanisms in the Invertebrata. *Tab. biol.,* **21,** 46–68.

ZIEGELMEIER, E. 1958. Zur Lokomotion bei Naticiden (Gastropoda Prosobranchia). (Kurze Mitteilung über Schwimmbewegung bei *Polynices josephinus* Risso.) *Helgol. wiss. Meeresunters.,* **6,** 202–206.

ZIEGELMEIER, E. 1961. Zur Fortpflanzungsbiologie der Naticiden (Gastropoda Prosobranchia). *Helgol. wiss. Meeresunters.,* **87,** 94–118.

INDEX

Accessory boring organ (ABO) —
  muricids, 232–3, 621f
  naticids, 234–8, 625
Age —
  Nucella, 720
  Patella, 467, 696–7
  top shells, 702
Albumen, 340, 356–78 *passim,* 662–7
Albumen gland, 27, 49, 302, 339–51 *passim,* 687
Alimentary system, 148–229, 592f
Aorta —
  anterior, 28, 36, 137, 198, 204, 251, 260–1
  posterior, 28, 253, 260–1
Aphallism, 328, 333, 666
Archaeogastropods, 9, 13, 724–36
Architaenioglossans, systematic position, 733
Attachment of muscle to shell, 136

Bacteria —
  in *Nucella,* 604
  in *Pomatias,* 564
Bellerophonts, 8, 729
Blood —
  pigment, 727
  pores, 656
  pressure, 655
Boring of shells —
  muricids, 199, 621f
  naticids, 106, 231, 234–8, 625, 712
  Nucella, 196, 199, 231, 234, 621, 718
  tonnoideans, 626
Brackish and fresh water —
  effect on excretion, osmoregulation, 647f
  effect of reproduction, 662
  effect on shell, 51, 367
  prosobranchs, 367, 541–61, 731
Breeding —
  capacity of littorinids, 689
  migrations, 356, 663–4, 685
  periods —
    littorinids, 688–9
    Hydrobia ulvae, 708
    Nucella, 720
    Patella, 695
Buccal —
  cavity, 29, 142, 147, 240, 271, 532
  conoidean, 615
  eversion in feeding, 613, 621
  limits, 595f
  marginellid, 617

mass, 131, 133, 171–94, 271, 273, 277
muscles, 172f
Burrowing —
  foot in, 588, 715
  naticid, 588, 715
  shell in, 568, 589, 715
  Turritella, 568
Bursa copulatrix, 48–9, 295, 303–4, 330, 333, 349–52,
  548–9
Bursicles, 727–30

Caecum of stomach, 212, 225, 726–7, 734
Cannibalism, embryonic, 373, 381, 667
Capsule gland, 27, 49, 303, 350–2, 367, 549
Carbon dioxide and shell, 575
Carnivorous habit, 230–50 *passim,* 476f, 518f, 528–9,
  610–26, 704, 712–4
Cartilages, buccal or radular, 170, 186, 193–4, 407
Cephalic vesicle, 362, 429, 563
Cephalization, 730
Chalaza, 316, 368, 737
Chink shells, 509, 690–2
Chemical sense, 282–3, 610f, 695
Ciliary —
  currents —
    ctenidium, 89–90
    mantle cavity, 89, 452, 456–7, 535
    oesophagus, 200, 420
    veliger, 419, 671–2, 675
  feeding, 97–107, 554, 557–8, 626f, 675
  locomotion, 140, 579
Classification, 12–16, 724f
Cocculiniform limpet characters, 727
Coelomoducts, 634
Columella, 21, 57–8, 72, 404, 493, 560
Columellar muscle, 17, 57, 81–2, 131, 134, 180–1, 186,
  190, 388, 404, 448, 489
Conchiolin, 80, 82, 125–6, 363, 366, 371, 432, 437
Copulation, 49, 97, 301, 318, 320, 324, 358, 509, 548,
  562, 728, 734
Crab predation, 567, 571, 688, 694
Crystal sac, 367
Crystalline style —
  occurrence, 210, 511, 549, 564, 596, 735
  origin, 217, 227
Ctenidium, 4, 23, 86, 93–6, 98–101, 125, 131, 256, 272,
  275–6, 456, 516, 531, 548–9
  aspidobranch (bipectinate), 13, 93, 451, 534, 725–6,
  728, 739
  blood circulation, 41, 88, 256, 259–60, 700